DISCOVERING THE UNIVERSE

Discovering the Universe

Ninth Edition

Neil F. Comins
University of Maine

William J. Kaufmann III
San Diego State University

W. H. Freeman and Company
New York

In memory of my grandparents, Hyman and Mamie Finkelstein and Nathan and Estelle Comins.

Executive Editor: Anthony Palmiotto
Developmental Editor: Tony Petrites
Project Editor: Kerry O'Shaughnessy
Senior Media and Supplements Editor: Amy Thorne
Marketing Manager: Alicia Brady
Editorial Assistant: Nicholas Ciani
Design Manager: Blake Logan
Production Coordinator: Paul W. Rohloff
Senior Illustration Coordinator: Bill Page
Illustration Coordinator: Janice Donnola
Illustrations: George Kelvin, Emily Cooper, Dragonfly Media Group
Photo Editor: Bianca Moscatelli
Photo Researchers: Dena Digilio Betz
Composition: Prepare Inc.
Printing and Binding: Quad Graphics

Library of Congress Cataloging-in-Publication Control Number: 2010943282

ISBN-13: 978-1-4292-5520-2 / ISBN-10: 1-4292-5520-X
ISBN-13: 978-1-4292-8451-6 / ISBN-10: 1-4292-8451-X (NASTA-spec version)

Printed in the United States of America

Second Printing

W. H. Freeman and Company
41 Madison Avenue
New York, NY 10010
Houndmills, Basingstoke
RG21 6XS, England
www.whfreeman.com

CONTENTS OVERVIEW

CONTENTS

PREFACE

To confine our attention to terrestrial matters would be to limit the human spirit. *–Stephen Hawking*

Our knowledge about the universe is expanding at a phenomenal rate. New discoveries are being made in many astronomical realms: astronomers are probing the extent of water on the Moon, developing new theories of how the solar system formed, finding planets orbiting distant alien stars, discovering stars and stellar remnants with unexpected properties, among myriad other things.

Many of these scientific updates are included in this edition. I am also pleased to include a wide variety of modern learning techniques and new features in the ninth edition of *Discovering the Universe* while still providing the wide range of factual topics that are the hallmark of the text.

In the realm of astronomy education, educators continue to develop methods to help students understand how to think like scientists and grasp the core concepts, even when scientific theories are at odds with students' prior beliefs and misconceptions. In-class interactivities, where students respond to questions with "clickers," enrich the classroom experience. Online materials provide tutorials and practice questions that turn students from passive into active learners.

The ninth edition of *Discovering the Universe* continues to present concepts clearly and accurately to students, while strengthening the pedagogical tools to make the learning process even more worthwhile. The pedagogy includes

- presenting the observations and underlying physical *concepts* needed to connect astronomical observations to theories that explain them coherently and meaningfully.
- using both textual and graphical information to present concepts for students who learn in different ways.
- addressing student misconceptions in a respectful but rigorous manner, helping readers to understand why modern scientific views are correct.
- using analogies from everyday life to make cosmic phenomena more concrete.
- providing visually rich timelines that connect astronomical discoveries with other events throughout history.
- expanding student perspectives and confronting misconceptions by exploring plausible alternative situations (asking "What if…?" questions).
- pointing students towards cutting-edge research in "Frontiers yet to be discovered" sections and *Scientific American* articles.
- linking material presented in the book with enhanced material offered electronically.

MANY FEATURES BRING THE UNIVERSE INTO CLEARER FOCUS

What If…? margin questions about important concepts stretch students' thinking using hypothetical situations. These questions help to correct misconceptions by explaining to students what strange effects and consequences would result if their initial misconceptions were true.

> **WHAT IF… Earth's changing distance from the Sun caused the seasons (and how do we know it doesn't)?**
> Earth's orbit around the Sun is elliptical (we will discuss this oval shape in detail in Chapter 2). If the seasons *were* caused by the changing distance between Earth and the Sun, all parts of Earth should have the same seasons at the same time. In fact, the northern and southern hemispheres have exactly opposite seasons. Furthermore, Earth is closest to the Sun on or around January 3 of each year—the dead of winter in the northern hemisphere!

WHAT IS A PLANET?

SCIENTIFIC AMERICAN

The controversial new official definition of "planet," which banished Pluto, has its flaws but by and large captures essential scientific principles.

Adapted from an article by Steven Soter

Most of us grew up with the conventional definition of a planet as a body that orbits a star, shines by reflecting the star's light and is larger than an asteroid. Although imprecise, the definition categorized the bodies we knew at the time. But a series of discoveries in the 1990s rendered it untenable. Beyond Neptune's orbit, astronomers found hundreds of icy worlds occupying a doughnut-shaped region called the Kuiper belt. Around scores of other stars, they found other planets whose orbits look nothing like those in our solar system. They also discovered brown dwarfs, blurring the distinction between planet and star, and found planet-like objects drifting through interstellar space.

These findings ignited debate, prompting the International Astronomical Union (IAU) to redefine a planet as an object orbiting a star that is large enough to have settled into a round shape, and "has cleared the neighborhood around its orbit." This new definition stripped Pluto of its status as a planet. Some astronomers refused to use it and organized a protest petition.

When Earth Became a Planet

Astronomers' reevaluation of the nature of planets has deep historical roots. The ancient Greeks recognized seven lights in the sky that moved against the background pattern of stars: the sun, the moon, Mercury, Venus, Mars, Jupiter, and Saturn. They called them *planetes*, or wanderers. Note that Earth was not on this list. For most of human history, Earth was regarded not as a planet but as the center of the universe. After Nicolaus Copernicus proved the Sun lies at the center, astronomers redefined planets as objects orbiting the Sun, adding Earth to the list and deleting the Sun and the Moon. Telescope observers added Uranus in 1781 and Neptune in 1846.

Ceres, discovered in 1801, was initially welcomed as the missing planet between Mars and Jupiter. But astronomers began to have doubts when they found Pallas in a similar orbit the following year. Unlike the classical planets, which telescopes revealed as little disks, these bodies appeared as pinpricks of light. English astronomer William Herschel named them "asteroids." By 1851, 15 had been seen and astronomers began listing them by order of discovery rather than distance from the Sun, as for planets, making them a distinct population. If we still counted asteroids as planets, our solar system would now have more than 135,000 planets.

Pluto's story is similar. When Clyde Tombaugh discovered it in 1930, astronomers considered it the long-sought "Planet X" whose gravity would account for unexplained peculiarities in Neptune's orbit. But Pluto turned out to be smaller than all eight planets and seven of their satellites, including Earth's moon, and Neptune's orbital peculiarities proved to be nonexistent. For six decades, Pluto was a unique anomaly at the outer edge of the planetary system.

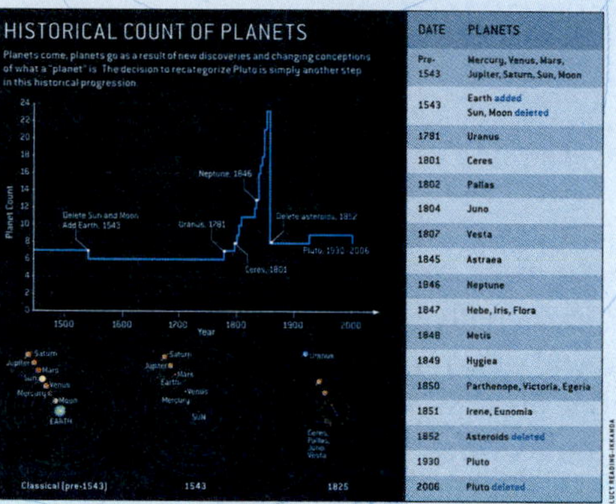

DATE	PLANETS
Pre-1543	Mercury, Venus, Mars, Jupiter, Saturn, Sun, Moon
1543	Earth added Sun, Moon deleted
1781	Uranus
1801	Ceres
1802	Pallas
1804	Juno
1807	Vesta
1845	Astraea
1846	Neptune
1847	Hebe, Iris, Flora
1848	Metis
1849	Hygiea
1850	Parthenope, Victoria, Egeria
1851	Irene, Eunomia
1852	Asteroids deleted
1930	Pluto
2006	Pluto deleted

When astronomers discovered that Pluto was one of many Kuiper belt objects (KBOs), they began to reconsider whether it should still be called a planet. Historically, revoking its status would not be unprecedented; the ranks of ex-planets include the Sun, the Moon, and asteroids. Nevertheless, many people have argued for continuing to call Pluto a planet, because almost everyone has grown accustomed to thinking of it as one.

The discovery in 2005 of Eris (formerly known as 2003 UB313 or Xena), a KBO even larger than Pluto, brought the issue to a head. If Pluto is a planet, then Eris must also be one, along with other large KBOs. On what objective grounds could astronomers decide?

Clearing the Air

To avoid a proliferation of planets, astronomers considered defining a planet as a body smaller than a star, but large enough for its gravity to form it into a round shape. Most bodies larger than several hundred kilometers are round; smaller ones often resemble giant boulders. Under these guidelines, Pluto, Ceres, and potentially dozens of KBOs would be considered planets.

But it is very difficult to observe the shapes of distant KBOs, and both asteroids and KBOs span an almost continuous spectrum of sizes and shapes. Ultimately, categorizing planets by degree of roundness seemed arbitrary.

An additional criterion provided a clear way to distinguish planets: They must be massive enough to dominate their orbital zone by flinging smaller bodies away, sweeping them up in direct collisions, or holding them in stable orbits. Lesser bodies occupy transient, unstable orbits or have a heavyweight guardian that stabilizes their orbits. For example, Earth controls bodies like near-Earth asteroids that stray too close and holds its moon in orbit. Each of the four giant planets protects many orbiting satellites. Jupiter and Neptune also maintain families of asteroids and KBOs (called Trojans and Plutinos, respectively) in special orbits called stable resonances, where an orbital synchrony prevents collisions with the planets.

The eight planets from Mercury through Neptune are thousands of times more likely to sweep up or deflect small neighboring bodies than even the largest asteroids and KBOs, which include Ceres, Pluto, and Eris. Asteroids, comets, and KBOs—including Pluto—live amid swarms of comparable bodies.

Planets grew from a flattened disk of gas and dust orbiting the primordial Sun. In the competition for limited raw material, some bodies won out, with a single large body dominating each orbital zone. The smaller bodies were swept up by the larger ones, ejected from the solar system or swallowed by the Sun, and the survivors became the planets we see today. The asteroids and comets, including the KBOs, are the leftover debris.

Endgame

The revised IAU definition may require additional guidelines to delineate what degree of clearing makes a celestial body a planet. But it removes the need for an upper mass limit to distinguish planets from both stars and brown dwarfs. It also classifies rare brown dwarf companions that orbit close to stars as planets.

The difference between planets and nonplanets is quantifiable in theory and by observation. All the planets in our solar system have enough mass to have swept up or scattered away most of the original planetesimals from their orbital zones. Today each planet contains at least 5,000 times more mass than all the debris in its vicinity. The asteroids, comets, and KBOs, including Pluto, live amid swarms of comparable bodies.

One objection to this definition is that astronomical objects should be classified by properties such as size, shape, or composition, not by location or dynamical context. This argument overlooks the fact that astronomers classify objects that orbit planets as "moons." Context and location are important. Distance from the Sun determined that close-in bodies became small, rocky planets and farther ones became gas and ice-rich giant planets. Planets dynamically dominate a large volume of orbital space—while asteroids, KBOs, and ejected planetary embryos do not. The eight planets are the dominant end products of disk accretion and differ recognizably from the vast populations of asteroids and KBOs.

The historical definition of nine planets retains a sentimental attraction. For 76 years, our schools taught that Pluto was a planet. While some argue that culture and tradition suffice to leave it that way, science cannot remain bound to misconceptions and must be derived from the natural world. The evolving definition of a planet reflects new discoveries and profound changes in our understanding of the solar system.

156 CHAPTER 5

FORMATION OF THE SOLAR SYSTEM AND OTHER PLANETARY SYSTEMS 157

Articles adapted from *Scientific American* explore current research in the field. Students follow renowned astronomy experts to the farthest frontiers of astronomy research. Dynamic visuals enhance student interest, while providing a glimpse into the world of science outside the textbook. The following articles were selected and edited exclusively to compliment the conceptual presentation and level of *Discovering the Universe*:

- What Is a Planet? (Chapter 5)
- Earth's Missing Ingredient (Chapter 6)
- The Red Planet's Watery Past (Chapter 7)
- The Strangest Satellites in the Solar System (Chapter 8)
- Bracing for a Solar Superstorm (Chapter 10)
- Cloudy with a Chance of Stars (Chapter 12)
- The Ghosts of Galaxies Past (Chapter 16)
- The First Few Microseconds (Chapter 18)
- The Color of Plants on Other Worlds (Chapter 19)

New theory of planet formation Chapter 5 now presents the Nice (pronounced NIECE) theory of solar system formation, walking students through a fully up-to-date picture of how the interactions between the Sun and planets evolved.

Margin questions about important concepts are presented in most sections of the book. These questions encourage students to frequently test themselves and correct their beliefs before errors accumulate. For example, after learning about inertia in Section 2-7, students are asked how, while driving in a car, they can show that their bodies have inertia. (They could put on the brakes and feel themselves being restrained by the seatbelts.) Answers to approximately one-third of the margin questions appear at the end of this text.

Margin photos provide a connection between the concepts being presented and their applications in everyday life.

Margin charts show the location in the sky of important astronomical objects cited in the text. Sufficient detail in the margin charts allows students to locate the objects

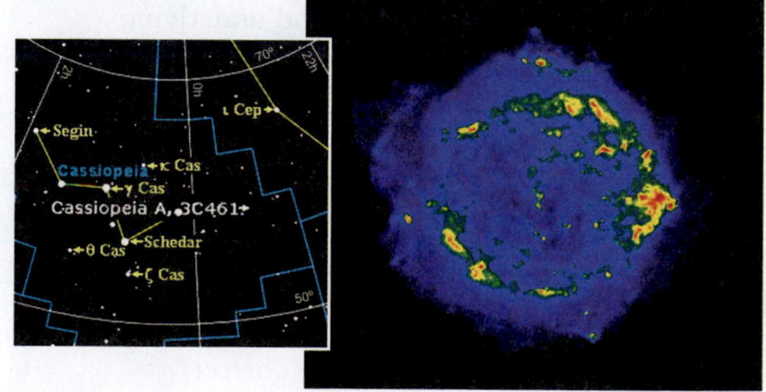

with either the unaided eye or a small telescope, as appropriate. In this example, a star chart of the constellation Cassiopeia is shown with a photo of the supernova remnant Cassiopeia A.

Revised coverage of planet classification The categories of *planets, dwarf planets,* and *small solar system objects* are explained and reconciled with the existing classes of objects, including planets, moons, asteroids, meteoroids, and comets. Also explained is how Pluto fits more comfortably with the dwarf planets than with the eight planets.

Dynamic art Summary figures appear throughout the book to show either the interactions between important concepts or the evolution of important objects. For example, the location of the Sun in the sky, which varies over the seasons, as does the corresponding intensity of the light and the appropriate ground cover, is shown in a sequence of drawings combined into one figure.

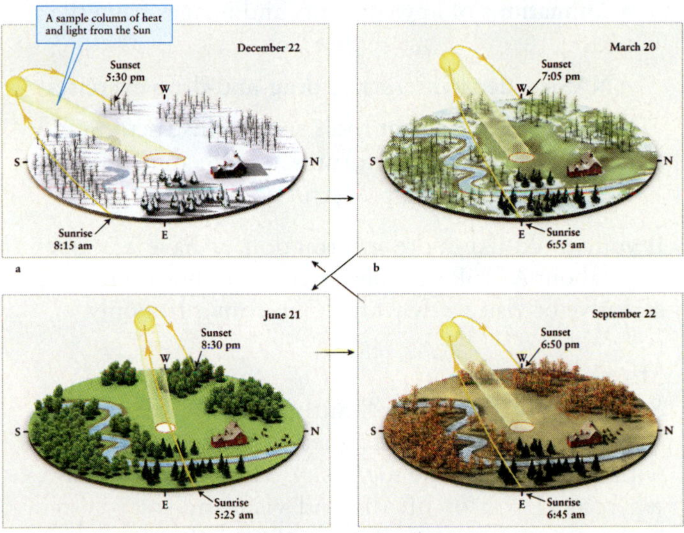

FIGURE 1-17 The Sun's Daily Path and the Energy It Deposits Here (a) On the winter solstice—the first day of winter—the Sun rises farthest south of east, is lowest in the noontime sky, stays up the shortest time, and its light and heat are least intense (most spread out) of any day of the year in the northern hemisphere. (b) On the vernal equinox—the first day of spring—the Sun rises precisely in the east and sets precisely in the west. Its light and heat have been growing more intense, as shown by the brighter oval of light than in (a). (c) On the summer solstice—the first day of summer—the Sun rises farthest north of east of any day in the year, is highest in the noontime sky, stays up the longest time, and its light and heat are most intense of any day in the northern hemisphere. (d) On the autumnal equinox, the same astronomical conditions exist as on the vernal equinox.

PROVEN FEATURES SUPPORT LEARNING

What Do You Think? and **What Did You Think?** questions in each chapter ask students to consider their present beliefs and actively compare them with the correct science presented in the book. Numbered icons mark the places in the text where each concept is discussed. Encouraging students to think about what they believe is true and then work through the correct science step-by-step has proved to be an effective teaching technique, especially when time constraints prevent instructors from working with students individually or in small groups as they try to reconcile incorrect beliefs with proper science.

Chapter opening narratives introduce and launch the chapter's topics and provide students with a context for understanding the material.

Learning objectives underscore the key chapter concepts.

Section headings are brief sentences that summarize section content and serve as a quick study guide to the chapter when reread.

Icons link the text to Web material

- ***Starry Night* ™** icons link the text to a specific interactivity in the ***Starry Night* ™** observing programs.

- Video icons link the text to relevant video clips available on the textbook's Web site.

- Animation icons link the text to animated figures available on the textbook's Web site.

Guided Discovery boxes offer hands-on experience with astronomy. Several use ***Starry Night* ™** software.

An Astronomer's Toolbox introduces equations used in astronomy. Most of the material in the book is descriptive, so essential equations are set off in numbered boxes to maintain the flow of the material. The toolboxes also contain worked examples, additional explanations, and practice doing calculations; answers are given at the end of the book. All the equations are summarized in Appendix C.

Insight into Science boxes are brief asides that relate topics to the nature of scientific inquiry and encourage critical thinking.

Wavelength tabs with photographic images show whether an image was made with radio waves (R), infrared radiation (I), visible light (V), ultraviolet light (U), X rays (X), or gamma rays (G).

Review and practice material

- **Summary of Key Ideas** is a bulleted list of key concepts.
- **What Did You Think?** questions at the end of each chapter answers the What Do You Think? questions posed at the beginning of the chapter.
- **Key Words, Review Questions,** and **Advanced Questions** help the students understand the chapter material.
- **Discussion Questions** offer interesting topics to spark lively, insightful debate.
- **Web Questions** take students to the Web for further study.
- **Observing Projects,** featuring *Starry Night* activities, ask students to be astronomers themselves.

What If... Selected chapters conclude with a "What If..." essay that encourages critical thinking by speculating on how changes in the universe could have profound effects on Earth.

An Astronomer's Almanac, a dynamic timeline that relates discoveries in astronomy to other historical events, opens each Part. The almanac provides strong context for the information presented.

MEDIA AND SUPPLEMENTS PACKAGE

FOR STUDENTS

The eBook: Affordable, Innovative, Customizable

The eBook for the ninth edition of *Discovering the Universe* is a complete online version of the textbook. It provides a rich learning experience by integrating a wealth of multimedia resources and adding unique features.

- Animations, videos, activities, and other visualization aids are delivered within the relevant sections of the electronic textbook, encouraging students to access the resources as the concepts are explained.
- Access from any Internet-connected computer is easy via a standard Web browser.
- Navigation to any section or subsection or to any printed book page number is quick and intuitive.
- Links to all glossary entries are provided.
- Students can highlight, take notes, and bookmark pages.
- Full-text search is built in, with an option to also search the complete glossary.

The eBook can be purchased online at http://ebooks.bfwpub.com.

AstroPortal

W. H. Freeman's AstroPortal combines a fully customizable eBook, tutorials, conceptual resources, and homework management in one affordable and easy-to-use learning management system. With its wide range of resources, AstroPortal can be used as a complete course management system or as an independent student study resource.

One Location, One Log-In AstroPortal integrates the Interactive eBook, Animations, Automatically Graded Homework, Course Management, Communication, and Gradebook:

- AstroPortal is completely **customizable** in terms of look, feel, functionality, and content.
- **eBook integration** gives students purchasing options: When used without a printed text, AstroPortal saves your students money. AstroPortal is approximately half the price of the printed text and includes supplements and tools that save students even more.

Features of the AstroPortal

- **Self-quizzes** gauge student understanding. Quiz results generate a student-specific set of additional exercises and eBook sections keyed to the questions students missed.
- **Tutorials** engage students in the scientific process of discovery and interpretation. Dozens of these concept-driven, experiential walkthroughs allow students to make observations, draw conclusions, and apply their knowledge. Interwoven with multimedia, activities, and questions, students receive a deep, self-guided exploration of the concepts.
- **Animations** of key concepts and images from the text.
- **NASA videos, interactive drag-and-drop exercises**
- **Course management tools** such as instructor and class blogs, discussion forums, e-mail, announcements, calendar, and powerful gradebook.

If you would like more information or have any questions about AstroPortal please contact your local representative or visit www.whfreeman.com/astronomy.

Starry Night

Starry Night™ is a brilliantly realistic planetarium software package. It is designed for easy use by anyone with an interest in the night sky. See the sky from anywhere on Earth or lift off and visit any solar system body or any location up to 20,000 light-years away. View 2,500,000 stars along with more than 170 deep-space objects such as galaxies, star clusters, and nebulae. Travel 15,000 years in time, check out the view from the International Space Station, and see planets up close from any one of their moons. Included are stunning OpenGL graphics. Handy star charts can be printed to explore outdoors. A download code for *Starry Night* is available with the text upon request. Use ISBN 1-4292-9447-7.

Observing Projects Using *Starry Night*™

by T. Alan Clark and William J. F. Wilson, University of Calgary, and Marcel Bergman
ISBN 1-4292-7806-4
Available for packaging with the text and compatible with both PC and Mac, this workbook contains a variety of comprehensive lab activities for use with *Starry Night*. The Observing Projects workbook can also be packaged with the *Starry Night* software.

Student Companion Web Site

The Companion Web site at www.whfreeman.com/dtu9e features a variety of study and review resources designed to help students understand the concepts. The open-access Web site includes the following:

- **Online quizzing** offers questions and answers with instant feedback to help students study, review, and prepare for exams. Instructors can access results through an online database or they can have results e-mailed directly to their accounts.
- **Animations and videos** both original and NASA-created, are keyed to specific chapters.
- **Interactive drag-and-drop exercises** based on text illustrations help students grasp the vocabulary in context
- **Flashcards** offer help with vocabulary and definitions

FOR INSTRUCTORS

Instructor's eBook

The eBook for the ninth edition of *Discovering the Universe* offers instructors flexibility and customization options not previously possible with any printed textbook. Instructors have access to:

- **eBook Customization** Instructors can choose the chapters that correspond with their syllabus, and students will get a custom version of the eBook with the selected chapters only.
- **Instructor Notes** Instructors can create an annotated version of the eBook by adding notes to any page. When students registered to this instructor's course log in, they see the notes.
- **Custom Content** Instructor content can include text, Web links, and images, allowing instructors to place any content they choose exactly where they want it. This feature can be used to augment, edit, or enrich the textbook. Instructor media resources are easily integrated, and eBook pages from other sites can be directly linked.

Instructor's Resources CD-ROM

ISBN 1-4292-7136-1

To help instructors create lecture presentations, course Web sites, and other resources, this CD-ROM contains the following resources:

- All images from the text
- Instructor's Manual
- Test Bank
- Online resources from the textbook's companion Web site

Test Bank CD-ROM

Windows and Mac versions on one disc ISBN 1-4292-8404-8

More than 3,500 multiple-choice questions are section-referenced. The easy-to-use CD-ROM allows instructors to add, edit, resequence, and print questions to suit their needs.

Online Course Materials (WebCT, Blackboard, Angel, Desire2Learn)

As a service for adopters, we will provide content files in the appropriate online course format, including the instructor and student resources for the textbook. The files can be used as is or can be customized to fit specific needs. Prebuilt quizzes, links, and activities, among other materials, are included.

Classroom Presentation and Interactivity

A set of online lecture presentations created in PowerPoint allows instructors to tailor their lectures to suit their own needs using images and notes from the textbook. These presentations are available on the instructor portion of the companion Web site.

Clicker Questions

Written by Neil Comins, these questions can be used as lecture launchers with or without a classroom response system such as iClicker. Each chapter includes questions relating to figures from the text and common misconceptions, as well as writing questions for instructors who would like to add a writing or class discussion element to their lectures.

ACKNOWLEDGMENTS

I am deeply grateful to the astronomers and teachers who reviewed the manuscript of this edition. This is a stronger and better book because of their conscientious efforts:

John Anderson, *University of North Florida*
Nadine G. Barlow, *Northern Arizona University*
John C. (Jack) Brandt, *University of New Mexico*
Stephen Eikenberry, *University of Florida*
Rica French, *Mira Costa College*
Thomasanna Hail, *Parkland College*
Javier Hasbun, *University of West Georgia*
Scott Hildreth, *Chabot College*
Mark Hollabaugh, *Normandale Community College*
Francine Jackson, *Framingham State College*
Fred Jacquin, *Onondaga Community College*
Katie Jore, *University of Wisconsin–Steven's Point*
Agnes Kim, *Georgia College & State University*
Rob Klinger, *Parkland College*
Charles Nelson, *Drake University*

Ron Olowin, *St. Mary's College of California*
Norm Siems, *Juniata College*
Gerard Williger, *University of Louisville*
Wayne Wooten, *Pensacola Junior College*

I would also like to thank the many people whose advice on previous editions has had an ongoing influence:

Kurt S. J. Anderson, *New Mexico State University*
Gordan Baird, *University of Arizona*
Nadine G. Barlow, *Northern Arizona University*
Henry E. Bass, *University of Mississippi*
J. David Batchelor, *Community College of Southern Nevada*
Jill Bechtold, *University of Arizona*
Peter A. Becker, *George Mason University*
Ralph L. Benbow, *Northern Illinois University*
John Bieging, *University of Arizona*
Greg Black, *University of Virginia*
John C. Brandt, *University of New Mexico*
James S. Brooks, *Florida State University*
John W. Burns, *Mt. San Antonio College*
Gene Byrd, *University of Alabama*
Eugene R. Capriotti, *Michigan State University*
Eric D. Carlson, *Wake Forest University*
Jennifer L. Cash, *South Carolina State University*
Michael W. Castelaz, *Pisgah Astronomical Research Institute*
Gerald Cecil, *University of North Carolina*
David S. Chandler, *Porterville College*
David Chernoff, *Cornell University*
Erik N. Christensen, *South Florida Community College*
Tom Christensen, *University of Colorado, Colorado Springs*
Chris Clemens, *University of North Carolina*
Christine Clement, *University of Toronto*
Halden Cohn, *Indiana University*
John Cowan, *University of Oklahoma*
Antoinette Cowie, *University of Hawaii*
Charles Curry, *University of Waterloo*
James J. D'Amario, *Harford Community College*
Purnas Das, *Purdue University*
Peter Dawson, *Trent University*
John M. Dickey, *University of Minnesota, Twin Cities*
John D. Eggert, *Daytona Beach Community College*
Stephen S. Eikenberry, *University of Florida*
Bernd Enders, *College of Marin*
Robert Frostick, *West Virginia State College*
Martin Gaskell, *University of Nebraska*
Richard E. Griffiths, *Carnegie Mellon University*
Bruce Gronich, *University of Texas, El Paso*
Siegbert Hagmann, *Kansas State University*
Thomasanna C. Hail, *University of Mississippi*
Javier Hasbun, *University of West Georgia*
David Hedin, *Northern Illinois University*
Chuck Higgins, *Penn State University*
Scott S. Hildreth, *Chabot College*
Thomas Hockey, *University of Northern Iowa*
Mark Hollabaugh, *Normandale Community College*
J. Christopher Hunt, *Prince George's Community College*
James L. Hunt, *University of Guelph*
Nathan Israeloff, *Northeastern College*
Kenneth Janes, *Boston University*
William C. Keel, *University of Alabama*
William Keller, *St. Petersburg Junior College*
Marvin D. Kemple, *Indiana University–Purdue University Indianapolis (IUPUI)*
Pushpa Khare, *University of Illinois at Chicago*
F. W. Kleinhaus, *Indiana University–Purdue University Indianapolis (IUPUI)*
Rob Klinger, *Parkland College*
H. S. La, *Clayton State University*
Patick M. Len, *Cuesta College*
John Patrick Lestrade, *Mississippi State University*
C. L. Littler, *University of North Texas*
M. A. K. Lohdi, *Texas Tech University*
Michael C. LoPresto, *Henry Ford Community College*
Phyllis Lugger, *Indiana University*
R. M. MacQueen, *Rhodes College*
Robert Manning, *Davidson College*
Paul Mason, *University of Texas, El Paso*
P. L. Matheson, *Salt Lake Community College*
Steve May, *Walla Walla Community College*
Rahul Mehta, *University of Central Arkansas*
Ken Menningen, *University of Wisconsin–Stevens Point*
J. Scott Miller, *University of Louisville*
Scott Miller, *Pennsylvania State University*
L. D. Mitchell, *Cambria County Area Community College*
J. Ward Moody, *Brigham Young University*
Siobahn M. Morgan, *University of Northern Iowa*
Steven Mutz, *Scottsdale Community College*
Gerald H. Newsom, *Ohio State University*
Bob O'Connell, *College of the Redwoods*
William C. Oelfke, *Valencia Community College*
Richard P. Olenick, *University of Dallas*
John P. Oliver, *University of Florida*
Melvyn Jay Oremland, *Pace University*
Jerome A. Orosz, *San Diego State University*
David Patton, *Trent University*
Jon Pedicino, *College of the Redwoods*
Sidney Perkowitz, *Emory University*
Lawrence Pinsky, *University of Houston*
Eric Preston, *Indiana State University*
David D. Reid, *Wayne State University*
Adam W. Rengstorf, *Indiana University*
James A. Roberts, *University of North Texas*
Henry Robinson, *Montgomery College*
Dwight P. Russell, *University of Texas, El Paso*
Barbara Ryden, *Ohio State University*
Itai Seggev, *University of Mississippi*
Larry Sessions, *Metropolitan State University*

C. Ian Short, Florida *Atlantic University*
John D. Silva, *University of Massachusetts at Dartmouth*
Michael L. Sitko, *University of Cincinnati*
Earl F. Skelton, George F. Smoot, *University of California at Berkeley*
Alex G. Smith, *University of Florida*
Jack Sulentic, *University of Alabama*
David Sturm, *University of Maine, Orono*
Paula Szkody, *University of Washington*
Michael T. Vaughan, *Northeastern University*
Andreas Veh, *Kenai Peninsula College*
John Wallin, *George Mason University*
William F. Welsh, *San Diego State University*
R. M. Williamon, *Emory University*
J. Wayne Wooten, *Pensacola Junior College*
Edward L. (Ned) Wright, *University of California at Los Angeles*
Jeff S. Wright, *Elon College*
Nicolle E. B. Zellner, *Rensselaer Polytechnic Institute*

I would like to add my special thanks to the wonderfully supportive staff at W. H. Freeman and Company who make the revision process so enjoyable. Among others, these people include Tony Petrites, Kerry O'Shaughnessy, Amy Thorne, Blake Logan, Bianca Moscatelli, Bill Page, Janice Donnola, Paul Rohloff, and Nicholas Ciani. Thanks also go to my copyeditor and indexer, Louise Ketz, and proofreader, Anna Paganelli. Thank you to T. Alan Clark, Marcel W. Bergman, and William J. F. Wilson for their work on the *Starry Night*™ material, and to Sharon Gudyup and Helen Hill at *Scientific American*. Warm thanks also to David Sturm, University of Maine, Orono, for his help in collecting current data on objects in the solar system; to my UM astronomy colleague David Batuski; and to my wife, Sue, for her patience and support while preparing this book.

Neil F. Comins
neil.comins@umit.maine.edu

ABOUT THE AUTHORS

Professor **Neil F. Comins** is on the faculty of the University of Maine. Born in 1951 in New York City, he grew up in New York and New England. He earned a bachelor's degree in engineering physics at Cornell University, a master's degree in physics at the University of Maryland, and a Ph.D. in astrophysics from University College, Cardiff, Wales, under the guidance of Bernard F. Schutz. Dr. Comins's work for his doctorate, on general relativity, was cited in Subramanyan Chandrasekhar's Nobel laureate speech. He has done theoretical and experimental research in general relativity, observational astronomy, computer simulations of galaxy evolution, and science education. The fourth edition of *Discovering the Universe* was the first book in this series that Dr. Comins wrote, having taken over following the death of Bill Kaufmann in 1994. He is also the author of four trade books, *What If the Moon Didn't Exist?*, *Heavenly Errors, The Hazards of Space Travel,* and *What if the Earth Had Two Moons? What If the Moon Didn't Exist?* has been made into planetarium shows, been excerpted for television and radio, translated into several languages, and was the theme for the Mitsubishi Pavilion at the World Expo 2005 in Aichi, Japan. *Heavenly Errors* explores misconceptions people have about astronomy, why such misconceptions are so common, and how to correct them. Dr. Comins has appeared on numerous television and radio shows and gives many public talks. Although he has jumped out of airplanes while in the military, today his activities are a little more sedate: He is a licensed pilot and avid sailor, having once competed against Prince Philip, Duke of Edinburgh.

William J. Kaufmann III was the author of the first three editions of *Discovering the Universe*. Born in New York City on December 27, 1942, he often visited the magnificent Hayden Planetarium as he was growing up. Dr. Kaufmann earned his bachelor's degree magna cum laude in physics from Adelphi University in 1963, a master's degree in physics from Rutgers in 1965, and a Ph.D. in astrophysics from Indiana University in 1968. At 27 he became the youngest director of any major observatory in the United States when he took the helm of the Griffith Observatory in Los Angeles. During his career he also held positions at San Diego State University, UCLA, Caltech, and the University of Illinois. Throughout his professional life as a scientist and educator, Dr. Kaufmann worked to bridge the gap between the scientific community and the general public to help the public share in the advances of astronomy. A prolific author, his many books include *Black Holes and Warped Spacetime, Relativity and Cosmology, The Cosmic Frontiers of General Relativity, Exploration of the Solar System, Planets and Moons, Stars and Nebulas, Galaxies and Quasars,* and *Supercomputing and the Transformation of Science*. Dr. Kaufmann died in 1994.

PART I

Understanding Astronomy

Telescopes enhance our views of the cosmos.
(David Parker/Photo Researchers, Inc.)

AN ASTRONOMER'S ALMANAC

2136 B.C. Chinese astronomers record solar eclipse.

586 B.C. Thales of Miletus predicts solar eclipse.

350 B.C. Aristotle proposes spherical Earth, geocentric cosmology.

ca. 270 B.C. Aristarchus of Samos proposes heliocentric cosmology.

Greek Golden Age

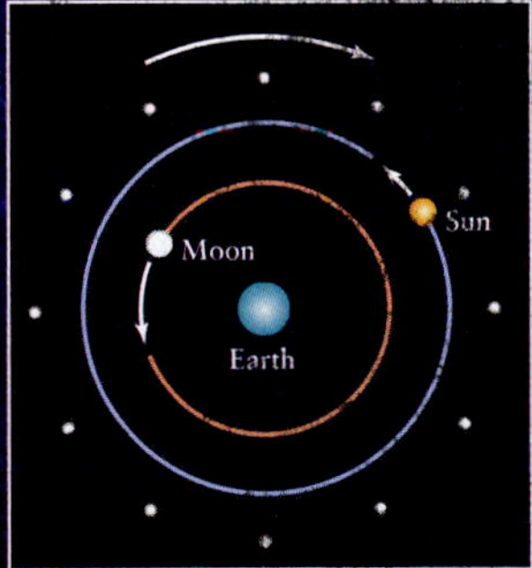

ca. A.D. 125 Claudius Ptolemy refines and details geocentric cosmology in his *Almagest*.

1512-1543 Nicolas Copernicus proposes heliocentric cosmology in his *Commentariolus* and *De Revolutionibus Orbium Coelestium*.

1576-1601 Tycho Brahe makes precise observations of stars and planets.

1589-1609 Galileo Galilei proposes that all objects fall with the same acceleration, independent of their masses; builds his first telescope, a refractor.

European Renaissance

1609-1610 Johannes Kepler publishes his three laws of planetary motion.

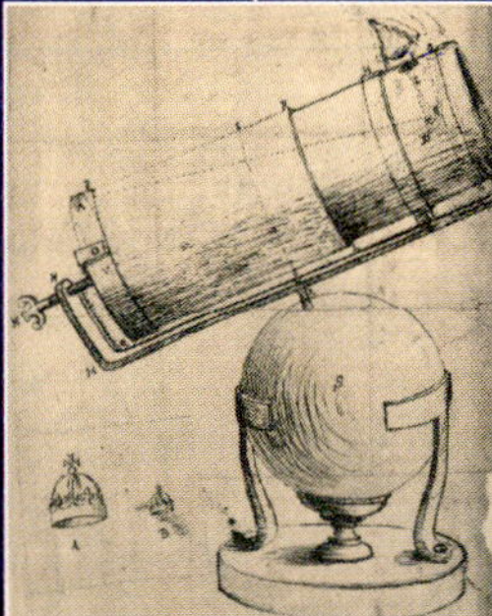

1665-1704 Isaac Newton deduces gravitational force from the orbit of the Moon; builds first reflecting telescope; proves that the planets obey Kepler's laws because they move under the influence of the gravitational force; and publishes compendium on light, *Opticks*.

1715 Edmond Halley calculates shadow path of a solar eclipse over Earth's surface.

1766 Henry Cavendish discovers hydrogen.

1800-1803 William Herschel discovers infrared radiation from the Sun. Thomas Young demonstrates wave nature of light. John Dalton proposes that matter is composed of atoms of different masses.

DISCOVERING ASTRONOMY

Industrial Revolution

1840-1849 J. W. Draper invents astronomical photography; takes first photographs of the Moon. Christian Doppler proposes that wavelength is affected by motion. Lord Rosse completes 60-in. reflecting telescope at Birr Castle in Ireland. Armand Fizeau and Jean-Bernard Foucault measure speed of light accurately.

1871-1873 Dimitri Mendeleyev develops periodic table of the elements. Henry Draper develops spectroscopy. James Clerk Maxwell asserts that light is an electromagnetic phenomenon.

1885-1888 Johann Balmer expresses spectral lines of hydrogen mathematically. Heinrich Hertz detects radio waves.

1895-1897 Wilhelm Roentgen discovers X rays. Joseph Thomson detects the electron. Yerkes 40-in. optical refracting telescope completed.

1900 Max Planck explains blackbody radiation. Paul Villard discovers gamma rays.

1913 Niels Bohr proposes quantum theory of the atom.

1930-1934 Karl Jansky builds first radio telescope. James Chadwick discovers the neutron. Bernhard Schmidt builds his Schmidt optical reflecting telescope.

1942-1949 J. S. Hey detects radio waves from the Sun. First astronomical telescope launched into space. Herbert Friedman detects X rays from the Sun. 200-in. optical reflecting telescope begins operation on Mt. Palomar, California.

Information Age

1963-1967 Largest single-dish radio telescope, 300 m across, begins operation at Arecibo, Puerto Rico. First Very Long Baseline Interferometer (VLBI) images.

1975 First charge-coupled device (CCD) astronomical observations.

1980 Very Large Array (VLA) radio observatory completed, Socorro, New Mexico.

1990-1996 Hubble Space Telescope launched. Keck 10-m optical/infrared telescopes begin operation at Mauna Kea, Hawaii. SOHO solar observatory launched.

1999 Chandra X-ray Telescope launched.

2004-present Two rovers travel on Mars, detectors search for gravitational radiation.

Stars appear to rotate around Polaris, the North Star (top), in this time exposure, taken January 26, 2006. Below Polaris is the 4-m telescope dome at Kitt Peak National Observatory near Tucson, Arizona. The image is composed of 114 thirty-second exposures of the night sky combined to make the equivalent of a nearly 1-h exposure in which Earth's rotation causes the stars to appear to move across the night sky. The orange glow on the horizon is from the city of Phoenix, 160 km (100 mi) away. (STAN HONDA/AFP/Getty Images)

Chapter 1

Discovering the Night Sky

WHAT DO YOU THINK?

1. Is the North Star—Polaris—the brightest star in the night sky?
2. What do astronomers define as constellations?
3. What causes the seasons?
4. When is Earth closest to the Sun?
5. How many zodiac constellations are there?
6. Does the Moon have a dark side that we never see from Earth?
7. Is the Moon ever visible during the daytime?
8. What causes lunar and solar eclipses?

Answers to these questions appear in the text beside the corresponding numbers in the margins and at the end of the chapter.

You have picked an exciting time to study astronomy. Our knowledge about the cosmos (or the *universe*) is growing as never before. Current telescope technology makes it possible for astronomers to observe objects that were invisible to them just a few years ago. These new observations have deepened our understanding of virtually every aspect of the universe. We can now watch it expand and see stars explode in distant galaxies; we have discovered planets orbiting nearby stars; we have seen newly-formed stars enshrouded in clouds of gas and dust; and we have identified black holes and other remnants of stellar evolution. Many of these objects are so far away that the light we see from them began its journey to Earth millions or even billions of years ago. Thus, as we look farther and farther out into the universe—defined as everything that we can ever detect or that can ever affect us—we also see farther and farther back into time.

Telescopes are not the only means by which we are deepening our understanding of what lies beyond Earth's atmosphere. We have also begun the process of physically exploring our neighborhood in space. In just the past half century, humans have walked on the Moon, and space probes have roamed over parts of Mars and dug into its soil, while others have crashed into a comet, landed on an asteroid, brought back debris from a comet, discovered active volcanoes and barren ice fields on the moons of Jupiter, landed on murky Titan, traveled through the shimmering rings of Saturn, and sped beyond the realm of the planets in our solar system, to mention just a few accomplishments. We are also witnessing the dawn of space tourism, with people buying trips to the International Space Station.

In the best locations, the night sky is truly breathtaking (Figure 1-1a). Even if you can't see the thousands of stars visible in clear locations (Figure 1-1b), software such as *Starry Night*™ can show them to you. The night sky can draw you out of yourself, inviting you to understand what is happening beyond Earth and inspiring you to think about our place in the universe.

Hundreds of years ago, the explanations of what people saw in the sky were based on beliefs that had to be accepted on faith—there was no way to test ideas of what the stars are, or whether the Moon really has liquid water oceans, or how the planets move, or why the Sun shines. Times have changed. We are fortunate to be living in an era when science has answers to many of the questions inspired by the cosmos.

Beautiful, intriguing, and practical, astronomy has something for everyone. This course and this book will help you better understand the universe by sharing what we have learned about some of these questions. As you proceed through this book, I hope that you will gain a new appreciation of the awesome power of the human mind to reach out, to observe, to explore, and to comprehend. One of the great lessons of modern astronomy is that by gaining, sharing, and passing on knowledge, we transcend the limitations of our bodies and the brevity of human life.

a

b

R I V U X G

ANIMATION 1.1

FIGURE 1-1 The Night Sky Without and With Light Pollution (a) Sunlight is a curtain that hides virtually everything behind it. As the Sun sets, places with little smog or light pollution treat viewers to beautiful panoramas of stars that can inspire the artist or scientist in many of us. This photograph shows the night sky in Goodwood, Ontario, Canada, during a power outage. (b) This photograph shows the same sky with normal city lighting. (© Todd Carlson/SkyNews Magazine)

In this chapter you will discover

- how astronomers organize the night sky to help them locate objects in it
- that Earth's spin on its axis causes day and night
- how the tilt of Earth's axis of rotation and Earth's motion around the Sun combine to create the seasons
- that the Moon's orbit around Earth creates the phases of the Moon and lunar and solar eclipses
- how the year is defined and how the calendar was developed

SCALES OF THE UNIVERSE

In learning a new field it is often useful to see the "big picture" before exploring the details. For this reason, we begin by surveying the major types of objects in the universe, along with their ranges of size and the scale of distances between them.

1-1 Astronomical distances are, well, astronomical

One of the thrills and challenges of studying astronomy is becoming familiar with and comfortable with the vast range of sizes that occur in it. In our everyday lives we typically deal with distances ranging from millimeters to thousands of kilometers. (The metric system of units is standard in science and will be used throughout this book; however, we will often provide the equivalent in U.S. customary units. Appendix E-10 lists conversion factors between the two sets of units.) A hundredth of a meter or a thousand kilometers are numbers that are easy to visualize and write. In astronomy, we deal with particles as small as a millionth of a billionth of a meter and systems of stars as large as a thousand billion billion kilometers across. Similarly, the speeds of some things, like light, are so large as to be cumbersome if you have to write them out in words each time. To deal with numbers much larger and much smaller than 1, we use a shorthand throughout this book called **scientific notation** or "powers of ten." (I suggest that you read Appendix A if you are unfamiliar with scientific notation.)

The size of the universe that we can observe and the range of sizes of the objects in it are truly staggering. Figure 1-2 summarizes the range of sizes from atomic particles up to the diameter of the entire universe visible

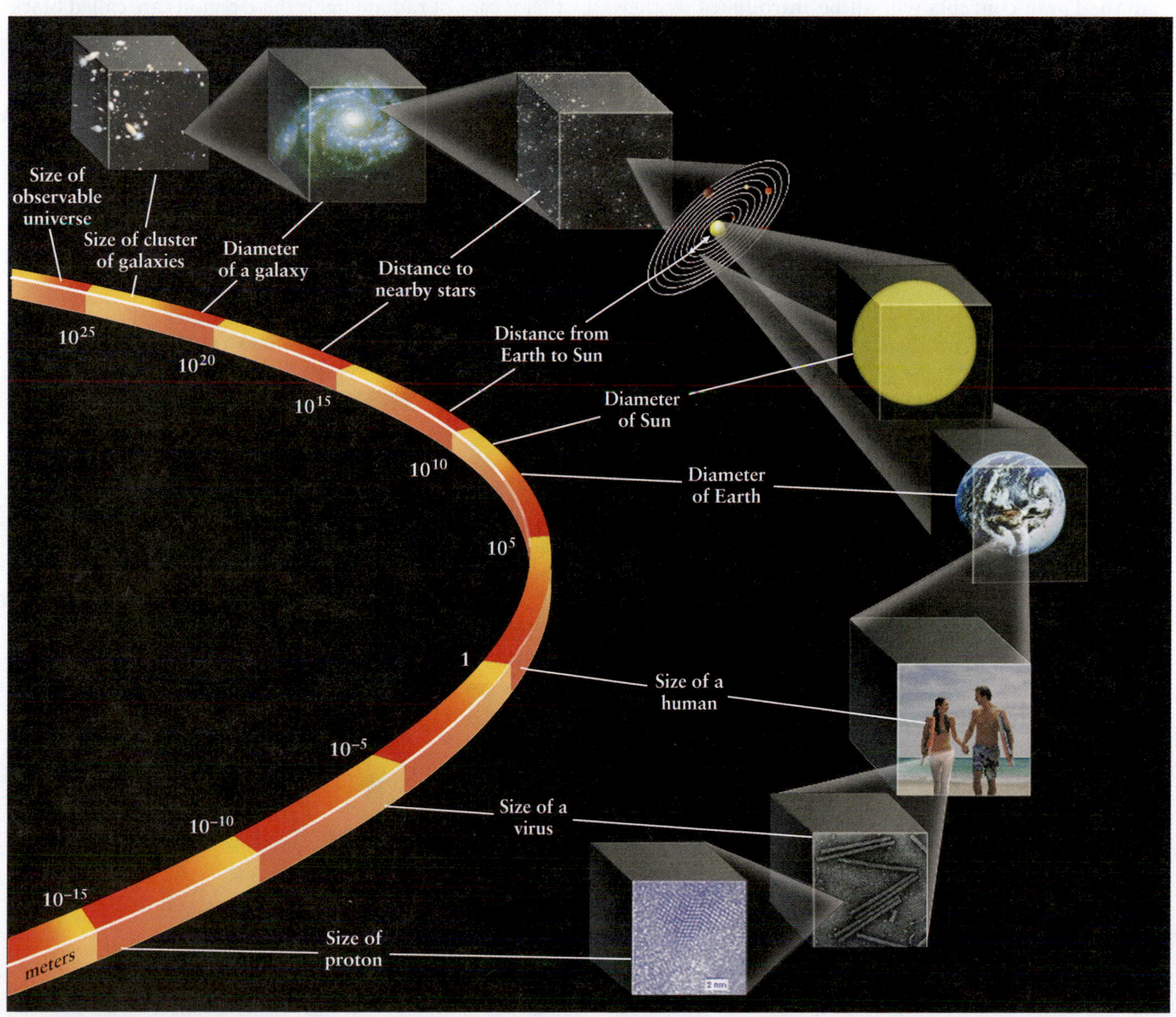

FIGURE 1-2 The Scales of the Universe This curve gives the sizes of objects in meters, ranging from subatomic particles at the bottom to the entire observable universe at the top. Every 0.5 cm up along the arc represents a factor of 10 larger. (Top to bottom: R. Williams and the Hubble Deep Field Team [STScI] and NASA; AAT; L. Golub, Naval Observatory, IBM Research, NASA; Richard Bickel/Corbis; Scientific American Books; Jose Luis Pelaez/Getty Images; Rothamsted Research Centre for Bioimaging; Courtesy of Florian Banhart/University of Mainz)

to us. Unlike linear intervals measured on a ruler, moving up 0.5×10^{-2} m (0.5 cm) along the arc of this figure brings you to objects 10 times larger. Because of this, going from the size of a proton (roughly 10^{-15} m) up to the size of an atom (roughly 10^{-10} m) takes about the same space along the arc as going from the distance between Earth and the Sun to the distance between Earth and the nearby stars.

This wide range of sizes underscores the fact that astronomy *synthesizes* or brings together information from many other fields of science. We will need to understand what atoms are composed of and how they behave; the nature and properties of light; the response of matter and energy to the force of gravity; the generation of energy by fusing particles together in stars; the ability of carbon—and carbon alone—to serve as the foundation of life; as well as other information. These concepts will all be introduced as they are needed.

What, then, have astronomers seen of the universe? Figure 1-3 presents examples of the types of objects we will explore in this text. An increasing number of planets like Jupiter, rich in hydrogen and helium (Figure 1-3a), as well as rocky planets not much larger than Earth, have been discovered orbiting other stars. Much smaller pieces of space debris—some of rock and metal called **asteroids** or **meteoroids** (Figure 1-3b), and others of rock and ice called **comets** (Figure 1-3c)—orbit the Sun (Figure 1-3d) and other stars. Vast stores of interstellar gas and dust are found in many galaxies; these "clouds" are often the incubators of new generations of stars (Figure 1-3e). Stars by the millions, billions, or even trillions are held together in galaxies by the force of gravity (Figure 1-3f). Galaxies like our own Milky Way often contain large amounts of that interstellar gas and dust, as well as regions of space where matter is so dense that it cannot radiate light; these regions are called **black holes** (Figure 1-3g). Groups of galaxies are held together

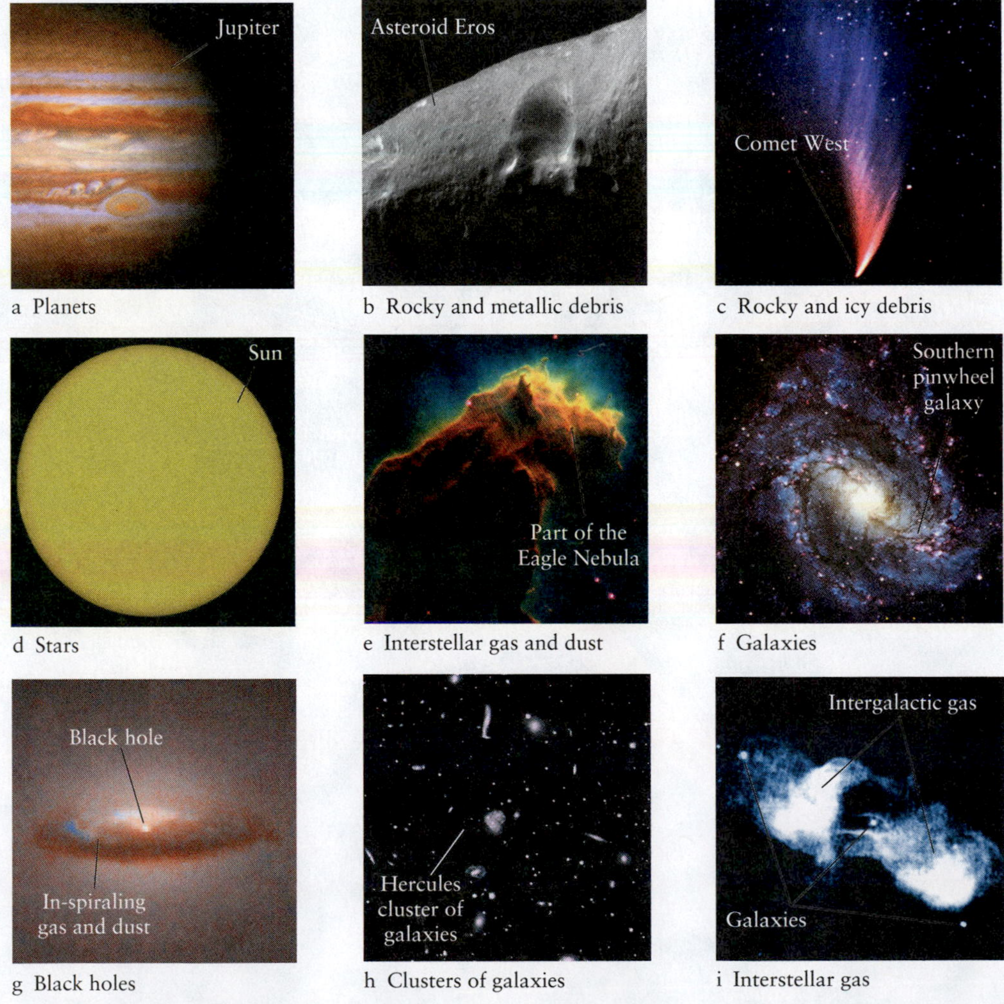

FIGURE 1-3 Inventory of the Universe Pictured here are examples of the major categories of objects that have been found throughout the universe. You will discover more about each type in the chapters that follow. (a: NASA/Hubblesite; b: NASA; c: Peter Stättmayer/European Southern Observatory; d: Big Bear Observatory; e: NASA/Jeff Hester & Paul Scowen; f: Anglo-Australian Observatory; g: NOAO; h: NASA; i: N. F. Comins and F. N. Owen/NRAO)

by gravity in clusters (Figure 1-3h), and clusters of galaxies are held together by gravity in superclusters. Huge quantities of intergalactic gas are often found between galaxies (Figure 1-3i).

Every object in astronomy is constantly changing—each has an origin, an active period you might consider as its "life," and each will have an end. In addition to examining the objects that fill the universe, we will also study the processes that cause them to change. After all is said and done, you will discover that all the matter and energy that astronomers have detected are but the tip of the cosmic iceberg—there is much more in the universe, but astronomers do not yet know its nature.

HI & LOIS © 1992 KING FEATURES SYNDICATE

PATTERNS OF STARS

When you gaze at the sky on a clear night where the air is free of pollution and there is not too much light from cities or other sources, there seem to be millions of stars twinkling overhead. In reality, the unaided human eye can detect only about 6000 stars over the entire sky. At any one time, you can see roughly 3000 stars in dark skies, because only half of the stars are above the *horizon*—the boundary between Earth and the sky. In very smoggy or light-polluted cities, you may see only a tenth of that number, or less (Figure 1-1).

In any event, you probably have noticed patterns, technically called *asterisms,* formed by bright stars, and you are familiar with some common names for these patterns, such as the ladle-shaped Big Dipper and broad-shouldered Orion. These recognizable patterns of stars are informally called *constellations* in everyday conversation, and they often have names derived from ancient legends (Figure 1-4a).

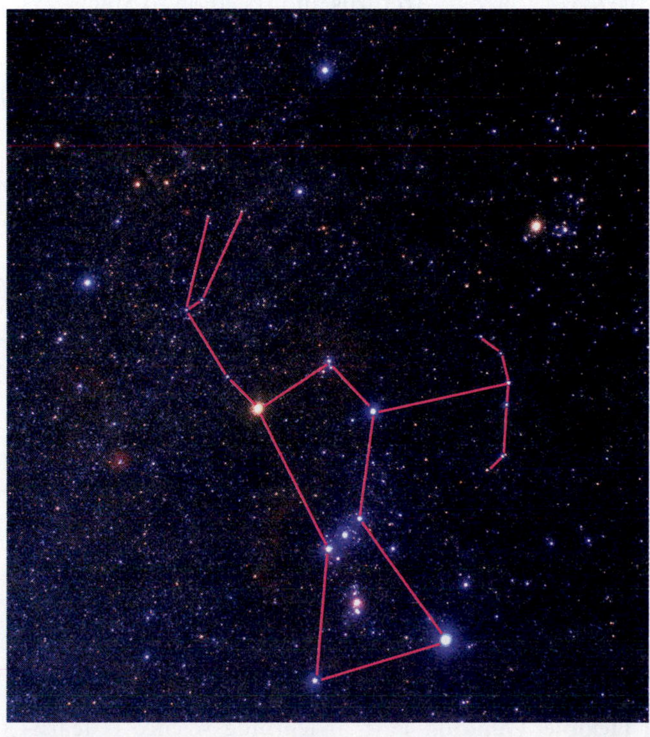

a

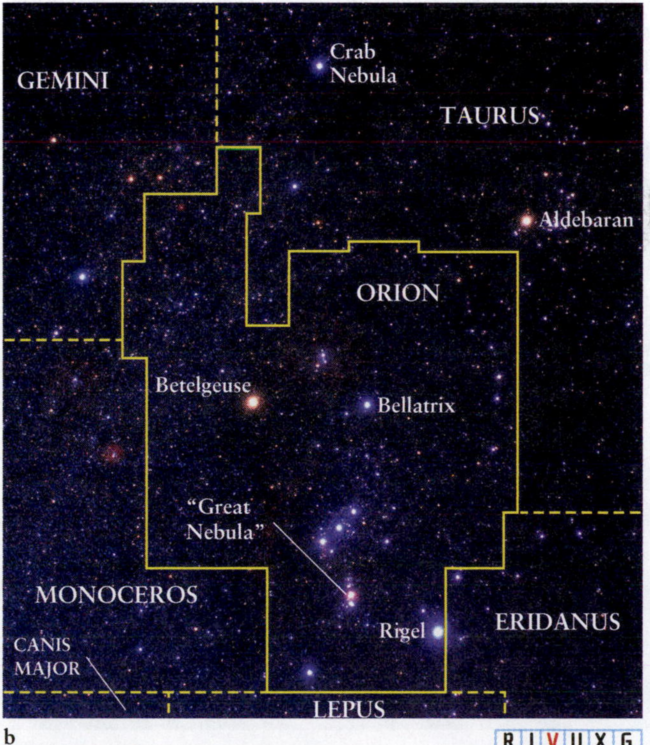

b

R I V U X G

FIGURE 1-4 The Constellation Orion (a) The pattern of stars (asterism) called Orion is prominent in the winter sky. From the northern hemisphere, it is easily seen high above the southern horizon from December through March. You can see in this photograph that the various stars have different colors—something to watch for when you observe the night sky. (b) Technically, constellations are entire regions of the sky. The constellation called Orion and parts of other nearby constellations are depicted in this photograph. All the stars inside the boundary of Orion are members of that constellation. The celestial sphere is covered by 88 constellations of differing sizes and shapes. (© 2004 Jerry Lodriguss/www.astropix.com)

1-2 Constellations make locating stars easy

1 You can orient yourself on Earth with the help of easily recognized constellations. For instance, if you live in the northern hemisphere, you can use the Big Dipper to find the direction north. To do this, locate the Big Dipper and imagine that its bowl is resting on a table (Figure 1-5). If you see the dipper upside down in the sky, as you frequently will, imagine the dipper resting on an upside-down table above it. Locate the two stars of the bowl farthest from the Big Dipper's handle. These are called the *pointer stars*. Draw a mental line through these stars in the direction away from the table, as shown in Figure 1-5. The first *moderately bright* star you then encounter is Polaris, also called the North Star because it is located almost directly over Earth's North Pole. So, while Polaris is not even among the 20 brightest stars (see Appendix E-6), it is easy to locate. Whenever you face Polaris, you are facing north. East is then on your right, south is behind you, and west is on your left. (There is no equivalent star over the South Pole.)

The Big Dipper also illustrates the fact that being familiar with just a few constellations makes it easy to locate stars and other constellations. The most effective way to do this is to use vivid visual connections, especially those of your own devising. For example, imagine gripping the handle of the Big Dipper and slamming its bowl straight down onto the head of Leo (the Lion). Leo comprises the first group of bright stars your dipper encounters. As shown in Figure 1-5, the brightest star in this group is Regulus, the dot of the backward question mark that traces the lion's mane. As another example, follow the arc of the Big Dipper's handle away from its bowl. The first bright star you encounter along that arc beyond the handle is Arcturus in Boötes (the Herdsman). Follow the same arc farther to the prominent bluish star Spica in Virgo (the Virgin). Spotting these stars and remembering their names is easy if you remember the saying "Arc to Arcturus and speed on to Spica."

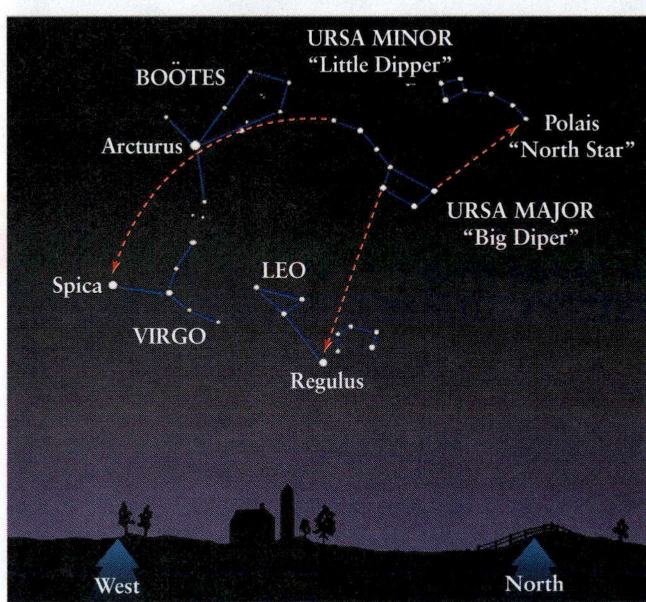

FIGURE 1-5 The Big Dipper as a Guide In the northern hemisphere, the Big Dipper is an easily recognized pattern of seven bright stars. This star chart shows how the Big Dipper can be used to locate the North Star as well as the brightest stars in three other constellations. While the Big Dipper appears right side up in this drawing of the sky shortly before sunrise, at other times of the night it appears upside down.

Insight Into Science

Compare What You Believe with What Nature and Science Show You If you thought Polaris is the brightest star in the night sky, how did you learn that? There is likely to be a lot of information you learned from a variety of sources that you will have to unlearn while learning astronomy. While this process is often uncomfortable, it is essential if you want to understand how science works and how scientists view the world. Be careful: It is human nature to change new information to fit our current beliefs rather than change what we believe. Fight that urge if science can show you that your beliefs are incorrect!

Different constellations are visible at night during different times of the year. During the winter months in the northern hemisphere, you can see some of the brightest stars in the sky. Many of them are in the vicinity of the "winter triangle," which connects bright stars in the constellations of Orion (the Hunter), Canis Major (the Larger Dog), and Canis Minor (the Smaller Dog), as shown in Figure 1-6. The winter triangle passes high in the sky at night during the middle of winter. It is easy to find Sirius, the brightest star in the night sky, by locating the belt of Orion and following a straight mental line from it to the left (as you face Orion). The first bright star that you encounter is Sirius.

The "summer triangle," which graces the summer sky as shown in Figure 1-7, connects the bright stars Vega in Lyra (the Lyre), Deneb in Cygnus (the Swan), and Altair in Aquila (the Eagle). A conspicuous portion of the Milky Way forms a beautiful background

Insight Into Science

Flexible Thinking Part of learning science is learning to look at things from different perspectives. For example, when learning to identify prominent asterisms, be sure to view them from different orientations (that is, with the star chart rotated at different angles), so that you can find them at different times of the night and the year.

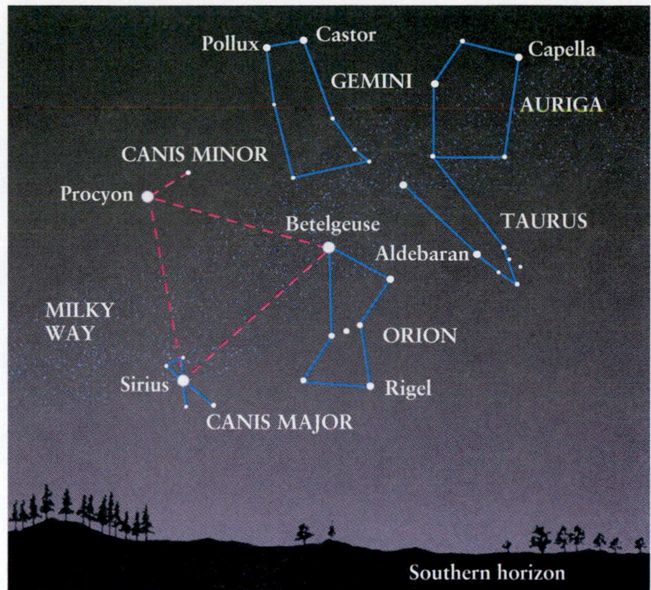

FIGURE 1-6 The Winter Triangle This star chart shows the southern sky as it appears during the evening in December. Three of the brightest stars in the sky make up the winter triangle. In addition to the constellations involved in the triangle, Gemini (the Twins), Auriga (the Charioteer), and Taurus (the Bull) are also shown.

for these constellations, which are nearly overhead during the middle of summer at midnight. For more on the constellations, see *Guided Discovery: The Stars and Constellations*.

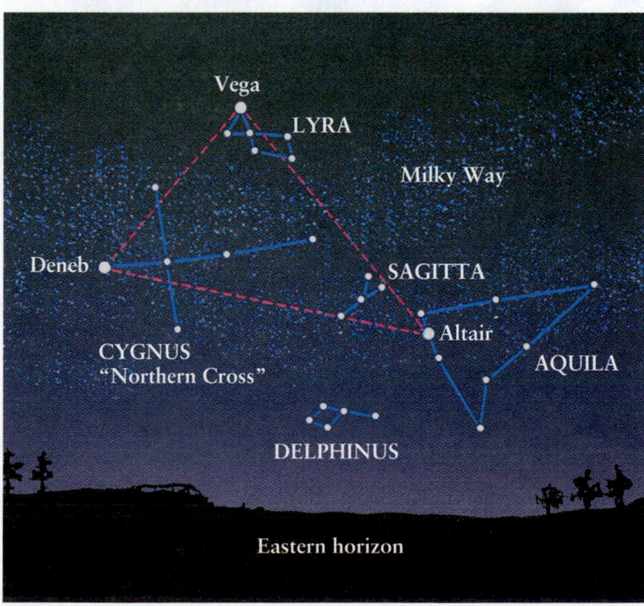

FIGURE 1-7 The Summer Triangle This star chart shows the northeastern sky as it appears in the evening in June. In addition to the three constellations involved in the summer triangle, the faint stars in the constellations Sagitta (the Arrow) and Delphinus (the Dolphin) are also shown.

Astronomers require more accuracy in locating dim objects than is possible simply by moving from constellation to constellation. They have therefore created a sky map, called the **celestial sphere,** and applied a coordinate system to it, analogous to the coordinate system of north-south latitude and east-west longitude used to navigate on Earth. If you know a star's celestial coordinates, you can locate it quickly. For such a sky map to be useful in finding stars, the stars must be fixed on it, just as cities are fixed on maps of Earth.

Referring to the first star map at the end of this book, create a story with which you can remember the connection between the constellations Sagittarius and Scorpius.

1-3 The celestial sphere aids in navigating the sky

ANIMATION 1.1

If you look at the night sky year after year, you will see that the stars do indeed appear fixed relative to one another. Furthermore, throughout each night the entire pattern of stars appears to rigidly orbit Earth. We employ this artificial, Earth-based view of the heavens to make celestial maps by pretending that the stars are attached to the inside of an enormous hollow shell, the celestial sphere, with Earth at its center (Figure 1-8). Visualized another way, you can imagine the half of the celestial sphere that is visible at night as a giant bowl covering Earth.

Students using a celestial sphere. (JupiterImages/Thinkstock/Alamy)

Whereas asterisms such as the Big Dipper are often called "constellations" in normal conversation, astronomers use the word **constellation** to describe an entire area of the celestial sphere and all the objects in it (see Figure 1-4b). The celestial sphere is divided into 88 constellations of differing sizes and shapes. (Keep in mind that most constellations and their asterisms—the recognizable star patterns—have the same name. For example, Orion is the name of a constellation, as well as the associated asterism, which sometimes complicates conversation.) The boundaries of the constellations are all straight lines that meet at right angles (see Figure 1-4b). Some constellations, like Ursa Major (the Large Bear), are very large, while others, like Sagitta (the Arrow), are relatively small. To describe a star's location, we might say "Albireo in the constellation Cygnus (the Swan)," much as we would refer to "Chicago in the state of Illinois," "Melbourne in the state of Victoria," or "Ottawa in the province of Ontario." 2

Guided Discovery

The Stars and Constellations

Many students take an introductory astronomy course expecting to be taught the familiar stars and constellations. Often, instructors just do not have time to cover this material. You can, however, learn them on your own in several ways.

First, use the two steps for memorizing the night sky discussed in the text:

1. Observe easily identified constellations.
2. Associate these constellations with nearby, lesser-known stars and constellations.

For example, you may not remember where the star Aldebaran in Taurus is in the sky, but if you remember that Orion is fighting Taurus, you can locate Aldebaran by following a line defined by the belt of Orion to the right (away from Sirius). The first bright star you encounter is Aldebaran.

Make your own connections between the constellations. Have fun while you're doing it. Chances are that you will remember the phrase "slam the Big Dipper's bowl downward to hit Leo the Lion on the head" rather than "the first bright group of stars directly below the Big Dipper is Leo."

Second, study print or software sky materials. A very efficient way to learn the constellations is to use the board and card game *Stellar 28* or the computer version called *Skytracker Constellation Game*. In addition, astronomy computer programs such as *Starry Night*™ are available. *Starry Night*™ shows many things in addition to what constellations are up at night, including the motion, location, and phases of the planets; the location of deep sky objects, such as nebulae; and the sky as seen from any location on Earth on any date.

A third way to familiarize yourself with the night sky is to go out on clear nights armed with star charts to see which constellations are up. You will find a set of such charts from the *Griffith Observer* magazine at the back of this book. To use the charts, first select the one that best corresponds to the date and time of your observation. Hold the chart vertically and turn it so that the direction you are facing outdoors shows at the bottom. The patterns and connections between stars and constellations will soon become evident. Using a flashlight with a red plastic coating over the light will make it easier to read the chart without ruining your night vision. You will find stargazing a truly enjoyable experience.

The stars seem fixed on the celestial sphere only because of their remoteness. In reality, they are at widely varying distances from Earth, and they do move relative to one another. But we neither see their motion nor perceive their relative distances because the stars are so far from us. You can understand this by imagining a jet plane just 1 kilometer overhead traveling at 1000 km (620 mi) per hour across the sky. Its motion is unmistakable. However, a plane moving at the same speed and altitude, but appearing along the horizon, seems to be moving about a hundred times more slowly. And an object that is at the distance of the Sun, traveling at the same speed and moving across the sky, would appear to be going nearly 100 million times more slowly than the plane overhead.

The stars (other than the Sun) are all more than 40 trillion km (25 trillion mi) from us. Therefore, although the patterns of stars in the sky do change, their great distances prevent us from seeing those changes over the course of a human lifetime. Thus, as unrealistic as it is, the celestial sphere is so useful for navigating the heavens that it is used by astronomers even at the most sophisticated observatories around the world.

As shown in Figure 1-8, we can project key geographic features from Earth out into space to establish directions and bearings. If we expand Earth's equator

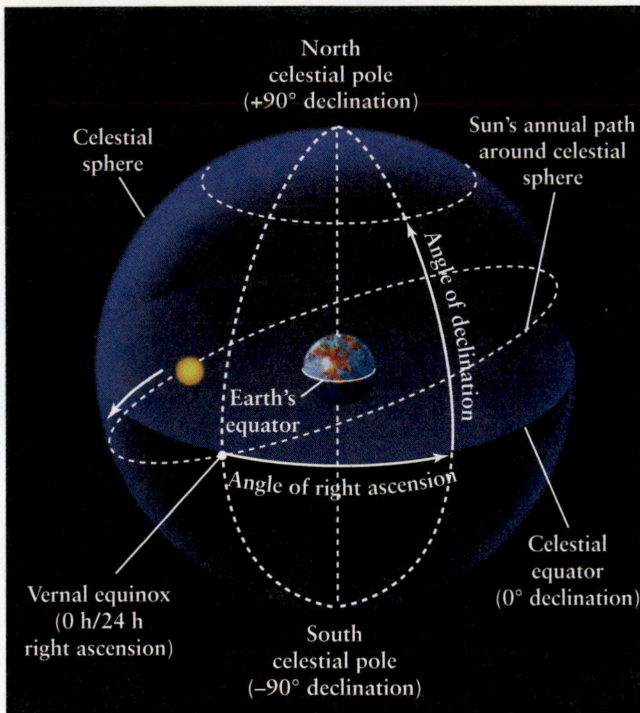

FIGURE 1-8 The Celestial Sphere The celestial sphere is the apparent "bowl" or hollow sphere of the sky. The celestial equator and poles are projections of Earth's equator and axis of rotation onto the celestial sphere. The north celestial pole is therefore located directly over Earth's North Pole, while the south celestial pole is directly above Earth's South Pole. Analogous to longitude and latitude, the coordinates in space are right ascension (R.A.) and declination (Decl.), respectively. The star in this figure has the indicated R.A. and Decl.

onto the celestial sphere, we obtain the **celestial equator,** which divides the sky into northern and southern hemispheres, just as Earth's equator divides Earth into two hemispheres. We can also imagine projecting Earth's North Pole and South Pole out into space along Earth's axis of rotation. Doing so gives us the **north celestial pole** and the **south celestial pole,** also shown in Figure 1-8.

Using the celestial equator and poles as reference features, astronomers divide up the surface of the celestial sphere in precisely the same way that the latitude and longitude grid divides Earth. The equivalent to latitude on Earth is called **declination (Decl.)** on the celestial sphere. It is measured from 0° to 90° north or south of the celestial equator. The equivalent of longitude on Earth is called **right ascension (R.A.)** on the celestial sphere, measured from 0 h to 24 h around the celestial equator (see Figure 1-8). The boundaries of the constellations, introduced above, run along lines of constant right ascension and constant declination.

Just as the location of Greenwich, England, defines the *prime meridian,* or zero of longitude on Earth, we need to establish a zero of right ascension. This zero value is defined as one of the places where the Sun's annual path across the celestial sphere intersects the celestial equator. (We will explore later in this chapter why the Sun appears to move in a circle around the celestial sphere during the course of a year.) The celestial equator and the Sun's path intersect at two points. The equivalent on the celestial sphere of Earth's prime meridian is where the Sun crosses the celestial equator moving northward. Angles of right ascension are measured from this point, called the **vernal equinox** (Figure 1-8).

> The planets move through constellations, while the stars do not. What two possible properties of the planets could allow them to do that?

In navigating on the celestial sphere, astronomers measure the distance between objects in terms of angles. If you are not familiar with angular separation, read *An Astronomer's Toolbox 1-1*.

EARTHLY CYCLES

The Sun systematically rises and sets at different times and in different places throughout the year. Similarly, the Moon rises and sets at different times each day and repeats its cycle roughly once every 29½ days. Furthermore, as noted in Section 1-2, we do not see the same constellations up in the sky every night of the year, but the cycle of constellations we can see at night repeats each year. The daily and annual rhythms of the sky, Earth, and all life on it arise from three celestial motions: Earth's spinning, which causes day and night, as well as causing the apparent daily motion of the celestial sphere and all the objects on it; Earth's orbit around the Sun, which creates the seasons, the year, and the change in times at which constellations are up at night; and the Moon's orbit around Earth, which creates the lunar phases, the cycle of tides, and the spectacular phenomena we call eclipses.

1-4 Earth's rotation creates the day-night cycle and its revolution defines a year

Earth spins on its axis. Such motion is called **rotation.** We do not feel Earth's rotation because our planet is so physically large compared to us that Earth's gravitational attraction holds us firmly on its surface. Earth's rotation causes the stars—as well as the Sun, the Moon, and planets—to appear to rise on the eastern horizon, move across the sky, and set on the western horizon. Earth's daily rotation, causing the Sun to rise and set, thereby creates day and night. The **diurnal motion,** or daily motion, of the celestial bodies is apparent in time-exposure photographs, such as that shown in Figure 1-9 and the one opening this chapter.

Take a friend outside on a clear, warm night to observe the diurnal motion of the stars for yourselves. Soon after dark, find a spot away from bright lights, and

Observational Measurements Using Angles

ANIMATION 1.3

Ancient mathematicians invented a system of angles and angular measure that is still used today to denote the relative positions and apparent sizes of objects in the sky. To locate stars on the celestial sphere, for example, we do not need to know their distances from Earth. All we need to know is the angle from one star to another in the sky, a property that remains fixed over our lifetimes because the stars are all so far away.

An **arc angle,** often just called an **angle,** is the opening between two lines that meet at a point. Angular measure is a method of describing the size of an angle. The basic unit of angular measure is the **degree,** designated by the symbol°. A full circle is divided into 360°. A right angle measures 90°. As shown in the figure below, the angle between the two "pointer stars" in the Big Dipper is about 5°.

Astronomers also use angular measure to describe the apparent sizes of celestial objects. For example, imagine the full Moon. As seen from Earth, the angle across the Moon's diameter is nearly ½°. We therefore say that the **angular diameter,** or **angular size,** of the Moon is ½°. Alternatively, astronomers say that the Moon "subtends" an angle of ½°. In this context, subtend means "to extend across."

An adult human hand held up at arm's length provides a means of estimating angles. For example, five fingers extended at arm's length cover an angle of some 8–10°, whereas the tip of your finger is about 1½° wide. Various segments of your index finger extended to arm's length can be similarly used to estimate angles a few degrees across, as shown in the figure above on the right.

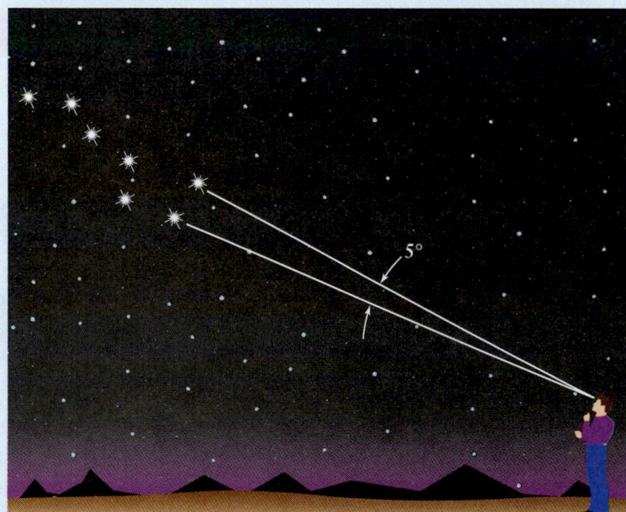

The Big Dipper The angular distance between the two "pointer stars" at the front of the Big Dipper is about 5°. For comparison, the angular diameter of the Moon is about ½°.

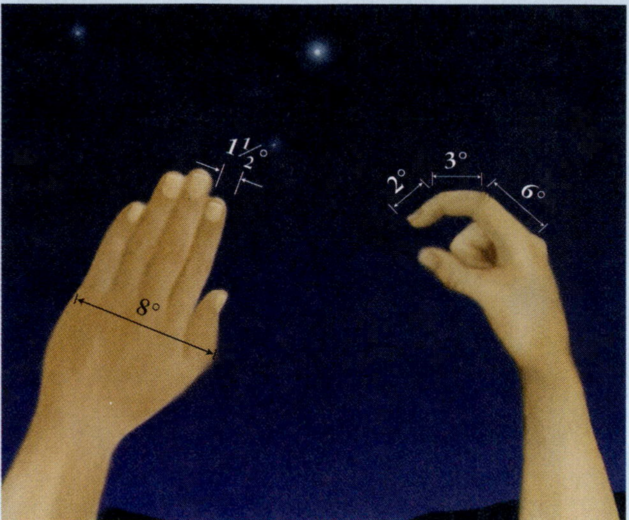

Estimating Angles with the Human Hand Various parts of the adult human hand extended to arm's length can be used to estimate angular distances and sizes in the sky. (These angles vary slightly from person to person because hands and arms are different sizes.)

To talk about smaller angles, we subdivide the degree into 60 arcminutes (abbreviated 60 arcmin or 60′). An arcminute is further subdivided into 60 arcseconds (abbreviated 60 arcsec or 60″). A dime viewed face-on from a distance of 1.6 km (1 mile) has an angular diameter of about 2″.

From everyday experience, we know that an object looks bigger when it is nearby than when it is far away. The angular size of an object therefore does not necessarily tell you anything about its actual physical size. For example, the fact that the Moon's angular diameter is ½° does not tell you how big the Moon really is. But if you also happen to know the distance to the Moon, then you can calculate the Moon's physical diameter. In general, the physical diameter of an object can be calculated from the equation:

$$\text{physical diameter} = \text{distance} \times \tan(\text{angular diameter})$$

where tan (angular diameter) means the tangent of the angle denoted "angular diameter." In the Moon's case, using a measured distance (see Appendix E-3) of 384,400 km and an angular diameter of ½°, we find the diameter to be roughly 3350 km. The difference between this value and the exact diameter of 3476 km is due primarily to the approximate value of ½° that we have used.

Try these questions: The Sun is 1.5×10^8 km away and has a diameter of 1.4×10^6 km. How large an angle does it make in our sky? How does that compare to the arc angle of the Moon in our sky? What arc angle would the Moon make in our sky if it were twice as far away? Half as far?

(Answers appear at the end of the book.)

FIGURE 1-9 Circumpolar Star Trails This long exposure, taken from Australia's Siding Spring Mountain and aimed at the south celestial pole, shows the rotation of the sky. The stars that pass between the pole and the ground are all circumpolar stars. Note the lack of a bright, short star arc near the celestial pole compared to that of Polaris in the chapter opening photograph. (Anglo-Australian Observatory/David Malin Images)

note the constellations in the sky relative to some prominent landmarks near you on Earth. A few hours later, check again from the same place. You will find that the entire pattern of stars (as well any visible planets and the Moon, if it is up) has shifted. New constellations will have risen above the eastern horizon, while other constellations will have disappeared below the western horizon. If you check again just before dawn, you will find the stars that were just rising in the east when the night began are now low in the western sky.

Different constellations are visible at night during different times of the year. This occurs because Earth orbits, or revolves, around the Sun. **Revolution** is the motion of any astronomical object around another astronomical object. Earth takes 1 year, or about 365¼ days, to travel once around the Sun. A year on Earth is measured by the motion of our planet relative to the stars. For example, draw a straight line from the Sun through Earth to some star on the opposite side of Earth from the Sun. As Earth revolves, that line inscribes a straight path on the celestial sphere and returns to the original star 365¼ days later. The length of any cycle of motion, such as Earth's orbit around the Sun, that is measured with respect to the stars is called a **sidereal period.**

If Earth were spinning over a fixed place on the Sun, rather than revolving around it, then every star would rise and set at the same times throughout the year. As a result of Earth's motion around the Sun, however, the stars rise approximately 4 min earlier each day than they did the day (or night) before. This effect accumulates, bringing different constellations up at night throughout the year. Figure 1-10 summarizes this motion. When the Sun is within the boundaries of (colloquially, "in") Virgo (September 18–November 1), for example, the hemisphere containing the Sun and the constellations around Virgo are in daylight (see Figure 1-10a). When the Sun is

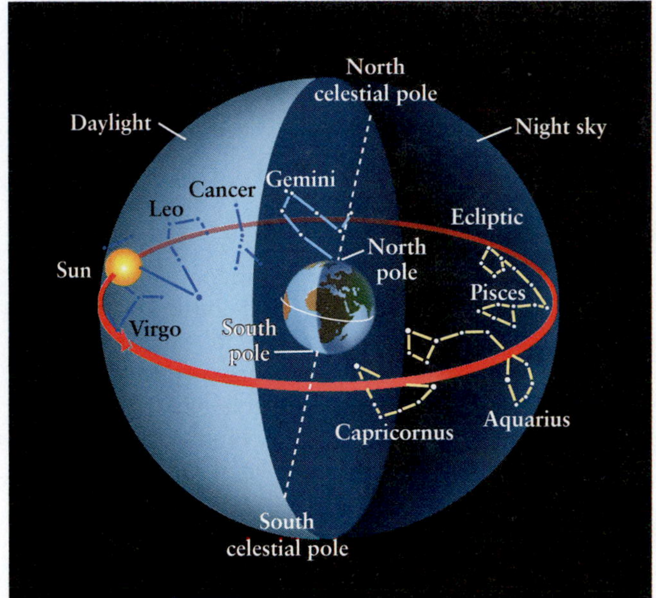

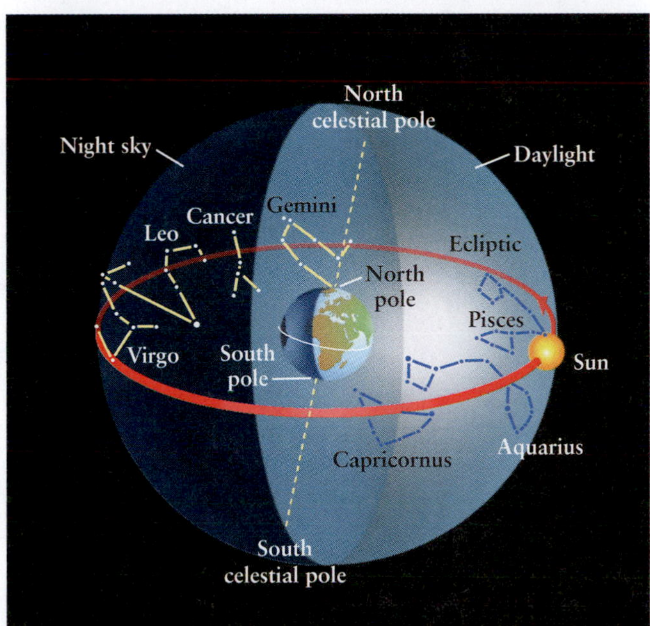

FIGURE 1-10 Why Different Constellations Are Visible at Different Times of the Year (a) On the autumnal equinox each year, the Sun is in the constellation Virgo. As seen from Earth, that part of the sky is then daylight and we see stars only on the other half of the sky, centered around the constellation Pisces. (b) Six months later, the Sun is in Pisces. This side of the sky is then bright, while the side centered on Virgo is in darkness.

Why are asterisms like the Big Dipper sometimes seen upright and sometimes upside down?

up, so are Virgo and the surrounding constellations, so we cannot see them. During that time of year, the constellations on the other side of the celestial sphere, centered on the constellation Pisces, are in darkness. Thus, when the Sun is "in" Virgo, Pisces and the constellations around it are high in our sky at night.

Six months later, when the Sun is "in" Pisces, that half of the sky is filled with daylight, while Virgo and the constellations around it are high in the night sky (see Figure 1-10b). These arguments apply everywhere on Earth at the same time because the Sun moves along its path very slowly as seen from Earth, taking a year to make one complete circuit.

We spoke earlier of stars rising on the eastern horizon and setting on the western horizon. Depending on your latitude, some of the stars and constellations never disappear below the horizon. Instead, they trace complete circles in the sky over the course of each night (see Figure 1-9). To understand why this happens, imagine that you are standing on Earth's North Pole at night. Looking straight up, you see Polaris. Because Earth is spinning around its axis directly under your feet, all the stars appear to move from left to right (counterclockwise) in horizontal rings above you. The exception is Polaris, which always remains at the North Pole's **zenith.** (Every place has a different zenith, which is the point directly overhead anywhere on Earth.) As seen from the North Pole, no stars rise or set (Figure 1-11). They just seem to revolve around Polaris in horizontal circles. Stars and constellations that never go below the horizon are called **circumpolar**. While there is no bright South Pole star equivalent to Polaris, all stars seen from the South Pole are also circumpolar and move from right to left (clockwise).

If you live in the northern hemisphere, Polaris is always located above your northern horizon at an angle equal to your latitude. Only the stars and constellations that pass between Polaris and the land directly below it are circumpolar. As you go farther south in the northern hemisphere, the number of stars and constellations that are circumpolar decreases. Likewise, as you go farther north in the southern hemisphere, the number of stars and constellations that are circumpolar decreases.

Now visualize yourself at the equator. All the stars appear to rise straight up in the eastern sky and set straight down in the western sky (Figure 1-12). Polaris is barely visible on the northern horizon. While Polaris never sets, all the other stars do, and therefore none of the stars are circumpolar as seen from the equator.

As you can see from these last two mental exercises, the angle at which the stars rise and set depends on your viewing latitude. Figure 1-13 shows stars setting at 35° north latitude. Polaris is fixed at 35° above the horizon to the right of this figure, not at the zenith, as it is at the North Pole, nor on the horizon, as seen from the equator. As another example, except for those stars in the corners, all the stars whose paths are shown in Figure 1-9 are circumpolar because they are visible all night, every night.

FIGURE 1-11 Motion of Stars at the Poles Because Earth rotates around the axis through its poles, stars seen from these locations appear to move in huge, horizontal circles. This is the same effect you would get by standing up in a room and spinning around; everything would appear to move in circles around you. At the North Pole, stars move left to right, while at the South Pole they move right to left.

FIGURE 1-12 Rising and Setting of Stars at the Equator Standing on the equator, you are perpendicular to the axis around which Earth rotates. As seen from there, the stars rise straight up on the eastern horizon and set straight down on the western horizon. This is the same effect you get when driving straight over the crest of a hill; the objects on the other side of the hill appear to move straight upward as you descend.

FIGURE 1-13 Rising and Setting of Stars at Middle North Latitudes Unlike the motion of the stars at the poles (see Figure 1-11), the stars at all other latitudes do change angle above the ground throughout the night. This time-lapse photograph shows stars setting. The latitude determines the angle at which the stars rise and set. (David Miller/DMI)

1-5 The seasons result from the tilt of Earth's rotation axis combined with Earth's revolution around the Sun

Imagine that you could see the stars even during the day, so that you could follow the Sun's apparent motion against the background constellations throughout the year. (The Sun appears to move among the stars because Earth orbits around it.) From day to day, the Sun traces a straight path on the celestial sphere. This path is called the **ecliptic**. As you can see in Figure 1-14a, the ecliptic makes a closed circle bisecting the celestial sphere. The ecliptic is precisely the loop labeled "Sun's annual path around celestial sphere" in Figure 1-8.

The term "ecliptic" has a second use in astronomy. Earth orbits the Sun in a plane also called the ecliptic. The two ecliptics exactly coincide: Imagine yourself on the Sun watching Earth move day by day. The path of Earth on the celestial sphere as seen from the Sun is precisely the same as the path of the Sun as seen from Earth (Figure 1-14b).

Insight Into Science

Define Your Terms As with interpersonal communication, using the correct words is very important in science. Usually, scientific words each have a specific meaning, such as "rotation" denoting spin and "revolution" meaning one object orbiting another. Be especially careful to understand the context in which words with more than one meaning, such as "ecliptic" and "constellation," are used.

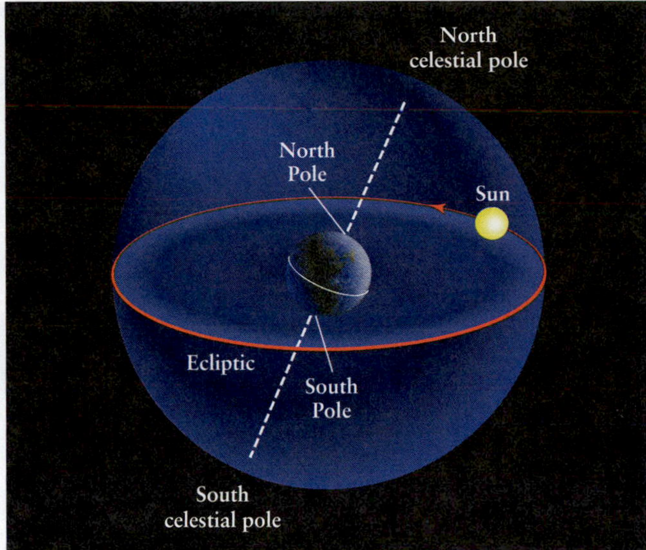

a

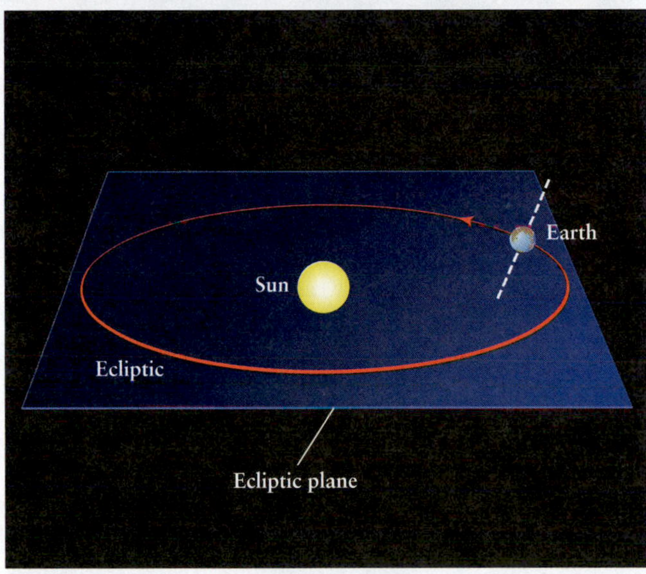

b

FIGURE 1-14 The Ecliptic (a) The ecliptic is the apparent annual path of the Sun on the celestial sphere. (b) The ecliptic is also the plane described by Earth's path around the Sun. The planes created by the two ecliptics exactly coincide. As in (a), the rotation axis of Earth is shown here tilted 23½° from being perpendicular to the ecliptic.

Recall also the discussion of the line running from the Sun through Earth to the celestial sphere in Section 1-4.

Equinoxes and Solstices The ecliptic and the celestial equator are different circles tilted 23½° with respect to each other on the celestial sphere. This occurs because Earth's rotation axis is tilted 23½° away from a line perpendicular to the ecliptic (Figures 1-14 and 1-15). These two circles intersect at only two points, which are exactly opposite each other on the celestial sphere (Figure 1-16). Each of these two points is called an **equinox** (from the Latin words meaning "equal night"),

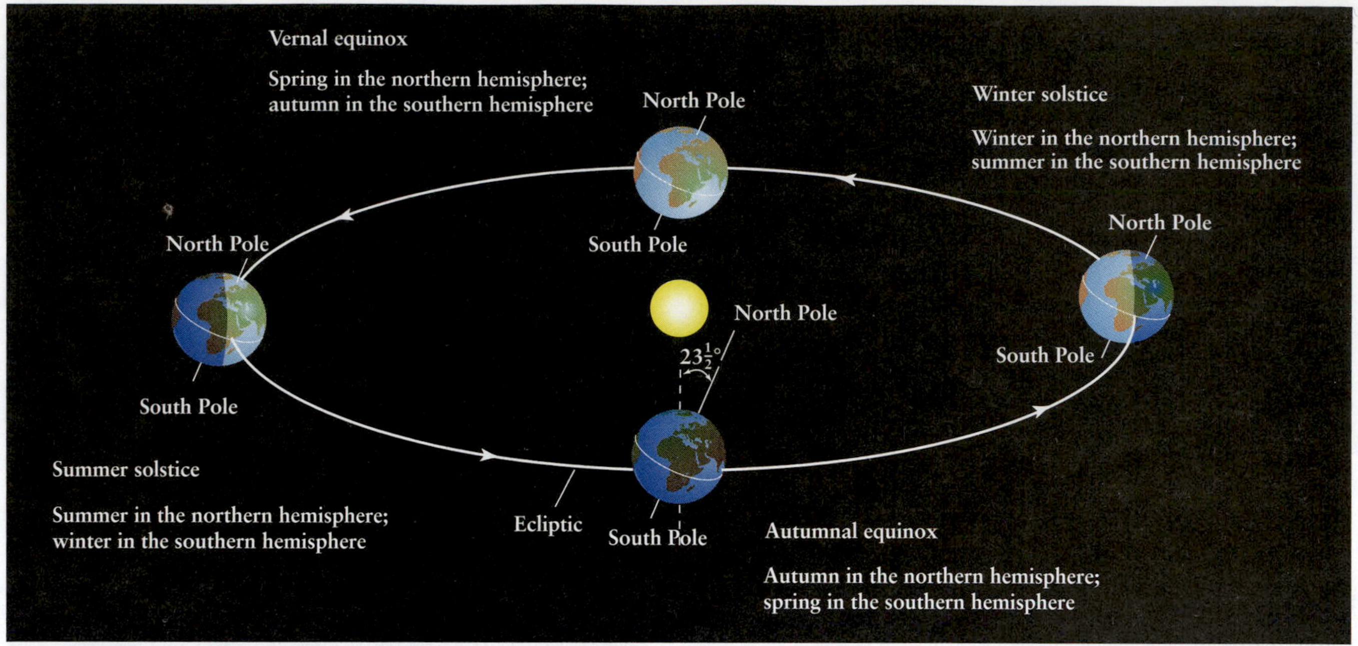

FIGURE 1-15 The Tilt of Earth's Axis Earth's axis of rotation is tilted 23½° from being perpendicular to the plane of Earth's orbit. Earth maintains this orientation (with its North Pole aimed at the north celestial pole near the star Polaris) throughout the year as it orbits the Sun. Consequently, the amount of solar illumination and the number of daylight hours at any location on Earth vary in a regular fashion with the seasons.

because when the Sun appears at either point, it is directly over Earth's equator, resulting in 12 h of daytime and 12 h of nighttime everywhere on Earth on that day.

Except for tiny changes each year, Earth maintains this tilted orientation as it orbits the Sun. Therefore,

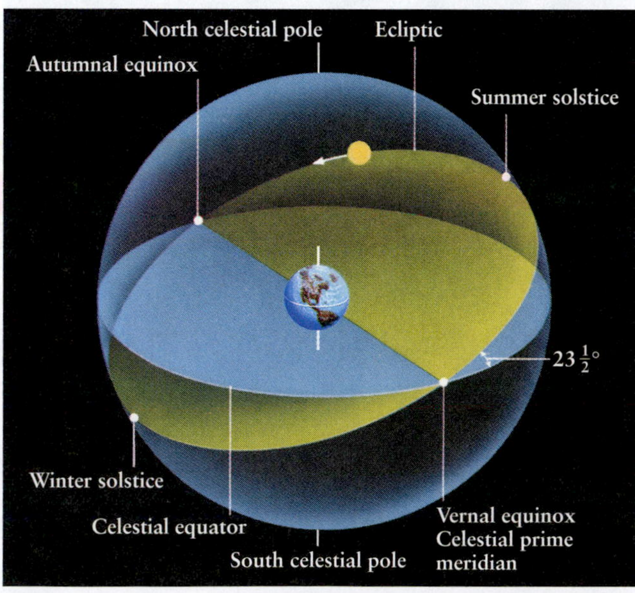

FIGURE 1-16 The Seasons Are Linked to Equinoxes and Solstices The ecliptic is inclined to the celestial equator by 23½° because of the tilt of Earth's axis of rotation. The ecliptic and the celestial equator intersect at two points called the equinoxes. The northernmost point on the ecliptic is called the summer solstice; the southernmost point is called the winter solstice.

Polaris is above the North Pole throughout the year. For half the year, the northern hemisphere is tilted toward the Sun, and as a result, the Sun rises higher in the northern hemisphere's sky than it does during the other half of the year (Figure 1-17). Equivalently, when the southern hemisphere is tilted toward the Sun, the Sun rises higher in the southern hemisphere's sky.

Consider the location of the Sun throughout the year as seen from the northern hemisphere. The day that the Sun rises farthest south of east is around December 22 each year (Figure 1-17) and is called the **winter solstice.** The winter solstice is the point on the ecliptic farthest south of the celestial equator (see Figure 1-16). It is also the day when the Sun rises to the lowest height at noon (Figure 1-17a), and it signals the day of the year in the northern hemisphere with the fewest number of daylight hours.

As the Sun moves along the ecliptic after the winter solstice, it rises earlier, is more northerly on the eastern horizon, and it passes higher in the sky at midday than it did on preceding days. Three months later, around March 20, the Sun crosses the celestial equator heading northward. This is called the vernal equinox and is one of the two days on which the Sun rises due east and sets due west (Figure 1-17b). The vernal equinox is the "prime meridian" of the celestial sphere, as discussed earlier. Three months after the vernal equinox, around June 21, the Sun rises farthest north of east and passes highest in the sky (Figure 1-17c). This is the **summer solstice** (see Figure 1-16), the day of the year in the northern hemisphere with the most daylight.

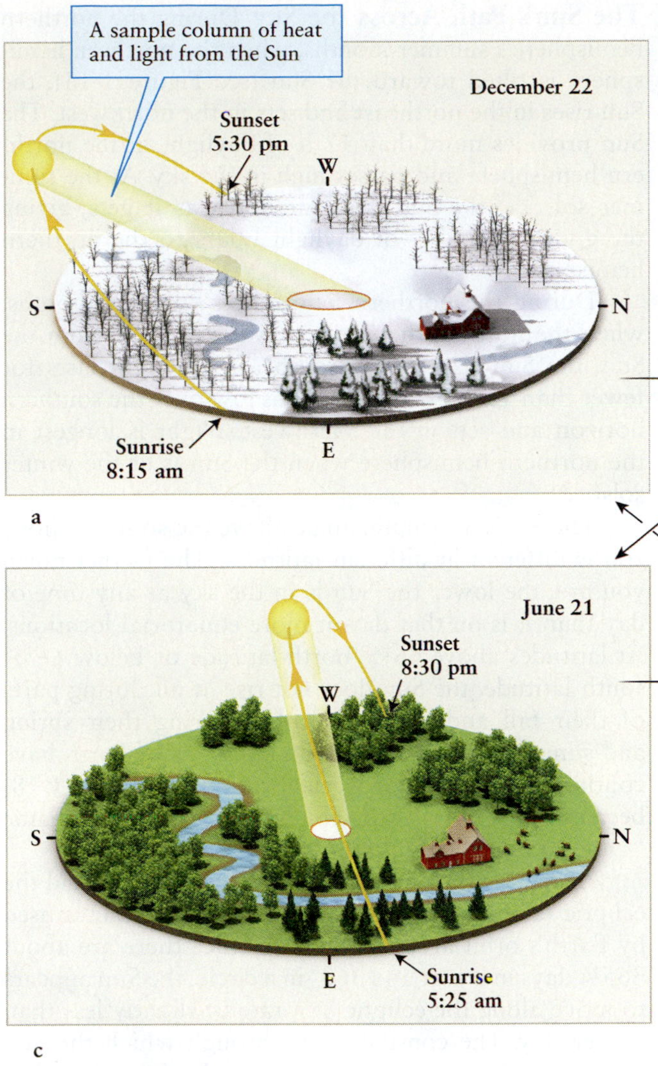

FIGURE 1-17 The Sun's Daily Path and the Energy It Deposits Here (a) On the winter solstice—the first day of winter—the Sun rises farthest south of east, is lowest in the noontime sky, stays up the shortest time, and its light and heat are least intense (most spread out) of any day of the year in the northern hemisphere. (b) On the vernal equinox—the first day of spring—the Sun rises precisely in the east and sets precisely in the west. Its light and heat have been growing more intense, as shown by the brighter oval of light than in (a). (c) On the summer solstice—the first day of summer—the Sun rises farthest north of east of any day in the year, is highest in the noontime sky, stays up the longest time, and its light and heat are most intense of any day in the northern hemisphere. (d) On the autumnal equinox, the same astronomical conditions exist as on the vernal equinox.

From June 21 through December 21, the Sun rises farther south than it did the preceding day. Its highest point in the sky is lower each succeeding day—the cycle of the previous 6 months reverses. The **autumnal equinox** occurs around September 22 (Figure 1-17d), with the Sun heading southward across the celestial equator, as seen from Earth.

The Seasons The higher the Sun rises during the day, the more daylight hours there are. During the days with longer periods of daylight, more light and heat from the Sun strike that hemisphere. Furthermore, when the Sun is higher in the sky, its energy is more concentrated on Earth's surface (see the footprint that the "cylinder of light" from the Sun makes in Figure 1-17a–d). Thus, during these days more energy is deposited on each square meter of the surface, thereby warming the surface more than when the Sun is lower in the sky. *The temperature and, hence, the seasons are determined by the duration of daylight at any place and the height of the Sun in the sky there.* (Bear in mind that winds and clouds greatly affect the weather throughout the year—we ignore these effects here.)

Explain why Figure 1-13 must have been taken facing west. *Hint*: Examine Figure 1-17.

To summarize, the Sun is lowest in the northern 
sky on the winter solstice. This marks the beginning of

winter in the northern hemisphere. As the Sun moves northward, the amount of daylight and heat deposited increases daily. The vernal equinox marks a midpoint in the amount of light and heat from the Sun onto the northern hemisphere and is the beginning of spring. When the Sun reaches the summer solstice, it is highest in the northern sky and is above the horizon for the most hours of any day of the year. This is the beginning of summer. Returning southward, the Sun crosses the celestial equator once again on the autumnal equinox, the beginning of fall.

4 While the distance between Earth and the Sun changes by 5 million km (3 million mi) throughout the year, this variation has only a minor effect on the seasons. The variation in Earth's distance from the Sun over the year would have a greater effect if it were not for the fact that the southern hemisphere has more area covered by oceans than does the northern hemisphere. As a result, when Earth is closer to the Sun (and the Sun is high in the southern hemisphere's sky), the southern oceans scatter more light and heat directly back into space than occurs when the Sun is higher over the northern hemisphere during the other half of the year. Had the extra energy sent back into space when we are closer to the Sun been absorbed by our planet, Earth would indeed heat more during this time than when the Sun is over the northern hemisphere.

WHAT IF... Earth's changing distance from the Sun caused the seasons (and how do we know it doesn't)?
Earth's orbit around the Sun is elliptical (we will discuss this oval shape in detail in Chapter 2). If the seasons *were* caused by the changing distance between Earth and the Sun, all parts of Earth should have the same seasons at the same time. In fact, the northern and southern hemispheres have exactly opposite seasons. Furthermore, Earth is closest to the Sun on or around January 3 of each year—the dead of winter in the northern hemisphere!

Insight Into Science

Expect the Unexpected
The process of science requires that we question the obvious, that is, what we think we know. Many phenomena in the universe defy commonsense explanations. The fact that the changing distance from Earth to the Sun has a minimal effect on the seasons is an excellent example.

The Sun's Path Across the Sky During the northern hemisphere's summer months, when the northern hemisphere is tilted toward the Sun (see Figure 1-15), the Sun rises in the northeast and sets in the northwest. The Sun provides more than 12 h of daylight in the northern hemisphere and passes high in the sky. At the summer solstice, the Sun is as far north as it gets, giving the greatest number of daylight hours to the northern hemisphere.

During the northern hemisphere's winter months, when the northern hemisphere is tilted away from the Sun, the Sun rises in the southeast. Daylight lasts for fewer than 12 h, as the Sun skims low over the southern horizon and sets in the southwest. Night is longest in the northern hemisphere when the Sun is at the winter solstice.

The Sun's maximum angle above the southern horizon is different at different latitudes. The farther north you are, the lower the Sun is in the sky at any time of day than it is on that day at more equatorial locations. At latitudes above 66½° north latitude or below 66½° south latitude, the Sun does not rise at all during parts of their fall and winter months. During their spring and summer months, those same regions of Earth have continuous sunlight for weeks or months (Figure 1-18) because the Sun is then circumpolar; hence the name "Land of the Midnight Sun."

The Sun takes 1 year to complete a trip around the 5
ecliptic (as noted above, this motion is actually caused by Earth's orbit around the Sun). Since there are about 365¼ days in a year and 360° in a circle, the Sun appears to move along the ecliptic at a rate of slightly less than 1° per day. The constellations through which the Sun moves throughout the year as it travels along the ecliptic are called the **zodiac** constellations. We cannot see

FIGURE 1-18 The Midnight Sun This time-lapse photograph was taken on July 19, 1985, at 69° north latitude in northeastern Alaska. At that latitude, the Sun is above the horizon continuously from mid-May until the end of July. (Doug Plummer/Science Photo Library)

Table 1-1 The 13 Constellations of the Zodiac

Constellation	Dates of the Sun's passage through each
Pisces	March 13–April 20
Aries	April 20–May 13
Taurus	May 13–June 21
Gemini	June 21–July 20
Cancer	July 20–August 11
Leo	August 11–September 18
Virgo	September 18–November 1
Libra	November 1–November 22
Scorpius	November 22–December 1
Ophiuchus	December 1–December 19
Sagittarius	December 19–January 19
Capricorn	January 19–February 18
Aquarius	February 18–March 13

the stars of these constellations when the Sun is among them, of course, but we can plot the Sun's path on the celestial sphere to determine through which constellations it moves. Traditionally, there were 12 zodiac constellations whose borders were set in antiquity. In 1930, the boundaries were redefined by astronomers, and the Sun now moves through 13 constellations throughout the year. (The thirteenth zodiac constellation is Ophiuchus, the Serpent Holder. The Sun passes through Ophiuchus from December 1 to December 19 each year.) Table 1-1 lists all the zodiac constellations and the dates the Sun passes through them. You may not have the "sign" that you think you have.

1-6 Clock times based on the Sun's location created scheduling nightmares

The Sun's daily motion through the sky provided our distant ancestors' earliest reference for time because the Sun's location determines whether it is day or night and roughly whether it is before or after midday. The Sun's motion through the sky came to determine the length of the **solar day**, upon which our 24-h day is based. Ideally, this is the interval of time between when the Sun is highest in the sky on one day until the time it is highest in the sky on the next day. However, the length of the solar day actually varies throughout the year. This occurs because Earth's orbit around the Sun is not perfectly circular—our planet's speed along the ecliptic increases as we approach the Sun and it decreases as we move away from it—and because Earth's rotation axis is tilted 23½° from being perpendicular to the ecliptic. These two effects change the apparent speed of the Sun across the sky from day to day. The *average* time interval between consecutive noontimes throughout the year is 24 h, which determines the time we use on our clocks. This is called the *mean* (or *average*) solar day.

What region of Earth has the smallest range of seasonal temperature changes and why?

Traditionally, noontime was taken to be the instant when the Sun is highest in the sky. But as we have just seen, the interval from one noontime to the next is not exactly 24 h, so the Sun is not always highest at noon. The difference between clock noontime and astronomical noontime (when the Sun is highest) is as much as 16 min.

Because Earth is rotating eastward, the Sun is highest at different longitudes (measured east and west along lines running between Earth's poles) at different times. For example, astronomical noon in New York City occurs earlier than it does a little farther west, in Philadelphia. Before the advent of time zones, local time was based on astronomical noon. To travel west from New York to Philadelphia by train, for example, you had to know the departure time at New York, using New York time, as well as the arrival time in Philadelphia (say, if someone was going to meet you) in Philadelphia time. Such time considerations became very confusing and burdensome as society became more complex. **Time zones** were established in the late nineteenth century to alleviate this problem. In a time zone, everyone agrees to set their clocks to the same time. Time zones are based on the time at 0° longitude in Greenwich, England, a location called the *prime meridian,* as mentioned earlier. With some variations due to geopolitical boundaries, every 15° of longitude around the globe begins a new time zone. The resulting 24 time zones are shown in Figure 1-19. Going from one time zone to the next usually requires you to change the time on your wristwatch by exactly 1 h.

Along with the solar day, there is also a *sidereal day,* the length of time from when a star is in one place in the sky until it is next in the same place. The solar and sidereal days differ from each other in length because while Earth rotates, it also revolves around the Sun. This motion of Earth in its orbit day by day changes the location of the stars, bringing them back to their original positions 4 min earlier each day. Therefore, the sidereal day is 23 h, 56 min long, while the solar day is 24 h long.

Why would a calendar based on sidereal days not be satisfactory?

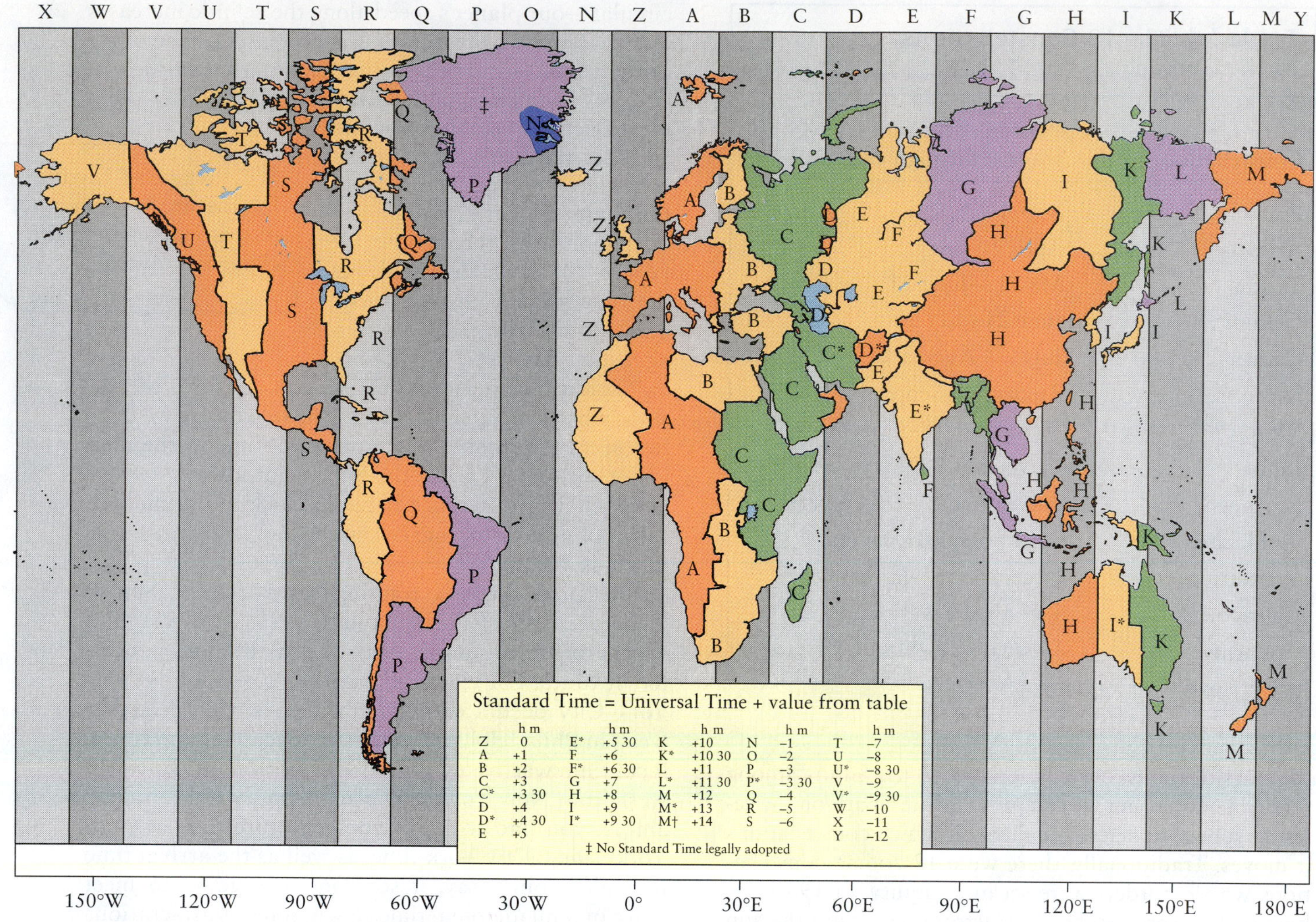

FIGURE 1-19 Time Zones of the World For convenience, Earth's 360° circumference is divided into 24 time zones. Ideally, each time zone would run due north-south. However, political considerations make many zones irregular. Indeed, there are even a few zones only a half-hour wide.

1-7 Calendars based on equal-length years also created scheduling problems

Just as the day is caused by Earth's rotation, the year is the unit of time based on Earth's revolution about the Sun. Earth does not take exactly 365 days to orbit the Sun, so the year is not exactly 365 days long. Basing the year on a 365-day cycle led to important events (such as holidays) occurring on the wrong day. To resolve this problem, Roman statesman Julius Caesar implemented a new calendar in the year 46 B.C. Because measurement revealed to ancient astronomers that the length of a year is approximately 365¼ days, this "Julian" calendar established a system of leap years to accommodate the extra quarter of a day. By adding an extra day to the calendar every 4 years, Caesar hoped to ensure that seasonal astronomical events, such as the beginning of spring, would occur on the same date year after year.

The Julian calendar would have worked just fine if a year were exactly 365¼ days long and if Earth's rotation axis (currently pointing toward Polaris, as discussed earlier) never changed direction. However, neither assumption is correct. Thus, over time, a discrepancy accumulated between the calendar and the actual time—astronomical and cultural events continued to fall on different dates each year. To straighten things out, a committee established by Pope Gregory XIII recommended a refinement, thus creating the Gregorian calendar in 1582. Pope Gregory began by dropping 10 days (October 5, 1582, was proclaimed to be October 15, 1582), which brought the first day of spring back to March 20. Next, he modified Caesar's system of leap years. Caesar had added February 29 to every calendar year that is evenly divisible by four. In the Julian system, for example, 1700, 1800, 1900, and 2000 would all have been leap years with 366 days. But this system produces an error of about 3 days every four centuries. To solve the problem, Pope Gregory decreed that century years would be leap years only if evenly divisible by 400. For example, the years 1700, 1800, and 1900 were not leap years under the improved Gregorian system. But the year 2000—which can be divided evenly by 400—was a leap year. We use the Gregorian system today. It assumes that the year is 365.2425 mean solar days long, which is very close to the length of the *tropical year,* defined as the time interval from one vernal equinox to the next. In fact, the error is only 1 day in every 3300 years. That won't cause any problems for a long time.

1-8 Precession is a slow, circular motion of Earth's axis of rotation

As noted just above, the orientation of Earth's axis of rotation changes slightly with respect to the celestial sphere (that is, it "points" in a slightly different direction). While this change is small over our lifetimes, it eventually causes the north celestial pole to drift away from Polaris. This major change in orientation is caused by gravitational forces from the Moon and the Sun pulling on the slight equatorial bulge at Earth's equator created by Earth's rotation—our planet's diameter is about 43 km (27 mi) greater at the equator than from pole to pole. **Gravitation (gravity)** is the universal force of attraction between all matter. The strength of the gravitational force between two objects depends on the amounts of mass the objects have and the distance between them, as we explain in more detail in Chapter 2.

Because of Earth's tilted axis of rotation, the Sun and Moon are usually not located directly over Earth's equator. As a result, their gravitational attractions on Earth's equatorial bulge provide forces that pull the bulge toward them (Figure 1-20a). However, Earth does not respond to these forces from the Sun and Moon by tilting so that its equator is closest to them. Instead, Earth maintains the same tilt (23½° from the ecliptic), but the direction in which its axis of rotation points on the celestial sphere changes—a motion called **precession**. This is exactly the same behavior exhibited by a spinning top (Figure 1-20b). If the top were not spinning, gravity would pull it over on its side. But when it is spinning, the combined actions of gravity and rotation cause the top's axis of rotation to precess or wobble in a circular path. As with the toy top, the combined actions of gravity from the Sun and the Moon plus rotation cause Earth's axis to trace a circle in the sky while remaining tilted about 23½° away from the ecliptic.

In the mid-1990s, astronomers simulated the behavior of Earth and discovered that without a large moon, Earth would not keep to a 23½° tilt, but rather would change its angle relative to the ecliptic dramatically over millions of years. Therefore, the Moon has played a crucial role in stabilizing Earth and maintaining the seasons as we know them.

Earth's rate of precession is slow compared to human time scales. It takes about 26,000 years for the north celestial pole to trace out a complete circle around the sky, as shown in Figure 1-20c. (The south celestial pole executes a similar circle in the southern sky.) At the present time, Earth's north axis of rotation points within 1° of the star Polaris. In 3000 B.C., it pointed near the star Thuban in the constellation Draco (the Dragon). In A.D. 14,000, the pole star will be near Vega in Lyra. As Earth's axis of rotation precesses, its equatorial plane also moves. Because Earth's equatorial plane defines the location of the celestial equator in the sky, the celestial equator also changes over time. Recall that the intersections of the celestial equator and the ecliptic define the equinoxes, so these key locations in the sky shift slowly from year to year. This entire phenomenon is often called the **precession of the equinoxes**. This change was discovered by the great Greek astronomer Hipparchus in the second century B.C. Today, the vernal equinox is located in the constellation Pisces (the Fish). Two thousand years ago, it was located in Aries (the Ram). Around the year A.D. 2600, the vernal equinox will move into Aquarius (the Water Bearer).

1-9 The phases of the Moon originally inspired the concept of the month

While the Moon's precessional effect on Earth unfolds over millennia, many lunar effects are noticeable every day. As the Moon orbits Earth, it moves from west to east (right to left) on the celestial sphere, changing position among the background stars. Its position relative to the Sun also changes, and as a result, we see different **lunar phases**.

The Sun illuminates half of the Moon at all times (Figure 1-21). The Moon's phase that we see depends on how much of its sunlit hemisphere is facing Earth. When the Moon is closest to the Sun in the sky, its dark hemisphere faces Earth. This phase, during which the Moon is at most a tiny crescent, is called the *new* Moon. During the 7 days following the new phase, more of the Moon's illuminated hemisphere becomes exposed to our view, resulting in a phase called the *waxing crescent* Moon. At the *first quarter* Moon, we see half of the illuminated hemisphere and half of the dark hemisphere. "Quarter Moon" refers to how far in its cycle the Moon has gone, rather than what fraction of the Moon appears lit by sunlight.

During the next week, still more of the illuminated hemisphere can be seen from Earth, giving us the phase called the *waxing gibbous*

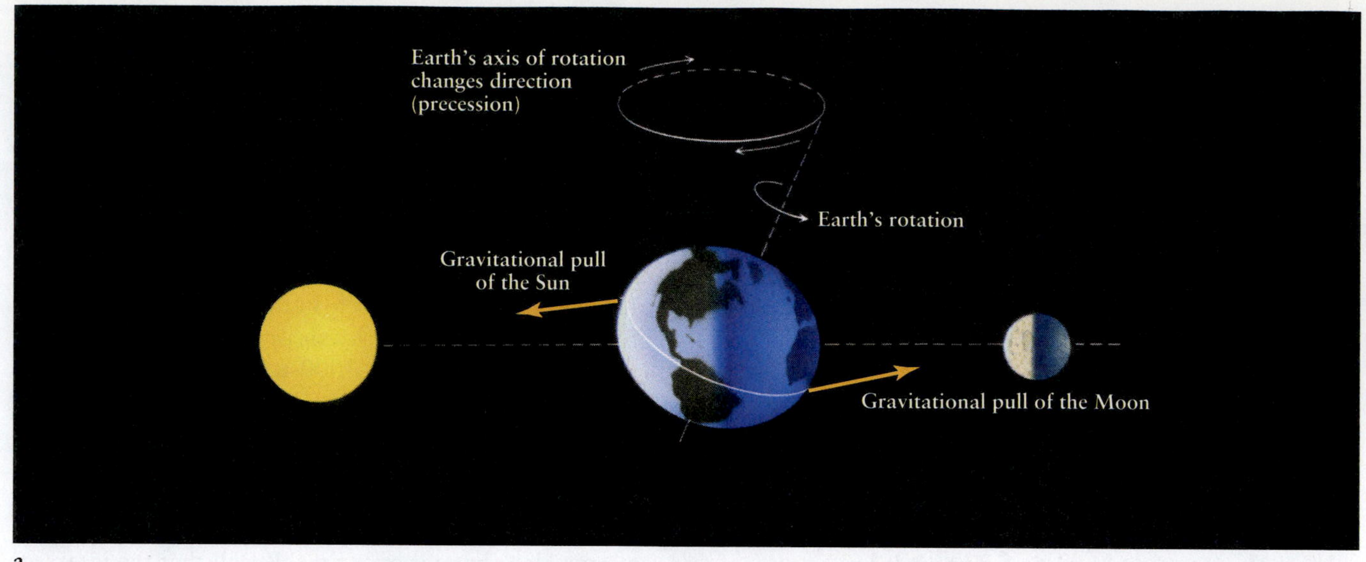

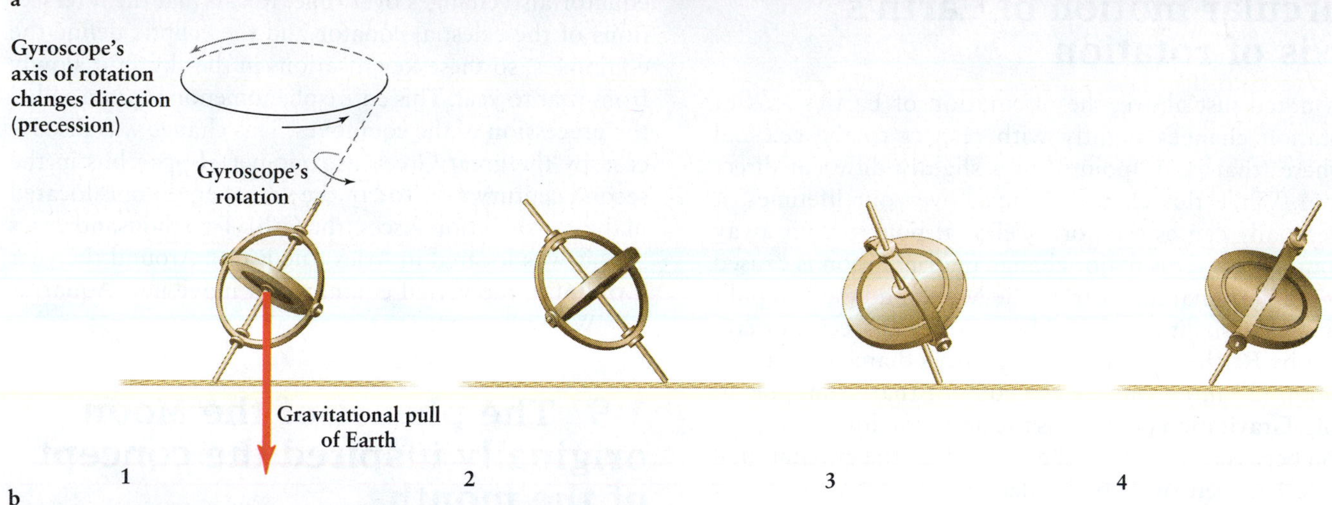

FIGURE 1-20 Precession and the Path of the North Celestial Pole (a) The gravitational pulls of the Moon and the Sun on Earth's equatorial bulge cause Earth to precess. (b) The situation is analogous to the motion of a gyroscope. The top of the gyroscope shows the motion of Earth's North Pole or South Pole, while the point on which the gyroscope spins represents the center of Earth. As the gyroscope spins, Earth's gravitational pull causes the gyroscope's axis of rotation to move in a circle—to precess. (c) As Earth precesses, the north celestial pole slowly traces out a circle among the northern constellations. At the present time, the north celestial pole is near the moderately bright star Polaris, which serves as the pole star. The total precession period is about 26,000 years.

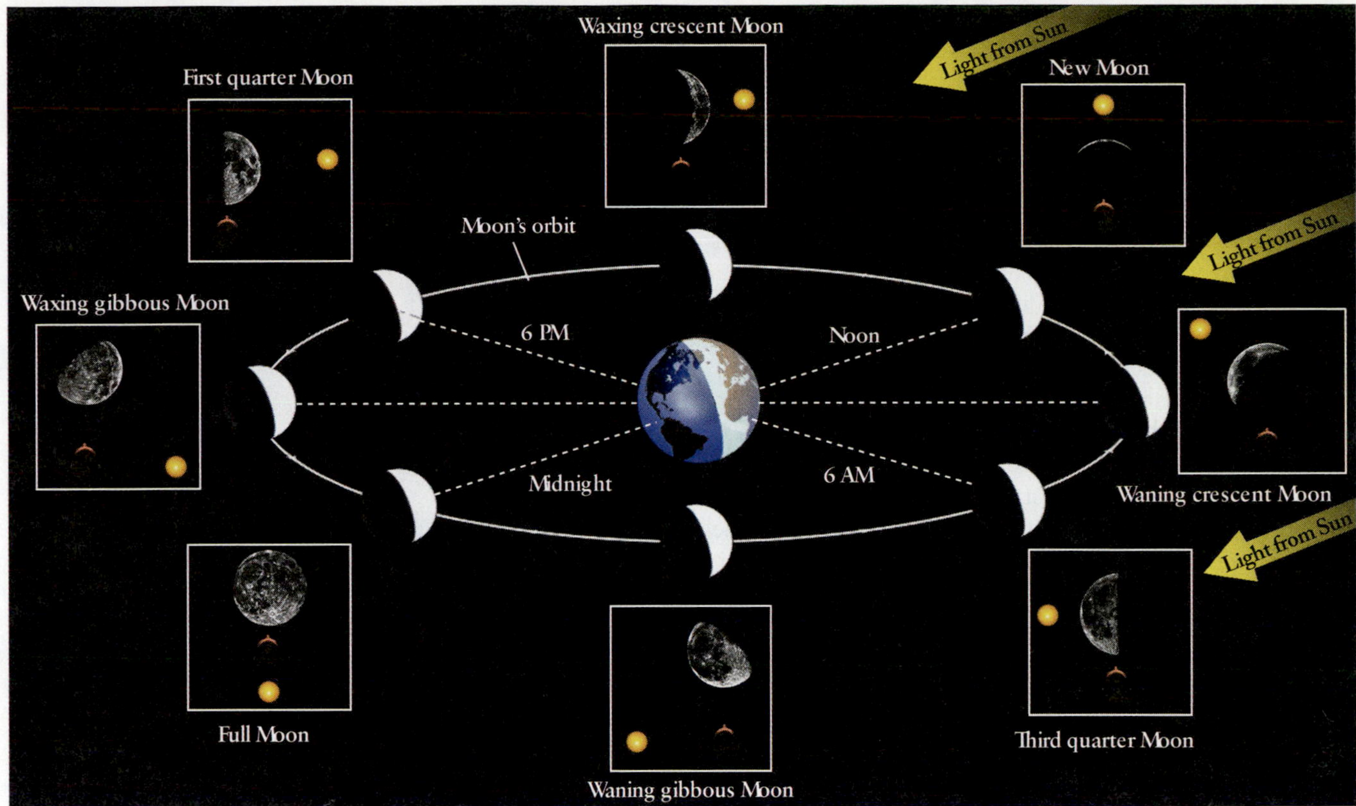

FIGURE 1-21 The Phases of the Moon The diagram shows the Moon at eight locations on its orbit as viewed from far above Earth's North Pole. Light from the Sun illuminates one-half of the Moon at all times, while the other half is dark. It takes about 29½ days for the Moon to go through all its phases. The inset drawings with photographs show the resulting lunar phases *as seen from Earth.* (Yerkes Observatory and Lick Observatory)

Moon. "Gibbous" means "rounded on both sides." When the Moon arrives on the opposite side of Earth from the Sun, we see virtually all of the fully illuminated hemisphere. This phase is the full Moon. Over the following two weeks, we see less and less of the illuminated hemisphere as the Moon continues along its orbit. This movement produces the phases called the *waning gibbous* Moon, the *third quarter* Moon, and the *waning crescent* Moon. The Moon completes a full cycle of phases in 29½ days.

What phase does the Moon really have in this cartoon?

THAT'S A QUARTER MOON.

9-24

"I can buy it for you, Mommy. I have a quarter!"

WHAT IF... Mark Twain were right?
Mark Twain once wrote, "Everyone is a moon and has a dark side which he never shows to anybody." Twain implies that the side of the Moon we don't see is its dark side. But the dark side of an astronomical body is the side facing away from the Sun—the nighttime side of that world. If we never saw the Moon's dark side, we would always see its sunlit side. That is, we would always see a full Moon. In any phase other than full, we see part or all of the Moon's dark side.

6 Confusion often occurs over the terms "far side" and "dark side" of the Moon. The far side is the side of the Moon facing away from Earth. The dark side is the side of the Moon on which the Sun is not shining. By examining the photographs in Figure 1-21, you can see that the same physical side of the Moon faces Earth all the time (that is, we always see the same craters). The half of the Moon that never faces Earth is the far side. However, the far side is not always the dark side, because we see part of the dark side whenever we see less than a full Moon. (What did Pink Floyd have to say about the dark side of the Moon?)

Figure 1-21 shows the Moon at various positions in its orbit. Remember that the bright side of the Moon is on the right (west) side of the waxing Moon, while the bright side is on the left (east) side of the waning Moon. This information can tell you at a glance whether the Moon is waxing or waning. When looking at the Moon through a telescope, the best place to see surface details is where the shadows are longest. This occurs at the boundary between the bright and dark regions, called the **terminator.**

7 Figure 1-21a also shows local time around the globe, from noon, when the Sun is highest in the sky, to midnight, when it is on the opposite side of Earth. These time markings roughly indicate when the Moon is highest in the sky. For example, at first quarter, the Moon is 90° east of the Sun in the sky; hence, the Moon is highest at sunset. At full Moon, the Moon is opposite the Sun in the sky; thus, the Moon is highest at midnight. Using this information, you can see that the Moon is visible during the daytime (Figure 1-22) for a part of most days of the year.

Since the dawn of civilization, people have sought accurate timekeeping systems. Ancient Egyptians wanted to know when the Nile would flood. Farmers everywhere needed to know when to plant crops. Migratory tribes wanted to know when the weather would change. Religious leaders scheduled observances in accordance with celestial events. Thus, astronomers have traditionally been responsible for telling time. Indeed, of the four ways in which time cycles are set, three are astronomical in origin: Time is determined by the positions of the Moon, Sun, or stars or, in our own age, by technological means, such as atomic clocks.

Is the Moon waning or waxing in Figure 1-22?

The approximately 4 weeks that the Moon takes to complete one cycle of its phases inspired our ancestors to invent the concept of a month. Astronomers find it useful to define two types of months, depending on whether the Moon's motion is measured relative to the stars or to the Sun. Neither type corresponds exactly to the months of our usual calendar, which have different (and in the case of February, varying) lengths.

The **sidereal month** is the time it takes the Moon to complete one full orbit of 360° around Earth (Figure 1-23). As with the sidereal day, the length of the sidereal month is determined by the location of the Moon in its orbit around Earth as measured with respect to the stars. Equivalently, this is the time it takes the Moon to start at one place on the celestial sphere and return to exactly the same place again. The sidereal orbital period of the Moon takes approximately 27.3 days. The **synodic month**, or **lunar month**, is the time it takes the Moon to complete one 29½-day cycle of phases (for example, from new Moon to new Moon or from full Moon to full Moon) and thus is measured with respect to the Sun rather than the stars.

FIGURE 1-22 The Moon During the Day The Moon is visible at some time during daylight hours virtually every day. The time of day or night it is up in our sky depends on its phase. (Richard Cummins/SuperStock)

The synodic month is longer than the sidereal month because Earth is orbiting the Sun while the Moon goes through its phases. As shown in Figure 1-23, the Moon must travel *more* than 360° along its orbit to complete a cycle of phases (for example, from one new Moon to the next), which takes about 2.2 days longer than the sidereal month.

Both the sidereal month and synodic month vary somewhat, because the gravitational pull of the Sun on the Moon affects the Moon's speed as it orbits Earth. The sidereal month can vary by as much as 7 h, while the synodic month can change by as much as 12 h.

The terms *synodic* and *sidereal* are also used in discussing the motion of the other bodies in the solar system. The synodic period of a planet is the time between consecutive straight alignments of the Sun, Earth, and that planet (during which time interval the planet also goes through a cycle of phases, as seen from Earth). Recall that any orbit measured with respect to the stars is called "sidereal," including orbits of the planets around the Sun, as well as orbits of moons around their planets. Earth's sidereal year is 365.2564 days. From year to year, our sidereal year differs from the time between consecutive vernal equinoxes (called a *tropical year*) primarily due to Earth's precession (see Section 1-8).

ECLIPSES

8 Eclipses are among the most spectacular natural phenomena. During a **lunar eclipse,** the brilliant full Moon often darkens to a deep red. A lunar eclipse occurs when the Moon passes through Earth's shadow. This can happen only when the Sun, Earth, and the Moon are in a straight line at full Moon. During a **solar eclipse,** broad daylight is transformed into an eerie twilight, as the Sun seems to be blotted from the sky. A solar eclipse occurs when the Moon's shadow moves across Earth's surface. As seen from Earth, the Moon moves in front of the Sun—a new Moon.

1-10 Eclipses don't occur during every new or full phase

At first glance, it would seem that eclipses should happen at every new and full Moon, but in fact, they occur much less often because the Moon's orbit is tilted 5° from the plane of the ecliptic (Figure 1-24). Consequently, the new

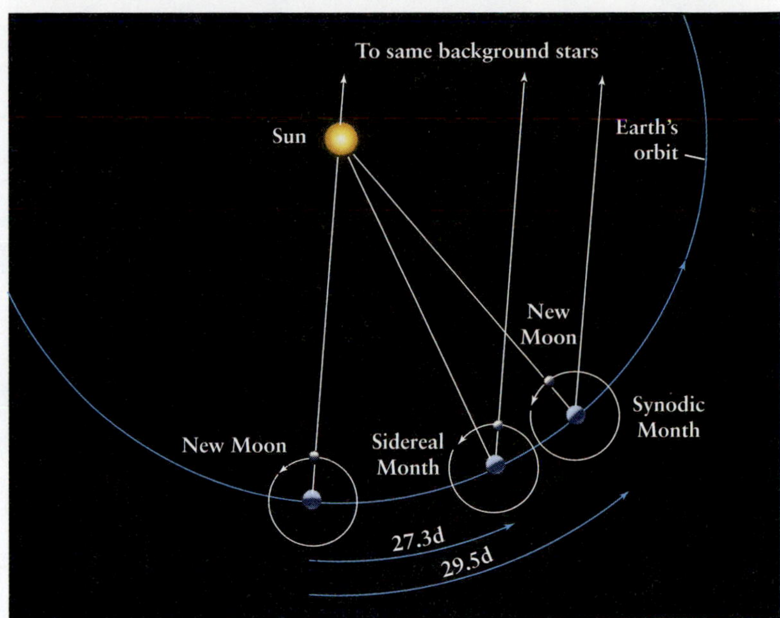

FIGURE 1-23 The Sidereal and Synodic Months The sidereal month is the time it takes the Moon to complete one revolution with respect to the background stars, about 27.3 days. However, because Earth is constantly moving in its orbit about the Sun, the Moon must travel through more than 360° to get from one new Moon to the next. The synodic month is the time between consecutive new Moons or consecutive full Moons, about 29½ days.

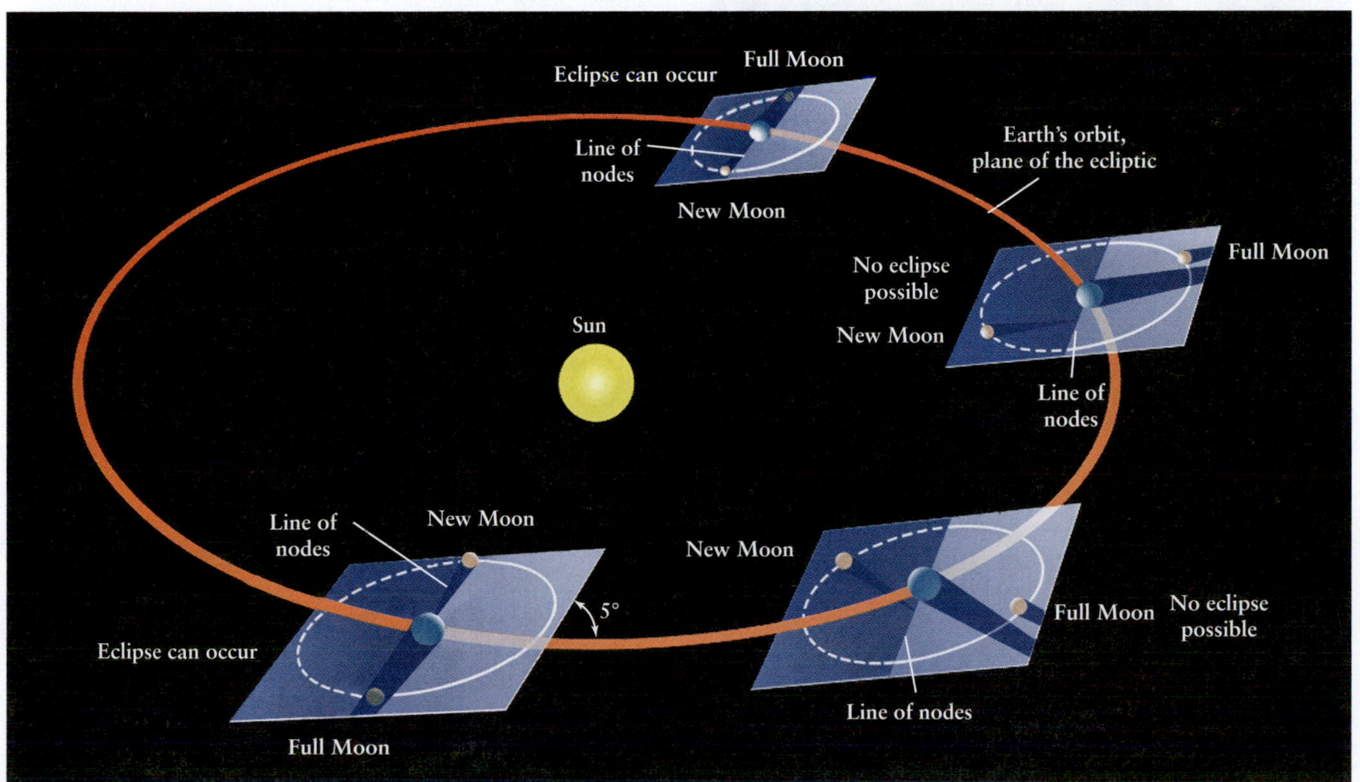

FIGURE 1-24 Conditions for Eclipses The Moon must be very nearly on the ecliptic at new Moon for a solar eclipse to occur. A lunar eclipse occurs only if the Moon is very nearly on the ecliptic at full Moon. When new Moon or full Moon phases occur away from the ecliptic, no eclipse is seen because the Moon and Earth do not pass through each other's shadows.

Moon and full Moon usually occur when the Moon is either above or below the plane of Earth's orbit. In such positions, a true alignment between the Sun, Moon, and Earth is not possible and so an eclipse cannot occur.

Indeed, because its orbit is tilted 5° from the ecliptic, the Moon is usually above or below the plane of our orbit around the Sun. The Moon crosses the ecliptic at what is called the **line of nodes** (see Figure 1-24). When the Moon crosses the plane of the ecliptic during its new or full phase, an eclipse takes place. By calculating the number of times a new Moon takes place on the line of nodes, we find that at least two and no more than five solar eclipses occur each year. Lunar eclipses occur just about as frequently as solar eclipses, with the maximum number of eclipses (solar plus lunar) possible in a year being seven.

1-11 Three types of lunar eclipses occur

Earth's shadow has two distinct parts, as shown in Figure 1-25a. The **umbra** is the part of the shadow where all direct sunlight is blocked by Earth. If you were in Earth's umbra looking at Earth, you would not see the Sun behind it at all. The **penumbra** of the shadow is where Earth blocks only some of the sunlight. If you were in Earth's penumbra looking at Earth, you would see a crescent Sun behind it. The Moon has an analogous umbra and penumbra.

Depending on how the Moon travels through Earth's shadow, three kinds of lunar eclipses may occur. A **penumbral eclipse**, when the Moon passes through only Earth's penumbra, is easy to miss. The Moon still looks full, just a little dimmer than usual and sometimes slightly reddish in color (path 1 in Figure 1-25a).

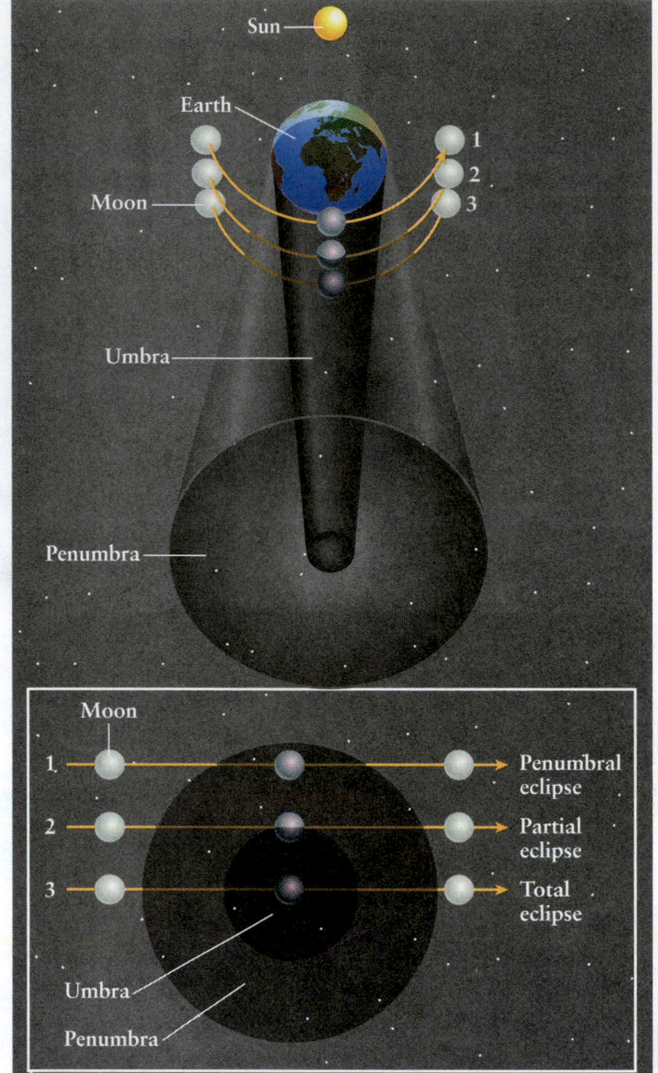

a

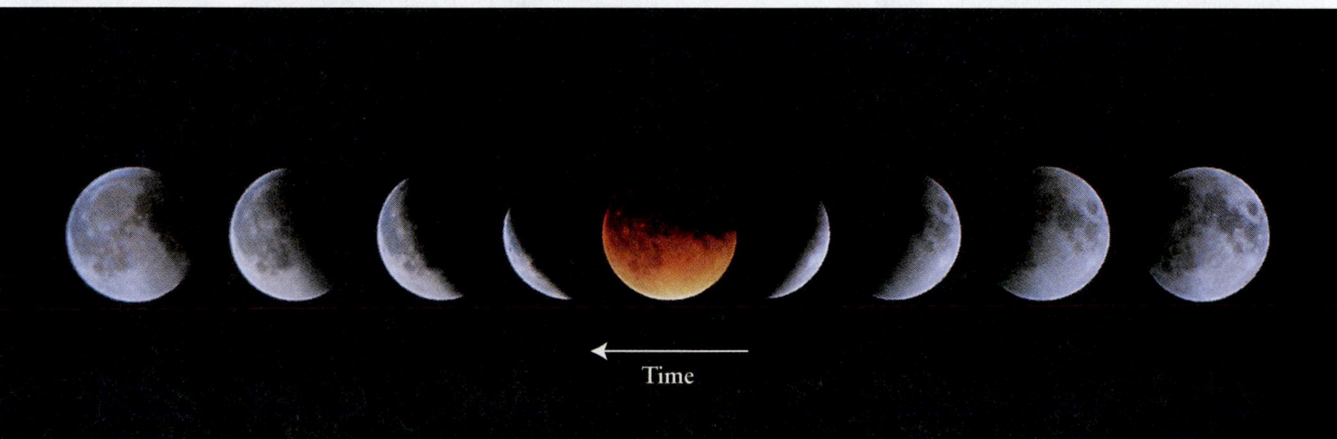

b

R I V U X G

FIGURE 1-25 Three Types of Lunar Eclipses (a) People on the nighttime side of Earth see a lunar eclipse when the Moon moves through Earth's shadow. The umbra is the darkest part of the shadow. In the penumbra, only part of the Sun is covered by Earth. The inset shows the various lunar eclipses that occur, depending on the Moon's path through Earth's shadow. (b) This sequence of nine photographs was taken over a 3-h period during the total lunar eclipse of January 20, 2000. During the total phase, the Moon has a distinctly reddish color. (Fred Espenak, NASA/Goddard Space Flight Center; © Fred Espenak, MrEclipse.com)

INTERACTIVE EXERCISE 1.4

Table 1-2 Lunar Eclipses, 2011–2014

Date	Visible from	Type	Duration of totality (h:min)
2011 June 15	South America, Europe, Africa, Asia, Australia	Total	1:41
2011 Dec 10	Europe, eastern Africa, Asia, Australia, North America	Total	:52
2012 Jun 04	Asia, Australia, Pacific, Americas	Partial	
2012 Nov 28	Europe, eastern Africa, Asia, Australia, Pacific, North America	Penumbral	
2013 Apr 25	Europe, Africa, Asia, Australia	Partial	
2013 May 25	Americas, Africa	Penumbral	
2013 Oct 18	Americas, Europe, Africa, Asia	Penumbral	
2014 Apr 15	Australia, Pacific, Americas	Total	1:18
2014 Oct 08	Asia, Australia, Pacific, Americas	Total	:59

When just part of the lunar surface passes through the umbra, a bite seems to be taken out of the Moon, and we see a **partial eclipse** (path 2 in Figure 1-25a). When the Moon travels completely into the umbra, we see a **total eclipse** of the Moon (path 3 in Figure 1-25a). Total lunar eclipses with the maximum duration, lasting for up to 1 h and 47 min, occur when the Moon is closest to Earth and is traveling directly through the center of the umbra. Table 1-2 lists all the total and partial lunar eclipses from 2011 through 2014.

Even during a total eclipse, the Moon does not completely disappear. A small amount of sunlight passing through Earth's atmosphere is bent into Earth's umbra. The light deflected into the umbra is primarily red and orange, and thus the darkened Moon glows faintly in rust-colored hues (Figure 1-25b). At sunrise and sunset, the sky appears red or orange for the same reason, because at those times more of Earth's atmosphere deflects red and orange light from the Sun toward you.

Everyone on the side of Earth over which a lunar eclipse occurs can see it, provided that clouds do not obscure the event. Lunar eclipses are perfectly safe to watch with the naked eye.

1-12 Three types of solar eclipses also occur

Because of their different distances from Earth, the Sun and the Moon have nearly the same angular diameter as seen from Earth—about ½°. When the Moon completely covers the Sun, the result is a total solar eclipse. You must be at a location within the Moon's umbra to see a total solar eclipse. During those few precious moments, hot gases (the **solar corona**) surrounding the Sun can be observed and photographed (Figure 1-26). By studying the light from the outer layers of the Sun, astronomers have been able to learn about its atmospheric temperature, chemistry, and gas activity.

You can see in Figure 1-27 that only the tip of the Moon's umbra ever reaches Earth's surface. As Earth turns and the Moon orbits, the tip traces an **eclipse path**

Why does the new Moon sometimes appear as a crescent?

FIGURE 1-26 A Total Eclipse of the Sun During a total solar eclipse, the Moon completely covers the Sun's disk, and the solar corona can be photographed. This halo of hot gases extends for millions of kilometers into space. This gorgeous image was taken in southwestern Mongolia during the August 1, 2008, solar eclipse. (© 2008 Miloslav Druckmüller, Martin Dietzel, Peter Aniol, Vojtech Rušin)

FIGURE 1-27 The Geometry of a Total Solar Eclipse During a total solar eclipse, the tip of the Moon's umbra traces an eclipse path across Earth's surface. People inside the eclipse path see a total solar eclipse, whereas people inside the penumbra see only a partial eclipse. The photograph in this figure shows the Moon's shadow on Earth. It was taken from the *Mir* space station during the August 11, 1999, total solar eclipse. The Moon's umbra appears as the very dark spot on the eastern coast of the United States. The umbra is surrounded by the penumbra. (Jean-Pierre Haigneré, Centre National d'Etudes Spatiales, France/GSFS)

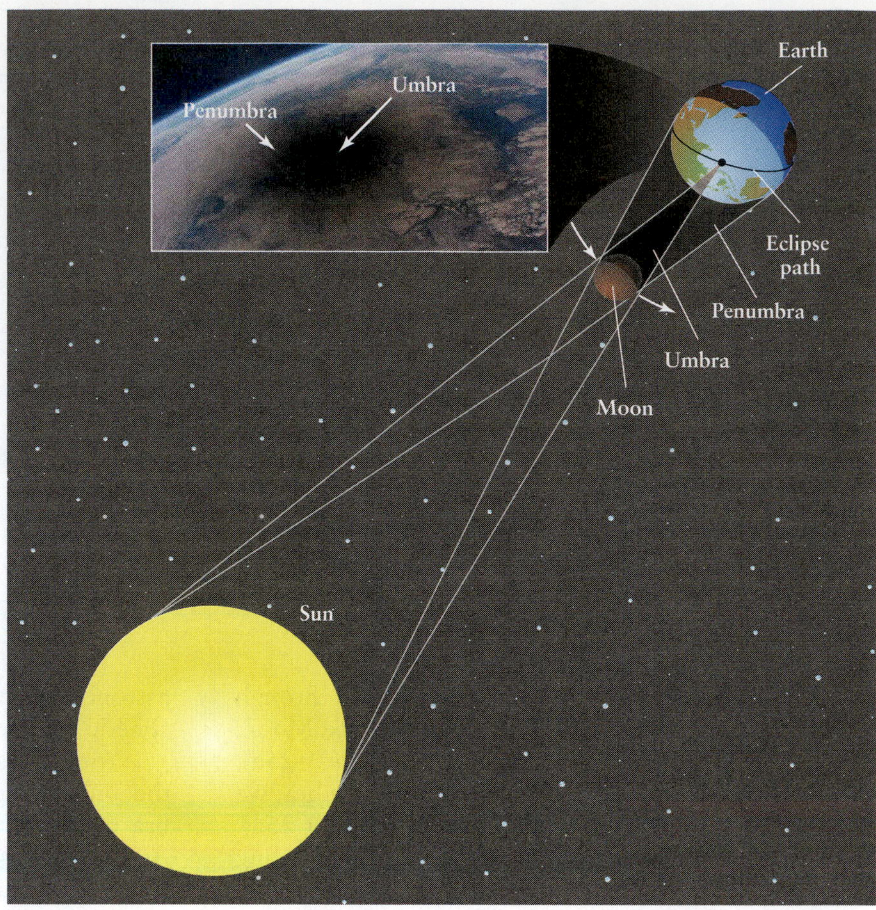

Looking at a solar eclipse through a filter (left) and a filtered telescope. (AP Photo/Mahmoud Tawil)

Looking at a solar eclipse projected by a telescope. (Guillermo Gonzalez/Visuals Unlimited)

across our planet. Figure 1-28 shows the paths over Earth of all the total eclipses through 2020. Only people within these areas are treated to the spectacle of a total solar eclipse, which is why cruise lines love solar eclipses. ***Viewing the Sun directly for more than a moment at any time without an approved filter causes permanent eye damage.*** It is only safe to look at a total solar eclipse without a filter during the brief time when the Moon completely blocks the Sun. At all other times during the eclipse, you must view it either through an approved filter or in the image made on a flat surface by a telescope or a pinhole camera (see margin photographs).

Earth's rotation and the orbital motion of the Moon cause the umbra to race along the eclipse path at speeds in excess of 1700 km/h (1050 mi/h). For this reason, complete blockage of the Sun (totality) during a total eclipse never lasts for more than 7½ min at any one location on the eclipse path, and it usually lasts for only a few moments.

The Moon's umbra is also surrounded by a penumbra (see Figure 1-27). The photograph in Figure 1-27 shows the dark spot produced by the Moon's umbra and the less dark penumbral ring surrounding it on Earth's surface during a total solar eclipse. During a solar eclipse, the Moon's penumbra extends over a large portion of Earth's surface. When *only* the penumbra sweeps across Earth's surface, as happens in high latitude regions, the Sun is only partly covered by the Moon. This circumstance results in a *partial eclipse of the Sun.* Similarly, people in the penumbra of a total eclipse see a partial eclipse. In either case, the Sun looks crescent-shaped as seen from Earth. ***Never look directly at a partial solar eclipse.***

The Moon's orbit around Earth is not quite a perfect circle. The distance between Earth and the Moon, which averages 384,400 km (238,900 mi), varies by a few percent as the Moon moves around Earth. The width of the eclipse path depends primarily on the Earth-Moon distance during an eclipse. The eclipse path is widest—up to 270 km (170 mi)—when the new Moon happens to be at the point in its orbit nearest Earth. Usually, however, the path is much narrower.

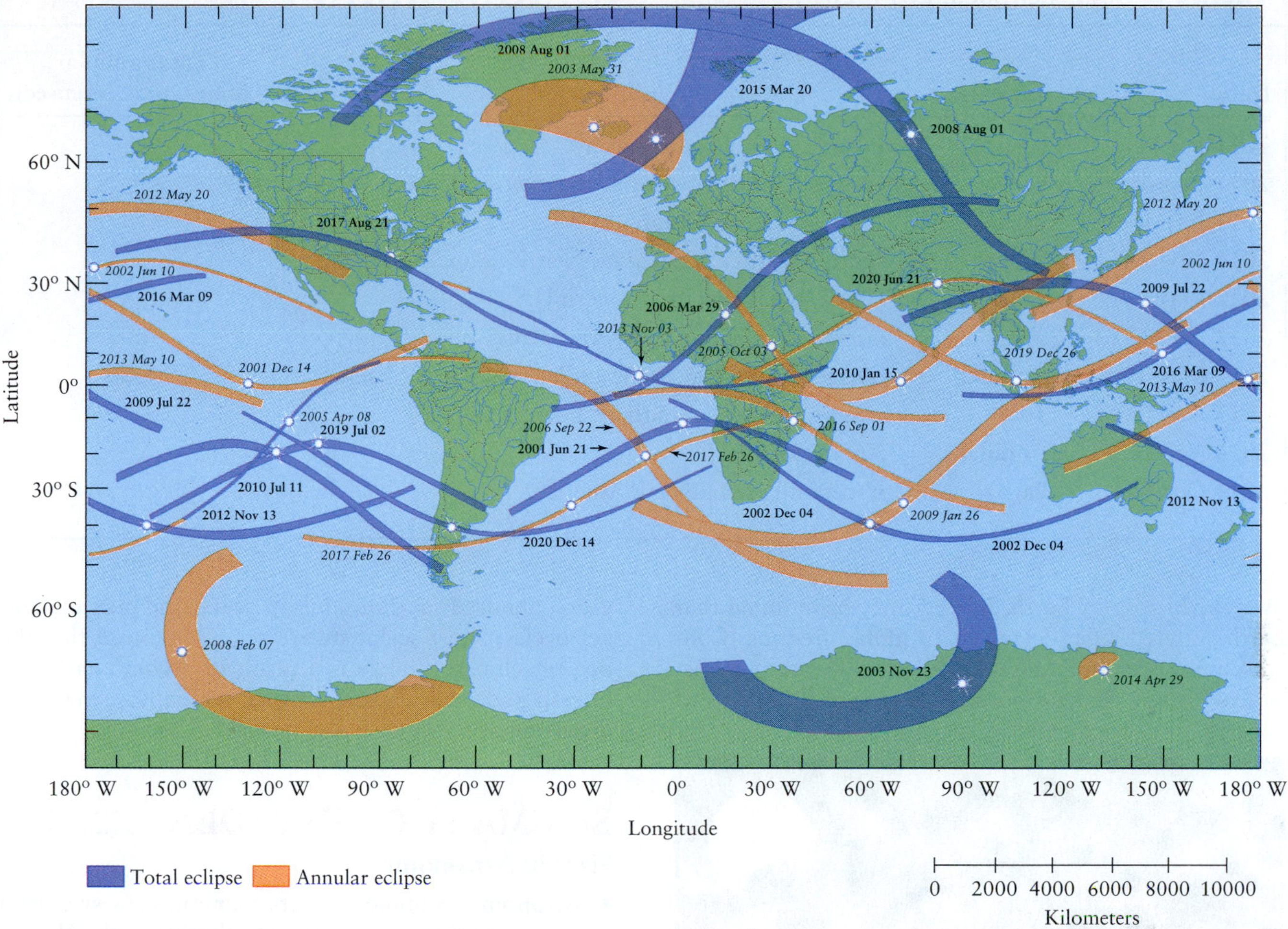

FIGURE 1-28 Eclipse Paths for Total and Annular Eclipses 2001–2020 This map shows the eclipse paths for the 14 total solar and 13 annular eclipses that occur between 2001 and 2020. In each eclipse, the Moon's shadow travels along the eclipse path in a generally eastward direction across Earth's surface. (Courtesy of Fred Espenak, NASA/Goddard Space Flight Center)

If a solar eclipse occurs when the Moon is farthest from Earth, then the Moon's umbra falls short of Earth and no one sees a total eclipse. From Earth's surface, the Moon then appears too small to cover the Sun completely, and a thin ring or "annulus" of light is seen around the edge of the Moon at mid-eclipse. This type of eclipse is called an **annular eclipse** (Figure 1-29). The length of the Moon's umbra is nearly 5000 km (3100 mi) shorter than the average distance between the Moon and Earth's surface. Thus, the Moon's shadow often fails to reach Earth, making annular solar eclipses slightly more common than total solar eclipses. Table 1-3 lists all the total, partial, and annular solar eclipses from 2011 through 2014. It is *never* safe to look directly at a partial or annular eclipse.

A total solar eclipse is a dramatic event. The sky begins to darken, the air temperature falls, and the winds increase as the Moon's umbra races toward you. All nature responds: Birds go to roost, flowers close their petals, and crickets begin to chirp as if evening had arrived. As the moment when the Sun becomes totally eclipsed approaches, the landscape is bathed in shimmering bands of light and dark as the last few rays of sunlight peek out from behind the edge of the Moon. Finally, the corona blazes forth in a star-studded daytime sky. It is an awesome sight, but always take care in viewing it.

1-13 Frontiers yet to be discovered

In this chapter we have examined several factors that explain the changing seasons on Earth. Geologists have discovered a variety of cycles of global temperature change that have occurred

> Why aren't there any annular lunar eclipses?
>
> Why are some of the paths in Figure 1-28 wider than others?

Table 1-3 Solar Eclipses, 2011–2014

Date	Type	Visible from	Total/annular eclipse time (min:sec)
2011 Jan 11	Partial	Europe, Africa, central Asia	
2011 June 1	Partial	eastern Asia, northern North America, Iceland	
2011 July 1	Partial	southern Indian Ocean	
2011 Nov 25	Partial	southern Africa, Tasmania, New Zealand	
2012 May 20	Annular	China, Japan, Pacific, western U.S.	5:46
2012 Nov 13	Total	northern Australia, southern Pacific	4:02
2013 May 10	Annular	northern Australia, Solomon Islands, central Pacific	6:03
2013 Nov 03	Total	eastern Americas, southern Europe, Africa	1:40
2014 Apr 29	Annular	Antarctica	
2014 Oct 23	Partial	northern Pacific, North America	

over the history of Earth. Indeed, they have found that Earth suffered several periods of global freezing. Conversely, we are now undergoing a global warming. Historically, these changes occur over tens of thousands of years, hundreds of thousands of years, and possibly longer cycles. Most scientists are now convinced that the present climate change is a result of a combination of human and other causes, some of which have yet to be discovered.

R I V U X G

FIGURE 1-29 An Annular Eclipse of the Sun This composite of five exposures taken at sunrise in Costa Rica shows the progress of an annular eclipse of the Sun that occurred on December 24, 1974. Note that at mid-eclipse the edge of the Sun is visible around the Moon. (Dennis Di Cicco)

SUMMARY OF KEY IDEAS

Sizes in Astronomy

- Astronomy examines objects that range in size from the parts of an atom ($\sim 10^{-15}$ m) to the size of the observable universe ($\sim 10^{26}$ m).
- Scientific notation is a convenient shorthand for writing very large and very small numbers.

Patterns of Stars

- The surface of the celestial sphere is divided into 88 unequal areas called constellations.
- The boundaries of the constellations run along lines of constant right ascension or declination.

Earthly Cycles

- The celestial sphere appears to revolve around Earth once in each day-night cycle. In fact, it is actually Earth's rotation that causes this apparent motion.
- The poles and equator of the celestial sphere are determined by extending the axis of rotation and the equatorial plane of Earth out onto the celestial sphere.
- Earth's axis of rotation is tilted at an angle of 23½° from a line perpendicular to the plane of Earth's orbit (the plane of the ecliptic). This tilt causes the seasons.
- Equinoxes and solstices are significant points along Earth's orbit that are determined by the relationship between the Sun's path on the celestial sphere (the ecliptic) and the celestial equator.

• Earth's axis of rotation slowly changes direction relative to the stars over thousands of years, a phenomenon called precession. Precession is caused by the gravitational pull of the Sun and the Moon on Earth's equatorial bulge.

• The length of the day is based upon Earth's rotation rate and the average motion of Earth around the Sun. These effects combine to produce the 24-h day upon which our clocks are based.

• The phases of the Moon are caused by the relative positions of Earth, the Moon, and the Sun. The Moon completes one cycle of phases in a synodic month, which averages 29½ days.

• The Moon completes one orbit around Earth with respect to the stars in a sidereal month, which averages 27.3 days.

Eclipses

• The shadow of an object has two parts: the umbra, where direct light from the source is completely blocked; and the penumbra, where the light source is only partially obscured.

• A lunar eclipse occurs when the Moon moves through Earth's shadow. During a lunar eclipse, the Sun, Earth, and the Moon are in alignment, with Earth between the Sun and the Moon, and the Moon is in the plane of the ecliptic.

• A solar eclipse occurs when a strip of Earth passes through the Moon's shadow. During a solar eclipse, the Sun, Earth, and the Moon are in alignment, with the Moon between Earth and the Sun, and the Moon is in the plane of the ecliptic.

• Depending on the relative positions of the Sun, the Moon, and Earth, lunar eclipses may be penumbral, partial, or total, and solar eclipses may be annular, partial, or total.

A WHAT DID YOU THINK?

1 *Is the North Star—Polaris—the brightest star in the night sky?* No. Polaris is a star of medium brightness compared with other stars visible to the naked eye.

2 *What do astronomers define as constellations?* Astronomers sometimes use the common definition of a constellation as a pattern of stars. Formally, however, a constellation is an entire area of the celestial sphere and all the stars and other objects in it. Viewed from Earth, the entire sky is covered by 88 different-sized constellations. If there is any room for confusion, astronomers refer to the patterns as asterisms.

3 *What causes the seasons?* The tilt of Earth's rotation axis with respect to the ecliptic causes the seasons. They are not caused by the changing distance from Earth to the Sun that results from the shape of Earth's orbit.

4 *When is Earth closest to the Sun?* On or around January 3 of each year.

5 *How many zodiac constellations are there?* There are 13 zodiac constellations, the least-known one being Ophiuchus.

6 *Does the Moon have a dark side that we never see from Earth?* Half of the Moon is always dark. Whenever we see less than a full Moon, we are seeing part of the Moon's dark side. So, the dark side of the Moon is not the same as the far side of the Moon, which we never see from Earth.

7 *Is the Moon ever visible during the daytime?* The Moon is visible at some time during daylight hours almost every day of the year. Different phases are visible during different times of the day.

8 *What causes lunar and solar eclipses?* When the Moon is crossing the ecliptic in the full or new phase, the shadows of Earth or the Moon, respectively, then fall on the Moon or Earth. These shadows on the respective surfaces are eclipses.

Key Terms for Review

angle, 14
angular diameter (angular size), 14
annular eclipse, 31
arc angle, 14
asteroid, 8
autumnal equinox, 19
black hole, 8
celestial equator, 12
celestial sphere, 11
circumpolar star, 16
comet, 8
constellation, 11
declination, 13
degree (°), 14
diurnal motion, 13
eclipse path, 29
ecliptic, 17
equinox, 17
gravitation, 23
line of nodes, 28
lunar eclipse, 27
lunar phase, 23
meteoroid, 8
north celestial pole, 13
partial eclipse, 29
penumbra, 28
penumbral eclipse, 28
precession, 23
precession of the equinoxes, 23
revolution, 15
right ascension, 13
rotation, 13
scientific notation, 7
sidereal month, 26
sidereal period, 15
solar corona, 29
solar day, 21
solar eclipse, 27
south celestial pole, 13
summer solstice, 18
synodic month (lunar month), 26
terminator, 26
time zone, 21
total eclipse, 29
umbra, 28
vernal equinox, 13
winter solstice, 18
zenith, 16
zodiac, 20

Review Questions

1. Where is the horizon? **a.** directly overhead, **b.** along the celestial equator, **c.** the boundary between land and sky, **d.** along the path that the Sun follows throughout the day, **e.** the line running from due north, directly overhead, ending due south.

2. How many constellations are there? **a.** 2, **b.** 12, **c.** 13, **d.** 56, **e.** 88.

3. Which of the following lies on the celestial sphere directly over Earth's equator? **a.** ecliptic, **b.** celestial equator, **c.** north celestial pole, **d.** south celestial pole, **e.** horizon.

4. The length of time it takes Earth to orbit the Sun is **a.** an hour, **b.** a day, **c.** a month, **d.** a year, **e.** a century.

5. In Figure 1-8, what is another name for the "Sun's annual path around the celestial sphere"?

6. How are constellations useful to astronomers?

7. What is the celestial sphere, and why is this ancient concept still useful today?

8. What is the celestial equator, and how is it related to Earth's equator? How are the north and south celestial poles related to Earth's axis of rotation?

9. What is the ecliptic, and why is it tilted with respect to the celestial equator?

10. By about how many degrees does the Sun move along the ecliptic each day?

11. Through how many constellations does the Sun move every day?

12. Through how many constellations does the Sun move every year?

13. Why does the tilt of Earth's axis relative to its orbit cause the seasons as Earth revolves around the Sun? Draw a diagram to illustrate your answer.

14. What are the vernal and autumnal equinoxes? What are the summer and winter solstices? How are these four events related to the ecliptic and the celestial equator?

15. What is precession, and how does it affect our view of the heavens?

16. How does the daily path of the Sun across the sky change with the seasons?

17. Why is it warmer in the summer than in the winter?

18. Why is it convenient to divide Earth into time zones?

19. Why does the Moon exhibit phases?

20. What is the difference between a sidereal month and a synodic month? Which is longer? Why?

21. What is the line of nodes, and how is it related to solar and lunar eclipses?

22. What is the difference between the umbra and the penumbra of a shadow?

23. What is a penumbral eclipse of the Moon? Why is it easy to overlook such an eclipse?

24. Which type of eclipse—lunar or solar—have most people seen? Why?

25. How is an annular eclipse of the Sun different from a total eclipse of the Sun? What causes this difference?

26. When is the next leap year?

27. At which phase(s) of the Moon does a solar eclipse occur? A lunar eclipse?

28. Is it safe to watch a solar eclipse without eye protection? A lunar eclipse?

29. During what phase is the Moon "up" least in the daytime?

Advanced Questions

The answers to all computational problems, which are preceded by an asterisk (*), appear at the end of the book.

30. During which phase(s) does the Moon rise after sunrise and before sunset? After sunset and before sunrise? At sunset? At sunrise?

31. Why can't a person in Australia use the Big Dipper to find north?

32. Are there any stars in the sky that are not members of a constellation? If so, which ones?

33. At what places on Earth is Polaris seen on the horizon?

34. Where do you have to be on Earth to see the Sun at your zenith? If you stay at one such location for a full year, on how many days will the Sun pass through the zenith?

35. Where do you have to be on Earth to see the south celestial pole at your zenith? What is the maximum possible elevation (angle) of the Sun above the horizon at that location? On what date is this maximum elevation achieved?

36. Where on the horizon does the Sun rise at the time of the vernal equinox?

37. Consult a star map that shows the precession path of the south celestial pole and determine which, if any, of the bright southern stars could someday become south celestial pole stars.

38. Are there stars in the sky that never set where you live? Are there stars that never rise where you live? Does your answer depend on your location on Earth? Why or why not?

39. Using a diagram, demonstrate that your latitude on Earth is equal to the altitude of the north (or south) celestial pole above your northern (or southern) horizon.

40. Using a star map, determine which bright stars, if any, could someday mark the location of the vernal equinox. Give the approximate years when this should happen.

41. What is the phase of the Moon if it **a.** rises at 3 A.M.? **b.** sets at 9 P.M.? At what time does **c.** the full Moon set? **d.** the first quarter Moon rise?

42. What is the phase of the Moon if, on the first day of spring, the Moon is located at the position of **a.** the vernal equinox, **b.** the summer solstice, **c.** the autumnal equinox, **d.** the winter solstice?

*43. How many more sidereal months than synodic months are there in a year? Why?

44. How do we know that the phases of the Moon are not due to the Moon moving through Earth's shadow?

45. Do the paths of total solar eclipses fall more frequently on oceans or on land? Explain.

46. Can one ever observe an annular eclipse of the Moon? Why or why not?

47. During a lunar eclipse, does the Moon enter Earth's shadow from the east or the west? Explain your answer.

48. How long was the exposure for the photograph of circumpolar star trails in Figure 1-9?

49. Determine which stars whose paths are shown in Figure 1-9 are *not* circumpolar. An easy way to write the answer is in terms of distance on the photograph from the location of the south celestial pole.

50. How does Figure 1-9 show that there is no bright "South Pole" star?

51. Do we see all of the Moon's surface from Earth? Why or why not? *Hint:* Carefully examine the photographs in Figure 1-21.

52. Explain why the waning gibbous Moon, third quarter Moon, and waning crescent Moon photographs in Figure 1-21 are correctly oriented as seen from Earth.

*53. Assuming that the Sun makes an angle of $1/2°$ in our sky and is at a distance of 1.496×10^{11} m, what is the Sun's diameter? Divide this value by 2 to find the Sun's radius and explain why this result is slightly different from the value given in Appendix E-8 at the back of the book.

54. Make a drawing of an annular solar eclipse as seen from space similar to Figure 1-27. Be sure to make clear how it differs from a total solar eclipse.

55. Which of the five images of the Sun in Figure 1-29 is the first and which is the last in the sequence shown? Justify your answer.

56. Why is a small crescent of light often observed on the Moon when it is exactly in the new phase?

Discussion Questions

57. Examine the list of the 88 constellations in Appendix E-7. Are there any constellations whose names obviously date from modern times? Where are these constellations located? Why do they not have ancient names?

58. In his novel *King Solomon's Mines,* H. Rider Haggard described a total solar eclipse that was seen in both South Africa and the British Isles. Is such an eclipse possible? Why or why not?

59. Describe how a lunar eclipse would look if Earth had no atmosphere.

60. Examine a listing of total solar eclipses over the next several decades. What are the chances that you might be able to travel to one of the eclipse paths? Do you think you might go through your entire life without ever seeing a total eclipse of the Sun?

What if ...

61. The Moon moved about Earth in an orbit perpendicular to the plane of Earth's orbit? What would the cycle of lunar phases be? Would solar and lunar eclipses be possible under these circumstances?

62. Earth's axis of rotation were tilted at a different angle? What would the seasons be like where you are now if the axis of rotation were **a.** 0° and **b.** 45° to its orbital plane? What would be different about the seasons and the day-night cycle if you lived at one of Earth's poles in these two situations?

63. You watched Earth from the Moon? What would you see for Earth's **a.** daily motion, **b.** motion along the celestial sphere, and **c.** cycle of phases?

64. The Moon didn't rotate? Describe how its surface *features* would appear from Earth—that is, would we see all sides of it over time? Why or why not? (Ignore the change in phases when discussing its appearance.) *Carefully* study the photographs in Figure 1-21 and state whether the same features are visible at all times or whether we see different features over time. (Again, ignore the phases.) What can you conclude about whether the Moon actually rotates?

Web Questions

65. Search the Web and identify at least four cultures whose stories were used to name modern constellations. Briefly relate the stories of a constellation from each culture.

66. Search the Web for information about the Great Nebula of Orion (see label on Figure 1-4b; it is also called the Orion Nebula). Can the Great Nebula be seen with the naked eye? Does it exist alone in space or is it part of a larger system of interstellar material?

What has been learned by examining the Great Nebula with telescopes sensitive to infrared light?

67. Search the Web for information about the national flags of Australia, New Zealand, and Brazil, and the state flag of Alaska. Which stars are depicted on these flags? Explain any similarities or differences among these flags.

68. Search the Web for the English meaning of the Japanese word *subaru*. Make a drawing of the Subaru car's emblem and explain it.

69. Use the U.S. Naval Observatory's Web site to determine the times of sunset and sunrise on **a.** your birthday and **b.** today's date. Are the times the same? Explain why or why not.

70. Use data in this book or search the Web for information about the next total solar eclipse. Through which major cities, if any, does the path of totality pass? What is the maximum duration of totality? Find a location where this maximum duration is observed. Will the eclipse be visible (even as a partial eclipse) from your present location?

71. Search the Web for information about the next total lunar eclipse. Will the total phase of the eclipse be visible from your present location? If not, will the penumbral phase be visible? Draw a picture of the Sun, Earth, and the Moon at totality and indicate your location on the drawing of Earth.

ANIMATION 1.5 **72.** Access the animation "The Moon's Phases" in Chapter 1 of the *Discovering the Universe* Web site. This shows the Earth-Moon system as seen from a vantage point above Earth's North Pole. **a.** Describe where you would be on the diagram if you were on the equator and the time were 6:00 P.M. **b.** If it were 6:00 P.M. and you were standing on Earth's equator, would a third quarter Moon be visible? Why or why not? If the Moon would be visible, describe its appearance.

Observing Projects

STARRY NIGHT Many of the following projects are based on the planetarium program, *Starry Night*™. If your teacher selected the textbook with this software included or you acquired this software independently, install this program on a suitable computer and run it. You can learn to use the program with the help of the User's Guide, which can be accessed under **Help > User's Guide** on the main menu. There is also a comprehensive tutorial as part of the Student Exercises included with this program. Open the **Sky-Guide** pane and click on **Student Exercises > Tutorial.** You can also experiment on your own or use the online help.

73. On a clear, cloud-free night, use the star charts at the end of this book to see how many constellations of the zodiac you can identify. Which ones are easy to find? Which are difficult?

74. As a guide to the above nighttime observing, use *Starry Night*™ to help you identify the constellations of the zodiac that are visible in your sky this evening. Start the program and stop the advance of time by clicking the **Stop** button in the toolbar; set the time to just after sunset; and use the **S** key on your keyboard to view the southern horizon if you are in the northern hemisphere. (Use the **N** key to view the northern horizon if you are in the southern hemisphere.) Select **View > Constellations > Zodiac** to outline these constellations with stick figures. You can check your identification by right-clicking over the region of interest if you are using a PC (Ctrl-click if you are using a Mac) to display the name of the constellation or a name of a star in it (often with a form of the name of the constellation on the end of the star's name) on the top line of the contextual menu. If you still have trouble identifying specific constellations, select **Labels > Constellations** to label every zodiac constellation. Advance time by a few hours and use the hand tool to "grab" and move the sky to allow you to recognize more of the zodiacal constellations. The zodiac contains the path of the Sun across the sky in the course of a year. You can verify that the Sun passes through zodiacal constellations by displaying the Sun's path, known as the *ecliptic,* by selecting **View > Ecliptic Guides > The Ecliptic** from the menu. You can also display classical images of constellations that indicate where these star patterns acquired their names in antiquity by clicking on **View > Constellations > Auto Identify,** after which the constellation name and classical image will appear whenever you center a constellation on the screen. **a.** Adjust time and date to determine the neighboring zodiacal constellations to the constellation Leo. **b.** Click the **Home** button in the toolbar to return to your home location and to the present time and click the **Stop** button to stop the advance of time again. You can determine which zodiac constellation the Sun is passing through today by changing the time in the toolbar to 12:00:00 P.M. and using the hand tool to find the Sun in the sky. To remove the blue daylight sky, select **View > Hide Daylight** from the menu. You can display the modern boundaries of constellations by clicking **View > Constellations > Boundaries** and thereby be more specific about the location of the Sun within a specific constellation. Change the time and date to track the Sun completely round the zodiac. Make a list of the names of these constellations, adding the dates of boundary crossing between the names of each constellation. Where is the Sun from your location at the present time and date? **c.** Will this location be different when viewed from the opposite hemisphere from the one in which you are located?

STARRY NIGHT

75. Use *Starry Night™* to observe the diurnal motion of the sky. **a.** Click the **Home** button in the toolbar and use the hand tool (or press the **N** key on the keyboard) to center your field of view on the northern horizon (if you live in the northern hemisphere) or the southern horizon (press the **S** key) if you live in the southern hemisphere. Adjust the view so that the horizon appears approximately flat. Select **View > Hide Daylight** from the menu in order to see the background stars. Set the **Time Flow Rate** to 3 minutes and click on the **Run time forward** button in the toolbar to start time flow. (Adjust the rate of the sky motion appropriately to follow the general motion of the stars.) Do the stars appear to revolve clockwise or counterclockwise? Explain this in terms of Earth's rotation. Are any of the stars circumpolar? Explain why or why not. **b.** Click the **Stop** button in the toolbar to stop time flow and re-center your field of view on the southern horizon (if you live in the northern hemisphere) or the northern horizon (if you live in the southern hemisphere). Click **Play** to set the sky in motion again. Explain what you see. Are any of these stars circumpolar? Explain.

76. Examine the star charts that are published monthly in such popular astronomy magazines as *Sky & Telescope* and *Astronomy*. How do they differ from the star charts within the covers of this book? On a clear, cloud-free night, use a star chart to locate the celestial equator and the ecliptic on the night sky. Note the inclination of the Milky Way to the ecliptic and celestial equator. What do your observations tell you about the orientation of Earth and its orbit around the Sun relative to the rest of the Galaxy?

77. Observe the Moon on each clear night over the course of a month. Note the Moon's location among the constellations and record that location on a star chart that also shows the ecliptic. After a few weeks, your observations will begin to trace the Moon's orbit. Identify the orientation of the line of nodes by marking the points where the Moon's orbit and the ecliptic intersect. On what dates is the Sun near to the nodes marked on your star chart? Compare these dates with the dates of the next solar and lunar eclipses.

STARRY NIGHT

78. Use *Starry Night™* to study the Moon's path in the sky and the phenomenon of eclipses, both lunar and solar. Select **Favourites > Discovering the Universe > Moon Motion** to display the sky as seen from the center of a transparent Earth on August 30, 2010, with daylight removed and the Moon centered in the view. As a guide to the positions and motions in the sky, the celestial equator and the associated equatorial grid are shown. The ecliptic, representing the plane of the Earth's orbit extended onto the sky, is also shown. This line also represents the apparent path of the Sun across our sky in the course of a year. The Time Flow Rate is set to 1 hour but this can be adjusted if necessary to follow the Moon across the sky. **a.** Is the Moon on the ecliptic at this time? **b.** With the view locked onto the Moon, click the **Play** button. In which direction does the Moon move against the background stars? Keep in mind that the celestial equator runs in the east-west direction, with east on the left. Note also that the ecliptic intersects this equator at an angle of 23½°. Does the Moon ever change its direction of motion relative to the stars? Why or why not? Describe its path relative to the ecliptic. **c.** Determine how many days elapse between successive times that the Moon is on the ecliptic. **d.** Set the time and date to 11 P.M. on July 21, 2009, and set the Time Flow Rate to 1 minute. Note that the Moon is almost on the ecliptic and that it is close to the position of the Sun in the sky. Run time forward at this 1-minute rate and stop time when the Moon is closest to the Sun. Zoom in to a field of view of about 3° and adjust the time in single steps to move the Moon to the position of maximum eclipse. Sketch the Sun and Moon on a piece of paper. Even though the Moon may not be directly over the Sun on the screen, an eclipse is occurring. What type of eclipse is it? Why is the Moon not directly over the Sun at this time? **e.** Click on the **Events** tab, expand the **Event Filters** layer and deselect all but the **Lunar and Solar Eclipse Events.** Expand the **Events Browser** layer, set the **Start** and **End** dates to cover the period from Jan 21, 2009 to Jan 21, 2010, and click **Find Events.** Right-click on each solar eclipse in turn, choose **View Event,** zoom out to about 5°, and run time forward and backward at a rate of 1 minute until the Moon is again closest to the Sun. Draw the Sun and Moon. Are they separated by the same amount as in part d? If not, why not? If there is a difference, what effect does it have on the eclipse?

79. A lunar eclipse might occur while you are taking this course. If so, look up on the Internet the precise time that it will happen in either the current issue of a reference from the U.S. Naval Observatory, such as the *Astronomical Almanac* or *Astronomical Phenomena,* or in a magazine such as *Sky & Telescope* or *Astronomy,* which generally run articles about eclipses a month or two before they happen. Make arrangements to observe this eclipse. Note the times at which the Moon enters and exits Earth's umbra.

STARRY NIGHT

80. You can use *Starry Night™* to determine when the Sun rises and sets today. Similarly, you can find out when the Moon rises and sets today. Because of the scenery, you may have to approximate these times. Click the **Home** button to return to your home location and set the time to just after midnight this morning. Use **Run time forward** to determine the sunrise and sunset times and then rerun this time period again to note the period of the day or night over which the Moon is visible today.

WHAT IF... Earth's Axis Lay on the Ecliptic?

Imagine Earth is tilted so that its axis of rotation lies in the plane of its orbit about the Sun (see the accompanying figure). We'll call it NeoEarth. NeoEarth still rotates once every 24 h. We arbitrarily fix its north pole to point toward the star Tau Tauri, a star nearly on the ecliptic, just north of the bright star Aldebaran in the constellation Taurus (see Figure 1-4b or Figure 1-6).

NeoEarth's Seasons Let's see how the seasons unfold on NeoEarth. It is March 20, the date of the spring equinox. The Sun is directly over the equator, as shown in the figure below. For the next 3 months, the Sun rises higher in the northern hemisphere. Unlike on our Earth, the Sun does not stop moving northward when it is over 23½° north latitude. Rather, the Sun rises farther and farther north day by day until it appears over the north pole of NeoEarth at the summer solstice, around June 22. Three months later, at the autumnal equinox, the Sun again rises over the equator, and day and night have the same length everywhere.

The Sun appears over the south pole of NeoEarth at the winter solstice, around December 22. The Sun's apparent motion through NeoEarth's sky completes the seasonal cycle by moving north, appearing over the equator once again around March 20.

NeoEarth's Climate Day and night take on new meanings for inhabitants of NeoEarth. On our Earth, the regions above the Arctic Circle and below the Antarctic Circle have days or weeks of continual light in summer and days or weeks of constant darkness in winter. But on NeoEarth, *every place* has extended winter periods of constant darkness followed by extended summer periods of constant daylight. Spring and fall on NeoEarth have daily cycles of daylight and darkness, which separate the periods of continual daylight and darkness.

A Neo Day At the latitude of Atlanta, Georgia (33°46′), on NeoEarth, the day-night cycle occurs for only 7½ months out of the year. During the other 4½ months, there is continuous day or continuous night, coupled with harsh summers and winters. With variations, this sequence of events occurs everywhere on NeoEarth.

The seasonal cycle on NeoEarth prevents the formation of permanent polar ice caps. In summer, polar regions experience the same tropical heating and high temperatures as the equatorial regions of our Earth. The polar regions of NeoEarth in winter are exceptionally cold, so seasonal polar ice caps may form. Because the southern polar cap resting on Antarctica is not permanent, the oceans and the shorelines on the continents, are higher than those on our Earth.

If seasonal polar ice caps form, the dominant force controlling weather may shift from the jet streams that circle our Earth along lines of latitude to a pole-to-pole flow. Thermal flows created by intense heating at one location and cooling at others may replace our Earth's trade winds and other east-west winds. How else would NeoEarth differ from our world?

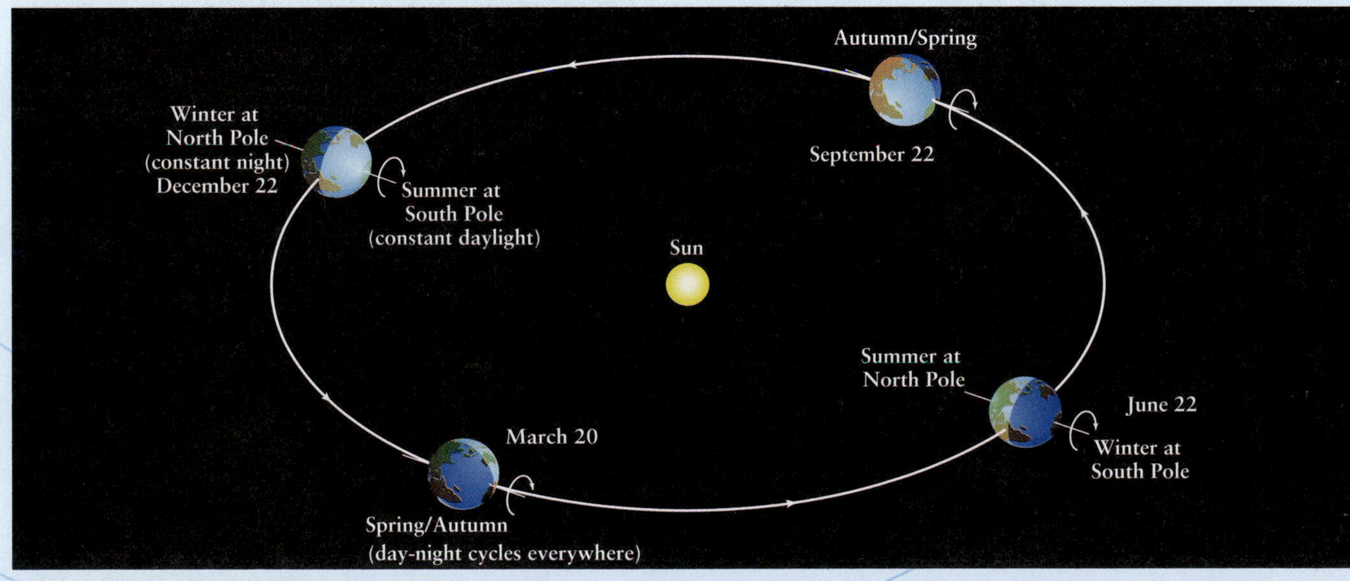

WHAT DO YOU THINK?

1. What makes a theory scientific?
2. What is the shape of Earth's orbit around the Sun?
3. Do the planets orbit the Sun at constant speeds?
4. Do all of the planets orbit the Sun at the same speed?
5. How much force does it take to keep an object moving in a straight line at a constant speed?
6. How does an object's mass differ when measured on Earth and on the Moon?
7. Do astronauts orbiting Earth feel the force of gravity from our planet?

Answers to these questions appear in the text beside the corresponding numbers in the margins and at the end of the chapter.

Gravity is the only universal force of attraction in the universe. Despite the magnificent solitude that astronaut Bruce McCandless II experienced floating a football-field length from the space shuttle *Challenger,* he was being held in orbit around Earth with almost the same gravitational attraction that our planet has on you. He is falling toward Earth, but continually missing it. Why? (STS-41B, NASA)

Chapter 2 Gravitation and the Motion of the Planets

Science enables us to understand and manipulate an awesome range of nature's properties. Scientists are a lot like detectives, and the process they follow when trying to explain scientific phenomena has a good deal in common with the activities of sleuths as they try to solve mysteries. For example, an investigator might suddenly realize that his prime suspect, the person around whom his whole case has revolved, could not have committed the crime because he was actually baking a soufflé at the time it occurred. Similarly, the early natural philosophers (as investigators into natural phenomena were then called) who studied the motions of heavenly bodies made their biggest leap forward when the evidence forced them to look beyond the suspect they believed to be at the center of everything—Earth.

This chapter traces how we moved from an Earth-centered view of the universe to a Sun-centered one and how we came to understand the motion of astronomical bodies under the influence of the gravitational force. The process of this discovery initially involved the efforts of a few determined people. To unravel the mysteries that puzzled them—such as why planets appear at times to change direction on the celestial sphere or why the assumption that Earth is the center of the universe fails to predict the locations of certain bodies—they used careful observations and a willingness to question their own and others' assumptions.

The groundwork for modern science was set down by Greek natural philosophers beginning around 2500 years ago, when Pythagoras and his followers began using mathematics to describe natural phenomena. About 200 years later, Aristotle asserted that the universe is comprehensible: It is governed by regular laws. The Greeks typically did not, however, perform experiments to test their ideas, an essential part of the scientific method used today. Nevertheless, they were among the first to leave a written record of their ideas, allowing succeeding generations to develop, criticize, and test their conclusions.

These Greek ideas, rediscovered in the seventeenth century, led to the development of the scientific method of examining, understanding, and predicting how things work: Science provides explanations for activities and events, and it makes predictions about things that have not yet happened or that have not yet been observed. These predictions are incredibly powerful tools that enable us to understand what we see without having to accept events on faith, or to fear that phenomena like the force of gravity will change on a whim. Science simplifies life and takes some of the uncertainty out of the world.

Consider the topic of this chapter, our understanding of gravity. Until the seventeenth century, when Isaac Newton made the conceptual leap that the force holding Earth in orbit around the Sun is the same force that holds us onto Earth, these two effects were considered to be separate and unrelated. Once Newton connected these forces and then wrote an equation to describe the behavior of Earth and of falling objects, people had, for the first time, the ability to reliably predict the future motion of projectiles and other falling objects. In the early twentieth century, Albert Einstein (Chapter 14) spearheaded discoveries that plumbed the depths of gravitational behavior even further, leading to equations that make even more accurate predictions about the motion of objects. The fundamentals of gravity we study here lead to our understanding of gravity's effects on the universe, including how it causes stars and planets to form (Chapter 5), enables stars to shine (Chapter 10), and holds galaxies together (Chapter 15), among many other things.

In this chapter you will discover

- what makes a theory scientific
- the scientific discoveries that revealed that Earth is not at the center of the universe, as previously believed
- Copernicus's argument that the planets orbit the Sun
- why the direction of motion of each planet on the celestial sphere sometimes appears to change
- that Kepler's determination of the shapes and other properties of planetary orbits depended on the careful observations of his mentor Tycho Brahe
- how Isaac Newton formulated an equation to describe the force of gravity and how he thereby explained why the planets and moons remain in orbit

SCIENCE: KEY TO COMPREHENDING THE COSMOS

Understanding how nature works enables us to manipulate the matter and energy that comprise our environment and thereby to create new things to make our lives better. Improvements in technology lead, in turn, to better research equipment, enabling us to make even deeper discoveries about space, time, matter, energy, and the relationships between them.

This spiral of understanding and application began centuries ago. In this chapter we begin by examining the nature of science and then use science to discover how gravity keeps planets and other objects orbiting the Sun and how gravity keeps moons orbiting their respective planets.

2-1 Science is both a body of knowledge *and* a process of learning about nature

Science is actually two things. First, it is a body of knowledge that we acquire through observations and experiments. The details of the motions of the Moon, the planets, and the Sun on the celestial sphere, described in Chapter 1, are examples of that knowledge. While nature can be discussed descriptively, as it is for the most part in this book, science also provides mathematical equations that quantify the effects being studied.

Second, science is a process for gaining more knowledge in a way that ensures that the information can be tested and thereby accepted by everyone. Science as a process is also called the **scientific method**, which describes how scientists ideally go about observing, explaining, and predicting physical reality. The scientific method (Figure 2-1) can begin in a variety of places, but most often it starts by people making observations or doing experiments. For example, the observation that some objects (for example, planets) move along the celestial sphere, while others (stars) remain fixed on it, demanded explanation. If a theory exists that purports to explain previous observations, new observations or experimental results are compared with the predictions of the theory. If the new data and old theory are not consistent, then a *hypothesis* that modifies or replaces the existing explanation is proposed. (If no theory explains observations or experimental results, a new hypothesis is proposed to explain them.) Hypotheses on related topics that make accurate predictions are incorporated together as a **scientific theory** (often just called a **theory**).

In everyday conversation, a theory is an idea based on common sense, intuition, or deep-seated personal beliefs. Such theories neither originate in equations nor lead to rigorous predictions. The word *theory* in science has a very different meaning. A scientific theory is an explanation of observations or experimental results that can be described quantitatively and tested formally. The mathematical description used in a scientific theory is considered a **model** of the real system. For example, Newton's *theory* (or, in earlier usage, *law*) of gravitation is written as an equation that predicts how bodies attract each other. (The word *gravity* is often used as shorthand for *gravitation,* and both are used in this book.)

As just noted, to be considered scientific, a theory 1
must make *testable* predictions that can be verified by making new observations and doing new experiments. Testing is a crucial aspect of the scientific method, which requires that the theory accurately forecast the results of new observations in its realm of validity. Newton's law of gravitation predicts that the Sun's gravitational force makes the planets move in elliptical orbits, and it predicts how long it should take each planet to orbit the Sun. As we will see shortly, observations have confirmed most of these predictions.

Scientists who develop new or more accurate models are going where no person has gone before. Many of them find this process of discovery as satisfying as an artist does in creating a masterpiece, an athlete breaking a world record, or an astronaut going into space. Scientists who make observations or do experiments that reveal previously unknown facts about nature often have similar reactions to their discoveries.

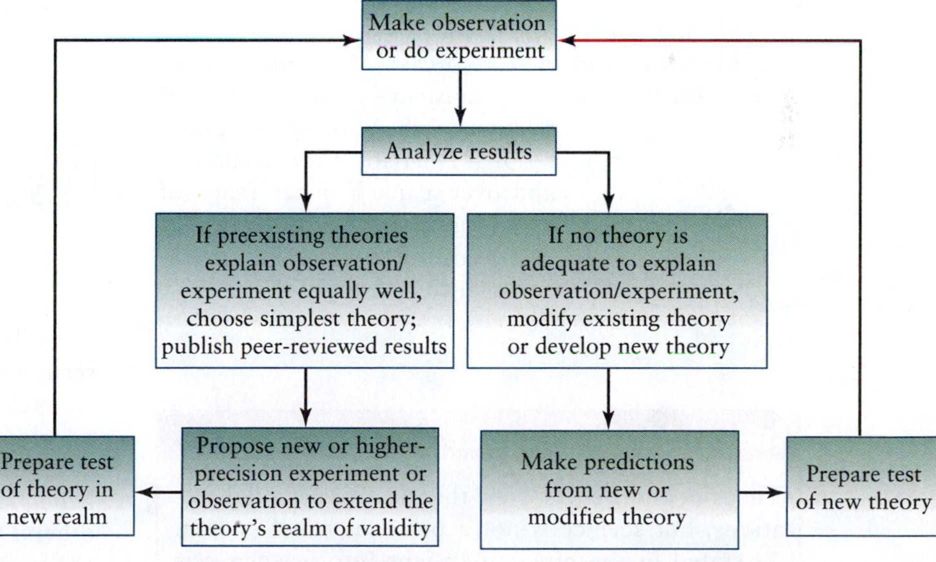

FIGURE 2-1 The Scientific Method This flow chart shows the basic steps in the process by which scientists study nature and develop new scientific theories. Different scientists start at different places on this chart, including making observations or doing experiments, creating or modifying scientific theories, or making predictions from theories. Anyone interested in some aspect of science and willing to learn the tools of the trade can participate in the adventure. (© Neil F. Comins)

Insight Into Science

Science Is Inclusive In principle, a scientific theory can be created, modified, or tested by anyone inclined to do so. In practice, however, being involved in the scientific enterprise requires that you understand the mathematical tools of science. Assuring that theories are written in terms of equations so that they can be carefully analyzed and tested by others is part of the process intended to prevent the scientific method from being derailed.

For a theory to be considered scientific, it must be potentially possible to disprove it. For example, Newton's law of gravitation can be tested and potentially disproved by observations and thus qualifies as a scientific theory. The idea that Earth was created in 6 days cannot be tested, much less disproved. It is not a scientific theory, but rather a matter of faith.

If the predictions of a theory are inconsistent with observations, the theory is modified, applied in more limited circumstances, or discarded in favor of a more accurate explanation. For example, Newton's law of gravitation is entirely adequate for describing the motion of an apple falling to Earth, the flight of a soccer ball, or the path of Earth orbiting the Sun; however, it is inaccurate in describing the orbit of Mercury around the Sun or the behavior of matter in the vicinity of a black hole. In these two cases, Newton's law of gravitation is replaced by Einstein's theory of general relativity, which describes gravitational behavior more accurately and over a much wider range of conditions than Newton's law, but at the cost of much greater mathematical complexity.

Give an example of one scientific hypothesis and one nonscientific hypothesis.

Insight Into Science

Theories and Beliefs New theories are personal creations, but science is not a personal belief system. As stated in the previous Insight Into Science, scientific theories make predictions that can be tested independently. If everyone who performs tests of the theory's predictions gets results consistent with the theory, the theory is considered valid in that realm. In comparison, belief systems—such as which sports team or political system is best—are personal matters. People will always hold differing opinions about such issues.

Science also strives to explain as many things as possible with as few theories as possible. We see billions upon billions of objects in the universe. It would be virtually impossible to study all of them separately so that we could come up with detailed descriptions of each one. Fortunately, individual theories explaining each object are not necessary. Scientists overcome this problem by noting that many of the bodies in space appear similar to each other. By categorizing them suitably and then applying the scientific method to these groups of objects, we form a few theories that describe many objects and how they have evolved. These few theories can then be tested and refined as necessary. Such groupings of objects have proven invaluable, and they give us insights into the structure and organization of billions of stars and galaxies that are, indeed, very similar to one another.

Insight Into Science

Keep It Simple When several competing theories describe the same concepts with the same accuracy, scientists choose the simplest one—namely, the theory that contains the fewest unproven assumptions. That basic tenet, formally expressed by William of Occam in the fourteenth century, is known as **Occam's razor.** Indeed, the original form of the Sun-centered cosmology, which we are about to explore, was appealing because it made the same predictions within a simpler model than did the Earth-centered cosmology. *Remember Occam's razor.*

While the vast majority of scientists carefully and scrupulously follow the rules of scientific research, we acknowledge that some experiments are run poorly. Some scientists have ignored experimental data or observations that do not mesh with cherished beliefs or have even fudged data or stolen data from others. Virtually all of these oversights and misdeeds are eventually discovered because most theories and their predictions are tested by several independent researchers.

The scientific method can be summarized in six words: *observe, hypothesize, predict, test, modify, economize.* I urge you to watch for applications of the scientific method throughout this book. Our first encounter with it is the discovery that Earth orbits the Sun.

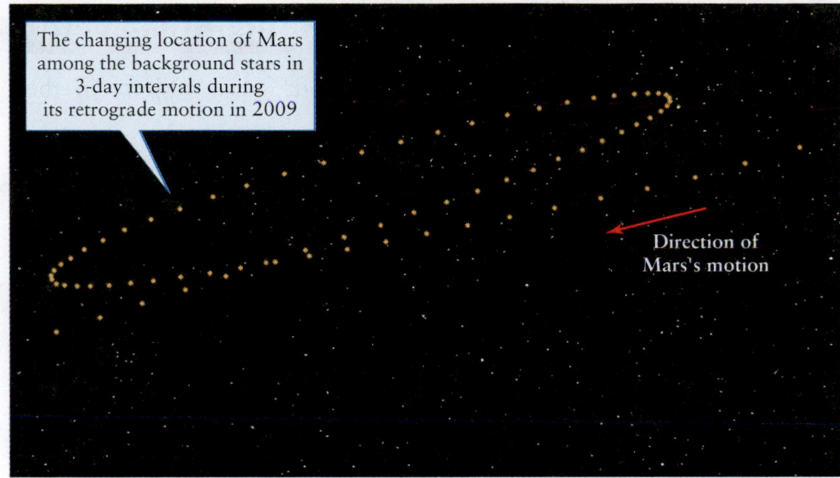

a

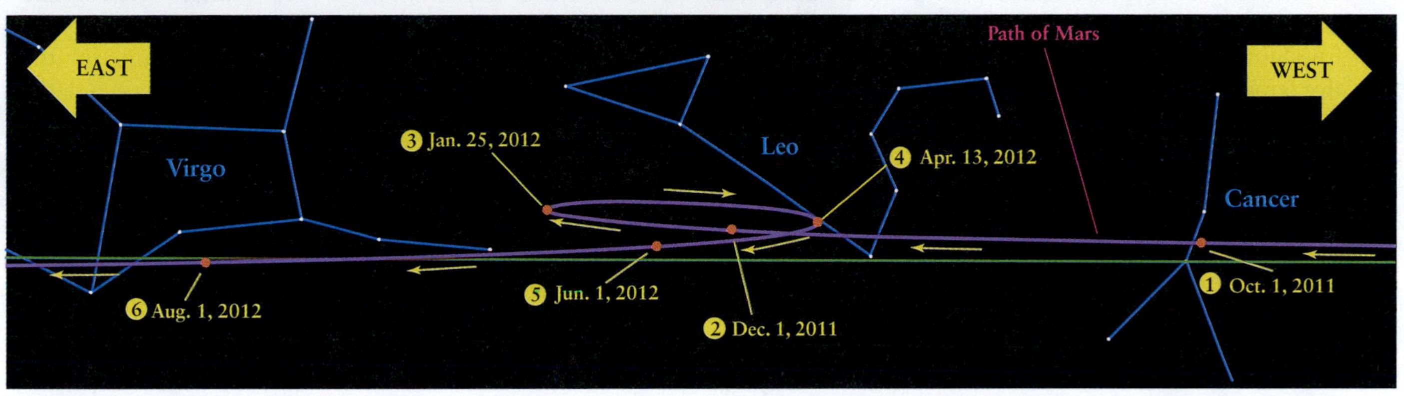

b

ANIMATION 2.1 **FIGURE 2-2 Paths of Mars** (a) The retrograde motion of Mars as it would be seen in a series of images taken on the same photographic plate. (b) To help visualize this motion on the celestial sphere, astronomers often plot the position of Mars (or other body in retrograde motion) on a star chart. From January 25, 2012, through April 13, 2012, Mars undergoes retrograde motion as seen from Earth. The retrograde path is sometimes a loop north (shown here) or south of the normal path, and sometimes an S-shaped path above or below it.

CHANGING OUR EARTH-CENTERED VIEW OF THE UNIVERSE

Early Greek astronomers tried to explain the motion of the five then-known planets: Mercury, Venus, Mars, Jupiter, and Saturn. Most people at that time held a *geocentric* view of the universe: Based on the observed motion of the celestial sphere, they believed that the Sun, the Moon, the stars, and the planets revolve around Earth. A theory of the overall structure and evolution of the universe is called a **cosmology**, so the prevailing Earth-centered cosmology was called *geocentric*. Geocentric cosmology, consistent as it is with casual observation of the sky, held sway until the sixteenth century, when Copernicus revolutionized our understanding of the cosmos.

2-2 The belief in a Sun-centered cosmology formed slowly

Explaining the motions of the five planets in a geocentric universe was one of the main challenges facing the astronomers of antiquity. The Greeks knew that the positions of the planets slowly shift relative to the "fixed" stars in the constellations. In fact, the word *planet* comes from a Greek term meaning "wanderer." They also observed that planets do not move at uniform rates through the constellations. From night to night, as viewed in the northern hemisphere, the planets usually move slowly to the left (eastward) relative to the background stars. This movement is called **direct motion.** Occasionally, however, a planet seems to stop and then back up for several weeks or months. This reverse movement (to the right or westward relative to the background stars) is called **retrograde motion.** Both direct and retrograde motions are best observed by photographing (Figure 2-2a) or plotting (Figure 2-2b) the nightly position of a planet against the background stars over a long period.

All planetary motions on the celestial sphere are much slower than the apparent daily movement of the entire sky caused by Earth's rotation. Throughout a single night it is very difficult to detect motion of the planets among the stars. Therefore, the planets always rise with the stars in the eastern half of the sky and set with them in the western half.

Referring to Figure 2-2b, why do you think Mars is seen sometimes above the ecliptic and sometimes below it?

Earth-Centered Universe

As we move through the twenty-first century, most of us find it hard to understand why anyone would believe that the Sun, planets, and stars orbit Earth. After all, we know that Earth spins on its axis. We know that the gravitational force from the Sun holds the planets in orbit, just as Earth's gravitational force holds the Moon in orbit. These facts have become part of our understanding of the motions of the heavenly bodies and we are taught these things from the time we are children.

Psychologists call this background information that we use to help explain things a *conceptual framework*. Any conceptual framework contains all of the information we take for granted. For example, when the Sun rises, moves across the sky, and sets, we take for granted that it is Earth's rotation that causes the Sun's apparent motion.

Our ancestors possessed a different conceptual framework for understanding the cosmos. They did not know that Earth rotates. They did not know that the then-mysterious force that held them to the ground is the same force that attracts Earth to the Sun and the Moon to Earth. They did not know that the Sun is a star, just like the fixed points of light in the sky, and they did not know any of the other laws of motion that we take for granted.

Because they did not feel Earth move under their feet, or see any other indication that Earth is in motion, our forebears sensed nothing to support the belief that we are in orbit around the Sun. The obvious conclusion for one who has a prescientific conceptual framework, even today, is that Earth stays put while objects in the heavens move around it.

This prescientific conceptual framework for understanding the motions of the heavenly bodies was based on the senses and on common sense. That is, people observed motions and drew "obvious," commonsense conclusions. Today, we incorporate the known and tested laws of physics in our understanding of the natural world. Many of these realities are utterly counterintuitive, and, therefore, the conceptual frameworks that we possess are less consistent with common sense than those held in the past. Studying science helps us develop intuition that is consistent with the actual workings of nature.

Geocentric Explanation of the Planets' Retrograde Motion The early Greeks, working in a geocentric cosmology, developed many theories to account for the occasional retrograde motion of the planets and the resulting loops that the planets trace out against the background stars. One of the most successful ideas was expounded by the last of the great ancient Greek astronomers, Ptolemy, who lived in Alexandria, Egypt, 1900 years ago. His basic concepts are sketched in the accompanying figure. Each planet is assumed to move in a small circle called an *epicycle*, the center of which moves in a larger circle called a *deferent*, whose center is offset from Earth. As viewed from Earth, the epicycle moves eastward along the deferent, and both it and the planet on it revolve in the same direction (counterclockwise in the figure below).

Most of the time, the motion of the planet on its epicycle adds to the eastward motion of the epicycle on the deferent. Thus, the planet is seen from Earth to be in direct

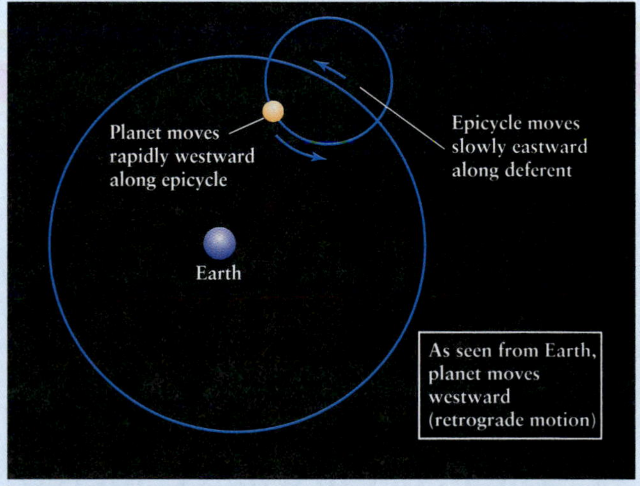

ANIMATION 2.2 **Figure GD2-1 A Geocentric Explanation of Planetary Motion** Each planet revolves around an epicycle, which in turn revolves around a deferent centered approximately on Earth. As seen from Earth, the speed of the planet on the epicycle alternately (a) adds to or (b) subtracts from the speed of the epicycle on the deferent, thus producing alternating periods of direct and retrograde motions.

motion (to the left or eastward) against the background stars throughout most of the year (Figure GD2-1a). However, when the planet is on the part of its epicycle nearest Earth, its motion along the epicycle subtracts from the motion of the epicycle along the deferent. The planet thus appears to slow and then halt its usual movement to the left (eastward motion) among the constellations, and then seems to move to the right (westward) among the stars for a few weeks or months (Figure GD2-1b). This concept of epicycles and deferents enabled Greek astronomers to explain the retrograde loops of the planets.

Using the wealth of astronomical data in the Library of Alexandria, including records of planetary positions covering hundreds of years, Ptolemy deduced the sizes of the epicycles and deferents and the rates of revolution needed to produce the recorded paths of the planets. After years of arduous work, Ptolemy assembled his calculations in the *Almagest*, in which the positions and paths of the Sun, the Moon, and the planets were described with unprecedented accuracy. In fact, the *Almagest* was so successful that it became the astronomer's bible. For more than 1000 years, Ptolemy's cosmology endured as a useful description of the workings of the heavens.

Eventually, however, the commonsense explanation of the Earth-centered cosmology began to go awry. Errors and inaccuracies that were unnoticeable in Ptolemy's day compounded and multiplied over the years, especially errors due to precession, the slow change in the direction of Earth's axis of rotation. Fifteenth-century astronomers made some cosmetic adjustments to the Ptolemaic system. However, the system became less and less satisfactory as more fanciful and arbitrary details were added to keep it consistent with the observed motions of the planets. After Newton's time, scientists knew that orbital motion required a force to be acting on the body. However, nothing in Ptolemy's epicycle theory produced such a force.

The effort to understand planetary motion—and especially to explain retrograde motion—in a geocentric cosmology resulted in an increasingly contrived and complex model (see *Guided Discovery: Earth-Centered Universe*). The ancient Greek astronomer Aristarchus of Samos proposed a more straightforward explanation of planetary motion, namely, that all of the planets, including Earth, revolve around the Sun. The retrograde motion of Mars in this **heliocentric** (Sun-centered) **cosmology** occurs because the faster-moving Earth overtakes and passes the Red Planet (Figure 2-3). The occasional retrograde movement of a planet is

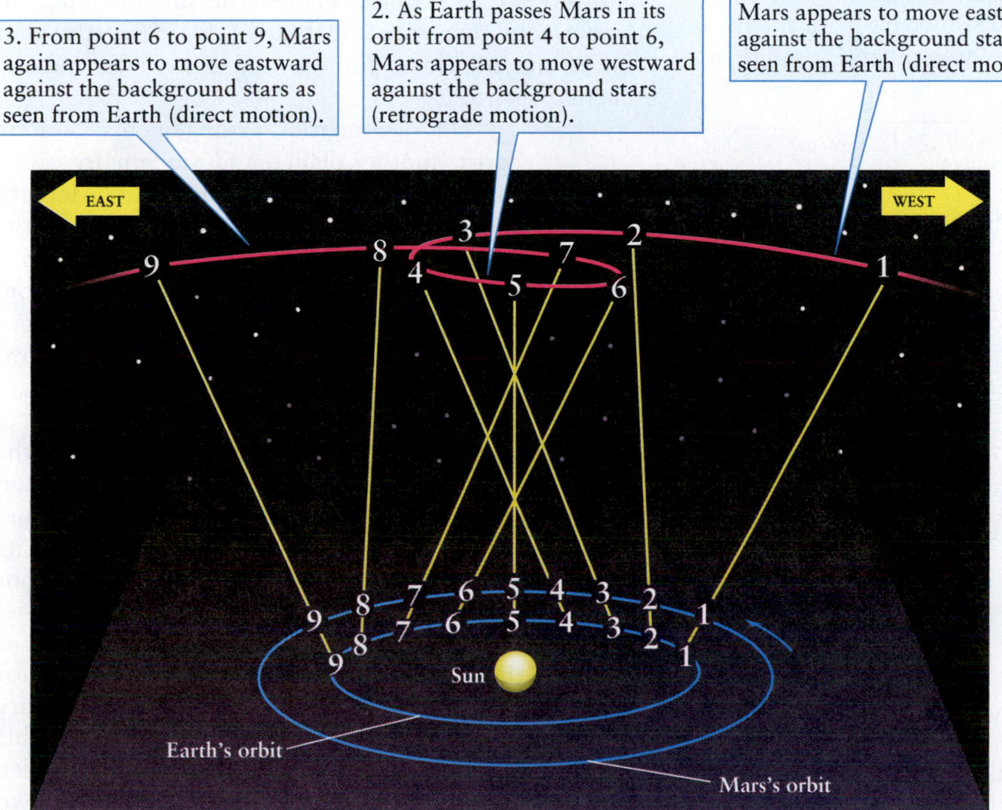

ANIMATION 2.3

FIGURE 2-3 A Heliocentric Explanation of Retrograde Motion Earth travels around the Sun more rapidly than does Mars. Consequently, as Earth overtakes and passes this slower-moving planet, Mars appears (from points 4 through 6) to move backward among the background stars for a few months.

merely the result of our changing viewpoint as we orbit the Sun—an idea that is beautifully simple compared to the geocentric system with all of its complex planetary motions. (The word *heliocentric* is misleading. Although the local planets, moons, and small pieces of space debris do orbit the Sun, the stars and innumerable other objects in space do not. In fact, the Sun and the bodies that orbit it all orbit the center of our Milky Way Galaxy.)

2-3 Copernicus devised the first comprehensive heliocentric cosmology

Over the centuries following Aristarchus, increasingly accurate observations of the planets' locations revealed errors in the predictions of geocentric cosmology. To reconcile that cosmology with the data, more and more complex motions were attributed to the planets. By the mid-1500s, the geocentric cosmology had become truly unwieldy in its efforts to predict the motions of the planets accurately. It was then that the Polish mathematician, lawyer, physician, economist, cleric, and artist Nicolaus Copernicus resurrected Aristarchus's theory. Copernicus (see *Guided Discovery: Astronomy's Foundation Builders*) was motivated by an effort to simplify the celestial scheme.

Because simplicity and accuracy are hallmarks of science, the complex geocentric model eventually gave way to the simpler, more elegant heliocentric cosmology, but it took over 1800 years from the time of Aristarchus before that fundamental change in cosmology was accepted by most astronomers. Contrary to popular lore today, the belief that Earth is the center of the universe was not a primarily religious construct (that is, it did not proceed from the idea that we humans are special because we are "at the center of everything"). The geocentric model began from the observation that everything in space appears to go around Earth and that like us today, our ancestors didn't feel Earth moving. Religions did not particularly feed that theory because all the good stuff was alleged to be going on "in heaven," not on Earth. Indeed, regardless of our seeming to be at the center of all that our ancestors could see, the Earth was often considered the lowest place (in terms of quality) in the universe—Earth at the center was considered by some theologians and philosophers as the garbage heap of the universe down to which all the bad stuff fell. Philosophers, such as Aristotle, also perceived the heavenly bodies beyond the Moon as perfect, while Earth and the Moon were considered to be corrupted.

After assuming that the planets orbit the Sun rather than Earth, Copernicus determined by observations which planets are closer to the Sun than Earth and which are farther away. Because Mercury and Venus are always observed fairly near the Sun, he correctly concluded that their orbits must lie inside Earth's. The other planets visible to Copernicus—Mars, Jupiter, and Saturn—can sometimes be seen high in the sky in the middle of the night, when the Sun is far below the horizon. This placement can occur only if Earth comes between the Sun and a planet. Copernicus therefore concluded (also correctly) that the orbits of Mars, Jupiter, and Saturn lie outside Earth's orbit.

The geometric arrangements among Earth, another planet, and the Sun are called **configurations**. For example, when Mercury or Venus is directly between Earth and the Sun (Figure 2-4), we say the planet is in a configuration called an **inferior conjunction**; when either of these planets is on the opposite side of the Sun from Earth, its configuration is called a **superior conjunction.**

The angle between the Sun and a planet as viewed from Earth is called the planet's **elongation**. A planet's elongation varies from zero degrees to a maximum value, depending upon where we see it in its orbit around the Sun. At *greatest eastern* or *greatest western elongation*, Mercury and Venus are as far from the Sun in angle as they can be. This position is about 28° for Mercury and about 47° for Venus. When either Mercury or Venus rises before the Sun, the planet is visible in the eastern sky as a bright "star" and is often called the "morning star." Similarly, when either of these two

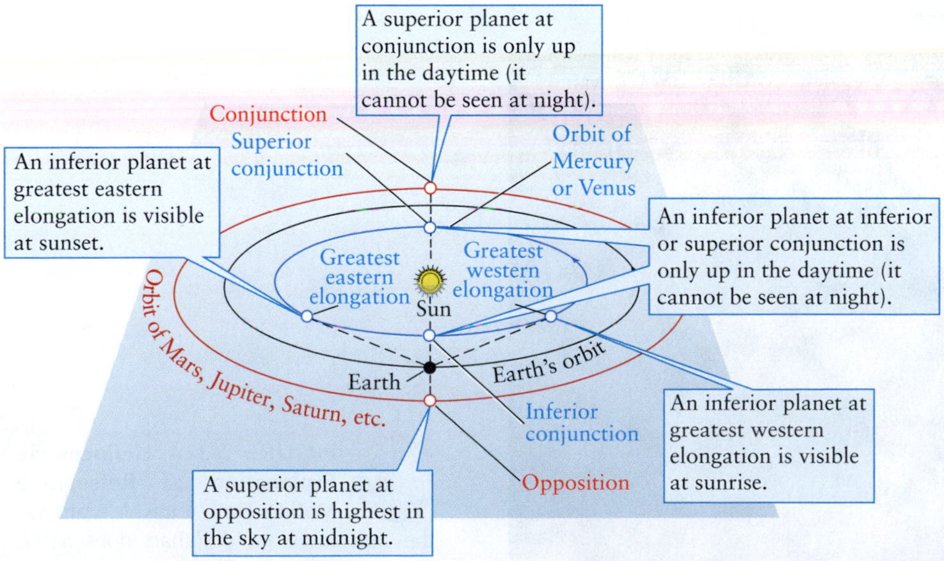

INTERACTIVE EXERCISE 2.1

FIGURE 2-4 Planetary Configurations Key points along a planet's orbit have names, as shown. These points identify specific geometric arrangements between Earth, another planet, and the Sun.

Astronomy's Foundation Builders

In the two centuries between 1500 and 1700, human understanding of the motion of celestial bodies and the nature of the gravitational force that keeps them in orbit surged forward as never before. Theories related to this subject were developed by brilliant thinkers, whose work established and verified the heliocentric model of the solar system and the role of gravity.

(World History Archive/Alamy)

Nicolaus Copernicus (1473–1543) Copernicus, the youngest of four children, was born in Torun, Poland. He pursued his higher education in Italy, where he received a doctorate in canon law and studied medicine. Copernicus developed a heliocentric theory of the known universe and just before his death in 1543 published this work under the title *De Revolutionibus Orbium Coelestium.* His revolutionary theory was flawed in that he assumed that the planets had circular orbits around the Sun. This was corrected by Johannes Kepler.

(INTERFOTO/Alamy)

Tycho Brahe (1546–1601) and Johannes Kepler (1571–1630) Tycho (depicted here and within the portrait of Kepler) was born to nobility in the Danish city of Knudstrup, which is now part of Sweden. At age 20, he lost part of his nose in a duel and wore a metal replacement thereafter. In 1576, the Danish king Frederick II built Tycho an astronomical observatory that Tycho named Uraniborg (after Urania, Greek muse of astronomy). Tycho rejected both Copernicus's heliocentric theory and the Ptolemaic geocentric system. He devised a halfway theory called the *Tychonic system.* According to Tycho's theory, Earth is stationary, with the Sun and the Moon revolving around it, while all the other planets revolve around the Sun.

(Painting by Jean-Leon Huens, courtesy of National Geographic Society)

Kepler was educated in Germany, where he spent 3 years studying mathematics, philosophy, and theology. In 1596, Kepler published a booklet in which he attempted to mathematically predict the planetary orbits. Although his theory was altogether wrong, its boldness and originality attracted the attention of Tycho Brahe, whose staff Kepler joined in 1600. Kepler deduced his three laws from Tycho's observations.

(Stock Montage, Inc./Alamy)

Galileo Galilei (1564–1642) Born in Pisa, Italy, Galileo studied medicine and philosophy at the University of Pisa. He abandoned medicine in favor of mathematics. He held the chair of mathematics at the University of Padua, and eventually returned to the University of Pisa as a professor of mathematics. There Galileo formulated his famous law of falling bodies: All objects fall with the same acceleration regardless of their weight. In 1609 he constructed a telescope and made a host of discoveries that contradicted the teachings of Aristotle and the Roman Catholic Church. He summed up his life's work on motion, acceleration, and gravity in the book *Dialogues Concerning the Two Chief World Systems,* published in 1632.

(Lebrecht Music and Arts Photo Library/Alamy)

Isaac Newton (1642–1727) Newton delighted in constructing mechanical devices, such as sundials, model windmills, a water clock, and a mechanical carriage. He received a bachelor's degree in 1665 from the University of Cambridge. While there, he began developing the mathematics that later became calculus (developed independently by the German Gottfried Leibniz). While pursuing experiments in optics, Newton constructed a reflecting telescope and also discovered that white light is actually a mixture of all colors. His major work on forces and gravitation was the tome *Philosophiae Naturalis Principia Mathematica*, which appeared in 1687. In 1704, Newton published his second great treatise, *Opticks*, in which he described his experiments and theories about light and color. Upon his death in 1727, Newton was buried in Westminster Abbey, the first scientist to be so honored.

planets sets after the Sun, the planet is visible in the western sky and is then called the "evening star." Because these two planets are not always at their greatest elongations, they are often very close in angle to the Sun. This positioning is especially true of Mercury, often making it hard to see from Earth. Venus is often nearly halfway up the sky at sunrise or sunset and therefore quite noticeable during much of its orbit. Because they are so bright and sometimes appear to change color due to the motion of Earth's atmosphere, Venus and Mercury are often mistaken for UFOs. (The same motion of the air causes the road in front of your car to shimmer on a hot day.)

Planets farther from the Sun than Earth have different configurations. When one of these planets is located behind the Sun, as seen from Earth, that planet is said to be in **conjunction**. When a planet is opposite the Sun in the sky, that planet is at **opposition**. It is not difficult to determine when a planet happens to be located at one of the key positions in Figure 2-4. For example, when Mars is at opposition, it appears high and bright in the sky at midnight.

It is relatively easy to follow a planet as it moves from one configuration to another. However, these observations alone do not tell us the planet's actual orbit, because Earth, from which we make the observations, is also moving. Copernicus was therefore careful to distinguish between two characteristic time intervals, or *periods*, of each planet.

Recall from your study of the Moon in Chapter 1 that the sidereal period of an orbit is the true orbital period of any astronomical body. A planet's **sidereal period** is the time it takes that body to make one complete orbit of the Sun. In other words, an observer fixed at the Sun's location watching that planet would see it move through the background stars to go from one point on the celestial sphere, around the sphere, and back to that same point again. The sidereal period is the length of a year for each planet.

The other useful time interval that Copernicus used is the **synodic period.** The synodic period is the time that elapses between two successive identical configurations as seen from Earth. This period can be from one opposition to the next, for example, or from one conjunction to the next (Figure 2-5). The synodic period tells us, among other things, when to expect a planet to be closest to Earth and, therefore, most easily studied.

Insight Into Science

Take a Fresh Look When a scientific concept is hard to visualize, try another perspective. For example, a planet's sidereal period of orbit is easy to understand when viewed from the Sun but more complicated as seen from Earth. The synodic period of each planet, on the other hand, is easily determined from Earth. As we will see, especially when we study Einstein's theories of relativity, each of these perspectives is called a *frame of reference*.

Thus, nearly 500 years ago, Copernicus was able to obtain the first six entries shown in Table 2-1 (the others are contemporary results included for completeness). Copernicus was then able to devise a straightforward geometric method for determining the distances of the planets from the Sun. His answers turned out to be remarkably close to the modern values, as shown in Table 2-2. From these two tables it is apparent that the farther a planet is from the Sun, the longer it takes to complete its orbit.

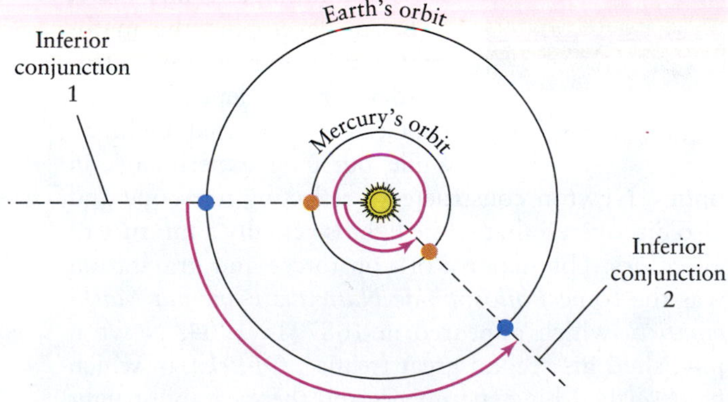

FIGURE 2-5 Synodic Period The time between consecutive conjunctions of Earth and Mercury is 116 days. As is typical of synodic periods for all planets, the location of Earth is different at the beginning and end of the period. You can visualize the synodic periods of the outer planets by putting Earth in Mercury's place in this figure and putting one of the outer planets in Earth's place.

Table 2-1 Synodic and Sidereal Periods of the Planets (in Earth Years)

	Synodic (year)	Sidereal (year)
Mercury	0.318	0.241
Venus	1.599	0.616
Earth	—	1.0
Mars	2.136	1.9
Jupiter	1.092	11.9
Saturn	1.035	29.5
Uranus	1.013	84.0
Neptune	1.008	164.8

Table 2-2 Average Distances of the Planets from the Sun

	Measurement (AU)	
	By Copernicus	Modern
Mercury	0.38	0.39
Venus	0.72	0.72
Earth	1.00	1.00
Mars	1.52	1.52
Jupiter	5.22	5.20
Saturn	9.07	9.54
Uranus	Unknown	19.19
Neptune	Unknown	30.06

Copernicus presented his heliocentric cosmology, including supporting observations and calculations, in a book entitled *De Revolutionibus Orbium Coelestium* (On the Revolutions of the Celestial Spheres), which was published in 1543, the year of his death. His great insight was the conceptual simplicity of a heliocentric cosmology compared to geocentric views, especially in explaining retrograde motion. However, Copernicus incorrectly assumed that the planets travel along circular paths around the Sun. Without using epicycles similar to those used in geocentric theory (see Figure GD2-1), many of his predictions were no more accurate than those of the earlier theory! As we will see shortly, by changing the shape of the orbits, Kepler was able to do away with epicycles and make even more accurate predictions than either the geocentric or original Copernican theories.

2-4 Tycho Brahe made astronomical observations that disproved ancient ideas about the heavens

In November 1572, a bright star suddenly appeared in the constellation Cassiopeia. At first, it was even brighter than Venus, but then it began to grow dim. After 18 months the star faded from view.

Modern astronomers recognize this event as a supernova explosion, the violent death of a certain type of star (see Chapter 13). In the sixteenth century, however, the prevailing opinion was quite different. Teachings dating back to Aristotle and Plato argued that the heavens were permanent and unalterable. From that perspective, the "new star" of 1572 could not really be a star at all, because the heavens do not change. Many astronomers and theologians of the day argued that the sighting must be some sort of bright object quite near Earth, perhaps not much farther away than the clouds overhead. A 25-year-old Danish astronomer named Tycho Brahe (see *Guided Discovery: Astronomy's Foundation Builders*) realized that straightforward observations might reveal the distance to this object.

Consider what happens when two people look at a nearby object from different places—they see it in different positions relative to the things behind it. Furthermore, their heads face at different angles when looking at it. This variation in angle that occurs when viewing a nearby object from different locations is called **parallax** (Figure 2-6).

FIGURE 2-6 Parallax Nearby objects are viewed at different angles from different places. These objects also appear to be in different places with respect to more distant objects when viewed at the same time by observers located at different positions. Both effects are called parallax, and they are used by astronomers, surveyors, and sailors to determine distances. (Tobi Zausner)

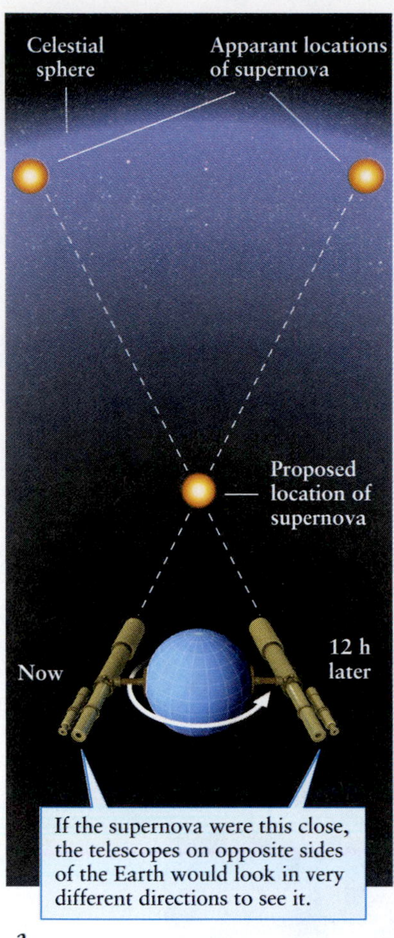

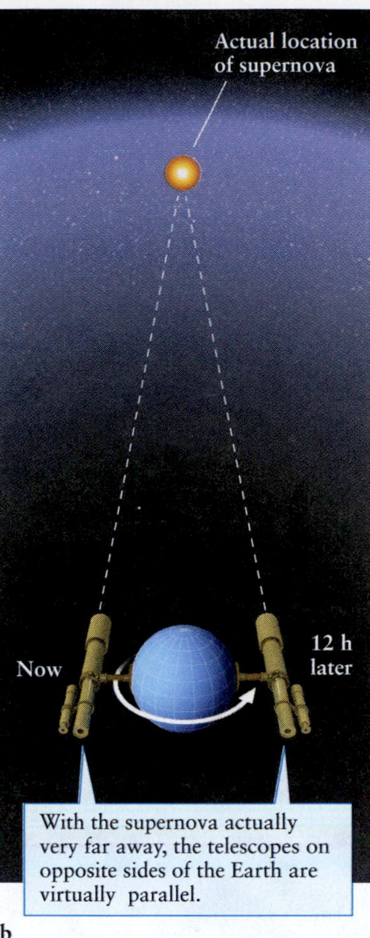

FIGURE 2-7 The Parallax of a Nearby Object in Space Tycho thought that Earth does not rotate and that the stars revolve around it. From our modern perspective, the changing position of the supernova would be due to Earth's rotation as shown. (a) Tycho argued that if an object is near Earth, its position relative to the background stars should change over the course of a night. (b) Tycho failed to measure such changes for the supernova in 1572. This is illustrated in (b) by the two telescopes being parallel to each other. He, therefore, concluded that the object was far from Earth.

> Describe a simple experiment to demonstrate that your eyes (and, implicitly, your brain) use parallax to determine distances.

Tycho reasoned as follows: If the new star is nearby, its position should shift against the background stars over the course of a night (Figure 2-7a). His careful observations, done in the spirit of the scientific method, failed to disclose any parallax, and so the new star had to be far away, farther from Earth than anyone had imagined (Figure 2-7b). Tycho summarized his findings in a small book, *De Stella Nova* (On the New Star), published in 1573.

Tycho's astronomical records were soon to play an important role in the development of a heliocentric cosmology. From 1576 to 1597, Tycho made comprehensive observations, measuring planetary positions with an accuracy of 1 arcmin, about as precise as is possible with nontelescopic instruments. (Arcminute [arcmin] is defined in *An Astronomer's Toolbox 1-1*.) Upon Tycho's death in 1601, most of these invaluable records were given to his gifted assistant, Johannes Kepler (see *Guided Discovery: Astronomy's Foundation Builders*).

KEPLER'S AND NEWTON'S LAWS

Until Kepler's time, astronomers had assumed that heavenly objects move in circles. For philosophical and aesthetic reasons, circles were considered the most perfect and most harmonious of all geometric shapes. However, using circular orbits failed to yield accurate predictions for the positions of the planets. For years, Kepler tried to find a shape for orbits that would fit Tycho's observations of the planets' positions against the background of distant stars. Finally, he began working with a geometric form called an **ellipse**.

2-5 Kepler's laws describe orbital shapes, changing speeds, and the lengths of planetary years

You can draw an ellipse as shown in Figure 2-8a. Each thumbtack is at a **focus** (plural **foci**). The longest diameter (major axis) across an ellipse passes through both foci. Half of that distance is called the **semimajor axis**. In astronomy, the length of the semimajor axis is also the average distance between a planet and the Sun.

To Kepler's delight, the ellipse turned out to be the 2
curve for which he had been searching. Predictions of the locations of planets based on elliptical paths were in very close agreement with where the planets actually were. Keep in mind that the following three laws that Kepler discovered merely quantified (gave equations for) observations that Tycho had made—Kepler did not have a theory to explain them. That would come nearly 80 years later, with the work of Isaac Newton.

Kepler published his discovery of elliptical orbits in 1609 in a book known today as *New Astronomy*. This important discovery is now considered the first of **Kepler's laws**:

Kepler's First Law: *The orbit of a planet around the Sun is an ellipse with the Sun at one focus.*

The shapes of ellipses have two extremes. The roundest ellipse, occurring when the two foci merge, is a circle. The most elongated ellipse approaches being a straight line. The shape of a planet's orbit around the Sun is described by its *orbital eccentricity*, designated

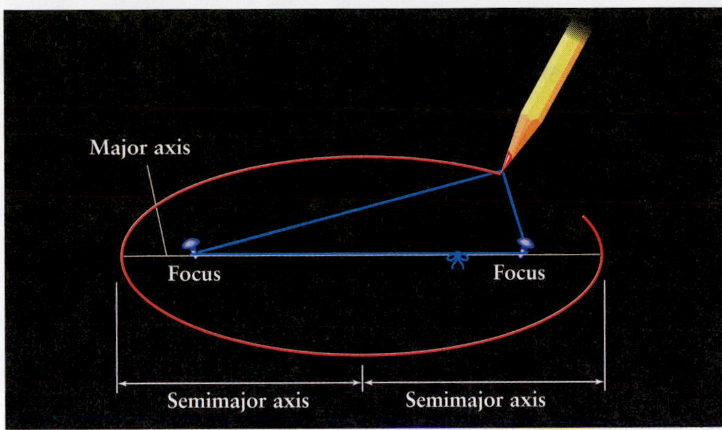

a The geometry of an ellipse

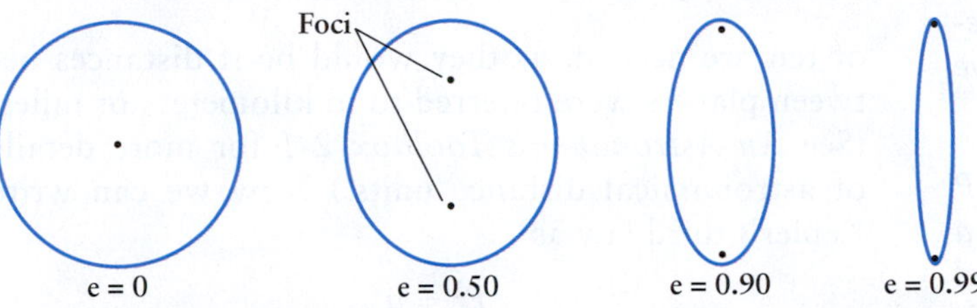

b Ellipses with different eccentricities

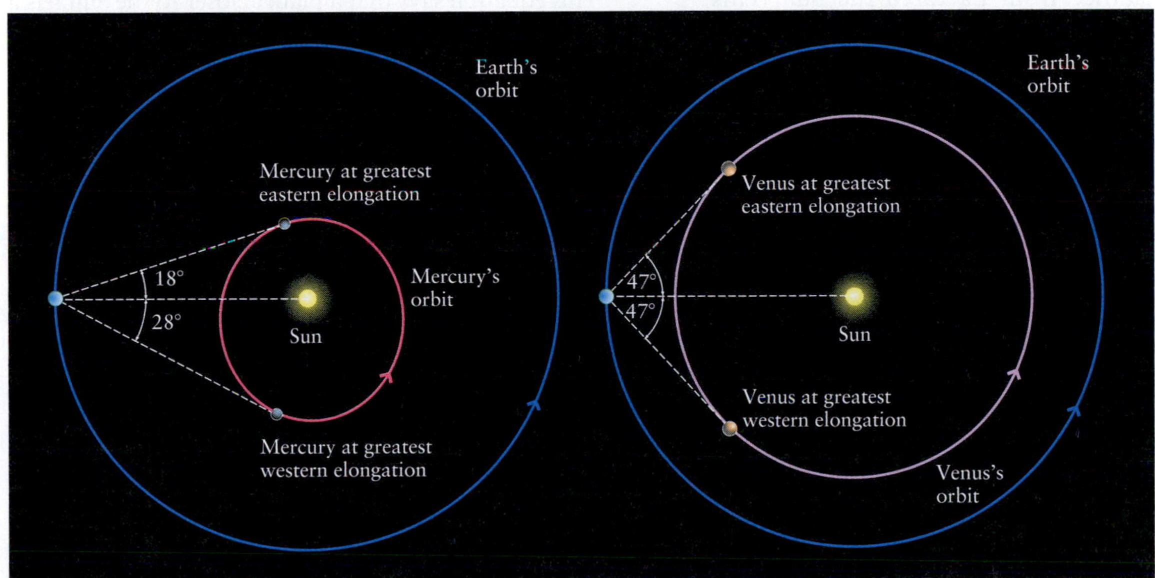

c

FIGURE 2-8 Ellipses (a) The construction of an ellipse: At all places along an ellipse, the sum of the distances to the two foci is a constant. An ellipse can be drawn with a pencil, a loop of string, and two thumbtacks, as shown. If the string is kept taut, the pencil traces out an ellipse. The two thumbtacks are located at the two foci of the ellipse. (b) A series of ellipses with different eccentricities (*e*). Eccentricities range between 0 (a circle) to just under 1.0 (almost a straight line). Note that all eight planets have eccentricities of less than 0.21. (c) Mercury has an especially eccentric orbit around the Sun. As seen from Earth, the angle of Mercury at greatest elongation ranges from 18° to 28°. In contrast, Venus's orbit is nearly circular, with both greatest elongations of 47°.

by the letter *e*, which ranges from 0 (a circular orbit) to just under 1.0 (nearly a straight line). Figure 2-8b shows a sequence of ellipses and their associated eccentricities. Observations have revealed that there is no object at the second focus of each elliptical planetary orbit. Figure 2-8c shows the effect of orbital eccentricity. For example, the maximum elongation of Mercury ($e = 0.21$) seen from Earth varies by 10°, while the maximum elongation of Venus ($e = 0.01$) varies by less than 1°.

3 Tycho's observations also showed Kepler that planets do not move at uniform speeds along their orbits. Rather, a planet moves most rapidly when it is nearest the Sun, a point on its orbit called **perihelion.** Conversely, a planet moves most slowly when it is farthest from the Sun, called its **aphelion.**

After much trial and error, Kepler discovered a way to describe how fast a planet moves anywhere along its orbit. This discovery, also published in *New Astronomy*, is illustrated in Figure 2-9. Suppose that it takes 30 days for a planet to go from point A to point B. During that time, the line joining the Sun and the planet sweeps out a nearly triangular area (shaded in Figure 2-9). Kepler discovered that the line joining the Sun and the planet sweeps out the same area during any other 30-day interval. In other words, if the planet also takes 30 days to go from point C to point D, then the two shaded segments in Figure 2-9 are equal in area. Kepler's second law, also called the **law of equal areas**, can be stated thus:

Kepler's Second Law: *A line joining a planet and the Sun sweeps out equal areas in equal intervals of time.*

A consequence of Kepler's second law is that each planet's speed decreases as it moves from perihelion to aphelion. The speed then increases as the planet moves from aphelion toward perihelion.

Kepler was also able to relate a planet's year to its distance from the Sun. This discovery, published in 1619, is Kepler's third law. This relationship predicts the planet's sidereal period if we know the length of the semimajor axis of the planet's orbit:

Kepler's Third Law: *The square of a planet's sidereal period around the Sun is directly proportional to the cube of the length of its orbit's semimajor axis.*

(© 1980 Sidney Harris)

The relationship is easiest to use if we let P represent the sidereal period in Earth years and a represent the length of the semimajor axis measured in **astronomical units (AU)**. One astronomical unit is the average distance from Earth to the Sun. This unit is commonly used when measuring distances to various objects in the solar system, because no powers of ten are needed, as they would be if distances between planets were referred to in kilometers or miles. (See *An Astronomer's Toolbox 2-1* for more details of astronomical distance units.) Now we can write Kepler's third law as

$$P^2 = a^3$$

This equation says that a planet closer to the Sun has a 4
shorter year than does a planet farther from the Sun. Using this equation with Kepler's second law reveals that planets closer to the Sun move more rapidly than those farther away. Using data from Tables 2-1 and 2-2, we can demonstrate Kepler's third law as shown in Table 2-3.

When Newton derived Kepler's third law using the law of gravitation, discussed later in this chapter, he discovered that the mass of a planet affects the period of its orbit around the Sun. The **mass** of an object is a measure of the total number of particles that it contains and is expressed in units of kilograms. For example, the mass of the Sun is 2×10^{30} kg, the mass of a hydrogen atom is 1.7×10^{-27} kg, and the mass of the author of this book is 83 kg. At rest, the Sun, a hydrogen atom, and I have these same masses regardless of where we happen to be in the universe. It is important not to confuse the concept of mass with the concept of weight. Your **weight** is the force with which you push down on a scale due to the gravitational attraction of the world on which you stand.

However, the effect of the planet's mass on the period of its orbit is exceedingly small for all the planets in the solar system, which is why the equation for Kepler's third law, as shown in Table 2-3, gives such good results for the planets' orbits even though it does not take their masses into account. When calculating the motion of pairs of stars orbiting each other, the effects of the masses must be taken into account, as described in *An Astronomer's Toolbox 11-4*.

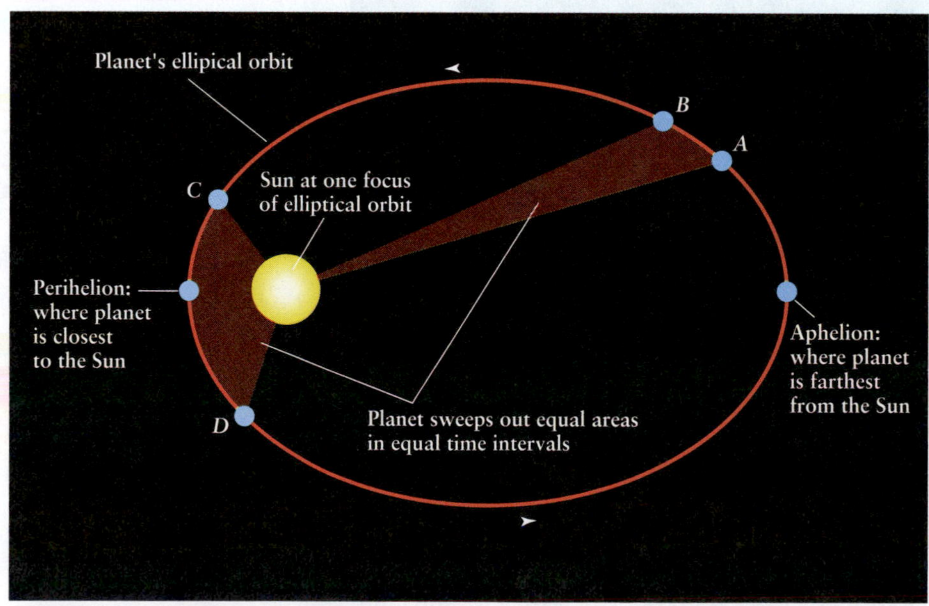

ANIMATION 2.4 **FIGURE 2-9 Kepler's First and Second Laws** According to Kepler's first law, every planet travels around the Sun along an elliptical orbit with the Sun at one focus. According to his second law, the line joining the planet and the Sun sweeps out equal areas (the burgundy-colored regions) in equal intervals of time (time from A to B equals time from C to D). *Note:* This drawing shows a highly elliptical orbit, with $e = 0.74$. Even though this is a much greater eccentricity than that of any planet in the solar system, the concept still applies to all planets and other orbiting bodies.

Table 2-3 A Demonstration of Kepler's Third Law

	Sidereal period P (year)	Semimajor axis a (AU)	P^2	=	a^3
Mercury	0.24	0.39	0.06		0.06
Venus	0.61	0.72	0.37		0.37
Earth	1.00	1.00	1.00		1.00
Mars	1.88	1.52	3.53		3.51
Jupiter	11.86	5.20	140.7		140.6
Saturn	29.46	9.54	867.9		868.3
Uranus	84.01	19.19	7058		7067
Neptune	164.79	30.06	27,160		27,160

WHAT IF... Earth were at the center of the universe? Let's return to the geocentric universe for a moment and consider a consequence of mass in the scenario where everything orbits around Earth. In 1905, Albert Einstein developed his special theory of relativity in which he showed that the faster objects move, the greater mass they acquire. If accelerated to the speed of light, 300,000 km/s (186,000 mi/s), any object that starts with even a tiny mass (such as a feather) would have more mass than everything else in the entire universe, which would make the object impossible to move. Thus, the speed of light is the ultimate speed limit. If Earth were at the center of the universe and everything orbited around us, objects farther away than 1.5×10^{12} km would each have to be traveling faster than the speed of light if they were to travel around Earth once each day. Therefore, these distant stars would have infinite mass and violate the universe's speed limit. The star closest to Earth other than the Sun is 40×10^{12} km from us. Enough said.

Kepler's three laws apply not only to the planets orbiting the Sun, but also to any object orbiting another under the influence of their mutual gravitational attraction. Thus, Kepler's laws apply to moons orbiting planets, artificial satellites orbiting Earth, and even (with the above caveat) two stars revolving around each other.

2-6 Galileo's discoveries strongly supported a heliocentric cosmology

While Kepler was in central Europe working on the laws of planetary orbits, an Italian physicist was making dramatic observations in southern Europe. Galileo Galilei did not invent the telescope, but he was one of the first people to point the new device toward the sky and publish his observations. He saw things that no one had ever imagined—mountains on the Moon and spots on the Sun. He also discovered that the apparent size of Venus as seen through his telescope was related to the planet's phase (Figure 2-10).

We saw in Chapter 1 that the Moon's orbit around Earth is not circular. Where in its orbit is the Moon moving fastest, and where is it moving slowest?

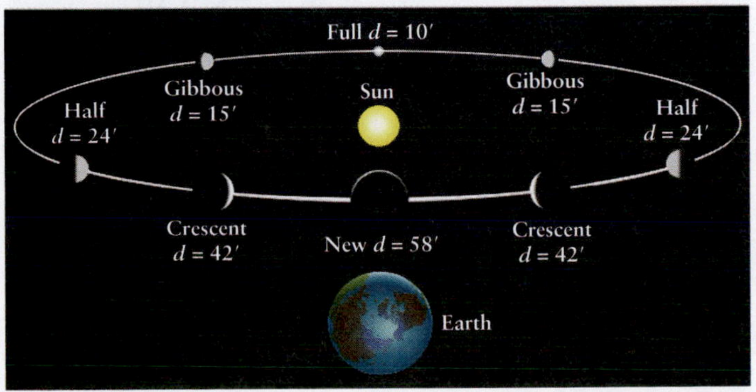

FIGURE 2-10 The Changing Appearance of Venus This figure shows how the appearance (phase) of Venus changes as it moves along its orbit. The number below each view is the angular diameter (d) of the planet as seen from Earth, in arcseconds. Note that the phases correlate with the planet's angular size and its angular distance from the Sun, both as seen from Earth. These observations clearly support the idea that Venus orbits the Sun.

An Astronomer's Toolbox 2-1

Units of Astronomical Distance

ANIMATION 2.5

Throughout this book we will find that some of our traditional units of measure become cumbersome. It is fine to use kilometers to measure the diameters of craters on the Moon or the heights of volcanoes on Mars. However, it is as awkward to use kilometers to express the large distances to planets, stars, or galaxies as it is to talk about the distance from New York City to San Francisco or Sydney to Perth in millimeters. Astronomers have therefore devised new units of measure.

When discussing distances across the solar system, astronomers use a unit of length called the astronomical unit (AU), which is the average distance between Earth and the Sun:

$$1 \text{ AU} \approx 1.5 \times 10^8 \text{ km} \approx 9.3 \times 10^7 \text{ mi}$$

Jupiter, for example, is an average of 5.2 times farther from the Sun than is Earth. Thus, the average distance between the Sun and Jupiter can be conveniently stated as 5.2 AU. This value can be converted into kilometers or miles using the previous relationship.

When talking about distances to the stars, astronomers choose between two different units of length. One is the **light-year (ly)**. A light-year is the *distance* that light travels in a year through a vacuum (that is, in the absence of air, glass, or other medium). Do keep in mind that the word *year* in this unit helps describe a separation between two objects rather than representing a unit of time.

$$1 \text{ ly} \approx 9.46 \times 10^{12} \text{ km} \approx 63{,}000 \text{ AU}$$

The spaces between the planets, stars, and galaxies are nearly ideal vacuums. One light-year is roughly equal to 6 trillion miles. Proxima Centauri, the closest star to Earth, other than the Sun, is just over 4.2 ly from us.

The second commonly used unit of length is the **parsec (pc)**, the distance at which two objects separated by 1 AU make an angle of 1 arcsec. Imagine taking a journey far into space, beyond the orbits of the outer planets. Watching the solar system as you move away, the angle between the Sun and Earth becomes smaller and smaller. When they are side by side from your perspective, and you measure the angle between them as 1/3600° (called 1 arcsec), you have reached a distance that astronomers call 1 parsec, as shown in the figure below. The parsec turns out to be longer than the light-year, specifically,

$$1 \text{ pc} \approx 3.09 \times 10^{13} \text{ km} \approx 3.26 \text{ ly}$$

Thus, the distance to the nearest star can be stated as 1.3 pc as well as 4.2 ly. Whether one uses light-years or parsecs is a matter of personal taste.

For larger distances, *kilolight* years (kly), *megalight* years (Mly), *kiloparsecs* (kpc), and *megaparsecs* (Mpc) are used. The prefixes "kilo" and "mega" simply mean "thousand" and "million," respectively:

$$1 \text{ kly} = 10^3 \text{ ly}$$
$$1 \text{ Mly} = 10^6 \text{ ly}$$
$$1 \text{ kpc} = 10^3 \text{ pc}$$
$$1 \text{ Mpc} = 10^6 \text{ pc}$$

For example, the distance from Earth to the center of our Milky Way Galaxy is about 8.6 kpc, and the rich cluster of galaxies in the direction of the constellation Virgo is 20 Mpc away.

Try these questions: The nearest star (other than the Sun) is 4.2 ly away. How many miles away is it? How many kilometers?

(Answers appear at the end of the book.)

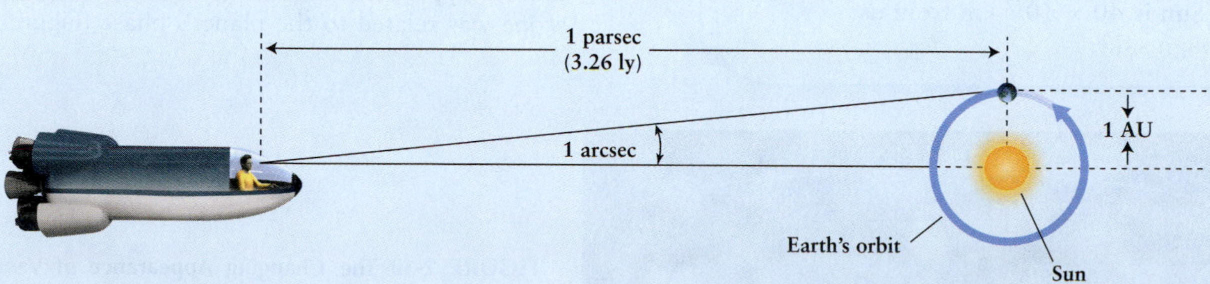

A Parsec The parsec, a unit of length commonly used by astronomers, is equal to 3.26 ly. The parsec is defined as the distance at which 1 AU perpendicular to the observer's line of sight makes an angle of 1 arcsec.

Venus appears smallest at full phase and largest at crescent phase. These observations were a big chink in the geocentric cosmology's armor, as that model could not explain why Venus has phases or changes size, while a heliocentric cosmology explains both. Galileo's observations, therefore, supported the conclusion that Venus orbits the Sun, not Earth.

In 1610, Galileo (see *Guided Discovery: Astronomy's Foundation Builders*) also discovered four moons near Jupiter. Today, in honor of their discoverer, these are called the **Galilean moons** (or **satellites**, another term for moon). Galileo concluded that the moons orbit Jupiter because he saw them move in straight lines from one side of the planet to the other. (He did not see them move in elliptical orbits because from Earth we see their orbits from edge-on.) Confirming observations were made in 1620 (Figure 2-11). These observations all provided further proof that Earth is not at the center of the universe. Like Earth in orbit around the Sun, Jupiter's four moons obey Kepler's third law: The square of a moon's orbital period around Jupiter is directly proportional to the cube of its average distance from the planet.

Galileo's telescopic observations constituted the first fundamentally new astronomical data since humans began recording what they saw in the sky. In contradiction to then-prevailing opinions, these discoveries strongly supported a heliocentric view of the universe. Because Galileo's ideas could not be reconciled with certain passages in the Bible or with the writings of Aristotle and Plato, the Roman Catholic Church condemned him, and he was forced to spend his later years under house arrest "for vehement suspicion of heresy." In 1992, Pope John Paul II stated that the church erred in this condemnation.

What is the shape of the International Space Station's orbit around Earth?

A major stumbling block prevented seventeenth-century thinkers from accepting Kepler's laws and Galileo's conclusions about the heliocentric cosmology. Once anything on Earth is put in motion, it quickly comes to rest. So why don't the planets orbiting the Sun stop, too?

The scientific method clarified most of the issues surrounding planetary orbits, leading to the equations and laws developed by the brilliant and eccentric (for example, he believed in alchemy) scientist Isaac Newton, who was born on Christmas Day in 1642, less than a year after Galileo died. In the decades that followed, Newton revolutionized science more profoundly than any person before him, and in doing so, he found physical and mathematical proofs of the heliocentric cosmology.

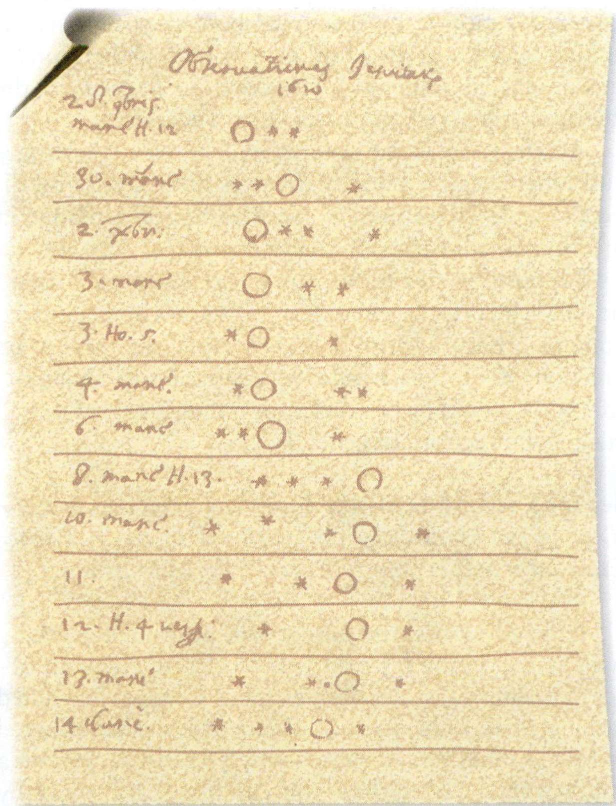

a

b

R I V U X G

FIGURE 2-11 Jupiter and Its Largest Moons In 1610, Galileo discovered four "stars" that move back and forth across Jupiter. He concluded that they are four moons that orbit Jupiter just as our Moon orbits Earth. (a) Observations made by Jesuits in 1620 of Jupiter and its four visible moons. (b) Photograph of the four Galilean satellites alongside an overexposed image of Jupiter. Each satellite would be bright enough to be seen with the unaided eye were it not overwhelmed by the glare of Jupiter. (Rev. Ronald Royer/Photo Researchers, Inc.)

Insight Into Science

Theories and Explanations Scientific theories (or laws) based on observations can be useful for making predictions even if the reasons that these theories work are unknown. The explanation for Kepler's laws came decades after Kepler deduced them, when Newton derived them in 1665 with his mathematical expression for gravitation, the force that holds the planets in their orbits.

2-7 Newton formulated three laws that describe fundamental properties of physical reality

Until the mid-seventeenth century, virtually all mathematical astronomy was done empirically. That is, astronomers from Ptolemy to Kepler created equations directly from data and observations.

How did people deduce that there is no air (and hence no air friction to slow planets down) in space before airplanes or even people-carrying balloons were invented?

Isaac Newton (see *Guided Discovery: Astronomy's Foundation Builders*) introduced a new approach. He began with three physical assumptions, now called **Newton's laws of motion**, which led to equations that have since been tested and shown to be correct in many everyday situations. He also found a formula for the force of **gravity** (or *gravitation*), the attraction between all objects due to their masses. Putting the assumptions into mathematical form and combining them with the equation for gravity, Newton was able to derive Kepler's three laws and use them to predict the orbits of bodies such as comets and other objects in the solar system. Newton also was able to use these same equations to predict the motions of bodies on and near Earth, such as the path of a projectile or the speed of a falling object.

5 **Newton's first law—the law of inertia:** *Inertia is the property of matter that keeps an object at rest or moving in a straight line at a constant speed unless acted upon by a net external force.*

If all of the external forces acting on an object do not cancel each other out, then there is a net external force acting on the object. Equivalently, we say that there is an unbalanced external force. For example, if you put a soccer ball between your hands and press on it so that it doesn't move, your hands represent a balanced pair of forces acting on the ball. In that case, you are exerting no net external force on the ball. Conversely, when your foot hits a soccer ball and the ball sails away, your foot *has* exerted a net external force on the ball.

At first, this law might seem to conflict with your everyday experience. For example, if you shove a chair, it does not move at a constant speed forever but comes to rest after sliding only a short distance. From Newton's viewpoint, however, a "net external force" does indeed act on the moving chair—namely, friction between the chair's legs and the floor. Without friction, the chair would continue in a straight path at a constant speed. A net external force changes the motion of an object.

Newton's first law tells us why the planets keep moving in orbit around the Sun. First, they do not come to rest because there is virtually no air in space and hence no force from, for example, air friction opposing their motion. Second, they do not move in straight lines because there is an outside force acting on the planets to continually change their directions and keep them in orbit. As we shall see, that force is the Sun's gravity.

Newton's second law describes quantitatively how a force changes the motion of an object. To better appreciate the concepts of force and motion, we must first understand two related quantities: velocity and acceleration.

Imagine an object motionless in space. Push on it and it begins to move. At any moment, you can describe the object's motion by specifying both its speed and direction. Speed and direction of motion together constitute an object's **velocity**. If you continue to push on the object, its speed will increase—it will accelerate.

Acceleration is the rate at which velocity changes with time. Because velocity involves both speed and direction, a slowing down, a speeding up, or a change in direction are all forms of acceleration.

Suppose, for example, an object revolved around the Sun in a perfectly circular orbit. As this object moved along its orbit, its speed would remain constant, but its direction of motion would be continuously changing. This body would have acceleration that involved only a change of direction. In general:

Newton's second law—the force law: *The acceleration of an object is directly proportional to the net force acting on it and is inversely proportional to its mass.*

In other words, the harder you push on something that can move, the faster it will accelerate. Also, an object of greater mass accelerates more slowly when acted on by a force than does an object of lesser mass acted on by the same force. That is why you can accelerate a child's wagon faster than you can accelerate a car by pushing on them equally hard.

(ScienceCartoonsPlus.com)

Newton's second law can be succinctly stated as an equation. If a **force** acts on an object, the object will experience an acceleration such that

$$\text{Force} = \text{mass} \times \text{acceleration}$$

6 Force is usually expressed in pounds or newtons. For example, the force with which I am pressing down on the ground is 814 newtons (183 lb). But I weigh 814 newtons only on Earth. I would weigh 136 newtons (30.5 lb) on the Moon, which has less mass and so pulls me down with less gravitational force. Orbiting in the space shuttle, my apparent weight (measured by standing on a scale in the shuttle) would be 0, but my mass would be the same as when I am on Earth. Because I still have inertia in the shuttle, I would have to push against something in order to float across the cabin. Whenever we describe the properties of planets, stars, or galaxies, we speak of their masses, never of their weights.

Newton's final assumption, called *Newton's third law,* is the law of action and reaction.

Newton's third law—the law of action and reaction: *Whenever one object exerts a force on a second object, the second object exerts an equal and opposite force on the first object.*

For example, I weigh 183 lb on Earth, and so I press down on the floor with a force of 183 lb. Newton's third law says that the floor is also pushing up against me with an equal force of 183 lb. (If it were less, I would fall through the floor, and if it were more, I would be lifted upward.) In the same way, Newton realized that because the Sun is exerting a force on each planet to keep it in orbit, each planet must also be exerting an equal and opposite force on the Sun. As each planet accelerates toward the Sun, the Sun in turn accelerates toward each planet.

Because the Sun is pulling on the planets, why don't they fall onto it? Conservation of angular momentum provides the answer. Angular momentum is a measure of how much energy is stored in an object due to its rotation and revolution. The details of momentum are presented in *An Astronomer's Toolbox 2-2*. As the orbiting planets fall toward the Sun, their angular momentum provides them with motion perpendicular to that infall, meaning that the planets continually fall toward the Sun, but they continually miss it. Because their angular momentum is conserved, planets neither spiral into the Sun nor fly away from it. Angular momentum remains constant unless acted on by an external torque (also defined in *An Astronomer's Toolbox 2-2*).

> Sitting in a moving car, how can you experimentally verify that your body has inertia?

Angular momentum depends on three things: how fast an object rotates or revolves, how much mass it has, and how spread out that mass is. The greater an object's angular motion or mass, or the more the mass is spread out, the greater its angular momentum. Consider, for example, a twirling ice skater. She rotates with a constant mass, practically free of outside forces. When she wishes to rotate more rapidly, she decreases the spread of her mass distribution by pulling her arms and outstretched leg in closer to her body (Figure 2-12). According to conservation of angular momentum, as the spread of mass decreases, the rotation rate must increase. In astronomy, we encounter many instances of the same law, as giant objects, such as stars, contract.

> If you are on a freely spinning merry-go-round, what will happen to it as you move toward the center?

a

b

FIGURE 2-12 Conservation of Angular Momentum As this skater brings her arms and outstretched leg in, she must spin faster to conserve her angular momentum. (Getty Images)

An Astronomer's Toolbox 2-2

Energy and Momentum

Scientists identify two types of energy that are available to any object. The first, called **kinetic energy**, is associated with the object's motion. For speeds much less than the speed of light, we can write the amount of kinetic energy, KE, of an object as

$$KE = \tfrac{1}{2} mv^2$$

where m is the object's mass and v is its speed. Kinetic energy is a measure of how much work the object can do on the outside world or, equivalently, how much work the outside world has done to give the object this speed.

Work is also a rigorously defined concept that often is at odds with our intuition. Work is defined as the product of the force, F, acting on an object and the distance, d, over which the object moves in the direction of the force:

$$W = Fd$$

For example, if I exert a horizontal force of 50 N (N is the unit newtons and is the metric unit of force) and thereby move an object 10 m in the direction I push it, then I have done 50 N × 10 m = 500 J of work. (I have used the relationship that 1 newton × 1 meter = 1 joule.)

The second type of energy is called **potential energy**, the energy available to an object as a result of its location in space. For example, if you hold a pencil above the ground, the pencil has potential energy that can be converted into kinetic energy by Earth's gravitational force. How does that conversion get underway? Just let go of the pencil.

There are various kinds of potential energy, such as the potential energy stored in a battery and the potential energy stored in objects under the influence of gravity. We will focus on *gravitational potential energy.* Far from extremely massive objects, like stars, or extremely dense objects, like black holes, gravitational potential energy, PE, can be written as

$$PE = \frac{GmM}{r}$$

where the constant $G = 6.6683 \times 10^{-11}$ N m²/kg², m is the mass of the object whose gravitational potential energy you are measuring, M is the mass of the object generating the gravitational attraction, and r is the distance between the centers of mass of these two objects.

Near the surface of Earth, this equation simplifies to

$$PE = mgh$$

where $g = 9.8$ m/s² (32 ft/s²) is the gravitational acceleration at Earth's surface, and h is the height of the object above Earth's surface.

Potential energy can be converted into kinetic energy and vice versa. After you drop a pencil, its gravitational potential energy begins to decrease while its kinetic energy begins to increase at the *same rate*. The pencil's total energy is conserved. Conversely, if you throw a pencil up in the air, the kinetic energy you give it will immediately begin to decrease, while its potential energy increases at the same rate.

Related to the motion of an object, and hence to its kinetic energy, are the concepts of *linear momentum*, usually just called **momentum**, and **angular momentum**. Momentum, p, is described by the equation,

$$\mathbf{p} = m\mathbf{v}$$

where v is the velocity of the object. Both $\mathbf{p}$ and $\mathbf{v}$ are in boldface to indicate that they both represent motion in some direction or another, as well as a numeric value. Simple algebra reveals that kinetic energy and momentum are related by

$$KE = \frac{p^2}{2m}$$

Linear momentum, then, indicates how much energy is available to an object because of its motion in a straight line (linear motion).

Angular momentum, L, can be expressed mathematically as

$$\mathbf{L} = I\boldsymbol{\omega}$$

where I is the **moment of inertia** of an object, and ω (lowercase Greek omega) is the angular speed and direction of the rotating object. Just as an object's mass indicates how hard it is to change an object's straight-line motion, the moment of inertia indicates how hard it is to change the rate at which an object rotates or revolves. The moment of inertia depends on an object's mass and shape. Kinetic energy due to angular motion can be written as

$$KE = \frac{L^2}{2I}$$

Newton's first law can also be expressed in terms of **conservation of linear momentum:**

A body maintains its linear momentum unless acted upon by a net external force.

Equivalently, for angular motion we can write the **conservation of angular momentum:**

An Astronomer's Toolbox 2-2 (continued)

A body maintains its angular momentum unless acted upon by a net external torque.

Torques are created when a force acts on an object in some direction other than toward the center of the object's angular motion, as shown in the accompanying figure. Earth has angular momentum from two sources, namely from spinning on its rotational axis and from orbiting the Sun. Likewise, the Moon has angular momentum because it spins on its rotation axis and it orbits Earth. Virtually all objects in astronomy have angular momentum, and it is probably fair to say that conservation of angular momentum is among the most important laws in the cosmos. After all, conservation of angular momentum is what keeps the planets in orbit around the Sun, the moons in orbit around the planets, and astronomical bodies rotating at relatively constant rates, as well as causing many other rotation-related effects that we will encounter throughout this book.

Try these questions: How does tripling the linear momentum of an object change its kinetic energy? How does halving the angular momentum of an object change its kinetic energy? How much work would you do if you pushed on a desk with a force of 100 N, while it moved 20 m? How much work would you do if you pushed on a desk with a force of 500 N, and it moved 0 m? What two things can you vary to change the angular momentum of an object?

(Answers appear at the end of the book.)

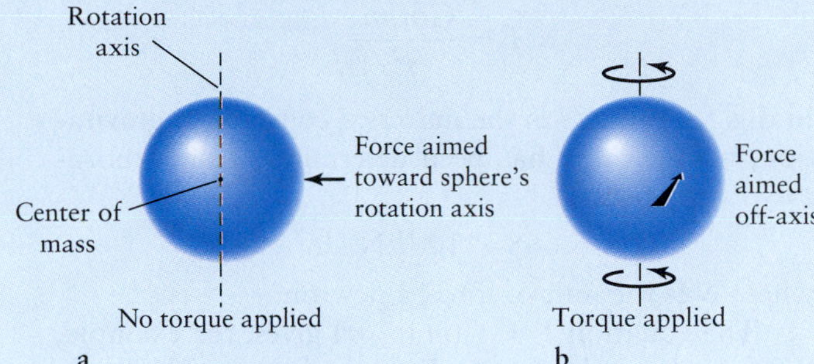

Angular Momentum and Torque (a) When a force acts through an object's rotation axis or toward its center of mass, the force does not exert a torque on the object. (b) When a force acts in some other direction, then it exerts a torque, causing the body's angular momentum to change. If the object can spin around a fixed axis, like a globe, then the rotation axis is the rod running through it. If the object is not held in place, then the rotation axis is in a line through a point called object's *center of mass*. The center of mass of any object is the point that follows a smooth, elliptical path as the object moves in response to a gravitational field. All other points in the spinning object wobble as it moves.

We have now reconstructed the central relationships between matter and motion. Scientific explanation of the heliocentric cosmology still requires a force to hold the planets in orbit around the Sun and the moons in orbit around the planets. Newton identified that, too.

2-8 Newton's description of gravity accounts for Kepler's laws

Isaac Newton did not invent the idea of gravity. An observant seventeenth-century person would understand that some force pulls things down to the ground. It was Newton, however, who gave us a quantitative description of the action of gravity, or *gravitation*, as it is more properly called. Using his first two laws, Newton proved mathematically that the force acting on each of the planets is directed toward the Sun. He expanded this result to the idea that the nature of the force pulling a falling apple straight down to the ground is the same as the nature of the force on the planets from the Sun. More generally, the gravitational force from every object acts to pull every other object directly toward it.

Newton succeeded in formulating a mathematical model that describes the behavior of the gravitational force that keeps the planets in their orbits (presented in *An Astronomer's Toolbox 2-3*).

Newton's law of universal gravitation: *Two objects attract each other with a force that is directly proportional to the product of their masses and inversely proportional to the square of the distance between them.*

In other words, gravitational force decreases with distance. Move twice as far away from an object and you feel only one-quarter of the force from it that you felt before. Despite its weakening, the force of gravity from each object extends throughout the universe. Also, an object with twice the mass of another object exerts twice the gravitational force as the less massive object.

Using his law of gravity along with his three laws stated earlier, Newton found that he could mathematically explain Kepler's three laws. For example, whereas Kepler discovered by trial and error that the period of orbit, P, and average distance between the Sun and planet, a, are related by $P^2 = a^3$, Newton mathematically derived this equation (corrected by including a tiny contribution due to the mass of the planet, as mentioned earlier). Bodies in elliptical orbits are bound by the force of gravity to remain in orbit.

An Astronomer's Toolbox 2-3

Gravitational Force

From Newton's law of gravitation, if two objects that have masses m_1 and m_2 are separated by a distance r, then the gravitational force, F, between them is

$$F = \frac{Gm_1m_2}{r^2}$$

In this formula, G is the **universal constant of gravitation**, whose value has been determined from laboratory experiments:

$$G = 6.668 \times 10^{-11}\ \mathrm{N \cdot m^2 \cdot kg^{-2}}$$

where N is the unit of force, a newton.

The equation $F = G(m_1m_2/r^2)$ gives, for example, the force from the Sun on Earth and, equivalently, the force from Earth on the Sun. If m_1 is the mass of Earth (6.0×10^{24} kg), m_2 is the mass of the Sun (2.0×10^{30} kg), and r is the distance from the center of Earth to the center of the Sun (1.5×10^{11} m):

$$F = 3.6 \times 10^{22}\ \mathrm{N}$$

This number can then be used in Newton's second law, $F = ma$, to find the acceleration of Earth due to the Sun. This yields

$$a_{\mathrm{Earth}} = F/m_1 = 6.0 \times 10^{-3}\ \mathrm{m/s^2}$$

Newton's third law says that Earth exerts the same force on the Sun, so the Sun's acceleration due to Earth's gravitational force is

$$a_{\mathrm{Sun}} = F/m_2 = 1.8 \times 10^{-8}\ \mathrm{m/s^2}$$

In other words, Earth pulls on the Sun, causing the Sun to move toward it. Because of the Sun's greater mass, however, the amount that the Sun accelerates Earth is more than 300,000 times greater than the amount that Earth accelerates the Sun.

Try these questions: Earth's radius is 6.4×10^6 m and 1 kg is a mass equivalent to a weight of 9.8 N (or 2.2 lb) on Earth. What is the force that Earth exerts on you in newtons and pounds? What is the force that you exert on Earth in these units? What would the Sun's force be on Earth if our planet were twice as far from the Sun as it is? How does that force compare to the force from the Sun at our present location?

(Answers appear at the end of the book.)

7 It seems plausible that astronauts in the International Space Station do not feel any force of gravity from Earth, but they do. Orbiting 330 km (approximately 200 mi) above Earth's surface, they feel 90% as much gravitational force from the planet as we do standing on it. They are weightless, however, because as they fall earthward, their angular momentum carries them around the planet at just the right rate to continually miss it.

Newton also discovered that some objects orbiting the Sun can follow nonelliptical paths. His equations led him to conclude that orbits can also be **parabolas** or **hyperbolas** (Figure 2-13). In both cases, such bodies would make only one pass close to the Sun and then travel out of the solar system, never to return. To date, all of the objects observed in the solar system began their existence in elliptical orbits, but some comets (small bodies of rock and ice) have received enough energy from being pulled by planets or from expelling jets of gas to develop parabolic or hyperbolic orbits.

Using the equations Newton derived, the orbits of the planets and their satellites could be calculated with unprecedented precision. Using his laws, mathemati-

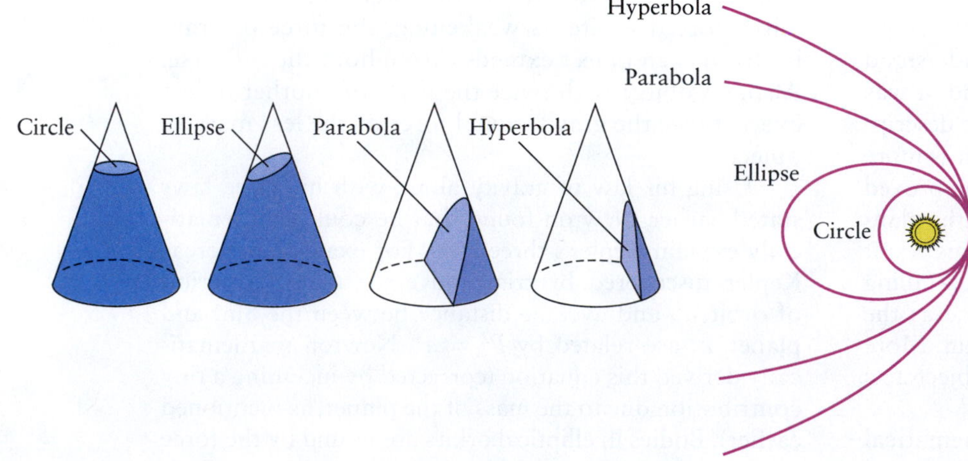

INTERACTIVE EXERCISE 2.2

FIGURE 2-13 Conic Sections A conic section is any one of a family of curves obtained by slicing a cone with a plane, as shown. The orbit of one body around another can be an ellipse, a parabola, or a hyperbola. Circular orbits are possible because a circle is just an ellipse for which both foci are at the same point.

WHAT IF... The International Space Station and the Astronauts in It Didn't Feel Any Gravity from Earth?

Newton's first law says that the astronauts and the space station will move in a straight line unless acted upon by a net external force. Without gravity, there would be no external force acting on either the astronauts or the station, so they would all move together in a straight line—instead of orbiting Earth, the space station would move away from us.

cians proved that Earth's axis of rotation must precess because of the gravitational pull of the Moon and the Sun on Earth's equatorial bulge (recall Figure 1-20). In the spirit of the scientific method, Newton's laws and mathematical techniques were used to predict new phenomena. For example, Edmond Halley was intrigued by historical records of a comet that was sighted about every 76 years. Using his friend Newton's methods, Halley worked out the details of the comet's orbit and predicted its return in 1758. It was first sighted on Christmas night of that year, and to this day the comet bears Halley's name (Figure 2-14).

Perhaps the most dramatic early use of the scientific method with Newton's ideas was their role in the discovery of the eighth planet in our solar system.

FIGURE 2-14 Halley's Comet Halley's Comet orbits the Sun with an average period of about 76 years. During the twentieth century, the comet passed near the Sun twice—once in 1910 and again, as shown here, in 1986. The comet will pass close to the Sun again in 2061. During its last visit, the comet spread more than 5° across the sky, or 10 times the diameter of the Moon. (W. Liller/Large Scale Phenomena Network/NSSDC/NASA)

The seventh planet, Uranus, had been discovered by William Herschel in 1781 during a systematic telescopic survey of the sky. Fifty years later, however, it was clear that Uranus was not following the orbit predicted by Newton's laws. Two mathematicians, John Couch Adams in England and Urbain-Jean-Joseph Leverrier in France, independently calculated that the deviations of Uranus from its predicted orbit could be explained by the gravitational pull of a then-unknown, more distant planet. Each man predicted that the planet would be found at a certain location in the constellation of Aquarius in September 1846. A telescopic search on September 23, 1846, by German astronomer Johann Galle, revealed Neptune less than 1° from its calculated position. Although sighted with a telescope, Neptune was really discovered with pencil and paper.

Insight Into Science

Quantify Predictions Mathematics provides a language that enables science to make quantitative predictions that can be checked by anyone. For example, in this chapter, we have seen how Kepler's third law and Newton's universal law of gravitation correctly predict the motion of objects under the influence of the Sun's gravitational attraction.

ANIMATION 2.6

It is a testament to Newton's genius that his three laws were precisely the basic ideas needed to understand so much about the natural world. Newton's process of deriving Kepler's laws and the universal law of gravitation helped secure the scientific method as an invaluable tool in our process of understanding the universe. Figure 2-15 shows some of the effects of gravity at the scales of planets, stars, and galaxies.

2-9 Frontiers yet to be discovered

The science related to forces and orbits described in this chapter was well established by the beginning of the nineteenth century. However, questions remain. What in nature gives matter mass? Why are the inertial mass (defined in Newton's second law) and the mass used in his law of gravitation identical? Experiments are underway at high-energy particle accelerators such as CERN, near Geneva, Switzerland, and the Fermi National Accelerator near Chicago, Illinois, to explain this. In another vein, careful observations have revealed that Newton's law of gravitation gives very slightly inaccurate predictions for the orbital path of Mercury. We will explore this issue further in Chapter 14.

FIGURE 2-15 Gravity Works at All Scales This figure shows a few of the effects of gravity here on Earth, in the solar system, in our Milky Way Galaxy, and beyond. Top: Space station (NASA); Couple holding hands (Paul Burns/Digital Vision/Getty Images); Center: Black hole (NASA); Bottom: Galaxy cluster (ESA, NASA, J.-P. Kneib [Caltech/Observatoire Midi-Pyrénées] and R. Ellis [Caltech])

SUMMARY OF KEY IDEAS

Science: Key to Comprehending the Cosmos

- The ancient Greeks laid the groundwork for progress in science by stating that the universe is comprehensible.
- The scientific method is a procedure for formulating theories that correctly predict how the universe behaves.
- A scientific theory must be testable, that is, capable of being disproved.
- Theories are tested and verified by observation or experimentation and result in a process that often leads to their refinement or replacement and to the progress of science.
- Observations of the cosmos have led astronomers to discover some fundamental physical laws of the universe.

Origins of a Sun-Centered Universe

- Common sense (for example, Earth doesn't appear to be moving) led early natural philosophers to devise a geocentric cosmology, which placed Earth at the center of the universe.
- Copernicus's heliocentric (Sun-centered) theory simplified the general explanation of planetary motions compared to the geocentric theory.
- The heliocentric cosmology refers to motion of planets and smaller debris orbiting the Sun. Other stars do not orbit the Sun.
- The sidereal orbital period of a planet is measured with respect to the stars, and determines the length of the planet's year. A planet's synodic period is measured with respect to the Sun as seen from the moving Earth (for example, from one opposition to the next).

Kepler's and Newton's Laws

- Ellipses describe the paths of the planets around the Sun much more accurately than do the circles used in previous theories. Kepler's three laws give important details about elliptical orbits.
- The invention of the telescope led Galileo to new discoveries, such as the phases of Venus and the moons of Jupiter, that supported a heliocentric view of the universe.
- Newton based his explanation of the universe on three assumptions, now called Newton's laws of motion. These laws and his law of universal gravitation can be used to deduce Kepler's laws and to describe most planetary motions with extreme accuracy.
- The mass of an object is a measure of the amount of matter in it; weight is a measure of the force with which the gravity of a world pulls on an object's mass when the two objects are at rest with respect to each other (or, equivalently, how much the object pushes down on a scale).
- The path of one astronomical object around another, such as that of a comet around the Sun, is an ellipse, a parabola, or a hyperbola. Ellipses are bound orbits, while objects with parabolic and hyperbolic orbits fly away, never to return.

A WHAT DID YOU THINK?

1 *What makes a theory scientific?* A theory is an idea or set of ideas proposed to explain something about the natural world. A theory is scientific if it makes predictions that can be objectively tested and potentially disproved.

2 *What is the shape of Earth's orbit around the Sun?* All planets have elliptical orbits around the Sun.

3 *Do the planets orbit the Sun at constant speeds?* No. The closer a planet is to the Sun in its elliptical orbit, the faster it is moving. The planet moves fastest at perihelion and slowest at aphelion.

4 *Do all of the planets orbit the Sun at the same speed?* No. A planet's speed depends on its average distance from the Sun. The closest planet moves fastest, while the most distant planet moves slowest.

5 *How much force does it take to keep an object moving in a straight line at a constant speed?* Unless an object is subject to an outside force, like friction, it takes no force at all to keep it moving in a straight line at a constant speed.

6 *How does an object's mass differ when measured on Earth and on the Moon?* Assuming the object doesn't shed or collect pieces, its mass remains constant whether on Earth or on the Moon. Its weight, however, is less on the Moon.

7 *Do astronauts orbiting Earth feel the force of gravity from our planet?* Yes. They are continually pulled earthward by gravity, but they continually miss it because of their motion around it. Because they are continually in free fall, they feel weightless.

Key Terms for Review

acceleration, 56
angular momentum, 58
aphelion, 51
astronomical unit, 52
configuration (of a planet), 46
conjunction, 48
conservation of angular momentum, 58
conservation of linear momentum, 58
cosmology, 43
direct motion, 43
ellipse, 50
elongation, 46
focus (of an ellipse), 50
force, 57
Galilean moons (satellites), 55
gravity, 56
heliocentric cosmology, 45
hyperbola, 60

inferior conjunction, 46
Kepler's laws, 50
kinetic energy, 58
law of equal areas, 52
law of inertia, 56
law of universal gravitation, 59
light-year, 54
mass, 52
model, 41
moment of inertia, 58
momentum, 58
Newton's laws of motion, 56
Occam's razor, 42
opposition, 48
parabola, 60
parallax, 49
parsec, 54
perihelion, 51
potential energy, 58
retrograde motion, 43
scientific method, 41
scientific theory, 41
semimajor axis (of an ellipse), 50
sidereal period, 48
superior conjunction, 46
synodic period, 48
theory, 41
universal constant of gravitation, 60
velocity, 56
weight, 52
work, 58

Review Questions

1. Who wrote down the equation for the law of gravitation? **a.** Copernicus **b.** Brahe **c.** Newton **d.** Galileo **e.** Kepler.

2. Which of the following most accurately describes the shape of Earth's orbit around the Sun? **a.** circle **b.** ellipse **c.** parabola **d.** hyperbola **e.** square.

3. Of the following planets, which takes the longest time to orbit the Sun? **a.** Earth **b.** Uranus **c.** Mercury **d.** Jupiter **e.** Venus.

4. What is a Sun-centered model of the solar system called?

5. How long does it take Earth to complete a sidereal orbit of the Sun?

6. How did Copernicus explain the retrograde motions of the planets?

7. Which planets can never be seen at opposition? Which planets never pass through inferior conjunction?

8. At what configuration (superior conjunction, greatest eastern elongation, etc.) would it be best to observe Mercury or Venus with an Earth-based telescope? At what configuration would it be best to observe Mars, Jupiter, or Saturn? Explain your answers.

9. What are the synodic and sidereal periods of a planet?

10. What are Kepler's three laws? Why are they important?

11. In what ways did the astronomical observations of Galileo support a heliocentric cosmology?

12. How did Newton's approach to understanding planetary motions differ from that of his predecessors?

13. What is the difference between mass and weight?

14. Why was the discovery of Neptune a major confirmation of Newton's universal law of gravitation?

15. Why does an astronaut have to exert a force on a weightless object to move it?

Advanced Questions

The answers to all computational problems, which are preceded by an asterisk (*), appear at the end of the book.

16. From the definition KE = $\frac{1}{2}mv^2$, derive the equation KE = $p^2/2m$, as discussed in *An Astronomer's Toolbox 2-2*.

***17.** Convert: **a.** 8.3 pc (parsec) into light-years **b.** 6.52 ly into parsecs **c.** 8450 AU into kilometers **d.** 2.7×10^3 Mpc into kiloparsecs.

18. Is it possible for an object in the solar system to have a synodic period of exactly 1 year? Explain your answer.

19. Describe why there is a systematic decrease in the synodic periods of the planets from Mars outward, as shown in Table 2-1.

***20.** A line joining the Sun and an asteroid was found to sweep out 5.2 square astronomical units of space in all of 2006. How much area was swept out in 2007? In the 5 years from 2002 to 2007?

***21.** A comet moves in a highly elongated orbit ($e \approx 0.95$) around the Sun, with a period of 1000 years. What is the length of the semimajor axis of the comet's orbit? Referring to Figure 2-8b, estimate the farthest distance that the comet can get from the Sun.

***22.** The orbit of a spacecraft around the Sun has a perihelion distance of 0.5 AU and an aphelion distance of 3.5 AU. What is the spacecraft's orbital period?

23. Make diagrams of Jupiter's phases as seen from Earth and as seen from Saturn.

24. Make a drawing of a plausible S-shaped path of Mars undergoing retrograde motion. Make another drawing showing Mars undergoing retrograde motion in a backward-S shaped path.

25. In what direction (left or right, eastward or westward) across the celestial sphere do the planets normally appear to move as seen from Australia? In what direction is retrograde motion as seen from there?

26. The dictionary defines astrology as "the study that assumes and attempts to interpret the influence of the heavenly bodies on human affairs." Based on what you know about scientific theory, is astrology a science? Why or why not? Feel free to further explore astrology, if you wish, before answering this question.

Discussion Questions

27. Which planet would you expect to exhibit the greatest variation in apparent brightness as seen from Earth? Explain your answer.

28. Use two thumbtacks (or pieces of tape), a loop of string, and a pencil to draw several ellipses. Describe how the shapes of the ellipses vary as you change the distance between the thumbtacks.

What if…

*29. Earth were 2 AU from the Sun? What would be the length of the year? Assuming that such physical properties as rotation rate were as they are today, what else would be different here?

*30. Earth were $^1/_2$ AU from the Sun? What would be the length of the year? Assuming that such physical properties as rotation rate were as they are today, what else would be different here?

*31. Earth were 10 AU from the Sun? How much stronger or weaker would the Sun's gravitational pull be than it is on Earth today?

*32. Earth had twice its present mass? Assume that all other properties of Earth and its orbit remain the same. What would be the acceleration of the more massive Earth due to the Sun compared to the present acceleration of Earth from the Sun? *Hint:* Try combining $F = m_1 a$ and the force equation in *An Astronomer's Toolbox 2-3*, where m_1 is the mass of Earth in both equations. Given that acceleration determines the period of the planet's orbit, how would the year on the more massive Earth compare to a year today?

33. The Sun suddenly disappeared? What would Earth's path in space be in response to such an event? Describe how Earth would change, as a result, and how humans might survive on a Sunless planet.

34. The skies of Earth were perpetually cloudy? How might that have changed the history of our understanding of the cosmos, and how might humans under such conditions eventually learn what is really "out there"?

35. Scientists remained believers in the first theory of the cosmos that they decided was correct? How might that change the dynamics by which science evolves in the face of new data that conflict with earlier theories?

Web Questions

36. Search the Web for information about Galileo. What were his contributions to physics? Which of Galileo's new ideas were later used by Newton to construct his laws of motion? What incorrect beliefs about astronomy did Galileo hold?

37. Search the Web for information about Kepler. Before he realized that the planets move on elliptical paths, what other models of planetary motion did he consider? What was Kepler's idea of the "music of the spheres"?

38. Search the Web for information about Newton. What were some of the contributions that he made to physics other than developing his laws of motion? What contributions did he make to mathematics?

Observing Projects

STARRY NIGHT

39. It is quite probable that, within a few weeks of your reading this chapter, one of the planets visible to the unaided eye (all the planets out as far as Saturn) will be in opposition or at greatest eastern elongation, making it visible in the evening sky from your location. Using *Starry Night*™, the Internet, or a reference book, such as the current issue of the *Observer's Handbook* of the Royal Astronomical Society of Canada, the *Astronomical Almanac*, or the pamphlet *Astronomical Phenomena* (both of the latter published by the U.S. Naval Observatory), select a planet that is at or near such a configuration. Launch the *Starry Night*™ program and click the **Sunset** button in the toolbar. Open the **Options** side pane and expand the **Solar System** layer. Click on both boxes in the **Planets-Moons** row to show and label the planets in the view and click on the left-hand boxes for all other objects to remove them from view. Use the hand tool to search the sky for planets as far away as Saturn. If none are in your sky, change the date by a few days and search again. Repeat until you have found one or two visible planets up at night. Plan to make your observations at that time. Make a sketch to show the positions of planets with respect to nearby stars, noting the time and date of this observation. If possible, confirm that your observation is a planet by observing a few days later and showing that the object has moved. (This step is vital in confirming discoveries of new bodies in the solar system.) If a small telescope or spotting scope is available, make a drawing of the view of the planet through it, being careful to include planet phases, moons, surface features, and nearby stars.

40. Use the *Starry Night*™ software to observe the moons of Jupiter. Select **Favourites > Discovering the Universe > Galilean Moons. a.** Note the positions of the moons. Step time forward in increments of 6 hours and draw the positions of the moons at each step. **b.** From your drawings, can you tell which moon orbits closest to Jupiter and which orbits farthest away? Explain your reasoning. **c.** Determine the periods of orbits of these moons (change the Time Flow Rate if necessary). **d.** Are there times when you see fewer than four Galilean moons? What has happened to the other moons at those times?

41. Use *Starry Night*™ to observe the phases of Venus. Select **Favourites > Discovering the Universe > Phases of Venus** and click the **Now** button in the toolbar. **a.** Draw the current shape (phase) of Venus. With the **Time Flow Rate** set to 30 days, step time forward, drawing Venus to scale at each step. Make a total of 20 time steps and drawings. **b.** From your drawings, determine when the planet is nearer or farther from the Earth than is the Sun. **c.** Deduce from your drawings when Venus is coming toward us or is moving away from us. **d.** Explain why Venus goes through this particular cycle of phases.

42. Select **Favourites > Discovering the Universe > Phases of Mars** and repeat the procedure of the previous observing project using Mars instead of Venus. Compare your results for the two planets. Why are the cycles of phases as seen from Earth different for the two planets?

43. In this exercise, you will observe which planets have the greatest orbital eccentricities. Using *Starry Night*™, select **Favourites > Discovering the Universe > Inner Solar System.** This will show a face-on view of the orbits of the inner four planets. Using the fact that Venus's orbital eccentricity is 0.007 (very nearly circular), note which two of the inner four planets have the least circular orbits. Now select **Favourites > Discovering the Universe > Outer Solar System.** Which outer planet has the most eccentric orbit? Run time forward to observe the relative speeds of these distant planets in their respective orbits.

Three of the four buildings containing the Very Large Telescope optical telescopes at the Paranal Observatory, Atacama, Chile. (ESO)

Chapter 3

Light and Telescopes

WHAT DO YOU THINK?

1. **What is light?**
2. **Which type of electromagnetic radiation is most dangerous to life?**
3. **What is the main purpose of a telescope?**
4. **Why do all research telescopes use mirrors, rather than lenses, to collect light?**
5. **Why do stars twinkle?**

Answers to these questions appear in the text beside the corresponding question numbers in the margins and at the end of the chapter.

With our eyes alone we can see visible light from several thousand stars. Until the seventeenth century, few people even dreamed that there were more of them. It was then that telescopes revolutionized human understanding of the universe, showing for the first time how little of the cosmos we normally see. The process of discovery continues, as telescopes reveal new things in space nearly every day. We can think of the radiation emitted by objects out there as the medium of natural cosmic communication, and telescopes as the means by which we gather and read those cosmic messages.

We have also discovered that visible light is only a tiny fraction of the energy emitted by objects in space. Indeed, such phenomena as interstellar clouds of gas and dust, the bodies lying behind these clouds, newly forming stars, intergalactic gas clouds, and a variety of exotic objects, such as black holes and neutron stars, are nearly invisible to even our best optical telescopes. However, many of these objects strongly emit a variety of nonvisible radiations (namely radio waves, microwaves, infrared and ultraviolet radiations, X rays, and gamma rays) that we now have the technology to detect. Today, we use telescopes on the ground, floating in the air, orbiting Earth, or traveling elsewhere in the solar system to see these myriad "stealth" objects in space.

In this chapter you will discover

- the connection between visible light, radio waves, X rays, and other types of electromagnetic radiation
- the debate in past centuries over what light is and how Einstein resolved this question
- how telescopes collect and focus light
- why different types of telescopes are used for different types of research
- the limitations of telescopes, especially those that use lenses to collect light
- what the new generations of land-based and space-based high-technology telescopes being developed can do
- how astronomers use the entire spectrum of electromagnetic radiation to observe the stars and other astronomical objects and events

THE NATURE OF LIGHT

So far in this text we have used the word *light* in its everyday sense—the stuff to which our eyes are sensitive. This is more properly called *visible light*, and it is a form of **electromagnetic radiation**. Perhaps contrary to one's intuition, this radiation is composed of particles, called *photons*, that have properties of waves. Detecting electromagnetic radiation with telescopes is the essence of observational astronomy. Although human perception of objects here on Earth and in space comes primarily from the visible light that our eyes detect, visible light is only a tiny fraction of all the electromagnetic radiation emitted by objects in the universe. The rest of this radiation, invisible to our eyes, is detected by high-tech sensors attached to specially designed telescopes. In this chapter, we examine telescopes for all types of electromagnetic radiation. To understand how telescopes work, we begin by exploring the properties of the electromagnetic radiation that they collect.

3-1 Newton discovered that white is not a fundamental color and proposed that light is composed of particles

From the time of Aristotle until the late seventeenth century, most people believed that white is the fundamental color of light. The colors of the rainbow (or, equivalently, the colors created by light passing through a prism) were believed to be added or created somehow as white light went from one medium through another. Isaac Newton performed experiments during the late 1600s that disproved these beliefs. He started by passing a beam of sunlight through a glass prism, which spread the light out into the colors of the rainbow (Figure 3-1a). The change in direction as light travels from one medium into another is called **refraction**, and the resulting spread of colors (complete or with colors missing) is called a **spectrum** (plural, **spectra**).

Then Newton selected a single color and sent it through a second prism (Figure 3-1b). The light that emerged from the first prism was refracted by the second prism, but *it remained the same color.* The fact that individual colors of refracted light were unchanged by the second prism led Newton to conclude that the colors of a full spectrum (shades of red, orange, yellow, green, blue, and violet; indigo is not considered a separate color in astronomy) were in fact properties of the light itself, and that white light is a mixture of colors. To prove this last point, he recombined all the spectrum colors, thereby recreating white light. Thus, different colors of the spectrum are different entities. But what, Newton wondered, is the nature of light that it could break apart into distinct colors and then be completely reconstituted?

Back in the mid-1600s, the Dutch scientist Christiaan Huygens proposed that light travels in the form of waves. Newton, on the other hand, performed many experiments in optics that convinced him that light is composed of tiny particles of energy. It turns out that both ideas were right.

In 1801, the English physicist Thomas Young demonstrated that light is indeed composed of waves.

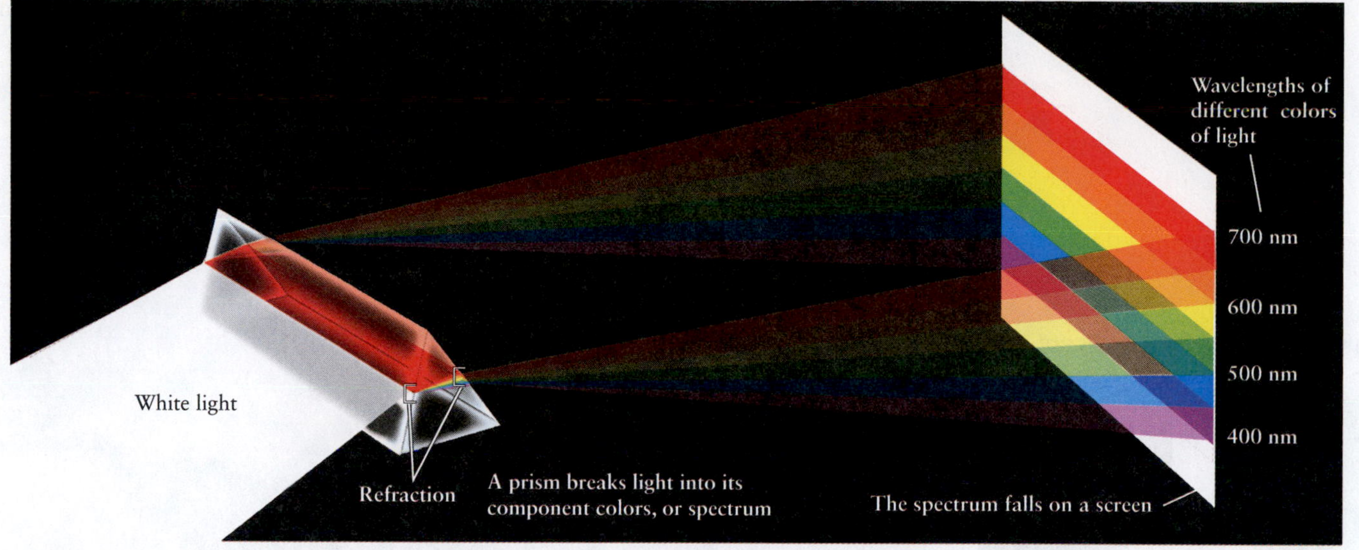

a

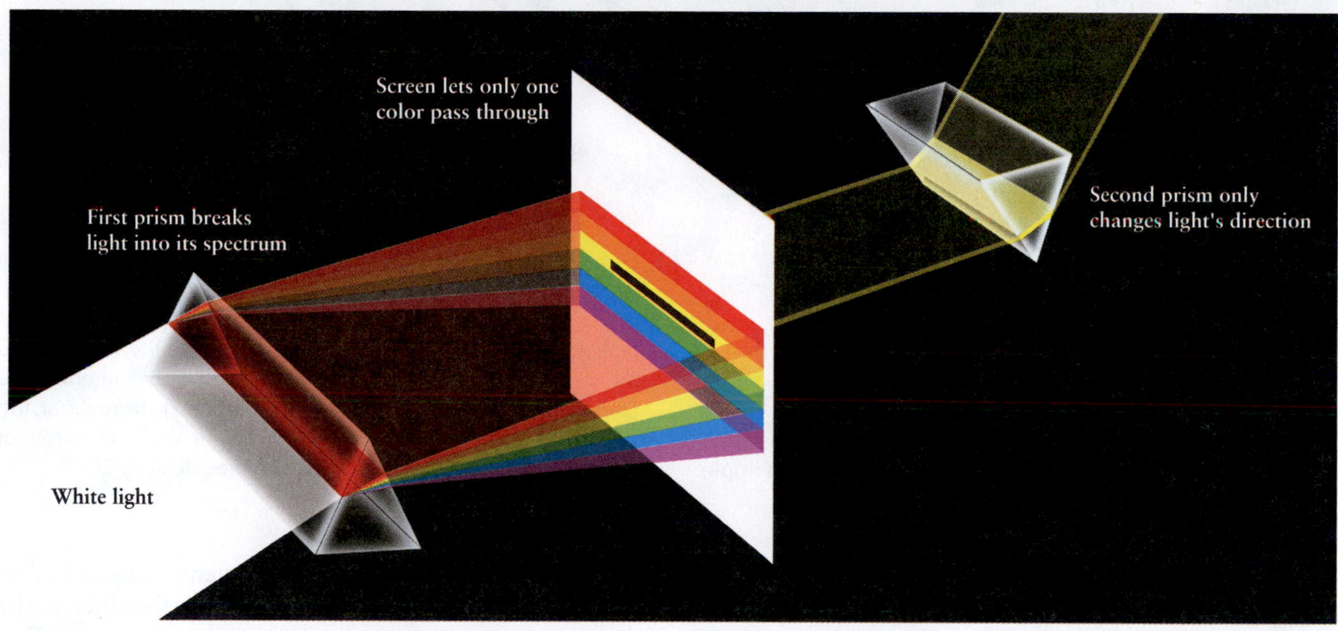

b

FIGURE 3-1 Prisms and a Spectrum (a) When a beam of white light passes through a glass prism, the light is separated or refracted into a rainbow-colored band called a spectrum. The numbers on the right side of the spectrum indicate wavelengths in nanometers (1 nm = 10^{-9} m). (b) This drawing of Newton's experiment illustrates that glass does not add to the color of light, only changes its direction. Because color is not added, this experiment shows that color is an intrinsic property of light.

Young sent light of a single color through two parallel slits (Figure 3-2a). He reasoned that if the light were waves, then these waves would behave the way waves on the surface of water behave, flowing through similar gaps (Figure 3-2b). In particular, he theorized that the light waves from each slit would interact with light waves from the other. For example, when two light waves meet, with one wave going up and the other going down, they would interfere with each other and partially or totally cancel each other out, leaving dark regions on the screen. When two waves that are both moving up or both moving down meet, they would reinforce each other and create bright regions. As you can see in Figure 3-2, this is precisely what happened. (If light were just random particles going through the slits, they would not interfere with each other in this way, and the pattern on the screen would just yield two bright regions, one behind each slit.) The analogy with water waves ends here. For example, light waves do not have to travel through a medium, unlike water waves, which travel through the liquid medium of water.

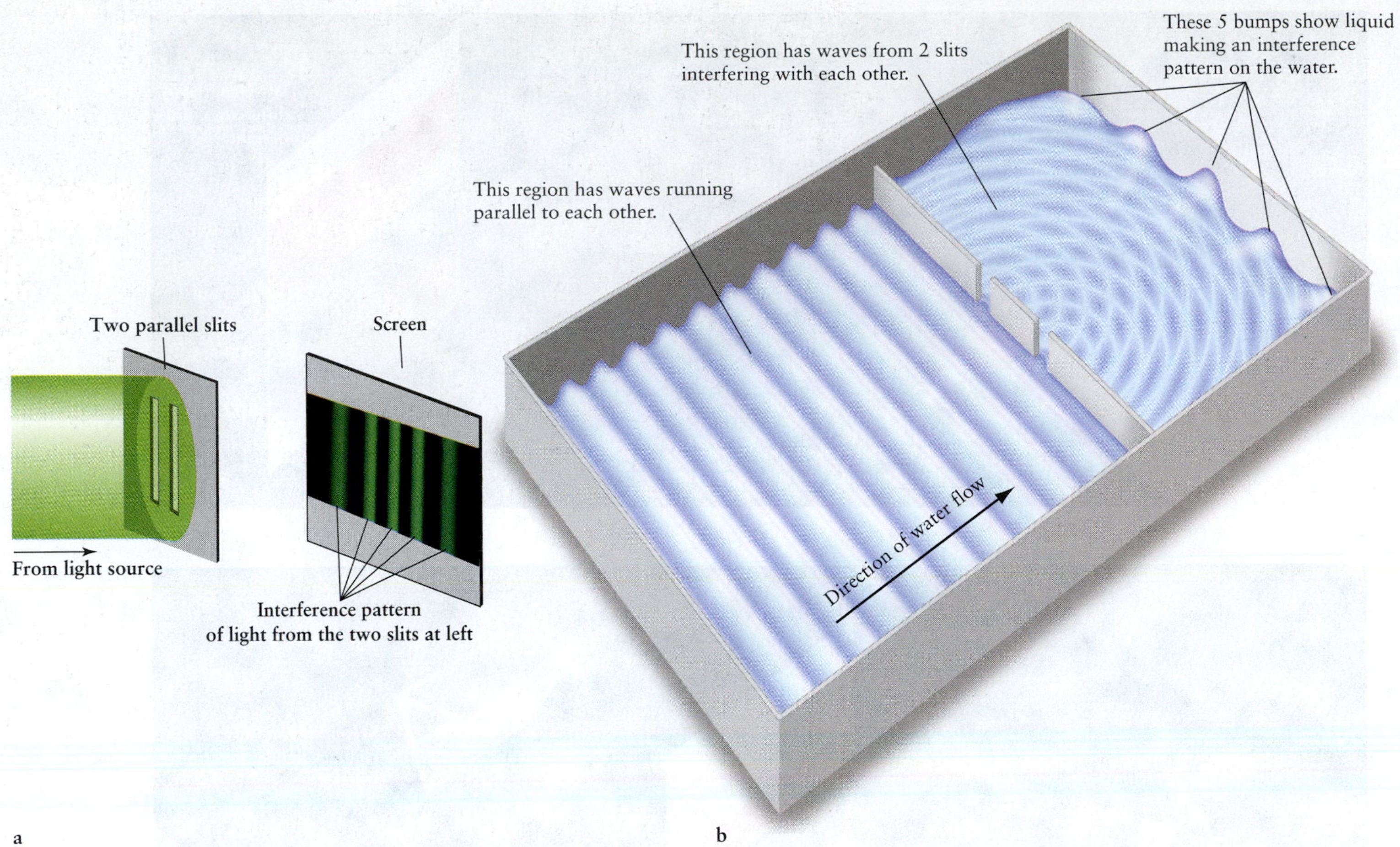

FIGURE 3-2 Wave Travel (a) Electromagnetic radiation travels as waves. Thomas Young's interference experiment shows that light of a single color passing through a barrier with two slits behaves as waves that create alternating light and dark patterns on a screen. (b) Water waves passing through two slits in a ripple tank create interference patterns. As with light, the water waves interfere with each other, creating constructive interference (crests) and destructive interference (troughs) throughout the right side of the tank and on the far right wall. (University of Colorado, Center for Integrated Plasma Studies, Boulder, CO)

Further insight into the wave character of light came from calculations by the Scottish physicist James Clerk Maxwell in the 1860s. Maxwell unified the descriptions of the basic properties of electricity and magnetism into four equations. By combining these equations, he demonstrated that electric and magnetic effects should travel through space together in the form of coupled waves (Figure 3-3) that have equal amplitudes. Maxwell's suggestion that some of these waves, now called *electromagnetic radiation*, are observed as visible light was soon confirmed by a variety of experiments. Despite the name electromagnetic radiation, neither visible light, nor any other type of electromagnetic radiation is electrically charged.

Newton showed that sunlight is composed of all the colors of the rainbow. Young, Maxwell, and others showed that light travels as waves. What makes the colors of the rainbow distinct from each other? The answer is surprisingly simple: Different colors are waves with different wavelengths. A **wavelength**, usually designated by λ, the lowercase Greek letter lambda, is the distance between two successive wave crests (see Figure 3-3).

What color is refracted least?

The wavelengths of all colors are extremely small, less than a thousandth of a millimeter. To express these tiny distances conveniently, scientists use a unit of length called the *nanometer* (nm), where 1 nm = 10^{-9} m. Another unit you might encounter when talking to astronomers is the *angstrom* (Å), where 1 Å = 0.1 nm = 10^{-10} m. Experiments demonstrate that visible light has wavelengths ranging from about 400 nm for the shortest wavelength of violet light to about 700 nm for the longest wavelength of red light. Intermediate colors of the rainbow fall between these wavelengths (see Figure 3-1a). The complete spectrum of colors from the longest wavelength to the shortest is red, orange, yellow, green, blue, and violet. Referring to Figure 3-1, you can also see that the amount of refraction that different colors undergo depends on their wavelengths: *The shorter the wavelength, the more the light is refracted.*

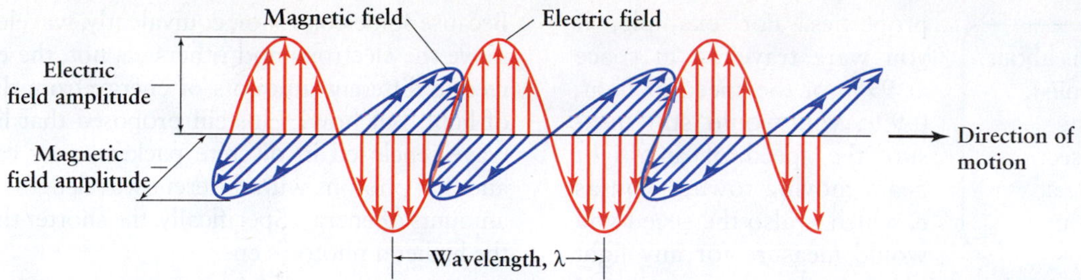

FIGURE 3-3 Electromagnetic Radiation All forms of electromagnetic radiation (radio waves, microwaves, infrared radiation, visible light, ultraviolet radiation, X rays, and gamma rays) consist of electric and magnetic fields oscillating perpendicular to each other and to the direction in which they move. In empty space this radiation travels at a speed of 3×10^5 km/s. These fields are the mathematical description of the electric and magnetic effects. The distance between two successive crests, denoted by λ, is called the wavelength of the light.

3-2 Light travels at a finite, but incredibly fast, speed

The fact that we see lightning before we hear the accompanying thunderclap tells us that light travels faster than sound. But does that mean that light travels instantaneously from one place to another, or does it move with a measurable speed?

The first evidence for the finite speed of light came in 1675, when Ole Rømer, a Danish astronomer, carefully timed eclipses of Jupiter's moons (Figure 3-4). Rømer discovered that the moment at which a moon enters Jupiter's shadow depends on the distance between Earth and Jupiter. When Jupiter is in opposition—that is, when Jupiter and Earth are on the same side of the Sun—the Earth–Jupiter distance is relatively short compared to when Jupiter is near conjunction. At opposition, Rømer found that eclipses occur slightly earlier than predicted by Kepler's laws, and they occur slightly later than predicted when Jupiter is near conjunction.

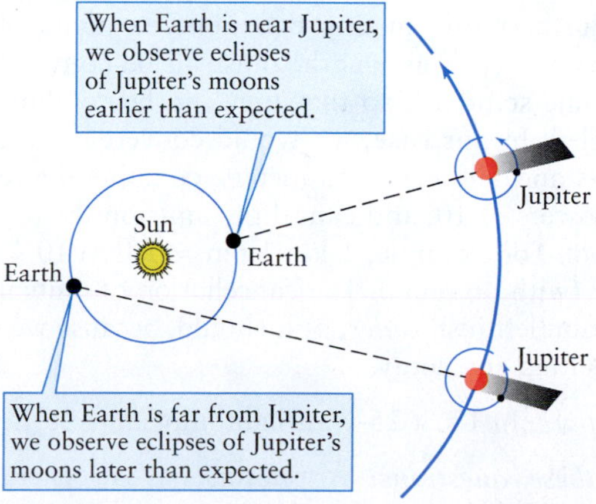

FIGURE 3-4 Evidence that Light Travels at a Finite Speed The times of the eclipses of Jupiter's moons as seen from Earth depend on the relative positions of Jupiter, Earth, and the Sun. Rømer correctly attributed the variations in these times to the variations in the time that it takes light from these events to reach Earth.

Rømer correctly concluded that light travels at a finite speed, and so it takes more time to travel longer distances across space. The greater the distance to Jupiter, the longer the image of an eclipse takes to reach our eyes. From his timing measurements, Rømer concluded that it takes 162 min for visible light to traverse the diameter of Earth's orbit (2 AU). Incidentally, Rømer's interpretation of the data requires a heliocentric cosmology—that both Earth and Jupiter orbit the Sun.

Rømer's subsequent calculation of the speed of light was off by 25% because the value for the astronomical unit (the average distance from Earth to the Sun) that existed at that time was highly inaccurate. Nevertheless, he proved his main point—light travels at a finite speed. The first accurate laboratory measurements of the speed of visible light were performed in the mid-1800s.

Maxwell's equations also reveal that light of all wavelengths travels at the same speed in a vacuum (a region that contains no matter), and, despite a few atoms per cubic meter, the space between planets and stars is a very good vacuum. The constant speed of light in a vacuum, usually designated by the letter c, has been measured to be 299,792.458 km/s, which we generally round to

$$c = 3.0 \times 10^5 \text{ km/s} = 1.86 \times 10^5 \text{ mi/s}$$

(Standard abbreviations for units of speed, such as *km/s* for kilometers per second and *mi/s* for miles per second, will be used throughout the rest of this book.) Light that travels through air, water, glass, or any other substance always moves more slowly than it does in a vacuum.

The value c is a fundamental property of the universe. The speed of light appears in equations that describe, among other things, atoms, gravity, electricity, magnetism, distance, and time. Light has extraordinary

The speed of sound is about 0.34 km/s (or 0.21 mi/s). How can this and the information in this section be used to determine a person's distance from a lightning strike?

properties. For example, if you were traveling in space at 99% of the speed of light, 0.99*c*, you would still measure the speed of any light beam moving toward you as *c*, which is also the speed you would measure for any light beam moving away from you!

3-3 Einstein showed that light sometimes behaves as particles that carry energy

1 By 1905, scientists were comfortable with the wave nature of light. However, in that year, Albert Einstein threw a monkey wrench into that theory when he proposed that light is composed of particles that have wave properties, creating what is now called the *wave-particle duality.* He used this idea to explain the *photoelectric effect.* Physicists knew that electrons are bound onto a metal's surface by electric forces and that it takes energy to overcome those forces. Shorter wavelengths of light can knock some electrons off the surfaces of metals, while longer wavelengths of light cannot, no matter how intense the beam of long-wavelength light. Because some colors (or, equivalently, wavelengths) can remove the electrons and others cannot, the electrons must receive different amounts of energy from different colors of light. But how? Einstein proposed that light travels as waves enclosed in discrete packets, now called **photons,** and that photons with different wavelengths have different amounts of energy. Specifically, the shorter the wavelength, the higher a photon's energy.

$$\text{Photon energy} = \frac{\text{Planck's constant} \times \text{the speed of light}}{\text{Wavelength}}$$

where Planck's constant (named for the German physicist Max Planck) has the value 6.67×10^{-34} J · s, where J is the unit of energy called a *joule* (see *An Astronomer's Toolbox 3-1*), and the wavelength, the distance between wave crests or troughs, is shown in Figure 3-3. (J · s stands for joules multiplied by seconds, or *joule-seconds.*) Einstein's concept of light, confirmed in numerous experiments, means that light can act both as waves (as when passing through slits) and as particles (as when striking matter).

The waves shown in Figure 3-3 are moving to the right. If you count the number of wave crests that pass a given point per second, you have found the **frequency** of the photon. The unit of frequency is *hertz*, named

An Astronomer's Toolbox 3-1

Photon Energies, Wavelengths, and Frequencies

All photons with the same energy are identical. This energy depends solely on the photon's wavelength or, equivalently, its frequency. The wavelength, λ, and frequency, f, of a photon are related by the simple equation

$$c = f\lambda \text{ or, equivalently, } f = c/\lambda$$

where c is the speed of light. Knowing either the wavelength or frequency, you can calculate the other value with these equations. Also, if you know either of them, you also know the photon's energy, as introduced in the equation from Section 3-3,

$$E = hc/\lambda \quad \text{or} \quad E = hf$$

Planck's constant, h, is 6.67×10^{-34} J · s and the speed of light, c, is 300,000 km/s.

Example: A photon of red light has a wavelength 700 nm. What is its energy? *Note:* All distances in the following equation must be converted to the same units, such as meters.

$$E_{\text{red}} = \frac{(6.67 \times 10^{-34}\ \text{J} \cdot \text{s})(300{,}000\ \text{km/s})}{700\ \text{nm}}$$

Simplifying, we find $E_{\text{red}} = 2.86 \times 10^{-19}$ J, meaning that each photon of red light with wavelength 700 nm has an energy of 2.86×10^{-19} J.

An aside on units: Just as the numbers in the numerator of an equation multiply together, the numbers in the denominator multiply together, and the resulting numerator is divided by the denominator, so are units combined in the same way. The same units found in both the numerator and denominator cancel. For example, a result such as 7 km s/s would have the seconds cancel on the top and bottom, leaving 7 km as the answer.

Furthermore, in any equation, the units of the same type (such as length) must all be converted to the same standard, so that they can be combined or canceled. In our case, we would convert both nanometers and kilometers to meters, combine the resulting powers of 10, and cancel the units on the top and bottom. For example, 1 km/1 nm = 10^3 m/10^{-9} m = 10^{+12} (with no units). The cancellation of units in the equation left just *joules*, as it should, because we were calculating an energy.

Compare: In 1 s, a 25-watt lightbulb emits 25 J.

Try these questions: A photon has an energy of 4.90×10^{-19} J. Calculate its wavelength in nanometers. Referring to Figure 3-1: What is this photon's color? What is the wavelength of a photon with twice this energy? What is the energy of a green photon?

(Answers appear at the end of the book.)

in honor of the German physicist Heinrich Hertz. One hertz means that 1 cycle per second—or that 1 wave crest per second—passes any point. A thousand hertz means a thousand cycles per second, and so on. The frequency is used, among many other things, to identify radio stations. For example, WCPE radio in Wake Forest, North Carolina, has a frequency of 89.7 megahertz (a megahertz is a million hertz or a million cycles per second). Frequency is discussed further in *An Astronomer's Toolbox 3-1*.

All photons with the same wavelength are identical to each other, and, therefore, every photon of a given wavelength carries the same amount of energy as every other photon with that wavelength. The energy delivered by a photon is either enough to eject an electron from the surface of the metal or it is not; there is no middle ground. Extensive testing in the twentieth century confirmed both the wave and particle properties of light.

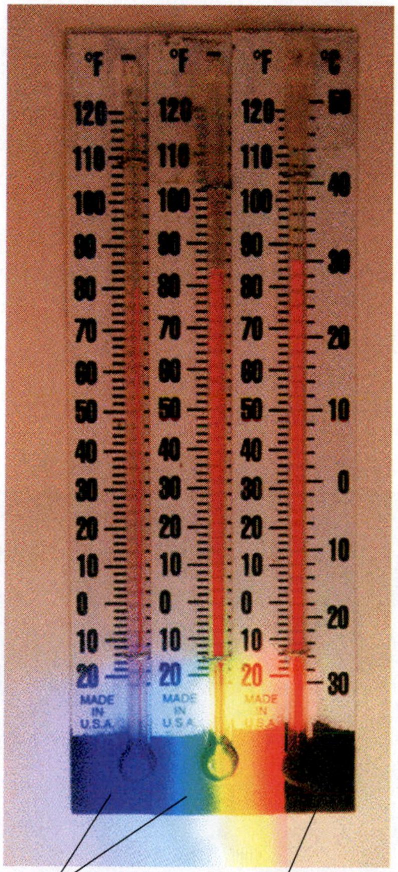

How are the frequencies of ocean waves determined?

While the energy of a single photon is fixed by its wavelength, the total number of photons passing per second from that source with a given energy determines the intensity of the electromagnetic radiation at that wavelength (that is, how bright the object appears to be). The more photons detected, the higher the intensity, and vice versa. However, the intensity of light does not change the energy per photon. If one photon is unable to eject an electron from a metal, then billions of photons with that energy will still be unable to remove that electron.

Visible light heats these thermometers

Infrared radiation heats this thermometer

R I V U X G

FIGURE 3-5 Experimental Evidence for Infrared Radiation This photograph shows the visible colors separated by a prism. The two thermometers in the region illuminated by visible light have temperatures less than the thermometer to the right of red. Therefore, there must be more radiation energizing (that is, heating) the warmest thermometer. This energy is what we call infrared radiation—invisible to the human eye, but detectable as heat. (NASA/JPL-Caltech)

3-4 Visible light is only one type of electromagnetic radiation

We have said that visible light has a narrow range of wavelengths, from about 400 to 700 nm. However, Maxwell's equations place no length restrictions on the wavelengths of electromagnetic radiation. What lies on either side of this interval? Around 1800, the British astronomer William Herschel discovered **infrared radiation** in an experiment with a prism. When he held a thermometer just beyond the red end of the visible spectrum, the thermometer registered a temperature increase, indicating that it was being heated by an invisible form of energy (Figure 3-5). Infrared radiation, discovered before Maxwell's equations were formulated, was later identified as electromagnetic radiation with wavelengths slightly longer than red light. Our bodies detect infrared radiation as heat.

In experiments with electric sparks in 1888, Hertz succeeded in producing electromagnetic radiation a few centimeters in wavelength, now known as **radio waves**.

At wavelengths shorter than those of visible light, **ultraviolet (UV) radiation** extends from about 400 nm to 10 nm. In 1895, Wilhelm Roentgen invented a machine that produces electromagnetic radiation with wavelengths shorter than 10 nm, now called **X rays**. Modern versions of Roentgen's machine are found in medical and dental offices and airport security checkpoints. X rays have wavelengths between about 10 and 0.01 nm. Photons with even shorter wavelengths are called **gamma rays**. These boundaries are all arbitrary and primarily used as convenient divisions in the electromagnetic spectrum, which is actually continuous.

We now know that visible light occupies only a tiny fraction of the full range of possible wavelengths, collectively called the **electromagnetic spectrum**. As shown in Figure 3-6, the electromagnetic spectrum stretches from the longest-wavelength radio waves, through microwaves, infrared radiation, visible light, ultraviolet radiation, and X rays, to the shortest-wavelength photons, gamma rays.

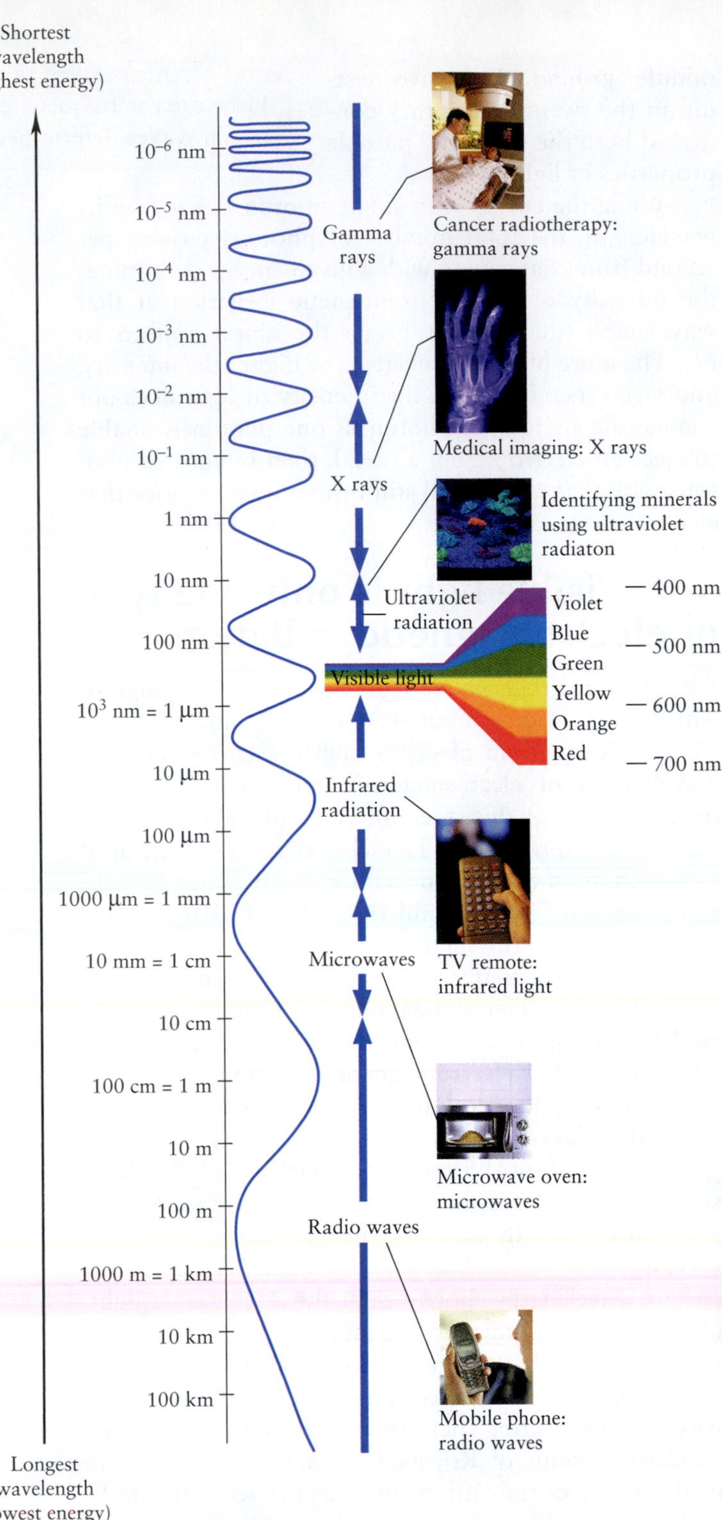

FIGURE 3-6 The Electromagnetic Spectrum The full array of all types of electromagnetic radiation is called the electromagnetic spectrum. It extends from the longest-wavelength radio waves to the shortest-wavelength gamma rays. Visible light forms only a tiny portion of the full electromagnetic spectrum. Note that 1 μm (micrometer) is 10^{-6} m, and 1 nm (nanometer) is 10^{-9} m. The insets show how we are now using all parts of the electromagnetic spectrum here on Earth. (From top: Will and Deni McIntyre/Science Photo Library; Edward Kinsman/Photo Researchers, Inc.; Chris Martin-Bahr/Science Photo Library; Bill Lush/Taxi/Getty; Michael Porsche/Corbis; Ian Britton, Royalty-Free/Corbis)

On the long-wavelength side of the visible spectrum, infrared radiation covers the range from about 700 nm to 1 mm. Astronomers interested in infrared radiation often express wavelength in *micrometers* or *microns* (abbreviated μm), where 1 μm = 1000 nm = 10^{-6} m. From roughly 1 mm to 10 cm is the range of microwaves. Microwaves are sometimes considered as a separate class of photons and sometimes categorized as infrared radiation or radio waves. Formally, radio waves are all electromagnetic waves longer than 10 cm.

The various types of electromagnetic radiation share many basic properties. For example, they are all photons, they all travel at the same speed, and they all sometimes behave as particles and sometimes as waves. But, because of their different wavelengths (and therefore different energies), they interact very differently with matter. For example, X rays penetrate deeply into your body tissues, while visible light is mostly stopped and scattered by the surface layer of skin; your eyes respond to visible light but not to infrared radiation; and your radio detects radio waves but not ultraviolet radiation.

Earth's atmosphere is relatively transparent to visible light, radio waves, microwaves, short-wavelength infrared, and long-wavelength ultraviolet. As a result, these radiations pass through the atmosphere without much loss and can be detected by ground-based telescopes sensitive to them. Astronomers say that the atmosphere has *windows* for these parts of the electromagnetic spectrum (Figure 3-7).

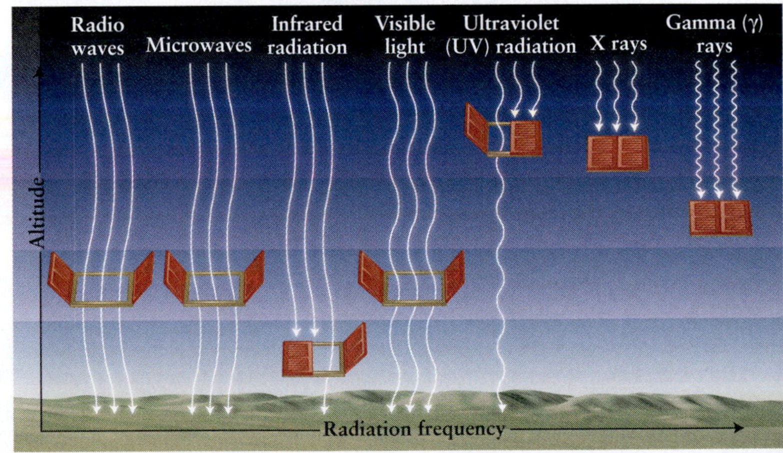

FIGURE 3-7 "Windows" Through the Atmosphere Different types of electromagnetic radiation penetrate into Earth's atmosphere in varying amounts. Visible light, radio waves, microwaves, short-wavelength infrared, and long-wavelength ultraviolet reach all the way to Earth's surface. The other types of radiation are absorbed or scattered by the gases in the air at different characteristic altitudes (indicated by heights of windows). Although the atmosphere does not have actual "windows," astronomers use the term to characterize the passage of radiation through it.

The longest-wavelength ultraviolet radiation, called UVA, causes tanning and sunburns. Ozone (O_3) in Earth's atmosphere normally screens out intermediate-wavelength ultraviolet radiation, or UVB. Until recently, the ozone in the *ozone layer* high in the atmosphere was being depleted by human-made chemicals, such as chlorofluorocarbons (CFCs) and bromine-rich gases. As a result, more UVB is reaching Earth's surface, and these highly energetic photons severely damage living tissue, causing skin cancer and glaucoma, among other diseases.

2 Earth's atmosphere is completely opaque to the other types of electromagnetic radiation, meaning that they do not reach Earth's surface. (This opacity is a good thing, because short-wavelength ultraviolet radiation [UVC], X rays, and gamma rays are devastating to living tissue. Gamma rays, packing the highest energies, are the deadliest.) Direct observations of these wavelengths must be performed high in the atmosphere or, ideally, from space.

As noted earlier, photons with different energies interact with matter in different ways. Higher-energy photons will pass through or rip apart material from which lower-energy photons will bounce off. Telescope designs for collecting and focusing radiation, therefore, differ depending on the energies or, equivalently, the wavelengths of interest. Knowing the energies of photons and their effects on matter enables astronomers to design both the telescopes to collect them and the devices used to record their presence. In the next section we will consider the lengths to which astronomers have gone to capture visible and invisible (or nonoptical) electromagnetic radiation.

OPTICS AND TELESCOPES

3 Since the time of Galileo, astronomers have been designing instruments to collect more light than the human eye can gather on its own. Collecting more light enables us to see things more brightly, in more detail, and at a greater distance. There are two basic types of telescopes—those that collect light through lenses, or **refracting telescopes**, and those that collect it from mirrors, or **reflecting telescopes** (**reflectors**). The earliest telescopes, such as Galileo's, used lenses, which have a variety of shortcomings as light-gathering devices. Consequently, all modern research telescopes use mirrors to collect light. Lenses are still used in the eyepieces of all home telescopes to straighten the gathered light, so when we look through these telescopes directly, our brains can accurately interpret what we see. Lenses are also used to collect light in binoculars and cameras. Reflecting and refracting telescopes have the same main purpose—to collect as many photons as possible so we can better observe the sky (see Section 3-6). Because astronomers exclusively use reflecting telescopes, we will begin exploring telescopes by discussing how these telescopes work. Then we will consider how lenses collect light, how refracting telescopes work, and, finally, how astronomers have developed telescopes to see nonvisible electromagnetic radiation.

3-5 Reflecting telescopes use mirrors to concentrate incoming starlight

The first reflecting telescope was built in the seventeenth century by Isaac Newton (Figure 3-8). To understand how these telescopes work, consider a flow of photons, more commonly called a *light ray*, moving toward a flat mirror. In Figure 3-9a, a light ray strikes the mirror, and we imagine a perpendicular line coming out of the mirror at that point. According to the principle of **reflection**, the angle between the incoming light ray and the perpendicular (dashed line) is always equal to the angle between the outgoing, reflected light ray and the perpendicular. This principle is often stated as, "The angle of incidence equals the angle of reflection." This rule also applies if the mirror is curved (Figure 3-9b).

Using this principle, Newton determined that a concave (hollowed-out) mirror, ground in the shape of a parabola, causes all parallel incoming light rays that

FIGURE 3-8 Replica of Newton's Reflecting Telescope Built in 1672, this reflecting telescope has a spherical primary mirror 3 cm (1.3 in.) in diameter. Its magnification was 40×. (Royal Greenwich Observatory/Science Photo Library)

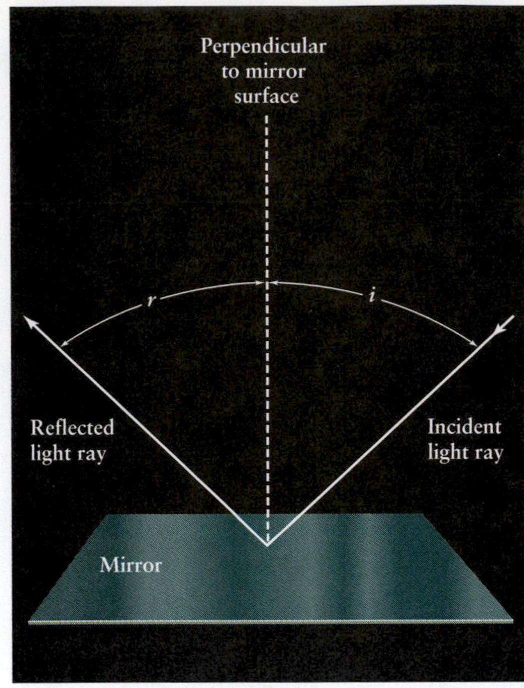

a

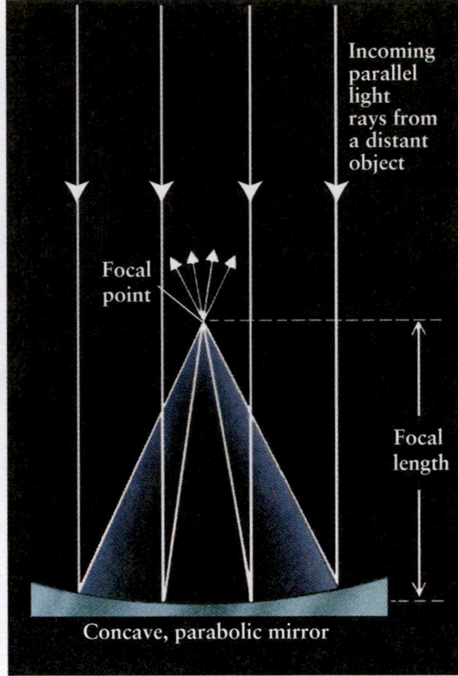

b

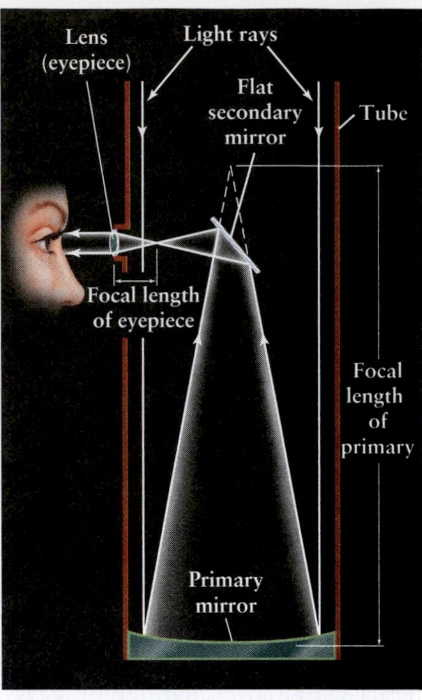

c

FIGURE 3-9 Reflection (a) The angle at which a beam of light strikes a mirror (the angle of incidence, *i*) is always equal to the angle at which the beam is reflected from the mirror (the angle of reflection, *r*). (b) A concave, parabolic mirror causes parallel light rays to converge and meet at the focal point. The distance between the mirror and focal point is the focal length. (c) A Newtonian telescope uses a flat mirror, called the secondary mirror, to send light toward the side of the telescope. The light rays are made parallel again by passing through a lens, called the eyepiece. The dashed line shows where the focal point of this primary mirror would be if the secondary mirror were not in the way.

Which has more energy, an infrared photon or an ultraviolet photon?

strike the mirror to converge to a **focal point** (Figure 3-9b). The distance between this **primary mirror** and the focal point, where the image of the distant object is formed, is called the **focal length** of the mirror. Focal points exist for light from sources that are extremely far away, like the stars. (Figure 3-10 shows why stars can be considered "far away.") If the object is larger than a point, like the Moon or a planet, then the light will converge to a plane, called the **focal plane**, located at the distance of the focal length. Likewise, point objects like stars that are near each other on the celestial sphere also come into focus near each other on the focal plane.

To view the image, Newton placed a small, flat mirror at a 45° angle between the primary mirror and the focal point, as sketched in Figure 3-9c and Figure 3-11a. This **secondary mirror** reflects the light rays to one side of the telescope, and the viewer observes the image through an **eyepiece lens**. A telescope with this optical design is still called a **Newtonian reflector**. We will discuss how this lens works in Section 3-8. Suffice it to say for now that research telescopes do not use eyepieces. As we will see shortly, light-sensitive detectors are placed in their focal planes instead.

Newtonian telescopes are popular with amateur astronomers because they are convenient to use while the observer is standing up. However, they are not used in research observatories because they are lopsided. If astronomers attach their often heavy and bulky research equipment onto the side of a Newtonian telescope, the telescope twists and distorts the image in unpredictable ways.

Three basic designs exist for the reflecting telescopes used in research. In the first, a hole is drilled directly through the center of the primary mirror. A convex (outwardly curved) secondary mirror placed between the primary mirror and its focal point reflects the light rays back through the hole (Figure 3-11b). This design is called the **Cassegrain focus** (after a 1672 design by Laurent Cassegrain). This secondary mirror extends the telescope's focal length. Compact, relatively low-weight equipment is bolted to the bottom of the telescope, and the light is brought into focus in it. The advantage of this design over Newtonian telescopes is that the attached equipment is balanced and does not distort the telescope frame and, hence, the image.

The second design that astronomers use has two variations, both of which use a third mirror to direct light out the side of the telescope, at the place where it pivots (Figure 3-11c). Heavier or bulkier optical equipment that requires firmer mounting can be located at the

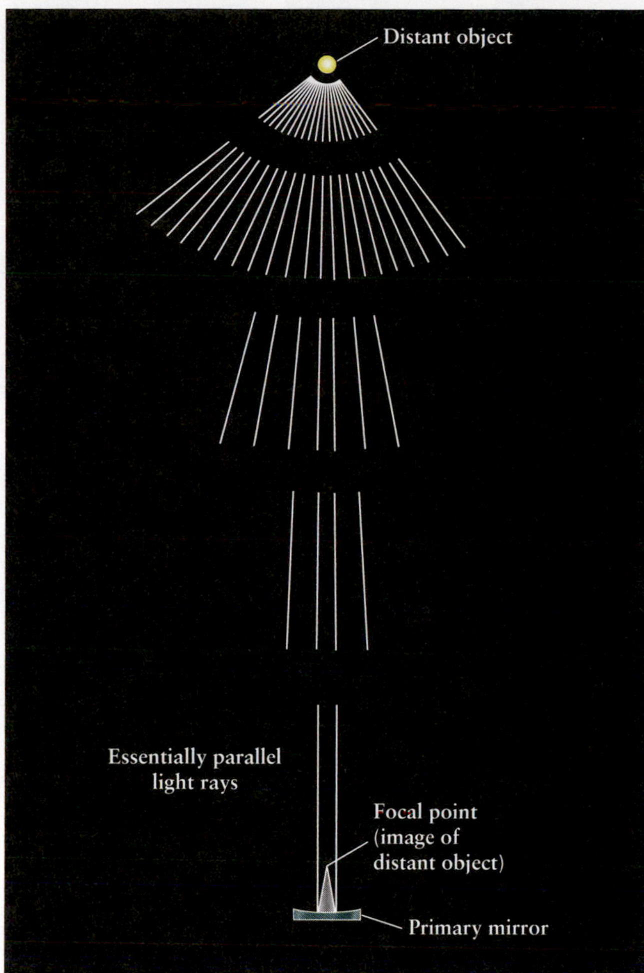

FIGURE 3-10 Parallel Light Rays from Distant Objects As light travels away from any object, the light rays, all moving in straight lines, separate. By the time light has traveled trillions of kilometers, only the light rays moving in virtually parallel tracks are still near each other.

Nasmyth focus (after Scottish engineer James Nasmyth, who developed it). If an extremely long focal length is desired, a **coudé focus** (named after a French word meaning "bent like an elbow") is used. Equipment that profits from the use of the Nasmyth focus and coudé focus includes a variety of *spectrographs*, instruments that separate light from objects into its individual colors to determine the objects' chemistries, surface temperatures, and motions toward or away from us.

In the third design, an observing device is located at the undeflected focal point, directly in front of the primary mirror. This arrangement is called the **prime focus** (Figure 3-11d). This design has the advantages of making the brightest image (for a given exposure time) and having the fewest reflections that can cause light loss and distortion (that is, no secondary mirror).

Where in Figure 3-9c does the light from the primary mirror actually come to a focal point?

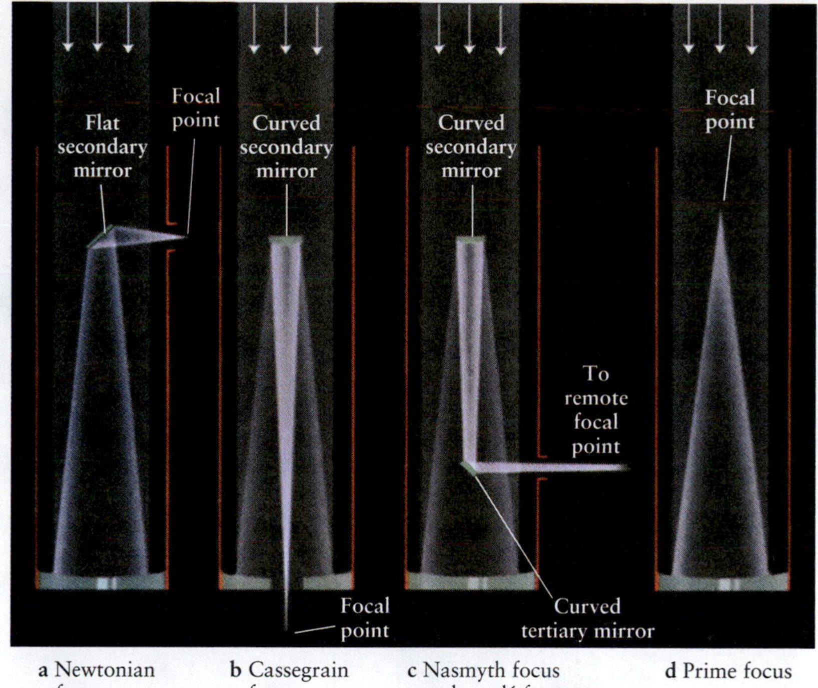

ANIMATION 3.1

FIGURE 3-11 Reflecting Telescopes Four of the most common optical designs for reflecting tele-scopes: (a) Newtonian focus (popular among amateur astronomers) and the three major designs used by researchers—(b) Cassegrain focus, (c) Nasmyth focus and coudé focus, and (d) prime focus.

3-6 Telescopes brighten, resolve, and magnify

As mentioned earlier, a telescope's most important function is to provide astronomers with as bright an image as possible. The brighter an object appears, the more information about it we can extract. The observed brightness of any object depends on the total number of photons collected from it, which, in turn, depends on the area of the telescope's primary mirror. Analogously, the pupils in our eyes get larger in dark environs to allow more photons to strike the retina and create a brighter image than otherwise. For exposures of equal times, a telescope with a large primary mirror produces brighter images and detects fainter objects than a telescope with a smaller primary mirror (Figure 3-12).

Insight Into Science

Costs and Benefits Science now relies heavily on technology to conduct experiments or to make observations. The cost of cutting-edge astronomical observations may run to hundreds of millions of dollars or more. The return on such investments is a better understanding of how the universe works, how we can harness its resources, and our place in it.

FIGURE 3-12 Light-Gathering Power Because a large primary mirror collects more starlight than does a smaller one, a larger telescope produces a brighter image than a smaller one, all other things being equal. The same principle applies to telescopes that collect light using an objective lens rather than a primary mirror. The two photographs of the Andromeda Galaxy were taken through telescopes with different diameters and were exposed for equal lengths of time at equal magnification. (AURA)

The **light-gathering power** of a telescope is directly related to the area of the telescope's primary mirror. Recall that the area and diameter of a circle are related by the formula

$$\text{Area} = \frac{\pi d^2}{4}$$

where d is the diameter of the mirror and π (pi) is about 3.14. Consequently, *a mirror with twice the diameter of another mirror has 4 times the area of the smaller mirror and, therefore, collects 4 times as much light as does the smaller one in the same amount of time*. For example, a 36-cm-diameter mirror has 4 times the area of an 18-cm-diameter mirror. Therefore, the 36-cm telescope has 4 times the light-gathering power of a telescope half its size.

Another vital function of any telescope is to reveal greater detail of objects that are more than just points of light. Such *extended objects* include the Moon, the Sun, planets, galaxies, interstellar gas clouds, and clusters of stars, among other things. A large telescope increases the sharpness of the image and the degree of detail that can be seen. **Angular resolution** (often called just *resolution*) measures the clarity of images (Figure 3-13). The angular resolution of a telescope is measured as the arc angle between two adjacent stars whose images can just barely be distinguished by the telescope. The smaller the angle, the sharper the image. Large, modern telescopes, like the Keck telescopes in Hawaii, have angular resolutions better than 0.1 arcsec. As a general rule, *a telescope with a primary mirror twice the diameter of another telescope's primary will be able to see twice as much detail as the smaller telescope*.

The final function of a telescope is to make objects appear larger. This property is called **magnification.** Magnification is associated with resolution, because the larger the image, the more detail of the image you can potentially see (Figure 3-14). The magnification of a reflecting telescope is equal to the focal length of the primary mirror divided by the focal length of the eyepiece lens

$$\text{Magnification} = \frac{\text{Focal length of primary}}{\text{Focal length of eyepiece}}$$

a

b

FIGURE 3-13 Resolution The larger the diameter of a telescope's primary mirror, the finer the detail the telescope can resolve. These two images of the Andromeda Galaxy, taken through telescopes with different diameters, show this effect. (a) Shows a lower-resolution image taken through a smaller telescope. In this photograph most individual stars blur together to make the galaxies look like fuzzy blobs. (b) Shows the same field of view through a larger-diameter telescope. Many more individual stars and interstellar gas clouds are visible here than in (a). Increasing the exposure time of the smaller-diameter telescope (a), will only brighten the image, not improve the resolution. (AURA)

For example, if the primary mirror of a telescope has a focal length of 100 cm and the eyepiece has a focal length of 0.5 cm, then the magnifying power of the telescope is

$$\text{Magnification} = \frac{100 \text{ cm}}{0.5 \text{ cm}} = 200$$

This property is usually expressed as 200×.

There is a limit to the magnification of any telescope. Try to magnify beyond that limit, and the image becomes distorted. As a rule, *a telescope with a primary mirror twice the diameter of another telescope's primary will have twice the maximum magnification of the smaller telescope.*

Normal eyeglasses or contact lenses are designed to improve which aspect of vision presented in this section?

a

b

FIGURE 3-14 Magnification The same telescope can magnify by different amounts, depending on the focal length of the eyepiece. (a) A low-magnification image of the Moon. (b) An image of the Moon taken with magnification 4 times greater than image (a). Note in this case that the increased magnification leads to increased resolution (that is, more detail can be seen in the larger image). (NASA)

FIGURE 3-15 Mosaic of Charge-Coupled Devices (CCDs) (a) These 40 CCDs combine to provide up to 378 million light-sensitive pixels that store images collected by the CFHT telescope on the dormant volcano Mauna Kea in Hawaii. Electronic circuits transfer the data to a waiting computer. (b) This image of the Rosette Nebula, a region of star formation 5000 ly away in the constellation Monoceros (the Unicorn), was taken with the CCD in (a). The image shows the incredible detail that can be recorded by large telescopes and high-resolution CCDs. (J. C. Cuillandre/Canada-France-Hawaii Telescope)

3-7 Storing and analyzing light from space is key to understanding the cosmos

The invention of photography during the nineteenth century was a boon for astronomy. By taking a long exposure with a camera mounted at the focal plane of a telescope, an astronomer could record extremely faint features that could not be seen by just looking through the telescope. The reason photography is better is that our nervous system refreshes (that is, replaces) the images we receive from our eyes several times a second, whereas film adds up the intensity of all the photons that affect its emulsion. By keeping telescopes aimed precisely, so that images do not blur, exposures of an hour or more became quite routine. Astrophotographs and their modern electronic equivalents, discussed in the next few paragraphs, make all objects appear brighter and reveal greater detail in extended objects—such as galaxies, star clusters, and planets—than we could otherwise see.

Astronomers have long known, however, that a photographic plate is an inefficient detector of light because it depends on a chemical reaction to produce an image. Typically, only 2% of the light striking film triggers a reaction in the photosensitive material. Thus, roughly 98% of the light falling onto a photographic plate is wasted.

Rapidly evolving technology has changed all that. We have replaced photographic film with highly efficient electronic light detectors called **charge-coupled devices (CCDs)**. Very good CCDs respond to between 50 and 75% of the light falling on them; great CCDs are sensitive to more than 90% of the photons that strike them. Clear CCD resolution is much better than that of film, and CCDs respond more uniformly to light of different colors. Divided into an array of small, light-sensitive squares called picture elements or, more commonly, **pixels**, each CCD is a square or rectangle a few centimeters on a side. Several of them are often used together to create a larger image (Figure 3-15). Digital cameras and picture-taking cell phones use this same CCD technology. The largest grouping of CCDs used on a telescope has 378 megapixels (a megapixel is 1 million pixels), compared to most digital cameras, which typically have between 6 and 18 megapixels.

When an image from a telescope is focused on the CCD, an electric charge builds up in each pixel in direct proportion to the intensity of the light (number of photons) falling on that pixel. When the exposure is finished, the charge on each pixel is read into a computer. Figure 3-16 shows one photograph and two CCD images of the same region of the sky, all taken with the same telescope. You can see how many details visible in the CCD images are absent in the ordinary photograph.

a

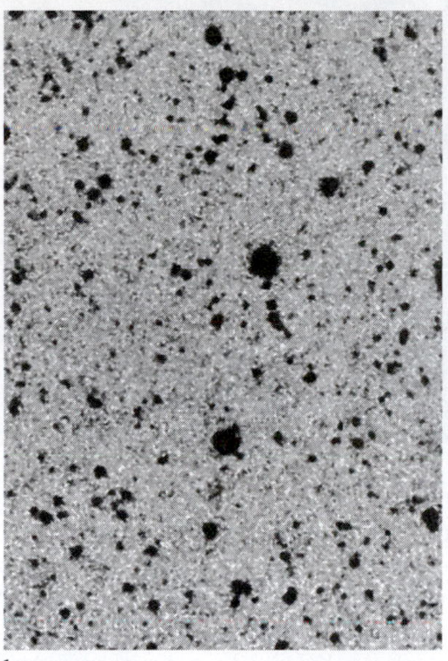

b

c

FIGURE 3-16 Photography versus CCD Images These three views of the same part of the sky, each taken with the same 4-m telescope, compare CCDs to photographic plates. (a) A negative print (black stars and white sky) of a photographic image. (b) A negative CCD image. Notice that many faint stars and galaxies that are invisible in the ordinary photograph can be seen clearly in this CCD image. (c) This (positive) color view was produced by combining a series of CCD images taken through colored filters. (Patrick Seitzer, NOAO)

3-8 Eyepieces, refracting telescopes, and binoculars use lenses to focus incoming light

Although light travels at about 300,000 km/s in a vacuum, it moves more slowly through any medium, such as glass. As light enters the glass, it slows abruptly, much like a person walking from a boardwalk onto a sandy beach. The person's pace suddenly decreases as he or she steps from the smooth, hard pavement onto the sand; and just as he or she steps back onto the boardwalk and resumes the original pace, light that exits a piece of glass resumes its original speed.

As a result of changing speed, light also changes direction as it passes from one transparent medium into another. As noted earlier, this latter change is called *refraction.* You see refraction every day when looking through windows. Imagine a stream of photons from a star that enters a window, as shown in Figure 3-17a. As a light ray goes from the air into the glass, the light ray's direction changes so that it is more perpendicular to the surface of the glass than it was before entering. Once inside the glass, the light ray travels in a straight line. Upon emerging from the other side, the light ray bends once again, resuming its original direction and speed. The net effect is only a slight, uniform displacement of the objects beyond the glass.

Unlike windows, lenses have surfaces of varying thickness (Figure 3-17b). These curved surfaces force the light rays to emerge from the lens in different directions than they had before entering the lens. Lenses that are thicker at their centers than at their edges are called *convex* lenses and force the light that enters them to converge as it passes through. Conversely, lenses thinner at their centers than around their edges are called *concave* lenses. They cause light rays to diverge. Large convex lenses called **objective lenses** were used in telescopes (and still are in many built for home use) instead of primary mirrors, and small lenses are still used as eyepieces to bring the light rays back to being parallel, so that when we view directly through the telescope, our eyes can make sense of what we see.

Why do the human eye and brain clear the images that they receive many times per second?

Parallel light rays that enter an objective lens converge and meet at the focal point of the lens. The distance between the focal point and the lens is the lens's focal length (see Figure 3-17b). As with reflecting telescopes, if the object is close enough or large enough to be more than just a dot as seen through the telescope, all the light from it does not converge at the focal point, but rather focuses along the focal plane (Figure 3-18).

A refracting telescope or **refractor** (Figure 3-19), is an arrangement of two lenses used to gather light. The objective lens at the top of the telescope has a large diameter and long focal length. Like a primary mirror, its purpose is to collect as much light as possible.

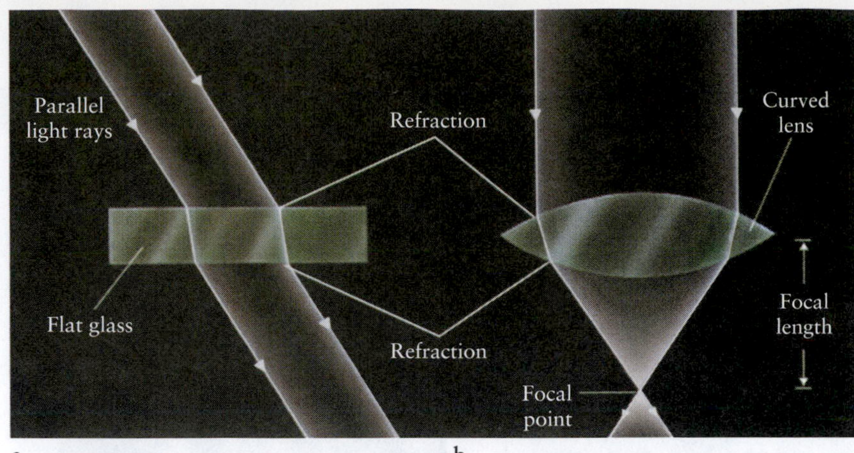

FIGURE 3-17 Refraction Through Uniform and Variable Thickness Glasses (a) Refraction is the change in direction of a light ray when it passes into or out of a transparent medium such as glass. A light ray that enters a denser medium, such as moving from air into water or glass, is bent or refracted to an angle more perpendicular to the surface than the angle at which it was originally traveling. If the glass is flat, then the light leaving it is refracted back to the direction it had before entering the glass. There is no overall change in the direction in which the light travels. (b) If the glass is in the shape of a suitable convex lens, parallel light rays converge to a focus at the focal point. As with parabolic mirrors, the distance from the lens to the focal point is called the focal length of the lens. (c) The straw as seen through the side of the liquid is magnified and offset from the straw above the liquid because the liquid is given a curved shape by the side of the glass. The straw, as seen through the top of the liquid, is refracted but does not appear magnified because the surface of the water is flat and the beaker has uniform thickness. (c: Ray Moller/Dorling Kindersley)

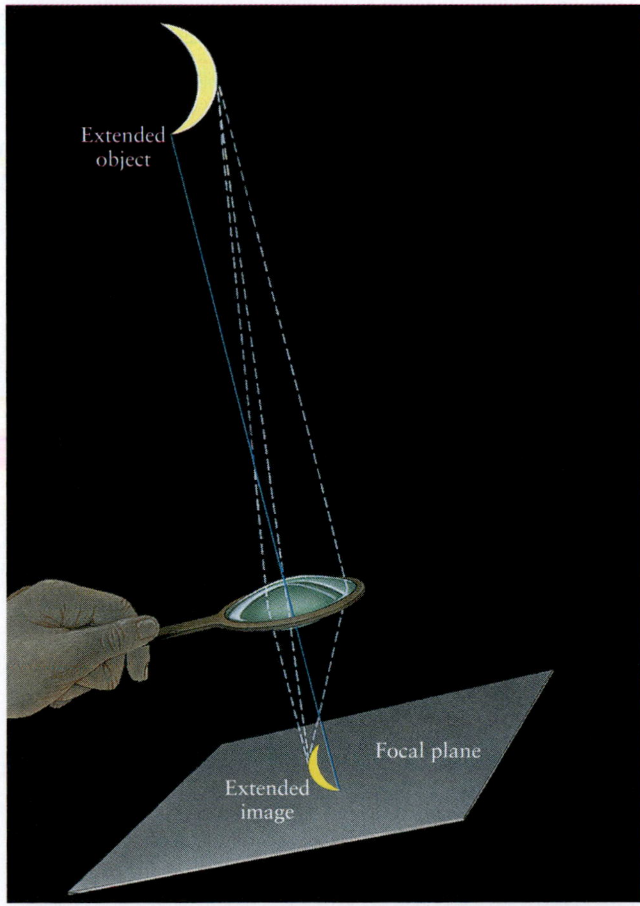

FIGURE 3-18 Extended Objects Create a Focal Plane Light from objects larger than points in the sky does not all converge to the focal point of a lens. Rather, an image of the object is created at the focal length in what is called the focal plane.

The eyepiece lens, at the bottom of the telescope, is smaller and has a short focal length. The mathematics of magnification for a refracting telescope is the same as for a reflecting telescope; however, the focal length of the objective lens substitutes for the focal length of the primary mirror. Likewise, all of the rules for the limits on telescopes are the same for refractors as for reflectors.

The largest refracting telescope in the world, completed in 1897 and located at the Yerkes Observatory in Williams Bay, Wisconsin, near Chicago (Figure 3-20), has an objective lens that is 102 cm (40 in.) in diameter with a focal length of $19^1/_3$ m ($63^1/_2$ ft). The lens was ground by the premier American lensmaker of the nineteenth century, Alvan Clark. The second-largest refracting telescope, located at Lick Observatory near San Jose, California, has an objective lens of 91 cm (36 in.) in diameter. No major refracting telescopes were constructed in the twentieth century or are planned for this century.

Refracting telescopes suffer from a variety of problems that have limited their use as research instruments. These problems include the following:

- It is hard to grind a lens to the very complicated shape necessary to have all parallel light rays of any color passing through it converge to the same focal point. When, for example, the lens is given a spherical shape (relatively easy to manufacture), the light at different distances from the center of the lens has different focal lengths (Figure 3-21). This phenomenon is called **spherical aberration.** Such aberration does not occur for reflecting telescopes because the shape of a mirror

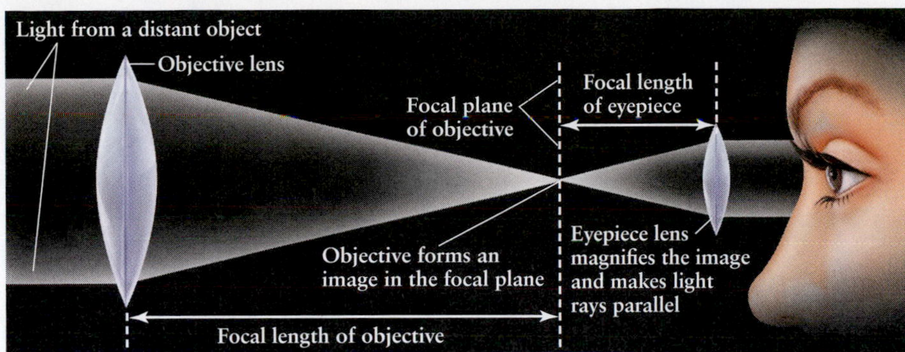

FIGURE 3-19 Essentials of a Refracting Telescope A refracting telescope consists of a large, long-focal-length objective lens that collects and focuses light rays and a small, short-focal-length eyepiece lens that restraightens the light rays. The lenses work together to brighten, resolve, and magnify the image formed at the focal plane of the objective lens.

needed to focus light from every part of the mirror to the same focal point is a simple parabola, which is easy to manufacture (see Section 3-10).

- Even when light passes through an ideally shaped lens, different colors of light are refracted by different amounts, so they have different focal lengths. This phenomenon is called **chromatic aberration** (Figure 3-22a).

As a result of spherical and chromatic aberration, objects look blurred (Figure 3-22b). Chromatic and spherical aberration can be corrected (Figure 3-23) by adding a second lens with a different shape and made of a different type of glass (which refracts light by a different amount than does the first lens). Chromatic aberration is not a problem for reflectors because the reflecting surface is on the top of the mirror so that light never enters the glass.

- A lens must be supported only around its edges to avoid blocking the light. The weight of a large lens can cause it to sag and, thereby, distort the image. This distortion is not a problem with reflectors because the entire underside of the mirror can be supported, as necessary.
- Air bubbles in the glass cause unwanted refractions and, hence, distorted images.
- Glass does not allow all wavelengths to pass through it equally.

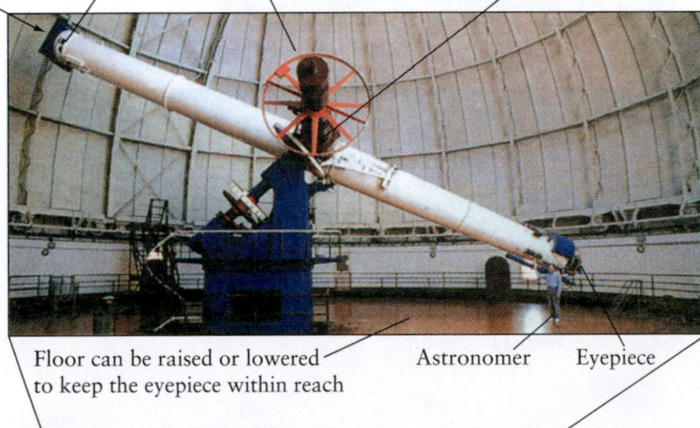

FIGURE 3-20 The Largest Refracting Telescope This giant refracting telescope, built in the late 1800s, is housed at Yerkes Observatory near Chicago. The objective lens is 102 cm (40 in.) in diameter, and the telescope tube is $19\frac{1}{3}$ m ($63\frac{1}{2}$ ft) long. (Yerkes Observatory)

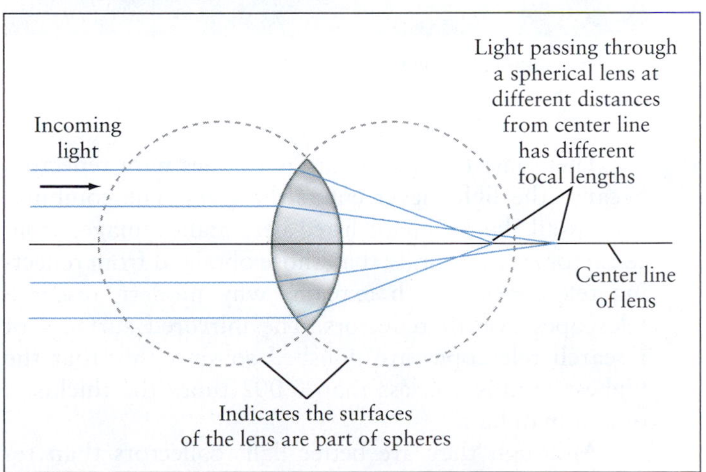

FIGURE 3-21 The Geometry of a Spherical Lens If both sides of a lens are spherical surfaces, as shown here, then light rays of the same color passing through at different distances from the center of the lens are refracted by different amounts. Therefore, spherical lenses have different focal lengths for these different light rays and so they give blurry images.

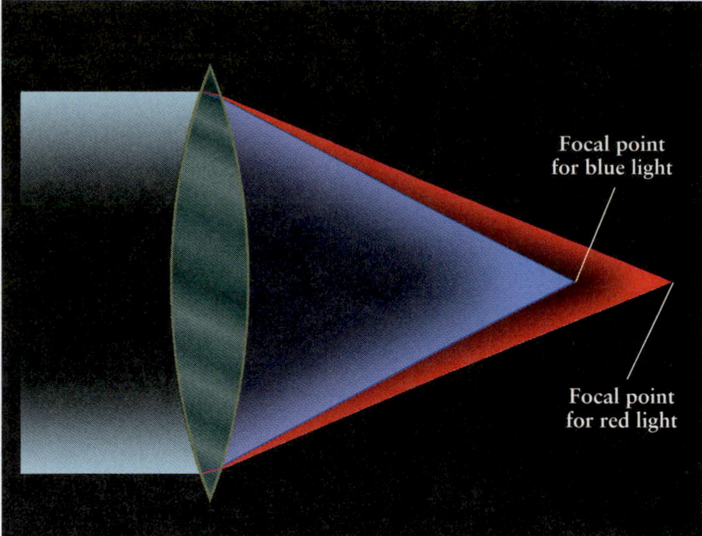

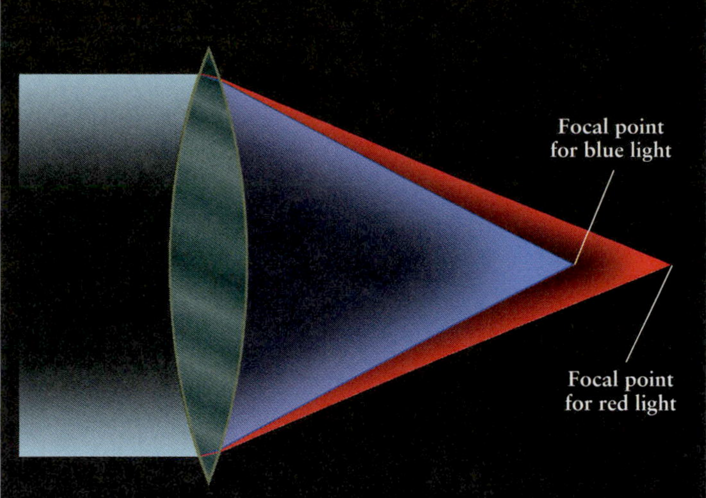

a The problem: Chromatic aberration

b RIVUXG

FIGURE 3-22 Chromatic Aberration (a) Light of different wavelengths is refracted by different amounts when passing through a medium such as glass. Therefore, single lenses such as this one have different focal lengths for light of different colors passing through them. (b) Image showing chromatic aberration. Note the rainbows caused by light passing through a lens. (Stan Zurek, CC-BY-SA-3.0/Wikimedia Commons)

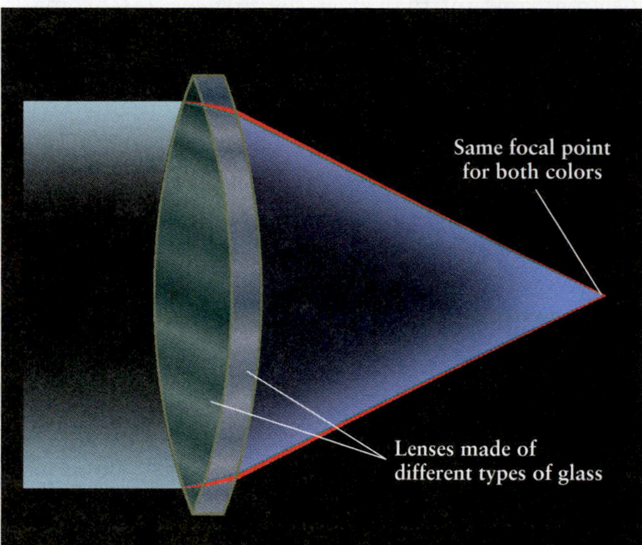

a The solution: Use two lenses

b RIVUXG

FIGURE 3-23 Achromatic Lens (a) By using two differently-shaped lenses (often of different types of glass), light of different wavelengths can be brought into focus at the same focal length. Such achromatic lenses are used in cameras and many telescopes. (b) Same object as in Figure 3-22b imaged through an achromatic lens. Note that the rainbows seen in Figure 3-22b do not occur here. (Stan Zurek, CC-BY-SA-3.0/Wikimedia Commons)

4 These last two points are not issues with reflectors because the light never enters the glass. The combination of all the problems listed here makes images from refractors less accurate than those obtained from reflecting telescopes, which explains why modern research telescopes are all reflectors. The mirrored surfaces of research telescopes are polished so smoothly that the highest bumps are less than 0.002 times the thickness of a human hair.

Although they are better light collectors than refracting telescopes, reflecting telescopes are not perfect; there are several prices to pay for the advantages they offer over refractors. Two of the most important prices to pay are blocked light and spherical aberration. We consider these problems in the next two sections.

How do human eyes focus light?

3-9 Secondary mirrors dim objects but do not create holes in them

You have probably noticed (see Figure 3-11) that the secondary mirror of a reflector blocks some incoming light—one unavoidable price that astronomers must pay. Typically, a secondary mirror prevents about 10% of the incoming light from reaching the primary mirror. This problem is addressed by constructing primary mirrors with sufficiently large surface areas to compensate for the loss of light. You might also think that, because light is missing from the center of the telescope due to blockage by the secondary mirror, a corresponding central "hole" appears in the images.

However, this problem does not occur, because light from all parts of each object being observed enters all parts of the telescope (Figure 3-24). Indeed, covering any part of the primary mirror or objective lens darkens the image but does not limit which parts of the object you can see through the telescope.

3-10 Shaping telescope mirrors and lenses is an evolving science

To make a reflector, an optician traditionally grinds and polishes a large slab of glass into a concave, spherical surface. Before computer control, grinding a spherical surface was much easier than grinding the ideal parabolic surface. However, light that enters a spherical telescope mirror at different distances from the mirror's center comes into focus at different focal lengths. Images taken directly from such telescopes (Figure 3-25a) have the same spherical aberration as the refractors described above.

We can avoid spherical aberration in reflecting telescopes by making the mirror parabolic (Figure 3-25b) or by using a thin correcting lens, called a **Schmidt corrector plate,** with a spherical mirror. Designed by the Estonian-Swedish optician Bernhard Schmidt, the corrector plate is located at the top of the telescope (Figure 3-25c). The light coming into the telescope is refracted by the plate just enough to compensate for spherical aberration and to bring all of the light into focus at the same focal length. These correctors have the added benefit of focusing light from a larger angle in the sky than would be in focus without the plate. A Schmidt corrector plate enables astronomers to map large areas of the sky with relatively few photographs at moderately high magnification. In other words, the Schmidt corrector plate acts like a wide-angle lens on a camera. Schmidt corrector plates are often used in conjunction with Cassegrain

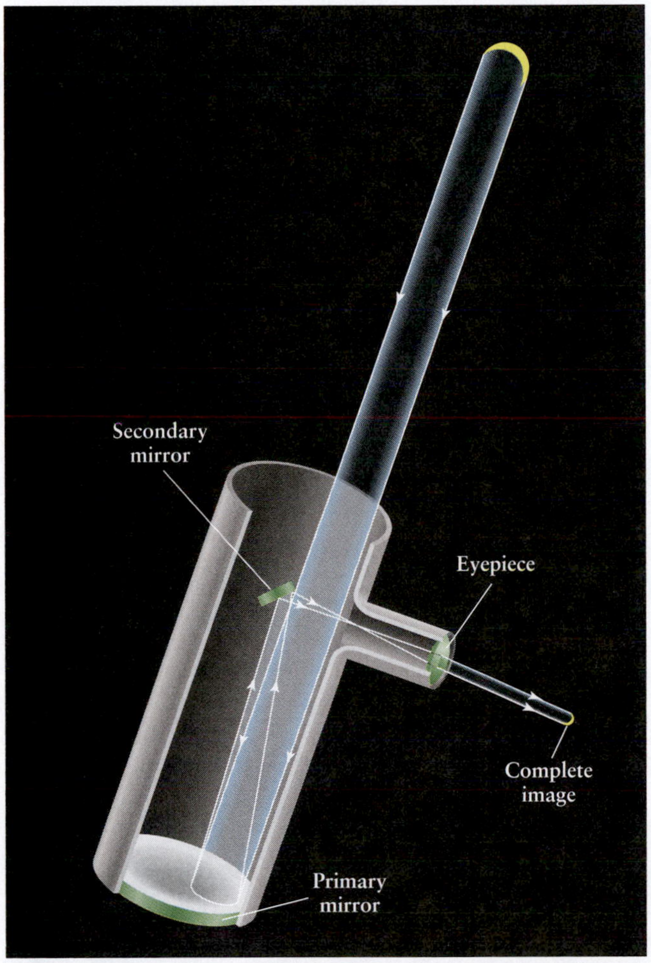

FIGURE 3-24 The Secondary Mirror Does Not Create a Hole in the Image Because the light rays from distant objects are parallel, light from the entire object reflects off all parts of the mirror. Therefore, every part of the object sends photons to the eyepiece. This figure shows the reconstruction of the entire Moon from light passing through just part of this telescope. The same drawing applies everywhere on the primary mirror that is not blocked by the secondary mirror.

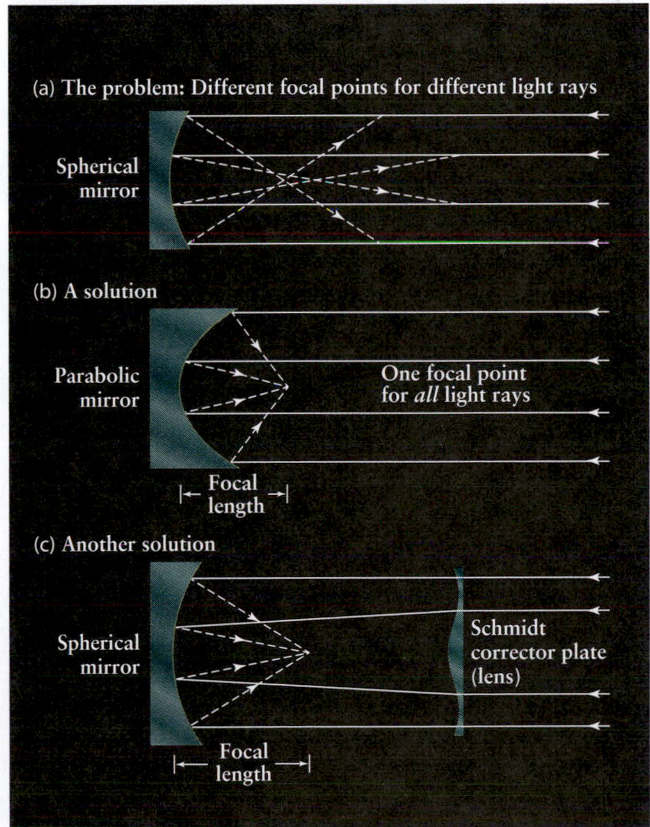

FIGURE 3-25 Spherical Aberration (a) Different parts of a spherically concave mirror reflect light to slightly different focal points. This effect, spherical aberration, causes image blurring. This problem can be overcome by (b) using a parabolic mirror or (c) using a Schmidt corrector plate (a specially curved lens) in front of the telescope.

Buying a Telescope

If you feel elated by discovering treasures, you may want to consider viewing the night sky through powerful binoculars or a telescope. Both types of instrument enable you to find many beautiful, exciting objects in space. Binoculars are nice because you can quickly scan the heavens to see craters on the Moon, star clusters, the Great Nebula of Orion, and even the Andromeda Galaxy. Telescopes open new vistas and even enable amateur astronomers to discover new comets, among other things, adding to our scientific database and immortalizing the discoverers' names.

There are four basic types of telescope mounts that steer telescopes around the sky: the Fork Equatorial Mount, the German Equatorial Mount, the Altitude-Azimuth (Alt-Azimuth) Mount, and the Dobsonian Mount (see figures). The Telescope Mounts table presents the advantages and disadvantages of each mount.

Reflecting and refracting telescopes both have pros and cons, as discussed in the text. Refractors and Newtonian reflectors tend to be longer and bulkier than any other types. Newtonians are the least expensive and simplest telescopes to build. As a result, they are often used with the inexpensive Dobsonian mounts. If you plan to take photographs, a Cassegrain telescope is the best instrument because it remains balanced as the telescope tracks objects across the sky.

If you want to see especially large areas of the sky, you may want to purchase a telescope with a Schmidt corrector plate. The most common wide-field telescopes are of the Schmidt-Cassegrain design. These telescopes have a Schmidt corrector plate on top and a Cassegrain mirror on the bottom, with their eyepieces located underneath.

If you build or buy a telescope with the eyepiece positioned so that you have to crawl under it or climb a ladder to see through it, you will find the experience less satisfying than if you can observe in a comfortable position. If the eyepiece is not easily accessible, you can buy a *diagonal mirror* (a right-angle mirror that goes between the telescope body and the eyepiece) to help correct this problem.

You will also want to get a few different eyepieces with different focal lengths (three is a good number to start with), so that you can look at large areas of the sky under low magnification and details of small areas under high magnification. As discussed in the text, the magnification of a telescope depends on the focal lengths of the objective lens (or primary mirror) and the eyepiece. The objective lens or primary mirror is fixed in the telescope. However, all telescopes come with removable eyepieces. Change the eyepiece to another of different focal length, and you can change the telescope's magnification. Keep in mind, however, that increasing the magnification decreases the area of the sky that you see.

Unless your telescope is computerized, it is also essential to have a *finder scope*, which is a small, low-magnification, very-wide-field telescope, with cross-hairs, attached directly onto your main telescope. If the finder and your main scope are well aligned, you can quickly zero in on an object.

If you plan to take photographs with a CCD camera or other instrument on your telescope or to show the cosmos to large groups of people, you will need a *tracking motor* to enable the telescope to follow the stars automatically. Tracking motors require their own power from batteries, a 12-volt outlet in a car, or a

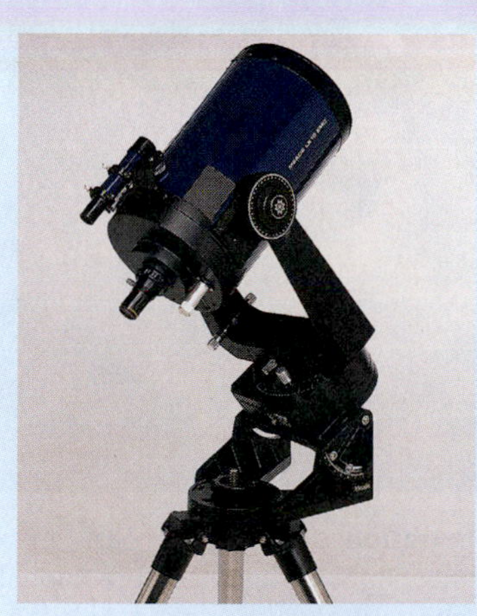

a Fork Equatorial Mount
(Andy Crawford/Dorling Kindersley)

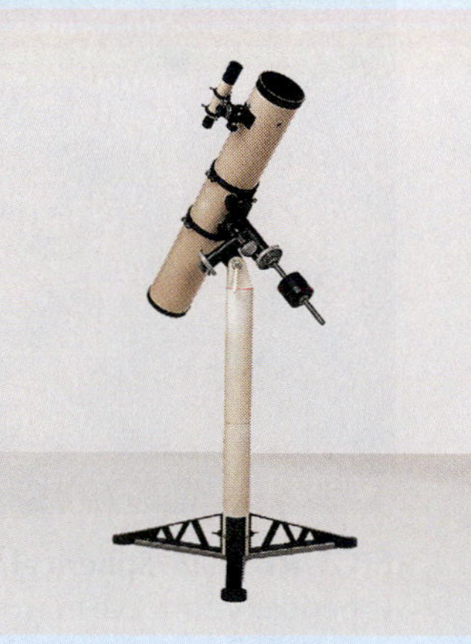

b German Equatorial Mount
(Andy Crawford/Dorling Kindersley)

c Alt-Azimuth Mount
(Celestron Images)

d Dobsonian Mount
(Andy Crawford/Dorling Kindersley)

110-volt outlet. The motor should also be able to run at high speed so you can slew (turn the telescope rapidly) from one object to the next. For photography, you will also need a *camera adapter.*

Viewing the Sun is very exciting, especially when you can observe sunspots. If you want to observe the Sun, it is essential that you buy a good *Sun filter.* **Never, ever look at the Sun directly, either through a telescope or with your naked eye. Doing so for even a second can lead to partial blindness!**

Finally, you will need a flashlight with a red plastic film over the lens or one with a red LED light. Because red light does not cause the pupils of your eyes to contract, they will remain dilated (wide open) while you use the red flashlight to inspect your equipment and your star charts.

Telescope Mounts

Type of Telescope	Pros	Cons
Fork Equatorial	Can track objects in the sky with a clock drive Useful for taking CCD or film photographs Not too heavy or cumbersome	Eyepiece is in an uncomfortable position near celestial poles
German Equatorial	Can track objects in the sky with a clock drive Useful for taking CCD or film photographs Easy to change direction telescope points	Hard at first to learn to set up Takes a long time to set up Very heavy compared to other types of mounts
Alt-Azimuth	Compact, easy to set up, store, and carry Eyepiece is convenient for viewing	Must be computerized to track objects in the sky
Dobsonian	Least expensive for a given diameter primary mirror Easy to set up Easy to use	Eyepiece is often in an inconvenient position Cannot track objects in the sky or take photographs without specialized equipment

mirrors to create *Schmidt-Cassegrain* telescopes, which are popular for home use because they are compact and give relatively wide-angled images. However, the plate does not allow for as much magnification as a telescope with a parabolic mirror. (If you are interested in buying a telescope, you might find helpful the *Guided Discovery: Buying a Telescope.*)

Although some parabolic primary mirrors have been meticulously hand-ground over the past century, the advent of computer-controlled grinding and rotating furnaces (Figure 3-26), in which the liquid glass is actually spun into a parabolic shape, has now made it economical to cast parabolic mirrors with diameters of several meters. That spinning a liquid causes it to develop a parabolic surface was discovered by Isaac Newton in 1689, when he created the effect in a bucket of spinning water.

3-11 Earth's atmosphere hinders astronomical research

Earth's atmosphere affects the light from objects in space before it reaches ground-based telescopes. Even for those wavelengths that have windows through the atmosphere (see Section 3-4), some light is absorbed in the atmosphere, making all objects appear dimmer than they would appear from space. Also, the air is turbulent and filled with varying amounts of impurities and moisture. You have probably seen turbulence while driving in a car on a hot day, when the road ahead appears to shimmer. Blobs of air, heated by Earth, move upward to create this effect. Light passing through such a blob is refracted, because each hot air mass has a different density than the cooler air around it. Because each blob behaves like a lens, images of objects beyond them appear distorted.

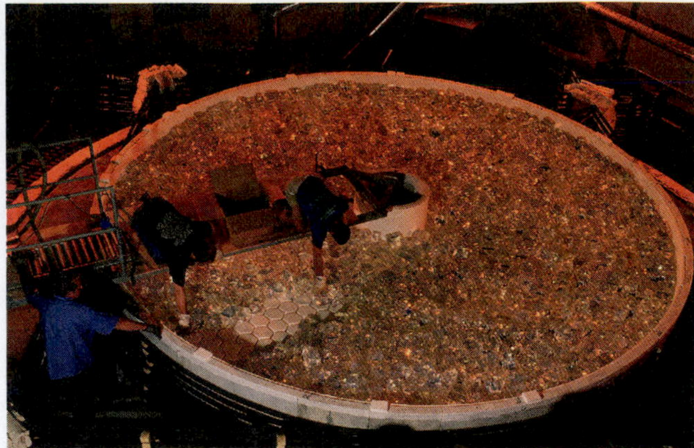

a

b

R I V U X G

FIGURE 3-26 Rotating Furnace for Making Parabolic Telescope Mirrors (a) To make each 8.4-m primary mirror for the Large Binocular Telescope II on Mount Graham in Arizona, 40,000 lb of glass are loaded into a rotating furnace and heated to 1450 K (2150°F). This image shows glass fragments loaded into the cylindrical furnace. (b) After melting, spinning, and cooling, the mirror's parabolic surface is ready for final smoothing and coating with a highly reflective material. (a: Roger Ressmeyer/Corbis; b: University of Arizona, Steward Observatory)

5 The atmosphere over our heads is similarly moving and changing density, and the starlight passing through it is similarly refracted. Because air density changes rapidly, the resulting changes in refraction make the stars appear to change brightness and position rapidly, an effect we see as **twinkling**. When photographed from Earth for more than a few seconds, twinkling smears out a star's image, causing it to look like a disk rather than a pinpoint of light (Figure 3-27a). Astronomers use the expression "seeing" to describe how steady the atmosphere is; when the seeing is bad, much twinkling is occurring and, therefore, telescopic images are spread out and blurry.

The angular diameter of a star's smeared-out image, called the **seeing disk**, is a realistic measure of the best possible resolution. The size of the seeing disk varies from one observatory site to another. At Palomar Observatory in California, the seeing disk is roughly 1 arcsec (1″). The best conditions on Earth have been recorded at the observatory on the 14,000-ft summit of Mauna Kea, the tallest volcano on the island of Hawaii. The seeing disk there has been as small as 0.2″.

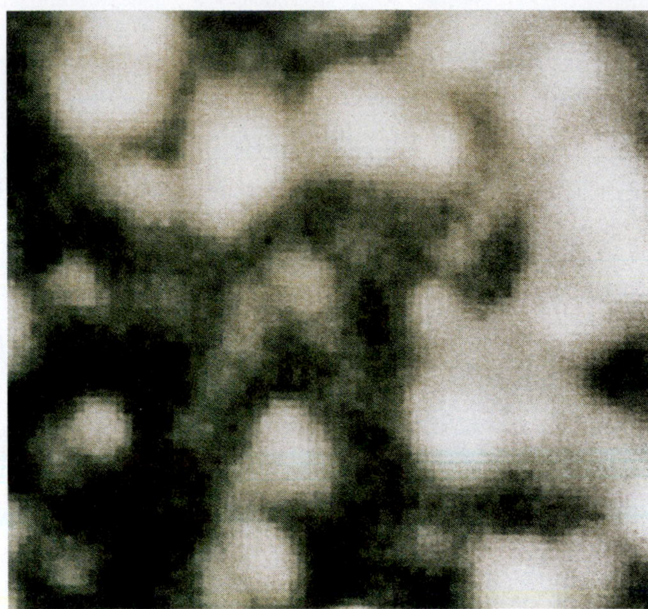

a

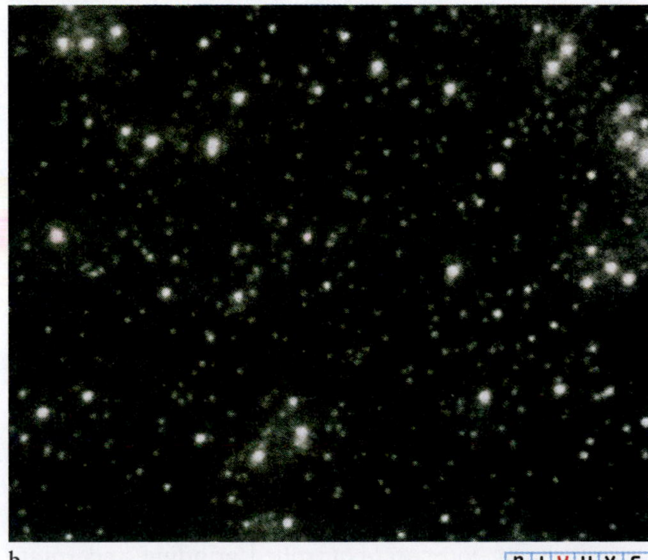

b

R I V U X G

FIGURE 3-27 Effects of Twinkling The same star field photographed with (a) a ground-based telescope, which is subject to poor seeing conditions that result in stars twinkling, and (b) the Hubble Space Telescope, which is free from the effects of twinkling. (NASA/ESA)

Insight Into Science

Research Requires Patience Seeing conditions—indeed, most observing situations in science—are rarely ideal. Besides such natural phenomena, which are beyond their control, scientists must also contend with equipment failures, late deliveries of parts, and design flaws. Furthermore, because travel time is so long, some missions (like the robotic spacecraft roving on Mars or the *New Horizons* voyage to Pluto and the Kuiper belt) take years or even decades to complete.

Without the effects of Earth's atmosphere, stars do not twinkle. As a result, photographs taken from telescopes in space reveal stars as much finer points (Figure 3-27b), and extended objects, such as planets and galaxies, appear in greater detail.

Light pollution from cities poses another problem for Earth-based telescopes (Figure 3-28). Keep in mind that the larger the primary mirror, the more light it gathers and, therefore, the more information astronomers can obtain from its images or spectra. The 5-m (200-in.) telescope at the Palomar Observatory between San Diego and Los Angeles is the first truly great large telescope, providing astronomers with invaluable insights into the universe for decades. However, light pollution from the two cities now fills the night sky, seriously reducing the ability of that telescope to collect light from objects in space. Not surprisingly, the best observing sites in the world are high on mountaintops—above smog, water vapor, and clouds—and far from city lights. An even better location, astronomers have discovered, is to observe space from space, eliminating the interference of both the lights of civilization and Earth's atmosphere. Removing these effects is why Hubble and other orbiting telescopes achieve such magnificent resolution in the images they take.

WHAT IF... Stars actually twinkled?
Stars appear to significantly change their brightness (twinkle) in fractions of a second. If they actually did vary in brightness that quickly, they would do so by changing size—bigger is brighter. If this expansion and contraction occurred, the rapid motion of their gases would cause stars to blow apart in a matter of seconds.

R I V U X G

FIGURE 3-28 Light Pollution These two images of Tucson, Arizona, were taken from the Kitt Peak National Observatory, which is 38 linear miles away. They show the dramatic growth in ground light output between 1959 (top) and 1989 (bottom). Since 1972, light pollution, a problem for many observatories around the world, has been partially controlled by local ordinances passed by cities. (NOAO/AURA/NSF/Galaxy)

3-12 The Hubble Space Telescope provides stunning details about the universe

For decades, astronomers dreamed of observatories in space. Such facilities would eliminate the image distortion created by twinkling and by poor atmospheric transparency due to pollution, volcanic debris, and water vapor. These telescopes could operate 24 hours per day and over a wide range of wavelengths—from the infrared through the visible range and far out into the gamma-ray part of the spectrum. Over the past 30 years, NASA and other space agencies have launched a variety of space telescopes, including four of what NASA calls its Great Observatories. The first Great Observatory to go up was the Hubble Space Telescope (HST).

Soon after HST was placed in orbit from the space shuttle in 1990, astronomers discovered that the telescope's 2.4-m primary mirror had been ground to the incorrect shape. Therefore, it suffered from spherical aberration, which caused its

Where in a typical house can the seeing problems described in Section 3-11 be observed?

FIGURE 3-29 The Hubble Space Telescope (HST) This photograph of HST hovering above the space shuttle's cargo bay was taken in 1993, at completion of the first servicing mission. HST studies the heavens at infrared, visible light, and ultraviolet wavelengths. (NASA)

images to be surrounded by a hazy glow. During a repair mission in December 1993, astronauts installed corrective optics that eliminated the problem (Figure 3-29). The telescope was further upgraded in 1997, 1999, 2002, and 2009. Now HST has a resolution of better than 0.1″, which is better than can be obtained by telescopes on Earth's surface without the use of advanced technology (see Section 3-13).

The observations taken by HST continue to stagger the imagination. Hubble has made discoveries related to the planets in our solar system, planetary systems around other stars, other galaxies and the distances to them, black holes, quasars, the formation of the earliest galaxies, the age of the universe, and many other topics. The Hubble Space Telescope developed technical difficulties in 2002. A final repair mission to it occurred in October 2008, bringing it back to pristine condition. Hubble's success, coupled with rapidly improving technology, has spurred scientists and engineers to develop the 6.5-m-diameter James Webb Space Telescope to replace the Hubble. This new telescope is scheduled for launch in 2014.

3-13 Advanced technology is spawning a new generation of superb ground-based telescopes

The clarity of images taken by the HST and the "seeing" issues described earlier may suggest that ground-based observational astronomy is a dying practice. However, two exciting techniques, called *active optics* and *adaptive optics*, enable telescopes on the ground to match the quality of Hubble—or better it!

Historically, primary mirrors have been thick and heavy, to help keep them rigid and perfectly shaped. Nevertheless, these mirrors were not ideal because they would warp slightly as the telescope changed angle on the sky and the mirrors could not compensate for the seeing conditions, which causes the objects to move and therefore blur. However, in the 1980s, astronomers discovered that changing the shape of the primary would enable them to compensate for these effects. By making thinner mirrors, the shape of the primary could be changed by pistons called *actuators* located under the mirror. Thus, the field of **active optics** was created. This technology finds the best orientation for the primary mirror in response to changes in temperature and the shape of the telescope mount. Actuators adjust the mirror every few seconds to help keep the telescope optimally aimed at its target. With active optics, the New Technology Telescope in Chile and the Keck telescopes in Hawaii routinely achieve resolutions as fine as 0.3″, whereas the resolution is much worse for telescopes without active optics at the same sites.

Even better resolution can be achieved with **adaptive optics**, which uses sensors to determine the amount of twinkling created by atmospheric turbulence. The stellar motion due to this twinkling is neutralized by computer-activated, motorized supports that actually reshape a smaller mirror installed farther down the optical path of the telescope. Adaptive optics effectively eliminate atmospheric distortion and produce remarkably clear images with resolution as fine as 0.03″ (Figure 3-30). Many large ground-based telescopes now use adaptive optics on many observing runs, resulting in images comparable to those from HST (Figure 3-30c).

Until the 1980s, telescopes with primary mirrors of between 2 and 6 m were the largest in the world. Now, new technologies in mirror building and computer control allow us to construct much larger telescopes. At least 72 reflectors around the world today have primary mirrors measuring 2 m or more in diameter. Among these are 13 reflectors that have mirrors between 8 and 10 m in diameter, with at least 8 other very large telescopes under construction. Appendix I is a list of the telescopes larger than 2 m in diameter that are in operation.

a

b

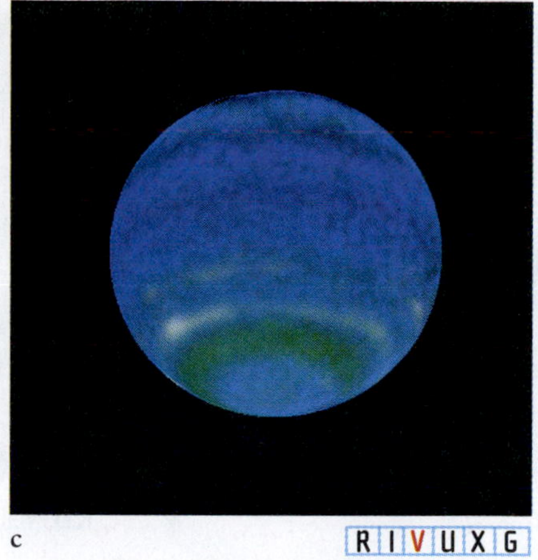
c

FIGURE 3-30 Images from Earth and Space (a) Image of Neptune from an Earth-based telescope without adaptive optics. (b) Image of Neptune from the same Earth-based telescope with adaptive optics. (c) Image of Neptune from the Hubble Space Telescope, which does not incorporate adaptive optics technology. (a & b: Courtesy of Center for Adaptive Optics, University of California; c: NASA, L. Sromovksy, and P. Fry, University of Wisconsin-Madison)

FIGURE 3-31 The 10-m Keck Telescopes Located on the dormant (and hopefully extinct) Mauna Kea volcano in Hawaii, these huge twin telescopes each consist of 36 hexagonal mirrors measuring 1.8 m (5.9 ft) across. Each Keck telescope has the light-gathering, resolving, and magnifying ability of a single mirror 10 m in diameter. Inset: View down the Keck I telescope. The hexagonal apparatus near the top of the photograph shows the housing for the 1.4-m secondary mirror. (W. M. Keck Observatory, Courtesy of Richard J. Wainscoat)

Because the cost of building very large mirrors is enormous, astronomers have devised less expensive ways to collect the same amount of light. One approach is to make a large mirror out of smaller pieces, fitted together like floor tiles. The largest examples of this segmented-mirror technique are the 10.4-m Gran Telescopio Canarias in the Canary Islands, and the 10-m (400-in.) Keck I and Keck II telescopes on the summit of Mauna Kea. In each of these telescopes, 36 hexagonal mirrors are mounted side by side to collect the same amount of light as a single primary mirror of 10.4 or 10 m, respectively (Figure 3-31). Another method, called **interferometry**, combines images from different telescopes. For example, used together to observe the same object, the two Keck telescopes have the resolving power of a single 85-m telescope. The four 8.2-m reflectors of the Very Large Telescope at the Paranal Observatory (see the image at the beginning of this chapter) combine to create the resolution that would come from a single telescope 200 m in diameter. This observatory has an ideal resolution of about 0.002″.

Why are military and spy agencies around the world so interested in cutting-edge telescope technologies?

NONOPTICAL ASTRONOMY

Looking back at Figure 3-6, you can see that visible light represents a very tiny fraction of the electromagnetic spectrum. As late as the 1940s, astronomers had no idea how much nonvisible radiation is emitted by objects in space. However, today we know that many objects in space emit undetectable amounts of visible light, but release large amounts of nonvisible radiation. Specially designed telescopes gather electromagnetic energy in all of the nonvisible parts of the spectrum. From

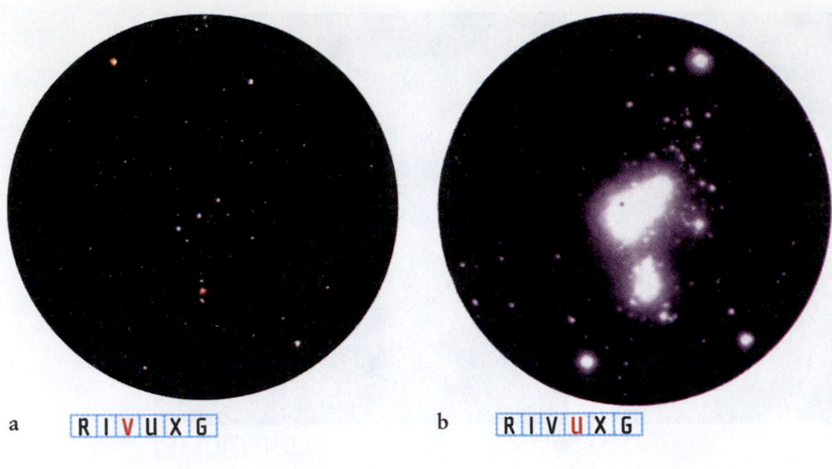

FIGURE 3-32 Orion as Seen in Visible, Ultraviolet, and Infrared Wavelengths (a) An ordinary optical photograph of the constellation Orion. (b) An ultraviolet image of Orion obtained during a brief rocket flight on December 5, 1975. The 100-s exposure captured wavelengths ranging from 125 to 200 nm. (c) A false-color view from the Infrared Astronomical Satellite uses color to display the intensity of infrared wavelengths. (a: R. C. Mitchell, Central Washington University; b: G. R. Carruthers, NRL; c: NASA/JPL-Caltech)

the radio waves and infrared radiation being emitted by vast interstellar gas clouds to the ultraviolet radiation and X rays from the remnants of stars to bursts of gamma rays of extraordinary power from merging black holes and other sources, our growing ability to see the entire electromagnetic spectrum is revealing intriguing phenomena. To get a sense of how much we do not see in the visible part of the spectrum, look at Figure 3-32, which shows ultraviolet, infrared, and visible images of the familiar constellation Orion.

3-14 A radio telescope uses a large concave dish to collect radio waves

The first evidence of nonvisible radiation from outer space came from the work of a young radio engineer, Karl Jansky, at Bell Laboratories. Using radio antennas, Jansky was investigating the sources of static that affect short-wavelength radiotelephone communication. In 1932, he realized that a certain kind of radio noise is strongest when the constellation Sagittarius is high in the sky. Because the center of our Milky Way Galaxy is located in the direction of Sagittarius, Jansky correctly concluded that he was detecting radio waves from elsewhere in the Galaxy. Jansky's accidental discovery of radio signals from space led to a worldwide effort to "see" what nonvisible wavelengths could teach us about the cosmos.

Insight Into Science

Think "Outside the Box" Observations and experiments often require making connections between seemingly unrelated concepts. Jansky's proposal that some radio waves originate in space is an example of connecting apparently disparate fields—astronomy and radio engineering.

Radio telescopes record radio signals from the sky. Just as a mirror reflects visible light, the metal surfaces of radio telescopes reflect radio waves. Each radio telescope has a large concave dish (Figure 3-33) that focuses radio photons in the same way that an optical telescope mirror focuses visible photons. A small antenna tuned to the desired wavelength is located at the telescope's prime focus or Cassegrain focus. The incoming signal is relayed to amplifiers and recording instruments.

Radio waves have the longest wavelengths of all electromagnetic radiation. Because the angular resolution of any telescope decreases as the wavelengths it collects increases, a radio telescope gives a fuzzier picture than any other type of telescope with the same diameter. Indeed, the first small radio telescopes produced blurry, indistinct images.

VIDEO 3.1

Very large radio telescopes create sharper radio images because, as with optical telescopes, the bigger the dish, the better the angular resolution. For this reason, most modern radio telescopes have

FIGURE 3-33 A Radio Telescope Recall that the secondary mirror or prime focus on most telescopes blocks incoming light or other radiation. This radio telescope at the National Radio Astronomy Observatory in Green Bank, West Virginia, has its prime focus hardware located off-center from the telescope's 100-m by 110-m oval reflector. By using this new design, there is no such loss of signal. Such configurations are also common on microwave dishes used to receive satellite transmissions for home televisions. (NRAO/AUI/NSF)

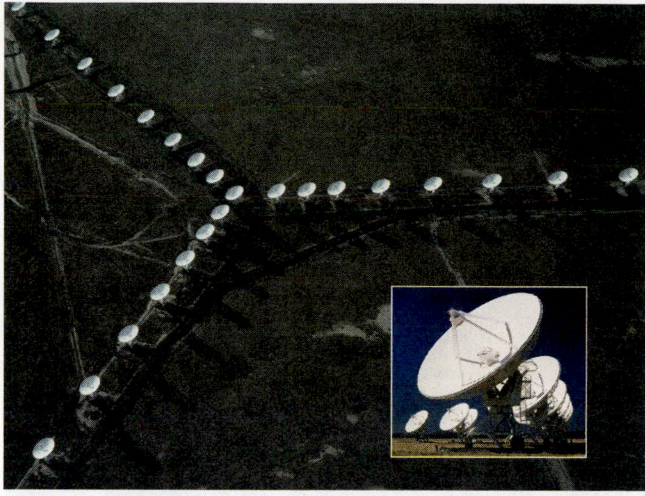

FIGURE 3-34 The Very Large Array (VLA) The 27 radio telescopes of the VLA system are arranged along the arms of a Y in central New Mexico. Besides being able to change the angles at which they observe the sky, astronomers can also move these telescopes by train cars so that the array can detect either wide areas of the sky (when they are close together, as in this photograph) or small areas with higher resolution (when they are farther apart). The inset shows the traditional secondary mirror assembly in the center of each of these antennas. (Jim Sugar/Corbis; inset: David Nunuk/Science Photo Library/Photo Researchers)

reflectors more than 25 m in diameter. Nevertheless, even the largest radio dish in existence (305-m-diameter in Arecibo, Puerto Rico) cannot come close to the resolution of the best optical telescopes. For example, a 6-m optical telescope has 2000 times better resolution than a 6-m radio telescope that detects radio waves of 1-mm wavelength.

To overcome the limitation on resolution set by telescope diameter, radio astronomers often use interferometry to produce high-resolution radio images. Recall from Section 3-13 that interferometry combines the data received simultaneously by two or more telescopes. The telescopes can be kilometers, continents, or even worlds apart. The radio signals received by all of the dishes are made to "interfere," or blend together, and with suitable computer-aided processing, the combined image of the source is sharp and clear. The results are impressive: The resolution of such a system is equivalent to that of one gigantic dish with a diameter equal to the distance between the farthest telescopes in the array.

Interferometry, exploited for the first time in the late 1940s, gave astronomers their first detailed views of "radio objects" in the sky. One of the most complex systems of radio telescopes used in this way began operating in 1980 on the plains of San Agustin near Socorro, New Mexico. Called the Very Large Array (VLA), this system consists of 27 concave dishes, each 26 m (85 ft) in diameter (Figure 3-34). The 27 telescopes are positioned along the three arms of a gigantic Y that can span a distance of 36 km (22 mi). Working together, they can create radio images with 0.1″ resolution. This system produces radio views of the sky with resolutions comparable to those of the very best optical telescopes.

More recently, radio telescopes separated by thousands of kilometers have been linked together in **very-long-baseline interferometry (VLBI)**. The Very Long Baseline Array (VLBA) consists of ten 25-m radio telescopes located across the United States, from Hawaii to New Hampshire. With a maximum baseline of 8000 km, VLBA has a resolving power of 0.001″.

The best angular resolution on Earth is obtained by combining radio data from telescopes on opposite sides of our planet. In that case, features as small as 0.00001″ can be distinguished at radio wavelengths—10,000 times sharper than the best views obtainable from single optical telescopes. Radio telescopes are also being put into space and used in even longer-baseline interferometers.

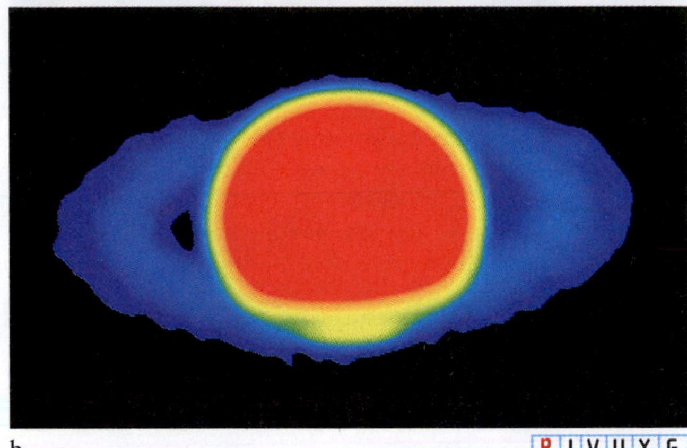

FIGURE 3-35 Visible and Radio Views of Saturn (a) This picture was taken by a camera on board a spacecraft as it approached Saturn. The view was produced by sunlight scattered from the planet's cloudtops and rings. (b) This false-color picture, taken by the VLA, shows radio emission from Saturn at a wavelength of 2 cm. (a: NASA; b: Image courtesy of NRAO/AUI)

To make radio images more comprehensible, radio astronomers often use gray scales (see Figure 1-3i) or false colors (Figure 3-35). The most intense radio emission is shown in red, the least intense in blue. Intermediate colors of the rainbow represent intermediate levels of radio intensity. Black indicates that there is no detectable radio radiation. Astronomers who work at other nonvisible-wavelength ranges also frequently use false-color techniques to display views obtained from their instruments.

Consider an example of how nonoptical astronomy, such as is done with these radio telescopes, can overcome limitations of optical telescopes. Late in the eighteenth century, the astronomer Sir William Herschel observed regions of our Milky Way from which no visible light was emitted. "Surely, there is a hole in the heavens," he reported on seeing the first such region. (He was not referring to a modern "black hole," but rather just to an area that was especially dark to his eyes.) In the twentieth century, astronomers discovered that these regions are actually clouds of interstellar gas and dust that prevent visible light from stars beyond them from reaching us, just as thick clouds of water vapor in our atmosphere can obscure the Sun and darken the sky. However, radio waves and some other nonvisible wavelengths pass through interstellar clouds. With the advent of radio telescopes, astronomers got their first glimpses of the variety of objects that lie beyond these clouds.

On the boundary between radio and infrared radiation, several microwave telescopes have been orbited around Earth. Notable among these are the Cosmic Background Explorer (COBE), launched in 1989, and the Wilkinson Microwave Anisotropy Probe (WMAP), launched in 2001. These telescopes have been used to measure the blackbody radiation in space and thereby take the temperature of the universe, results that we will discuss in Chapter 18.

3-15 Infrared and ultraviolet telescopes also use reflectors to collect their electromagnetic radiation

As with radio and microwave telescopes, infrared and ultraviolet telescopes are all reflectors. Indeed, using suitable CCDs, optical reflecting telescopes can also detect infrared and ultraviolet photons with wavelengths near visible light. The most sensitive infrared-detecting CCDs must be cooled to prevent the heat (and, hence, infrared photons) of the telescope from overwhelming the infrared radiation received from objects in space.

Because water vapor is the main absorber of infrared radiation from space, locating infrared observatories at sites of low humidity can overcome much of

Infrared radiation can travel through media that block visible light. (NASA/JPL-Caltech)

a

b

FIGURE 3-36 Spitzer Space Telescope (a) The mirror assembly for the Spitzer Space Telescope showing the 85-cm objective mirror. (b) Launched in 2003, this Great Observatory is taking images and spectra of planets, comets, gas, and dust around other stars and in interstellar space, galaxies, and the large-scale distribution of matter in the universe. Inset: An infrared image of a region of star formation invisible to optical telescopes. (a: Balz/SIRTF Science Center; b: NASA/JPL-Caltech; inset: NASA/JPL)

the atmosphere's hindrance. For example, the summit of Mauna Kea is exceptionally dry (most of the moisture in the air is below the height of this volcano). Infrared observations are made on the summit by the primarily optical-wavelength Subaru and two Keck telescopes, as well as by NASA's 3-m Infrared Telescope Facility (IRTF).

The best way to avoid the obscuring effects of water vapor is to place a telescope in orbit around Earth. The 1983 Infrared Astronomical Satellite (IRAS), Hubble Space Telescope's Near Infrared Camera and Multi-Object Spectrometer (NICMOS), the Infrared Space Observatory (ISO) launched in 1995, and the Spitzer Space Telescope launched in 2003 (Figure 3-36), have done much to reveal the full richness and variety of the infrared sky. Spitzer is the infrared equivalent to HST, one of NASA's Great Observatories. In 2010, NASA began flying a telescope called SOFIA (Stratospheric Observatory for Infrared Astronomy) in a modified Boeing 747. At altitude, a giant door in the side of the aircraft slides open and the telescope observes the night sky high above water vapor in the atmosphere.

Infrared telescopes detect radiation from small bodies in the solar system, as well as from the bands of dust in our galaxy, newly formed stars, the dust disks around stars at various stages of their evolution, and the most distant galaxies, whose radiation arrives here primarily at infrared wavelengths, among other infrared-emitting objects. Astronomers have located more than a half-million infrared sources in the sky, most of which are invisible at optical wavelengths. As with radio waves, some infrared radiation passes through interstellar clouds of dust and gas, allowing astronomers to see, for example, individual stars at the center of our galaxy some 245 trillion km (153 trillion mi) away (Figure 3-37).

Where in a typical house would infrared detectors indicate most activity?

b

R I V U X G

FIGURE 3-37 Views of the Milky Way's Central Regions (a) An optical image in the direction of Sagittarius, toward the Milky Way's center. The dark regions are interstellar gas and dust clouds that prevent visible light from beyond them from reaching us. (b) An infrared image of the same area of the sky, showing many more distant stars whose infrared radiation passes through the clouds and is collected by our telescopes. (Howard McCallon and Gene Kopan/2 MASS Project)

During the early 1970s, ultraviolet astronomy got off the ground—literally and figuratively. Both *Apollo* and *Skylab* astronauts used small telescopes above Earth's atmosphere to give us some of our first views of the ultraviolet sources in space. Small rockets have also been used to place ultraviolet telescopes briefly above Earth's atmosphere. A typical ultraviolet view is shown in Figure 3-32b.

Some of the finest early ultraviolet astronomy was accomplished by the International Ultraviolet Explorer (IUE), which was launched in 1978 and functioned until 1996. The space shuttle was transformed into an orbiting observatory twice in the 1990s, carrying aloft and then returning to Earth three ultraviolet telescopes. In 1992 the Extreme Ultraviolet Explorer (EUVE) was launched. As with infrared observations, ultraviolet images reveal sights previously invisible and often unexpected. Many objects in space emit ultraviolet radiation that astronomers can use to study the chemistries of these cosmic bodies. Therefore, in 1999, astronomers launched the Far Ultraviolet Spectroscopic Explorer (FUSE). Through 2007 this telescope provided us with information about such things as how much deuterium (hydrogen nuclei with one neutron) was created when the universe formed, the location of particularly hot interstellar gas and dust, and the chemical evolution of galaxies.

Telescopes dedicated to studying the Sun have observed it from the ground since 1941, and solar telescopes have studied it continuously from space since 1984. Since 1995, the Solar and Heliospheric Observatory (SOHO) has investigated the Sun from space with ultraviolet detectors. More recent orbiting observatories include STEREO (Solar TErrestrial RElations Observatory), launched in 2006, and SDO (the Solar Dynamic Observatory), launched in 2010. These observatories observe the Sun's outer layers rising and sinking, along with a wide range of energetic activity emanating from the Sun, including particles and radiation racing outward from it, among other phenomena. These telescopes show us breathtaking ultraviolet images and movies, some in 3D, of our star. In Chapter 10 we will discuss these observations and their physical origins.

3-16 X-ray and gamma-ray telescopes cannot use normal reflectors to gather information

X rays and gamma rays from space interact strongly with the particles in Earth's atmosphere, preventing these dangerous radiations from reaching our planet's surface. Therefore, direct observations of astronomical sources that emit these extremely short wavelengths must be made from space. Astronomers got their first look at X-ray sources during brief rocket flights in the late 1940s. Several small satellites, launched during the early 1970s, viewed the sources of both X rays and gamma rays in space, revealing hundreds of previously unknown objects, including several black holes (see Chapter 14). X-ray telescopes have also been carried on space shuttle missions. The insets in Figure 3-38 show how different our Sun appears when seen through X-ray and visible-light "eyes."

FIGURE 3-38 Nonvisible and Visible Radiation (a) The McMath-Pierce Solar Telescope at Kitt Peak Observatory near Tucson, Arizona (the inverted V-shaped structure), takes visible-light photographs of the Sun, such as the one shown in the inset. (b) This X-ray telescope was carried aloft in 1994 by the space shuttle. The inset shows an X-ray image of the Sun. Comparing the images in the two insets reveals how important observing nonvisible radiation from astronomical phenomena is to furthering our understanding of how the universe operates. (a: NOAO; b: Solar X-ray image from the Yohkoh mission of ISAS, Japan; the X-ray telescope was prepared by the Lockheed-Martin Solar and Astrophysics Laboratory, the National Astronomical Observatory of Japan, and the University of Tokyo with the support of NASA and ISAS; insets: L. Golub, Naval Observatory, IBM Research, NASA)

X-ray photons are tricky to collect. Because of their high energies, X rays penetrate even highly polished surfaces that they meet nearly head on. Therefore, normal reflecting mirrors cannot be used to focus them. Instead, X-ray telescopes are designed to deflect photons at a fairly shallow angle because X rays that are barely skimming or grazing a surface can be reflected and thereby focused. This process is analogous to skipping a flat rock on water; throw it at a steep angle and it immediately sinks; throw it at a shallow angle and it will skip off the surface. Figure 3-39 shows the design of such "grazing incidence" X-ray telescopes.

After being focused, X rays are detected in several ways. As at infrared, visible, and ultraviolet wavelengths, CCDs can detect these photons. Other devices, called *scintillators*, detect the visible light created as X rays pass through them. Still other instruments, called *calorimeters*, detect the heat generated by X rays passing through the detector.

At least seven X-ray telescopes are orbiting Earth today. Among them, the Chandra X-ray Observatory, a NASA Great Observatory named after the Nobel laureate astronomer Subrahmanyan Chandrasekhar, provides images with better than 1″ resolution, and the European Space Agency's XMM-Newton has 6″ resolution. (Your textbook author's early work in general relativity was cited in Chandra's Nobel Prize essay.) Other X-ray telescopes are carried high into the atmosphere by balloons. The balloons float at altitudes up to 40 km (25 mi) above Earth's surface and often stay up for several weeks. Whereas low-energy X rays do not penetrate that far into the atmosphere, higher-energy X rays do, giving these balloon-borne telescopes a wide range of objects to study.

Over 10,000 X-ray sources have been discovered all across the sky. Among these are planetary atmospheres, stars (see the Sun in Figure 3-38), stellar remnants, vast clouds of intergalactic gas, jets of gas emitted by

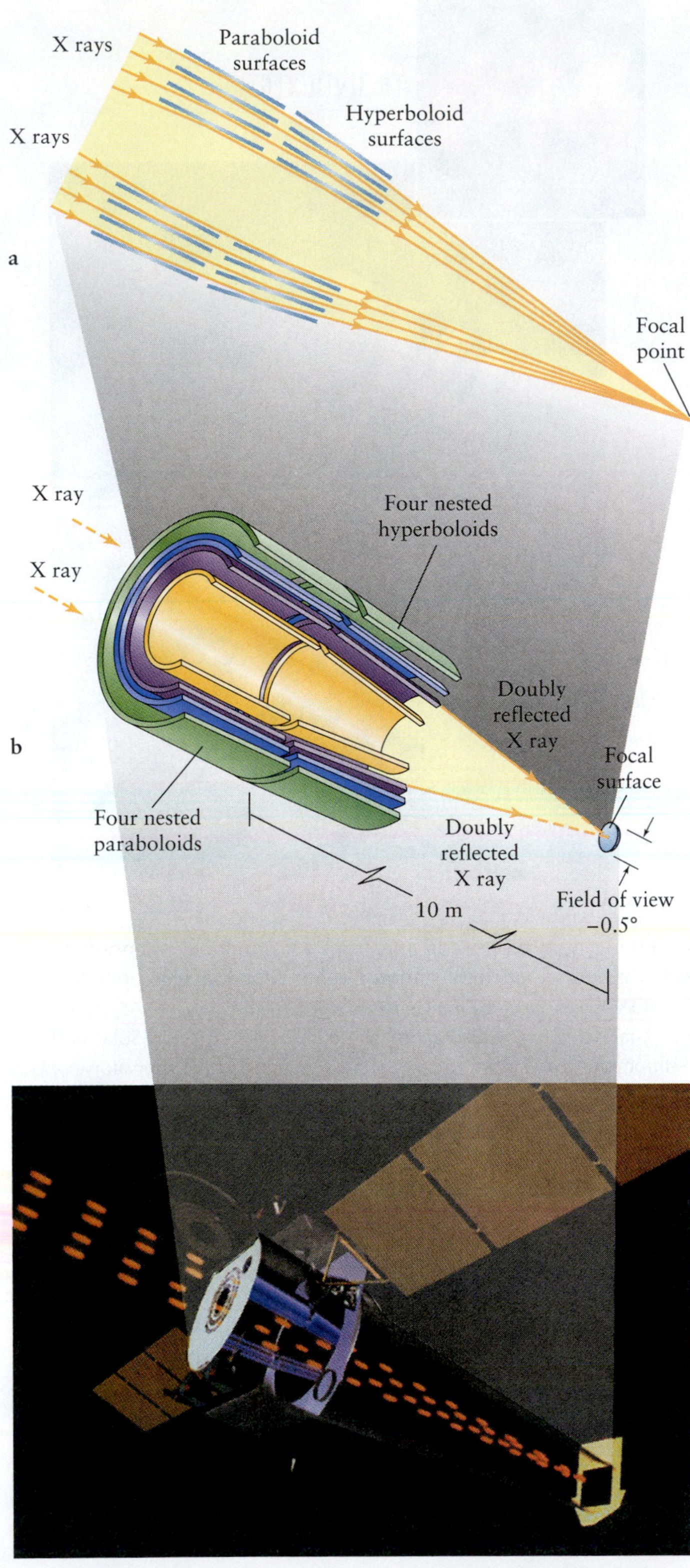

FIGURE 3-39 Grazing Incidence X-ray Telescopes (a) X rays penetrate objects they strike head-on. In order to be focused, X rays have to be gently nudged by skimming off cylindrical "mirrors." The shapes of the mirrors optimize the focus. (b) The diagram shows how X rays are focused in the Chandra X-ray Telescope. (a & b: NASA/JPL-Caltech; c: NASA/Chandra X-ray Observatory Center/Smithsonian Astrophysical Observatory)

galaxies, black holes, quasars, clusters of galaxies, and a diffuse X-ray glow that fills the universe.

The electromagnetic radiation with the shortest wavelengths and the most energy are gamma rays. In 1991, the Compton Gamma-ray Observatory, also a NASA Great Observatory, was carried into orbit by the space shuttle. Named in honor of Arthur Holly Compton, an American physicist and Nobel laureate who made important discoveries about gamma rays, this orbiting observatory carried four instruments that performed a variety of observations, giving us tantalizing views of the gamma-ray sky until the telescope failed in 2000. At least 10 gamma-ray telescopes, including the *SWIFT* Gamma-Ray Burst Mission and the Fermi Large Area Telescope, are presently in orbit, and several have been flown on balloon missions.

Several thousand gamma-ray sources have been discovered. However, gamma rays are too powerful even for grazing incidence telescopes. Therefore, astronomers have devised other methods of detecting them and determining their origin. These include absorbing them in crystals; allowing them to pass through tiny holes called *collimators* whose directions are well determined; and using chambers in which the gamma rays transform into electrons and positrons (positively charged electrons), leaving a track whose direction can be determined. These techniques are not nearly as precise as those used in other parts of the spectrum, and the best resolution gamma-ray instruments are only accurate to about 5′.

Astronomers are also taking advantage of the interaction between very high energy gamma rays and Earth's atmosphere to build telescopes on the ground that indirectly detect gamma rays. When these photons strike the atmosphere, they often cause particles in the air to move faster than the speed of light in air (but still slower than the speed of light in empty space, the ultimate speed limit). These particles quickly slow down, emitting short bursts of blue light that travel in the same direction as the incoming gamma rays; by detecting this light, it is possible to determine the direction from which the gamma rays came.

ANIMATION 3.3

We now have telescopes with which we can see the energy in the universe from virtually all parts of the electromagnetic spectrum (Figure 3-40). Telescopes provide us with more than just stunning images of objects in space. They also provide information about the chemistry of stars and interstellar gas and dust, the motion of objects toward or away from us, whether stars are rotating, and whether they are alone in space or orbiting a companion, among other things. We will explore how we get this information when we study the various objects in space.

Until recently, photons were the only sources of detailed astronomical information that we had. However, in the past four decades, physicists have built detectors for waves and particles whose nature is not electromagnetic. Chapters 10 and 14 discuss two of these devices: neutrino detectors and gravity wave detectors.

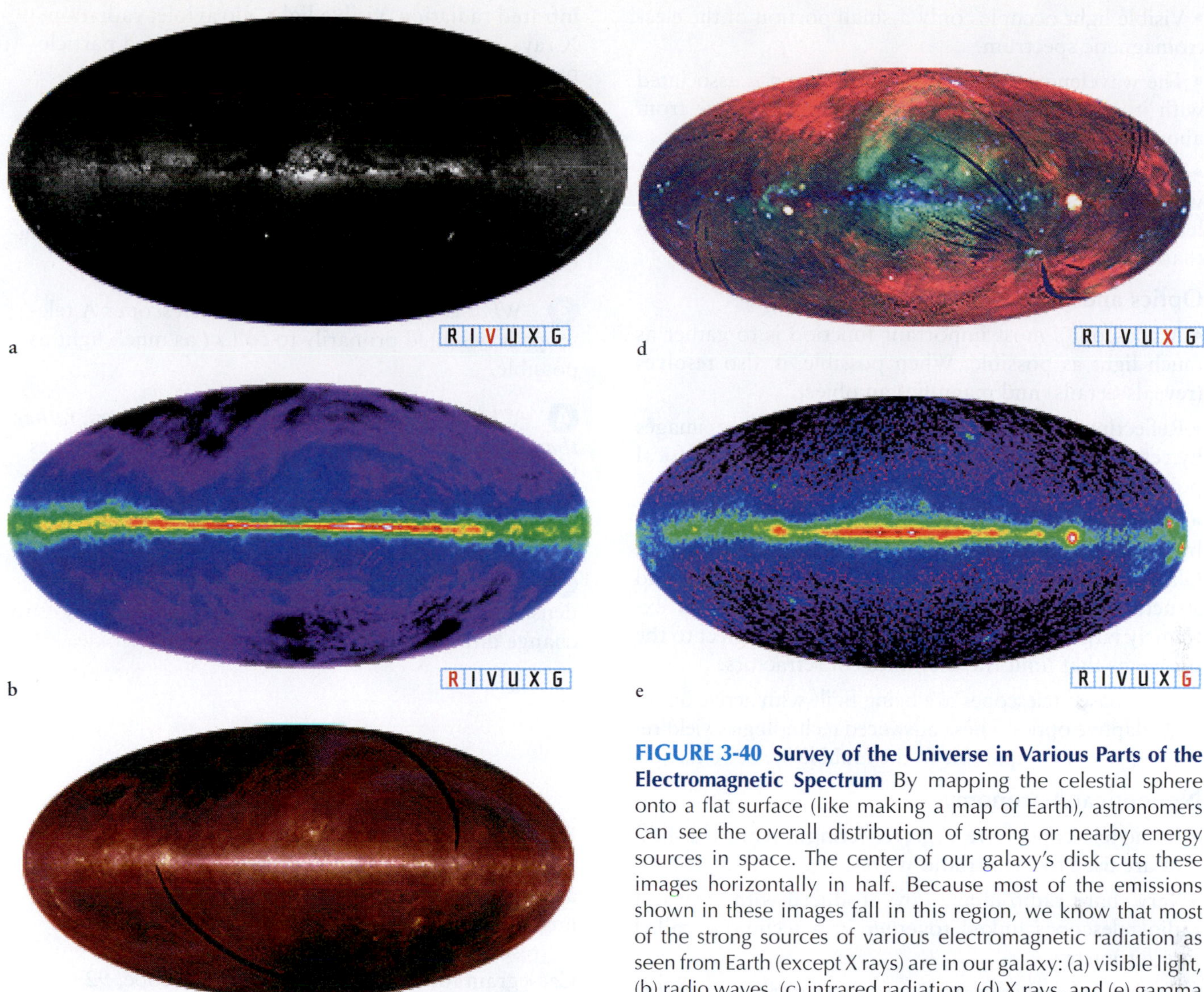

FIGURE 3-40 Survey of the Universe in Various Parts of the Electromagnetic Spectrum By mapping the celestial sphere onto a flat surface (like making a map of Earth), astronomers can see the overall distribution of strong or nearby energy sources in space. The center of our galaxy's disk cuts these images horizontally in half. Because most of the emissions shown in these images fall in this region, we know that most of the strong sources of various electromagnetic radiation as seen from Earth (except X rays) are in our galaxy: (a) visible light, (b) radio waves, (c) infrared radiation, (d) X rays, and (e) gamma rays. (GFSC/NASA)

3-17 Frontiers yet to be discovered

Before the twentieth century, astronomers were like the blind men in the fable who are trying to describe an elephant. Their perceptions were piecemeal: We could describe parts of the universe but not the whole. Our ancestors did not have the technology that could enable them to see the big picture. However, we are beginning to see it, and, as you will learn in the chapters that follow, our understanding of the cosmos is therefore increasing dramatically. A vast amount of observational information remains to be gathered. Indeed, literally every planet, moon, piece of interplanetary debris, star, stellar remnant, gas cloud, galaxy, quasar, cluster of galaxies, and supercluster of galaxies have stories to tell. Observational astronomy is so new an activity that we are still making new and often unexpected discoveries almost daily. There is still a lot to discover about the elephant.

SUMMARY OF KEY IDEAS

The Nature of Light

• Photons, units of vibrating electric and magnetic fields, all carry energy through space at the same speed, the speed of light (300,000 km/s in a vacuum, slower in any medium).

• Radio waves, microwaves, infrared radiation, visible light, ultraviolet radiation, X rays, and gamma rays are the forms of electromagnetic radiation. They travel as photons, sometimes behaving as particles, sometimes as waves.

• Visible light occupies only a small portion of the electromagnetic spectrum.

• The wavelength of a visible-light photon is associated with its color. Wavelengths of visible light range from about 400 nm for violet light to 700 nm for red light.

• Infrared radiation, microwaves, and radio waves have wavelengths longer than those of visible light. Ultraviolet radiation, X rays, and gamma rays have wavelengths that are shorter.

Optics and Telescopes

• A telescope's most important function is to gather as much light as possible. When possible, it also resolves (reveals details) and magnifies an object.

• Reflecting telescopes, or reflectors, produce images by reflecting light rays from concave mirrors to a focal point or focal plane.

• Refracting telescopes, or refractors, produce images by bending light rays as they pass through glass lenses. Glass impurity, opacity to certain wavelengths, and structural difficulties make it inadvisable to build extremely large refractors. Reflectors are not subject to the problems that limit the usefulness of refractors.

• Earth-based telescopes are being built with active optics and adaptive optics. These advanced technologies yield resolving power comparable to the Hubble Space Telescope.

Nonoptical Astronomy

• Radio telescopes have large, reflecting antennas (dishes) that are used to focus radio waves.

• Very sharp radio images are produced with arrays of radio telescopes linked together in a technique called interferometry.

• Earth's atmosphere is fairly transparent to most visible light and radio waves, along with some infrared and ultraviolet radiation arriving from space, but it absorbs much of the electromagnetic radiation at other wavelengths.

• For observations at other wavelengths, astronomers mostly depend upon telescopes carried above the atmosphere by rockets. Satellite-based observatories are giving us a wealth of new information about the universe and permitting coordinated observation of the sky at all wavelengths.

• Charge-coupled devices (CCDs) record images on many telescopes used between infrared and X-ray wavelengths.

A WHAT DID YOU THINK?

1 *What is light?* Light—more properly "visible light"—is one form of electromagnetic radiation. All electromagnetic radiation (radio waves, microwaves, infrared radiation, visible light, ultraviolet radiation, X rays, and gamma rays) has both wave and particle properties.

2 *Which type of electromagnetic radiation is most dangerous to life?* Gamma rays have the highest energies of all photons, so they are the most dangerous to life. However, ultraviolet radiation from the Sun is the most common everyday form of dangerous electromagnetic radiation that we encounter.

3 *What is the main purpose of a telescope?* A telescope is designed primarily to collect as much light as possible.

4 *Why do all research telescopes use mirrors, rather than lenses, to collect light?* Telescopes that use lenses have more problems, such as chromatic aberration, internal defects, complex shapes, and distortion from sagging, than do telescopes that use mirrors.

5 *Why do stars twinkle?* Rapid changes in the density of Earth's atmosphere cause passing starlight to change direction, making stars appear to twinkle.

Key Terms for Review

active optics, 100
adaptive optics, 100
angular resolution (resolution), 78
Cassegrain focus, 76
charge-coupled device (CCD), 80
chromatic aberration, 83
coudé focus, 77
electromagnetic radiation, 68
electromagnetic spectrum, 73
eyepiece lens, 76
focal length, 76
focal plane, 76
focal point, 76
frequency, 72
gamma ray, 73
infrared radiation, 73
interferometry, 100
light-gathering power, 78
magnification, 78
Newtonian reflector, 76
objective lens, 81
photon, 72
pixel, 80
primary mirror, 76
prime focus, 77
radio telescope, 92
radio wave, 73
reflecting telescope (reflector), 75
reflection, 75
refracting telescope, 75
refraction, 68
refractor, 81
Schmidt corrector plate, 85
secondary mirror, 76
seeing disk, 100
spectrum (*plural* spectra), 68
spherical aberration, 82
twinkling, 88
ultraviolet (UV) radiation, 73
very-long-baseline interferometry (VLBI), 93
wavelength (λ), 70
X ray, 73

Review Questions

The answers to all computational problems, which are preceded by an asterisk (*), appear at the end of the book.

1. Describe reflection and refraction. How do these processes enable astronomers to build telescopes?

2. Give everyday examples of refraction and reflection.

3. Describe a reflecting telescope by doing Interactive Exercise 3.1, and transcribe the drawing and correct labels to paper, if requested.

***4.** How much more light does a 3-m-diameter telescope collect than a 1-m-diameter telescope?

5. Explain some of the advantages of reflecting telescopes over refracting telescopes.

6. What are the three major functions of a telescope?

7. What is meant by the angular resolution of a telescope?

8. What limits the ability of the 5-m telescope at Palomar Observatory to collect starlight? There are several correct answers to this question.

9. Why will many of the very large telescopes of the future make use of multiple mirrors?

10. What is meant by adaptive optics? What problem does adaptive optics overcome?

11. Compare an optical reflecting telescope to a radio telescope. What do they have in common? How are they different?

12. Why can radio astronomers observe at any time of the day or night whereas optical astronomers are mostly limited to observing at night?

13. Why must astronomers use satellites and Earth-orbiting observatories to study the heavens at X-ray wavelengths?

14. What are NASA's four Great Observatories, and in what parts of the electromagnetic spectrum do (or did) they observe?

15. Why did Rømer's observations of the eclipses of Jupiter's moons support the heliocentric, but not the geocentric, cosmology?

Advanced Questions

16. Advertisements for home telescopes frequently give a magnification for the instrument. Is this a good criterion for evaluating such telescopes? Explain your answer.

***17.** The observing cage in which an astronomer sits at the prime focus of the 5-m telescope at Palomar Observatory is about 1 m in diameter. What fraction of the incoming starlight is blocked by the cage? *Hint:* The area of a circle of diameter d is $\pi d^2/4$, where $\pi \approx 3.14$.

***18.** Compare the light-gathering power of the Palomar Observatory's 5-m telescope to that of the fully dark-adapted human eye, which has a pupil diameter of about 5 mm.

19. Show by means of a diagram why the image formed by a simple refracting telescope is "upside down."

***20.** Suppose your Newtonian reflector has a mirror with a diameter of 20 cm and a focal length of 2 m. What magnification do you get with an eyepiece whose focal length is **a.** 9 mm, **b.** 20 mm, and **c.** 55 mm?

21. Why does no major observatory have a Newtonian reflector as its primary instrument, whereas Newtonian reflectors are popular among amateur astronomers?

22. From the ground, how can astronomers detect gamma-ray sources in space?

23. Why will many of the very large telescopes of the future have ultrathin primary mirrors?

Discussion Questions

24. Discuss the advantages and disadvantages of using a relatively small visible-light telescope in Earth's orbit (for example, the 2.4-m Hubble Space Telescope) versus a large visible-light telescope on a mountaintop (for example, the 8.3-m Subaru telescope on Mauna Kea, Hawaii).

25. If you were in charge of selecting a site for a new observatory, what factors would you consider?

26. Consider two identical Cassegrain telescope mirrors. One is set up as a prime focus telescope, whereas the other is used in a Cassegrain telescope. **a.** Sketch both telescopes. **b.** What are the differences between the two that make each useful in different observing situations?

What If...

27. Telescopes were first invented today? What objects or areas of the sky would you recommend that astronomers explore first? Why?

28. We had eyes sensitive to radio waves? How would our bodies be different, and how would our visual perceptions of the world be different?

29. Humans were unable to detect any electromagnetic radiation? How would that change our lives, and what alternatives might evolve (some species indeed have done this) to provide information about distant objects?

Web Questions

30. Several telescope manufacturers build Schmidt-Cassegrain telescopes. These devices use a correcting lens in an arrangement like that shown in Figure 3-25c. Consult advertisements on the Web and list the dimensions, weights, and costs of some of these telescopes. Why are they popular among amateur astronomers?

31. Discuss the advantages and disadvantages of setting up an observatory on the Moon. *Hint:* To get a broad perspective on this question, you might find it useful to explore the Web for the challenges of living on the Moon.

32. The Large Zenith Telescope (LZT) in British Columbia, Canada, uses a 6-m liquid mirror made of mercury. Use the Web to investigate this technology. How can a liquid metal be formed into the necessary shape of a telescope mirror? What are the advantages and disadvantages of a liquid mirror?

Observing Projects

33. During the daytime, obtain a telescope and several eyepieces of differing focal lengths. If you can determine the focal length of the telescope's objective lens or mirror (often printed on it), calculate and record the magnifying power for each eyepiece. Focus the telescope on some familiar object, such as a distant lamppost or tree. **DO NOT FOCUS ON THE SUN! Looking directly at the Sun through a telescope will cause blindness.** Describe the image you see through the telescope. Is it upside down? How does the image move as you slowly and gently shift the telescope left and right or up and down? Examine the distant objects under different magnifications. How does the field of view change as you go from low magnification to high magnification?

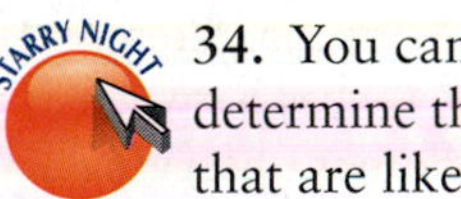

34. You can use *Starry Night*™ to help you to determine the brightness of the faintest stars that are likely to be visible to the naked eye from your location under your present sky conditions and to estimate the fraction of possible stars that you can see with the unaided eye. This exercise is best done outdoors with a laptop computer, or through a window from a very dark room, on a dark, clear night. Click on **Home** to restore the view to your present location and time, if you have not already done so. You can now adjust the amount of light pollution on the view to match your present sky, where only stars above a limiting brightness are discernable. On the Menu, select **View > Hide Daylight**. Open the **Options** side pane and expand the **Local View** layer. Place the cursor over the words **Local Light Pollution** and click the **Local Light Pollution Options…** button that appears. This will open the **Local View Options** dialog window. Move this dialog window to the side of the view. Click the checkbox to the left of the **Local Light Pollution** option to turn this feature on. You can now slide the bar to adjust the local light pollution level until the view matches your night sky. When you are satisfied that the view matches your sky, click on **OK** to dismiss the **Local View Options** dialog window. Move the cursor over some of the stars that appear on your screen to display their properties in the **HUD (Heads-up Display)**. (If necessary, open the **Preferences** dialog from the **File** (Windows) or **Starry Night** (Macintosh) menu and add the **Apparent magnitude** option to the **Cursor Tracking (HUD)** options.) **a.** Make a note of the apparent magnitudes of some of the faintest stars in this view and compare these values to the faintest apparent magnitude of about +6 that can be seen under ideal conditions by the human eye. **b.** The second goal is to estimate what fraction of the visible stars you can see under these conditions, compared to the total number of stars you might see under ideal conditions. Carefully do the following on the screen: Select a small, square section of the sky on the screen (maybe 8 cm [about 3 in, or roughly half the length of a typical pen] on each side) and count and record the number of stars in this square with your present setting. (To help with this, you might consider cutting a suitable hole in a sheet of paper to use as a mask). Now open the **Local View Options** dialog window again and set the **Local Light Pollution** to **Less** (which essentially gives you ideal conditions) and repeat the count of visible stars. Divide the first number you count by the second. What fraction of the stars that would be visible under ideal conditions were you seeing? **c.** To see how much light pollution occurs in large cities, adjust the slide bar for **Local Light Pollution** in the **Local View Options** dialog window to the far right for maximum light pollution and repeat the star counting within the same limited sky region. Compared to viewing under ideal conditions, what fraction of the stars are observers in large urban centers seeing?

35. Observe the stars when the Moon is either full, new, or in a quarter phase. Record the phase of the Moon and note, qualitatively, whether you see many more stars than just those visible in relatively bright asterisms such as the Big Dipper or Orion. Compare these qualitative observations of the number of visible stars with those made on a clear night about a week later, when the Moon's phase has changed and its contribution to the light of the night sky is different, noting the phase of the Moon again. **a.** Compare the numbers of stars on the two nights. On which night did you see more stars? Why? **b.** During which three phases of the Moon do you expect that astronomers prefer to make observations of faint objects?

36. On a clear night, view the Moon, a planet, and a star through a telescope using eyepieces of various focal lengths. (You can use *Starry Night*™ or consult a source on the Internet or such magazines as *Sky & Telescope* or *Astronomy* to determine the phase of the Moon and the locations of the planets.) **a.** How do the images change as you view with increasing magnification? **b.** Do they become distorted at any level of magnification?

STARRY NIGHT

37. In this exercise, you can use *Starry Night*™ to explore the appearance of a nearby galaxy, M31, the Andromeda Galaxy, under conditions that simulate the use of binoculars or a telescope to enhance the unaided-eye view. This will allow you to evaluate the effects of greater light-gathering power and improved resolution that come from the use of these optical aids. Select **Favourites > Discovering the Universe > Andromeda** to open a view of the sky as seen from New York City at just after midnight on the first day of summer, June 21, 2011, looking toward the northwest. This 100° field of view of the sky over Central Park is centered upon M31. You can use this view to show you where to look for this faint object in the sky by selecting **View > Constellations > Astronomical** to show this distant galaxy with respect to the Square of Pegasus and the V-shaped constellation of Andromeda. You can label these constellations by clicking on **Labels > Constellations** if you like. This is the approximate view that one has of this galaxy with the unaided eye from a dark-sky site. **a.** Sketch the constellation patterns of bright stars around M31 and include its position, for reference when you view the real sky. Remove the constellations and labels. **b.** Now use the **Zoom** facility on the toolbar to adjust the field of view to about 24° to match that of a pair of ordinary binoculars. (To display a more precise indication of the field of view of your own binoculars, click on the **FOV** tab and the **Edit** button to the right of **Binoculars**, open the **From List** box and select the appropriate binoculars from this list. You can now click on your binocular characteristics in this **FOV > Binoculars** list). Again, sketch and describe the galaxy's appearance. **c.** Finally, adjust the field of view to about 2°, to match that of a small telescope. Here, the galaxy will extend across the full field. Sketch and describe this galaxy again, noting particularly any details now apparent at this higher magnification. Locate and identify two other galaxies, M32 and M110, in this image. **d.** Is the *resolution* of this telescope image better, the same, or worse than that seen through binoculars or with the unaided eye?

WHAT IF... Humans Had Infrared-Sensitive Eyes?

Our eyes are sensitive to less than a trillionth of 1% of the electromagnetic spectrum—what we call visible light. But this minuscule resource provides an awe-inspiring amount of information about the universe. We interpret visible-light photons as the six colors of the rainbow—red, orange, yellow, green, blue, and violet. These colors combine to form all of the others that make our visual world so rich. But the Sun actually emits photons of all wavelengths. So, what would happen if our eyes had evolved to sense another part of the spectrum?

A Darker Vision Gamma rays, X rays, and most ultraviolet radiation do not pass through Earth's atmosphere. Because the world is illuminated by sunlight, Earth and the sky would look dark, indeed, if our eyes were sensitive only to these wavelengths. Radio waves, in contrast, easily pass through our atmosphere. But to see the same detail from radio waves that we now see from visible-light photons, our eyes would require a diameter 10,000 times larger. Each would be the size of a baseball infield!

What about infrared radiation? Although not all incoming infrared photons get through our atmosphere, short-wavelength ("near") infrared radiation passes easily through air. Depending on wavelength, the Sun emits between one-half and a ten-billionth as many infrared photons as red-light photons. Fortunately, most of these are in the near infrared.

Heat-Sensitive Vision To see infrared photons, human eyes would need to be only 5 to 10 times larger. Some snakes have evolved infrared vision. Portable infrared "night vision" cameras and goggles are available to us humans. Because everything that emits heat emits infrared photons, infrared sight would be very useful. Also, not everything we see with infrared sight would be due just to reflected sunlight—hotter objects would be intrinsically brighter than cool ones. For example, seeing infrared would allow us to observe changes in a person's emotional state. Someone who is excited or angry often has more blood near the skin and, thus, releases more infrared radiation (heat) than normal. Conversely, someone who is scared has less blood near the skin and, thus, emits less heat.

Night Vision The night sky would be a spectacular sight through infrared-sensitive eyes. Gas and dust clouds in the Milky Way absorb visible light, thus preventing the light of distant stars from getting to Earth. However, because most infrared radiation passes through these clouds, unaided, we would be able to see distant stars that we cannot see today. On the other hand, the white glow of the Milky Way, which is caused by the scattering of starlight by interstellar clouds, would be dimmer, because the gas and dust clouds do not scatter infrared light as much as they do visible light. (The haze created by the Milky Way would not vanish, however, because when gas and dust clouds are heated by starlight, they emit their own infrared radiation.)

Our concept of stars would be different, too. Many stars, especially young, hot ones, are surrounded by cocoons of gas and dust that emit infrared radiation. This dust is heated by the nearby stars. Instead of appearing as pinpoints, many stars would appear to be surrounded by wild strokes of color, and we would have an Impressionist sky.

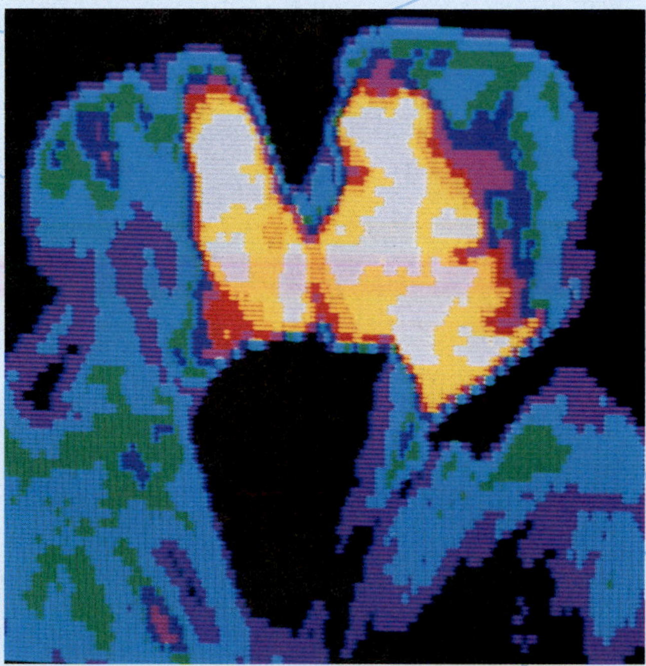

Kissing Is Hot The infrared (heat) from this kissing couple has been converted into visible light colors so that we can interpret the invisible radiation. The hottest regions are white, with successively cooler areas shown in yellow, orange, red, green, sky blue, dark blue, and violet. (D. Montrose/Custom Medical Stock Photo)

Visible Spectrum of the Sun (N.A. Sharp, NOAO/AURA/NSF)

Chapter 4

Atomic Physics and Spectra

WHAT DO YOU THINK?

1. **Which is hotter, a "red-hot" or a "blue-hot" object?**
2. **What color does the Sun emit most brightly?**
3. **How can we determine the age of space debris found on Earth?**

Answers to these questions appear in the text beside the corresponding numbers in the margins and at the end of the chapter.

"A star, extrasolar planet, galaxy, or interstellar gas cloud was discovered yesterday," is a statement that is true virtually every day of the year. The first thing that astronomers want to know about that newly discovered object is—everything. Much of this information is gained by focusing electromagnetic radiation from the object through a prism or similar device that separates its wavelengths, creating a spectrum. For astronomers, visible light is only the proverbial tip of the electromagnetic radiation iceberg. While probing other wavelengths, scientists in the past two centuries have discovered that most objects in space emit varying amounts of all types of electromagnetic radiation. This spectral analysis, along with information gleaned from the shapes and locations of objects, provides us with virtually all of the information we have about the cosmos. Furthermore, these observations provide the data needed to confirm the models of astronomical objects that scientists have developed.

In particular, you will see in this chapter that the resulting *spectrum* enables us to learn a variety of the physical properties of the object, including its surface temperature, chemical composition, mass, size, rotation rate, motion through space, and sometimes its distance from Earth.

We begin this ongoing adventure of discovery by discussing the nature of matter and how it interacts with electromagnetic radiation. In subsequent chapters, we consider related issues, such as why stars shine (that is, emit electromagnetic radiation), where different elements found on Earth and throughout the universe come from, and why objects have the motions they do, among many other questions.

In this chapter you will discover

- the origins of electromagnetic radiation
- the structure of atoms
- that stars with different surface temperatures emit different intensities of electromagnetic radiation
- that astronomers can determine the chemical compositions of stars and interstellar clouds by studying the wavelengths of electromagnetic radiation that they absorb or emit
- how to tell whether an object in space is moving toward or away from Earth

BLACKBODY RADIATION

We begin our study of why we see what we see with a familiar property of matter—temperature. We will discover that for many astronomical objects, their temperatures determine the relative numbers of photons that they emit at all different wavelengths.

4-1 An object's peak color shifts to shorter wavelengths as it is heated

Imagine that you have mounted an iron rod in a vise in a completely darkened room. You cannot see the rod because it emits too few visible photons, and you cannot feel its warmth without touching it because it is at the same temperature as the air in the room. Now imagine that there is also a propane torch in the room. You light it and heat the iron rod for several seconds, until it begins to glow. At the same time, you can feel the rod warm up, meaning that it is also emitting more infrared radiation.

The rod's first visible color is red—it glows "red 1
hot" (Figure 4-1a). Heated a little more, the rod appears orange and brighter, and feels hotter (Figure 4-1b). Heated more, it appears yellow and brighter still, and feels even hotter (Figure 4-1c). After even more heating, the rod appears white-hot and brighter yet. If it does not melt, further heating will make the rod appear blue and even brighter, and feel even hotter. As confirmed by the photographs in Figure 4-1, this experiment shows how the amount of electromagnetic radiation that any object emits changes with its temperature. A "red-hot" object is the coolest of all glowing bodies.

Insight Into Science

Many Phenomena Result from Several Effects Recall from Chapter 3 that white is not a fundamental color. Why then do objects at certain temperatures appear "white-hot," whereas cooler ones are red, orange, or yellow, and hotter ones appear blue or violet? The answer is that when blackbodies peak in the green part of the spectrum, all the color-sensitive cones in our eyes are overloaded, an effect that the brain interprets as "white."

If you were to put the light emitted by a heated object, like the iron rod, through a prism or diffraction grating (which separates the colors of light, just like a prism), you would discover that all wavelengths are present and that one is brightest. This peak wavelength is the color that the object appears to have. The experiment with the rod reveals two principles about the electromagnetic radiation emitted by objects as their temperatures change:

1. As an object heats up, it gets brighter, emitting more electromagnetic radiation at all wavelengths.
2. The brightest color (most intense wavelength) of the emitted radiation changes with temperature.

Most objects big enough to be seen with the naked eye continuously emit electromagnetic radiation as just

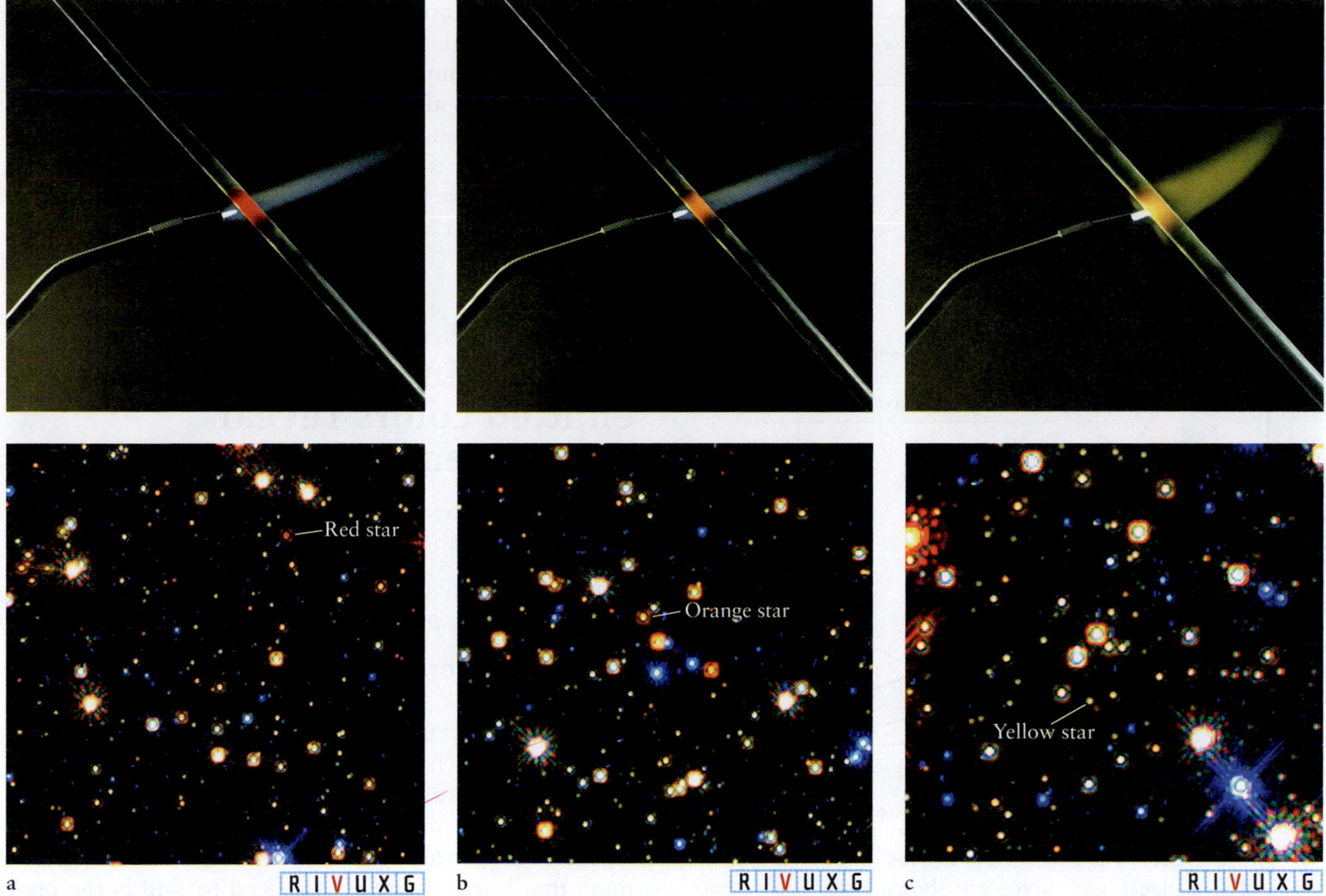

FIGURE 4-1 Objects at Different Temperatures Have Different Colors and Brightnesses This sequence of photographs shows the changing appearance of a piece of iron as it is heated. As the temperature increases, the amount of energy radiated by the bar increases, and so it appears brighter. The apparent color of the bar also changes because, as the temperature increases, the peak wavelength of light emitted by the bar decreases. The stars shown have roughly the same temperatures as the bars above them. (top photos: © 1984 Richard Megna Fundamental Photographs; bottom star photos: NASA)

described. The peak wavelength of any object's radiation, that is, where it emits most intensely, is denoted λ_{max}. When the object is relatively cool, such as a rock or an animal, λ_{max} is a radio, microwave, or infrared wavelength. When it is hot enough, like a fire or the Sun, λ_{max} is in the range of visible light, giving a hot object its characteristic color. Exceptionally hot objects, such as some stars, emit λ_{max} in the ultraviolet part of the electromagnetic spectrum.

Stars, molten rock, and iron bars are approximations of an important class of objects that astrophysicists call **blackbodies**. An ideal blackbody absorbs all of the electromagnetic radiation that strikes it. The incoming radiation heats up the blackbody, which then reemits the energy it has absorbed, but with different intensities at each wavelength than it received. Furthermore, the pattern of radiation emitted by blackbodies is independent of their chemical compositions. By measuring the intensity of radiation emitted by a blackbody at several wavelengths, it is possible to plot a curve of its emissions over all wavelengths (Figure 4-2). Ideal blackbodies have smooth **blackbody curves**, whereas objects that approximate blackbodies, such as the Sun, have more jagged curves whose variations from the ideal blackbody are caused by other physics (see Figure 4-3).

The total amount of radiation emitted by a blackbody at each wavelength depends only on the object's temperature and how much surface area it has. The bigger an object is, the brighter it is at all wavelengths. However, the *relative* amounts of different wavelengths (for example, the intensity of light at 750 nm compared to the intensity at 425 nm) depend on just the body's temperature. So, by examining the relative intensities of an object's blackbody curve, we are able to determine its temperature, regardless of how big or how far away it is. This process is analogous to how a thermometer tells your temperature no matter how big you are. (Temperatures

If a blackbody's peak wavelength is in the infrared, the object is often visible to our eyes. What color will it have?

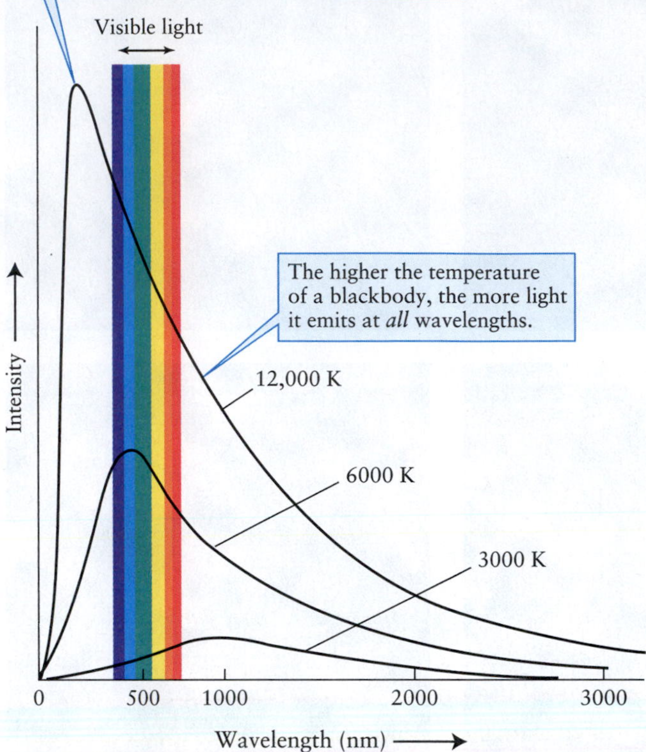

FIGURE 4-2 Blackbody Curves Three representative blackbody curves are shown here. Each curve shows the intensity of radiation over a wide range of wavelengths emitted by a blackbody at a particular temperature.

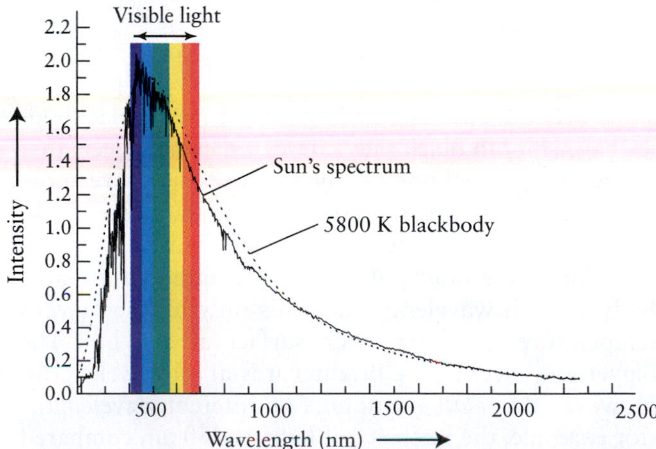

FIGURE 4-3 The Sun as a Blackbody This graph shows the similarity between the intensity of sunlight and the intensity of radiation from a 5800 K blackbody over a wide range of wavelengths. Earth's atmosphere scatters shorter-wavelength photons more than it scatters longer-wavelength ones. As seen from Earth, this scattering lowers the intensity of sunlight and shifts the blackbody peak to the right from where it would occur normally. To avoid these effects, the measurements of the Sun's intensity in this figure were made in space.

throughout this book are expressed in kelvins. If you are not familiar with the Kelvin temperature scale, review Appendix D: Temperature Scales.)

Rather than absorb and reradiate light from an outside source like our iron rod, stars radiate electromagnetic radiation that is generated inside them. Even so, stars behave nearly like blackbodies, and the self-generated radiation they emit follows the idealized blackbody curves fairly well.

4-2 The intensities of different emitted colors reveal a star's temperature

In 1893, the German physicist and Nobel laureate Wilhelm Wien showed that

The peak wavelength of radiation emitted by a blackbody is inversely proportional to its temperature.

In other words, the hotter any object becomes, the shorter its λ_{max}, and vice versa. **Wien's law** is the mathematical relationship between the location of the peak for each curve in Figure 4-2 and that blackbody's temperature.

Wien's law proves very useful in computing the surface temperature of a star. It says that to determine that temperature all we need to find is the peak wavelength of its electromagnetic radiation—we do not need to know the star's size, distance, or any other physical property.

In 1879, the Austrian physicist Josef Stefan observed that

An object emits energy per unit area at a rate proportional to the fourth power of its temperature in kelvins.

Think of the electric heating element on a stove. As this device becomes hotter, it gets brighter and emits more energy. Stefan's result says that if you double the temperature of an object (for example, from 500 to 1000 K), the energy emitted each second from each square meter of the object's surface increases by a factor of 2^4, or 16 times. If you triple the temperature (for example, from 500 to 1500 K), the rate at which energy is emitted increases by a factor of 3^4, or 81 times. Stefan's experimental results were supported by Ludwig Boltzmann's theory in 1885. In their honor, the intensity-temperature relationship for blackbodies is named the **Stefan-Boltzmann law.**

The Stefan-Boltzmann law and Wien's law are powerful tools that describe two basic properties of blackbody radiation, namely, the relative rate at which photons of all different wavelengths are emitted and the wavelength that is emitted most intensely. If you would like to learn how to use these two radiation laws to make predictions, see *An Astronomer's Toolbox 4-1*.

An Astronomer's Toolbox 4-1

The Radiation Laws

Wien's law can be stated as a simple equation. If λ_{max}, the wavelength of maximum intensity, is measured in meters and T is the temperature of the blackbody measured in kelvins, then

$$\lambda_{max} = 2.93 \times 10^{-3}/T$$

Example: The Sun's maximum intensity is at a wavelength of about 500 nm, or 5×10^{-7} m. From Wien's law, we can calculate the Sun's surface temperature as

$$T = \frac{2.93 \times 10^{-3}}{5 \times 10^{-7}}\ \text{K} = 5800\ \text{K}$$

The total energy emitted from each square meter of an object's surface each second is called the **energy flux**, F. In this context, flux means "rate of flow." The energy flux is related to the temperature, T, by the equation

$$F = \sigma T^4$$

where σ (lowercase Greek sigma) is called the Stefan-Boltzmann constant. It has a value of

$$\sigma = 5.67 \times 10^{-8}\ \text{J}/(\text{m}^2 \cdot \text{K}^4 \cdot \text{s})$$

with J being the energy unit joules, m the unit meters, K the unit kelvins, and s the unit seconds.

Because F is the energy emitted per second from each square meter of an object, multiplying F by the surface area, $4\pi r^2$, where r is the object's radius, yields the total energy emitted by a spherical object each second. This quantity, denoted L, is called **luminosity**:

$$L = F \cdot 4\pi r^2 = \sigma T^4 4\pi r^2$$

Luminosity is the total energy emitted per second by an entire object.

Compare! We consider one application of the Stefan-Boltzmann law. Suppose you observe two stars of equal size—one with a surface temperature of 10,000 K and the other with the same surface temperature as the Sun (5800 K). You can use the Stefan-Boltzmann law to determine how much brighter the hotter star is:

$$\frac{F_{hotter}}{F_{colder}} = \frac{\sigma \times 10{,}000^4}{\sigma \times 5800^4} \approx 8.8$$

So, the hotter star emits about 8.8 times as much energy from each square meter of its surface than does the cooler star. Because the stars are the same size, the hotter one is therefore 8.8 times brighter than the cooler one.

Try these questions: The color yellow is centered around 550 nm. If the Sun were actually yellow, what would be its surface temperature? (To find out why it is not yellow, see *Guided Discovery: The Color of the Sun.*) What is the peak wavelength in nanometers given off by a blackbody at room temperature of 300 K? Referring to Figure 3-6, what part of the electromagnetic spectrum is that in?

(Answers appear at the end of the book.)

Blackbody curves, such as those shown in Figure 4-2, provide the same information as Wien's law and the Stefan-Boltzmann law, as well as show the intensity of each blackbody at all wavelengths. Note that the blackbody curves for objects with different temperatures never touch or cross. Therefore, when astronomers measure the intensity of radiation from a star at a few wavelengths, they can determine the object's temperature by fitting these measurements to the appropriate, unique blackbody curve.

Although intensity curves of stars closely "follow" blackbody curves, they are not ideal blackbodies. The *differences* between an ideal blackbody curve and the curves seen from actual stars reveal stellar chemistries, the presence of companion stars and planets too dim to see, and the motions of stars toward or away from us, among other information.

2 Figure 4-3 shows how the intensity of sunlight varies with wavelength. The blackbody curve for a body with a temperature of 5800 K is also plotted in this figure. Note how the observed intensity curve for the Sun nearly reproduces the ideal blackbody curve at most wavelengths. The differences in the two curves are due to photons removed from the Sun's blackbody radiation by gases between the Sun's visible surface and our eyes. We discuss this phenomenon further shortly. Because the observed intensity curves for most stars and the idealized blackbody curves are so closely correlated, the laws of blackbody radiation can be applied to starlight. The peak of the intensity curve for the Sun is at a wavelength of about 500 nm, which is in the blue-green part of the visible spectrum.

The shapes of ideal blackbody curves were first derived mathematically in 1900 by the German physicist and Nobel laureate Max Planck. To determine these curves, he had to assume that objects receive electromagnetic radiation as individual packets of energy. This surprising assumption was reinforced in 1905 by Albert Einstein, who suggested that light—a wave—also behaves as particles, as discussed in Section 3-3.

We can now explain our experiment with the iron bar at the beginning of this chapter in terms of the energies and intensities of photons. Recall from Section 3-3 that the energy carried by a photon is inversely proportional

The Color of the Sun

Different people perceive the Sun to have different colors. To some it appears white, to others yellow. Still others, who notice it at sunrise or sunset, believe it to be orange or even red. But we have seen that the Sun actually emits all colors. Moreover, the peak in the Sun's spectrum falls between blue and green. Why doesn't the Sun appear turquoise? Several factors affect our perception of its color.

Before reaching our eyes, visible sunlight passes through Earth's atmosphere. All wavelengths are absorbed and reemitted by the molecules in the air, a process called *scattering*. Violet light is scattered most strongly, followed in decreasing order by blue, green, yellow, orange, and red. Thus, more violet, blue, and green photons are scattered by Earth's atmosphere than are yellow photons. The intense scattering of violet, blue, and green has the effect of shifting the peak of the Sun's intensity entering our eyes from blue-green toward yellow. (The sky is blue because of the strong scattering of blue light—it isn't violet because the Sun emits many fewer violet photons than blue ones.)

The perception of a yellow Sun is further enhanced by our eyes themselves. Our eyes do not see all colors equally well. Rather, the light-sensitive cones in our eyes each respond to one of three ranges of colors, which are centered on red, yellow, and blue wavelengths. None of the cones is especially sensitive to blue-green photons. By adding together the color intensities detected by the three types of cones, our brains *re-create* color. After combining all of the light that it can, the eye is most sensitive to the yellow-green part of the spectrum. We see blue and orange less well and violet and red most poorly. Although the eye sees yellow and green light about equally well, the Sun appears yellow because the air scatters more green light than yellow and our eyes are relatively insensitive to blue-green. Therefore, a casual glance at the Sun often leaves the impression of a yellow object.

Above Earth's atmosphere, the Sun's light is so intense that it saturates the color-sensitive cones in our eyes. The Sun appears "blindingly white" in space, as an astronaut told this author.

At sunrise and sunset we see an orange or red Sun. This coloration occurs because close to the horizon the Sun's violet, blue, green, and even yellow photons are all strongly scattered by the thick layer of atmosphere through which they travel, leaving the Sun looking redder and redder as it sets.

(Tim Henry and the OSPAN team, U.S. Air Force Research Laboratory) R I V U X G

(ImageGap/Alamy) R I V U X G

(Roger Ressmeyer/Corbis) R I V U X G

Table 4-1 Some Properties of Electromagnetic Radiation

	Wavelength (nm)	Photon energy (eV)*	Blackbody temperature (K)
Radio	$>10^7$	$<10^{-4}$	<0.03
Microwave**	10^7 to 4×10^5	10^{-4} to 3×10^{-3}	0.03 to 30
Infrared	4×10^5 to 7×10^2	3×10^{-3} to 2	30 to 4100
Visible	7×10^2 to 4×10^2	2 to 3	4100 to 7300
Ultraviolet	4×10^2 to 10^1	3 to 10^3	7300 to 3×10^6
X ray	10^1 to 10^{-2}	10^3 to 10^5	3×10^6 to 3×10^8
Gamma ray	$<10^{-2}$	$>10^5$	$>3 \times 10^8$

Note: > means greater than; <means less than.
*1 eV (electron volt) = 1.6×10^{-19} J.
**Microwaves, listed here separately, are often classified as radio waves or infrared radiation.

to its wavelength. In other words, long-wavelength photons, such as radio waves, carry little energy, while short-wavelength photons, like X rays and gamma rays, each carry much more energy. When the rod was cool, it mostly emitted lower-energy infrared and radio photons; when it was hot enough to glow, it emitted mostly visible photons and more of all types of photons, which is why it looked brighter and felt hotter. The relationship between the energy of a photon and its wavelength is called **Planck's law**, as explored in *An Astronomer's Toolbox 3-1*.

Together, Planck's law and Wien's law relate the temperature of an object to the energy of the photons that it emits (Table 4-1). A cool object emits primarily long-wavelength photons that carry little energy, while a hot object emits mostly short-wavelength photons that carry much more energy. In later chapters, we will find these relationships invaluable for understanding how stars of various temperatures interact with gas and dust in space.

WHAT IF... The Sun *Did* Have Its Peak Emission in the Yellow Part of the Spectrum?
In that case, the Sun's surface temperature would be 5100 K, rather than 5800 K, and Earth (in its present orbit) would be colder overall than it is today.

IDENTIFYING THE ELEMENTS BY ANALYZING THEIR UNIQUE SPECTRA

In 1814, the German optician Joseph von Fraunhofer repeated Newton's classic experiment of shining a beam of sunlight through a prism (recall Figure 3-1). However, Fraunhofer magnified the resulting rainbow-colored spectrum. He discovered that the solar spectrum contains hundreds of fine dark lines, which became known as **absorption lines** because the light of these colors has been absorbed by gases between the Sun and the viewer on Earth. Fraunhofer counted more than 600 such lines, and today physicists have detected more than 30,000 of them. Thousands of spectral lines are visible in the photograph of the Sun's spectrum shown in Figure 4-4.

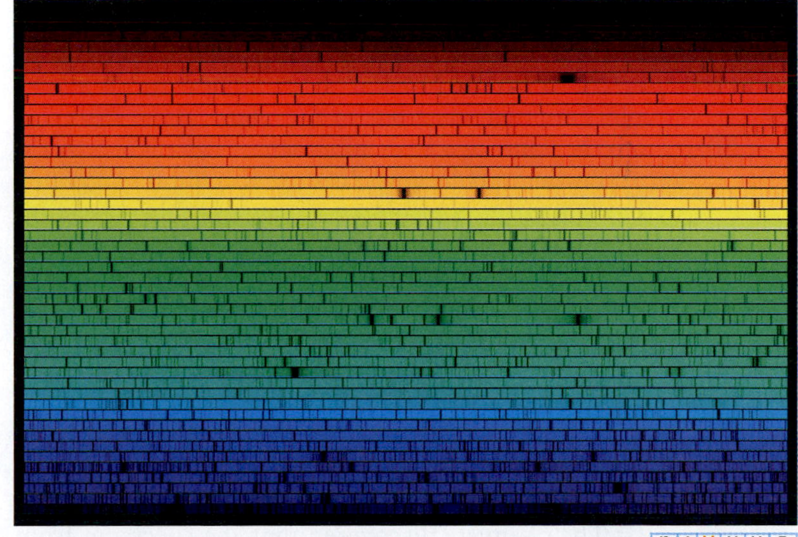

FIGURE 4-4 Solar Spectrum Starlight passing through a prism or diffraction grating spreads into its component colors. From such spectra we can learn an incredible amount about stars, including their masses, surface temperatures, diameters, chemical compositions, rotation rates, and motions toward or away from us. The above image, called a *spectrogram*, shows the spectrum of the Sun sliced and stacked to fit on this page. In it you can see thousands of absorption lines. As you study spectra, also note the distinct differences between the intensities of the various colors emitted by each star and by different stars. (N. A. Sharp, NOAO/AURA/NSF)

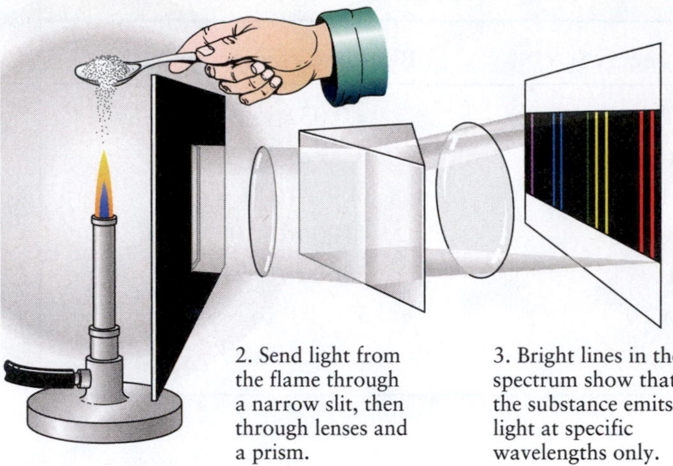

FIGURE 4-5 Early Spectroscope In the mid-1850s, Kirchhoff and Bunsen discovered that when a chemical substance is heated and vaporized, the resulting spectrum exhibits a series of bright spectral lines when passed through a slit, lenses, and a prism. This device is called a spectroscope. In addition, they found that each chemical element produces its own characteristic pattern of spectral lines. The lenses focus and magnify the spectrum.

By the mid-1800s, chemists discovered that they could produce spectral lines in the laboratory. Around 1857, the German chemist Robert Bunsen invented a special gas burner that produces a clean, constant flame. Certain chemicals are easy to identify by the distinctive colors emitted when bits of the chemical are sprinkled into the flame of a Bunsen burner.

Bunsen's colleague Gustav Kirchhoff suggested that light from the colored flames could best be studied by passing it through a prism. By separating the colors, the chemists could see exactly which ones were present. The process of analyzing the spectrum of an object is called *spectroscopy*. Bunsen and Kirchhoff collaborated in designing and constructing the first **spectroscope**. This device consists of a narrow slit, a prism, and several lenses that straighten the light rays and magnify the spectrum so that it can be closely examined (Figure 4-5).

The chemists discovered that the spectrum from a chemical heated by a flame consists of a pattern of thin, bright spectral lines, called **emission lines**, against a dark background (not a blackbody spectrum!). They next found that *the number of lines produced and their colors are unique to the element or compound heated to produce them.* Thus, in 1859, the technique of **spectral analysis**, the identification of chemical substances by their spectral lines, was born.

Where are radio waves relative to the horizontal axis of Figure 4-2?

4-3 Each chemical element produces its own unique set of spectral lines

A chemical **element** is a fundamental substance that cannot be broken down into more basic units while still retaining its properties. By the mid-1800s, chemists had already identified such familiar elements as hydrogen, oxygen, carbon, iron, gold, and silver. Spectral analysis promptly led to the discovery of additional elements, many of which are quite rare.

After Bunsen and Kirchhoff recorded the prominent spectral lines of all of the known elements, they began to discover other spectral lines in mineral samples. In 1860, for example, they found a new line in the blue portion of the spectrum of mineral water. After chemically isolating the previously unknown element responsible for the line, they named it *cesium* (from the Latin *caesius*, meaning "gray-blue"). The next year, a new spectral line in the red portion of the spectrum of a mineral sample led to the discovery of the element rubidium (from *rubidus*, for "red").

During a solar eclipse in 1868, astronomers found a new spectral line in the light coming from the upper atmosphere of the Sun when the main body of the Sun was hidden by the Moon. This line was attributed to a new element, which was named *helium* (from the Greek *helios*, meaning "Sun"). Helium was not actually discovered on Earth until 1895, when it was identified in gases obtained from a uranium compound.

The chemical elements are most conveniently displayed in the form of a **periodic table** (Appendix H). In addition to the 92 naturally occurring elements, there are additional, artificially produced ones. All of the human-made elements are heavier than uranium (U), and all are highly radioactive, meaning that they spontaneously decay into lighter elements shortly after being created in the laboratory.

Insight Into Science

Seek Relationships Finding how properties of different things relate to one another often provides invaluable insights into how they work. The periodic table, for example, enables scientists to determine the properties of similar chemical elements. Keep an eye out for other such relationships throughout this book.

During some experiments, Bunsen and Kirchhoff examined the blackbody spectra of hot sources whose light they passed through cooler gas. Instead of seeing a continuum of wavelengths, as they would have seen had the cool gas not been there, they saw dark *absorption lines* among the colors of the rainbow created by

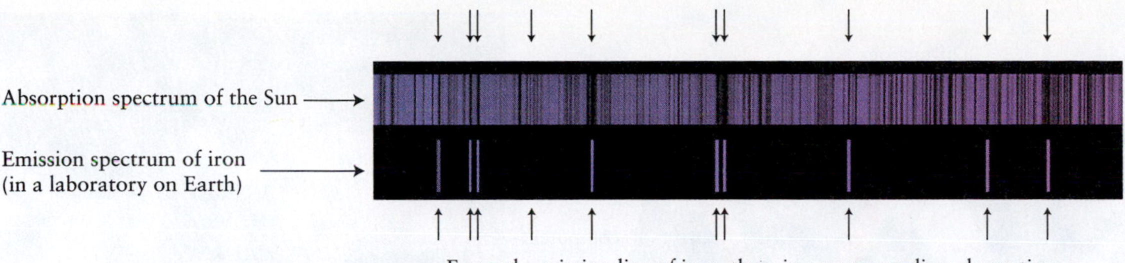

FIGURE 4-6 Iron in the Sun's Atmosphere The upper (absorption) spectrum is a portion of the Sun's spectrum from 425 to 430 nm. Numerous dark spectral lines are visible. The lower (emission) spectrum is a corresponding portion of the spectrum of vaporized iron. Several emission lines can be seen against the black background. The iron lines coincide with some of the solar absorption lines, proving that there is some iron (albeit a very tiny amount) in the Sun's atmosphere. (Carnegie Observatories)

the hot source. (This phenomenon is what Fraunhofer had seen earlier from the Sun—the cooler gases in his results were the atmospheres of Earth and the Sun.) In other experiments, when looking just at gases without hot sources behind them, they saw bright *emission lines* against otherwise dark backgrounds. Combining the results from both sets of experiments, they discovered that *the emission lines of a particular gas occur at exactly the same wavelengths as the absorption lines of that gas.*

Because each chemical element produces its own unique pattern of spectral lines, scientists can determine the chemical composition of a remote astronomical object by identifying the lines in its spectrum. For example, Figure 4-6 shows a portion of the Sun's absorption spectrum in the blue-violet part of the spectrum, along with the emission spectrum of an iron sample taken here on Earth. The spectral lines of iron also appear in the Sun's spectrum, so we can reasonably conclude that the Sun's atmosphere contains some vaporized iron. This gaseous iron absorbs and then scatters certain wavelengths from the continuum emitted below it.

After photography was invented, scientists preferred to do spectroscopy by making a permanent photographic record of spectra. A device for photographing a spectrum is called a **spectrograph**, and this instrument, in its numerous variations, is the astronomer's most important tool.

In its earliest form, a spectrograph was attached directly to a telescope. It consisted of a slit, two lenses, and a prism arranged to focus the spectrum of an object on film, similar to the device sketched in Figure 4-5. Although conceptually straightforward, this early type of spectrograph had severe drawbacks. A prism does not spread colors evenly: The blue and violet portions of the spectrum are spread out more than the red portion. In addition, because the blue and violet wavelengths must pass through more glass than the red wavelengths (see Figure 3-1), light is absorbed unevenly across the spectrum. Indeed, a glass prism is opaque to ultraviolet wavelengths.

Practical spectrographs used in research today (shown conceptually in Figure 4-7a) separate light from objects in space into the colors of the rainbow using a **diffraction grating**, which is a piece of glass on which thousands of closely spaced parallel grooves are cut. Some of the finest diffraction gratings have more than 10,000 grooves per centimeter. The spacing of the grooves must be very regular. Light waves are diffracted by the grooves in the diffraction grating, just as light or water waves are diffracted when passing through slits (see Figure 3-2). Different wavelengths are diffracted by different amounts. A spectrum is produced by combining this spreading of the colors in the diffraction grating with the interference between the waves (see Figure 3-2b).

Figure 4-7a shows the design of a modern diffraction grating spectrograph with a CCD (charge-coupled device; see Figure 3-15) that has replaced film to record the spectra. This optical device typically mounts at the Nasmyth, coudé, or Cassegrain focal point of a telescope. The image of the object to be examined is focused on the slit. After the spectrum of a star, galaxy, or other object has been recorded, the CCD collects an emission spectrum from a known source, such as a gas composed of helium, neon, and argon, that is also focused on the slit. This "comparison spectrum" is placed next to the spectrum of the object, as in Figure 4-6. Because the wavelengths of the spectral lines in the comparison spectrum are already known from laboratory experiments, these lines can be used to identify and measure the wavelengths of the lines in the spectrum of the star or galaxy under study. Figures 4-7b and 4-7c show diffraction spectra occurring on Earth.

From what we have just discussed, what do you think causes fireworks to have their distinctive colors?

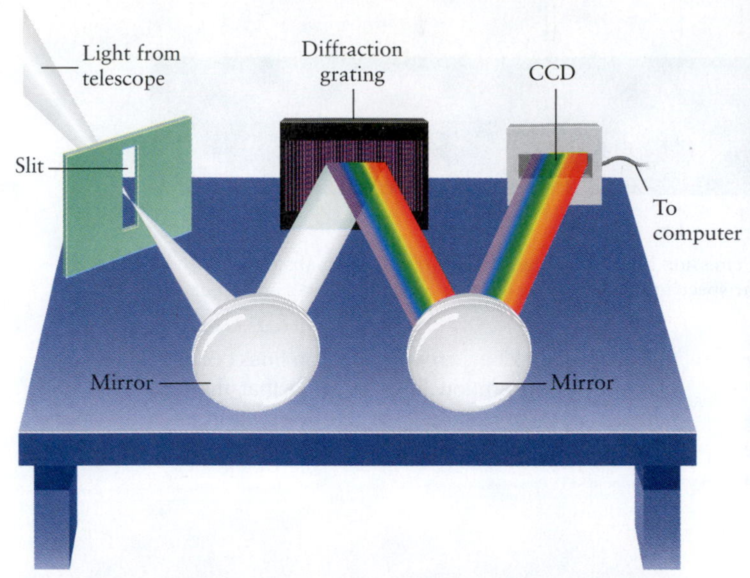

FIGURE 4-7 A Grating Spectrograph (a) The diffraction grating in a grating spectrograph has many parallel lines on its surface that reflect light of different colors in different directions. This separation of colors (wavelengths) allows the object's spectrum to be analyzed. (b) This peacock feather contains numerous natural diffraction gratings. The role of the parallel lines etched in a human-made diffraction grating is played by parallel rods of the protein melanin in the feathers. (c) CDs and DVDs store information on closely spaced bumps located on a set of nearly parallel tracks. Light striking these tracks systematically reflects different colors in different directions—a CD or a DVD behaves like a diffraction grating. (b: Paul Silverman/Fundamental Photos; c: W. Cody/Corbis)

The spectral data are converted by computer to a graph that plots light intensity against wavelength. Dark lines in the rainbow-colored spectrum appear as depressions or valleys on the graph, while bright lines in the spectrum appear as peaks. For example, Figure 4-8 shows both an absorption spectrum and an emission spectrum for hydrogen. The five absorption lines (Figures 4-8a, b) occur at precisely the same wavelengths as the five emission lines for hydrogen (Figures 4-8c, d).

4-4 The various brightness levels of spectral lines depend on conditions in the spectrum's source

INTERACTIVE EXERCISE 4.1

By the early 1860s, Kirchhoff had discovered the conditions under which different types of spectra are observed. His description is summarized today as **Kirchhoff's laws:**

Law 1 A solid, liquid, or dense gas produces a **continuous spectrum** *(also called a* **continuum***)—a complete rainbow of colors without any spectral lines. This is a blackbody spectrum.*

The light emitted by a hot iron rod and by an incandescent lightbulb are examples of continuous spectra that are bright enough for us to see.

Law 2 A rarefied (opposite of dense) gas produces an **emission line spectrum***—a series of bright spectral lines against a dark background.*

The light emitted by neon lights and by low-pressure sodium vapor lights are examples of emission line spectra. The neon has bright-red emission lines, while the low-pressure sodium has bright-yellow lines.

Law 3 The light from an object with a continuous spectrum that passes through a cool gas produces an **absorption line spectrum***—a series of dark spectral lines among the colors of the rainbow.*

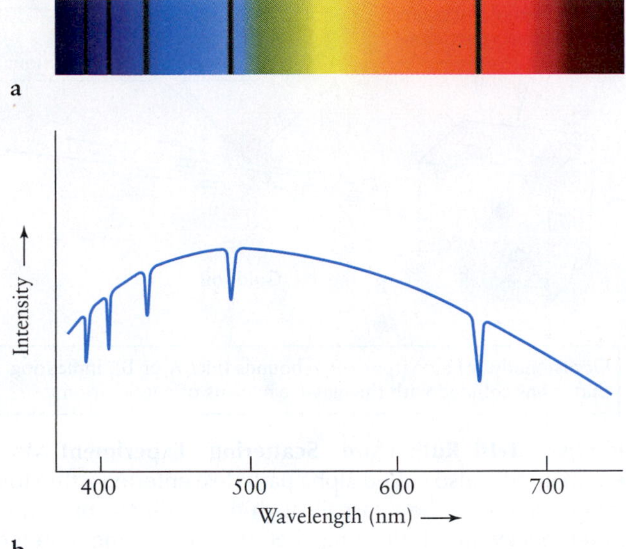

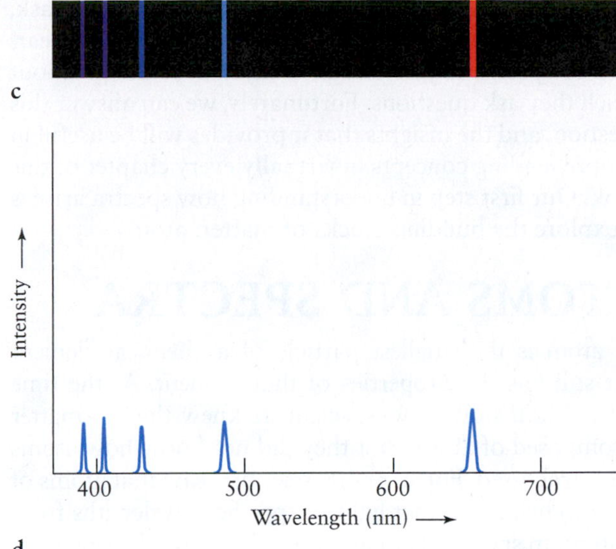

FIGURE 4-8 Spectrum of Hydrogen Gas (a) When a CCD is placed at the focus of a spectrograph, the spectrum is recorded. In this case, the spectrum is from a hot object whose light shines through hydrogen gas. (b) This spectrum is converted by computer into a graph of intensity versus wavelength. Note that the absorption lines appear as dips in the intensity-versus-wavelength curve. Conversely, when a gas emits only a few wavelengths, (c) its emission spectrum appears as a series of bright lines, which are converted into peaks (d) on a graph of intensity versus wavelength.

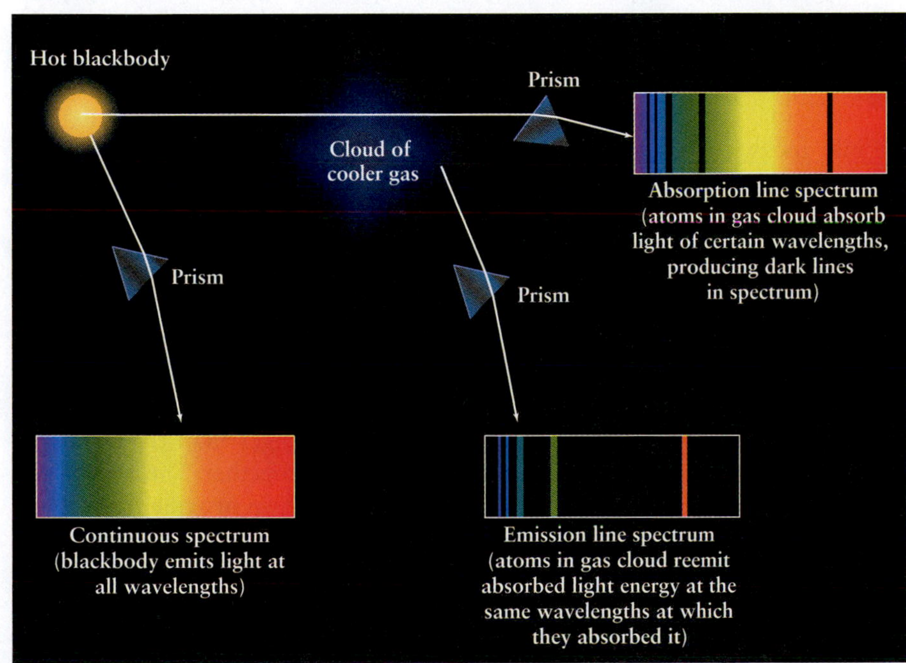

ANIMATION 4.2 **FIGURE 4-9 Continuous, Absorption Line, and Emission Line Spectra** This schematic diagram summarizes how different types of spectra are produced. The prisms are added for conceptual clarity, but in real telescopes diffraction gratings are used to separate the colors. A hot, glowing object emits a continuous spectrum. If this source of light is viewed through a cool gas, dark absorption lines appear in the resulting spectrum. When the same gas is viewed against a cold, dark background, its spectrum consists of just bright emission lines. The linked Animation shows the details of the activity drawn here.

Sunlight and the light from other stars pass through several cooler gases on their way to us. Hence, light from stars produces absorption line spectra.

In summary (Figure 4-9), a continuum is seen if there is no gas between a blackbody and the observer; absorption lines are seen if the background object is hotter than the gas between the observer and the object; and emission lines are seen if the background is cooler than such a gas.

Consider, for example, the spectrum of the Sun. We know that the Sun's surface emits a continuous blackbody spectrum, but here on Earth many absorption lines are seen in it (see Figure 4-4). Kirchhoff's third law explains why. There must be a cooler gas between the surface of the Sun and Earth. In fact, as noted earlier, there are two such gases: the Sun's lower atmosphere and Earth's entire atmosphere. Using the catalogs of spectra of different elements, the absorption lines in the Sun's spectrum tell us the chemical composition of the Sun's lower atmosphere and that of Earth's atmosphere.

> Diffraction gratings can either reflect or transmit light to create spectra. Are the peacock feathers in Figure 4-7b reflection or transmission diffraction gratings?

If you have younger siblings or are around children a lot, you are likely to hear them go through a stage of development in which they frequently ask, "Why?" Scientists retain that curiosity, and the appearance of spectra is just one of a myriad of things about which they ask questions. Fortunately, we can answer this question, and the insights that it provides will be useful in comprehending concepts in virtually every chapter of this book. Our first step in understanding how spectra arise is to explore the building blocks of matter: atoms.

ATOMS AND SPECTRA

An **atom** is the smallest particle of a chemical element that still has the properties of that element. At the time of Kirchhoff's discoveries, scientists knew that all matter is composed of atoms, but they did not know how atoms were structured. Furthermore, scientists saw that atoms of a gas somehow extract light of specific wavelengths from continuum spectra that pass through the gas, leaving dark absorption lines, and they perceived that the atoms then radiate light of precisely the same wavelengths—the bright emission lines (see Figure 4-6). But traditional theories of electromagnetism could not explain these phenomena. The answer came early in the twentieth century, with the development of nuclear physics and quantum mechanics.

4-5 An atom consists of a small, dense nucleus surrounded by electrons

The internal structure of atoms first came into focus in 1908 when New Zealand native and Nobel laureate Ernest Rutherford and his colleagues at the University of Manchester in England were investigating radioactivity. This phenomenon was studied earlier by Polish physicist (and *double* Nobel laureate) Marie Curie, among others. Over time, a **radioactive** element naturally and spontaneously transforms into another element by emitting particles. Certain radioactive elements, such as uranium and radium, were known to emit such particles with considerable speed. It seemed plausible that a beam of these high-speed particles would penetrate a thin sheet of gold. Rutherford and his associates found that almost all of the particles did pass through the gold sheet with little or no deflection. To their surprise, however, an occasional particle bounced back (Figure 4-10). It must have struck something very dense, indeed.

Within a few decades, the structure of atoms became evident. That dense "something" is now called the **nucleus** of the atom, and it consists of particles called **protons** and **neutrons**. Surrounding the nucleus, one or more **electrons** normally orbit. Newton's second law (see Section 2-7) explains that electrons move much more than the nuclei they orbit because the electrons have only 1/2000 the mass of a proton or neutron.

Unlike the planets, whose gravitational interaction keeps them orbiting the Sun, electrons orbit nuclei because electrons and protons have a property called

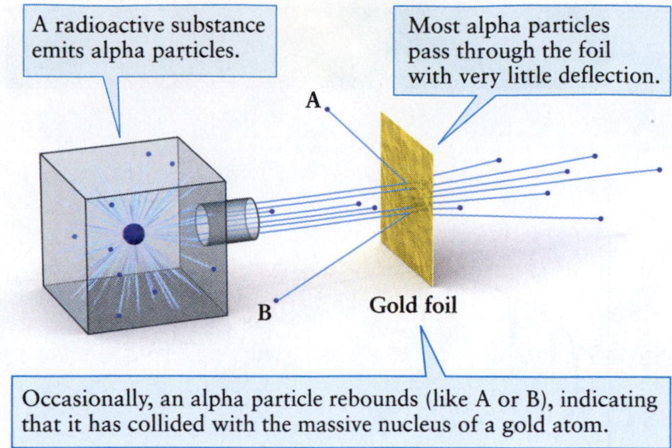

FIGURE 4-10 Rutherford Scattering Experiment Most helium nuclei (also called alpha particles) entering a thin foil scatter slightly as they pass through the medium, but some scatter backward, indicating that they have encountered very dense, compact objects. Such experiments were the first evidence that most matter is concentrated in what are now called atomic nuclei.

electric charge. All protons have the exact same positive charge, while all electrons have a negative charge equal in strength to the proton's charge. The terms *positive* and *negative* are arbitrary and just indicate that the charges are opposite to each other. Particles with opposite charges attract each other. Therefore, protons attract electrons, and it is this attraction that keeps electrons in orbit around nuclei. The interaction between charged particles is a result of the **electromagnetic force**, the second of four fundamental forces in nature. (Gravitation is the first fundamental force we have discussed.)

Particles with the same type of charge, such as a pair of protons or a pair of electrons, repel each other. For atoms that have more than one proton in their nuclei (and many do), the protons are pushing away from each other. For nuclei to have more than one proton, there must be an attractive force stronger than the repulsion of protons to keep them glued together. The electrically neutral neutrons in the nucleus help provide that attractive force, called the **strong nuclear force**. This is the third of the four fundamental forces.

Many scientists find it fascinating that all of the interactions between matter and energy in nature occur as a result of just four forces, the above three plus the **weak nuclear force**. The weak nuclear force is involved in some radioactive decays, such as when a neutron transforms into a proton. In Chapter 18, we will further explore the weak nuclear force in our study of the evolution of the whole universe. The properties of all four fundamental forces are summarized in Table 4-2.

The number of protons in an atom's nucleus, called the **atomic number**, determines the element of that atom. All hydrogen nuclei have 1 proton, all helium nuclei have 2, and so forth. There are 92 different types of elements that form naturally. Uranium is the most massive, with 92

Table 4-2 The Four Fundamental Forces of Nature

Name	Strength (compared to the strong force)	Range of effect (from each object)
Strong force	1	Inside atomic nuclei
Electromagnetic force	1/137	Throughout the universe
Weak force	10^{-5}	Inside atomic nuclei
Gravitational force	6×10^{-39}	Throughout the universe

protons (see the periodic table in Appendix F). As we will explore in Chapter 13, the fact that neutrons can transform into protons by emitting an electron and that protons can transform into neutrons by absorbing an electron is vital for the formation of elements with more than 26 protons.

In contrast, the nuclei of most elements can have different numbers of neutrons. For example, hydrogen always has 1 proton, but it can have 0, 1, or 2 neutrons; oxygen, with an atomic number of 8, always has 8 protons, but it can have 8, 9, or 10 neutrons. Each different combination of protons and neutrons is called an **isotope.** There are three isotopes of hydrogen and three isotopes of oxygen. Hydrogen with no neutrons is the most common hydrogen isotope, while oxygen with 8 neutrons is by far the most abundant isotope of oxygen.

Some isotopes are stable, meaning that the numbers of protons and neutrons in their nuclei do not change. However, many elements have isotopes that are unstable and come apart spontaneously. These isotopes are called radioactive. For example, carbon with 6 neutrons, ^{12}C ("carbon twelve"), is stable, while carbon with 8 neutrons, ^{14}C ("carbon fourteen"), is unstable. ^{14}C decays into nitrogen with 7 neutrons, ^{14}N. To learn how radioactive decay is used to determine the ages of different objects, see *An Astronomer's Toolbox* 4-2. Using radioactive age-dating techniques on the oldest rocks brought back from the Moon, astronomers are able to determine roughly when the Moon formed (4.5 billion years ago). Applying the same technique to space debris found on Earth (meteorites), we can determine that the solar system, consisting of the Sun and everything orbiting it, formed some 4.62 billion years ago.

The number of electrons that orbit an atom is normally equal to the number of protons in its nucleus, thus making the atom electrically neutral. Astronomers denote neutral atoms by writing the atomic symbol followed by the Roman numeral I. For example, neutral hydrogen is written as H I and neutral iron is Fe I.

When an atom contains a different number of electrons than protons, the atom is called an **ion.** The process of creating an ion is called **ionization.** Ions are denoted by the atomic symbol followed by a Roman numeral that is one greater than the number of missing electrons. Positively ionized hydrogen (missing its one electron) is denoted H II, while positively ionized iron with seven electrons missing is denoted Fe VIII. Note that negative ions also exist, in which nuclei have more electrons orbiting than they have protons.

Atoms can share electrons and, by doing so, become bound together. Such groups of atoms are called **molecules.** Molecules are the essential building blocks of all complex structures, including life.

4-6 Spectra occur because electrons absorb and emit photons with only certain wavelengths

Because protons and neutrons have masses about 2000 times greater than the mass of an electron, over 99.95% of the mass of any atom is concentrated in its nucleus. The electron orbits are far from the nucleus, typically 10,000 times farther away than the radius of the nucleus. This immense distance is why you may have heard the statement that matter is mostly empty space.

Be careful not to imagine electrons as miniature planets orbiting the nucleus as a "miniature Sun." Protons, neutrons, and electrons are not tiny solid bodies. Rather, like photons, they all have both wave and particle properties. The science that accurately describes their complex behavior is called **quantum mechanics.**

Quantum mechanics explains that electrons in atoms can exist in only certain *allowed orbits* around their nuclei, except when they are making a **transition** from one allowed orbit to another. These orbital conditions are completely unlike planets, which can exist at any distance (in any orbit) around the Sun. Each allowed electron orbit has a well-defined energy associated with it, and every different type of atom and molecule has a unique set of allowed orbits. These orbits and the transitions between them are the key to understanding the spectra that we have been discussing.

Consider an atom of the simplest hydrogen isotope, which contains just a single proton in its nucleus orbited by one electron. (This discussion generalizes directly to all of the other atoms and isotopes, made more complex only because they have more than one electron in orbit.) Figure 4-11 shows this hydrogen atom's lowest energy levels.

An Astronomer's Toolbox 4-2

Radioactivity and the Ages of Objects

The isotopes of many elements are radioactive, meaning that the elements spontaneously transform into other elements. Each radioactive isotope has a distinctive half-life, which is the time that it takes half of the initial concentration of the isotope to transform into another element. After two half-lives, a radioactive isotope is reduced to ½ × ½ or ¼ of its initial concentration (see the accompanying figure). Among the most important radioactive elements for determining the age of objects in astronomy is the isotope of uranium with 146 neutrons, ^{238}U ("U two-thirty-eight"). The half-life of ^{238}U decaying into lead is 4.5 billion years.

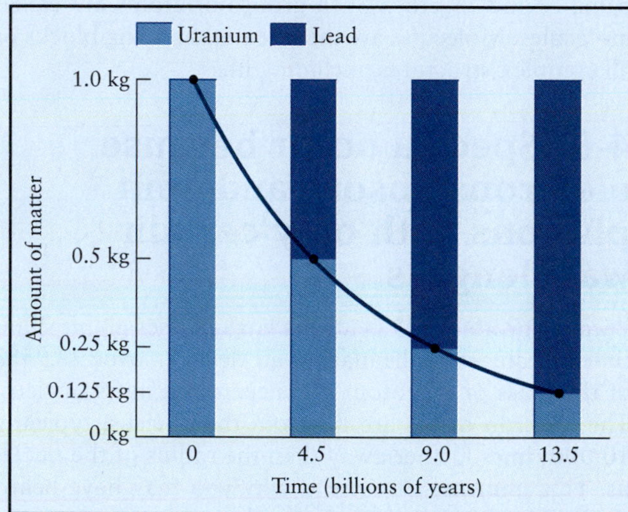

The Transformation of Uranium into Lead This figure shows the rate that 1 kg of uranium decays into lead, as described in the text. The 1-kg sample contains 0.125 kg of uranium after 13.5 billion years.

To determine the time since an object, such as a piece of space debris discovered on Earth, solidified, scientists estimate how much lead the object had when it formed. Then they measure the amounts of uranium and lead that it contains now. Subtracting the amount of original lead, they use the amount of uranium and lead, together with the graph of radioactive decay, to determine the object's age.

Example: Suppose a piece of space debris discovered 3
on Earth was determined to have equal amounts of lead and uranium. How long ago did this debris form? Assuming that it originally had no lead, we see from the chart that a 1-to-1 mix of lead to uranium occurs 4.5 billion years after the object formed. This period is one half-life of uranium.

Compare! This process works with any radioactive isotope. However, some isotopes have such short half-lives that they are not useful in astronomy. For example, the well-known carbon dating used to determine the ages of ancient artifacts on Earth is of little use in astronomy because ^{14}C has a half-life of only 5730 years. Because so much of the carbon in a sample has decayed away by then, carbon dating is useful only for time intervals shorter than 100,000 years, usually a period over which little of astronomical importance occurs.

Try these questions: What fraction of a kilogram of radioactive material remains after three half-lives have passed? Using the accompanying figure, estimate how much uranium will remain after 6 billion years. Approximately how many half-lives of ^{14}C pass in 100,000 years?

(Answers appear at the end of the book.)

Normally, the electron is in the lowest-energy allowed orbit or energy level, commonly called the **ground state**; this is labeled $n = 1$ in Figure 4-11. Each allowed orbit with successively higher energy is labeled $n = 2, 3, 4$, and so on. When an electron is in an orbit with more energy than the lowest energy state available to it, it is said to be in an **excited state.**

Electrons change orbits by absorbing or emitting photons. However, electrons cannot absorb just any photon that they encounter. Electrons can absorb only those photons with energies exactly enough to boost them up to a higher-energy allowed orbit. That is, the photons that are absorbed are those with energies equal to the difference between the energies of two allowed orbits. All photons that do not satisfy this condition pass straight through the atom. If the electron starts in its ground state, then it must get exactly the energy necessary to move it to an excited state. If it is already in an excited state, then by absorbing a photon, it must transition to a higher-energy excited state.

For example, referring to Figures 4-11 and 4-12, transitions between the $n = 2$ and the $n = 3$ energy levels require the electron to absorb a photon with energy equal to 12.1 eV − 10.2 eV = 1.9 eV (electron volt, a measure of energy). Recall from *An Astronomer's Toolbox 3-1* that a photon's energy corresponds to a certain wavelength. In this case, the photon absorbed by the electron has a wavelength of 656.3 nm (Figure 4-12a); it is a red photon (recall Figure 3-6).

Although some absorption occurs at optical wavelengths, it also happens in many other parts of the electromagnetic spectrum. Indeed, most of the transitions in hydrogen, shown in Figure 4-11 as the Lyman series (up from and down to $n = 1$) and the Paschen series

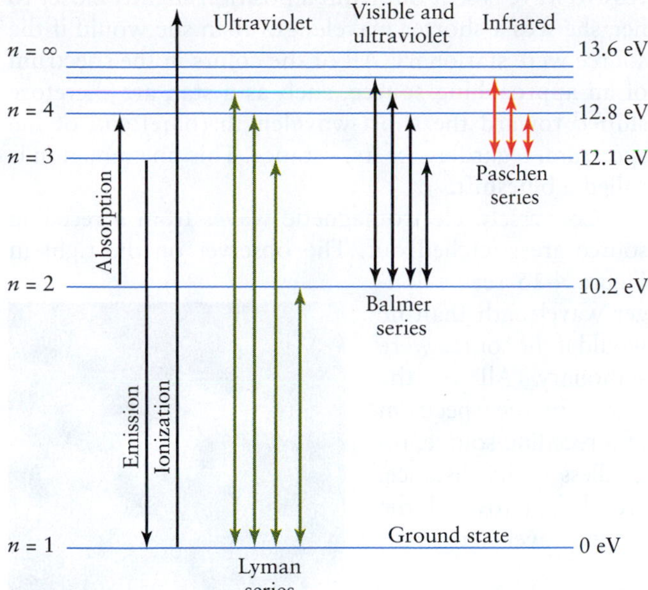

INTERACTIVE EXERCISE 4.2

FIGURE 4-11 Energy Level Diagram of Hydrogen The activity of a hydrogen atom's electron is conveniently displayed in a diagram showing some of the energy levels at which the electron can exist. A variety of electron jumps, or transitions, are also shown, including those that produce the most prominent lines in the hydrogen spectrum. For another perspective on the same information, you can try the linked Interactive Exercise.

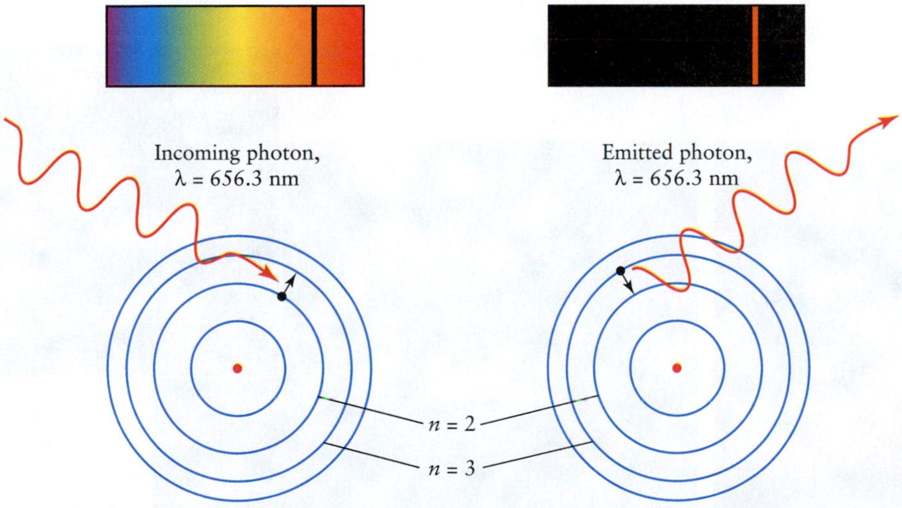

Incoming photon, $\lambda = 656.3$ nm

Emitted photon, $\lambda = 656.3$ nm

$n = 2$

$n = 3$

a Atom absorbs a 656.3-nm photon; absorbed energy causes electron to jump from the $n = 2$ orbit up to the $n = 3$ orbit

b Electron falls from the $n = 3$ orbit to the $n = 2$ orbit; energy lost by atom goes into emitting a 656.3-nm photon

ANIMATION 4.3

FIGURE 4-12 The Absorption and Emission of an H_α Photon This schematic diagram of hydrogen's four lowest allowed orbits shows what happens when a hydrogen atom absorbs or emits an H_α photon, which is red and has a wavelength of 656.3 nm. The spectral lines by such events are also shown. (a) A photon is absorbed by the electron, causing the electron to transition from orbit $n = 2$ up to orbit $n = 3$. (b) A photon is emitted as the electron makes a transition from orbit $n = 3$ down to orbit $n = 2$.

(up from and down to $n = 3$), are nonvisible photons. Only a few of the Balmer series (up from and down to $n = 2$) transitions are visible, with the rest being ultraviolet. We will refer to Balmer transitions when we study stars. They are named after the Swiss physicist Johann Balmer, who first calculated their wavelengths. The longest wavelength Balmer line is called H_α (*H alpha*), the second H_β (*H beta*), the third H_γ (*H gamma*), and so forth, ending with the shortest wavelength Balmer line, H_∞ (*H infinity*). (The first dozen lines of the series have Greek-letter subscripts; the remainder are identified by numerical subscripts.) Part of the spectrum of a star having Balmer absorption lines H_α through H_θ (*H theta*) is shown in Figure 4-13.

Absorption lines are therefore caused by photons being taken out of the stream of light by electrons, which thereby move into higher-energy allowed orbits. An observer on the right in Figure 4-9 looking at the hot blackbody through the cloud of cooler gas sees dark absorption lines where photons have been absorbed by the cooler gas.

Electrons in excited states are unstable. They lose energy by bumping into other particles and suddenly having either too much or too little energy than is necessary to be in that state. When this happens, the electron is forced to descend to a lower-energy allowed state, either the ground state or some allowed state between where it starts and the ground state. In either case, to make this transition down, the electron must *emit* a photon with energy once again equal to the difference between the energies of the starting and final allowed orbits (see Figure 4-12b).

The emitted photons have the same set of wavelengths as the absorbed photons. Furthermore, the emitted photons are sent out (that is, scattered) in all directions, which creates the glowing emission line spectra of hot objects. The color of a gas depends on the atoms and molecules that it contains. Hydrogen-rich gas clouds glow red because of the H_α emission (Figures 4-13 and 4-14a). Oxygen emits many green photons, so oxygen-rich gas clouds glow green (Figure 4-14b).

If an electron orbiting in a hydrogen atom encounters a photon with more than 13.6 eV, that photon is absorbed and knocks the electron completely out of orbit and away from the atom. This process is called *photoionization*. Each type of atom and molecule has a different photoionization energy, above which all electrons are kicked out of the atom or molecule. Photoionization occurs in stars and

Which Balmer line in Figure 4-11 is H_α?

interstellar nebulae (Figure 4-14b). Indeed, many spectral lines from stars and nebulae are those of ionized atoms that retain at least one electron.

4-7 Spectral lines shift due to the relative motion between the source and the observer

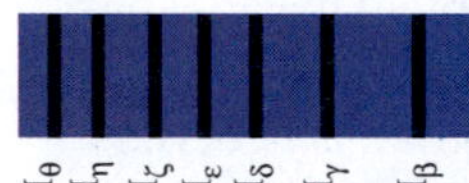

In 1842, Christian Doppler, a professor of mathematics in Prague, deduced that wavelength, hence color, is affected by motion. As objects move toward or away from you, they change color, in analogy to how the pitch of a siren changes as it passes you. As shown for the observer on the left in Figure 4-15, the wavelengths of electromagnetic radiation from an approaching source are compressed. The circles represent consecutive wave peaks (S_1 through S_4) emitted in all directions as the source moves along. Because each successive wave is sent out from a position slightly closer to her, she sees a shorter wavelength than she would if the source were stationary. All of the colors in the spectrum of an approaching source, such as a star, are therefore shifted toward the short-wavelength (blue) end of the spectrum, regardless of its distance. This phenomenon is called a **blueshift.**

Conversely, electromagnetic waves from a receding source are stretched out. The observer on the right in Figure 4-15 sees a longer wavelength than he would if the source were stationary. All of the colors in the spectrum of a receding source, regardless of its distance, are shifted toward the longer-wavelength (red)

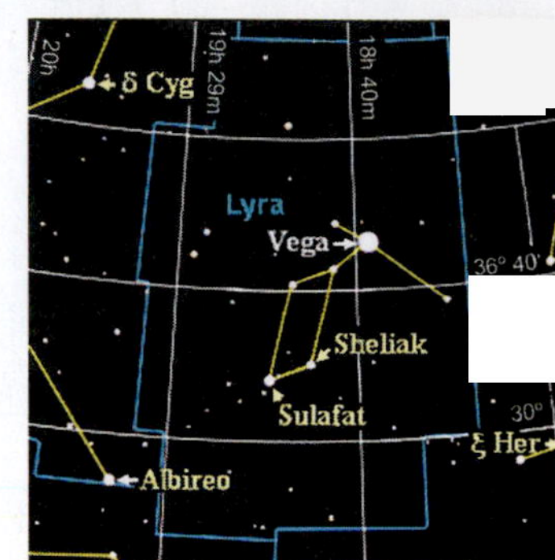

Shorter wavelength

H_θ H_η H_ζ H_ϵ H_δ H_γ H_β H_α

R I V U X G

FIGURE 4-13 Balmer Lines in the Spectrum of a Star This portion of the spectrum of the star Vega shows eight Balmer lines, from H_α at 656.3 nm through H_θ at 388.9 nm. (NOAO)

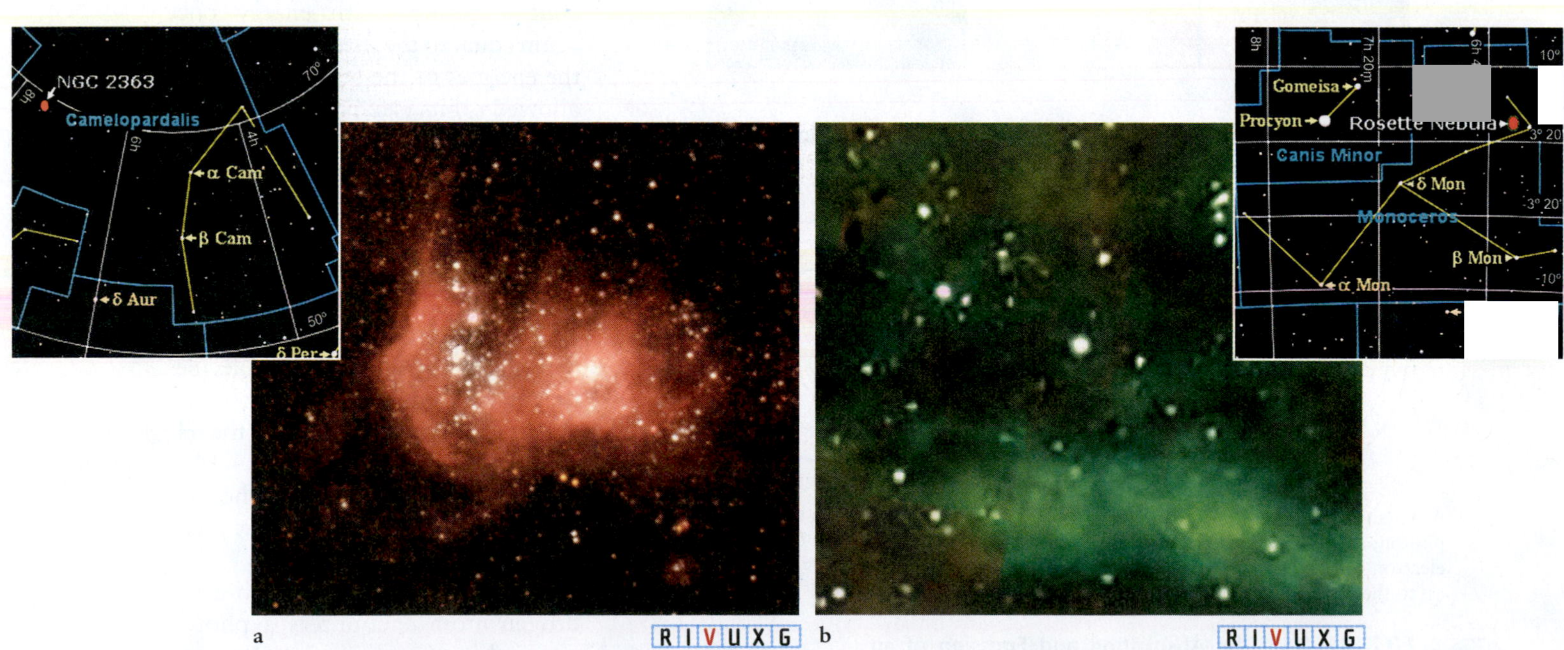

FIGURE 4-14 Emission Spectra from Interstellar Gas Clouds (a) Stars in this interstellar gas cloud (NGC 2363 in the constellation Camelopardus, the Giraffe) emit blackbody spectra. Electrons in the cloud's hydrogen gas absorb and reemit the red light from these stars. NGC 2363 is located some 10 million ly away. (b) Part of the Rosette Nebula (NGC 2237), an interstellar gas cloud in the constellation Monoceros (the Unicorn). The green glow is generated by doubly ionized oxygen atoms (O III; oxygen atoms missing two electrons) in the cloud that emit 501-nm photons. The Rosette is 3000 ly away. (a: L. Drissen, J.-R. Roy, and C. Robert/Département de Physique and Observatoire du Mont Mégantic, Université Laval; and NASA; b: T. A. Rector, B. Wolpa, M. Hanna, AURA/NOAO/NSF)

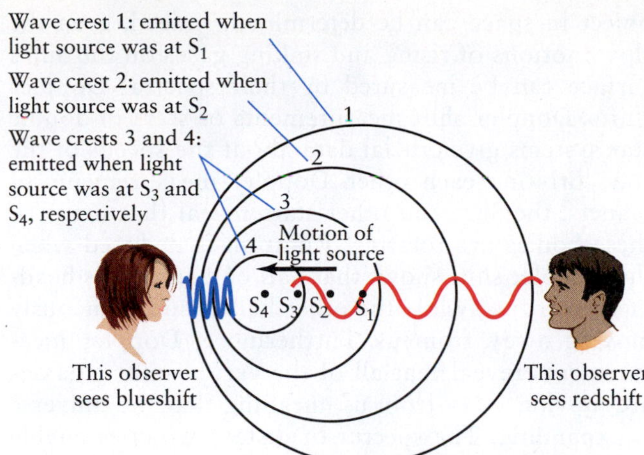

FIGURE 4-15 The Doppler Shift Wavelength is affected by motion between the light source and the observer. A source of light is moving toward the left. The four circles (numbered 1 through 4) indicate the location of light waves that were emitted by the moving source when it was at points S_1 through S_4, respectively. Note that the waves are compressed in front of the source but stretched out behind it. Consequently, wavelengths appear shortened (blueshifted) if the source is moving toward the observer and lengthened (redshifted) if the source is moving away from the observer. Motion perpendicular to the observer's line of sight does not affect wavelength.

end of the spectrum, producing a **redshift.** A blueshift or a redshift is also called a **Doppler shift.** The amount of Doppler shift varies directly with approaching or receding speed: When the speeds are small compared to the speed of light, an object that approaches twice as fast as another has all of its colors (or wavelengths) blueshifted twice as much as does the slower-moving object. An object moving away twice as fast as another object has its colors redshifted twice as much as the slower-moving object. Doppler shift applies to spectral lines as well. See *An Astronomer's Toolbox 4-3* for mathematical details of the Doppler shift.

The speed of an object toward or away from us is called its **radial velocity** (Figure 4-16) because the motion is along our line of sight or, put another way, along the "radius" drawn from Earth to the object as seen on the celestial sphere. Of course, the star or other object may also have a velocity perpendicular or transverse to our line of sight, like a car passing a pedestrian. Both the radial velocity and the **transverse velocity** are measured in kilometers per second.

In three dimensions, what shape will the circular wave crests (as shown in Figure 4-15) have?

Astronomers cannot measure the transverse velocity directly. Instead they measure the angle that the object moves among the stars on the celestial sphere as seen from Earth. This angle is called its **proper motion** and is often measured in arcseconds per year or arcseconds per century. The star with the greatest proper motion as seen from Earth is Barnard's star (Figure 4-17). Proper motion does not affect the perceived wavelength and cannot be determined by Doppler shift. This angle must be measured by comparing the locations of objects with higher proper motion to the

Does a police siren approaching you sound higher or lower in pitch than the siren at rest relative to you?

An Astronomer's Toolbox 4-3

The Doppler Shift

Suppose that λ_0 (lambda naught) is the wavelength of a spectral line from a stationary source. This is the wavelength that a reference book would list or a laboratory experiment would yield. If the source is moving toward or away from you, this particular spectral line is shifted to a different wavelength, λ. The size of the wavelength shift is usually written as $\Delta\lambda$ (delta lambda), where $\Delta\lambda = \lambda - \lambda_0$. This $\Delta\lambda$ is the difference between the wavelength that you actually observe in the spectrum of a star or galaxy and the wavelength listed in astronomy reference books.

Doppler proved that the wavelength shift is governed by the simple equation

$$\Delta\lambda / \lambda_0 = v/c$$

where v is the speed of the source toward or away from the observer. As usual, c is the speed of light (3×10^5 km/s).

Example: The spectral lines of hydrogen appear in the spectrum of the bright star Vega, as shown in Figure 4-13. The hydrogen line H_α has a normal wavelength of 656.285 nm, but in Vega's spectrum the line is located at 656.255 nm. The wavelength shift is −0.030 nm. The minus sign indicates that the star is moving toward us, so the star is approaching us with a speed of −14 km/s.

Try these questions: How fast would a star be moving toward us if all of its wavelengths shifted to $^9/_{10}$ of their rest wavelengths? If a star is moving away from us at 100 km/s, what is the change, $\Delta\lambda/\lambda_0$? If a star is moving toward us at $0.1c$, what is the change, $\Delta\lambda/\lambda_0$, in its wavelengths?

(Answers appear at the end of the book.)

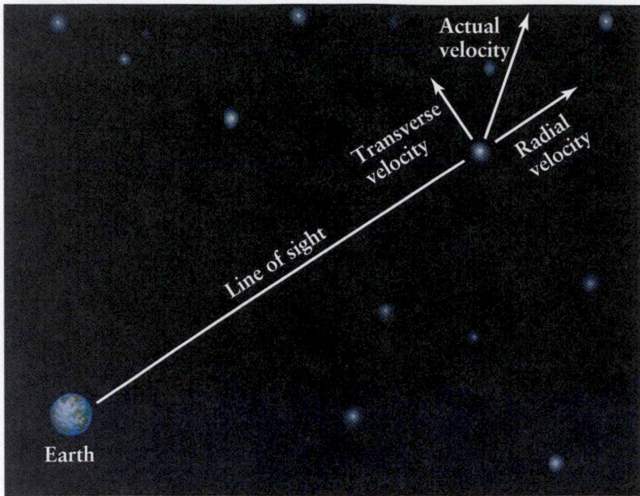

FIGURE 4-16 Radial and Transverse Velocities of a Star The actual velocity of a star (or other object) can be separated into radial and transverse motions. The speed of the star toward or away from Earth in kilometers per hour is its radial velocity. This motion creates a Doppler shift in the star's spectrum. The speed in kilometers per hour perpendicular to the radial velocity is called its transverse velocity, which does not affect the star's spectrum. The *angular* motion of the star across the celestial sphere (that is, in the direction of its transverse velocity) is called proper motion.

positions of objects with lower proper motions. Planets moving among the relatively fixed stars on the celestial sphere are examples of bodies that have large proper motions.

Stellar proper motions are so small that they can be measured for only relatively nearby stars in our Galaxy. However, the radial velocity of virtually every object in space can be determined. Indeed, even the slow motions of rising and sinking gases on the Sun's surface can be measured by their spectral Doppler shifts. Doppler shift measurements of stars in double star systems give crucial data about the speeds of the stars orbiting each other. Doppler measurements of planets, the Sun, and other stars reveal that many of these bodies are rotating. Rotation is deduced when the Doppler shift shows that half of a surface is heading toward us, while the other half is simultaneously moving away from us. Furthermore, Doppler measurements reveal that all of the very distant galaxies are moving away from us, meaning that the universe is expanding. The spectra of distant galaxies enable us to determine the rate of that expansion. In later chapters, we will refer to the Doppler shift whenever we need to convert an observed wavelength shift into a speed toward or away from us.

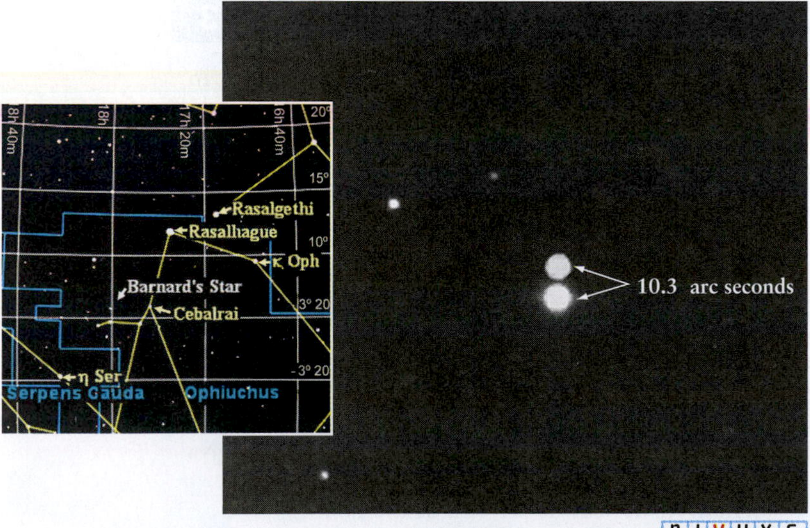

FIGURE 4-17 Proper Motion of Barnard's Star Two images of Barnard's star, taken a year apart in 2000 and 2001, show the proper motion of the star during that time. In addition to having the largest known proper motion (10.3″ per year), Barnard's star is one of the closest stars to Earth.

Insight Into Science

Remote Science Astronomical objects are so remote and their activities are often so complex that astrophysicists must use a tremendous amount of physics to interpret observations. For example, a single spectrum can contain information about stars, extrasolar planets, interstellar gas, Earth's motion and atmosphere, and the performance of the observing telescope and the equipment attached to it. All of these factors must be understood theoretically and accounted for.

With knowledge of spectroscopy, atomic and nuclear physics, and the Doppler shift, we can now list many of the vital properties of matter in space that are available to us. These properties include the chemical compositions of stars, interstellar gas clouds, and other objects; the temperatures of these objects; the rotation of objects in space; the motion of objects toward or away from Earth; the presence of planets and dim companion stars; and the rate at which the universe is expanding. Knowing these properties, astrophysicists have developed models that provide numerous insights into the ages, masses, distances, internal activity, rotation rates, and companion objects of stars, among other things.

Modern physics was born when Newton set out to understand the motions of the planets. Two and a half centuries later, scientists—including Maxwell, Planck, Einstein, Curie, Rutherford, and Doppler, among many others—discovered the basic properties of electromagnetic radiation and the structures of atoms. As we will see in the following chapters, the fruits of their labors have important implications for astronomy even today.

4-8 Frontiers yet to be discovered

Spectra provide much of what we know about the cosmos. Spectra of objects in space continue to yield new insights into the chemical composition of the planets, moons, local space debris, stars, interstellar gas and dust, and galaxies. Furthermore, as we obtain more accurate Doppler measurements of objects in our Galaxy, in other nearby galaxies, and, indeed, of entire galaxies, we are better able to determine the rate at which the objects in the universe are moving relative to one another. The more we understand of the chemistries and motions of objects in space, the more accurately we are able to explain the evolution of the solar system, stars, galaxies, and the universe.

SUMMARY OF KEY IDEAS

• By studying the wavelengths of electromagnetic radiation emitted and absorbed by an astronomical object, astronomers can learn about the object's temperature, chemical composition, rotation rate, companion objects, and movement through space.

Blackbody Radiation

• A blackbody is a hypothetical object that perfectly absorbs electromagnetic radiation at all wavelengths. The relative intensities of radiation that it emits at different wavelengths depend only on its temperature. Stars closely approximate blackbodies.

• Wien's law states that the peak wavelength of radiation emitted by a blackbody is inversely proportional to its temperature—the higher its temperature, the shorter the peak wavelength. The intensities of radiation emitted at various wavelengths by a blackbody at a given temperature are shown as a blackbody curve.

• The Stefan-Boltzmann law shows that a hotter blackbody emits more radiation at every wavelength than does a cooler blackbody.

Discovering Spectra

• Spectroscopy—the study of electromagnetic spectra—provides important information about the chemical composition of remote astronomical objects.

• Kirchhoff's three laws of spectral analysis describe the conditions under which absorption lines, emission lines, and a continuous spectrum can be observed.

• Spectral lines serve as distinctive "fingerprints" that identify the chemical elements and compounds comprising a light source.

Atoms and Spectra

• An atom consists of a small, dense nucleus (composed of protons and neutrons) surrounded by electrons. Atoms of different elements have different numbers of protons, while different isotopes have different numbers of neutrons.

• Quantum mechanics describes the behavior of particles and shows that electrons can only be in certain allowed orbits around the nucleus.

• The nuclei of some atoms are stable, while others (radioactive ones) spontaneously split into pieces.

• The spectral lines of atoms of a particular element correspond to the various electron transitions between allowed orbits of that element with different energy levels of those atoms. When an electron shifts from one energy level to another, a photon of the appropriate energy (and hence a specific wavelength) is absorbed or emitted by the atom.

• The spectrum of hydrogen at visible wavelengths consists of part of the Balmer series, which arises from electron transitions between the second energy level of the hydrogen atom and higher levels.

• Every different element, isotope, and molecule has a different set of spectral lines.

• When a neutral atom loses or gains one or more electrons, it is said to be charged. An atom loses an electron when the electron absorbs a sufficiently energetic photon, which rips it away from the nucleus.

• The motion of an object toward or away from an observer causes the observer to see all of the colors from the object blueshifted or redshifted, respectively. This effect is generically called a Doppler shift.

• The equation that describes the Doppler effect states that the size of a wavelength shift is proportional to the radial velocity between the light source and the observer.

A WHAT DID YOU THINK?

1 *Which is hotter, a "red-hot" or a "blue-hot" object?* Of all objects that glow visibly from heat generated or energy stored inside them, those that glow red are the coolest.

2 *What color does the Sun emit most brightly?* The Sun emits all wavelengths of electromagnetic radiation. The colors it emits most intensely are in the blue-green part of the spectrum. Because the human eye is less sensitive to blue-green than to yellow, and because Earth's atmosphere scatters blue-green wavelengths more readily than longer wavelengths, we normally see the Sun as yellow.

3 *How can we determine the age of space debris found on Earth?* We measure how much the long-lived radioactive elements, such as ^{238}U, have decayed in the object. Carbon dating is only reliable for organic materials that formed within the past 100,000 years.

It cannot be used for determining the age of rocks and minerals on Earth or from space. These substances were formed more than 4.5 billion years ago. Radioactive carbon in them has long since decayed to stable isotopes.

Key Terms for Review

absorption line, 111
absorption line spectrum, 114
atom, 116
atomic number, 116
blackbody, 107
blackbody curve, 107
blueshift, 120
continuous spectrum (continuum), 114
diffraction grating, 113
Doppler shift, 121
electromagnetic force, 116
electron, 116
element, 112
emission line, 112
emission line spectrum, 114
energy flux, 109
excited state, 118
ground state, 118
ion, 117
ionization, 117
isotope, 117
Kirchhoff's laws, 114
luminosity, 109
molecule, 117
neutron, 116
nucleus (of an atom), 116
periodic table, 112
Planck's law, 111
proper motion, 121
proton, 116
quantum mechanics, 117
radial velocity, 121
radioactive, 116
redshift, 124
spectral analysis, 112
spectrograph, 113
spectroscope, 124
Stefan-Boltzmann law, 108
strong nuclear force, 116
transition (of an electron), 117
transverse velocity, 121
weak nuclear force, 116
Wien's law, 108

Review Questions

1. A blackbody glowing with which of the following colors is hottest? **a.** yellow, **b.** red, **c.** orange, **d.** violet, **e.** blue

2. Of the following photons, which has the lowest energy? **a.** infrared, **b.** gamma ray, **c.** visible light, **d.** ultraviolet, **e.** X ray

3. The spectrum of which of the following objects will show a blueshift? **a.** an object moving just eastward on the celestial sphere, **b.** an object moving just northward on the celestial sphere, **c.** an object moving directly toward Earth, **d.** an object moving directly away from Earth, **e.** an object that is not moving relative to Earth

4. What is a blackbody? What does it mean to say that a star appears almost like a blackbody? If stars appear to be like blackbodies, why are they not black?

5. What is Wien's law? How could you use it to determine the temperature of a star's surface?

6. What is the Stefan-Boltzmann law? How do astronomers use it?

7. Using Wien's law and the Stefan-Boltzmann law, state the changes in color and intensity that are observed as the temperature of a hot, glowing object increases.

8. What color will an interstellar gas cloud composed of hydrogen glow, and why?

9. What is an element? List the names of five different elements, and briefly explain what makes them different from each other.

10. How are the three isotopes of hydrogen different from each other?

11. Explain how the spectrum of hydrogen is related to the structure of the hydrogen atom.

12. Why do different elements have different patterns of lines in their spectra?

13. Explain why the Doppler shift tells us only about the motion directly along the line of sight between a light source and an observer, but not about motion across the celestial sphere.

14. What is the Doppler shift, and why is it important to astronomers?

Advanced Questions

The answers to all computational problems, which are preceded by an asterisk (*), appear at the end of the book.

***15.** Approximately how many times around the world could a beam of light travel in 1 s?

***16.** The bright star Regulus in the constellation of Leo (the Lion) has a surface temperature of 12,200 K. Approximately what is the dominant wavelength (λ_{max}) of the light that it emits?

***17.** The bright star Procyon in the constellation of Canis Minor (the Little Dog) emits the greatest intensity of radiation at a wavelength $\lambda_{max} = 445$ nm. Approximately what is the surface temperature of the star in kelvins?

***18.** As observed from Earth, the wavelength of H_α in the spectrum of the star Megrez in the Big Dipper is 486.112 nm. Laboratory measurements demonstrate that the normal wavelength of this spectral line is 486.133 nm. Is the star coming toward us or moving away from us? At what speed?

***19.** In the spectrum of the bright star Rigel, H_α is observed to have a wavelength of 656.331 nm. Is the star coming toward us or moving away from us? How fast?

***20.** Imagine driving down a street toward a traffic light. How fast would you have to go so that the red light (700 nm) would appear green (500 nm)? What fraction of the speed of light c ($c = 300{,}000$ km/s) is this velocity?

Discussion Questions

21. Compare the technique of identifying chemicals by their spectral line patterns with that of identifying people by their fingerprints.

22. Suppose you look up at the night sky and observe some of the brightest stars with your naked eye. Is there any way of telling which stars are hotter and which are cooler? Explain your answer.

23. How could we exclude Earth's atmosphere as the source of the iron absorption lines in Figure 4-6?

What If...

*24. The Sun were twice its actual diameter but still had the same surface temperature? At what wavelength would that new Sun emit its radiation most intensely? How many times brighter would the new Sun be than the present Sun? How might things on Earth be different under the bigger Sun?

25. All of the nearby stars were observed to have redshifted spectra? What conclusions could we draw? What other measurement would be useful to have for each star, and why?

26. No stars had any Doppler shift? What would that say about stellar motions relative to the solar system? Is that possible? *Hint:* Consider Newton's law of gravitation from Chapter 3.

27. The spectrum of the Sun (Figure 4-4) had several absorption lines missing when taken from above Earth's atmosphere compared to its spectrum taken at Earth's surface? What would this indicate about the chemistries of the Sun and of Earth's atmosphere?

Web Questions

*28. **Blackbody Peak Colors** Search the Web for an interactive blackbody calculator that uses Wien's equation to give the peak wavelength as a function of temperature (try the keywords *blackbody spectrum calculator*). Noting that the visible spectrum runs from 400 nm (violet) to 700 nm (red), determine whether any of the following stars have a peak wavelength in the visible spectrum: Rigel, $T = 14{,}000$ K; Deneb, $T = 9500$ K; Arcturus, $T = 4500$ K; Vega, $T = 11{,}500$ K; Betelgeuse, $T = 3100$ K. *Hint:* You may have to convert length scales. Recall that 1 nm = 10^{-9} m.

Observing Projects

STARRY NIGHT

29. Use the *Starry Night*™ program to examine the temperatures of several relatively nearby stars. Select **Favourites > Discovering the Universe > Atlas.** Use the **Find** pane to locate each of the following stars: (i) Altair, (ii) Procyon, (iii) Epsilon Indi, (iv) Tau Ceti, (v) Epsilon Eridani, (vi) Lalande 21185. First, click the magnifying glass icon in the left margin of the edit box at the top of the **Find** pane and select **Star** from the drop-down menu. Next, type the name of the star in the edit box and then press the **Enter** (**Return**) key. Information for each star can then be found by clicking on the **Info** tab at the far left of the *Starry Night*™ window. For each star, record its temperature (listed in the **Info** pane under **Other Data**). Then answer the following questions: (a) Which of the stars has a longer wavelength of maximum emission λ_{max} than the Sun? (b) Which of the stars has a shorter λ_{max} than the Sun? (c) Which of the stars has a reddish color?

STARRY NIGHT

30. You can use *Starry Night*™ to examine the celestial objects listed below. Select **Favourites > Discovering the Universe > Atlas** to show the whole sky as would be seen from the center of a transparent Earth. Ensure that deep space objects are displayed by selecting **View > Deep Space > Messier Objects** and **View > Deep Space > Bright NGC Objects** from the menu. Also, select **View > Deep Space > Hubble Images** to turn this option off. Open the **Find** pane and then type the name of the object in the edit box followed by the **Enter** (**Return**) key. The object will be centered in the view. For each object, use the zoom controls to adjust your view until you can see the object in detail. For each object, decide whether you think it will have a continuous spectrum, an absorption line spectrum, or an emission line spectrum, and explain your reasoning. The list of objects to observe includes:

(a) The Lagoon Nebula in Sagittarius (With a field of view of about 6° × 4°, you can compare and contrast the appearance of the Lagoon Nebula with the Trifid Nebula just to the north of it.)

(b) M31, the great galaxy in the constellation Andromeda (*Hint:* The light coming from this galaxy is the combined light of hundreds of billions of individual stars.)

(c) The Moon (*Hint:* Moonlight is simply reflected sunlight.)

STARRY NIGHT

31. Use the *Starry Night*™ program to compare the brightness of two similar-size stars in the constellation Auriga. Select **Favourites > Discovering the Universe > Auriga.** The two stars, Capella and Delta Aurigae, are labeled in this view. Select **Preferences** from the **File** menu (Windows) or **Starry Night** menu (Macintosh) and set the HUD so that it includes **Temperature** and **Radius** in the display. You will notice that these two stars have the same radius but differ in temperature. From these data, which of these stars is intrinsically brighter and by what proportion?

PART II

Understanding the Solar System

R I V U X G A montage of the planets (plus Pluto, a dwarf planet) in our solar system presented in correct relative sizes. The orbits in the background are also drawn to scale. (NASA)

AN ASTRONOMER'S ALMANAC

Moveable Type in Use — American Revolution — Telegraph in Use — Telephone

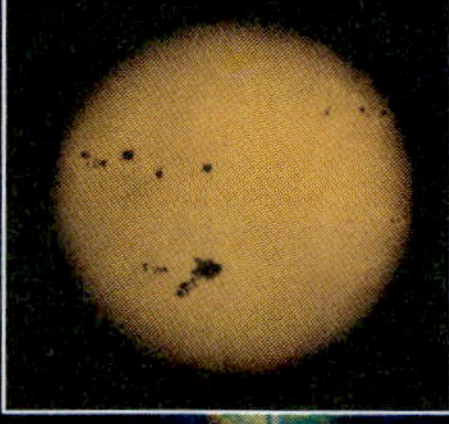

1610–1613 Galileo Galilei observes phases of Venus and major moons of Jupiter –– Io, Europa, Ganymede, and Callisto–– and uses observations of sunspots to demonstrate the Sun's rotation.

1655–1656 Jupiter's Great Red Spot discovered by Giovanni Cassini. Christiaan Huygens identifies Saturn's rings as rings, and discovers its largest moon, Titan.

1671 Giovanni Cassini calculates first relatively accurate estimate of distance between Earth and Sun.

1705–1757 Edmond Halley makes the first prediction of the period of a comet's orbit, calculating that the comet we now call Halley's would return in 1758. Comet's return toward inner solar system first observed by Johann Palitzsch.

1781–1784 William Herschel first to observe Uranus and to detect clouds around Mars.

1801 Giuseppi Piazzi discovers the asteroid Ceres.

1814 Joseph von Fraunhofer studies the Sun's spectrum and catalogs over 600 spectral lines.

1821 Alexis Bouvard detects irregularities in Uranus's orbit.

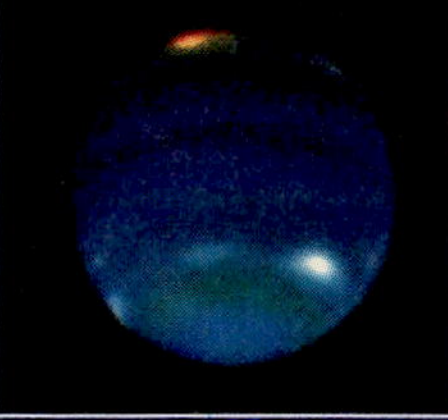

1843–1849 Sunspot cycle discovered. Neptune's presence predicted independently by John Adams and Urbain Leverrier. Neptune first observed by Johann Galle. Triton discovered by William Lassell. Edouard Roche shows that Saturn's rings did not form from a preexisting moon.

1856–1859 James Clerk Maxwell proves that Saturn's rings are not solid. Richard Carrington discovers solar flares.

1861–1866 Gustav Kirchhoff uses spectral analysis to determine the chemical composition of the Sun's atmosphere. Richard Carrington discovers the Sun's differential rotation. Giovanni Schiaparelli determines that meteor showers are caused by Earth passing through comet debris.

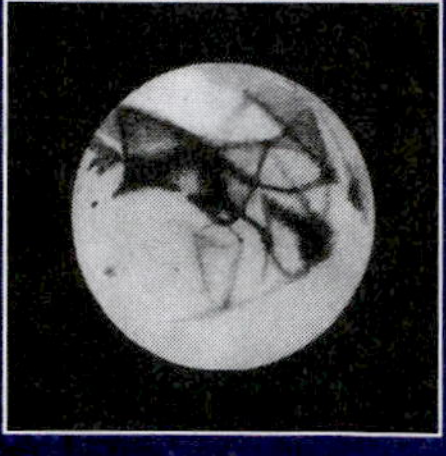

1877 Phobos and Deimos discovered by Asaph Hall. Giovanni Schiaparelli observes "canali" (Italian for "channels") on Mars.

1906 First Trojan asteroid, Achilles, discovered by Max Wolf. Limb darkening explained by Karl Schwarzschild.

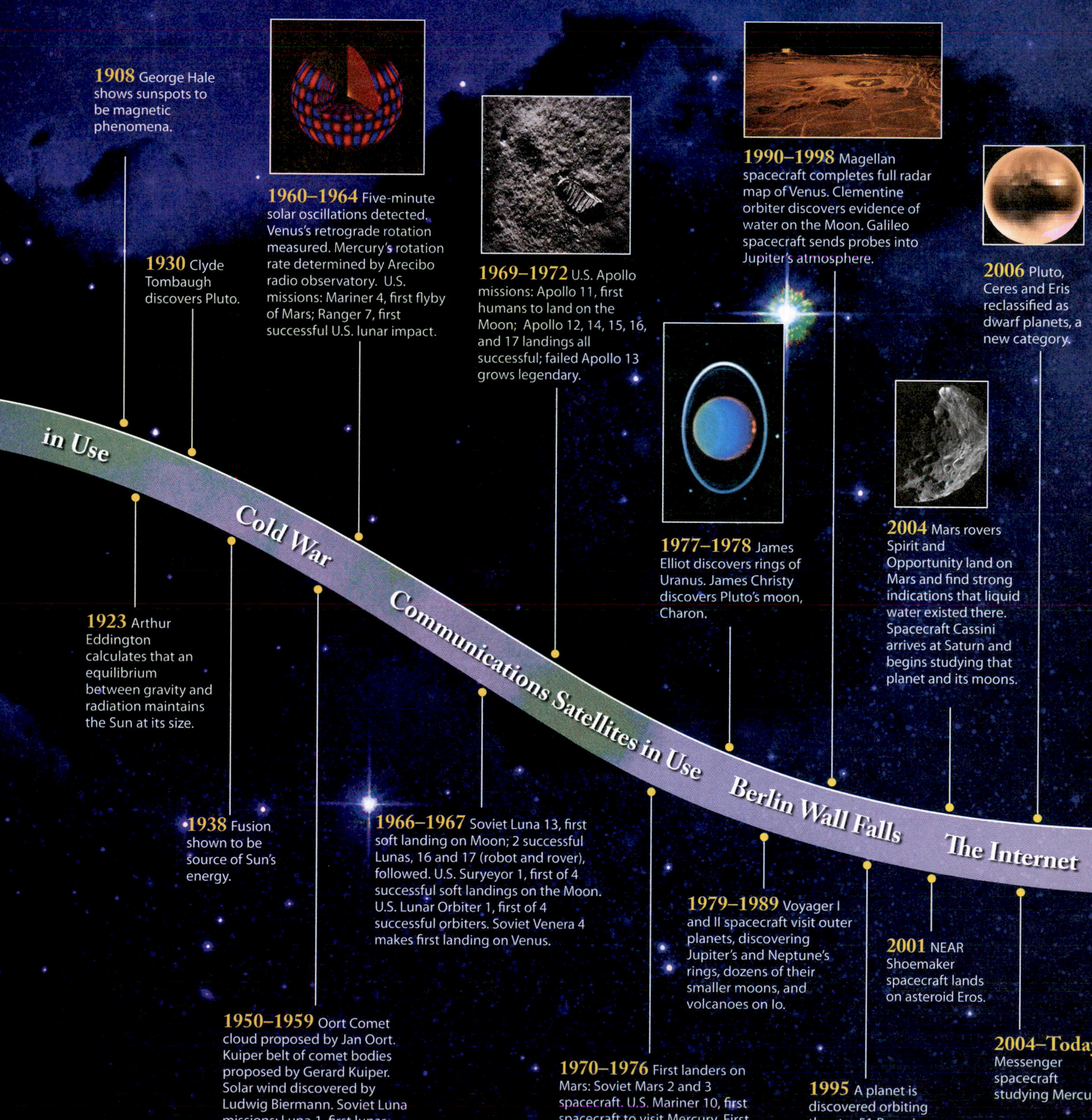

in Use
Cold War
Communications Satellites in Use
Berlin Wall Falls
The Internet
1908 George Hale shows sunspots to be magnetic phenomena.
1923 Arthur Eddington calculates that an equilibrium between gravity and radiation maintains the Sun at its size.
1930 Clyde Tombaugh discovers Pluto.
1938 Fusion shown to be source of Sun's energy.
1950–1959 Oort Comet cloud proposed by Jan Oort. Kuiper belt of comet bodies proposed by Gerard Kuiper. Solar wind discovered by Ludwig Biermann. Soviet Luna missions: Luna 1, first lunar flyby; Luna 2, first lunar impact; Luna 3, first photo of lunar far side.
1960–1964 Five-minute solar oscillations detected. Venus's retrograde rotation measured. Mercury's rotation rate determined by Arecibo radio observatory. U.S. missions: Mariner 4, first flyby of Mars; Ranger 7, first successful U.S. lunar impact.
1966–1967 Soviet Luna 13, first soft landing on Moon; 2 successful Lunas, 16 and 17 (robot and rover), followed. U.S. Surveyor 1, first of 4 successful soft landings on the Moon. U.S. Lunar Orbiter 1, first of 4 successful orbiters. Soviet Venera 4 makes first landing on Venus.
1969–1972 U.S. Apollo missions: Apollo 11, first humans to land on the Moon; Apollo 12, 14, 15, 16, and 17 landings all successful; failed Apollo 13 grows legendary.
1970–1976 First landers on Mars: Soviet Mars 2 and 3 spacecraft. U.S. Mariner 10, first spacecraft to visit Mercury. First U.S. landings on Mars by Viking 1 and 2 spacecraft.
1977–1978 James Elliot discovers rings of Uranus. James Christy discovers Pluto's moon, Charon.
1979–1989 Voyager I and II spacecraft visit outer planets, discovering Jupiter's and Neptune's rings, dozens of their smaller moons, and volcanoes on Io.
1990–1998 Magellan spacecraft completes full radar map of Venus. Clementine orbiter discovers evidence of water on the Moon. Galileo spacecraft sends probes into Jupiter's atmosphere.
1995 A planet is discovered orbiting the star 51 Pegasi, the first of over 500 extra solar planets detected so far.
2001 NEAR Shoemaker spacecraft lands on asteroid Eros.
2004 Mars rovers Spirit and Opportunity land on Mars and find strong indications that liquid water existed there. Spacecraft Cassini arrives at Saturn and begins studying that planet and its moons.
2004–Today Messenger spacecraft studying Mercury.
2006 Pluto, Ceres and Eris reclassified as dwarf planets, a new category.

WHAT DO YOU THINK?

1 How many stars are there in the solar system?

2 Were the Sun and planets among the first generation of objects created in the universe?

3 How long has Earth existed, and how do we know this?

4 What typical shape(s) do moons have, and why?

5 Have any Earthlike planets been discovered orbiting Sunlike stars?

Answers to these questions appear in the text beside the corresponding numbers in the margins and at the end of the chapter.

R I V U X G The Orion Nebula, 1500 ly away, as imaged by the Hubble Space Telescope. Part of the much larger Orion Molecular Cloud, the Orion Nebula contains over 3000 stars and many more in the process of forming. The bright central region is the home of the Trapezium, four young, bright stars that cause the surrounding gas and dust to glow. The bright region in the upper left, denoted M43, is glowing due to the radiation from one young, massive star. (NASA, ESA, M. Robberto (STScI/ESA), and the HST Orion Treasury Project Team)

Chapter 5

Formation of the Solar System and Other Planetary Systems

1 The Sun and all bodies that orbit it—planets, moons, and various kinds of debris (called asteroids, comets, and meteoroids)—make up the **solar system**, the formation of which would make quite a dramatic movie. The story began quietly, with lots of small encounters between vast numbers of bit players (tiny gas and dust particles). Most of them quickly plunged into the center of the theater and together they formed the Sun, the star of the show. Plot lines developed as the remaining matter interacted. Things started to heat up as matter came together in progressively more violent collisions and near misses. Particles clumped into pebbles, rocks, and mountain-sized pieces of rubble. Sometimes big chunks of debris smashed each other to bits. Eventually, the heavyweights (in this case, bodies destined to become planets and large moons) managed to wipe out most of the smaller players. The drama also led to the evolution of life, a subplot featuring the third rock from the Sun.

Yes, *The Making of the Solar System* would be a great film—except that it would be in "slo-mo." Just to get to the part where the planets are well established in their roles, you would have to watch solar system formation for about a hundred million years. In addition, because astronomers and geologists are continually making new discoveries about our solar system, the script would have to be continually revised and scenes reshot. In fact, several times over the years, astronomers have decided that some bodies thought to be major actors really just play minor roles after all. Perhaps the most interesting way to view such a movie would be to compare it to films that show the formation of planetary systems around other stars, an area of intense research today.

In this chapter you will discover

- how the solar system formed
- why the environment of the early solar system was much more violent than it is today
- how astronomers define the various types of objects in the solar system
- the relationships between planets, dwarf planets, small solar system bodies, and other classifications of objects in the solar system
- how and why the planets are grouped as they are
- how the moons formed throughout the solar system
- what the debris of the solar system is made of
- that disks of gas and dust, as well as planets, have been observed around a growing number of stars
- that newly forming stars and planetary systems are being discovered every year

THE SOLAR SYSTEM CONTAINS HEAVY ELEMENTS, FORMED FROM AN EARLIER GENERATION OF STARS

How did the solar system form? How were its varied building blocks of rock, metal, ice, and gas created? How has the solar system changed since its formation, and what does its history tell us about the planets we see today? Within the past few decades, telescopes and space probes, along with the theories of modern science and computer simulations of these theories, have finally provided some of the answers to these age-old questions.

5-1 Stars transform matter from lighter elements into heavier ones

Our study of the formation of the solar system begins with an inconsistency, namely that Earth, the Moon, and many other objects that orbit the Sun are composed primarily of heavy elements including (among many others) oxygen, silicon, aluminum, iron, carbon, and calcium. (Note that we capitalize the word *moon* only when we are referring to Earth's Moon.) Furthermore, most of these bodies contain extremely little hydrogen and helium. However, observations of the spectra of the Sun, other stars, and interstellar clouds reveal that hydrogen and helium are by far the most abundant elements in the universe. These two elements account for about 99.9% of all observed atoms (or, equivalently, 98% of observed mass). All of the other elements combined account for about 0.1% of the observed atoms (and, hence, 2% of the observed mass of the universe). How is it that Earth, the Moon, Mars, Venus, Mercury, and many smaller bodies in the solar system typically contain less than 0.15% hydrogen and helium? Somehow, these familiar astronomical objects formed from matter that had been enriched with the heavier elements and depleted in hydrogen and helium.

There is a good reason for the overwhelming abundance of hydrogen and helium throughout the universe. Astronomers believe that the universe formed about 13.75 billion years ago in a violent event called the *Big Bang*. This event created not only all the matter and energy that exists, but also space and time. (We explore these issues in Chapter 18.) Only the lightest elements—hydrogen, helium, and a tiny amount of lithium—were created in the process. These first elements came into existence as gases. The first stars, composed only of these three elements, condensed out of this primeval matter

FIGURE 5-1 How Stars Lose Mass (a) The brightest star in Scorpius, Antares, is nearing the end of its existence. Strong winds from its surface are expelling large quantities of gas and dust, creating this nebula reminiscent of an Impressionist painting. The scattering of starlight off this material makes it appear especially bright, even at a distance of 604 ly. (b) The planetary nebula Abell 39 is 7000 ly from Earth. With a relatively gentle emission of matter, the central star shed its outer layers of gas and dust in an expanding spherical shell now about 6 ly across. (c) A supernova is the most powerful known mechanism for a star to shed mass. The Crab Nebula, even though it is about 6000 ly from Earth, was visible during the day for 3 weeks during 1054. (a: David Malin/Anglo-Australian Observatory; b: George Jacoby [WIYN Obs.] et al., WIYN, AURA, NOAO, NSF; c: Malin/Pasachoff/Caltech)

within a few hundred million years after the Big Bang occurred.

2 The difference in chemical composition between the gas and stars in the early universe and the solar system today shows that the solar system did not form as a *direct* result of the Big Bang. Rather, as we shall see shortly, the solar system formed some 9 billion years after the universe came into existence. Let us briefly explore the chemical transformations that led to our being here, which began deep inside those first stars. Once stars form, gravity compresses the matter in their central *cores* so much that the hydrogen there is transformed into helium in a process called *fusion.* Hydrogen fusion has the interesting property that some of the hydrogen's mass is converted into electromagnetic energy (photons), a portion of which eventually leaks out through the star's surface, enabling it to shine. The core's helium subsequently fuses to create carbon, also converting some mass into energy. If the star has enough pressure in its core, the carbon transforms into even heavier elements, such as nitrogen, oxygen, neon, silicon, and iron. We explore details of the creation of these heavier elements in Chapters 12 and 13.

Stars also shed matter into space (Figure 5-1). Some of the heavy elements formed inside them are eventually ejected, along with most of the hydrogen and helium remaining in their outer layers. The outer layers of stars are expelled at different rates, ranging from continuous outflows called *stellar winds* (Figure 5-1a), to the more energetic *planetary nebulae* (so named not because they have any direct connections to real planets, but rather because they looked like planets in early, low-resolution telescopes) (Figure 5-1b), to spectacular detonations called *supernovae* (Figure 5-1c). Supernovae are such powerful events that fusion occurs during the explosion, creating even denser elements that are also expelled from the star. Planetary nebulae and supernovae leave only tiny, dim, but very hot stellar cores, remnants of once-mighty stars.

Of the matter ejected from stars, some leaves as small particles of dust, such as the soot formed in a fire or in diesel exhaust. The dust is typically a millionth of a meter (a micron) in size. Over time, enough gas and dust were emitted by enough stars to form interstellar clouds rich in hydrogen and helium, but which also contained small amounts of metals. (In a curious alternative use of a common word, astronomers define **metals** to be all of the elements in the universe other than hydrogen and helium.) Some of the atoms in these clouds also combine to form molecules of, for example, water, carbon dioxide, methane, and ammonia. We will turn next to these clouds in order to understand the formation of the solar system and why dense planets like Earth are so rich in metals and poor in light elements.

5-2 Gravity, rotation, collisions, and heat shaped the young solar system

Observations of clouds in which stars are forming today, such as the nearby Orion Molecular Cloud (Figure 5-2 and the image that opened this chapter), reveal that most stars form inside interstellar clouds of gas and dust. Furthermore, typically stars form in groups called *open clusters* that contain between a few and a few thousand stars. It is likely that the solar system formed in such a cluster. (Stars in open clusters eventually drift apart, which is why even the closest stars to the solar system are light-years away from us.)

Fragments of such a cloud do not spontaneously collapse to form stars because the particles in the cloud are moving fast enough (meaning they are warm enough) to avoid being pulled together by their mutual gravitational attraction. Another way to say this is that the gas comprising the cloud has a high enough internal pressure to prevent the collapse. Pressure is the force that the gas exerts on any area of either itself or anything else (think of the pressure that the gas in a balloon exerts on the surface of the balloon to keep it inflated).

The conditions of temperature, pressure, and density under which a fragment of a gas cloud is forced to collapse under the influence of its own gravitational attraction are referred to as **Jeans instability,** after the British physicist Sir James Jeans, who calculated them in 1902. Pieces of interstellar clouds become "Jeans unstable," allowing stars and their hosts of orbiting bodies to form, for at least three reasons. First, winds from nearby stars compress gas and dust in the cloud. Second, the explosive force of a nearby supernova can compress regions of gas and dust. Third, pairs of clouds collide and compress each other. Each collapsing fragment destined to become a star (or a pair of stars orbiting each other called a *binary star system*) and possibly planets is called a **dense core.**

Some chemical evidence in space debris that we have analyzed on Earth suggests that the dense core from which our solar system formed may have become Jeans unstable when its gas and dust was hit by the shock wave from a supernova. We call that dense core the **solar nebula** (also called the proto-solar or pre-solar nebula).

Many of the concepts in what follows comprise the **Nice model** (pronounced like "niece" and named after the city in France where it was first developed) of the formation of the solar system. *We are in an epoch of*

FIGURE 5-2 Dusty Regions of Star Formation (a) These three bright young stars in the constellation Monoceros are still surrounded by much of the gas and dust from which they formed. These stars and dust make up a tiny part of a much larger cloud, known as the Cone Nebula. Astronomers hypothesize that the solar system formed from a similar fragment of an interstellar gas and dust cloud. (b) Newly formed stars in the Orion Nebula. Although visible light from many of the stars is blocked by the nebula, their infrared emission travels through the gas and dust to us. (a: ACS Science & Engineering Team, NASA; b: NASA; K. L. Luhman/Harvard-Smithsonian Center for Astrophysics and G. Schneider, E. Young, G. Rieke, A. Cotera, H. Chen, M. Rieke, R. Thomson/Steward Observatory, University of Arizona, Tucson, AZ)

astronomical research in which insights into the process by which the solar nebula transformed into the Sun and its host of orbiting bodies are coming at an unprecedented rate. The details of this theory are tentative and are likely to evolve as simulations become more realistic. Therefore, the following scenario is a prime example of how science is a work in progress, rather than a presentation of complete understanding of nature.

Essentially, the Nice model proposes that in the outer solar system Jupiter formed first, followed by Saturn, and then by Neptune and Uranus. The latter two planets were then flung out to their present orbits by gravitational forces from Jupiter and Saturn. This model is based on computer simulations of the evolution of the outer regions of the solar nebula.

Insight Into Science

Computer-Aided Analysis Many equations are so complex that they require computer analysis to enable us to understand their implications. For example, the study of the physics of the formation of the planets and the Sun began with observations, which then led to refined computer models.

The solar nebula initially had a diameter of at least 1000 AU (1000 times the average distance from Earth to the Sun) and a total mass about 2 to 3 times the mass of the Sun (Figure 5-3). At first, that nebula was a very cold collection of gas and dust—well below the freezing point of water. Although most of the solar nebula was hydrogen and helium, ice and ice-coated dust grains composed of heavy elements were scattered across this vast volume. Deep inside the nebula, gravitational attraction caused gas and dust to fall rapidly toward its center. (In terms of the physics presented in Chapter 2, the cloud's gravitational potential energy was being converted into kinetic energy—energy of motion.) As a result, the density, pressure, and temperature at the center of the nebula began to increase, producing a concentration of matter called the **protosun.**

As the protosun continued to increase in mass and to contract, atoms within it collided with increasing speed and frequency. Such collisions created heat, causing the protosun's temperature to soar. Therefore, the first heat and light emitted by our Sun came from colliding gas, not from nuclear fusion, as it does today. At the same time, there was also considerable activity in the outer regions of the disk.

Rotation played a key role in the formation of the solar system. When it first developed, the solar nebula was a very slowly rotating ensemble of particles. This rotation occurred because the formation of interstellar clouds is turbulent, so the clouds are created with many slowly swirling regions, like smoke rising from a fire. As a result of this rotation, the solar nebula had angular momentum (see *An Astronomer's Toolbox 2-2*), which prevented its outer regions from collapsing into the protosun. (A nonswirling collapse would have created a star without any planets or other orbiting matter.) Because angular momentum is conserved (see Section 2-7), rotating matter falling inward in the solar nebula revolved faster and faster, like the skater in Figure 2-12 when she pulls her arms in. Even the debris that formed the protosun was orbiting as it fell inward, like water spiraling before it goes down a drain. Therefore, the protosun was spinning and the Sun that formed from it rotates. We will discuss the Sun's rotation further in Chapter 10.

The outer parts of the solar nebula collapsed inward more slowly than the central matter that formed the protosun. This outer region was probably very ragged, but mathematical studies show that the combined effects of gravity, collisions, and rotation would transform even an irregular dense core into a rotating disk with a warm center and outer cold edge about 30 AU from the protosun (see Figure 5-3). You can see an analogous effect of rotation in the way the ice skater's dress in Figure 2-12 is forced into a plane as she spins.

Although we cannot see our solar system as it was before planets formed, astronomers have found what we think are similar disks of gas and dust surrounding other young stars, including those shown in Figure 5-4. Called **protoplanetary disks** or **proplyds**, these systems are undergoing the same initial stage of evolution that we have described here for our solar system.

As occurs with all gases, the protosun's temperature increased as it became denser. As the protosun radiated more and more heat, the temperature around it began to increase. The protosun's increasing temperature vaporized all common icy substances (including carbon dioxide, water, methane, and ammonia) in the inner region of the disk. These gases then blew away, along with the abundant hydrogen and helium gas that also initially existed there. Within a hundred thousand years or so of the disk forming, no gases or ices orbited closer than roughly 3 AU, between what are now the orbits of Mars and Jupiter. Temperatures were low enough beyond this distance that gases, primarily hydrogen and helium, and ices remained in orbit around the protosun. The boundary beyond which these gases and ices of water, carbon dioxide, methane, and ammonia, persisted is called the **snow line** (or *frost line*). Within a million years of when the Jeans instability occurred, the solar system had differentiated into an inner region containing just dust grains (that are not easily vaporized) and an outer region rich in hydrogen gas, helium gas, and ice-covered dust particles.

At first, neighboring gases and tiny dust grains, colliding with each other beyond the snow line, orbited the Sun in tandem. In other words, both the gas and dust particles dramatically changed directions as they collided, some moving faster, others slower as a result of these collisions, but on average both groups of particles

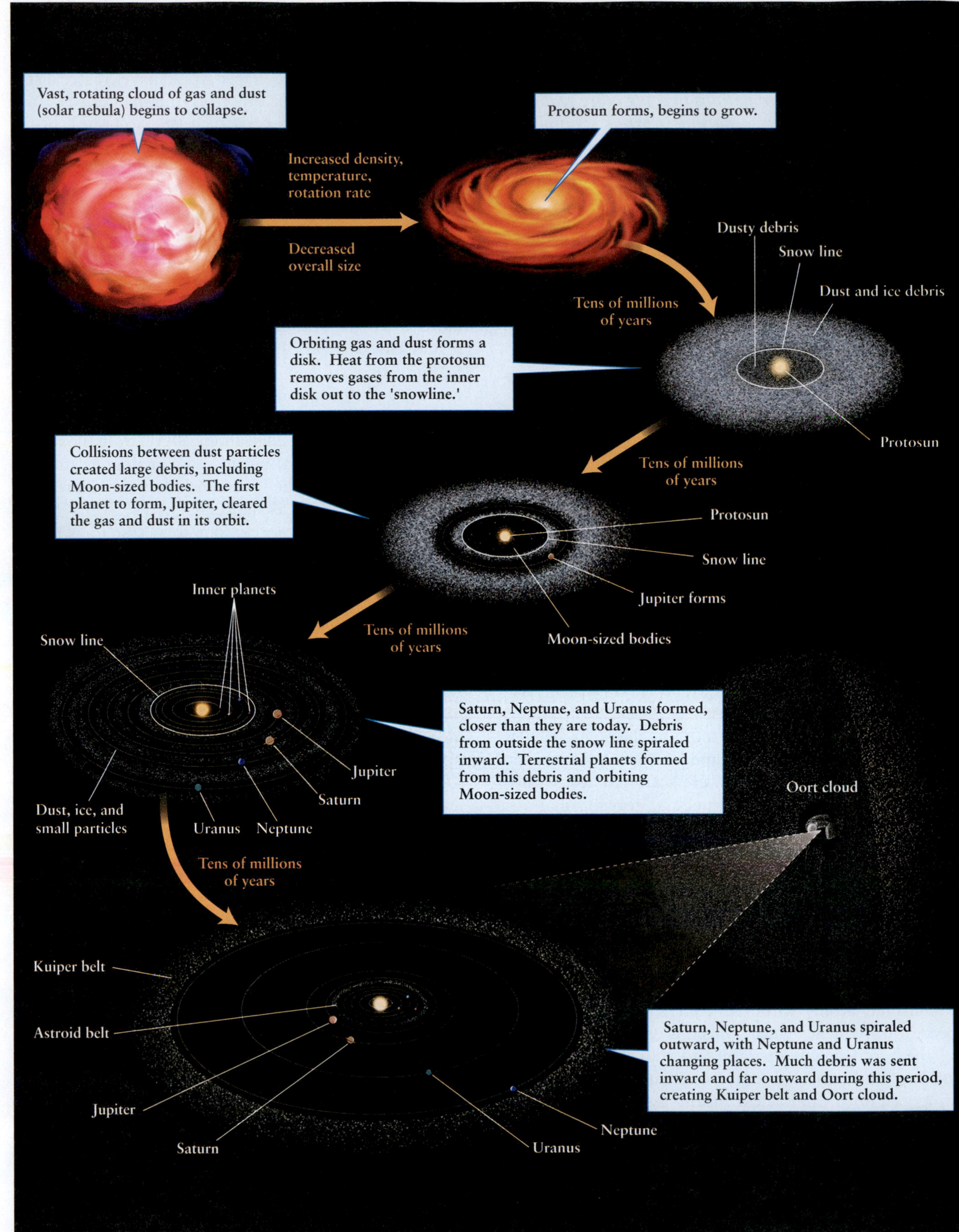

FIGURE 5-3 The Formation of the Solar System This sequence of drawings shows stages in the formation of the solar system.

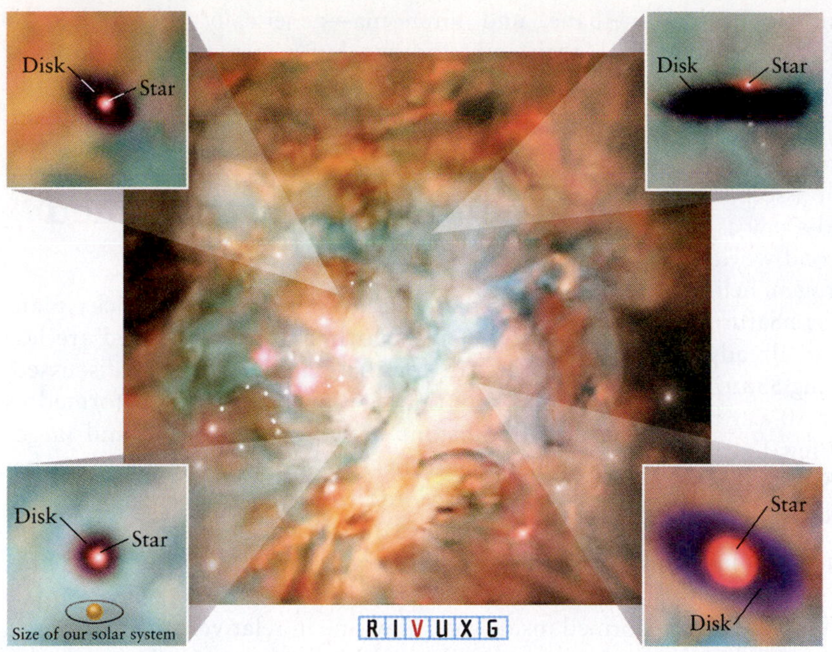

VIDEO 5.1 **FIGURE 5-4 Young Circumstellar Disks of Matter** The heart of the Orion Nebula as seen through the Hubble Space Telescope. The four insets are false-color images of protoplanetary disks within the nebula. A recently formed star is at the center of each disk. The disk in the upper right is seen nearly edge-on. Our solar system is drawn to scale in the lower left image. (C. R. O'Dell and S. K. Wong, Rice University; NASA)

orbited at the same speed. These orbits changed, however, as the dust grains collided with each other, coalesced, and thereby grew larger and larger. When the dust particles reached sizes of a few millimeters, they began plowing through the surrounding gas. As the dust particles pushed aside the gas, the gas took energy away from the dust particles (much as you lose energy walking into a headwind). As a result, this dust spiraled inward toward the protosun. The collision and **accretion**—the coming together of smaller pieces of matter to form larger ones—of dust particles continued as they spiraled inward. Through accretion, the dust grew into pieces of debris meters across. Calculations show that a piece of debris a meter across could have spiraled inward to half its original distance from the protosun in just 1000 years! Collisions destroyed some of these bigger bodies, while some of them merged and grew even larger. Eventually, a few billion of these boulders reached dimensions of a kilometer or so, large enough for their mutual gravitational attractions to enhance the rate at which they collided. Such pieces of debris are called **planetesimals**.

The formation of planetesimals was a turning point in the young solar system's evolution. They collided with countless dust particles as well as with each other. Sometimes colliding planetesimals destroyed each other, and returned to myriad pieces of dust. Eventually, however, these collisions led to the growth of several hundred substantial bodies that persisted and eventually reached the size and mass of our Moon. At this size, the gravitational attraction of the planetesimals was strong enough for them to either sweep up the surrounding rubble or to fling it far away, thereby clearing out nearby space and limiting their own growth.

As some Moon-sized planetesimals merged, they began to pull onto themselves the remaining gas beyond the snow line. In this way, the giant planets formed—and the most gigantic, Jupiter, formed first.

What force ensured that once the solar nebula began collapsing, it continued to do so?

5-3 The giant planets formed in sequence

Jupiter Many planetesimals were formed in substantially noncircular orbits. As a result, they crossed paths and a few of them collided and merged with each other, creating larger bodies. One planetesimal located 1 or 2 AU beyond the snow line accreted enough of its companions to become an Earthlike body containing perhaps 10 Earth masses of rock and metal. This body had enough gravitational attraction to pull onto itself vast amounts of the hydrogen, helium, and ices that existed near its orbit. Within only a few thousand years of forming, this object grew into the planet Jupiter. Like the protosun, infalling gases heated Jupiter up; in fact, the planet was heated so much that for a short time it outshone the protosun.

As it grew, Jupiter stopped migrating inward toward the protosun, and its gravity stirred up the surrounding gas enough so that some of the gas was sent inward toward the protosun and some sent outward. In this way, Jupiter "cleared its neighborhood," and its growth halted. By this point, Jupiter had accumulated about 300 Earth masses of hydrogen and helium, plus several Earth masses of water and other simple molecules. From the inside out, Jupiter was a rocky world, surrounded by a layer of water, which was surrounded in turn by a much thicker layer of predominantly hydrogen and helium. Jupiter rotates today because it formed from swirling debris in the solar nebula. Likewise, all the other planets in our solar system rotate.

Saturn A few million years after Jupiter formed, collisions of planetesimals led to the formation of a second rocky world of about 10 Earth masses located a few astronomical units out beyond Jupiter. This second world underwent the same process of accreting hydrogen, helium, and ice, thereby creating Saturn. By the time Saturn formed, however, a lot of gas in its vicinity had already been stirred up and dissipated by Jupiter, leaving Saturn with less matter to accrete. Saturn has about 80 Earth masses of hydrogen and helium, along with a few Earth masses of water. Note that when it formed, Saturn was closer to the protosun than it is to the Sun today.

Neptune and Uranus The gravitational effects of Jupiter and Saturn forced many planetesimals to become concentrated out beyond Saturn's orbit, leading to the formation of two more rocky planets destined to become the cores of Neptune and Uranus. The debris that far from the protosun had been spiraling inward for several million years; thus, these two planets are believed to have formed substantially closer to Saturn than they are today. Furthermore, computer simulations suggest that at the time of their formation, Neptune was closer to the protosun than Uranus (the opposite of their positions now!). While substantial ice remained for Neptune and Uranus to collect, they formed so late that Jupiter and Saturn had dissipated most of the hydrogen and helium in their neighborhoods. Therefore, Neptune and Uranus became rich in water, along with carbon dioxide, methane, and ammonia—generically called "ices"—from their surroundings, but accreted only a couple of Earth masses of hydrogen and helium. While Jupiter and Saturn are called *gas giants,* Neptune and Uranus are called *ice giants.*

What two elements comprise most of Saturn's outer layer?

5-4 The inner planets formed primarily from collisions

The details explaining how the inner four rocky planets, Mercury, Venus, Earth, and Mars, formed are less well known than for the giant planets just discussed. It appears likely that the inner planets also formed as a result of collisions between Moon-sized and larger planetesimals. Within 100,000 years of the dust beginning to coalesce and to spiral inward, as described above, bodies with 0.1 Earth masses were able to form in it and collide with each other inside the snow line. It is unknown whether most of these planetesimals formed inside the snow line in relatively circular orbits and then collided to build planets in relatively circular orbits, or whether the four inner worlds formed from collisions of planetesimals that came from beyond the snow line in highly elliptical orbits that then somehow became circular. Whether the inner planets formed in orbits near their present average distances from the Sun or whether they migrated inward, like the planetesimals outside the snow line described above, is also unknown.

In the end, however, the collisions of hundreds of Moon-sized planetesimals led to the existence of four rocky inner planets: Mercury, Venus, Earth, and Mars (Figure 5-5). Because these planets were located inside the snow line, they did not acquire coatings of hydrogen and helium.

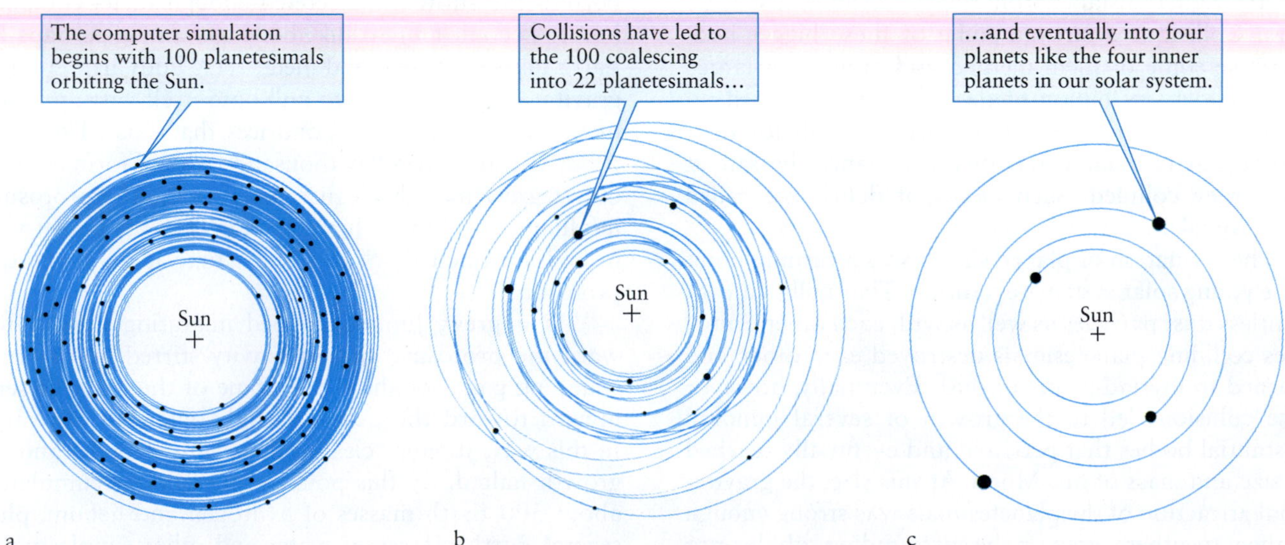

FIGURE 5-5 Accretion of the Inner Planets This computer simulation shows the formation of the inner planets over time.

Because Mercury, Venus, and Mars have similar densities as Earth (in Latin, *terra*), they are composed of similar chemicals. Consequently, all four bodies are called **terrestrial** planets.

The large quantities of water that exist on Earth today, that existed on Mars, and that are likely to have existed briefly on Venus came in part from ices in the planetesimals that created these worlds and in part from impacts of water-rich comets. Thus, this theory can show how our world came to have so much metal and so little hydrogen and helium, as proposed earlier in the chapter. And you can see from the formation process of Earth that *we are literally made of stardust!*

One planetesimal impact on the young Earth was particularly notable. When the young Earth had between 80% and 90% of its present mass, it was struck by a planetesimal with about the mass of Mars (10% of Earth's present mass). The impact likely tilted Earth's rotation axis and splashed a lot of debris off Earth's surface and into orbit. This event occurred about 4.5 billion years ago, within 100 million years after Earth started forming. The orbiting debris created a short-lived ring, which clumped together (in a process analogous to how planetesimals grew from collisions) and formed our Moon.

5-5 The changing orbits of the giant planets spread debris throughout the solar system

The Nice model also provides interesting insights into the evolution of the outer solar system after the four giant planets were formed. These planets were so much more massive than the surviving planetesimals with which they orbited that they were able to gravitationally fling some of this remaining debris toward the inner solar system. To do so while conserving energy and angular momentum, the planets had to spiral outward in response. Lower-mass Saturn spiraled outward faster than Jupiter. Recall from Kepler's third law (Section 2-5) that as planets move farther from the Sun, the periods of their orbits lengthen. Eventually, Saturn reached a distance where it orbited the Sun once every time Jupiter orbited twice. This situation, where the periods of two orbits are related by whole numbers, is called a *resonance* because the resulting gravitational interactions between the two bodies reinforce each other. The process is analogous to pushing someone on a swing—if you push at the right time (that is, in resonance with the person's motion), they will go higher and higher.

As Saturn passed through this distance, resonant tugs from Jupiter caused Saturn to accelerate outward. Rapidly approaching Neptune and Uranus, Saturn's gravitational attraction in turn gave these bodies extra strong tugs, forcing them much farther away from the Sun than they were originally. Recall that Neptune formed closer to the young Sun than Uranus. Because it was closer to Saturn, Neptune received stronger gravitational tugs from Saturn (and Jupiter) than did Uranus. As a result, Neptune was flung out past Uranus.

At that time there were still billions of planetesimals in the realm of the giant planets. The gravitational effect of the outwardly moving planets caused most of these planetesimals to change orbits. Some were sent inward, but most were hurled into orbits much farther away from the Sun than they had been. The outward flying planetesimals spread way beyond the original disk (which had a radius of 30 AU, as noted above). Billions of them were flung into a bagel-shaped volume beyond the orbit of Neptune that is now called the **Kuiper belt** (Figure 5-6), named after Gerard Kuiper, who first proposed the existence of such a region in 1951. Many more planetesimals were sent out even further than the Kuiper belt, creating a spherical distribution of debris called the **Oort comet cloud.**

Kuiper belt objects, or **KBOs**, were first observed in 1992. Most of the Kuiper belt objects are ice-rich comet nuclei. Some of them have developed highly elongated orbits that occasionally bring them closer to the Sun than Earth orbits. When this happens, the Sun's radiation vaporizes some of their ices, producing the long, flowing tails that we associate with comets.

Kuiper belt objects are distributed from about 30 AU to about 50 AU from the Sun. The smallest Kuiper belt object observed so far, discovered using the Hubble Space Telescope in 2009, is only 1 km across. Astronomers believe that there are many Kuiper belt objects much smaller than this. Even nonscientists, however, are familiar with one of several large bodies that the giant planets flung out—Pluto. Pluto became locked in a 2:3 resonant orbit with Neptune, meaning that it orbits the Sun twice in the same time that Neptune orbits 3 times. The gravitational effects of resonances are so pronounced that thousands of other bodies, called *plutinos,* have the same type of resonant orbit around the Sun (two orbits of each plutino for every three orbits of Neptune) as does Pluto. Astronomers have observed several other bodies similar in size to Pluto that were also flung far into the Kuiper belt. All bodies orbiting beyond Neptune that have enough mass to pull themselves into spherical shape are called *plutoids*. Pluto is one such body.

Pluto takes 248 years to orbit the Sun once. Are the orbital periods of the plutinos greater than, less than, or equal to 248 years? Why?

The Oort cloud extends out perhaps 50,000 AU, a quarter of the way to the nearest star (see Figure 5-6). Most bodies in that distribution are believed to be a few kilometers in size or smaller. One candidate Oort cloud object, Sedna, is in an extremely elliptical orbit (Figure 5-6a), taking it between 76 and 976 AU. *All* the objects that orbit the Sun farther than Neptune are called **trans-Neptunian objects**, or **TNOs**. Plutinos and plutoids are examples of TNOs.

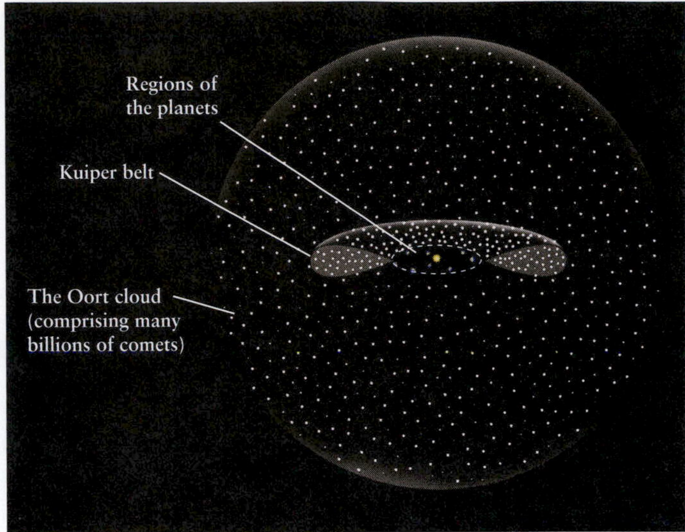

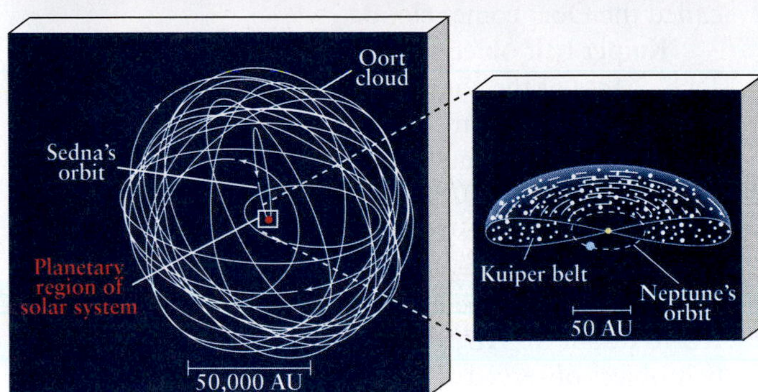

a Orbits of some Oort and Kuiper belt objects

b

FIGURE 5-6 The Kuiper Belt and Oort Cloud (a) The classical Kuiper belt of comets spreads from Neptune out 50 AU from the Sun. Most of the estimated 200 million belt comets are believed to orbit in or near the plane of the ecliptic. The spherical Oort cloud extends from beyond the Kuiper belt. (b) Orbits of bodies in the Oort cloud and Kuiper belt. (b: NASA and A. Field/Space Telescope Science Institute)

5-6 The asteroid belt is leftover debris

Now we turn to the major reservoir of leftover debris remaining in the *inner* solar system, the **asteroid belt,** located between Mars and Jupiter. Orbiting in this region are millions of planetesimals and smaller pieces of debris. The total mass of debris in the asteroid belt is about 5% of our Moon's mass, much less than the mass of a planet. There were more bodies in the asteroid belt in the distant past, but impacts and the gravitational tug of Jupiter pulled some bodies away and kept the remaining debris spread out, thereby preventing it from ever coming together to form a larger body.

The largest asteroid, Ceres, has a diameter of about 900 km (compared to Earth's diameter of 12,756 km). The next largest, Pallas and Vesta, are each about 500 km in diameter. Still smaller asteroids are increasingly numerous. Many thousands of kilometer-sized asteroids have been observed. A close-up picture of the asteroid Gaspra is shown in Figure 5-7. Some asteroids have their own moons, such as the asteroid Ida, which is orbited by heavily cratered Dactyl.

Which elements on Earth may have been unchanged since the universe began?

Some asteroids in the inner solar system orbit between the asteroid belt and the Sun. Many of these objects have highly elliptical orbits

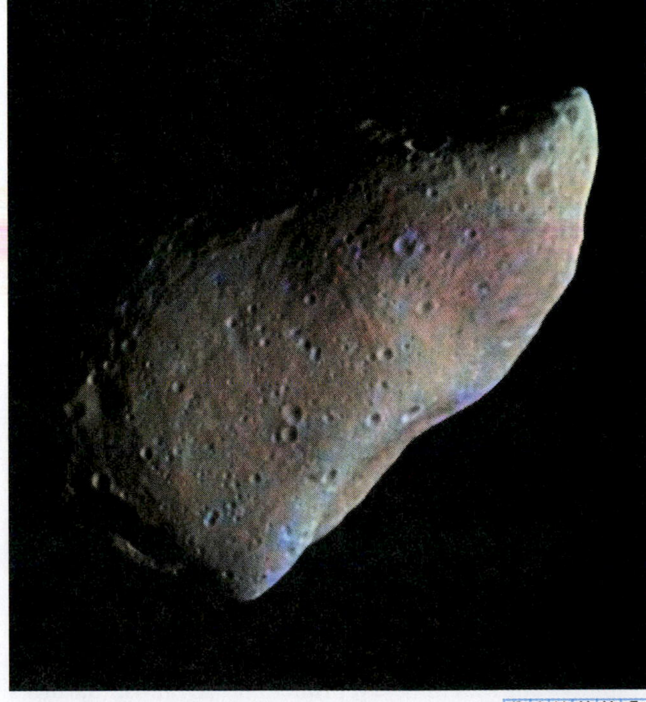

FIGURE 5-7 An Asteroid This picture of the asteroid Gaspra was taken in 1991 by the *Galileo* spacecraft on its way to Jupiter. The asteroid measures 12 × 20 × 11 km. Millions of similar chunks of rock orbit the Sun between the orbits of Mars and Jupiter. (NASA)

that take them across the paths of some of the planets, including Earth.

5-7 The infalling debris from the giant planets led to the Late Heavy Bombardment

Young Earth, along with every other body that orbited the Sun, was struck by countless pieces of space debris throughout its life. Indeed, these impacts still occur today. Depending on the speed and angle at which they strike land, impacting bodies larger than about 0.5 m create scars, called **craters.** Most of the craters on our planet's surface have been erased, as we will study in the next chapter, but we can still see numerous ones on our Moon, as well as on all the other bodies with solid surfaces in the solar system (Figure 5-8).

During the first 400 million years that the planets existed, the rate of impacts declined steadily. But then, about 4.1 billion years ago, the impact rate skyrocketed as planetesimals in the asteroid belt were knocked out of orbit as debris flung inward by the giant planets (see Section 5-5) struck them. Numerous asteroids were pushed into orbits that crossed the paths of the inner planets. Thus began the *Late Heavy Bombardment,* which ended as these errant asteroids were destroyed by either striking something, falling into the Sun, or eventually being flung out of the inner solar system.

The Moon was struck by so many objects during the Late Heavy Bombardment that large regions of its surface were completely obliterated, creating gigantic craters, which remained on the surface for hundreds of millions of years. These craters were eventually filled in when magma (molten rock) from inside leaked out through the thin crust below the craters, thereby forming the maria (dark plains) that we see today (Figure 5-8). Mercury, which is particularly dense when compared to the other inner planets, is believed to have lost much of its lower-density outer layers in a collision that may have also occurred during this time.

R I V U X G

FIGURE 5-8 Our Moon This photograph, taken by astronauts in 1972, shows thousands of craters produced by impacts of leftover rocky debris from the formation of the solar system. Age-dating of lunar rocks brought back by the astronauts indicates that the Moon is about 4.5 billion years old. Most of the lunar craters were formed during the Moon's first 700 million years of existence, when the rate of bombardment was much greater than it is now. (NASA)

The Moon's virtually airless environment has preserved important information about the early history of the solar system. Radioactive dating of Moon rocks brought back by the Apollo astronauts indicates that the rate of impacts declined dramatically about 3.8 billion years ago. That time is taken to be the end of the Late Heavy Bombardment. Since that time, impact cratering has continued, but at a very low level.

5-8 The categories of solar system objects have evolved

While it makes sense that the objects in the asteroid belt are called asteroids, the naming of bodies there and throughout the solar system is more complicated than that. Traditionally, an **asteroid** is a piece of space debris composed primarily of rock and metal that is larger than about 10 m (no exact minimum size has been agreed upon). Bodies smaller than this and composed primarily of rock and metal are called **meteoroids.** All debris that is a roughly equal mix of ice and rock are called **comets.** More precisely, the solid body that exists even when a comet has no tail is called a *comet nucleus.* The asteroid belt contains all three types of debris. Indeed, all the remaining planetesimals have become today's asteroids, meteoroids, and comets.

Another layer of terminology about bodies orbiting the Sun (and by inference, any other stars) was introduced in 2006, when the International Astronomical Union redefined the term *planet.* According to the most recently accepted definition, a **planet** is a celestial body that (1) orbits the Sun; (2) has enough mass so that its own gravitational attraction causes it to be essentially spherical, rather than being a large, irregular rock in space; and (3) has enough gravitational attraction to clear its neighborhood of other orbiting debris. Planets "clear their neighborhoods" by either pulling surrounding debris onto themselves or, in some cases, flinging debris far away from their orbits. There are thus eight planets in the solar system: Mercury, Venus, Earth, Mars, Jupiter, Saturn, Uranus, and Neptune.

A **dwarf planet** satisfies conditions (1) and (2), but not condition (3). These objects are smaller than planets, with less mass and hence less gravitational attraction. This definition is the reason that Pluto was demoted from planet status. So far, of the myriad candidate bodies, astronomers have agreed on five dwarf planets: Pluto, Ceres, Haumea, Makemake (pronounced *mah-kay mah-kay*), and Eris.

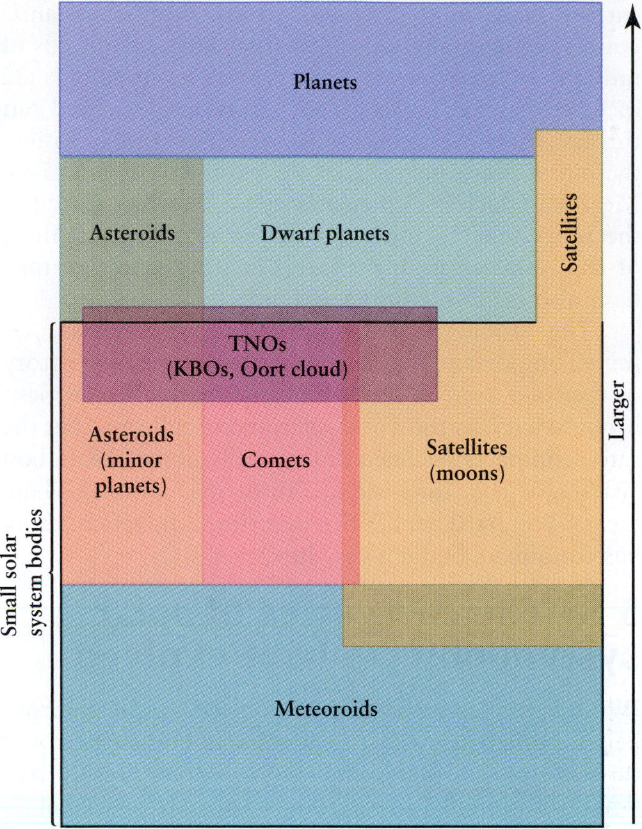

FIGURE 5-9 Different Classifications of Solar System Objects Some of the definitions of the different types of objects in the solar system overlap. For example, the largest asteroids are also classified as dwarf planets; various trans-Neptunian objects (TNOs) are asteroids or comets; some comets are satellites of Jupiter; some Kuiper belt objects (KBOs) are satellites of other KBOs. Furthermore, TNOs exist in two groups: KBOs and Oort cloud bodies. Some moons are as large as dwarf planets. Indeed, some moons are as large as small planets!

A **moon** or **natural satellite** is an object orbiting a larger body that is, in turn, orbiting the Sun. Unlike ours, most moons are potato-shaped rocks. These moons are planetesimals that were captured by the various planets, dwarf planets, and asteroids.

Everything in the solar system that is not a planet, dwarf planet, or moon is called a **small solar system body** (SSSB). The tricky part is that the largest asteroids are dwarf planets, while the vast majority of asteroids, as well as all meteoroids and all comets are small solar system bodies. Figure 5-9 summarizes these relationships.

> How might some of the asteroids and some of the meteoroids in the solar system be related?

5-9 The Sun developed while the planets matured

During the 100 million years that the inner planets were completing their formation, the protosun was also evolving. Throughout this time, the temperature and pressure at the center of the contracting protosun continued to increase. Finally, its core reached about 10^7 K, which is hot enough to initiate hydrogen fusion, in which hydrogen is converted into helium. Such fusion also generated energy in the form of gamma rays, which in turn provided enough pressure from the Sun's central region to stop the protosun's collapse. At that point, the Sun became a star. Hydrogen fusion continues in the Sun's core today and the energy emitted by the Sun over the past 4.5 billion years has enabled Earth to sustain life.

Sunlike stars take approximately 100 million years to form from a nebula and settle down, which means that the Sun probably became a full-fledged star around the time the accretion of the inner planets was completed. Radiation from the Sun heated and dispersed the remaining gases in the solar system, thereby helping to limit the sizes of the planets.

Because the planets formed in a disk, they should all orbit in more or less the same plane as Earth (that is, nearly in the plane of the ecliptic). Observations bear out this conclusion. The angles of the orbital planes of the other planets with respect to the ecliptic are called their **orbital inclinations**, and all are 7° or less (Figure 5-10).

Furthermore, according to the Nice and other models of solar system formation, all of the planets should orbit the Sun in the same direction as does Earth. Imagine traveling far above Earth's North Pole and looking down at Earth's orbit. You would see that our planet is going counterclockwise around the Sun. All the other planets also orbit counterclockwise as observed from the same vantage point.

To find out more about the solar system's origins, 3
astronomers study the leftover interplanetary debris: asteroids, meteoroids, and comets. Many of these bodies are thought to be virtually unchanged remnants of the solar system's formation. According to radioactive dating of the oldest debris ever discovered from space (see *An Astronomer's Toolbox* 4-2), the solar system was roughly in the form we know today just under 4.6 bil-

> **WHAT IF... The Solar System Had More than One Star?**
> We saw in Chapter 2 that Earth has a low-eccentricity orbit around the Sun. The low eccentricity keeps our planet's temperature in a relatively narrow range suitable for the formation and evolution of life. Under most circumstances, the gravitational tugs of two or more "Suns" would have the effect of forcing Earth into an extremely elliptical orbit, making the surface frequently too hot or too cold for most life-forms that live on Earth today.

lion years ago. As we will see throughout the book, a lot happened between the formation of the universe some 13.75 billion years ago and the formation of the solar system, some 9 billion years later.

While most of the moons in the solar system are planetesimals captured by the gravitational attractions of the planets, the larger ones that orbit Jupiter and Saturn probably formed from disks of gas and dust that orbited those worlds in analogy to the disk that orbited the Sun in which the planets coalesced. Of all the moons orbiting full-fledged planets, only one, our Moon, was created as the result of a collision between its planet and a substantial body crossing the planet's path.

On the young Earth, one of the conditions that allowed for the beginning of biological evolution was the presence of liquid water. Strong evidence indicates that liquid water existed on Mars and may still exist inside it and inside Jupiter's moons Europa and Ganymede, as well as Saturn's moon Titan, raising the hotly debated question of whether simple life has evolved on those worlds as well.

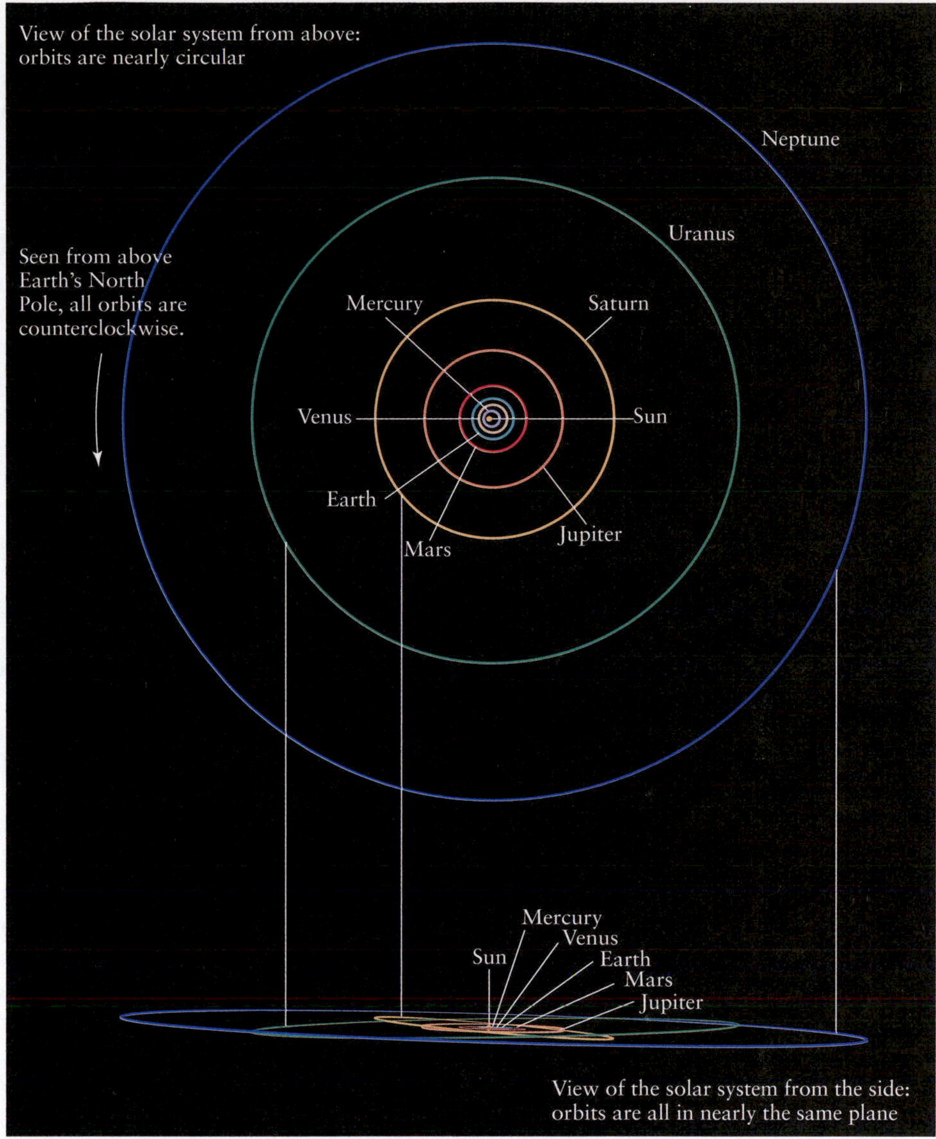

ANIMATION 5.1 **FIGURE 5-10 The Solar System** This scale drawing shows the distribution of planetary orbits around the Sun. All orbits are counterclockwise because the view is from above Earth's North Pole. The four terrestrial planets are located close to the Sun; the four giant planets orbit at much greater distances. Seen from above the disk of the solar system, most of the orbits appear nearly circular. Mercury has the most elliptical orbit of any planet.

COMPARATIVE PLANETOLOGY

There are two ways of studying objects in the solar system. One approach is to compare and contrast one feature of all similar objects, such as all of the atmospheres or all of the surfaces of planets, and then move on to the next feature. This method has the advantage of showing the "big picture" for each feature as it arises. The other approach is to take an object and explore all of its properties before moving on to the next object. This method has the advantage of connecting all of the properties of each object directly to each other, so that you do not inadvertently connect, say, the atmosphere of Venus with the surface of Mercury. Because both approaches are valuable, we do both in this book.

5-10 Comparisons among the eight planets show distinct similarities and significant differences

The planets that emerged from the accretion of planetesimals in the young solar system were Mercury, Venus, Earth, Mars, Jupiter, Saturn, Uranus, and Neptune. Before exploring the properties of the individual planets in the upcoming chapters, it is instructive to compare a variety of their orbital and physical properties.

Orbits Table 5-1 lists some orbital characteristics of the eight planets. As shown in Figure 5-10, the four inner planets—Mercury, Venus, Earth, and Mars—are crowded close to the Sun. In contrast, the orbits of the four large outer planets—Jupiter, Saturn, Uranus, and Neptune—are widely spaced at greater distances from it. Kepler's laws (see Chapter 2) showed us that all of the planets have elliptical orbits. Even so, most of their

Table 5-1 Orbital Characteristics of the Planets

	Average distance from Sun		Orbital period
	(AU)	(10^6 km)	(year)
Mercury	0.39	58	0.24
Venus	0.72	108	0.62
Earth	1.00	150	1.00
Mars	1.52	228	1.88
Jupiter	5.20	778	11.86
Saturn	9.54	1427	29.46
Uranus	19.19	2871	84.01
Neptune	30.06	4497	164.79

orbits are nearly circular. The exception is Mercury, whose orbit is noticeably oval.

Size The planets fall into three size groups (Figure 5-11). The four inner planets form one group, Jupiter and Saturn form the second group, and Uranus and Neptune compose the third, intermediate-size group. Jupiter is the largest planet, with a diameter about 11 times bigger than that of Earth, whereas Mercury, with about a third of Earth's diameter, is the smallest planet. Indeed, two moons, Ganymede and Titan, are larger than Mercury. Neptune and Uranus are each about 4 times larger in diameter than Earth, the largest of the four inner planets. The diameters of the planets are given in Table 5-2.

Mass Mass, a measure of the total amount of matter an object contains, is another characteristic that distinguishes the inner planets from the outer planets. Planetary masses are determined by measuring the periods of moons' orbits or, for planets without moons, the deflection by the planets of passing objects. This information is used in Kepler's third law (or similar equations) to determine the planets' masses. The four inner planets have small masses compared to the giant outer ones. Again, first place goes to Jupiter, whose mass is 318 times greater than Earth's (see Table 5-2). Uranus and Neptune, ice giants, have masses intermediate between the terrestrial and the gas giant planets.

Density Size and mass can be combined in a useful way to provide information about the chemical composition of a planet (or any other object). Matter composed of heavy elements, such as iron or lead, has more particles (protons and neutrons) and hence more mass packed into the same volume than does matter composed of light elements, such as hydrogen, helium, or carbon. Therefore, objects made primarily of heavier elements have a greater average density than objects composed primarily of lighter elements, where average density is given by the equation

$$\text{Average density} = \frac{\text{Total mass}}{\text{Total volume}}$$

The chemical composition (kinds of elements) of any object, therefore, determines its **average density**—how much mass the object has in a unit of volume. For example, a kilogram of copper takes up less space (is denser) than a kilogram of water, even though both have the same mass (Figure 5-12). Average density is expressed in kilograms per cubic meter. To help us grasp this concept, we often compare the average densities of the planets to the density of something familiar, namely liquid water, which has a density of 1000 kg/m^3.

The four inner planets have high average densities compared to water (see Table 5-2). In particular, the average density of Earth is 5520 kg/m^3. Because the density of typical surface rock is only about 3000 kg/m^3, this high density implies that Earth contains a large amount of material inside it, namely iron and nickel, that is much denser than surface rock. We will explore the terrestrial planets in Chapters 6 and 7.

ANIMATION 5.2

FIGURE 5-11 The Sun and the Planets This figure shows the eight planets drawn to size scale in order of their distances from the Sun (distances are not to scale). The four planets that orbit nearest the Sun (Mercury, Venus, Earth, and Mars) are small and made of rock and metal. The next two planets (Jupiter and Saturn) are large and composed primarily of hydrogen and helium. Uranus and Neptune are intermediate in size and contain roughly equal amounts of ices, hydrogen and helium, and terrestrial material. (Calvin J. Hamilton and NASA/JPL)

Table 5-2 Physical Characteristics of the Planets

	Diameter		Mass		Average density
	(km)	(Earth = 1)	(kg)	(Earth = 1)	(kg/m^3)
Mercury	4878	0.38	3.3×10^{23}	0.06	5430
Venus	12,100	0.95	4.9×10^{24}	0.81	5250
Earth	12,756	1.00	6.0×10^{24}	1.00	5520
Mars	6786	0.53	6.4×10^{23}	0.11	3950
Jupiter	142,984	11.21	1.9×10^{27}	317.94	1330
Saturn	120,536	9.45	5.7×10^{26}	95.18	690
Uranus	51,118	4.01	8.7×10^{25}	14.53	1290
Neptune	49,528	3.88	1.0×10^{26}	17.14	1640

In sharp contrast, Jupiter, Saturn, Uranus, and Neptune have relatively low average densities (see Table 5-2). Indeed, Saturn's average density is less than that of water. The low densities of the giant outer planets suggest that they contain significant amounts of the lightest elements, hydrogen and helium, as discussed earlier in this chapter. As we saw in Section 5-3, Jupiter and Saturn share similar compositions, being primarily hydrogen and helium, while Uranus and Neptune both contain vast quantities of water and ammonia, as well as much hydrogen and helium. These four worlds are sometimes called *Jovian planets* (the Roman god Jupiter was also known as Jove) because, like Jupiter, they all have relatively low densities and high masses compared to Earth. However, the differences in the chemistries of Jupiter and Saturn compared to those of Uranus and Neptune make the name *giant planets* (giant compared to Earth) more appropriate in collectively describing these four. As noted earlier, Jupiter and Saturn are specifically called gas giants, while Uranus and Neptune are ice giants. We will explore the giant planets as two groups in Chapter 8.

FIGURE 5-12 The Volumes of Objects with Different Densities All of the objects in this image have the same mass. However, the chemicals from which they form have different densities (mass per volume), so they all take up different amounts of space (volume). (Richard Megna/Fundamentals Photographs, NYC)

Spectra Further information about the chemical composition of bodies in the solar system is obtained from their spectra. For solar system objects, spectra are provided primarily by sunlight scattered off the surface or clouds that surround each object. As we saw in Chapter 4, spectra provide us with details of an object's surface (or atmospheric chemical composition) and rotation rate. From spectra, we can confirm that the outer layers of the giant planets are primarily hydrogen and helium, and that the surface of Mars is rich in iron oxides, among many other things.

Insight Into Science

Astronomical Measurement Astronomers use the laws of physics to infer things we cannot measure directly. For example, we can get an overall idea of the chemical compositions of planets from their masses and volumes. Kepler's laws enable us to determine the mass of a planet from the period of its moons' orbits. The measured diameter of each planet (determined by its distance from Earth and its angular size in the sky) yields its volume. As shown in the equation earlier in this section, dividing total mass by total volume yields the average density. Comparing this density to the densities of known substances gives us information about the planets' interior chemistries.

Venus and Mercury do not have any moons. What objects do you think astronomers observed being deflected in order to find these planets' mass?

Albedo The surfaces or the upper cloud layers of the various planets scatter (send in all directions) different amounts of light. The fraction of incoming light returning directly into space is called a body's **albedo.** An object that scatters no light has an albedo of 0.0; for example, powdered charcoal has an albedo of nearly 0.0. An object that scatters all of the light that strikes it (a high-quality mirror comes close) has an albedo of 1.0. Multiplying the albedo of an object by 100 gives the percentage of light directly scattered off that body. Three planets (Mercury, Earth, and Mars) have albedos of 0.37 or less. Such albedos result from dark, dry surfaces, or from a mixture of light (water and clouds) and dark surfaces (continents) exposed to space. In contrast, Venus, Jupiter, Saturn, Uranus, and Neptune have albedos of 0.47 or more. High albedos like these imply bright materials exposed to space. The albedos of Venus, Jupiter, Saturn, Uranus, and Neptune are high because these planets are all completely enshrouded by clouds, which are very good reflectors of sunlight.

Why is Earth's albedo continually changing?

4 **Moons** Every planet, except Mercury and Venus, has moons. At least 168 planetary moons are known to exist in the solar system (up from 99 known in 2001), and more are still being discovered. Table 5-3 lists the numbers of moons orbiting each planet. Unlike our Moon, most moons are irregularly shaped, and look more like potatoes than spheres. We will discover that there is as much variety among moons as there is among planets.

Table 5-3 Number of Moons Orbiting Each Planet

Planet	Moons
Mercury	0
Venus	0
Earth	1
Mars	2
Jupiter	63
Saturn	62
Uranus	27
Neptune	13

PLANETS OUTSIDE OUR SOLAR SYSTEM

Understanding the process by which the solar system was formed is being aided by observations of different stages of planet formation occurring around other stars. Although astronomers have been searching for planets outside the solar system for centuries, it was only in the 1980s that observations leading to their discovery were first made.

5-11 Planets and smaller debris that orbit other stars have been discovered

Stellar evolution and star formation are ongoing processes. You can see a cloud in the process of forming stars today with your naked eye, namely the middle "star" in Orion's sword (see the figure opening this chapter), which looks like a fuzzy blob in the night sky. Called the Orion Nebula or Great Nebula of Orion, this "star" is actually part of a gigantic interstellar cloud region called the Orion Molecular Cloud in which new stars and planets are being created. Recall from Chapter 3 that visible light provides limited information; you can see in Figure 3-32 and Figure 5-2b that infrared images reveal even more of this region. Likewise, ultraviolet images provide more insight than do visible ones.

As indicated earlier in the chapter, astronomers have discovered stars in early stages of formation that are surrounded by proplyds—disks of gas and dust. These disks are similar to what the solar nebula was believed to have been like over four-and-a-half billion years ago. Indirect observational evidence for the existence of planets outside the solar system, called *exoplanets* or *extrasolar planets,* first came from distortions of proplyd disks.

If one or more planets orbit in a proplyd disk around a young star, their gravitational pull will affect the disk of gas and dust around the star, causing the disk to warp or become off-center from its star. An example of serious warping can be seen in the edge-on disk surrounding the star Beta Pictoris. This star and the material that orbits it formed only 20 million years ago. In 2006, astronomers discovered that Beta Pictoris actually has two disks (Figure 5-13), tilted slightly with respect to each other. Computer models show that a Jupiterlike planet orbiting out of the plane of the big disk would attract debris from that disk, thereby forming the smaller disk in the plane of its orbit. Beta Pictoris also has millions of comet nuclei, which were discovered in 2001. By studying such systems as Beta Pictoris, we can also begin addressing such questions as whether Earth acquired its water from comets in the young solar system.

This interaction between planets and disks is seen in other systems. The nearby star Fomalhaut has an off-centered disk of gas and dust. This star and its entourage are only some 200 million years old. Early observations

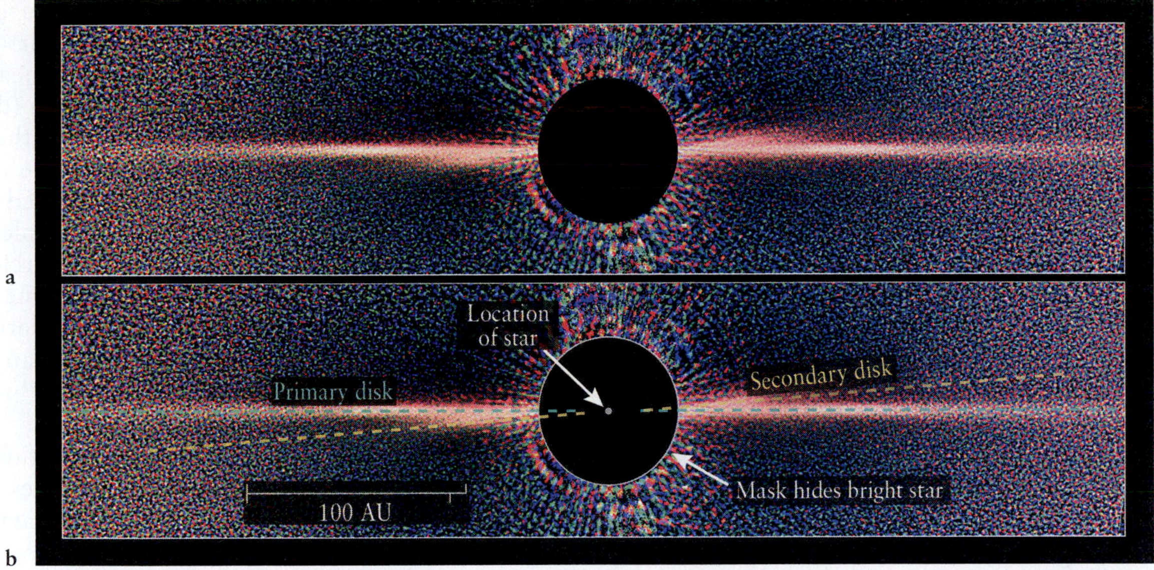

FIGURE 5-13 A Circumstellar Disk of Matter (a) Hubble view of Beta Pictoris, an edge-on disk of material 225 billion km (140 billion mi) across that orbits the star Beta Pictoris (blocked out in this image) 50 ly from Earth. Twenty million years old, this disk is believed to be composed primarily of iceberglike bodies that orbit the star. The smaller disk is believed to have been formed by the gravitational pull of a roughly Jupiter-mass planet in that orbit. Because the secondary disk is so dim, the labeling for this image is added in (b). (NASA, ESA, and D. Golimawski)

suggested that this distortion is caused by a planet tugging on the gas as it orbits. That planet, Fomalhaut b, was first observed in 2004 (Figure 5-14). This exoplanet is roughly 10 times further from its star than Saturn is from the Sun. The pull of the planet is so great that the center of Fomalhaut's disk is displaced 15 AU from the star.

Along with disks of gas and dust, large numbers of comets, and exoplanets, astronomers have also discovered asteroids orbiting other stars. The first system discovered with asteroids is Zeta Leporis, a young star just 70 ly from Earth. That asteroid belt, located about the same distance from its star as is our asteroid belt, is estimated to contain about 200 times as much mass as ours. The pieces of asteroid debris have not been directly observed, but were inferred from the existence of hot (room-temperature) dust surrounding Zeta Leporis. The most likely explanation for this high-temperature dust is that it is being generated by the collisions of asteroids. As noted in Section 5-6, our solar system has one asteroid belt, located between Mars and Jupiter. Observations reveal that the star Epsilon Eridani has two asteroid belts, along with planets and a ring of icy debris.

Observing exoplanets directly as they orbit other stars is extremely challenging because even high-albedo planets like Jupiter are typically very dim compared to the stars that they orbit. An orbiting body labeled 2M1207b (the 2M designates it as listed in the 2Mass catalog of point objects in space) *almost* qualifies as an observed exoplanet. This object (Figure 5-15), first seen in 2004, has a mass between 5 and 8 times the mass of

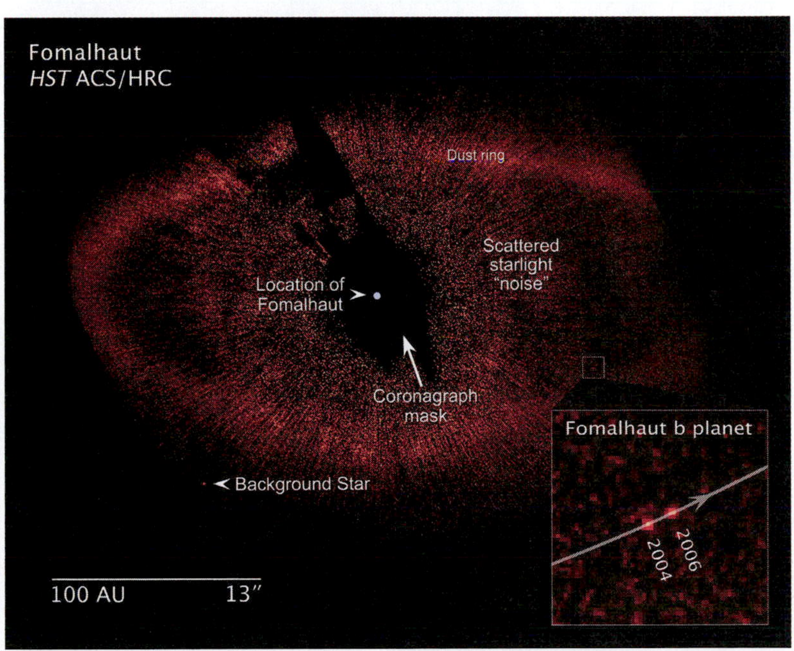

FIGURE 5-14 Visible Image of an Exoplanet The star Fomalhaut, blocked out so that its light does not obscure the disk, is surrounded by gas and dust in a ring whose center is separated from the star by 15 AU, nearly as far as Uranus is from the Sun. This offset is due to the gravitational effects of giant planet Fomalhaut b orbiting the star. This system is 25 ly from Earth. The dimmer debris in that system and between it and Earth scatters light that is called "noise" in this image. (NASA, ESA, and Z. Levay (STScl)

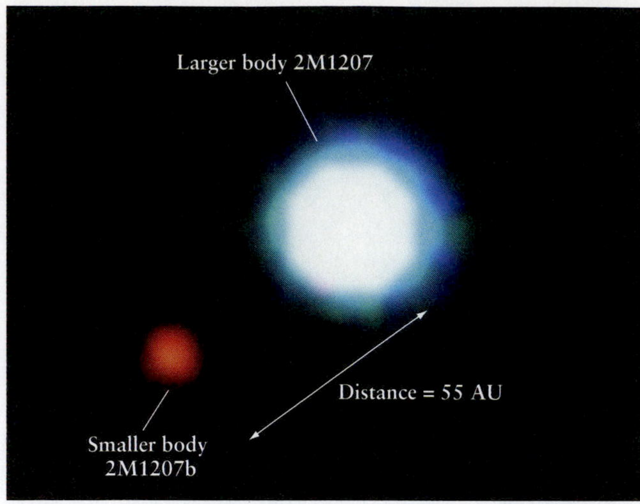

FIGURE 5-15 Image of an Almost-Extrasolar Planet This infrared image, taken at the European Southern Observatory, shows the two bodies 2M1207 and 2M1207b. Neither is quite large enough nor massive enough to be a star, and evidence suggests that 2M1207b did not form from a disk of gas and dust surrounding the larger body; hence, it is not a planet. This system is about 170 ly from our solar system in the constellation Hydra. (ESO/VLT/NACO)

Jupiter. It orbits a slightly bigger object, which itself is too small to sustain fusion in its core. Hence, neither object shines very brightly. Observations indicate that both objects were formed together, rather than 2M1207b forming from a disk of gas and dust orbiting the bigger object. Bodies that form simultaneously, even if they have masses too small to be stars, are not considered planets!

Since 1995, most exoplanets have been discovered by their effects on the stars they orbit (Figure 5-16a). Some planets are discovered by measuring changes in the radial velocities of their stars (defined in Section 4-7), which are calculated from the Doppler shifts of the stars' spectra (Figure 5-16b). This motion toward or away from Earth is created by the gravitational pull of the planet. The Doppler shift changes cyclically as the star and planet circle their common center of mass, and the length of time of one cycle is the period of the planet's orbit. Modern spectroscopic techniques are so good that we can detect stars approaching or receding at speeds as low as 4 km/h (2.5 mi/h).

Other planets are discovered from variations in a star's proper motion, or motion among the background stars (also defined in Section 4-7). A planet's gravitational attraction causes its star to deviate from motion in a straight line. This *astrometric method* of discovery looks for just such a wobble (Figure 5-16c).

Yet other planets are located by observing changes in the brightnesses of their stars. As a planet passes or transits between us and its star, it partially eclipses the star (Figure 5-16d). In 2003, astronomers used this *transit method* to detect a planet only one-sixteenth as far from its star as Mercury is from the Sun. That planet orbits once every 28 h, 33 min.

By observing transits, astronomers can also study the spectrum of a planet's atmosphere. When a gas giant planet labeled HD 209458b passed in front of the star HD 209458 (Henry Draper catalog star number 209458) some 150 ly from Earth, hydrogen, sodium, oxygen, carbon, and carbon monoxide in the planet's

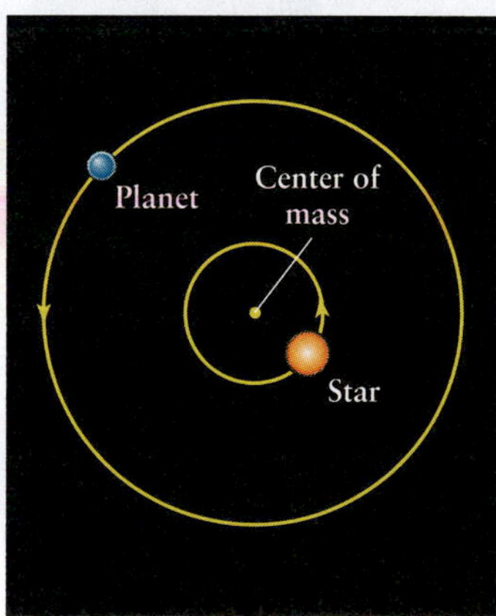

a A star and its planet

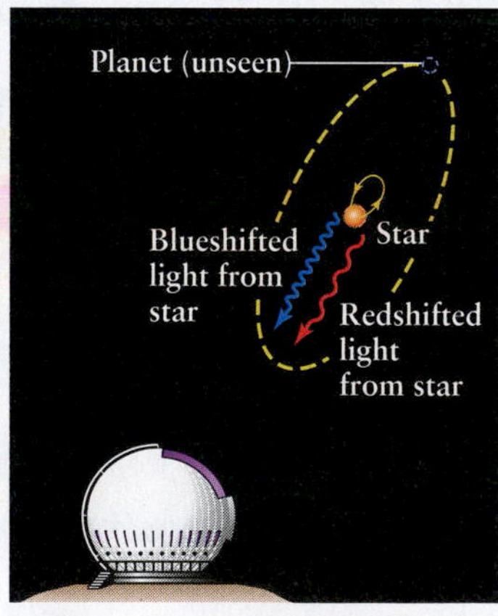

b The radial velocity method

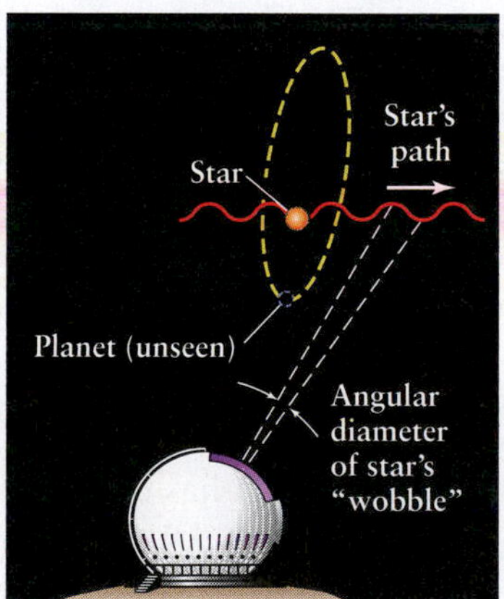

c The astrometric method

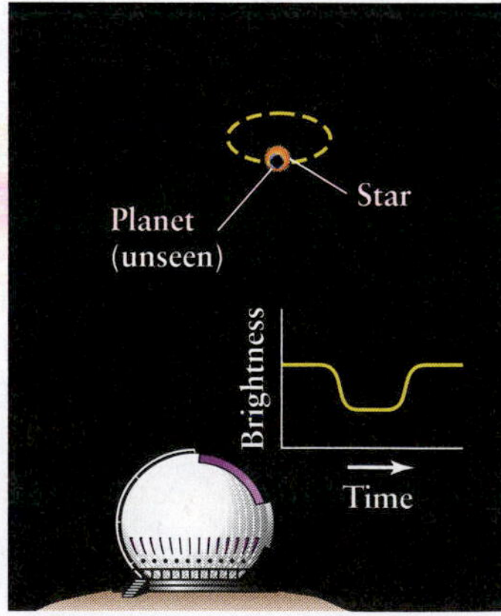

d The transit method

ANIMATION 5.3

FIGURE 5-16 Three Traditional Methods of Detecting Exoplanets (a) A planet and its star both orbit around their common center of mass, always staying on opposite sides of that point. The star's motion around the center of mass often provides astronomers with the information that a planet is present. (b) As a planet moves toward or away from us, its star moves in the opposite direction. Using spectroscopy, we can measure the Doppler shift of the star's spectrum, which reveals the effects of the unseen planet or planets. (c) If a star and its planet are moving across the sky, the motion of the planet causes the star to orbit its center of mass. This motion appears as a wobbling of the star across the celestial sphere. (d) If a planet happens to move in a plane that takes it across its star (that is, the planet transits the star), as seen from Earth, then the planet will hide some of the starlight, causing the star to dim. This change in brightness will occur periodically and can reveal the presence of a planet.

atmosphere absorbed certain wavelengths of starlight. Therefore, the star's spectrum has extra absorption lines (introduced in Section 4-6). The Hubble Space Telescope further revealed that the outer layers of this planet are being heated so much that at least 10,000 tons of hydrogen are evaporating into space from it every second. In 2010, a tail created by this departing gas was observed. Based on this rate of mass loss, the planet has lost at least 0.1% of its mass over its lifetime of 5 billion years. Furthermore, observations in 2010 revealed winds in the outer cloud layers of this planet reaching 7000 km/h (4300 mi/h), more than 3 times faster than any winds known to occur in our solar system. The atmospheres of other exoplanets also include carbon dioxide and water vapor.

The transit method has also revealed that some exoplanets have much more high-density material in their cores than do our giant planets. The changing intensity of light from the star HD 149026 showed that the planet passing in front of it, which has a mass 115 times that of Earth, is much smaller than other planets of that mass. Calculations reveal that about 70 Earth masses (roughly two-thirds) of that planet is rocky material in its core. For comparison, Jupiter's terrestrial core is about 4% of its total mass.

As you can see in Figure 5-17, many giant planets orbit surprisingly close to their stars, much closer than Earth is to the Sun. If the Nice model of how giant planets form is correct, then these planets are spiraling inward after having formed farther away from their stars than they are now.

WHAT IF... Jupiter Were Closer to the Sun than Earth?

The mass of Jupiter is so great that if it had always been closer to the Sun than Earth, the gravitational pull of the gas giant would pull Earth out of its present orbit and cause us to orbit where the temperature would be too cold to sustain most life on Earth. If the Nice model is correct and Jupiter got closer to the Sun than we are by migrating inward from its initial, more distant orbit, then Jupiter is likely to have pulled Earth into itself or flung Earth far from its present orbit as it spiraled in past our planet.

A dramatic consequence of the inward spiraling of planets was discovered in 2001, when the remnants of at least one planet were discovered in the atmosphere of star HD 82943 that still has at least two other planets in orbit around it. This star's atmosphere contained a rare form of lithium that is found in planets, but which is destroyed in stars within 30 million years after they form. The presence of this isotope, ^{6}Li, means that at least one planet spiraled so close to the star that it was vaporized. In 2008, a roughly Jupiter-mass planet, WASP-12b, was discovered so close to its star, WASP-12, that tides from the star have made the planet egg-shaped. WASP-12b's atmosphere is now being pulled onto its star at a rate of about 2×10^{17} tons per year. The planet will be devoured within 10 million years.

In 2010, observations of exoplanets transiting their stars revealed that some of the planets are orbiting in the opposite direction to which their stars rotate (or spin). This motion is intriguing because it seems to run counter to the predictions of the Nice model, in which planets should orbit in the same direction as the stellar rotation, as occurs here in our solar system.

The power of the transit method is demonstrated even further by its ability to reveal when a star has several planets orbiting it. To do this, astronomers watch the largest planet transit a star several consecutive times. The interval between transits is the length of the year on that planet. When there are other planets in a star system, their gravitational tugs on the observed planet change the length of time it takes to orbit the star. So if the length of a year for a transiting planet changes, then astronomers can deduce that there must be other planets in that system.

In 2004, astronomers began finding exoplanets by using a property of space that Albert Einstein discovered. His theory of general relativity, which we explore in detail in Chapter 14, shows that matter warps its surrounding space, causing, among other effects, passing light to change direction. This phenomenon is quite analogous to how light passing through a lens changes direction and is focused—light can also be focused by gravity, an effect called **microlensing** (or *gravitational microlensing*). Figure 5-18 shows how a star with a planet, passing between Earth and a distant star, can focus the light from the distant star, causing it to appear to change brightness. This change occurs twice, once as the distant star's light is focused toward us by the closer star and again when the light is focused toward us by the planet and the star together.

The vast majority of planets that have been discovered orbiting stars that shine like the Sun have masses ranging between a half and a few times the mass of Jupiter (see Figure 5-17). The smallest-mass planet known with confidence has about twice the mass of our Moon. However, this exoplanet, along with at least two other planets, orbits the remnant of a star that exploded, destroying any life that may have existed on its planets. The two other planets in that system each have about 4 times Earth's mass.

In 1999, astronomers began observing some stars wobbling in ways too complicated to be caused by a single planet, implying the presence of several planets. In a multiple-planet system, each planet contributes a tug on the star. By combining the effects of two or more planets, the observed pattern of the star's motion (Doppler or astrometric) can be reproduced. The first star discovered with multiple planets was Upsilon Andromedae

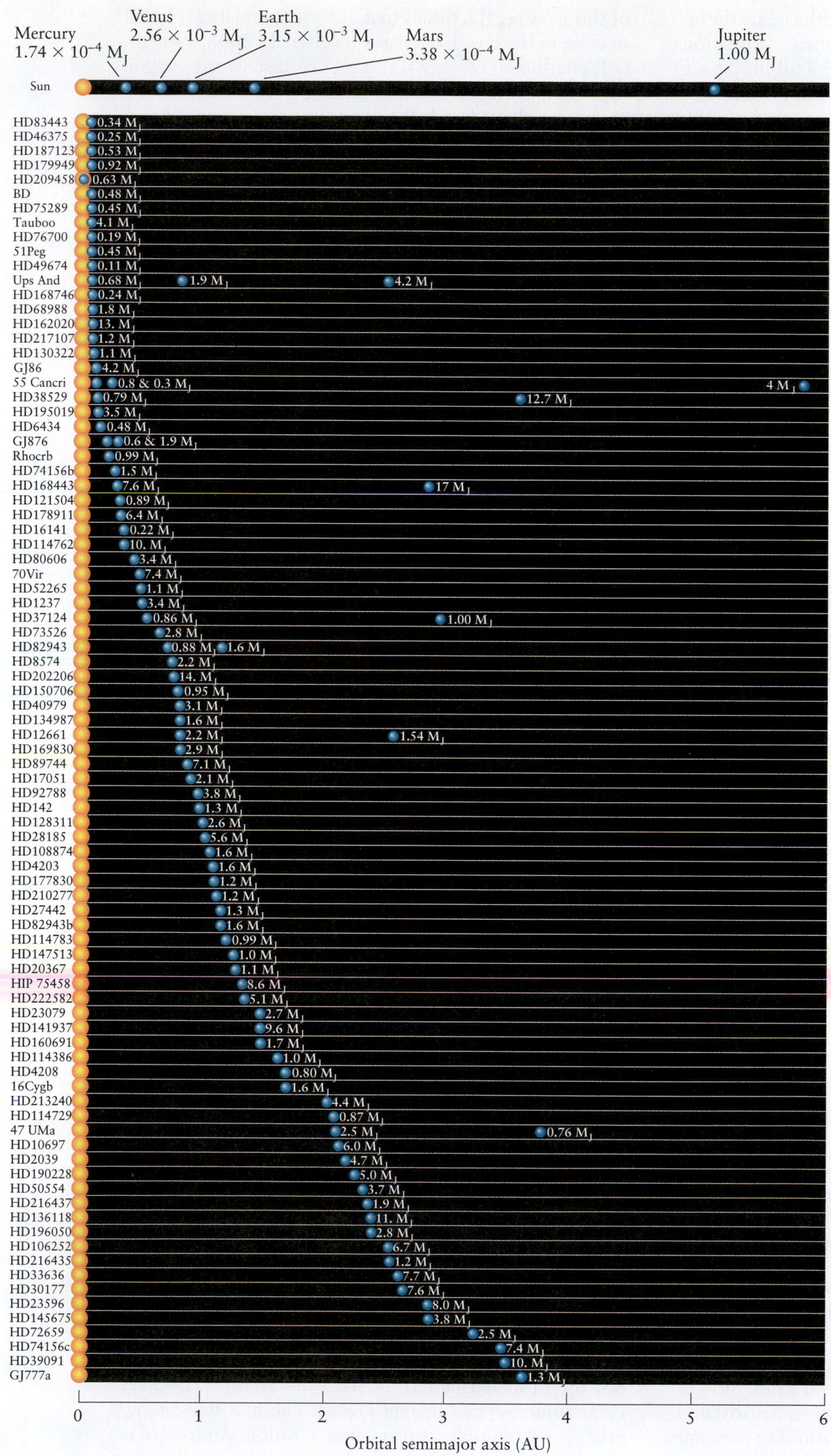

FIGURE 5-17 Planets and Their Stars This figure shows the separations between exoplanets and their stars. The corresponding star names are given on the left of each line. Note that many systems have giant planets that orbit much closer than 1 AU from their stars. (M_J is shorthand for the mass of Jupiter.) For comparison, the solar system is shown at top. (California and Carnegie Planet Search)

(Figure 5-19). One of its planets, labeled Upsilon Andromedae B, is 10 times closer to its star than Mercury is to the Sun. Using the infrared-sensitive Spitzer Space Telescope, astronomers have determined that the difference in temperature between the daytime and nighttime sides of Upsilon Andromedae B is 1400 K. The star 55 Cancri A, in the constellation Cancer, has at least five planets.

5 In 2005, microlensing was used to detect a terrestrial planet labeled Gliese 876d orbiting a shining star, Gliese 876. (The Gliese catalogs list nearby stars and their known planets.) This planet has 6.8 times Earth's mass. At least four (and possibly six) similar-mass planets, Gliese 581b, c, d, e, and possibly f and g, have been discovered orbiting another Gliese catalog star, Gliese 581, since 2007. Most, if not all, of these planets are likely to be composed primarily of rock and metal, with water on or near their surfaces. The stars they orbit are much cooler than our Sun. All of these planets orbit their stars closer than Mercury orbits the Sun, and, hence, they are likely to be in synchronous orbits, as our Moon is in orbit around Earth. Therefore, one side of each planet faces its star all the time. This side is swelteringly hot, while the other side is ice cold. So far, 45 stars have been observed with two or more planets that orbit them; several are listed in Figure 5-17.

The number of exoplanets that are being discovered is in the process of exploding with the activation of NASA's *Kepler* orbiting observatory, which was designed to search for Earthlike planets. It began discovering them in 2009, the year of its launch. Within six weeks *Kepler* had identified five previously unknown exoplanets. In one year of operation it had found more than 700 other candidate exoplanets. These discoveries are now being confirmed. Compare this rate to the discovery by other technologies of just 506 exoplanets between 1992 and 2010.

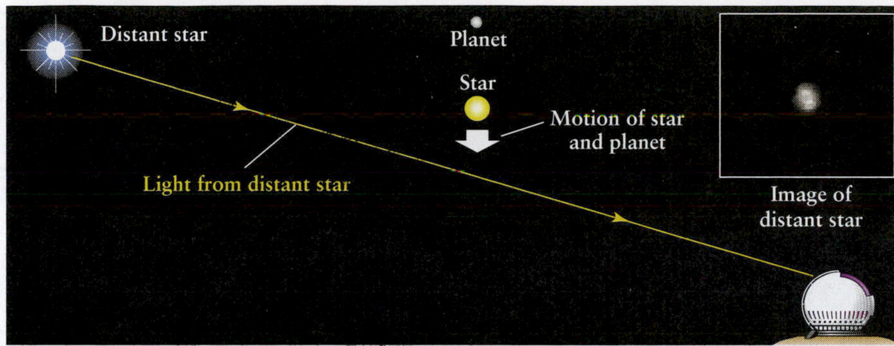

a No microlensing

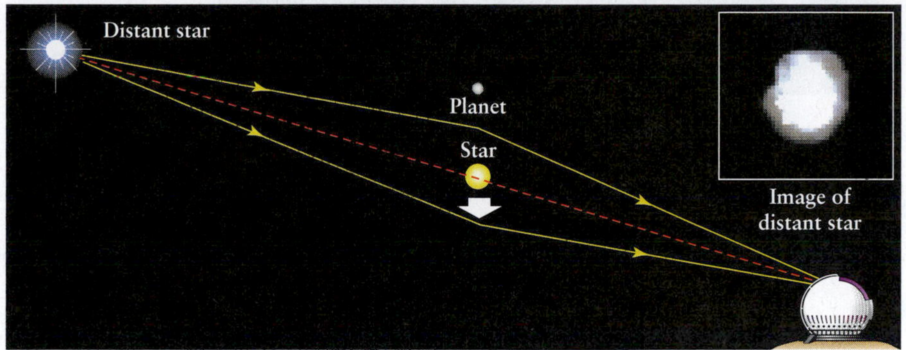

b Microlensing by star

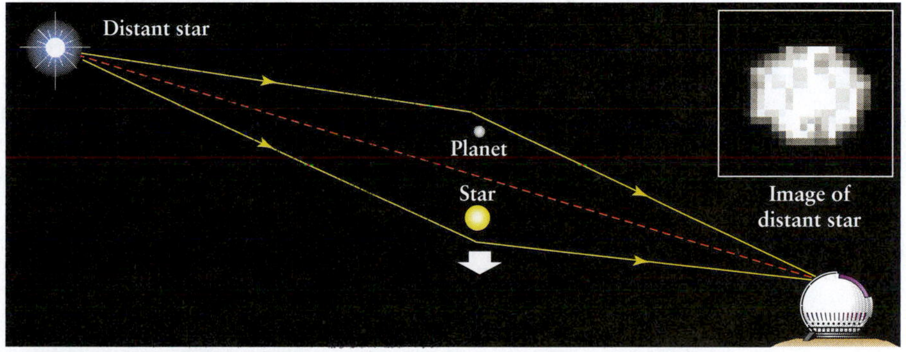

c Microlensing by star and planet

FIGURE 5-18 Microlensing Reveals an Exoplanet Gravitational fields cause light to change direction. As a star with a planet passes between Earth and a more distant star (b), the light from the distant star is focused toward us, making the distant star appear brighter. The focusing of the distant star's light occurs twice, once by the closer star and once by the closer star and its planet (c), making the distant star change brightness. For these simulations, the closer star and planet are 17,000 ly away, while the distant star is 24,000 ly away.

5-12 Dust and exoplanets orbit a breathtaking variety of stars

While some planets are orbiting stars similar to the Sun, many planets orbit burned-out remnants of stars and even orbit pairs or trios of stars. Most of the planets we have found are around stars in the disk of the Milky Way Galaxy (which is where our solar system resides), but a few have been found in large clumps of stars outside our Galaxy's disk. (The word *galaxy* when used alone is capitalized only when referring to our Milky Way.) These clumps, called *globular clusters,* contain very old stars that formed shortly after the universe began. Because it is in a globular cluster, the oldest known planet dates back to the cluster's formation, 13 billion years ago.

The vast majority of the more than 500 exoplanets discovered so far have highly elliptical orbits, with eccentricities up to at least $e = 0.71$ (see Section 2-5). Star systems with massive planets in highly eccentric orbits are unlikely to have life-sustaining Earthlike planets, because the changing location of the massive planet is likely to prevent smaller planets from staying in stable orbits.

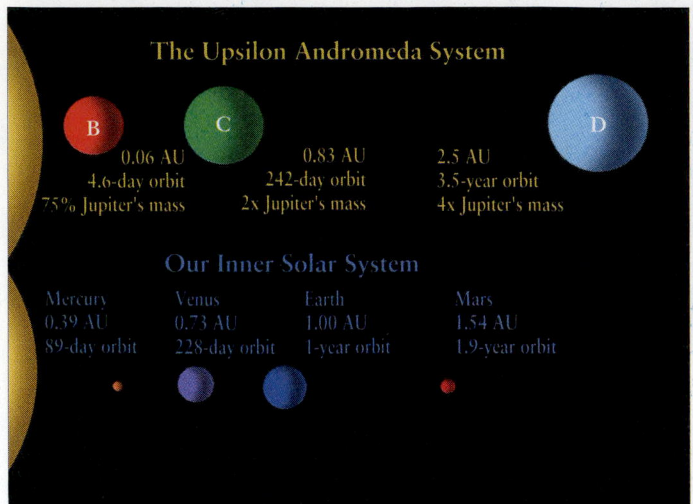

a

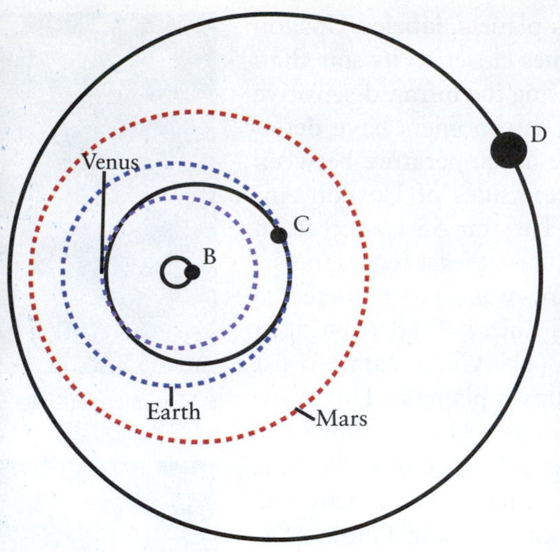

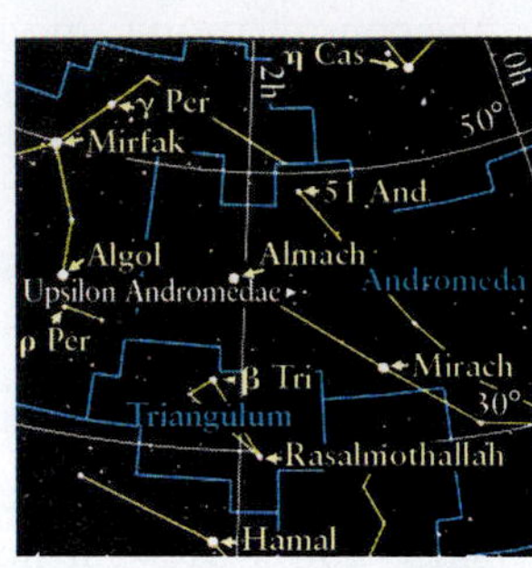

b

FIGURE 5-19 A Star with Three Planets (a) The star Upsilon Andromedae has at least three planets, discovered by measuring the complex Doppler shift of its starlight. This star system is located 44 ly from Earth, and the planets all have masses similar to Jupiter's. (b) The orbital paths of the planets, labeled B, C, and D, along with the orbits of Venus, Earth, and Mars, are drawn for comparison. (Adam Contos, Harvard-Smithsonian Center for Astrophysics)

Given the growing number of exoplanets that are being discovered orbiting a wide range of stars, astronomers are beginning to make estimates for the number of Jupiterlike planets and the number of them orbiting Sunlike stars in our Milky Way Galaxy. These numbers range from between 1 billion and 30 billion planets, of which up to 2 billion are believed to be in orbit around Sunlike stars. All of these numbers will be refined as better estimates are determined for the percentage of star systems with planets.

Although astronomers had expected to find disks of gas and dust orbiting young stars, consistent with the theory explaining the formation of the solar system, they have also found disks of dusty debris orbiting stars as old as the Sun (so they are over 4 billion years old). These stars are all known to have planets, and the rubble surrounding them was most likely formed by the collisions of asteroids and comets with each other and perhaps with the planets in these systems.

We should start finding Earthlike planets that orbit Sunlike stars in the coming years. The problem, until recently, has been the sensitivity of our technologies. The difficulty in finding such planets is primarily that they are so dim, small, and low in mass compared to their stars. Their dimness makes it exceedingly hard for current telescopes to see the starlight they scatter compared to the monumental brightness of the stars themselves. When these planets eclipse their stars, the small sizes of the orbiting planets make the change in the brightness of their star also very hard to detect. Likewise, the low masses of the planets compared to their stars means that they cause very small Doppler shifts of just a few meters per second, which we are only now able to detect.

When we do start detecting terrestrial planets orbiting other stars, which will be the best candidates for supporting life? We are most likely to discover extraterrestrial life (of the sort we are familiar with on Earth) on exoplanets orbiting at about 1 AU from stars similar in mass (within about a factor of 2) to the Sun. At this distance, such stars provide enough heat to keep water liquid, and these stars shine long enough for life to evolve. We discuss the issues related to other life in the universe in Chapter 19.

"ANOTHER ONE UNINHABITED. THAT'S THREE DOWN AND SEVERAL HUNDRED BILLION TO GO."

(ScienceCartoonsPlus.com)

5-13 Frontiers yet to be discovered

The formations of the solar system and other planetary systems are so complex that we are continually making refinements to our models of how these processes started and evolved. Our understanding of the development of the solar system is likely to be greatly advanced by further study of proplyds and planets that orbit other stars. For example, observations of these systems should help us pin down the timescales involved in forming planetary systems. Is the Nice model of planet formation correct? Did the collision of Moon-sized bodies create the terrestrial planets in the solar system?

Another major challenge in solar system astronomy is identifying and determining the properties of the myriad Kuiper belt and Oort cloud objects. What are their compositions? Are there more Kuiper belt objects like Pluto? Although the most poorly understood bodies in the solar system to date are the KBOs, every other object in the solar system also has secrets to be uncovered.

The exoplanets and their environs present many mysteries, such as how Jupiter-mass planets came to be closer to their planets than any such bodies in our solar system and how some planets ended up orbiting in the opposite direction to their stars' rotation. How common are Earthlike planets, and around what kinds of stars do they typically form? Although none of the currently known exoplanets is a candidate to sustain life as we know it, we are likely to eventually discover such worlds.

We begin a detailed exploration of the solar system in Chapter 6 by examining the two bodies we know best—Earth and its Moon. By understanding them, we will be better able to make some sense of the remarkably alien neighboring worlds we encounter in the following three chapters.

SUMMARY OF KEY IDEAS

Formation of the Solar System

• Hydrogen, helium, and traces of lithium, the three lightest elements, were formed shortly after the formation of the universe. The heavier elements were produced much later by stars and are cast into space when stars die. By mass, 98% of the observed matter in the universe is hydrogen and helium.

• The solar system formed 4.6 billion years ago from a swirling, disk-shaped cloud of gas, ice, and dust, called the solar nebula.

• The planets and other debris in the solar system today formed from gas, ice, and dust in the solar nebula orbiting the protosun.

• The outer solar system, beyond the snow line, had both dust and ice (including hydrogen and helium), while inside the snow line, such ices were vaporized by the protosun.

• Jupiter and Saturn were initially worlds of rock and metal that pulled onto themselves large amounts of hydrogen and helium, along with some water.

• Uranus and Neptune were also initially worlds of rock and metal, but they attracted more water and less hydrogen and helium than the other giant planets.

• The Nice model of solar system formation proposes that in the outer solar system Jupiter formed first, followed by Saturn, and then by Neptune and Uranus, which were then flung out to their present orbits by gravitational forces from Jupiter and Saturn.

• The four inner planets formed through the collisions of Moon-sized bodies, probably after the outer four planets were formed.

• The Sun formed at the center of the solar nebula. After about 100 million years, the temperature at the protosun's center was high enough to ignite thermonuclear fusion reactions.

• For 800 million years after the Sun formed, impacts of asteroidlike objects on the young planets dominated the history of the solar system.

Categories of Solar System Objects

• Astronomical objects smaller than the eight planets are classified as dwarf planets or small solar system bodies (SSSBs).

• A variety of other names, including asteroids, comets, meteoroids, trans-Neptunian objects, plutinos, plutoids, Kuiper belt objects (KBOs), and Oort cloud objects, overlap with the designations "dwarf planet" and "SSSB."

• KBOs and Oort cloud objects are trans-Neptunian objects—they orbit farther from the Sun than the outermost planet.

• To date, five objects—Pluto, Ceres, Eris, Haumea, and Makemake—have been classified as dwarf planets.

• Other objects orbit the Sun beyond Neptune. At least 1500 KBOs have been observed. A few potential Oort cloud objects have also been identified.

Comparative Planetology

• The four inner planets of the solar system share many characteristics and are distinctly different from the four giant outer planets.

• The four inner, terrestrial planets are relatively small, have high average densities, and are composed primarily of rock and metal.

• Jupiter and Saturn have large diameters and low densities and are composed primarily of hydrogen and helium. Uranus and Neptune have large quantities of water as well as much hydrogen and helium.

• Pluto, once considered the smallest planet, has a size, density, and composition consistent with other large Kuiper belt objects.

• Asteroids are rocky and metallic debris in the solar system, are larger than about 10 m in diameter, and are found primarily between the orbits of Mars and Jupiter. Meteoroids are smaller pieces of such debris. Comets are debris that contain both ice and rock.

Planets Outside Our Solar System

- Astronomers have observed disks of gas and dust orbiting young stars.
- At least 506 exoplanets have been discovered orbiting other stars.
- Most of the exoplanets that have been discovered have masses roughly equal to the mass of Jupiter.
- Exoplanets are discovered indirectly as a result of their effects on the stars they orbit.

WHAT DID YOU THINK?

1 *How many stars are there in the solar system?* One, the Sun.

2 *Were the Sun and planets among the first generation of objects created in the universe?* No. All matter and energy were created by the Big Bang. However, much of the material that exists in our solar system was processed inside stars that evolved before the solar system existed. The solar system formed billions of years after the Big Bang occurred.

3 *How long has Earth existed, and how do we know this?* Earth formed along with the rest of the solar system, about 4.6 billion years ago. The age is determined from the amount of radioactive decay that has occurred on Earth.

4 *What typical shape(s) do moons have, and why?* Although some moons are spherical, most look roughly like potatoes. Those that are spherical are held together by the force of gravity, which pulls down high regions. Those that are potato-shaped are held together by the electromagnetic interaction between atoms, just like rocks. These latter moons are too small to be reshaped by gravity.

5 *Have any Earthlike planets been discovered orbiting Sunlike stars?* Not really. Most exoplanets are Jupiterlike gas giants. The planets similar in mass and size to Earth are either orbiting remnants of stars that exploded or, in the case of Gliese 581, a star much less massive and much cooler than the Sun.

Key Terms for Review

accretion, 137
albedo, 146
asteroid, 141
asteroid belt, 140
average density, 144
comet, 141
crater, 141
dense core, 134
dwarf planet, 141
Jeans instability, 134
Kuiper belt, 139
Kuiper belt object (KBO), 139
meteoroid, 141
metals, 134
microlensing, 149
moon (natural satellite), 142
Nice model, 134
Oort cloud, 139
orbital inclination, 142
planet, 141
planetesimal, 137
protoplanetary disks (proplyds), 135
protosun, 135
small solar system body (SSSB), 142
snow line, 135
solar nebula, 134
solar system, 132
terrestrial planet, 144
trans-Neptunian objects (TNOs), 139

Review Questions

1. About how long after the universe came into existence did our solar system form? **a.** 0 years (they formed together), **b.** a million years, **c.** 10 million years, **d.** a billion years, **e.** 10 billion years

2. Pluto is most similar in composition to which of the following objects? **a.** Eris **b.** Jupiter **c.** our Moon **d.** Earth **e.** the Sun

3. Which planets are terrestrial and which are giant?

4. To test your understanding of the formation of the solar system, do Interactive Exercise 5.1 on the Web site. Explain what is happening in each figure. You can print out your results, if required.

5. Describe four methods for discovering exoplanets.

6. Which giant planet formed first?

7. According to the Nice theory, where did the Kuiper belt objects and Oort cloud objects come from?

8. What created most of the craters in the solar system? What else could create craters? (*Hint:* Think of craters on Earth.)

9. Jeans instability is responsible for what event in the life of the solar system?

10. Name and briefly describe one small solar system body.

11. Name one dwarf planet (other than Pluto) and state where it is located.

12. Asteroids, meteoroids, and comets are also classified as what (one) kind of object today?

Advanced Questions

13. What two properties of a planet must be known to determine its average density? How are these properties determined?

14. How can Neptune have more mass than Uranus but a smaller diameter? *Hint:* See Table 5-2.

15. How would a planet orbiting one of the first-generation stars be different than planets formed today? What chemical elements would that planet contain? Could we humans exist on that world? Why or why not?

16. What is significant about the snow line described in this chapter?

What If...

17. Earth had a highly elliptical orbit, like Mercury? What would be different about Earth and life on it?

18. The accretion process of planet formation were still going on in the solar system? How would life on Earth be different, in general, and how might humans be different?

19. The solar system passed through a cloud of gas and dust that was beginning to collapse to form a new star and planet system? What might happen to Earth, and how would the passage affect the appearance of the sky?

Web Questions

20. Search the Web for information about recent observations of protoplanetary disks. What insights about the formation of the solar system have astronomers gained from these observations? Explain the evidence astronomers have, from these observations, that planets are forming in these disks.

21. In 2000, exoplanets with masses comparable to the mass of Saturn were first detected around the stars HD 16141 (also called 79 Ceti) and HD 46375. Search the Web for information about these planets. How do their masses compare to previously discovered exoplanets? Do these planets move around their stars in the same kind of orbit as Saturn follows around the Sun? If not, explain the differences.

22. Search the Web for information about the planet that orbits the star HD 209458. This planet has been detected using two methods. What are they? What have astronomers learned about this planet?

23. **Determining Terrestrial Planet Orbital Periods.** Access the animation "Planetary Orbits" in Chapter 5 of the *Discovering the Universe* Web site. Focus on the motions of the inner planets during the last half of the animation. Using the stop and start buttons, determine how many days it takes Mars, Venus, and Mercury to orbit the Sun once if it takes Earth approximately 365 days.

24. Search the Web for information about the cool, dim star Gliese 581. What is its mass compared to our Sun's? How bright is it (in other words, what is its luminosity) compared to the Sun? How many planets are known to be orbiting it? How do their masses compare to Earth's? What is special about the planet Gliese 581c?

Observing Projects

25. Use the *Starry Night*™ planetarium software to get an introductory overview of the solar system. Launch *Starry Night*™ and select **View > Ecliptic Guides > The Ecliptic** and **Labels > Planets-Moons** from the menu. Click the **Sunset** button in the toolbar to set the time to sunset today at your home location. **a.** Use the hand tool or gaze controls to examine the sky near the ecliptic and make a list of the planets that are in your sky. **b.** Advance the time in the toolbar by 1 hour and search along the ecliptic again for planets and add them to your list.
c. Repeat this procedure until you have reached sunrise.
d. Set the date for 6 months from now and repeat **a–c.** How do the lists compare? Explain their similarities and differences.

26. Many young stars are still embedded in the clouds of gas and dust from which they, and any associated planets, were formed. For example, from the northern hemisphere, the famous Orion Nebula is visible in the winter evening sky, while the Lagoon, Omega, and Trifid nebulae are visible in the summer night sky. Examine some of these nebulae with a telescope. Describe their appearance. As you look at each nebula, try to determine which stars in your telescope's field of view are actually associated with the nebulosity. Use the **Find** pane in *Starry Night*™ to help you to find these nebulae. The table below gives their coordinates (right ascension and declination).

Nebula	Right ascension	Declination
Lagoon	$18^h\ 03.8^m$	−24° 23′
Omega	$18^h\ 20.8^m$	−16° 11′
Trifid	$18^h\ 02.3^m$	−23° 02′
Orion	$5^h\ 35.4^m$	−5° 27′

27. Use the *Starry Night*™ program to examine magnified images of the terrestrial planets Mercury, Venus, Earth, and Mars. Select the appropriate view from the **Discovering the Universe** folder in the **Favourites** pane. **a.** Describe each planet's appearance. From what you observe in each case, is there any way of knowing whether you are looking at a planet's surface or at the tops of complete cloud cover over the planet? **b.** Which planet or planets have clouds? **c.** If a planet has clouds, open its contextual menu and choose **Surface Image/ Model > Default** to remove these clouds and use the location scroller to examine the planet's surface. Which planet shows the heaviest cratering? **d.** Which of these planets show evidence of liquid water? **e.** What do you notice about Venus's rotation compared to the other planets?

28. Use the *Starry Night*™ program to investigate stars in our near neighborhood that have planets orbiting them. First, click the **Home** button in the toolbar. Open the **Options** pane and use the checkboxes in the **Local View** layer to turn off **Daylight** and the **Local Horizon.** Then use the **Find** pane to find and center upon each of the following stars: (i) 47 Ursae Majoris; (ii) 51 Pegasi; (iii) 70 Virginis; (iv) Rho Coronae Borealis. To do this, click the magnifying glass icon on the left side of the edit box at the top of the **Find** pane and select **Star** from the drop-down menu; then type the name of the star in the edit box and press the Enter or Return key on the keyboard. Click on the **Info** tab for full information about each star. Expand the **Other Data** layer and note the luminosity of each of these stars. **a.** Which stars are more luminous than the Sun? **b.** Which are less luminous? **c.** How do you think these differences would have affected temperatures in the nebula in which each star's planets formed?

WHAT IS A PLANET?

The controversial new official definition of "planet," which banished Pluto, has its flaws but by and large captures essential scientific principles.

Adapted from an article by Steven Soter

Most of us grew up with the conventional definition of a planet as a body that orbits a star, shines by reflecting the star's light and is larger than an asteroid. Although imprecise, the definition categorized the bodies we knew at the time. But a series of discoveries in the 1990s rendered it untenable. Beyond Neptune's orbit, astronomers found hundreds of icy worlds occupying a doughnut-shaped region called the Kuiper belt. Around scores of other stars, they found other planets whose orbits look nothing like those in our solar system. They also discovered brown dwarfs, blurring the distinction between planet and star, and found planet-like objects drifting through interstellar space.

These findings ignited debate, prompting the International Astronomical Union (IAU) to redefine a planet as an object orbiting a star that is large enough to have settled into a round shape, and "has cleared the neighborhood around its orbit." This new definition stripped Pluto of its status as a planet. Some astronomers refused to use it and organized a protest petition.

When Earth Became a Planet

Astronomers' reevaluation of the nature of planets has deep historical roots. The ancient Greeks recognized seven lights in the sky that moved against the background pattern of stars: the sun, the moon, Mercury, Venus, Mars, Jupiter, and Saturn. They called them *planetes,* or wanderers. Note that Earth was not on this list. For most of human history, Earth was regarded not as a planet but as the center of the universe. After Nicolaus Copernicus proved the Sun lies at the center, astronomers redefined planets as objects orbiting the Sun, adding Earth to the list and deleting the Sun and the Moon. Telescope observers added Uranus in 1781 and Neptune in 1846.

Ceres, discovered in 1801, was initially welcomed as the missing planet between Mars and Jupiter. But astronomers began to have doubts when they found Pallas in a similar orbit the following year. Unlike the classical planets, which telescopes revealed as little disks, these bodies appeared as pinpricks of light. English astronomer William Herschel named them "asteroids." By 1851, 15 had been seen and astronomers began listing them by order of discovery rather than distance from the Sun, as for planets, making them a distinct population. If we still counted asteroids as planets, our solar system would now have more than 135,000 planets.

Pluto's story is similar. When Clyde Tombaugh discovered it in 1930, astronomers considered it the long-sought "Planet X" whose gravity would account for unexplained peculiarities in Neptune's orbit. But Pluto turned out to be smaller than all eight planets and seven of their satellites, including Earth's moon, and Neptune's orbital peculiarities proved to be non-

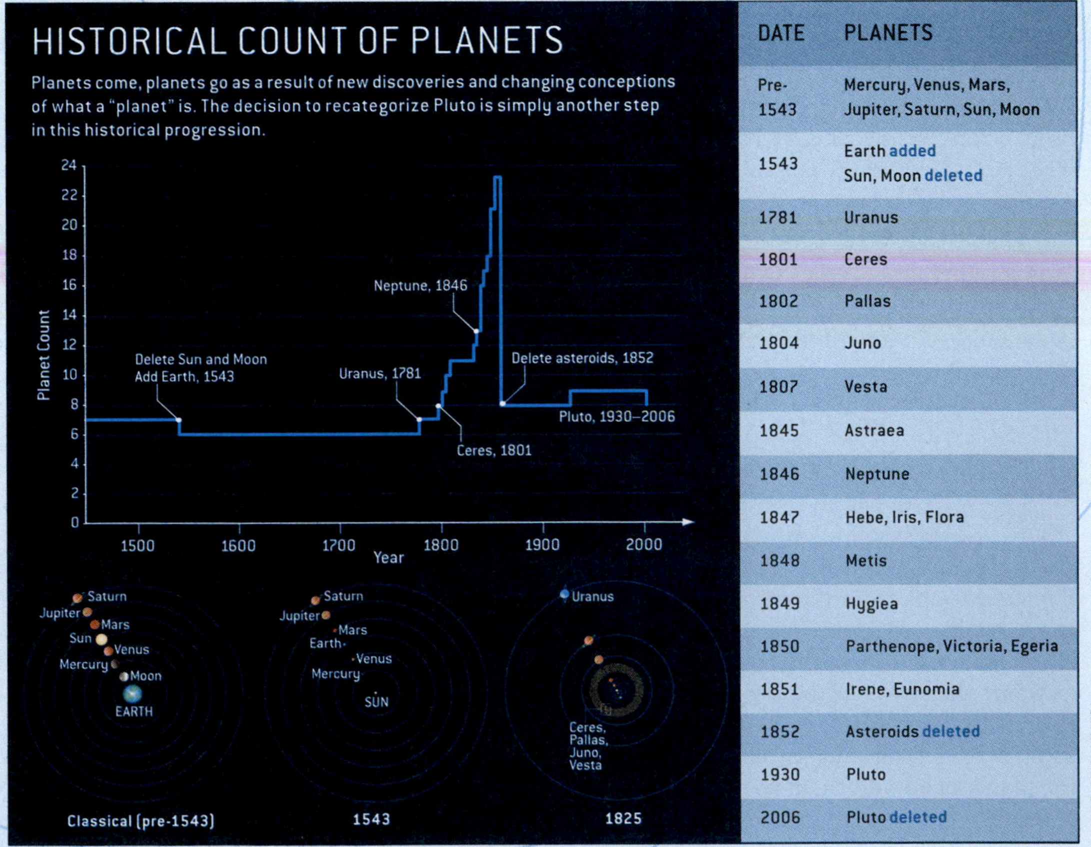

DATE	PLANETS
Pre-1543	Mercury, Venus, Mars, Jupiter, Saturn, Sun, Moon
1543	Earth added Sun, Moon deleted
1781	Uranus
1801	Ceres
1802	Pallas
1804	Juno
1807	Vesta
1845	Astraea
1846	Neptune
1847	Hebe, Iris, Flora
1848	Metis
1849	Hygiea
1850	Parthenope, Victoria, Egeria
1851	Irene, Eunomia
1852	Asteroids deleted
1930	Pluto
2006	Pluto deleted

LUCY READING-IKKANDA

existent. For six decades, Pluto was a unique anomaly at the outer edge of the planetary system.

When astronomers discovered that Pluto was one of many Kuiper belt objects (KBOs), they began to reconsider whether it should still be called a planet. Historically, revoking its status would not be unprecedented; the ranks of ex-planets include the Sun, the Moon, and asteroids. Nevertheless, many people have argued for continuing to call Pluto a planet, because almost everyone has grown accustomed to thinking of it as one.

The discovery in 2005 of Eris (formerly known as 2003 UB313 or Xena), a KBO even larger than Pluto, brought the issue to a head. If Pluto is a planet, then Eris must also be one, along with other large KBOs. On what objective grounds could astronomers decide?

Clearing the Air

To avoid a proliferation of planets, astronomers considered defining a planet as a body smaller than a star, but large enough for its gravity to form it into a round shape. Most bodies larger than several hundred kilometers are round; smaller ones often resemble giant boulders. Under these guidelines, Pluto, Ceres, and potentially dozens of KBOs would be considered planets.

But it is very difficult to observe the shapes of distant KBOs, and both asteroids and KBOs span an almost continuous spectrum of sizes and shapes. Ultimately, categorizing planets by degree of roundness seemed arbitrary.

An additional criterion provided a clear way to distinguish planets: They must be massive enough to dominate their orbital zone by flinging smaller bodies away, sweeping them up in direct collisions, or holding them in stable orbits. Lesser bodies occupy transient, unstable orbits or have a heavyweight guardian that stabilizes their orbits. For example, Earth controls bodies like near-Earth asteroids that stray too close and holds its moon in orbit. Each of the four giant planets protects many orbiting satellites. Jupiter and Neptune also maintain families of asteroids and KBOs (called Trojans and Plutinos, respectively) in special orbits called stable resonances, where an orbital synchrony prevents collisions with the planets.

The eight planets from Mercury through Neptune are thousands of times more likely to sweep up or deflect small neighboring bodies than even the largest asteroids and KBOs, which include Ceres, Pluto, and Eris. Asteroids, comets, and KBOs—including Pluto—live amid swarms of comparable bodies.

Planets grew from a flattened disk of gas and dust orbiting the primordial Sun. In the competition for limited raw material, some bodies won out, with a single large body dominating each orbital zone. The smaller bodies were swept up by the larger ones, ejected from the solar system or swallowed by the Sun, and the survivors became the planets we see today. The asteroids and comets, including the KBOs, are the leftover debris.

Endgame

The revised IAU definition may require additional guidelines to delineate what degree of clearing makes a celestial body a planet. But it removes the need for an upper mass limit to distinguish planets from both stars and brown dwarfs. It also classifies rare brown dwarf companions that orbit close to stars as planets.

The difference between planets and nonplanets is quantifiable in theory and by observation. All the planets in our solar system have enough mass to have swept up or scattered away most of the original planetesimals from their orbital zones. Today each planet contains at least 5,000 times more mass than all the debris in its vicinity. The asteroids, comets, and KBOs, including Pluto, live amid swarms of comparable bodies.

One objection to this definition is that astronomical objects should be classified by properties such as size, shape, or composition, not by location or dynamical context. This argument overlooks the fact that astronomers classify objects that orbit planets as "moons." Context and location are important. Distance from the Sun determined that close-in bodies became small, rocky planets and farther ones became gas and ice-rich giant planets. Planets dynamically dominate a large volume of orbital space—while asteroids, KBOs, and ejected planetary embryos do not. The eight planets are the dominant end products of disk accretion and differ recognizably from the vast populations of asteroids and KBOs.

The historical definition of nine planets retains a sentimental attraction. For 76 years, our schools taught that Pluto was a planet. While some argue that culture and tradition suffice to leave it that way, science cannot remain bound to misconceptions and must be derived from the natural world. The evolving definition of a planet reflects new discoveries and profound changes in our understanding of the solar system.

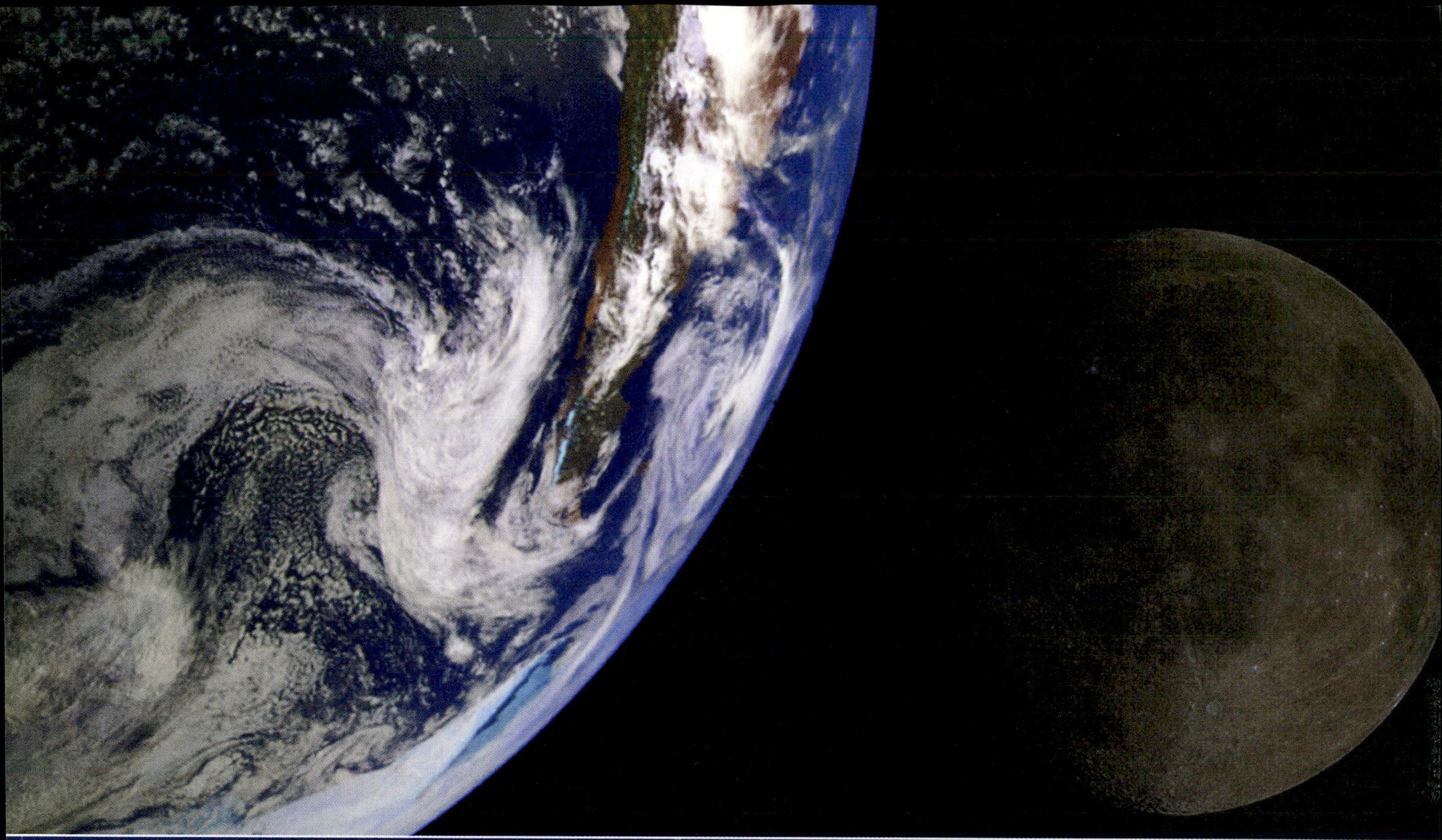

Chapter 6
Earth and the Moon

RIVUXG Earth and the Moon as seen from space. (NASA)

WHAT DO YOU THINK?

1. Can Earth's ozone layer, which has been partially depleted, be naturally replenished?
2. Who was the first person to walk on the Moon, and when did this event occur?
3. Do we see all parts of the Moon's surface at some time throughout the lunar cycle?
4. Does the Moon rotate and, if so, how fast?
5. What causes the ocean tides?
6. When does the spring tide occur?

Answers to these questions appear in the text beside the corresponding numbers in the margins and at the end of the chapter.

VIDEO 6.1

Imagine you are entering our solar system from a distant star system, looking for habitable worlds. Your spacecraft approaches the inner solar system on the opposite side of the Sun from Earth. You encounter Mars, with its thin, chilled, unbreathable atmosphere and barren desert landscapes. Not very promising. Moving much closer to the Sun, the next planet you happen to encounter, Venus, is enshrouded in corrosive clouds hiding a menacingly hot surface. Finally, moving around the Sun and heading back outward you spy Earth, orbiting our star between these two relatively forbidding worlds—one too cold, the other too hot. Ever-changing white clouds pirouette above the browns and blues of Earth's continents and oceans. Its close companion, the Moon, pockmarked by countless craters, causes those oceans to move up and down in a ceaseless rhythm. Is Earth what you are looking for?

In this chapter, we will explore Earth and the Moon, noting what makes Earth suitable for life. Our detailed and rapidly developing knowledge of Earth allows us to present material on topics that range from the magnetic fields surrounding it, to its atmosphere and climate, to the depth and composition of its crust, and to the properties extending to its metal-rich core. As you will see, our knowledge about the Moon is much more limited. We will explore in this chapter and the next one other terrestrial worlds with somewhat different properties from Earth, leading to vastly different environments from our own.

In this chapter you will discover

- why Earth is such a satisfactory environment for life
- that Earth is constantly in motion inside and out
- how Earth's magnetic field helps protect us
- what made the craters on the Moon
- how the Sun and the Moon cause Earth's tides
- that Earth and the Moon each have two (different) major types of surface features
- that water ice has been found on the Moon

R I V U X G

FIGURE 6-1 Views of Earth's Surface An oasis in the forbidding void of space, Earth is a world of unsurpassed beauty and variety. Ever-changing cloud patterns drift through its skies. More than two-thirds of its surface is covered with oceans. This liquid water, in combination with a huge variety of chemicals from its lands, led to the formation and evolution of life over most of the planet's surface. (NASA)

EARTH: A DYNAMIC, VITAL WORLD

Looking at Earth from outer space (Figure 6-1), one first notices that its clouds, water, and land combine to scatter, on average, 37% of the sunlight it receives right back into space. That amount of scattered light (an albedo of 0.37) and its variation throughout the year are unique among the planets in our solar system. In contrast, cloud-covered worlds, such as Venus, scatter much more light, while bone-dry worlds, such as Mercury, Mars, or our Moon, scatter much less. Furthermore, Earth's albedo varies, consistent with changing cloud and ice layers, as well as with different amounts of land and water facing the Sun. Water, the nearly universal solvent in which life on Earth first formed, covers about 71% of our planet's surface. (A solvent is a liquid in which various substances can dissolve.)

Earth is a geologically active, ever-changing world. Earthquakes shake many regions of it. Volcanoes pour huge quantities of molten rock from just under the crust (the outer layer) onto the surface, and gas released from this lava vents into the oceans and atmosphere. Some mountains are still rising, while others are wearing away. Flowing water erodes topsoil and carves river valleys. Rain and snow help rid the atmosphere of dust particles—and life teems virtually everywhere, making Earth unique in the solar system. Figure 6-2 details Earth's important physical and orbital properties. The symbol ⊕ in Figure 6-2 and hereafter is astronomers' shorthand for Earth. The other planets have different symbols that we will encounter later.

6-1 Earth's atmosphere has evolved over billions of years

Earth's atmosphere is a key part of our planet's dynamic activity and its ability to sustain life. Essential to the existence of complex life is air that contains oxygen mole-

EARTH: VITAL STATISTICS	
Planet symbol:	⊕
Average distance from the Sun:	1.000 AU = 1.496×10^8 km
Maximum distance from the Sun:	1.017 AU = 1.521×10^8 km
Minimum distance from the Sun:	0.983 AU = 1.471×10^8 km
Orbital eccentricity:	0.017
Average orbital speed:	29.79 km/s
Sidereal period of revolution (year):	365.26 days (1.00 year)
Sidereal rotation period:	0.997 days
Sidereal rotation period (day):	1.00 day = 24.0 h
Inclination of equator to orbit:	23.5°
Diameter (equatorial):	12,756 km
Mass:	5.974×10^{24} kg ($1M_\oplus$)
Average density:	5520 kg/m^3
Albedo (average):	0.31
Escape speed:	11.2 km/s
Surface temperature range:	Maximum 60°C = 140°F = 333K Mean: 14°C = 57°F = 287 K Minimum: −90°C = −130°F = 183K
Atmospheric composition (by number of molecules):	78.08% nitrogen (N_2), 20.95% oxygen (O_2), 0.035% carbon dioxide (CO_2), about 1% water vapor

R I V U X G

FIGURE 6-2 Earth's Vital Statistics The planet symbol for Earth is ⊕. Other planets have different symbols; these are often used as shorthand for information about each world. For example, the mass of Earth is often denoted $M_\oplus$. This figure provides information about Earth's physical and orbital properties. Astronauts spend many hours watching our world as they orbit it. The image here gives you an idea of why they find it so fascinating. (NASA)

cules. The air we breathe is predominantly a 4-to-1 mixture of nitrogen to oxygen, two gases that are found only in small amounts in the atmospheres of other planets.

The theory of planet formation and the models that attempt to reproduce the early history of Earth indicate that the present atmosphere is the third one to envelop our planet. Earth's earlier atmospheres contained no free oxygen. Composed of trace remnants of hydrogen and helium left over from the formation of the solar system, the first atmosphere did not last very long. The atoms and molecules in these gases were too light to stay near Earth. Heated by sunlight (meaning that they moved faster and faster), these particles gained enough kinetic energy to overcome Earth's gravitational pull and fly away into space.

The gases of the second atmosphere came from inside Earth, vented through volcanoes and cracks in Earth's surface. That atmosphere was composed primarily of carbon dioxide and water vapor, along with some nitrogen. These gases consist of more massive atoms than the hydrogen and helium in the first atmosphere. Therefore, the speeds imparted to the gases of the second (and third) atmospheres by sunlight were not high enough to enable them to leave the Earth's gravitational bonds. That is why Earth developed a permanent atmosphere.

There was roughly 100 times as much gas in that second atmosphere as there is in the air today. Carbon dioxide and water store a great deal of heat, helping to create what is called the *greenhouse effect,* about which we will say more shortly. The high concentration of these gases early in Earth's life stored more heat than the atmosphere does today. Earth back then would have been a serious hothouse had the Sun reached its full intensity at that time. However, the Sun was dimmer and cooler back then, so the high concentrations of carbon dioxide and water in the air actually prevented the young Earth from freezing.

Oceans began forming when water precipitated out of this atmosphere, as well as from impacts of water-rich space debris, such as the comet nuclei mentioned in Chapter 5. These events occurred within 300 million years of Earth's formation. Indeed, geologists have discovered evidence for liquid water on Earth's surface 4.3 billion years ago. The oceans eventually soaked up about half the carbon dioxide in that second atmosphere, as the rain absorbed it from the air and carried it earthward. (Water holds a great deal of carbon dioxide, as you can see by the amount of carbon dioxide fizz in a can of soda.) As life evolved in the oceans, much of that dissolved carbon dioxide was transformed into the shells of many organisms and then, as the organisms died, their carbon-rich shells sank to the ocean bottoms. Shells piled up and compressed the shells underneath them into rock, such as limestone.

Early plant life in the oceans and then on the shores removed most of the remaining carbon dioxide in the air by converting it into oxygen. The most efficient of such conversion mechanisms is photosynthesis, which today helps maintain the balance between carbon dioxide and oxygen in the air. The molecules that remained in the air as the carbon dioxide was being removed were mostly nitrogen, argon, and water vapor. Argon interacts very little with other atoms—it is an inert gas.

Oxygen that entered the air early in Earth's history came from oxygen-rich molecules, like carbon dioxide, that were broken apart by the Sun's ultraviolet radiation. Then came oxygen released as a by-product of ocean plant life activity. In both cases, this oxygen did not stay in the air for long. Oxygen is highly reactive, and early atmospheric oxygen combined quickly with many elements on Earth's surface, notably iron, to form new compounds, such as iron oxides (otherwise known as rust).

About 2 billion years ago, after all of the surface minerals that could combine with oxygen had done so, the atmosphere began to fill with this gas. With the carbon dioxide reduced to a very low level, the air became the primarily nitrogen-oxygen mixture that we breathe today. Although nitrogen was only a tiny fraction of that earlier atmosphere, it now dominates the thinner air of today.

Because air has mass, it is pulled down toward Earth by gravity, thereby creating a pressure on the surface. Recall that we define pressure as a force acting over some area:

$$\text{Pressure} = \frac{\text{Force}}{\text{Area}}$$

The weight of the air pushing down creates a pressure at sea level of 14.7 pounds per square inch, which is commonly denoted as 1 atmosphere (atm). That value is the pressure acting on your skin. The atmospheric pressure decreases with increasing altitude, falling by roughly half with every gain of 5.5 km of altitude. Seventy-five percent of the mass of the atmosphere lies within 11 km (approximately 7 mi, or 36,000 ft) of Earth's surface.

Lower Layers of the Atmosphere All of Earth's weather—clouds, rain, sleet, and snow—occurs in the lowest layer of the atmosphere, called the **troposphere.** Commercial jets generally fly at the top of this layer to avoid having to plow through the denser air below. If Earth were the size of a typical classroom globe, the troposphere would be only as thick as a sheet of paper wrapped around its surface.

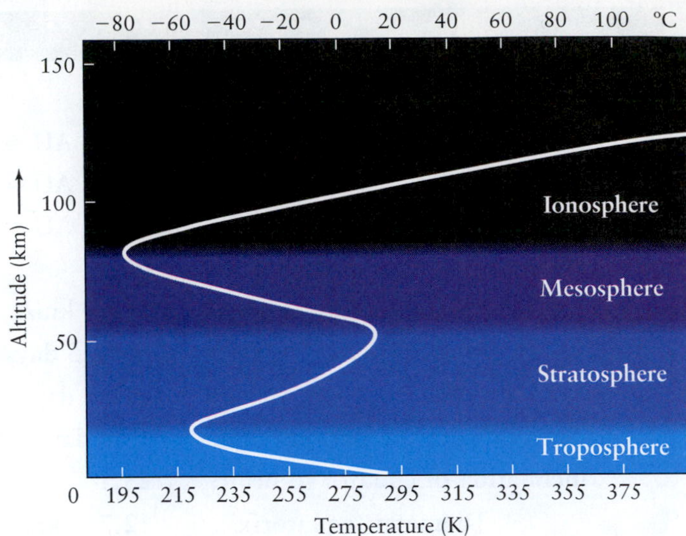

FIGURE 6-3 Temperature Profile of Earth's Atmosphere The atmospheric temperature changes with altitude because of the way sunlight and heat from Earth's surface interact with various gases at different heights.

Starting at Earth's surface, atmospheric temperature initially decreases as you go upward, reaching a minimum of 218 K (−67°F) at the top of the troposphere. Above this level is the region called the **stratosphere,** which extends from 11 to 50 km (approximately 7 to 31 mi) above Earth's surface. This region is the realm of the **ozone layer** (Figure 6-3).

Insight Into Science

What's in a Word? The word "layer" in the term *ozone layer* suggests a narrow region, perhaps a few meters thick. Not so. The ozone layer extends in altitude over some 40 km of the atmosphere!

The Ozone Layer The ozone layer has been much in the news in recent years. What is it, and how threatened is it? Ozone molecules (three oxygen atoms bound together, O_3) are created in the stratosphere when short-wavelength ultraviolet radiation from the Sun strikes and splits normal oxygen molecules, O_2, into two separate oxygen atoms. Each resulting oxygen atom then combines with different O_2 molecules to create ozone, O_3. Once created, ozone efficiently absorbs intermediate-wavelength solar ultraviolet rays, thereby heating the air in this layer while preventing much of this lethal radiation from reaching Earth's surface.

The troposphere (the lowest part of the atmosphere) is heated primarily by Earth, which is why the temperature in this part of the atmosphere decreases with altitude—the farther away from the heat source of Earth, the cooler the air gets. However, because the ozone in the stratosphere stores significant heat from the Sun above it, the temperature in the stratosphere actually rises with altitude to about 285 K (50°F) at its top (see Figure 6-3).

Not much ozone actually exists in the ozone layer. If all of the ozone were compressed to the density of the air we breathe, it would be a layer only a few millimeters (about $^1/_8$ in.) thick. This amount of ozone is enough, however, to protect us from most of the Sun's lethal intermediate-wavelength ultraviolet radiation.

Ultraviolet radiation has been a double-edged sword in the history of Earth. Because Earth's early atmosphere lacked ozone, more ultraviolet radiation penetrated to the planet's surface, where it provided much of the energy necessary for life to form and evolve in the oceans. After aquatic life developed, however, the formation of the ozone layer by the same ultraviolet radiation was essential in lessening the amount of ultraviolet radiation that reached the surface, thereby allowing life-forms to leave the oceans and survive on land.

Today, the ozone layer is still vital, because the energy it absorbs would otherwise cause skin cancer and various eye diseases. In sufficiently high doses, ultraviolet radiation even damages the human immune system by destroying some of the molecules that normally repair cells in our bodies. Unfortunately, over the past few decades, the ozone layer has been partially depleted by human-made chemicals rich in bromine or chlorine, such as chlorofluorocarbons (CFCs). Until the past few years, ozone levels in the midlatitudes of the northern hemisphere have been decreasing by about 4% per decade. Much more rapid and dramatic drops in the ozone layer over Earth's poles, especially over the South Pole, have been observed for more than two decades. These seasonal *ozone holes* typically extend over 24 million km^2, or 4.6% of Earth's surface area. There are some hints that today the ozone layer is decreasing more slowly and that the ozone holes are diminishing, but these trends are still unclear.

1 Fortunately, as explained earlier, sunlight creates ozone. If current international efforts aimed at banning ozone-depleting compounds (such as some previously common refrigerants) continue, the ozone layer will naturally replenish itself in a century.

Upper Layers of the Atmosphere Note in Figure 6-3 that above the stratosphere lies the **mesosphere.** Atmospheric temperature again declines as you move up through the mesosphere, reaching a minimum of about 200 K (−103°F) at an altitude of about 80 km (50 mi). This minimum marks the bottom of the **ionosphere,** also called the **thermosphere,** above which the Sun's ultraviolet radiation heats and ionizes atoms, producing charged particles that reflect radio waves.

In this era of burgeoning satellite usage and human space travel, the effects of the Sun's electromagnetic radiation, along with the flow of particles from it and from other sources, on Earth's atmosphere and on astronauts are increasingly important. As a result, *space weather* satellites are now continually monitoring our neighborhood in space and warning about solar events that could imperil humans and machines out there.

FIGURE 6-4 The Greenhouse Effect in a Car The glass windows in this car allow visible light to enter but prevent the infrared radiation released by the car's interior from escaping. The infrared radiation, therefore, heats the air in the car to a higher temperature than the outside air. This action also occurs in greenhouses and is called the greenhouse effect. (Tobi Zausner)

Carbon Dioxide and the Greenhouse Effect Today

Perhaps you have had the experience of parking your car in the sunshine on a warm summer day. You roll up the windows, lock the doors, and go on an errand. After a few hours, you return to discover that the interior of your automobile has become considerably hotter than the outside air temperature (Figure 6-4).

What happened to make your car so warm? First, sunlight entered it through the windows. This radiation was absorbed by the dashboard and the upholstery, increasing their temperatures. Because they become warm, your dashboard and upholstery emit infrared radiation, which you detect as heat and which your car windows do not permit to escape. This energy is therefore trapped inside your car and absorbed by the air and interior surfaces. As more sunlight comes through the windows and is trapped, the temperature continues to increase. The same trapping of sunlight warms greenhouses, as well as the entire Earth.

ANIMATION 6.1 The greenhouse effect created by the carbon dioxide (CO_2) in the second atmosphere strongly heated the young Earth. When the carbon dioxide was dramatically reduced, that effect also decreased, but it never went away completely.

The greenhouse effect in our atmosphere works like this: All wavelengths of electromagnetic radiation from the Sun approach Earth's atmosphere. Some of this energy is scattered by clouds, land, and water right back into

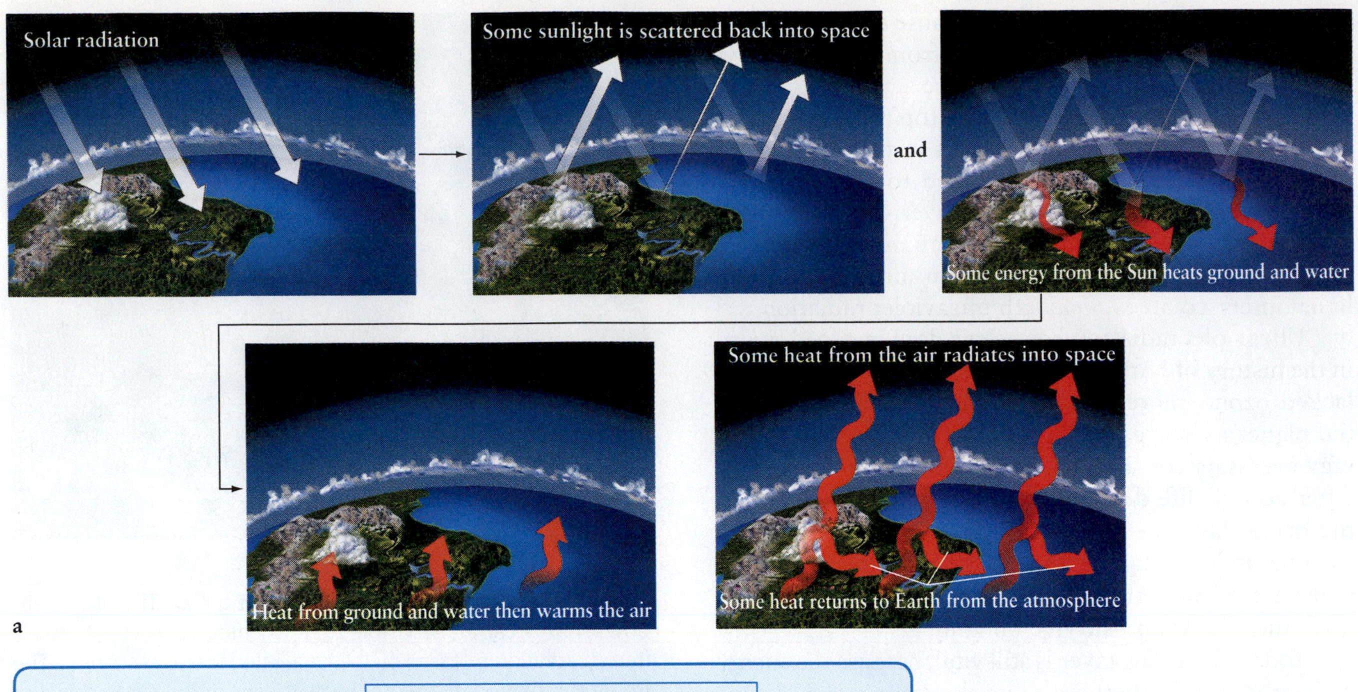

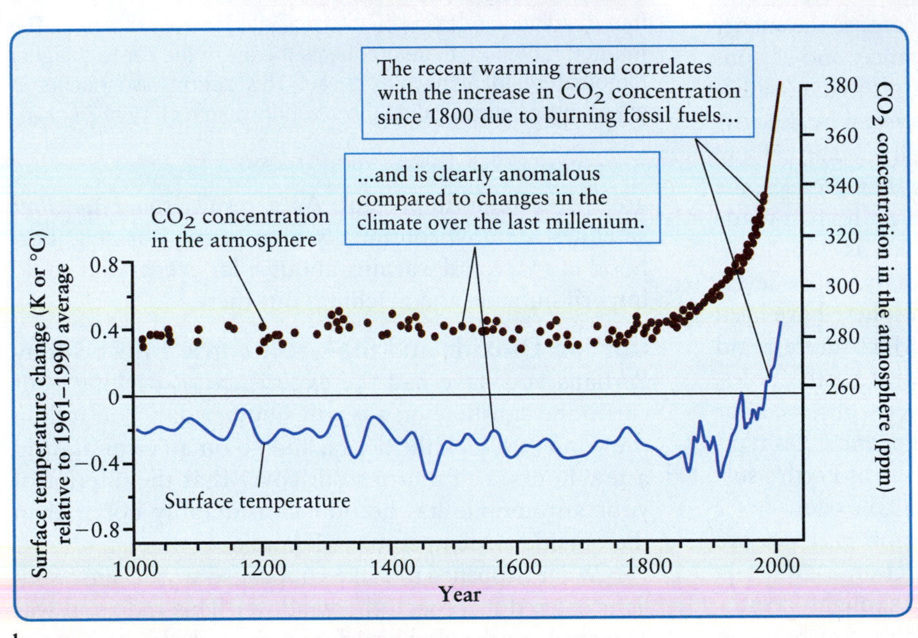

FIGURE 6-5 The Greenhouse Effect (a) Sunlight and heat from Earth's interior warm Earth's surface, which in turn radiates energy, mostly as infrared radiation. Much of this radiation is absorbed by atmospheric carbon dioxide and water, heating the air, which in turn increases Earth's temperature even further. In equilibrium, Earth radiates as much energy as it receives. (b) The amount of carbon dioxide in our atmosphere since 1000 A.D. has been determined. The increase in carbon dioxide since 1800 due to burning fossil fuels and deforestation have caused a dramatic temperature increase.

space (Figure 6-5a), while some is absorbed in the atmosphere, thereby heating it. Some visible light, along with some ultraviolet radiation and radio waves, penetrate to Earth's surface and is absorbed by it. This energy helps heat the planet (along with heat from the air and from inside the planet) until the surface material can store no further energy. Behaving nearly like a blackbody, Earth radiates the excess energy, primarily as infrared radiation (where its blackbody spectrum peaks).

Unlike most visible light, which passes relatively easily through the gases in our atmosphere, infrared radiation is absorbed by certain gases, especially water, carbon dioxide, and ozone in the air. The energy from these photons further heats the atmosphere. Carbon dioxide levels are increasing in Earth's atmosphere, primarily as a result of human activity, such as the burning of fossil fuels. Increased atmospheric carbon dioxide stores more heat in the air, and the air becomes warmer. Normally, carbon dioxide is removed from the air (and oxygen added) by the process of photosynthesis in green leaves. With increased deforestation and the expansion of cities, fewer plants and trees are left, so the carbon dioxide in the air remains there longer, overheating the air and creating *global warming*. Figure 6-5b shows how the CO_2 concentration has changed over the past millennium.

The consequences of the recent global warming trend are quite complex and serious. They include the melting of glaciers and polar ice caps, rising sea levels, changes in ocean heating and circulation patterns, changes in weather patterns worldwide, and increases in undesirable weather events, such as hurricanes, tornadoes, droughts, floods, heat waves, and shoreline erosion. With scientific data on the human effects that add to global warming in hand, many nations of the world are working together to decrease its causes.

6-2 Plate tectonics produce major changes on Earth's surface

Large landmasses called continents protrude through Earth's oceans. These landmasses are regions of Earth's outermost layer, or **crust,** that are composed of relatively low-density rock (compared to the rock at the bottoms of the oceans). By studying the various kinds of rocks at a particular location on a continent, geologists can deduce the history of that site. For example, whether an area was once covered by an ancient sea or flooded by lava from volcanoes is readily apparent from the kinds of rocks currently found there.

The crust has distinct boundaries (an example is shown in Figure 6-6), many occurring beneath Earth's oceans. These divisions suggest that the continents are separate bodies. Earthquakes and volcanoes often occur near the boundaries, hinting at activity taking place below Earth's crust.

Anyone who carefully examines the shapes of Earth's continents might conclude that landmasses move. Eastern South America, for example, looks like it would fit snugly against western Africa. Between 1912 and 1915, the German scientist Alfred Wegener published his observations on this remarkable fit between landmasses on either side of the Atlantic Ocean, an idea proposed by Isaac Newton two centuries earlier. Wegener was inspired to advocate the hypothesis of **continental drift**—the idea that the continents on either side of the Atlantic Ocean have moved apart.

An increase in the amount of vegetation on Earth will have what effect on the carbon dioxide level in the atmosphere?

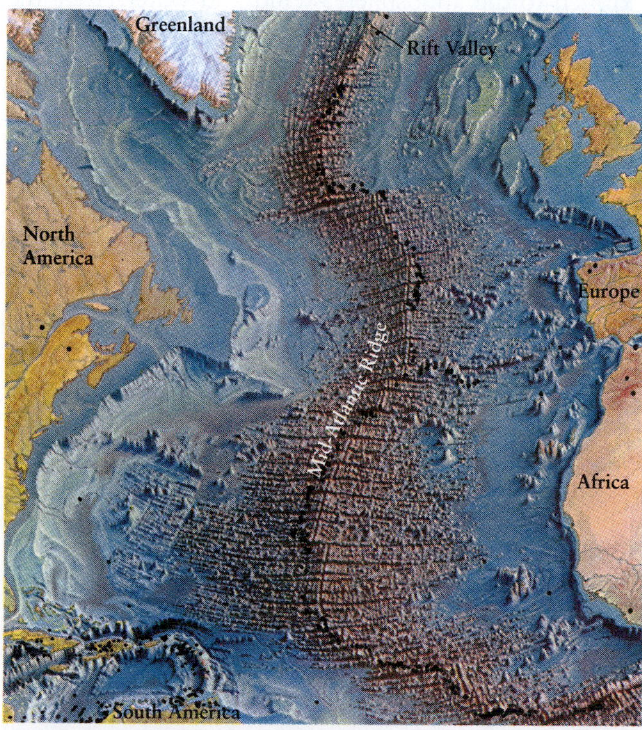

FIGURE 6-6 The Mid-Atlantic Ridge This artist's rendition of the bottom of the North Atlantic Ocean shows an unusual mountain range in the middle of the ocean floor. Called the Mid-Atlantic Ridge, these mountains are created by lava seeping up from Earth's interior along a rift that extends from Iceland to Antarctica. The black dots indicate locations of earthquakes. (Courtesy of M. Tharp and B. C. Heezen)

Wegener argued that a single, gigantic supercontinent called *Pangaea* (pronounced pan-gee-yah, a Greek word for "all lands") began to break up and drift apart some 200 million years ago. Initially, most geologists ridiculed Wegener's ideas. Although it was generally accepted that the continents do float on denser rock beneath them, few geologists could accept that entire continents move across Earth's face at speeds as great as several centimeters per year. Wegener and other "continental drifters" could not explain what forces would shove the massive continents around.

Then, in the mid-1950s, geologists discovered long, underwater mountain ranges. The Mid-Atlantic Ridge, for example, stretches from Iceland to Antarctica (see Figure 6-6). Careful examination revealed that molten rock from Earth's interior is being forced upward there as some mechanism causes the ocean floor to slide apart on either side of the ridge. This **seafloor spreading** is moving the Americas away from Europe and Africa at a speed of roughly 3 cm per year. The relative motion of the plates worldwide ranges from about 1.5 to 7 cm per year.

Insight Into Science

Before they are accepted, scientific theories may undergo years or decades of testing and replication. Newton speculated about continental drift in the late seventeenth century. Wegener had proposed plate tectonics by 1912. His hypothesis was not verified until the 1950s and was not widely accepted until the late 1960s.

Seafloor spreading provided the physical mechanism that allowed for Wegener's hypothesis of crustal motion to be developed into the modern theory of **plate tectonics.**

Earth's surface is made up of about a dozen major thick rafts of rock called *tectonic plates* that move relative to one another. These plates float on the slowly moving, denser matter below them. The plates have not always been separate, as they are today. Sometimes they merge into a single mass called a *supercontinent.* In recent years, geologists have uncovered evidence that

a 237 million years ago: the supercontinent Pangaea

b 152 million years ago: the breakup of Pangaea

c The continents today

FIGURE 6-7 The Supercontinent Pangaea (a) The present continents are pieces of what was once a larger, united body called Pangaea. (b) Geologists believe that Pangaea must have first split into two smaller supercontinents, which they call Laurasia and Gondwana. (c) These bodies later separated into the continents of today. Gondwana split into Africa, South America, Australia, and Antarctica, while Laurasia divided to become North America and Eurasia.

points to a whole succession of supercontinents that broke apart and reassembled. Pangaea is only the most recent supercontinent in this cycle, which repeats on average every 500 million years (Figure 6-7).

The boundaries between plates are the sites of some of the most impressive geological activity on our planet. Earthquakes and volcanoes tend to occur at the boundaries of Earth's crustal plates, where the plates are colliding (at convergent boundaries), separating (at divergent boundaries), or grinding against each other (at slip-strike fault boundaries). For example, the Eyjafjallajökull (pronounced *ā-yuh-flah-lah-yolk*) volcano in Iceland, which erupted in 2010, is located on a divergent boundary. If the epicenters of earthquakes are plotted on a map, the boundaries between tectonic plates (Figure 6-8) stand out clearly. For example, the San Andreas fault, running along the West Coast of the United States, is the slip-strike fault that occurs where the North American and Pacific plates are rubbing against each other (Figure 6-8b). The Red Sea and Gulf of Suez are expanding as Africa and the Middle East move apart (Figure 6-8c). Great mountain ranges are thrust up by collisions between pairs of continent-bearing plates composed of low-density rock. During such collisions, the colliding boundaries of both plates move up. For example, the Himalayas are being formed today as the Indian-Australian and Eurasian plates collide (Figure 6-8d). When a continent-bearing plate collides with a plate of higher density, ocean-bottom rock, the denser plate moves downward (subducts) into Earth under the other. This is currently occurring off the coasts of Japan and Chile.

Tectonic plate motion helps explain why Earth's oceans have so few craters today: Countless impacts occurred in the oceans, which cover most of our planet. Some of the impacting bodies were slowed or obliterated before cratering the ocean floor, but others were large enough or fast enough to hit the floor. The resulting craters in the oceans have been drawn underground by tectonic activity and thereby erased. Impact craters created on the dry landmasses (continents and islands) are exposed to the atmosphere. Wind and water erode the rest, with silt from rivers helping to cover any remnants. Only the youngest craters on Earth remain visible.

What powers all of this activity? Some enormous energy source must be shifting landmasses, and it must have been doing so for billions of years. The answer to that question leads us to a scientific model of Earth's interior.

6-3 Earth's interior consists of a rocky mantle and an iron-rich core

Geologists have calculated that Earth was composed entirely of liquid rock and metal soon after its formation, about 4.6 billion years ago. The violent impact of space debris—along with energy released by the breakup of radioactive elements due to *radioactive decay* and *nuclear fission*—heated, melted, and kept the young Earth molten. In radioactive decay, radioactive nuclei emit small particles, such as electrons, protons, neutrons, or helium nuclei. In nuclear fission, the atomic nuclei break into a substantial pieces, such as when a uranium atom breaks into a rubidium atom and a cesium atom, while also releasing other particles and energy. Fission also creates the heat used to generate electricity in nuclear power plants. That process differs from how energy is generated by *nuclear fusion* in the Sun, wherein lighter atoms are fused together, as mentioned in Chapter 5 and discussed in detail in Chapter 10.

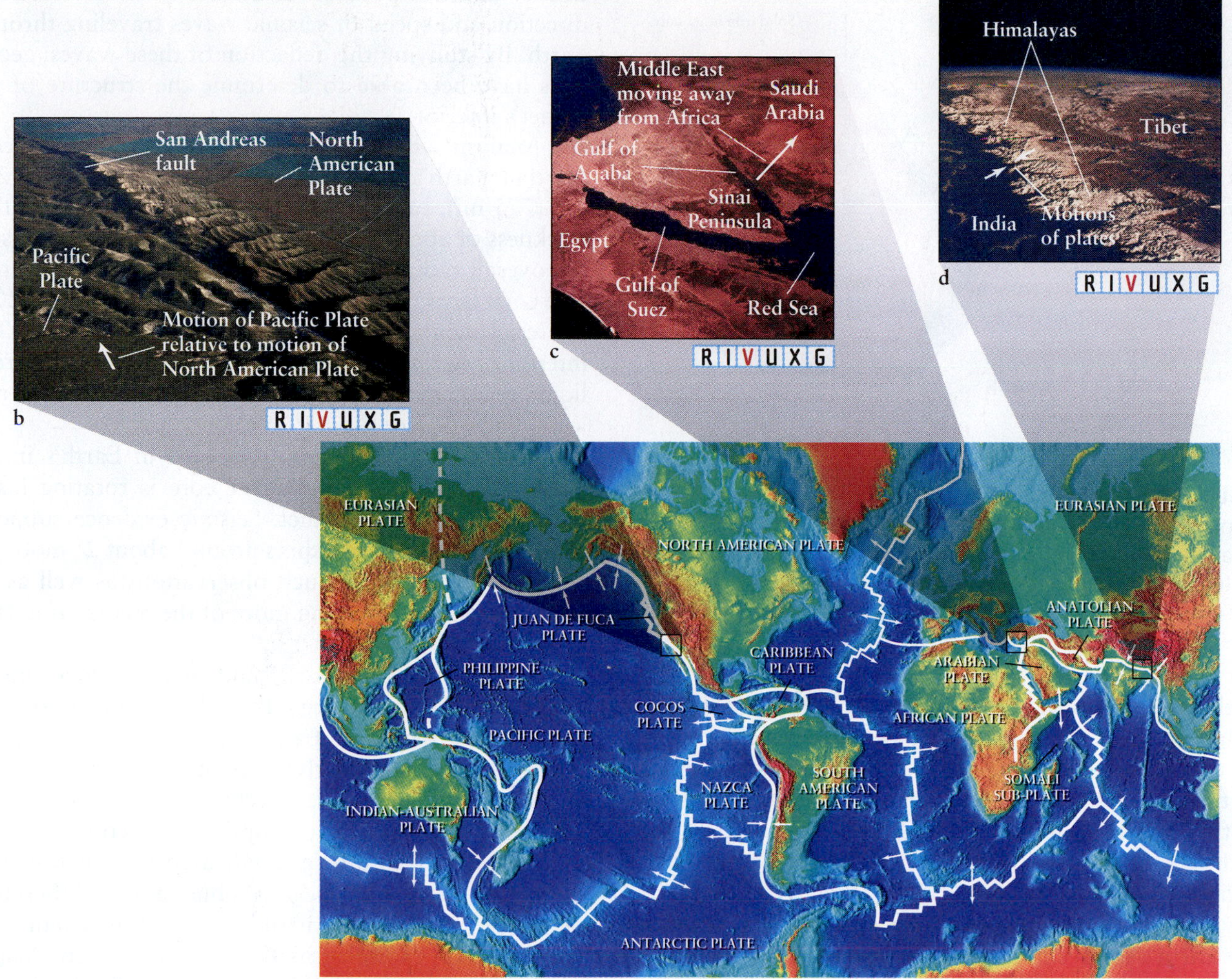

FIGURE 6-8 Earth's Major Tectonic Plates (a) Earth's surface is divided into a dozen or so rigid plates that move relative to one another. The boundaries of the plates are the scenes of violent seismic and geologic activity, such as earthquakes, volcanoes, rising mountain ranges, and sinking seafloors. The arrows indicate whether plates are moving apart (←→), together (→←), or sliding past one another ↑↓. (b) **Rubbing of Two Plates.** The San Andreas fault, running up the west coast of North America, formed because the Pacific Plate is moving northwest along the North American Plate. (c) **Separation of Two Plates.** The plates that carry Egypt and Saudi Arabia are moving apart, leaving the trench that contains the Red Sea. (d) **Collision of Two Plates.** The plates that carry India and China are colliding. As a result, the Himalayas are being thrust upward. In this photograph, taken by astronauts in 1968, Mount Everest is one of the snow-covered peaks near the center. (a: Digital image by Peter W. Sloss, NOAA-NESDIS-NGDC; b: Craig Aurness/Corbis; c: *Gemini 12*, NASA; d: *Apollo 7*, NASA)

Most of the iron and other dense elements sank toward the center of the young, liquid Earth, just as a rock sinks in a pond. At the same time, most of the less dense materials were forced upward toward the surface (Figure 6-9a). This process, called **planetary differentiation,** produced a layered structure within Earth: a very dense central **core** surrounded by a **mantle** of less dense minerals, which, in turn, was surrounded by a thin crust of relatively light minerals (Figure 6-9b).

Differentiation explains why most of the rocks you find on the ground are composed of lower-density elements like silicon and aluminum. Much of the iron, gold, lead, and other denser elements found on Earth today actually had to return to the surface, emerging through volcanoes and other lava flows. Despite this return of heavier elements to the surface, the average density of crustal rocks is 3000 kg/m^3, considerably less than the average density of Earth as a whole (5520 kg/m^3).

What type of boundary exists between Africa and the Middle East?

To understand Earth's interior, geologists have created a

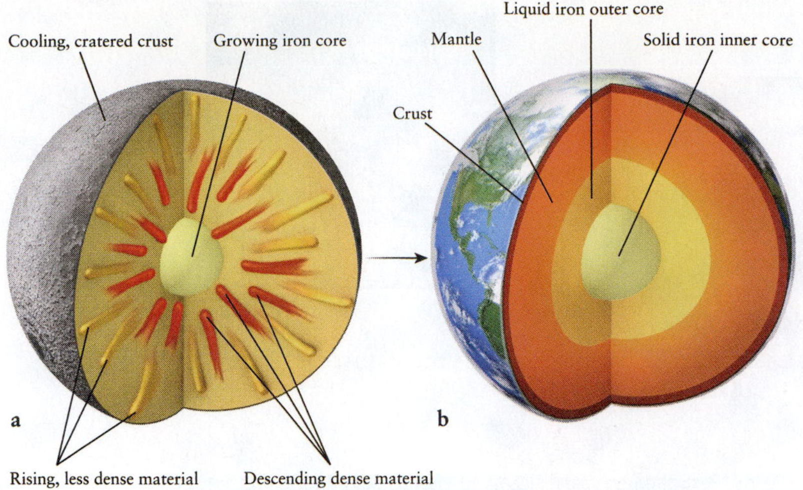

FIGURE 6-9 Cutaway Model of Earth Early Earth was initially a homogeneous mixture of elements with no continents or oceans. (a) Molten iron sank to the center and light material floated upward to form a crust. (b) As a result, Earth has a dense iron core and a crust of light rock, with a mantle of intermediate density between them.

mathematical model that includes the effects of differentiation, gravity, radioactivity, chemistry, heat transfer, rotation, and other factors. The model predicts that Earth's temperature increases steadily from about 290 K on the surface to nearly 5000 K (9000°F) at the center. Radioactivity in Earth's interior helps replenish heat lost through the planet's surface. In addition to temperature, pressure also increases with increasing depth below Earth's surface because the deeper you go, the more mass presses down on you. These factors shape Earth's inner layers.

You might think that a temperature of 5000 K would melt virtually any substance. The mantle, in particular, is composed largely of minerals rich in iron and magnesium, both of which have melting points of only slightly more than 1250 K at Earth's surface. However, the melting point of any substance also depends on the pressure to which it is subjected: the higher the pressure, the higher a material's melting point. The high pressures within Earth (1.4 million atm at the bottom of the mantle) keep the mantle solid to a depth of about 2900 km and also allow for a solid iron core in the center of the planet. In other words, the inner core's pressure is so high that it forces the very hot atoms there into a solid. Surrounding the solid inner core is a molten outer core, where the pressure is too low for that hot material to become solid.

Geologists have tested their models of Earth's interior by studying the response of the planet to earthquakes. These events produce a variety of **seismic waves,** vibrations that travel through Earth either as ripples, like ocean waves, or by compressing matter, like sound waves. Geologists use sensitive **seismographs** to detect and record these vibratory motions. The varying density and composition of Earth's interior affects the direction and speed of seismic waves traveling through Earth. By studying the deflection of these waves, geologists have been able to determine the structure of the planet's interior.

Analysis of seismic recordings led to the discovery that Earth's solid core has a radius of about 1250 km (775 mi), surrounded by a molten iron core with a thickness of about 2250 km (1375 mi). For comparison, the overall radius of our planet is 6378 km (3963 mi). The cores are composed of roughly 80% iron and 20% lighter elements, such as nickel, oxygen, and sulfur. The interior of our planet therefore has a curious structure: a liquid region sandwiched between a solid inner core and a solid mantle (see Figure 6-9b).

A sophisticated computer model of Earth's interior predicts that the solid inner core is rotating faster than the rest of the planet. Seismic evidence supports the model that the core spins around about 2° more per year than the surface. Such observations as well as refined models are revealing more of the secrets of Earth's interior.

The tremendous heat and molten rock inside Earth provide the energy that drives the motion of the tectonic plates in Earth's crust. That heat, from the pressure of the overlying rock and from the decay of radioactive elements, affects Earth's surface today. This heat transport is accomplished by **convection.** In this process a liquid or gas is heated from below and, as a result, it expands, becomes buoyant, and therefore rises, carrying the heat it received with it. Eventually, the liquid or gas releases the heat, cools, condenses, and sinks back down. This motion of fluid causes a circular current, called a *convection current,* to arise. You witness convection every time you see liquid simmering in a pot. Heated from below, blobs of hot fluid move upward. The rising liquid releases its heat at the surface (Figure 6-10a).

Earth's upper mantle is rigid. It comprises the bottoms of the tectonic plates. The rock of the mantle below the plates, although solid, is weak and hot enough to have an oozing, plastic flow (like stretching Silly Putty). As sketched in Figure 6-10b, molten rock, or *magma,* seeps upward along cracks or rifts in the ocean floor, where convection currents of mantle rock are carrying plates apart. Crustal rock sinks back down into Earth where plates collide. The continents ride on top of the plates as they are pushed around by the convection currents circulating beneath them.

Supercontinents like Pangaea apparently sow the seeds of their own destruction by blocking the flow of heat from Earth's interior. As soon as a supercontinent forms, temperatures beneath it increase. As heat accumulates, the supercontinent domes upward and cracks. Overheated molten rock wells up to fill the resulting rifts, which continue to widen as pieces of the fragmenting supercontinent are pushed apart.

WHAT IF . . . Earth Did Not Have Tectonic Plates? Heat from Earth's interior drives the motion of the tectonic plates, which in turn allow some of that heat to escape near plate boundaries through volcanoes and mid-ocean ridges. Without those ever-changing boundaries, the heat within Earth would have no easy way of escaping. As a result, Earth's interior would get hotter and hotter until it melted large regions of Earth's surface. This process is actually believed to occur on Venus, which lacks plate tectonics.

Earth's molten, iron-rich interior affects the planet's exterior in another important way, one that leaves fewer obvious signs than tectonic plate motion. This effect is the creation of a magnetic field that extends through Earth's surface and out into space. As we will see next, that magnetic field results from the convection of molten metal inside Earth combined with the planet's rotation.

6-4 Earth's magnetic field shields us from the solar wind

If you have ever used a compass, you have seen the effect of Earth's magnetic field. This planetary field is quite similar to the field that surrounds a bar magnet (Figure 6-11a). Magnetic fields are created by electric charges in motion. Thus, the motion of your compass is caused by electric charges moving inside Earth. The electric wires in your house also carry moving charges called electric currents that, in turn, create magnetic fields of their own.

Many geologists believe that convection of molten iron in Earth's outer core, combined with our planet's rotation, creates the electric currents, which, in turn, create Earth's magnetic field. The details of this so-called **dynamo theory** of Earth's magnetic field are still being developed.

Heated from below, gases also rise. What does this suggest about Earth's atmosphere?

The magnetic fields near Earth's south rotation pole loop out tens of thousands of kilometers into space and then return near Earth's north rotation pole. The places where the magnetic fields pierce Earth's surface most intensely are called the north and south magnetic poles. These poles move continuously. Evidence in solidified lava reveals that Earth's magnetic field completely reverses or flips (the north magnetic pole becomes the south magnetic pole, and vice versa) on an irregular schedule, ranging upward from a few tens of thousands of years. Each flip takes a few thousand years to complete. The

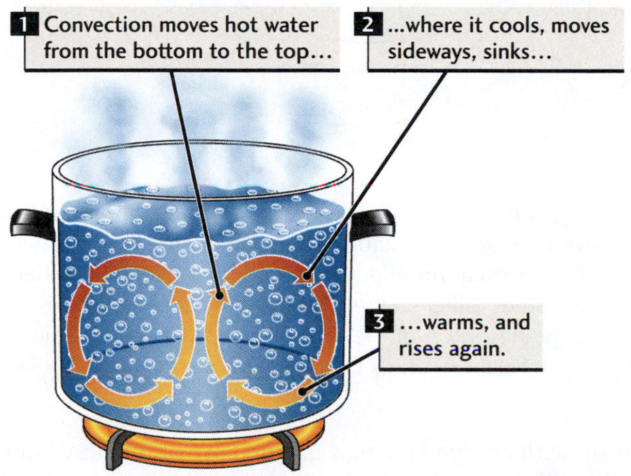

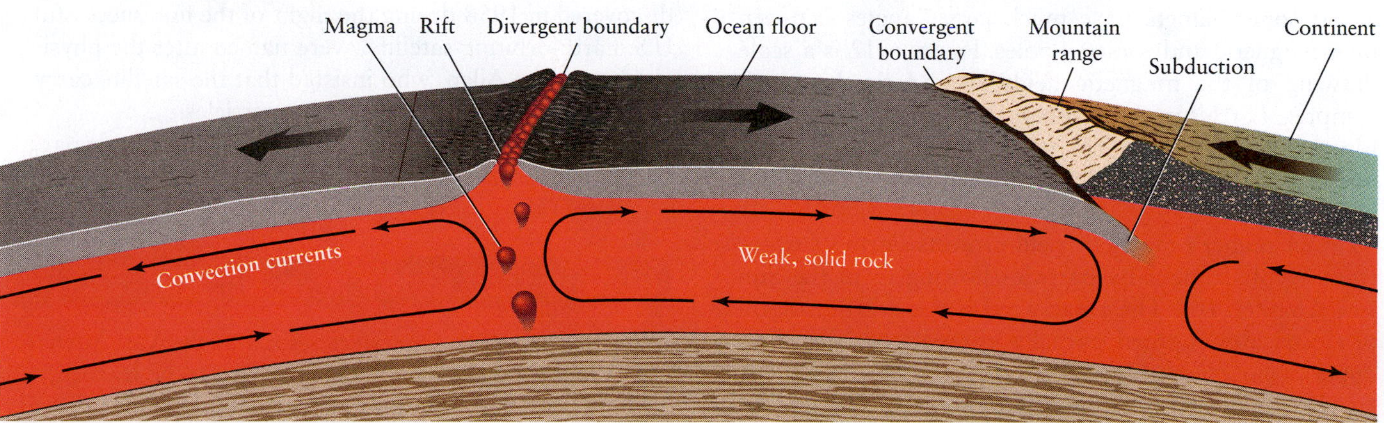

ANIMATION 6.4 **FIGURE 6-10 Convection and the Mechanism of Plate Tectonics** (a) Heat supplied by a heating coil warms the water at the bottom of a pot. The heated water consequently expands, decreasing its density. This lower-density water rises (like bubbles in soda) and transfers its heat to the cooler surroundings. When the hot rising water gets to the top of the pot, it loses a lot of heat into the room, becomes denser, and sinks back to the bottom of the pot to repeat the process. (b) Convection currents in Earth's interior are responsible for pushing around rigid plates on its crust. New crust forms in oceanic rifts, where magma oozes upward between separating plates. Mountain ranges and deep oceanic trenches are formed where plates collide and crust sometimes sinks back into the interior. Note that not all tectonic plates move together or apart—some scrape against each other.

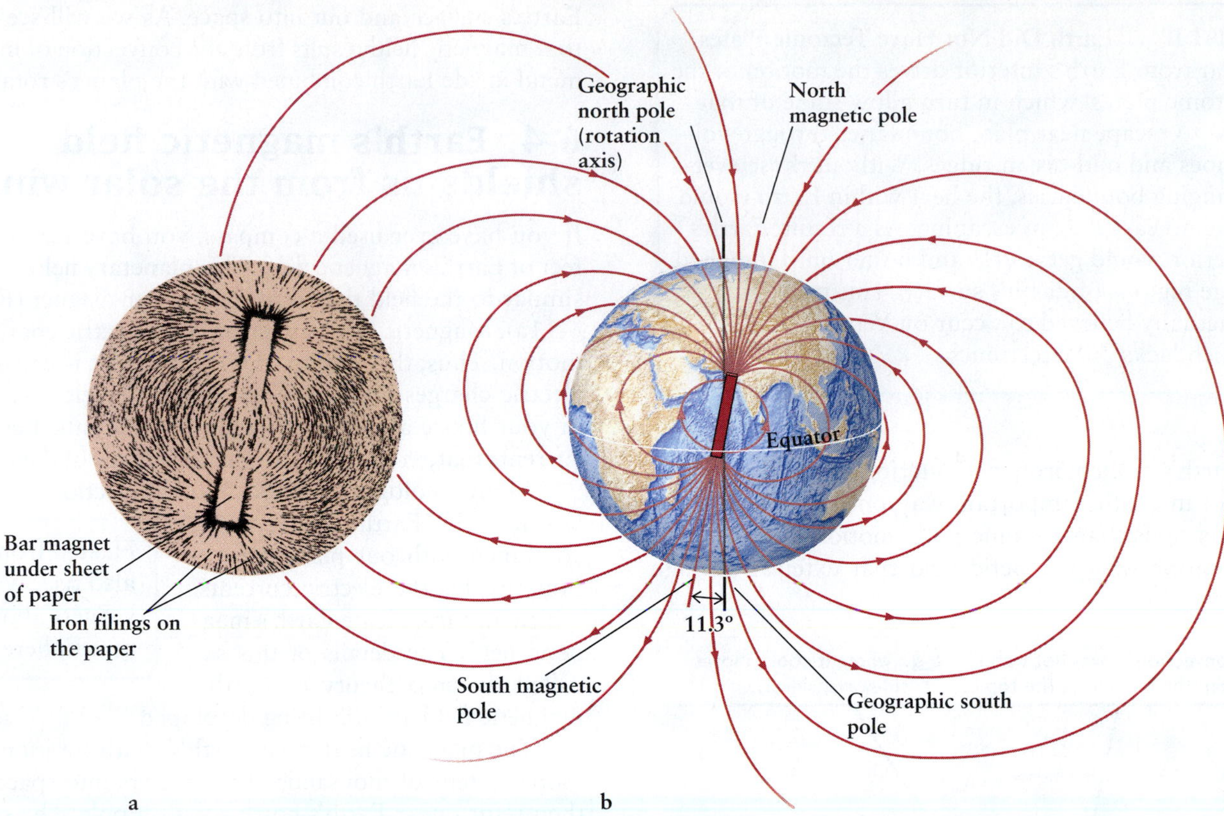

FIGURE 6-11 Earth's Magnetic Field (a) The magnetic field of a bar magnet is revealed by the alignment of iron filings on paper. (b) Although Earth does not contain a bar magnet, its rotation, combined with moving electric charges in its core, creates an equivalent field. Note that the field is not aligned with Earth's rotation axis. By convention, the magnetic pole near Earth's north rotation axis is called the *north magnetic pole* even though it is actually the south pole on a magnet! We will see similar misalignments and flipped magnetic fields when we study other planets. Note that this figure does not include the distortion of the field by the solar wind, which is depicted in Figure 6-12. (a: Jules Bucher/Photo Researchers)

last reversal was about 600,000 years ago. Today, the magnetic and rotation poles of Earth are about 11.3° apart; some planets have much larger angles between their magnetic and rotation poles. Figure 6-12 is a scale drawing of the magnetic fields around Earth, which comprise Earth's magnetosphere. Magnetic fields always form complete loops. (The unconnected lines in Figure 6-11b actually are completed outside of the boundaries of the figure.)

Our planet's magnetic field protects Earth's surface from bombardment by particles flowing from the Sun (see Chapter 10). The **solar wind** is the erratic stream of electrically charged particles ejected from the Sun's upper atmosphere. Near Earth, the particles in the solar wind move at speeds of approximately 400 km/s, or about a million miles per hour.

Magnetic fields can change the direction of moving charged particles. Thus, solar wind particles heading toward Earth are deflected by our planet's magnetic field. Many particles, therefore, pass around Earth. However, the field also traps some of these charged particles in two huge, doughnut-shaped rings, called the **Van Allen radiation belts** (hereafter *Van Allen belts*), which surround Earth (see Figure 6-12). The region between the two belts fills up with charged particles, which are drained by lightning on Earth, making this region unstable. These belts, discovered in 1958 during the flight of the first successful U.S. Earth-orbiting satellite, were named after the physicist James Van Allen, who insisted that the satellite carry a Geiger counter to detect charged particles.

Sometimes the Van Allen belts overload with particles. The particles then leak through the magnetic

Insight Into Science

Pace of Discovery Discoveries such as the Van Allen belts remind us of how much remains to be learned, even about our own planet. Such fundamental discoveries continue to be made, for many reasons. For example, the technology necessary to make certain observations may just have been developed, or our equations that describe various phenomena may become more complete. Sometimes these equations are so complex that their implications are not immediately clear. Progress in science requires unquenchable curiosity and unfailing open-mindedness.

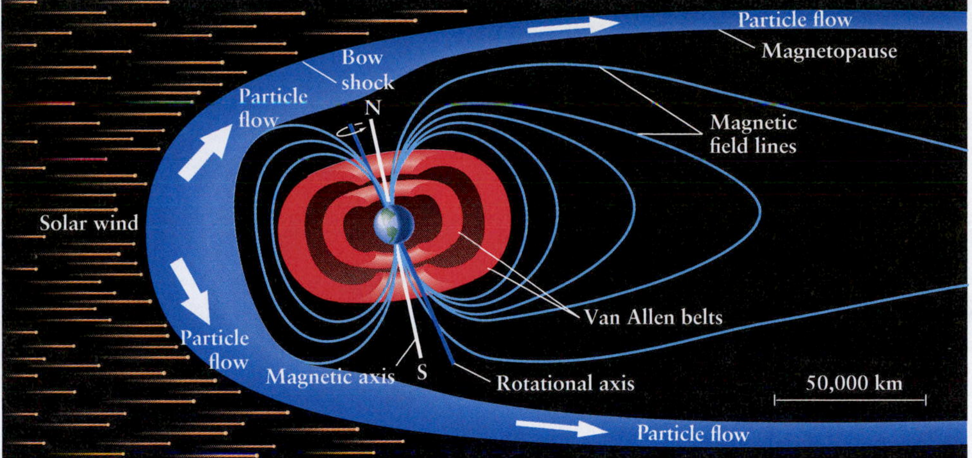

FIGURE 6-12 Earth's Magnetosphere A slice through Earth's magnetic field, which surrounds the entire planet, carves out a cavity in space that excludes most of the charged particles ejected from the Sun, called the solar wind. Most of the particles of the solar wind are deflected around Earth by the fields in a turbulent region colored blue in this drawing. Because of the strength of Earth's magnetic field, our planet traps some charged particles in two huge, doughnut-shaped rings called the Van Allen belts (in red).

VIDEO 6.2

fields near the poles and cascade down into Earth's upper atmosphere, usually in a ring-shaped pattern (Figure 6-13a). As these high-speed, charged particles collide with gases in the upper atmosphere, the gases fluoresce (give off light) like the gases in a neon sign. The result is a beautiful, shimmering display, called the **northern lights (aurora borealis)** or the **southern lights (aurora australis)**, depending on the hemisphere in which the phenomenon is observed (Figure 6-13b, c).

Occasionally, a violent event on the Sun's surface, called a **coronal mass ejection,** sends a burst of protons and electrons straight through the Van Allen belts and into the atmosphere. These events partially deplete the ozone layer and cause spectacular auroral displays that can often be seen over a wide range of latitudes and longitudes. Coronal mass ejections also disturb radio transmissions and can damage communications satellites and electric transmission lines.

What happens to the Van Allen belts when Earth's magnetic field is flipping?

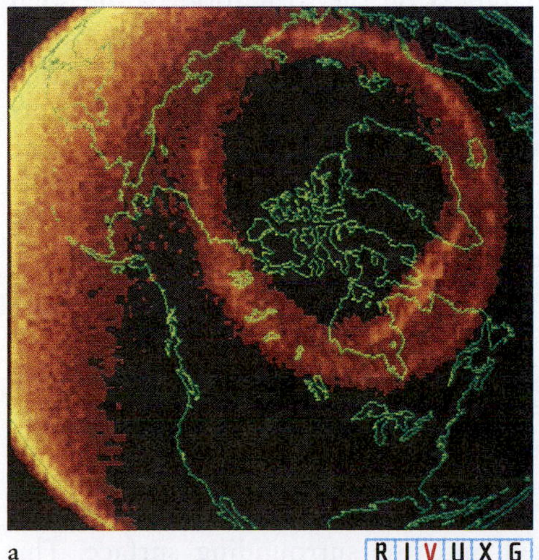

a R I V U X G

b R I V U X G

c R I V U X G

FIGURE 6-13 The Northern Lights (Aurora Borealis) A deluge of charged particles from the Sun can overload the Van Allen belts and cascade toward Earth, producing auroras that can be seen over a wide range of latitudes. (a) View of an aurora from the deep space satellite *Dynamics Explorer 1*. The green lines show the locations of landmasses under the aurora. Auroras typically occur 100 to 400 km above Earth's surface. (b) View of aurora from space. Part of a space shuttle is visible at the bottom left of the picture. (c) Aurora borealis in Alaska. The gorgeous aurora seen here is mostly glowing green due to emission by oxygen atoms in our atmosphere. Some auroras remain stationary for hours, while others shimmer, like curtains blowing in the wind. (a: L.A. Frank and J.D. Crave, University of Iowa; b: MTU/Geological & Mining Engineering & Sciences; c: J. Finch/Photo Researchers, Inc.)

THE MOON: VITAL STATISTICS

Distance from Earth (center to center):	Average: 384,400 km Maximum (apogee): 405,500 km Minimum (perigee): 363,300 km
Orbital eccentricity:	0.055
Synodic period of revolution (lunar phases):	29.53 days
Sidereal rotation period (around Earth):	27.32 days
Inclination of equator to orbit:	6.68°
Inclination of orbit to ecliptic:	5.15°
Diameter (equatorial):	3476 km = 0.272 Earth diameter
Mass:	7.35×10^{22} kg = 0.012 Earth mass
Average density:	3340 kg/m^3
Albedo (average):	0.12
Escape speed:	2.4 km/s
Average surface temperatures:	Day: 130°C = 266°F = 403 K Night: −180°C = −292°F = 93K
Atmosphere:	Traces of argon (Ar), helium (He), sodium (Na), and potassium (K)
Surface gravity (Earth = 1):	0.17

R I V U X G

FIGURE 6-14 The Moon and Its Vital Statistics Our Moon is one of seven large satellites in the solar system. The Moon's diameter of 3476 km (2158 mi) is slightly less than the distance from New York to San Francisco. (NASA)

THE MOON AND TIDES

Earth's only natural satellite provides one of the most dramatic sights in the nighttime sky. The Moon is so large and so close to us that some of its surface features are readily visible to the naked eye. Even without a telescope, you can easily see dark gray and light gray areas that cover vast expanses of the Moon (Figure 6-14). People have dreamed of visiting our nearest neighbor in space for millennia. One of the first movies ever made (1902) was *A Trip to the Moon,* based on Jules Verne's book *From the Earth to the Moon.* The wonder of actually sending people there for the first time was one of the few things that has ever caused humanity to stop its quarreling long enough to participate, even vicariously, in a great adventure. To date, 12 people have walked on the Moon. Our exploration of it begins with a study of its surface.

6-5 The Moon's surface is covered with craters, plains, and mountains

Perhaps the most familiar and characteristic features on the Moon are its craters (Figure 6-15a). As discussed in Chapter 5 on the formation of the solar system, all lunar craters that we have examined are the result of bombardment by meteoritic material (rather than volcanic activity). Nearly all of these craters are circular, indicating that they were not merely gouged out by moderate-speed rocks, which would typically create oval craters if they struck the surface at an angle. Instead, the high-speed (upward of 180,000 km/h or 112,000 mi/h) collisions with the Moon's surface vaporized the rapidly moving debris coming in at any angle, often with the power of a large nuclear bomb. The resulting explosions of hot gas and rubble produced the round-rimmed craters that we observe today (Figures 6-15b, c, and d).

Meteoritic impact causes material from the crater site to be ejected onto the surrounding surface. This pulverized rock is called an **ejecta blanket.** You can see some of the more recent ejecta blankets, which are lighter-colored than the older ones, in Figures 6-14 and 6-15. Ejecta blankets darken with time as their surfaces roughen from impacts by the solar wind and other particles from space. Therefore, the lightness of a crater's ejecta is one clue astronomers use to determine how long ago a crater formed.

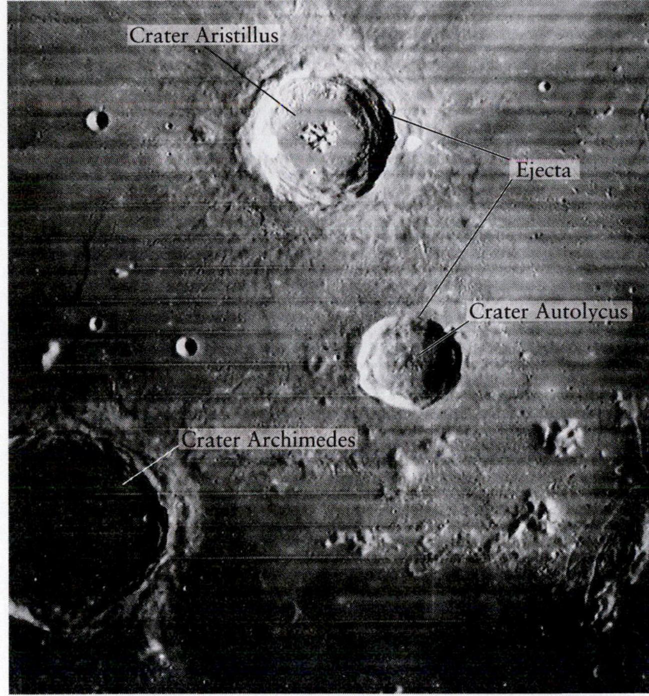

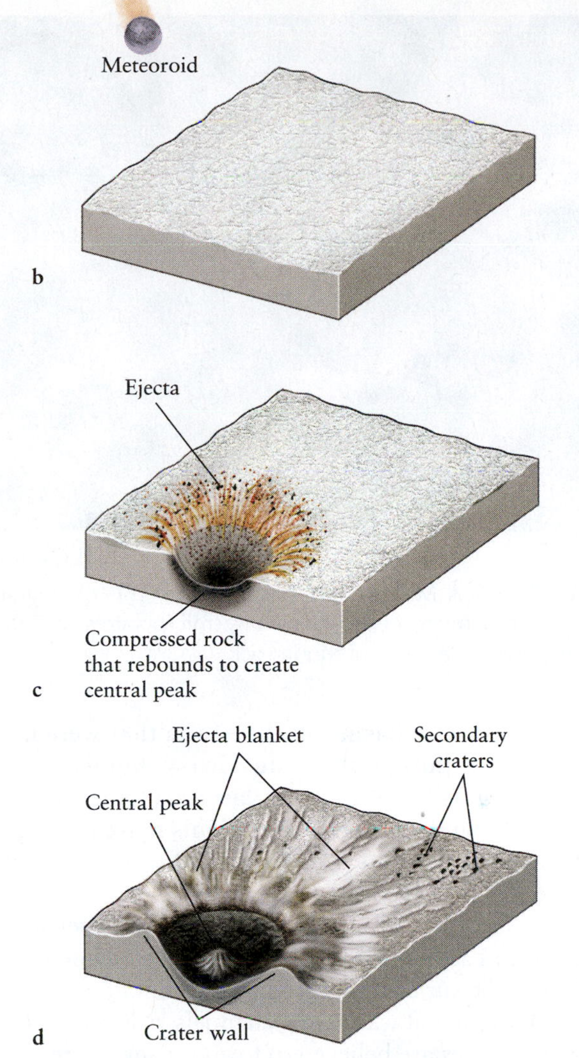

FIGURE 6-15 Lunar Craters (a) This photograph, taken from lunar orbit by astronauts, includes the crater Aristillus. Note the crater's central peaks; collapsed, terraced crater wall; and ejecta blanket. Numerous smaller craters resulting from the same impact pockmark the surrounding lunar surface. The following three drawings show the crater formation process: (b) An incoming meteoroid, (c) upon impact, is pulverized and the surface explodes outward and downward. (d) After the impact, the ground rebounds, creating the central peak and causing the crater walls to collapse. The lighter region is the ejecta blanket. (a: 2004 Lunar and Planetary Institute/Universities Space Research Association)

Craters larger than about 20 km often form *central peaks* (see Figure 6-15). These features occur because the impact compresses the crater floor so much that, afterward, the crater rebounds and pushes the peak upward. As the peak goes up, the crater walls collapse and form *terraces*.

Through an Earth-based telescope, some 30,000 craters, with diameters ranging from 1 km to more than 100 km, are visible. Following a tradition established in the seventeenth century, the most prominent craters are named after astronomers, physicists, mathematicians, and philosophers, such as Kepler, Copernicus, Pythagoras, Plato, and Aristotle. Close-up photographs from lunar orbit reveal millions of craters too small to be seen with Earth-based telescopes. Indeed, extreme close-up photos of the Moon's surface reveal countless microscopic craters (Figure 6-16).

Earth, by comparison, has only 178 known craters caused by impacts. Many craters on Earth have been pulled inside our planet by plate tectonic motion and obliterated. Many other craters of a few kilometers in diameter and smaller have been worn away (eroded) by weathering effects of wind and water. Earth's atmosphere vaporized countless space rocks that would otherwise have created small craters here. Moreover, those that do hit are slowed down by the atmosphere so that they produce smaller craters than they otherwise would have. We will discuss Earth's craters further in Chapter 9.

Besides its craters, the most obvious characteristic of the Moon visible from Earth is that its surface is various shades of gray (see Figure 6-14). Most prominent are the large, dark gray plains called **maria** (pronounced *mar-ee-uh*). The singular form of this term, **mare** (pronounced *mar-ay*), means "sea" in Latin and was introduced in the seventeenth century when observers using early telescopes thought that these features were large bodies of water. We know now that no liquid water exists on our satellite. Nevertheless, we retain these poetic, fanciful names, including Mare Tranquillitatis (Sea of Tranquility), Mare Nubium (Sea of Clouds), Mare Nectaris (Sea of Nectar), and Mare Serenitatis (Sea of Serenity). As introduced in Chapter 5 and as we will discuss further

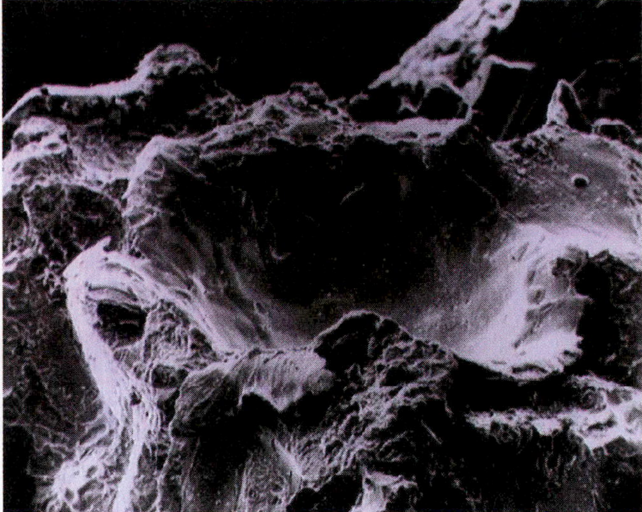

FIGURE 6-16 A Microscopic Lunar Crater This photograph made with a microscope shows tiny microcraters less than 1 mm across on a piece of Moon rock. (NASA)

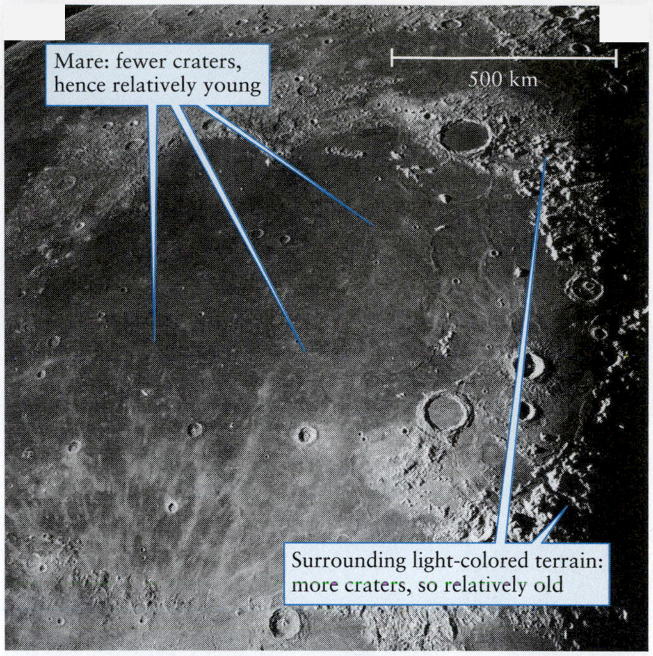

FIGURE 6-17 Mare Imbrium and the Surrounding Highlands Mare Imbrium, the largest of the 14 dark plains that dominate the Earth-facing side of the Moon, is ringed by lighter-colored highlands strewn with craters and towering mountains. The highlands were created by asteroid impacts pushing land together. (NASA)

shortly, maria are basins on the Moon that were filled in with lava during the Late Heavy Bombardment (see Section 5-7). One of the largest of the maria is Mare Imbrium (Sea of Showers). This mare is roughly circular and measures 1100 km (700 mi) in diameter (Figure 6-17).

Although the maria seem quite smooth in telescopic views from Earth, close-up photographs from lunar orbit and from the surface reveal that they contain small craters and occasional wandering channels, called **rilles** (Figure 6-18). Rilles are believed to form for several reasons: When underground tubes of molten rock stop flowing and the magma in them either drains out or solidifies in place and thereby condenses, the ceilings of the tubes are no longer supported, hence they collapse. Also, if lava rivers flowed on the Moon's surface and then stopped running, they would have drained away, leaving beds that we see as maria. Support for the existence of lava tubes on the young Moon came in 2009, when a Japanese lunar orbiter, *SELENE*, observed a nearly circular hole about 65 m in diameter and 85 m deep on the Moon's surface. This width-to-height ratio is inconsistent with it being an impact crater. Rather, it is believed to be the opening of one such tube (Figure 6-18c).

As we will discuss further in Section 6-8, the same side of the Moon always faces Earth. One of the surprises stemming from the earliest lunar exploration is that the far side is much more heavily cratered than the near side. A few tiny maria, including Mare Moscoviense, Mare Ingenii, and Mare Australe, plus one moderate-sized mare, Orientale, grace the Moon's far side, the side we never see from Earth (Figure 6-19). Unlike the mare on our side, those on the far side are only partially filled with lava. Otherwise, craters and mountains cover the entire far side of the Moon.

Which of the three named craters in Figure 6-15a is the youngest?

Detailed measurements by astronauts in lunar orbit demonstrated that the maria on the Moon's Earth-facing side are 2 to 5 km *below* the average lunar elevation on that side, whereas the ones on the far side are 4 to 5 km *above* the average lunar elevation. The flat, low-lying, dark gray maria cover only 17% of the entire lunar surface. The remaining 83% is composed of light gray, heavily cratered mountainous regions, called **highlands** (see Figure 6-14). The difference in heights of the maria on the two sides of the Moon apparently stems from the way they were formed. Those maria on the near side were created by flows of lava that filled low-lying basins (enormous craters) surrounded by highlands. These mountains were pushed up by large impacts during the Late Heavy Bombardment. The small maria on the far side were more likely lava flows that pushed up and out relatively high in the mountains, rather than into low-lying basins.

The relative ease with which maria on the near side of the Moon formed, compared to the far side, is explained by the observation that the crust on the near side is thinner than on the far side. The impacts that created what became the maria, formed craters that thinned the near side crust even further, making eventual seepage of lava easier on our side.

a

b RIVUXG

c

FIGURE 6-18 Details of a Lunar Mare (a) Close-up views of the lunar surface reveal rilles and numerous small craters on the maria. Astronauts in lunar orbit took this photograph of Mare Tranquillitatis (Sea of Tranquility) in 1969 while searching for potential landing sites for the first human landing. At 1100 km (700 mi) across, this mare is the same size as the distance from London to Rome or from Chicago to Philadelphia. (b) Astronaut David Scott on Hadley's rille during the *Apollo 15* mission to the Moon. (c) Possible lava tube outlet on the Moon. (a: NASA; b: Kennedy Space Center/NASA; c: NASA)

Analyses of Moon rocks and observations from lunar orbit indicate that violent impacts of huge asteroids thrust up the surrounding land to create the lunar highlands. One such mountain range can be seen around the edge of Mare Imbrium (see Figure 6-17). Although they take their names from famous terrestrial ranges, such as the Alps, the Apennines, and the Carpathians, the lunar mountains were not formed through plate tectonics, as mountains on Earth were.

6-6 Visits to the Moon yielded invaluable information about its history

VIDEO 6.3

Reaching the Moon became national obsessions for the United States and the Soviet Union in the early 1960s. A series of successful space missions by the Soviet Union spurred

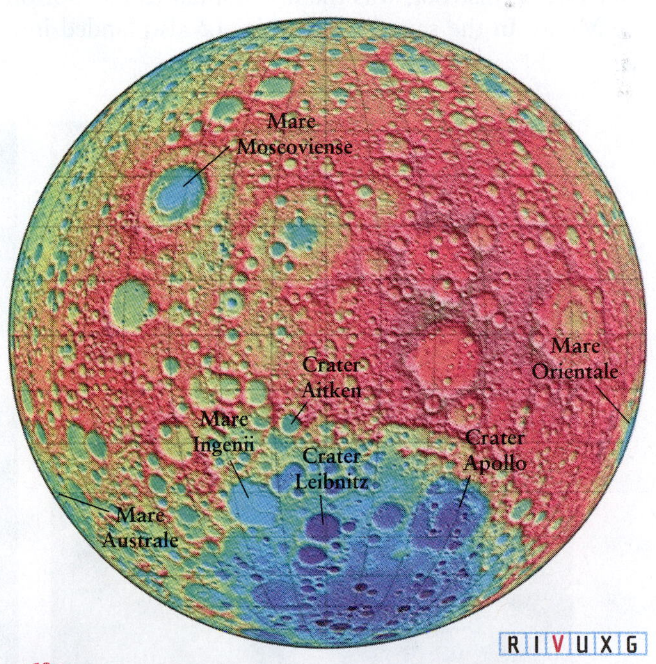

VIDEO 6.4

FIGURE 6-19 The Far Side of the Moon Using a laser mounted on the *Lunar Reconnaissance Orbiter,* this detailed image of the lunar far side was made in 2010. Going by the colors of the rainbow, violet indicates lowest terrain, while red indicates highest. (NASA)

VIDEO 6.5 **FIGURE 6-20 An Apollo Astronaut on the Moon** *Apollo 17* astronaut Harrison Schmitt enters the Taurus-Littow Valley on the Moon. The enormous boulder seen here slid down a mountain to the right of this image, fracturing on the way. This final Apollo mission landed in the most rugged terrain of any Apollo flight. (NASA)

President John F. Kennedy's goal of landing a person on the Moon by the end of that decade.

2 The first of six manned lunar landings, *Apollo 11*, set down on Mare Tranquillitatis on July 20, 1969. Astronaut Neil Armstrong was the first human to set foot on the Moon. In the same year, *Apollo 12* also landed in a mare. Human voyages into space encountered a setback when *Apollo 13* experienced a nearly fatal explosion en route to the Moon. Fortunately, a heroic effort by the astronauts and ground personnel brought it safely home. *Apollo 14* through *Apollo 17* took on progressively more challenging lunar terrain. The period of human visitation to our Moon lasted less than 4 years, culminating (and ending) in 1972, when *Apollo 17* landed amid mountains just east of Mare Serenitatis (Figure 6-20).

Astronauts discovered that the lunar surface is covered with fine powder (Figure 6-21a) and rock fragments (Figure 6-21b). The powdered rock, ranging in thickness from 1 to 20 m, is called the **regolith** (from the Greek, meaning "blanket of stone"). Regolith formed during 4.5 billion years of relentless micrometeorite bombardment that pulverized the surface rock. We do not refer to this layer as soil, because the term "soil" suggests the presence of decayed biological matter, which is not found on the Moon's surface.

A major goal of the Apollo program was to study the geology of the Moon and help determine its history. The astronauts brought back a total of 382 kg (842 lb) of lunar rocks, which have provided important information about the early history of the Moon. By carefully measuring trace amounts of radioactive elements, geologists determined that lunar rock from the maria is solidified lava that dates back only 3.1 to 3.8 billion years. This indicates that the maria developed about a billion years after the Moon formed and that the maria were formed from molten rock, consistent with this material having poured out of the Moon after the Late Heavy Bombardment. These Moon rocks are composed mostly of the same minerals that are found in volcanic rocks in

FIGURE 6-21 The Regolith The Moon's surface is covered by a layer of (a) powdered rock and (b) small pieces of rock. Called *regolith*, the powdered rock was created over billions of years as a result of bombardment by space debris; it sticks together like wet sand, as illustrated by this *Apollo 11* astronaut bootprint. (NASA)

FIGURE 6-22 Mare Basalt This 1.53-kg (3.38-lb) specimen of mare basalt was brought back by *Apollo 15* astronauts in 1971. Small holes that cover about a third of its surface suggest that gas was dissolved in the lava from which this rock solidified. When the lava reached the airless lunar surface, bubbles formed as the pressure dropped and the gas expanded. Some of the bubbles were frozen in place as the rock cooled. (NASA)

Hawaii and Iceland. The rock of these low-lying lunar plains is called **mare basalt** (Figure 6-22). About 17% of the Moon's surface rock is basalt.

In contrast to the dark mare basalt, the lunar highlands are covered with **anorthosite,** dark rock that is somewhat lighter than basalt (Figure 6-23). On Earth, anorthositic rock is found only in very old mountain ranges, such as the Adirondacks in the eastern United States. Compared to the mare basalts, which have more of the heavier elements like iron, manganese, and titanium, anorthosite is rich in calcium and aluminum. Anorthosite is, therefore, less dense than basalt. Typical anorthositic specimens from the highlands are between 4.0 and 4.3 billion years old. These ancient specimens of highland rock probably represent samples of the Moon's original (pre–Late Heavy Bombardment) crust. Many highland rocks brought back to Earth are **impact breccias** (Figure 6-24), which are composites of different rock fused together as a result of meteorite impacts.

The rough, powdered regolith, composed of intrinsically dark basaltic and anorthositic rock, absorbs most of the light incident on it. The Moon's albedo is only 0.07, meaning that it scatters only 7% of the light striking it.

Four of the six Apollo missions that landed on the Moon set up scientific packages containing seismic detectors (Figure 6-25). These devices are sensitive to moonquakes, motions of the Moon's surface analogous to earthquakes. Between 1969 and 1977, they detected more than 12,500 events, including a few quakes strong enough to knock books off shelves (if there had been shelves or books there at the time). Moonquakes occur for any of four reasons: impacts, motion of the lunar surface due to changing tides created by Earth's gravity, expansion and contraction due to heating and cooling of the Moon's surface, and other activity inside the Moon that is still under investigation.

With several seismic detectors working simultaneously, geologists were able to deduce information about the Moon's interior from the timing and strength of the vibrations that were detected. The data suggested that the Moon has a small iron core, possibly a layer of hot rock above it, and a thick layer of solid rock above that, surrounded by the crust. The presence of the iron core, accounting for only 1 to 4% of the Moon's total mass, was confirmed by data from the *Lunar Prospector* satellite in 1998. This is a small proportion compared to

FIGURE 6-23 Anorthosite The lunar highlands are covered with this ancient type of rock, which is believed to be the material of the original lunar crust. This sample's dimensions are 18 cm × 16 cm × 7 cm. Although this sample is medium gray, other anorthosites retrieved from the Moon have been white, while others are darker gray than this one. This rock was brought back by *Apollo 16* astronauts. (NASA)

FIGURE 6-24 Impact Breccias These rocks are created from shattered debris fused together under high temperature and pressure. Such conditions prevail immediately following impacts of space debris on the Moon's surface. (NASA)

FIGURE 6-25 ***Apollo 11*** **Landing Site** On the Moon's Sea of Tranquility, astronaut Buzz Aldrin stands next to the package of equipment containing the seismic detector. The corner reflectors are used, even today, to determine the distance from Earth to the Moon. The stereo camera took pairs of images of the Moon's surface. Seeing them through special glasses gives a 3-D close-up view of the Moon's surface. (Figure 6-21b is one of such a pair.) The bottom half of the lander is still on the Moon's surface. The top half brought astronauts Neil Armstrong and Buzz Aldrin back into lunar orbit, where they transferred to the command module to fly home. (NASA)

Earth's core, which has about 30% of our planet's total mass.

Like Earth, the Moon's interior was initially molten. As the Moon aged, heat flowed up and out through the surface, as still happens there and on Earth today. Because the Moon is smaller than Earth, it initially had less heat stored inside and it lost that heat more rapidly than does our planet. This is why less, if any, of the Moon's interior is molten today.

As the Moon's interior cooled and solidified, it became more compact; virtually all substances (except water) shrink when they become solid. Therefore, as the Moon's interior solidified, gravity pulled its solid surface inward. This process could not happen uniformly (think of an apple that is drying on the inside). Therefore, parts of the Moon settled inward faster than other parts. As a result, sharp cliffs called **scarps** formed where one region of the surface moved down before adjacent regions could (Figure 6-26a). In 2010, astronomers using a camera on the *Lunar Reconnaissance Orbiter* spacecraft observed the existence of scarps over much of the Moon's surface.

Since spacecraft began orbiting the Moon, scientists have noted that their orbits do not follow the elliptical paths predicted by Kepler's laws. Every once in awhile, usually over a circular mare, the spacecraft momentarily dip Moonward, as they are pulled by a higher-than-normal concentration of mass. These *mass concentrations*, or **mascons**, are local regions of relatively dense rock and metal near the Moon's surface. The mascons found in maria are believed to be the magma that filled the bottom of the mare after the basins were formed by impacts of very large bodies during the Late Heavy Bombardment. The magma solidified after emerging from inside the Moon. Before the mascons were well understood, a small satellite put in orbit in 1972 by *Apollo 16* astronauts came under their influence, and their stronger-than-average gravitational tug caused the satellite to crash into the Moon. The satellite, intended to orbit for years, lasted only 35 days. New mascons continue to be discovered, including seven found by the *Lunar Prospector* spacecraft in 1998.

In 1994, the lunar-orbiting *Clementine* spacecraft sent radio signals to the Moon's surface. These signals scattered off the materials there, and then were received and analyzed on Earth. The results were intriguing, suggesting that frozen water exists in craters near the Moon's south pole. The presence of ice on the Moon was confirmed by the Lunar Crater Observing and Sensing Satellite (LCROSS), the Indian *Chandrayaan-1* lunar orbiter, and the *Lunar Prospector* lunar orbiter. These craters alone, ranging in size from 2 to 15 km across, contain an estimated 600 million metric tons of water ice. This ice, in polar craters that are never exposed to sunlight, is not in danger of being evaporated (Figure 6-26b). The ice was probably deposited on the Moon during comet and asteroid impacts.

While water ice at the poles, blocked from the Sun's heat by crater walls, doesn't surprise astronomers and geologists, the discovery by *Chandrayaan-1* in 2010 of water (in the form of OH^-, called the hydroxyl radical, and H, hydrogen atoms) in the Moon's regolith was unexpected. Compared to Earth, the Moon is bone dry. However, if, as now appears likely, water-bearing rock is distributed over most of the Moon's surface, that body is estimated to contain as much water as is in the Great Lakes of North America, which have a volume of about 5400 cubic miles.

The presence of water ice would greatly simplify human settlement of the Moon, because ice can provide us not only with liquid water, but also with rocket fuel and breathable oxygen. The space shuttle's primary rocket engine was fueled by liquid oxygen and liquid hydrogen. Separating the Moon's water into liquid oxygen and liquid hydrogen would create the same fuel. An existing water reservoir would make establishing a lunar colony much less expensive and more feasible than lugging all that water from Earth or dragging a water-rich comet to the Moon.

Astronauts found no traces of life on the Moon. Life as we know it requires liquid water, of which the Moon has none, to form and evolve. The Moon has an atmosphere of helium, argon, sodium, and potassium that is so thin as to be considered an excellent vacuum compared to the density of the air we breathe. An atmosphere is held around a planet or moon by that body's

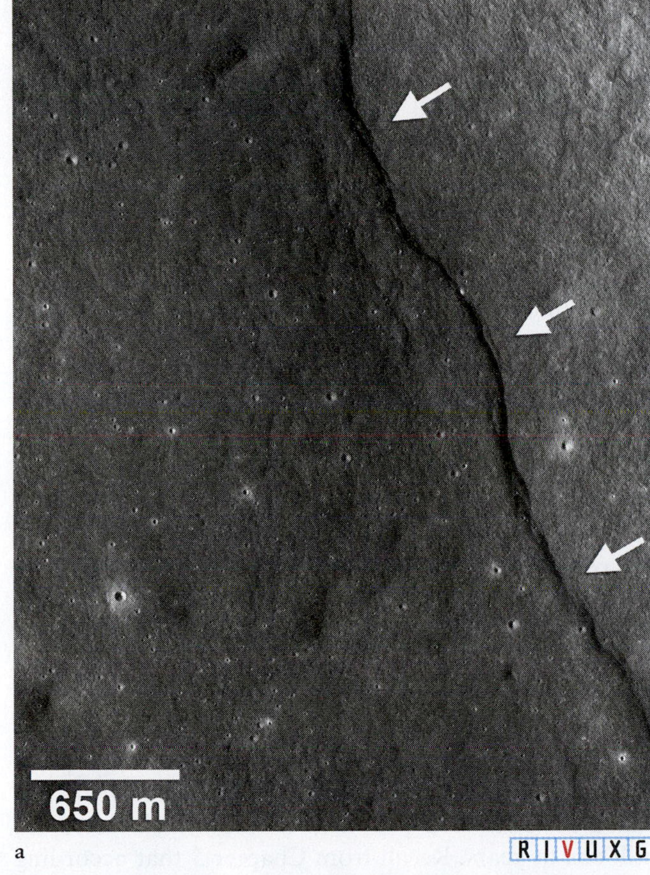

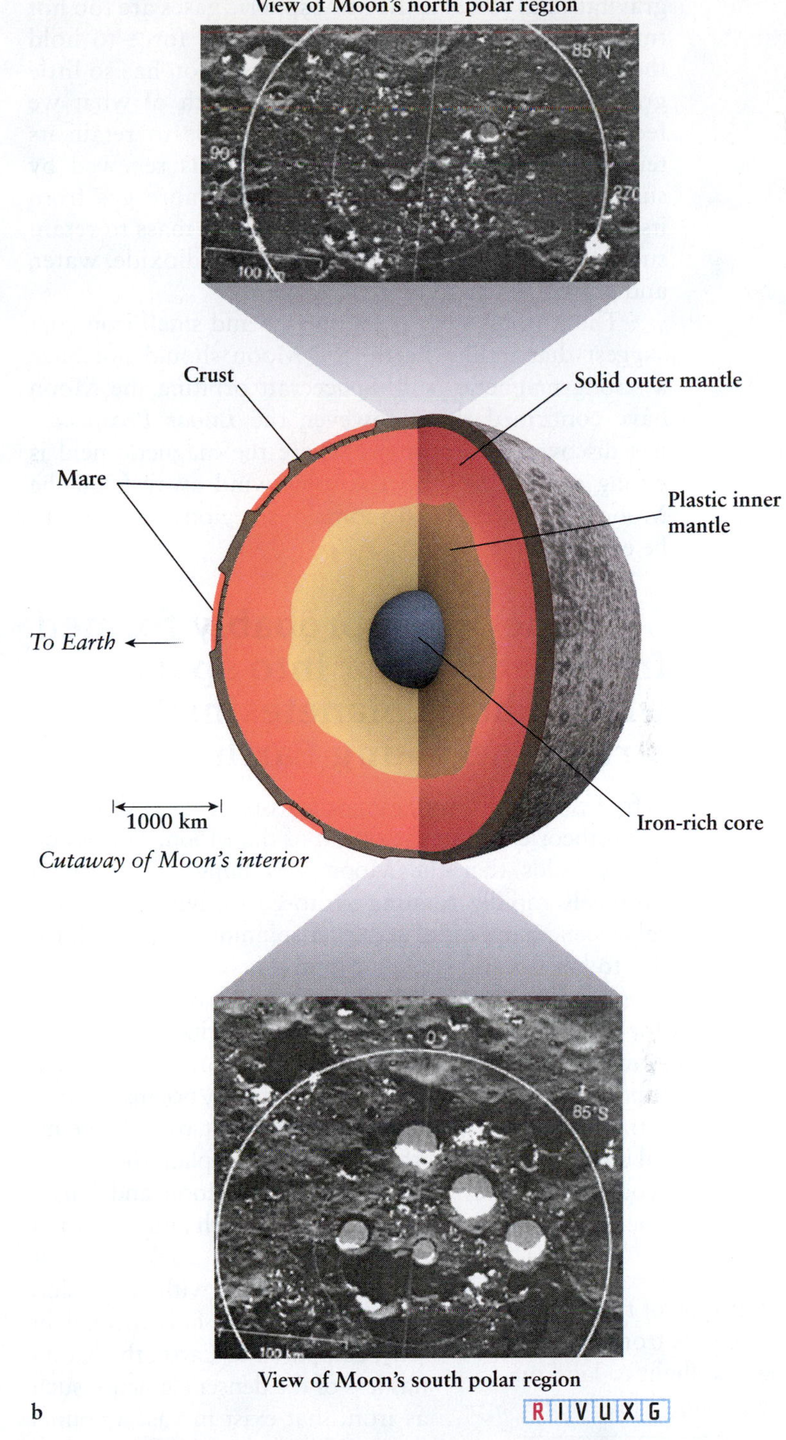

FIGURE 6-26 The Moon's Interior and Its Effects on the Moon's Surface (a) As the Moon's interior shrank, the surface settled irregularly, creating long lines of cliffs called scarps. (b) Based on seismic experiments left on the Moon by Apollo astronauts and the studies done by *Lunar Prospector,* we know that the Moon has a crust, a mantle, and a core. The lunar crust has an average thickness of about 60 km on its Earth-facing side and about 100 km on the far side. The crust and solid upper mantle extend inward to about 800 km, where the nonrigid inner mantle begins. The Moon's core has a radius of between 220 and 450 km. Although the main features of the Moon's interior are analogous to those of Earth, the proportions and details are quite different. The *Clementine* spacecraft revealed that the south polar region on the far side of the Moon has a significant basin where the crust was apparently stripped away by an impact. Top inset: A radar image of the Moon's north polar region. The areas computer-colored in white and light gray are regions where the Sun never shines and that contain water. Bottom inset: A radar image of the Moon's south polar region, also showing regions of permanent or near-permanent darkness. (a: NASA/Goddard/Arizona State University/Smithsonian; b (insets): NASA/Galaxy)

gravitational force. If the atmospheric gases are too hot (moving too fast) for the gravitational force to hold them, they will leak into space. Our Moon has so little gravitational attraction (about one-sixth of what we feel at Earth's surface) that it is unable to retain its tenuous atmosphere. The atmosphere is renewed by sunlight striking the Moon, releasing more gas from its rocks. Earth, in contrast, has enough mass to retain such gases as nitrogen, oxygen, carbon dioxide, water, and argon.

The Moon's slow rotation rate and small iron core suggest that unlike Earth, the Moon should not have a strong magnetic field. Spacecraft orbiting the Moon have confirmed this. However, the *Lunar Prospector* has discovered local areas where the magnetic field is strong enough to keep the solar wind away from the Moon's surface. These anomalous regions have yet to be explained.

6-7 The Moon probably formed from debris cast into space when a huge planetesimal struck the young Earth

Before the Apollo program, scientists debated three different theories about the origin of the Moon. The **fission theory** holds that the Moon was flung out from an extremely rapidly rotating proto-Earth, but no geological evidence supports Earth ever spinning fast enough for this to have occurred. The **capture theory** posits that the Moon was formed elsewhere in the solar system and then drawn into orbit around Earth by gravitational forces. However, it is physically very difficult for a planet to capture such a large moon. Furthermore, because bodies formed in different places have different overall chemical compositions, this theory does not explain the similar geochemistries of the surfaces of the Moon and Earth. The **cocreation theory** proposes that Earth and the Moon were formed near each other at the same time. As with the fission theory, this theory fails to explain why, compared to Earth, the Moon has less of the denser elements, such as iron, that exist in vast amounts deep within the Earth.

What type of rock did astronauts from *Apollo 11*, the first flight to land on the Moon, bring back?

A fourth theory, now held to be correct by most astronomers, was first proposed in the 1980s. It applies the established fact of age-old collisions in the solar system to the formation of the Moon. Called the **collision-ejection theory**, or *large impact theory*, it proposes that the newly formed Earth was struck at an angle by a Mars-sized planetesimal that literally splashed some of Earth's surface layers into orbit around the young planet. Evidence from Moon rocks places this event as occurring between 60 and 100 million years after the solar system began forming nearly 4.6 billion years ago. Earth had differentiated (see Section 6-3) before these times, so the material ejected from Earth that formed the Moon came from Earth's crust and mantle, but not its iron-rich core. A computer simulation of this cataclysm is shown in Figure 6-27. This orbiting matter became a short-lived ring that eventually condensed into the Moon, just as planetesimals condensed to form Earth.

The collision-ejection theory is consistent with the known facts about the Moon. For example, rock vaporized by the impact would have been depleted of *volatile* (easily evaporated) elements and water, leaving the Moon relatively low in water (compared to Earth), as we now know to be true. Also, the Moon has little iron-rich matter because that material had sunk deep into Earth before the planetesimal struck. The material from Earth that splashed into orbit and formed the Moon was mostly the lighter rock floating on Earth's surface. Furthermore, most of the debris from the collision would orbit near the plane of the ecliptic as long as the orbit of the impacting planetesimal had been near that plane. Also, the impact of an object large enough to create the Moon could have tipped Earth's axis of rotation, inaugurating the seasons (Figure 6-28). The fact that this theory is both consistent with observations and explains two seemingly separate things (the existence of the Moon and the tilt of Earth) makes it especially appealing to scientists.

The surface of the newborn Moon probably remained molten for millennia, due to heat released during the impact of rock fragments falling onto the young satellite and the decay of short-lived radioactive isotopes inside it. As the Moon gradually cooled, low-density lava floating on its surface began to solidify into the anorthositic crust that exists today in the highlands.

By correlating the ages of Moon rocks with the density of craters at the sites where they were collected, geologists have determined how the rate of impacts on the Moon changed over the ages. In summary, the ancient, heavily cratered lunar highlands are evidence of an intense bombardment of asteroids that dominated the Moon's early history. This barrage began 4.5 billion years ago, when the Moon's surface first solidified, and ended about 3.8 billion years ago. This period of impacts was punctuated by a severe pounding 4 billion years ago called the Late Heavy Bombardment, which lasted for millions of years. Recall from Chapter 5 that according to the Nice model, this bombardment was caused by the migration of the giant planets.

The frequency of impacts gradually tapered off as meteoroids and planetesimals were swept up by the newly formed planets, as discussed in Chapter 5. Recorded among the final scars at the end of this crater-making era are the impacts of more than a dozen objects, each measuring at least 100 km across. As these huge rocks slammed into the young Moon, they blasted gigantic craters in regions that would later fill with lava to become the maria. Meanwhile, heat from the decay of long-lived radioactive elements, like uranium and

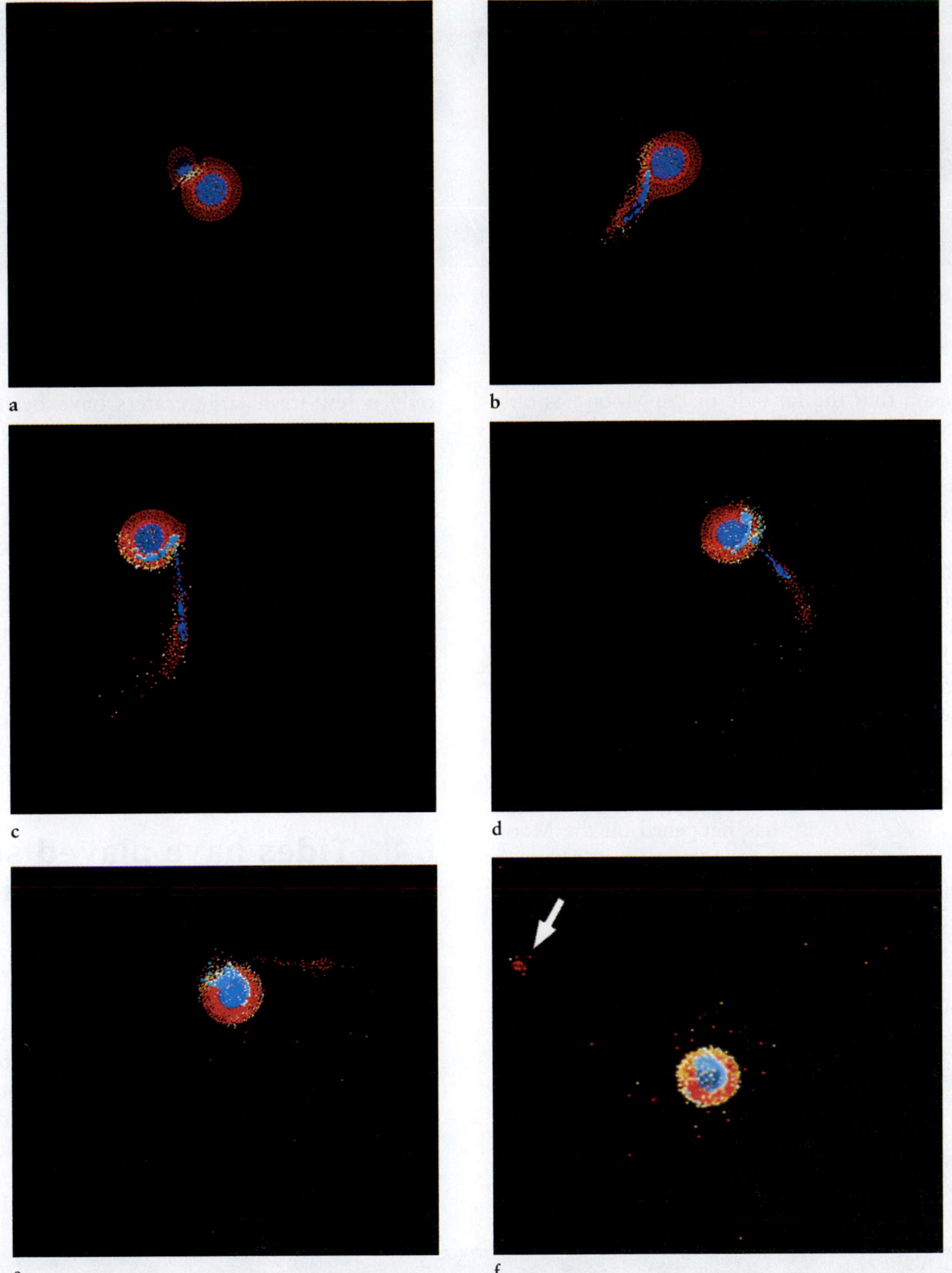

ANIMATION 6.5 **FIGURE 6-27 The Moon's Creation** This computer simulation models the creation of the Moon from material ejected by the impact of a large planetlike body with the young Earth. Successive views (a–f) show the impact, splash, and formation of the Moon. The debris probably formed a short-lived ring before coalescing to become the Moon. Blue and green indicate iron from the cores of Earth and the asteroid; red, orange, and yellow indicate rocky mantle material. In this simulation, the impact ejects both mantle and core material, but most of the dense iron falls back onto Earth. The surviving ejected rocky matter coalesces to form the Moon (indicated by arrow). (W. Benz)

thorium, began to remelt the inside of the Moon. Then, from 3.8 to 3.1 billion years ago, great floods of molten rock from just below the surface (not from the Moon's core) gushed up from the lunar interior through the weak spots in the crust created by these large impacts. Lava filled the impact basins and created the maria we see today. Because most of the bombardment was over by then, maria have few craters.

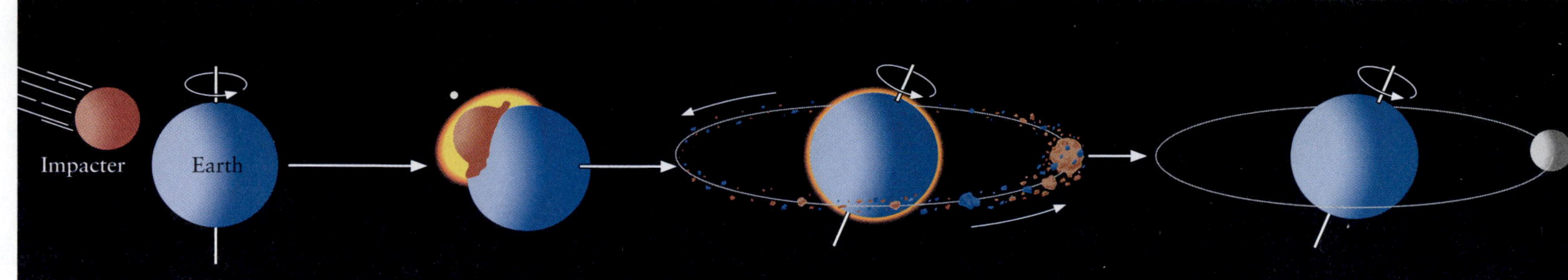

FIGURE 6-28 Tilting Earth's Axis The collision that created the Moon could have also knocked Earth's rotation axis over so that today it has a 23½° tilt, thereby creating the seasons.

The reason that the far side of the Moon has only one significant mare apparently stems from the Moon's crust on the side facing Earth being thinner than the crust on the far side. This difference was created when the Moon was young and molten. At that time, the Moon was much closer to Earth (we will explore this shortly). The Moon's rapid orbit and the tides created on it by Earth caused much of the low-density, crustal lava to flow away from Earth to the lunar far side, where it remained when the crust solidified. It was therefore easier for the molten rock inside the Moon to escape through the large craters on the thin crust of our side than through those on the thicker-crusted far side.

Why are we fortunate on Earth that the Moon is rarely being struck today?

Relatively little large-scale activity has happened on the Moon since those ancient times. Although only a few fresh large craters have been formed, the small-sized debris still plentiful in the solar system continues to create small ones. Back in 1953, an impact was observed that has been connected with the formation of a 1.5-km (0.93-mi) diameter crater. Smaller impacts are observed nearly every year when large groups of rocky space debris, left in orbit from comets as they vaporize, cross the Moon's path. By and large, however, the astronauts visited a world that has remained largely unchanged for more than 3 billion years. Unlike Earth, which is still quite geologically active, any molten rock inside the Moon has apparently stopped directly affecting that world's surface. Astronomers hope to learn more about the Moon's interior when modern seismographs are eventually placed on its surface.

6-8 Tides have played several important roles in the history of Earth and the Moon

Gravitational attraction keeps Earth and the Moon together, while the angular momentum (see *An Astronomer's Toolbox 2-2*) given to the Moon when it was formed keeps the Moon from being pulled back onto Earth. Contrary to appearances, the Moon does not orbit Earth. Rather, both bodies are in motion around their center of mass, called their *barycenter* (Figure 6-29a), just as a pair of objects tied together and sent spinning across an air table move around a similar common point (Figure 6-29b).

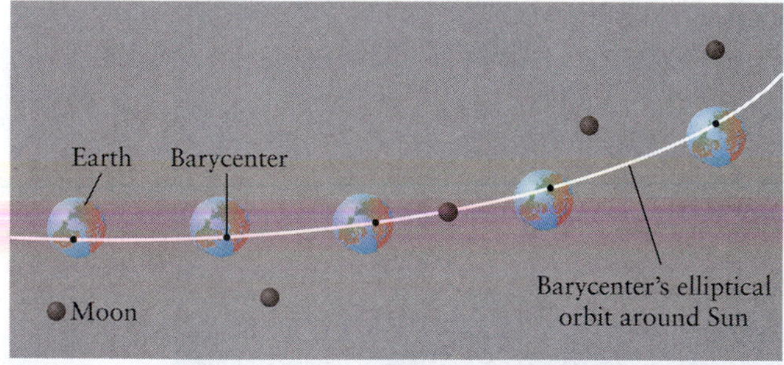

a

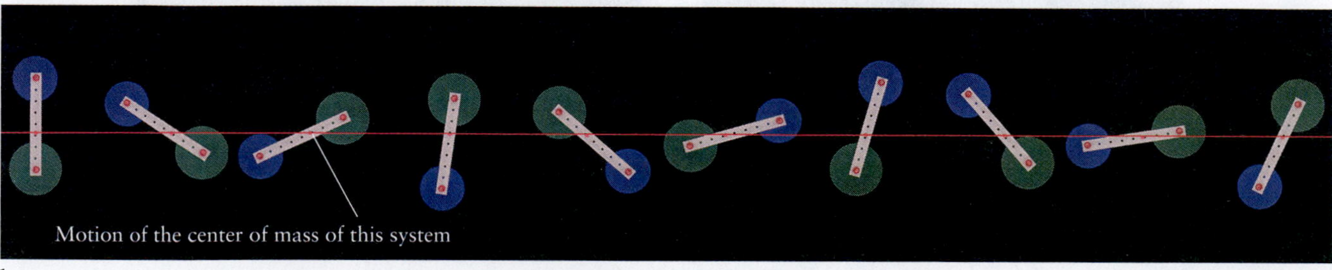

b

FIGURE 6-29 Motion of Earth-Moon System (a) The paths of Earth and the Moon as their barycenter follows an elliptical orbit around the Sun. (b) Analogously, this time-lapse image shows a pair of different-mass disks connected by a piece of wood sliding across a table. Note that the center of mass of this collection of objects moves in a straight line, while its other parts follow curved paths. (© Tom Patagnes)

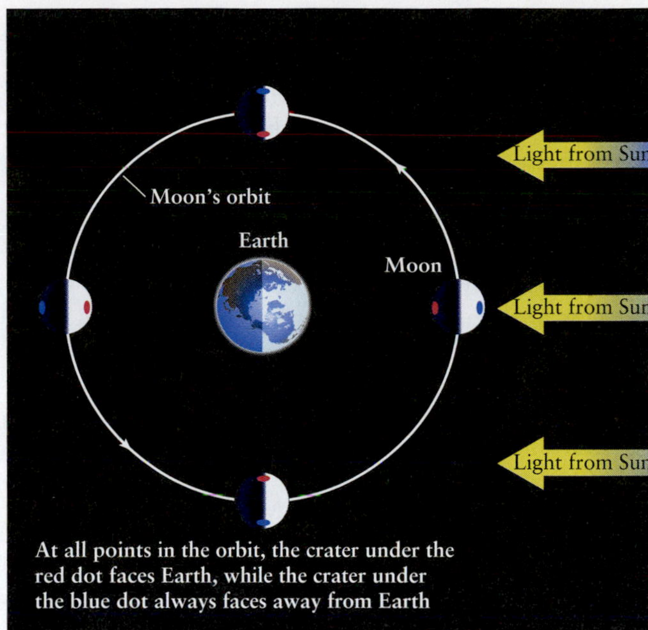

ANIMATION 6.6 **FIGURE 6-30 Synchronous Rotation of the Moon** The motion of the Moon around Earth as seen from above Earth's north polar region (ignoring Earth's orbit around the Earth-Moon barycenter). For the Moon to keep the same side facing Earth as it orbits our planet, the Moon must rotate on its axis at precisely the same rate that it revolves around Earth.

3 4 Except for slight variations as described below, the same surface features on the Moon always face Earth, regardless of what phase the Moon is in (see Figure 1-21). This phenomena occurs because the Moon is in **synchronous rotation** around Earth: The Moon rotates on its axis at the same rate that it and Earth orbit their barycenter (Figure 6-30). To see why the Moon must be rotating if it keeps the same side facing us, imagine standing up, holding your arm out in front of you with your palm facing you, and twirling yourself around. You are Earth, while your palm represents the side of the Moon facing Earth. The only way your palm can always face you (be in synchronous rotation) is if your hand (the Moon) rotates at exactly the same rate that it swings (revolves). So the Moon must be rotating.

If the Moon's orbit were perfectly circular, if the Moon's rotation axis were perpendicular to the plane of its orbit around the Earth, if the Moon were orbiting in the plane of the ecliptic, and if we always saw the Moon at the same distance from the Earth throughout the day, then we would see precisely 50% of the Moon's surface. In fact, none of these statements is precisely true. As a result, we actually see a little beyond the poles and around the sides of the Moon. We can see about 59% percent of the Moon's total surface at different times throughout the year. These changes in what we see of the Moon from Earth are called *librations*.

We have seen the Moon's complete "far side" only from spacecraft (see Figure 6-19). The far side of the Moon is often, and incorrectly, called its "dark side." The dark side of the Moon is exactly like the dark side of Earth—it is the side where the Sun is below the horizon. Whenever we view less than a full Moon, we see part of its dark side, while at the same time some sunlight is then striking part of its far side.

The origin of the Moon's synchronous rotation is in the tidal force that acts between Earth and the Moon. This tidal force also creates ocean tides on Earth. To understand how the Moon came to have synchronous rotation, we investigate the forces acting between Earth and the Moon. The discussion in *Guided Discovery: Tides* focuses on how the Moon creates the tides on Earth. We will use this information to see how Earth once created tides on the Moon that changed the Moon's original rotation rate to synchronous rotation.

Two-thirds of the tide height on Earth results from 5
the gravitational tug-of-war between Earth and the Moon as the two bodies orbit the barycenter. The Sun generates most of the remaining third of the tides, with Jupiter and the other planets contributing a tiny fraction.

As explained in *Guided Discovery: Tides*, two high  6
tides always occur on opposite sides of Earth (see Figure GD6-1). When the Sun, the Moon, and Earth are aligned, the Sun and the Moon create pairs of high ocean tides in the same directions, and the resulting combined tides are the highest high tides of the lunar cycle. These **spring tides** (so-called not because they only occur in the spring—they occur in all seasons—but because the term derives from the German word *springen,* meaning to "spring up") occur at every new and full Moon (see Figure GD6-2a). Note that it does not matter whether the Sun and Moon are on the same or opposite sides of Earth when generating the spring tides, because the Sun and the Moon both create high tides on opposite sides of Earth.

Why haven't the ocean tides on Earth put our planet into synchronous rotation with respect to the Moon?

At first quarter and third quarter, the Sun and the Moon form a right angle as seen from Earth. The gravitational forces they exert compete with each other, so the tidal distortion is the least pronounced. The especially small tidal shifts on these days are called **neap tides** (Figure GD6-2b).

Finally, we can explain the Moon's synchronous rotation: When the Moon was young, it was molten and Earth's gravity created huge tides of molten rock up to 18 m (60 ft) high there. Friction generated by these tides between the liquid rock and the mantle rock below it caused the Moon to speed up or slow down (we don't know which) until one of the high tides became fixed directly between the Moon's center and the center of Earth. As the surface solidified, it locked onto the interior, causing the entire Moon to rotate at the same rate as its surface. The entire Moon was then in synchronous rotation.

Tides

Tides result from a combination of Earth's motion around its barycenter with the Moon (or Sun) and the changing strength of the gravitational force from the Moon (or Sun) across Earth. Ignoring Earth's rotation and focusing on the Moon for the moment, let's first consider the motion of Earth and the Moon around their barycenter (see Figure 6-29a). The barycenter is located 1720 km (1068 mi) below Earth's surface on the line between the centers of Earth and the Moon. As a result of its motion around the barycenter, every point on Earth feels an equally strong force away from the Moon, unlike the outward force that you feel while riding on a merry-go-round, which increases with distance from the center. The force away from the Moon is represented at three locations by the force labeled F_{out} in Figure GD6-1a.

Opposing this force away from the Moon, Earth also has a gravitational attraction to its satellite. Recall from Chapter 2 (*An Astronomer's Toolbox 2-3*) that the gravitational force exerted by one body on another decreases with distance. Therefore, the Moon's gravitational attraction on Earth is greatest at the point on Earth's surface closest to the Moon. This attraction is greater than the force felt at the center of Earth, which in turn is greater than the force felt on the opposite side of Earth from the Moon. The gravitational force in these places is labeled F_{grav} in Figure GD6-1a.

Because forces in the same or opposite directions simply add or subtract, respectively, we can subtract the two forces F_{out} and F_{grav} at any point on Earth, due to the presence of the Moon. The resulting net force is labeled F_{tide} in Figure GD6-1a, and is the tidal force acting at these places. For example, at the point on Earth directly below the Moon, the tidal force is very strong and points toward the Moon. At the center of Earth, the two forces cancel each other completely. At the point on the opposite side of Earth from the Moon, the tidal force is equal in strength to the tidal force directly under the Moon, but is directed away from the Moon. The strength of the tidal force at various places is summarized in Figure GD6-1b.

Consider the point on Earth closest to the Moon. The strength of the tidal force, F_{tide}, seems to imply that the Moon's gravitational force lifts the water there to create the high tide. However, the force of gravitation from the Moon is not strong enough to raise that water more than a few centimeters. Rather, the high tide closest to the Moon occurs because ocean waters from nearly halfway around to the opposite side of Earth are pulled Moonward by the gravitational force acting on them. This water slides toward the Moon and fills the ocean directly under it, creating a tidal bulge or high tide. Likewise, the net outward force acting on the opposite side of Earth from the Moon pushes the waters that are just over halfway around from the Moon to the side directly away from it, creating a simultaneous, second high tide of equal magnitude on the opposite side of Earth from the Moon. Where the water has been drained away in this process, low tides occur.

Let's spin Earth back up. As it rotates, the geographic locations of the high and low tides continually change. Figure GD6-1b indicates that high tides occur when the Moon is high in the sky or below our feet, while low tides occur when the Moon is near either horizon. This timing means that there are two cycles of high and low tides at most places on Earth each day. These cycles do not span exactly 24 h, however, because the Moon is moving along the celestial sphere,

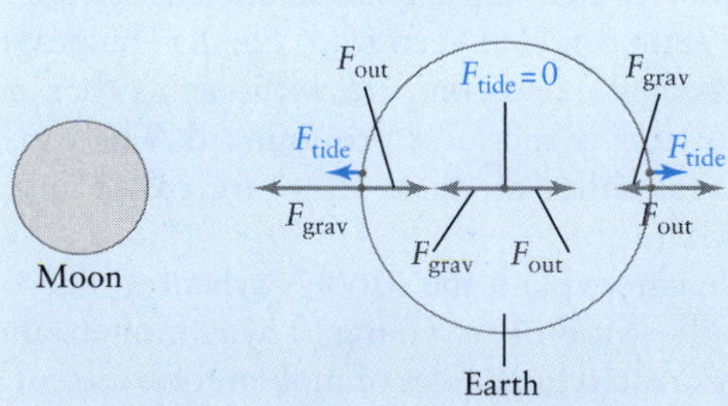

ANIMATION 6.7 **FIGURE GD6-1 Tidal Forces** (a) The Moon induces tidal forces, F_{tide}, on Earth. At each point, this force is the difference between the force, F_{out}, created by the orbital motion of the two bodies around their barycenter, and the Moon's gravitational force, F_{grav}, at that point. The magnitude and direction of each arrow represent the strength and direction of each force. (b) Water slides along Earth to create the tides. Ignoring Earth's rotation and the effects of the continents for now, this figure shows how two high tides are created on Earth by the Moon's gravitational pull. The Sun has a weaker, but otherwise identical, effect.

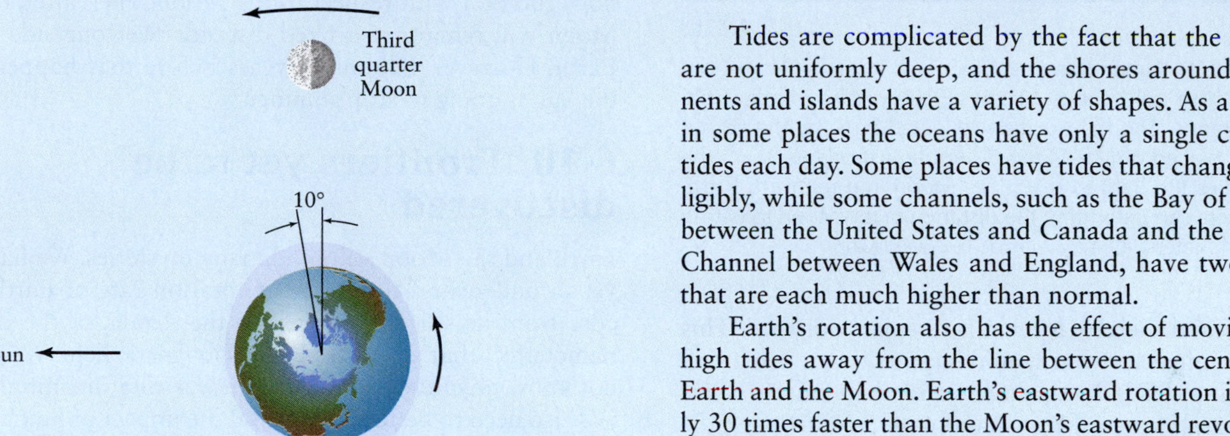

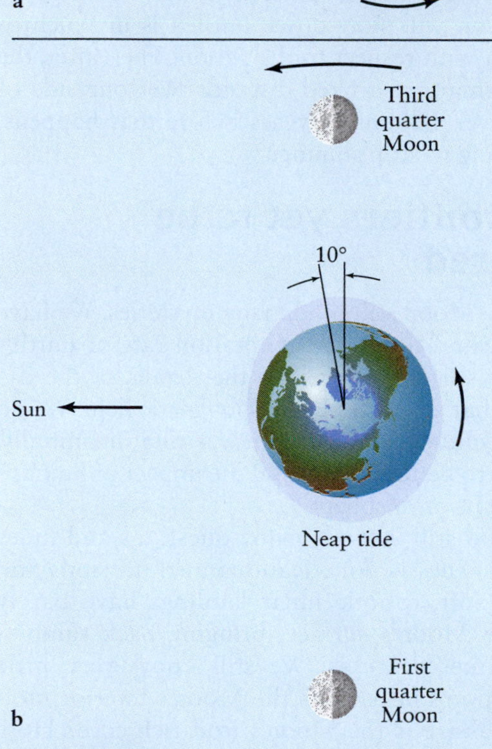

FIGURE GD6-2 Tides on Earth The gravitational forces of the Moon and the Sun deform the oceans. Due to Earth's rapid rotation, the high tide closest to the Moon leads it around Earth by about 10°. (a) The greatest deformation (spring tides) occurs when the Sun, Earth, and the Moon are aligned with the Sun and the Moon either on the same or opposite sides of Earth. (b) The least deformation (neap tides) occurs when the Sun, Earth, and the Moon form a right angle.

changing the time at which high or low tide occurs at any location from day to day. As a result, the time between consecutive high tides is 12 h, 25 min.

Tides are complicated by the fact that the oceans are not uniformly deep, and the shores around continents and islands have a variety of shapes. As a result, in some places the oceans have only a single cycle of tides each day. Some places have tides that change negligibly, while some channels, such as the Bay of Fundy between the United States and Canada and the Bristol Channel between Wales and England, have two tides that are each much higher than normal.

Earth's rotation also has the effect of moving the high tides away from the line between the centers of Earth and the Moon. Earth's eastward rotation is nearly 30 times faster than the Moon's eastward revolution around Earth. Therefore, as it rotates, Earth drags the high tides eastward with it. As a result, the high tide that should be directly between the centers of Earth and the Moon is actually 10° ahead of the Moon in its orbit around Earth (Figure GD6-2). This high tide stays at roughly this position because the Moon's gravitational force pulls the tide westward while it is being dragged eastward by Earth's rotation; the two effects balance each other out. A similar argument applies to the tide on the opposite side of Earth.

We can repeat this analysis for the Earth-Sun system and get the same qualitative results.

WHAT IF . . . The Moon Were Not in Synchronous Rotation with Earth?
We would, of course, be able to see all parts of the Moon's surface. More interestingly, the Moon's rotation would cause land tides to travel along the Moon as it rotated and revolved around the barycenter. These tides would cause the rock of the Moon to rub against itself, creating heat, and causing more of the Moon's interior to be molten than may be the case today. That molten rock would escape the Moon's surface in volcanoes and through cracks in its surface, creating a very colorful and dynamic Moon.

6-9 The Moon is moving away from Earth

In 1897, Sir George Darwin (son of the evolutionist Charles Darwin) proposed that the Moon is moving away from Earth. To understand why, read *Guided Discovery: Tides* and see Figure GD6-2, which shows that the high tide nearest the Moon is actually 10° ahead of the Moon in its orbit around Earth. This offset occurs because friction between the ocean and the ocean floor causes the rapidly spinning Earth to drag the tidal bulge ahead of the line between the center of Earth and the center of the Moon. The gravitational force from the water in the high ocean tide nearest the Moon acts back on the Moon, pulling it ahead in its orbit, giving

FIGURE 6-31 Lunar Ranging Beams of laser light are fired through three telescopes at the Observatoire de la Côte d'Azur, France. The light is then reflected back by the corner reflectors placed on the Moon by Apollo astronauts. From the time it takes the light to reach the Moon and return to Earth, astronomers can determine the distance to the Moon to within a few millimeters. (Observatoire de la Côte d'Azur)

it energy and thereby forcing it to spiral outward. This effect is similar to what would happen if you tied a ball on a string, let it hang down, and then started spinning yourself around. Your hand would act through the string, giving the ball energy and starting it to spin and move farther and farther away from your body.

To test the concept that the Moon is moving away from Earth, *Apollo 11, 14,* and *15* astronauts placed sets of reflectors on the Moon (see Figure 6-25), similar to the orange and red ones found on cars. These reflectors are specially designed so that light striking them from any direction is reflected back toward its source. Pulses of laser light were fired at the Moon from Earth (Figure 6-31), and the time it took the light to reach the Moon, bounce off a reflector array, and return to Earth was carefully recorded. Knowing the speed of light and the time it takes for the light pulse to make the round trip, astronomers can calculate the distance to the Moon with an error of only a few millimeters. From such measurements made since 1969, astronomers have established that the Moon is currently spiraling away from Earth at a rate of 3.8 cm per year.

This means, of course, that the Moon used to be closer to Earth. Although the initial distance of the Moon when it coalesced is not known, its present rate of recession tells us that it was at least half its present distance; more accurate calculations suggest one-tenth. At that distance, the tides on Earth were a thousand times higher than they are today.

What effects would the original tides on Earth have had on the geology of our young planet?

Where does the Moon get the energy necessary to spiral away from Earth? The answer is from Earth itself. As the tides move westward to try to stay under the Moon, they encounter the eastward-moving landmasses, which disrupt the water's motion. Because the ocean tides move westward, while the planet rotates eastward, the oceans actually push on the continents opposite to Earth's direction of rotation. The result is that the tides are slowing Earth's rotation. The energy lost by Earth as it spins down is gained by the Moon. At present, the day is slowing down by about one-thousandth of a second each century. Based on fossil and other evidence, geologists calculate that when Earth first formed, the day was only between 5 and 10 h long.

Will the Moon ever leave Earth completely as a result of these dynamics? In fact, it will not. Rather, Earth's rotation will slow down until it is in synchronous rotation with respect to the Moon. Thereafter, the Moon will remain at a fixed distance over one side of Earth. However, billions of years before that happens, the Sun is going to stop shining.

6-10 Frontiers yet to be discovered

Earth and the Moon still hold many mysteries. We have yet to understand the different rotation rate of Earth's core from its surface, as well as the details of the dynamo effect that generates Earth's magnetic field. We do not know precisely how fast Earth was rotating initially. We also need to better understand the impact of humans on Earth and its atmosphere.

The Moon still inspires many questions and hides many secrets. The six American-manned missions and three Soviet soft, robotic lunar landings have barely scratched the Moon's surface, bringing back samples from only nine locations. We still know very little about the Moon's far side. Is the Moon's interior molten? How massive is the Moon's iron-rich core? How old are the youngest lunar rocks? Did lava flows occur over western Oceanus Procellarum only 2 billion years ago, as crater densities there suggest? Is the Moon really geologically dead, or does it just appear dead because our examination of the lunar surface has been so cursory? Such questions can be answered only by sending new and improved science experiments, such as the seismometers mentioned earlier in the chapter, to the Moon.

SUMMARY OF KEY IDEAS

Earth: A Dynamic, Vital World

- Earth's atmosphere is about four-fifths nitrogen and one-fifth oxygen. This abundance of oxygen is due to the biological processes of life-forms on the planet.
- Earth's atmosphere is divided into layers named the troposphere, stratosphere, mesosphere, and ionosphere. Ozone molecules in the stratosphere absorb ultraviolet light rays.
- The outermost layer, or crust, of Earth offers clues to the history of our planet.

• Earth's surface is divided into huge plates that move over the upper mantle. Movement of these plates, a process called *plate tectonics,* is caused by convection in the mantle. Also, upwelling of molten material along cracks in the ocean floor occurs during seafloor spreading. Plate tectonics is responsible for most of the major features of Earth's surface, including mountain ranges, volcanoes, and the shapes of the continents and oceans.

• Study of seismic waves (vibrations produced by earthquakes) shows that Earth has a small, solid inner core surrounded by a liquid outer core. The outer core is surrounded by the dense mantle, which in turn is surrounded by the thin, low-density crust on which we live. Earth's inner and outer cores are composed primarily of iron. The mantle is composed of iron-rich minerals.

• Earth's magnetic field produces a magnetosphere that surrounds the planet and deflects the solar wind.

• Some charged particles from the solar wind are trapped in two huge, doughnut-shaped rings called the Van Allen belts. An Earthward deluge of particles from a coronal mass ejection on the Sun can produce exceptional auroras.

The Moon and Tides

• The Moon has heavily cratered highlands and relatively smooth-surfaced maria. Powdered into regolith, the anorthosite rock of the highland is brighter than then powdered basalts of the maria.

• Many lunar rock samples are solidified lava formed largely of minerals also found in Earth rocks.

• Anorthositic rock in the lunar highlands was formed between 4.0 and 4.3 billion years ago, whereas the mare basalts solidified between 3.1 and 3.8 billion years ago. The Moon's surface has undergone very little geologic change over the past 3 billion years.

• Impacts have been the only significant "weathering" agent on the Moon; the Moon's regolith (pulverized rock layer) was formed by meteoritic action. Lunar rocks brought back to Earth contain no water and are depleted of volatile elements.

• Frozen water has been discovered in numerous places just below the Moon's surface.

• The collision-ejection theory of the Moon's origin, accepted by most astronomers, holds that the young Earth was struck by a huge planetesimal, and debris from this collision coalesced to form the Moon.

• The Moon was molten in its early stages, and the anorthositic crust solidified from low-density magma that floated to the lunar surface. The mare basins were created later by the impact of planetesimals and were then filled with lava from the lunar interior.

• Gravitational interactions between Earth and the Moon produce tides in the oceans of Earth and set the Moon into synchronous rotation. The Moon is moving away from Earth, and, consequently, Earth's rotation rate is decreasing.

A WHAT DID YOU THINK?

1 *Can Earth's ozone layer, which has been partially depleted, be naturally replenished?* Yes. Ozone is created continuously from normal oxygen molecules by their interaction with the Sun's ultraviolet radiation.

2 *Who was the first person to walk on the Moon, and when did this event occur?* Neil Armstrong was the first person to set foot on the Moon. He and Buzz Aldrin flew on the *Apollo 11* spacecraft piloted by Michael Collins. Armstrong and Aldrin set down the Eagle Lander on the Moon on July 20, 1969.

3 *Do we see every part of the Moon's surface at some time during the lunar cycle?* No. Because the Moon's rotation around Earth is synchronous, we always see the same side. The far side of the Moon has been seen only from spacecraft that pass or orbit it.

4 *Does the Moon rotate and, if so, how fast?* The Moon rotates at the same rate that it revolves around Earth, once every 27.3 Earth days. If the Moon did not rotate, then, as it revolved, we would see its entire surface from Earth, which we do not.

5 *What causes the ocean tides?* The tides are created by gravitational forces from the Moon and Sun combined with Earth's motion around the barycenter.

6 *When does the spring tide occur?* Spring tides occur during each full and new Moon.

Key Terms for Review

anorthosite, 177
capture theory, 180
cocreation theory, 180
collision-ejection theory, 180
continental drift, 165
convection, 168
core, 167
coronal mass ejection, 171
crust, 165
dynamo theory, 169
ejecta blanket, 172
fission theory, 180
highlands, 174
impact breccias, 177
ionosphere (thermosphere), 163
mantle, 167
mare (plural *maria*), 187
mare basalt, 177
mascons, 178
mesosphere, 163
neap tide, 183
northern lights (aurora borealis), 171
ozone layer, 162
planetary differentiation, 167
plate tectonics, 165
regolith, 176
rille, 174
scarps, 178
seafloor spreading, 165
seismic waves,168
seismograph, 168
solar wind, 170
southern lights (aurora australis), 171
spring tide, 183
stratosphere, 162
synchronous rotation, 183
troposphere, 162
Van Allen radiation belts, 170

Review Questions

1. The Moon's surface is best described as: **a.** fine-grained powder, **b.** solid rock, **c.** rocky rubble, **d.** liquid water oceans and dry land, **e.** molten rock.

2. What type of chemical or molecule is most common in Earth's atmosphere? **a.** carbon dioxide, **b.** oxygen, **c.** water, **d.** nitrogen, **e.** hydrogen.

3. Why does Earth's albedo change daily? Seasonally?

4. Why is Earth's surface not riddled with craters as is that of the Moon?

5. List the layers of Earth's atmosphere.

6. Describe the process of plate tectonics. Give specific examples of geographic features created by plate tectonics.

7. To review Earth's tectonic plates, do Interactive Exercise 6.1 and print out or sketch the completed diagram.

8. To review the mechanism of plate tectonics, do Interactive Exercise 6.2 and print out or sketch the completed diagram.

9. How do we know about Earth's interior, given that the deepest wells and mines extend only a few kilometers into its crust?

10. Describe the interior structure of Earth.

11. Why is the center of Earth not molten?

12. Describe Earth's magnetosphere.

13. To review Earth's magnetosphere, do Interactive Exercise 6.3 and print out or sketch the completed diagram.

14. What are the Van Allen belts?

15. What kind of features can you see on the Moon with a small telescope?

16. Explain why the maria appear darker than the lunar highlands.

17. Why are there so few craters on the maria?

18. To review the crater formation history in the solar system, do Interactive Exercise 6.4 and print out or sketch the completed diagram.

19. Briefly describe the main differences and similarities between Moon rocks and Earth rocks.

20. How do we know that the maria were formed after the lunar highlands?

21. What is a tidal force? How do tidal forces produce tides in Earth's oceans?

22. What is the difference between spring tides and neap tides? During which phase(s) of the Moon do each occur?

23. Why do most scientists support the collision-ejection theory for the Moon's formation?

24. Why are virtually all of the craters on the Moon circular, even though many impacts there were not head-on?

25. Why hasn't the water ice near the Moon's poles been evaporated by sunlight?

Advanced Questions

26. Explain how the outward flow of energy from Earth's interior drives the process of plate tectonics.

27. Why do some geologists believe that Pangaea was the most recent in a succession of supercontinents?

28. Why are active volcanoes, such as Mount St. Helens, usually located in mountain ranges along the boundaries of tectonic plates?

29. Why is more lunar detail visible through a telescope when the Moon is near quarter phase than when it is at full phase?

30. Why are the Moon rocks retrieved by astronauts so much older than typical Earth rocks, even though both worlds formed at nearly the same time?

31. Some people who supported the fission theory proposed that the Pacific Ocean basin is the scar left when the Moon pulled away from Earth. Explain why this idea is wrong.

32. Apollo astronauts left seismometers on the Moon that radioed seismic data back to Earth. The data showed that moonquakes occur more frequently when the Moon is at perigee (closest to Earth) than at other locations along its orbit. Give an explanation for this finding.

33. Why do you think that no Apollo missions landed on the far side of the Moon?

34. How might studying albedo help astronomers locate habitable worlds orbiting other stars?

Discussion Questions

35. If Earth did not have a magnetic field, do you think auroras would be more common or less common than they are today? Explain.

36. Comment on the idea that without the Moon's presence, life would have developed far more slowly.

37. Identify and compare the advantages and disadvantages of lunar exploration by astronauts as opposed to mobile, unmanned instrument packages and robots.

38. When was the last earthquake near your hometown? How far is your hometown from a plate boundary? What kinds of topography (for example, mountains, plains, seashore) dominate the geography of your hometown area? Does that topography and the frequency of earthquakes seem to be consistent with your hometown's proximity to a plate boundary?

39. The ice on the Moon is believed to be mixed with rock just under the Moon's surface. How might that water be economically extracted and purified?

What If...

40. Earth had two moons? Assume one is at our Moon's distance and the other is at half that distance. Describe the motion of the two moons in the sky and how they might appear to us. What would be different here on Earth?

41. The Moon orbited Earth in the opposite direction from Earth's rotation, rather than in the same direction? What would be different about Earth and life on it? Assume that today such a counter-revolving Moon would be at the same distance as our Moon.

42. Our Moon were one-tenth of its actual distance from Earth? What would be different about Earth and life on it? *Hint:* The heights of tides vary as $1/r^3$, where r is the distance between the centers of Earth and the Moon.

43. The Moon, located in its current orbit, were as massive as Earth? What would be different about Earth, life on Earth, and the Moon?

44. Earth's interior were entirely solid, rather than partly molten? What would be different about Earth and life on it?

45. Earth were now in synchronous rotation with the Moon? That is, suppose that Earth rotated at the same rate that the Moon orbits Earth. What would be different about Earth and life on it?

Web Questions

46. Use the Web to learn more about plate tectonics. What global changes might accompany the formation and breakup of a supercontinent? How might these changes affect the evolution of life? What life-forms dominated Earth when Pangaea existed some 200 million years ago, and also when fragments of the preceding supercontinent were as dispersed as today's continents?

47. Search the Web for information about "Pangaea Ultima," a supercontinent that may form in the distant future. When is it expected to form? How will it compare to the Pangaea of 200 million years ago (see Figure 6-7)?

48. Use the Web to determine the status of the Antarctic and Arctic ozone holes. How has the situation changed over the past few years? Explain why most scientists who study this issue blame chemicals called CFCs for the existence of the ozone holes.

49. In 1989, representatives of many countries signed a treaty called the *Montreal Protocol* to protect the ozone layer. Use the Web to learn the current status of this treaty. How many nations have signed it? Has the treaty been amended? If so, how? List when various substances that destroy the ozone layer are being phased out.

50. Search the Web for current information about upcoming lunar science missions. When is each scheduled to be launched? What new investigations are planned for each mission? What existing scientific issues may these missions resolve?

51. Search the Web for current information about tourist trips into space. For when are these scheduled? How long are the trips? What are their costs? What preparation is involved?

52. Refinements to the modeled mass and impact angle of the planetesimal that struck Earth and created the Moon have recently been made. Search the Web for this information and explain the justification for these new parameters.

53. Some astronomers have observed changes in brightness and color on the Moon, called *lunar transient phenomena.* Search the Web and explain what these events are and what their origins are believed to be.

Observing Projects

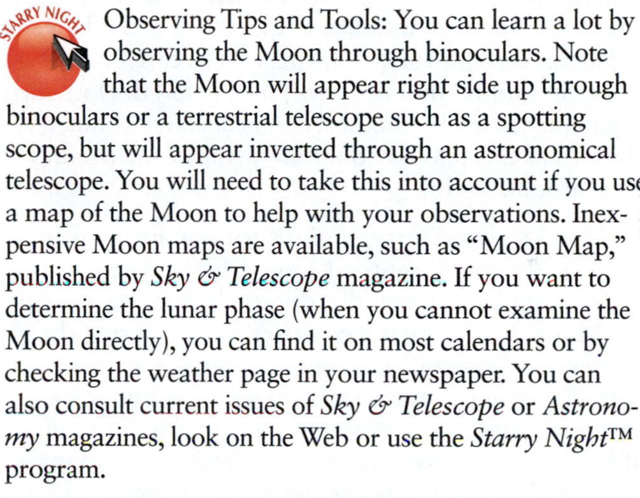

Observing Tips and Tools: You can learn a lot by observing the Moon through binoculars. Note that the Moon will appear right side up through binoculars or a terrestrial telescope such as a spotting scope, but will appear inverted through an astronomical telescope. You will need to take this into account if you use a map of the Moon to help with your observations. Inexpensive Moon maps are available, such as "Moon Map," published by *Sky & Telescope* magazine. If you want to determine the lunar phase (when you cannot examine the Moon directly), you can find it on most calendars or by checking the weather page in your newspaper. You can also consult current issues of *Sky & Telescope* or *Astronomy* magazines, look on the Web or use the *Starry Night*™ program.

54. Use a telescope or binoculars to observe the Moon. Compare the texture of the lunar surface that you see on the maria with that of the lunar highlands. How does the visibility of details vary with distance from the terminator (the boundary between day and night on the Moon)? Why?

55. If you live near the ocean, observe the tides to see how the times of high and low tides are correlated with the position of the Moon in the sky.

56. Observe the Moon through a telescope or binoculars for a few nights over a period of 2 weeks between new Moon and full Moon. Make sketches of various surface features, such as craters, mountain ranges, and maria. How does the appearance of these features change with the Moon's phase? Which features are most easily seen at a low angle of illumination (near the terminator)? Which features show up best at full Moon, when the Moon's surface is directly illuminated by the Sun (that is, far from the terminator)?

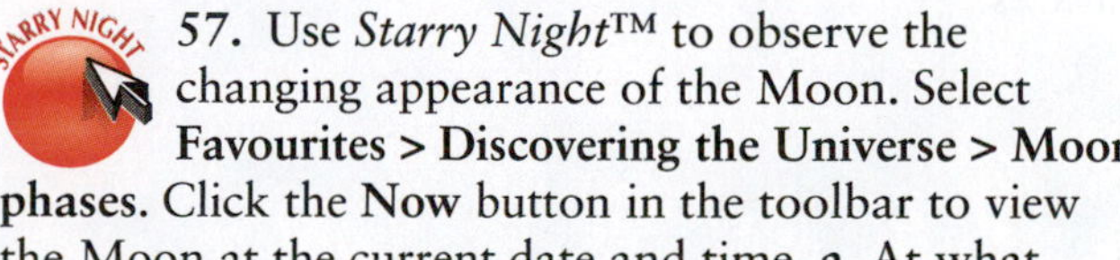

57. Use *Starry Night*™ to observe the changing appearance of the Moon. Select **Favourites > Discovering the Universe > Moon phases**. Click the **Now** button in the toolbar to view the Moon at the current date and time. **a.** At what

phase is the Moon at the present time? To see the changing phases of the Moon, set the **Time Flow Rate** to between 1 and 6 hours, depending on your computer's speed, and click the **Play** button to start time flow. **b.** Describe how the phase of the Moon changes. From your viewpoint, which way does the shadow move across the Moon and how does the shape of the illuminated part of the Moon (that is, its phase) change? **c.** Choose **Favourites > Discovering the Universe > Atlas.** Now open the **Find** pane, type Moon into the search box and press the Return (Enter) key on the keyboard in order to center the view upon the Moon. Zoom in to a field of view of 1°. Set the **Time Flow Rate** to **1 day** and, while clicking the **Step time forward** button, carefully observe the lunar features. Do these features appear to remain stationary in the view? (They *would* remain stationary if the Moon's synchronous rotation were the *only* motion that the Moon had relative to Earth.) If not, explain. (*Hint*: A Web search on the word "libration" [not libation!] may help.) **d.** Does the apparent size of the Moon remain constant? If not, explain what this tells us about the shape of the Moon's orbit around Earth.

STARRY NIGHT

58. Use *Starry Night*™ to view Earth as it appears from the Moon. Change your viewing location to the *Apollo 11* landing site on the Moon by selecting **Favourites > Discovering the Universe > Tranquility Base.** (You can see an astronaut deploying instrumentation near the Lunar Exploration Module in this view, along with the shadow of his colleague. Note the halo around the head in this shadow, caused by the opposition effect, in which dust particles hide their own shadows in the anti-sun direction.) Click the **Now** button in the toolbar and use the **Find** pane to locate Earth in the lunar sky (double-click the listing for Earth in the **Find** pane). If Earth is near the "new" phase, it might be hard to see, but you can show its position by displaying its label if you check the box to the left of "Earth" in the **Find** pane. With the cursor over the image of Earth in the view, click the right mouse button (**Ctrl–click** on a Mac) and select **Magnify** from the contextual menu. With the **Time Flow Rate** set to 10 minutes, click the **Play** button. **a.** Do you see changes in the positions of surface features of Earth as time advances? If so, explain why, particularly when compared to the observations of the Moon from Earth in the previous project, where no significant change was seen in the face of the Moon as time advanced. **b.** Run time for a much longer period and describe the phases of Earth, ignoring any changes in the positions of surface features.

STARRY NIGHT

59. In this exercise, you can use the *Starry Night*™ program to examine Earth as seen from space. Select **Favourites > Discovering the Universe > Earth.** You can use the **Zoom** buttons in the toolbar to zoom in and out on the Earth image. **a.** Use the **Location Scroller** to move around Earth and attempt to position the Sun behind Earth, noting the phase of Earth as you do so. What is the observed phase of Earth when the Sun is in the same region of the observer's sky as Earth? From the relationship between Earth's phase and the Sun's position in the observer's view, can you suggest a technique whereby watching the phase of a planet while moving around it using the location scroller would allow you to locate the Sun in the observer's sky? You can use this sun-finding technique from a position above Mars. Open the **Find** pane and remove any text from the search box at the top of the pane so that a list of Solar System items is shown. Go to Mars by clicking the down arrow to the left of the listing for this planet and select **Go There** from the menu that appears. This places you at about 17,000 km above the planet's surface. Use the location scroller to move around the planet at this distance and use the above method of observing the planet's phase to locate the Sun in your sky. **b.** Return to Earth by selecting **File > Revert** and then use the location scroller to move Earth around until South America is visible. Set the **Time Flow Rate** to **3000x** and allow time to flow until the terminator (the dividing line between the day and night sides of Earth) is over Eastern Brazil, at the eastern edge of South America, and then click the **Stop** button. As seen from eastern Brazil, is the sun rising or setting? Explain.

STARRY NIGHT

60. In this exercise, you can use *Starry Night*™ to examine the Moon. Select **Favourites > Discovering the Universe > Moon Phases.** Click **Play** and then **Stop** time flow when the Moon's phase is full. **a.** Based only on what you can see in the image, what evidence can you find that the Moon is geologically inactive? Explain. **b.** Spreading outward from some of the largest craters on the Moon are straight lines of slightly lighter-colored material, called rays, which were caused by material ejected outward by the impact that caused the crater. Use the zoom controls and the location scroller to get a better view of these lunar features. Can you see whether this ejected material disturbed crater walls during the violent impacts that caused these craters? Place the cursor over these craters and open the contextual menu (**right-click** on a PC, **Ctrl-click** on a Mac). The name of the crater will be found in the command "Mark *crater name* on surface." List at least three craters on the lunar surface that have rays.

WHAT IF... The Moon Didn't Exist?

Throughout history, people have woven myths about the Moon and its effects on everything from childbirth to stock market activity. Countless romances have begun under a full Moon, and entire nations have dedicated themselves to reaching it.

What if the Moon never existed? Would life have even developed on Earth? If so, how would it be different? Would a self-aware species like ourselves have evolved?

Delayed Origins It would have been much more difficult for life to start evolving on a Moonless Earth. In the first place, the minerals from which life developed in our Earth's young oceans were swept down there, in large part, by gigantic tides, a thousand times higher than the tides of today, created by the Moon when it was some 10 times closer to Earth than at present. The newly-formed Earth, spinning more than 4 times faster than at present, caused these tides to move miles inland and back out to sea every 1½ h or so.

A Moonless Earth would still have tides, caused by the Sun. However, these tides would never be more than one-third as high as Earth's present tides. Minerals would wash into the oceans incredibly slowly from the flow of rivers. As a result, it would have likely taken much longer, hundreds of millions or even billions of years longer, for enough minerals to be dissolved for life to firmly establish itself.

Harsh Conditions Animal life's transition from oceans to land would also be much harder because of continuous winds, between 50 and 150 mi/h, created by the planet's rapid rotation. The resulting waves (wind causes waves) would be enormous and perpetual. Fish with legs near shores would almost certainly be pounded to a pulp rather than being able to sedately move onto land and then back into the water, as apparently happened on Earth.

But we see on Earth how life develops in the most incredible forms and places. It seems plausible, then, that sea life would find a way to make the transfer to dry land and continue to evolve there.

Allowing then for diverse terrestrial life on the Moonless Earth, what would be different about that life compared to life on our Earth? For one thing, creatures evolving there would have to withstand the perpetual pummeling from winds and the debris they carry. Turtle-like shells are one solution.

Obviously, there would be no eclipses or moonlight—all clear nights would be equally dark and star-filled. Therefore, nocturnal animals would be less successful at hunting, foraging, and traveling. Instead, these animals might evolve more enhanced senses to compensate for their inability to see visible light well at night.

Rush Hour Clearly, naked apes do not seem a likely bet on a Moonless Earth, or birds battling the ever-present winds. But given enough time, complex, even self-aware life likely would evolve. After all, we evolved because of the challenges faced by our ancestors, and Earth without a Moon would clearly provide challenges of its own.

The physiology of life on Earth evolved based on a 24-h day. This is most evident in our biological clocks, or circadian rhythms, which are the internal mechanisms that regulate sleeping, waking, eating, and other cyclic activities. Faced with a 6-h day, these circadian rhythms would be hopelessly out of sync with the natural world. To function on such a world, all of its creatures would have to evolve biological clocks based on 6-h cycles, which certainly could have occurred. Considering all that we have to do now, imagine what life would be like with only 6 h in each day!

Wobbling Earth Finally, rapidly rotating terrestrial planets without large moons are unstable, and calculations show that they dramatically change the direction of their rotation axes. If Earth had no Moon, we would have seasons that vary dramatically over the millennia, from those like we have today to times of no seasons to times when the Sun passes over every place on Earth, including the poles! Our Moon stabilizes Earth's rotation, preventing such phenomena. It would be incomparably harder for complex life to persist on a Moonless Earth than it is on our world.

THE EARTH'S MISSING INGREDIENT

The discovery of a novel high-density mineral means that Earth's mantle is a more restless place than scientists suspected—and offers new clues to the planet's history

Adapted from an article by Kei Hirose

The deepest hole that humans have dug reaches 12 km below Russia's Kola Peninsula. Although we have a spacecraft headed for Pluto—about six billion kilometers from the Sun—we cannot send a probe into the deep Earth. For practical purposes, the center of the planet, 6380 km below, is farther away than the edge of our solar system.

We know Earth's core, mantle, and crust form concentric layers, much like an onion. The mantle constitutes about 85 percent of the planet's volume, and its slow stirring drives the geologic cataclysms—earthquakes and volcanoes—of the crust. This middle domain is a mix of silicon, iron, oxygen, and magnesium, in roughly equal concentrations, plus smaller amounts of other elements. These elements combine into different mineral layers at different depths.

Although the nature and composition of most mantle layers have been understood for decades, the lowermost layer remained a puzzle. But the mystery was solved in 2002 with the synthesis of a novel, dense mineral that forms at high heat and pressure in the mantle's bottom 300 km: postperovskite.

Rock Bottom

Geophysicists map Earth's structure by measuring seismic waves that travel through the planet. These waves may be refracted or reflected when they cross the boundaries between different materials. When those waves cross boundaries between any of the mantle's five layers, they accelerate because of changes in rock structure.

The minerals that make up rock are composed of atoms arranged into a particular geometric pattern, or crystal, each with its own composition, physical properties, and color. Below certain depths in the mantle, enormous heat and pressure forces elements to rearrange into new crystal structures, undergoing a so-called "phase transition."

To study these structures, early geologists searched for mantle rocks brought to the surface by magma. Because the diamonds that often stud these rocks form in the uppermost mantle about 150 km down, their host rocks probably come from that depth. Scientists rarely get specimens from below 200 km.

As researchers learned to generate high pressures and temperatures in laboratories, they began synthesizing minerals believed to make up the lower mantle. Starting at a depth of 660 km, a dense magnesium/silicon mix becomes rock's main component. It belongs to a vast family of crystals called perovskites that is used to manufacture electronics and superconductors.

Scientists first synthesized magnesium silicate perovskite in 1974. For the next 30 years, experts believed that this mineral probably existed throughout the mantle unchanged. But in the 1960s, scientists discovered a new seismic anomaly in the lower mantle at around 2600 km below Earth's surface. In 1983, that boundary was attributed to a change in the relative quantity of elements, not to a phase-transition. This assumption was made partly because perovskite is an "ideal" crystal structure, so experts doubted the tightly-packed atoms could be compressed any tighter. But a change in element quantities was surprising. In theory, convection should mix the lower mantle's contents with those of overlying layers, creating consistency in the kinds and ratios of elements throughout.

To clarify the situation, we used a diamond anvil cell, squeezing samples of mantle-like materials to high pressure between gem-quality diamonds and heating them with a laser. At great pressure, even diamond—the hardest known material—starts to deform. To push the pressure higher, we needed to optimize the shape of the diamond anvil's tips so they wouldn't break, which they did many times. In 2001, we were the first lab in the world to study the effects of intense pressure on perovskite.

Crystal Clear

To monitor changes in the crystal structure of our samples under extreme conditions, we made high-quality images at 1-second intervals using hair-thin, intense X-ray beams. We repeated the experiments many times, observing the new diffraction pattern through the crystals each time. When we reheated a sample at low pressure, it reverted to perovskite. If the transition was reversible, it ruled out a change in the sample's chemical composition. We had transformed magnesium silicate perovskite into a new structure.

Next, we found that the transition actually happened at the temperature and pressure where the jump in seismic-wave velocity occurred, 2600 km down. We had discovered a new phase transition and a new material which must be predominant in the lower mantle and could impact its dynamics.

We named the new phase postperovskite. Its structure is essentially identical to that of two known crystals, uranium ferrous sulfate and calcium iridiate. The density of postperovskite is 1 to 1.5 percent higher than perovskite.

[THE NEW PICTURE]

A More Complex Planet

Earth is structured like an onion, with different materials appearing in each concentric layer. The discovery of a new, high-density material, called postperovskite, implies the existence of a new layer of that onion and explains puzzling behavior by seismic waves traveling through the planet.

CRUST (6–35 km)
DEPTH (km)
410
520
660
2,600
2,900
5,100
6,400
MANTLE
UPPER MANTLE
Olivine
Modified spinel
Spinel
LOWER MANTLE
Perovskite
Postperovskite
CORE
Outer core
Inner core

CRUST (UP TO 35 KILOMETERS OF DEPTH)

The continents, which are in part submerged by the oceans, are made of diverse rock that is up to several billion years old and relatively light. Thus, they float on the denser mantle underneath. The heavy basaltic rock that forms the bulk of the oceanic crust originates from mantle magma that erupts at underwater ridges and eventually sinks back into the mantle, typically within 100 million years.

MANTLE

Mantle rock consists primarily of oxygen, silicon and magnesium. Despite being mostly solid, it does deform on geologic timescales. In fact, the rock slowly flows as convective currents stir the entire mantle. That flow dissipates Earth's inner heat and propels continental drift.

UPPER MANTLE (35–660 KM)

As greater depths bring higher pressures and temperatures, the mantle's elemental components arrange into different crystal structures (minerals), forming layers. Three minerals—olivine, modified spinel and spinel—give the layers of the upper mantle their respective names.

LOWER MANTLE (660–2,900 KM)

The lower mantle was for decades thought to be relatively uniform in structure. But seismological data suggested that something different was happening at the bottom.

- **Perovskite layer**

The most prevalent mineral here (70 percent by weight) is a magnesium silicate ($MgSiO_3$) belonging to the family of crystal structures called perovskites. In this densely packed structure, magnesium ions (*yellow*) are surrounded by octahedral silicon-oxygen groups (*blue double-pyramid shapes*). Until recently, scientists thought that no denser crystal arrangement of these elements could exist.

- **Postperovskite layer**

At the pressures and temperatures of the bottom 300 km of the mantle, perovskite transforms into a new structure: the magnesium ions and the silicon-oxygen groups arrange themselves into separate layers. The transition releases heat and reduces volume by roughly 1.5 percent—a small difference, but one with dramatic effects on the entire planet.

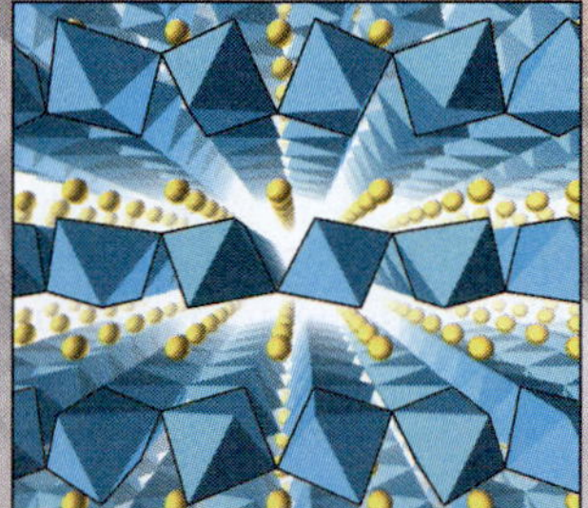

CORE (2,900–6,400 KM)

The deepest part of Earth consists predominantly of iron, which is liquid in the outer core and solid in the inner core. Convection stirs the outer core just as it stirs the mantle, but because the core is much denser, little mixing occurs between the mantle and the core. Core convection is thought to produce the planet's magnetic field.

EITH KASNOT (*cutaway*); KEN EWARD (*minerals*)

[THE DISCOVERY]

DEEP EARTH IN THE LAB

The author's team re-created the conditions of the lower mantle using a diamond-anvil cell. The cylindrical steel cell is designed so that tightening its screws—usually done by hand—concentrates pressure onto a surface just micrometers wide between the tips of two diamonds. The team placed samples between the tips and heated them with a laser. At the same time, an X-ray beam revealed how the samples' crystal structure changed.

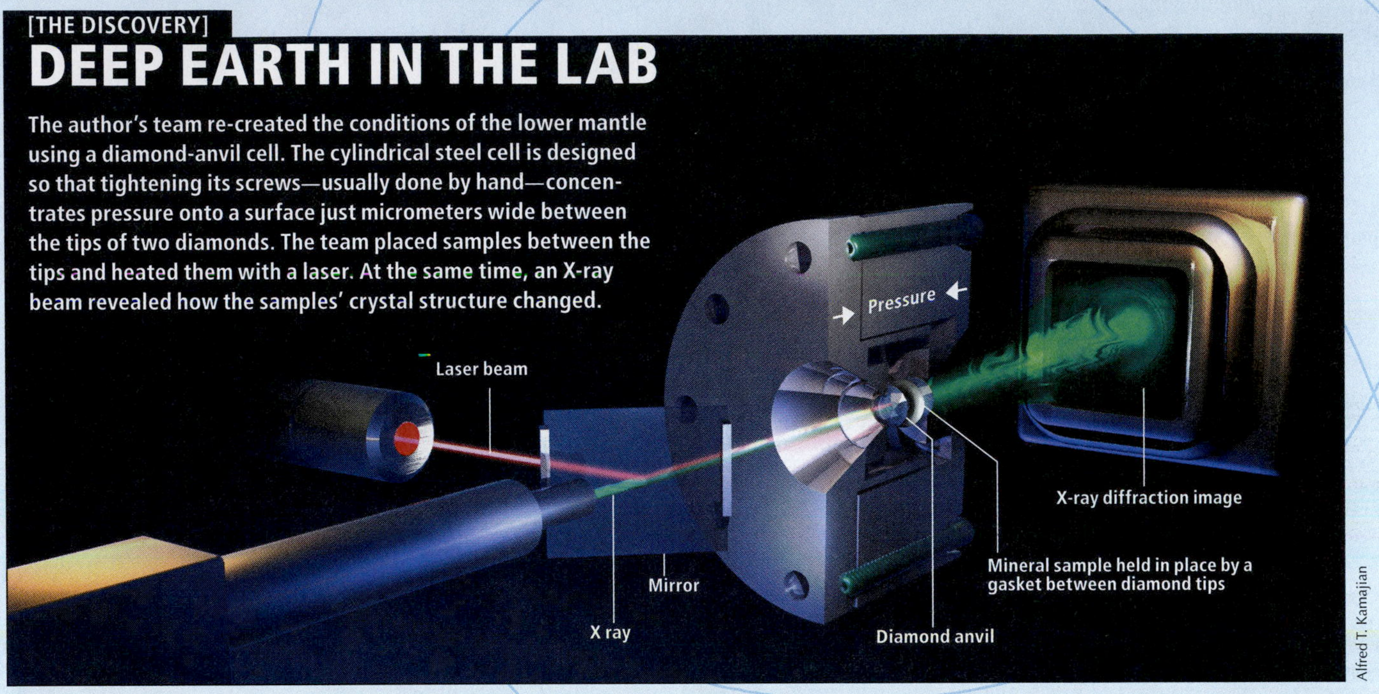

Taking the Heat

Since then, researchers have built on our findings to craft an exciting new picture of the processes occurring within Earth. Our discovery cast light on the amount of heat flowing from the core to the mantle. The core is mostly iron, making it twice as dense as the mantle. So virtually no mixing of elements occurs at the core-mantle boundary and heat is exchanged mostly by conduction. The mantle is rich in radioactive isotopes, but the core is probably not, implying that its heat (perhaps 4000 to 5000 K) is mostly left over from the formation of Earth. Since then, the core has cooled with time as heat transferred into the mantle at the core-mantle boundary.

By approximating the conductivity of materials in the lower mantle, we estimated that heat flows from the core into the mantle at a rate comparable to the combined output of all the world's power stations: 5 to 10 terawatts. This rate is a larger flow of energy than previously thought, producing a faster rate of core cooling. This means that Earth's core must have started at a higher temperature than was previously thought.

Inside the young Earth, the core was liquid. But at some point in the planet's history, the inner core began to crystallize, forming an inner, solid core and an outer, liquid core. The faster cooling rate and current size suggest the inner core may be less than a billion years old, which is young compared with Earth's age of 4.6 billion years: Otherwise the inner core would be much larger than we observe at present.

The inner core's formation and its influence on geomagnetism may have made it possible for life to leave the seas. Scientists believe that convection of liquid metal in the outer core generates the planet's magnetic field, which is amplified by the solid inner core's regulation of that convection. The geomagnetic field shields Earth from solar wind and cosmic rays, which cause genetic mutations and is especially dangerous for land-living creatures.

As the planet cooled some 2.3 billion years ago, some perovskite turned into postperovskite, boosting heat flow from the core and raising temperatures throughout the mantle. By speeding up mantle convection, postperovskite raises the temperature of the upper mantle by hundreds of degrees, which increases volcanic activity. Researchers estimate that faster plate motion and increased volcanism may have caused the continents to grow twice as fast during the past 2.3 billion years than they had previously, though this debate continues.

A postperovskite layer would also enhance the exchange of angular momentum between the liquid core and solid mantle whenever the core's flow-pattern changes. Simulations show that this exchange would alter Earth's rotational speed in a way that matches millisecond variations in the length of the day on decade-length time scales. Postperovskite's conductivity and its effects could also help explain the periodic change in the orientation of Earth's axis.

Although postperovskite is present only at the bottom of Earth's mantle, it could make up larger portions of other planets. Postperovskite may be a main component of the rocky cores of Uranus and Neptune. But the thick hydrogen layers that envelop the rocky cores of Jupiter and Saturn make pressures and temperatures too high to stabilize postperovskite.

Postperovskite could also be the most abundant constituent of some exoplanets. By analyzing optical spectrum data of other stars and their planets, astronomers have deduced that some, particularly "super-Earths" (those smaller than the size of 10 Earths) may share similar chemical composition with our planet.

To Be Continued

Questions remain about the structure of Earth's postperovskite-rich lower mantle. Large anomalies in seismic-wave speeds hint that it may be composed of two distinct masses that may be denser than the surrounding rock but still light enough to float on the outer core—similar to the way continents float on the outer mantle. These two "hidden continents," located roughly under Africa and under the Pacific Ocean, could affect the convection patterns in the entire mantle and plate tectonics at the surface. How did these masses form? Are they growing? Could the one under the Pacific Ocean have had something to do with the mantle plume that formed Hawaii?

The discovery of postperovskite explains a great deal about Earth's lowermost mantle, but the iron-rich metallic core remains enigmatic. We've known since 1952 that the liquid outer core is about 10 percent less dense than pure iron or iron-nickel alloy. One or more lighter elements, perhaps sulfur, silicon, oxygen, carbon, or hydrogen must therefore be present. The core's temperature is best estimated from the melting temperature of iron alloys at a pressure corresponding to that of the core's solid-liquid boundary. But current estimates range over more than 2000 K: Melting temperature depends strongly on composition, which remains unknown. The crystal structure of iron at inner-core conditions is also unknown, making it difficult to interpret seismological observations. Very recently we have produced diamond anvils that can reach the full range of pressures and temperatures of Earth's core, opening the door to addressing these unsolved mysteries about the center of our planet.

RIVUXG Photographed by NASA's *Mars Reconnaissance Orbiter*, this shaft descends at least 78 m (255 ft), and probably much deeper, into the side of volcano Arsia Mons. Similar features form on Earth when deep underground caves collapse. (NASA/JPL/University of Arizona)

Chapter 7

The Other Terrestrial Planets

Q WHAT DO YOU THINK?

1. Which terrestrial planet—Mercury, Venus, Earth, or Mars—has the coolest surface temperature?
2. Which planet is most similar in size to Earth?
3. Which terrestrial planet—Mercury, Venus, Earth, or Mars—has the hottest surface temperature?
4. What is the composition of the clouds that surround Venus?
5. Does Mars have liquid water on its surface today? Did it have liquid surface water in the past?
6. Is life known to exist on Mars today?

Answers to these questions appear in the text beside the corresponding numbers in the margins and at the end of the chapter.

Every human being is unique: Each person has his or her own genetic makeup and personal history. On the other hand, people have many similarities: we are either male or female; we all breathe air; and, under normal circumstances, we have common characteristics, such as two eyes, two hands, and ten toes, among other traits. To understand humans fully, biologists and physicians study our common features and then our individual peculiarities, just as psychiatrists and psychologists study our common behaviors as well as our differences.

The planets in our solar system also have similarities and differences that astronomers are learning to understand. For example, there are three basic groups of planets: terrestrial worlds (Mercury, Venus, Earth, and Mars), all roughly similar in size and chemistry to Earth; the much larger gas giant, or Jovian, worlds Jupiter and Saturn; and the planets intermediate in both size and chemistry, the ice giant worlds Uranus and Neptune.

In this first of two chapters on the rest of the planets, we explore the terrestrial worlds both individually and in comparison to each other and to Earth. The remaining four outer planets are presented in the next chapter.

In this chapter you will discover

- Mercury, a Sun-scorched planet with dormant volcanoes, a heavily cratered surface, and a substantial iron core
- Venus, perpetually shrouded in thick, poisonous clouds and mostly covered by gently rolling hills
- Mars, a red, dusty planet that once had running water on its surface and may still have liquid water underground

MERCURY

ANIMATION 7.1

The closest planet to the Sun, Mercury is a truly inhospitable world of temperature extremes. Its incredibly thin atmosphere and weak magnetic field leave it virtually unprotected from countless impacts and a continuous bath of deadly solar radiation. Despite having a surface nearly as dark as coal, Mercury sometimes appears as one of the brightest objects in our sky, due to its proximity to the Sun. We can see it from Earth only for a few hours before sunrise or a few hours after sunset because its angle from the Sun (its elongation) is always less than 28°.

7-1 Photographs from *Mariner 10* and *Messenger* spacecraft reveal Mercury's lunarlike surface

Until late last century, Mercury (Figure 7-1) was a complete mystery. Earth-based telescopic studies provided virtually no information about the planet's surface. The three visits by *Mariner 10* between March 1974 and March 1975 photographed only 45% of Mercury's surface. Between 2008 and 2011 the *Messenger* spacecraft imaged the rest of it. (*Messenger* is short for MErcury Surface, Space, ENvironment, GEochemistry, and Ranging.)

First impressions of Mercury evoke a lunar landscape, and there are several similarities. Craters in the Moon's highlands are densely packed, as are some of the craters on Mercury (Figure 7-2). Astronomers conclude that most of the craters on both Mercury and the Moon were produced by impacts in the first 800 million years after these bodies condensed from the solar nebula. The impact craters on both worlds have similar features, such as ejecta blankets and central peaks, as discussed for the Moon in Section 6-5. Unlike our Moon, however, Mercury also has a variety of volcanic craters out of which poured large amounts of lava that filled smaller impact craters and created smooth regions between larger ones. Like the Moon, Mercury has an exceptionally low albedo of 0.12 (its surface scatters 12% of incoming light). It is very bright as seen from Earth only because the sunlight scattering from Mercury is so intense.

The most impressive feature discovered by *Mariner 10* is a huge circular region called *Caloris Basin*. The name of this feature derives from *caloris,* Latin for "heat," and in fact this feature is the closest part of Mercury to the Sun at the planet's perihelion (closest approach to the Sun). Caloris Basin (Figure 7-3a) measures 1550 km (963 mi) in diameter, meaning that it has nearly 3 times the area of Texas. When seen lying along the *terminator* (the border between day and night) where the shadows are longest and details clearest, the basin (Figure 7-3b) is surrounded by a 2-km-high ring of mountains, beyond which are relatively smooth plains. Like the lunar maria, Caloris Basin was probably gouged out by the impact of an asteroid that penetrated the planet's crust. *Messenger* revealed small impact craters inside the Caloris Basin that have been filled in by lava from volcanoes that were active long after both the basin and those craters were formed. A giant volcano 95 km in diameter (bigger than the state of Delaware) has been seen in the basin.

The Caloris impact was a tumultuous event that shook the entire planet. Indeed, the collision affected the side of Mercury directly opposite of Caloris Basin. That area (Figure 7-3c) has a jumbled, hilly surface covering nearly half a million square kilometers, about twice the

MERCURY: VITAL STATISTICS

Planet symbol:	☿
Average distance from the Sun:	0.387 AU = 5.79×10^7 km
Maximum distance from the Sun:	0.467 AU = 6.98×10^7 km
Minimum distance from the Sun:	0.307 AU = 4.60×10^7 km
Orbital eccentricity:	0.21
Average orbital speed:	47.9 km/s
Sidereal period of revolution (year):	88.0 Earth days = 0.24 Earth year
Sidereal rotation period:	58.6 Earth days
Solar rotation period (day):	176 Earth days
Inclination of equator to orbit:	0.5°
Inclination of orbit to ecliptic:	7° 00′ 16″
Diameter (equatorial):	4880 km = 0.382 Earth diameter
Mass:	3.30×10^{23} kg = 0.0553 Earth mass
Average density:	5430 kg/m^3 = 0.984 Earth density
Escape speed:	4.3 km/s
Surface gravity (Earth = 1):	0.38
Albedo:	0.12
Average surface temperatures:	Day: 350°C = 662°F = 623 K Night: −170°C = −274°F = 103 K
Atmosphere:	Very thin, transient hydrogen (H_2), helium (He), potassium (K), sodium (Na), and oxygen (O_2)

FIGURE 7-1 Mercury's Vital Statistics Heavily cratered Mercury was visited 3 times by the *Mariner 10* spacecraft in 1974 and 1975 and 3 times by *Messenger* since 2008. This image was taken from an altitude of 27,000 km (17,000 mi). *Messenger* will settle into orbit around Mercury on March 18, 2011. (NASA/John Hopkins University Applied Physics Laboratory/Carnegie Institution of Washington)

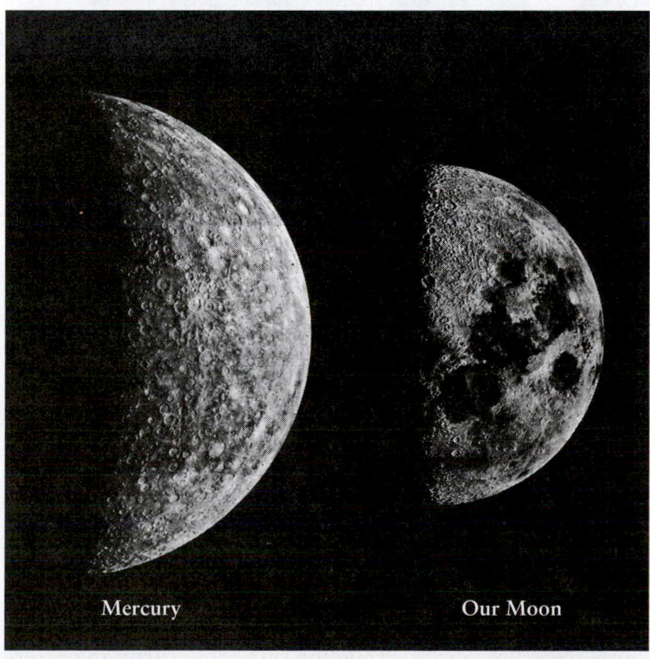

FIGURE 7-2 Mercury and Our Moon Mercury and our Moon are shown here to the same size scale. Mercury's radius is 2439 km and the Moon's is 1738 km. For comparison, the distance from New York to Los Angeles is 3944 km (2451 mi). Mercury's surface is more uniformly cratered than that of the Moon. Daytime temperatures at the equator on Mercury reach 700 K (800°F), hot enough to melt lead or tin. (NASA, UCO/Lick Observatory)

size of Wyoming. The hills, which appear as tiny wrinkles that cover most of the photograph, are about 5 to 10 km wide and between 100 and 1800 m high. Geologists believe that energy from the Caloris impact traveled through the planet and became focused, like light through a lens, as it passed through Mercury. As this concentrated energy reached the far surface of the planet, jumbled hills were pushed up. A similar pair of phenomena to Caloris and the jumbled terrain on the other side of Mercury can be seen in our Moon's Mare Orientale and chaotic hills on the opposite side of

> The rings of mountains frozen into Mercury's and the Moon's surfaces have what *transient* analog on Earth? (*Hint:* Think about impacts.)

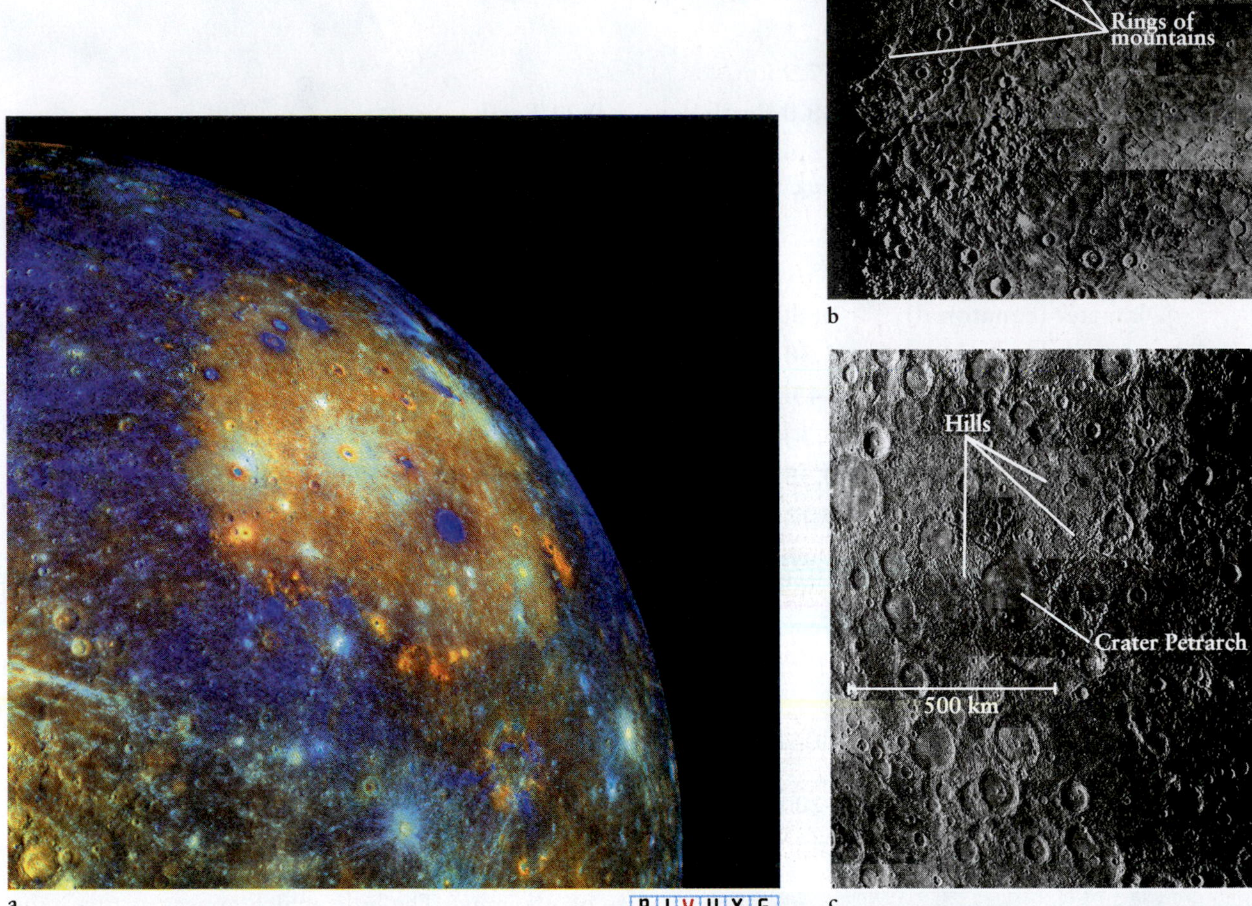

FIGURE 7-3 Major Impacts on Mercury and on Our Moon (a) **Caloris Basin**. *Messenger* sent back this view of a huge impact basin on Mercury's equator. The basin is the entire orange-colored portion of the figure. This image has been color enhanced to show the different surface compositions of Mercury. The orange regions around the edge of the basin are believed to be volcanic features. (b) Only about half the Caloris Basin appears here because it happened to lie on the terminator when the *Mariner 10* spacecraft sped past the planet. Although the center of the impact basin is hidden in the shadows (just beyond the left side of the picture), several semicircular rings of mountains reveal its extent. (c) **Unusual, Hilly Terrain.** What look like tiny, fine-grained wrinkles on this picture are actually closely spaced hills, part of a jumbled terrain that covers nearly 500,000 square km on the opposite side of Mercury from the Caloris Basin. The large, smooth-floored crater, Petrarch, has a diameter of 170 km (106 mi). This impact crater was produced more recently than Caloris Basin. (a: NASA/Johns Hopkins University Applied Physics Laboratory/Arizona State University/Carnegie Institution of Washington, image reproduced courtesy of Science/AAAS; b and c: NASA)

that world. Finding such a similar pair of phenomena on a different world supports the connection between the impact and the jumbled surface.

Looking at Mercury in more detail, we start to see a variety of differences from the features on the Moon. Unlike the Moon and its maria, Mercury lacks extensive craterless regions. Instead of maria, Mercury has plains with relatively small craters. Such plains are not found on the Moon. Figure 7-4 shows a typical close-up view of Mercury. As we learned in Section 6-7, the lunar maria were produced by extensive lava flows that occurred between 3.1 and 3.8 billion years ago. Ancient lava flows also formed the Mercurian plains. Some of this material came from volcanoes, while the rest came through craters

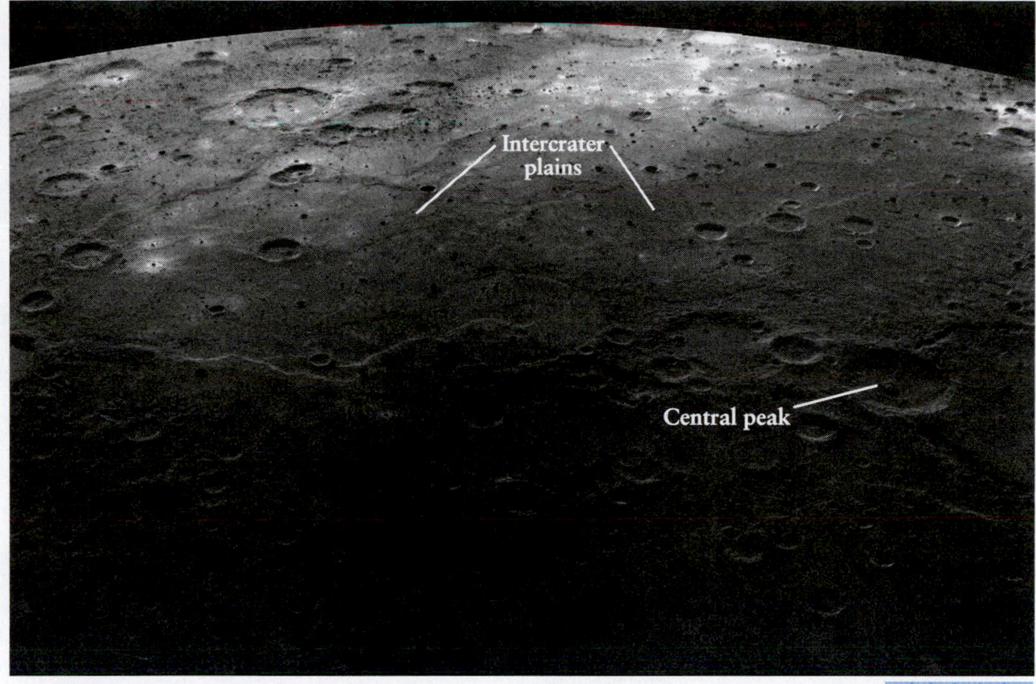

R I V U X G

VIDEO 7.1

FIGURE 7-4 Mercury's Craters and Plains This view of Mercury's northern hemisphere was taken by *Messenger* as it sped past the planet in 2009. Numerous craters on the top left and bottom half of the image are separated by a broad intercrater plain. The image is about 880 km (550 mi) across. (NASA/Johns Hopkins University Applied Physics Laboratory/Carnegie Institution of Washington)

created as large meteorites punctured the planet's thin, newly formed crust, allowing lava to well up from the molten interior to flood low-lying areas. The existence of craters pitting Mercury's plains suggests that these plains formed just over 3.8 billion years ago, near the end of the Late Heavy Bombardment. Mercury's plains are therefore older than most of the lunar maria, leaving more time for cratering to eradicate marialike features. Today, 40% of Mercury's surface is covered by these lightly cratered plains, while the rest is more heavily cratered.

Dormant volcanoes were discovered on Mercury in 2008 (Figure 7-5) by the *Messenger* spacecraft, providing further sources of lava to fill in low-lying areas, as well as to cover over many of the planet's impact craters.

Mariner 10 and *Messenger* also revealed numerous, long cliffs, called scarps, meandering across Mercury's surface (Figure 7-6). They have also been observed on our Moon (see Figure 6-26). Scarps are believed to have developed as Mercury cooled. Almost everything that cools, contracts. Therefore, as Mercury's mantle and molten iron core cooled and contracted, its surface moved inward. Because it was already solid, Mercury's crust could not collapse uniformly. Instead, the crust wrinkled as it contracted, forming the scarps. These features and the lack of recent volcanic activity suggest that the planet's interior is solid to a significant depth. Otherwise, lava would have leaked out as the scarps formed.

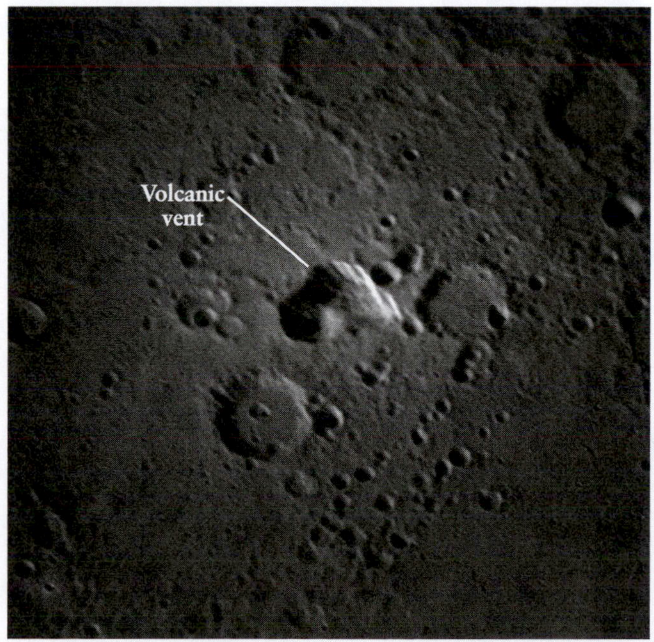

R I V U X G

FIGURE 7-5 Possible Volcanic Vent The central indentation in this *Messenger* image from 2009 is believed to be a caldera (sunken vent) of an explosive volcano on Mercury. It is unlikely to be an impact crater, as it completely lacks a raised crater wall. (NASA/Johns Hopkins University Applied Physics Laboratory/Carnegie Institution of Washington)

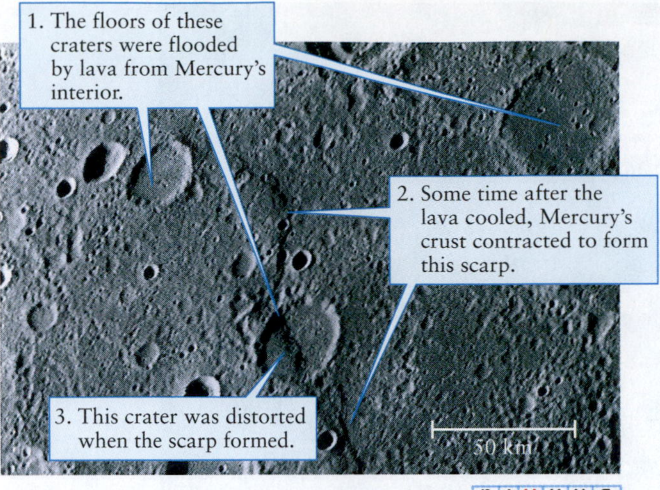

ANIMATION 7.2

FIGURE 7-6 Scarps on Mercury A long, meandering cliff, called Santa Maria Rupes, runs from north to south across this *Mariner 10* image of a region near Mercury's equator. This cliff, called a *scarp* by geologists, is more than 1 km high and runs for several hundred kilometers. Note how the crater in the center of the image was distorted vertically when the scarp formed. (NASA)

Messenger continues to provide intriguing images of Mercury. Figure 7-7 is one of the most interesting photos taken so far. The features displayed in this photograph include lines of secondary craters, created by ejecta from impacts; craters filled with volcanic lava; a large (210-km-diameter) impact crater and concentric ring created as a ripple of surface rock by the impact, both of which are filled in with lava; a scarp; and a crater with as-yet unexplained light rock revealed inside it. The large double-ringed crater is a smaller version of Caloris Basin.

7-2 Mercury has a higher percentage of iron than does Earth

Mercury's average density of 5430 kg/m^3 is quite similar to Earth's (5520 kg/m^3). As we saw in Section 6-3, typical rocks on Earth's surface have a density of only about 3000 kg/m^3 because they are composed primarily of lightweight elements. The high average densities of both Mercury and our planet are caused by their dense interiors.

Because Mercury is less dense than Earth, you might conclude that Mercury has a lower percentage of iron, a common heavy element, than does our planet. Indeed, *Messenger* revealed that Mercury's surface has a low iron content. However, in terms of the percentage of its total mass that is iron, Mercury is the most iron-rich planet in the solar system. Only Earth's greater mass, pressing inward and thereby compressing our planet's inner parts, makes it denser than Mercury.

FIGURE 7-7 Geology on Mercury This *Messenger* image shows a variety of interesting features, including lava-filled craters, secondary craters created by debris that splashed out of an impact crater, a scarp cutting through an old crater, and as-yet unidentified, light-colored rock at the bottom of a crater. (NASA/Johns Hopkins University Applied Physics Laboratory/Carnegie Institution of Washington)

Figure 7-8 shows a scale drawing of Mercury's interior, where an iron core contains nearly 60% of the planet's mass and 75% of its radius. Surrounding the core is a 600-km-thick rocky mantle. For comparison, Earth's iron core occupies only 17% of the planet's

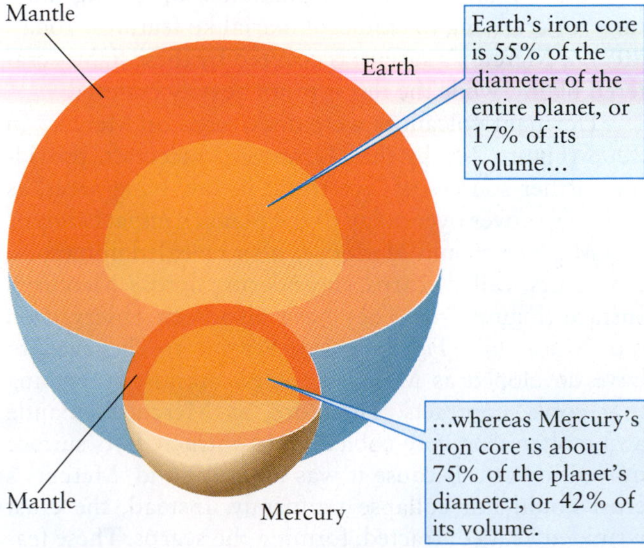

FIGURE 7-8 The Interiors of Earth and Mercury Mercury has the highest percentage of iron of any planet in the solar system. Consequently, its iron core occupies an exceptionally large fraction of its interior.

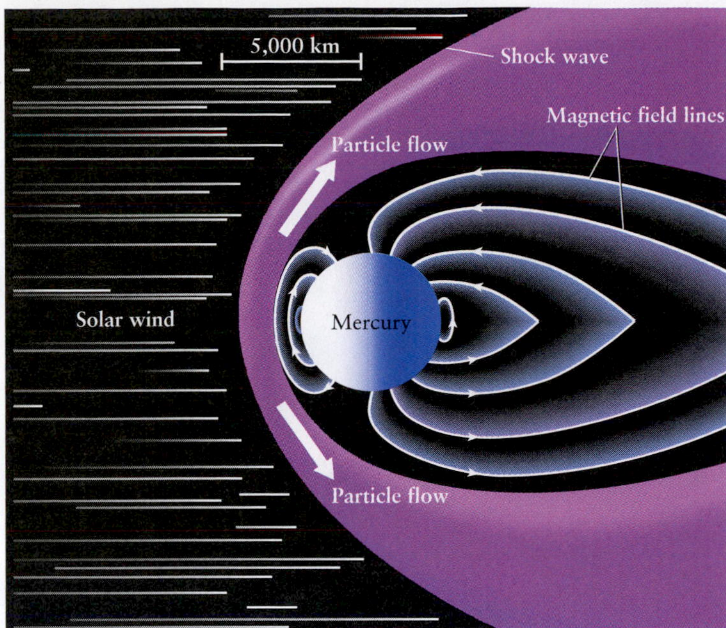

FIGURE 7-9 Mercury's Magnetosphere Mercury's magnetic field is about 1% as strong as Earth's field, just strong enough to deflect the solar wind.

volume. *Mariner 10* discovered that Mercury has a varying magnetic field that interacts with the solar wind (Figure 7-9) sometimes more weakly than does the Earth's field, sometimes more strongly (recall Section 6-4). Indeed, scientists have observed Mercury's field changing significantly over just a few years. The cause of the change is still unknown. Until 2008, astronomers didn't even know what generated Mercury's magnetic field. In that year, the *Messenger* spacecraft passing by Mercury revealed that the planet has a liquid iron core that generates the field. As we saw in Section 6-4, electric currents in a planet's liquid iron core create a planetwide magnetic field. Earth creates its electric current by rotating once a day. Mercury rotates 59 times more slowly. That slower rotation and the lower volume of molten iron inside Mercury (compared to Earth) explain why Mercury's magnetic field is so weak.

Events early in Mercury's history must somehow account for its high iron content. We know that the inner regions of the primordial solar nebula were incredibly hot. Perhaps only iron-rich minerals were able to withstand the solar heat there, and these subsequently formed iron-rich Mercury. According to another theory, an especially intense outflow of particles from the young Sun stripped Mercury of its low-density mantle shortly after the Sun formed. A third possibility is that, during the final stages of planet formation, Mercury was struck by a large planetesimal (debris formed as the solar system came into being), just as Earth was struck by a Mars-sized body that led to the formation of our Moon. Computer simulations show that this cataclysmic collision would have ejected much of Mercury's lighter mantle (Figure 7-10).

7-3 Mercury's rotation and revolution are coupled

Mercury has one of the most unusual orbits in the solar system. Recall from Section 6-8 that when Earth was young, its gravitational force created tides on the Moon, thereby forcing the Moon into synchronous rotation. Something similar (but not identical) happened

If Mercury were struck by a large planetesimal, why would this collision not produce a moon, as happened when Earth was struck early in its history?

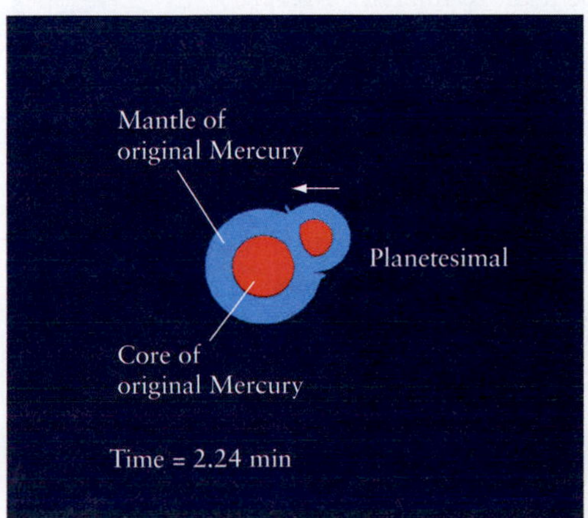

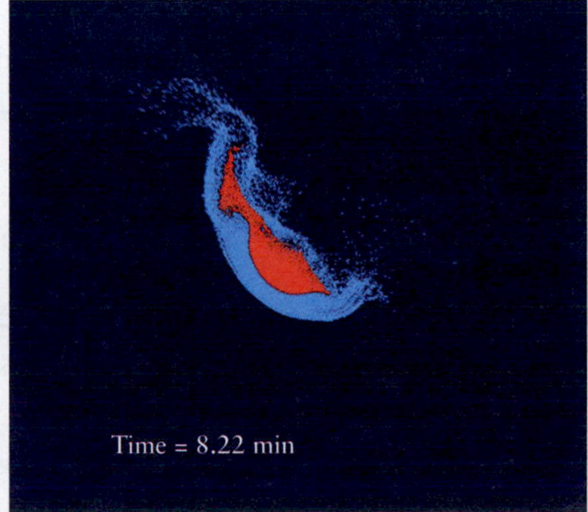

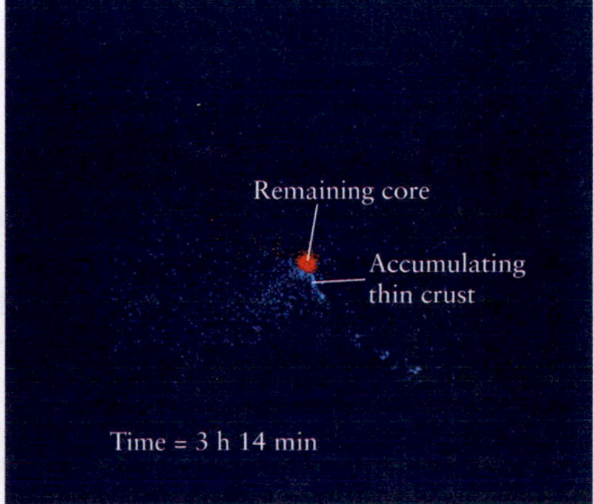

FIGURE 7-10 The Stripping of Mercury's Mantle To account for Mercury's high iron content, one theory proposes that a collision with a massive planetesimal stripped Mercury of most of its rocky mantle. These three images show a computer simulation of a nearly head-on collision between proto-Mercury and a body one-sixth its mass. Both worlds are shattered by the impact, which vaporizes much of their rocky mantles. Mercury eventually reforms from the remaining iron-rich debris. The rest of the original Mercury and the impactor leave this area of the solar system. (Royal Astronomical Society)

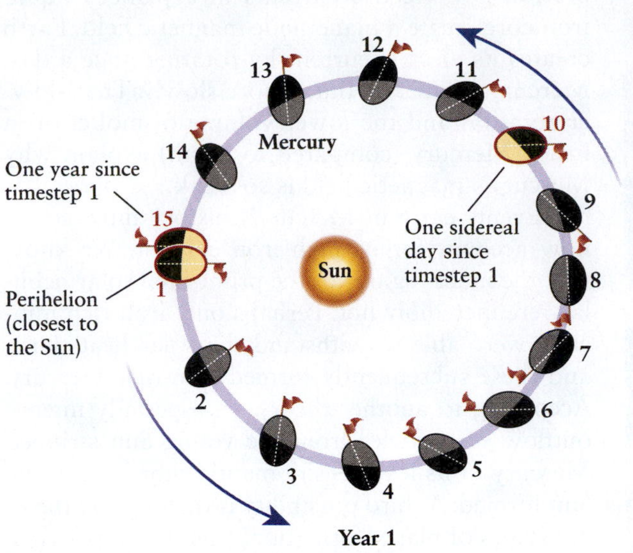

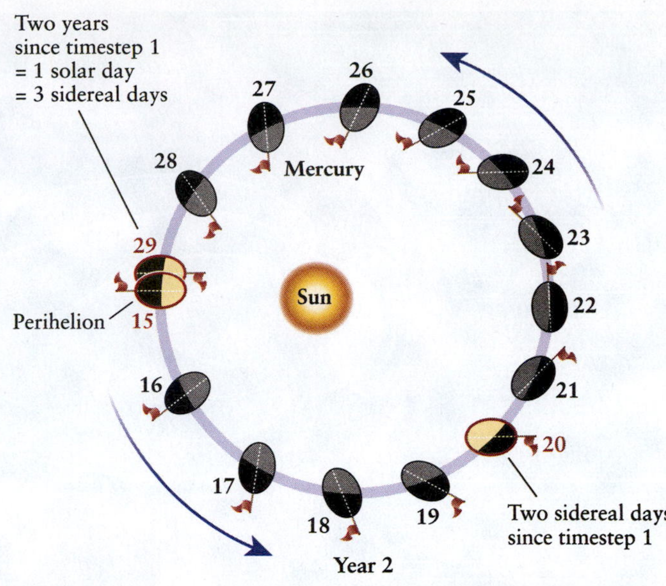

ANIMATION 7.3

FIGURE 7-11 3-to-2 Spin-Orbit Coupling Mercury undergoes three sidereal rotations every 2 years. You can see this by following the flag on Mercury (a) from timestep 1 to timestep 15 and then (b) from timestep 15 to timestep 29. The flag points to the right four times during this interval (at times 1, 10, 20, and, finally, at 29). This means that Mercury has rotated three times in exactly 2 sidereal Mercurian years. Because a sidereal year is 88 Earth days long, a sidereal day on Mercury is 58.7 Earth days long. During the same interval, however, the Sun is at noon as seen from the flag's location only twice: at timestep 1 and timestep 29. Therefore, a solar day is 176 Earth days long.

with the Sun playing the role of Earth and Mercury acting in the place of our Moon. To begin with, the Sun's gravity produced two significant tidal bulges on opposite sides of young, molten Mercury that locked into place when the planet solidified. Mercury is so close to the Sun (average separation: 0.387 AU) that the Sun's gravitational force on Mercury's tidal bulges changed the planet's rotation rate. However, Mercury's highly eccentric orbit ($e = 0.21$) prevented the planet from being locked into synchronous orbit like our Moon. Instead, Mercury developed what is called a **3-to-2 spin-orbit coupling.** This coupling means that Mercury undergoes three sidereal rotations (rotations measured with respect to the distant stars, not the Sun), while undergoing two revolutions around the Sun (Figure 7-11).

The major consequence of spin-orbit coupling is that one or the other of Mercury's regions of high tide is facing the Sun whenever the planet is at perihelion (see Figure 7-11).

A Day on Mercury Is 2 Mercury Years Long The motion of the Sun across Mercury's sky is unique in the solar system. First, a solar day there (noon to noon) is 176 Earth days long, twice the length of a year on Mercury! Furthermore, if you were to set up a camera at the location of high tide on Mercury to take a series of pictures, you would see the Sun start to rise in the east, stop high in the sky, move back toward the east, stop again, and then resume its westward journey. This is analogous to the retrograde motion of the planets that we observe from Earth (see Section 2-2).

Why isn't Mercury in synchronous rotation with respect to the Sun?

WHAT IF . . . Mercury Were as Massive as Earth? A more massive Mercury would have a molten interior similar to that of Earth. As such a planet orbited the Sun in the highly elliptical orbit that Mercury has, the tidal force from the Sun on that planet's land would cause Mercury to change shape. Mercury's crust would move (rise and fall), creating great friction between adjacent rocks. Such friction generates heat. Combining this heat with that from its interior, these tides would cause a more massive Mercury to have continually active volcanoes and possibly seas of molten rock.

7-4 Mercury's atmosphere is the thinnest of all terrestrial planets

Mercury's mass is only 5.5% that of Earth. Like our Moon, the force of gravity is too weak on Mercury to hold a permanent atmosphere, but trace amounts of seven different gases—hydrogen, helium, sodium, calcium, magnesium, potassium, and oxygen—have been detected around Mercury. Scientists think the Sun is the source of the hydrogen and helium gas near Mercury, while sodium, magnesium, calcium, and potassium gas escape from rocks inside the planet (a process called *outgassing*, which also occurs on Earth). Different concentrations of these gases were observed at different locations over the planet. Oxygen observed in Mercury's atmosphere may come from polar ice that is slowly evaporating. All of these gases drift into space and are continually replenished in the atmosphere from their respective sources. This atmosphere is at least 10^{17} times less dense than the air we breathe.

1 **Mercury's Temperature Range Is the Most Extreme in the Solar System** Earth's thick atmosphere stores heat, which helps explain why it stays warm on cloudy days and at night. Nevertheless, Earth continually loses heat into space. Because of Earth's rapid daily rotation, this lost heat is quickly replaced by the Sun during daylight hours, so the average temperature change between day and night on Earth is only about 11 K (20°F). Mercury's slow rotation and minimal atmosphere cause the difference in temperature between day and night there to be far greater than on Earth. At noon on Mercury, the surface temperature is 700 K (800°F). At the terminator, where day meets night, the temperature is about 425 K (305°F). On the night side, shortly before sunrise, the temperature falls to as low as 100 K (−280°F), the coldest temperature on any terrestrial planet! The resulting daily range of temperature is therefore 600 K (1080°F) on Mercury.

Just as astronomers use scattered radio waves to search for water on the Moon (see Section 6-6), they have also sent radio waves to Mercury. In 1992, researchers made an extraordinary discovery—evidence for ice near Mercury's poles, in craters that are permanently in shadow. Dozens of circular regions, presumably craters, sent back signals with characteristics distinct to ice. Confirmation of the presence of this ice will have to wait until the *Messenger* spacecraft begins orbiting Mercury in 2011. If the ice is there, its origin—whether from cometary impacts, from gases rising from inside the planet and then freezing, or from both sources—also remains to be determined.

Insight Into Science

Model Building In modeling real situations, scientists must consider several different effects simultaneously. Omit any crucial property and you get inaccurate results. For example, consider how long ice can remain at Mercury's poles. Astronomers must take into account the planet's distance from the Sun, the tilt of its axis of rotation, its rotation rate, its surface features, its chemical composition, and whether the ice on it is exposed or mixed with other material.

VENUS

2 One of the few things that Mercury and Venus have in common is that neither has a moon. Venus and Earth, on the other hand, have much in common. They have almost the same mass, the same diameter, and the same average density. Indeed, if Venus were located at the same distance from the Sun as is Earth, then it, too, might well have evolved life. However, Venus is 30% closer to the Sun than Earth, and this one difference between the two planets leads to a host of others, making Venus inhospitable to life.

7-5 The surface of Venus is completely hidden beneath a permanent cloud cover

At nearly twice the distance from the Sun as Mercury, Venus is often easy to view without interference from the Sun's glare. At its greatest western elongation, Venus is seen high above the western horizon after sunset, where, like Mercury, it is often called the "evening star." Conversely, high in the eastern sky before sunrise, it is called the "morning star."

Venus is easy to identify because it is usually one of the brightest objects in the night sky. Only the Sun and the Moon outshine Venus at its greatest brilliance. Venus is often mistaken for a UFO (Figure 7-12) because when it is low on the horizon, its bright light is strongly refracted by Earth's atmosphere, making it appear to rapidly change color and position. You can see the appearance of the inner planets using *Starry Night*™ (see also *Guided Discovery: The Inner Solar System*).

FIGURE 7-12 UFO? Just Venus Venus is bright and is often seen near the horizon where rising and sinking gases in Earth's atmosphere make it appear to move and change color, like alleged UFOs. (Joe Orman)

Unlike Mercury, Venus is intrinsically bright because it is completely surrounded by light-colored, highly reflective clouds (Figure 7-13). Because visible light telescopes cannot penetrate this thick, unbroken layer of clouds, we did not even know how fast Venus rotates until 1962. In the 1960s, however, both the United States and the Soviet Union began sending probes there. The Americans sent fragile, lightweight spacecraft into orbit near the planet. The Soviets, who had more powerful rockets, sent more durable spacecraft directly into the Venusian atmosphere.

Building spacecraft whose technology could survive the descent into Venus's atmosphere proved to be more frustrating than anyone had expected. Finally, in 1970, the Soviet probe *Venera* (Russian for "Venus") 7 managed to transmit data for 23 min directly from the Venusian surface. Soviet missions continued until 1985, measuring a surface temperature of 750 K (900°F) and a surface air pressure of 90 atm, among other things. This value is the same pressure you would feel if you were swimming 0.82 km (2700 ft) underwater on Earth. 3

In contrast to Earth's present nitrogen- and oxygen-rich atmosphere (see Section 6-1), Venus's thick atmosphere

Guided Discovery

The Inner Solar System

Here, you will learn more about the appearances of the inner planets as seen from Earth, using the *Starry Night*™ program.

Mercury The first thing to demonstrate is that Mercury always appears close to the horizon when it is visible in our night sky from Earth because its orbit places it close to the Sun in the sky. Click the **Home** button in the toolbar to move to your home location. Open the **Find** pane and remove any text from the search box at the top of the **Find** pane so that the list of solar system items is shown. Double-click the listing for **Mercury** to center this planet in the view.

1. If Mercury is below the horizon, the **Mercury is not currently visible** warning dialog will come up. In this case, click the **Best Time** button. If Mercury is to the west of the Sun (to the right if your location is in the northern hemisphere and the gaze direction is south, or to the left if your location is in the southern hemisphere and the gaze direction is toward the north), go to 2a. If Mercury is to the east of the Sun, go to 2b.

2a. If Mercury is already up in the sky, select **View > Show Daylight** or use the **Daylight** button in the button bar to be sure that the display of daylight is turned on. If Mercury is west of the Sun, set up the view with Mercury as the "morning star." To do this, click the **Sunrise** button in the toolbar and run time forward or backward at about 300× until Mercury is just above the eastern horizon and the Sun is below the horizon. This shows Mercury as the "morning star." Now, run time forward and you will see the white dot of Mercury fade. (If it is too close to the Sun, it will not be visible at all. Change the date and try again.) Now go to step 3.

2b. If Mercury is east (to the left) of the Sun in the sky, click the **Sunset** button in the toolbar. If necessary, select **View > Show Daylight**, or use the button bar to turn on the display of daylight in the view. Run time forward or backward at about 300× until the Sun is about to set. Step forward in time by 1-min intervals and watch Mercury first appear as the "evening star" and then set on the western horizon. (If it is too close to the Sun, it will not be visible at all. Change the date and try again.)

3. You can now explore Mercury's orbit. Select **Favourites > Discovering the Universe > Atlas** and open the **Find** pane. Double-click the listing for Mercury to center this planet in the view. You may also wish to click the checkbox to the left of the listing for the Sun to label it in the view. **Zoom** in to a field of view of about 70°. Set the **Time Flow Rate** to **6 hours** (more or less, depending on your computer speed) and click the **Play** button. Describe and explain the motion of the Sun relative to Mercury. You can learn more about Mercury as seen from Earth by trying Observing Project 51 at the end of the chapter.

Venus Repeat the same procedures for Venus as are described above for Mercury, starting at step 1. What similarities and differences do you see?

Mars Select **Favourites > Discovering the Universe > Atlas** and set the **Date** in the toolbar to February 23, 2010, and the **Time** to midnight. Open the **Find** pane and double-click the listing for Mars to center this planet in the view. Click the checkbox to the left of the listing for the Sun in the **Find** pane to label it in the view. Now **Zoom** in on Mars until you see its surface features. What is the phase of Mars? **Zoom** out and use the hand tool or cursor keys to adjust the view so that both the Sun and Mars are visible on the screen. Use the angular separation tool to measure the angular distance between the Sun and Mars on this date. Does the Sun ever get this far in angle from Mercury or Venus, as seen from Earth? Explain.

VENUS: VITAL STATISTICS

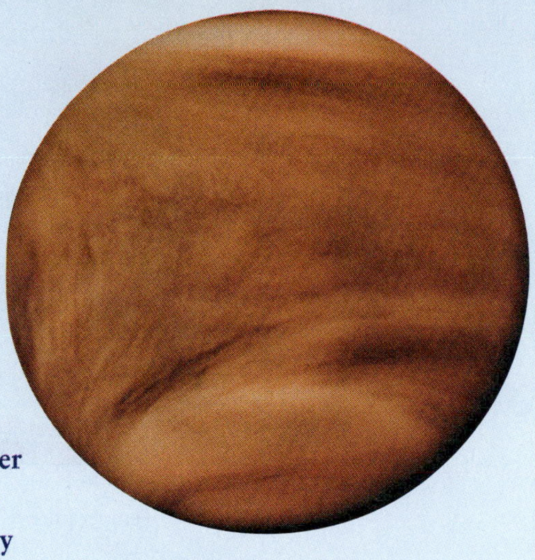

Planet symbol:	♀
Average distance from the Sun:	0.723 AU = 1.082×10^8 km
Maximum distance from the Sun:	0.728 AU = 1.089×10^8 km
Minimum distance from the Sun:	0.718 AU = 1.075×10^8 km
Orbital eccentricity:	0.007
Average orbital speed:	35.0 km/s
Sidereal period of revolution (year):	224.7 days = 0.615 Earth year
Sidereal rotation period:	243.0 days (retrograde)
Solar rotation period (day):	116.8 Earth days
Inclination of equator to orbit:	177.4°
Inclination of orbit to ecliptic:	3.39°
Diameter (equatorial):	12,104 km = 0.949 Earth diameter
Mass:	4.87×10^{24} kg = 0.815 Earth mass
Average density:	5240 kg/m^3 = 0.949 Earth density
Escape speed:	10.4 km/s
Surface gravity (Earth = 1):	0.91
Albedo:	0.59
Average surface temperature:	460°C = 860°F = 733 K
Atmospheric composition (by number of molecules):	96.5% carbon dioxide (CO_2), 3.5% nitrogen (N_2), 0.003% water vapor (H_2O)

R I V U X G

FIGURE 7-13 Venus's Vital Statistics Venus's thick cloud cover efficiently traps heat from the Sun, resulting in a surface temperature even hotter than that on Mercury. Unlike Earth's clouds, which are made of water droplets, Venus's clouds are very dry and contain droplets of concentrated sulfuric acid. This ultraviolet image was taken by the *Pioneer Venus Orbiter* in 1979. (NASA)

is 96% carbon dioxide, with the remaining 4% mostly nitrogen. This atmosphere is remarkably similar to Earth's early carbon dioxide–rich atmosphere. These gases were vented from inside Venus through volcanoes and other openings. Unlike Earth, however, Venus has no liquid oceans to dissolve the carbon dioxide and no living organisms to convert it into oxygen and carbon compounds, so it remains in the atmosphere today.

Soviet spacecraft also discovered that Venus's clouds are confined to a 20-km-thick layer located 48 to 68 km above the planet's surface. Above and below the clouds are 20-km-thick layers of haze. Beneath the lower level of haze, the Venusian atmosphere is clear all the way down to the surface.

The results of the early probing of Venus's atmosphere are summarized in Figure 7-14. Both pressure and temperature decrease smoothly with increasing altitude. From the changes in temperature and pressure above a planet's surface, we can begin to understand the structure of its atmosphere.

4 Unlike the clouds on Earth, which appear white from above, the cloud tops of Venus appear yellowish or yellow-orange to the human eye. These colors are typical of sulfur and its compounds. Indeed, spacecraft found substantial amounts of sulfur dust in Venus's upper atmosphere and sulfur dioxide and hydrogen sulfide at lower elevations. The clouds themselves are composed of droplets of concentrated sulfuric acid! Because of the tremendous atmospheric pressure at Venus's surface, the droplets remain suspended as a thick mist, rather than falling as rain. Standing on the surface of Venus, you would experience a perpetually cloudy, yellow day.

All of the major chemical compounds spewed into our air by Earth's volcanoes have been detected in Venus's atmosphere. Among these molecules are large quantities of sulfur compounds. Because many of these substances are very short-lived, they must be constantly replenished by new eruptions. In fact, the abundance of sulfur compounds in Venus's atmosphere does vary. These data suggest the possibility that the sulfurous compounds in Venus's atmosphere come from active volcanoes. This idea is supported by the discovery, in 2007, of lightning on Venus. Lightning is a frequent event around active volcanoes here on Earth, such as occurred when the Eyjafjallajökull volcano erupted in 2010 in Iceland

These discharges create short-lived magnetic fields that were detected by the *Venus Express* satellite.

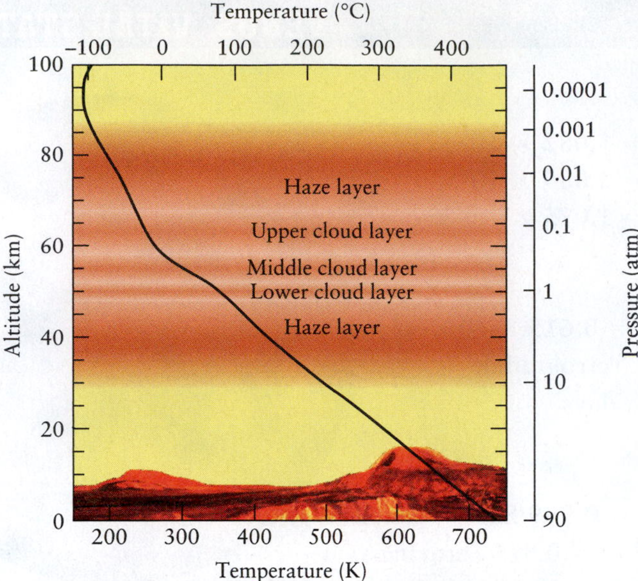

FIGURE 7-14 Temperature and Pressure in the Venusian Atmosphere The pressure at the Venusian surface is a crushing 90 atm (1296 lb/in^2). Above the surface, atmospheric pressure and temperature decrease smoothly with increasing altitude. The temperature decreases from a maximum of nearly 750 K (900°F) on the ground to a minimum of about 173 K (−150°F) at an altitude of 100 km. (NASA/Magellan Images [JPL])

Why does the day have a yellow hue on Venus?

(Figure 7-15). The belief that lightning occurs on Venus comes from the detection of variations in the magnetic field around the planet caused by the electrical discharges that are lightning.

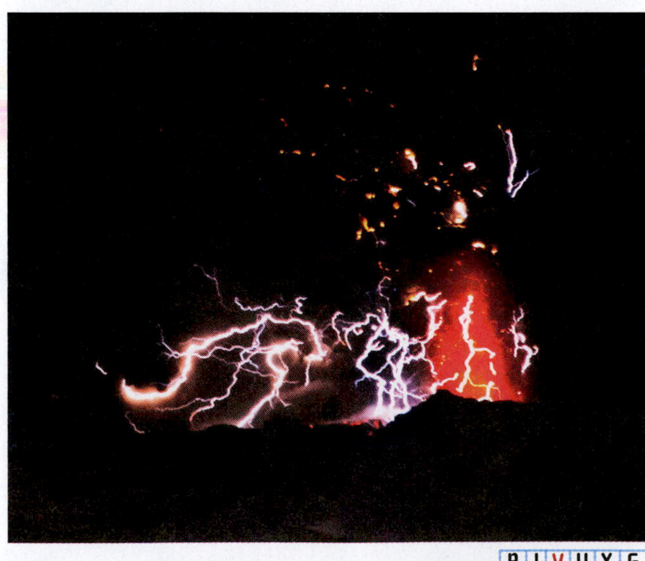

FIGURE 7-15 Lightning on Earth This photograph of lightning generated by the Eyjafjallajökull volcano in Iceland in 2010 shows the phenomena that may have recently been observed on Venus. (David Jon/NordicPhotos/Getty Images)

7-6 The greenhouse effect heats Venus's surface

At first, no one could believe reports that the surface temperature on Venus was higher than the surface temperature on Mercury, which, after all, is closer to the Sun. Setting aside this initial skepticism, astronomers found a straightforward explanation—the **greenhouse effect** (Figure 7-16).

Carbon dioxide is responsible for the warming of Venus's atmosphere. Like your car windows, carbon dioxide is transparent to visible light but absorbs infrared radiation. Although most of the visible sunlight striking the Venusian cloud tops is reflected back into space, enough light reaches the Venusian surface to heat it. The warmed surface, which is also heated from inside the planet, in turn, emits infrared radiation, which is absorbed by Venus's carbon dioxide–rich atmosphere. Some of this trapped radiation returns to the planet's surface, producing the high temperatures found there. The rest of the heat is radiated into space.

Without the greenhouse effect, the surface of Venus would have a noontime temperature of 465 K. But with the greenhouse effect, that temperature is actually a swel-

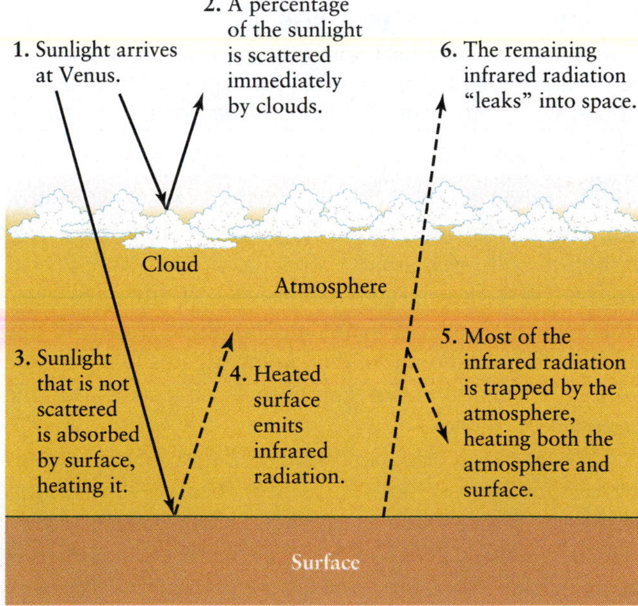

FIGURE 7-16 The Greenhouse Effect A portion of the sunlight penetrates through the clouds and atmosphere of Venus, heating its surface. The surface in turn emits infrared radiation, much of which is absorbed by carbon dioxide (and to a much lesser degree, water vapor). The trapped radiation helps increase the average temperatures of the surface and atmosphere. Some infrared radiation does penetrate the atmosphere and leaks into space. In a state of equilibrium, the rate at which the planet loses energy to space in this way is equal to the rate at which it absorbs energy from the Sun.

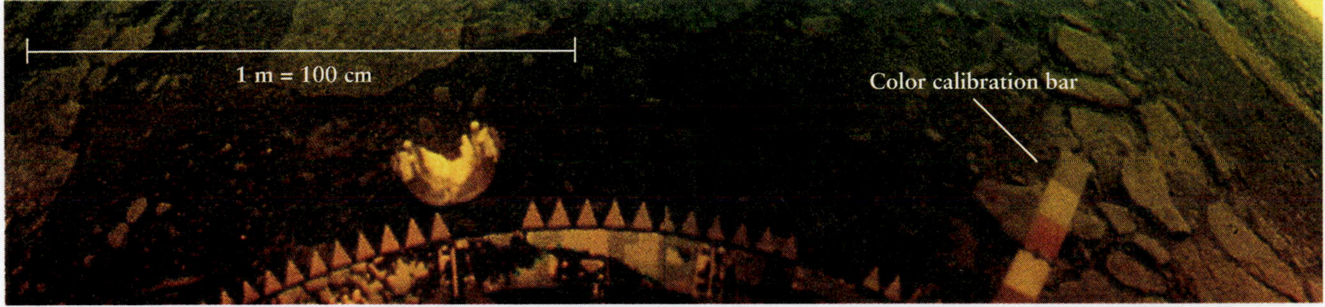

a Image from *Venera 13*

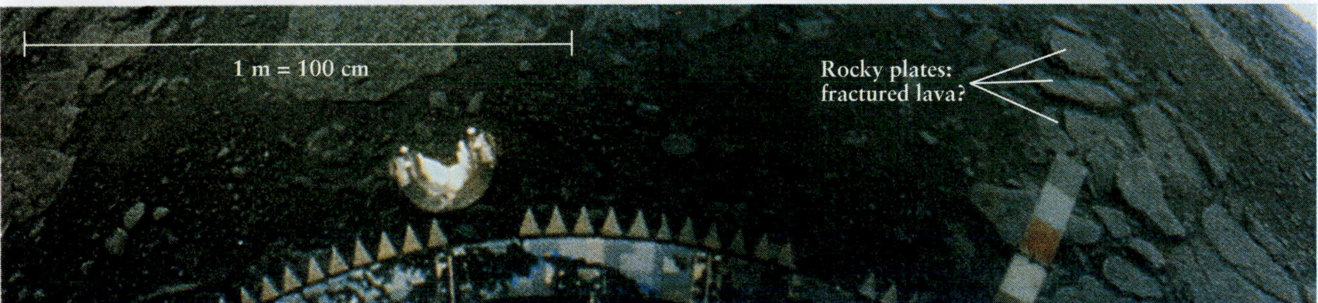

b Color-corrected image

R I V U X G

c

FIGURE 7-17 The Venusian Surface (a) This color photograph, taken by a Soviet spacecraft, shows rocks that appear orange because the light was filtered through the thick, sulfur-rich clouds. (b) By comparing the apparent color of the spacecraft to the color it was known to be, computers can correct for the sulfurous light. The actual color of the rocks is gray. In this view, the rocky plates that cover the ground may be fractured segments of a thin layer of lava. The toothed wheel in each image is part of the landing mechanism that keeps the spherical spacecraft from rolling. (c) The horizon and yellow sky of Venus imaged by the *Venera* lander. (a: Courtesy of C. M. Pieters and the USSR Academy of Sciences; b and c: Don Mitchell)

tering 750 K, hotter than the hottest spot on Mercury! Furthermore, the thick atmosphere keeps the night side of Venus at nearly the same temperature, unlike the night side of Mercury, where the temperature drops precipitously.

This high temperature prevents liquid water from existing on Venus. Just like Earth, Venus was likely to have initially had water both in the debris from which it formed and from the impacts of ice-bearing bodies, such as comets. The high atmospheric temperature has evaporated any water that came to Venus's surface, which is now bone dry. Today the total amount of water in Venus's atmosphere would uniformly cover the planet to a depth of only 3 cm. The equivalent depth of a uniform ocean on Earth today would be 2.8 km (1.7 mi). In 2007, using the *Venus Express* spacecraft, astronomers determined from the chemistry of its present atmosphere that Venus once had at least 150 times as much water. The atmospheric heat has caused most of the water that was once in its air to drift into space.

Where on Earth (besides in a greenhouse) does the greenhouse effect occur?

7-7 Venus is covered with gently rolling hills, two "continents," and numerous volcanoes

Soviet spacecraft that landed on Venus provided us with intriguing close-up images of the planet's arid surface. Figure 7-17 is a view of Venus's regolith taken by *Venera 13* in 1982. Russian scientists believe that this region was covered with a thin layer of lava that contracted and fractured upon cooling to create the rounded, interlocking shapes seen in the photographs.

Indeed, measurements by several spacecraft indicate that Venusian rock is quite similar to lava rocks, called *basalt,* which are common on Earth and the Moon.

The entire Venusian surface was mapped by the *Magellan* spacecraft, which arrived at Venus in 1990. While in orbit about the planet, *Magellan* sent radar signals through the clouds surrounding Venus. Like radar used by police, some of the radar signals bounced off Venus and returned to the spacecraft. By measuring the time delay of the radar echo, scientists determined the heights and depths of Venus's hills and valleys. As a result, astronomers have been able to construct a three-dimensional map of the planet. The detail (or *resolution*) visible on this map is about 75 m. You could see a football stadium on Venus if there were any. (None were detected.)

Venus is remarkably flat compared to Earth. More than 80% of Venus's surface is covered with volcanic plains and gently rolling hills created by numerous lava flows. Figure 7-18 is an image using the *Magellan* radar data to create a view of a Venusian landscape.

Global radar images revealed just two large highlands, or "continents," rising well above the generally level surface of the planet (Figure 7-19). The continent at high latitudes in the northern hemisphere is called Ishtar Terra (after the Babylonian goddess of love) and is approximately the same size as Australia. Ishtar Terra is dominated by a high plateau, Lakshmi Planum, ringed by towering mountains. The highest mountain is Maxwell Montes, whose summit rises to an altitude of 11 km (6.8 mi) above the average surface. For comparison, Mount Everest on Earth rises 9 km above sea level.

The larger Venusian continent, Aphrodite Terra (named after the Greek counterpart of the Roman goddess Venus), is a belt of highlands that straddles the equator. Aphrodite Terra is 16,000 km (10,000 mi) in length and 2000 km (1200 mi) wide, giving it an area about one-half that of Africa. The global view of the surface of Venus in Figure 7-20 shows that most of Aphrodite Terra is covered by vast networks of faults and fractures.

Venus has more than 1600 major volcanoes and volcanic features and fewer than a thousand impact craters (Figure 7-21), as compared to the hundreds of thousands of impact craters seen on the Moon and Mercury. Today, Venus's thick atmosphere heats, and thereby vaporizes, much of the infalling debris that would otherwise create craters. Although pieces larger than around

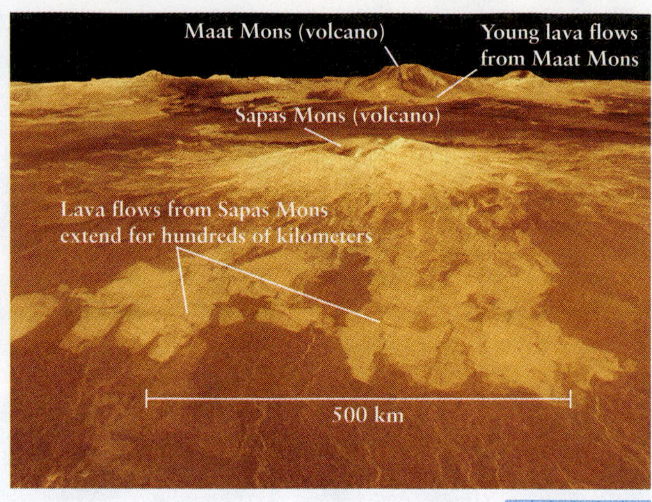

FIGURE 7-18 A Venusian Landscape A computer combined radio images to yield this perspective view of Venus as you would see it from an altitude of 4 km (2.5 mi). The color results from light being filtered through Venus's thick clouds. The brighter color of the extensive lava flows indicates that these flows reflect radio waves more strongly. The vertical scale has been exaggerated 10 times to show the gentle slopes of Sapas Mons and Maat Mons, volcanoes named for ancient Phoenician and Egyptian goddesses, respectively. (NASA, JPL Multimission Image Processing Laboratory)

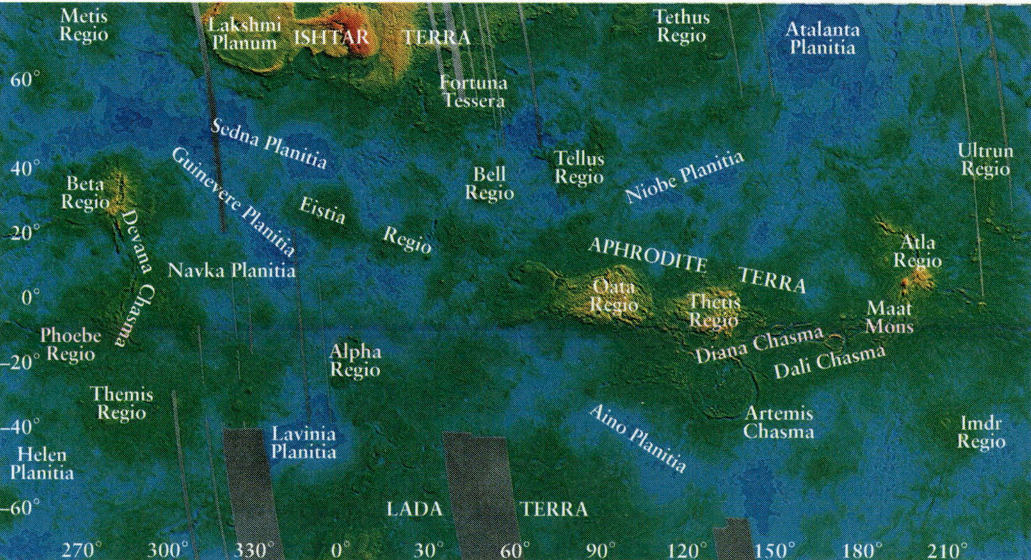

FIGURE 7-19 A Map of Venus This false-color radar map of Venus, analogous to a topographic map of Earth, shows the large-scale surface features of the planet. The equator extends horizontally across the middle of the map. Color indicates elevation—red for highest, followed by orange, yellow, green, and blue for lowest. The planet's highest mountain is Maxwell Montes on Ishtar Terra. Scorpion-shaped Aphrodite Terra, a continentlike highland, contains several spectacular volcanoes. Do not confuse the blue and green for oceans and land. (Peter Ford/MIT, NASA/JPL)

R I V U X G

FIGURE 7-20 A "Global" View of Venus A computer using numerous *Magellan* images creates a simulated globe. Color is used to enhance small-scale structures. Extensive lava flows and lava plains cover about 80% of Venus's relatively flat surface. The bright band running almost east-west is the continentlike highland region Aphrodite Terra. (NASA)

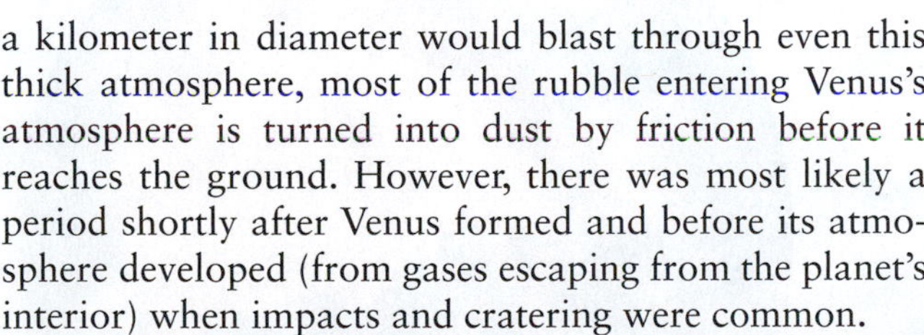

a kilometer in diameter would blast through even this thick atmosphere, most of the rubble entering Venus's atmosphere is turned into dust by friction before it reaches the ground. However, there was most likely a period shortly after Venus formed and before its atmosphere developed (from gases escaping from the planet's interior) when impacts and cratering were common.

If we accept that the young Venus was cratered like every other solid object in the solar system, we still need to explain the low number of craters on it today. The problem is that unlike Earth, Venus lacks the large-scale tectonic plates that divide up and refresh Earth's surface. Put another way, Venus is a one-plate planet. Without plates moving relative to each other, how can the craters be erased? Several mechanisms to explain this are being explored. Most theories posit that the lack of surface motion has forced Venus's crust to become thicker than Earth's crust. In one theory, heat from radioactive elements inside the planet causes mantle convection without tectonic plate motion. The temperatures inside the planet eventually become high enough in the regions of rising mantle to create massive volcanoes out of which pour enormous amounts of magma that cover large areas of the surface. A variation of this theory proposes that the thickening crust is suddenly thinned from below by the horizontal convective motion of the mantle, like stripping layers off of plywood. When so thinned, heat is able to rise through the crust and cause many volcanoes to form.

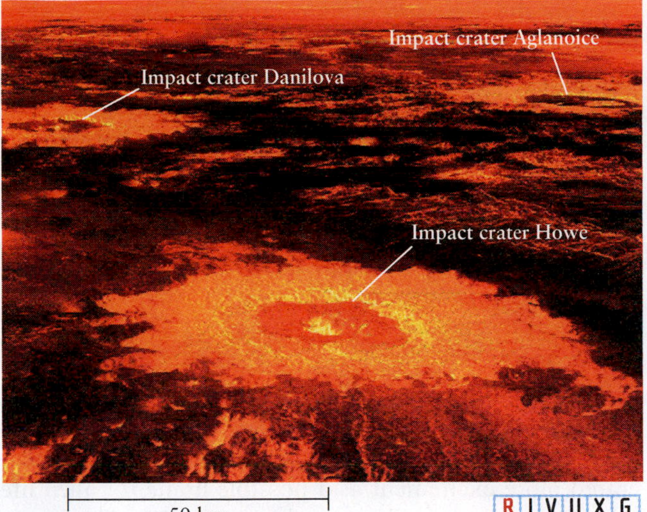

R I V U X G

FIGURE 7-21 Craters on Venus These three impact craters, with extensive ejecta surrounding each, are located on Venus's southern hemisphere. They were imaged using radar by the *Magellan* spacecraft. The colors are based on the *Venera* images (see Figure 7-17). (NASA/Magellan Images [JPL])

Another theory proposes that, episodically, the one-plate surface of Venus is heated below enough to crack and begin Earthlike tectonic plate motion until it cools down and again becomes a single plate. During the period of plate motion, the surface is erased and replaced.

In yet another theory, the mantle becomes so hot that large sections of the crust melt more or less simultaneously, destroying the old crust and its craters. Indeed, that theory has been expanded to periodic global meltdowns occurring every 700 million years or so. For all of these models, when enough heat is released from the planet's interior, its surface once again solidifies, and with the possible exception of local volcanic activity, the crust rethickens.

Why does little cratering occur on Venus today, even compared to the present low rate of cratering on our Moon?

The sulfur content of the air and the traces of active volcanoes on the surface are compelling indicators that, like Earth, Venus has a molten interior. Because the average density of Venus is similar to that of Earth, its core is predominantly iron. Currents in the molten iron should generate a magnetic field; however, none of the spacecraft sent to Venus has detected one. In general, magnetic fields are created by rotation of currents (see Section 6-4). Therefore, Venus's apparent lack of a magnetic field is plausible only if it rotates exceptionally slowly. In fact, the planet takes 116.8 Earth days to get from one sunrise to the next (if you could see the Sun from Venus's surface).

Unlike Earth and Mercury, Venus exhibits **retrograde rotation,** in which the direction of Venus's orbit around the Sun (counterclockwise as seen from space far above Earth's North Pole) is *opposite* the direction of its rotation (clockwise as seen from the same vantage point). In other words, sunrise on Venus occurs in the west. Venus's rotation axis is tilted more than 177°, compared to Earth's 23½° tilt. Because Venus's axis is within 3° of being perpendicular to the plane of its orbit around the Sun, the planet has no seasons. Although we do not know the cause of Venus's retrograde rotation, one likely explanation is that a monumental impact flipped the rotation axis early in the planet's existence.

MARS

Mars's distinctive rust-colored hue (Figure 7-22) makes it stand out in the night sky. For centuries Mars has generated more excitement as a possible home for alien life than any other world in our solar system. The idea began following the first well-documented telescopic observations of Mars in 1659 by the Dutch physicist Christiaan Huygens. Huygens identified a prominent, dark surface feature that reemerged about every 24 hours, suggesting a rate of rotation similar to that of Earth.

Belief that advanced life exists on Mars skyrocketed at the end of the nineteenth century after Giovanni Virginio Schiaparelli, an Italian astronomer, reported in 1877 seeing 40 lines crisscrossing the Martian surface (Figure 7-23). He called these dark features *canali,* an Italian term meaning "channels." It was soon mistranslated into English as *canals,* implying the existence on Mars of intelligent creatures capable of engineering feats. This speculation led Percival Lowell, scion of a wealthy Boston family, to finance a major new observatory near Flagstaff, Arizona. By the end of the nineteenth century, Lowell had allegedly observed 160 Martian "canals."

By the beginning of the twentieth century, it was fashionable to speculate that the Martian canals formed an enormous, planetwide irrigation network to transport water from melting polar ice caps to vegetation near the equator. In view of the planet's reddish, desert-like appearance, Mars was thought to be a dying planet whose inhabitants must go to great lengths to irrigate their farmlands. No doubt the Martians would readily abandon their arid ancestral homeland and invade Earth for its abundant resources.

MARS: VITAL STATISTICS

Planet symbol:	♂
Average distance from the Sun:	1.52 AU = 2.28×10^8 km
Maximum distance from the Sun:	1.67 AU = 2.49×10^8 km
Minimum distance from the Sun:	1.38 AU = 2.10×10^8 km
Orbital eccentricity:	0.093
Average orbital speed:	24.1 km/s
Sidereal period of revolution (year):	687 Earth days = 1.88 Earth years
Sidereal rotation period:	24 h 37 m 22 s
Solar rotation period (day):	24 h 39 m 35 s
Inclination of equator to orbit:	25.19°
Inclination of orbit to ecliptic:	1.85°
Diameter (equatorial):	6787 km = 0.53 Earth diameter
Mass:	6.42×10^{23} kg = 0.107 Earth mass
Average density:	3950 kg/m^3 = 0.716 Earth density
Escape speed:	5.0 km/s
Surface gravity (Earth = 1):	0.38
Albedo:	0.16
Surface temperatures:	Maximum: 20°C = 70°F = 293 K Mean: −53°C = −63°F = 220 K Minimum: −140°C = −220°F = 133 K
Atmospheric composition (by number of molecules):	95.3% carbon dioxide (CO_2) 2.7% nitrogen (N_2) 0.03% water vapor (H_2O) 2% other gases

R I V U X G

FIGURE 7-22 Mars Vital Statistics The lighter orange region in the center of the photograph in this figure is called Arabia Terrae, after the Arabian peninsula on Earth. (NASA/USGS)

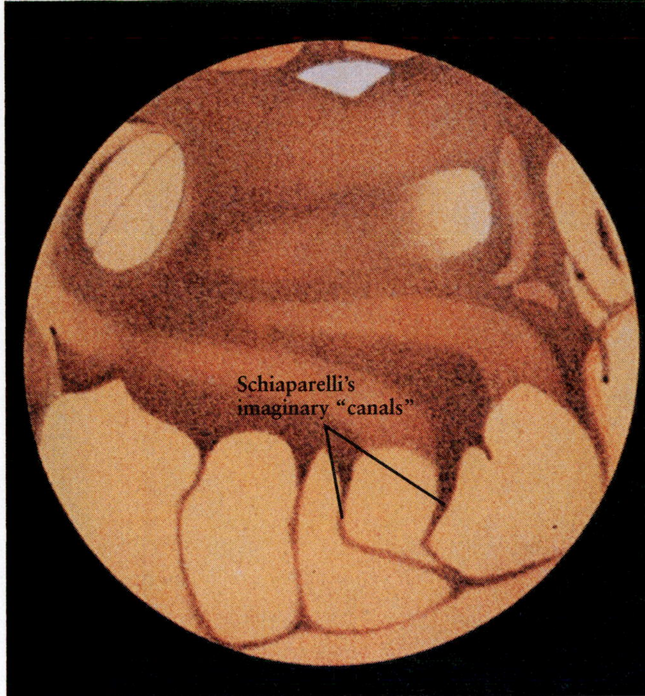

FIGURE 7-23 The Illusion of Martian Canals Giovanni Schiaparelli examined Mars through a telescope 20 cm (8 in.) in diameter, the same size used by many amateur astronomers today. His drawings of the red planet showed features perceived by Percival Lowell and others as irrigation canals. Higher-resolution images from Earth and spacecraft visiting Mars failed to show the same features. (Michael Hoskin, ed., *The Cambridge Illustrated History of Astronomy*, Cambridge University Press, 1997, p. 286. Illustration by G. V. Schiaparelli. Courtesy of Institute of Astronomy, University of Cambridge, UK)

Insight Into Science

Perception and Reality Our tendency to see patterns, even where they do not exist, makes it easy for us to leap to unjustified conclusions. Scientific images and photographs require objective analysis. This is why scientific discoveries are not considered to be authentic until they have been reproduced and verified by independent researchers.

7-8 Mars's global features include plains, canyons, craters, and volcanoes

Spacecraft journeying to Mars have sent back pictures since the 1970s. They show that Mars's surface is completely dry and that it has broad plains, shallow craters, enormous inactive volcanoes, and vast canyons (Figure 7-24). The broad northern plain is called Vastitas Borealis (the **northern vastness** or **northern lowlands**) and is shown in blue on Figure 7-24. This plain is some 5 km (3 mi) below the much more heavily cratered and hilly plains in the south, called the **southern highlands.** The cause of the significant difference in heights of the two hemispheres is not known, but it may stem from either early tectonic activity (see Section 6-2 for a discussion of tectonics) or from a powerful early impact.

Between the two hemispheres, *Mariner 9* discovered a vast canyon now called Valles Marineris, which runs roughly parallel to the Martian equator (Figure 7-25). Valles Marineris stretches over 4000 km, about one-fifth the circumference of Mars. If this canyon were located on Earth, it would stretch from New York to Los Angeles. The individual canyons that make up Valles Marineris are up to 6 km (4 mi) deep and 190 km (120 mi) wide. Valles Marineris begins with heavily fractured terrain in the west and ends with ancient-cratered terrain in the east. Geologists believe that Valles Marineris is a large crack that formed as the planet cooled, and was enhanced by nearby rising crust to its west and widened further by water-driven erosion. Some of the eastern parts of this system appear to have been formed almost entirely by water flow, similar to Earth's Grand Canyon. Although remote observations such as those made by the *Mariner* and myriad other spacecraft have strongly suggested water has long existed on Mars, the chemical verification of water there, in the form of ice, was first made in 2008 by the *Phoenix Mars Lander*.

Most impact craters are located on Mars's southern hemisphere (Figure 7-26). This suggests that the northern vastness has been resurfaced by some process that eradicated ancient lowland craters, consistent with either of the earlier explanations of the differences in hemispheres.

Most of the volcanoes on Mars are located in the northern hemisphere. The largest volcano, Olympus Mons (Figure 7-27a), covers an area as big as the state of Missouri and rises 26 km (16 mi) above the surrounding plains—nearly 3 times the height of Mount Everest. The highest volcano on Earth, Mauna Loa in the Hawaiian Islands, has a summit only 17 km above the ocean floor. The clouds surrounding the volcano in this image are composed of carbon dioxide.

Around Olympus Mons and other volcanoes, orbiting spacecraft have observed numerous cratered, cone-shaped features that are typically the size of a football field (Figure 7-27b). These cones are believed to have been created by ice below the Martian surface that was liquefied and vaporized by lava from the nearby volcanoes. The expanding water pushed its way up to the surface, deforming the land before exploding out the centers of these "ice" volcanoes. Similar clusters of cones on Earth, formed by the same mechanism, are found in Iceland.

Astronomers have discovered impact craters on Mars that formed in the past few years. In one case, an image showed no craters, while an image taken 67 days later of the same area revealed a cluster of new, small craters. Careful study of these new craters shows that they uncover ice that had been buried only a meter or two under the planet's surface before the meteorite strikes.

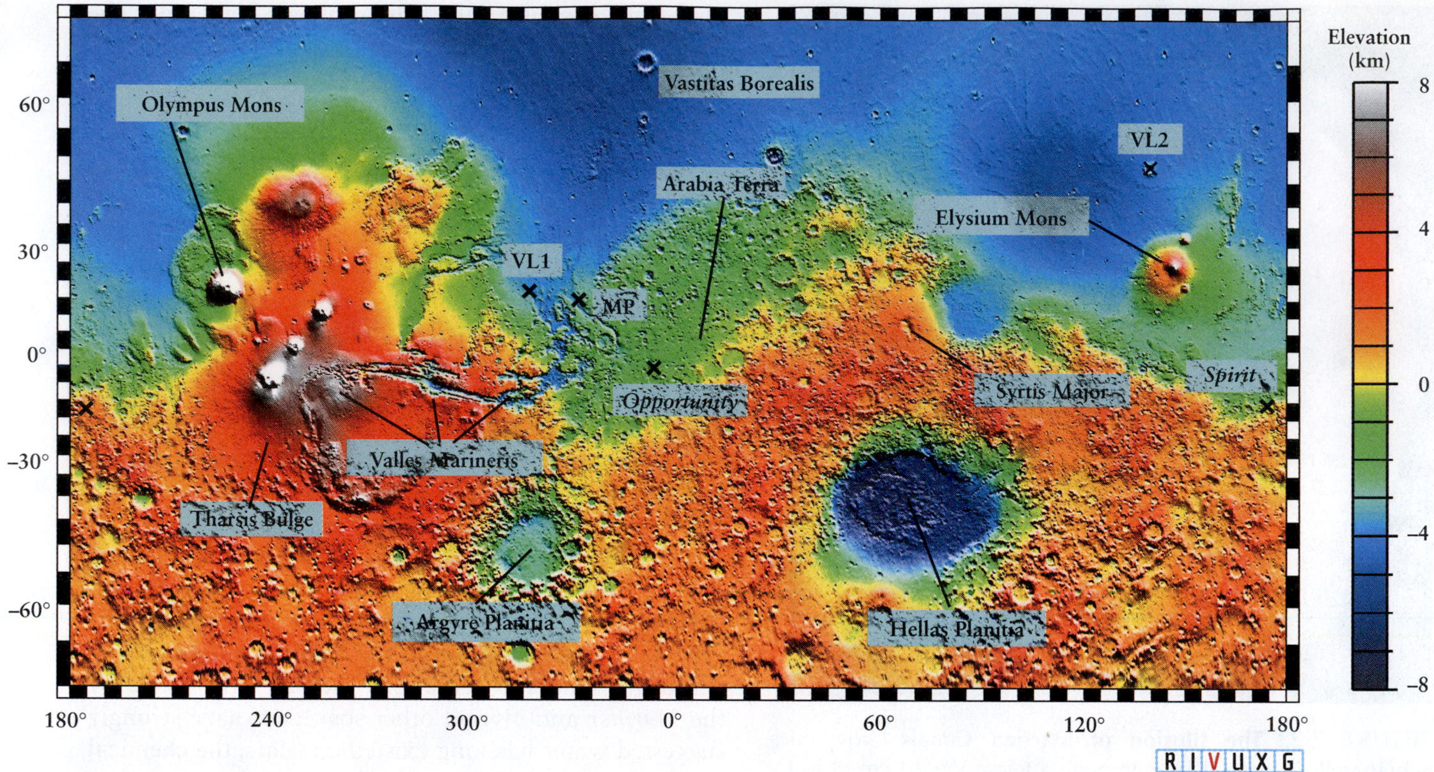

ANIMATION 7.6

FIGURE 7-24 The Topography of Mars The color coding on this map of Mars shows elevations above (positive numbers) or below (negative numbers) the planet's average radius. To produce this map, an instrument on board *Mars Global Surveyor* fired pulses of laser light at the planet's surface, then measured how long it took each reflected pulse to return to the spacecraft. The *Viking 1 Lander* (VL1), *Viking 2 Lander* (VL2), *Mars Pathfinder* (MP), *Opportunity*, and *Spirit* landing sites are each marked with an x. The volcanoes depicted in white are higher than the scale on the right can show. (MOLA Science Team, NASA/GSFC)

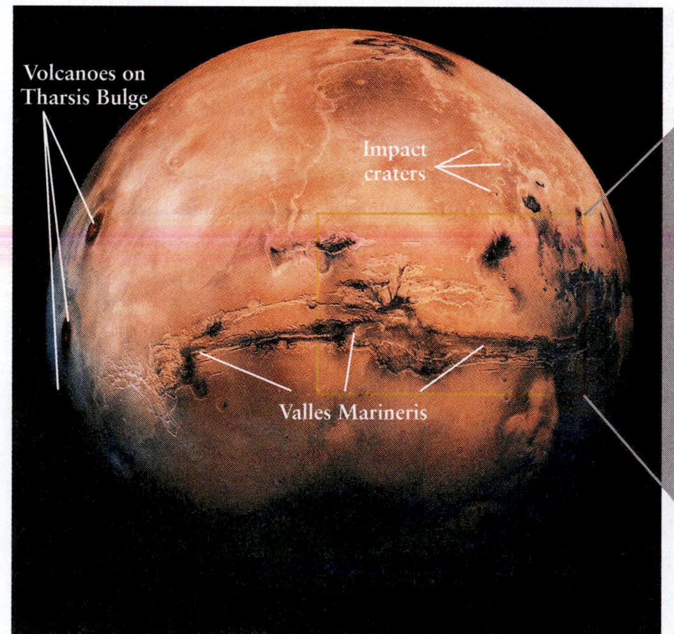

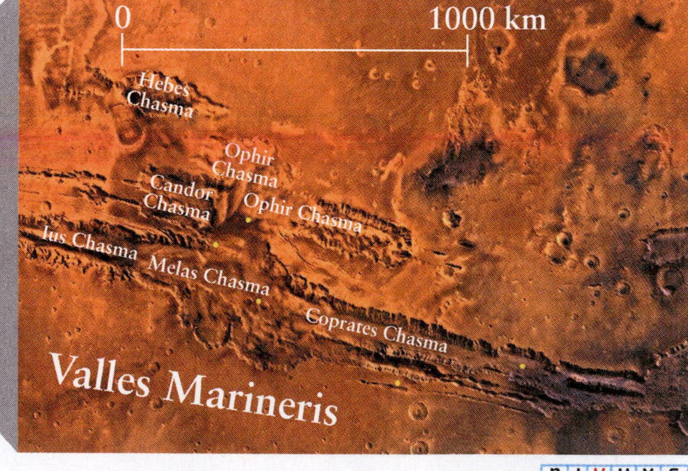

FIGURE 7-25 Martian Terrain This high-altitude photograph shows a variety of the features on Mars, including enormous volcanoes (left) on the highland, called Tharsis Bulge; impact craters (upper right); and vast, windswept plains. The enormous Valles Marineris canyon system crosses horizontally just below the center of the image. Inset: Details of the Valles Marineris, which is about 100 km (60 mi) wide. The canyon floor has two major levels. The northern (upper) canyon floor is 8 km (5 mi) beneath the surrounding plateau, whereas the southern canyon floor is only 5 km (3 mi) below the plateau. (USGS/NASA; inset: NASA/GSFP/LTP)

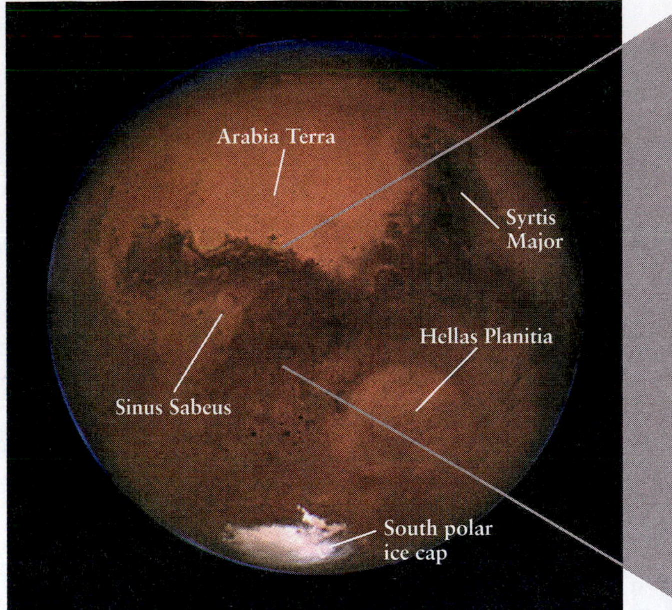

Mars from the Hubble Space Telescope

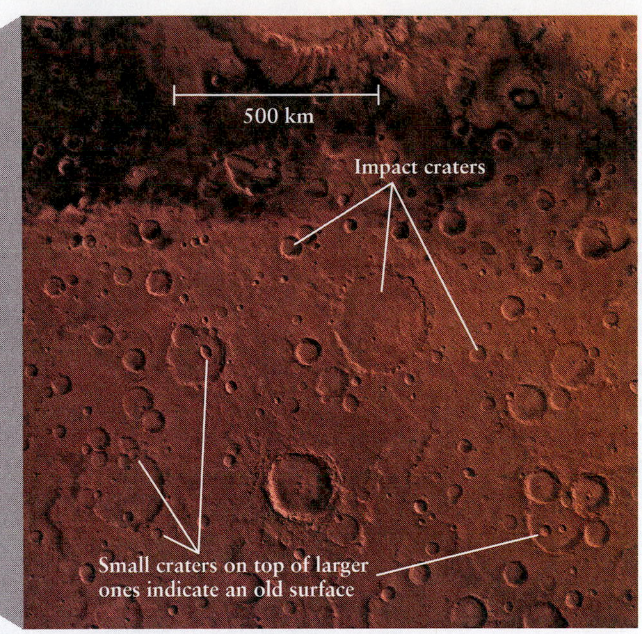

Closeup of Sinus Sabeus region

R I V U X G

FIGURE 7-26 Craters on Mars This image taken by the Hubble Space Telescope was made during the opposition of 2003. Arabia Terra looks like the Arabian peninsula on Earth. It is an old, highly eroded region dotted with numerous flat-bottom craters. Lava-covered Syrtis Major was first identified by Christiaan Huygens in 1659. A single impact carved out Hellas Planitia, which is 5 times the size of Texas. Inset: This mosaic of images from the *Viking 1* and *2* orbiter spacecraft shows an extensively cratered region located south of the Martian equator. Note how worn down these craters are compared to those on Mercury and our Moon. (NASA; J. Bell, Cornell University; and M. Wolff, SSI; inset: USGS)

a

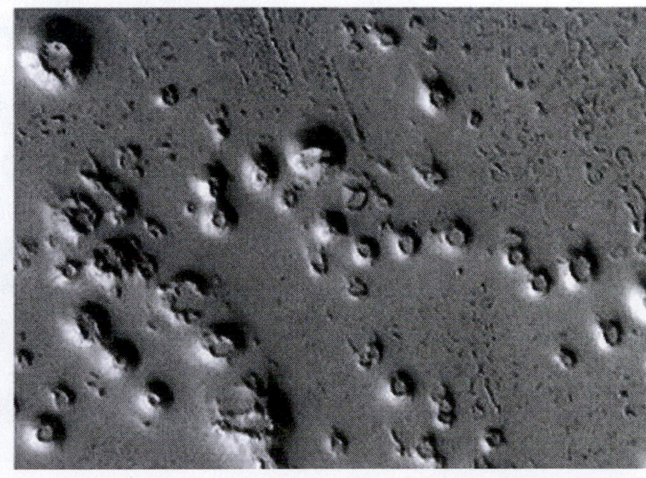
b

R I V U X G

FIGURE 7-27 The Olympus Caldera (a) This view of the summit of Olympus Mons is based on a mosaic of six pictures taken by one of the *Viking* orbiters. The caldera consists of overlapping, volcanic craters and measures about 70 km across. The volcano is wreathed in mid-morning clouds brought upslope by cool air currents. The cloud tops are about 8 km below the volcano's peak. (b) These cones on Mars may have been created when lava from Olympus Mons heated underground ice, causing the resulting water and vapor to expand, raise the planet's surface, and burst out. (Peter Lanagan [LPL, U. Arizona] et al., MOC, MGS, NASA)

Planetwide, high-resolution photographs of Mars over the past 40 years have failed to show one canal of a size consistent with those allegedly seen from Earth (see Figure 7-23). We now know for certain that Schiaparelli's *canali* were optical illusions, and the science fiction writers who wrote about advanced cultures there were completely off the mark.

Insight Into Science

Science Fact and Fiction Science fiction may foreshadow real scientific discoveries. (Can you think of some examples?) But science also reins in the fantasies of the science fiction writer. A century ago, millions of people, including top scientists, believed that technologically advanced life existed on Mars. However, scientific evidence says no.

7-9 Although no canals exist on Mars, it does have some curious natural features

Mars not only lacks canals, it also has no cities, roads, or other signs of civilization. However, astronomers have photographed several surface features that, at first glance, could have been crafted by intelligent life-forms. In 1976, a hundred years after Schiaparelli's alleged discovery of canali, the *Viking 1* orbiter spacecraft photographed a feature that appeared to be a humanlike face (Figure 7-28a). However, when the *Mars Global Surveyor* photographed the surface in greater detail and from a different angle in 1998 (Figure 7-28b), it found no facial features. If anything, the same spot looked more like a giant heel print.

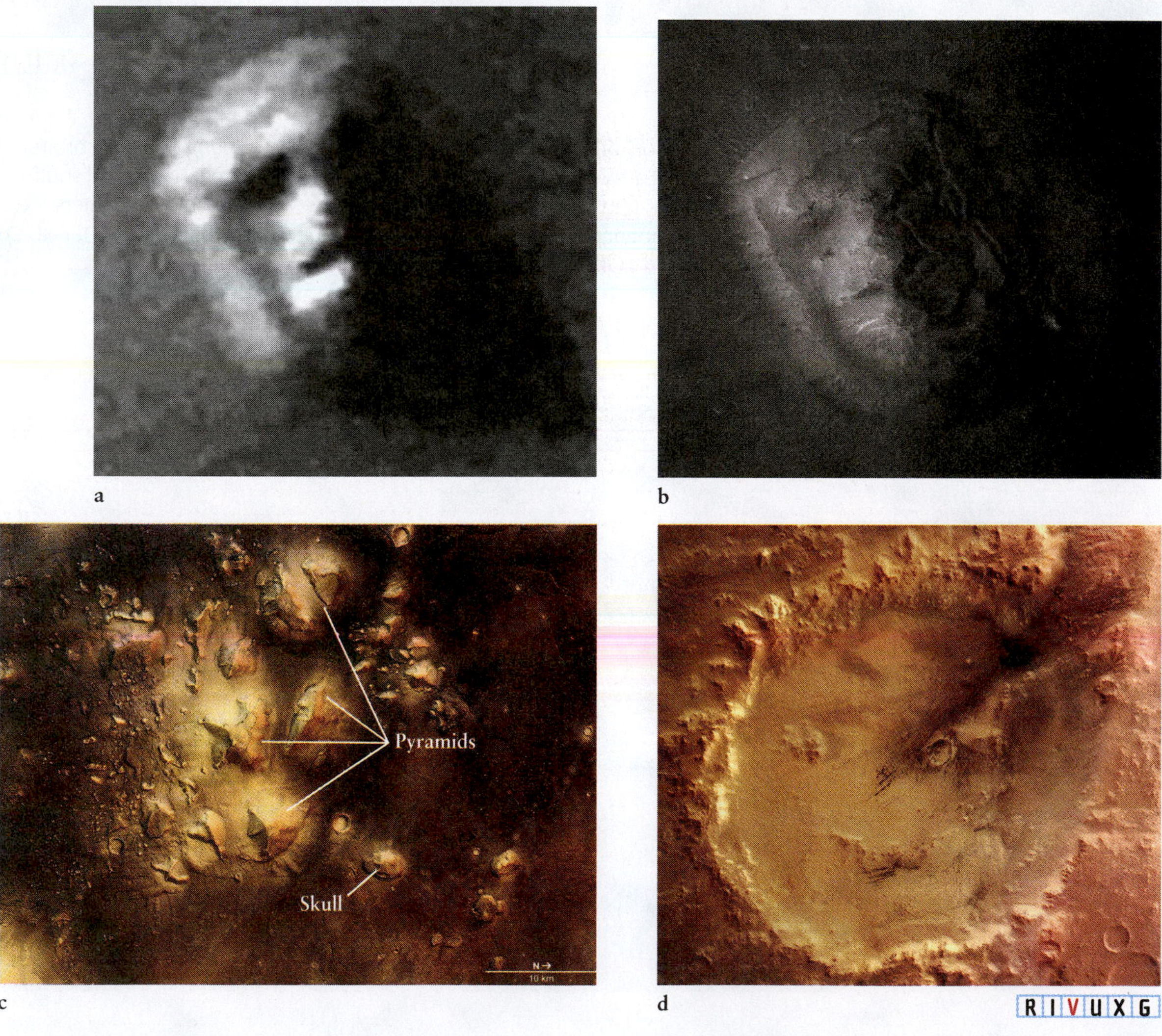

FIGURE 7-28 In the Eye of the Beholder Images (a) and (b) are of the same site on Mars, taken 22 years apart. They show how, at high resolution, the apparent face in (a) changed to a more "natural-looking" feature in (b). This change in appearance was due to weathering of the site, improved camera technology, and the difference in angle at which the photographs were taken. (c) In the same region of Mars, other erosion features also appear to be pyramids and skulls. (d) The Galle crater and its interior features combine to give the impression of a "happy face." (a and b: NSSDC/NASA, NASA/JPL/Malin Space Science Systems; c and d: ©ESA/DLR/FU Berlin [G. Neu Kum])

Insight Into Science

Keep It Simple II Was the Martian "face" made by intelligent life or by wind-sculpted hills? When two or more alternative interpretations present themselves, scientists apply the principle of Occam's razor (Section 2-1) and always choose the explanation that requires the fewest unproven assumptions.

Other photographs show a collection of pyramids, a skull (Figure 7-28c), and a "happy face" (Figure 7-28d). Are any of these features relics of an advanced civilization? Scientists universally believe that they are all naturally formed features. The first face is a region of hard rock around which softer debris is being worn away. Similarly, the pyramids and skull are consistent with winds eroding softer rock around harder rock pushed upward from inside Mars billions of years ago. The same thing happens on Earth. The happy face is made from a large impact crater and several small natural features inside it.

7-10 Mars's interior is less molten than the inside of Earth

Part of the Valles Marineris was apparently formed when large regions of Mars's surface moved apart due to tectonic plate motion. This activity is generated by the flow of rock under the planet's surface (see Section 6-2 for more on plate tectonics). However, there is no evidence that plate tectonic activity occurs on Mars today. This lack of activity indicates that more of Mars's interior is cooled, solid rock than exists inside Earth. This conclusion is supported by the observation that Mars lacks a global magnetic field (see Section 6-4 for discussion of the field-generating dynamo effect). However, observations made in 2003 revealed that the Sun creates tiny tides (less than a centimeter) on the solid body of Mars. As the land rises and sinks ever so slightly through the day on Mars, friction created by this motion in the planet's interior provides enough heat to keep some of the interior molten. (Rub your hands vigorously together to see how friction creates heat.)

ANIMATION 7.7

Although a global magnetic field on Mars is absent today, there are remnants of such a field. In 1997 the *Mars Global Surveyor* (which operated until 2006) discovered patterns of local surface magnetic fields in nine places on Mars, similar to regions on Earth where tectonic plates separate and lock Earth's magnetic field in the rock (see Section 6-4). Geologists propose that when Mars was young, its internal "dynamo" created a strong magnetic field and that tectonic plates helped craft its surface while locking traces of its changing magnetic field in the then-molten rock. However, the planet quickly cooled, its global magnetic field vanished, and tectonic-plate activity apparently ceased nearly 4 billion years ago.

What two major activities created Valles Marinaris?

A more molten interior early in its existence is also consistent with the enormous volcanoes that exist on Mars. Volcanoes are places where molten rock oozes or bursts from a body's interior. The volcanoes on Mars are the type where molten rock oozed out, called *shield volcanoes,* like the ones that created the Hawaiian Islands. Such volcanoes are very different from Mount Saint Helens in Washington state or Mount Etna in Sicily, where the material from inside Earth is ejected violently. On Earth, the Hawaiian Islands are only the most recent additions to a long chain of volcanoes. These volcanoes resulted from **hot-spot volcanism,** a process by which molten rock rises to the surface from a fixed hot region far below. The Pacific tectonic plate is slowly moving northwest at a rate of several centimeters per year. As a result, new volcanoes are created above the hot spot, while older ones move off and become extinct, eventually disappearing beneath the ocean.

Have you ever seen something that you interpreted as one thing, but it turned out to be something else?

Because Mars's surface is apparently frozen in place due to the lack of tectonic plate motion, a hot spot can keep pumping lava upward through the same vent for millions of years. One result is Olympus Mons—a single giant volcano rather than a long chain of smaller ones (see Figure 7-27a). This volcano's summit has collapsed to form a volcanic crater, called a **caldera,** large enough to contain the state of Rhode Island. Because Mars's solid crust and mantle (the region directly below the surface) are now so thick, they insulate the surface, preventing any more molten rock from escaping. Therefore, we do not expect to see any more large-scale volcanic activity there.

In 2008, radar on the *Mars Reconnaissance Orbiter* circling Mars observed that the surface of Mars does not buckle under the weight of its north polar ice cap, whereas the Earth does warp under similar ice layers. The fact that Mars doesn't change shape implies that the upper layers of the planet, the crust and mantle, must be more rigid than the same regions on Earth. As a consequence of this discovery, astronomers have concluded that liquid water, if it exists inside Mars, is deeper inside the planet than previously thought. If life on Mars does exist, then it likely lies deep inside the planet as well.

The Martian crust is at least 40 km thick under the northern lowlands and 70 km thick under the southern highlands. However, the boundary between the thin and thick crusts does not line up with the boundary between high and low terrain. Current models of Mars suggest that it has a core about 3400 km in diameter and that at least some of that core is molten. A detailed understanding of Mars's interior waits for the placing of seismic detectors on the planet's surface.

7-11 Martian air is thin and often filled with dust

To understand more about the activity on Mars's surface, we must examine its atmosphere, which has only 0.6% the pressure of Earth's atmosphere. As on Venus, some 95% of Mars's thin atmosphere is composed of carbon dioxide. The remaining 5% consists of nitrogen, argon, and some traces of oxygen. Unlike the gravitational attraction of tiny Mercury, which is too weak to hold any gases as a permanent atmosphere, Mars's gravitational force is just strong enough to prevent carbon dioxide, nitrogen, argon, and oxygen from escaping into space, but not strong enough to hold down water vapor.

If the Hawaiian Islands were fixed over their hot spot, rather than moving, what would happen to them?

A growing body of evidence indicates that there was once a vast water-filled ocean in the northern hemisphere of Mars. This evidence includes layers of rock deposited by water and channels in which water from the south flowed northward. Observations made in 2008 suggest that liquid water existed on the surface of Mars for 2½ billion years—in other words, until about 2 billion years ago. Much of that water has evaporated and drifted into space, never to return. Today, the concentration of water vapor in Mars's atmosphere is 30 times lower than it is in Earth's air. If all of the water vapor could somehow be squeezed out of the Martian atmosphere, it would not fill even one of the five Great Lakes of North America. The remainder of the water that was on Mars's surface has turned to ice, much of which remains there today. Since 2002, satellites orbiting Mars have discovered enough water ice under Mars's south pole to cover the entire planet to a depth of 11 m (36 ft). In 2008, the *Phoenix Mars Lander* dug into the Martian regolith and observed ice just under the surface. Chemical tests made by it confirmed that the ice was frozen water. Water ice is being discovered at many latitudes from the polar regions to the equator. In 2008, the *Mars Reconnaissance Orbiter* discovered buried glaciers extending dozens of kilometers from mountains and cliffs. It is certain that much more water remains to be found inside Mars.

Despite the low density of Mars's air, the red planet experiences Earthlike seasons because of a striking coincidence, first noted in the late 1700s by the German-born English astronomer William Herschel. Just as Earth's equatorial plane is tilted 23½° from the plane of its orbit, Mars's equator makes an angle of about 25° from the plane of its orbit. If, as is now suspected, the northern hemisphere of Mars is lower than the southern hemisphere because the north was struck by a large body, that impact could have also tilted the planet and initiated the seasons there. The Martian seasons last nearly twice as long as Earth's, because Mars takes nearly 2 Earth years to orbit the Sun. When Mars is near opposition, even current telescopes for home use reveal its daily changes. Dark seasonal markings on the Martian surface can be seen to vary, and prominent polar caps of frozen carbon dioxide shrink noticeably during the spring and summer months (Figure 7-29).

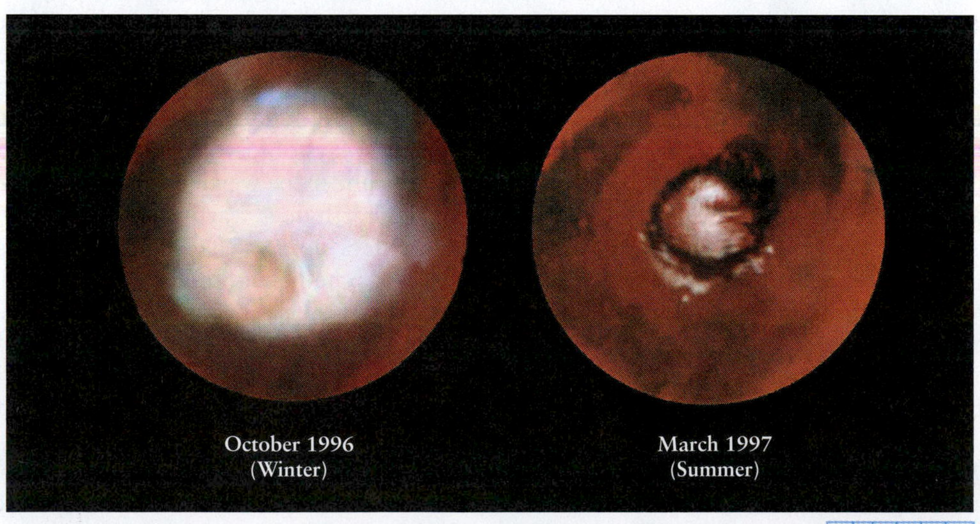

FIGURE 7-29 Changing Seasons on Mars During the Martian winter, the temperature decreases so much that carbon dioxide freezes out of the Martian atmosphere. A thin coating of carbon dioxide frost covers a broad region around Mars's north pole. During the summer in the northern hemisphere, the range of this north polar carbon dioxide cap decreases dramatically. During the summer, a ring of dark sand dunes is exposed around Mars's north pole. (S. Lee/J. Bell/M. Wolff/Space Science Institute/NASA)

a RIVUXG

b RIVUXG

FIGURE 7-30 The Atmosphere of Mars (a) When Mars's sky is relatively free of dust, it appears similar in color to our sky, as shown in this sunset photo taken by the rover *Opportunity*. The darker, brown color in which the Sun is immersed is due to lingering dust in the sky. When less dust is present, the Sun looks almost white during Martian sunsets. Most of the images we have of Mars's sky show colors like those seen in (b). (b) Taken by the *Mars Pathfinder*, this photograph shows the *Sojourner* rover snuggled against a rock named Moe on the Ares Vallis to run tests on the rock. At the top of the image, the pink color of the Martian sky is evident. (NASA/Lunar and Planetary Institute)

In 1994, astronomers using computer simulations discovered that the axis tilts of rapidly rotating planets, like Mars and Earth, are prone to change over millions of years unless such a planet is stabilized by the presence of a massive moon. Earth has such a moon, hence our planet maintains its present 23½° tilt, but the two tiny moons of Mars do not have enough gravitational attraction to keep the orbital inclination of Mars from changing dramatically. Observations of the underground polar regions by the *Mars Reconnaissance Orbiter* reveal seasonal changes on Mars over periods of millions of years consistent with Mars's rotation axis changing far from its present tilt of 25°.

Although sometimes blue (Figure 7-30a), Mars's atmosphere is often pastel red, sometimes turning shades of pink and russet (Figure 7-30b). All of these latter colors are due to fine dust blown from the planet's desert-like surface during windstorms. The dust is iron oxide, familiar here on Earth as rust. Mars's sky also changes color because the amount of dust in the air varies with the season. During the winter, carbon dioxide ice adheres to the dust particles and drags them to the ground. This helps clear and lighten the air. In the summer months, the carbon dioxide is not frozen, and the dust blown by surface winds remains aloft longer.

The sky color also changes over periods of many years. In 1995, for example, the amount of dust was observed to have dropped dramatically compared to that observed in the 1970s. The reason for such long-term changes is still under investigation.

As on Earth, the temperature changes with the seasons on Mars. For example, one day at the landing site of the rover *Spirit*, near the equator, the temperature ranged from 188 K to 243 K, whereas 100 days later it ranged from 200 K to 263 K (see Figure 7-24 for the site location). During the afternoons, heat from the planet's surface warms the air and sometimes creates whirlwinds, called **dust devils** (Figure 7-31). A similar phenomenon occurs in dry or desert terrain on Earth. Martian dust devils reach altitudes of 6 km (20,000 ft). Spacecraft on Mars detect drops in air pressure as dust devils sweep past, just as on Earth. Some dust devils are large enough to be seen by orbiting spacecraft.

The winds on Mars help explain why its surface is eroding. Over the past two centuries, astronomers have seen faint surface markings disappear under a reddish-orange haze as thin Martian winds have stirred up finely powdered dust from the Martian regolith. Some storms obscure the entire planet, as happened in 2001. Over the ages, winds have worn down crater walls, and deposits of dust have filled in the crater bottoms (Figure 7-32). But the Martian atmosphere is so thin that 3 billion years of sporadic storms have not carried enough of Mars's extremely fine-grained powder to eradicate the craters completely.

Why do you think it is easier to stand up in a dust devil on Mars than in the same equivalent-speed event here on Earth?

a

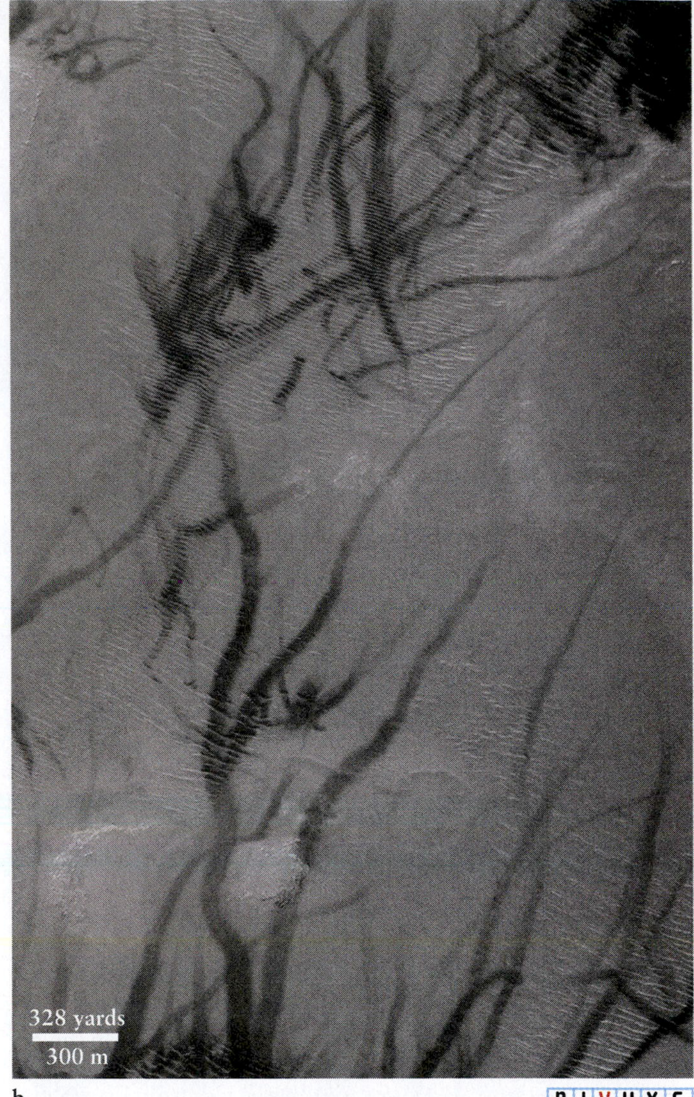

b

FIGURE 7-31 Martian Dust Devil (a) This *Spirit* image is one in a movie sequence of a dust devil moving left to right across the surface of Mars. The rovers have filmed several such events. (b) These dark streaks are the paths of dust devils on the Argyre Planitia of Mars. The tracks cross hills, sand dunes, and boulder fields, among other features on the planet's surface. (a: NASA/JPL; b: NASA/JPL/Malin Space Science Systems)

7-12 Surface features indicate that water once flowed on Mars

5 Despite disproving the theory that Mars has broad canals or any other liquid surface water, Mars-orbiting spacecraft did reveal many dried-up riverbeds (Figure 7-33a); lakes (Figure 7-34a); river deltas (Figure 7-34), where water and debris emptied into lakes and oceans; sedimentation laid down by water flow (Figure 7-35); and other water-related features. Some of the riverbeds include intricate branched patterns and delicate channels meandering among flat-bottomed craters. Rivers on Earth invariably follow similarly winding courses (Figure 7-33b).

Surface rovers *Spirit* and *Opportunity* found strong evidence that water did indeed create many of these features (Figure 7-36), including the presence of water-formed hematite—iron-rich rocks that look like blueberries—and rocks with extremely high levels of salts that had previously been dissolved in liquid water. Such salts take millions of years to seep into rocks in the quantities they have been found, implying that liquid water was on the red planet's surface for at least that length of time. The origin of at least some of the water that was on Mars is believed to have been ice-rich bodies that struck the surface, releasing their water into the atmosphere, where it later rained down on the surface. Although water existed on Mars's surface for long periods, there are indications in several places that large

FIGURE 7-32 Crater Endurance Photographed by the rover *Opportunity*, this crater on Mars is about 130 m (430 ft) across. Rocks are visible in the crater and in vertical cliffs along its walls. By studying such rocks, astronomers hope to understand more of the history of water on Mars. (JPL/NASA)

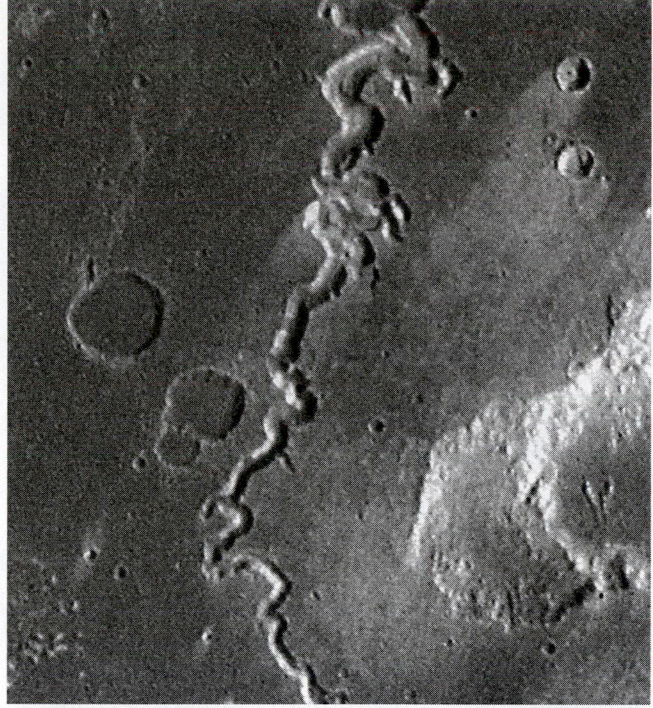

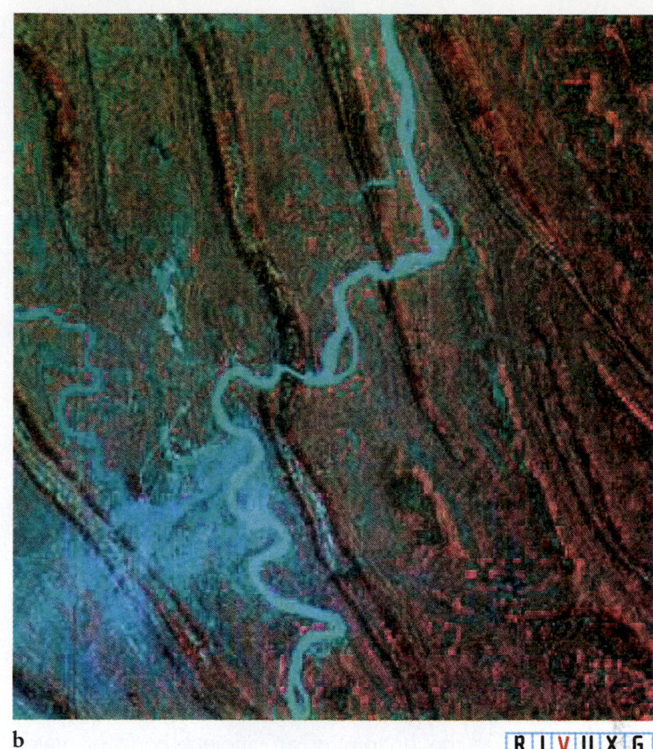

FIGURE 7-33 Rivers on Mars and Earth (a) Winding canyons on Mars, such as the one in this *Viking 1* orbiter image, appear to have been formed by sustained water flow. This theory is supported by the terraces seen on the canyon walls in high-resolution *Mars Global Surveyor* images. Long periods of water flow require that the planet's atmosphere was once thicker and its climate more Earthlike. (b) The Yangtze River near Chongqing, China. Typical of rivers on Earth, it shows the same snakelike curve as the river channels on Mars. (a: NASA; b: TMSC/NASA)

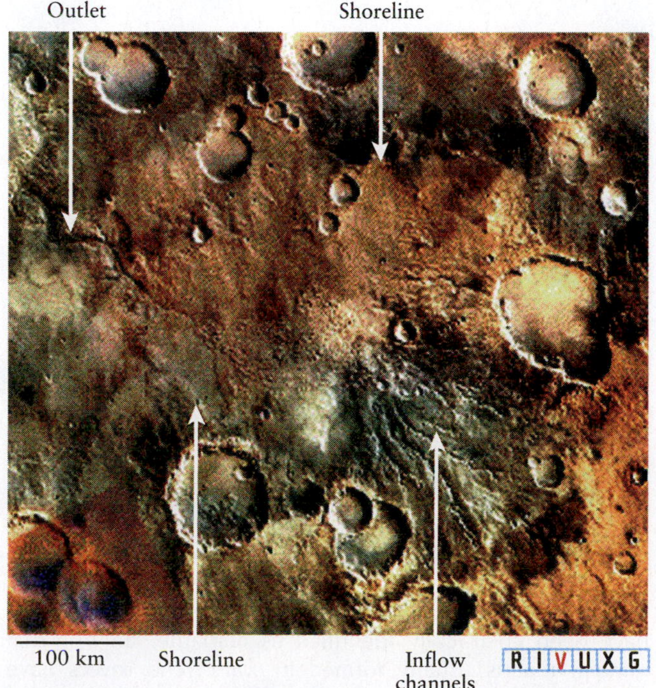

FIGURE 7-34 Evidence of Water on Mars (a) This dry Martian lake, photographed by the *Mars Global Surveyor* with a resolution of 1.5 m, is an excellent example of how geology and astronomy overlap. The features of this dry lake are consistent with those found on lakebeds on Earth. (b) This network of dry riverbeds is located on Mars's cratered southern hemisphere. (a: NASA/JPL; b: NASA)

VIDEO 7.5

FIGURE 7-35 Ancient Oceans and Lakes on Mars (a) This *Mars Global Surveyor* image of a portion of Valles Marineris reveals terrain with "stair-step layers." Such terrain is likely to have been created by sedimentation at the bottom of an ancient body of water. (b) This image of Burns Cliff, photographed by the rover *Opportunity*, shows a close-up of layers of rock laid down on a body of water on Mars that went through wet and dry periods. (a: Malin Space Science Systems/JPL/NASA; b: NASA/JPL/Cornell)

bodies of water have been gone for billions of years. The evidence for this is the discovery by the rovers of lava rocks that contain olivine and pyroxene, rocks that dissolve quickly in water.

The surface of Mars continues to change. For instance, images taken by *Mars Global Surveyor* of one crater wall show that sometime between 1999 and 2005, there was a significant landslide on it (Figure 7-37a, b). While this new feature was initially attributed to liquid water gushing out and initiating the landslide, subsequent observations of similar activity suggest that the crater wall is so steep that dry sandy material could have

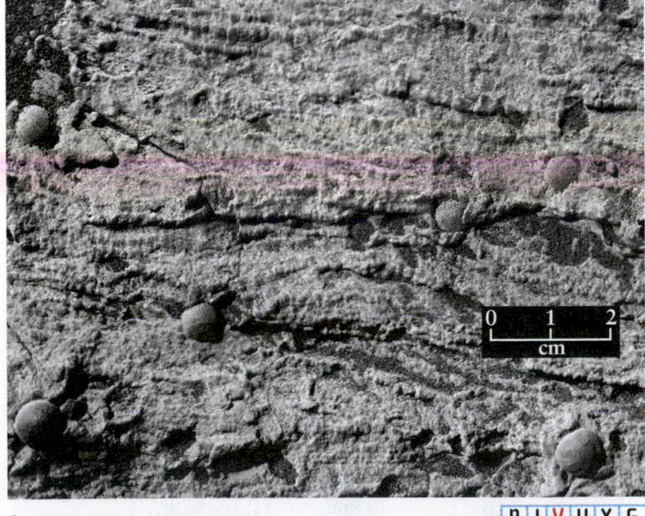

FIGURE 7-36 Layers of Rock Laid Down by Water (a) This close-up image taken by the rover *Opportunity* shows a small section of rock layers in a location called the Dells. The angled and curved layering seen here is created only on Earth by water flow, strongly suggesting that this sediment was also deposited by water. The nearly spherical rocks, called "blueberries" because they are dark, have been chemically identified as hematite, an iron-rich mineral that is usually formed in water. The rovers have found blueberries strewn in a wide variety of locations. (b) This iron meteorite, dubbed the "Heat Shield Rock" because it was discovered behind the Mars rover *Sojourner*'s heat shield, is surrounded by hematite blueberries. (a: NASA/JPL/USGS; b: NASA/JPL/Cornell)

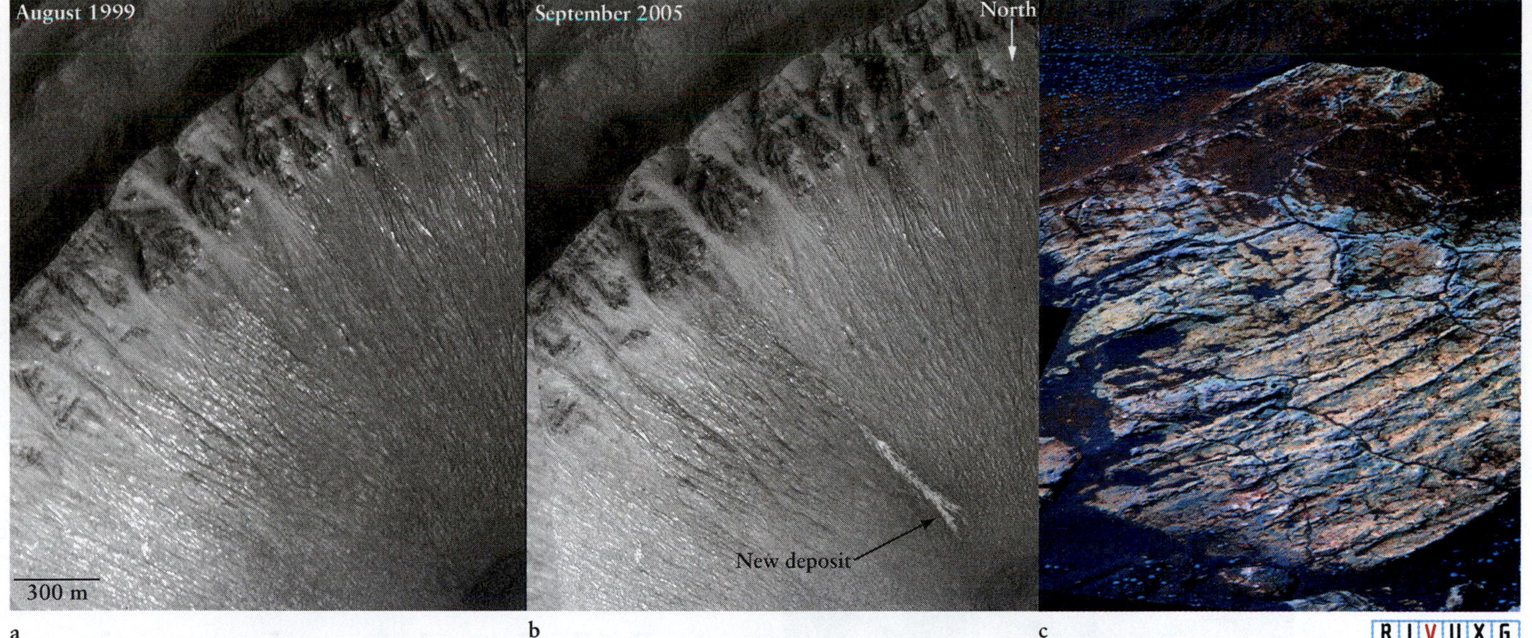

FIGURE 7-37 Martian Gullies (a, b) Two images of the same southern hemisphere crater on Mars taken 6 years apart. Whether the new deposit was initiated by ground vibration or by a short period of liquid water flow from inside Mars is still under investigation. (c) Polygonal cracks are visible in this *Opportunity* image of Escher Rock. They were believed to have formed when this area was flooded. Water seeped into the rock, which cracked as the water froze, expanded, and then evaporated away. (a and b: NASA/JPL/Malin Space Science Systems; c: NASA/JPL/Cornell)

begun sliding due to a small Marsquake (earthquake on Mars) or a tremor created by a meteorite impact. In 2008, astronomers using the *Mars Reconnaissance Orbiter* observed avalanches in progress close to Mars's north pole, where liquid water is even less likely to exist near the surface today (Figure 7-38).

In 2004, the rover *Opportunity* discovered Martian rock (Figure 7-37c) with polygon-shaped cracks similar to dried mud flats found on Earth. The best explanation for these features on Mars is that long after the lakes and oceans dried there, an impact released water from inside the planet that briefly reflooded the surface rock. The water seeped into the rock and upon freezing, the water expanded and cracked the rock. This scenario would require vast amounts of water inside Mars for several reasons. First, most of the liberated water heated by the impact would have left as vapor. Second, liquid water on the surface would not persist very long because of the low air pressure. When air pressure is very low, molecules easily escape from a liquid's surface, which would cause water to vaporize. Because the pressure of Mars's atmosphere is only 0.6% that of Earth's atmosphere, any liquid water on Mars today would quickly transform into ice or furiously boil and evaporate into the thin Martian air, and then be lost into space. Thus, if liquid water did cause the rock formations seen by *Opportunity*, there must have been a lot of it to start with, in order for there to be enough to cover the large areas of the Martian surface that have such cracks.

The total amount of water that once existed on Mars's surface is not known. However, by studying flood channels that exist there, geologists estimate that there was once enough water to cover the planet to a depth of 500 m (1500 ft). As noted earlier, Earth has enough water to cover our planet to a depth of 2.8 km (1.7 mi), assuming that Earth's surface was everywhere a uniform height.

Further evidence that water once flowed on Mars comes from so-called *SNC meteorites* found on Earth (Figure 7-39a). These space rocks are believed to have once been pieces of Mars because their chemistries are consistent with those of rocks studied on Mars's surface and because they contain trace gases in relative amounts found only in the current Martian atmosphere. The meteorites were ejected into space during especially powerful impacts on that planet's surface. They also contain water-soaked clay, which is not expected to be found on any objects in the solar system besides Mars, Earth, and Jupiter's moons Europa, Ganymede, and Callisto. At least 57 meteorites from Mars have been identified.

Some of the water that was on Mars is still near the surface today, frozen in several places on the red planet from its poles to its equator. Recent satellite measurements indicate that as much as 90% of the ice at the poles is water ice, the remainder being dry ice (frozen carbon dioxide). The temperature at Mars's poles, typically 160 K (−70°F), keeps the water permanently frozen, like the permafrost found in Siberia and Antarctica. Seasonal variations at Mars's poles (see Figure 7-29) are created by the freezing and evaporating of dry ice, which reaches thickness of 2 m. Indeed, during the

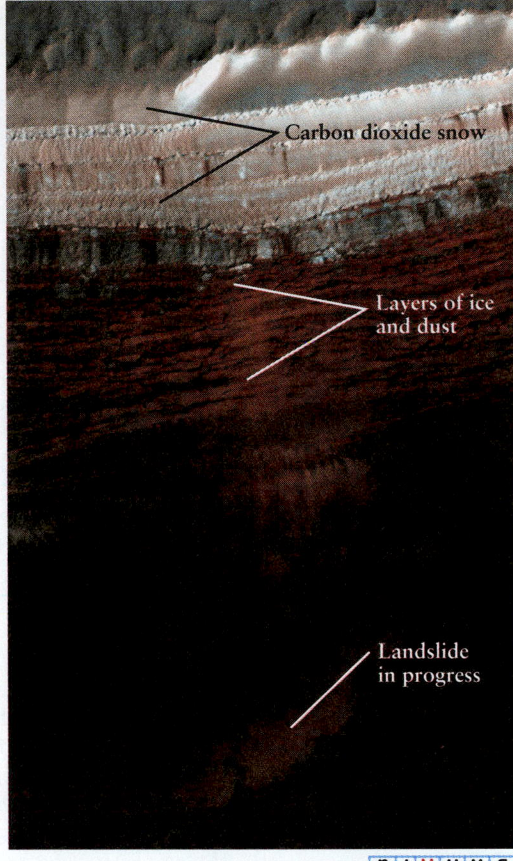

FIGURE 7-38 An Active Avalanche on Mars This image, taken by the *Mars Reconnaissance Orbiter*, shows an active landslide of falling ice, dust, and rock. This debris first hit the cliff wall about halfway down and then continued falling down the less steep slope, as seen in the bottom of the image. The cloud of debris near the bottom is about 180 m (590 ft) wide. (NASA/JPL-Caltech/University of Arizona)

winter, about one-third of the carbon dioxide in Mars's atmosphere precipitates down as dry ice.

The 1997 visit by *Pathfinder* and its little rover (see Figure 7-30b) to the Ares Vallis revealed evidence of several floods at the mouth of a dried flood plain. This evidence includes layers of sediment, clumps of rock and sand stuck together like similar groupings created by water on Earth, rounded rocks worn down as they were dragged by the water, and rocks aligned by water flow. The rover performed some 20 chemical analyses of rocks in the landing area. The soil at the site is a rusty color, laden with sulfur. The source of this sulfur is a mystery, as is the large amount of silicon-rich rock the rover explored.

There is observational evidence of an ancient shoreline encircling the northern lowlands of Mars. The belief that the northern lowlands were an ocean long ago is supported by the analysis of a 1.2-billion-year-old meteorite from Mars. This rock has remnants of salt believed to have been deposited there when the Martian ocean dried up.

7-13 Search for microscopic life on Mars continues

The discovery in the 1960s that water once existed on the surface of Mars rekindled speculation about Martian life. Although it was clear that Mars has neither civilizations nor fields of plants, microbial life-forms still seemed possible. Searching for signs of organic matter was one of the main objectives of the ambitious and highly successful *Viking* missions.

The two *Viking* spacecraft were launched during the summer of 1975. Each consisted of two modules—an orbiter and a lander. Almost one year later, both *Viking* landers set down on rocky plains north of the Martian equator.

The landers confirmed the long-held suspicion that the red color of the planet is due to large quantities of

a

b

FIGURE 7-39 A Piece of Mars on Earth Mars's thin atmosphere does little to protect it from impacts. Some of the debris ejected from impact craters there apparently traveled to Earth. (a) An SNC meteorite recovered in Antarctica. The 4.1-billion-year-old meteorite, designated ALH84001, shows strong evidence of having been exposed to liquid water on Mars, perhaps for hundreds of years. (b) Possible fossil remains of primitive bacterial life on Mars, although theories of nonbiological origins have also been presented. (a: NASA; b: NASA/JPL) iron in its soil. Despite the high iron content of its crust, Mars has a lower average density (3950 kg/m^3) than that of other terrestrial planets (this value is more than 5000 kg/m^3 for Mercury, Venus, and Earth). Mars must therefore contain overall a smaller percentage of iron than these other planets.

Each *Viking* lander was able to dig into the Martian regolith and retrieve rock samples for analysis, which showed that the rocks at both sites are rich in iron, silicon, and sulfur. The Martian regolith (Figure 7-40) ranges in consistency from clay to sand, although its chemistry is different from clay and sand on Earth. The *Viking* landers each carried a compact biological laboratory designed to test for microorganisms in the Martian soil. Three biological experiments were conducted, each based on the idea that living things alter their environment: They eat, they breathe, and they release waste products. During each experiment, a sample of the Martian regolith was placed in a closed container, with or without a nutrient substance. The container was then examined for any changes in its contents.

The first data returned by the *Viking* biological experiments caused great excitement. In almost every case, rapid and extensive changes were detected inside the sealed containers. However, further analysis showed that these changes were due solely to nonbiological chemical processes. Apparently, the Martian regolith (Figure 7-40) is rich in chemicals that effervesce (fizz) when moistened. A large amount of oxygen is tied up in the regolith in the form of unstable chemicals called *peroxides* and *superoxides,* which break down in the presence of water to release oxygen gas.

The chemical reactivity of the Martian regolith comes from the Sun's ultraviolet radiation and from electric activity inside dust devils. Ultraviolet photons easily break apart molecules of carbon dioxide (CO_2) and water vapor (H_2O) by knocking off oxygen atoms, which then become loosely attached to chemicals in the regolith. Ultraviolet photons also produce highly reactive ozone (O_3) and hydrogen peroxide (H_2O_2) near the planet's surface, which become incorporated in the regolith. Dust devils suck up lots of dust from the planet's surface and transform it into a variety of compounds.

Which ice on Mars changes more actively today, carbon dioxide ice or water ice?

Here on Earth, hydrogen peroxide is commonly used as an antiseptic. When you pour this liquid on a wound, it fizzes and froths as the loosely attached oxygen atoms chemically combine with organic material and destroy germs. The *Viking* landers may have failed to detect any organic compounds on Mars because the superoxides and peroxides in the Martian regolith act as antiseptics there, making it sterile.

In 2003, astronomers began detecting methane in 6
Mars's atmosphere. This discovery is intriguing, because sunlight breaks down methane within a few centuries. Therefore, the gas needs to be continuously replenished. Indeed, plumes of methane emerging from the planet's interior have now been observed. One of these plumes was observed to contain 19,000 tons of the gas. This simple compound, CH_4, is often a by-product of biological activity (cows come to mind in this regard). Methane is also trapped inside planets, like Earth and possibly Mars, during their formation and thereafter leaks out. Although the discovery of methane on Mars hints that simple life might be active under the planet's surface today, it is by no means positive proof that such life exists.

Possible signs of ancient life on Mars have been discovered here on Earth. When cut open, several of the Martian meteorites have shown microscopic features that could be fossils of Martian bacterial life and their excretions (see Figure 7-39b). These features, no larger than 500 nm (1/100 the diameter of a human hair), are 30 times smaller than bacteria found on Earth. Detailed analysis of the meteorites in 2001 revealed the presence of several organically created features, including tiny spheres similar to those found with some Earth bacteria, and magnetite crystals, which are also used by some bacteria on Earth as compasses to find food. The debate as to whether the features in these meteorites are fossil evidence of life on Mars has raged unabated for several years, and it is not yet resolved. (It is worth noting that similar claims of fossils in meteorites from Mars in the 1960s have been disproved.)

The search for microscopic life on Mars is not over. If there is liquid water under the Martian surface, then it is entirely possible that life has evolved in it and may still exist. This belief is based on the incredibly wide range of places that life has developed in water on Earth, from volcanic vents on the bottoms of oceans, to geysers, to lakes under the Antarctic polar ice cap.

Sulfur salts

R I V U X G

FIGURE 7-40 Regolith of Mars The rover *Spirit's* wheels churned up the regolith of Mars, revealing sulfur-based salts just below the surface. (NASA/JPL/Cornell)

7-14 Mars's two moons look more like potatoes than spheres

Two tiny, irregularly-shaped moons orbit close to Mars's surface. *Phobos* (Greek, meaning "fear") and *Deimos* ("panic") are so small that they were not discovered until 1877. In mythology, these names are either Mars's sons by Aphrodite or the two horses that pulled Mars's war chariot—take your pick. Potato-shaped Phobos is the inner and larger of the two moons (Figure 7-41a). It is heavily cratered, and its surface has been transformed into dust at least 1 m thick by countless tiny impacts over the eons. Football-shaped Deimos is less cratered than Phobos (Figure 7-41b).

> What does the observation that so much ultraviolet radiation from the Sun reaches the surface of Mars indicate about the Martian atmosphere?

The gravitational deflections by Phobos of spacecraft passing near it have been used to determine the moon's total mass. Measurements of Phobos's mass and volume made in 2010 suggest that it may be very porous, with up to 30% of its interior composed of empty space.

Both Phobos and Deimos orbit in the same direction that Mars rotates, just as our Moon orbits Earth. However, Phobos is so close to the planet that this moon goes around 3 times in 1 Martian day. This rapid revolution creates the unusual situation wherein Phobos rises in the west, races across the sky in only 5½ hours as seen from Mars's equator, and sets in the east. As seen from Mars, Deimos rises in the east and takes about 3 Earth days to creep from one horizon to the other. Phobos creates tiny tides on Mars, which in turn pull that rapidly orbiting moon inward (just the opposite effect of Earth on our more slowly orbiting Moon). Phobos, descending at a rate of 1.8 m/Earth year, will either strike Mars or be pulled apart and become a ring within 50 million years.

Phobos and Deimos were not formed like our Moon, which condensed from debris that "splashed" into orbit as the result of a collision between our planet and another large body. For decades Phobos and Deimos were both believed to be captured asteroids (which may still be true for Deimos). While some asteroids have densities as low as that of Phobos, astronomers have trouble showing how Phobos could have been captured and put into its relatively circular orbit without being pulled apart. A viable explanation of Phobos's presence is still under development. Both moons are in synchronous rotation as they orbit Mars.

Insight Into Science

Imagine the Moon The definitions we use for words based on our everyday experience often fail us in astronomy. For example, the word *moon* usually creates an image of a spherical body, like our Moon. In reality, most of the moons in the solar system have very irregular shapes, like Phobos and Deimos.

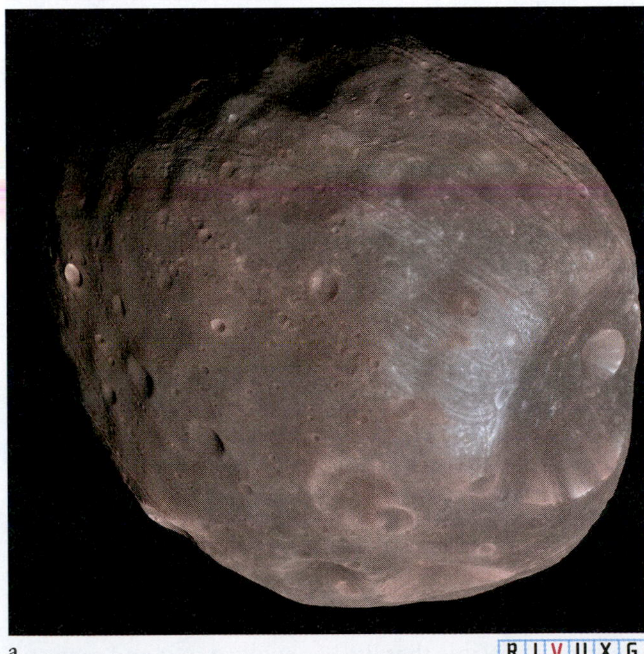

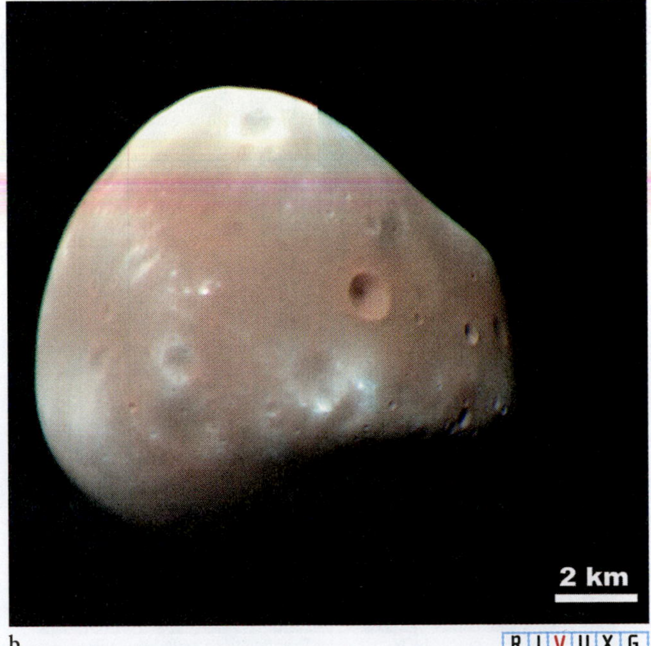

FIGURE 7-41 Phobos and Deimos (a) Phobos, the larger of Mars's two moons, is potato-shaped and measures approximately 28 × 23 × 20 km. It is dominated by crater Stickney (on the right of the image), named for discoverer Asaph Hall's wife (Angeline Stickney). (b) Deimos is less cratered than Phobos and measures roughly 16 × 12 × 10 km. (a: NASA/JPL-Caltech/University of Arizona; b: NASA/JPL-Caltech/University of Arizona)

7-15 Comparisons of planetary features provide new insights

Now that we have examined the terrestrial planets individually, it can be useful to see how their various features compare to each other and to Earth. Table 7-1 (The Inner Planets: A Comparison) summarizes much of this material.

Size and Mass Earth is the largest and most massive of all four terrestrial planets. In this regard, Venus is almost the sister planet to Earth, with nearly 95% of Earth's diameter and 82% of Earth's mass. Although Mars is most similar to Earth in other ways, such as its history of surface water and rotation rate, it is only about half (53%) of Earth's diameter and 11% of Earth's mass. Mercury, with 38% of Earth's diameter and a scant 5.5% of Earth's mass, is much closer in size to Earth's Moon than it is to the other terrestrial planets; Mercury is only 1.4 times bigger than the Moon.

Surface Features Although all four terrestrial planets have craters, only Mercury has them in large numbers, similar to what we see on our Moon. Venus has erased most of its craters by melting them or covering them with magma. Mars has removed many of its craters as a result of erosion and weather. Earth has removed most of its craters by tectonic plate motion and weathering. Like Earth, Venus has continentlike plateaus and ocean bottomlike lowlands. Mars has a lower northern cap than the rest of the planet, but no real continentlike regions. Mercury has relatively uniform height. While all four terrestrial planets have volcanoes, those of Mercury and Mars are extinct. Water erosion occurred just on Mars and Earth.

Interior The interior chemistries of the four terrestrial planets are similar, with cores consisting primarily of iron surrounded by rock. Mercury, Venus, Earth, and Mars all have partially molten cores. Although Mercury is the smallest terrestrial planet, it has the highest density, meaning that it has the highest percentage of iron of these (and, in fact, of any) planets. This high percentage is probably caused by Mercury losing more of its outer, rocky layer, probably as a result of impacts, than did any other terrestrial planet.

Water Earth contains by far the highest percentage of water of the terrestrial planets. Mars contains water frozen near its surface, and we have yet to determine whether it has any liquid water deep inside. Venus contains very little water compared to either Earth or Mars because Venus is so hot that it has evaporated surface water into its atmosphere, and water in its interior has probably been mostly ejected through volcanoes or when the surface periodically melts. Mercury apparently has some water (very little compared to Earth or Mars) frozen at its poles, the result of collisions with water-rich comets. Because its interior is so iron rich, the water-bearing layers were probably blasted into space by impacts early in Mercury's existence.

> Would our Moon have to be closer or farther away to orbit in the same direction, but rise in the west?

Atmosphere Venus has by far the densest atmosphere of the terrestrial planets, with about 90 times as much gas as the air we breathe. Furthermore, Venus's atmosphere is composed primarily of carbon dioxide, with a minor component of nitrogen. The thick atmosphere has protected Venus's surface from all but the most massive infalling space debris. Its thick atmosphere and the possibility that Venus's surface periodically melts, explains why Venus has few impact craters. Venus's atmosphere is most similar in composition to the air around Mars, although the density of Mars's air is only 0.6% as dense as the air we breathe. Although Mars's thin atmosphere has enabled many pieces of space debris to strike the planet and form craters, these craters are continually being eroded, primarily by wind.

Earth's atmosphere, once very similar to that of Venus, was transformed by water and life into the nitrogen-oxygen atmosphere we have today. As with Venus, Earth's atmosphere protects the surface from most impacts. Many craters have been removed by our planet's plate tectonic motion, which apparently does not occur today on any other planet. Mercury's gravity is too low to hold any gases as a permanent atmosphere. That planet is surrounded by a very thin atmosphere of transient gases from the Sun, the planet's interior, and possibly from polar ice. As these gases drift into space, they are replaced by fresh gas. Because its atmosphere is so thin and its surface has been unchanged by internal activity for billions of years, Mercury is the most heavily cratered of the terrestrial planets.

Temperature Some of the temperatures on the terrestrial worlds are surprising at first glance. Mercury, which is closest to the Sun, has a hot daytime surface of about 700 K (800°F). Its lack of atmosphere allows a lot of this heat to escape at night, bringing its nighttime temperature down to a frigid 100 K (−280°F), much colder than on any other terrestrial planet. Venus's thick atmosphere creates a greenhouse effect that keeps that planet at 750 K (890°F), even hotter than Mercury. Earth's surface temperature ranges from about 330 K (140°F) to 180 K (−130°F), and Mars is only slightly colder, with temperatures ranging from 280 K (45°F) down to 160 K (−170°F).

Rotation and Magnetic Fields All of the terrestrial planets rotate, with Earth's solar day being shortest at 24 h. This motion, combined with Earth's liquid iron core, creates a strong magnetic field that surrounds our planet. Mars has virtually the same solar day of 24 h 39 min, but its molten iron core is much smaller than that of Earth and it has no global magnetic field, only local magnetic fields. The solar days of Venus (117 Earth

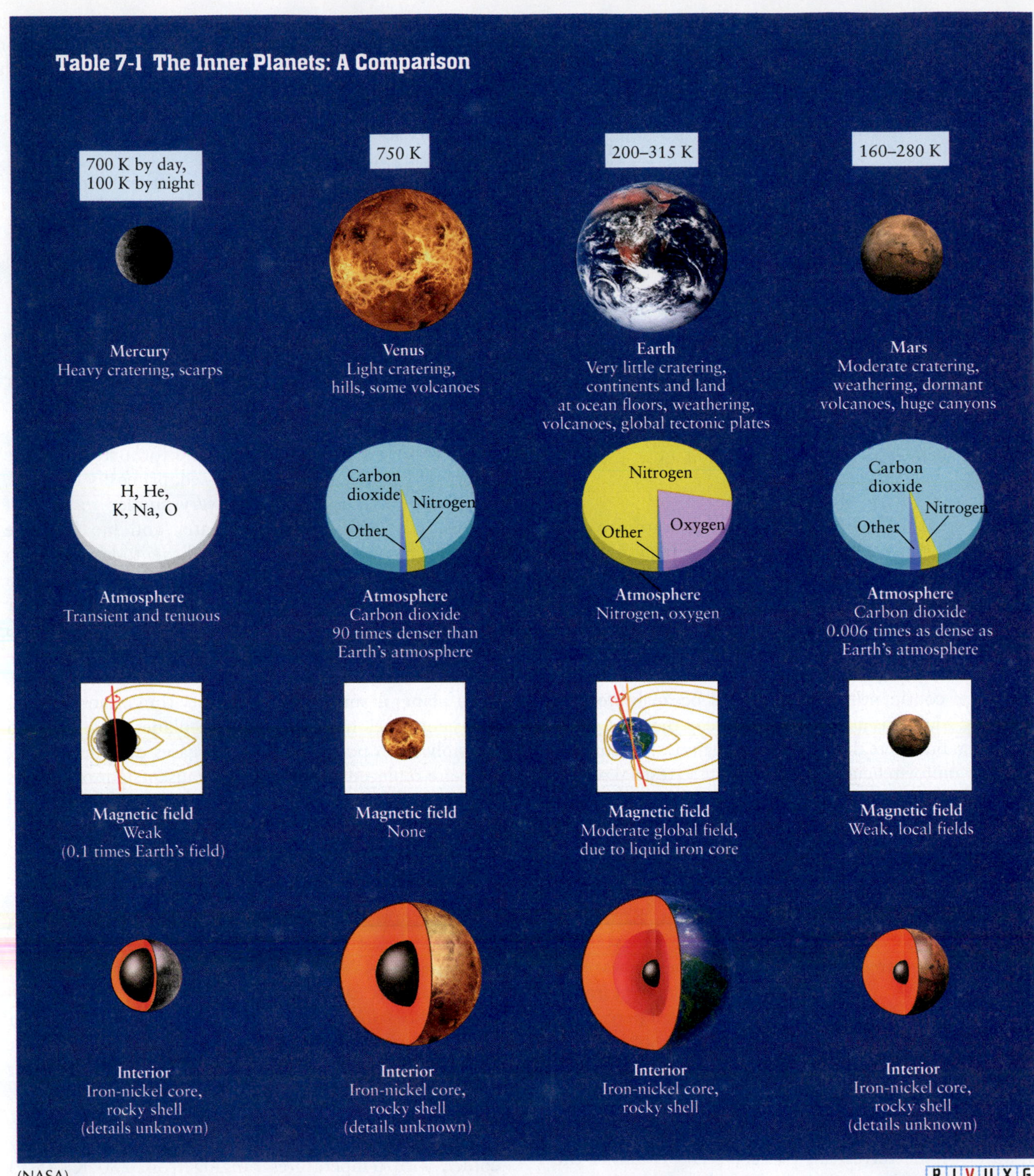

(NASA)

R I V U X G

days) and Mercury (176 Earth days) are both extremely long by Earth standards, and, indeed, a solar day on Mercury is 2 Mercurian years long! No magnetic field has been detected around Venus. Mercury has a weak global field that apparently results from the extremely high amount of iron it contains.

What does the surface of Venus lack, without which its atmosphere cannot transform into one like Earth's?

7-16 Frontiers yet to be discovered

All three terrestrial planets orbiting the Sun with us have much to reveal. What is the chemical composition of Mercury's surface rocks? What is the planet's cooling history, and how much of its core is molten? Is there really ice at its poles? If so, how did it get there? What will the planet's internal structure reveal?

Which volcanoes on Venus are active? Are they the source of sulfur compounds in Venus's atmosphere? Can we find observational evidence to support the conjecture that Venus's rotation axis has flipped over? Likewise, can we find further evidence that its surface periodically undergoes significant recovering?

Some of the most intriguing questions about our solar system focus on Mars. Did life ever exist there? Is there liquid water under the red planet's surface today? If so, how far along did life evolve in that water? What are the similarities and differences between such life and life on Earth? Significant similarities might imply a common origin. Furthermore, what is the surface water history of Mars? What causes its local magnetic fields? What does its interior look like? Where did its moons come from? It is likely that we will have answers to many of these questions in the coming decades.

SUMMARY OF KEY IDEAS

All four inner planets are composed primarily of rock and metal, and thus they are classified as terrestrial.

Mercury

- Even at its greatest orbital elongations, Mercury can be seen from Earth only briefly after sunset or before sunrise.
- The Mercurian surface is pocked with craters like the Moon's, but extensive, smooth plains lie between these craters. Long cliffs meander across the surface of Mercury. These scarps probably formed as the planet cooled, solidified, and shrank.
- The long-ago impact of a large object formed the huge Caloris Basin on Mercury and shoved up jumbled hills on the opposite side of the planet. Several similar, but smaller, impact features also exist on Mercury.
- Mercury has an iron core, which fills more of its interior than Earth's core fills Earth.
- Mercury has a weak, global magnetic field that partially shields it from the solar wind.

Venus

- Venus is similar to Earth in size, mass, and average density, but it is covered by unbroken, highly reflective clouds that conceal its other features from observers using visible-light telescopes.
- Although most of Venus's atmosphere is carbon dioxide, its dense clouds contain droplets of concentrated sulfuric acid mixed with yellowish sulfur dust. Active volcanoes on Venus are likely to be a constant source of this sulfurous veil.
- Venus's exceptionally high temperature is caused by the greenhouse effect, as the dense carbon dioxide atmosphere traps and retains heat emitted by the planet. The surface pressure on Venus is 90 atm, and the surface temperature is 750 K. Both temperature and pressure decrease as altitude increases.
- The surface of Venus is surprisingly flat and mostly covered with gently rolling hills. There are two major "continents" and several large volcanoes. The surface of Venus shows evidence of local tectonic activity but not the large-scale motions that play a major role in continually reshaping Earth's surface.

Mars

- Earth-based observers found that the Martian solar day is nearly the same as that of Earth, that Mars has polar ice caps of carbon dioxide snow that expand and shrink with the seasons, and that the Martian surface undergoes seasonal color changes.
- A century ago, observers reported networks of linear features that many perceived as canals. These observations led to speculation about self-aware life on Mars.
- The Martian surface has many flat-bottomed craters, several huge volcanoes, a vast equatorial canyon, and dried-up riverbeds—but no canals formed by intelligent life. River deltas and dry riverbeds on the Martian surface indicate that large amounts of water once flowed there.
- Liquid water would quickly boil away in Mars's thin present-day atmosphere, but the planet's polar ice caps contain significant quantities of frozen water, and a layer of permafrost exists beneath parts of the regolith.
- The Martian atmosphere is composed mostly of carbon dioxide. The surface pressure is less than 0.01 atm.
- Chemical reactions in the regolith, together with ultraviolet radiation from the Sun, apparently act to sterilize the Martian surface.
- Mars has no global magnetic fields, but local fields pierce its surface in at least nine places.
- Mars has two irregularly shaped moons, Phobos and Deimos. Both are in synchronous rotation with Mars. How they came into orbit is still under investigation.

A WHAT DID YOU THINK?

1. *Which terrestrial planet—Mercury, Venus, Earth, or Mars—has the coolest surface temperature?* The nighttime side of Mercury, closest planet to the Sun, is the coldest surface of any terrestrial planet.

2. *Which planet is most similar in size to Earth?* Venus is most similar to Earth in size.

3. *Which terrestrial planet—Mercury, Venus, Earth, or Mars—has the hottest surface temperature?* Venus is hottest, its temperature raised above that of Mercury because of the greenhouse effect in Venus's atmosphere.

4. *What is the composition of the clouds that surround Venus?* The clouds are made primarily of sulfuric acid.

5 *Does Mars have liquid water on its surface today? Did it have liquid surface water in the past?* Mars has no liquid surface water today, but there are very strong indications that it had liquid water on its surface in the past.

6 *Is life known to exist on Mars today?* No current life has yet been discovered on Mars, but it may exist in underground water oceans.

Key Terms for Review

3-to-2 spin-orbit coupling, 204
caldera, 217
dust devil, 219
greenhouse effect, 208
hot-spot volcanism, 217
northern vastness (northern lowlands), 213
retrograde rotation, 211
southern highlands, 213

Review Questions

1. Which pair of planets have atmospheres with the most similar chemical compositions? **a.** Earth and Venus, **b.** Earth and Mars, **c.** Venus and Mars, **d.** Mercury and Mars, **e.** Mercury and Venus

2. Which planet is least likely to have water ice on or just under its surface? **a.** Earth, **b.** Mercury, **c.** Mars, **d.** Venus

3. Which object is most similar to Venus in mass and diameter? **a.** Earth, **b.** Mars, **c.** Mercury, **d.** our Moon

4. Why is Mercury so difficult to observe? When is the best time to see the planet? (*Hint: Guided Discovery: The Inner Solar System* in Section 7-5 can help.)

5. Compare the surfaces of Mercury and our Moon. How are they similar? How are they different?

6. Compare the interiors of Mercury and Earth. How are they similar? How are they different?

7. To better understand the interiors of Mercury and Earth, do Interactive Exercise 7.1 on the Web. You can print out the result, if requested.

8. What are the longest features found on Mercury? Why are the examples of this feature probably much older than tectonic features on Earth?

9. Briefly describe a scientific theory explaining why Mercury has such a large iron core.

10. Astronomers often refer to Venus as Earth's twin. What physical properties do the two planets have in common? In what ways are the two planets dissimilar?

11. Why is it hotter on Venus than on Mercury?

12. What evidence exists for active volcanoes on Venus?

13. Describe the Venusian surface. What kinds of geologic features would you see if you could travel around the planet?

14. Why do astronomers believe that Venus's surface was not molded by the kind of continuous tectonic activity that shaped Earth's surface?

15. Why is Mars red?

16. What is the greenhouse effect? What role does it play in the atmospheres of Venus and Earth?

17. During which configuration of planets is it easiest to observe Mars from Earth? (*Hint: Guided Discovery: The Inner Solar System* in Section 7-5 can help.)

18. Compare the cratered regions of Mercury, the Moon, and Mars. Assuming that the craters on all three worlds originally had equally sharp rims, what can you conclude about the environmental histories of these worlds?

19. How would you tell which craters on Mars were formed by meteoritic impacts and which by volcanic activity?

20. What geologic features indicate that plate tectonic activity once occurred on Mars? What features created by tectonic activity on Earth are not found on Mars?

21. To better understand the surface features of Mars, do Interactive Exercise 7.2 on the Web. You can print out your results, if requested.

22. To compare the surfaces of Mercury, Venus, and Mars, do Interactive Exercise 7.3 on the Web. You can print out your results, if requested.

23. What evidence have astronomers accumulated that liquid water once existed in large quantities on Mars's surface? What evidence is there that water still exists, under its surface?

Advanced Questions

24. What evidence do we have that the surface features on Mercury were not formed during recent geologic history?

25. Venus takes 440 days to move from greatest western elongation to greatest eastern elongation, but it needs only 144 days to go from greatest eastern elongation to greatest western elongation. With the aid of a diagram, explain why.

26. As seen from Earth, the brightness of Venus changes as it moves along its orbit. Describe the main factors that determine Venus's variations in brightness as seen from Earth. (*Hint:* See the discussion of Venus in Chapter 2.)

27. How might Venus's cloud cover change if all of Venus's volcanic activity suddenly stopped? How might these changes affect the overall Venusian environment?

28. Compare Venus's continents with those on Earth. What do they have in common? How are they different?

29. Explain why Mars has the longest synodic period of all of the planets, although its sidereal period is only 687 days.

30. With carbon dioxide accounting for about 95% of the atmospheres of both Mars and Venus, why is there little greenhouse effect on Mars today?

Discussion Questions

31. If you were planning a new *Messenger* mission to Mercury, what features and observations would be of particular interest to you and why?

32. If you were designing a space vehicle to land on Venus, what special features would be necessary? In what ways would this mission and landing craft differ from a spacecraft designed for a similar mission to Mercury?

33. Suppose someone told you that the *Viking* mission failed to detect life on Mars simply because the tests were designed to detect terrestrial life-forms, not Martian life-forms. How would you respond?

34. Compare the scientific opportunities for long-term exploration offered by the Moon and Mars. What difficulties would there be in establishing a permanent base or colony on each of these two worlds?

35. Imagine you are an astronaut living at a base on Mars. Describe your day's activities, what you see, the weather, the spacesuit you are wearing, and so on.

What If...

36. Mercury had synchronous rotation? How would the temperatures on such a planet be different than they are today? Where would humans set up camp on such a world?

37. Venus could support self-aware life, but it still had permanent cloud cover? In what ways would life there be different than it is here? How would their perceptions of the cosmos be different from ours?

38. Mars had the same mass, surface features, and atmosphere as Earth? In what ways would life there be different than it is here?

39. Mars rotated once every 20 Earth days rather than once every 1.026 Earth days? What would be different on the alternate Mars?

40. The carbon dioxide content of Earth's atmosphere increased? What would happen to Earth? (This is not a completely hypothetical question, because carbon dioxide levels are increasing today.)

Web Questions

41. Search the Web for the latest information about the *Messenger* mission to Mercury. Exactly when was it launched? What scientific experiments does it carry? What scientific issues are these instruments intended to resolve? How long is it expected to function in orbit around Mercury?

ANIMATION 7.1

42. Elongations of Mercury Access the animation "Elongations of Mercury" in Chapter 7 of the *Discovering the Universe* Web site. **a.** Note the dates of the greatest eastern and western elongations in the animation. Which time interval is greater: from a greatest eastern elongation to a greatest western elongation, or vice versa? **b.** Based on what you observe in the animation, draw a diagram to explain your answer to the question in **a.**

43. Search the Web for the latest information about proposed future missions to Venus. What scientific experiments will they carry? What scientific issues are these instruments intended to resolve?

44. In 1999, two NASA spacecraft—*Mars Climate Orbiter* and *Mars Polar Orbiter*—failed to reach their destinations. Search the Web for information on these missions. What were their scientific goals? How and why did the missions fail? Which current and future missions, if any, are intended to replace these missions?

45. Search the Web for information about possible manned missions to Mars. How long would such a mission take? How expensive would they be? What are some advantages and disadvantages of a manned mission compared to an unmanned one?

ANIMATION 7.8

46. Conjunctions of Mars Access and view the animation "The Orbits of Earth and Mars" in Chapter 7 of the *Discovering the Universe* Web site. **a.** The animation highlights three dates when Mars is in opposition, so that Earth lies directly between Mars and the Sun. By using the **Stop** and **Play** buttons in the animation, find 2 times during the animation when Mars is in *conjunction*, so that the Sun lies directly between Mars and Earth (see Figure 2-4). For each conjunction, make a drawing showing the positions of the Sun, Earth, and Mars, and record the month and year when the conjunction occurs. **b.** When Mars is in conjunction, at approximately what time of day does it rise as seen from Earth? At what time of day does it set? Is Mars suitably placed for telescopic observations when it is in conjunction?

Observing Projects

STARRY NIGHT

47. Mercury can be seen in the eastern sky at dawn for a few days around greatest western elongation. It can also be seen in the western sky at dusk for a few days around greatest eastern elongation. The following table shows the dates of the next few greatest elongations of Mercury. Magazines such as *Sky & Telescope* or *Astronomy* will tell you which, if any, of these greatest elongations will be especially favorable for viewing the planet. Alternatively, you can use *Starry Night*™ to examine these elongations. Select **Favourites > Discovering the Universe > A fast little world.** This view shows Mercury and its orbital path around the Sun. (You can remove the blue sky by clicking on **View > Hide Daylight** to show this motion against the background stars.) Click the **Stop** button to stop Mercury's orbital motion. Note how close Mercury stays to the Sun during its orbital journey. **a.** Use the **Angular Separation** tool to measure the approximate angular separation between the Sun and the maximum positions of Mercury from the Sun along its orbit. **b.** The table below lists times of

Mercury's greatest elongation for the next few years. You can use *Starry Night*™ to decide which of these times are favorable for observation, where the planet is at its highest above the horizon. For each of the times listed in the table below for Mercury's **Greatest western elongation**, move to the appropriate **Date** and click the **Sunrise** button in the toolbar to place the Sun on the horizon. Right-click over Mercury and center on the planet. You can tabulate Mercury's altitude, as presented in the **Gaze** box in the upper right of the window, for each of these listed times of greatest elongation and decide which of these will be favorable for viewing. Which of these times of greatest eastern elongation are most preferable in the next 4 years? **c.** Why is Mercury at different altitudes (angles above the horizon) on different dates at these sunrise times? **d.** Now, adjust the **Date** to each of the listed dates for **Greatest eastern elongation**, click the **Sunset** button, and tabulate Mercury's altitude to determine which of these events are favorable dates for observing Mercury in the evening. Set the **Time Flow Rate** to 1 minute at any of these times and click **Play** to observe Mercury's descending motion with respect to the horizon. What are the favorable dates for observing Mercury at greatest eastern elongation in the next 4 years?

If an elongation is favorable, make plans to observe this innermost planet of the solar system, a planet that is rarely seen by anyone but an astronomer. Set aside several evenings (or mornings) around the date of the favorable elongation to reduce the chances of being "clouded out." Select an observing site that has a clear, unobstructed view of the horizon where the Sun sets (or rises). Make arrangements to have a telescope at your disposal, if possible. Search for the planet on the dates you have selected and make a drawing of its appearance through your telescope.

Greatest Elongations of Mercury,* 2011–2014

Year	Greatest western (morning) elongations	Greatest eastern (evening) elongations
2011	January 9, May 7, September 3, December 23	March 23, July 20, November 14
2012	April 18, August 16, December 4	March 5, July 1, October 26
2013	March 31, July 30, November 18	February 16, June 12, October 9
2014	March 14, July 12, November 1	January 31, May 5, September 21

*Mercury can be seen in the eastern sky at dawn for a few days around greatest western elongation. It can be seen in the western sky at dusk for a few days around greatest eastern elongation.

48. Venus can be seen in the eastern sky in the hours before dawn for several months around greatest western elongation. It can also be seen in the western sky in the hours after dusk for several months around greatest eastern elongation. Refer to the following table to see if Venus is currently near a position of greatest elongation. If so, view the planet through a telescope. Make a sketch of the planet's appearance. From your sketch, can you determine whether Venus is closer to us or farther from us than the Sun?

Greatest Elongations of Venus,* 2011–2014

Year	Greatest western elongations	Greatest eastern elongations
2011	January 8	(none)
2012	March 27	August 15
2013	(none)	November 1
2014	March 22	(none)

*Venus can be seen in the eastern sky in the hours before dawn for several months around greatest western elongation. It can be seen in the western sky in the hours after dusk for several months around greatest eastern elongation.

49. Observe Venus through a telescope once a week for a month and make a sketch of the planet's appearance on each occasion. From your sketches, can you determine whether Venus is approaching us or moving away from us?

50. Consult such magazines as *Sky & Telescope* and *Astronomy*, or use *Starry Night*™ to determine Mars's location among the constellations. If Mars is suitably placed for observation, arrange to view the planet through a telescope and sketch this planet, noting particularly its phase. What magnifying power seems to give you the best image? Can you distinguish any surface features? Can you see a polar cap or dark markings? If not, can you offer an explanation for Mars's bland appearance?

51. You can use *Starry Night*™ to follow Mercury's orbit as it orbits the Sun. Select **Favourites > Discovering the Universe > Mercury's Motion**. The view is from the center of a transparent Earth and is centered upon the Sun. **a.** With the **Time Flow Rate** at 3 hours, click the **Play** button in the toolbar to watch the motion of the planet Mercury against the background stars as time progresses and describe this motion. Pay particular attention to the relative speeds when Mercury is passing near to the Sun on east-to-west and west-to-east passes, respectively. **b.** Click the **Stop** button and change the **Time Flow Rate** to **5 sidereal days**. Right-click over Mercury and select **Centre** from the contextual menu to place Mercury at the center of the view. Single-step time and follow the planet's motion around the Sun, stopping occasionally to zoom in and determine Mercury's phase. Describe the pattern of phase changes and relate these changes to the position of Mercury with respect to the Sun.

STARRY NIGHT **52.** You can use *Starry Night*™ to monitor the motion of Mars as seen from Earth, as the red planet moves against the background stars. Select **Favourites > Discovering the Universe > Motion of Mars**. This view from the center of a transparent Earth is locked upon Mars in the center of the view. With the **Time Flow Rate** at 1 sidereal day, click the **Play** button in the toolbar to watch Mars move against the background stars. You will see that Mars begins to slow down against the background stars as time progresses. Click the **Stop** button when Mars appears almost to stop in its motion. Right-click on a nearby star and select **Centre** from the contextual menu to lock the view upon this background star. Click the **Play** button and describe how Mars now moves relative to the background stars. Run time backward until Mars is at its stationary point furthest to the east (leftward) position and stop time again. Change the **Time Flow Rate** to 5 sidereal days and single-step time forward, tracing Mars's motion on tracing paper or transparent film with a soft felt pen. What do we call this odd motion of Mars? What would you expect its phase to be at the midpoint between the two stationary points at which East-West motion ceases for a short time? Use the time controls in the toolbar to move to the time at which Mars is approximately at this mid-position, center upon this planet, and check your conclusion by zooming in on Mars.

STARRY NIGHT **53.** In this exercise, you can use *Starry Night*™ to examine the apparent size of the Sun in the sky from the positions of all of the inner planets, starting with Mercury. In order to record your measurements, draw up a table with the following columns: Planet Name, Angular Radius of the Sun in arcminutes and arcseconds, Angular Radius of the Sun in arcseconds, Ratio of Sun Angular Radius to That as Seen from Earth, and Semi-Major Axis of Orbit. (The Semi-Major Axis values can be obtained by clicking on **Find**, right-clicking over each planet's name, and clicking on **Show Info** in the list that appears. The planet's average orbital radius is shown as **Orbit size**.) Open the **Find** pane and ensure that the **Query** box is empty in order to display a list of the planets. Stop time advance. **a.** Click on the **Down** arrow to the left of Mercury in the **Find** pane to open the drop-down list and select **Go There** to go to the surface of this planet. Then double-click the listing for the **Sun** in the **Find** pane to center the view on the Sun. Now **Zoom** in (+) until the Sun fills the view. Use the angular separation tool to measure the angular radius of the Sun from this viewpoint. (Place the angular separation tool over the Sun's center, hold down the mouse button and move the cursor to the edge of the Sun.) Record the value of this radius, displayed in arcminutes and arcseconds, in your table. **b.** Repeat the procedure in part **a** for Venus, Earth, and Mars in turn, noting the apparent angular size of the Sun from each planet. Qualitatively, how does the apparent size of the Sun change when going from Mercury to Mars? **c.** You can now find a formal relationship between the angles you measured and the distances of the planets from the Sun. Convert the angular solar radii you have recorded into arcseconds and enter these values in the appropriate column of your table (1 arcminute = 60 arcseconds). Divide the solar angular radius as measured from Earth by its radius as measured from Mercury and record this number. Similarly, divide the Sun's radius at Earth by the Sun's radius at Venus, then divide the Sun's radius at Earth by the Sun's radius at Earth (this should be 1, of course), and finally, divide the Sun's radius at Earth by the Sun's radius at Mars. How do these ratios compare with the orbital radii of the planets? (Can you verify that these ratios are nearly equal to the values of semi-major axes in your table for each of the planets?) If the measurements are not exactly the semi-major axes, explain why. (*Hint*: Draw triangles with the Sun's radius as one side and the angle from the planet as the opposite vertex and use simple trigonometry.)

STARRY NIGHT **54.** In this exercise, you can use *Starry Night*™ to compare the orbits of Mercury and the Earth. Open **Favourites > Discovering the Universe > Mercury and Earth**. The view, from a location about 2.3 AU from the Sun, shows the orbits of Mercury and Earth. Use the location scroller to examine and compare the orbits of Mercury and Earth. **a.** Which of the orbits of Earth and Mercury is closest to circular? **b.** Are the orbits of Earth and Mercury in the same plane? If not, measure the angle between these two orbital planes by zooming in on the Sun to a field of view about 10° wide and then tracing the two orbital planes onto tracing paper or transparent film with a soft felt-tipped pen.

THE RED PLANET'S WATERY PAST

New observations by rovers and orbiters indicate that liquid water not only existed on Mars, it once covered large parts of the planet's surface, perhaps for more than a billion years.

Adapted from an article by Jim Bell

By February 2005, the Mars Exploration Rover *Spirit* had spent over a year in Gusev Crater, a 2-km-deep, Connecticut-sized hole in the red planet's surface. Because Gusev lies at the end of an ancient, dry river valley longer than the Grand Canyon, many of us on the rover's mission team expected to find evidence of water from billions of years ago. But on the plains where it landed, the rover found no sign of lakes or flowing water: *Spirit*'s photographs documented dust, sand, and dry volcanic rock.

But everything changed once the rover reached the Columbia Hills, 2.6 km from the landing site. *Spirit* dug deep tracks into the soil as it climbed the western slope of Husband Hill. At a slippery patch dubbed Paso Robles, the wheels uncovered whitish deposits and the mission team slammed on the brakes, pulling a U-turn. It was unlike anything seen in Gusev.

The deposits were hydrated sulfate minerals, rich in iron and magnesium, concentrated just below the dusty surface. On Earth, these kinds of deposits are found where saltwater has evaporated or where groundwater interacts with volcanic gases or fluids. (Though no active volcanoes have been found on Mars, eruptions occurred earlier in its history.) We realized these buried sulfate salts could be remnants of a watery environment, just one of a number of surprising rover discoveries.

SNAPSHOTS FROM THE HUNT FOR WATER

Spirit and *Opportunity*, the Mars Exploration Rovers that have been operating on the Red Planet since January 2004, have revealed some of the best evidence of a warm, wet past.

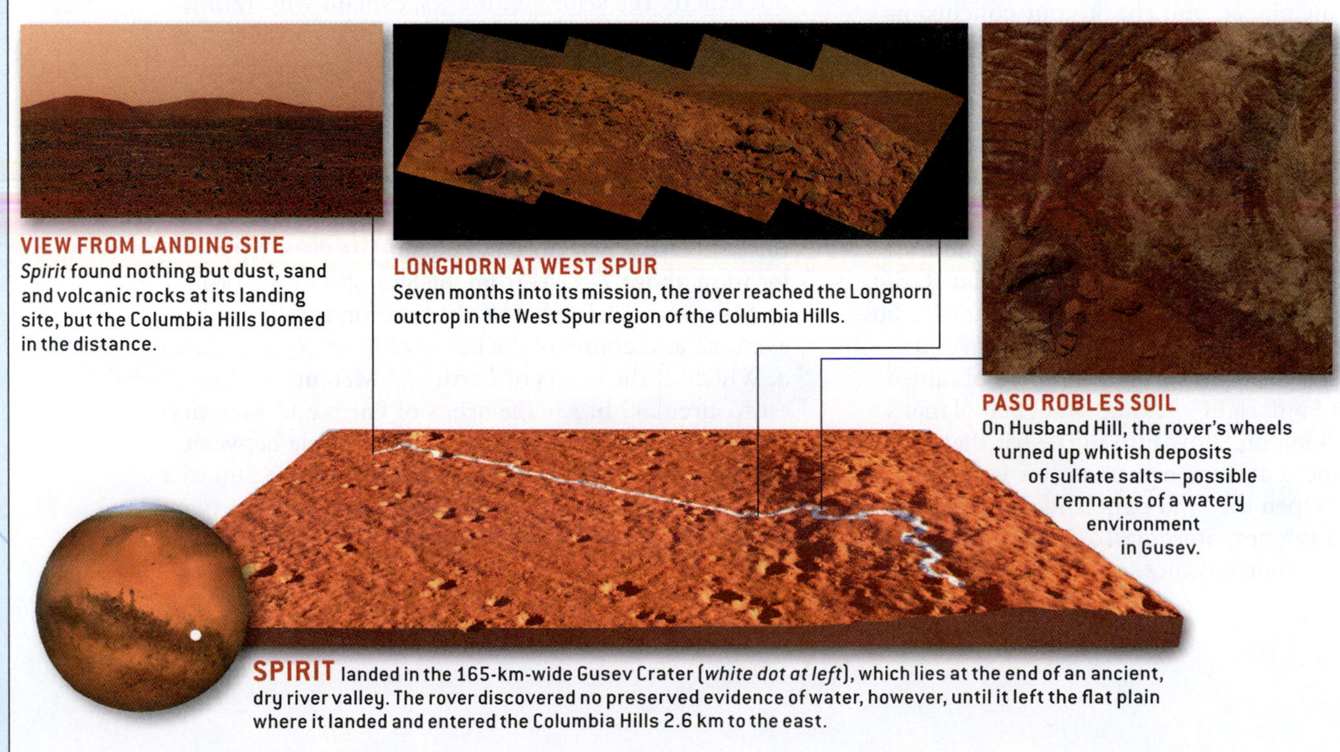

VIEW FROM LANDING SITE *Spirit* found nothing but dust, sand and volcanic rocks at its landing site, but the Columbia Hills loomed in the distance.

LONGHORN AT WEST SPUR Seven months into its mission, the rover reached the Longhorn outcrop in the West Spur region of the Columbia Hills.

PASO ROBLES SOIL On Husband Hill, the rover's wheels turned up whitish deposits of sulfate salts—possible remnants of a watery environment in Gusev.

SPIRIT landed in the 165-km-wide Gusev Crater (*white dot at left*), which lies at the end of an ancient, dry river valley. The rover discovered no preserved evidence of water, however, until it left the flat plain where it landed and entered the Columbia Hills 2.6 km to the east.

Spirit's serendipitous find was consistent with discoveries made by *Opportunity* (the rover investigating the planet's other side) and satellites photographing the surface. For decades, scientists believed Mars had always been cold, dry, and inhospitable. Signs of occasional floods and water-altered minerals were considered brief deviations dating from just after Mars formed 4.6 billion years ago. But new rover, orbital, and meteorite studies paint a drastically different picture. Water apparently covered large parts of the Martian surface for long periods.

New satellite images have provided compelling evidence that stable, Earth-like conditions prevailed on Mars for long periods. The most exciting discoveries are features resembling river deltas. The largest example, photographed by the *Mars Global Surveyor*, is at the end of a valley network that drains into Eberswalde Crater, southeast of the Valles Marineris canyon system. It terminates in a 10-km-wide fan, much like a delta formed at the mouth of a river flowing into a shallow lake.

The structure of the Eberswalde fan suggests it grew and changed shape many times, most likely responding to changes in an ancient river. If the Eberswalde fan is an ancient river-delta deposit, it implies that water persistently flowed on Mars, eroding sedimentary materials and transporting them downstream. Orbital images revealed similar fans elsewhere on Mars, but only 5 percent of the surface has been photographed at the resolution needed to identify these features.

Further orbital studies may allow researchers to test the river delta hypothesis. But to determine how long water flowed

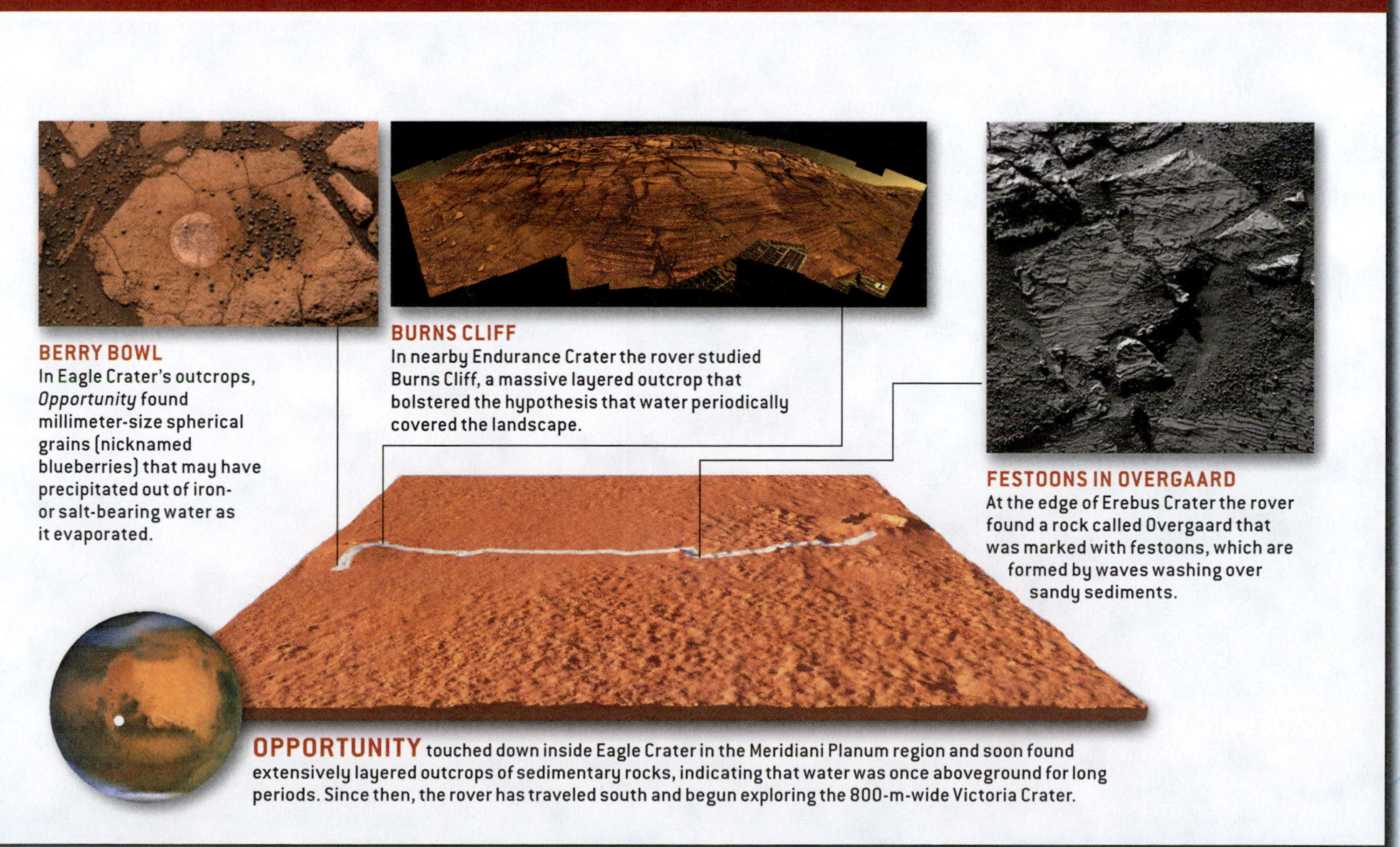

BERRY BOWL In Eagle Crater's outcrops, *Opportunity* found millimeter-size spherical grains (nicknamed blueberries) that may have precipitated out of iron- or salt-bearing water as it evaporated.

BURNS CLIFF In nearby Endurance Crater the rover studied Burns Cliff, a massive layered outcrop that bolstered the hypothesis that water periodically covered the landscape.

FESTOONS IN OVERGAARD At the edge of Erebus Crater the rover found a rock called Overgaard that was marked with festoons, which are formed by waves washing over sandy sediments.

OPPORTUNITY touched down inside Eagle Crater in the Meridiani Planum region and soon found extensively layered outcrops of sedimentary rocks, indicating that water was once aboveground for long periods. Since then, the rover has traveled south and begun exploring the 800-m-wide Victoria Crater.

NASA/STScI/JIM BELL (*global images*); RON MILLER (*traverse maps*); NASA/JPL/CORNELL (*landing site, Longhorn, Paso Robles*); NASA/JPL/CORNELL (*Berry Bowl, Burns Cliff*); NASA/JPL-CALTECH/CORNELL (*festoons*)

to create the fans, rock samples must be sent to Earth to measure the ages of the landforms—or must be examined by rovers that perform radioisotope dating.

The Emerging Paradigm

The emerging story is that Mars had an extensive watery past: puddles, ponds, lakes, and/or seas that existed for long periods in a thicker, warmer atmosphere. During its first billion or so years, the red planet was much more Earthlike, probably hospitable to life as we know it. But as sulfur built up, presumably because of Mars's volcanic history, the waters became acidic and the planet's geologic activity waned. Clays gave way to sulfates as acid rain altered the volcanic rocks and broke down any carbonates that may have formed. The atmosphere thinned; perhaps it was lost to space when the magnetic field shut off, or maybe it was blown off by catastrophic impacts or sequestered in the crust. Eventually, Mars became cold and arid. This sequence of events would explain why the volcanic rocks from the past few billion years remain unweathered. The older rocks underneath—exposed by impacts, erosion, or rovers—hold the key to the past.

Did life ever form on Mars, and if so, was it able to evolve as the environment changed? The answer mostly depends on the duration of Earthlike conditions—and an answer that current images and data can't give. It may ultimately prove impossible to use impact craters to establish ages on a surface that has seen so many episodes of massive burial and erosion. A better method would involve bringing samples to Earth for radioisotope dating or sending miniature age-dating instruments on Martian missions. Until then, orbital spacecraft will hunt for key mineral deposits and identify optimal sites for future exploration, which may someday reveal the duration of the red planet's watery era. The past decade's findings may be only a small taste of an even more exciting century of discoveries.

A MARTIAN CHRONICLE

Based on new evidence from recent missions to Mars, scientists have proposed a timeline positing an extensive watery past (dates are approximate).

4.6 billion to 4.2 billion years ago

ERA OF GIANT IMPACTS
After Mars's formation, asteroids and comets bombard the planet, forming huge impact basins and triggering intense volcanism. Oceans of magma (liquid rock) flow across the surface.

4.2 billion to 3.5 billion years ago

EPISODES OF EARTH-LIKE CONDITIONS
As the impacts lessen, liquid water fills some of the basins and carves enormous river valleys. The water weathers the underlying rock, producing clays and other hydrated silicates.

MICHAEL CARROLL (*above and opposite page*)

EBERSWALDE FAN (*left*), photographed from orbit by the Mars Global Surveyor, lies at the end of a valley network leading into Eberswalde Crater. The meandering, overlapping channels in the 10-kilometer-wide fan suggest that it was once a river delta draining into a shallow lake that may have filled much of the crater. The artist's rendering (*right*) portrays this delta as it might have looked billions of years ago.

3.5 billion to 2.5 billion years ago

DRYING OUT AND COOLING DOWN
Sulfur from Mars's volcanoes dissolves in the pools of water, turning them acidic and destroying the clays. Surface water begins to freeze, but sporadic floods create large outflow channels.

2.5 billion years ago to present

ARID AND INHOSPITABLE
Volcanic activity wanes, and dust covers much of the planet. But liquid water may persist underground and occasionally burst to the surface, forming gullies in the walls of canyons and craters.

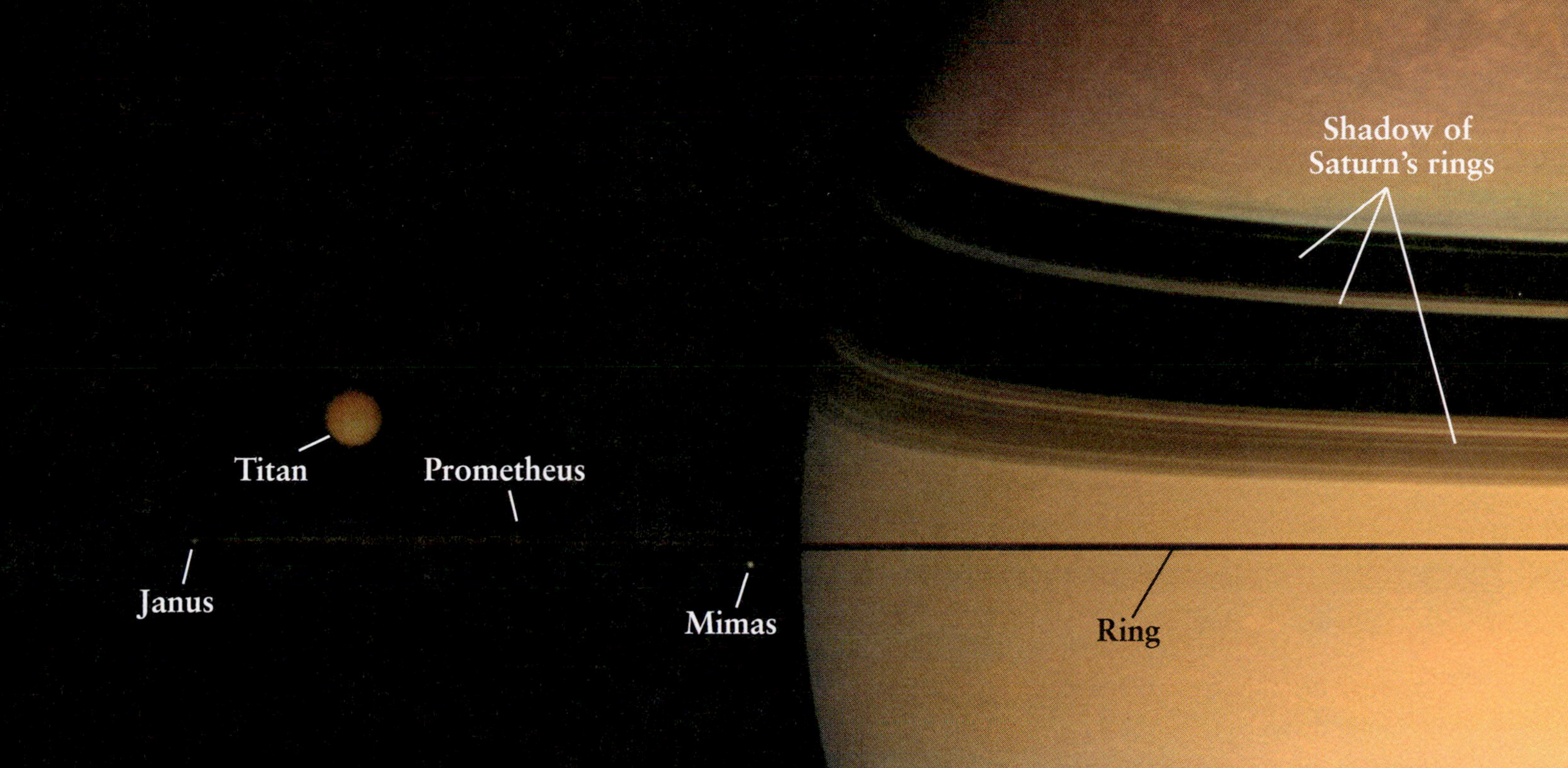

RIVUXG Saturn, Four of Its Moons, and Its Rings. This *Cassini* image gives you a feel for how thin the rings of Saturn are. (NASA/JPL/Space Science Institute)

Chapter 8
The Outer Planets

WHAT DO YOU THINK?

1. **Is Jupiter a "failed star"? Why or why not?**
2. **What is Jupiter's Great Red Spot?**
3. **Does Jupiter have continents and oceans?**
4. **Is Saturn the only planet with rings?**
5. **Are the rings of Saturn solid ribbons?**

Answers to these questions appear in the text beside the corresponding numbers in the margins and at the end of the chapter.

We can pretty easily imagine a trip to Mars, exploring its vast canyons and icy polar regions. Even Venus and Mercury lend themselves to being compared to Earth and the Moon. However, we find few similarities between Earth and the giant outer planets: Jupiter, Saturn, Uranus, and Neptune. Until recently, Pluto was also considered one of the outer planets. That has changed as described in Section 5-8. Now Pluto is categorized as a dwarf planet; we will explore it in Chapter 9.

The four giant worlds lack solid surfaces and are so much larger, rotate so much faster, and have such different chemical compositions from our planet that upon seeing them, one knows, to paraphrase Dorothy, "You're not on Earth anymore." There is nothing on Earth remotely similar to the swirling red and brown clouds of Jupiter, the ever-changing ring system of Saturn, or the blue-green clouds of Uranus and Neptune. The outer four worlds collectively have at least 164 moons, which have an amazing range of shapes, sizes, and surfaces. Some of these moons have water interiors; some have volcanoes; some have surface ice; one, Titan, has a thick atmosphere; and one, Hyperion, is arguably the most bizarre-looking object in the solar system. We begin our exploration of the outer planets with magnificent Jupiter.

In this chapter you will discover

- Jupiter, an active, vibrant, multicolored world more massive than all of the other planets combined
- Jupiter's diverse system of moons
- Saturn, with its spectacular system of thin, flat rings and numerous moons, including bizarre Enceladus and Titan
- Uranus and Neptune, ice giants similar to each other and different from Jupiter and Saturn

JUPITER: VITAL STATISTICS

Planet symbol:	♃
Average distance from the Sun:	**5.20 AU = 7.78×10^8 km**
Maximum distance from the Sun:	**5.46 AU = 8.16×10^8 km**
Minimum distance from the Sun:	**4.95 AU = 7.41×10^8 km**
Orbital eccentricity:	**0.048**
Average orbital speed:	**13.1 km/s**
Orbital period:	**11.86 years**
Rotation period:	**9 h 50 m 28 s (equatorial) 9 h 55 m 30 s (internal)**
Inclination of equator to orbit:	**3.12°**
Inclination of orbit to ecliptic:	**1.30°**
Diameter:	**142,984 km = 11.21 Earth diameters (equatorial) 33,700 km = 10.48 Earth diameters (polar)**
Mass:	**1.90×10^{27} kg = 318 Earth masses**
Average density:	**1330 kg/m^3 = 0.241 Earth density**
Escape speed:	**60.2 km/s**
Surface gravity (Earth = 1):	**2.36**
Albedo:	**0.52**
Average temperature at cloud tops:	**−108°C = −162°F = 165 K**
Atmospheric composition (by number of molecules):	**89.8% hydrogen (H_2), 10.2% helium (He), and traces of methane (CH_4), ammonia (NH_3), water vapor (H_2O), and other gases**

Jupiter   Earth

R I V U X G

VIDEO 8.1

FIGURE 8-1 Jupiter and Its Vital Statistics This view was sent back from *Voyager 1* in 1979. Features as small as 600 km (370 mi) across can be seen in the turbulent cloud tops of this giant planet. The complex cloud motions that surround the Great Red Spot are clearly visible. Also, clouds at different latitudes have different rotation rates. The inset image of Earth shows its size relative to Jupiter. (NASA/JPL; inset: NASA)

JUPITER

Jupiter's multicolored bands, dotted with ovals of white and brown (Figure 8-1), give it the appearance of a world unlike any of the terrestrial planets. Viewed even through a small telescope (see Figure 2-11b), you can also see up to four of its moons. As the high-resolution images in this chapter reveal, Jupiter is a world of breathtaking beauty.

1 Jupiter is the largest planet in the solar system: More than 1300 Earths could be packed into its volume. Using the orbital periods of its moons in Kepler's laws, astronomers have determined that Jupiter is 318 times more massive than Earth. Indeed, Jupiter has more than 2½ times as much mass as all of the other planets combined. This huge mass has created the myth that Jupiter is a failed star, meaning that it has almost enough matter to shine on its own, like the Sun, which we will study in Chapter 10. In fact, Jupiter would have to be 75 times more massive than it is to generate energy like the Sun, and therefore to be classified as a star. Nevertheless, Jupiter emits approximately twice as much energy as it receives from the Sun. Its extra energy comes from radioactive elements in its core and from an overall contraction amounting to less than 10 cm per century.

8-1 Jupiter's outer layer is a dynamic area of storms and turbulent gases

Jupiter is permanently covered with clouds (see Figure 8-1). Because it rotates about once every 10 hours—the fastest of any planet—Jupiter's clouds are in perpetual motion and are confined to narrow bands of latitude. Even through a small telescope, you can see Jupiter's dark, reddish bands called **belts,** alternating with light-colored bands called **zones.** The zones are white due to ammonia vapor at their tops, while the sources of the colors of the belts are still under investigation. These belts and zones are gases flowing east or west, with very little north-south motion. In contrast, winds on slower-rotating Earth wander over vast ranges of latitude (compare Figure 6-1).

Jupiter's belts and zones provide a framework for turbulent swirling cloud patterns, as well as rotating storms similar in structure to hurricanes or cyclones on Earth. These storms are known as *white ovals* and *brown ovals* (Figure 8-2). The white ovals are observed to be cool clouds higher than the average clouds in Jupiter's atmosphere. The brown ovals are warmer and lower clouds, seen through holes in the normal cloud layer. The various oval features last from hours to centuries.

FIGURE 8-2 Close-ups of Jupiter's Atmosphere The dynamic winds, rapid rotation, internal heating, and complex chemical composition of Jupiter's atmosphere create its beautiful and complex banded pattern. (a) A *Voyager 2* southern hemisphere image showing a white oval that has existed for over 40 years. (b) A *Voyager 2* northern hemisphere image showing a brown oval. The white feature overlapping the oval is a high cloud. (NASA)

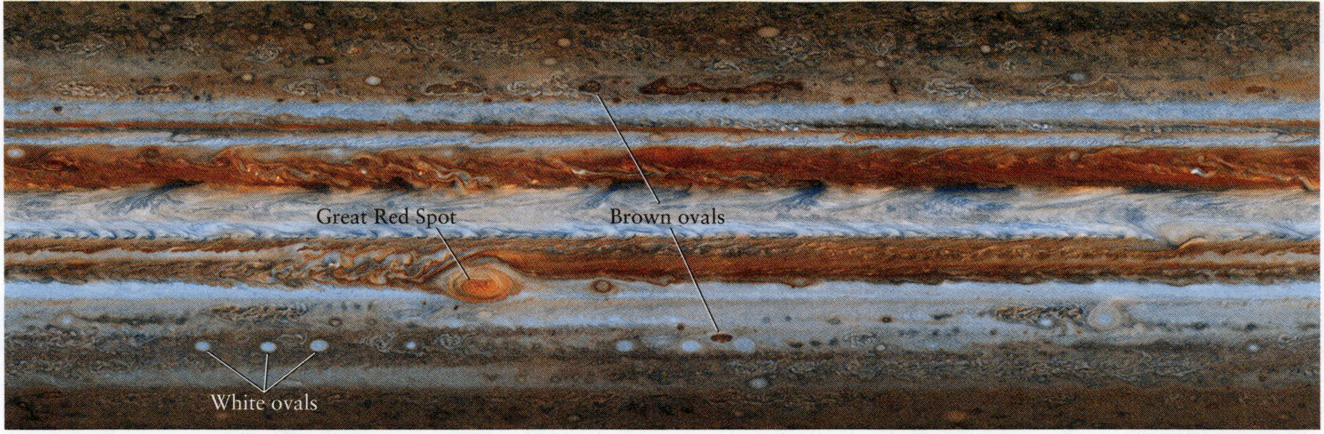

FIGURE 8-3 Jupiter Unwrapped *Cassini* images of Jupiter were combined and opened to give a maplike representation of the planet. The banded structure is absent near the poles. (Courtesy NASA/JPL-Caltech)

Computers show us how the cloud features on Jupiter would look if the planet's atmosphere were unwrapped and laid flat like a map of Earth (Figure 8-3).

Jupiter's most striking feature is its **Great Red Spot** (Figure 8-4), which is so large that it can be seen through a small telescope. This feature changes dimensions, and at present is about 25,000 km long by 12,000 km wide—large enough so that two Earths could easily fit side by side inside it. The Great Red Spot was first observed around 1656, either by the English scientist Robert Hooke or the Italian astronomer Giovanni Cassini. Because earlier telescopes were unlikely to have been able to see it, the Great Red Spot could well have formed long before that time.

The Great Red Spot is a hurricane- or typhoon- 
like storm of swirling gases. Heat welling upward from inside Jupiter has maintained it for more than three-and-a-half centuries. (Consider what life would be like for us if Earth sustained storms for such long periods!) Between 1998 and 2000, three smaller white storms on Jupiter merged (Figure 8-5a–d), creating a larger white storm that became red in 2006 (Figure 8-5e inset). Called *Red Spot Jr.* (technically named Oval BA), it is similar to, but somewhat smaller than, the Great Red Spot and located at nearly the same latitude (Figure 8-5e). As with the Great Red Spot, the cause of the red color is still being studied.

In 1690, Cassini noticed that the speeds of Jupiter's clouds vary with latitude, an effect called **differential rotation.** Near the poles, the rotation period of Jupiter's atmosphere, 9 h 55 min 30 s, is 5 min longer than at the equator. Furthermore, clouds at different latitudes circulate in opposite directions—some eastward, some westward. At their boundaries, the clouds rub against each other, creating beautiful swirling patterns (see Figure 8-2). The interactions of clouds at different latitudes also help provide stability for storms like the Great Red Spot.

Astronomers first determined Jupiter's overall 3
chemical composition from its average density—only 1330 kg/m^3. Recall from Section 5-10 that average density is mass divided by volume. We determine Jupiter's mass using Kepler's third law and the orbital periods of its moons, while the trigonometry of Jupiter's

Turbulence downwind of the Great Red Spot

Winds on the north side of the Great Red Spot flow westward.

Winds within the Great Red Spot circulate counterclockwise.

Winds on the south side of the Great Red Spot flow eastward.

20,000 km

Earth's diameter

R I V U X G

FIGURE 8-4 The Great Red Spot This image of the Great Red Spot shows the counterclockwise circulation of gas in the Great Red Spot that takes about 6 days to make one rotation. The clouds that encounter the spot are forced to pass around it, and when other oval features are near it, the entire system becomes particularly turbulent, like batter in a two-bladed blender. (NASA/JPL)

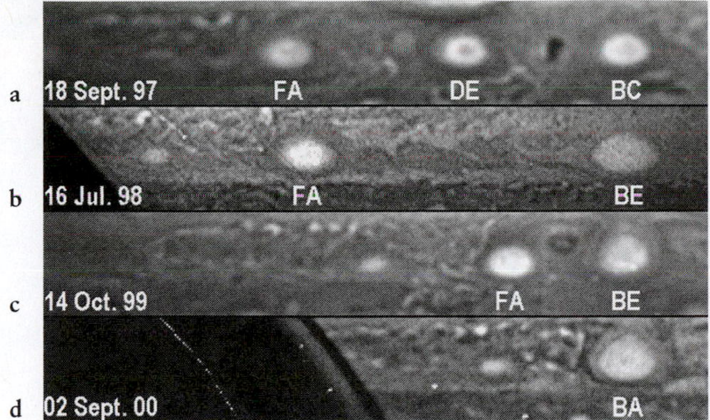

FIGURE 8-5 Creating Red Spot Jr. (a–d) For 60 years prior to 1998, the three white ovals labeled FA, DE, and BC traveled together at the same latitude on Jupiter. Between 1998 and 2000, they combined into one white oval, labeled BA, which (e) became a red spot, named Red Spot Jr., in 2006. (a–d: NASA/JPL/WFPC2; e: NASA, ESA, A. Simon-Miller [NASA/GSFC], and I. de Pater [University of California Berkeley])

distance from Earth and angular size in our sky reveal its diameter and hence its volume. This low density implies that Jupiter is composed of primarily the lightweight elements hydrogen and helium surrounding a relatively small central volume thousands of kilometers below its cloud tops containing water, metal, and rock. The planet has no solid continents, islands, or water oceans on its surface.

Spectra from Earth-based telescopes and from the *Galileo* probe sent into Jupiter's upper atmosphere in 1995 give more detail about the chemistry of this giant world's upper level. About 86% of its atoms are hydrogen and 13% are helium. The remainder consists of molecular compounds, such as methane (CH_4), ammonia (NH_3), and water vapor (H_2O). Keeping in mind that different elements have different masses, we can convert these percentages of atoms into the masses of various substances in Jupiter's atmosphere: 75% hydrogen, 24% helium, and 1% other substances. Because the interior contains heavier elements than the surface, the overall mass distribution in Jupiter has been calculated to be 71% hydrogen, 24% helium, and 5% heavier elements.

Observations from spacecraft visiting Jupiter and its moons, combined with the scientific model of Jupiter's atmosphere developed to explain the observations of Jupiter's clouds and chemistry, indicate that it has three major cloud layers (Figure 8-6a). The uppermost Jovian cloud layer is composed of crystals of frozen ammonia. These crystals and the frozen water in Jupiter's clouds are white, so what chemicals create the subtle tones of brown, red, and orange? The answer is as yet unknown. Some scientists think that sulfur compounds, which can assume many different colors, depending on their temperature, play an important role; others think that phosphorus is involved, especially in the Great Red Spot. The middle cloud layer is primarily ammonium hydrosulfide, and the bottom cloud layer is mostly composed of water vapor.

FIGURE 8-6 Jupiter's and Saturn's Upper Layers These graphs display temperature profiles of (a) Jupiter's and (b) Saturn's upper regions, as deduced from measurements at radio and infrared wavelengths. Three major cloud layers are shown in each, along with the colors that predominate at various depths. Data from the *Galileo* spacecraft indicate that Jupiter's cloud layers are not found at all locations around the planet; there are some relatively clear, cloud-free areas.

The descent of a probe from the *Galileo* spacecraft into Jupiter's atmosphere in 1995 revealed wind speeds of up to 600 km/h (375 mi/h), higher-than-expected air density and temperature, and lower-than-expected concentrations of water, helium, neon, carbon, oxygen, and sulfur. These seemingly abnormal observations probably occurred because the probe descended into a particularly arid region of the atmosphere called a *hot spot*, akin to the air over a desert on Earth.

Below its cloud layer, Jupiter's mantle is entirely liquid. Here on Earth, the distinction between the gaseous air and the liquid oceans is very clear—jump off a diving board and you know when you hit the water. However, the conditions on Jupiter under which hydrogen liquefies are different from anything we normally experience. Solid, liquid, and gas are called *phases of matter.* Going from one phase to another in everyday life is called a *phase transition*. Under conditions of extreme pressure and temperature, phase transitions don't occur. As a result, on Jupiter there is no definite boundary between the planet's gaseous atmosphere and its liquid mantle. The hydrogen gradually gets denser until, 1000 km below the cloud tops, the pressure is high enough for the hydrogen to be what we would consider a liquid.

In introducing the solar system, we noted that the young planets heated up as they coalesced. After they formed, radioactive elements continued to heat their interiors. On Earth, this heat leaks out of the surface through volcanoes and other vents. Jupiter loses heat everywhere on its surface, because, unlike Earth, it has no surface landmasses to block the heat loss.

Still heated from deep within Jupiter, blobs of liquid hydrogen and helium move upward inside the planet. When these blobs reach the cloud tops, they release their heat and descend back into the interior. (The same process, *convection,* drives the motion of Earth's mantle and its tectonic plates, as well as liquid simmering on a stove; see Figure 6-10.) Jupiter's rapid, differential rotation draws the convective gases into bands of winds moving eastward and westward at different speeds around the planet.

What are the two most common elements in Jupiter's upper layers?

Astronomers believe that Jupiter's belts and zones result from the combined actions of the planet's convection and rapid differential rotation. Until recently, they also believed that the light-colored zones are regions of hotter, rising gas, while the dark belts are regions of cooler, descending gas (Figure 8-7). However,

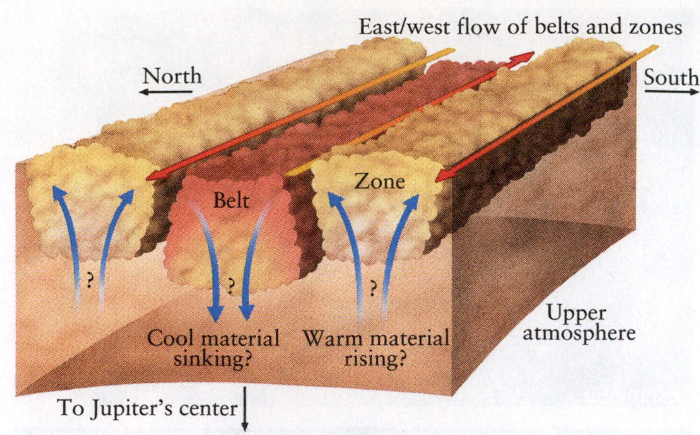

FIGURE 8-7 Original Model of Jupiter's Belts and Zones The light-colored zones and dark-colored belts in Jupiter's atmosphere were believed until recently to be regions of rising and descending gases, respectively. In the zones, gases warmed by heat from Jupiter's interior were thought to rise upward and cool, forming high-altitude clouds. In the belts, cooled gases were thought to descend and undergo an increase in temperature; the cloud layers seen there are at lower altitudes than in the zones. Observations by the *Cassini* spacecraft on its way to Saturn suggest that just the opposite may be correct (stay tuned)! In either case, Jupiter's rapid differential rotation shapes the rising and descending gas into bands of winds parallel to the planet's equator.

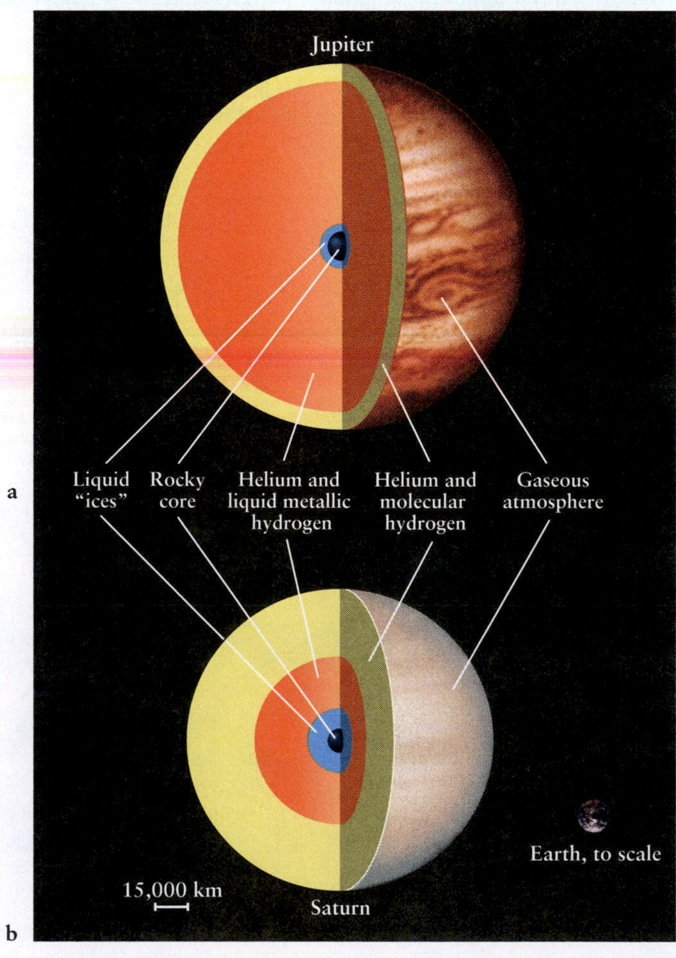

FIGURE 8-8 Cutaways of Jupiter and Saturn The interiors of both Jupiter and Saturn are believed to have four regions: a terrestrial rocky core, a liquid "ice" shell, a liquid metallic hydrogen shell, and a normal liquid hydrogen mantle. Their atmospheres are thin layers above the normal hydrogen, which boils upward, creating the belts and zones.

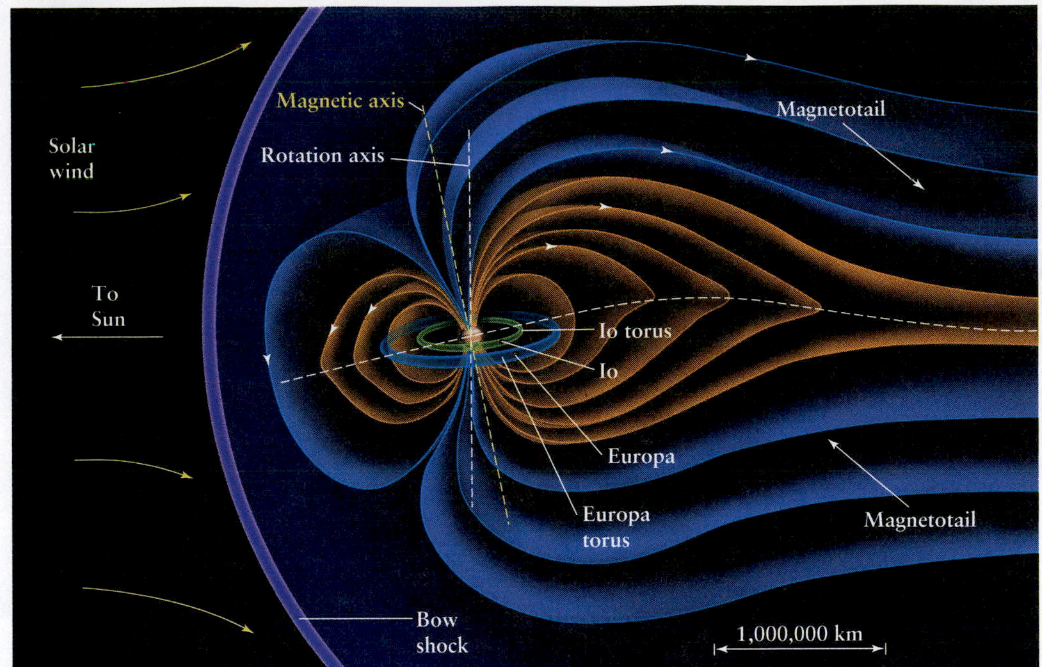

FIGURE 8-9 Jupiter's Magnetosphere (a) Created by the planet's rotation, the ion-trapping regions of Jupiter's magnetosphere (in orange, analogous to the Van Allen belts) extend into the realm of the Galilean moons. Gases from Io and Europa form tori (doughnut-shaped regions) in the magnetosphere. Some of Io's particles are pulled by the field onto the planet. Pushed outward by the solar wind, the magnetosphere has a "magnetotail" pointing away from the Sun. The magnetotail is often over 500 million km long and sometimes it reaches all the way to Saturn. (b) High-energy particles, trapped in Jupiter's magnetosphere, excite gases in the planet's upper atmosphere, causing them to glow as aurorae. The magnetosphere and cloud motion also lead to lightning on Jupiter. (b: J. Clarke, University of Michigan, and NASA)

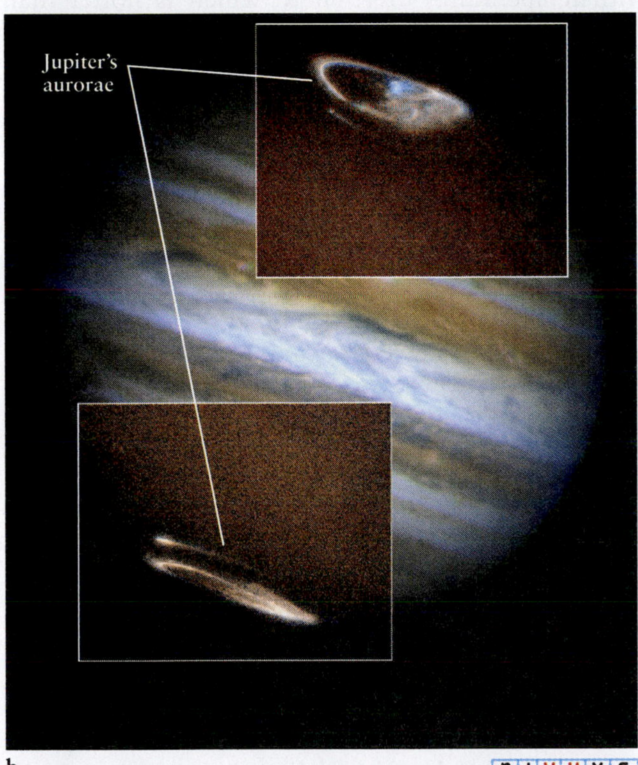

high-resolution observations by the passing *Cassini* spacecraft on its way to Saturn revealed numerous white clouds that are too small to be seen from Earth rising in the dark belts of Jupiter, while the zones are sinking gas—just the opposite of the original model! The correct explanation for the behavior of gases in the belts and zones may be provided by *Cassini,* when it studies the analogous belts and zones found on Saturn (see Section 8-9) in greater detail in the next few years.

8-2 Jupiter's interior has four distinct regions

Descending below Jupiter's clouds, we first encounter liquid molecular hydrogen and helium. As in Earth's oceans, the pressure in this fluid increases with depth. The gravitational force created by Jupiter's enormous mass compresses and heats its interior so much that 20,000 km (12,500 mi) below the cloud tops, the pressure is 3 million atm. Below this depth, the pressure is high enough to transform hydrogen into **liquid metallic hydrogen** (Figure 8-8). In this state, hydrogen acts like a metal (similar to the copper wiring in a house) in its ability to conduct electricity and heat. Electric currents that run through this rotating, metallic region of Jupiter generate a powerful planetary magnetic field (Figure 8-9). At the cloud level of Jupiter, this field is 14 times stronger per square meter than Earth's field is at our planet's surface. Jupiter's magnetic field has a region, like Earth's Van Allen belts, where charged particles are stored (depicted in orange in Figure 8-9a) and its tail (called the *magnetotail*) streams outward to Saturn, over 700 million km away.

For Earth-based astronomers, the first evidence of Jupiter's magnetosphere is a hiss of radio static, which varies cyclically over a period of 9 h 55 min 30 s, the planet's internal rotation rate. In recent years, astronomers have also been able to observe from Earth (and from space) permanent aurorae surrounding Jupiter's poles (Figure 8-9b). These aurorae have their distinctive ring shapes as a result of the glowing gases being trapped by the planet's magnetic field. Beginning with the two *Pioneer* and two *Voyager* spacecraft that journeyed past Jupiter in the 1970s, and continuing with the recent *Galileo* and *Cassini* spacecraft, the awesome dimensions of Jupiter's magnetosphere have been revealed: It is nearly 30 million km across, and envelops the orbits of many of its moons. If Jupiter's magnetosphere were visible from Earth, it would cover an area of the sky 16 times larger than the Moon as seen from here.

Based on the Nice theory of giant planet formation (Section 5-3), it appears that a terrestrial (rock and metal) protoplanet formed at Jupiter's orbit and attracted hydrogen and helium to form its outer layers. This terrestrial matter is now Jupiter's core. This core is only 4% of Jupiter's mass, which still amounts to nearly 13 times the mass of the entire Earth. Water, carbon dioxide, methane, and ammonia are likely to have existed as ice in Jupiter's terrestrial protoplanet, and they were also attracted onto the young and growing planet. When astronomers talk in general terms about "ice," they are referring to any or all of these four compounds.

The tremendous crushing weight of the bulk of Jupiter above the core—equal to the mass of 305 Earths—compresses the terrestrial core down to a sphere only 10,000 km in radius (see Figure 8-8). By comparison, Earth's diameter is 12,756 km. At the same time, the pressure forced the lighter ices out of the rock and metal, thereby forming a shell of these compounds between the solid core and the liquid metallic hydrogen layer. Calculations reveal that the temperature and pressure inside Jupiter should make these "ices" liquid down there! The pressure at Jupiter's very center is calculated to be about 70 million atm, and the temperature there is about 25,000 K, nearly 4 times hotter than the surface of the Sun. The interior of Jupiter is summarized in Figure 8-8a.

Interestingly, Jupiter gives off about twice as much energy as it receives from the Sun. Emitting this extra energy causes the planet to cool ever so slightly. As a planet cools, its gravity causes it to contract. Therefore, Jupiter is presently shrinking about a centimeter per decade. This process of cooling and contracting, called the *Kelvin-Helmholtz mechanism,* was proposed by Lord Kelvin and Hermann von Helmholtz in the late nineteenth century.

8-3 Impacts provide probes into Jupiter's atmosphere

On July 7, 1992, a comet nucleus (a clump of rock and ice a few kilometers across) passed so close to Jupiter that the planet's gravitational tidal force ripped it into at least 21 pieces. The debris from this comet was first observed in March 1993 by comet hunters Gene and Carolyn Shoemaker and David Levy. (Because it was the ninth comet they had found together, it was named Shoemaker-Levy 9 in their honor.) Shoemaker-Levy 9 was an unusual comet that actually orbited Jupiter, rather than just orbiting the Sun. Calculations of the comet's orbit showed that the pieces (Figure 8-10a) would return to strike Jupiter between July 16 and July 22, 1994.

Recall from Chapter 5 that impacts were extremely common in the first 800 million years of the solar system's existence. From more recent times, two chains of impact craters have been discovered on Earth, consistent with pieces of comets having hit our planet within the past 300 million years. However, it is very uncommon for pieces of space debris as large as several kilometers in diameter to collide with planets today. Therefore, the discovery that Shoemaker-Levy 9 would hit Jupiter created great excitement in the astronomical community. Seeing how a planet and a comet respond to such an impact would allow astronomers to deduce information about the planet's atmosphere and interior as well as about the striking body's properties.

The impacts occurred as predicted, with most of Earth's major telescopes—as well as those on several spacecraft—watching closely (Figure 8-10b). At least 20 fragments from Shoemaker-Levy 9 struck Jupiter, and 15 of them had detectable impact sites. The impacts resulted in fireballs some 10 km in diameter with temperatures of 7500 K, which is hotter than the surface of the Sun. Indeed, the largest comet fragment gave off as much energy as 600 million megatons of TNT, far more than the combined energy that could be released by all the nuclear weapons currently stockpiled on Earth. Impacts were followed by crescent-shaped ejecta that contained a variety of chemical compounds. Ripples or waves spread out from the impact sites through Jupiter's clouds in splotches that lasted for months.

The observations suggested that the pieces of comet did not penetrate very far into Jupiter's upper cloud layer. This fact, in turn, supports the belief that the pieces were not much larger than a kilometer in diameter. The ejecta from each impact included a dark plume that rose high into Jupiter's atmosphere. The darkness was apparently due to carbon compounds that vaporized from the cometary bodies. Also detected from the comet collisions were water, sulfur compounds, silicon, magnesium, and iron.

On July 19, 2009, an amateur astronomer discovered a new dark patch near Jupiter's south pole. Further observations revealed that this feature was created by another impact that had occurred about 4 hours before it was actually observed. Unlike the impact of 1994, this collision occurred when a small asteroid, about 0.5 km in diameter, struck Jupiter. Astronomers were able to distinguish between the two causes (comet or asteroid) by noting that gases and dust drifting along with the pieces of comet that struck Jupiter created halos around the impact sites, while the asteroid, lacking the surrounding gas and dust as it slammed into Jupiter, had no halo. Based on the 1994 and 2009 impacts, it appears that Jupiter is struck by pieces of debris 0.5 to 1 km in diameter every 10 to 15 years.

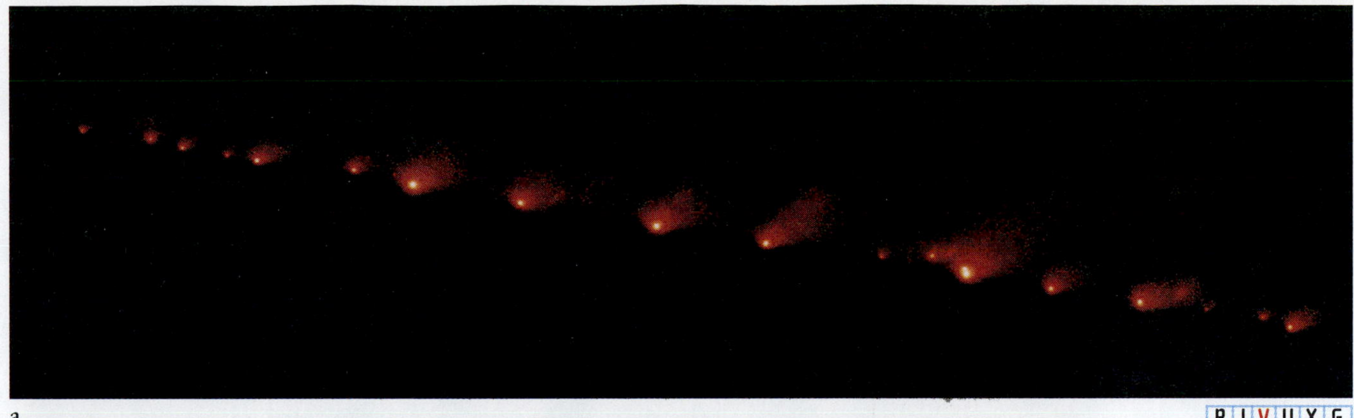

FIGURE 8-10 Comet Shoemaker-Levy 9 and Its Encounter with Jupiter (a) This comet, originally orbiting Jupiter, was torn apart by the planet's gravitational force on July 7, 1992, fracturing into at least 21 pieces. Its returning debris, shown here in May 1994, struck the planet between July 16 and July 22, 1994. (b) Shown here are visible (left) and ultraviolet (right) images of Jupiter taken by the Hubble Space Telescope after three pieces of Comet Shoemaker-Levy 9 struck the planet. Astronomers had expected white remnants (the color of condensing ammonia or water vapor); the darkness of the impact sites may have come from carbon compounds in the comet debris. Note the aurorae in the ultraviolet image. (a: H. A. Weaver, T. E. Smith, STScI and NASA; b: NASA)

JUPITER'S MOONS AND RINGS

Jupiter hosts at least 63 moons. The four largest—Io, Europa, Ganymede, and Callisto—may have formed at the same time as Jupiter from debris orbiting the planet, just as the planets formed from debris orbiting the Sun (see Chapter 5). Jupiter's other moons are probably captured planetesimals and smaller pieces of space debris.

Galileo was the first person to observe the four largest moons, in 1610, seen through his meager telescope as pinpoints of light. He called them the *Medicean stars* to attract the attention of the Medicis, rulers of Florence and wealthy patrons of the arts and sciences. To Galileo, the moons provided evidence supporting the then-controversial Copernican cosmology; at that time, Western theologians asserted that all cosmic bodies orbited Earth. The fact that the Medicean stars orbited Jupiter raised grave concerns in some circles.

To the modern astronomer, these moons are four extraordinary worlds, different both from the rocky terrestrial planets and from hydrogen-rich Jupiter. Now called collectively the **Galilean moons** or **Galilean satellites,** they are named after the mythical lovers and companions of the Greek god Zeus (called Jupiter by the Romans). From the innermost moon outward, they are Io, Europa, Ganymede, and Callisto.

These four worlds were photographed extensively by the *Voyager 1*, *Voyager 2*, and *Cassini* flybys and by the *Galileo* spacecraft that orbited Jupiter (Figure 8-11). The two inner Galilean satellites, Io and Europa, are about the same size as our Moon. The two outer satellites, Ganymede and Callisto, are roughly the size of Mercury. Figure 8-11 presents comparative information about these six bodies.

Why did the comet impacts of 1994 not create craters on Jupiter?

THE GALILEAN MOONS, MERCURY, AND EARTH'S MOON: VITAL STATISTICS

	Mean distance from Jupiter (km)	Sidereal period (day)	Diameter (km)	Mass		Mean density (kg/m^3)
				(kg)	(Moon = 1)	
Io	421,600	1.77	3630	8.94×10^{22}	1.22	3570
Europa	670,900	3.55	3138	4.80×10^{22}	0.65	2970
Ganymede	1,070,000	7.16	262	1.48×10^{23}	2.01	1940
Callisto	1,883,000	16.69	4800	1.08×10^{23}	1.47	1860
Mercury	—	—	4878	3.30×10^{23}	4.49	5430
Moon	—	—	3476	7.35×10^{22}	1.00	3340

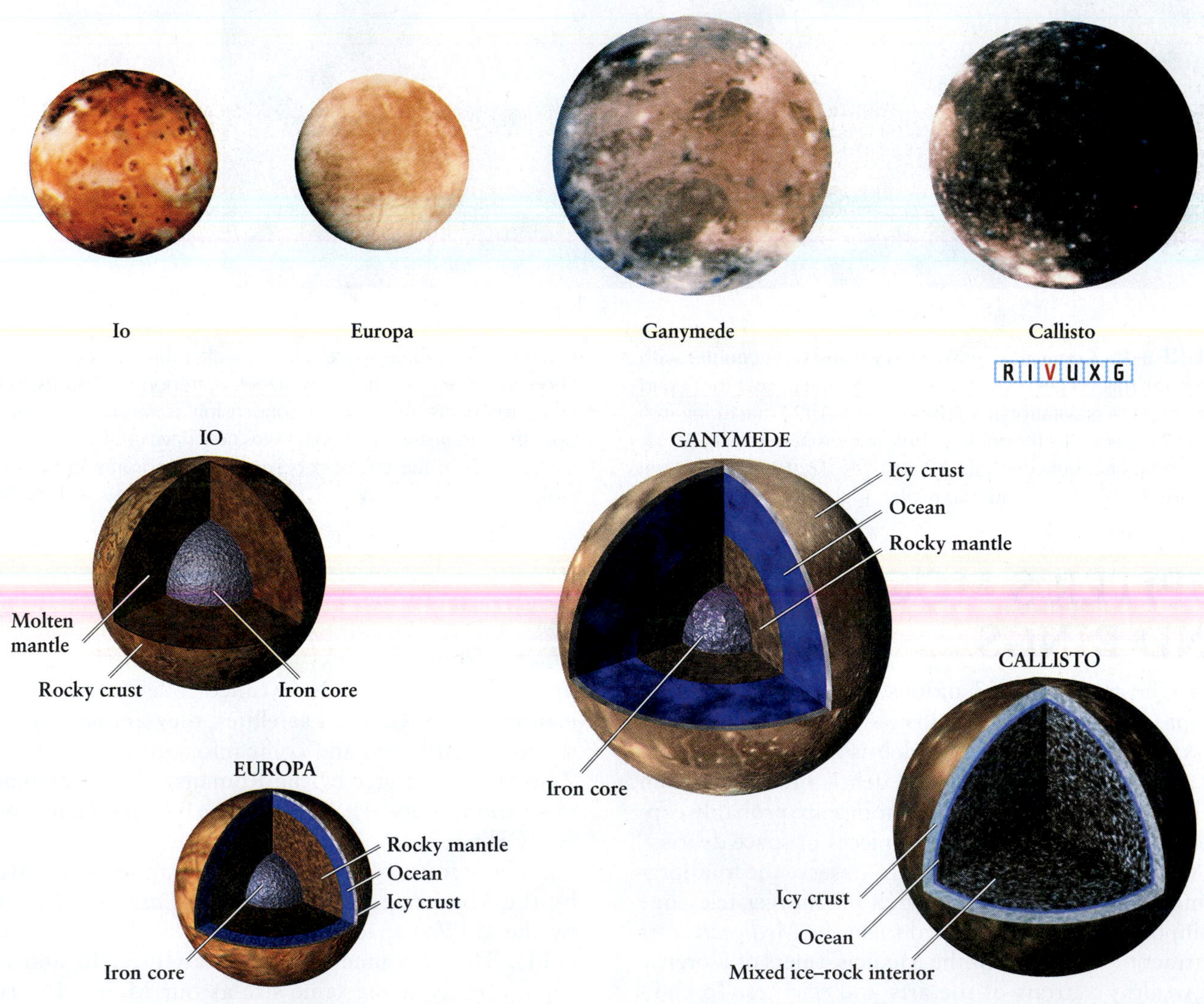

FIGURE 8-11 The Galilean Satellites The four Galilean satellites are shown here to the same scale. Io and Europa have diameters and densities comparable to our Moon and are composed primarily of rocky material. Ganymede and Callisto are roughly as big as Mercury, but their low average densities indicate that each contains a thick layer of water and ice. The cross-sectional diagrams of the interiors of the four Galilean moons show the probable internal structures of the moons, based on their average densities and on information from the *Galileo* mission. (NASA and NASA/JPL)

8-4 Io's surface is sculpted by volcanic activity

Sulfurous Io is among the most exotic moons in our solar system (Figure 8-12a). At first glance, this moon's density of 3570 kg/m^3 seems to place its chemical composition between that of the terrestrial and Jovian planets. However, its density has that value in large measure because its mass is so small that it cannot compress its interior nearly as much as can Earth or more massive planets. Allowing for its lower compression, astronomers calculate that Io is mostly rock and iron, like Earth, rather than being composed of lighter elements, as is Jupiter.

Io zooms through its orbit of Jupiter once every 1.8 days. Like our Moon, it is in synchronous rotation with its planet. Images of Io reveal giant plumes emitted by about 300 active volcanoes (Figure 8-12c) and geysers (similar to Old Faithful on Earth). Most of this ejected material falls back onto Io's surface; the rest is moving fast enough to escape into space. Observations indicate that Io's volcanoes emit 10 trillion tons of matter each year in plumes up to 500 km high. That amount of material is enough to resurface Io to a depth of 1 m each century (Figure 8-12b). One eruption in 2002 occurred over an area the size of London, some 1600 sq km (620 sq mi). The volcanoes also emit basaltic lava flows rich in magnesium and iron. Io's volcanoes are named after gods and goddesses associated with fire in Greek, Norse, Hawaiian, and other mythologies. Io also has numerous black "dots" on its surface, which, apparently, are

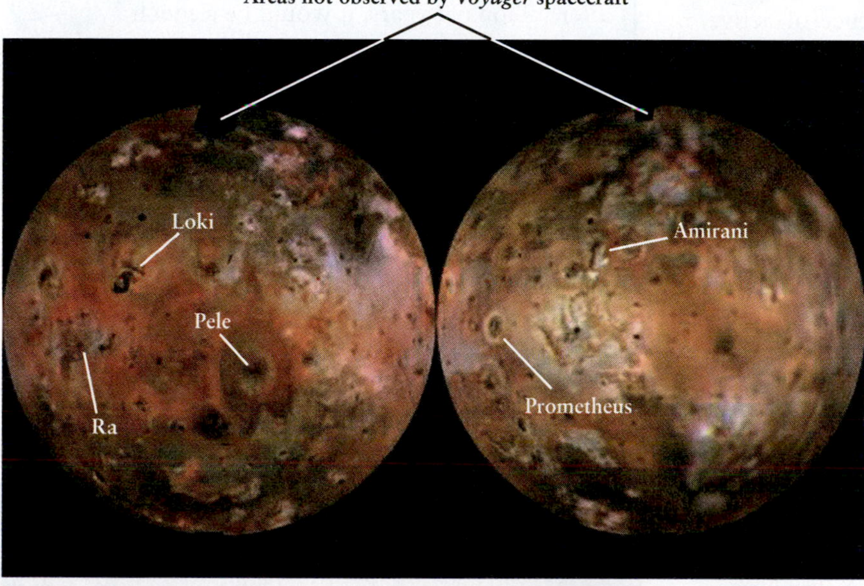

VIDEO 8.4

FIGURE 8-12 Io (a) These *Voyager* images show both sides of Io. The range of colors results from surface deposits of sulfur ejected from Io's numerous volcanoes. Plumes from the volcano Prometheus rise up 100 km. Prometheus has been active in every image taken of Io since the *Voyager* flybys of 1979. (b) Photographed in 1999 and then 2000 (*shown here*), the ongoing lava flow from this volcanic eruption at Tvashtar Catena has considerably altered this region of Io's surface. (c) *Galileo* image of an eruption of Pilan Patera on Io. (a: NASA/JPL; b: University of Arizona/JPL/NASA; c: Galileo Project, JPL, NASA) dormant volcanic vents. Old lava flows radiate from many of these locations, which are typically 10 to 50 km in diameter and cover 5% of Io's surface.

Insight Into Science

What's in a Name? Following up on the preconceptions that common words create (see *Insight Into Science: Imagine the Moon* in Chapter 7), objects with familiar names often have different characteristics than we expect. We tend to envision "moons" as inert, airless, lifeless, dry places similar to our Moon. However, as we will see throughout this chapter, starting with Io, some moons have active volcanoes, atmospheres, and vast underground liquid water oceans.

Just before their discovery, the existence of active volcanoes on Io was predicted from analysis of the gravitational forces to which that moon is subjected. As it rapidly orbits Jupiter, Io repeatedly passes between Jupiter and one or another of the remaining Galilean satellites. These moons pull on Io, causing it to slightly change its distance from Jupiter. This change in distance creates a change in the tidal forces (see Section 6-8) acting on it from the planet. As the distance between Io and Jupiter varies, the resulting tidal stresses alternately squeeze and flex the moon. This ongoing motion of Io's interior generates heat through friction, creating as much energy inside Io as the detonation of 2400 tons of TNT every second. Gas and molten rock from inside Io eventually make their way to the moon's surface, where they are ejected through the volcanoes and other vents.

What creates most of the heat inside Io that causes it to have volcanoes and geysers?

Satellite instruments have identified sulfur and sulfur dioxide in the material erupting from Io's volcanoes. Sulfur is normally bright yellow. If heated and suddenly cooled, however, it forms molecules that assume a range of colors, from orange and red to black, which accounts for Io's tremendous range of colors (see Figure 8-12). Sulfur dioxide (SO_2) is an acrid gas commonly discharged from volcanic vents here on Earth and, apparently, on Venus. When eruptions on Io release this gas into the cold vacuum of space, it crystallizes into white flakes, which fall onto the surface and account for the moon's whitish deposits.

The *Galileo* spacecraft detected an atmosphere around Io. Composed of oxygen, sulfur, and sulfur dioxide, it is only one-billionth as dense as the air we breathe. Io's atmosphere can sometimes be seen to glow blue, red, or green, depending on the gases involved. Gases ejected from Io's volcanoes have also been observed extending out into space and forming a doughnut-shaped region around Jupiter called the *Io torus,* about which we will say more shortly.

WHAT IF . . . Io Were in a More Elliptical Orbit? Io is in an exceedingly circular orbit, with an orbital eccentricity (defined in Section 2-5) of $e = 0.0041$. If Io's orbital eccentricity were the same as that of our Moon ($e = 0.055$), it would be even more dynamic than it is today. Currently, more than 95% of Io's surface is solid, with the remaining surface either active volcanoes or molten lava. With a more eccentric orbit, the moon would change shape significantly more than it does today, leading to much greater internal stresses, friction, and heating. Io would experience so much more volcanic activity and lava flow that perhaps half of its surface would be molten or otherwise active at any moment. Io would also be ejecting more lava into space and hence, after billions of years, it would be a much smaller moon.

8-5 Europa apparently harbors liquid water below its surface

Images of Europa's ice and rock surface from *Voyager 2* and the *Galileo* spacecraft suggest that Jupiter's second-closest Galilean moon contains liquid water (Figure 8-13). Europa orbits Jupiter every 3½ days and, like Io, is in synchronous rotation. The changing gravitational tug

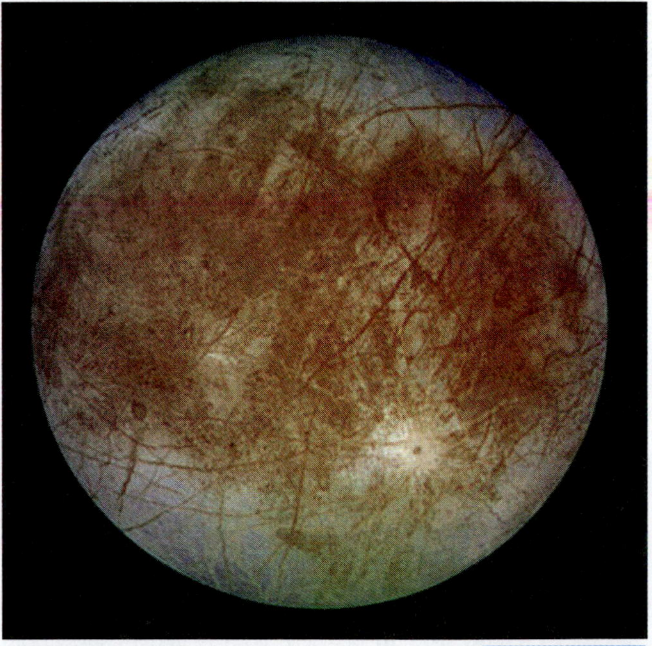

FIGURE 8-13 Europa Imaged by the *Galileo* spacecraft, Europa's ice surface is covered by numerous streaks and cracks that give the satellite a fractured appearance. The streaks are typically 20 to 40 km wide. (NASA/JPL)

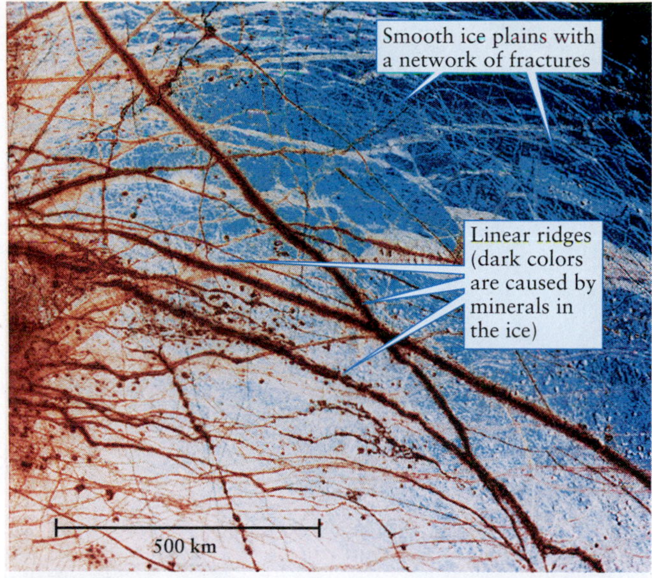

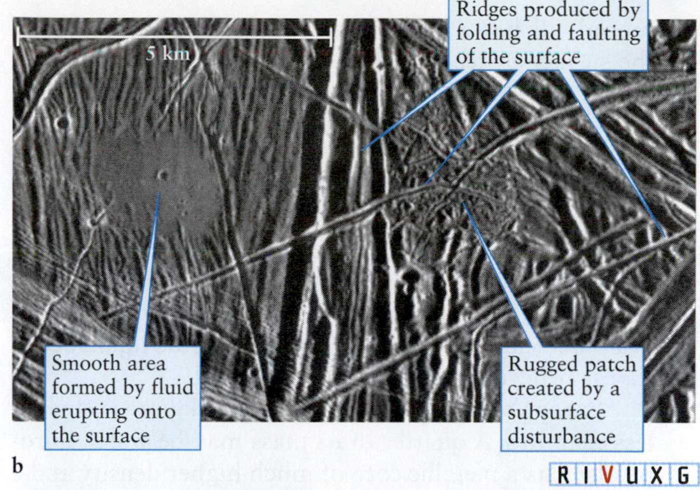

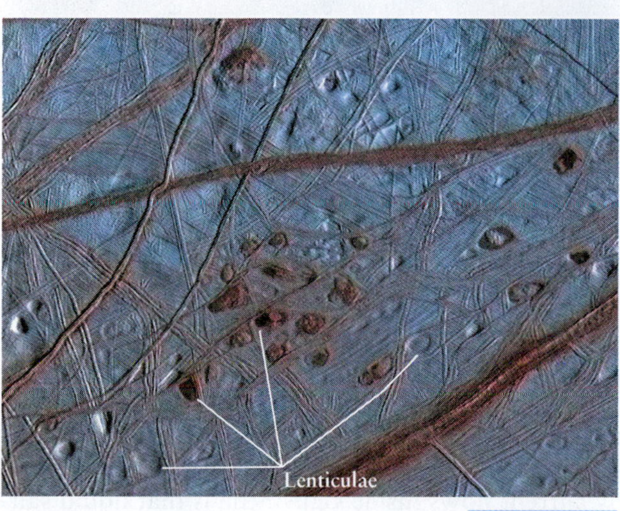

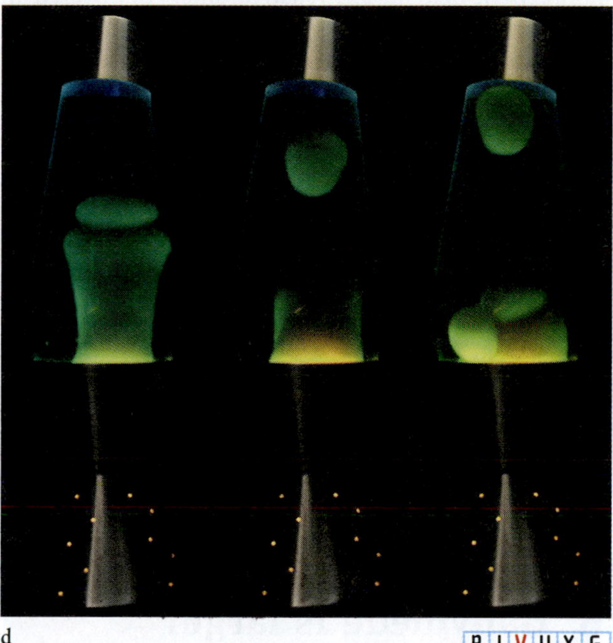

FIGURE 8-14 Surface Features on Europa (a) This false color *Galileo* image of Europa combining visible and infrared observations shows smooth plains of ice, mineral ridges deposited by upwelling water, and numerous fractures believed to be caused by tidal stresses. (b) This region of Europa's surface shows the jumbled, stressed features common to the surface, as well as direct indications of liquid water activity underground. (c) Lenticulae attributed to rising warmed ice and debris travel up from the moon's interior by convection, arriving at and then leaking out at the surface. The white domes are likely to be rising material that has not yet reached the surface. (d) A lava lamp in which warmed material rises through cooler liquid. The rising material in Europa is analogous to the rising motion of the blobs in a lava lamp, except that on Europa, the motion is through ice. (a, b: NASA/JPL; c: NASA/JPL/University of Arizona/University of Colorado; d: Bianca Moscatelli)

of Io creates stress inside Europa similar to the magma-generating distortion that Io undergoes. These stresses apparently cause numerous cracks or fractures seen on Europa's surface (Figure 8-14a). Furthermore, astronomers think this stress creates enough heat inside Europa to keep the water in a liquid state just a few kilometers below the moon's ice and rock surface.

Besides the cracks, numerous other features on Europa support the idea that it has liquid water inside. Just as in the Arctic region on Earth, the movement of surface features on Europa creates ice floes, along with swirls, strips, and ridges, driven by circulating water underneath the moon's surface (Figure 8-14b). Also indicative of liquid water is the moon's reddish color. The coloring may be due to salt deposits left after liquid water rose to the surface and evaporated (Figures 8-14a, c). *Galileo* spacecraft images suggest that some of Europa's features have moved within the past few million years, and perhaps are still in motion. Indeed, the chaotic surface revealed by *Galileo* is interpreted as having formed as a result of water volcanism on Europa, strengthening the belief that this

What region on Earth does the surface of Europa most resemble?

moon still has a liquid water layer. Replenishment of the surface by tectonic plate motion would explain why only a few small impact craters have survived.

Galileo also photographed red and white domes called *lenticulae* on Europa (Figure 8-14c) that geologists think are rising warmed ice mixed with other material, possibly including organic matter. These lenticulae are typically 100 m high and 10 km (6 mi) wide. The rising material behaves like the blobs in lava lamps (Figure 8-14d), although the lenticulae are calculated to take 100,000 years to reach the surface from the liquid ocean that is believed to exist inside the moon.

Europa's average density of 2970 kg/m^3 is slightly less than Io's. A quarter of its mass may be water. Europa also has a metallic core of much higher density and a weak magnetic field. In 1995, astronomers discovered an extremely thin atmosphere containing molecular oxygen surrounding Europa. This gas is about 10^{-11} times less dense than the air we breathe. The oxygen may come from water molecules on the moon's surface that were broken up by ultraviolet radiation from the Sun.

The underground ocean of Europa is especially interesting to scientists because virtually all liquid water locations on Earth support life, and life may well have evolved in Europa's oceans. If it exists, and if it evolved differently from life on Earth, then life-forms from Earth would likely contaminate it, even wiping it out. This process would be analogous to the deadly impact of diseases on Earth, such as the introduction of smallpox to the Americas from Europe. To prevent the expired spacecraft *Galileo* from contaminating any life that may exist on Europa, NASA sent that spacecraft into Jupiter's thick atmosphere, where the machine was vaporized.

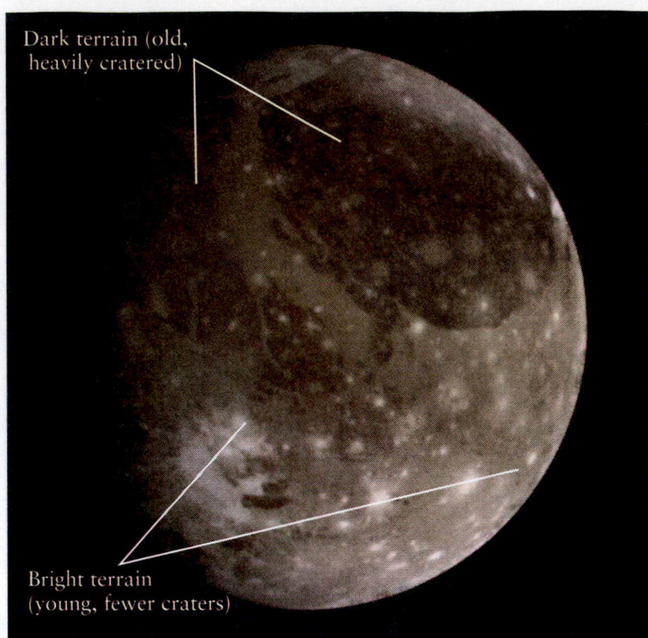

VIDEO 8.5

FIGURE 8-15 Ganymede This side of Ganymede is dominated by the huge, dark, circular region called Galileo Regio, which is the largest remnant of Ganymede's ancient crust. Darker areas of the moon are older; lighter areas are younger, tectonically deformed regions. The light white areas in and around some craters indicate the presence of water ice. Large impacts create white craters, filled in by ice from below the surface. (NASA/JPL)

8-6 Ganymede is larger than Mercury

Ganymede is the largest satellite in the solar system (Figure 8-15). Its diameter is greater than Mercury's, but its density of 1940 kg/m^3 is much less than that of Mercury. This moon also has a permanent magnetic field that is twice as strong as Mercury's field. Ganymede orbits Jupiter in synchronous rotation once every 7.2 days. Like its neighbor Europa, Ganymede has an iron-rich core, a rocky mantle, an underground liquid water ocean, a thin atmosphere, and a covering of dirty ice.

The existence of an underground ocean is implied by the discovery of a second magnetic field around Ganymede, which continually varies. This field has a different origin than the permanent one just mentioned; the second field is generated by Jupiter's magnetic field. As Ganymede orbits Jupiter, the planet's powerful magnetic field creates an electric current inside the moon, which, in turn, creates Ganymede's varying magnetic field. (The same process is used to create electric currents in power stations here on Earth, where falling water or steam is used to rotate powerful magnets at high speeds. Their magnetic fields cut across wires, thereby pushing electrons in the wire. These moving electrons are the electric current we use.) The best explanation of why current flows inside Ganymede is that liquid saltwater exists there, and saltwater is a good conductor of electricity. This implies, of course, the presence of a liquid ocean. Furthermore, salts have been observed on Ganymede's surface. They were apparently carried upward and deposited there as water leaked out and froze. As on Europa, liquid water may imply the presence of life.

Like our Moon, Ganymede has two very different kinds of terrain (Figures 8-15, 8-16). Dark, polygon-shaped regions are its oldest surface features, as judged by their numerous craters. Light-colored, heavily grooved terrain is found between the dark, angular islands. These lighter regions are much less cratered and therefore younger. Ganymede's grooved terrain consists of parallel mountain ridges up to 1 km high and spaced 10 to 15 km apart. These features suggest that the process of plate tectonics may have dominated Ganymede's early history. But, unlike Europa, where tectonic activity still occurs today, tectonics on Ganymede bogged down 3 billion years ago as the satellite's crust froze solid.

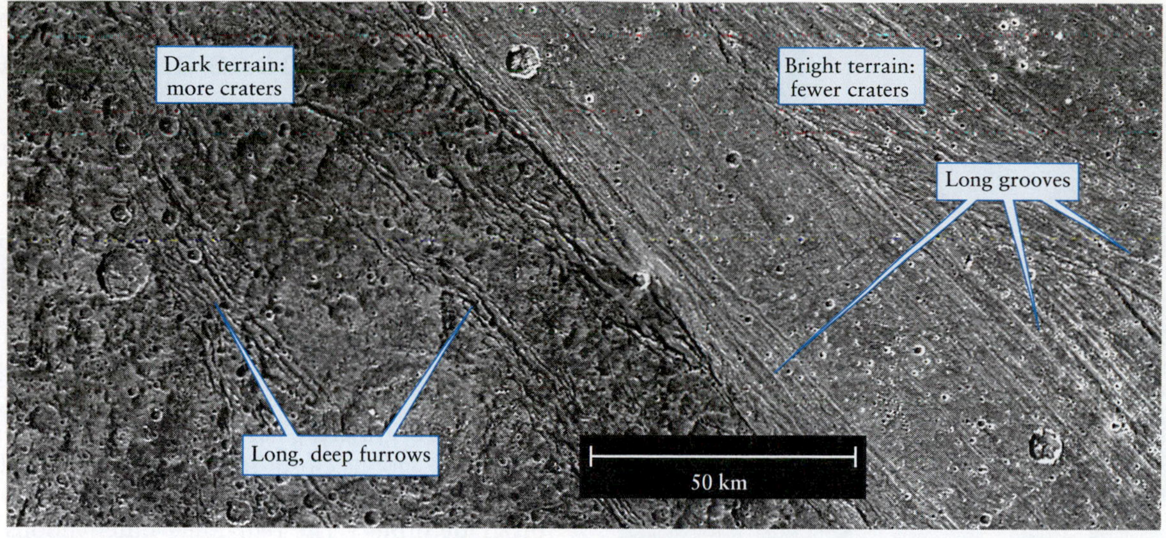

FIGURE 8-16 Two Surfaces of Ganymede The older, rougher, more heavily cratered parts of Ganymede are the dark terrain. These regions are surrounded by younger, smoother, less cratered bright terrain. The parallel ridges suggest that the bright terrain has been crafted by tectonic processes. (NASA)

Another mechanism that may have created Ganymede's large-scale features is a bizarre property of water. Unlike most liquids, which shrink upon solidifying, water expands when it freezes. Seeping up through cracks in Ganymede's original crust, water may have thus forced apart fragments of that crust. This process could have produced jagged, dark islands of old crust separated by bands of younger, light-colored, heavily grooved ice.

Ganymede has a thin atmosphere known to contain oxygen, and possibly other gases. As that moon passes through Jupiter's magnetic field, that oxygen is bombarded by electrons trapped by the field. These collisions cause the gas in Ganymede's atmosphere to give off radiation, creating auroras, which were first observed in 1998.

8-7 Callisto bears the scars of a huge asteroid impact

Callisto is Jupiter's outermost Galilean moon. It orbits Jupiter in 16.7 days, and, like the other Galilean moons, its rotation is synchronous. Callisto is 91% as big and 96% as dense as Ganymede, and has a thin atmosphere of hydrogen and carbon dioxide. Like Ganymede, Callisto apparently harbors a substantial liquid water ocean. As with Ganymede, this ocean's presence is inferred by Callisto's changing magnetic field. The heat that keeps the ocean liquid likely comes from energy released by radioactive decay inside the moon.

Although numerous large impact craters are scattered over Callisto's dark, ancient, icy crust, it has very few craters smaller than 100 m across (Figure 8-17). Astronomers speculate that the smaller craters have disintegrated. Unlike Ganymede and Europa, Callisto has no younger, grooved terrain. The absence of grooved terrain suggests that tectonic activity never began there, probably because the satellite simply froze too rapidly. It is bitterly cold on Callisto's surface. *Voyager* instruments measured a noontime temperature of 155 K (−180°F), and the nighttime temperature plunges to 80 K (−315°F).

Which planet is smaller than Ganymede?

Like Mercury, our Moon, and other bodies we will encounter shortly, Callisto carries the cold, hard evidence of what happens when one astronomical body strikes another. *Voyager 1* photographed a huge impact basin named Valhalla on Callisto (see Figure 8-17a). An asteroid-sized object produced Valhalla Basin, which is located on Callisto's Jupiter-facing hemisphere. Like throwing a rock into a calm lake, ripples ran out from the impact site along Callisto's surface, cracking the surface and freezing into place. The largest remnant rings surrounding the impact crater have diameters of 3000 km. In 2001, the *Galileo* spacecraft revealed spires 80–100 m high on Callisto (Figure 8-17b). These spires are also believed to have been created by an impact, perhaps the same one. One feature that is missing from Callisto is the jumbled terrain on the side opposite of Valhalla Basin, as is seen opposite of large impact sites on Mercury and our Moon. The smoothness of Callisto can be explained by a model that shows that a liquid water interior would dampen the impact shock and thereby prevent the opposite side from becoming disturbed. This theory supports the idea that Callisto has liquid water inside. The probable interiors of the Galilean moons are shown in Figure 8-11.

What features on our Moon and on Mercury are similar to Callisto's Valhalla impact basin?

8-8 Other debris orbits Jupiter as smaller moons and ringlets

Besides the four Galilean moons, Jupiter has at least 59 other moons, a set of tenuous *ringlets,* and two doughnut-shaped tori of electrically-charged gas particles. The non-Galilean moons are all irregular in shape and less than 275 km in diameter. Four of these moons (Figure 8-18) are inside Io's orbit; all of the other known moons are outside Callisto's orbit. The inner moon Amalthea (see Figure 8-18) is red-colored, has about the same density as water, and is apparently made of pieces of rock and ice barely held together by the moon's own gravity. The Galilean moons, along with the smaller moons closer to Jupiter and six of the outer moons, orbit in the same direction that Jupiter rotates (**prograde orbits**). The remaining outer moons revolve in the opposite direction (**retrograde orbits**). The outer ones appear to be individual captured planetesimals, while the inner ones are probably smaller pieces broken off a single larger body.

Jupiter is only the first of four planets with rings. Its 4
ring system is composed of several parts (Figure 8-19a). Debris knocked off the moons Adrastea and Metis is responsible for the main ring. The outer edge of this ring has three well-defined **ringlets** (Figure 8-19b) that are pieces of debris ranging in size from gravel to small

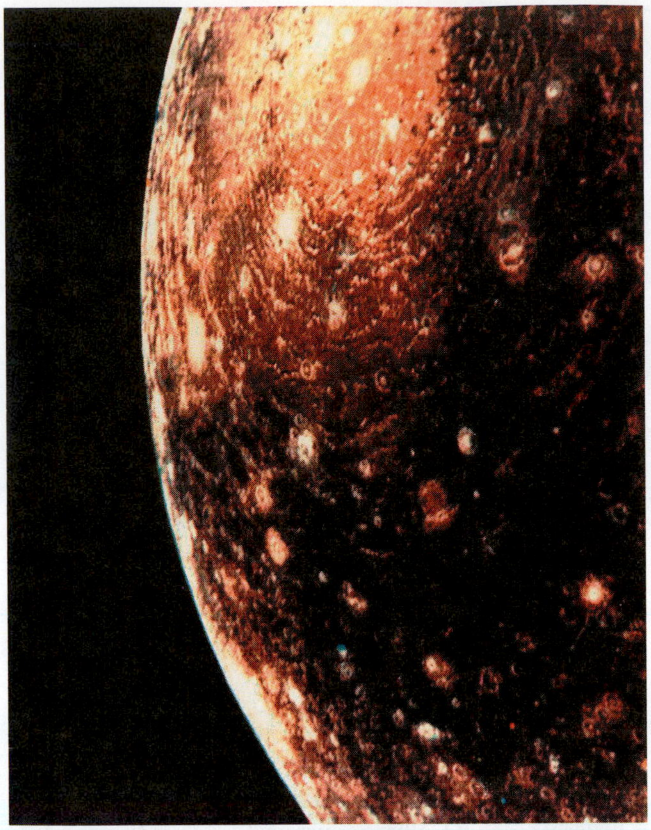

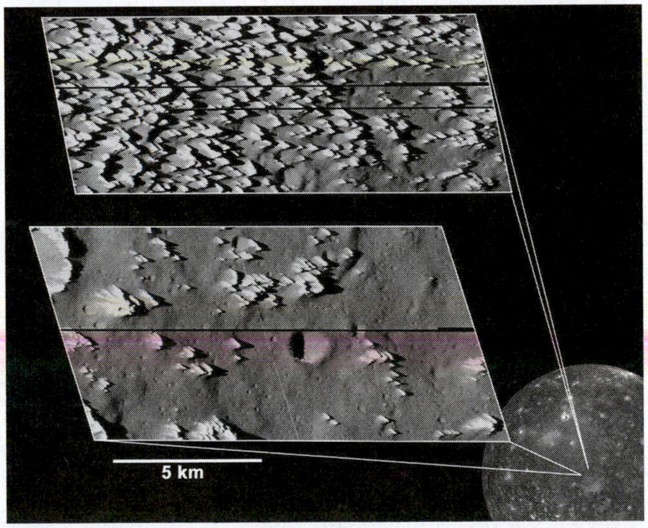

FIGURE 8-17 Callisto The outermost Galilean satellite is almost exactly the same size as Mercury. Numerous craters pockmark Callisto's icy surface. (a) The series of faint, concentric rings that cover much of this image is the result of a huge impact that created the impact basin Valhalla. Valhalla dominates the Jupiter-facing hemisphere of this frozen, geologically inactive world. (b) The two insets in this *Galileo* mission image show spires that contain both ice and some dark material. The spires were probably thrown upward as the result of an impact. The spires erode as dark material in them absorbs heat from the Sun. (a: Courtesy of NASA/JPL; b: NASA/JPL/Arizona State University)

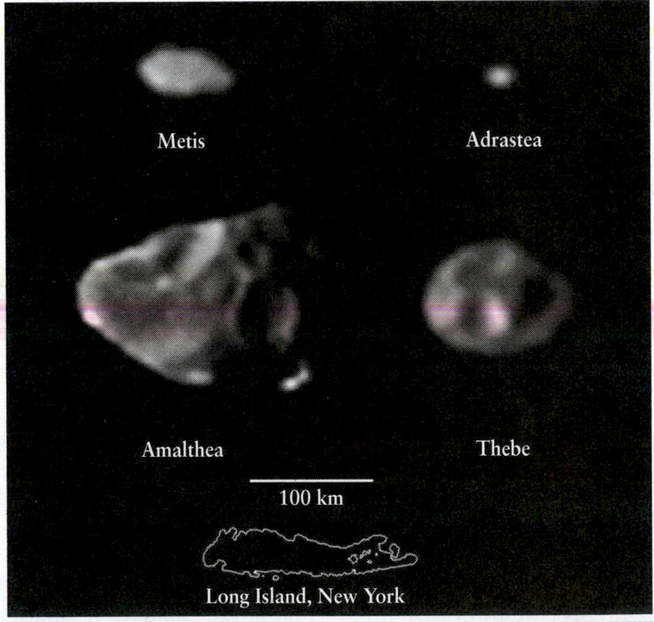

FIGURE 8-18 Irregularly-Shaped Inner Moons The four known inner moons of Jupiter are significantly different from the Galilean satellites. These bodies are roughly oval-shaped. Although craters have not yet been resolved on Adrastea and Metis, their irregular shapes strongly suggest that they are cratered. All four moons are named for characters in mythology relating to Jupiter (Zeus, in Greek mythology). (NASA/JPL, Cornell University)

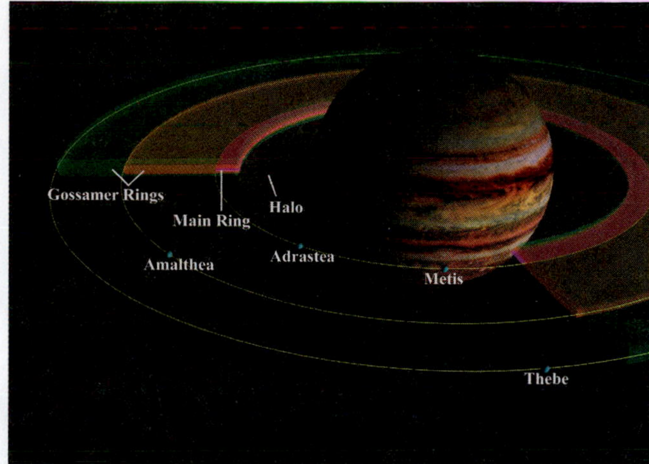

a

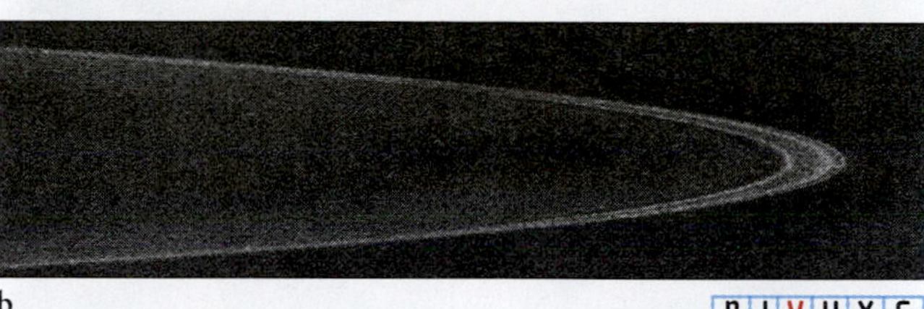

b

R I V U X G

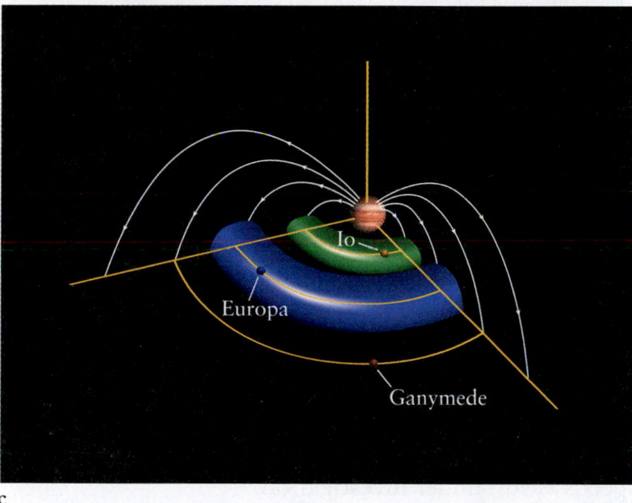

c

FIGURE 8-19 Jupiter's Ring and Torus (a) Cutaway diagram of Jupiter's rings, which are generated from debris blasted off the inner moons Adrastea, Metis, Amalthea, and Thebe. (b) A portion of Jupiter's faint ring system, photographed by the *New Horizons* spacecraft heading to Pluto. The outer three bright rings are composed of pebble- to rock-sized fragments. The rest is mostly dust. The brightest portion of the ring is about 6000 km wide. While the outer edge of the ring is sharply defined, the inner edge is somewhat fuzzy. A tenuous sheet of material extends from the ring's inner edge all the way down to the planet's cloud tops. (c) Quarter images of Io's and Europa's tori (also called *plasma tori* because the gas particles in them are charged, that is, the gases are plasmas). Some of Jupiter's magnetic field lines are also drawn in. Plasma from tori flow inward along these field lines toward Jupiter. (a: NASA/JPL/Cornell University; b: NASA/Johns Hopkins University Applied Physics Laboratory/Southwest Research Institute; c: NASA/JPL/Johns Hopkins University Applied Physics Laboratory)

boulders. The rest of the ring, extending about 10,000 km inward toward Jupiter, is composed of dust-sized particles. This ring is less than 50 km (30 mi) thick, which is exceptionally thin compared to its width. Apparently some of the smallest particles leak inward, creating the halo ring (see Figure 8-19a).

> What other objects that we have already studied in the solar system are similar to the smaller moons of Jupiter?

Debris ejected from the moons Amalthea and Thebe create thicker rings of very tiny dust particles called the *gossamer rings*. While some of the ring material is continually being ejected by collisions between particles and by radiation from the Sun, impacts freeing regolith from the moons are continually replenishing all the rings.

At least two doughnut-shaped regions (or tori) of electrically charged gas particles, called *plasmas,* orbit Jupiter. One is in the same orbit as Io (Figure 8-19c). The Io torus consists of sulfur and oxygen ions (charged atoms) along with free electrons. These particles were ejected by Io's geysers, and they are held in orbit by Jupiter's strong magnetic field. Guided by the field, some of this matter spirals in toward Jupiter, thereby creating the aurora seen there (see Figure 8-9b).

The other torus is in Europa's orbit (Figure 8-19c), and consists of hydrogen and oxygen ions and electrons created from water molecules kicked off Europa's surface by radiation from Jupiter. The mass of this ring of gas is calculated to be only about 5.4×10^4 kg (6×10^4 tons).

SATURN

Saturn, with its ethereal rings, presents the most spectacular image of the planets (Figure 8-20). Giant Saturn has 95 times as much mass as Earth, making it second in mass and size to Jupiter. Like Jupiter, Saturn has a thick, active atmosphere composed predominantly of hydrogen. Indeed, the percentage of hydrogen is higher in Saturn's atmosphere than in Jupiter's atmosphere. Saturn also has a strong magnetic field. Ultraviolet images from the Hubble Space Telescope reveal aurorae around Saturn differing from those of Jupiter in that Saturn's are spirals (see Figure 8-20), while Jupiter's are rings (see Figure 8-9b).

8-9 Saturn's atmosphere, surface, and interior are similar to those of Jupiter

Partly obscured by the thick, hazy atmosphere above them, Saturn's clouds lack the colorful contrast visible on Jupiter. Nevertheless, photographs do show faint stripes in Saturn's atmosphere (Figure 8-21a), similar to Jupiter's belts and zones. Their existence indicates that Saturn also has internal heat that transports the cloud

SATURN: VITAL STATISTICS

Planet symbol:	♄
Average distance from the Sun:	9.57 AU = 1.43×10^9 km
Maximum distance from the Sun:	10.1 AU = 1.51×10^9 km
Minimum distance from the Sun:	9.06 AU = 1.36×10^9 km
Orbital eccentricity:	0.056
Average orbital speed:	9.64 km/s
Orbital period:	29.4 years
Rotation period:	10 h 13 m 59 s (equatorial) 10 h 39 m 25 s (internal)
Inclination of equator to orbit:	26.7°
Inclination of orbit to ecliptic:	2.48°
Diameter:	120,536 km = 9.45 Earth diameters (equatorial) 108,680 km = 8.52 Earth diameters (polar)
Mass:	5.69×10^{26} kg = 95.2 Earth masses
Average density:	687 kg/m^3
Escape speed:	35.5 km/s
Surface gravity (Earth = 1):	0.92
Albedo:	0.46
Average temperature at cloud tops:	−180°C = −292°F = 93 K
Atmospheric composition (by number of molecules):	96.3% hydrogen (H_2), 3.3% helium (He), 0.4% methane (CH_4), ammonia (NH_3), water vapor (H_2O), and other gases

Spiral aurora

Saturn

Earth

R I V U X G

FIGURE 8-20 Saturn and Its Vital Statistics Combined visible and ultraviolet images from the Hubble Space Telescope reveal spiral arcs of auroras near Saturn's south pole. The inset image of Earth shows its size relative to Saturn. Note that there is much less contrast between Saturn's clouds than those of Jupiter. (NASA, ESA, J. Clarke [Boston University], and Z. Levay [STScI]; inset: NASA)

gases by convection. Changing features in the atmosphere show that Saturn, too, has differential rotation—ranging from about 10 h 14 min at the equator to 10 h 40 min at high latitudes. As on Jupiter, some of the belts and zones move eastward, while others move westward. In 2006, the *Cassini* spacecraft revealed a hexagonal boundary surrounding Saturn's north pole (Figure 8-21b). While belts and zones flow around it, this feature is nearly stationary. The mechanism that creates and maintains the boundary is still under investigation. Although Saturn lacks a long-lived spot like Jupiter's Great Red Spot, it does have storms, including a major one discovered by the Hubble Space Telescope in 1994 and a pair of storms that merged in 2004 (Figure 8-22). In 2004, *Cassini* detected lightning in Saturn's atmosphere.

Saturn's atmosphere is composed of the same gases as Jupiter's. However, because its mass is smaller than Jupiter's, Saturn's gravitational force on its atmosphere is less. Therefore, Saturn's atmosphere is more spread out than that of its larger neighbor (see Figure 8-6). Saturn has impressive surface winds. At its equator, storms have been clocked moving at speeds of 1600 km/h (1000 mph), which is some 10 times faster than Earth's jet streams and 3 times faster than the fastest winds on Jupiter. The reason for Saturn's extremely high wind speeds is still under investigation.

As noted in the introduction to Saturn above, the chemical composition of the planet's atmosphere is different from that of Jupiter. Saturn's atmosphere is about 96.3% hydrogen and 3.3% helium, while Jupiter has about 89.8% hydrogen and 10.2% helium. A related observation is that Saturn emits about 2.3 times as much energy as it receives from the Sun. Unlike Jupiter, whose extra energy is emitted primarily by the Kelvin-Helmholtz mechanism, much of Saturn's extra energy is believed to be generated by a mechanism called *helium rain*. Helium rain works like this: Because it is smaller than Jupiter and farther from the Sun, Saturn's atmosphere is cool enough for the helium in it to condense into a liquid—the helium rain—that then descends deep into the planet. As it strikes the interior, the rain's energy helps heat the planet, which in turn radiates that energy.

Astronomers infer that Saturn's interior structure resembles that of Jupiter. A layer of molecular hydrogen

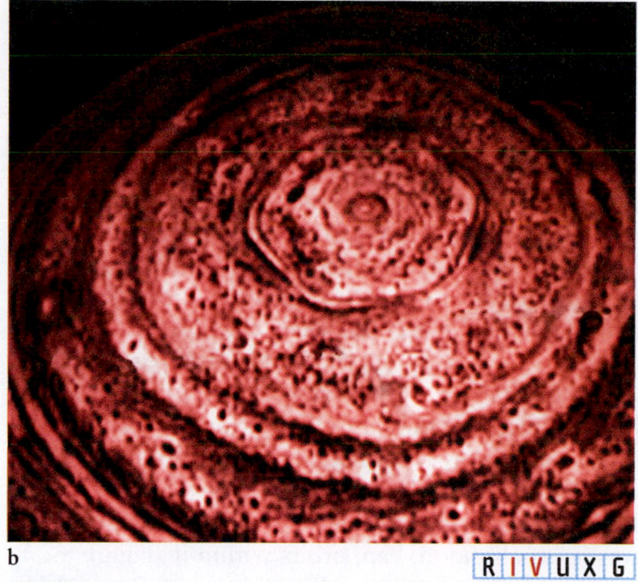

FIGURE 8-21 Belts and Zones on Saturn (a) *Cassini* took this extremely high resolution image of Saturn in 2007. Details as small is 53 km (33 mi) across can be seen. There is less swirling structure between belts and zones on Saturn than on Jupiter. (b) Combining infrared and visible images, *Cassini* took this view of a hexagonal pattern of clouds that rotates much more slowly than the surrounding belts and zones. The pattern's origin is still under investigation. (a: NASA/JPL/Space Science Institute; b: NASA/JPL/University of Arizona)

just below the clouds surrounds a layer of liquid metallic hydrogen, liquid "ices," and a rock and metal core (see Figure 8-8).

Because of its smaller mass, Saturn's interior is also less compressed than Jupiter's, so the pressure inside Saturn is insufficient to convert as much hydrogen into liquid metal. Saturn's rocky core is about 7400 km in radius, while its layer of liquid metallic hydrogen is 12,000 km thick (see Figure 8-8). To alchemists, Saturn was associated with the extremely dense element lead. This is wonderfully ironic in that at 687 kg/m^3, Saturn is, on average, the least dense body in the entire solar system; an object with that density on Earth floats in water.

8-10 Saturn's spectacular rings are composed of fragments of ice and ice-coated rock

Even when viewed through a telescope from the vicinity of Earth, Saturn's magnificent rings are among the most interesting objects in the solar system. They are tilted 27° from the perpendicular to Saturn's plane of orbit, so sometimes they are nearly edge-on to us, making them virtually impossible to see (Figure 8-23). In 1675, Giovanni Cassini discovered an intriguing feature—a dark division in the rings. This 5000-km-wide gap, called the **Cassini division,** separates the dimmer **A ring** from the brighter **B ring,** which lies closer to the planet. By the mid-1800s, astronomers using improved telescopes detected a faint **C ring** just inside the B ring. Figure 8-24 shows a modern perspective of these rings.

The Cassini division occurs because the gravitational force from Saturn's moon Mimas combines with the gravitational force from the planet to keep the region clear of debris. Whenever matter drifts into the Cassini division, Mimas (orbiting at a different rate than the matter in the Cassini division) periodically exerts a force on this matter, eventually pulling it out of the division. This effect is called a **resonance** and is similar to what happens when you repeatedly push someone on a swing

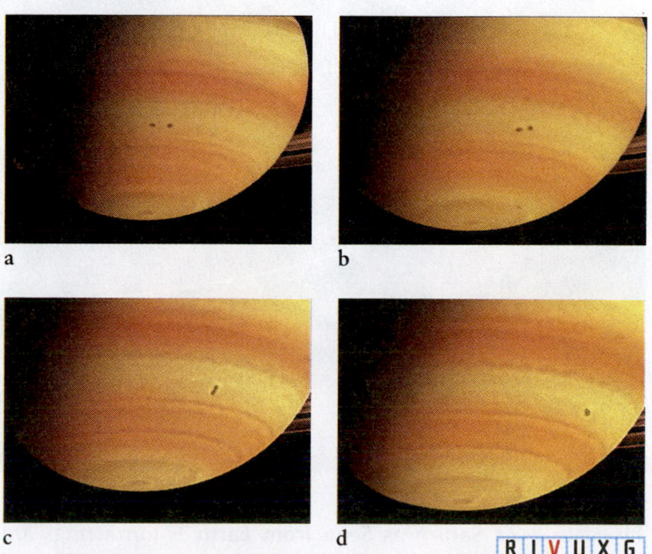

FIGURE 8-22 Merging Storms on Saturn This sequence of *Cassini* images shows two hurricanelike storms merging into one on Saturn in 2004. Each storm is about 1000 km (600 mi) across. (NASA/JPL/Space Science Institute)

What two effects cause Saturn's belt and zone system?

at the right time, giving them energy that enables them to go higher and higher.

A second gap exists in the outer portion of the A ring, named the **Encke division** (Figures 8-24, 8-25), after the German astronomer Johann Franz Encke, who allegedly saw it in 1838. (Many astronomers have argued that Encke's report was erroneous, because his telescope was inadequate to resolve such a narrow gap.) The first undisputed observation of the 270-km-wide division was made by the American astronomer James Keeler in the late 1880s, with the newly constructed 36-in. refractor at the Lick Observatory in California. A 40-km space near the outer edge of the bright A ring is now named the *Keeler gap* in honor of his work. Unlike the Cassini division, the Encke division is kept clear because a small moon, Pan, orbits within it (Figure 8-25). The gravitational tugs of Pan, Daphnis, Prometheus, and other moons also cause the rings to ripple, as shown in the inset of Figure 8-24.

The *Voyager* cameras sent back the first high-quality pictures of the **F ring,** a thin set of ringlets just beyond the outer edge of the A ring (Figure 8-26). Two tiny satellites, Prometheus and Pandora, have orbits on either side of the F ring and serve to keep it intact. The outer of the two satellites orbits Saturn at a slower speed than do the ice particles in the ring, as governed by Kepler's third law. As the ring particles pass near it, they receive a tiny, backward gravitational tug, which slows them down, causing them to fall into orbits a bit closer to Saturn. Meanwhile, the inner satellite orbits the planet faster than the F ring particles. This moon's gravitational force pulls them forward and nudges the particles back into a higher orbit. The combined effect of these two satellites is to focus the icy particles into a well-defined, narrow band about 100 km wide. In 2010, the *Cassini* spacecraft observed snowballs as large as 20 km (12 mi) in diameter forming in the F ring. These bodies form as small bits of the F ring debris swirl around and collide with each other, under the gravitational attractive force of the passing moon Prometheus, which tugs on them. At other times, large pieces collide and break into smaller ones.

Because of their confining influence, Prometheus and Pandora are called **shepherd satellites** or **shepherd moons.** Among the most curious features of the F ring is that the ringlets are sometimes braided or intertwined and sometimes separate. Although stabilized by the shepherd moons, the F ring does ripple and shows varying brightness. Astronomers theorize that the changes in brightness result from collisions between clumps of matter in the ring.

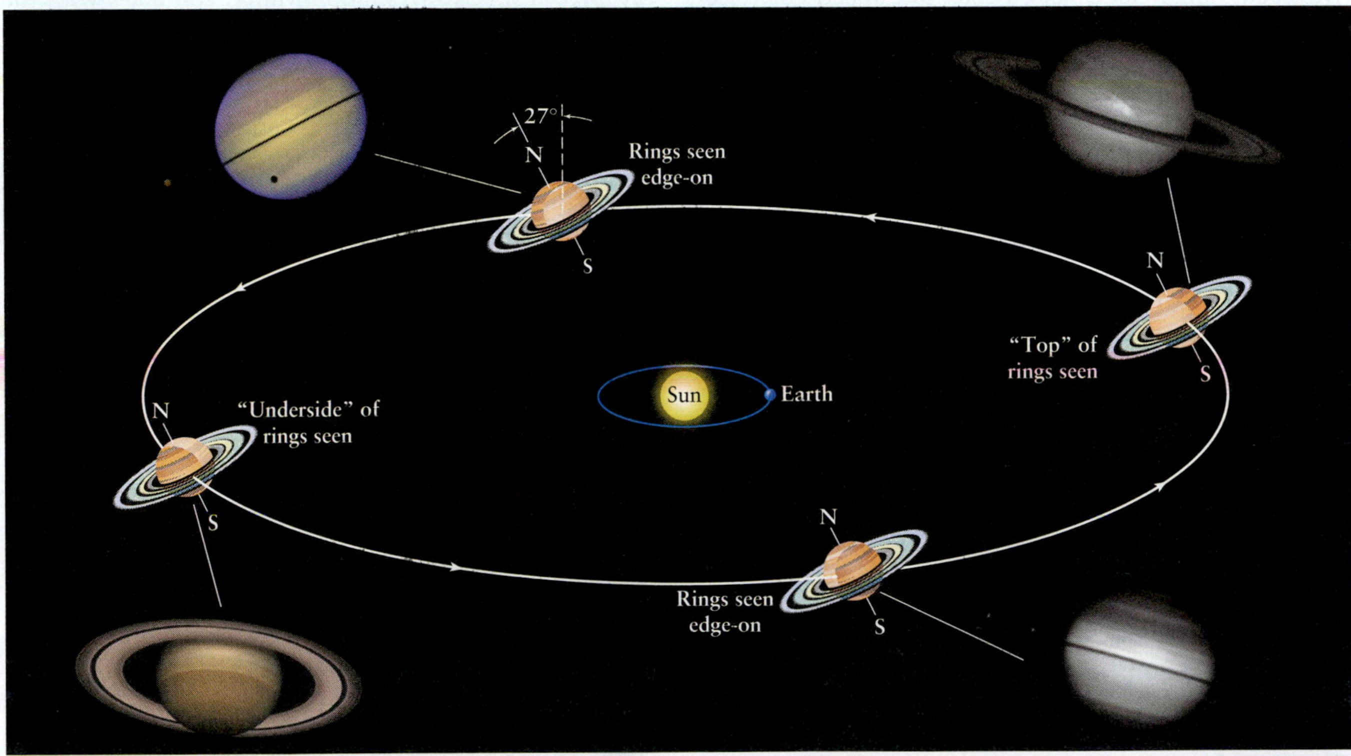

FIGURE 8-23 Saturn as Seen from Earth Saturn's rings are aligned with its equator, which is tilted 27° from the plane of Saturn's orbit around the Sun. Therefore, Earth-based observers see the rings at various angles as Saturn orbits the Sun. The plane of Saturn's rings and equator keeps the same orientation in space as the planet goes around its orbit, just as Earth keeps its 23½° tilt as it orbits the Sun. The accompanying Earth-based photographs show how the rings seem to disappear entirely about every 15 years. (Top left: E. Karkoschka/U. of Arizona Lunar and Planetary Lab and NASA; bottom left: AURA/STScI/NASA; bottom right and top right: A. Bosh/Lowell Obs. and NASA)

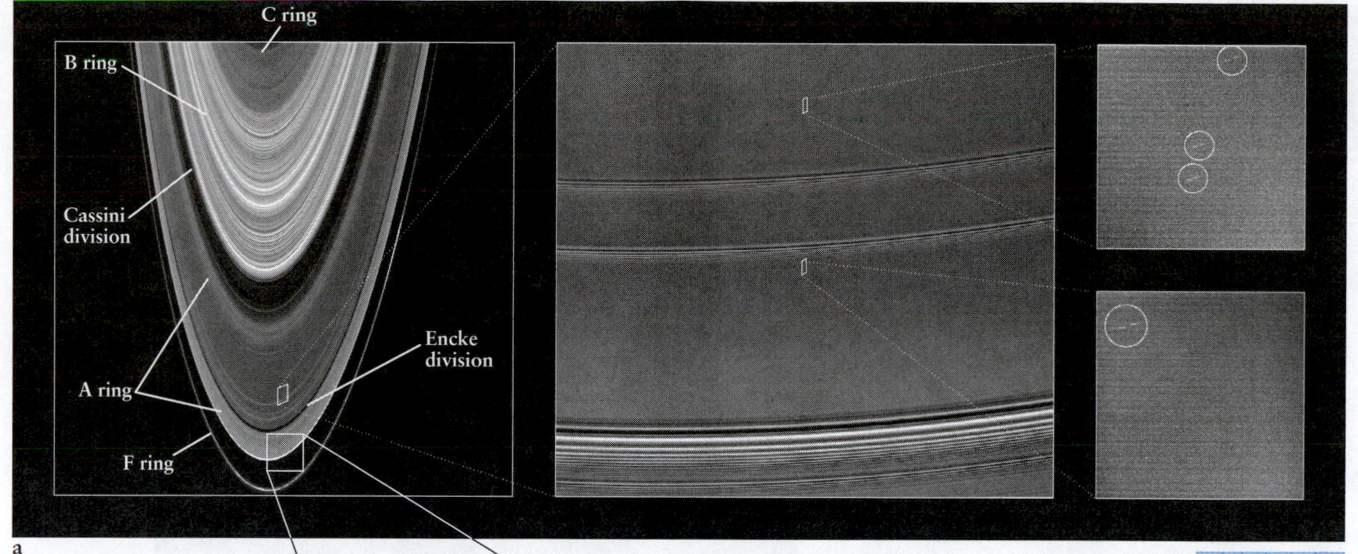

R I V U X G

VIDEO 8.6

VIDEO 8.7

FIGURE 8-24 Numerous Thin Ringlets Constitute Saturn's Inner Rings This *Cassini* image shows some of the structure of Saturn's rings, including some of the moonlets orbiting in them. Inset: As moons orbit near or between rings, they often cause the ring ices to develop ripples, like the grooves in a phonograph record. (NASA/JPL/Space Science Institute)

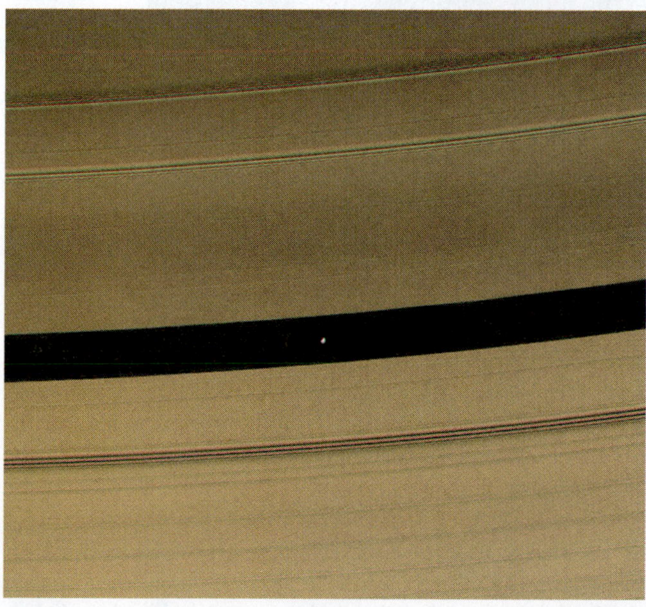

R I V U X G

FIGURE 8-25 The Moon Pan Orbiting in Encke's division, Pan is the "shepherd" moon that keeps the division clear of small debris. It is the innermost known moon of Saturn. (NASA)

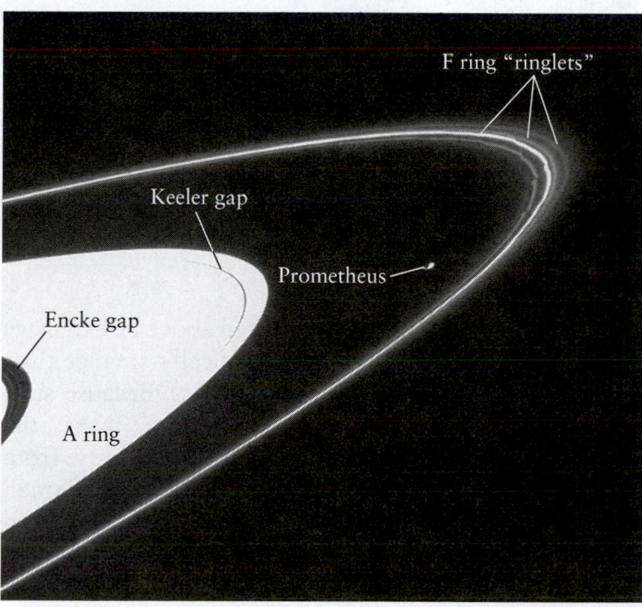

R I V U X G

FIGURE 8-26 The F Ring and One of Its Shepherds Two tiny satellites, Prometheus and Pandora, each measuring about 50 km across, orbit Saturn on either side of the F ring. Sometimes the ringlets are braided, sometimes parallel to each other. In any case, the passage of the shepherd moons causes ripples in the rings, which lead to the formation of large snowballs in them. The gravitational effects of these two shepherd satellites confine the particles in the F ring to a band about 100 km wide. (NASA)

FIGURE 8-27 Saturn's Ring System (a) Photographed with the Sun behind Saturn, the inner and intermediate regions of Saturn's ring system are shown to be very different from each other. Beyond the F ring, the particles are dust- and pebble-sized. (b) Another view of the inner and intermediate rings, where subtle color differences are indicated. (c) Superimposed on this *Cassini* image are labels that indicate how far the rings extend into the moon system of Saturn. Titan (off image on right) is 1.2 million km (750, 000 mi) from the center of Saturn. (a, b: NASA/JPL/Space Science Institute; c: NASA/JPL)

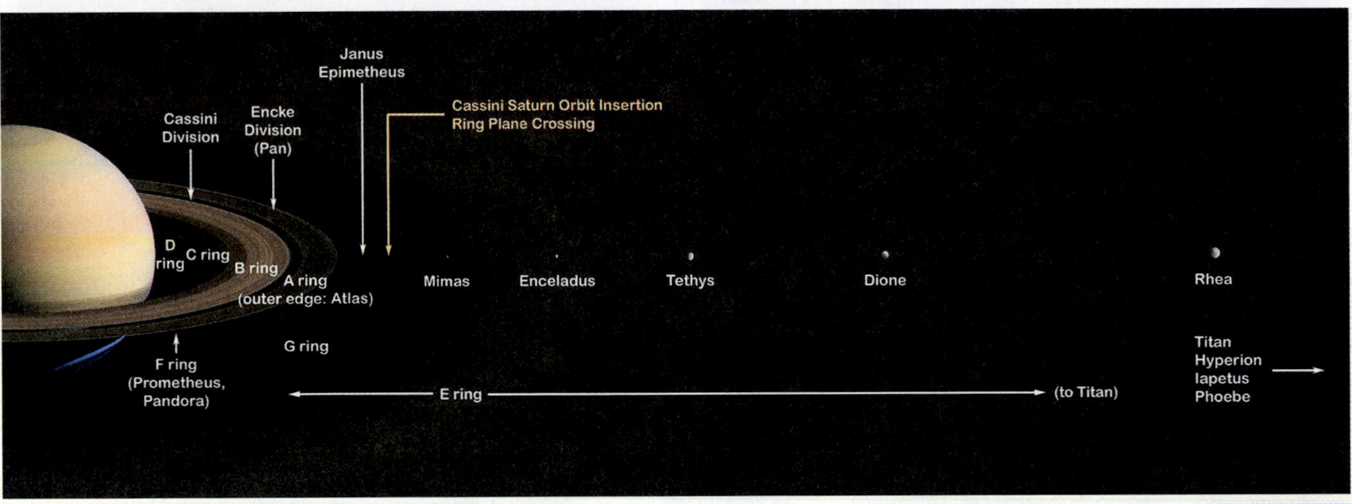

Beyond the major rings described above lie extensive rings of very fine dust particles, typically smaller than a millionth of a meter (a micron) across. Because such rings are very unstable (that is, the particles are drifting away from each other) there must be small moons that stabilize the rings and replace lost debris. The moons Janus, Epimetheus, and Pallene are associated with these rings. Their gravitational attractions slow the loss of the dust, while also providing replacement particles.

Why are there gaps in Saturn's rings?

Figure 8-27a shows a *Cassini* image of part of this system, including a ring discovered in the orbit of the moon Pallene. The *Cassini* image in Figure 8-27b gives us a more comprehensive view of Saturn's ring system. Nevertheless, the rings go out farther still. The ring system shown in Figure 8-27c actually extends out to Titan, 1.2 million km from Saturn. In 2009, astronomers using the Spitzer Space Telescope observed a separate ring in the same orbit as the moon Phoebe (Figure 8-28), 13 million km (8.1 million mi) from Saturn. Called the *Phoebe ring*, its particles are believed to come from that moon, probably ejected as a result of meteorite impacts. Unlike the main ring system, which orbits over Saturn's equator, the Phoebe ring is inclined 27° from the equator.

Returning to the inner rings, pictures from the *Cassini* spacecraft and the Hubble Space Telescope show that the bulk of Saturn's rings lie in a remarkably thin disk—about 10 m (33 ft) thick according to recent observations! This value is amazing when you consider that the total ring system has a width (from inner edge to outer edge) of over 380,000 km (236,000 mi), more than the distance from Earth to our Moon. In 2009, when the rings were seen edge on from Earth, astronomers discovered that small amounts of ring material extend up to 18,000 km above and below the plane of the rings, probably pulled out by the gravitational attraction of Saturn's moons and by the planet's magnetic field. In 2006, *Cassini* discovered a new class of small moons, called *moonlets,* orbiting in Saturn's rings. As many as 10 million of them may exist there. One theory is that they are pieces of a larger body (see Figure 8-24) that broke up to form the rings.

Because Saturn's rings are very bright (their albedo is 0.80), the particles that form them must be highly reflective. Astronomers had long suspected that the rings consist primarily of water ice, along with ice-coated rocks. Spectra taken by Earth-based observatories and by the spacecraft visiting Saturn have confirmed this suspicion. Because the temperature of the rings ranges from 93 K (−90°F) in the sunshine to less than 73 K (−330°F) in Saturn's shadow, frozen water in the rings is in no danger of melting or evaporating.

Saturn's rings are slightly salmon-colored. This coloring suggests that they contain traces of organic molecules, which often have such hues. It appears that the rings have gained this material by being bombarded with debris from the outer solar system.

To determine the size of the particles in Saturn's rings, scientists measured the brightness of different rings from many angles as the *Voyager* and *Cassini* spacecraft flew past the planet. Astronomers also measured changes in radio signals received from these spacecraft as they passed behind the rings. Different rings are composed of debris with different sizes, ranging from particles of the sizes found in wood smoke, through those a few centimeters across, to particles roughly 10 m in diameter. The outermost main ring, the A ring (see Figure 8-24), is composed of these larger clumps of rubble, which frequently collide with one another and thereby change sizes. This process is analogous to the planet-formation activity described in Chapter 5. Of the particles that would be visible to the naked eye (that is, excluding dust particles), the centimeter-sized particles are most common.

High-resolution images reveal that the ring structure seen from Earth actually consists of thousands of 
closely spaced ringlets (Figure 8-29). Even the spaces

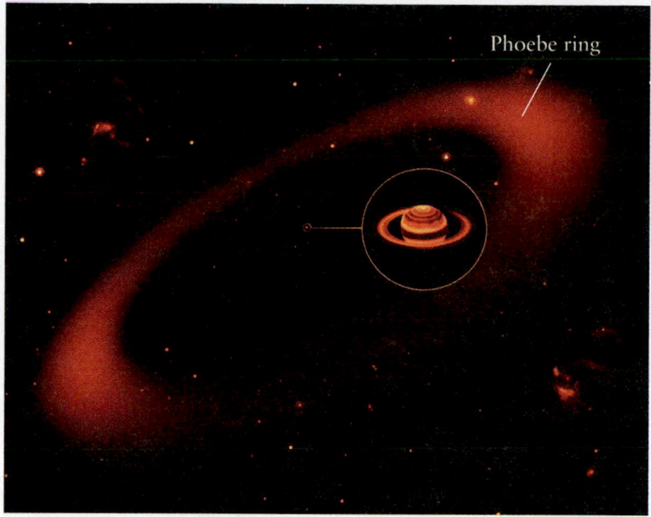

FIGURE 8-28 Saturn and Its Outer, Giant Ring This artist's rendition of Saturn's giant ring is drawn to scale with an infrared image of Saturn and the rings we normally see. The giant ring spans the region from 6 million km (3.7 million mi) to 18 million km (11.1 million mi) beyond Saturn. Put another way, the giant ring is as wide as 30 Saturns placed side by side. (NASA/JPL-Caltech/Keck)

FIGURE 8-29 Spokes in Saturn's Rings Believed to be caused by Saturn's magnetic field moving electrically charged particles that are lifted out of the ring plane, these dark regions move around the rings like the spokes on a rotating wheel. (NASA/JPL/Space Science Institute)

WHAT IF . . . A Comet Passed Through Saturn's Inner Ring System?

Comets are pieces of rocky and icy space debris orbiting the Sun. As we saw in Section 8-3, they sometimes pass close to, and even strike, planets. Suppose a comet with the diameter (10 km) and mass (about 10^{-5} times the total mass of Saturn's rings) of Halley's comet struck Saturn's inner ring system like a rock dropped into a pond. The ring debris at the impact site would become part of the comet, while spiral ripples of ring particles that surrounded the impact site would spread around Saturn. This debris would strike other ring particles. Clumps of debris would stick together, forming larger pieces that could stay in orbit, while much of the ring system would drift into space, never to return. In the end, Saturn would have a much less spectacular set of rings.

that we view from Earth as empty—namely the Cassini and Encke divisions—are filled with myriad smoke-sized dust particles. When spacecraft *Cassini* passed through a gap in the ring plane between the F and G rings (see Figure 8-27a), it was peppered with 100,000 impacts from these dust particles in less than 5 min.

Saturn's shell of liquid metallic hydrogen produces a planetwide magnetic field that apparently affects its rings. Due to Saturn's much smaller volume of liquid metallic hydrogen, the magnetic field at Saturn's surface is only about two-thirds as strong as the magnetic field here on Earth. Data from spacecraft show that Saturn's magnetosphere contains radiation belts similar to Earth's Van Allen belts. Furthermore, dark **spokes** move around Saturn's rings (see Figure 8-29); these are believed to be created by electric charges on the ring material interacting with the planet's magnetic field. The magnetic field lifts charged particles out of the plane in which the rings orbit. Spreading the particles out decreases the light scattered from them and therefore makes the rings appear darker. The magnetic field causes the regions of spread particles to change, making the spokes appear to revolve around the planet.

8-11 Titan has a thick atmosphere, clouds, and lakes filled with liquids

Only 7 of Saturn's 62 known moons are spherical. Four of them are shown in Figure 8-30a–c, e. The other 58 moons are oblong, suggesting that they are captured asteroids. These oblong satellites include Phoebe (Figure 8-30d) and Hyperion (Figure 8-30f). Twelve oblong moons discovered in 2001 move in clumps, suggesting that they are pieces of a larger body that was broken up by impacts.

a Mimas (diameter 392 km)

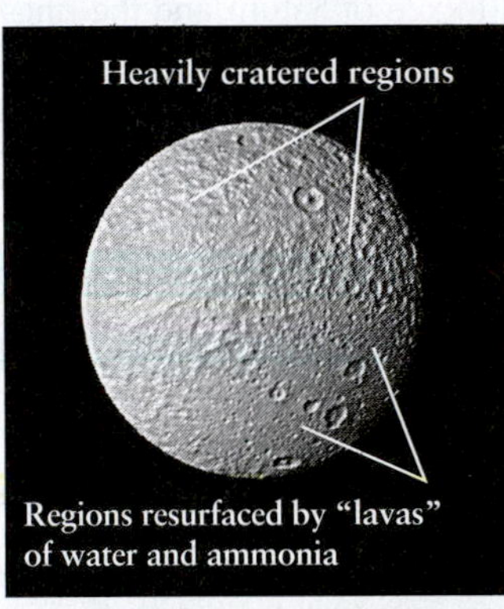

b Tethys (diameter 1060 km)

c Dione (diameter 1120 km)

d Phoebe (diameter 220 km)

RIVUXG

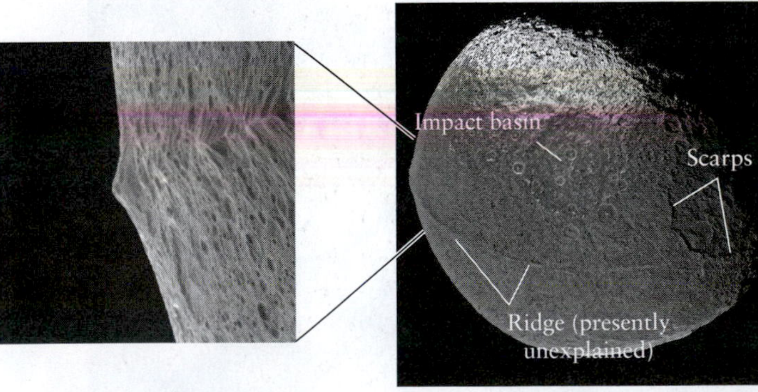

e Iapetus (diameter 1460 km)

f Hyperion (diameter 410 km)

FIGURE 8-30 Saturn's Diverse Moons (a–c) These *Voyager 1, Voyager 2,* and *Cassini* images show the variety of surface features seen on three of Saturn's seven spherical moons. They are not shown to scale (refer to the diameters given below each image). (d) In comparison, this *Cassini* image shows the nonspherical moon Phoebe, almost as dark as coal, carrying many craters, landslides, grooves, and ridges. Phoebe is barely held in orbit by Saturn. Astronomers believe that it was captured after wandering in from beyond the orbit of Neptune. Two particularly intriguing moons are (e) Iapetus and (f) Hyperion. The ridge running along the equator of Iapetus is believed to have developed as the moon formed. Apparently Iapetus cooled so rapidly that the ridge did not have time to settle away. Perhaps the most bizarre object photographed in the solar system, Hyperion, shows innumerable impact craters. These features are different from craters seen in other objects in that the crater walls here have not filled in the bottom of the craters. This moon's low gravity and the pull of nearby Titan may explain this unusual phenomenon. (NASA/JPL/Space Science Institute)

Saturn's largest moon, Titan, is second in size to Ganymede among the moons of the solar system and is the only moon in the solar system to have a dense atmosphere. Like many of the other moons of Saturn, Titan is in synchronous rotation around the planet. About 10 times more gas lies above each square meter of Titan's surface than lies above each square meter of Earth. As a result, the air pressure on the surface of Titan is about 1.5 times the air pressure we feel on Earth.

Christiaan Huygens discovered Titan in 1655, the same year he proposed that Saturn has rings. By the early 1900s, several scientists had begun to suspect that Titan might have an atmosphere. This moon is larger than (although less massive than) Mercury and far colder than that planet. Calculations reveal that Titan is cool enough and massive enough to retain heavy gases in its atmosphere. Because of this, Titan was a primary target for the *Voyager* missions (Figure 8-31a). Unexpectedly, its thick haze completely blocked any view of its surface. Observations also reveal that clouds of methane form and dissipate seasonally near both of Titan's poles and in the mid-southern latitudes. Although the causes for

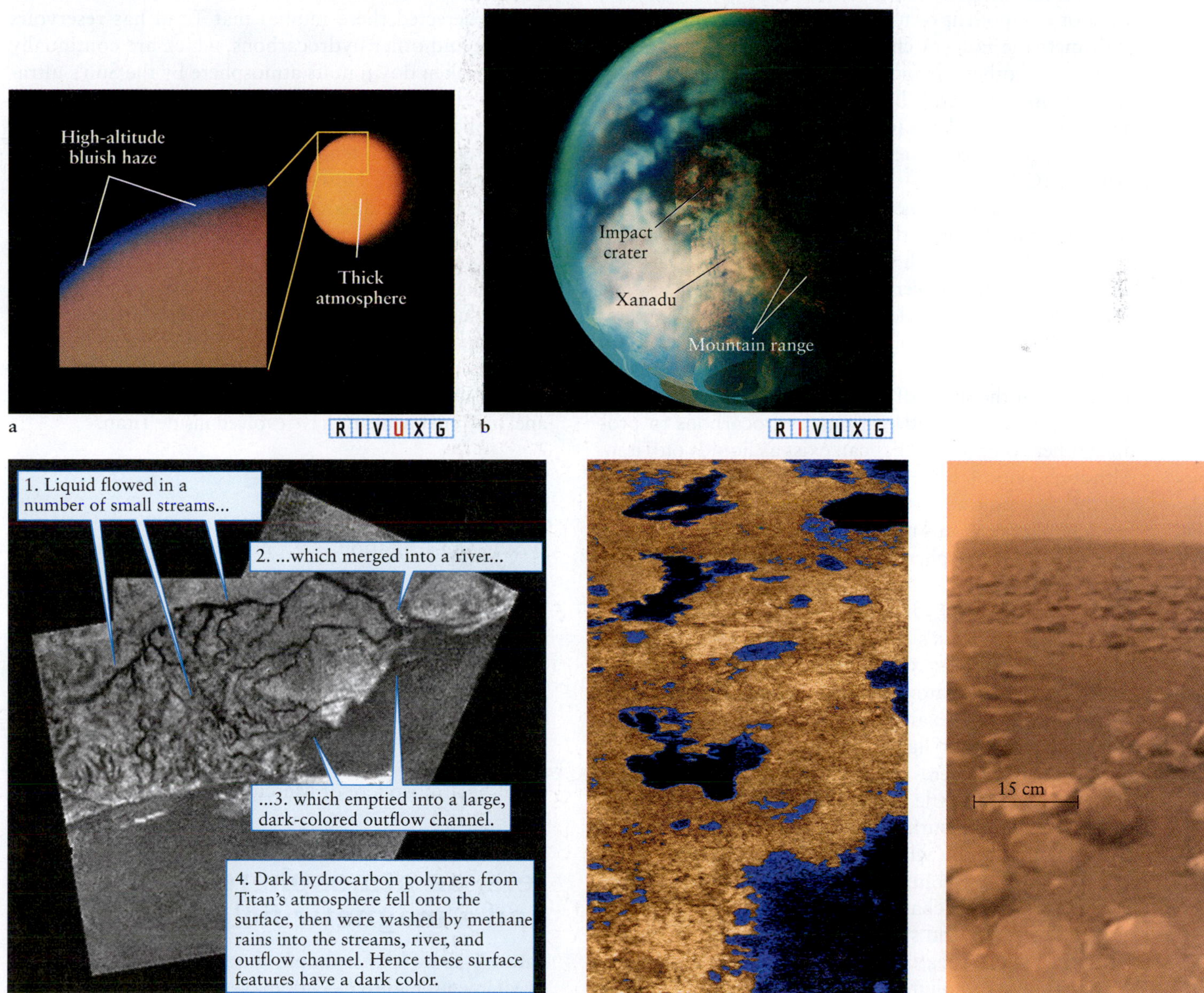

FIGURE 8-31 Surface Features on Titan (a) *Voyager* images of Titan's smoggy atmosphere. (b) *Cassini* image of Titan (diameter 5150 km) showing lighter highlands, called Xanadu, and dark, flat, lowlands that may be hydrocarbon seas. Resolution is 4.2 km (2.6 mi). (c) Riverbeds meandering across the Xanadu highlands of Titan. These are believed to have been formed by the flow of liquid methane and ethane. (d) Lakes filled with liquid methane and ethane. (e) The *Huygens* probe took this image at Titan's surface on January 14, 2005. What appear like boulders here are actually pebbles strewn around the landscape. The biggest ones are about 15 cm (6 in.) across. (a: NASA/JPL/Space Science Institute; b: NASA/JPL/University of Arizona; c, e: NASA/JPL/ESA/University of Arizona; d: Cassini Radar Mapper/JPL/ESA/NASA)

the appearance and disappearance of these clouds are not all known, volcanoes, geysers, and evaporation from Titan's lakes are all considered as possible factors.

Voyager data indicated that roughly 90% of Titan's atmosphere is nitrogen. Most of this gas probably formed from the breakdown of ammonia (NH_3) into hydrogen and nitrogen atoms under the influence of the Sun's ultraviolet radiation. Because Titan's gravity is too weak to retain hydrogen, this gas has escaped into space, leaving behind ample nitrogen. The unbreathable atmosphere is about 4 times as dense as that of Earth.

The second most abundant gas on Titan is methane, a major component of natural gas. Sunlight interacting with methane induces chemical reactions that produce a variety of other carbon-hydrogen compounds, or **hydrocarbons.** Spacecraft have detected small amounts of nearly 20 hydrocarbons in Titan's atmosphere, including ethane (C_2H_6), acetylene (C_2H_2), ethylene (C_2H_4), and propane (C_3H_8).

Methane and ethane condense into droplets in the cold air, making Titan the only other world in the solar system besides Earth where rain reaches the surface. The rain, however, is very different on the two worlds. Titan's raindrops are calculated to be 1 cm in diameter (roughly twice the size of large raindrops on Earth) and they fall 6 times slower than rain drops here—rain on Titan falls at the speed of slowly falling snowflakes here. Nitrogen combines with various hydrocarbons to produce other compounds that can exist as liquids on Titan. Although one of these compounds, hydrogen cyanide (HCN), is a poison, some of the others are the building blocks of life's organic molecules. Indeed, biologists who study Titan's atmosphere now think that its chemistry is very similar to that of the young Earth.

Many of the hydrocarbons and carbon-nitrogen compounds in Titan's atmosphere can form complex compounds called **polymers.** These molecules are long repeating chains of atoms; rubber, cellulose, and plastics are the best-known examples on Earth. Scientists hypothesize that droplets of lighter polymers remain suspended in Titan's atmosphere and form a mist, while heavier polymer particles settle down onto Titan's surface.

Images of the surface taken by the *Cassini* spacecraft (Figure 8-31b), which arrived in the Saturn system in 2004, show few craters, indicating that Titan's surface undergoes dynamic change. Despite the fact that activity in its atmosphere and surface continually covers craters and other surface features, Titan does have a (presumably long-lived) mountain range with peaks as high as 1.5 km (1 mi). These peaks are covered with organic matter and possible methane snow.

About 20% of Titan's surface has sand dunes 100 m high. The composition of the sand is not yet clear, although they may be frozen organic molecules. Using the Keck I telescope in 1999, astronomers observed infrared emissions from Titan's surface consistent with a partially liquid surface. Rain and rivers of methane have been observed (Figure 8-31c). In 2007, radar images from *Cassini* revealed a smooth dark surface of over 10^5 km^2 (39,000 mi^2), the largest lake of liquid methane and ethane seen on Titan. Other images, such as Figure 8-31d, reveal similar features elsewhere on that world, including islands in some lakes. Titan's lakes are all located near that world's poles.

Cassini launched a probe, named *Huygens,* that successfully parachuted onto Titan and photographed rocks (Figure 8-31e), highlands, and channels. Upon impact, *Huygens* vaporized methane that had been just centimeters below the surface. The quantity of methane and ethane detected there implies that Titan has reservoirs of these and other hydrocarbons, which are continually being broken down in its atmosphere by the Sun's ultraviolet radiation.

We have little reason to suspect that life exists on Titan's surface because the temperature there, 95 K (−288°F), is prohibitively cold. However, in 2008 astronomers using data from *Cassini* discovered that surface features, including lakes, canyons, even mountains, drift some 15 km per year. They concluded that this drift is possible because a layer of liquid water and ammonia exists 100 km under Titan's surface. That layer acts as a lubricant that allows the surface to move relative to the deep interior, which has been determined to be a mix of rock and ice. The combination of liquid water and organic compounds, such as methane, may enable life to have evolved inside Titan.

8-12 Rhea has ice

Rhea, Saturn's second-largest moon (Figure 8-32a), is in synchronous rotation. Astronomers have observed that its leading face has many more craters than its trailing face. This moon has a density consistent with it being composed mostly of water ice, with rocky debris mixed in. In 2008, *Cassini* observed white and blue ice features on its trailing side that are associated with scarps (long lines of cliffs), like those found on Mercury. In 2010, *Cassini* detected an atmosphere containing 70% oxygen and 30% carbon dioxide. It is about 100 times thicker than the atmosphere of Mercury.

8-13 Enceladus has water jets, an atmosphere, and a magnetic field

Saturn's sixth largest moon, Enceladus (500 km or 310 mi diameter, Figure 8-33a), moved from the sidelines toward center stage in 2005, as *Cassini* began studying it. This bright world has an icy, wrinkled surface similar in appearance to Jupiter's Europa and Ganymede. The fact that these latter two bodies contain liquid water suggests that Enceladus does, too. This theory is supported by observations that entire regions of Enceladus's southern hemisphere are free of craters. Instead, there are ice blocks the size of large houses, which have apparently

FIGURE 8-32 Rhea Rhea (diameter 1530 km) is heavily cratered. The bluish regions on the inset are believed to be ices uncovered as a result of impacts. (NASA/JPL/Space Science Institute; inset: NASA/JPL/Space Science Institute/Universities Space Research Association/Lunar & Planetary Institute)

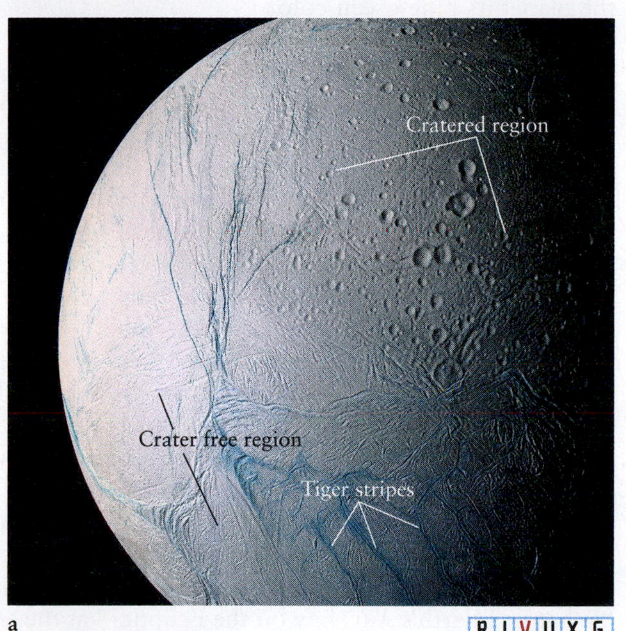

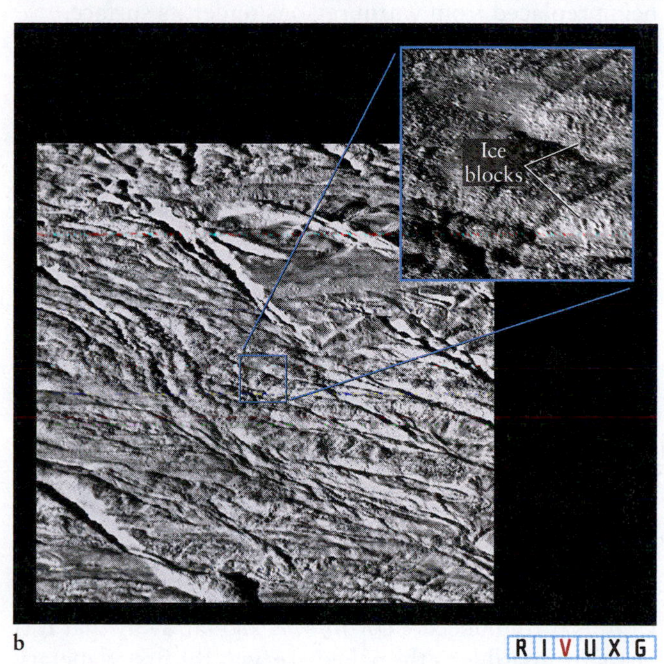

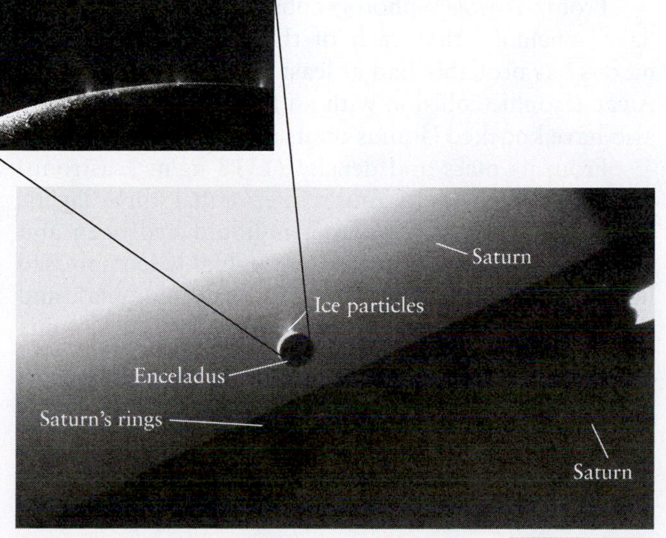

FIGURE 8-33 Enceladus (a) *Cassini* view of the two distinct landscapes on Enceladus, one heavily cratered, the other nearly crater-free. The blue "tiger stripes" are believed to be due to upwelling of liquid that froze at the surface. (b) The crater-free region near the south pole. The ridges are thought to be created by tectonic flows. Inset shows ice boulders. (c) Icy particles ejected from Enceladus may be continually coming out of the moon. (a–c: NASA/JPL/Space Science Institute; inset: NASA/JPL/SSI)

What liquids appear to exist on Titan?

erased many craters due to tectonic plate motion (Figure 8-33b). In addition, the spacecraft revealed that dark lines on Enceladus, dubbed *tiger stripes* (Figure 8-33a), are less than 1000 years old. These features are cracks through which water wells up from inside, freezing into a smooth, dark surface. The tiger stripes are typically 100 km long and 35 km apart.

Cassini's detectors also discovered that Enceladus has a thin atmosphere and a magnetic field (consistent with salty liquid water in its interior). Water vapor has been discovered over the moon's south pole. Water shooting out of Enceladus at speeds up to 2200 km/h (1400 mi/h) from the regions of the tiger stripes was first seen in 2007 (Figure 8-33c). Observations indicate that it leaves Enceladus in relatively steady jets, unlike geysers on Earth. Because this world lacks the gravitational attraction to keep the water it ejects, water is apparently being replaced from warm regions under its surface.

The most promising explanation of what powers the jets and what caused the recent formation of the tiger stripes is the tidal force from Saturn causing Enceladus to flex. Similar to how Jupiter and Europa's gravitational forces cause Io to change shape and thereby heat as it orbits, the gravitational forces from Saturn and its host of other moons cause Enceladus to change shape. This motion creates internal friction and heat that keeps some water inside Enceladus liquid, while also stressing the surface, allowing the tiger stripes to form.

Besides water vapor, the atmosphere of Enceladus is composed of hydrogen, carbon dioxide, nitrogen, carbon monoxide, and organic compounds. *Cassini* also detected dust and salt-rich ice grains surrounding Enceladus. This material is the source of Saturn's E ring.

URANUS

Uranus (pronounced *YUR-uh-nus*), so far away that it is virtually invisible to the naked eye, was the first planetary system discovered by a telescopic search of the heavens. The planet is 19.2 AU from the Sun and takes 84 Earth years to orbit it. Since its discovery in 1781, Uranus has orbited the Sun fewer than 3 times.

8-14 Uranus sports a hazy atmosphere and clouds

Until early 1996, observations of Uranus, the fourth most massive planet in our solar system, revealed few notable features in the visible part of the spectrum. Even the 1986 visit of *Voyager 2* showed a remarkably bland world. It took the Hubble Space Telescope's infrared camera to find what *Voyager*'s visible light camera could not: Uranus has a system of belts and zones. Its hydrogen atmosphere has traces of methane along with a high-altitude haze, under which are clear air and ever-changing methane clouds that dwarf the typical cumulus clouds we see on Earth. Uranus's clouds are towering and huge, each typically as large as Europe. Along with the rest of the atmosphere, the clouds go around the planet once every 16½ h, with variation due to differential rotation, as also occurs on Jupiter and Saturn.

What evidence do astronomers have that Enceladus has liquid water in its interior?

Uranus contains 14½ times as much mass as Earth and is 4 times bigger in diameter (see Figure 8-34). Its outer layers are composed predominantly of gaseous hydrogen and helium. The temperature in the upper atmosphere of the planet is so low (about 73 K, or −330°F) that the methane and water there condense to form clouds of ice crystals. Because methane freezes at a lower temperature than water, methane forms higher clouds over Uranus. Methane efficiently absorbs red light, while scattering blue and green, thereby giving the planet its blue-green color.

Uranus's rotation axis lies very nearly in the plane of its orbit around the Sun (Figure 8-35). Put another way, recall that Earth's rotation axis is tilted about 23½° from being perpendicular to the ecliptic. Uranus's rotation axis is inclined 98° from a line perpendicular to its plane of orbit (see lower left drawing in Figure 8-35). Therefore, its rotation axis lies within 8° of the plane of orbit.

Because its north pole points below the plane of its orbit around the Sun, Uranus is one of only two planets with retrograde rotation (Venus is the other). As Uranus orbits the Sun, its north and south poles alternately point almost directly toward or directly away from the Sun, producing exaggerated seasons (see Figure 8-35). In the summertime, near Uranus's north pole, the Sun is almost directly overhead for many Earth years, at which time southern latitudes are subjected to a continuous, frigid winter night. Forty-two Earth years later, the situation is reversed. For more on this, see the essay "What if… Earth's Axis Lay on the Ecliptic?" at the end of Chapter 1.

From *Voyager* photographs, planetary scientists have concluded that each of the five largest Uranian moons has probably had at least one shattering impact. A catastrophic collision with an Earth-sized object may also have knocked Uranus on its side, as we see it today.

From its mass and density (1318 kg/m^3), astronomers conclude that Uranus's interior has three layers. The outer 30% of the planet is liquid hydrogen and helium, the next 40% inward is highly compressed liquid water (with some methane and ammonia), and the inner 30% is a rocky core (Figure 8-36). Ammonia dissolves easily in water, so an ocean would explain the scarcity of this gas in the planet's atmosphere.

Uranus's surface magnetic field is about three-quarters that of Earth. That strength is reasonable, considering the planet's mass and rotation rate, but everything else about the magnetic field is extraordinary. The field is remarkably tilted—59° from its axis of rotation—and its

URANUS: VITAL STATISTICS

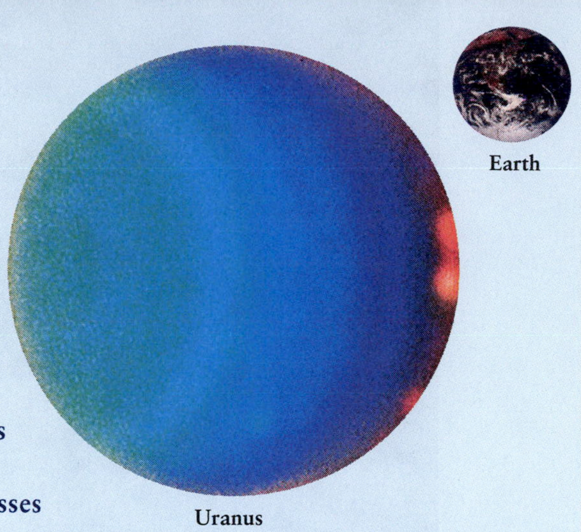

Planet symbol:	⛢
Average distance from the Sun:	19.2 AU = 2.87×10^9 km
Maximum distance from the Sun:	20.0 AU = 3.00×10^9 km
Minimum distance from the Sun:	18.4 AU = 2.75×10^9 km
Orbital eccentricity:	0.047
Average orbital speed:	6.83 km/s
Orbital period:	84.1 years
Rotation period (internal):	17.24 h
Inclination of equator to orbit:	97.9°
Inclination of orbit to ecliptic:	0.77°
Diameter:	57,118 km = 4.01 Earth diameters (equatorial)
Mass:	8.68×10^{25} kg = 14.5 Earth masses
Average density:	1318 kg/m^3
Escape speed:	21.3 km/s
Surface gravity (Earth = 1):	0.90
Albedo:	0.56
Average temperature at cloud tops:	−218°C = −360°F = 55 K
Atmospheric composition (by number of molecules):	82.5% hydrogen (H_2), 15.2% helium (He), 2.3% methane (CH_4)

RIVUXG

NEPTUNE: VITAL STATISTICS

Earth

Neptune

Planet symbol:	Ψ
Average distance from the Sun:	30.1 AU = 4.50×10^9 km
Maximum distance from the Sun:	30.4 AU = 4.54×10^9 km
Minimum distance from the Sun:	29.8 AU = 4.45×10^9 km
Orbital eccentricity:	0.009
Average orbital speed:	5.5 km/s
Orbital period:	164.8 years
Rotation period (internal):	16.11 h
Inclination of equator to orbit:	29.6°
Inclination of orbit to ecliptic:	1.77°
Diameter:	49,528 km = 3.88 Earth diameters (equatorial)
Mass:	1.024×10^{26} kg = 17.1 Earth masses
Average density:	1638 kg/m^3
Escape speed:	23.5 km/s
Surface gravity (Earth = 1):	1.1
Albedo:	0.41
Average temperature at cloud tops:	−218°C = −360°F = 55 K
Atmospheric composition (by number of molecules):	79% hydrogen (H_2), 18% helium (He), 3% methane (CH_4)

RIVUXG

FIGURE 8-34 Uranus, Earth, and Neptune These images of Uranus, Earth, and Neptune are to the same scale. Uranus and Neptune are quite similar in mass, size, and chemical composition. Both planets are surrounded by thin, dark rings, quite unlike Saturn's, which are broad and bright. The clouds on the right of Uranus (false-color pink) are each the size of Europe. (NASA)

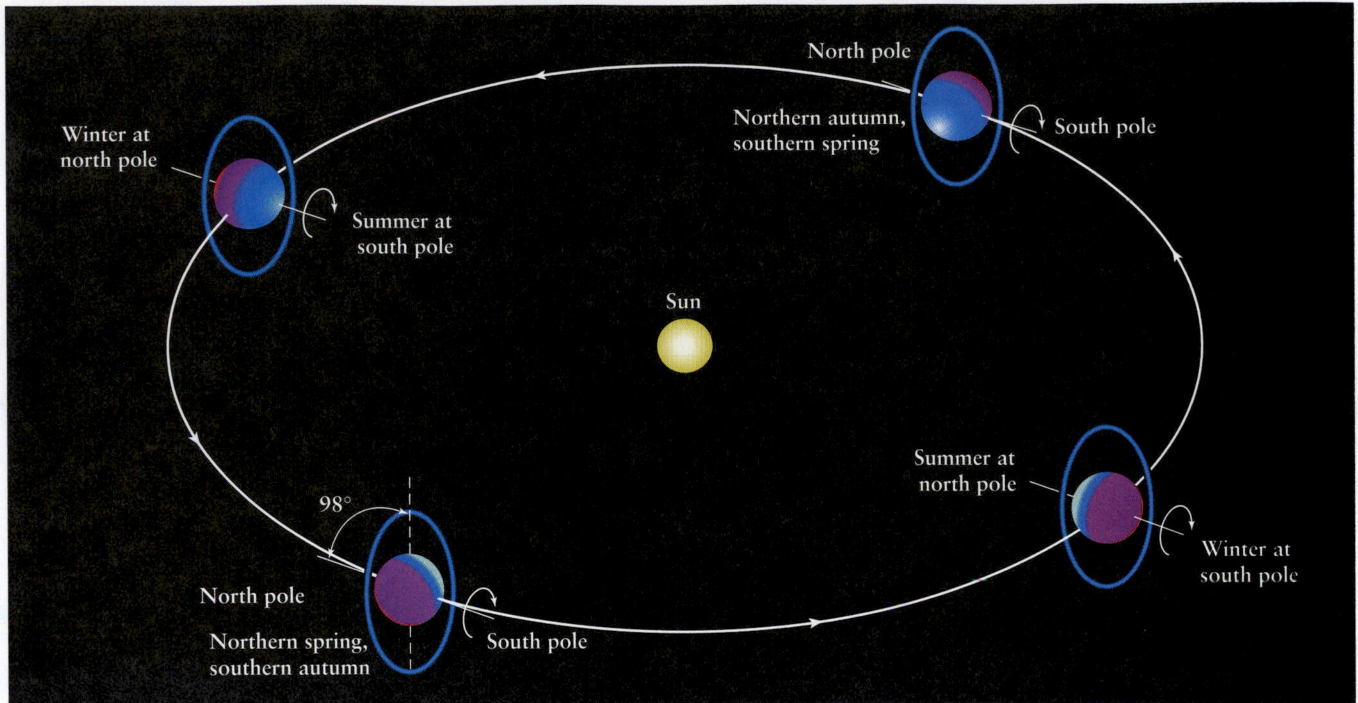

ANIMATION 8.2 **FIGURE 8-35 Exaggerated Seasons on Uranus** Uranus's axis of rotation is tilted so steeply that it lies nearly in the plane of its orbit. Seasonal changes on Uranus are thus greatly exaggerated. For example, during midsummer at Uranus's south pole, the Sun appears nearly overhead for many Earth years, during which time the planet's northern regions are subjected to a long, continuous winter night. Half an orbit later, the seasons are reversed.

> Besides a terrestrial world, what dominates the composition of Uranus's interior?

axis does not even pass through the center of the planet (Figure 8-37).

Because of the large angle between the magnetic field of Uranus and its rotation axis, the magnetosphere of Uranus wobbles considerably as the planet rotates. Astronomers use this wobble to determine the interior rotation speed of the planet, which rotates once every 17 h 14 min. The interior rotation rate is considerably slower than the rotation rate of the clouds on its surface. A rapidly changing magnetic field like this occurs elsewhere in astronomy. For example, it helps explain the behavior of pulsars, a type of star we will study in Chapter 13.

8-15 A system of rings and satellites revolves around Uranus

Uranus has 13 known rings, each divided into ringlets (Figure 8-38). The first nine of the thin, dark rings were discovered accidentally in 1977, when Uranus passed in front of a star. The star's light was momentarily blocked by each ring, thereby revealing their existence

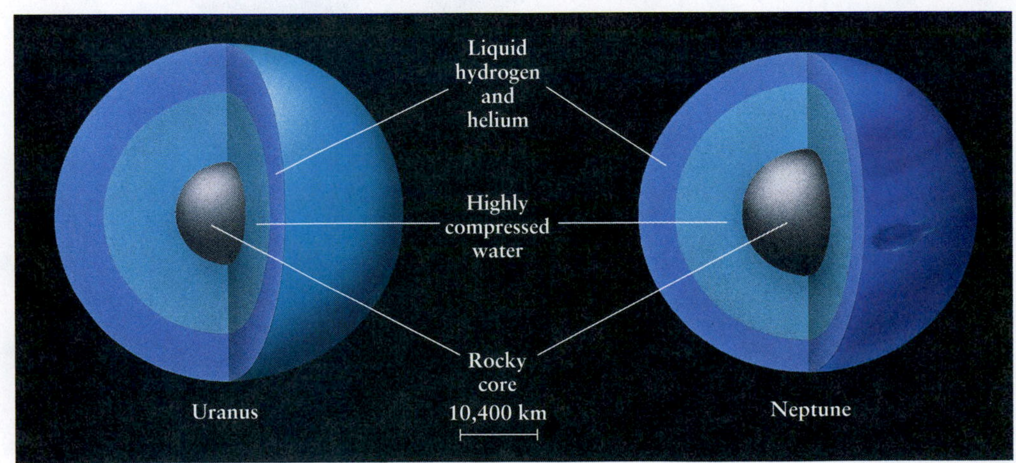

ANIMATION 8.3 **FIGURE 8-36 Cutaways of Uranus and Neptune** The interiors of both Uranus and Neptune are believed to have three regions: a terrestrial rocky core surrounded by a liquid water mantle, which is surrounded, in turn, by liquid hydrogen and helium. Their atmospheres are thin layers at the top of their hydrogen and helium layers.

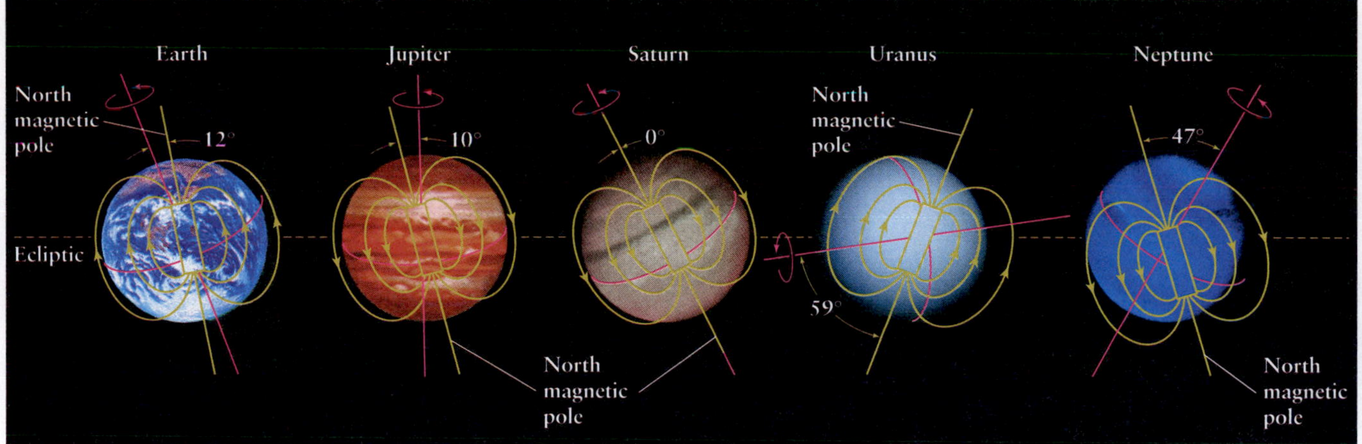

FIGURE 8-37 The Magnetic Fields of Five Planets This drawing shows how the magnetic fields of Earth, Jupiter, Saturn, Uranus, and Neptune are tilted relative to their rotation axes. Note that the magnetic fields of Uranus and Neptune are offset from the centers of the planets and steeply inclined to their rotation axes. Jupiter, Saturn, and Neptune have north magnetic poles on the hemisphere where Earth has its south magnetic pole.

to astronomers (Figure 8-39). Blocking of the light of a more distant object, such as the star here, by something between it and us, such as Uranus's rings, is called an **occultation.** The other rings were detected by *Voyager* and the Hubble Space Telescope. The rings are held in orbit by shepherd moons, including Cordelia and Ophelia.

Most of Uranus's 27 known moons, like its rings, orbit in the plane of the planet's equator. Five of these satellites, ranging in diameter from 480 to nearly 1600 km, were known before the *Voyager* mission. However, *Voyager*'s cameras discovered 10 additional satellites, each smaller than 50 km across. Still others have since been observed from Earth. Several of these tiny, irregularly shaped moons are among the shepherd satellites.

Comparing observations taken 30 years apart, between the 1970s and the first decade of the twenty-first

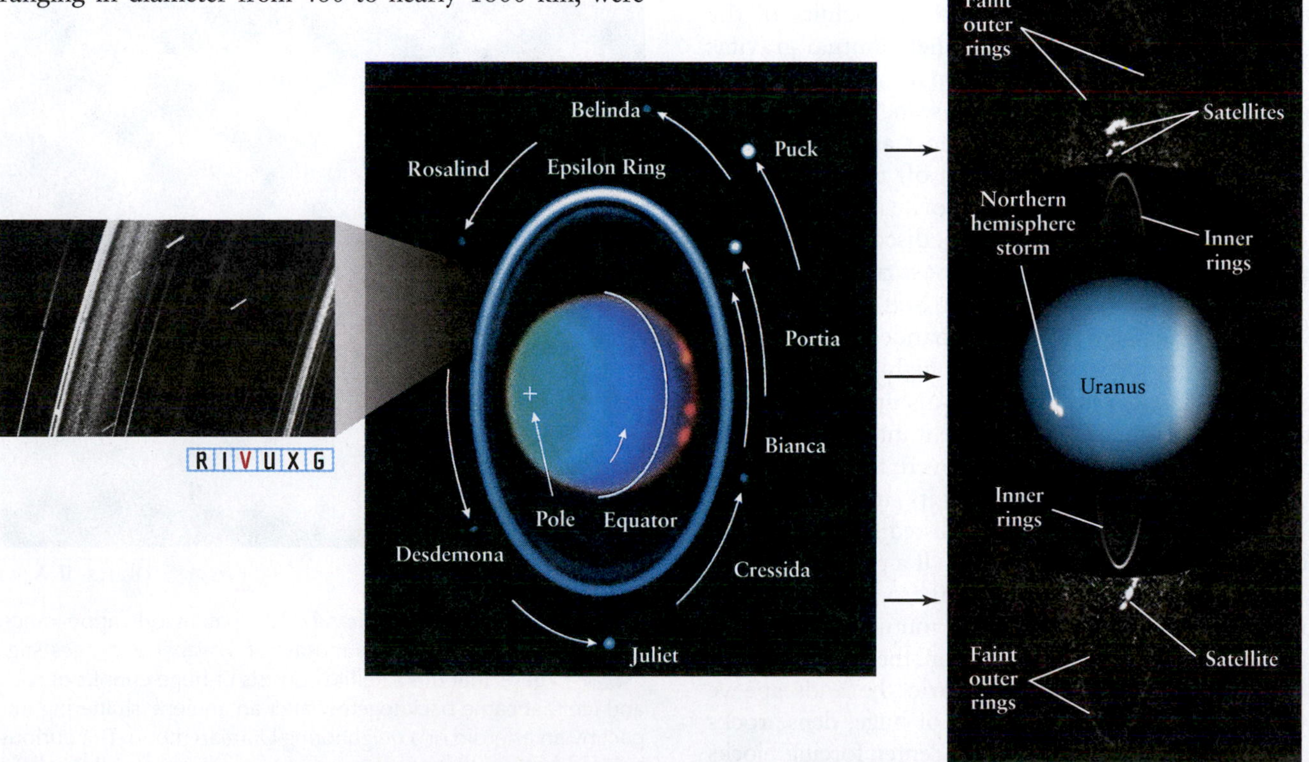

FIGURE 8-38 The Rings and Moons of Uranus (Right) Full-scale image of Uranus and its inner and outer rings. (Center) This image of Uranus, its rings, and eight of its moons was taken by the Hubble Space Telescope. (Left) Close-up of part of the ring system taken by *Voyager 2* when the spacecraft was in Uranus's shadow looking back toward the Sun. Numerous fine dust particles between the main rings gleam in the sunlight. The short streaks are star images blurred because of the spacecraft's motion during the exposure. (NASA; ESA; and M. Showalter, SETI Institute)

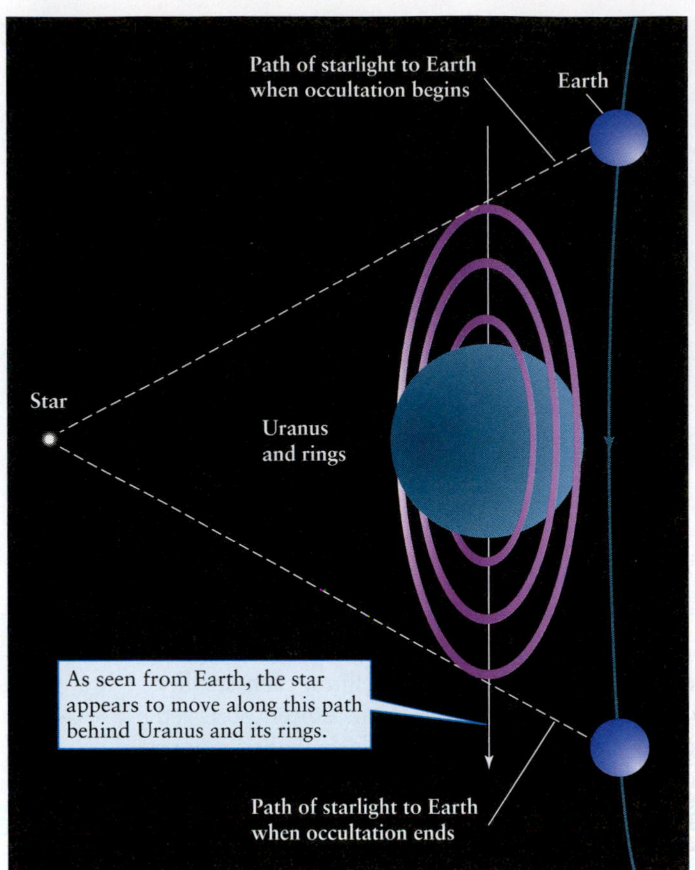

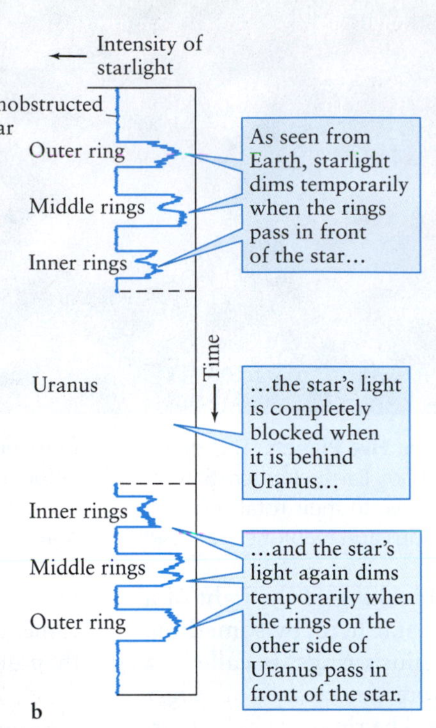

FIGURE 8-39 Discovery of the Rings of Uranus (a) Light from a star is reduced as the rings move in front of it. (b) With sensitive light detectors, astronomers can detect the variation in light intensity. Such dimming led to the discovery of Uranus's rings. Of course, the star vanishes completely when Uranus occults it.

century, reveals that the moons in the vicinity of the rings are changing orbits due to their mutual gravitational influences and the gravitational tugs of the rings. Furthermore, some of the ring debris is being lost and replaced by debris blasted off the moons as a result of meteorite impacts. Observations suggest that the moon Mab, discovered in 2003, is supplying the dust for the outermost ring.

Why do Uranus's rings remain in orbit?

Miranda is the most fascinating and bizarre of Uranus's moons. Unusual wrinkled and banded features cover Miranda's surface (Figure 8-40). Its highly varied terrain suggests that it was once seriously disturbed. Perhaps a shattering impact temporarily broke it into several pieces that then recoalesced, or perhaps severe tidal heating, as we saw on Io, moved large pieces of its surface.

Miranda's core originally consisted of dense rock, while its outer layers were mostly ice. If a powerful impact did occur, blocks of debris broke off from Miranda and then drifted back together through mutual gravitational attraction. Recolliding with that moon, they formed a chaotic mix of rock and ice. In this scenario, the landscape we see today on Miranda is the result of huge, dense rocks trying to settle toward the satellite's center, forcing blocks of less dense ice upward toward the surface.

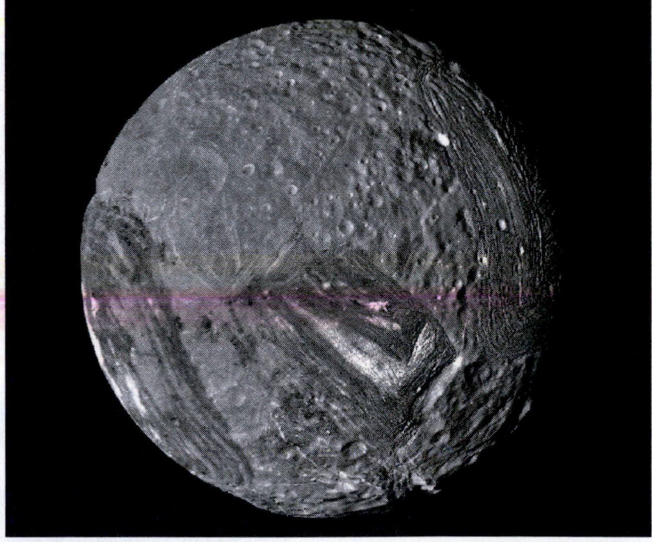

R I V U X G

VIDEO 8.8 **FIGURE 8-40 Miranda** The patchwork appearance of Miranda in this mosaic of *Voyager 2* images suggests that this satellite consists of huge chunks of rock and ice that came back together after an ancient, shattering impact by an asteroid or a neighboring Uranian moon. The curious banded features that cover much of Miranda are parallel valleys and ridges that may have formed as dense, rocky material sank toward the satellite's core. At the very bottom of the image—where a "bite" seems to have been taken out of the satellite—is a range of enormous cliffs that jut upward as high as 20 km, twice the height of Mount Everest. (NASA)

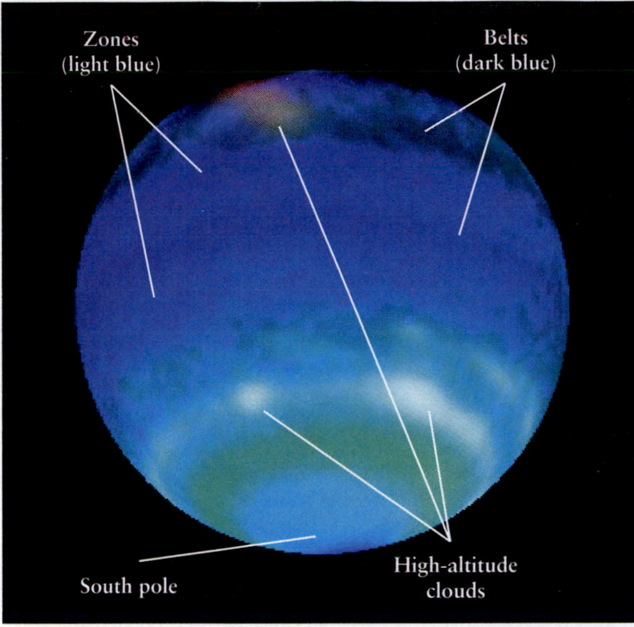

R I V U X G

FIGURE 8-41 Neptune's Banded Structure Several Hubble Space Telescope images at different wavelengths were combined to create this enhanced-color view of Neptune. The dark blue and light blue areas are the belts and zones, respectively. The slightly darker belt running across the middle of the image lies just south of Neptune's equator. White areas are high-altitude clouds, presumably of methane ice. The very highest clouds are shown in yellow-red, as seen at the very top of the image. The green belt near the south pole is a region where the atmosphere absorbs blue light, probably indicating some differences in chemical composition. (Lawrence Sromovsky, University of Wisconsin-Madison and STScI/NASA)

NEPTUNE

Neptune is physically similar to Uranus (review Figures 8-34 and 8-36). Neptune has 17.1 times Earth's mass, 3.88 times Earth's diameter, and a density of 1640 kg/m^3. Its rapid rotation draws its clouds into belts and zones. Unlike on Uranus, however, cloud features can readily be discerned on Neptune. Its whitish, cirruslike clouds consist of methane ice crystals. As on Uranus, the methane absorbs red light, leaving the planet's belts and zones with a banded, bluish appearance (Figure 8-41).

Insight Into Science

Process and Progress Astronomers were able to predict Neptune's position on the basis of theory before it was observed. Likewise, the volcanoes on Io were also predicted. In other cases in science, theories have even correctly predicted results before the technology existed to verify them. Once confirmed, valid theories lead to other predictions and further scientific understanding.

Like the other giant planets, the atmosphere of Neptune experiences differential rotation. The winds on Neptune blow at speeds of up to 2000 km/h—among the fastest known in the solar system. Neptune's rotation axis is tilted nearly 30° from the plane of its orbit around the Sun. Recall that the tilt of Earth's axis creates our seasons. Hubble Space Telescope observations of Neptune in 2003 showed seasonal changes near its poles.

8-16 Neptune was discovered because it had to be there

Neptune's discovery is storied because it illustrates a scientific prediction leading to an expected discovery. Recall from Section 2-8 that in 1781, the British astronomer William Herschel discovered Uranus. The planet's position was carefully plotted, and, by the 1840s, it was clear that even considering the gravitational effects of all of the known bodies in the solar system, Uranus was not following the path predicted by Newton's and Kepler's laws. Either the theories behind these laws were wrong, or there had to be another, yet-to-be-discovered body in the solar system pulling on Uranus.

Independent, nearly simultaneous calculations by an English mathematician, John Adams, and a French astronomer, Urbain Leverrier, predicted the same location for the alleged planet. That planet, Neptune, was located in 1846 by the German astronomer Johann Galle, within a degree or two of where it had to be to have the observed influence on Uranus.

In August 1989, nearly 150 years after Neptune's discovery, *Voyager 2* arrived at the planet to cap one of NASA's most ambitious space missions. Scientists were overjoyed at the detailed, close-up pictures and wealth of data about Neptune sent back to Earth by the spacecraft.

At the time *Voyager 2* passed it, a giant storm raged in Neptune's atmosphere. Called the **Great Dark Spot,** the storm was about half as large as Jupiter's Great Red Spot. The Great Dark Spot (Figure 8-42) was located at about the same latitude on Neptune and occupied a similar proportion of Neptune's surface as the Great Red Spot does on Jupiter. Although these similarities suggested that similar mechanisms created the spots, the Hubble Space Telescope in 1994 showed that the Great Dark Spot had disappeared. Then, in April 1995, another spot developed in the opposite hemisphere.

Neptune's interior is believed to be very similar in composition and structure to that of Uranus: a rocky core surrounded by ammonia- and methane-laden water (see Figure 8-36). Neptune's surface magnetic field is about 40% that of Earth. As with Uranus, Neptune's magnetic axis (the line connecting its north and south magnetic poles) is tilted sharply from its rotation axis. In this case, the tilt is 47°. Also like Uranus, Neptune's

The "spots" on Neptune are analogous to what features on Earth?

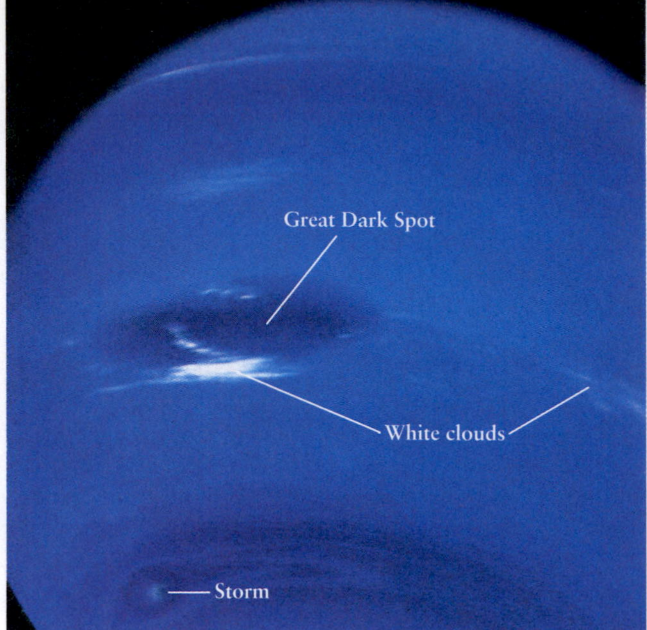

FIGURE 8-42 Neptune This view from *Voyager 2* looks down on the southern hemisphere of Neptune. The Great Dark Spot's longer dimension at the time was about the same size as Earth's diameter. It has since vanished. Note the white, wispy methane clouds. (NASA/JPL)

magnetic axis does not pass through the center of the planet (see Figure 8-37).

We saw in Section 8-2 that the magnetic fields of Jupiter and Saturn are believed to be generated by the motions of their liquid metallic hydrogen. However, Uranus and Neptune lack this material, and their magnetic fields have a different origin. These fields are believed to exist because molecules such as ammonia, which dissolve in their water layers, lose electrons (become ionized). These ions, moving with the planets' rotating, fluid interiors, create the same dynamo effect, and hence the magnetic field.

FIGURE 8-43 Neptune's Rings Two main rings are easily seen in this view alongside overexposed edges of Neptune. In taking this image, the bright planet was hidden so that the dim rings would be visible, hence the black rectangle running down the center of the figure. Careful examination also reveals a faint inner ring. A fainter-still sheet of particles, whose outer edge is located between the two main rings, extends inward toward the planet. (NASA)

8-17 Neptune has rings and has captured its moons

Like Uranus, Neptune is surrounded by a system of thin, dark rings (Figure 8-43). It is so cold at these distances from the Sun that both planets' ring particles retain methane ice. Scientists speculate that eons of radiation damage have converted this methane ice into darkish carbon compounds, thus accounting for the low reflectivity of the rings.

Neptune has 13 known moons. Twelve have irregular shapes and highly elliptical orbits, which suggest that Neptune captured them. Triton, discovered in 1846, is spherical and was quickly observed to have a nearly circular, retrograde orbit around Neptune. It is difficult to imagine how a planet rotating one way and a satellite revolving the other way could form together. Indeed, only the small outer satellites of Jupiter and Saturn have retrograde orbits, and these bodies are probably captured asteroids. Some scientists have therefore suggested that Triton was also captured 3 or 4 billion years ago by Neptune's gravity. Upon being captured, Triton was most likely in a highly elliptical orbit. However, the tides that the moon creates on Neptune's liquid surface would, in turn, have made Triton's orbit more circular.

Triton's average density of about 2100 kg/m^3 indicates that it is an equal mix of rock and ice. It has a thin nitrogen, methane, and carbon dioxide atmosphere, which changes density with the seasons on that world. When Triton was younger, the tidal force exerted on it by Neptune due to the moon's initially elliptical orbit caused Triton to stretch and flex, providing enough energy to melt much of its interior and obliterate its original surface features, including craters. Triton's south polar region is shown in Figure 8-44. Note that very few craters are visible. Calculations based on these observations indicate that Triton's present surface is only about 100 million years old.

Triton does exhibit some surface features seen on other icy worlds, such as long cracks resembling those on Europa and Ganymede. Other features unique to Triton are quite puzzling. For example, the top part of Figure 8-44 reveals a wrinkled terrain that resembles the skin of a cantaloupe. Triton also has a few frozen lakes like the one shown in Figure 8-45. Some scientists have speculated that these lakelike features are the calderas of extinct cryovolcanoes. A mixture of methane, ammonia, and water, which can have a melting point far below that of pure water, could have formed a kind of cold lava on Triton.

Voyager instruments measured a surface temperature of 36 K (−395°F), making Triton the coldest world that our probes have ever visited. Nevertheless, *Voyager*

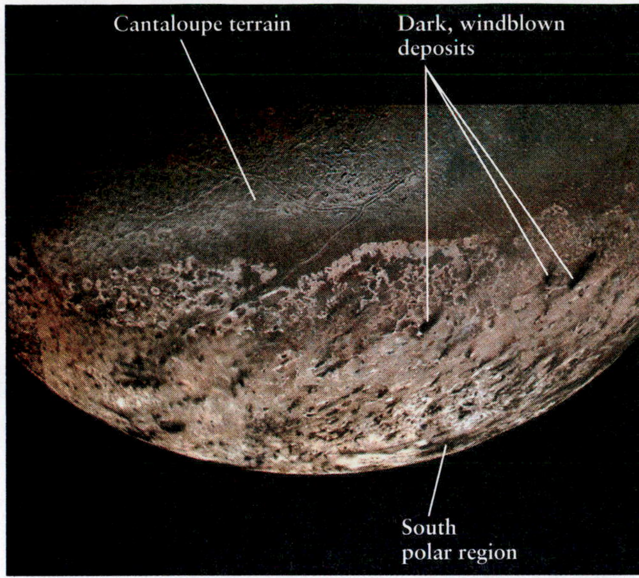

FIGURE 8-44 **Triton's South Polar Cap** Approximately a dozen high-resolution *Voyager 2* images were combined to produce this view of Triton's southern hemisphere. The pinkish polar cap is probably made of nitrogen frost. A notable scarcity of craters suggests that Triton's surface was either melted or flooded by icy lava after the era of bombardment that characterized the early history of the solar system. (NASA/JPL)

VIDEO 8.11

VIDEO 8.12

cameras did glimpse two towering plumes of gas extending up to 8 km above the satellite's surface. These are apparently jets of nitrogen gas warmed by interior radioactive decay and escaping through vents or fissures.

Triton continues to create tides on Neptune. Whereas the tides on Earth cause our Moon to spiral outward, the tides on Neptune cause Triton (in its retrograde orbit) to spiral inward. Within the next quarter of a billion years, Triton will reach the **Roche limit**, the distance at which a planet creates tides on its moon's solid surface high enough to pull its moon apart. Pieces of Triton will then literally float into space until the entire moon is demolished! By destroying Triton, Neptune will create a new ring system that will be much more substantial than its present one (Figure 8-46).

FIGURE 8-45 **A Frozen Lake on Triton** Scientists think that the feature in the center of this image is a basin filled with water ice. The flooded basin is about 200 km across. (NASA)

8-18 Comparative planetology of the outer planets

Now that we have examined the individual outer planets, it is instructive to compare their various properties. Table 8-1, *The Outer Planets: A Comparison*, summarizes much of this material.

Is our Moon inside or outside Earth's Roche limit?

Size and Mass The four giant planets, Jupiter, Saturn, Uranus, and Neptune, are all much larger (roughly 4 to 11 times Earth's

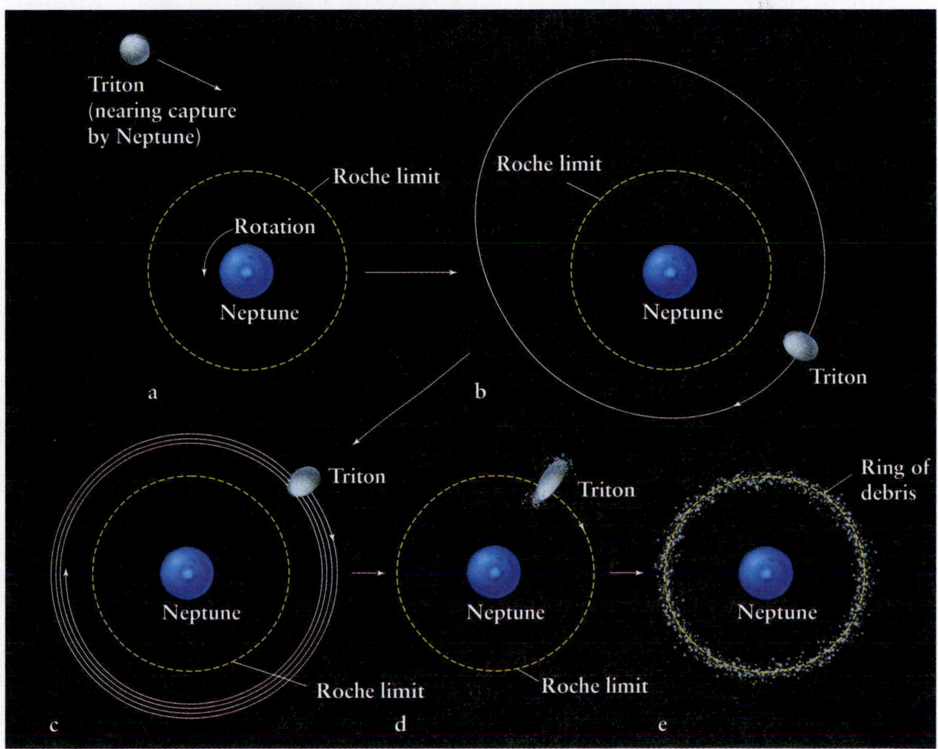

FIGURE 8-46 **The Capture and De-struction of Triton** This series of drawings depicts how (a) Triton was captured by Neptune in a retrograde orbit. (b) The tides that Triton then created on the planet caused that moon's orbit to become quite circular and (c) to spiral inward. (d) It will eventually reach Neptune's Roche limit and (e) be pulled apart to form a ring. (Stephen P. Meszaros/NASA, M. Buie, K. Horne, and D. Tholen)

Table 8-1 The Outer Planets: A Comparison

	Interior	Surface	Rings	Atmosphere	Magnetic Field*
Jupiter	Terrestrial core, liquid metallic hydrogen shell, liquid hydrogen mantle	No solid surface, atmosphere gradually thickens to liquid state, belt and zone structure, hurricanelike features	Yes	Primarily H, He	19,000 × Earth's total field; at its cloud layer, 14 × Earth's surface field
Saturn	Similar to Jupiter, with bigger terrestrial core and less metallic hydrogen	No solid surface, less distinct belt and zone structure than Jupiter	Yes	Primarily H, He	570 × Earth's total field; at its cloud layer, ⅔ × Earth's surface field
Uranus	Terrestrial core, liquid water shell, liquid hydrogen and helium mantle	No solid surface, weak belt and zone system, hurricanelike features, color from methane absorption of red, orange, yellow	Yes	Primarily H, He, some CH_4	50 × Earth's total field; at its cloud layer, 0.73 × Earth's surface field
Neptune	Similar to Uranus	Similar to Uranus	Yes	Primarily H, He, some CH_4	35 × Earth's total field; at its cloud layer, 0.4 × Earth's surface field

For detailed numerical comparisons between planets, see Appendix Tables E-1 and E-2.
*To see the orientations of these magnetic fields relative to the rotation axes of the planets, see Figure 8-37.

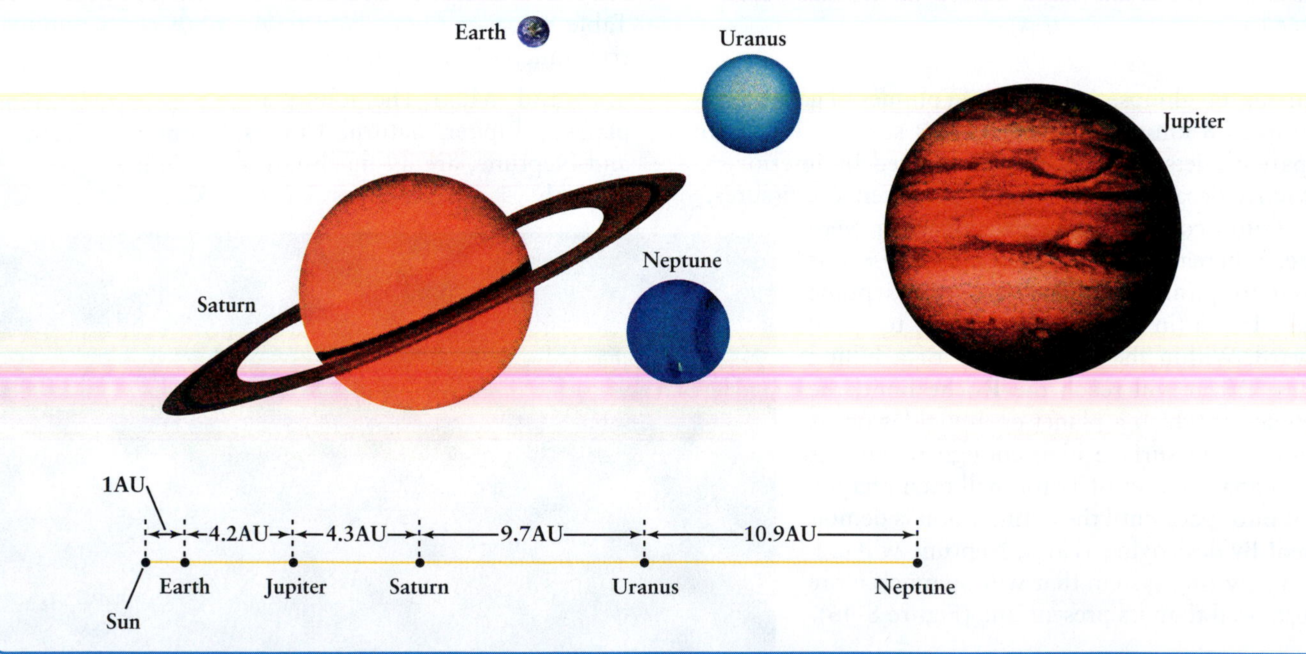

diameter) and more massive (roughly 14 to 318 times Earth's mass) than Earth.

How do the numbers and types of moons found orbiting the giant planets compare to those that orbit the terrestrial planets?

Atmospheres and Rotations The four outer planets have thick hydrogen- and helium-rich atmospheres that are permanently and completely covered with clouds. All of the giant planets rotate rapidly compared to Earth. Their sidereal rotation rates range from about 10 h to about 17¼ h. This rapid rotation draws their clouds into parallel bands, called *belts* and *zones*. Earth's slower rotation allows the clouds and winds here to roam over a much greater range of latitudes than on the giant planets.

Interiors All four giant planets have terrestrial bodies as their cores. That is where their internal similarities to Earth end. Jupiter and Saturn have small amounts of "ice" surrounding their cores. These ices are surrounded by thick liquid metallic hydrogen layers, which, in turn, are surrounded by layers of normal hydrogen and helium. In contrast, Uranus's and Neptune's terrestrial cores are surrounded by large amounts of water, which, in turn, are surrounded by liquid hydrogen and helium.

Magnetic Fields and Rings The four giant planets all have magnetic fields that store particles from the solar wind, similar to Earth's Van Allen belts. Like Earth's magnetic field, those of Jupiter, Uranus, and Neptune are tilted relative to their rotation axes. Saturn's magnetic field is along its rotation axis. The four giants all have rings. Saturn's rings are the most massive and distinctive, followed by those of Uranus, Jupiter, and Neptune.

Moons Among them, the giant planets have at least 164 moons. Each of the giants has between one and a few spherical moons, like our Moon. However, most of the moons in the solar system are much smaller, irregularly-shaped bodies. These objects are most likely captured space debris. Unlike our Moon, all of the satellites of the giants are at least a few hundred thousand times less massive than their planets. Indeed, most moons are millions or billions of times less massive than their planets. Recall that our Moon is only 81 times less massive than Earth.

8-19 Frontiers yet to be discovered

The outer planets hold countless new insights into the formation and evolution of the solar system. Considering that most of the known extrasolar planets are Jupiterlike gas giants, our outer planets also have a lot to tell us about planets throughout our Galaxy. Some specific issues include: What are the dynamics that create the belts and zones of Jupiter? How has the Great Red Spot persisted for so long? Is there life in the underground oceans of the outer Galilean moons? What, exactly, do the individual ring particles around each of the planets look like? How long ago did Saturn's ring system form? How long will it last? Are the moving spokes in Saturn's rings really caused by its magnetic field, and if they are why do the spokes appear as they do? What caused Miranda's surface to become so profoundly disturbed? These are but a few of the questions about the outer planets that are likely to be answered during this century.

SUMMARY OF KEY IDEAS

Jupiter and Saturn

- Jupiter is by far the largest and most massive planet in the solar system.
- Jupiter and Saturn probably have rocky cores surrounded by a thin layer of ice and a thick outer layer of liquid metallic hydrogen and an outer layer of ordinary liquid hydrogen and helium. Both planets have an overall chemical composition very similar to that of the Sun.
- The visible features of Jupiter exist in the outermost 100 km of its atmosphere. Saturn has similar features, but they are much fainter. Three cloud layers exist in the upper atmospheres of both Jupiter and Saturn. Because Saturn's cloud layers extend through a greater range of altitudes, the colors of the Saturnian atmosphere appear muted.
- The colored ovals visible in the Jovian atmosphere are gigantic storms, some of which (such as the Great Red Spot) are stable and persist for years or even centuries.
- Jupiter and Saturn have strong magnetic fields created by electric currents in their metallic hydrogen layers.
- Four large satellites orbit Jupiter. The two inner Galilean moons, Io and Europa, are roughly the same size as our Moon. The two outer moons, Ganymede and Callisto, are approximately the size of Mercury.
- Io is covered with a colorful layer of sulfur compounds deposited by frequent explosive eruptions from volcanic vents. Europa is covered with a smooth layer of frozen water crisscrossed by an intricate pattern of long cracks.
- The heavily cratered surface of Ganymede is composed of frozen water with large polygons of dark, ancient crust separated by regions of heavily grooved, lighter-colored, younger terrain. Callisto has a heavily cratered ancient crust of frozen water.
- Saturn is circled by a system of thin, broad rings lying in the plane of the planet's equator. Each major ring is composed of a great many narrow ringlets that consist of numerous fragments of ice and ice-coated rock. Jupiter has a much less substantial ring system.
- Titan has a thick atmosphere of nitrogen, methane, and other gases, as well as lakes of methane and ethane.
- Enceladus has areas with very different surface features: an older, heavily cratered region and a newer, nearly crater-free surface created by tectonic activity.

Uranus and Neptune

- Uranus and Neptune are quite similar in appearance, mass, size, and chemical composition. Each has a rocky core surrounded by a thick, watery mantle topped by a layer rich in hydrogen and helium; the axes of their magnetic fields are steeply inclined to their axes of rotation; and both planets are surrounded by systems of thin, dark rings.
- Uranus is unique in that its axis of rotation lies near the plane of its orbit, producing greatly exaggerated seasons on the planet.
- Uranus has five moderate-sized satellites, the most bizarre of which is Miranda.
- Triton, the largest satellite of Neptune, is an icy world with a tenuous nitrogen atmosphere. Triton moves in a retrograde orbit that suggests it was captured into orbit by Neptune's gravity. It is spiraling down toward Neptune and will eventually break up and form a ring system.

WHAT DID YOU THINK?

1 *Is Jupiter a "failed star"? Why or why not?* No. Jupiter has 75 times too little mass to shine as a star.

2 *What is Jupiter's Great Red Spot?* The Great Red Spot is a long-lived, oval cloud circulation, similar to a hurricane on Earth.

3 *Does Jupiter have continents and oceans?* No. Jupiter is surrounded by a thick atmosphere primarily composed of hydrogen and helium that gradually becomes liquid as you move inward. The only solid matter in Jupiter is its core.

4 *Is Saturn the only planet with rings?* No. All four giant planets (Jupiter, Saturn, Uranus, and Neptune) have rings.

5 *Are the rings of Saturn solid ribbons?* No. Saturn's rings are all composed of thin, closely spaced ringlets consisting of particles of ice and ice-coated rocks. If they were solid ribbons, Saturn's gravitational tidal force would tear them apart.

Key Terms for Review

A ring, 257
B ring, 257
belt, 241
C ring, 257
Cassini division, 257
differential rotation, 242
Encke division, 258
F ring, 258
Galilean moon (satellite), 247
Great Dark Spot, 271
Great Red Spot, 242
hydrocarbon, 264
liquid metallic hydrogen, 245
occultation, 269
polymer, 264
prograde orbit, 254
resonance, 257
retrograde orbit, 254
ringlet, 254
Roche limit, 273
shepherd satellite (moon), 258
spoke, 262
zone (atmospheric), 241

Review Questions

1. Which is the most massive planet in the solar system? **a.** Earth, **b.** Neptune, **c.** Saturn, **d.** Jupiter, **e.** Mercury

2. Which of the following planets does *not* have rings? Choose only one. **a.** Mars, **b.** Uranus, **c.** Neptune, **d.** Saturn, **e.** Jupiter

3. Which is the least massive planet in the solar system? **a.** Mercury, **b.** Mars, **c.** Uranus, **d.** Jupiter, **e.** Venus

4. Which planet is presently known to have the most moons? **a.** Mars, **b.** Saturn, **c.** Uranus, **d.** Jupiter, **e.** Neptune

5. Describe the appearance of Jupiter's atmosphere. Which features are long-lived and which are relatively fleeting?

6. To test your knowledge of Jupiter's belt and zone structure, do Interactive Exercise 8.1 on the Web. You can print out your results, if required.

7. What causes the belts and zones in Jupiter's atmosphere?

8. To test your knowledge of Jupiter's internal structure, do Interactive Exercise 8.2 on the Web. You can print out your results, if required.

9. What is liquid metallic hydrogen? Which planets contain this substance? What produces this form of hydrogen?

10. Compare and contrast the surface features of the four Galilean satellites, discussing their geologic activity and their evolution.

11. To test your knowledge of the Galilean moons, do Interactive Exercise 8.3 on the Web. You can print out your results, if required.

12. What energy source powers Io's volcanoes?

13. Why are numerous impact craters found on Ganymede and Callisto but not on Io or Europa?

14. Describe the structure of Saturn's rings. What are they made of?

15. To test your knowledge of Saturn's rings, do Interactive Exercise 8.4 on the Web. You can print out your results, if required.

16. Why do features in Saturn's atmosphere appear to be much fainter and more "washed out" than comparable features in Jupiter's atmosphere?

17. Explain how shepherd satellites affect some planetary rings. Is "shepherd satellite" an appropriate term for these objects? Explain your answer.

18. Describe Titan's atmosphere. What effect does sunlight have on it?

19. Describe Titan's surface.

20. Describe the seasons on Uranus. Why are the Uranian seasons different from those on any other planet?

21. Briefly describe the evidence that supports the idea that Uranus was struck by a large planetlike object several billion years ago.

22. Why are Uranus and Neptune distinctly bluer than Jupiter and Saturn?

23. Compare the ring systems of Saturn and Uranus. Why were Uranus's rings unnoticed until the 1970s?

24. How do the orientations of Uranus's and Neptune's magnetic axes differ from those of the other planets?

25. To test your knowledge of planetary magnetic fields, do Interactive Exercise 8.5 on the Web. You can print out your results, if required.

26. Explain why Triton will never collide with Neptune, even though Triton is spiraling toward that planet.

Advanced Questions

27. Consult the Internet or a magazine such as *Sky & Telescope* or *Astronomy* to determine which satellite missions are now under way. What data and pictures have they sent back that update information presented in this chapter?

28. Long before the *Voyager* flybys, Earth-based astronomers reported that Io appeared brighter than usual for a few hours after emerging from Jupiter's shadow. Explain this brief brightening of Io.

29. Compare and contrast Valhalla on Callisto with Caloris Basin on Mercury.

30. As seen by Earth-based observers, the intervals between successive edge-on presentations of Saturn's rings alternate between 13 years, 9 months, and 15 years, 9 months. Why are these two intervals not equal?

31. Compare and contrast the internal structures of Jupiter and Saturn with the internal structures of Uranus and Neptune. Can you propose an explanation for why the differences between these two pairs of planets occurred?

32. Neptune has more mass than Uranus, but Uranus is the larger of the two planets. Reconcile these two facts.

Discussion Questions

33. Suppose that you were planning a mission to Jupiter, using an airplanelike vehicle that would spend many days, even months, flying through the Jovian clouds. What observations, measurements, and analyses should this aircraft make? What dangers might it encounter, and what design problems would you have to overcome?

34. Discuss why astronomers believe that Europa, Ganymede, and Callisto may harbor some sort of marine life. Why do they not expect any life on the surfaces of these worlds?

35. Suppose you were planning separate missions to each of Jupiter's Galilean moons. What questions would you want these missions to answer, and what kinds of data would you want your spacecraft to send back? Given the different environments on the four satellites, how would the designs of the four spacecraft differ?

36. NASA and the Jet Propulsion Laboratory have tentative plans to place spacecraft in orbit about Uranus and Neptune in this century. What kinds of data should be collected, and what questions would you like to see answered by these missions?

What If...

37. Jupiter, at its present location, were a star? What would Earth be like? (*Hint:* Recall that to be a star, Jupiter would have to have 75 times more mass than it has today.)

38. Jupiter had formed at one-third its present distance of 5.2 AU from the Sun? What would Earth be like?

39. Io were struck by another object of similar size? (*Hint:* You can create a variety of different scenarios by imagining the impacting body striking from different directions, with different speeds, and at different angles.)

40. Jupiter were orbiting in the opposite direction than it actually is? What effects might this have on the other planets? Would this change affect Earth? If so, how?

Web Questions

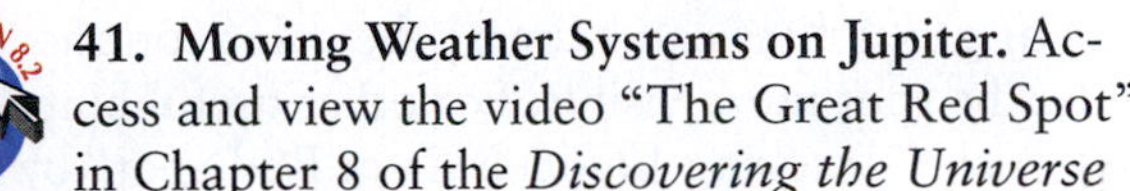

41. **Moving Weather Systems on Jupiter.** Access and view the video "The Great Red Spot" in Chapter 8 of the *Discovering the Universe* Web site. **a.** Near the bottom of the video window you will see a white oval moving from left to right. By stepping through the video one frame at a time, estimate how long it takes this oval to move a distance equal to its horizontal dimension. (*Hint:* You can keep track of time by noticing how many frames it takes a feature in the Great Red Spot, at the center of the video window, to move in a complete circle around the center of the spot. The actual time for this feature to complete a circle is about 6 days.) **b.** The horizontal dimension of the white oval is about 4000 km. At what approximate speed (in km/h) does the white oval move? (Remember that speed = distance/time.)

42. Search the Web, especially the Web sites at NASA's Jet Propulsion Laboratory and at the European Space Agency, for information about the current status of the *Cassini* mission. When did *Cassini* arrive at Saturn? Where in the Saturn system is it presently? What are the current plans for its tour of Saturn's satellites? What plans have been adopted for the *Cassini* extended mission, which began in 2008?

43. In 2004, astronomers reported the discovery of two new moons of Saturn. Search the Web for information about these. How were they discovered? Have the observations been confirmed? How large are these moons? What sort of orbits do they follow?

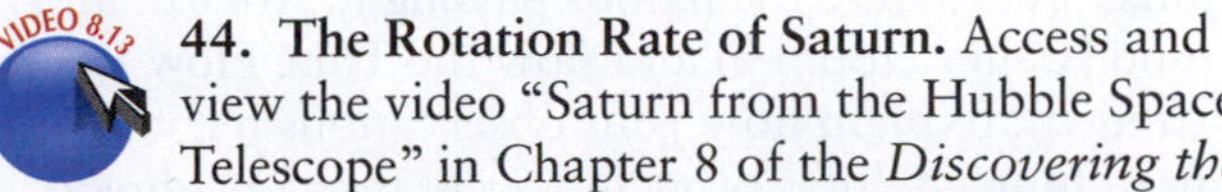

44. **The Rotation Rate of Saturn.** Access and view the video "Saturn from the Hubble Space Telescope" in Chapter 8 of the *Discovering the Universe* Web site. The total time that actually elapses in this video is 42.6 h. Using this information, identify and follow an atmospheric feature and determine the rotation period of Saturn.

Observing Projects

45. Consult such magazines as *Sky & Telescope* and *Astronomy* or your *Starry Night*™ software to determine whether Jupiter is currently visible in the night sky. If so, make arrangements to view the planet through a telescope. What magnifying power seems to give you the best view? Draw a picture of what you see. Can you see any belts and zones? How many? Can you see the Great Red Spot?

46. Make arrangements to view Jupiter's Great Red Spot through a telescope. Consult the *Sky & Telescope* Web site, which lists the times when the center of the Great Red Spot passes across Jupiter's central region, as seen from Earth. The Great Red Spot is well placed for viewing for 50 minutes before and after this time. You will need a refractor with an objective lens of at least 15 cm (6 in.) diameter or a reflector with an objective of at least 20 cm (8 in.) diameter. The use of a pale blue or green filter can increase the color contrast and make the spot more visible. For other useful hints, see the article "Tracking Jupiter's Great Red Spot" by Alan MacRobert (*Sky & Telescope*, September 1997). Sketch Jupiter based on your observations.

47. If Jupiter is visible at night, observe it through a pair of binoculars or a telescope. Can you see all four Galilean moons? Make a drawing of what you observe. Use *Starry Night*™ set to the time of your observations to identify the moons. Click the **Home** button in the toolbar and click the **Stop** button. Then use the **Find** pane to locate Jupiter. Close the **Find** pane and use the **Zoom** controls in the toolbar to set a field of view about 25′ wide. Select **Labels > Planets-Moons** from the menu. You should be able to see all the moons labeled in their present positions. Compare the locations of the moons during your observations with the equivalent view on *Starry Night*™ to identify these moons.

48. Use *Starry Night*™ to observe the motion of the Galilean moons of Jupiter. Select **Favourites > Discovering the Universe > Atlas** and click the **Now** button in the toolbar. Open the **Find** pane and double-click the entry for Jupiter. Use the **Zoom** controls to change the field of view to about 30′ wide. Change the **Time Flow Rate** to 20 minutes (adjust as needed) and click the **Play** button to start time flow. You will see the four Galilean moons orbiting Jupiter. **a.** Are all four moons ever on the same side of Jupiter? **b.** Observe the moons passing in front of and behind Jupiter (zoom in and slow the **Time Flow Rate** as needed). Explain how your observations tell you that all four satellites orbit Jupiter in the same direction. You can examine these moons more closely by centering and zooming in on each moon in turn.

49. Consult such magazines as *Sky & Telescope* and *Astronomy* or your *Starry Night*™ software to determine whether Saturn is currently visible in the night sky. If so, view Saturn through a small telescope. Make a sketch of what you see. Estimate the angle at which the rings are tilted to your line of sight. Can you see the Cassini division? Can you see any belts or zones in Saturn's clouds? Do you observe a faint, starlike object near Saturn that might be Titan? What observations could you perform to test whether this starlike object is a Saturnian satellite?

50. Use *Starry Night*™ to observe the changing appearance of Saturn. Select **Favourites > Discovering the Universe > Atlas** from the menu and click the **Now** button in the toolbar. Use the **Find** pane to locate and lock the view on Saturn (double-click the entry for Saturn). Zoom in until Saturn and its rings are clearly visible. Change the **Time Flow Rate** to 1 year. Now, use the single-step time control buttons to observe the changing aspect or orientation of the rings as time advances in 1-year steps. Describe qualitatively how the rings change orientation, as seen from Earth. **a.** During which of the next 30 years will we see the rings edge on? **b.** Why do we see the rings widen and narrow again in our view in a time of about 15 years?

51. Determine whether Uranus or Neptune is currently visible in the night sky. If so, make arrangements to view them through a telescope. To help you find these planets, use your *Starry Night*™ software or the star chart published each January in *Sky & Telescope,* showing the paths of Uranus and Neptune against the background stars. For a more detailed view of each planet than you will get through a telescope, locate them with your *Starry Night*™ software and zoom to high resolution. Also, locate Triton with the software.

52. This exercise allows you to use *Starry Night*™ to observe Saturn from a distance of about 132,000 km from the planet. Select **Favourites > Discovering the Universe > Saturn.** Click the **Now** button in the toolbar and click the **Stop Time** button. From the image of Saturn that appears, write down where you expect that the Sun is located (e.g., above, down to the right, etc.). Use the location scroller to change the view so that Saturn's rings are edge on and move around until the Sun appears in the same region of the sky as Saturn from your viewpoint. (You can label the Sun if necessary by clicking the box to the left of the Sun in the **Find** pane.) **a.** Is the Sun in the same plane as the rings of Saturn? **b.** What is the phase of Saturn from your viewpoint at this time? **c.** Does this phase agree with what you would expect, with the Sun in this position?

WHAT IF... We Lived on a Metal-Poor Earth?

Earth provides a wealth of building blocks necessary for the development and evolution of complex life-forms. More than 80 elements on or near Earth's surface combine in countless ways essential for its diversity of flora and fauna. Especially important for life on Earth are metals, such as iron. The human body typically contains more than 3 g of iron, mostly in the form of hemoglobin that helps transport oxygen through the bloodstream. A slight iron imbalance leads to anemia or toxicity, and without abundant metals, most of our technologies would never have developed.

But what if the solar system formed from an interstellar gas cloud containing fewer metals and high-mass elements—that is, a solar system with less iron, nickel, copper, and other metals crucial for life as we know it?

Metal-Rich versus Metal-Poor Earth formed with almost 6×10^{21} metric tons of star matter, almost one-third of it iron. If the solar system formed from interstellar gas containing relatively few heavy elements, Earth would contain a much lower fraction of such elements as uranium, lead, iron, and nickel. Earth would be composed of a correspondingly higher fraction of lower-mass elements, such as carbon, nitrogen, oxygen, silicon, and aluminum. That version of Earth—let's call it Lithia—would be profoundly different from our planet.

The Environment of Lithia Without heavy elements like iron, Lithia's density, gravity, and magnetic fields are much lower than Earth's. Let's assume that the density of Lithia is similar to that of the Moon. (We'll assume that the Moon doesn't change in its characteristics or its distance from Lithia.)

If Lithia has the same radius as Earth, the force of gravity on Lithia's surface is only 60% that on Earth. That is, you would weigh 40% less on Lithia. Mobile life-forms that evolve on the metal-poor planet require much less muscle strength to get about and less bone density to resist the planet's gravity.

With fewer radioactive elements to provide heat, the core of Lithia cooled off quickly compared with the core of Earth. As a result, Lithia has much less heat flowing upward through its mantle and crust, so most likely there is no plate tectonic activity. Lithia also has less volcanic activity. But without plate motion to spread the lava flows, volcanoes will grow to higher elevations than on Earth. The lack of crustal motions also means that pockets of high-density material will not rise from below, forming the metal ores mined on Earth.

The Moon's orbit and the tides that it induces are different, too. With Lithia having 60% of Earth's mass, the Moon at its present distance orbits once every 36 days instead of once every 29.5 days. Thus, the cycle of lunar phases is longer, but the cycle of lunar-induced tides is slightly shorter. Surprisingly, the tides have roughly the same height, because the tidal effects depend only on the Moon's gravity and not Earth's, and the distance between Earth and Lithia and their satellite is the same.

Life on Lithia The low abundance of iron and the resulting lack of a planetary magnetic field might preclude the development of complex life-forms on Lithia. And yet, the incredible diversity of life we find on Earth suggests that evolution on Lithia might well occur. If advanced life-forms evolve on Lithia before the Sun uses up its nuclear fuel and becomes a red giant, the chemical differences between Lithia and Earth dictate that they will differ profoundly from the complex life that we find on Earth.

Without the heat of radioactive elements to keep Lithia's core molten for billions of years and with fewer metals, Lithia will have a much weaker magnetic field—if any at all—than Earth. Inhabitants can expect auroras to grace the night sky continuously. On the other hand, high-energy particles and the radiation they create in the atmosphere will bombard any life-forms on the surface continuously and will also adversely affect the ozone layer. A larger amount of the Sun's harmful ultraviolet radiation will reach the surface. Without more protection, life as we know it could not survive.

A Barren Landscape With a thin atmosphere and an abundance of lighter elements, such as silicon (silicon dioxide is sand), Lithia might resemble a barren desert here on Earth. (Photri)

THE STRANGEST SATELLITES IN THE SOLAR SYSTEM

Found in stretched, slanted, loop-d-loop orbits, an odd breed of planetary satellites opens a window into the formation of the planets.

Adapted from an article by David Jewitt, Scott S. Sheppard, and Jan Kleyna

Between 2001 and 2006, our team found 62 undiscovered moons around our solar system's giant planets. Other groups found 24 more. Most of these "irregular" moons had escaped detection because they tend to be smaller and fainter than regular satellites and are distributed over a large region.

Most irregular moons have large, elliptical orbits and circle clockwise, opposite their host planet's rotation. "Regular" moons, such as Earth's Moon or Jupiter's large Galilean satellites, have circular, nearly equatorial, prograde orbits. Our Moon travels counterclockwise—the same direction Earth rotates on its axis and revolves around the Sun. All the planets move counterclockwise, a pattern probably mimicking the swirling gas and dust they emerged from 4.5 billion years ago. Regular moons likely share this motion because they coalesced from disks around their planets. The contrary behavior of the irregular moons reflects a different origin.

Because they roam so far from their host planet, they are tugged by planetary and solar gravity. Their rotation is so rapid that the moons trace out strange, looping trajectories.

These bodies are not well explained. It seems they are products of a time when the gravitational tug of the newly-formed planets scattered—or snatched—small bodies from their orbits.

What a Drag

The irregular moons' properties suggest they are leftover planetary building blocks that originally orbited the Sun. Understanding how they were captured is not easy. Asteroids and comets are routinely pulled into short-lived orbits around the giant planets, like leaves caught in a vortex on a windy day, swirling a few dozen times and then blowing away.

One of these, Comet D/Shoemaker-Levy 9 entered a temporary orbit around Jupiter in the twentieth century. Had it not crashed into the planet in 1994, it would have been ejected to again orbit the Sun within a few hundred years.

A body must lose energy and slow down to enter a permanent, stable orbit around a planet. Because energy is not dissipated efficiently in the solar system today, moon capture must have happened when the solar system had different properties. In the 1970s, scientists proposed three mechanisms, all surrounding planet formation, early in solar system history.

The first argues that the moons lost energy to friction as they passed through the extended atmospheres of the embryonic gas giants. Jupiter and Saturn probably formed when a core of rock and ice of roughly 10 times Earth's mass pulled in gas from the primordial disk surrounding the young Sun. Before settling into compact forms, the planets may have passed through a distended phase when their atmospheres extended hundreds of times farther.

Passing asteroids and comets met one of three fates. If they were small, they burned up in the atmosphere, like a meteor. If

HOW TO SNAG A MOON

The strange orbital properties of the irregular moons indicate that they started off in orbit around the Sun and later were captured by their current host planets. Astronomers have proposed three capture mechanisms.

For all three, the initial stage is the formation of asteroid-size bodies called planetesimals. Many agglomerate to form the rocky cores of the giant planets. The leftovers are vulnerable to being captured.

they were large, they plowed through and continued orbiting the Sun. Medium-sized bodies slowed and were captured.

However, the gas-drag model does not explain the irregular satellites around Uranus and Neptune. Because of their greater distance from the Sun and the lower density of material in the outer regions of the disk, their cores took longer to reach the mass needed to trigger gaseous collapse. Before that happened, the solar nebula had largely dissipated, so Uranus and Neptune never had extended atmospheres like Jupiter and Saturn. How can gas drag operate when there is not much gas?

Three's a Crowd

The second method also places capture during the planetary growth phase. As the gas giants' cores captured gas and grew, there was sudden growth in each planet's gravitational realm, or Hill sphere. Nearby objects were trapped in those extended gravity fields, a process dubbed "pull-down" capture.

But this mechanism does not account for the moons around Uranus and Neptune. Neither underwent runaway growth. Models indicate that they grew slowly, accumulating

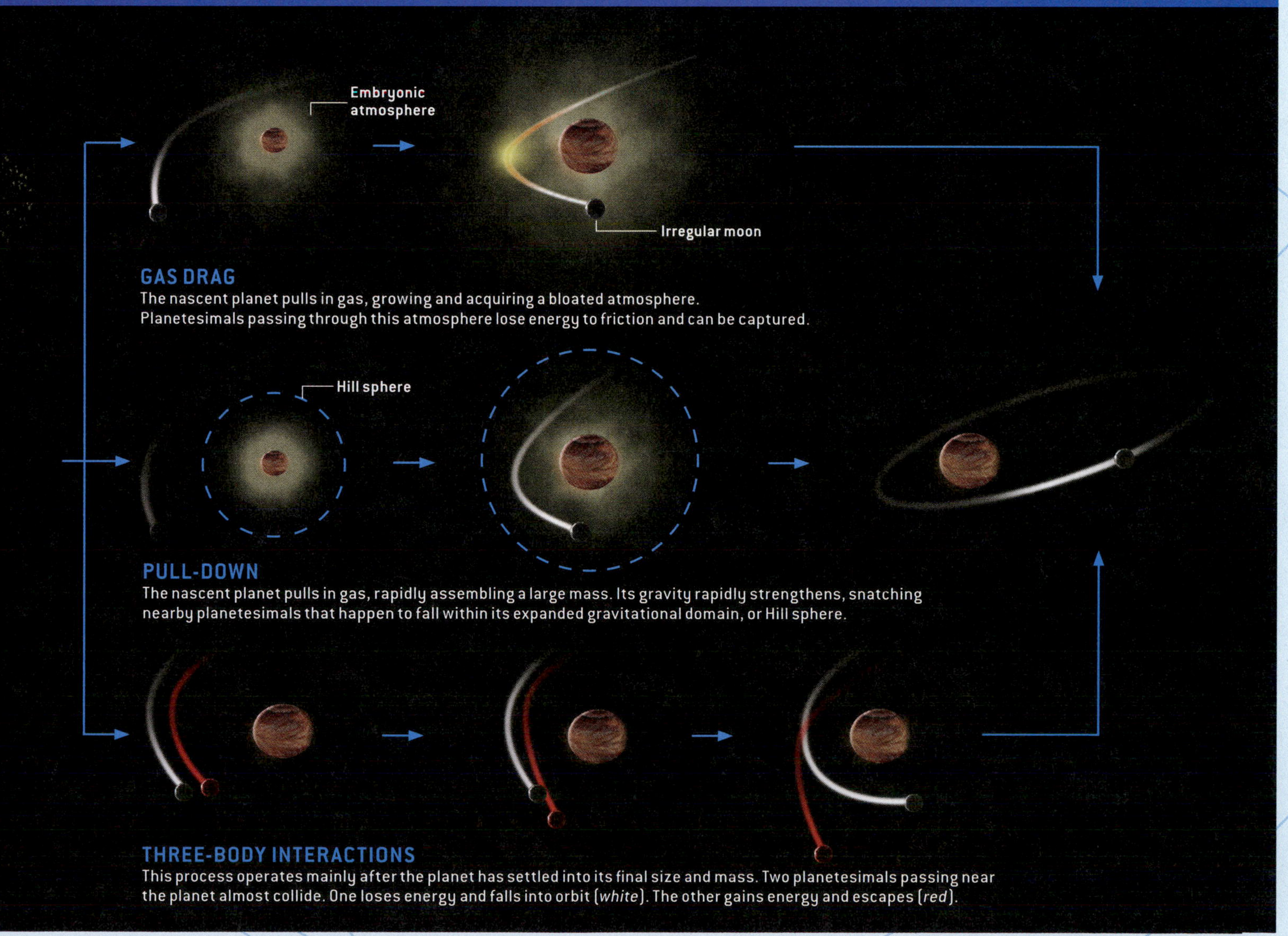

GAS DRAG
The nascent planet pulls in gas, growing and acquiring a bloated atmosphere. Planetesimals passing through this atmosphere lose energy to friction and can be captured.

PULL-DOWN
The nascent planet pulls in gas, rapidly assembling a large mass. Its gravity rapidly strengthens, snatching nearby planetesimals that happen to fall within its expanded gravitational domain, or Hill sphere.

THREE-BODY INTERACTIONS
This process operates mainly after the planet has settled into its final size and mass. Two planetesimals passing near the planet almost collide. One loses energy and falls into orbit (*white*). The other gains energy and escapes (*red*).

asteroid- and comet-size bodies over tens or hundreds of millions of years.

Another scenario, "three-body capture," emerged in 1971. It suggested that collisions between two bodies in a planet's Hill sphere could dissipate enough energy to allow one to be captured. Newer studies show that the three bodies need only interact gravitationally for one to gain energy at the expense of the others. Recently, astronomers suggested another form of three-body capture where an object is sheared apart by planetary gravity, ejecting one component and pulling the other into orbit.

Planetary Movements

Three-body capture works for both gas and ice giants, all of which may have 100 irregular moons larger than one kilometer in diameter, with many smaller ones. This theory does not require a massive gaseous envelope or runaway planetary growth, just a sufficient number of collisions or gravitational interactions. These would probably have occurred near the end of planet formation, after the Hill spheres grew to their present proportions and before leftover debris from planet formation was cleared out.

But where did the irregular moons come from? The moons could be asteroids and comets that massed near the planet that snatched them. While most of their cohorts were absorbed into the bodies of the planets or catapulted from the solar system, the irregular moons survived.

Another possible model dates back to 700 million years after planetary formation, in a solar system still choked with debris. Gravitational interactions between Jupiter and Saturn sparked oscillations that shook the system, scattering billions of asteroids and comets that were originally formed in the Kuiper belt. A fraction of them were captured.

Spectral measurements should eventually be able to test these hypotheses. If irregular moons of different planets have different compositions, it would support the hypothesis that moons formed near their host planets. If they have similar compositions, all moons likely formed together and dispersed, supporting the second theory.

Two things are evident: First, capture of irregular moons occurred early in the solar system's history, during planet formation or immediately afterward. Second, similarities among the irregular moon populations of all four outer planets suggest they arose by three-body interactions.

Like skid marks after a car crash, the irregular moons provide us with tantalizing clues about events that we never could have witnessed.

Comet Hale-Bopp. (Walter Pacholka, Astropics/Photo Researchers)

Chapter 9

Vagabonds of the Solar System

WHAT DO YOU THINK?

1. Are the asteroids a former planet that was somehow destroyed? Why or why not?
2. How far apart are the asteroids on average?
3. How are comet tails formed? Of what are they made?
4. In which directions do a comet's tails point?
5. What is a shooting star?

Answers to these questions appear in the text beside the corresponding numbers in the margins and at the end of the chapter.

Now that we have explored the properties of the planets and their moons in Chapters 7 and 8, we finish studying the objects orbiting the Sun in this chapter. In Section 5-8, we saw that astronomers have several overlapping ways of classifying the bodies orbiting the Sun. To briefly review, recall that a **planet** is a celestial body that (1) is in orbit around the Sun, (2) has sufficient mass for its self-gravity to pull itself into a nearly spherical shape (although rotation causes many planets to be wider at their equators than at their poles), and (3) also has enough mass for its gravitational force to clear the neighborhood around its orbit. A **dwarf planet** satisfies (1) and (2), but it does not have enough gravity to clear its orbital neighborhood of debris. (Note that the dwarf planet Haumea is egg-shaped, rather than being uniformly wider at its equator, like Earth. This is apparently due to Haumea's exceptionally rapid rotation.) All objects in the solar system that are not planets, dwarf planets, or moons are called **small solar system bodies** (SSSBs).

So far, Pluto, Ceres (also an asteroid), Eris (named after a Greek goddess), Makemake (pronounced *mah-key mah-key* and named after a god of the Rapa Nui people of Easter Island), and Haumea (named after a goddess of Hawaii) have earned the title of dwarf planet. The latter three bodies were discovered in 2005. Dozens of other solar system objects are being evaluated for inclusion in this class. Bear in mind that as the properties of more solar system objects are pinned down, the definitions of dwarf planets and SSSBs may be refined.

All dwarf planets that orbit beyond Neptune are called *plutoids,* which, besides Pluto, include Eris, Makemake, and Haumea. These latter three objects are also members of the Kuiper belt.

In this chapter you will discover

- the properties of dwarf planets and small solar system bodies
- asteroids and meteoroids—pieces of interplanetary rock and metal
- comets—objects containing large amounts of ice and rocky debris
- space debris that falls through Earth's atmosphere
- the asteroid belt and the Kuiper belt, both filled with a variety of debris, including orbiting pairs of objects
- the impacts from space 250 million and 65 million years ago that caused mass extinctions of life on Earth
- wayward asteroids that could again threaten life on Earth

DWARF PLANETS

9-1 Pluto and its moon, Charon, are about the same size

Clyde Tombaugh discovered Pluto (Figure 9-1) because it moved among the background stars from night to night (Figure 9-2), a discovery technique that was, and still is, standard for finding other dwarf planets and small solar system bodies.

Unlike the planets, Pluto's orbital eccentricity of $e = 0.25$ (see Section 2-5) is so great that it is sometimes closer to the Sun than Neptune, as it was from 1979 to 1999. Pluto is now farther away from the Sun than Neptune and will continue to be so for about the next 230 years (Figure 9-3). Pluto's orbit is also more tilted, with respect to the plane of the ecliptic, than that of any planet (Figure 9-3c).

Little was known about Pluto for half a century until astronomers noticed that its image sometimes appears oblong (Figure 9-4). This observation led to the discovery in 1978 of Pluto's largest moon, Charon (pronounced *KER-en,* after the mythical boatman who ferried souls across the River Styx to Hades, the domain ruled by the Roman god Pluto). The mass of an object in space can only be determined directly from its gravitational attraction on another body. We know the mass of the Sun because we can determine the length of Earth's year. We know the mass of Mars because we can determine the periods of its moons Phobos and Deimos. The mass of Pluto was not known until the orbit of Charon was accurately determined in the early 1980s.

From 1985 through 1990, the orbit of Charon was oriented so that Earth-based observers could watch it eclipse Pluto. Astronomers used observations of these eclipses to determine that Pluto is only twice as wide as its satellite: Pluto's diameter is 2380 km and Charon's is 1190 km. Their similarity in size has led to a debate as to whether they should be classified as a binary dwarf-planet system.

The average distance between Charon and Pluto is less than one-twentieth the distance between Earth and our Moon. Furthermore, both bodies always keep the same side facing each other. That is to say, they both have synchronous rotation with respect to each other. As seen from the satellite-facing side of Pluto, Charon neither rises nor sets, but instead hovers in the sky, perpetually suspended in the same place above the horizon.

The exceptional similarities between Pluto and Charon suggest that this binary system may have formed when Pluto collided with a body of similar size. Perhaps chunks of matter were stripped from this second body, leaving behind a mass, now called Charon, vulnerable to capture by Pluto's gravity. Alternatively, perhaps Pluto's gravity captured Charon into orbit during a close

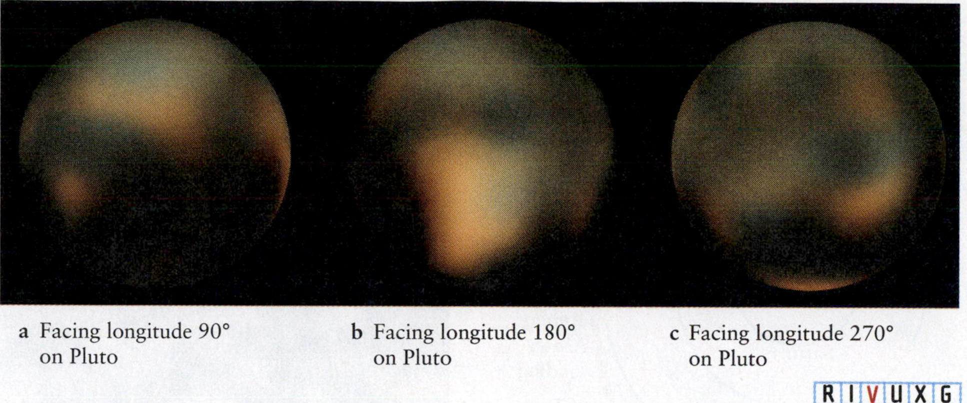

FIGURE 9-1 Pluto These three Hubble Space Telescope images of Pluto show little detail but indicate that the major features of Pluto's surface each cover large amounts of its area. Comparing these observations to previous ones reveals that the surface changes in color and brightness seasonally. (NASA, ESA, and M. Buie [Southwest Research Institute])

encounter between the two worlds. For either of these two scenarios to be feasible, many Pluto-sized objects must have existed in the outer regions of the young solar system. One astronomer estimates that there had to be at least a thousand Pluto-sized bodies for a collision or close encounter between two of them to have occurred at least once since the solar system formed 4.6 billion years ago.

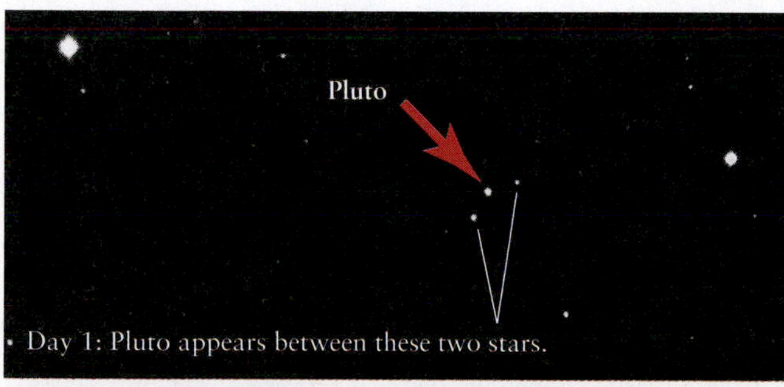

FIGURE 9-2 Discovery of Pluto Pluto was discovered in 1930 by searching for a dim, starlike object that slowly moved against the background stars. These two photographs were taken 1 day apart. (UC Regents/Lick Observatory)

In 2005, the Hubble Space Telescope detected two other moons orbiting Pluto (Figure 9-5). Called Nix (named after the goddess of the night and mother of Charon) and Hydra (named after the nine-headed serpent living at the gates of Hades), these moons are both much smaller than Charon.

Like Neptune's moon Triton, Pluto and Charon are probably both composed of nearly equal amounts of rock and ice. The surface of Pluto (see Figure 9-1) has 12 distinct regions—more large-scale features than any object in the solar system other than Earth. Such features on Earth include oceans and continents, while on Pluto they are likely to include regions dominated by rock, regions dominated by ice, major impact regions, and mountain ranges.

Studies of Pluto's surface show that different regions vary in temperature from one another by up to 25 K. Its spectrum shows that its surface contains frozen nitrogen, methane, and carbon monoxide. Hubble observations of Pluto show that its surface has several distinct features (see Figure 9-1). The orange tinge may be due to carbon-rich compounds on its surface, the result of methane there being chemically altered by the Sun's ultraviolet radiation. The brightest regions are believed to be covered with frozen carbon monoxide, while the darker regions probably contain rocky debris. The dwarf planet is also observed to have a very thin atmosphere of nitrogen and carbon monoxide. In contrast, Charon's surface appears to be covered predominantly with water ice. So intriguing, unusual, and unexplored are Pluto and its companions that NASA launched the *New Horizons* spacecraft in 2006 to visit them (scheduled to arrive there in July 2015) and then to continue deeper into the Kuiper belt.

> What objects are classified today as planets? What objects are classified as dwarf planets?

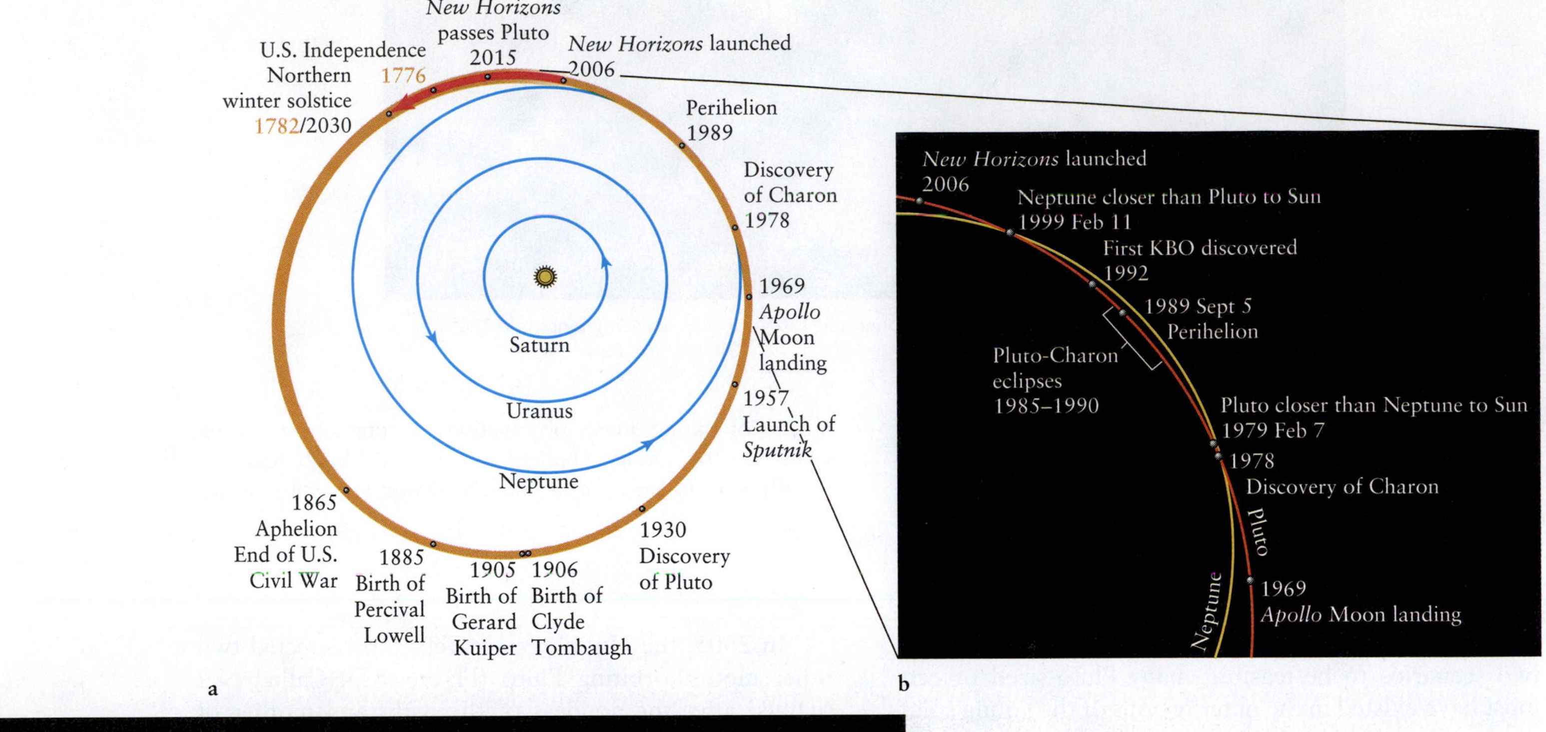

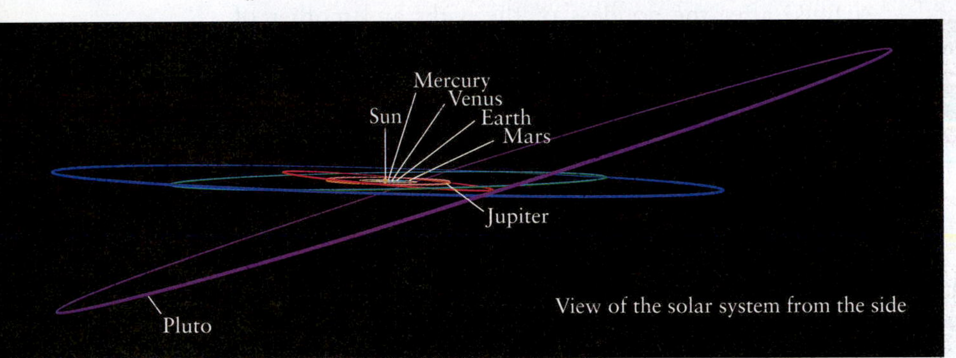

FIGURE 9-3 Orbit of Pluto (a) The high-eccentricity orbit of dwarf planet (and KBO) Pluto stands out compared to the orbits of the outer three planets. Notice how many significant events occurred on Earth during Pluto's present orbit of the Sun. (b) Details of Pluto's passage inside the orbit of Neptune. The two bodies will never collide. (c) A nearly edge-on view of the ecliptic and Pluto's orbit compared to it.

9-2 Ceres is a dwarf planet in the asteroid belt, while Eris, Haumea, and Makemake are dwarf planets in the Kuiper belt

On New Year's Day, 1801, the Sicilian astronomer Giuseppe Piazzi was carefully mapping faint stars in the constellation Taurus. He noticed a dim, previously uncharted "star" that shifted its position slightly over the next several nights. Uranus had been discovered in this way by William Herschel just 20 years before. Later that year, the orbit of this object was determined to lie between Mars and Jupiter. At Piazzi's request, it was named Ceres (pronounced *See-reez*), after the patron goddess of Sicily. Based on erroneous observations of Ceres's size, Piazzi's object was initially catalogued as a planet. In 1855, further observations revealed that all of the then-known planets—Mercury, Venus, Earth, Mars, Jupiter, Saturn, and Uranus—were all much larger than Ceres. We now know that Piazzi was the first astronomer to discover an asteroid.

Ceres is spherical like the planets (Figure 9-6), but its diameter is a scant 940 km (585 mi), only one-quarter the diameter of our Moon. In 2006, Ceres was classified as a dwarf planet. Recent calculations suggest that, like the planets, Ceres is differentiated, containing a rocky core and clad in an icy outer layer.

Eris was first observed in 2003, before the present definitions of planets, dwarf planets, and SSSBs were established. This body is presently located 97 AU from the Sun, 3 times farther from the Sun than Pluto. Furthermore, its 557-year orbit is so eccentric that for part of that time it is closer to the Sun than Pluto (Figure 9-7a). Eris has a diameter of about 2400 km (Pluto is 2300 km

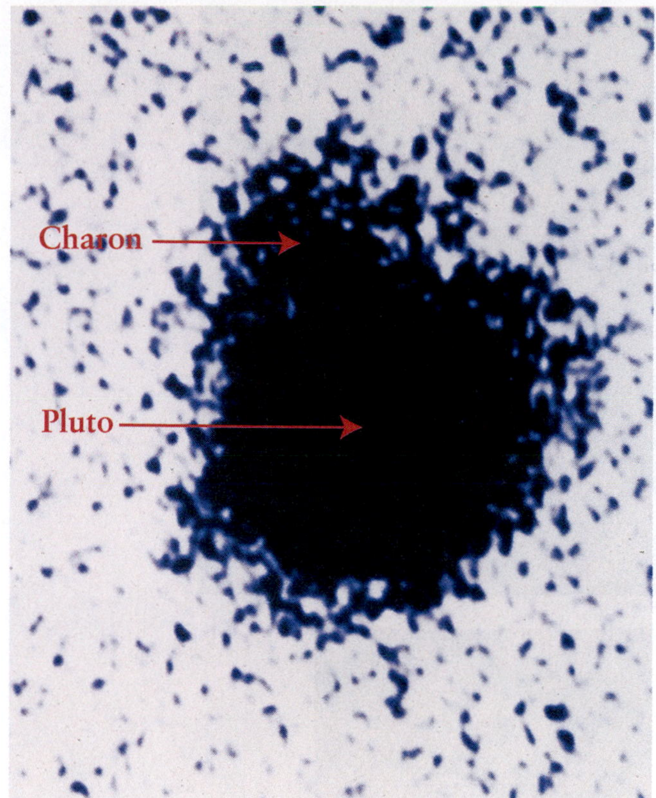

VIDEO 9.1 **FIGURE 9-4 Discovery of Charon** Long ignored as just a defect in the photographic emulsion, the bump on the upper left of this image of Pluto led astronomer James Christy to discover the moon Charon. (U.S. Naval Observatory)

VIDEO 9.2 **FIGURE 9-5 Pluto's Three Moons** Observations by the Hubble Space Telescope in 2005 revealed two small moons, each about 5000 times dimmer than Pluto. Named Nix and Hydra, they are between 2 and 3 times farther from Pluto than is its moon Charon. The lines radiating from Pluto and Charon are artifacts of the exposure. (NASA)

FIGURE 9-6 Comparison of Ceres with the Moon and Earth (a) Ceres, the Moon, and Earth are shown here to scale. Dwarf planet Ceres is the largest asteroid. This image of Ceres suggests that it has regions of ice and rock on its surface. The asteroid will be visited by the *Dawn* spacecraft in 2015. (NASA)

across), and it is very bright, with an albedo of about 0.86. Observations suggest that one-quarter of this world is water ice, a greater volume than all of the fresh water on Earth. A moon orbiting Eris, named Dysnomia (after the daughter of the Greek goddess Eris), was discovered in 2005 (Figure 9-7b). The moon is about 300 km across.

Haumea, discovered in 2005, rotates once every 3 h 55 min. In other words, a day on Haumea is less than 4 h long. Because of that rapid rotation, this world has distorted itself so that it looks like a watermelon. If it were not rotating, calculations reveal that it would be spherical, allowing it to be classified as a dwarf planet. Haumea has two known moons, H'iaka (about 310 km across) and Namaka (roughly 170 km across). Spectra of both moons reveal that unlike the rest of the bodies observed in the outer solar system, their surfaces are made of almost pure water ice. In contrast, most objects have surfaces of rock and ice, where the "ice" often contains molecules other than water.

Makemake, also discovered in 2005, rotates once every 7 h 45 min. It is about three-quarters the size of Pluto and it orbits the Sun once every 310 years. Unlike the other Kuiper belt dwarf planets, Makemake has no known moons.

While astronomers decide which other bodies will be designated dwarf planets, we turn to the small solar system bodies, which include the rest of the asteroids, all the comets, and the smallest rubble of all, the meteoroids.

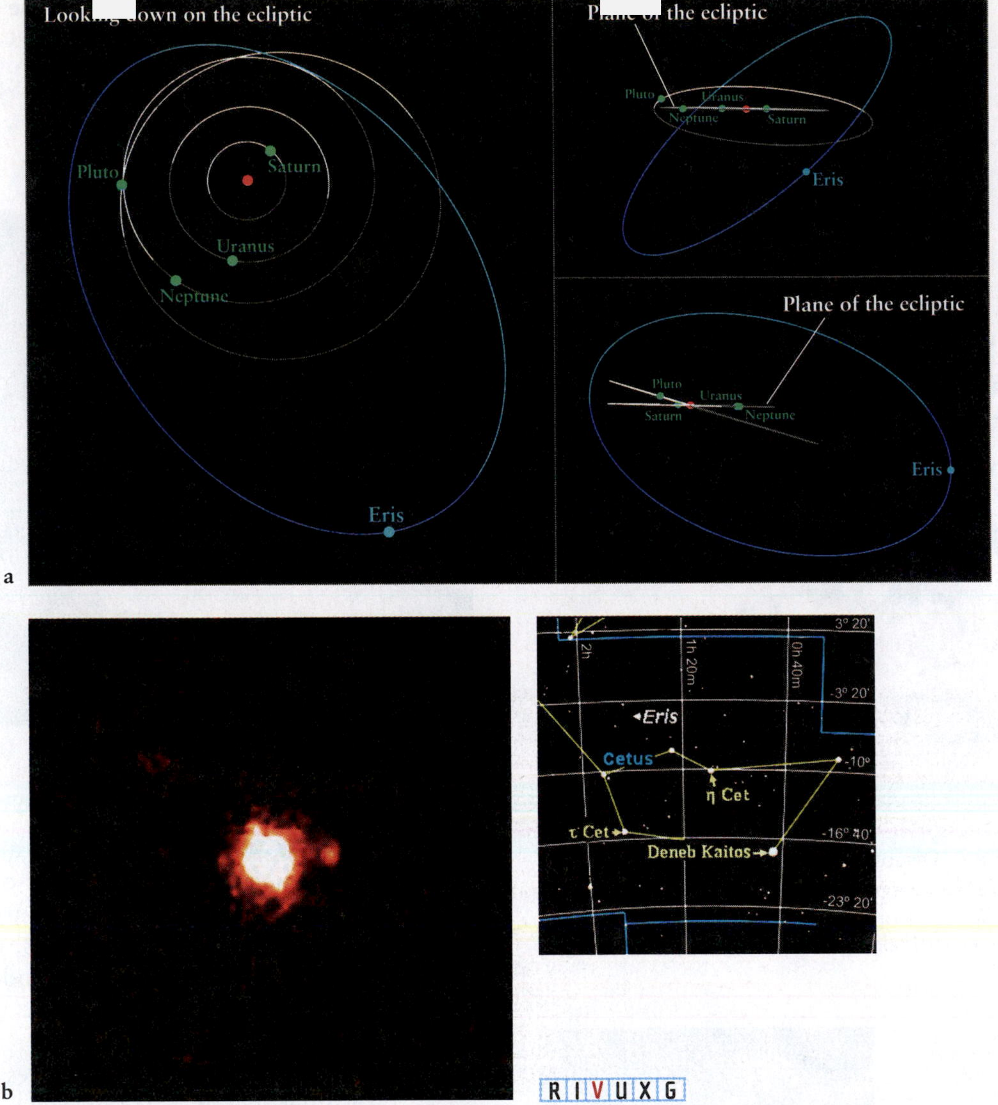

FIGURE 9-7 Dwarf Planet Eris (a) Three perpendicular views of the orbit of Eris and Dysnomia compared to the planets and Pluto. Eris and Dysnomia's orbit around the Sun ranges from 38 to 98 AU, and has an orbital eccentricity, $e = 0.44$, and an orbital inclination of 44°. (b) A Keck Telescope image of dwarf planet Eris and its moon Dysnomia. (a: Orionist; b: M.E. Brown, W. M. Keck Observatory)

SMALL SOLAR SYSTEM BODIES

We know that the solar system formed from a rotating disk of gas and dust. Matter in the disk that had too much angular momentum to fall onto the protosun coalesced at varying orbital distances into planetesimals. Many of these chunks of rock and metal eventually collided, forming the planets, dwarf planets, and larger moons of the solar system, as well as myriad pieces of dust, rock, and boulder-sized debris. Others were captured whole by various planets as small, irregularly-shaped moons, like Phobos and Deimos in orbit around Mars. Many planetesimals still orbit the Sun today in splendid isolation. These are the asteroids, sometimes called *minor planets* (not to be confused with dwarf planets).

ASTEROIDS

9-3 Most asteroids orbit the Sun between Mars and Jupiter

In 1802, less than two years after the asteroid (now also a dwarf planet) Ceres was discovered, the German astronomer Heinrich Olbers observed another faint, starlike object that moved against the background stars. He called it Pallas, after the Greek goddess of wisdom. Like Ceres, Pallas orbits the Sun in a nearly circular path between the orbits of Mars and Jupiter. Pallas is even dimmer and smaller than Ceres and has a diameter of only 600 km (375 mi).

Only two more of these asteroids—Juno and Vesta—were found until the mid-1800s, when telescopes

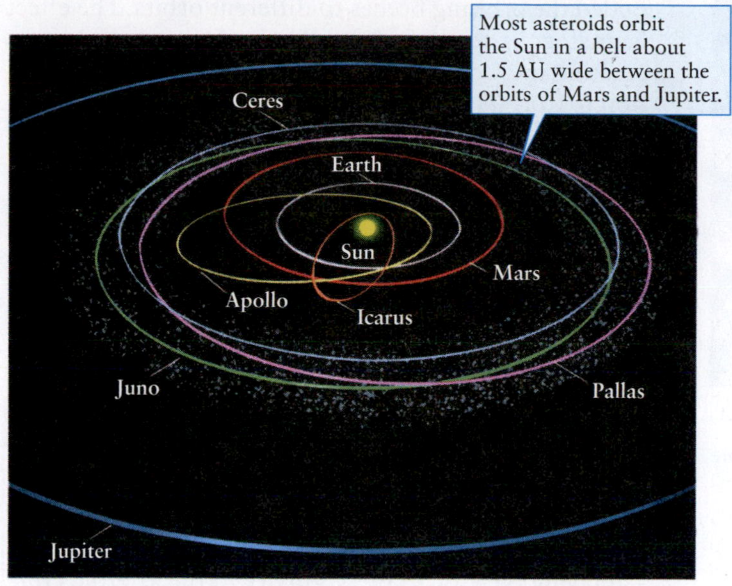

Trojan asteroids
Jupiter
Belt asteroids
Mars
Earth
Venus
Mercury
Trojan asteroids
b

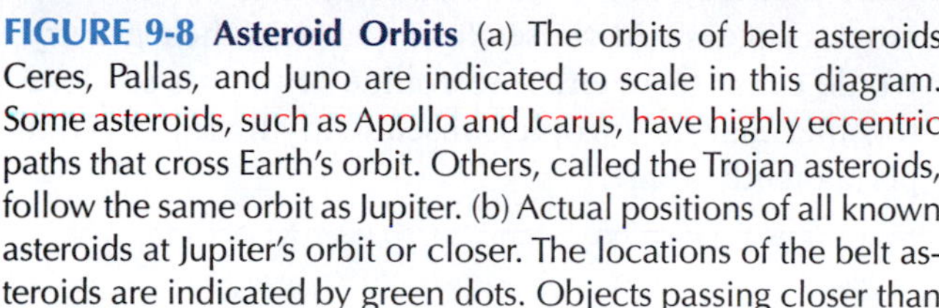

FIGURE 9-8 Asteroid Orbits (a) The orbits of belt asteroids Ceres, Pallas, and Juno are indicated to scale in this diagram. Some asteroids, such as Apollo and Icarus, have highly eccentric paths that cross Earth's orbit. Others, called the Trojan asteroids, follow the same orbit as Jupiter. (b) Actual positions of all known asteroids at Jupiter's orbit or closer. The locations of the belt asteroids are indicated by green dots. Objects passing closer than 1.3 AU to the Sun are shown by red circles. Objects observed at least twice are indicated by filled circles, and objects seen only once are indicated by outline circles. Jupiter's Trojan asteroids are deep blue squares. Comets are filled and unfilled light-blue squares. Although the asteroids appear packed together in this drawing, they are typically millions of kilometers apart. The small scale here is deceiving! (Minor Planet Center)

improved. Astronomers then began to stumble across many more asteroids orbiting the Sun at distances from 2 to 3½ AU, between the orbits of Mars and Jupiter. This region of the solar system is now called the **asteroid belt** (Figure 9-8). Asteroids whose orbits lie entirely within this region are called **belt asteroids.**

The spacecraft *Dawn* was sent in 2007 to explore Vesta and Ceres, neither of which can be well resolved from Earth and both of which hold clues to planet formation. The probe will arrive at Vesta in 2011 and Ceres in 2015.

Asteroids were discovered one by one throughout the nineteenth century. The next real breakthrough came in 1891, when the German astronomer Max Wolf applied photographic techniques to asteroid searching. A total of 300 asteroids had been found up to that time, each painstakingly discovered by scrutinizing the skies for faint, uncharted "stars" whose positions shifted slowly from one night to the next. With the advent of astrophotography, however, the floodgates were opened. Astronomers could simply aim a camera-equipped telescope at the stars and take long exposures. If an asteroid happened to be in the field of view, it left a distinctive trail on the photographic plate (Figure 9-9). Using this technique, Wolf alone discovered 228 asteroids.

Insight Into Science

Confirming Observations New findings in science must be confirmed or replicated by other competent scientists before discoveries or observations are accepted by the scientific community. Therefore, asteroid observations need to be repeated several times to determine exact orbits and eliminate the possibility that a sighting is of a known asteroid or other object.

Ceres, the largest asteroid, alone accounts for about 30% of the mass of all known asteroids combined. Only three asteroids—Ceres, Vesta, and Pallas—have diameters greater than 300 km. Records of all objects smaller than planets in our solar system are kept by the Minor Planet Center in Cambridge, Massachusetts. The center reports that as of January 2011, the existence of 257,455 asteroids had been confirmed, with more than 66,628,500 other observations awaiting confirmation as new asteroids.

Where is Eris located in the solar system?

The number of asteroids increases dramatically with decreasing size: Only 41 asteroids have

R I V U X G

FIGURE 9-9 Discovering Asteroids In 1998, the Hubble Space Telescope found this asteroid while observing objects in the constellation Centaurus. The exposure, tracking stars, shows the asteroid as a 19″ streak. This asteroid is about 2 km in diameter and was located about 140 million km (87 million mi) from Earth. (R. Evans and K. Stapelfeldt, Jet Propulsion Laboratory and NASA)

diameters between 200 and 300 km; 250 more are bigger than 100 km across; and there are thought to be tens of millions that are less than 1 km across.

1 You may have heard the common belief that the asteroids were formerly a single planet that was somehow destroyed. Indeed, it is plausible that the asteroid belt region once had many more objects in it—possibly several Earth masses worth of debris. However, Jupiter's gravitational force pulled many planetesimals out of this region before large numbers of them could meet, coalesce, and form a single planet-sized object.

If all of the present asteroids had once been part of a single body, it would have had a diameter of only 1500 km, or 12% of Earth's diameter. This is less than two-thirds the diameter of Pluto and half the diameter of our Moon. Like Pluto, such a body would not have enough gravitational attraction to clear the space around it of other bodies. Therefore, a single body that contains the mass of all of the asteroids would not qualify as a planet either. (Let's now consider Jupiter's gravitational effect on the asteroids in more detail.)

9-4 Jupiter's gravity creates gaps in the asteroid belt

In 1867, the American astronomer Daniel Kirkwood called attention to gaps in the asteroid belt. These features, called **Kirkwood gaps,** show the influence of Jupiter's gravitational attraction. Figure 9-10, a graph of asteroid orbital periods, has gaps at simple fractions ($\frac{1}{3}$, $\frac{2}{5}$, $\frac{3}{7}$, and $\frac{1}{2}$) of Jupiter's orbital period. The gravitational pull of Jupiter created these empty regions by giving objects in them periodic reinforcing tugs that pulled the orbiting bodies to different orbits. The effect is analogous to the gravitational resonance of Mimas creating the Cassini division in Saturn's rings (see Section 8-10).

It is likely that the asteroid belt contains tens of millions of asteroids, yet the typical separation between "neighboring" asteroids is an impressive 10 million km. This is quite unlike the image that has been created by innumerable popular movies and television shows of asteroids so close together that you must dodge them as you fly past.

Insight Into Science

Get Real Scientists continually apply the "laws of nature" to new situations, problems, observations, experiments, and even popular culture. For example, applying Newton's law of gravity to the asteroids reveals that they could never swarm, as science fiction movies suggest. At those close quarters, their gravity would cause them either to collide or to pass so close together that they would subsequently fly rapidly and permanently apart.

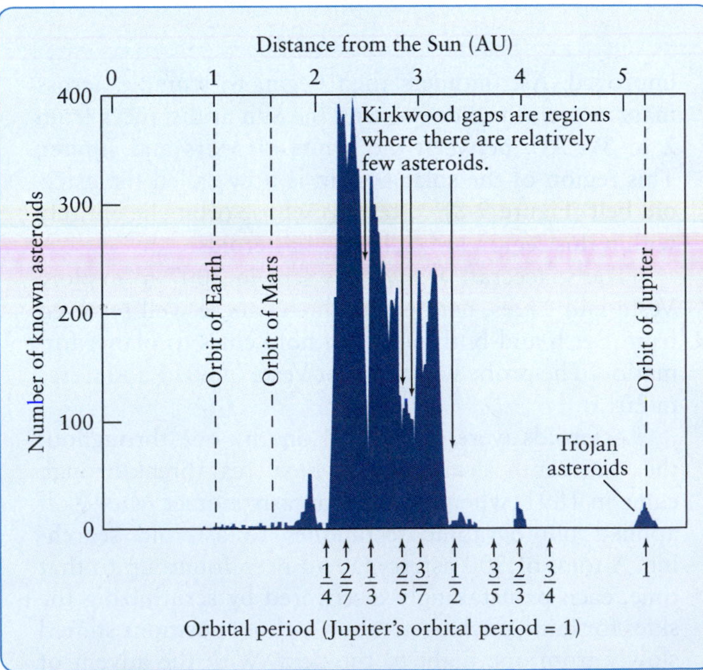

FIGURE 9-10 The Kirkwood Gaps This graph displays the number of asteroids at various distances from the Sun. Note that few asteroids have orbital periods that correspond to such simple fractions as $\frac{1}{3}$, $\frac{2}{5}$, $\frac{3}{7}$, and $\frac{1}{2}$ of Jupiter's orbital period. Resonant orbits with Jupiter have deflected asteroids away from these orbits. The Trojan asteroids accompany Jupiter as it orbits the Sun.

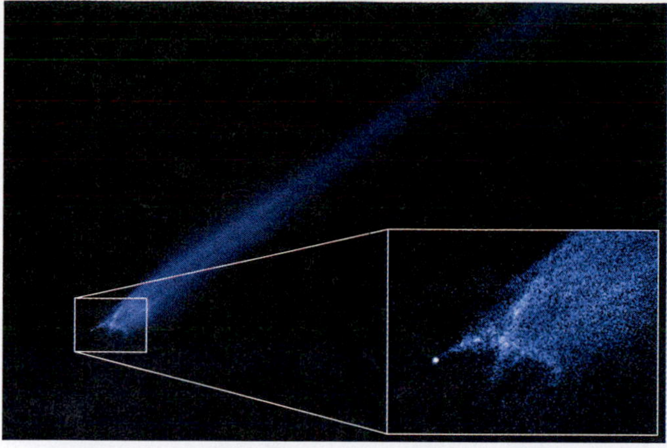

FIGURE 9-11 Collision Between Two Asteroids Observed in 2010, this X-shaped "object" (inset) is believed to be the collision of two small asteroids. The event created dust that was pushed away from the Sun, which is to the left and below this image. The collision occurred 2 AU from the Sun and 1 AU from Earth. (NASA, ESA, and D. Jewitt [UCLA])

Despite the large average separation between asteroids, the gravitational influences of Mars and Jupiter have sent some asteroids smashing into each other at various times over the past 4.6 billion years. In 2010, astronomers observed such an event (Figure 9-11). Typical collision velocities are estimated to be 3600 to 18,000 km/h (2000 to 11,000 mi/h), which is more than sufficient to shatter rock. In some collisions, the resulting fragments may not have enough speed to escape from one another's gravitational attraction, and they reassemble. The asteroid Toutatis appears to be composed of two similarly sized pieces connected to each other, as does the asteroid Castalia. These bodies are therefore likely to have been broken apart and reformed.

Alternatively, several large fragments, such as Ida and Dactyl (Figure 9-12), may end up orbiting each other. Dactyl is a pockmarked asteroid some 1.5 km in diameter that orbits Ida at a distance of 100 km. Petit-Prince is the satellite of Eugenia. Petit-Prince is 13 km across, orbiting 1200 km from the larger body. Asteroid 87 Sylvia has two known satellites, named Romulus and Remus. To date, 195 asteroid-satellite systems have been observed.

In 1918, the Japanese astronomer Kiyotsugu Hirayama drew attention to groups of asteroids that share nearly identical orbits. These fragments are pieces of parent asteroids. Pursuing this line of research, astronomers in 2002 discovered that two large asteroids collided only a few million years ago (very recently in solar system history) and created about 20 separate families or clusters of smaller asteroids. Asteroids in each of these families

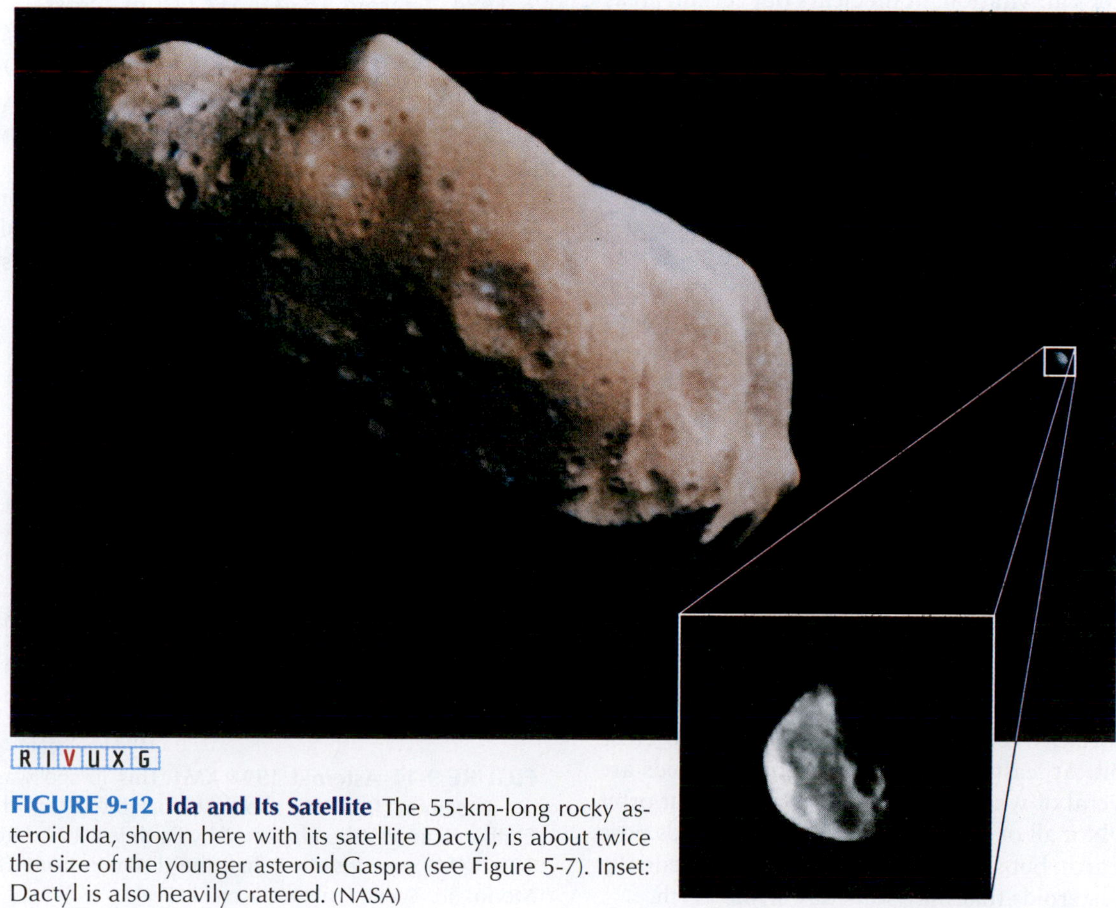

FIGURE 9-12 Ida and Its Satellite The 55-km-long rocky asteroid Ida, shown here with its satellite Dactyl, is about twice the size of the younger asteroid Gaspra (see Figure 5-7). Inset: Dactyl is also heavily cratered. (NASA)

orbit together. The largest remnant cluster of this impact is named Karin, after its largest member, an asteroid some 20 km (12 mi) across.

In the early 1990s, the Jupiter-bound *Galileo* spacecraft passed near two asteroids—Gaspra (see Figure 5-7) and Ida (Figure 9-12)—and sent back close-up views. Because Ida's surface is more heavily cratered than Gaspra's, we deduce that Ida is much older.

Why do astronomers doubt that the asteroid belt was once made up of a single planet?

The Sun is not the only star with an asteroid belt. In 2001, astronomers discovered that the star Zeta Leporis, 70 ly from Earth in the constellation Lepus (the Hare, see star chart on left), has a disk of debris that appears to contain asteroids. This star system is less than 0.5 billion years old; astronomers hope it will provide insights into the early evolution of the asteroids and other objects in the disk of gas and dust that surrounded the early Sun.

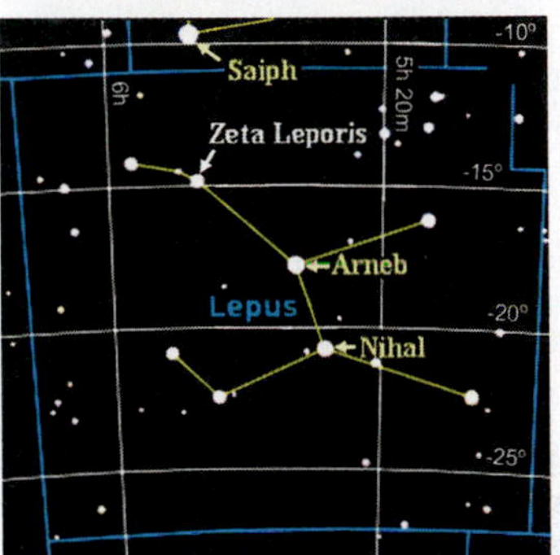

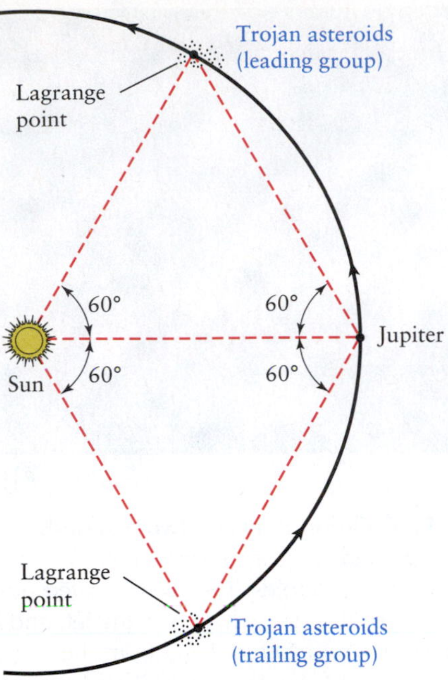

FIGURE 9-13 Jupiter's Trojan Asteroids Groups of asteroids orbit at the two stable Lagrange points along Jupiter's orbit, trapped by the combined gravitational forces of Jupiter and the Sun.

9-5 Asteroids also orbit outside the asteroid belt

While Jupiter's gravitational pull clears out certain areas within the asteroid belt, it actually captures asteroids at two locations in the path of its own orbit. The gravitational forces of the Sun and Jupiter work together to hold asteroids in orbit at these locations, called **stable Lagrange points**, in honor of the French mathematician Joseph Lagrange, whose calculations explained them. One Lagrange point is located 60° ahead of Jupiter, and the other is 60° behind, as shown in Figure 9-13 (see also Figure 9-8b).

The asteroids trapped at Jupiter's Lagrange points are called **Trojan asteroids**, each named after a hero of the Trojan War. As of January 2010, 4782 Trojan asteroids orbiting with Jupiter have been cataloged. Neptune has seven known Trojan asteroids, Mars has four, and astronomers continue to search for Trojans orbiting with other planets.

Some asteroids have highly elliptical orbits that bring them into the inner regions of the solar system (see the red dots in Figure 9-8b). Others have similarly elliptical orbits that extend from the asteroid belt out beyond the farthest reaches of Pluto's orbit. The Amor asteroids cross Mars's orbit, but do not get as close to the Sun as Earth's orbit, whereas the **Apollo asteroids** do cross Earth's orbit. At least 3796 *Earth-crossing* asteroids are known, several of which are pairs of asteroids that orbit each other. Not all of these Earth-crossing asteroids pose threats to Earth, but at least 1180 of these are potentially hazardous asteroids that may someday strike Earth.

Several close calls took place in recent years. In 1972, space debris was observed to skip off Earth's atmosphere and retreat back into space. On December 9, 1994, asteroid 1994 XM1 (10 m across—the size of a small bus; Figure 9-14) passed within 105,000 km (65,000 mi) of Earth. In 2004, asteroid 2004 FU162 (6 m in diameter) passed only 6500 km above Earth's surface. Near misses within the range of these two events happen virtually every year.

During these close encounters, astronomers can examine the details of asteroids. For example, an asteroid's brightness often varies as it rotates because

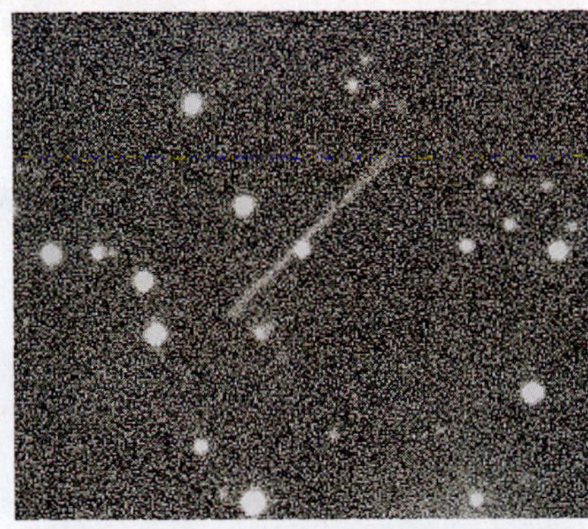

FIGURE 9-14 Asteroid 1994 XM1 This image was obtained on December 9, 1994, shortly before the asteroid arrived in Earth's vicinity. When it passed by Earth just 12 h later, asteroid 1994 XM1 was less than half the distance from Earth to the Moon. (Jim Scotti, Spacewatch on Kitt Peak)

different surface features scatter different amounts of light. Such data show that typical rotation periods for asteroids are between 5 and 20 h, although one asteroid, labeled 1998 KY26, rotates about once every 10.7 min. This body is the fastest rotating object known in the solar system.

The proximity of asteroids and the promise of learning more about these ancient and extremely varied members of our solar system prompted NASA to send the *Near Earth Asteroid Rendezvous (NEAR) Shoemaker* spacecraft to visit asteroids Mathilde (Figure 9-15a) and Eros. JAXA, the Japanese space agency, sent the *Hayabusa* spacecraft to the asteroid Itokawa (Figure 9-15b), where the spacecraft attempted to take samples. These maneuvers were only partially successful. The *Hayabusa* returned to Earth in June 2010 with its cargo. Dust particles were discovered in the spacecraft. In November 2010, some of the dust particles it carried were confirmed to have come from asteroid Itokawa. They are now being analyzed.

NEAR Shoemaker revealed that Mathilde is only 1.3 times denser than water and has an albedo of 0.04, making it darker than charcoal. This heavily cratered, carbon-rich body therefore has only half the density of the other asteroids astronomers have studied, such as Ida. Eros is about 3 times as dense as water and rotates once every 5¼ h. In February 2000, *NEAR Shoemaker* went into orbit around Eros, and for a year the spacecraft's cameras and other sensors sent back a wealth of information about it.

NEAR Shoemaker showed that Eros (Figure 9-16) is a chunk of rock and metal. Whether this asteroid is solid or a collection of loosely held pieces is still under investigation. Analyzing Eros's spectra reveals that it is probably much the same as it was when it coalesced 4.6 billion years ago. Thus, it was never hot enough to differentiate (separate rock from metal). Infrared observations reveal that like our Moon, Eros has regolith. It also has several substantial craters and is strewn with boulders (Figure 9-16c). Astronomers infer from the lack of high crater walls that recent impacts have caused the asteroid to vibrate, collapsing earlier craters. On February 12, 2001, NASA engineers landed *NEAR Shoemaker* on Eros so gently that the spacecraft continued to transmit data after landing. Details of Eros as small as 1.4 m across were imaged by *NEAR Shoemaker*.

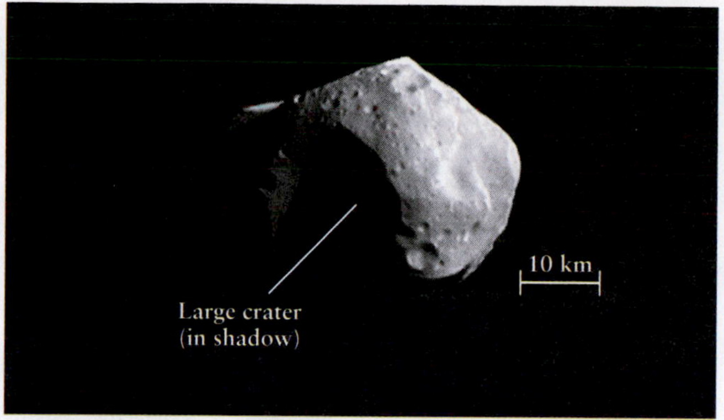

FIGURE 9-15 Asteroids (a) Reflecting only half as much light as a charcoal briquette, Mathilde is half as dense as typical stony asteroids. Slightly larger than Ida (see Figure 9-12), irregularly shaped Mathilde measures 66 km × 48 km × 46 km, rotates once every 17.4 days, and has a mass equivalent to 110 trillion tons. The part of the asteroid shown is about 59 km × 47 km. The large crater in shadow is about 20 km across. (b) The near-Earth asteroid Itokawa was visited by the Japanese space probe *Hayabusa*. Samples of dust particles from it are now being analyzed. (a: Johns Hopkins University, Applied Physics Laboratory; b: ISAS, JAXA)

WHAT IF . . . The Asteroids Were a Destroyed Planet? If the asteroids had been a planet with the mass of Mercury (the smallest known planet), we calculate that there would be 350 times as much debris in the asteroid belt as is known to exist there today. In that case, the rate at which asteroids would leave the asteroid belt due to collisions and near misses with other asteroids would be much greater. Higher rates of rocky debris streaming Earthward over billions of years would have almost certainly led to more mass extinctions and other nightmares for life on Earth.

Describe the appearance of small asteroids.

a

b

c

FIGURE 9-16 Asteroid Eros The *Near-Earth Asteroid Rendezvous (NEAR) Shoemaker* spacecraft took these images of asteroid Eros in February 1999. (a) The top of the figure is the asteroid's north polar region. Eros's dimensions are 33 km × 13 km × 13 km (21 mi × 8 mi × 8 mi) and it rotates every $5\frac{1}{4}$ h. Its density is 2700 kg/m^3, close to the average density of Earth's crust and twice as dense as asteroid Mathilde. (b) Looking into the large crater near the top of (a), which is 5.3 km (3.3 mi) across. (c) This is the penultimate image taken by *NEAR Shoemaker* before it gently landed on Eros. Taken from an altitude of 250 m (820 ft), the image is only 12 m across. You can see rocks and boulders buried to different depths in the regolith. (Johns Hopkins Applied Physics Laboratory)

COMETS

While asteroids consist primarily of rock and metal, other small solar system bodies are composed of frozen water, along with rock, metal, and ices of other compounds. We have seen in earlier chapters that water ice, along with carbon dioxide, methane, and ammonia ices, was locked up in planets and moons. Ices in the young solar system also condensed with roughly equal amounts of small rocky and metallic debris into bodies that still remain in orbit around the Sun. These dirty snowballs in space are the **comets.**

Comets are named after their discovers whenever possible. For example, Comet Shoemaker-Levy 9 was the ninth comet discovered by Carolyn and Eugene Shoemaker and separately, but simultaneously, by David Levy. Today, with large teams of astronomers using sophisticated observing equipment to locate comets, it is sometimes impossible to attribute their discoveries to just a few team members. Comets found under these conditions are named after the team or the research program that finds them.

9-6 Comets come from far out in the solar system

In 2001, astronomers were able to measure the temperature at which ammonia ice formed in Comet LINEAR (named after the Lincoln Near Earth Asteroid program that discovered it). This temperature determines the ammonia's solid structure, and by determining the structure of the ammonia in Comet LINEAR, scientists were able to calculate the temperature when it formed and therefore how far the comet was from the Sun at that time. That comet coalesced between the present orbits of Saturn and Uranus.

In the first few hundred million years of the solar system's existence, comets formed in its outer reaches at roughly the distances of Saturn, Uranus, and Neptune. In that region, water was plentiful and the temperature was low enough for the ices to condense into chunks several kilometers across. After they formed, gravitational tugs from Uranus and Neptune flung the comets in every direction, as we saw in Section 5-5.

The solar system is thought to contain two reservoirs of comets: the Kuiper belt and the Oort cloud (see Figure 5-6). Comets head inward as a result of near misses with other cometary bodies or due to gravitational disturbances from relatively nearby stars. The Kuiper belt (Figure 9-17) is centered on the plane of the ecliptic. Its main body of objects extends out some 50 AU from the Sun. Astronomers believe most comets entering the inner solar system come from the Kuiper belt because most of these comets have orbits relatively close to the plane of the ecliptic. If most comets came from the spherical Oort cloud, the planes of their orbits when near the Sun would be random, with most located far from the plane of the ecliptic.

The Kuiper belt objects were first discovered in 1992. These bodies fall into two groups. Those with roughly circular orbits are called *classical KBOs* and are found between 30 AU (Neptune's orbit) and 50 AU from the Sun. The second group, called *scattered KBOs,* have more elliptical orbits that range from 35 AU to at least 200 AU from the Sun. At least 1524 KBOs have been observed (Figure 9-17 and Figure 9-18). Although most of the

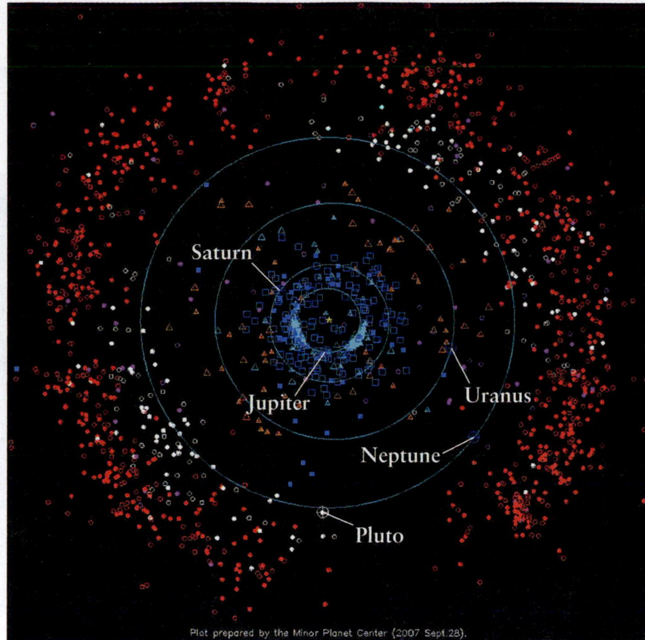

FIGURE 9-17 Current Positions of Known Dwarf Planets and SSSBs in the Outer Solar System Objects with unusually high eccentricity orbits are shown as cyan triangles. Objects roaming among the outer planets, called Centaur objects, are orange triangles. Plutinos are white circles. Miscellaneous objects are magenta circles, and classical KBOs are red circles. Objects observed only once are denoted by open symbols; objects with two separate observations are denoted by filled symbols. Comets are filled and unfilled light-blue squares. (Courtesy of Gareth Williams, Minor Planet Center)

observed KBOs are tens of kilometers across, astronomers have also begun to observe some KBOs only 10 to 100 m in diameter. Based on these latter observations, astronomers estimate that 10^{15} such objects exist out there.

At least 1% of the KBOs are orbiting pairs (Figure 9-18c), including Quaoar, which is 1250 km across, some 400 km larger than Ceres or about half the size of Pluto. At least 122 of the KBOs orbit the Sun in the same region as Pluto. These latter bodies are also called *plutinos* (Section 5-5). The period of Neptune's orbit around the sun is two-thirds as long as the orbital periods of Pluto and the plutinos. Indeed, it appears that Pluto and the plutinos were forced into their present orbit by Neptune's gravity. Thus, astronomers say that plutinos are locked into a *2:3 resonance* with Neptune.

The vast majority of comets are believed to lie even farther from the Sun. Unlike the Kuiper belt comets and the rest of the solar system, these bodies are believed to have a spherical distribution around the Sun, called the **Oort cloud** (Figure 5-6), named after the Dutch astronomer Jan Oort, who first proposed its existence in the 1950s.

In 2003, astronomers discovered an object larger than Quaoar. Called Sedna, this body is roughly 1800 km (1100 mi) across and is presently 86 AU from the Sun. Sedna has a highly elliptical orbit (Figure 9-19) that takes it from the outer realm of the main Kuiper belt and possibly into the Oort cloud. Because the distance from the Sun to the inner edge of the Oort cloud is not known, astronomers are still debating whether Sedna is another KBO or the first known Oort cloud object. Sedna has a sidereal rotation period of 10 h.

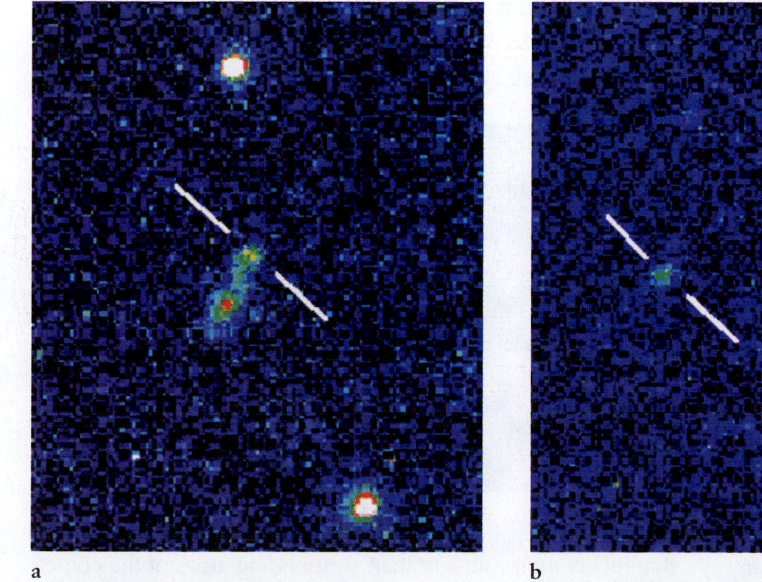
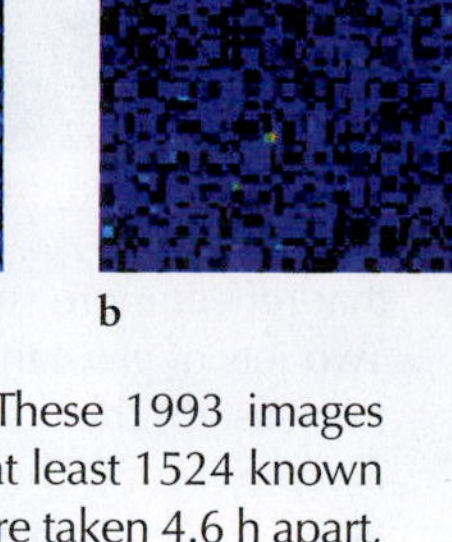

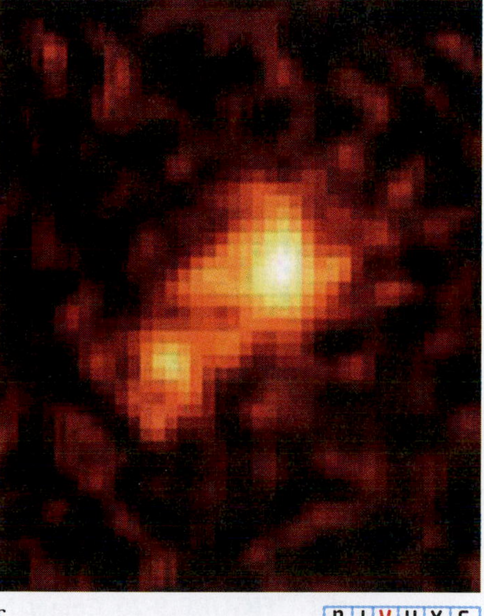

a b c

R I V U X G

FIGURE 9-18 Kuiper Belt Objects (a, b) These 1993 images show the discovery (white arrows) of one of at least 1524 known KBOs. These two images of KBO 1993 SC were taken 4.6 h apart, during which time the object moved against the background stars. (c) The KBO 1998 WW31 and its moon (lower left). (a and b: Alan Fitzsimmons, Queen's University of Belfast; c: C. Veillet/CFHT)

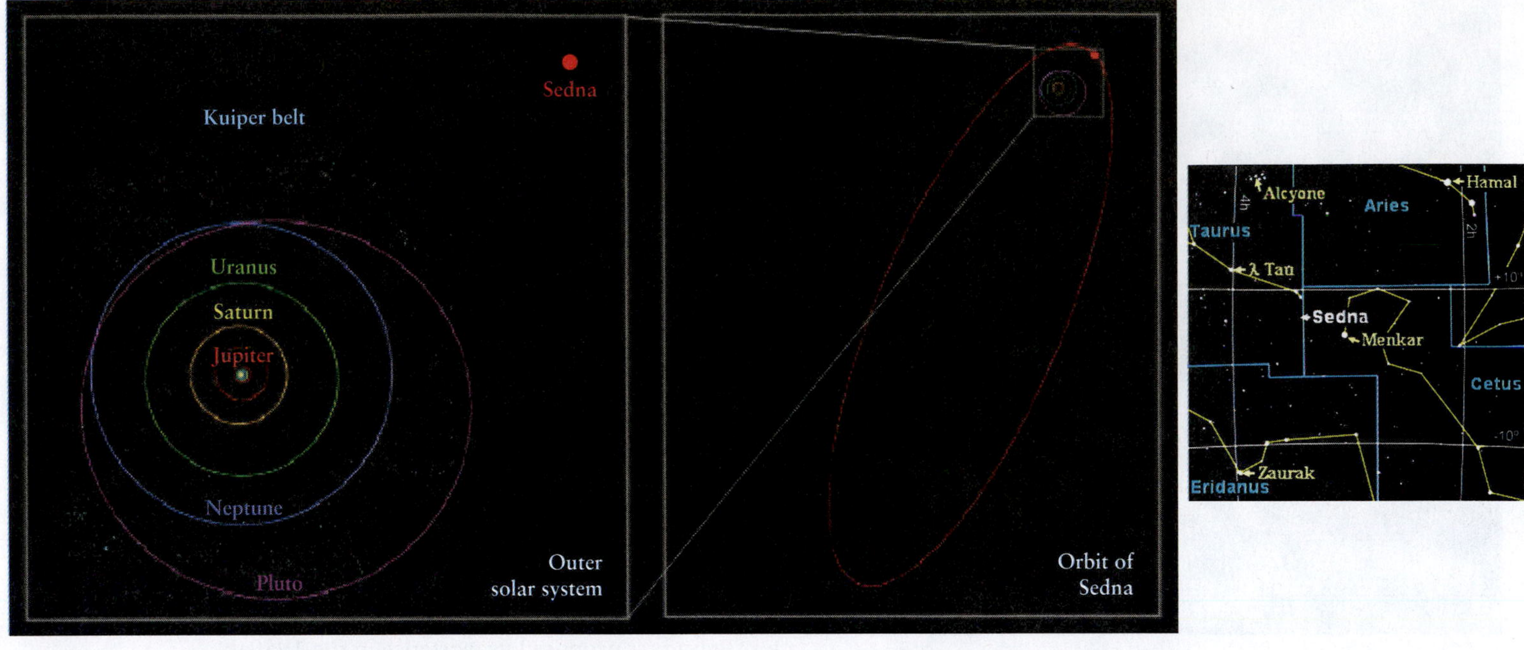

FIGURE 9-19 Sedna's Orbit (a) The farthest known body in the solar system is in a highly elliptical orbit (b) that ranges from the outer reaches of the Kuiper belt and possibly extends inward to the inner Oort cloud. (NASA/Caltech)

Astronomers calculate that the Oort cloud extends out at least 50,000 AU, one-fifth the distance to the nearest stars. Most of these comets have orbits so nearly circular that they never even get as close as Pluto is to the Sun. However, occasionally, a near miss between two comet nuclei or the gravitational tug from a passing star or passing giant molecular cloud nudges a distant comet toward the inner solar system. As a result of its inward plunge, comets from the Oort cloud, like Hale-Bopp and Hyakutake, have highly elliptical, parabolic, or even slightly hyperbolic orbits (see Section 2-8). Often Oort cloud comets also have orbits that are highly tilted, even perpendicular, to the plane of the ecliptic.

Because the comets in the Kuiper belt and Oort cloud are far from the Sun, they are completely frozen. Solid cometary bodies, called **nuclei** (singular, **nucleus**), are typically between 1 and 10 km across, although some, like that of Comet Halley, are 15 km or more in diameter. The first pictures of a comet's nucleus were obtained when a fleet of spacecraft flew past Comet Halley in 1986 (Figure 9-20a). Halley's potato-shaped nucleus is darker than coal, probably because of carbon-rich

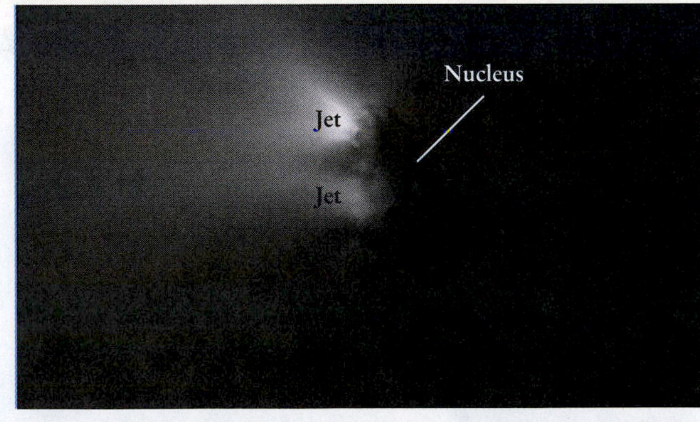

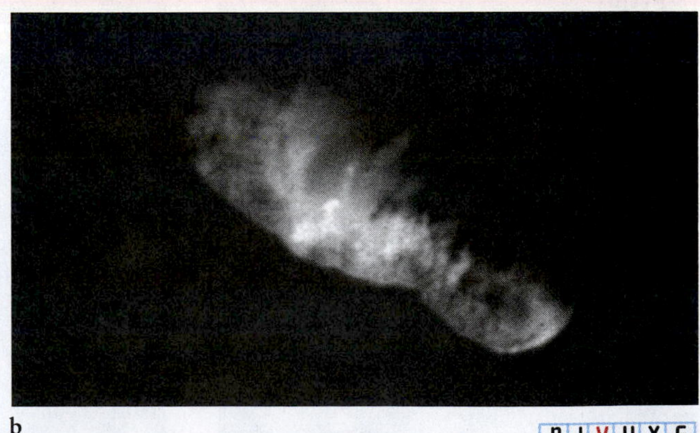

FIGURE 9-20 Comet Nuclei (a) This image, taken by the *Giotto* spacecraft, shows the potato-shaped nucleus of Comet Halley. Its dark nucleus measures 15 km in its longest dimension and about 8 km in its shortest. The numerous bright areas on the nucleus are icy outcroppings that reflect more sunlight than surrounding areas of the comet. Two jets of gas can be seen emanating from the left side of the nucleus. (b) The nucleus of Comet Borrelly, in an image taken by *Deep Space 1*. The nucleus is 8 km (5 mi) long. (a: Max-Planck-Institut für Aeronomie; b: *Deep Space 1* Team, JPL, NASA)

a

R I V U X G

b

R I V U X G

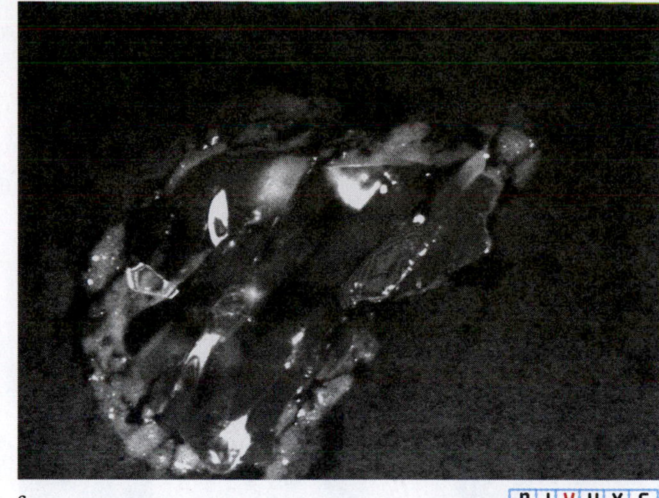

c

R I V U X G

FIGURE 9-21 Comet Wild 2 (a) This picture shows two images combined. One is a high-resolution photograph showing the surprisingly heavily cratered comet. The other image is a longer photograph showing gas and dust jetting from the comet. Its tails are millions of kilometers long. (b) A substance called aerogel was used to capture particles from Comet Wild 2's dust tail. A piece of space debris pierced the aluminum foil holding the aerogel and embedded in it, along with pieces of the foil. (c) A 2-μm piece of comet dust, composed of a mineral called forsterite. On Earth this mineral is used to make gems called peridot. (a, b: NASA/JPL; c: NASA/JPL-Caltech/ University of Washington.)

compounds left behind as ice evaporates. In 2001, the *Deep Space 1* spacecraft took high-resolution images of Comet Borrelly's nucleus (Figure 9-20b). The ends of the comet are very rugged, while the center has long, rolling plains, from which jets of gas appear to originate.

To better determine the composition of the gas emitted by comets, the *Stardust* probe visited Comet Wild (pronounced *vilt*) 2, where it collected samples and returned them to Earth in 2006 (Figure 9-21). Among the most interesting discoveries was the presence of **amino acids,** the building blocks of proteins upon which terrestrial life is based. Comet Wild 2 itself showed many unexpected features, including towering spires, many craters and cliffs, and more than 24 jets, rather than the two or three jets seen on other comets.

As a comet nucleus comes within 20 AU of the Sun, solar radiation begins to sublimate (change directly from solid to gas) the ices on its surface. The liberated gases form an atmosphere, or **coma,** around the nucleus. Because the coma scatters sunlight, it appears as a fuzzy, luminous ball. The largest coma ever measured was more than a million kilometers across—nearly as large as the Sun. Not visible to the human eye is the **hydrogen envelope,** a sphere of tenuous gas surrounding the comet's nucleus and measuring as much as 20 million km in diameter (Figure 9-22).

9-7 Comet tails develop from gases and dust pushed outward by the Sun

The most visible and inspiring features of comets are their long, flowing, diaphanous tails (Figure 9-23). 3 4 Comet tails develop from coma gases and dust pushed outward by radiation and particles from the Sun. Thus, comet tails do not trail behind the nucleus, as the exhaust from a jet plane does in Earth's atmosphere. Rather, *at the comet's nucleus, the tails always point away from the Sun* (Figure 9-24), regardless of the direction of the comet's motion. The implication that something from the Sun "blows" the comet's gases radially outward led Ludwig Biermann to predict the existence of the solar wind (see Section 6-4 and Section 10-3). This stream of particles from the Sun was actually discovered in 1962, a full decade after it was predicted, by instruments on the spacecraft *Mariner 2*. In 2001, NASA launched the *Genesis* spacecraft, which collected pristine solar wind particles for 2½ years. Although it crash-landed back on Earth in September 2004, some of the particles that it collected were able to be salvaged and studied.

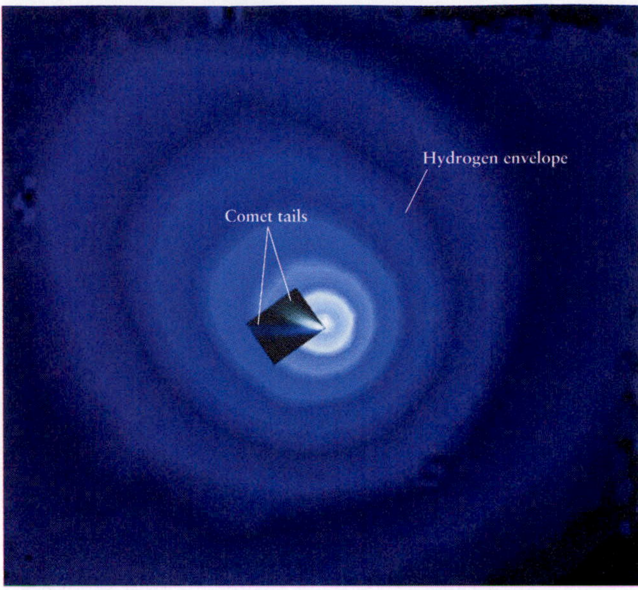

FIGURE 9-22 Comet Hale-Bopp In 1997, Comet Hale-Bopp had a hydrogen envelope 1 AU in diameter (blue ovals). This gas was observed in the ultraviolet. The visible light inset shows the scale of the visible tails (see also the image at the opening of this chapter). (Johns Hopkins University and Naval Research Laboratory; inset: Mike Combi)

Insight Into Science

Starting with Observations Science often advances by working from observations and experiments that require scientific explanations. For example, seeing comet tails and how they behave led Biermann to wonder what causes them. The simplest physical explanation is that matter from each cometary body is being evaporated and pushed upon by something from the Sun—the solar wind. Remember Occam's razor (Section 2-1).

FIGURE 9-23 Comet West Astronomer Michael M. West first noticed this comet on a photograph taken with a telescope in 1975. After passing near the Sun, Comet West became one of the brightest comets of the 1970s. This photograph shows the comet in the predawn sky in March 1976. (Hans Vehrenberg)

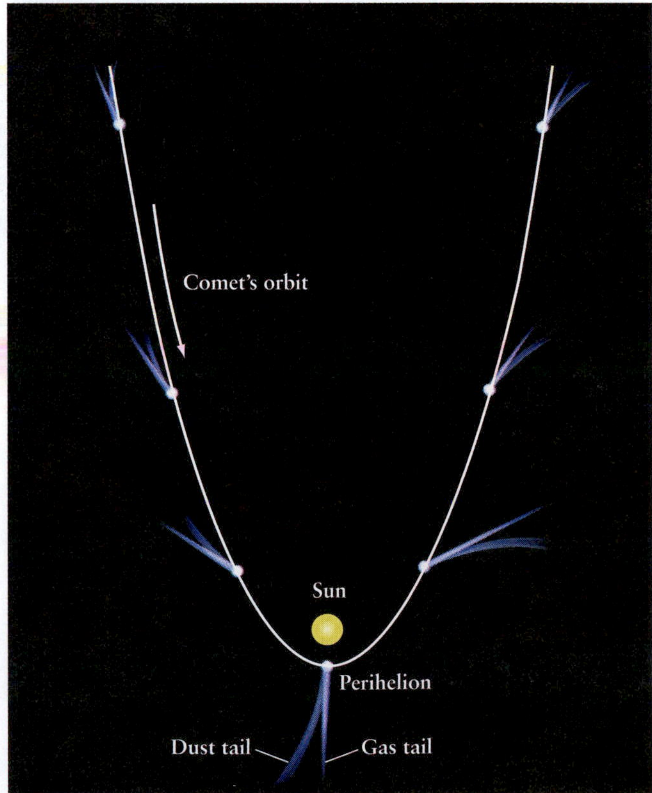

FIGURE 9-24 The Orbit and Tails of a Comet The sunlight and solar wind blow a comet's dust particles and ionized atoms away from the Sun. Consequently, comets' tails always point away from the Sun.

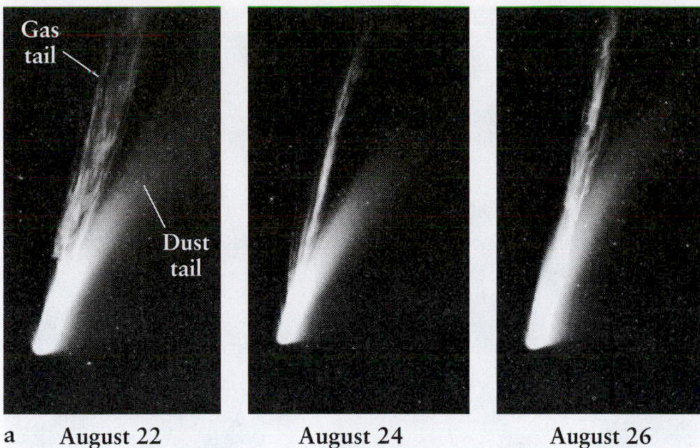

FIGURE 9-25 The Two Tails of Comet Mrkos (a) Comet Mrkos dominated the evening sky in August 1957. These three views, taken at 2-day intervals, show dramatic changes in the comet's gas tail. In contrast, the slightly curved dust tail remained fuzzy and featureless. (b) Wind blowing smoke from this forest fire causes the smoke column to change shape and direction, just like the solar wind and sunlight cause the tails of comets to change shape and direction. (a: Palomar Observatory; b: Paul Von Baich/The Image Works)

Outflowing particles and radiation from the Sun produce two comet tails: a **gas** (or **ion** or **plasma**) **tail** and a **dust tail** (Figure 9-25 and the figure opening this chapter). Positively charged ions (atoms missing one or more electrons) from the coma are swept directly away from the Sun by the solar wind to form the gas tail. Gas tails are composed primarily of water, carbon monoxide, carbon dioxide, methanol, formaldehyde, and ammonia. Such a tail often appears blue, because carbon monoxide ions in the gas strongly scatter blue light. The relatively straight gas tail can change dramatically from night to night (see Figure 9-25). These changes occur because the solar wind that pushes the gases changes rapidly and the gases themselves are very light and easily moved about, like the smoke from a fire (Figure 9-25b).

The dust tail is formed when sunlight strikes dust particles that have been freed from the comet's evaporating nucleus. Light exerts pressure on any object that absorbs or scatters it. While this pressure, called **radiation pressure** or **photon pressure,** is quite weak, fine-grained dust particles in a comet's coma are sufficiently light to be pushed from the vicinity of the comet, thus producing a dust tail. The dust tail is often white or light gray. The dust particles are massive enough not to flow straight away from the Sun; rather, the dust tail arches in a path that lies between the gas tail and the direction from which the comet came (Figure 9-25 and Figure 9-26). Even though gas tails of most comets appear significantly brighter and often longer than dust tails and other rocky debris, observations reveal that about 3 times as much mass leaves a typical comet in the form of dust and rocks than leaves as gas.

What is the solid part of a comet called?

Genesis spacecraft crash landed in Utah. (USAF 388th Range Squadron)

Figure 9-26 outlines the elements of comets, although many have bigger or smaller versions of these features.

FIGURE 9-26 The Structure of a Comet The solid part of a typical comet (the nucleus) is roughly 10 km in diameter. The coma can be as large as 10^5 to 10^6 km across, and the hydrogen envelope is typically 10^7 km in diameter. A comet's tail can be enormous—even longer than 1 AU. Comet Wild 2 (inset) is examined further in Figure 9-21. (This drawing is not to scale.) (Inset: NASA/JPL)

For example, some comets, like the one shown in Figure 9-27, have a large, bright coma but short, stubby tails. Others have an inconspicuous coma but one or more tails of astonishing length. The gas tail in Figure 9-28 stretches more than 150 million km (1 AU) in length.

9-8 Comets are fragile yet durable

To better understand the chemistry and structure of comets, astronomers sent the *Deep Impact* spacecraft, roughly the size of a small refrigerator, slamming into Comet Tempel 1 at 36,000 km/h (22,000 mi/h) in 2005 (Figure 9-29). Observations indicate that the projectile gouged a crater 100 m across and 30 m deep. This collision sent up a cloud of vapor and melted rock estimated to have had a mass of 10 million kg (22 million lb), still less than a ten-millionth of the comet's total mass. The particles ejected were typically smaller than 100 microns (μm).

The size of the impact crater and the volume of matter ejected provide strong evidence that this comet is composed of very weakly bound fragments of rock and ice, rather than solid pieces of each. Spectra of the debris thrown upward revealed many molecules containing carbon that had been parts of organic compounds. This finding supports the conjecture by some astronomers and geologists that comet impacts brought such compounds to Earth early in our planet's existence.

Even though this comet has an extremely low gravitational force with which to hold itself together, it managed to pull over 90% of the dust that it ejected back onto itself. The craters on this comet and others show that they have been hit by other pieces of space debris throughout their histories. The comet's ability to essentially reform itself after this and other collisions helps explain why such objects have not all been annihilated over time.

Which comet tail points directly away from the Sun, and why?

9-9 Comets do not last forever

Astronomers, many of them amateurs, typically discover at least a dozen new comets each year. Falling Sunward from the Kuiper belt, most are **long-period comets,** which have such eccentric orbits that they leave the inner solar system after one pass by the Sun and typically take tens of thousands up to millions of years to return. In 2006, astronomers observed the first comet whose orbit is so elliptical that it is believed to come from even farther out, from the Oort cloud. Called 2006 SQ372, this body has passed the orbit of Neptune on its way out to a maximum distance of 240 billion km from the Sun. This comet will return to the inner solar system in 22,500 years.

Some of these long-period comets can become trapped in the inner solar system. The transformation of their orbits can occur in two different ways. First, a comet can pass so close to a giant planet that the planet's

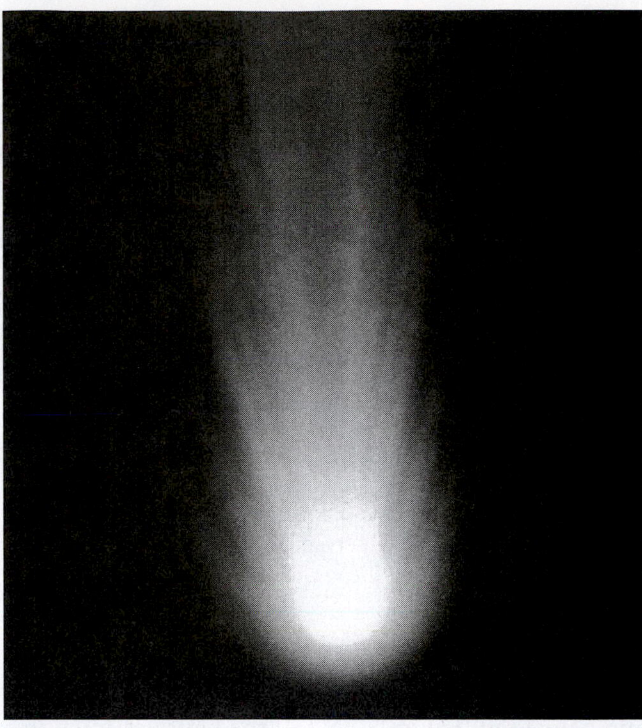

R I V U X G

FIGURE 9-27 The Head of Comet Brooks This comet had an exceptionally large, bright coma. Named after its discoverer, William R. Brooks, it dominated the night skies in October 1911. (UCO/Lick Observatory)

R I V U X G

FIGURE 9-28 The Tail of Comet Ikeya-Seki Named after its codiscoverers in Japan, this comet dominated the predawn skies in late October 1965. The yellow in the tail comes from emission by sodium atoms in the dust that was released by the comet. Although its coma was tiny, its tail spanned over 1 AU. (Roger Lynds/NOAO/AURA/NSF)

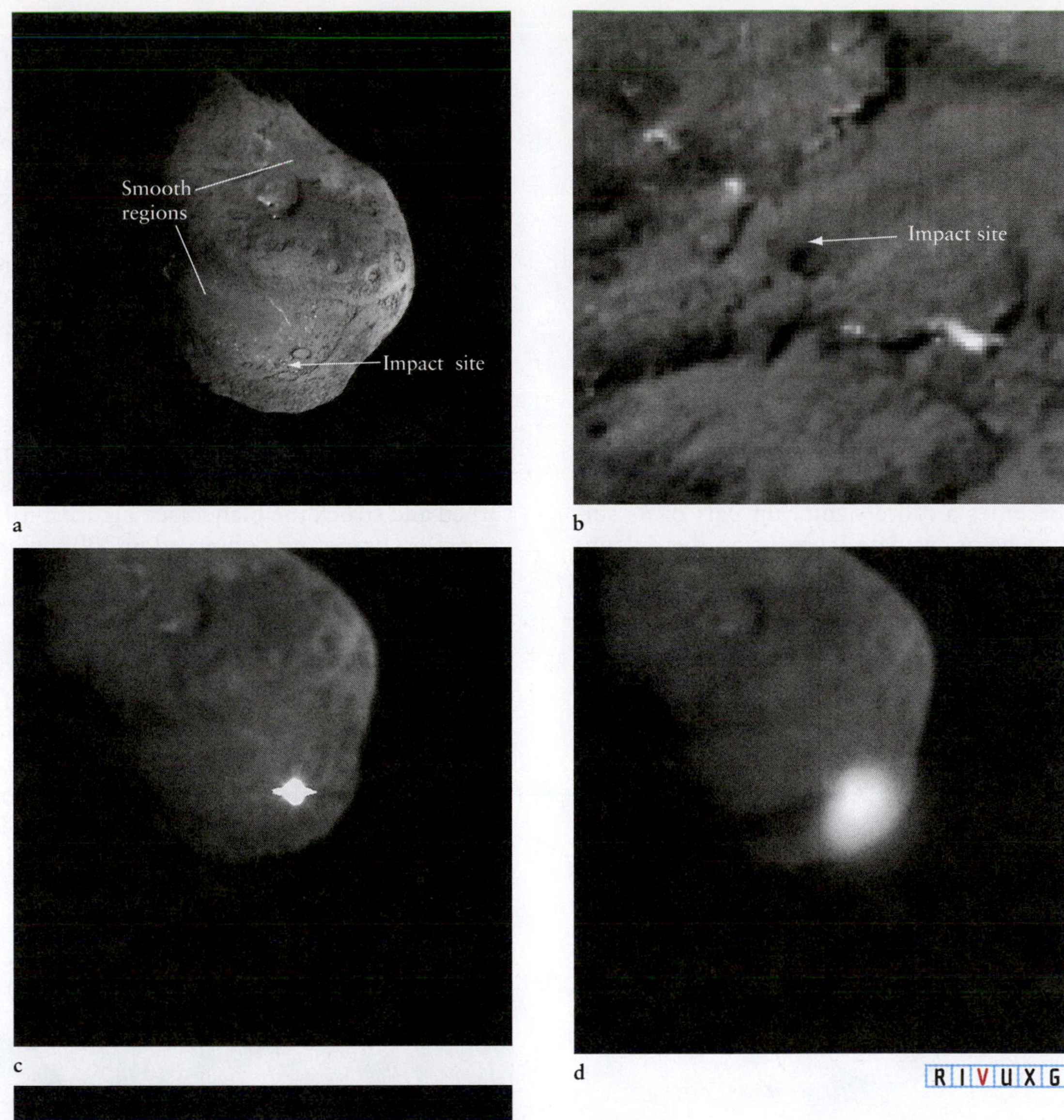

FIGURE 9-29 Comet Tempel 1 (a) This composite image of Comet Tempel 1 has higher resolution at the bottom, as the projectile from *Deep Impact* headed in that direction. The smooth regions on the comet have yet to be explained. (b) Thirty seconds before the projectile struck the comet. (c) Seconds after impact, hot debris explodes away from the comet nucleus. The white horizontal half-ellipses are areas where the CCDs were overloaded with light from the event. (d) Moments later, the gases and dust were expanding outward. (e) An image taken 67 s after impact. Within minutes, the cloud of debris became much larger than the entire nucleus. (a: NASA/JPL/UMD; b–e: NASA/JPL-Caltech/UMD)

gravitational force profoundly changes the comet's orbit, slowing it down and trapping it in the inner solar system (Figure 9-30). Second, the comet can undergo a number of less significant gravitational encounters with planets and other bodies whose cumulative effect is also to trap the comet in the inner solar system. In either case, the long-period comet becomes a **short-period comet,** orbiting the Sun in fewer than 200 years. Like Halley's comet, short-period comets appear again and again at predictable intervals. Halley's comet has a 76-year orbit and will next pass close to the Sun in 2062.

What is the composition of a comet?

Comets cannot survive an infinite number of passages near the Sun. A comet's first close encounter causes it to lose a large amount of its surface ice. In doing so, the dust it contains becomes a more significant part of its surface. This dust helps insulate it, reducing the amount of ice and dust the comet loses in subsequent close passages near the Sun. A typical comet is estimated to make between 60 and 100 passages to perihelion (closest approach to the Sun) before losing all its ice. Because the rock and metal debris in a comet are held together by the ices in which they are imbedded, when the ices are gone, the solid pieces drift apart, many of them remaining in orbit around the Sun. Halley's comet has been seen at least 27 times, meaning that it has fewer than 73 more close passes to the Sun before it dissipates. Because it returns every 76 years or so, it will survive for about $73 \times 76 = 5548$ more years. A comet can also be torn apart when it comes too close to a planet or the Sun, or it can be destroyed completely by striking a planet, a moon, or the Sun. A spectacular example was Comet Shoemaker-Levy 9. As discussed in Section 8-3, that comet fragmented under the tidal force from Jupiter in 1992. Two years later, with the world's astronomers watching carefully, the pieces returned and struck the planet (see Figure 8-10). Another impact on Jupiter was observed in 2009, when a single object several hundred meters across struck it.

Shoemaker-Levy 9 broke up because its nucleus was held together very weakly, like Comet Tempel 1. As discussed previously, Shoemaker-Levy 9 may have been composed of separate pieces that had stuck together until Jupiter's tidal force pulled them apart. Since then, astronomers have observed at least five other comets break up, in 1999, 2000 (Comet LINEAR), 2001, 2002 (Comet Toit-Neujmin-Delporte), and 2006 (Schwassmann-Wachmann 3). Figure 9-31

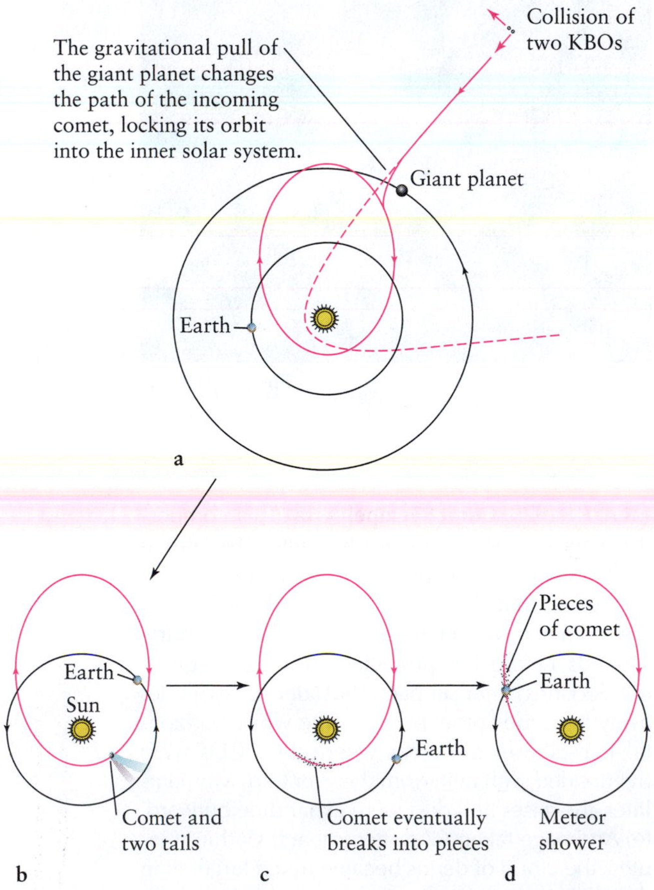

FIGURE 9-30 Transformation and Evolution of a Long-Period Comet (a) The gravitational force of a giant planet can change a comet's orbit. Comets initially on highly elliptical orbits are sometimes deflected into more circular paths that keep them in the inner solar system. (b–d) These figures show the evolution of a comet into gas, dust, and rubble, and why debris from some of these comets strikes Earth (see also Figure 9-37).

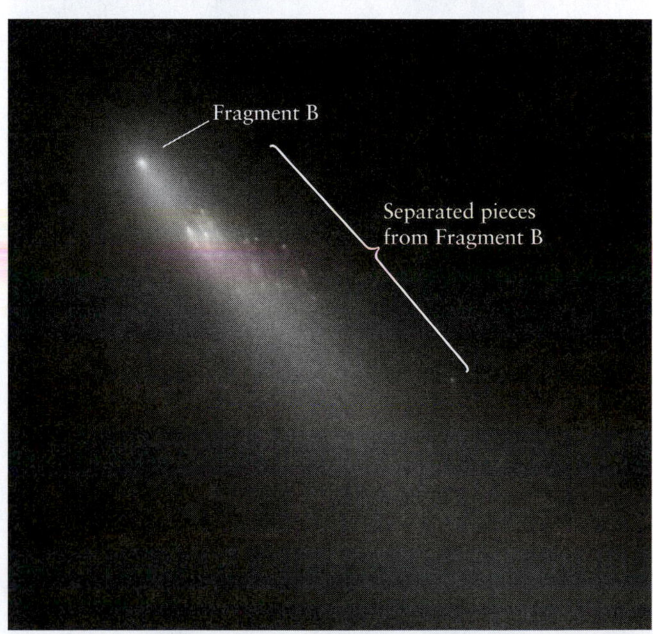

FIGURE 9-31 The Fragmentation of Comet Schwassmann-Wachmann 3 This comet, with a 5.4-year orbit, has been coming apart for decades. In 2006, it further fragmented after passing perihelion. One piece, Fragment B, shed at least 30 smaller pieces, shown here. (NASA; ESA; H. Weaver [APL/JHU]; M. Mutchler and Z. Levay [STLScI])

shows a piece of Comet Schwassmann-Wachmann 3 coming apart shortly after it passed perihelion. Soon thereafter, its remaining dust and rock fragments spread out in a loose collection of debris that continues to orbit the Sun along the comet's path.

Comets occasionally lose mass quickly by ejecting it in bursts. Several times, starting in September 1995, astronomers observed Comet Hale-Bopp eject 7 to 10 times more mass than usual (Figure 9-32). Astronomers think that these events resulted from surface ice and dust being heated by the Sun and then being rapidly ejected from a local region on the comet's surface, called a *vent*. Hale-Bopp rotates with a period of about 12 h. The ejected matter therefore formed pinwheel-shaped distributions spiraling from the comet's nucleus at a speed of 109 km/h (68 mi/h). Although the ejected matter did not represent a significant fraction of the comet's mass, its light-reflecting dust made it look like a huge comet fragment.

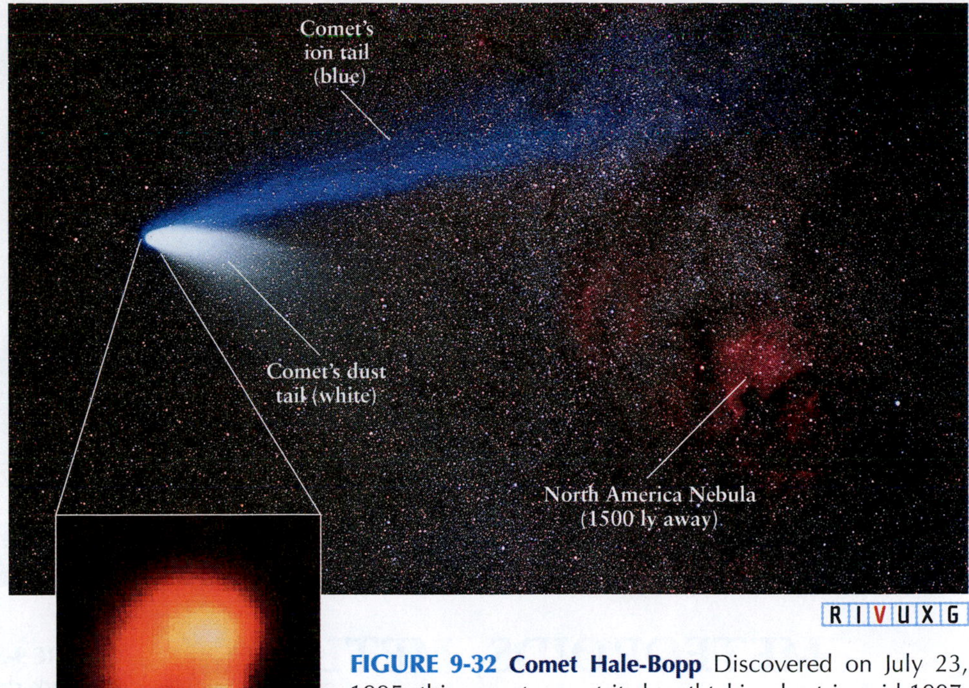

FIGURE 9-32 Comet Hale-Bopp Discovered on July 23, 1995, this comet was at its breathtaking best in mid-1997. Inset: Jets of gas and debris were observed shooting out from Comet Hale-Bopp several times. This image shows the comet nucleus (lower bright region), an ejected piece of the comet's surface (upper bright region), and a spiral tail. The ejected piece eventually disintegrated, following the same spiral pattern as the tail. (Tony and Daphne Hallas, Astrophotos)

For thousands of years, astronomers have observed comets that pass very closely to the Sun. This was difficult, of course, because of the Sun's brilliance, so only a few dozen had been carefully observed prior to the mid-1990s. Called *sungrazing comets*, they come within about 0.01 AU (1.46×10^6 km) of the Sun's surface and are typically sublimated and thereby separated into smaller pieces near perihelion. At the end of the nineteenth century, Heinrich Kreutz observed that sungrazers all have the same orbit, indicating that they are pieces of a single comet that broke up perhaps centuries ago. They are called *Kreutz comets* in his honor. In 1995, the Solar and Heliospheric Observatory (SOHO) began a dedicated mission of observing the Sun and, among other things, locating sungrazing comets. To date, it has observed over 1000 Kreutz comets (Figure 9-33).

What evidence do we have that Comet Shoemaker-Levy 9 was not one solid chunk of rock surrounded by ice?

The Sun is not the only star that comets orbit. A star labeled CW Leonis, about 500 ly from Earth, is presently enlarging in size and vaporizing billions of comets around it. Astronomers have observed the spectra of the water vapor from these cometary bodies.

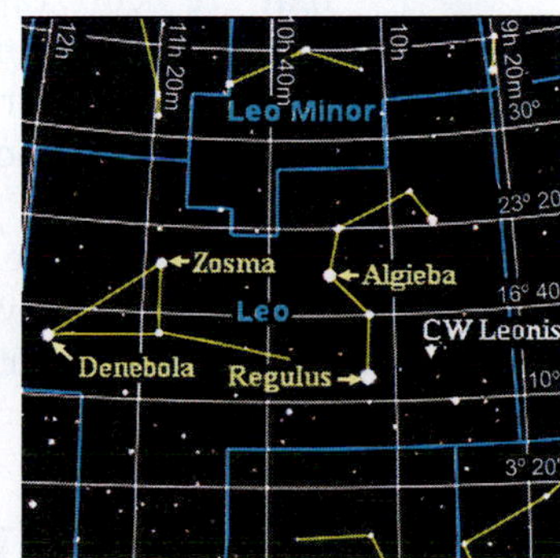

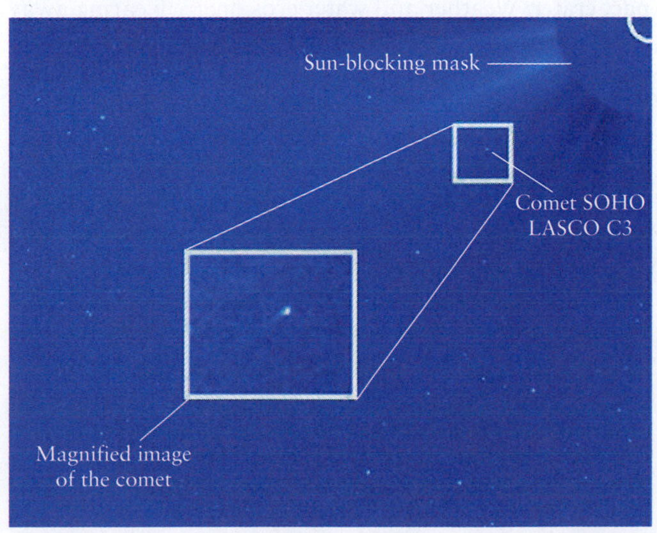

FIGURE 9-33 Sungrazing Comet Comet SOHO LASCO C3 is shown in the smaller box and magnified in the larger one. Discovered in March 2004, it was the 750th sungrazing comet discovered from the SOHO data. It completely sublimated near perihelion. (ESA/NASA-SOHO/LASCO)

WHAT IF . . . Different Comets Had Very Different Chemistries?
Water ice, as well as the others ices frozen in comets, have atoms of different isotopes (see Section 4-5). The amount of each isotope in a comet can be compared to the amount of each isotope of the same substances found on Earth. Very different amounts of isotopes on a comet compared to amounts found on Earth strongly suggest that the comet did not originate from the same gas and dust that created Earth. Indeed, such variations would (and in some cases do) suggest that some comets originated in orbit around other stars and were eventually captured by the Sun.

FIGURE 9-34 Meteor This brilliant meteor is seen lighting up the dark desert skies of the California desert area of Joshua Tree National Park. Just to the right of the meteor trail are the Pleiades. (Wally Pacholka, Astropics)

METEOROIDS, METEORS, AND METEORITES

As noted earlier, asteroids occasionally collide with each other, sending fragments into interplanetary space. These smaller pieces, along with rocky and metallic debris from evaporating comets, and material that never coalesced with larger bodies, are still orbiting the Sun. **Meteoroids** are rocky and metallic debris smaller than asteroids and are scattered throughout the solar system. Although no official size standard distinguishes the two, meteoroids are no more than about 10 m across, and the vast majority of them are smaller than 1 mm. **Meteors** are meteoroids that are being vaporized in the atmosphere. **Meteorites** are remnants of meteoroids that land intact.

Meteorite impact, Poughkeepsie, NY, 1992

By studying the chemistry of various asteroids, astronomers are now beginning to identify which ones were the origins for other asteroids and for meteoroids. For example, the asteroid Braille (named in honor of Louis Braille, developer of the alphabet for the blind) is a piece of debris 2 km long that was blasted off the 500-km-diameter asteroid Vesta.

9-10 Small, rocky debris peppers the solar system

5 Passing meteoroids are often pulled by gravity toward Earth's atmosphere. As they move through the atmosphere, they compress and heat the gases in front of them so much that the surrounding air glows. The glowing gas, accompanied by vaporized pieces of the infalling matter, creates the trail we see, and the meteoroid becomes a meteor (Figure 9-34). Common names for these dramatic streaks of light flashing across the sky include *shooting stars, fireballs,* and *bolides.* Fireballs are meteors at least as bright as Venus; bolides are bright meteors that explode in the air. Therefore, "shooting stars" are not stars of any kind or dying stars.

Most meteors vaporize completely before they can strike Earth. Their dust settles to the ground, often carried by raindrops. (This dust is not the source of acid rain, however, which comes from natural and human-made gases ejected into the atmosphere.) Pieces of a meteor that survive the fiery descent to Earth may leave an **impact crater.** Weather and water erosion are wearing away all of the nearly 200 known impact craters on Earth and thousands more have long since been drawn into Earth by the motion of Earth's tectonic plates. Indeed, the craters that exist today are all less than 500 million years old because of forces reshaping Earth's surface.

One of Earth's least-weathered impact craters is Meteor (or Barringer) Crater near Winslow, Arizona (Figure 9-35). Measuring 1.2 km across and 200 m deep, it formed approximately 50,000 years ago when an iron-rich meteoroid some 50 m across (about half the length of a football field) struck the ground at 40,000 km/h (25,000 mi/h). The blast was like the detonation of a 20-megaton hydrogen bomb.

On a typical clear night, you can expect to see a meteor about every 10 min. However, at predictable times throughout each year, Earth is inundated with them. These **meteor showers** occur when Earth passes

FIGURE 9-35 **Meteor Crater** An iron meteor measuring 50 m across struck the ground in Arizona 50,000 years ago. The result was this beautifully symmetric impact crater. (D. J. Roddy and K. Zeller/USGS)

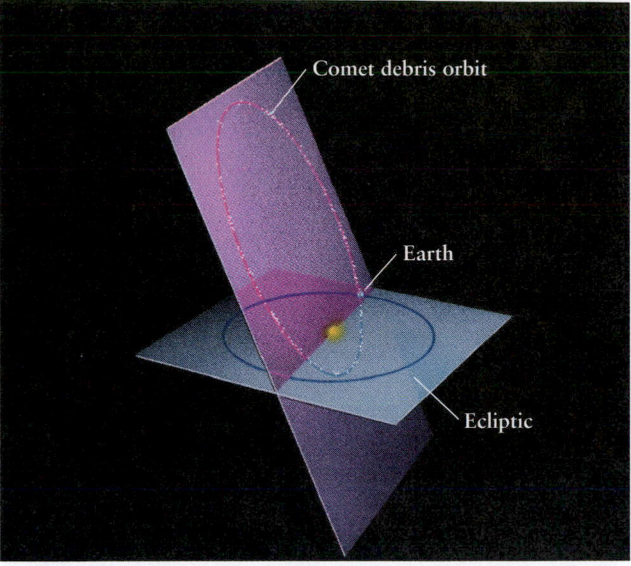

FIGURE 9-36 **The Origin of Meteor Showers** As comets dissipate, they leave debris behind that spreads out along their orbits. When Earth plows through such material, many meteors can be seen emanating from the same place within a very short time—a meteor shower. As shown in this diagram, many comets have high orbital inclinations.

through the orbit of debris left behind by a comet (Figures 9-36 and 9-37; also see Figure 9-30).

Some 30 meteor showers can be seen each year (see Table 9-1). Because the meteors in each shower appear to come from a fixed region of the sky, they are named after the constellation from which the meteors appear to radiate (Figure 9-38). For example, meteors in the Perseid shower appear to originate in the constellation Perseus. More than one meteor can be seen each minute at the peaks of such prodigious meteor showers as the Perseids, which take place in the summertime and are among the most exciting light shows in astronomy. Except for the Lyrids, meteor showers are best seen after midnight.

Table 9-1 Prominent Yearly Meteor Showers

Shower	Date of maximum intensity	Typical hourly rate	Constellation
Quadrantids	January 3	40	Boötes
Lyrids	April 22	15	Lyra
Eta Aquarids	May 4	20	Aquarius
Delta Aquarids	July 30	20	Aquarius
Perseids	August 12	80	Perseus
Orionids	October 21	20	Orion
Taurids	November 4	15	Taurus
Leonids	November 16	15	Leo Major
Geminids	December 13	50	Gemini
Ursids	December 22	15	Ursa Minor

Why are the times when meteor showers occur so predictable?

The Moon also moves through the debris that creates meteor showers on Earth. Whereas our atmosphere vaporizes most of the meteors before they strike Earth, the Moon's thin atmosphere provides no protection, so the Moon is struck by numerous meteoroids during each shower. The Leonid meteor shower of 1999 provided astronomers with an excellent opportunity to observe contemporary impacts on the Moon (see Figure 9-38). Some of these impacts were even bright enough to see with the naked eye. When a 10-kg meteorite strikes the Moon, the impact is as powerful as 10^4 pounds of TNT, and the resulting cloud is momentarily between 5×10^4 and 10×10^4 K, which is hotter than the surface of the Sun. Other meteor showers have provided further impact sightings on the Moon.

9-11 Meteorites are space debris that land intact

Although most meteors that enter Earth's atmosphere completely vaporize or are destroyed on impact, some land without totally disintegrating. Such pieces of debris are called meteorites, and they come from a variety of sources. Many are thought to have been broken off from asteroids by collisions; some are debris that were never part of larger bodies; still others come from the Moon, Mars, Vesta, or comets. We saw in Section 7-12, for example, how SNC meteorites offer clues to the chemistry of Mars. By studying the chemistry and age of one meteorite

FIGURE 9-37 Meteor Shower This time exposure, taken in 1998, shows meteors streaking away from the constellation Leo Major. They are part of the Leonid meteor shower. This shower occurs because Earth is moving through debris left by Comet Temple-Tuttle. (Jerry Lodriguss/Photo Researchers Inc.)

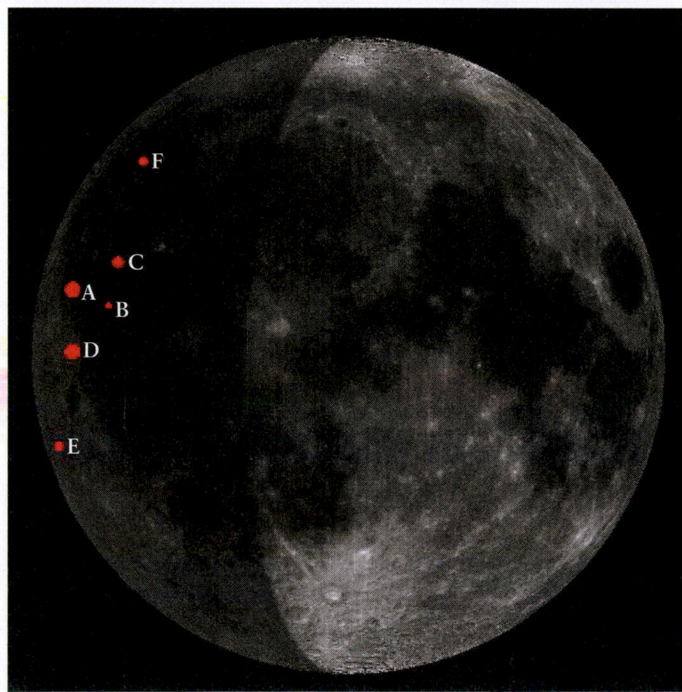

FIGURE 9-38 Recent Impacts on the Moon The locations A–F are places on the Moon where impacts were observed from Earth in 1999 during the Leonid meteor shower. The impacting bodies hit the Moon at around 260,000 km/h (160,000 mi/h) and had masses of between 1 and 10 kg. Each impact created a short-lived cloud that momentarily heated to between 5 × 10^4 and 10 × 10^4 K, much hotter than the surface of the Sun. (NASA)

from the Moon that was found on Earth, geologists have determined that it was ejected into space when the Mare Imbrium impact basin formed. Meteorites from the asteroid Vesta show signs of having been molten rock on that world. They were ejected when an impact occurred there, creating a crater 460 km (280 mi) wide and 13 km (8 mi) deep, which has been seen by the Hubble Space Telescope.

Meteorites tell us the age of the solar system. Measuring the relative amounts of various radioactive elements in them allows astronomers to determine how long ago they formed (see *An Astronomer's Toolbox 4-2*). The oldest known meteorites became solid bodies about 4.57 billion years ago, and impacts on Earth as early as 4 billion years ago have been identified, based on how they disturbed preexisting rock here. Therefore, the solar system is at least 4.57 billion years old, and we will use 4.6 billion years as its age throughout the book.

People have been picking up debris from space for thousands of years. The descriptions of meteorites in historical Chinese, Indian, Islamic, Greek, and Roman literature show that early peoples placed special significance on these "rocks from heaven." Meteorites also have a practical "impact": Infalling space debris is increasing Earth's mass by nearly 300 tons per day on average (Figure 9-39).

Stony Meteorites Often Look like Ordinary Rocks Meteorites are classified as stones (or stony), stony-irons, and irons. Most **stony meteorites** look much like ordinary rocks, although some are covered with a dark *fusion crust* (Figure 9-40a), created when the meteorite's outer layer melts during its fiery descent through the atmosphere (see Figure 9-37). When a stony

FIGURE 9-39 The Mass of Impacts on Earth The Vatican Obelisk is about 300 tons, the amount of mass that strikes Earth daily. As a result, Earth's mass increases by this amount every day. (John and Dallas Heaton/Corbis)

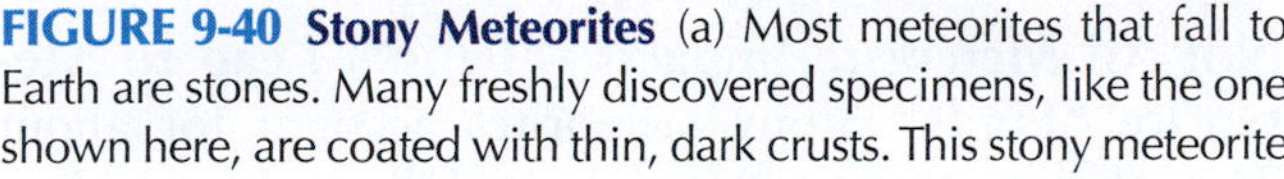

FIGURE 9-40 Stony Meteorites (a) Most meteorites that fall to Earth are stones. Many freshly discovered specimens, like the one shown here, are coated with thin, dark crusts. This stony meteorite fell in Morocco. (b) Some stony meteorites contain tiny specks of iron, which can be seen when the stones are cut and polished. This specimen was discovered in Ohio. (The R. N. Hartman Collection)

meteorite is cut in two and polished, tiny flecks of iron are sometimes found in the rock (Figure 9-40b).

Iron Meteorites Are Very Dense and Look Noticeably Different from Rocks **Iron meteorites** (Figure 9-41a) may also contain from 10% to 20% nickel by weight. Iron is moderately abundant in the universe as well as being one of the most common rock-forming elements, so it is not surprising that iron is an important constituent of asteroids and meteoroids. Another element, iridium, is common in the iron-rich minerals of meteorites but rare in ordinary rocks on Earth's surface. Iridium binds strongly to iron, so when Earth was first forming and iron was descending into our planet's core, it carried down most of the iridium that was associated with it. Measurements of iridium in Earth's crust can thus tell us the rate at which meteoritic material has been deposited over the ages.

FIGURE 9-41 Iron Meteorites (a) Irons are composed almost entirely of iron-nickel minerals. The surface of a typical iron is covered with thumbprint-like depressions created as the meteorite's outer layers vaporized during its high-speed descent through the atmosphere. This specimen was found in Argentina. (b) When cut, polished, and etched with a weak acid solution, most iron meteorites exhibit interlocking crystals in designs, called Widmanstätten patterns. This meteorite was found in Australia. (a: The R. N. Hartman Collection; b: R. A. Oriti)

Meteorites with a high iron content can be located with a metal detector. They also look unusual and hence are more likely to be noticed. Consequently, the easily found iron and stony-iron meteorites dominate most museum collections. In 1808, Count Alois von Widmanstätten, director of the Imperial Porcelain Works, in Vienna, discovered a conclusive test for the most common type of iron meteorite. Most iron meteorites have a unique structure of long nickel-iron crystals, called **Widmanstätten patterns,** which become visible when the meteorites are cut, polished, and briefly dipped into a dilute solution of acid (Figure 9-41b). Because nickel-iron crystals can grow to lengths of several centimeters only if the molten metal cools slowly over many millions of years, Widmanstätten patterns are never found in counterfeit meteorites (or "meteorwrongs").

Stony-Iron Meteorites Are the Most Exotic of All Space Debris on Earth The final category of meteorites are the **stony-iron meteorites,** which consist of roughly equal amounts of rock and iron. Figure 9-42, for example, shows the greenish mineral olivine suspended in a matrix of iron.

To understand why different types of meteorites exist, we consider their formation. Many meteorites were once pieces of asteroids. Heat from impacts and from the rapid decay of radioactive isotopes melted newly formed asteroid interiors. Over the next few million years, differentiation occurred, just as in young Earth. Iron sank toward the asteroid's center, while lighter rock floated up to its surface. Iron meteorites are fragments of asteroid cores, and stony-irons come from the boundary regions between the iron cores and stony crusts.

Stony meteorites have a variety of origins. Some come from the outer layers of asteroids. Other stony meteorites, ordinary **chondrites,** show no evidence of ever having been melted as parts of asteroids. They may therefore be primordial material from which our solar system was created. **Carbonaceous chondrites** are rare chondrites that contain small glass-rich beads called *chondrules.* These meteorites also contain complex carbon compounds, including simple sugars and glycerin. They also have as much as 20% water bound into their minerals. The organic compounds would have been broken down and the water driven out if these meteorites had been significantly heated. Undifferentiated asteroid Mathilde, shown in Figure 9-15a, has a very dark gray color and virtually the same spectrum as a carbonaceous chondrite meteorite, so it is likely composed of primordial material.

Amino acids are among the organic compounds occasionally found inside carbonaceous chondrites. By observing their spectra, we know that organic compounds exist on asteroids and smaller space debris. Indeed, some scientists suspect that amino acid–rich carbonaceous chondrites may have played a role in the origin of life on Earth.

Not All Meteorite Impacts ("Falls") Lead to Meteorite "Finds" Stony meteorites account for about 94% of all meteoritic material that falls on Earth. Most stony meteorites are not identified as meteorites because, well, they look like ordinary stones. Indeed, the percentages of the stony, iron, and stony-iron meteorites that are discovered, called *finds,* are quite different from the percentages that actually land, called *falls.* Nevertheless, astronomers and geologists have a good idea about how many of each type strike land. They obtain the correct percentages of impacts by carefully surveying areas in which only meteorites land, namely snow and ice-covered regions, such as Antarctica, or on deserts. By counting all of the debris under the surface using metal detectors and other technologies, the actual number of impacts of each type of meteorite is determined. Table 9-2 lists the relationship between the falls and finds.

R I V U X G

FIGURE 9-42 Stony-Iron Meteorite Stony-irons account for about 1% of all meteorites that fall to Earth. This specimen, a variety called a pallasite, was found in Antarctica. This meteorite has been thinly cut and appears to glow because of a light located behind it. (James C. Hartman)

9-12 The Allende meteorite and Tunguska mystery provide evidence of catastrophic explosions

A chance to study debris immediately after a meteorite impact came shortly after midnight on February 8, 1969, when a brilliant, blue-white light shot across the

Table 9–2 Percentages of Meteorite "Falls" and "Finds"

	Stony	Iron	Stony-Iron
Falls	94%	4.8%	1.2%
Finds	53%	42%	5%

night sky over Chihuahua, Mexico. Hundreds of people witnessed the dazzling display. The light disappeared in a spectacular and deafening explosion that dropped thousands of rocks and pebbles over the terrified onlookers. Within hours, teams of scientists were on their way to collect specimens of carbonaceous chondrites, collectively named the *Allende meteorite,* after the locality in which they fell (Figure 9-43).

One of the most significant discoveries to come from the Allende meteorite was evidence of the detonation of a nearby supernova 4.6 billion years ago (see Section 5-2). Among nature's most violent and spectacular phenomena, a *supernova explosion* occurs when a massive star blows apart in a cataclysm that hurls matter outward at tremendous speeds, as we will explore in Chapter 13. During this detonation, violent collisions between atomic nuclei produce a host of radioactive elements, including a short-lived radioactive isotope of aluminum. Based on its decay products, scientists found unmistakable evidence that this isotope once lay within the Allende meteorite. Some astronomers interpret this as evidence for a supernova in our vicinity at about the time the Sun was born. By compressing interstellar gas and dust, the supernova's shock wave may have helped stimulate the birth of our solar system.

Finding a meteorite in Antarctica. (NASA Johnson Space Center)

Why are stony-iron meteorites so rare compared to stony or iron meteorites?

Sixty years before the meteoritic event in Chihuahua, Mexico, at 7:14 A.M. local time on June 30, 1908, a much more spectacular explosion occurred, this one over the Tunguska region of Siberia. The blast, comparable to a nuclear detonation of several megatons, knocked a man off his porch some 60 km away and was audible more than 1000 km away. Millions of tons of dust were injected into the atmosphere, darkening the air as far away as California.

Preoccupied with wars, along with political and economic upheaval, neither Russia nor its successor, the Soviet Union, sent a scientific expedition to the site until 1927. At that time, Soviet researchers found that trees had been seared and felled radially outward in an area about 30 km in diameter (Figure 9-44). There was no

a

b

R I V U X G

FIGURE 9-43 Pieces of the Allende Meteorite (a) This carbonaceous chondrite fell near Chihuahua, Mexico, in February 1969. Note the meteorite's dark color, caused by a high abundance of carbon. Geologists believe that this meteorite is a specimen of primitive planetary material. The ruler is 15 cm long. (b) Sliced open, the Allende meteorite shows round, rocky inclusions called chondrules in a matrix of dark rock. (a: J. A. Wood; b: The R. N. Hartman Collection)

R I V U X G

FIGURE 9-44 Aftermath of the Tunguska Event In 1908, a stony asteroid traveling at supersonic speed struck Earth's atmosphere and exploded over the Tunguska region of Siberia. Trees were blown down for many kilometers in all directions from the impact site. (SOVFOTO) clear evidence of a crater. In fact, the trees at "ground zero" were left standing upright, although they were completely stripped of branches and leaves. Because no significant meteorite samples were found, for many years scientists assumed that a small comet had exploded in the atmosphere.

> What evidence do astronomers have from the Allende meteorite that a supernova explosion occurred at about the time the solar system formed?

Recently, however, several teams of astronomers have argued that a small comet, composed of primarily light elements and ice, would break up too high above the ground to cause significant damage. They argue that the Tunguska explosion was actually caused by a small asteroid or large meteoroid traveling at supersonic speed. The Tunguska event is consistent with an explosion of an asteroid about 80 m (260 ft) in diameter entering Earth's atmosphere at a shallow angle, moving at 79,000 km/h (50,000 mi/h) and exploding in the air as a result of becoming exceedingly hot. The resulting shock wave, slamming into the ground, would cause the damage seen without creating a crater. The debate on the cause of the Tunguska event continues.

FIGURE 9-45 Iridium-Rich Layer of Clay This photograph of strata in the Apennine Mountains of Italy shows a dark-colored layer of iridium-rich clay sandwiched between white limestone (bottom) from the late Mesozoic era and grayish limestone (top) from the early Cenozoic era. The coin is the size of a U.S. quarter. (W. Alvarez)

9-13 Asteroid impacts with Earth have caused mass extinctions

In the late 1970s, the geologist Walter Alvarez and his father, physicist Luis Alvarez, discovered the remnants of a far more powerful impact. Working at a site of exposed marine limestone in the Apennine Mountains in Italy that had been on Earth's surface 65 million years ago, the Alvarez team discovered an exceptionally high abundance of iridium in a dark-colored layer of clay between limestone strata (Figure 9-45). Recall from above that iridium that was present on the surface when Earth differentiated sank to the core. The iridium found by the Alvarez team had to have been deposited much later.

Since 1979, when this discovery was announced, a comparable layer of iridium-rich material has been uncovered at numerous sites around the world. In every case, geologic dating reveals that this apparently worldwide iridium-rich layer was deposited about 65 million years ago. Paleontologists were quick to realize the significance of this date, because it was back then that all of the dinosaurs rather suddenly became extinct. In fact, two-thirds of all animal species disappeared within a brief span of time 65 million years ago.

ANIMATION 9.3

The Alvarez discovery supported an astronomical explanation for the dramatic extinction of so much of the life that once inhabited our planet—an asteroid impact. There is no universal agreement on the disasters that befell Earth during this episode. One scenario has an asteroid 10 km in diameter slamming into Earth at high speed and throwing enough dust into the atmosphere to block out sunlight for several years. As the temperature dropped drastically and plants died for lack of sunshine, the dinosaurs would have perished, along with many other creatures in the food chain that were highly dependent on vegetation. The dust eventually settled, depositing an iridium-rich layer over Earth. Another theory posits that the impact created enough heat to cause planetwide fires, followed by changes in the oceans that killed many species of life in them. Whatever scenario turns out to be correct, small creatures capable of ferreting out seeds and nuts were among the animals that managed to survive this holocaust, setting the stage for the rise of mammals and, consequently, the evolution of humans.

In 1992, a team of geologists discovered that the hypothesized asteroid crashed into a site on the eastern edge of Mexico. They based this conclusion on glassy debris and violently shocked grains of rock ejected from the multiringed, 195-km-diameter Chicxulub (pronounced *Chih-chuh-lube*) Crater buried under the Yucatán Peninsula and western Caribbean Sea in Mexico (Figure 9-46). From the known rate at which radioactive potassium decays, the scientists have pinpointed the date when the asteroid struck—64.98 million years ago. In 1998, geologists digging on the Pacific Ocean floor discovered a piece of meteoritic debris with precisely the same age, apparently a piece of the offending asteroid. While some geologists and paleontologists are not yet convinced that an asteroid impact led to the extinction of the dinosaurs, most agree that this hypothesis better fits the available evidence than any other explanation that has been offered so far.

By comparing the chemical composition of the debris at Chicxulub with the chemical composition of meteorites, astronomers and geologists in 2007 determined that the asteroid that caused the mass extinction 65 million years ago came from a group of asteroids called Baptistinas. This group was created as a result of a larger body, some 170 km in diameter, breaking up after a collision 160 million years ago.

Another Impact May Have Led to an Earlier, More Devastating Mass Extinction The end of the dinosaurs' reign was not the only mass extinction caused by an impact. In 2001, evidence came to light concerning the Permian-Triassic boundary of 250 million years ago that suggests another devastating blow from space. That time is called the "Great Dying," a mass extinction during which some 80% of the species of life living on land and 90% of those living in the oceans perished. Rocks from that time discovered in places from Japan to Hungary show evidence, in the form of *fullerenes*—soccer ball–shaped molecules containing at least 60 carbon atoms—of an impact from space. Trapped inside these fullerenes were gases that could only have been forced into them by exploding stars. (This signature of gas-filled fullerenes from space has now also been discovered in the layer of rock that existed on Earth's surface 65 million years ago.)

A 250-million-year-old crater that may have been the relevant impact site was found in 2004 off the northwestern coast of modern-day Australia. Keep in mind that both this impact and the one that ended the reign of the dinosaurs occurred on a world whose surface is in continual motion. Therefore, their present locations are not the same as their initial impact sites. The Chicxulub impact actually occurred in the Pacific Ocean, whereas the Australian impact occurred before the continents were separated from Pangaea (see Section 6-2). The impact explanation of the Great Dying is not universally accepted. Another possible explanation is a massive amount of volcanic eruption that darkened the skies and changed the climate.

Could a catastrophic impact happen again? So many asteroids cross Earth's path that scientists agree that it is a matter of "when" rather than "if." NASA and other agencies are actively cataloging potentially dangerous asteroids. The good news is that studies of craters show that larger asteroids strike Earth significantly less often than do smaller ones. Although asteroids large enough to create Meteor Crater strike Earth about once every 10,000 years, killer asteroids, like the one that killed off the dinosaurs, collide with Earth only once every 100 million years. The threat of a catastrophic impact by an asteroid or comet in our lifetimes, thankfully, is remote.

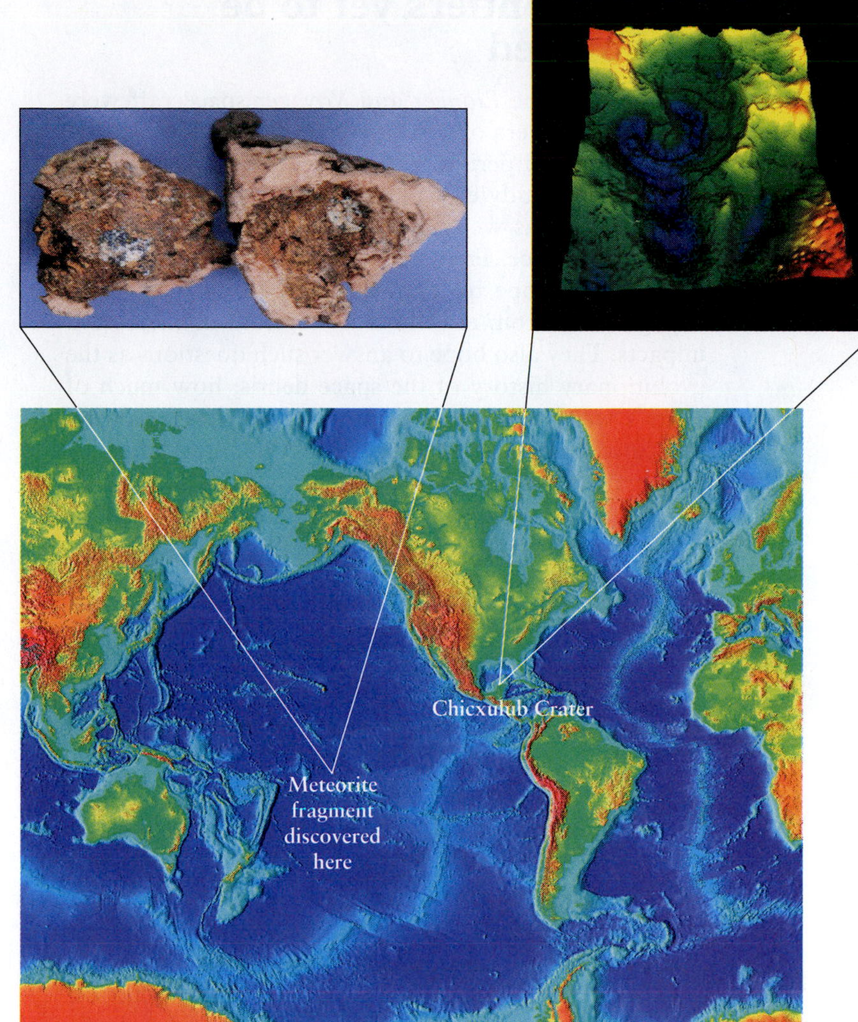

FIGURE 9-46 Confirming an Extinction-Level Impact Site By measuring slight variations in the gravitational attraction of different materials under Earth's surface, geologists create images of underground features. Concentric rings of the underground Chicxulub Crater (right inset) lie under a portion of the Yucatán Peninsula. This crater has been dated to 65 million years ago and is believed to be the site of the impact that led to the extinction of the dinosaurs. A piece of 65-million-year-old meteorite discovered in the middle of the Pacific Ocean in 1998 is believed to be a fragment of that meteorite. The fragment, about 0.3 cm (0.1 in.) long, was cut into two pieces for study (left inset). (Virgil L. Sharpton, Lunar and Planetary Institute; right inset: digital image by Peter W. Sloss, NOAA-NESDIS-NGDCD; left inset: Frank T. Kyte, UCLA)

How did the Chicxulub Crater move from the Pacific to the Caribbean?

9-14 Frontiers yet to be discovered

The years of the *Pioneer* and *Voyager* spacecraft were the first golden era of solar system research. We are now in another such period, with spacecraft either in development or already en route to planets, asteroids, comets, and even, as we will see shortly, out observing the weather in space. From studies of solar system debris, astronomers hope to learn whether life on Earth was brought here from elsewhere by asteroid or meteoroid impacts. They also hope to answer such questions as the evolutionary history of the space debris; how much of Earth's water came here during the planet's formation and how much landed afterward from comet impacts; whether the asteroids have sufficiently valuable compositions to justify mining them; whether comets can be harvested to supply water and other materials for people colonizing the solar system; and whether the Oort cloud really exists.

SUMMARY OF KEY IDEAS

Asteroids

• Pieces of solar system debris larger than about 10 m and composed primarily of rock and metal are called asteroids.

• Tens of thousands of belt asteroids with diameters larger than a kilometer are known to orbit the Sun between the orbits of Mars and Jupiter. The gravitational attraction of Jupiter depletes certain orbits within the asteroid belt. The resulting Kirkwood gaps occur at simple fractions of Jupiter's orbital period.

• Jupiter's and the Sun's gravity combine to capture Trojan asteroids in two locations, called stable Lagrange points, along Jupiter's orbit.

• The Apollo asteroids move in highly elliptical orbits that cross the orbit of Earth. Many of these asteroids will eventually strike the inner planets.

• A belt asteroid, Ceres, along with four KBOs (Pluto, Eris, Haumea, and Makemake) are classified as dwarf planets.

• Pluto, a KBO and dwarf planet, is an icy world that may well resemble the moon Triton.

Comets

• Comet nuclei are fragments of ice and rock often orbiting at a great inclination to the plane of the ecliptic. In the Kuiper belt and Oort cloud, comets have fairly circular orbits. When close to the Sun, they generally move in highly elliptical orbits.

• Many comet nuclei orbit the Sun in the Kuiper belt, a doughnut-shaped region beyond Neptune. Billions of cometary nuclei are also believed to exist in the spherical Oort cloud located far beyond the Kuiper belt.

• As an icy comet nucleus approaches the Sun, it develops a luminous coma surrounded by a vast hydrogen envelope. A gas (or ion) tail and a dust tail extend from the comet, pushed away from the Sun by the solar wind and radiation pressure.

Meteoroids, Meteors, and Meteorites

• Boulder-sized and smaller pieces of rock and metal in space are called meteoroids. When a meteoroid enters Earth's atmosphere, it produces a fiery trail, and it is then called a meteor. If part of the object survives the fall, the fragment that reaches Earth's surface is called a meteorite.

• Meteorites are grouped in three major classes according to their composition: iron, stony-iron, and stony meteorites. Rare stony meteorites, called carbonaceous chondrites, may be relatively unmodified material from the primordial solar nebula. These meteorites often contain organic hydrocarbon compounds, including amino acids.

• Fragments of rock from "burned-out" comets produce meteor showers.

• An analysis of the Allende meteorite suggests that a nearby supernova explosion may have been involved in the formation of the solar system some 4.6 billion years ago.

• An asteroid that struck Earth 65 million years ago probably contributed to the extinction of the dinosaurs and many other species. Another impact may have caused the Great Dying of life 250 million years ago. Such devastating impacts occur on average every 100 million years.

A WHAT DID YOU THINK?

1 *Are the asteroids a former planet that was somehow destroyed? Why or why not?* No. The gravitational pull from Jupiter prevented a planet from ever forming in the asteroid belt. Also, the total mass of the asteroids is much less than even the mass of tiny Pluto, a dwarf planet.

2 *How far apart are the asteroids on average?* The distance between asteroids averages 10 million km.

3 *How are comet tails formed? Of what are they made?* Ices in comet nuclei are turned into gas by absorbing energy from the Sun. Debris is released in this process. Sunlight and the solar wind push on the gas and dust, creating the tails.

4 *In which directions do a comet's tails point?* Comets' gas tails point directly away from the Sun; their dust tails make arcs pointing away from the Sun.

5 *What is a shooting star?* A shooting star is a piece of space debris plunging through Earth's atmosphere—a meteor. It is not a star.

Key Terms for Review

amino acid, 297
Apollo asteroid, 292
asteroid belt, 289
belt asteroid, 289
carbonaceous chondrite, 308
chondrites, 308
coma (of a comet), 297
comet, 294
dust tail, 299
dwarf planet, 284
gas (ion) tail, 299
hydrogen envelope, 297
impact crater, 304
iron meteorite, 307
Kirkwood gaps, 290
long-period comet, 300
meteor, 304
meteor shower, 304
meteorite, 304
meteoroid, 304
nucleus (of a comet), 296
Oort cloud, 295
planet, 284
radiation (photon) pressure, 299
short-period comet, 302
small solar system bodies (SSSBs), 284
stable Lagrange points, 292
stony meteorite, 306
stony-iron meteorite, 308
Trojan asteroid, 292
Widmanstätten patterns, 308

Review Questions

1. A piece of space debris that you pick up from the ground is called a(n): **a.** asteroid, **b.** meteoroid, **c.** meteor, **d.** meteorite, **e.** comet

2. Space debris that is a roughly equal mix of rock and ice is called a(n): **a.** asteroid, **b.** comet, **c.** meteoroid, **d.** meteorite, **e.** meteor

3. Which is the rarest type of meteorite found on Earth? **a.** iron, **b.** stony-iron, **c.** stony

4. Which part of a comet is solid? **a.** nucleus, **b.** halo, **c.** gas tail, **d.** dust tail, **e.** coma

5. Suppose you were standing on Pluto. Describe the motions of Charon relative to the horizon. Under what circumstances would you never see Charon?

6. Describe the circumstantial evidence that supports the idea that Pluto is one of a thousand similar icy worlds that once occupied the outer regions of the solar system.

7. What role did Charon play in enabling astronomers to determine Pluto's mass?

8. Why are asteroids, meteoroids, and comets of special interest to astronomers who want to understand the early history of the solar system?

9. Describe the objects in the asteroid belt, including their sizes, orbits, and separation.

10. To test your understanding of the asteroid belt, do Interactive Exercise 9.1 on the Web. You can print out your results, if required.

11. Why are there many small asteroids but only a few very large ones?

12. Describe the different chemistries of the two tails of a comet.

13. In what directions do comet tails point, and why?

14. What are the Kirkwood gaps, and what causes them?

15. What are the Trojan asteroids, and where are they located?

16. Describe the three main classifications of meteorites. How do astronomers believe that these different types of meteorites originated?

17. Why do astronomers believe that the debris that creates many isolated meteors comes from asteroids, whereas the debris that creates meteor showers is related to comets?

18. To test your understanding of comets, do Interactive Exercise 9.2 on the Web. You can print out your results, if required.

19. Why is the phrase "dirty snowball" an appropriate characterization of a comet's nucleus?

20. What and where is the Kuiper belt, and how is it related to debris left over from the formation of the solar system?

21. Why do scientists think the Tunguska event was caused by a large meteoroid and not a comet?

22. What evidence in Figure 9-32 supports the labeling of the gas and dust tails?

Advanced Questions

The answers to computational problems, which are preceded by an asterisk (*), appear at the end of the book.

23. Would you expect the surfaces of Pluto and Charon to be heavily cratered? Explain.

24. How did the regolith on asteroid Eros form?

25. Why are comets generally brighter after passing perihelion (closest approach to the Sun) than before reaching perihelion?

26. Can you think of another place in the solar system where a phenomenon similar to the Kirkwood gaps in the asteroid belt is likely to exist? Explain your answer.

27. Where on Earth might you find large numbers of stony meteorites that have not been significantly changed by weathering?

***28.** Assuming a constant rate of meteor infall, how much mass has Earth gained in the past 4.6 billion years?

Discussion Questions

29. Suppose it was discovered that the asteroid Hermes had been perturbed in such a way as to put it on a collision course with Earth. Describe what you would do to counter such a catastrophe using present technology.

30. From the abundance of craters on the Moon and Mercury, we know that numerous asteroids and meteoroids struck the inner planets during the very early history of the solar system. Is it reasonable to suppose that numerous comets also pelted the planets 4 to 4½ billion years ago? What effects would such a cometary bombardment have had, especially with regard to the evolution of the primordial atmospheres of the terrestrial planets and oceans on Earth?

What If ...

31. An ocean on Earth were struck by a comet nucleus a kilometer across? What physical effects would occur to Earth?

32. We passed through the tail of a comet? What would happen to Earth and life on it?

33. As some astronomers have recently argued, passage of the solar system through an interstellar cloud of gas could perturb the Oort cloud, causing many comets to deviate slightly from their original orbits? What might be the consequences for Earth?

34. All of the space debris (asteroids, meteoroids, and comets) had been cleared out of the solar system 3.8 billion years ago? What would have been different in the history of Earth and in the history of life on Earth?

Web Questions

35. The discovery of Charon (see Figure 9-4) was made by an astronomer at the U.S. Naval Observatory. Search the Web to find out why the U.S. Navy carries out work in astronomy.

36. Search the Web to find out why some scientists disagree with the idea that a tremendous impact led to the demise of the dinosaurs. (They do not dispute that the impact occurred, only what its consequences were.) What are their arguments? From what you have learned, what is your opinion?

37. Several scientific research programs are dedicated to the search for near-Earth objects (NEOs), especially those that might someday strike our planet. Search the Web for information about at least one of these programs. How does the program search for NEOs? How many NEOs are now known, and how many has this program found? Will any of these NEOs pose a threat in your lifetime?

38. Search the Web to learn whether there are any comets visible at present. List them. What constellations are they presently in? Are any visible to the naked eye? (Recall that the unaided human eye sees objects brighter than about sixth magnitude.)

Observing Projects

39. Use the *Starry Night*™ program to observe Pluto and Charon. Select **Favourites > Discovering the Universe > Pluto.** The view is centered upon Pluto from about 42,000 km above its surface. Charon, one of Pluto's moons, is also shown almost in front of the planet, along with its orbit. You can measure the period of revolution of Charon and compare it to the rotation period of Pluto. To help with this, the time has been adjusted so that the south pole stick on Charon lines up with the star Atria. Note that a red meridian line between the poles of Pluto is pointing directly toward the observer at this specific time. Write down the time and date shown in the toolbar. With the **Time Flow Rate** at 30,000×, click the **Play** button and observe the motion of Charon around Pluto and pay particular attention to the rotation of Pluto. Stop the time advance after Charon has completed a complete orbit and adjust the time to return Charon precisely to its aligned position with the star Atria and again note the date and time. **a.** What is the period of Charon's orbit? **b.** What did Pluto do during this revolution of Charon? What is Pluto's rotation period? **c.** How do these two periods compare? Explain this result.

40. Use the *Starry Night*™ program to explore some of the dwarf planets of the solar system. Select **Favourites > Discovering the Universe > Dwarf Planets.** This view, from a position in space about 97 AU from the Sun, shows the orbits of several dwarf planets, with Neptune's orbit as a reference. **a.** Do the dwarf planets revolve around the Sun in the same direction as the planets or in the opposite direction? **b.** Use the location scroller to change the view so that the plane of Neptune's orbit appears edge on in the view. How do the orbital planes of the dwarf planets compare to those of the planets?

41. In this exercise, you can use *Starry Night*™ to locate the Sun as seen from the dwarf planet Eris. Select **Favourites > Discovering the Universe > Eris.** The view is from a location 6125 km above the surface of Eris, with the constellations displayed in the view. Use the location scroller to move around Eris at this elevation. Do this until an object that you believe is the Sun lies just to one side of Eris. (*Hint:* Think about what Eris's phase will be when you are in this position, and move the location scroller accordingly.) Place the cursor over that object and use the heads-up display to see if you are correct. If not, move around with the location scroller until you find the Sun. With the Sun next to Eris in the view, make a simple drawing of Eris, the Sun, and a few constellations showing the Sun's location in Eris's sky. **a.** Locate the constellation of Orion. Does it look the same from Eris as it does from Earth? Explain why or why not. **b.** Move to the surface of Earth on this date. (*Hint:* Open **Options > Viewing Location...** and, in the **View from** box, enter the surface of Earth.) Then use the **Find** facility to **Centre** the Sun in the sky on this date. If the Sun is not visible from the current location, click the **Hide Horizon** button in the dialog window that appears. Is the Sun in the same constellation as seen from Eris as it is from Earth? Explain why or why not.

42. Make arrangements to view an asteroid. At opposition, some of the largest asteroids are bright enough to be seen through a modest telescope. You can use *Starry Night*™ to see which bright asteroids are visible. Click the **Home** button in the toolbar followed by the **Stop** button to stop time flow. Set the time in the toolbar to your expected observing time. Open the **Find** pane and clear any data in the search box to bring up a list of solar system objects. Click the "+" button to the left of the **Asteroids** entry to expand the list of asteroids. Asteroids that are visible from your location at the time of your planned observation will appear in black type while those that are not visible will appear in light gray type. Double-click on one of the asteroids that will be visible at the time of your planned observation. Zoom in on the asteroid progressively and print a chart of its position in the sky at various zoom factors. To print a chart, move the mouse over the asteroid in the view and open its contextual menu (right-click on a PC or Ctrl-click on a Mac), and select **Print Chart. . . .** In the **Print Settings** dialog, select the **3 Pane** layout, choose 3 different fields of view, such as 90°, 50°, and 5°, and select **Use current settings.** As an alternative to using *Starry Night*™, you can check the "Minor Planets" section of the current issue of the *Astronomical Almanac* to see if any bright asteroids are close to opposition, or check the current issue, as well as the most recent January issue, of *Sky & Telescope,* for star charts showing the paths of bright asteroids among the constellations. You will need an appropriate sky chart to distinguish the asteroid from background stars. Observe the asteroid on at least two occasions separated by a few days. On each night, draw a star chart of the objects in your telescope's field of view. Has the position of one star-like object shifted between observing sessions? Does the position of the moving object agree with the path plotted on your star charts? Do you feel confident that you have in fact observed an asteroid? Explain.

43. Make arrangements to view a comet through a telescope. Since astronomers discover about a dozen comets each year, a comet is usually visible somewhere in the sky. Unfortunately, most comets are quite dim and so you will need access to a moderately large telescope. You can use *Starry Night*™ to see which comets are visible. First, update comets online by opening the **LiveSky** menu at the top of the program and selecting **Update Data Files.** If the data files were updated, close and restart the program. Set the time in the toolbar to your expected observing time, stop time flow, and then find a comet that is visible by opening the **Find** pane, clearing the search box and expanding the **Comets** list. The names of those comets visible at your selected observing time will be in black type. To find the expected apparent magnitude of a comet, click the menu button at the extreme left of the listing for the comet and select **Show Info.** The comet's apparent magnitude will be listed under the **Other Data** layer of the **Info** pane. A comet brighter than about 5th magnitude will be visible with binoculars while a comet fainter than about 12th magnitude will not be easily visible, even through a telescope. If the brightness of the comet you have chosen is too low, try other comets. (*Hint:* Remember that the larger the apparent magnitude value, the lower the brightness.) Double-click and zoom in on a selected comet to see the direction of its tail. Of course, *Starry Night*™ can only depict the comet with a symbolic image and will not show the real appearance of any comet. You can print finder charts as described in the previous Observing Project on asteroids. Alternatively, consult the Web or recent issues of the *IAU Circular,* published by the International Astronomical Union's Central Bureau for Astronomical Telegrams, which contains predicted positions and the anticipated brightness of comets in the sky. Also, if there is an especially bright comet in the sky, the latest issue of *Sky & Telescope* might contain useful information. Observe the comet through a telescope. Can you distinguish it from background stars? Can you see its coma? How many tails do you see? (Note that a telescope with a low magnification, or a pair of binoculars, is more suitable for viewing a bright, nearby comet with an extended tail.)

44. Make arrangements to view a meteor shower. You can find a list of the major showers and their dates of maximum intensity in Table 9-1. The *Starry Night*™ database contains the positions of the radiant points from which major showers of meteors appear to come in the sky. Select an appropriate meteor shower from the list. Click the **Home** button and set the date to the listed date for this shower and the time to midnight. Select **View > Solar System > Meteor Showers** and **Labels > Meteor Showers** from the menu. Find the selected meteor shower in the view, move the cursor over its radiant point, right-click (Ctrl-click on a Mac) to open the contextual menu and select **Show Info.** Expand the top layer of the **Info** pane to show the **Dates** of maximum intensity and the approximate **Zenith Hourly Rate (ZHR)** for this shower. Often, a good display can be seen a day or two before or after the maximum date. Ideally, you need a clear, moonless sky and a relatively clear horizon. The Moon's glow in the sky will hide many of the faint meteor trails. Attempt to count the number of meteors that you see in 10-min intervals and compare these rates with those in the *Starry Night*™ database. You should remove these radiant points from your view before leaving this exercise.

45. Use *Starry Night*™ software to observe several comets. Change the date in the toolbar to March 22, 1997, and click the **Stop** button to stop time flow. Change the viewing location to either the North or South Pole by selecting **Options > Viewing Location. . .** from the menu. This is the date of the equinox and from either pole, the Sun is somewhere on the horizon. Open the **Find** pane and expand the **Comets** list. Locate Comet Hale-Bopp in the expanded list and double-click on it. Zoom in (if necessary) until you can see the comet and its tail. **a.** Predict in which direction the Sun is located relative to the comet and explain how you made your prediction. Find the Sun in the view to verify your hypothesis. **b.** Using the Comet list again, double-click on the name of another visible comet to center on it, but do not zoom in. Then use the hand tool or gaze controls to locate the Sun. From your observations, predict the direction of the comet's tail on the sky, and explain how you made your prediction. Center again on the comet and zoom in until you can see its tail. (If this comet does not have a visible tail, repeat this part with yet another comet.) Was your prediction correct?

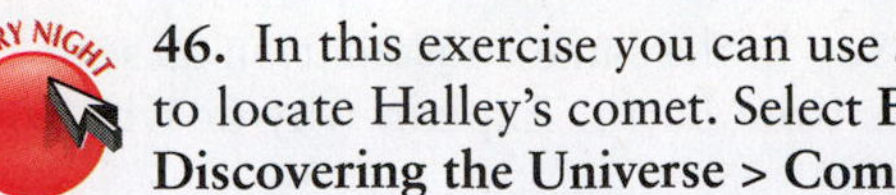

46. In this exercise you can use *Starry Night*™ to locate Halley's comet. Select **Favourites > Discovering the Universe > Comet Halley's Orbit** from the menu. (Note: The apparent curvature of the planes of orbits in this and other views is a consequence of the wide field of the display.) **a.** Explain the reason for the direction in which the comet's tail is pointed. **b.** From the direction of the tail, can you predict the direction of Halley's motion? If so, give its direction and explain how you deduced it. If not, explain why not. Click the **Play** button to start time flow and watch the orbital motion of Halley's comet. **c.** Does the comet orbit the Sun in the same or in the opposite sense to the planets visible in the view? Right-click over the Sun, select **Centre** and then use the location scroller to adjust the view until Earth's orbital plane is flat on the screen. Run time forward and back to watch Halley's comet as it crosses this plane in its close approach to the Sun, noting particularly the direction of its tail.

47. In this exercise you can use the *Starry Night*™ program to locate the largest asteroid, Ceres, now designated a dwarf planet. Select **Favourites > Discovering the Universe > Ceres and the Main Belt.** Use the location scroller to modify the view to determine whether Ceres moves in the plane of the ecliptic.

WHAT DO YOU THINK?

1. What percentage of the solar system's mass is in the Sun?
2. Does the Sun have a solid and liquid interior like Earth?
3. What is the surface of the Sun like?
4. Does the Sun rotate? If so, how fast?
5. What makes the Sun shine?
6. Are matter and energy conserved?

Answers to these questions appear in the text beside the corresponding numbers in the margins and at the end of the chapter.

RIVUXG This ultraviolet image of the Sun, taken in 2010 by the *Solar Dynamics Observatory*, shows a prominence (upper left) in which magnetic fields carry gases above the Sun's surface in a loop that can extend upward hundreds of thousands of kilometers. (NASA/GSFC/AIA)

Chapter 10

The Sun: Our Extraordinary Ordinary Star

Dawn. Light and heat from the Sun begin to arouse our senses and signal the start of daily activity. The earliest societies, realizing that the Sun is essential to the existence and maintenance of life on Earth, revered it. The same respect for the Sun's awesome power motivated astronomers in the nineteenth and twentieth centuries to figure out how it shines.

1 A quick glance (never longer) at the Sun shows a brilliant disk of light, apparently no larger than the Moon. Despite appearances, the Sun is a body of monumental size and mass compared to any object we have studied so far. If the Sun's surface were as close to Earth as the Moon's surface is now, the Sun would spread two-thirds of the way across the sky, not to mention that it would be so hot on Earth that our planet would quickly vaporize. The Sun contains as much mass as 333,000 Earths. In other words, the Sun contains 99.85% of the solar system's mass. All of the planets and other objects orbiting the Sun *combined* have only 0.15% of its mass.

The Sun is the closest star to Earth. By studying it, astronomers have come to understand how most stars throughout the universe work. As powerful and majestic as it is in our sky, we are going to discover in the next few chapters that the Sun is an ordinary star, neither among the most massive nor among the least massive. Similarly, it is neither among the brightest nor among the dimmest of stars. Figure 10-1 lists the Sun's properties.

In this chapter you will discover

- why the Sun is a typical star
- how today's technology has led to new understanding of solar phenomena, from sunspots to the powerful ejections of solar matter that sometimes enter our atmosphere
- that some features of the Sun generated by its varying magnetic field occur in cycles
- how the Sun generates the energy that makes it shine
- new insights into the nature of matter from solar neutrinos

THE SUN: VITAL STATISTICS

Sun symbol:	$\odot$
Mass (1 $M_\odot$):	1.989×10^{30} kg $= 3.33 \times 10^{5}$ Earth masses
Visual diameter (1 $R_\odot$):	1.392×10^{6} km $= 10^{9}$ Earth diameters
Luminosity (1 $L_\odot$):	3.827×10^{26} watts
Mean angular diameter:	32 arcmin
Rotation periods:	Equatorial: 25 days
	Polar: 35 days
Mean density:	1408 kg/m^3
Distances from Earth:	Mean (1 AU): 1.496×10^{8} km
	Maximum: 1.520×10^{8} km
	Minimum: 1.470×10^{8} km
Mean light travel time to Earth:	8.32 min
Mean temperatures:	Surface: 5800 K
	Center: 1.55×10^{7} K
Composition (by mass):	71.5% hydrogen (H), 27% helium (He), 1.5% other elements
Composition (by number of atoms):	91.3% hydrogen (H), 8.6% helium (He), 0.1% other elements
Distance from center of Galaxy:	26,000 ly = 8000 pc
Orbital period around center of Galaxy:	230 million years
Orbital speed around center of Galaxy:	220 km/s

R I V U X G

FIGURE 10-1 Our Star, the Sun The Sun emits most of its visible light from a thin layer of gas, called the photosphere, as shown. Although the Sun has no solid or even liquid region, we see the photosphere as its "surface." Astronomers always take great care when viewing the Sun by using extremely dark filters or by projecting the Sun's image onto a screen. (Celestron International)

THE SUN'S ATMOSPHERE

2 Although astronomers often speak of the solar "surface," the Sun is so hot that it has neither liquid nor solid matter anywhere inside it. Moving down through the Sun, we continually encounter ever denser and hotter gases.

10-1 The photosphere is the visible layer of the Sun

The Sun appears to have a surface only because most of its visible light comes from one specific gas layer (see Figure 10-1). This region, which is about 400 km thick, is appropriately called the **photosphere** ("sphere of light"). The density of the photosphere's gas is low by Earth standards, about 0.01% as thick as the air we breathe. The photosphere has a blackbody spectrum (recall Chapter 4) that corresponds to an average temperature of 5800 K (see Figure 4-3).

3 The photosphere is the innermost of the three layers that comprise the Sun's atmosphere. Because the upper two layers (discussed in the following two sections) are transparent to most wavelengths of visible light, we see through them down to the photosphere. We cannot, however, see through the shimmering gases of the photosphere, so everything below the photosphere is considered to be the Sun's interior.

As you can see in Figure 10-1, the photosphere appears darkest toward the edge, or **limb,** of the solar disk, a phenomenon called **limb darkening.** This occurs because we see regions of different temperatures at different depths in the photosphere. Here is how it works: Wherever we look on the Sun, we see light through roughly equal amounts of the Sun's atmosphere. Because the Sun is spherical, the light we see leaving the Sun at different places actually comes from different levels of the photosphere (Figure 10-2). Looking from Earth at the center of the Sun's disk (see Figure 10-2, bottom black line), we see farther into the Sun's atmosphere than when we look toward the limb (see Figure 10-2, top black line). The photosphere is hottest and brightest at its base, and so the center of the Sun's disk, where we see deepest into it, looks brighter and yellower than does its limb.

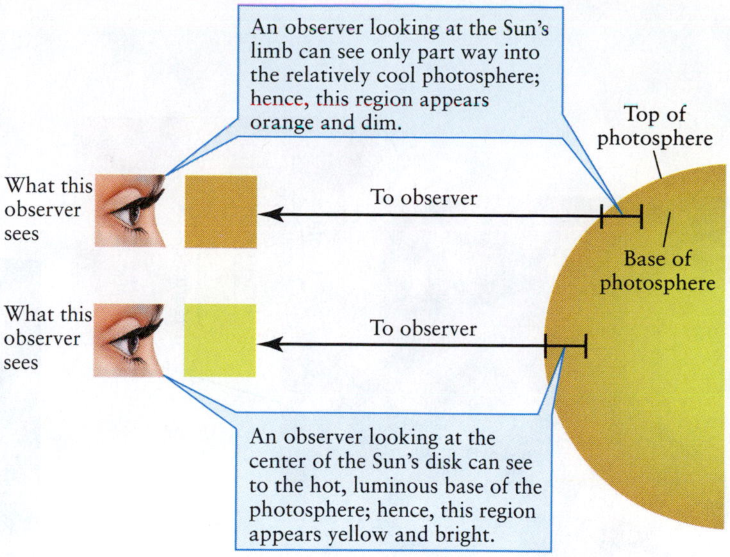

FIGURE 10-2 Limb Darkening The Sun's edge, or limb, appears distinctly darker and more orange than does its center, as seen from Earth (see Figure 10-1). This phenomenon occurs because we look through the same amount of solar atmosphere at all places. As a consequence, we see higher in the Sun's photosphere near its limb than when we look at its central regions. The higher photosphere is cooler and, because it is a blackbody, darker and more orange than the lower, hotter region of the photosphere.

Under good observing conditions and using a telescope with special dark filters for protection, you can see a blotchy pattern, called *granulation,* on the photosphere (Figure 10-3). The bright **granules** measure about 1000 km across and are surrounded by relatively dark boundaries. Time-lapse photography shows that granules form, disappear, and then reform in cycles that last several minutes. At any single moment, several million granules cover the solar surface.

Like liquid in a pot of simmering soup, the gases in granules rise and fall. In Section 4-7 we learned that radial motion of a light source affects the wavelengths of its spectral lines through the Doppler effect (see Figures 4-15 and 4-16 and *An Astronomer's Toolbox 4-3*). By carefully measuring the Doppler shifts of spectral lines in various parts of individual solar granules, astronomers have determined that hot gases move upward in the center of each granule and cooler gases cascade downward around its edges. This motion is caused by convection. We saw in Section 6-3 that convection moves the continents on Earth, and in Section 8-1 that it also helps form the belts and zones on the giant planets. Once the Sun's hot gases arrive at the photosphere, they radiate energy out into space. We see this energy as visible light and other electromagnetic radiation. Upon radiating, the gases cool, spill over the edges of the granules, and plunge back down into the Sun along the boundaries between granules (see Figure 10-3a inset).

Which part of the disk of the Sun appears brighter, the center or the edge? Why?

According to the Stefan-Boltzmann law (see Section 4-2), hotter regions emit more photons per square meter than do cooler regions. Note in Figure 10-3a that the centers of the granules are brighter than their edges. Observations indicate that a granule's center is typically 100 K hotter than its edge, thus explaining the observed brightness difference.

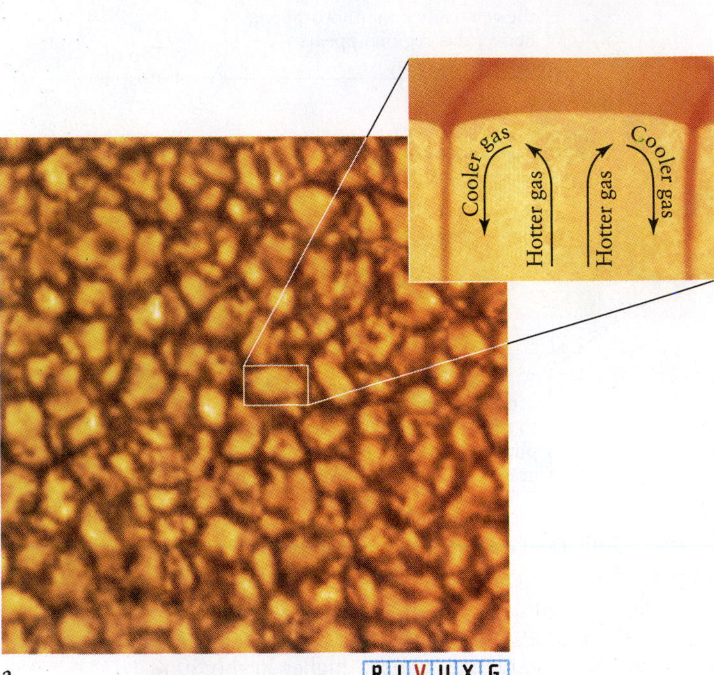

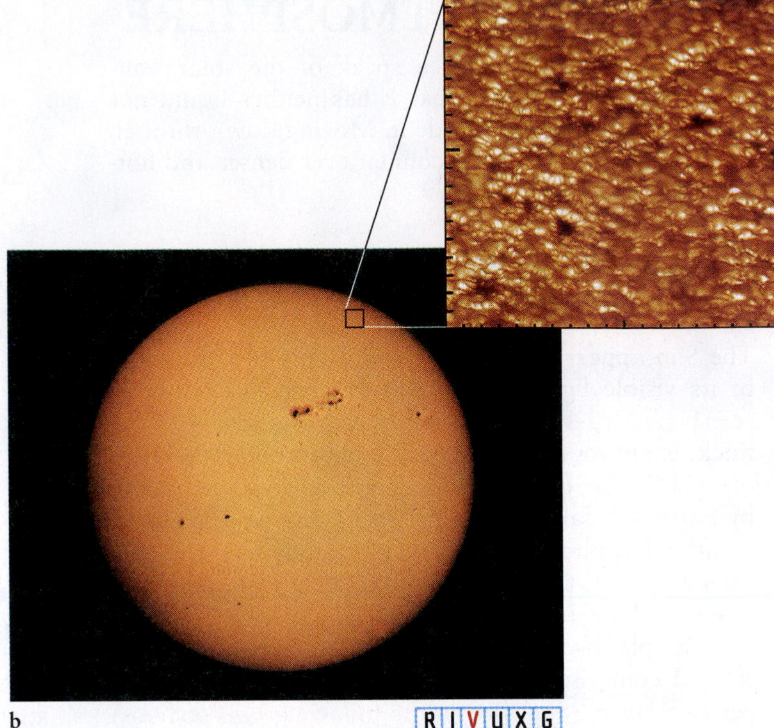

VIDEO 10.1

FIGURE 10-3 Solar Granulation (a) High-resolution photographs of the Sun's surface reveal a blotchy pattern, called granulation. Granules, which measure about 1000 km across, are convection cells in the Sun's photosphere. (Inset) Gas rising upward produces the bright granules. Cooler gas sinks downward along the darker, cooler boundaries between granules. This convective motion transports energy from the Sun's interior outward to the solar atmosphere. (b) At lower resolution, the Sun's surface appears relatively smooth (the dark regions will be discussed shortly). (Inset) Viewed near the Sun's limb, granules are seen to bulge upward at their centers as a result of the convection that creates them. (a: MSFC/NASA; inset: Goran Scharmer, Lund Observatory; b: Celestron International; inset: G. Scharmer [ISP, RSAS] et al., Lockheed-Martin Solar and Astrophysics Lab)

10-2 The chromosphere is characterized by spikes of gas called spicules

Immediately above the photosphere is a dim layer of less dense stellar gas, called the **chromosphere** ("sphere of color"). This unfortunate name suggests that it is the layer we normally see, but for centuries the chromosphere was visible only when the photosphere was blocked during a total solar eclipse.

During an eclipse, the chromosphere is visible as a pinkish strip some 2000 km thick around the edge of the dark Moon. Today, astronomers can also study the chromosphere through filters that pass light with specific wavelengths strongly emitted by it—but not by the photosphere—or through telescopes sensitive to nonvisible wavelengths that the chromosphere emits intensely (Figure 10-4).

High-resolution images of the chromosphere reveal numerous spikes, which are jets of gas called **spicules** (Figures 10-4 and 10-5a). A typical spicule rises for several minutes at the rate of 72,000 km/h (45,000 mi/h) to a height of nearly 10,000 km (Figure 10-5b). Then it collapses and fades away. At any one time, roughly a third of a million spicules cover a few percent of the Sun's chromosphere.

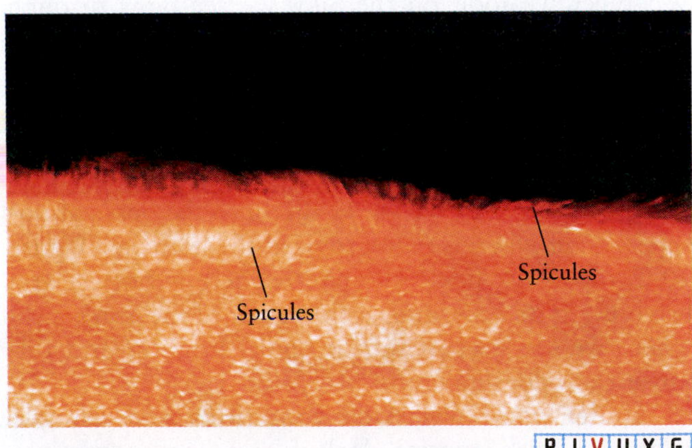

FIGURE 10-4 The Chromosphere This photograph of the chromosphere was taken by the *Hinode* (Japanese for "sunrise") satellite. The dark bumps are the tops of granules, and the light regions are hotter gases in spicules. The spicules on the edge or limb of the Sun give a sense of the height of these gas jets. (Hinode JAXA/NASA)

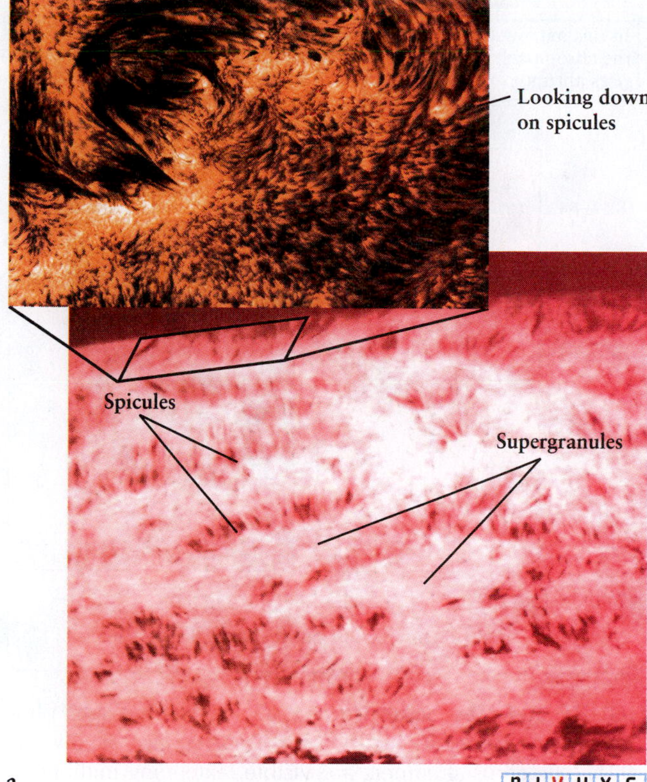

a

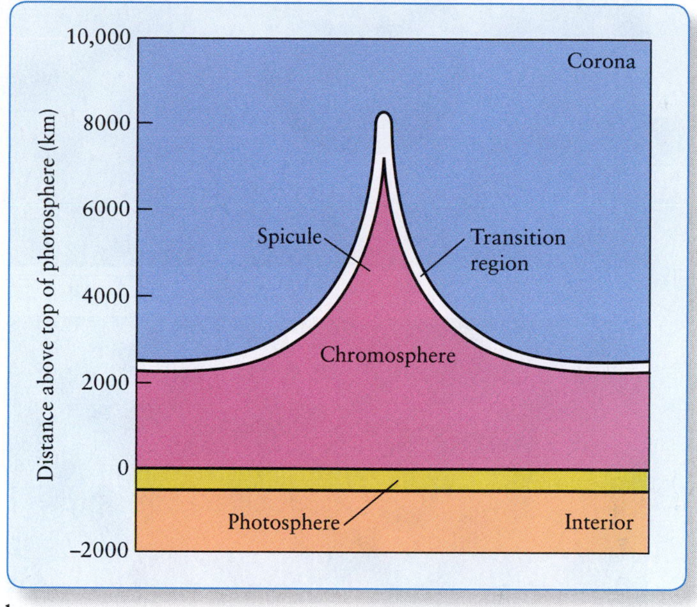

b

FIGURE 10-5 Spicules and Supergranules (a) Spicules appear in this photograph of the Sun's chromosphere. The Sun appears rose-colored in this image because the photo was taken through an H_α filter that passes red light from hydrogen and effectively blocks most of the photosphere's light. Surrounded by spicules, supergranules are regions of rising and falling gas in the chromosphere. Each supergranule spans hundreds of granules in the photosphere below. Inset: A view of spicules from above. (b) The spicules are jets of gas that surge upward into the Sun's outer atmosphere. This schematic diagram shows a spicule and its relationship to the solar atmosphere's layers. The photosphere is about 400 km thick. The chromosphere above it extends to an altitude of about 2000 km, with spicules jutting up to nearly 10,000 km above the photosphere. The outermost layer, the corona (discussed in Section 10-3), extends millions of kilometers above the photosphere. (a: NASA; inset: Swedish Solar Telescope/Royal Swedish Academy of Sciences, La Palma, Spain, by Bert De Pontieu, Lockheed Martin Solar and Astrophysics Lab)

Spicules are generally located on the boundaries of enormous regions of rising and falling chromospheric gas called **supergranules** (Figure 10-5a). A typical supergranule has a diameter slightly larger than Earth's and contains about 900 granules.

10-3 Temperatures increase higher in the Sun's atmosphere

It seems plausible that the temperature should decrease as you rise through the Sun's atmosphere. After all, by moving upward, you move farther from the Sun's internal heat source. Indeed, starting in the photosphere at 5800 K, the temperature drops to around 4000 K in the lower chromosphere. Surprisingly, however, the temperature then begins to increase as you ascend, reaching about 10,000 K at the top of the chromosphere. The outermost region of the Sun's atmosphere, the **corona,** extends several million kilometers from the top of the chromosphere (Figure 10-6). Between the chromosphere and corona is a **transition zone** in which the temperature skyrockets to about 1 million K (see Figure 10-6c).

Which part of the Sun do we normally see?

This unexpected increase in temperature was discovered around 1940 as a result of the high temperature's effect on the spectrum of the Sun's corona. The hotter a gas is, the more it is ionized (electrons are stripped off the atoms). Astronomers discovered that the corona contains the emission lines of several highly ionized (therefore, very hot) elements. For example, a prominent green line caused by the presence of Fe XIV, an iron atom stripped of 13 electrons, appears in the coronal spectrum. (Recall from Section 4-5 that Fe I is a neutral iron atom.) It is now known that coronal temperatures are typically in the range of 1 million to 2 million K, although some regions are even hotter.

Although its temperature is extremely high, the density of gas in the corona is very low, about 10 trillion times less dense than the air at sea level on Earth. The low density partly accounts for the dimness of

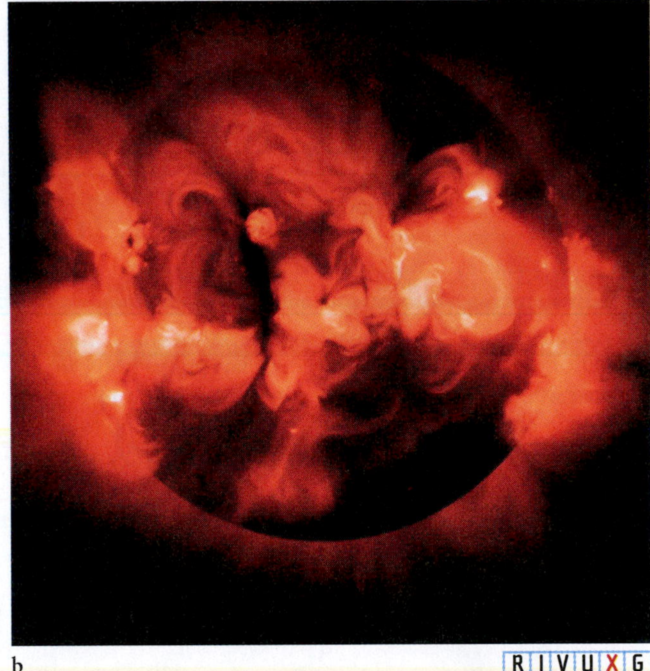

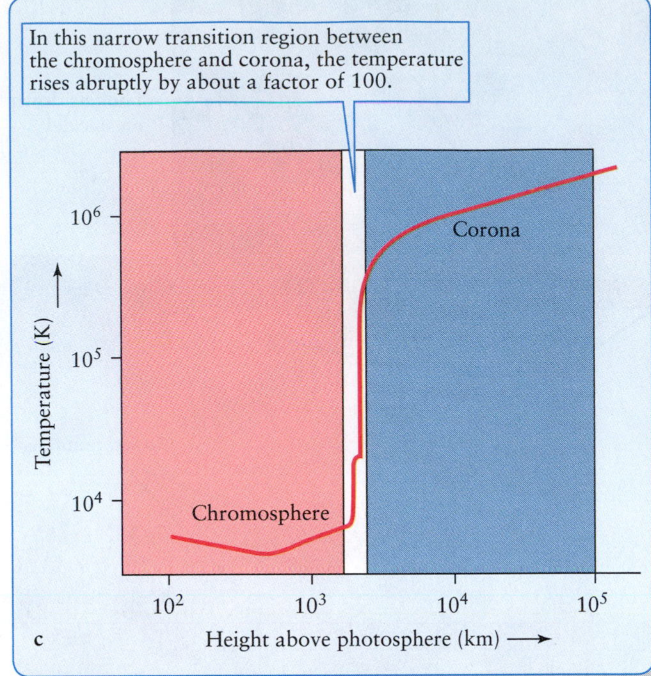

FIGURE 10-6 The Solar Corona (a) This visible-light photograph was taken during the total solar eclipse of July 11, 1991. Numerous streamers are visible, extending millions of kilometers above the solar surface. (b) This X-ray image of the Sun's corona, taken by the *Yohkoh* satellite in 1999, provides hints of the complex activity taking place on and in the Sun. The million-degree gases in the corona emit the X-rays visible here. (c) This graph shows how temperature varies with altitude in the Sun's chromosphere and corona and in the transition region between them (white). Note that both the height and temperature scales are nonlinear. (a: R. Christen and M. Christen, Astro-Physics Inc.; b: Greg Slater; c: adapted from A. Gabriel)

the corona, which otherwise would outshine the photosphere. Astronomers have observed that the corona is heated by energy carried aloft and released there by the Sun's complex magnetic fields, discussed in Sections 10-5 and 10-6.

The total amount of visible light that we receive from the solar corona is comparable to the brightness of the full Moon—or only about one-millionth as bright as the photosphere. As with the chromosphere, the corona can be seen only when the photosphere is blocked out or through special filters or at nonvisible wavelengths (such as ultraviolet and X-ray), at which the corona is especially bright compared to the photosphere. The photosphere is blocked naturally during a total eclipse or artificially with a specially designed telescope, called a *coronagraph*. Figure 10-6a is a photograph of the corona taken during a total eclipse (see also Figure 1-26). Figure 10-6b shows a stunning X-ray image of the corona.

Just as Earth's gravity prevents most of our atmosphere from escaping into space, so, too, does the Sun's gravity keep most of its outer layers from leaving. However, some of the gas is moving fast enough—around a million kilometers per hour—to escape the Sun's gravity forever and race into space. A portion of this gas comes directly from the corona, with the help of the Sun's magnetic fields, while some of it is funneled out from below the corona, also by magnetic fields. This outflow of particles is called the **solar wind.** (We saw its effects on Earth's magnetosphere in Section 6-4 and on comets in Section 9-7). The *heliosphere* is a bubble in space created by the solar wind that contains the Sun and planets. This outflow of gases from the Sun prevents most of the gases expelled in our direction by other stars from entering our solar system. During the years 2008–2010 the solar wind was the weakest it has been in recorded history. As a result, the heliosphere allowed in more particles from outside the solar system, called *galactic cosmic rays*.

In 2008, using data from the two venerable *Voyager* spacecraft that were launched in 1977 and that are now in the process of leaving the solar system, astronomers found that the heliosphere is not spherical. This discovery happened because the two spacecraft were sent out in different directions (and launched within 2 weeks of each other), and *Voyager 2* reached the inner edge of the heliosphere closer to the Sun than did *Voyager 1*. The information from the *Voyager* spacecraft provides only two data points about the location of the boundary of the heliosphere. To get a complete map of this boundary, NASA launched the *Interstellar Boundary Explorer* (IBEX) in 2008. While still in its early stages, IBEX has detected an intriguing feature of the boundary layer, namely a line or "ribbon" of extra hydrogen atoms running along parts of the boundary. Astronomers are working to explain this feature.

The Sun ejects around a million tons of matter each second as the solar wind. Even at this rate of emission, the mass loss due to the solar wind will amount to only a few tenths of a percent of the Sun's total mass throughout its lifetime. The solar wind particles reach speeds up to 2.9×10^6 km/h (1.8×10^6 mi/h). The wind achieves these high speeds, in part, by being accelerated by the Sun's magnetic field, discussed in Section 10-5. By the time the solar wind travels 1 AU and arrives in the vicinity of Earth, it has spread out so much that there are typically just 9 particles per cubic centimeter (cm^3), and they are moving around 1.1 to 1.4×10^6 km/h (6.8 to 8.5×10^5 mi/h). For comparison, there are 6×10^{19} molecules/cm^3 in the air we breathe, and the winds at Earth's surface are moving less than 240 km/h (150 mi/h).

Astronomers think that the Sun's surface chemical composition today is nearly identical to the composition of the solar nebula (see Section 5-2). Although electrons and hydrogen and helium nuclei comprise 99.9% of the solar wind, silicon, sulfur, calcium, chromium, nickel, neon, and argon ions have also been detected in it. To better understand the chemistry of the early solar system, the *Genesis* spacecraft was launched to collect pristine solar wind particles far outside Earth's magnetosphere (see Section 6-4) and to return them to Earth. By studying the amounts of the different chemicals in the solar wind, astronomers hope to better understand how the Sun works, as well as to learn the chemistries of the earlier generation of stars that exploded and whose gases eventually recondensed and formed the Sun and the other objects in the solar system. While attempting to land in 2004, *Genesis*'s parachute failed and the spacecraft crashed into Earth at 320 km/h (about 200 mi/h) (see Section 9-7 and the margin photo on page 297). Nevertheless, the collected samples have been salvaged and are being studied.

An important result from the *Genesis* mission concerns the chemistry of neon in the Sun. The amounts of different neon isotopes (atoms of neon that have different numbers of neutrons in their nuclei) in the Moon rocks brought back by Apollo astronauts suggested that the Sun was more active (hotter) in the past than it is today. The amounts of neon isotopes collected by *Genesis* were different from those in the Moon rocks. Astronomers deduce that the neon at different depths of the Moon was chemically altered by different amounts due to the radiation from space over billions of years. Taking this into account, they conclude that the Sun has maintained a more nearly constant temperature than previously believed.

Which of the Sun's three atmospheric layers is coolest? Which is densest?

THE ACTIVE SUN

Granules, supergranules, spicules, and the solar wind occur continuously. But the Sun's atmosphere is also periodically disrupted by magnetic fields that stir things up, creating a group of phenomena known collectively as the *active Sun*. The Sun's most obvious transient features are **sunspots,** regions of the photosphere that appear dark because they are cooler than the rest of the Sun's lower atmosphere. Sometimes sunspots occur in isolation (Figure 10-7a), but often they arise in clusters, called *sunspot groups* (Figure 10-7b).

Insight Into Science

Perception versus Reality Our senses (and often our technology) are limited, so we must be careful about interpreting what we perceive. For example, the sunspots in Figure 10-7 certainly look like black spots. However, they appear black only in contrast to the bright light around them. As we will discuss shortly, sunspots are actually red and orange.

10-4 Sunspots reveal the solar cycle and the Sun's rotation

Like other transient features of the active Sun, the average number and location of sunspots vary in fairly predictable cycles because of the changes in magnetic fields that produce them (as we will explore in Section 10-5). The sunspot cycle was discovered in 1843 by the German astronomer Samuel Schwabe. As shown in Figure 10-8a, the average sunspot cycle lasts approximately 11 years. Within an 11-year cycle, most sunspots appear at a **sunspot maximum** (Figure 10-8b). Sunspot maxima occurred most recently in 1979, 1989, and 2001. During a **sunspot minimum,** the Sun is often devoid of sunspots, as it was in 1986, 1997, and 2008 (Figure 10-8c). The next maximum in the cycle is projected to occur in 2013.

A typical sunspot is 10,000 km across and lasts between a few hours and a few months. Each sunspot has two parts: a dark, central region, called the *umbra,*

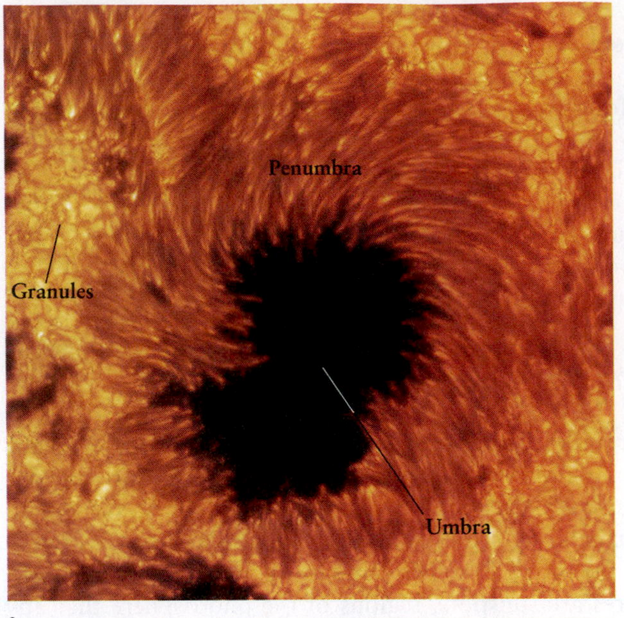

FIGURE 10-7 Sunspots (a) This dark region on the Sun is a typical isolated sunspot. Granulation is visible in the surrounding, undisturbed photosphere. (b) This high-resolution photograph shows a sunspot group in which several sunspots in a large group overlap. (a: Royal Swedish Academy of Science; b: National Solar Observatory)

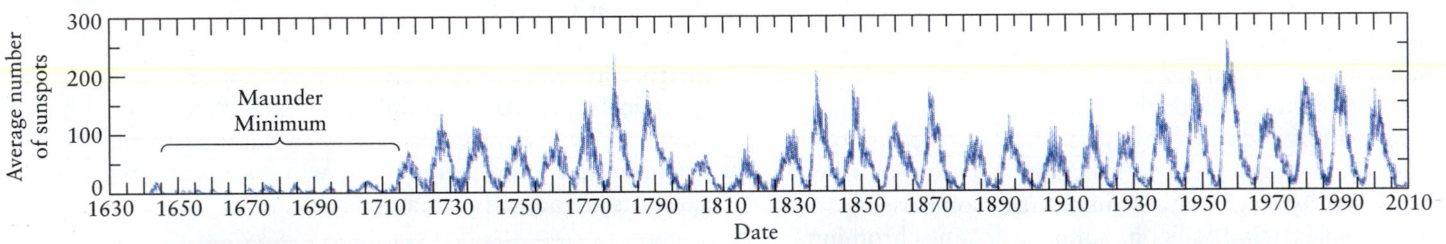

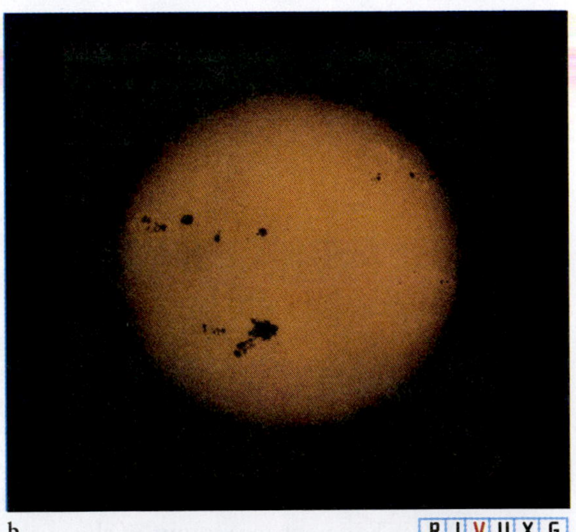

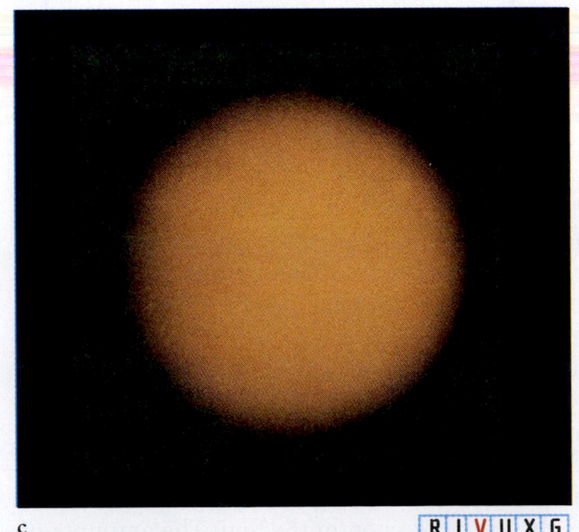

FIGURE 10-8 The Sunspot Cycle (a) The number of sunspots on the Sun varies with a period of about 11 years. The most recent sunspot maximum occurred in 2001, and the most recent sunspot minimum began in 2007 (and continued through 2010). (b) The active Sun has many sunspots (this photo was taken in 1979). (c) The Sun has many fewer sunspots when it is not active (this photo, showing a time with no sunspots, was taken in 1989). (b and c: NOAO)

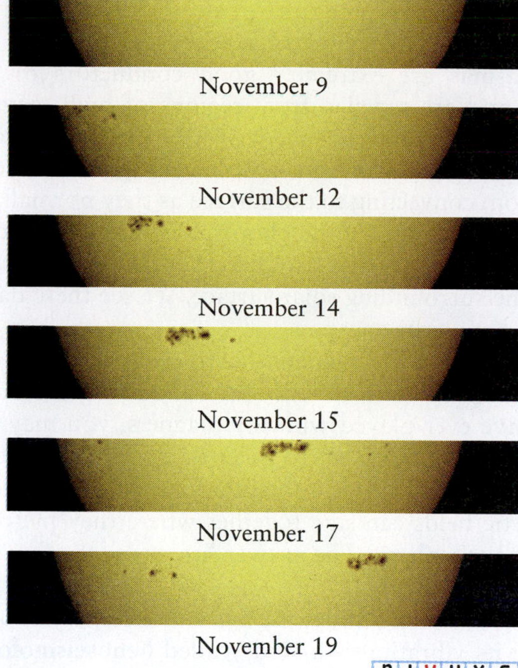

FIGURE 10-9 The Sun's Rotation This series of photographs taken in 1999 shows the same sunspot group over one-third of a solar rotation. Note how the sunspot groups changed over this time. By observing a group of sunspots from one day to the next in this same manner, Galileo found that the Sun rotates once in about 4 weeks. Sunspot activity also reveals the Sun's differential rotation: The equatorial regions rotate faster than the polar regions. (Carnegie Observatories)

and a brighter ring that surrounds the umbra, called the *penumbra,* visible in Figure 10-7. Although these are the same names as the blocked areas of an eclipse (see Section 1-11), we will see in the next section that their causes are completely different—another good example of how certain astronomical terms can have different meanings in different contexts.

Seen without the surrounding brilliant granules that outshine it, a sunspot's umbra appears red and its penumbra is orange. According to Wien's law (see Section 4-2), these colors indicate that the umbra is typically 4300 K and the penumbra is 5000 K, both cooler than the normal photosphere.

On rare occasions, a sunspot group is so large that it can be seen with the unaided eye. Chinese astronomers recorded such sightings 2000 years ago, and a huge sunspot group visible to the naked eye was seen in 2001. *Always use special dark filters or other means to protect your eyes when viewing the Sun. Looking directly at the Sun for more than a few moments causes eye damage!* Of course, a telescope gives a much better view, so it was not until Galileo used one that anyone examined sunspots in detail.

By following sunspots as they moved across the solar disk (Figure 10-9), Galileo discovered that *the Sun rotates once in about 4 weeks.* Galileo initially observed the Sun directly, undoubtedly contributing to his eventual blindness. Eventually, his protégé, Benedetto Castelli, developed the technique of projecting sunlight onto a screen to safely observe sunspots. Because a typical sunspot group lasts about 2 months, it can be followed for two solar rotations. Sunspot activity also reveals that like the giant planets, different latitudes of the Sun rotate at different rates, a phenomenon called **differential rotation:** The equatorial regions rotate more rapidly than the polar regions. A sunspot near the solar equator takes 25 days to go once around the Sun, whereas a sunspot at 30° north or south of the equator takes about 27 days. The rotation period at 75° north or south of the equator is about 33 days, and near the poles it is as long as 35 days. 4

The average latitude at which new sunspots appear changes throughout the sunspot cycle. At the beginning of each cycle, the sunspots appear mostly at about 30° north and south latitudes. Those sunspots that form later in the cycle typically occur closer to the equator. The "butterfly diagram" presented in Figure 10-10 shows that the range of latitudes on which sunspots are located and the average latitude at which they are found vary at the same 11-year rate as does the number of sunspots. Satellite measurements reveal that the Sun emits about 0.1% more energy at the peak of the sunspot cycle than at its minimum.

Which area of the photosphere is hottest, a sunspot's umbra, its penumbra, or a granule?

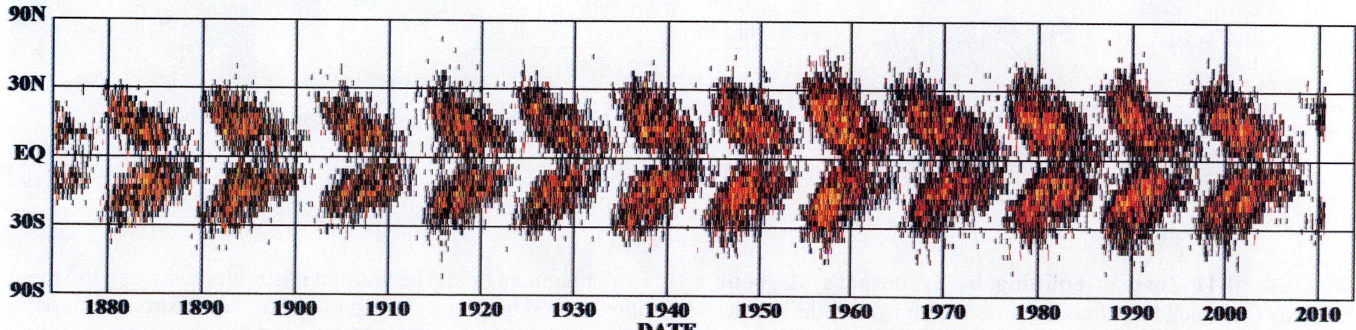

FIGURE 10-10 Locations of Sunspots Throughout the Sunspot Cycle (a) This "butterfly" diagram of sunspot locations shows that sunspots occur at changing latitudes throughout each cycle. From most common locations to least, the diagram is color coded yellow, orange, and black. (Hathaway/NASA/MSFC 2010/10)

10-5 The Sun's magnetic fields create sunspots

In 1908, the American astronomer George Ellery Hale discovered that sunspots are directly linked to intense magnetic fields on the Sun. When Hale focused a spectroscope on sunlight coming from a sunspot (Figure 10-11a), he found that each spectral line in the normal solar spectrum is flanked by additional, closely spaced spectral lines not usually observed (Figure 10-11b). This "splitting" of a single spectral line into two or more lines is called the **Zeeman effect.** The Dutch physicist Pieter Zeeman, who first observed it in the laboratory in 1896, showed that an intense magnetic field splits the spectral lines of a light source inside the field. The more intense the magnetic field, the more the split lines are separated.

Sunspots are areas where concentrated north or south magnetic fields project through the hot gases of the photosphere. But how do these fields create the spots? The answer lies in the interaction between the magnetic fields and the photosphere's gases. Because of the photosphere's high temperature, many atoms in it are *ionized:* One or more of their electrons have been stripped off by high-energy photons there. As a result, the photosphere is a mixture of electrically charged ions and electrons, called a **plasma.**

Plasmas are extremely good conductors of electricity and are repelled from regions of high magnetic field. Therefore, the magnetic field protruding through the photosphere prevents hot, ionized gases inside the Sun from convecting to the surface as they normally do. Thus, such regions of the photosphere are left relatively devoid of hot gas and are therefore cooler and darker than the surrounding solar surface. We see these darker, cooler regions as sunspots.

The fact that magnetic fields create sunspots presented a problem for astronomers studying the Sun. If you have ever played with toy magnets, you may have discovered that two north poles or two south poles repel each other. How is it then that bundles of same-pole magnetic fields can stay together where they pierce the Sun's photosphere? The answer lies under the Sun's visible layer.

To learn more about the Sun's interior, astronomers record its vibrations—a study called **helioseismology—** just as geologists use earthquakes to study Earth's interior structure. Although there are no true sunquakes, the Sun does vibrate in a variety of ways, somewhat like a vibrating drum head, which, when struck in different

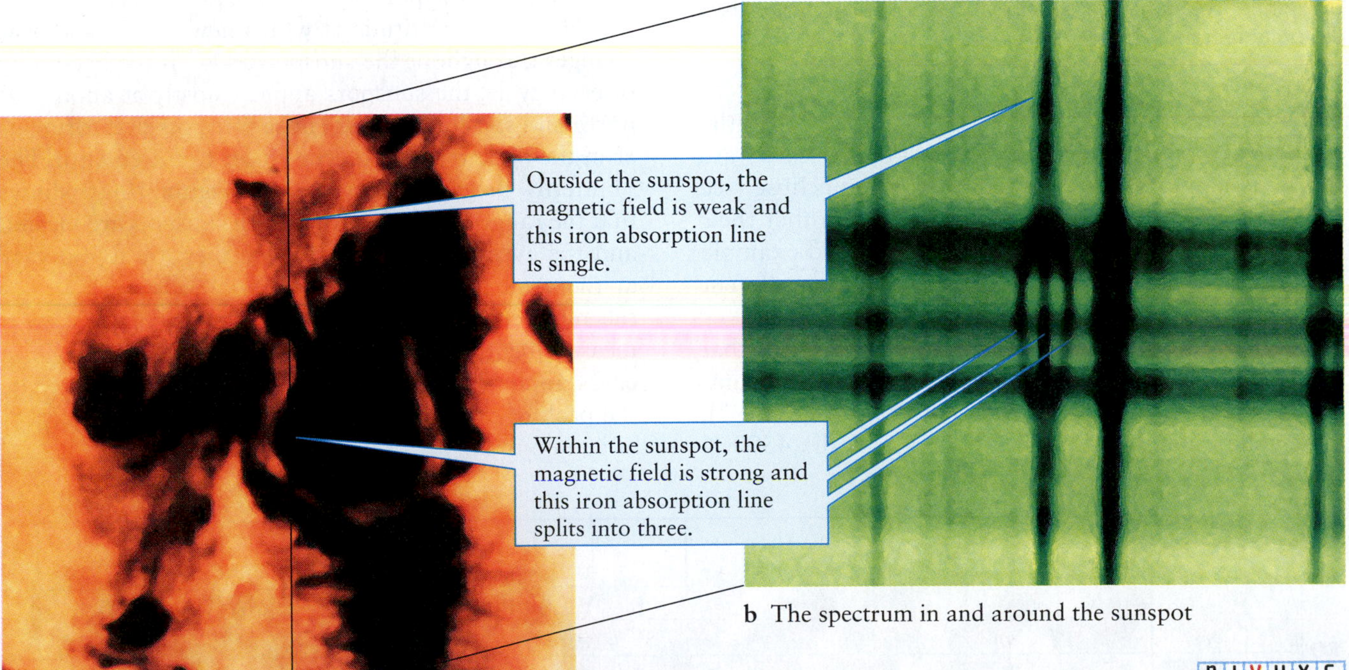

a A sunspot

b The spectrum in and around the sunspot

FIGURE 10-11 Zeeman Splitting by a Sunspot's Magnetic Field (a) The black line drawn across the sunspot indicates the location toward which the slit of the spectroscope was aimed. (b) In the resulting spectrogram, one line in the middle of the normal solar spectrum is split into three components by the Sun's magnetic field. The amount of splitting between the three lines is used to determine the magnetic field's strength. Typical sunspots have magnetic fields some 5000 times stronger than Earth's magnetic field. (a and b: NOAO)

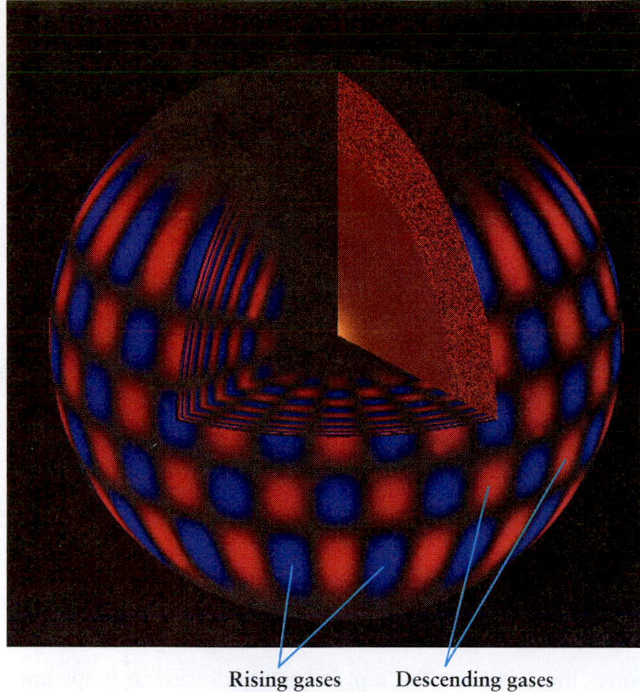

a

25 Days
35 Days

b

VIDEO 10.4

FIGURE 10-12 Helioseismology (a) This computer-generated image shows one of the myriad ways in which the Sun vibrates because of sound waves resonating in its interior. The regions that are moving outward are *blue;* those moving inward are *red.* The cutaway shows how deep these oscillations are believed to extend. (b) This cutaway picture of the Sun shows how the rate of solar rotation varies with depth and latitude. *Red* and *yellow* denote faster-than-average motion; *blue* regions move more slowly than average. The pattern of differential surface rotation, which varies from 25 days at the equator to 35 days near the poles, persists at least 19,000 km down into the Sun's convective layer. Sunspots preferentially occur on the boundaries between different rotating regions. Earthlike jet streams and other wind patterns have also been discovered in the Sun's atmosphere.

places, makes different sounds due to different modes of vibration. Solar vibrations, first noted in 1960, can be detected by using sensitive Doppler shift measurements. Astronomers have observed, for example, that portions of the Sun's surface move up and down by about 10 km every 5 min (Figure 10-12a).

Slower vibrations with periods ranging from 20 min to nearly an hour were discovered in the 1970s. Oscillations lasting several days have since been detected, as have pulses 16 months long. One important discovery from helioseismology is that deep inside, the Sun rotates like a rigid body, rather than with the differential rotation we see on its surface (Figure 10-12b).

While observing the vibrations of the Sun around sunspots, astronomers in 2001 apparently found the answer to the question of how sunspots persist for months despite the repulsion of the magnetic fields inside them. They discovered that below the photosphere, the gases that surround each sunspot whip around like hurricanes as large across as Earth's diameter. The circulation of charged gases around the magnetic fields holds sunspots in place.

Recall from Section 6-4 that magnetic field lines form complete loops, with each magnetic field having a north pole and a south pole. Likewise, when the Sun's magnetic field emerges through one sunspot or sunspot group, it forms a loop that reenters the Sun at another sunspot or sunspot group (Figure 10-13). We associate the name *north pole* or *south pole* with each sunspot or sunspot group, depending on whether the magnetic field there is pointing outward (a south pole) or pointing inward (a north pole). Sunspots and sunspot groups are connected in pairs—one where the magnetic field points out of the Sun, the other where it points into the Sun.

During his observations, Hale found that on one hemisphere of the Sun, sunspots with a magnetic north pole always come into view before the corresponding sunspots with a magnetic south pole. At the same time on the other hemisphere, the order is reversed (see Figure 10-13). Hale also found that this pattern reverses itself about every 11 years. The hemisphere where north magnetic poles come first during one 11-year cycle has south magnetic poles coming first during the next. Astronomers, therefore, speak of the 22-year **solar cycle,** the time it takes solar magnetic fields to return to their original orientation. The Sun is not alone in having its magnetic field reverse. In 2007 astronomers observed

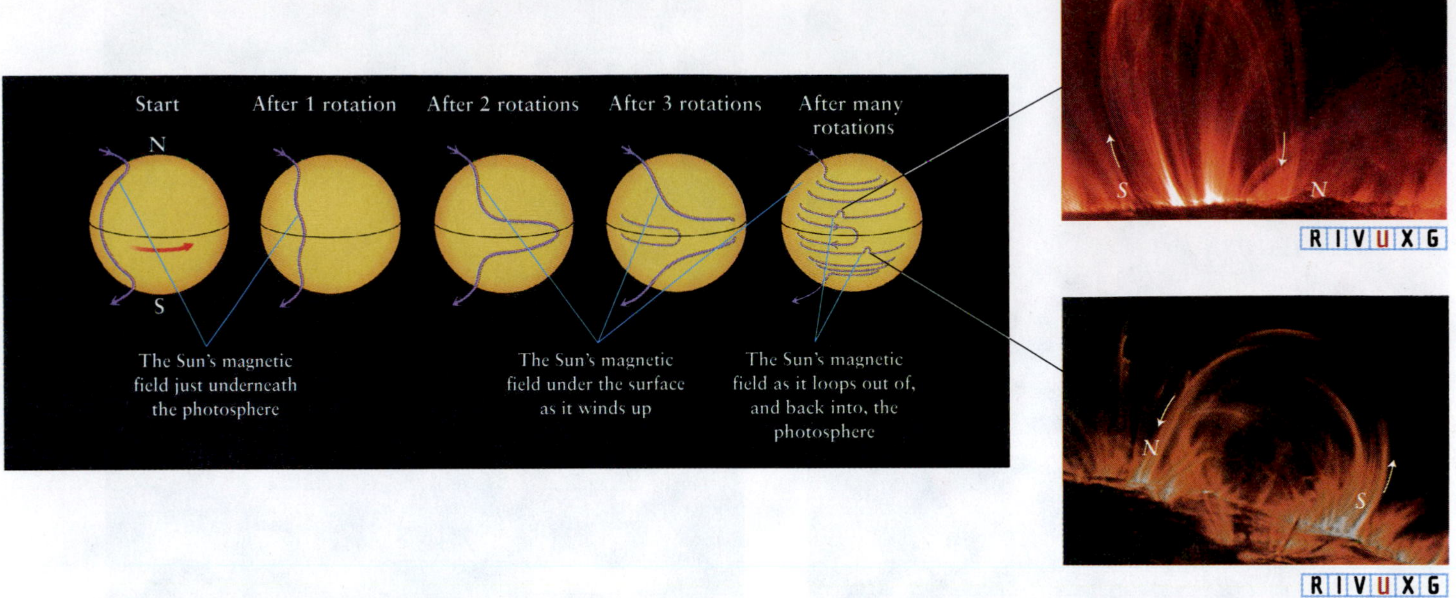

FIGURE 10-13 Babcock's Magnetic Dynamo In a plausible partial explanation for the sunspot cycle, differential rotation wraps a magnetic field around the Sun, just under its surface. Convection under the photosphere tangles the field, which becomes buoyant and rises through the photosphere, creating sunspots and sunspot groups. Insets: In each group, the sunspot that appears first as the Sun rotates (on the right side of each loop as you view them) has the same polarity as the Sun's magnetic pole in that hemisphere (*N* in the upper hemisphere and *S* in the lower hemisphere). To understand why, note that magnetic fields make complete loops: If a north pole enters the Sun, it leads to a south pole emerging, and vice versa. Follow the magnetic field in the northern hemisphere from where it enters the Sun (near the top) in counterclockwise loops until you encounter the bump in that hemisphere. Where the field emerges in the bump is a south pole and where it reenters the Sun is a north pole, as shown. Conversely, following the fields in the southern hemisphere, starting at the south pole, you encounter the southern bump from the left, emerging as a north pole and reentering the Sun as a south pole, as drawn. The Sun's magnetic fields are revealed by the radiation emitted from the gas they trap. These ultraviolet images show coronal loops up to 160,000 km (100,000 mi) high, with gases moving along the magnetic field lines at speeds of 100 km/s (60 mi/s). (Top inset: TRACE, Stanford-Lockheed Institute for Space Research, and NASA Small Explorer Program; bottom inset: NASA)

the magnetic field of the star Tau Bootis, located 50.9 ly from Earth, flip.

Based on these observations, we can now explain the Sun's magnetic field. This field is created as a result of the Sun's rotation and the resulting motion of the ionized particles found throughout it. This theory was proposed in 1960 by another American astronomer, Horace Babcock, as the **magnetic dynamo** model, in an effort to explain the 22-year solar cycle. When the Sun is quiet, its magnetic field lies just below the surface in the highly conducting plasma located there, unlike Earth's field, which passes through our planet's center.

As shown in Figure 10-13, the Sun's differential rotation causes its magnetic field to become increasingly stretched. Like expanding a rubber band, stretching the field this way causes it to store energy. At the same time, the magnetic field becomes tangled like the jumble of computer wires behind your desk as convection of the gases under the photosphere also causes the fields to rise and fall. Unlike a rubber band, however, magnetic field lines cannot break to release the energy stored in them. Rather, the fields must untangle themselves.

This untangling process begins as the mixed-up regions of magnetic field trap gases. This gas expands, thereby becoming buoyant and floating up through the solar surface, carrying the magnetic fields with them. Sunspots, with magnetic fields typically 5000 times stronger than Earth's magnetic field, form where loops or tangles of solar magnetic field leave and reenter the Sun. The gas bottled up in the loops of magnetic field eventually leaks out, and the fields untangle, interact with other parts of the Sun's magnetic field, and gradually settle back under the photosphere, at which point the sunspots associated with them disappear. This process of magnetic fields piercing the Sun's surface begins at high latitudes, and over the next 11 years it occurs closer and closer to the Sun's equator (see Figure 10-10).

Because of how these fields interact and vanish, every 11 years the Sun's entire magnetic field is reversed—the Sun's north magnetic pole becomes its south magnetic pole, and vice versa. After another 11-year cycle, the field is back to its original orientation. This is why the solar cycle is 22 years long. The most recent reversal of the Sun's magnetic field occurred in 2001.

Notable irregularities occur in the solar cycle. For example, the overall reversal of the Sun's magnetic field is often piecemeal and haphazard. More intriguing still is the strong historical evidence that all traces of sunspots and the sunspot cycle have vanished for decades at a time. For example, in 1893 the British astronomer E. Walter Maunder used historical observations to conclude that virtually no sunspots occurred from 1645 through 1715 (see Figure 10-8a). This period, called the *Maunder minimum,* coincides with a period of cold so extreme in Europe that it was called the *Little Ice Age.* At the same time, western North America was subject to severe drought.

Similar sunspot-free periods apparently occurred at irregular intervals in earlier times as well. Conversely, periods of increased sunspot activity in the eleventh and twelfth centuries coincided with periods of warmer-than-average temperatures. Interestingly, the minimum that began in 2007 has gone on longer than any minimum since the one of 1913. Furthermore, the peak of the present sunspot cycle, due in 2013, is predicted to have the fewest sunspots in 200 years. Whether these variations from the norm have or will have any bearing on global weather remains to be seen. Indeed, the question of whether there is a cause-and-effect link between the number of sunspots and periods of extreme temperatures on Earth is still open.

Insight Into Science

Cause and Effect While extremes of sunspot activity often occur at the same times as weather extremes on Earth, models of these phenomena must also take into account the possibility of purely terrestrial causes for these temperature changes. Remember, just because one event follows another does not mean that the first *causes* the second.

10-6 Solar magnetic fields also create other atmospheric phenomena

Figure 10-14 shows the active Sun's chromosphere and corona. The bright areas in this photograph are called **plages** (pronounced *plahzh,* from the French word for "beaches"). By studying the light emitted by calcium or hydrogen atoms in the plages, we know that they are hotter, and therefore brighter, than the surrounding chromosphere. Plages, which often appear just before nearby sunspots form, are thought to be created by the magnetic fields under the photosphere crowding upward just before they emerge through the photosphere. In pushing upward, the fields compress the gases of the upper Sun, causing this gas to become hotter and therefore to glow more brightly.

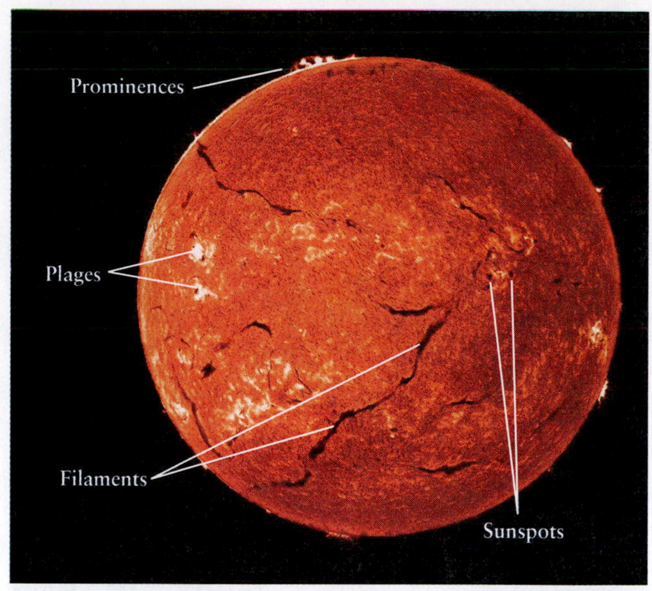

FIGURE 10-14 Active Sun in H_α This photograph shows the chromosphere and corona during a solar maximum, when sunspots are abundant. The image was taken through a filter that allowed only light from H_α emission to pass through. The hot, upper layers of the Sun's atmosphere are strong emitters of H_α photons. (See Section 4-6 for details of the Balmer series, H_α–H_∞.) A few large sunspots are evident. Most notable are features that do not appear at the solar minimum, such as the snakelike features shown here called filaments, bright areas called plages, and prominences (filaments seen edge on) observed at the solar limb. (NASA)

The dark streaks in Figure 10-14 are features in the corona called **filaments.** These features are huge volumes of gas lofted upward from the photosphere by the Sun's magnetic field. When viewed from the side rather than from above, filaments form gigantic loops or arches, called **prominences** (see Figures 10-14, 10-15a, and the figure that opens this chapter). The temperature of the gas in prominences can reach 50,000 K. These features are almost always associated with sunspots. Some prominences last for only a few hours, whereas others persist for months. The gases in the most energetic prominences escape the magnetic fields that confine them and surge out into space (Figure 10-15b).

X-ray photographs also reveal numerous coronal bright and dark spots that are hotter or colder, respectively, than the surrounding corona (Figure 10-16). Temperatures in the bright regions occasionally reach 4 million K. Many of the bright coronal hot spots are located over sunspots. The darker, cooler **coronal holes** act as conduits for gases to flow out of the Sun. Therefore, when a coronal hole on the rotating Sun faces Earth, the solar wind in our direction increases dramatically.

Violent, eruptive events on the Sun, called **solar flares,** release vast quantities of high-energy particles, as

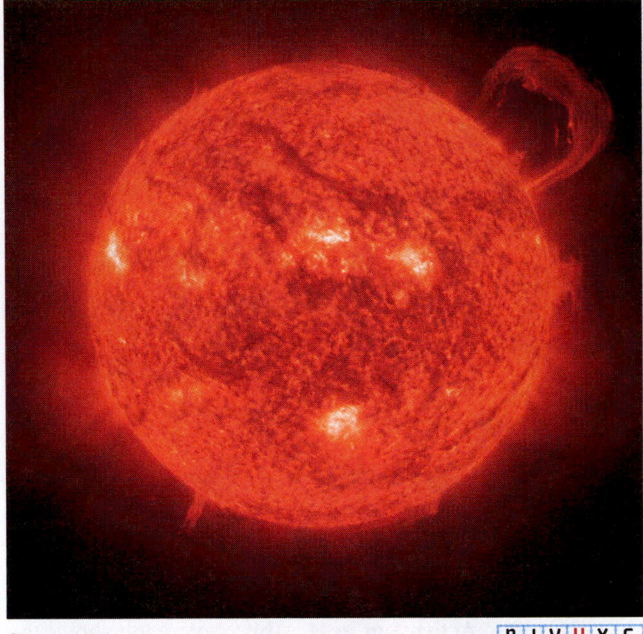

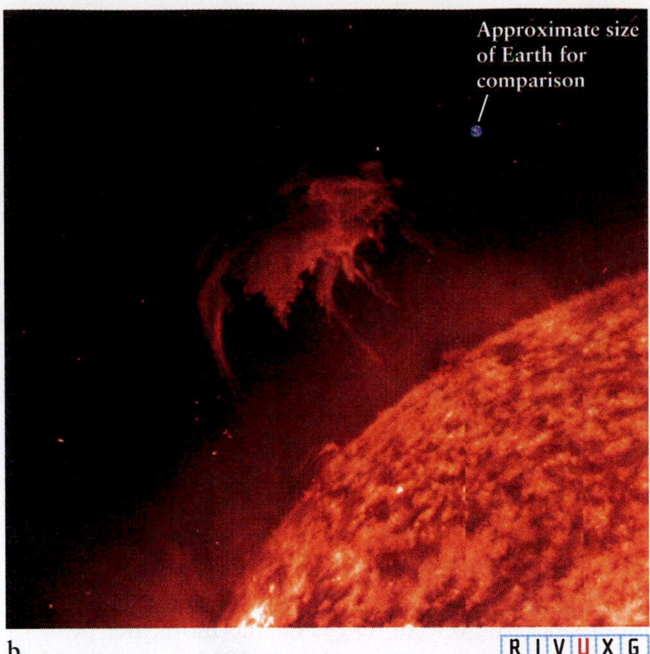

FIGURE 10-15 Prominences (a) A huge prominence arches above the solar surface in this SOHO image taken in 2001. The radiation that exposed this picture is from singly ionized helium at a wavelength of 30.4 nm, corresponding to a temperature of about 50,000 K. (b) The gas in these prominences was so energetic that it broke free from the magnetic fields that shape and confine it. This eruptive prominence occurred in 1999 (and did not strike Earth, which is shown for size). (a: ESA/NASA; b: Joseph B. Gurman, Solar Data Analysis Center, NASA)

well as X rays and ultraviolet radiation, from the Sun (Figure 10-17). Solar flares sometimes occur when sunspots collide (Figure 10-18). Typical flares each emit as much energy as is contained in all the fossil fuel ever stored inside the Earth. Flares are so powerful that they leave the region of the surface of the Sun in their vicinity quaking for an hour or more. At the maximum of the sunspot cycle, about 1100 flares occur per year.

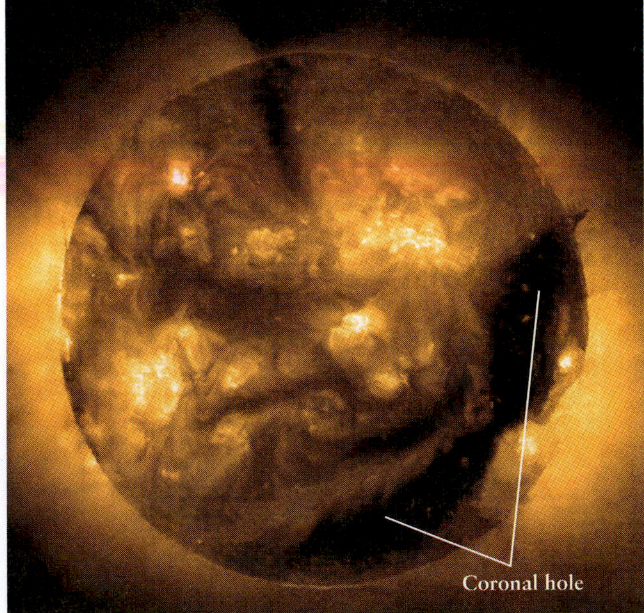

FIGURE 10-16 A Coronal Hole This X-ray picture of the Sun's corona was taken by the SOHO satellite. A huge coronal hole dominates the lower right side of the corona. The bright regions are emissions from sunspot groups. (SOHO/EIT/ESA/NASA)

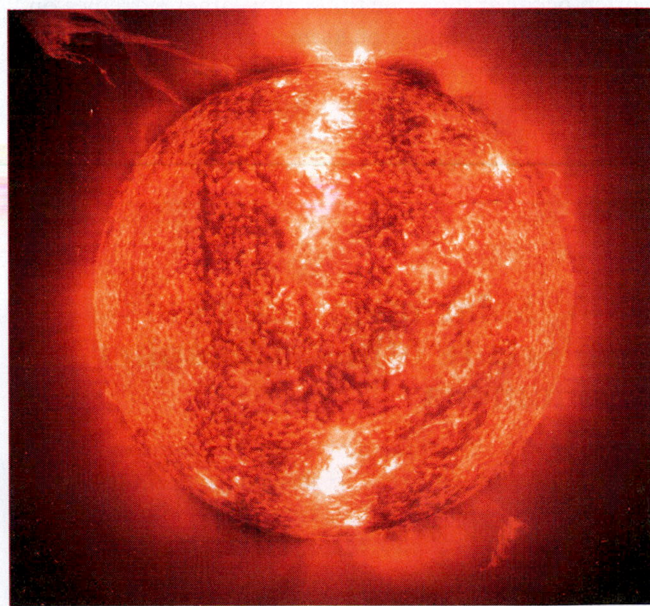

FIGURE 10-17 A Flare Solar flares, which are associated with sunspot groups, produce energetic emissions of particles from the Sun. This image, taken in 2000 by SOHO, shows a twisted flare (upper left part of figure) in which the Sun's magnetic field lines are still threaded through the region of emerging particles. (NASA)

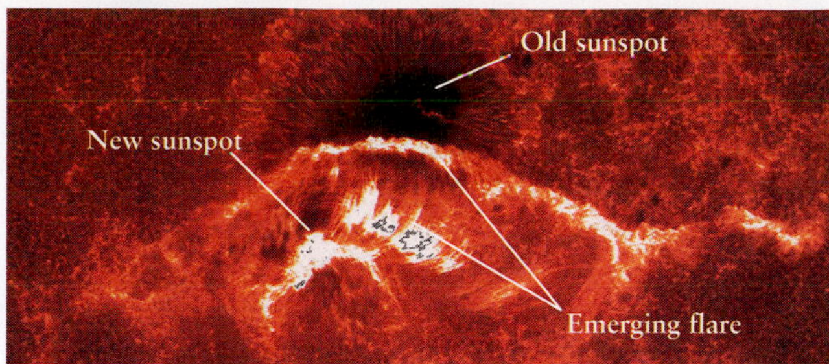

R I V U X G

FIGURE 10-18 Origin of a Solar Flare This *Hinode* satellite image of the Sun shows a developing sunspot colliding with a preexisting sunspot. The interacting magnetic fields funnel hot gases rapidly away from the Sun as a solar flare. (Hinode JAXA/NASA)

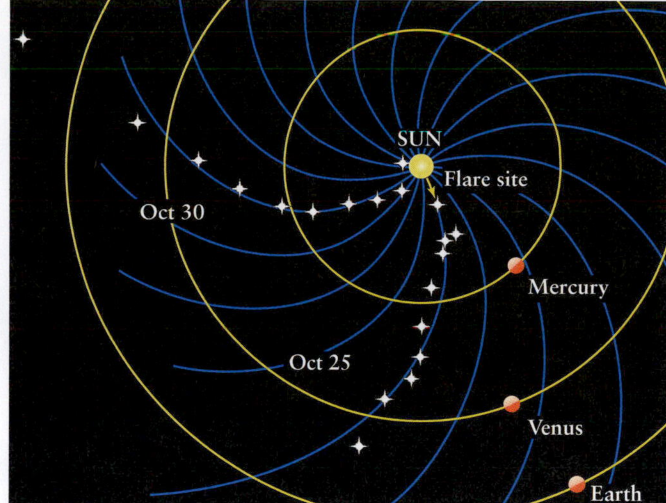

FIGURE 10-19 A Snapshot of the Sun's Global Magnetic Field By following the paths of particles emitted by a solar flare, astronomers have begun mapping the solar magnetic field outside the Sun. The field guides the outflowing particles, which, in turn, emit radio waves that indicate the position of the field. These data were collected by the *Ulysses* spacecraft in 1994. *Ulysses* was the first spacecraft to explore interplanetary space from high above the plane of the ecliptic.

Most of them last for less than an hour, but during that time, temperatures in the Sun's atmosphere soar to 5 million K. By following the paths of particles emitted by flares as they are deflected by the Sun's magnetic field, astronomers have, since the 1960s, been mapping that field as it extends out beyond Earth (Figure 10-19).

Astronomers have also observed huge, balloon-shaped volumes of high-energy gas being ejected from the corona. These **coronal mass ejections** typically expel 2 trillion tons of matter at 1.4×10^6 km/h, and each one lasts for up to a few hours (Figure 10-20). Coronal mass ejections have enough energy to break through the Sun's magnetic fields that normally contain them or even to carry fields outward. Flares have been observed to create shock waves that initiate some of the coronal mass ejections. The origins of other ejections are still under investigation. The numbers of plages, prominences, flares, and coronal mass ejections vary with the same 11-year cycle as the number of sunspots. Coronal mass ejections, the major source of hazardous particles from the Sun, occur with varying frequency throughout the sunspot cycle, but they never completely cease.

What holds the gas in loops above the Sun's surface?

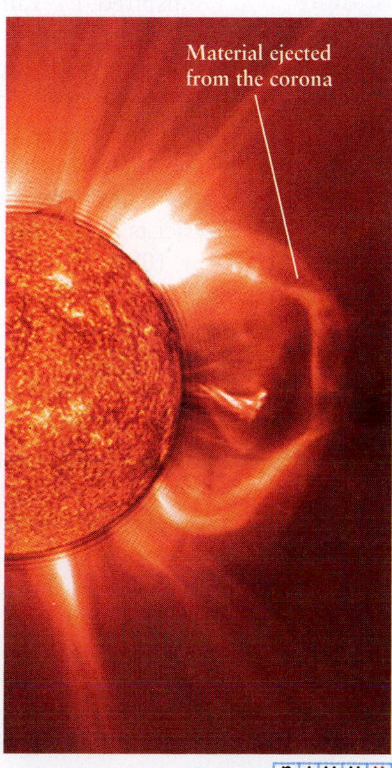

a A coronal mass ejection R I V U X G

b Two to four days later

FIGURE 10-20 A Coronal Mass Ejection (a) An X-ray image of a coronal mass ejection from the Sun taken by SOHO. (b) Two to 4 days later, the highest-energy gases from the ejection reach 1 AU. If they come our way, most particles are deflected by Earth's magnetic field (*blue*). However, as shown, some particles leak Earthward, causing aurorae, disrupting radio communications and electric power transmission, damaging satellites, and ejecting some of Earth's atmosphere into interplanetary space. (SOHO/EIT/LASCO/ESA/NASA)

Some coronal mass ejections, solar flares, and filaments head toward Earth. Their electromagnetic radiation gets here in about 8 min; their high-energy particles arrive a few days later. At times when these surges of particles are *not* coming to us, the normal solar wind particles are trapped by Earth's magnetic fields in the regions called the Van Allen belts (see Section 6-4 for more details). The additional particles from a coronal mass ejection or other solar event that *do* come our way overwhelm the Van Allen belts, enabling matter in them to cascade Earthward. One of the most spectacular of such events in recent years occurred on and around October 31, 2003 (Halloween in the United States), filling the night skies with aurorae in many places around the world. Perhaps you saw them.

Using the technology described in Chapter 3, several spacecraft now monitor the Sun and the space between it and Earth. These observers include the *Solar and Heliospheric Observatory* (SOHO), launched in 1995 by the European Space Agency (ESA) and NASA; the *Transition Region and Coronal Explorer* (TRACE), launched in 1998; the *Reuven Ramaty High Energy Solar Spectroscopic Imager* (RHESSI), launched in 2002; the *Solar Radiation and Climate Experiment* (SORCE), launched in 2003; *Hinode,* launched in 2006; and the *Solar Dynamics Observatory,* launched in 2010. These spacecraft observe gamma-ray, X-ray, ultraviolet, and visible parts of the Sun's spectrum, and they provide both scientific and space weather information.

Space weather can profoundly affect humans in space, as well as our assets out there and on the ground. Strong surges of high-energy particles from the Sun damage satellites, disrupt radio communications, short out power grids here, produce intense aurorae in Earth's atmosphere, cause some of Earth's atmosphere to gush into interplanetary space, and potentially harm or kill people in space. The satellites mentioned previously provide data that allow astronomers to accurately forecast several days in advance of the arrival of dangerous levels of solar particles. In the past few years, modeling the Sun's surface using helioseismology has become so sophisticated that we can even deduce what is happening on the side of the Sun facing away from Earth. As the Sun rotates, any hazardous activity there will sweep into view. By knowing what is coming even weeks ahead of time, astronauts and spacecraft can be better protected than ever before.

What two features on the Sun are actually the same thing, seen from different angles?

THE SUN'S INTERIOR

During the 1800s, geologists and biologists found convincing evidence that Earth must have existed in more or less its present form for at least hundreds of millions of years. This fact posed severe problems for astrophysicists, because at that time it seemed impossible to explain how the Sun could continue to shine for so long, radiating immense amounts of energy into space as it does. If the Sun were shining by burning coal or hydrogen gas, it would be ablaze for only 5000 years before consuming all of its fuel.

Everyday experience tells us that the Sun is the source of an enormous amount of energy. We have just seen that observations reveal hot gas, intense magnetic fields, and a variety of continuous and transient features on the Sun's surface. But the energy does not come from the surface gas or the magnetic fields—they have no mechanisms to create it. The origin of the Sun's electromagnetic radiation is deep within it. To understand why the Sun shines, we must understand where its energy originates and how that energy is transported to its surface.

10-7 Thermonuclear reactions in the core of the Sun produce its energy

In 1905, Albert Einstein provided an important clue to the source of the Sun's energy with his special theory of relativity. One of the implications of this theory is that matter and energy are related by the simple equation:

$$E = mc^2$$

In other words, a mass (m) can be converted into an amount of energy (E) equivalent to mc^2, where c is the speed of light. Because c^2 (that is, $c \times c$) is huge, namely 9×10^{10} km^2/s^2, a small amount of matter can be converted into an awesome amount of energy.

Inspired by Einstein's work, physicists discovered 5
that the Sun's energy output comes from the conversion of matter into energy. In the 1920s, the British astrophysicist Arthur Eddington proposed that the temperature at the center of the Sun, its **core,** is much greater than had ever been imagined. Calculations eventually revealed that the temperature there is about 15.5×10^6 K. Physicists also showed that at temperatures above about 10×10^6 K, hydrogen fuses into the element helium. In each such reaction, a tiny amount of mass is converted into energy in the form of gamma rays. Enough fusion occurs in its core to account for all of the energy emitted by the Sun.

The process of fusing nuclei at such extreme temperatures is called **thermonuclear fusion** (see *An Astronomer's Toolbox 10-1*). In particular, conversion of hydrogen into helium is called **hydrogen fusion.** The same process provides the devastating energy released by a hydrogen bomb. (We will encounter the thermonuclear fusion of helium and other elements in later chapters.)

Hydrogen fusion is also called *hydrogen burning,* even though nothing is burned in the conventional sense. The ordinary burning of wood, coal, or any flammable substance is a chemical process involving only the electrons orbiting the nuclei of the atoms. Thermonuclear fusion is a far more energetic process that involves violent collisions that change the atomic nuclei themselves. *An Astronomer's Toolbox 10-1* shows you just how energetic.

6 You may have heard statements like "matter is always conserved" or "energy is always conserved." We know that both of these concepts are incorrect, because mass can be converted into energy, and vice versa. What is true, however, is that the total amount of energy plus mass (multiplied by c^2) is conserved. So, the destruction of mass and the creation of energy by the Sun do not violate any laws of nature.

An Astronomer's Toolbox 10-1

Thermonuclear Fusion

What drives the thermonuclear fusion that powers the Sun? For nuclei to fuse, they must be brought together at incredibly high temperatures and pressures. That process is exactly what occurs in the Sun's core, where the entire mass of the Sun compresses inward. The core's temperature is 15.5 million K, its pressure is about 3.4×10^{11} atm, and its density is 160 times greater than that of water.

Normally, nuclei cannot contact one another because the positive electric charge on each proton prevents nearby protons from coming too closely together. (Remember that like charges repel each other.) But in the extreme heat and pressure of the Sun's center, the protons move so fast in such close proximity that they can stick, or fuse, together.

The nuclear transformations inside the Sun follow several routes, but each begins with the simplest atom, hydrogen (H). Most hydrogen nuclei consist of a single proton. The final outcome of fusion is the creation of a nucleus of the next simplest atom, helium (He), consisting of two protons and two neutrons. The fusion of hydrogen into helium takes several steps.

The accompanying diagram shows the most common path for hydrogen fusion in the Sun. This particular sequence is called the *proton-proton chain* (PP chain).

Note that proton-proton fusion releases positively charged electrons, e^+, called **positrons**, and neutral, nearly massless particles, called **neutrinos**, ν. When these positrons encounter regular electrons in the Sun's core, both particles are annihilated, and their mass is converted into energy in the form of gamma-ray photons. The final fusion, in which the helium forms, also returns two protons to the core, which are then available to fuse again.

Because the PP chain produces both neutrinos and the Sun's energy, we can summarize hydrogen fusion this way:

$$4\,{}^1\text{H} \rightarrow 1\,{}^4\text{He} + \text{neutrinos} + \text{energy}$$

Example: From our summary equation, we can calculate the energy released during a fusion reaction. We simply look at how much mass is converted into energy:

$$\begin{aligned} \text{Mass of 4 hydrogen atoms} &= 6.693 \times 10^{-27}\ \text{kg} \\ -\ \text{Mass of 1 helium atom} &= 6.645 \times 10^{-27}\ \text{kg} \\ \hline \text{Mass lost} &= 0.048 \times 10^{-27}\ \text{kg} \end{aligned}$$

Thus, a small fraction (0.7%) of the mass of the hydrogen going into the nuclear reactions does not show up in the mass of the helium. Ignoring the relatively small mass of the neutrino, this lost mass is converted into energy, as predicted by Einstein's famous equation:

$$\begin{aligned} E &= mc^2 \\ &= (0.048 \times 10^{-27}\ \text{kg}) \times (3 \times 10^8\ \text{m/s})^2 \\ &= 4.3 \times 10^{-12}\ \text{J} \end{aligned}$$

Compare! The energy released from the formation of a single helium atom would light a 10-watt lightbulb for almost one-half a trillionth of a second.

Now let us add up the energy emitted from the entire Sun. As noted earlier in this chapter, the Sun's mass, usually designated 1 $M_\odot$, is equal to 333,000 Earth masses. Its total energy output per second, called the **solar luminosity** and denoted 1 $L_\odot$, is 3.9×10^{26} watts. To produce this luminosity, the Sun converts 600 million metric tons of hydrogen into helium within its core each second. This value is twice the weight of the Empire State Building in New York City. This enormous rate is possible because the Sun contains a vast supply of hydrogen—enough to continue the present rate of energy output for another 5 billion years.

Try these questions: How many helium atoms would have to be created from hydrogen to release enough energy to light a 60-watt bulb for 12 h? How many joules of energy would be released if a U.S. penny (2.5×10^{-3} kg) were converted entirely into energy?

(Answers appear at the end of the book.)

(continued)

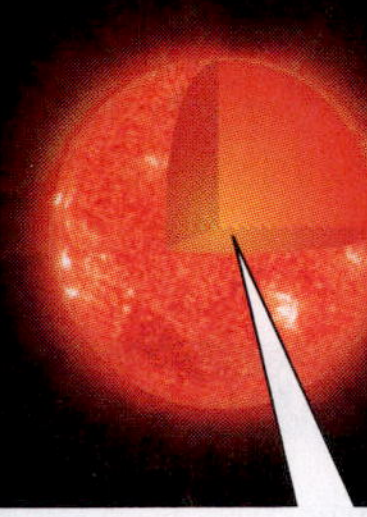

Hydrogen fusion in the Sun usually takes place in a sequence of steps called the proton-proton chain. Each of these steps releases energy that heats the Sun and gives it its luminosity.

STEP 1

(a) Two protons (hydrogen nuclei, ^{1}H) collide.

(b) One of the protons changes into a neutron (shown in blue). The proton and neutron form a hydrogen isotope (^{2}H).

(c) One by-product of converting a proton to a neutron is a neutral, nearly massless neutrino (ν). This escapes from the Sun.

(d) The other by-product of converting a proton to a neutron is a positively charged electron, or positron (e^+). This encounters an ordinary electron (e^-), annihilating both particles and converting them into gamma-ray photons (γ). The energy of these photons goes into sustaining the Sun's internal heat.

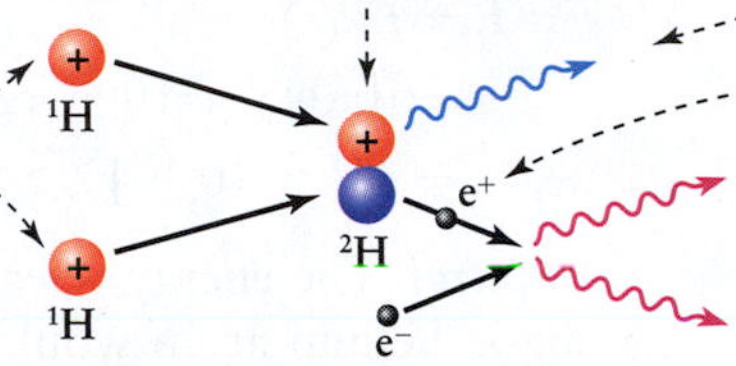

STEP 2

(a) The ^{2}H nucleus produced in Step 1 collides with a third proton (^{1}H).

(b) The result of the collision is a helium isotope (^{3}He) with two protons and one neutron.

(c) This nuclear reaction releases another gamma-ray photon (γ). Its energy also goes into sustaining the internal heat of the Sun.

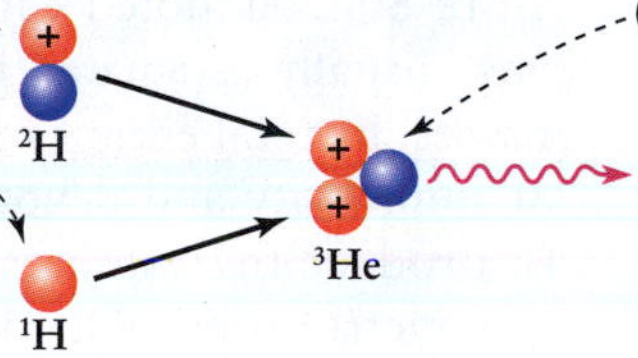

STEP 3

(a) The ^{3}He nucleus produced in Step 2 collides with another ^{3}He nucleus produced from three other protons.

(b) Two protons and two neutrons from the two ^{3}He nuclei rearrange themselves into a different helium isotope (^{4}He).

(c) The two remaining protons are released. The energy of their motion contributes to the Sun's internal heat.

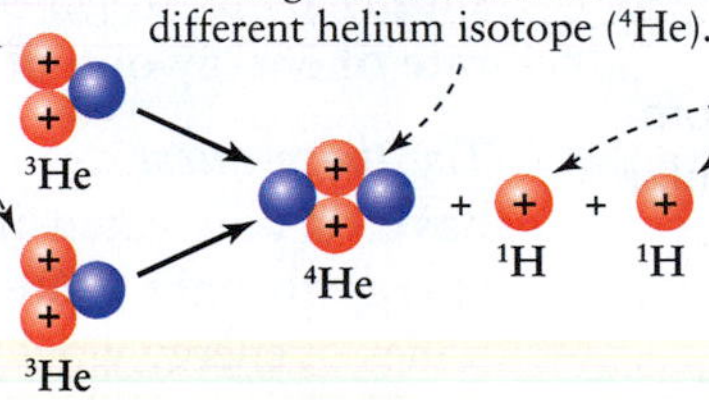

(d) Six ^{1}H nuclei went into producing the two ^{3}He nuclei, which combine to make one ^{4}He nucleus. Because two of the original ^{1}H nuclei are returned to their original state, we can summarize the three steps as:

$$4\ ^1\text{H} \longrightarrow\ ^4\text{He} + \text{energy} + \text{neutrinos}$$

(Courtesy of Wally Pacholka)

Hydrogen fusion also takes place in all of the stars visible to the naked eye. (Fusion follows a different sequence of steps in the most massive stars, but the net result is the same.)

Steps to Fuse Hydrogen into Helium by the Proton-Proton Chain.

WHAT IF... The Sun Had Been Formed Entirely from Helium Rather than Mostly from Hydrogen? As discussed above, the Sun shines because its core temperature of 15.5×10^6 K is high enough to fuse hydrogen into helium. The fusion temperature of helium, about 10^8 K, is above the present core temperature. This lack of fusion in a young helium "Sun" would allow its gravity to compress and heat its core more than occurs in our Sun. Eventually, the helium core would reach a hundred million K, allowing helium fusion to occur. The core helium in such a star would be used up in about 1 billion years, after which the star would shed its outer layers. Such a short stellar life would preclude the evolution of biological life on planets orbiting it.

10-8 The solar model describes how energy escapes from the Sun's core

A scientific theory describing the Sun's interior, called the **solar model,** explains how the energy from nuclear fusion in the Sun's core gets to its photosphere. The model begins with the inward force due to the Sun's gravity. This force increases the pressure and temperature in the Sun's core, thereby causing hydrogen fusion to occur there. Because the Sun is not shrinking today, however, there must be an outward force throughout its interior that counters the inward force of gravity. That outward force is created by the gamma-ray photons generated during fusion.

To begin their journeys outward, newly created gamma rays slam into nearby ions and electrons in the solar core. The photons then disappear, their energy being incorporated into the ions and electrons, which therefore rebound at very high speeds. Because they are densely packed together, these particles do not travel far before striking other particles above them. In each collision, the ions and electrons exert forces on one another. After each collision, the particles involved all rebound, emit photons, and collide with other particles. The result of these frequent interactions is pressure on the upper layers of the Sun sufficient to counterbalance gravity and prevent any part of the Sun from collapsing inward. The balance between the inward force of gravity and the outward force from the motion of the hot gas is called **hydrostatic equilibrium** (Figure 10-21).

Maintaining the Sun's hydrostatic equilibrium requires that the photons created in the Sun's core give up some of their energy. Photons collide with particles, which collide with other particles and thereby emit new photons. The emitted photons have slightly less energy than the photons that these particles had previously absorbed. The energy

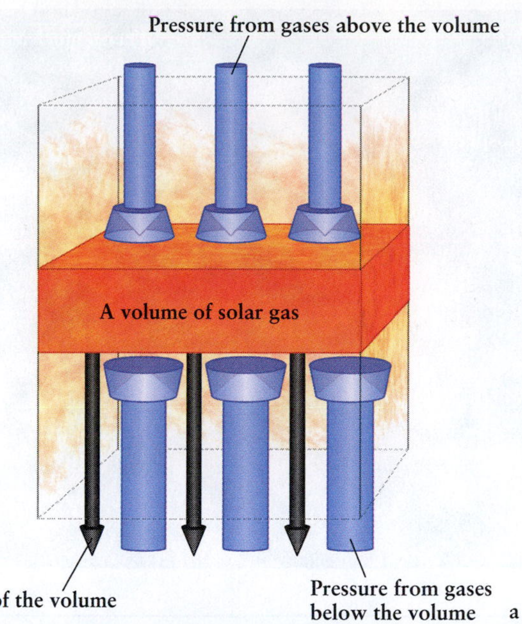

FIGURE 10-21 Hydrostatic Equilibrium (a) Matter deep inside the Sun is in hydrostatic equilibrium, meaning that upward and downward forces (or, equivalently, pressures) on the gases are balanced. (b) When the forces on the divers in water are in hydrostatic equilibrium, the divers neither sink nor rise. (JUPITERIMAGES/Comstock Images/Alamy)

the photons lose provides the outward force (and, hence, pressure) keeping the Sun in equilibrium. Photon energies slowly diminish as new photons are created closer and closer to the photosphere, an odyssey of photon absorption, particle collision, and photon emission that typically takes 170,000 years.

Where in the Sun is the energy that we see generated?

The outward movement of energy by photons hitting particles, which then bounce off other particles and thereby reemit photons, is called *radiative transport,* because individual photons are responsible for carrying energy from collision to collision. Calculations show that radiative transport is the dominant means of outward energy flow in the **radiative zone,** extending from the core to about 70% of the way out to the photosphere.

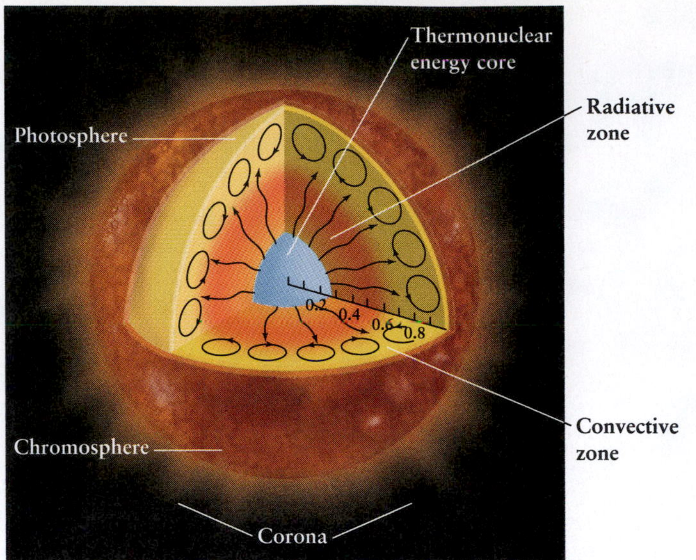

a

FIGURE 10-22 The Solar Model (a) Thermonuclear reactions occur in the Sun's core, which extends to a distance of 0.25 solar radius from the center. In this model, energy from the core radiates outward to a distance of 0.7 solar radius. Convection is responsible for energy transport in the Sun's outer layers. (b) The Sun's internal structure is displayed here with graphs that show how the luminosity, mass, temperature, and density vary with the distance from the Sun's center. A solar radius (the distance from the Sun's center to the photosphere) equals 696,000 km. (c) The nine most common elements in the Sun, by the numbers of atoms of each and by the percentage of the Sun's total mass they each comprise. (a: NASA)

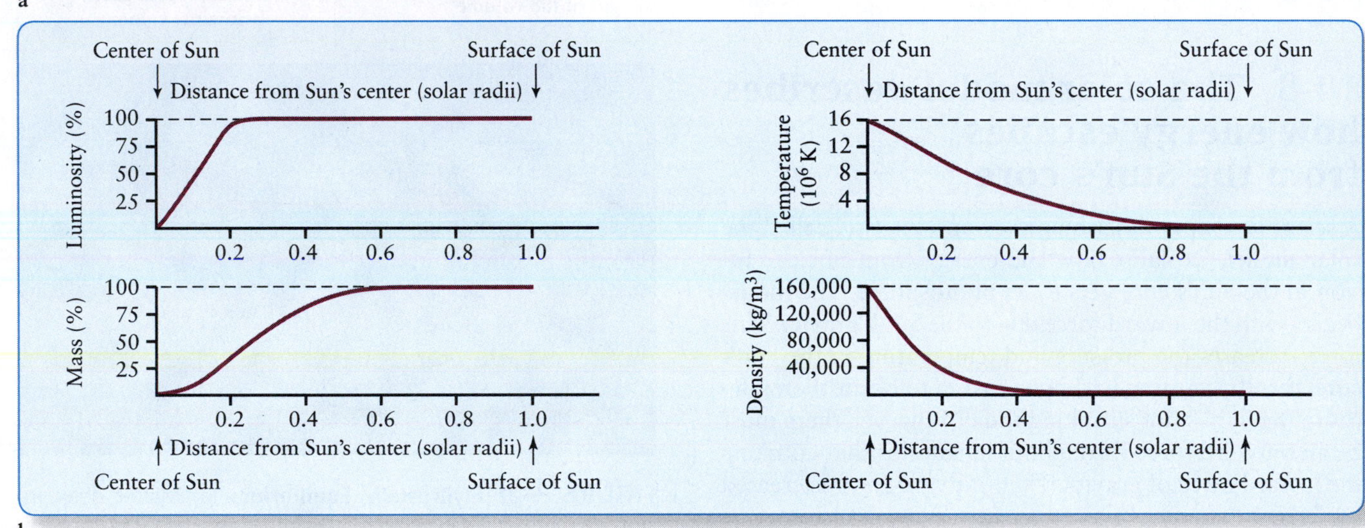

b

Element	Number of atoms (percent)	Percent of total mass
Hydrogen	91.3	71.5
Helium	8.61	27.0
Magnesium	0.074	0.039
Oxygen	0.048	0.60
Carbon	0.027	0.25
Neon	0.0082	0.13
Nitrogen	0.0067	0.073
Iron	0.0032	0.14
Silicon	0.0032	0.07

c

Near the photosphere, energy is carried the remaining distance to the surface by the bulk motion of the hot gas, rather than by the flying, energetic photons. As we discussed earlier, this circulation of gas blobs is called *convection*. We thus say that the Sun has a **convective zone.** Hot gas travels up to the top of this zone by convection and from there radiates photons into space. These photons are what we see as sunlight, and you can see that the photosphere, which is at the bottom of the solar atmosphere, is also at the top of the convective zone. Once the gases in the convective zone lose energy by emitting photons into space, they cool and settle back into the Sun. As noted earlier, this convective flow is also the origin of solar granules, and the departing energy is the visible light and other electromagnetic radiation that the Sun emits into space. Figure 10-22a is a sketch of our model of the internal structure of the Sun.

Different gamma rays created in the Sun's core lose different amounts of energy as they and their successors travel upward through the Sun. Therefore, the photons emitted from the photosphere have a wide range of energies and, hence, wavelengths. The most intense emission is in the visible part of the electromagnetic spectrum. This process of energy loss by photons traveling up from the Sun's core is the origin of the blackbody nature of the photosphere's spectrum.

Our model of the Sun's interior can be expressed as a set of mathematical equations, called the *equations of stellar structure*. The model quantitatively describes the Sun's internal characteristics, such as its pressure, temperature, and density at various depths. These equations also describe the conditions inside other stars as they evolve. Because the equations are so complex, astrophysicists today use computers to solve them. Figure 10-22b presents their results graphically. The four graphs show how the Sun's luminosity, mass, temperature, and density vary from the Sun's center to its photosphere. For example, the upper left graph gives the percentage of the Sun's luminosity created within that radius. The luminosity increases to 100% at about one-quarter of the way from the Sun's center to its photosphere. This trend tells us that all of the Sun's energy is produced within a volume extending out to $0.25R_{\odot}$, where $R_{\odot}$ is the radius of the Sun.

The mass curve rises to nearly 100% at about $0.6\ R_{\odot}$ from the Sun's center. Almost all of the Sun's mass is therefore confined to a volume extending only 60% of the distance from the Sun's center to its visible surface. This distribution of mass helps explain why the density of the gas in the photosphere is 10^4 times lower than the density of the air we breathe.

WHAT IF... The Sun Were More Massive?
The mass of the Sun determines how much its interior is compressed and thus how hot its core becomes: The more mass a star has, the hotter its core is. Hotter star cores produce more fusion activity, which leads to more energy leaving the star's surface, as well as a hotter surface. Therefore, the Sun would be brighter and, according to Wien's law, bluer than it is now. Using Kepler's third law (modified by the higher mass in this scenario), we can calculate that the length of our year would be shorter.

10-9 The mystery of the missing neutrinos inspired research into the fundamental nature of matter

As explained in *An Astronomer's Toolbox 10-1,* for every proton that changes into a neutron during thermonuclear fusion, a *neutrino* is released. Neutrinos have no electric charge, and they are extraordinarily difficult to detect because they rarely interact with ordinary matter. The Sun is largely transparent to neutrinos, allowing these particles to stream outward, unimpeded, from its core. Likewise, most of the neutrinos from the Sun that arrive at Earth pass right through it.

According to our model of fusion in the Sun, nearly 10^{38} *solar neutrinos* per second are produced in the Sun's core. This output is so huge that here on Earth roughly 100 billion solar neutrinos pass through every square centimeter of your body each second!

Occasionally, however, solar neutrinos strike neutrons and convert them into protons. If astronomers could detect even a few of these converted protons, it might be possible to build a "neutrino telescope" that could be used to detect the thermonuclear inferno in the Sun's core that is hidden from the view of telescopes that collect photons.

Inspired by such possibilities, the American chemist Raymond Davis designed and built a large neutrino detector. This device consisted of a huge tank that contained 100,000 gallons of perchloroethylene (C_2Cl_4, the fluid your local dry cleaner uses) located deep in the Homestake gold mine in Lead, South Dakota. All neutrino experiments are performed underground to help prevent them from being contaminated by other sources of energy, like high-energy protons from space. Because matter is virtually transparent to neutrinos, most of the solar neutrinos passed right through Davis's tank. On rare occasions, however, a solar neutrino hit the nucleus of one of the chlorine atoms in the cleaning fluid and converted one of its neutrons into a proton, creating a radioactive atom of argon. The rate at which argon was produced was therefore correlated with the number of solar neutrinos arriving at Earth.

On average, solar neutrinos created one radioactive argon atom every 3 days in Davis's tank. To the consternation of physicists and astronomers, this rate corresponds to only one-third of the neutrinos predicted to be created by the fusion in the Sun's core. There were three possible explanations for this unexpected result: The experiment could be faulty; the Sun might not be fusing at the expected rate; or our understanding of the properties of neutrinos could be in error. This experiment, which began in the mid-1960s, was repeated with extreme care by other researchers around the world who got the same results, suggesting that the experiment was not at fault. Calculations of how much energy the Sun must generate to shine as it does confirmed the number of neutrinos created per second. The problem had to lie in our understanding of the neutrino.

Why does the Sun not collapse under the influence of its own enormous gravitational attraction?

Neutrinos, like the planet Neptune, were discovered because, according to theory, they had to exist. Back in the first few decades of the twentieth century, physicists observed neutrons in nuclei spontaneously transforming into protons and electrons. The transformation is a result of the *weak nuclear force,* briefly introduced in Section 4-5. When these scientists calculated the energy and momenta of the neutron and the particles into which it changed, the numbers didn't match—conservation of energy and momentum (see Section 2-7) appeared to be violated. If this were indeed the case, it would have required a fundamental reworking of many of the laws of nature. A simpler solution was proposed in 1930 by the physicist Wolfgang Pauli, who posited the existence of a hard-to-detect particle that carried away the "lost" energy and momentum (this particle was named the *neutrino* in 1931 by physicist Enrico Fermi). Ray Davis shared the Nobel Prize in 2002 for his work in observing solar neutrinos.

Among the properties neutrinos had to have are very small masses compared to any other types of particles that have mass and very little interaction with other matter. This latter property had to be true because trillions upon trillions of them are being created in the Sun every second, with trillions passing through every cubic meter of Earth every second, and there were (and are) virtually no interactions between them and us. Indeed, neutrinos were not detected for 26 years, until 1956. Because electrons or positrons (identical to electrons, except with opposite electric charge) always accompany the formation of the neutrinos created in the Sun, they are often called *electron neutrinos*.

Two other kinds of neutrinos were subsequently proposed. One, the *muon neutrino*, is emitted in reactions in which elementary particles, called *muons* (or *antimuons*), are released, while the other, the *tau neutrino*, accompanies the formation of other elementary particles, the tau and antitau particles. The Sun emits neither muon nor tau neutrinos, and the first generation of detectors could only detect electron neutrinos. The solution to the neutrino rate dilemma lay in the existence of muon and tau neutrinos.

Initially, the three so-called "flavors" of neutrinos (electron, muon, and tau) were thought to be massless, and, if so, the equations revealed that they could not change from one flavor into another. But if neutrinos have even the slightest mass, then they can transform from one to another. The transformation of massive neutrinos would occur spontaneously, meaning that if an electron neutrino left the Sun's core and sped toward Earth, there would be a finite probability (about 65%, actually) that by the time it got here, it would have become a tau or muon neutrino. So, if neutrinos have mass, we should observe only one-third of them (which we do), because the rest would have changed flavor before they got to us.

In 1998, the super Kamiokande neutrino detector in Japan observed the other two types of neutrinos. The observations were repeated in 2002 at the Sudbury Neutrino Observatory (Figure 10-23) in Ontario, Canada. At least a dozen other neutrino detectors have begun operations or are being built to further study solar neutrinos.

To understand how the present generation of neutrino detectors work, let us consider one of the several interactions that occur between neutrinos and other matter. The Sudbury Neutrino Observatory is filled with water that contains a rare type of hydrogen nucleus, called *deuterium*, which consists of a proton and a neutron. When a solar neutrino is absorbed by a deuterium nucleus, the nucleus breaks apart into two protons and

FIGURE 10-23 A Solar Neutrino Experiment Located 2073 m (6800 ft) underground in the Creighton nickel mine in Sudbury, Canada, the Sudbury Neutrino Observatory is centered around a tank that contains 1000 tons of water. Occasionally, a neutrino entering the tank interacts with one or another of the particles already there. Such interactions create flashes of light, called Cerenkov radiation. Some 9600 light detectors sense this light. The numerous silver protrusions are the back sides of the light detectors prior to their being wired and connected to electronics in the lab (seen at the bottom of the photograph). (Ernest Orlando Lawrence/Berkeley National Laboratory)

an electron. As this electron rushes through the water, it emits a flash of light, called **Cerenkov radiation.** As the Russian physicist Pavel A. Cerenkov (pronounced *Che-REN-kov*) first observed, the flash occurs whenever a particle moves through a medium, such as water, faster than light can. Such motion does not violate the tenet that the speed of light *in a vacuum* (3×10^5 km/s) is the ultimate speed limit in the universe. Light is slowed considerably as it passes through water, and high-energy particles can exceed this reduced speed of light in a medium without violating the laws of nature.

Thus, scientists detect neutrinos by observing Cerenkov radiation flashes with light-sensitive devices, called *photomultipliers*, mounted in the water (see Figure 10-23). The three flavors of neutrinos have different interactions with water, all of which have been detected. This evidence that neutrinos can change *requires* that they have mass, and it explains the earlier low rate of solar neutrino observations.

10-10 Frontiers yet to be discovered

Despite centuries of observations and decades of modeling the Sun, it holds innumerable secrets yet to be uncovered. We have launched an armada of spacecraft specifically designed to observe it in different parts of the spectrum. Astronomers soon hope to understand more about such questions as how the Sun's magnetic field is generated and how it changes throughout the solar cycle; how the Sun's atmosphere heats so dramatically in the transition zone; why solar luminosity varies with time; just how the Sun's changing output affects the temperatures on Earth; the details of how flares and coronal mass ejections occur; and why the Sun rotates differentially.

SUMMARY OF KEY IDEAS

The Sun's Atmosphere

- The thin shell of the Sun's gases we see are from its photosphere, the lowest level of its atmosphere. The gases in this layer shine nearly as a blackbody. The photosphere's base is at the top of the convective zone.
- Convection of gas from below the photosphere produces features called granules.
- Above the photosphere is a layer of hotter, but less dense, gas called the chromosphere. Jets of gas, called spicules, rise up into the chromosphere along the boundaries of supergranules.
- The outermost layer of gases in the solar atmosphere, called the corona, extends outward to become the solar wind at great distances from the Sun. The gases of the corona are very hot, but they have extremely low densities.

The Active Sun

- Some surface features on the Sun vary periodically in an 11-year cycle. The magnetic fields that cause these changes actually vary over a 22-year cycle.
- Sunspots are relatively cool regions produced by local concentrations of the Sun's magnetic field protruding through the photosphere. The average number of sunspots and their average latitude vary in an 11-year cycle.
- A prominence is gas lifted into the Sun's corona by magnetic fields. A solar flare is a brief, but violent, eruption of hot, ionized gases from a sunspot group. Coronal mass ejections send out large quantities of gas from the Sun. Coronal mass ejections and flares that head in Earth's direction affect satellites, communication, and electric power, and cause aurorae.
- The magnetic dynamo model suggests that many transient features of the solar cycle are caused by the effects of differential rotation and convection on the Sun's magnetic field.

Why did the earlier neutrino detectors not detect the predicted number of neutrinos from the Sun?

The Sun's Interior

- The Sun's energy is produced by the thermonuclear process called hydrogen fusion, in which four hydrogen nuclei release energy when they fuse to produce a single helium nucleus.
- The energy released in a thermonuclear reaction comes from the conversion of matter into energy, according to Einstein's equation $E = mc^2$.
- The solar model is a theoretical description of the Sun's interior derived from calculations based on the laws of physics. The solar model reveals that hydrogen fusion occurs in a core that extends from the center to about a quarter of the Sun's visible radius.
- Throughout most of the Sun's interior, energy moves outward from the core by radiative diffusion. In the Sun's outer layers, energy is transported to the Sun's surface by convection.
- Neutrinos were originally believed to be massless. The electron neutrinos generated and emitted by the Sun were originally detected at a lower rate than is predicted by our model of thermonuclear fusion. The discrepancy occurred because electron neutrinos have mass, which causes many of them to change into other forms of neutrinos before they reach Earth. These alternative forms are now being detected.

A WHAT DID YOU THINK?

1 *What percentage of the solar system's mass is in the Sun?* The Sun contains about 99.85% of the solar system's mass.

2 *Does the Sun have a solid and liquid interior like Earth?* No. The entire Sun is composed of hot gases.

3 *What is the surface of the Sun like?* The Sun has no solid surface. Indeed, it has no solids or liquids anywhere. The level we see, the photosphere, is composed of hot, churning gases.

4 *Does the Sun rotate? If so, how fast?* The Sun's surface rotates differentially, varying between once every 35 days near its poles and once every 25 days at its equator.

5 *What makes the Sun shine?* Thermonuclear fusion in the Sun's core is the source of the Sun's energy.

6 *Are matter and energy conserved?* By themselves, they are not always conserved. Nuclear fusion converts matter into energy. Energy can also be converted into matter. The sum of the matter (multiplied by c^2) and energy *is* always conserved.

Key Terms for Review

Cerenkov radiation, 338
chromosphere, 320
convective zone, 336
core (of the Sun), 332
corona, 321
coronal hole, 329
coronal mass ejection, 331
differential rotation, 325
filament, 329
granule, 319
helioseismology, 326
hydrogen fusion, 332
hydrostatic equilibrium, 335
limb (of the Sun), 319
limb darkening, 319
magnetic dynamo, 328
neutrino, 333
photosphere, 319
plages, 329
plasma, 326
positron, 333
prominence, 329
radiative zone, 335
solar cycle, 327
solar flare, 329
solar luminosity ($L_\odot$), 333
solar model, 335
solar wind, 322
spicule, 320
sunspot, 323
sunspot maximum, 323
sunspot minimum, 323
supergranule, 321
thermonuclear fusion, 332
transition zone, 321
Zeeman effect, 326

Review Questions

The answers to computational problems, which are preceded by an asterisk (*), appear at the end of the book.

1. Describe the features of the Sun's atmosphere that are always present.

2. Describe the three main layers of the solar atmosphere and how you would best observe them.

3. Name and describe seven features of the active Sun. Which two are the same, seen from different angles?

4. Describe the three main layers of the Sun's interior.

***5.** When will the next sunspot minimum and sunspot maximum occur after the maximum in 2001 and the minimum in 2007? Explain your reasoning.

6. Why is the solar cycle said to have a period of 22 years, even though the sunspot cycle is only 11 years long?

7. How do astronomers detect the presence of a magnetic field in hot gases, such as the field in the solar photosphere?

8. Describe the dangers in attempting to observe the Sun. How have astronomers learned to circumvent these hazards?

9. Give an everyday example of hydrostatic equilibrium not presented in the book.

10. Give some everyday examples of heat transfer by convection and radiative transport.

11. What do astronomers mean by a "model of the Sun"?

12. Why do thermonuclear reactions in the Sun take place only in its core?

13. What is hydrogen fusion? This process is sometimes called "hydrogen burning." How is hydrogen burning fundamentally unlike the burning of a log in a fireplace?

14. Describe the Sun's interior, including the main physical processes that occur at various levels within the Sun.

15. What is a neutrino, and why are astronomers so interested in detecting neutrinos from the Sun?

Advanced Questions

The answers to computational problems, which are preceded by an asterisk (*), appear at the end of the book.

***16.** Using the mass and size of the Sun, calculate the Sun's average density. Compare your answer to the average densities of the outer planets. (*Hint:* The volume of a sphere of radius r is $\frac{4}{3}\pi r^3$.)

***17.** Assuming that the current rate of hydrogen fusion in the Sun remains constant, what fraction of the Sun's mass will be converted into helium over the next 5 billion years? How will this affect the chemical composition of the Sun?

***18.** Calculate the wavelengths at which the photosphere, chromosphere, and corona emit the most radiation. Explain how the results of your calculations suggest the best way to observe these regions of the solar atmosphere. (*Hint:* Use Wien's law and assume that the average temperatures of the photosphere, chromosphere, and corona are 5800 K, 50,000 K, and 1.5×10^6 K, respectively.)

19. When we are near a sunspot maximum, the Hubble Space Telescope must be moved to a higher orbit. Why? (*Hint:* Think about how the increased solar energy affects Earth's atmosphere.)

Discussion Questions

20. Discuss the extent to which cultures around the world have worshiped the Sun as a deity throughout history. Why do you think our star inspires such widespread veneration?

21. Discuss some of the difficulties of correlating solar activity with changes in the terrestrial climate.

22. Describe some advantages and disadvantages of observing the Sun **a.** from space and **b.** from Earth's South Pole. What kinds of phenomena and issues do solar astronomers want to explore from both Earth-orbiting and Antarctic observatories?

What If...

23. The Sun were not rotating? What about it would be different?

24. The typical solar wind were much stronger (say 100 times stronger) than it is now? What differences would there be in the solar system?

25. The typical solar wind were much weaker (say 100 times weaker) than it is now? What differences would there be in the solar system?

26. The Sun's brightness and heat output periodically changed, as they do in many stars? What differences would there be in the solar system? Unless your instructor gives you another time frame, assume that the light and heat vary together in a cycle that lasts 1 year.

Web Questions

27. Search the Web for all solar neutrino experiments. Make a list of them and indicate which are currently operating and which are still under construction. Summarize the results found by the active detectors.

28. Search the Web for information about features of the solar atmosphere called *sigmoids*. What are they? What causes them? How do sigmoids provide a way to predict coronal mass ejections?

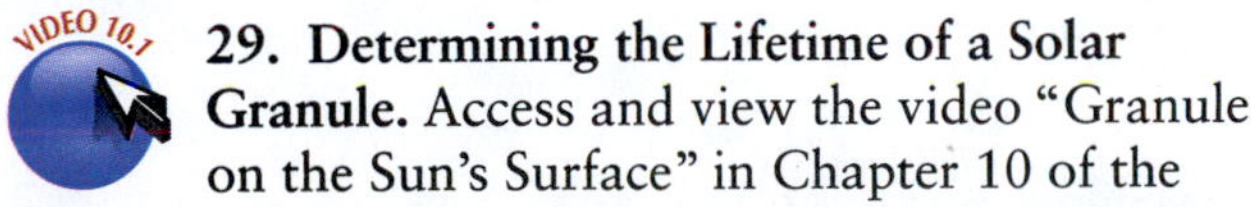

29. **Determining the Lifetime of a Solar Granule.** Access and view the video "Granules on the Sun's Surface" in Chapter 10 of the *Discovering the Universe* Web site. You will use it to determine the approximate lifetime of a solar granule. Select an area on the Sun's image and slowly and rhythmically repeat *Start, Stop, Start, Stop* until you can consistently predict the appearance and disappearance of granules. While keeping your rhythm, move to a different area of the video and continue monitoring the appearance and disappearance of granules. When you are confident that you have the timing right, move your eyes to the clock shown in the video (or work with a partner). Using your *Start-Stop* cycle, determine the length of time between the appearance and disappearance of the granules and record your answer.

Observing Projects

30. Use a telescope to view the Sun, but *only* when it is equipped with an appropriate and safe solar filter, or by projecting the Sun's image onto a screen or sheet of white paper. *Do not look directly at the Sun! Looking at the Sun causes blindness.* Do you see any sunspots? If so, sketch their appearance. Can you distinguish between the umbrae and penumbrae of the sunspots? Can you see limb darkening? Can you see granulation?

31. If you have access to an H_α filter attached to a telescope specially designed for viewing the Sun safely, use this instrument to examine the solar surface. How does the filtered appearance of the Sun differ from that in white light (see Observing Project 30)? What do sunspots look like in H_α? Can you see any prominences? Can you see any filaments? Are the filaments in the H_α image close to any sunspots seen in white light?

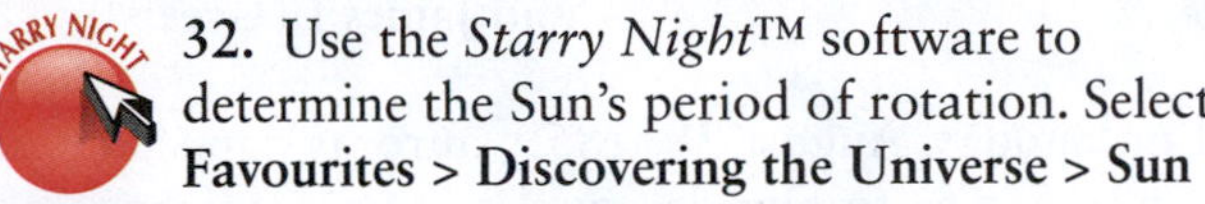

32. Use the *Starry Night*™ software to determine the Sun's period of rotation. Select **Favourites > Discovering the Universe > Sun Rotation.** From this position, 0.008 AU from the Sun on December 27, 2011, you can see numerous features on its surface. A line of solar longitude aligned with the eastern side of a major active region has been added to assist you in determining the average rotation period of the Sun. With the **Time Flow Rate** set to 1 day, step time forward to observe and measure the Sun's rotation by counting the 1-day steps to bring the longitude line back to its initial position. Compare your measurement with the quoted rotation information for the Sun presented in the text. In view of the fact that *Starry Night*™ does not show the Sun's differential rotation, which latitude region of the Sun is represented by your measured rotation period?

33. Use *Starry Night*™ to examine the solar chromosphere from a position just above the solar surface. Select **Favourites > Discovering the Universe > Solar Chromosphere** to view the Sun as it would appear at the wavelength of the red hydrogen H_α Balmer line that originates in the upper layers of the Sun's atmosphere. Use the hand tool to examine the various features that appear. **a.** At which compass directions on the horizon do you find quiescent prominences? **b.** At which compass directions do you find major sunspots with umbrae and penumbrae? **c.** In which direction do you see a bright flare on the Sun's surface? Examine particularly the way in which the intense magnetic fields around this latter feature have disturbed and aligned the gases in the solar atmosphere. Zoom in to a field of view of about 40° and scan around the horizon to examine the spicules that protrude above the Sun. These spicules make up a network along the boundaries of supergranular convection cells across the solar surface.

BRACING FOR A SOLAR SUPERSTORM

A recurrence of the 1859 solar superstorm would be a cosmic Katrina, causing billions of dollars of damage to satellites, power grids and radio communications

Adapted from an article by Sten F. Odenwald and James L. Green

As night fell on Sunday, August 28, 1859, auroras danced overhead. From Maine to Florida, vivid curtains of light took the skies. Startled Cubans saw them directly overhead; ships near the equator described crimson lights reaching halfway to the zenith. Many thought their cities had caught fire. Around the world, scientific instruments recording minute changes in Earth's magnetism suddenly shot off the scale, and electric currents surged into the world's telegraph systems. It took Baltimore telegraph operators from 8 P.M. until 10 A.M. the next day to transmit a 400-word report.

Just before noon on September 1st, astronomer Richard C. Carrington witnessed an intense white light flash from two enormous, dark sunspots. Seventeen hours later in the Americas, a second wave of auroras turned night to day as far south as Panama. People could read the newspaper by their crimson and green light. Gold miners in the Rocky Mountains ate breakfast at 1 A.M., thinking the Sun had risen on a cloudy day. Telegraph systems became unusable across Europe and North America.

The news media looked for researchers to explain the phenomena, but scientists scarcely understood auroral displays. Were they meteoritic matter, reflected light from polar icebergs—or high-altitude lightning?

The Great Aurora of 1859 ushered in new scientific models for space weather. The October 15th issue of *Scientific American* noted that "a connection between the northern lights and forces of electricity and magnetism is now fully established." Since then, scientists have established that auroras originate in violent events on the Sun, which fire off huge clouds of plasma and momentarily disrupt Earth's magnetic field.

The impact of the 1859 storm was muted only by the technological infancy of that time. Today, such a storm could severely damage satellites, disable radio communications, and cause continent-wide blackouts that would require weeks to repair. Although a storm of that magnitude occurs just once every 500 years, storms with half that intensity rage about every 50 years. The last one, on November 13, 1960, led to worldwide geomagnetic disturbances and radio outages. Without preparations, the costs of another superstorm could equal that of a major hurricane or earthquake.

The Big One

Solar magnetic activity waxes and wanes on an 11-year cycle. The current cycle began in January 2008; over the next five years, solar activity will increase. During the last cycle, 21,000 flares and 13,000 clouds of plasma exploded from the Sun's surface. These solar storms arise from churning solar gases. Flares are like lightning storms; they are bursts of energetic particles and intense X rays resulting from changes in the Sun's magnetic field. Coronal mass ejections (CMEs) are analogous to hurricanes; they are giant magnetic bubbles that hurl billion-ton plasma clouds into space at several million kilometers per hour. But these storms mostly send auroras dancing across polar skies.

No one living today has experienced a superstorm, but ice-core data from Greenland and Antarctica link jumps in concentrations of trapped nitrate gases with blasts of solar particles. A nitrate anomaly found for 1859 stands out as the biggest of the past 500 years, roughly equivalent to all major events of the past four decades combined. But as violent as it was, it does not appear to have been qualitatively different from lesser events.

Toasted Satellites, Lights Out

The most obvious victims of the next large geomagnetic storm will be satellites. Under ordinary conditions, cosmic-ray particles erode solar panels and reduce power generation by about two percent annually. Incoming particles also interfere with satellite electronics. Many communications satellites have been lost this way. A large solar storm can take one to three years off of a satellite's working lifetime and produce hundreds of glitches, from harmless commands to destructive electrostatic discharges.

To study satellite impact, we simulated 1000 ways a superstorm might unfold, with intensities ranging from the worst storm of the Space Age on October 20, 1989, to the 1859 superstorm. We found the storms would degrade solar panels. Total economic loss could range from about $20 billion to $70 billion, with the top end equaling approximately a year's revenue for all communications satellites. These figures do not include losses to satellite customers.

Fortunately, communications satellites fare well against once-a-decade events. Their life spans have grown from five years in 1980 to nearly 17 years today. To increase protection, engineers could thicken the shielding, lower solar panel voltages to lessen the risk of runaway electrostatic discharges, add extra backup systems, and make the software more robust to data corruption.

Impact of a Coronal Mass Ejection

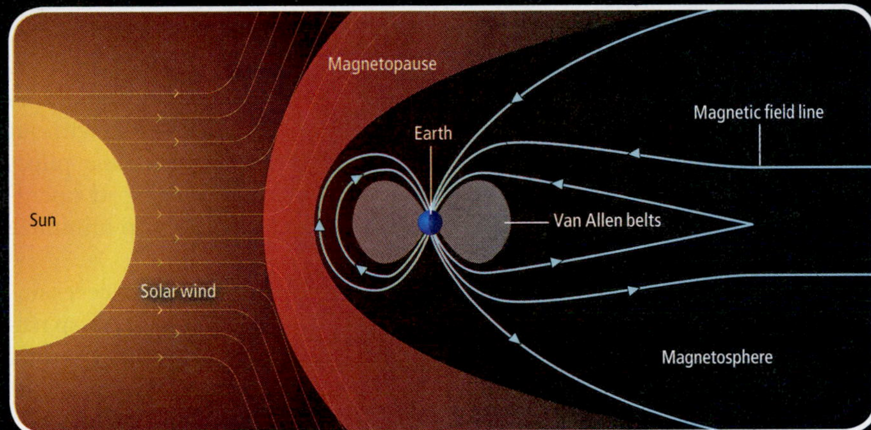

NORMAL CONDITIONS: Earth's magnetic field typically deflects the charged particles streaming out from the Sun, carving out a teardrop-shaped volume known as the magnetosphere. On the Sun-facing side, the boundary, or magnetopause, is about 60,000 kilometers from our planet. The field also traps particles in a doughnut-shaped region known as the Van Allen belts.

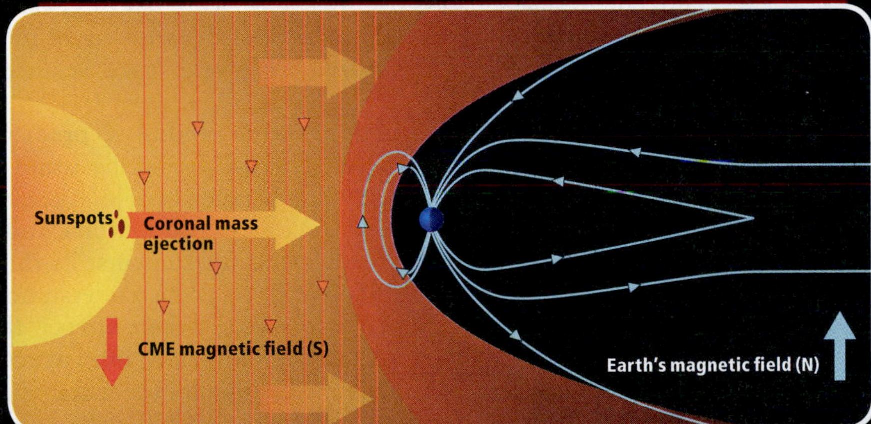

FIRST STAGES OF IMPACT: When the Sun fires off a coronal mass ejection (CME), this bubble of ionized gas greatly compresses the magnetosphere. In extreme cases such as superstorms, it can push the magnetopause into the Van Allen belts and wipe them out.

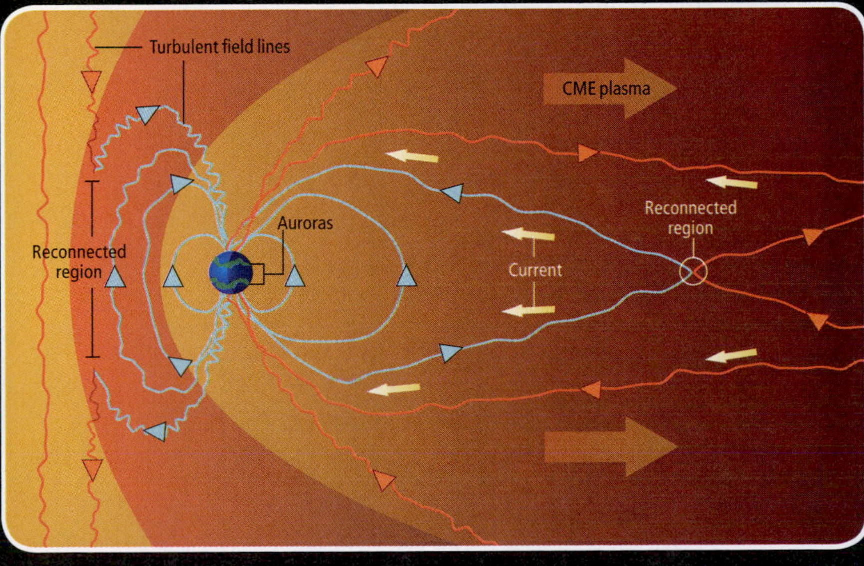

MAGNETIC RECONNECTION: The solar gas has its own magnetic field, and as it streams past our planet, it stirs up turbulence in Earth's magnetic field. If this field points in the opposite direction as Earth's, the two can link up, or reconnect—releasing magnetic energy that accelerates particles and thereby creates bright auroras and powerful electric currents.

To improve solar panels, engineers switched from silicon to gallium arsenide, making them smaller, increasing power production, and providing increased resistance to cosmic-ray damage.

It is harder to guard against other effects. X-ray energy would cause the atmosphere to expand, enhancing drag on military and commercial imaging and communications satellites that orbit below 600 km in altitude. Within months of a superstorm, low-orbiting satellites could burn up in the atmosphere.

While satellites have been designed to function amidst space weather, power grids are fragile in the best of times. During solar storms, large transformers could be literally fried by bursts of geomagnetically-induced direct current. Voltage regulation could collapse network-wide. The magnetic storm of May 15, 1921, would have caused a blackout affecting half of North America had it happened today, while a storm like that of 1859 could bring down the entire grid. North America faces greater danger than many other industrialized countries because of its proximity to the north magnetic pole.

A superstorm will also interfere with radio signals, rendering the Global Positioning System (GPS) useless for many military and civilian applications. High-energy particles will interfere with aircraft radio communications and could force the rerouting of hundreds of flights over the North Pole and across Canada and the northern U.S.

Getting Ready

Society's increasing vulnerability to solar storms has coincided with decreasing public awareness. Before 1950, magnetic storms, solar flares, and their effects often received lavish, front-page newspaper stories; since then, such stories have been buried on inside pages.

More reliable warnings of solar and geomagnetic storms would help. With adequate warning, satellite operators can defer critical maneuvers and watch for anomalies that, without quick action, could escalate into emergencies. Airline pilots could prepare for orderly flight diversions. Power grid operators could watch susceptible network components to minimize outages.

Agencies such as NASA and the National Science Foundation have worked over the past 20 years to develop space-weather forecasting capabilities. Currently National Oceanic and Atmospheric Administration's (NOAA) Space Weather Prediction Center provides daily space weather reports to more than 1000 businesses and government agencies. But this capability relies on a hodgepodge of satellites designed more for research purposes than for efficient, long-term space weather monitoring.

Some researchers feel that scientists' ability to predict space weather is where NOAA was in predicting atmospheric weather in the early 1950s. Inexpensive, long-term space buoys are needed to monitor weather conditions, and can be built using simple, off-the-shelf instruments. In the meantime, scientists have a long way to go to understand the physics of solar storms and to forecast their effects. If we really want to safeguard our technological infrastructure, we will have to redouble our investment in forecasting, modeling, and basic research to batten down for the next solar tempest.

PART III

Understanding the Stars

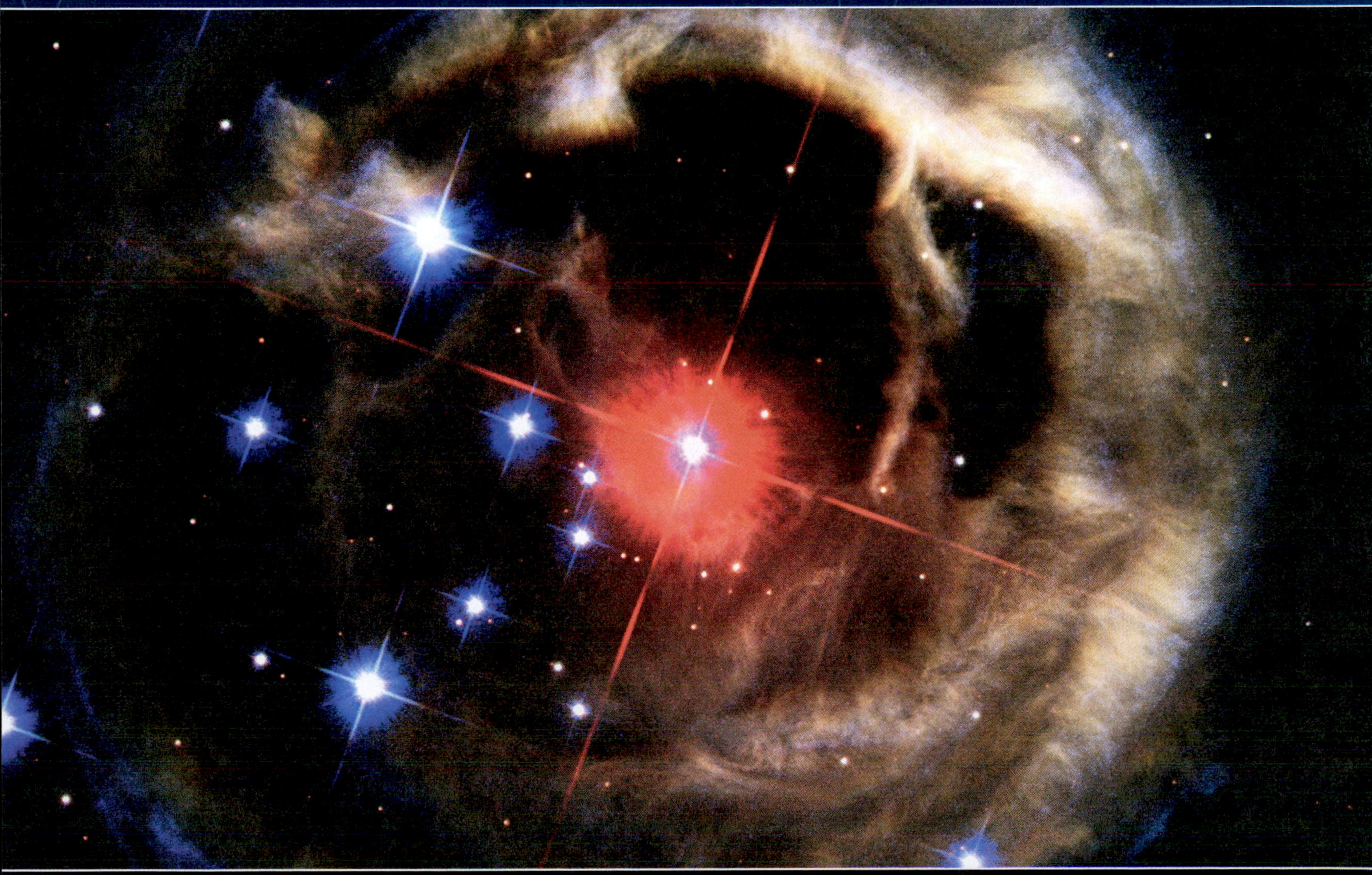

R I V U X G

Interstellar dust illuminated by a pulse of light emitted from the red giant star, V838 Monocerotis, in the center of the image. (NASA, ESA, and H. H. Bond [STScI])

AN ASTRONOMER'S ALMANAC

134 B.C. Hipparchus proposes first apparent magnitude scale.

1054 Supernova that becomes the Crab Nebula, observed worldwide, visible during the day for several weeks.

Gregorian calendar

1650 Giovanni Riccioli first observes an optical binary star system, Mizar A and Mizar B.

1718 Edmond Halley first measures proper motion of stars.

1781–1784 Charles Messier catalogs 107 nebulae, M1–M109 (two were spurious). Edward Piggot observes first Cepheid variable star.

1838 First stellar parallaxes measured by Friedrich Bessel.

1844 Sirius determined to be part of binary star system.

1856 Norman Pogson proposes modern apparent magnitude scale.

Darwin's theory of evolution

1862–1863 Sirius B, binary companion to Sirius A, is first observed by Alvan G. Clark. Angelo Secchi is first to classify stars by their spectra.

1889 E. C. Pickering first observes a spectroscopic binary, Mizar A and its dim companion.

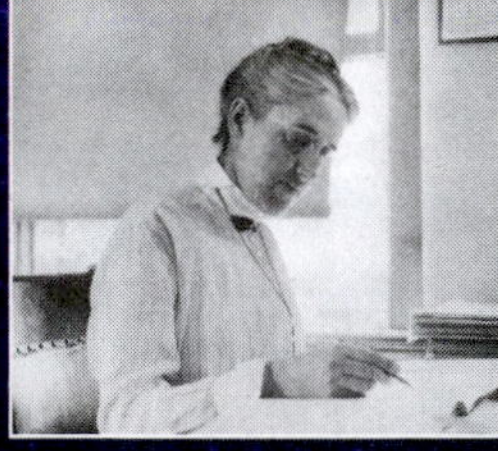

(AIP Emilio Segre Visual Archives)

1902–1908 James Jeans calculates criteria for the collapse of a gas cloud under the influence of gravity. Albert Einstein's special relativity describes the effect of motion on space, time, and matter. Henrietta Leavitt discovers Cepheid period-luminosity relationship.

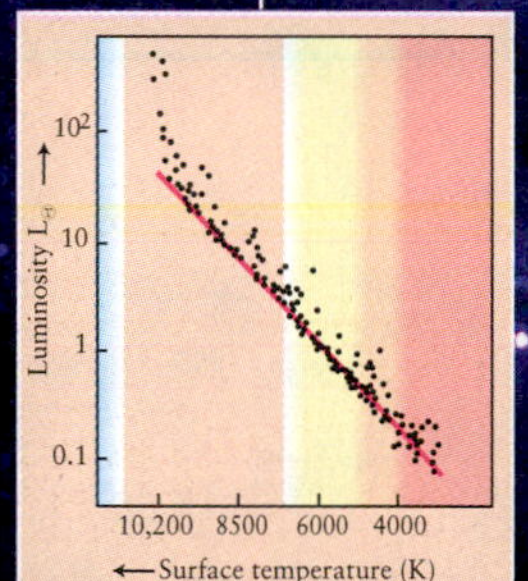

1911–1913 Relationship between stellar magnitudes and spectral types noted by Ejnar Hertzsprung and then by Henry Russell.

1915–1916 Einstein publishes his general theory of relativity. Karl Schwarzschild calculates nonrotating solution of Einstein's equations of general relativity (without electric charge).

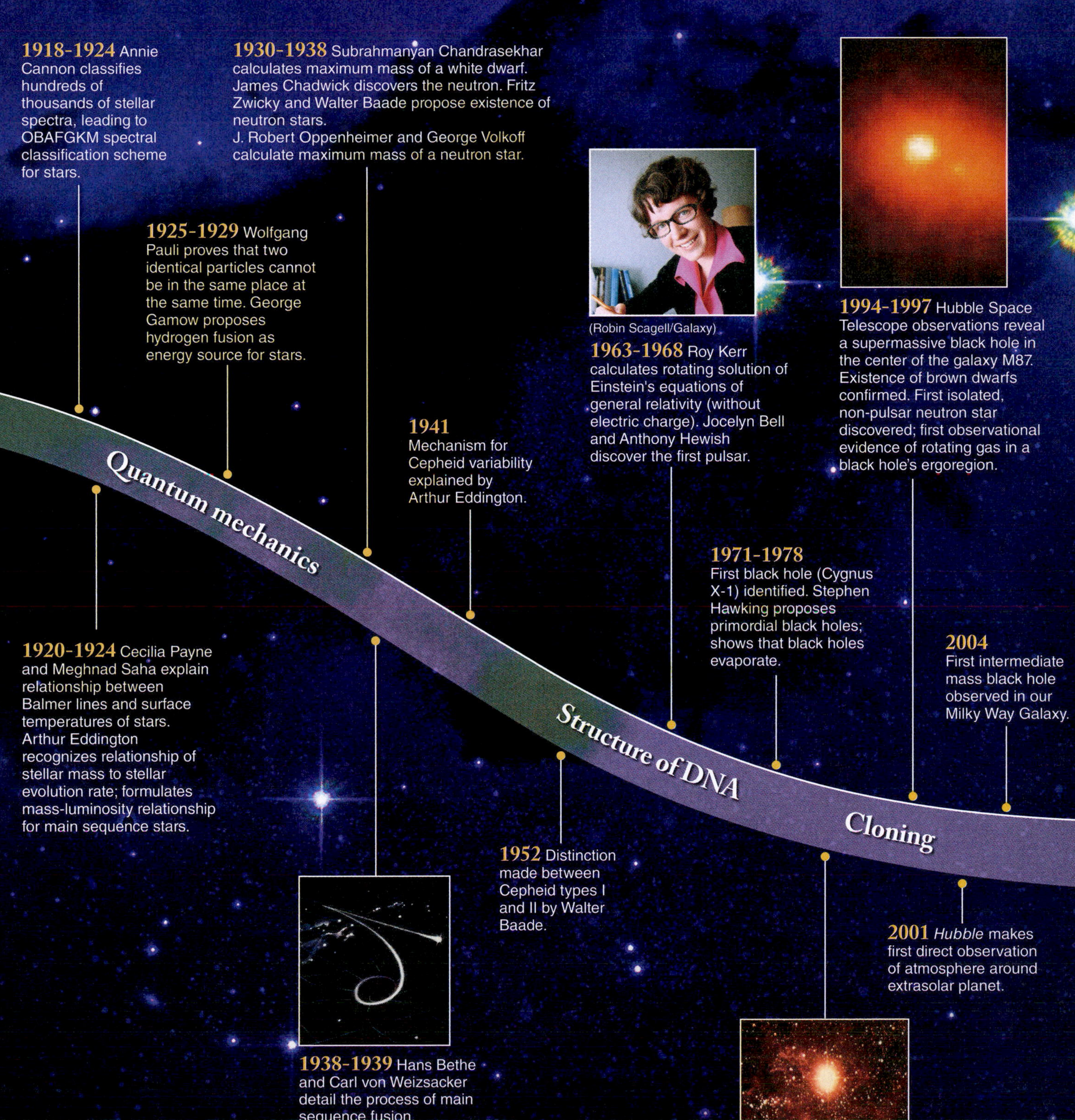

1918-1924 Annie Cannon classifies hundreds of thousands of stellar spectra, leading to OBAFGKM spectral classification scheme for stars.
1920-1924 Cecilia Payne and Meghnad Saha explain relationship between Balmer lines and surface temperatures of stars. Arthur Eddington recognizes relationship of stellar mass to stellar evolution rate; formulates mass-luminosity relationship for main sequence stars.
1925-1929 Wolfgang Pauli proves that two identical particles cannot be in the same place at the same time. George Gamow proposes hydrogen fusion as energy source for stars.
Quantum mechanics
1930-1938 Subrahmanyan Chandrasekhar calculates maximum mass of a white dwarf. James Chadwick discovers the neutron. Fritz Zwicky and Walter Baade propose existence of neutron stars.
J. Robert Oppenheimer and George Volkoff calculate maximum mass of a neutron star.
1938-1939 Hans Bethe and Carl von Weizsacker detail the process of main sequence fusion.
J. Robert Oppenheimer and Hartland Snyder demonstrate that collapsing, massive stars form black holes.
1941 Mechanism for Cepheid variability explained by Arthur Eddington.
1952 Distinction made between Cepheid types I and II by Walter Baade.
Structure of DNA
(Robin Scagell/Galaxy)
1963-1968 Roy Kerr calculates rotating solution of Einstein's equations of general relativity (without electric charge). Jocelyn Bell and Anthony Hewish discover the first pulsar.
1971-1978 First black hole (Cygnus X-1) identified. Stephen Hawking proposes primordial black holes; shows that black holes evaporate.
1987 Supernova 1987A observed as it brightened and then as it faded.
Cloning
1994-1997 Hubble Space Telescope observations reveal a supermassive black hole in the center of the galaxy M87. Existence of brown dwarfs confirmed. First isolated, non-pulsar neutron star discovered; first observational evidence of rotating gas in a black hole's ergoregion.
2001 Hubble makes first direct observation of atmosphere around extrasolar planet.
2004 First intermediate mass black hole observed in our Milky Way Galaxy.

WHAT DO YOU THINK?

1. How near to us is the closest star other than the Sun?
2. How luminous is the Sun compared with other stars?
3. What colors are stars, and why do they have these colors?
4. Are brighter stars hotter than dimmer stars?
5. Compared to the Sun, what sizes are other stars?
6. Are most stars isolated from other stars, as the Sun is?

Answers to these questions appear in the text beside the corresponding numbers in the margins and at the end of the chapter.

RIVUXG Stars Come in Many Colors. (Hubble Heritage Team/AURA/STScI/NASA)

Chapter 11
Characterizing Stars

Legend tells of a race of remarkable insects, the Ephemera, that inhabited a great forest. These noble creatures were blessed with great intelligence, but were cursed with tragically short life spans. To the Ephemera, the forest seemed eternal and unchanging. Members of each generation lived out their brief lives without ever noticing any changes in their leafy world. Nevertheless, careful observations and reasoning led some Ephemera to postulate that the forest was not static. They began to suspect that small green shoots grew to become huge trees and that mature trees eventually died, toppled over, and littered the forest with rotting logs, enriching the soil for future trees. Although unable to witness these transformations personally, the Ephemera predicted the existence of life processes stretching over the mind-boggling periods of many years.

To us, the cosmos seems eternal and unchanging; the views that greet us every night are virtually indistinguishable from those seen by our ancestors. This permanence, however, is an illusion. We see the stars because they emit vast amounts of radiation, created as a result of converting one element into another in their interiors. Because all stars contain finite amounts of the elements fueling these emissions, at different times in the life of each star, its various fuels are used up. This process causes the stars to change profoundly.

We do have one advantage over the Ephemera, who were so small that they could see only a few trees, leaves, shoots, and rocks. Astronomers are witnesses to literally billions of stars in various stages of evolution. Therefore, although we cannot see any one star go through more than a tiny fraction of the process, we can see every stage of **stellar evolution.** By understanding the changes that stars undergo, we gain insight into our place in the cosmos. We begin our journey to the stars by learning the properties of stars that astronomers used to develop models of why stars shine and how they evolve.

In this chapter you will discover

- that the distances to many nearby stars can be measured directly, whereas the distances to farther ones are determined indirectly
- the observed properties of stars on which astronomers base their models of stellar evolution
- how astronomers analyze starlight to determine a star's temperature and chemical composition
- how the total energy emitted by stars and their surface temperatures are related
- the different classes of stars
- the variety and importance of binary star systems
- how astronomers calculate stellar masses

11-1 Distances to nearby stars are determined by stellar parallax

Nothing in human experience prepares us for understanding the distances between the stars. On Earth, the greatest distance we read about in our daily lives is perhaps halfway around the globe, 20,000 km, which is about the distance from New York City to Perth, Australia. By observing the Sun, the Moon, and planets, we have some comprehension of hundreds of thousands, or even millions, of kilometers or miles.

How far away are the nearest stars? Ask 10 people 1
and you will probably hear answers that range from millions to billions of kilometers or miles. In fact, the closest star other than the Sun, Proxima Centauri, in the constellation Centaurus, is about 40 *trillion* km (4.0×10^{13} km or 25 trillion mi) away. Light takes more than 4 years to get here from there. Most of the stars you see in the night sky are many times farther away. (Centaurus is not visible from most of the northern hemisphere, as you can see by the negative declinations on the margin star chart on this page.) Given such enormous separations between astronomical bodies, how do we measure the distance to a nearby star?

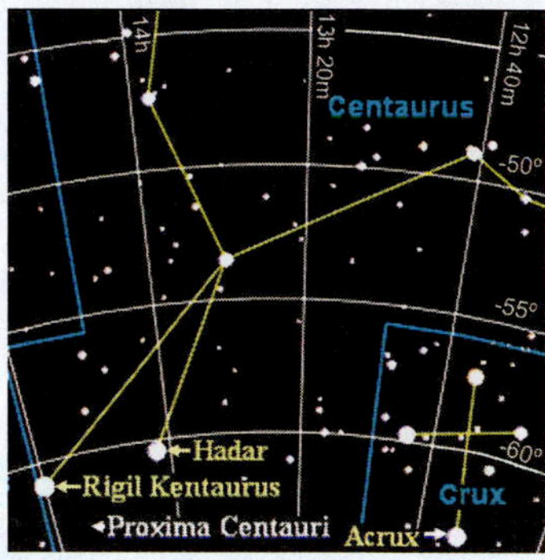

Because Proxima Centauri is closer to us than other stars, its position among the background stars changes as Earth orbits the Sun. We see precisely the same effect when, for example, Mars appears to have retrograde motion as we pass between it and the Sun (see Figures 2-2 and 2-3), and it is the effect that Tycho Brahe sought while observing the supernova of 1572 (see Figure 2-7). The apparent motion of nearby stars among the background of more distant stars, due to Earth's motion around the Sun, is called **stellar parallax.**

Parallax is an everyday phenomenon. We experience it when nearby objects appear to shift their positions against a distant background as we move (review Figure 2-6). We also experience it continuously when we are awake, because our eyes change angle when looking at objects at different distances. As you view a tree 10 m away, your eyes cross only slightly (Figure 11-1a). Looking at something closer, say this book, your eyes cross much more in order for both of them to focus on the same word (Figure 11-1b). The parallax angle formed between your eyes and the tree or the book lets your brain judge just how close these objects are to you. Working out distances of nearby stars is done similarly, but it requires painstaking measurements of tiny angles and a little geometry.

a

b

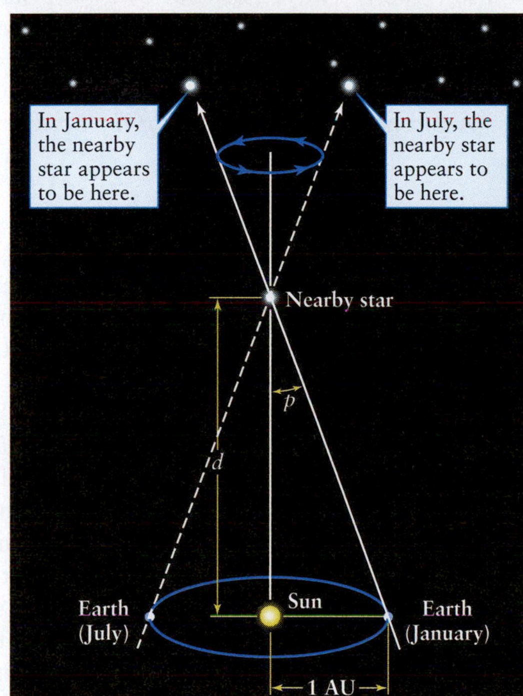

c Parallax of a nearby star

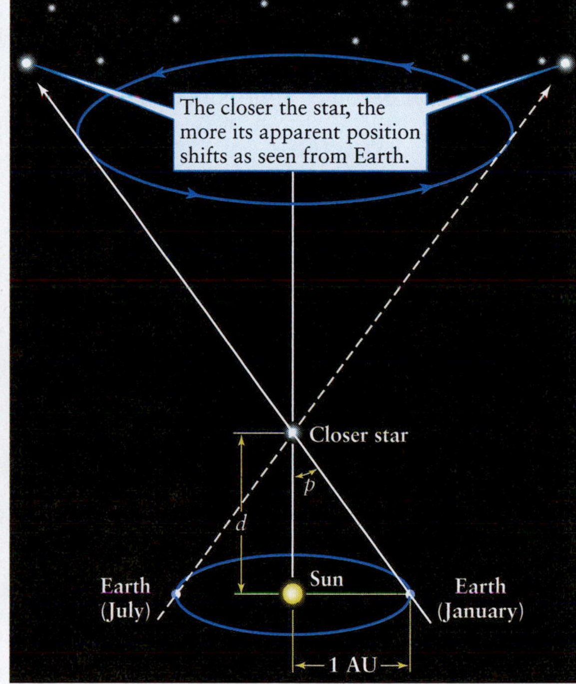

d Parallax of an even closer star

ANIMATION 11.1

FIGURE 11-1 Using Parallax to Determine Distance (a, b) Our eyes change the angle between their lines of sight as we look at things that are different distances away. Our eyes are adjusting for the parallax of the things we see. This change helps our brains determine the distances to objects and is analogous to how astronomers determine the distance to objects in space. (c) As Earth orbits the Sun, a nearby star appears to shift its position against the background of distant stars. The star's parallax angle (p) is equal to the angle between the Sun and Earth, as seen from the star. The stars on the scale of this drawing are shown much closer than they are in reality. If drawn to the correct scale, the closest star, other than the Sun, would be about 5 km (3.2 mi) away. (d) The closer the star is to us, the greater the parallax angle p. The distance to the star (in parsecs) is found by taking the inverse of the parallax angle p (in arcseconds), $d = 1/p$. (a, b: Mark Andersen/JupiterImages)

An Astronomer's Toolbox 11-1

Distances to Nearby Stars

Recall from *An Astronomer's Toolbox 2-1* that 1 parsec (1 pc) is the distance at which two objects 1 AU apart appear 1 arcsec apart. This distance is 3.09×10^{13} km, or 206,265 AU. The word *parsec* (from *par*allax *sec*ond) originated from the use of parallax to measure distance. Using parsecs, we can write down an especially simple equation for the distances to stars:

$$\text{Distance to a star in parsecs} = \frac{1}{\text{parallax angle of that star in arcseconds}}$$

or

$$d = \frac{1}{p}$$

where d is the distance to the star and p is the parallax angle of that star.

The equation is only this simple in these units, which is one of the main reasons why many astronomers discuss cosmic distances in parsecs rather than light-years. We will continue to primarily use light-years (ly) throughout this book, however, as they are more intuitive. In light-years, this same equation becomes approximately

$$\text{Distance to a star in light-years} \approx \frac{3.26}{\text{parallax angle of that star in arcseconds}}$$

or

$$d_{\text{ly}} \approx \frac{3.26}{p}$$

where d_{ly} is the distance to a star in light-years.

Example: The nearest star, Proxima Centauri, has a parallax angle of 0.77 arcsec, and so its distance is 1/0.77, or approximately 1.3 pc. Equivalently, Proxima Centauri is 4.24 ly away. The parallax of Proxima Centauri is comparable to the angular diameter of a dime seen from a distance of 3 km. The parallax angles of the 25 nearest stars are listed in Appendix Table E-5.

Try these questions: What is the average distance from Earth to the Sun in parsecs? What is the parallax angle of a star 100 ly away? How far away is a star with a parallax angle of 0.3850? Rigel is 773 ly away; how far away is it in parsecs? What is its parallax angle in arcseconds?

(Answers appear at the end of the book.)

As Earth moves from one side of its orbit around the Sun to the other, a nearby star's apparent position shifts among the more distant stars. Referring to Figure 11-1c, the parallax angle, p, is half the angle by which Earth shifts positions through the year as seen from that star, measured in arcseconds. The difference in parallax angles for stars at different distances can be seen by comparing Figures 11-1c and 11-1d. *An Astronomer's Toolbox 11-1* explores some details of how distances are determined from parallax angles.

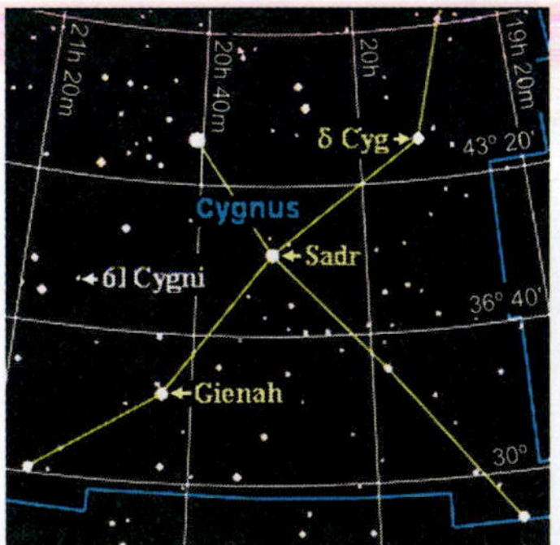

For which object—the Moon, Mars, or the star Sirius—is the parallax angle smallest from Earth?

The first stellar parallax measurement was made in 1838 by Friedrich Wilhelm Bessel, a German astronomer and mathematician. He found the parallax angle of the star 61 Cygni to be ⅓″, and so its distance is about 3 pc (see *An Astronomer's Toolbox 2-1* to review astronomical distance units). The precision of stellar parallax measurements is limited by the angular resolution of the telescope, as discussed in Section 3-6. Because parallax angles smaller than about 0.01″ are difficult to measure from Earth-based observatories, the stellar parallax method using Earth-based telescopes gives stellar distances only up to about 100 pc.

Telescopes in space are unhampered by our atmosphere and therefore have higher resolutions than Earth-based telescopes. Parallax measurements made in space thus enable astronomers to determine the distances to stars well beyond the reach of ground-based observations.

In 1989, the European Space Agency (ESA) launched a satellite, called *Hipparcos* (an acronym for *Hi*gh *P*recision *Par*allax *Co*llecting *S*atellite, and named for Hipparchus, an astronomer in ancient Greece who created an early classification system for stars). Although the satellite failed to achieve its proper orbit, astronomers used it to measure the distances to over 2.5 million of the nearest stars up to 500 ly (150 pc) away. The success of *Hipparcos* has led to plans for better satellites to collect parallax data from stars even farther away.

Despite the information gained from stellar parallax, astronomers need to know the distances to more remote stars for which parallax cannot yet be measured. Several methods of determining ever-greater distances will be introduced later in this chapter and in Chapters 12, 13, and 17.

Having established that different stars are at different distances from Earth, we now consider the brightnesses that stars appear to have as seen from our planet. Combining the distances and the varied brightnesses we observe will enable us to calculate how much light stars actually emit, and thereby to explore their evolution.

MAGNITUDE SCALES

Greek astronomers, from Hipparchus in the second century B.C. to Ptolemy in the second century A.D., undertook the classification of stars strictly by evaluating how bright they appear to be relative to each other. This comparison made sense because back then astronomers assumed the stars were all at the same distance from us, and, therefore, differences in brightness due to different stellar distances from Earth were not expected. Classification regardless of distance is still useful to help us navigate around the night sky, and so we will study it now. In Section 11-3 we will add the effects of different distances on how bright stars appear from Earth.

11-2 Apparent magnitude measures the brightness of stars as seen from Earth

The brightnesses of stars measured without regard to their distances from Earth are called **apparent magnitudes,** denoted by lowercase m. The brightest stars were originally said to be of first magnitude, and their apparent magnitudes were designated $m = +1$. Those stars that appeared to be about half as bright as first-magnitude stars were said to be second-magnitude stars (designated $m = +2$), and so forth, down to sixth-magnitude stars, the dimmest ones visible to the unaided eye. (Greek astronomers did not try to classify the Sun's dazzling brightness in this scheme.) Because stars do not appear with discrete levels of brightness, this system has noninteger magnitudes as well, such as +3.5 or +4.8.

More quantitative methods of classifying stars were developed in the mid-nineteenth century (see *An Astronomer's Toolbox 11-2*). While maintaining the basic idea that brighter objects have smaller numbers than dimmer ones, the scale used today gives the star Vega an apparent magnitude of 0.0. Because some bodies are brighter than Vega, astronomers assign *negative* numbers to the apparent magnitudes of the very *brightest* objects. Sirius, the brightest star in the night sky, has an apparent magnitude of $m = -1.44$. Figure 11-2a shows Sirius along with the apparent magnitudes of some of the stars in Orion. With this convention we can describe other bright objects in the sky, such as the Sun, the Moon, comets, and planets. At its brightest, Venus shines with an apparent magnitude of $m = -4.4$, the full Moon has an apparent magnitude of $m = -12.6$, and the Sun has an apparent magnitude of $m = -26.7$. *Remember:* Objects with negative apparent magnitudes appear brighter than those with positive apparent magnitudes—the more negative, the brighter.

Insight Into Science

Bigger Is Not Necessarily Brighter Numbering as well as naming schemes may be counterintuitive in science. You might expect brighter stars to have larger, more positive numbers than dimmer stars, but the apparent magnitude scheme is just the opposite. Similarly, we will see shortly that on the standard plot of stars used by astronomers, the Hertzsprung-Russell diagram, the hottest stars fall on the *left* and the coolest stars on the *right*.

An Astronomer's Toolbox 11-2

Details of the Magnitude Scales

The magnitude scales were created before accurate measurements of the relative brightnesses of stars could be made, and they have since been refined. Specifically, careful measurements reveal that the original first-magnitude stars were about 100 times brighter than the original sixth-magnitude stars. Therefore, astronomers chose the brightness factor of exactly 100 to *define* the range of brightness between modern first- and sixth-magnitude stars. In other words, it takes 100 stars of apparent magnitude $m = +6$ to provide as much light as we receive from a single star of apparent magnitude $m = +1$.

To find out how much brighter each magnitude is from the next dimmer one, we note that there are five integer magnitudes between first and sixth magnitude. Going from $m = +6$ to $m = +5$ increases (multiplies) the brightness we see by the same factor as going from $m = +5$ to $m = +4$, and so on. Going from $m = +6$ to $m = +1$ requires multiplying the same brightness factor from one magnitude to the next 5 times. The number we must multiply 5 times to get the range of brightness of 100 is $100^{1/5} \approx 2.512$, or, put another way, $2.512 \times 2.512 \times 2.512 \times 2.512 \times 2.512 \approx 100$. This mathematical statement means that *each successively brighter magnitude is approximately 2.512 times brighter than the preceding magnitude.*

Example: An $m = +3$ star is approximately 2.512 times brighter than an $m = +4$ star. Equivalently, it takes 2.512 fourth-magnitude stars to provide as much light as we receive from a single third-magnitude star.

Try these questions after reading Section 11-3: How much brighter is an $m = 0$ star than an $m = +4$ star? How much brighter is an $m = -2$ star than an $m = +5$ star? If one star is 7.93 times brighter than another and the brighter star has an absolute magnitude of $m = +3$, what is the absolute magnitude of the dimmer star? (*Hint for last question:* Recall that magnitudes need not be integers.)

(Answers appear at the end of the book.)

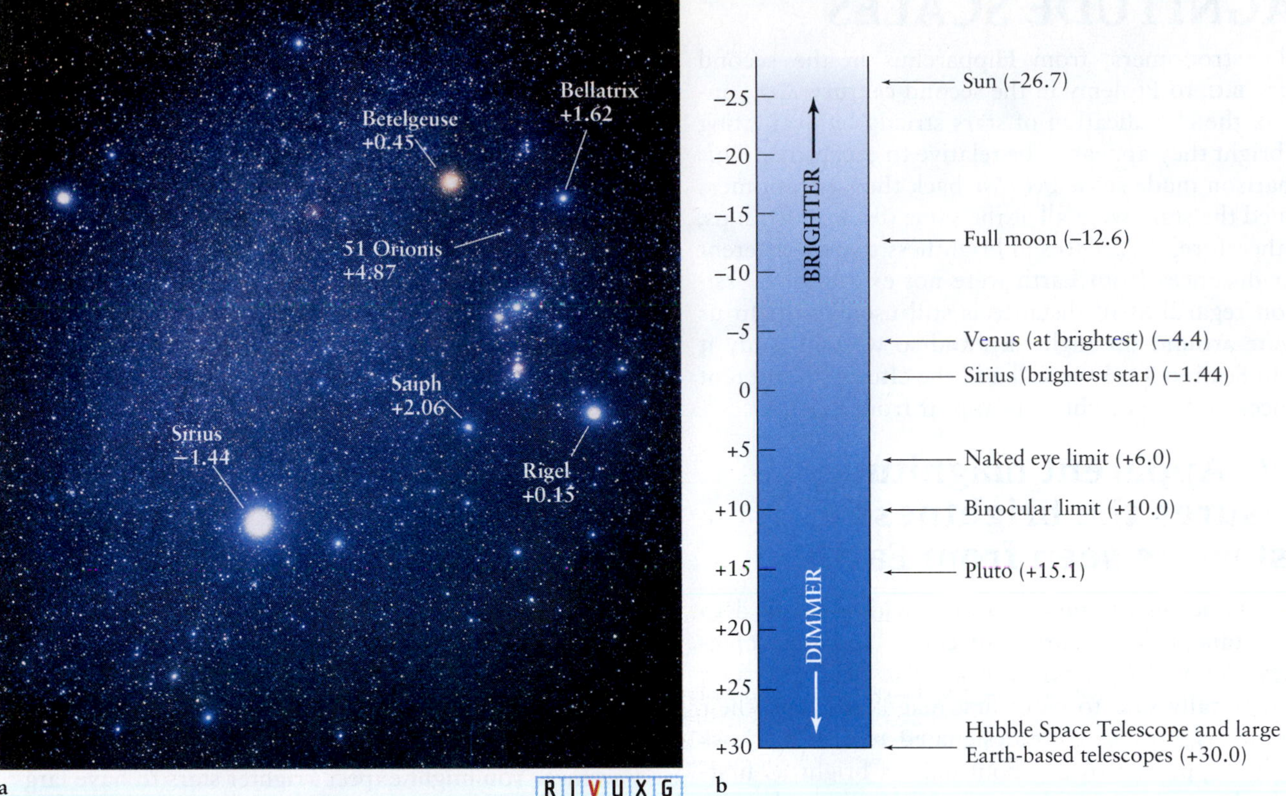

FIGURE 11-2 Apparent Magnitude Scale (a) Several stars in and around the constellation Orion, labeled with their names and apparent magnitudes. For a discussion of star names, see *Guided Discovery: Star Names.* (b) Astronomers denote the brightnesses of objects in the sky by their apparent magnitudes. Stars visible to the naked eye have magnitudes between $m = -1.44$ (Sirius) and about $m = +6.0$. However, CCD (charge-coupled device) photography through the Hubble Space Telescope or a large Earth-based telescope can reveal stars and other objects nearly as faint as magnitude $m = +30$. (a: Okiro Fujii, L'Astronomie)

Astronomers also extended the magnitude scale so that dimmer stars, visible only through telescopes, could be included in the magnitude system. For example, the dimmest stars visible through a pair of powerful binoculars have an apparent magnitude of about $m = +10$. Time-exposure photographs through telescopes reveal even dimmer stars. The Keck telescopes and the Hubble Space Telescope, among others, image stars nearly as dim as magnitude $m = +30$. Figure 11-2b illustrates the modern apparent magnitude scale. Similarly, apparent magnitudes from entire groups of stars, such as distant galaxies, can be measured.

Knowing, as we now do, that stars are at different distances from us, the apparent magnitudes do not directly reveal fundamental stellar properties. All other things being equal, the closer of two identical stars appears brighter to us (has a smaller apparent magnitude) than the farther star. We take the different distances into account with either of two measures: absolute magnitude and luminosity.

What might cause the closer of two identical stars to appear dimmer than the farther one?

11-3 Absolute magnitudes and luminosities do not depend on distance

To determine the total energy emitted by each star in space, astronomers need to take into account that different stars are at different distances from Earth. They do this by calculating the brightness each star would have if they were all at the same distance from here. Knowing how far away a star really is and how bright it appears (its apparent magnitude), we can calculate how bright it would be at any distance. **Absolute magnitude,** M, is the brightness each star would have at a distance of 10 pc. Unfortunately, absolute magnitudes have the same counterintuitive numeric scale as apparent magnitudes.

To understand the relationship between the apparent magnitude and the absolute magnitude, we need to know how the brightness of an object changes with distance. Suppose we observe two identical stars, one twice as far away as the other. How much dimmer will the farther one

Guided Discovery

Star Names

Many prominent stars received names from early Arabic or Greek stargazers. For example, Aldebaran is from the Arabic for "the follower" because it follows the prominent star cluster Pleiades around the sky. Betelgeuse, located in the shoulder of the constellation Orion, takes its name from the Arabic for "armpit of the central one." The name Sirius comes from the Greek for "sparkling," while Arcturus is from the Greek for "bear guard," because it is the brightest star in the constellation Boötes, the herdsman, next to the Great Bear (Ursa Major).

Unlike Betelgeuse or Aldebaran, most stars never received fanciful names. Indeed, most are so dim that they have been observed only through telescopes in the last two centuries. To study the stars yourself, you must be able to keep track of them. Astronomers have created a system of labels for all of them.

Up to the 24 most prominent stars in each constellation are assigned Greek lowercase letters:

α alpha	ι iota	ρ rho
β beta	κ kappa	σ sigma
γ gamma	λ lambda	τ tau
δ delta	μ mu	υ upsilon
ε epsilon	ν nu	φ phi
ζ zeta	ξ xi	χ chi
η eta	o omicron	ψ psi
θ theta	π pi	ω omega

A bright star's name is a Greek letter together with the Latin possessive (technically, genitive) form of the constellation name. The possessive forms of the names of the zodiac constellations are:

Constellation	Possessive
Aries	Arietis
Taurus	Tauri
Gemini	Geminorum
Cancer	Cancri
Leo	Leonis
Virgo	Virginis
Libra	Librae
Scorpius	Scorpii
Ophiuchus	Ophiuchi
Sagittarius	Sagittarii
Capricornus	Capricorni
Aquarius	Aquarii
Pisces	Piscium

In most cases, the brightest star in the constellation is α, the second brightest is β, the third is γ, and so on. For example, the brightest star in the constellation Leo is called α Leonis. This name is more informative than its common name, Regulus.

For the millions of stars extending beyond the 24 brightest in each constellation, a variety of catalogs list the stars numerically. For example, HDE 226868 is a bright blue star, the 226,868th star in the *Henry Draper Extended Catalogue* of stars.

Caveat emptor: The naming of stars is the responsibility of the International Astronomical Union, which does not ever recognize the commercial sale of star names. Companies that offer to name a star for a price do not have any official standing or recognition in the astronomical community, and neither do the names they promulgate. You can pay them to name a star for you, but the name is absolutely not official.

appear to be, as seen from Earth? As light moves outward from a source, it spreads out over increasingly larger areas of space and its brightness decreases. Thus, the farther away a source of light is, the dimmer it appears. The **inverse-square law** provides the rule for just how quickly the brightness of an object changes with distance.

Imagine light moving out from a star (Figure 11-3a). Start with a small square area of light that has moved out a distance $d = 1$. The light in that square has a certain brightness. When the same square of light has gone twice as far ($d = 2$), you can see on the figure that it has become 4 times larger. The light in each of the same-sized squares at $d = 2$ contains one-quarter of the photons that were in the single square at $d = 1$. Therefore, each small square at $d = 2$ is one-quarter as bright as the same-sized square at $d = 1$. Similarly, when the square of light has moved to $d = 3$, there are now nine squares the same size as the original, each one-ninth as bright as the original square at $d = 1$. For example, the Sun emits 3.83×10^{26} W (watts) of power (compared to 100 W for a bright home lightbulb). The Sun's power is spread over a wider area as it travels through space. When it passes Mercury, sunlight provides 9140 watts per square meter (W/m^2) of space. That same energy has spread out so much that by the time it passes us at 1 AU from the Sun, it provides only 1370 W/m^2. You see the same effect every day when you look at a car's headlights at different distances (Figure 11-3b).

This inverse-square law can be summarized mathematically as follows:

Apparent brightness decreases inversely with the square of the distance between the source and the observer.

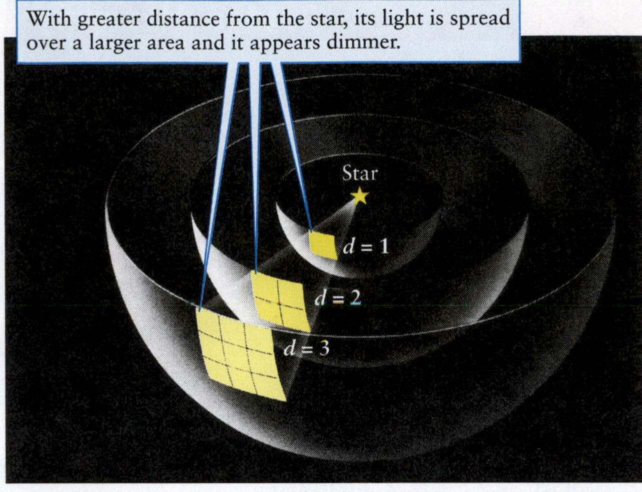

a

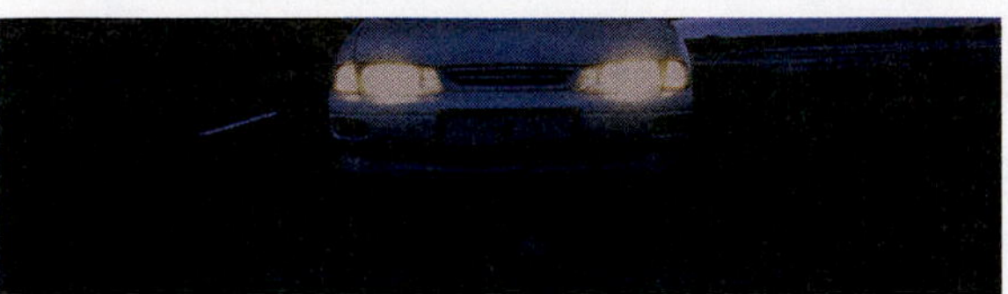

30 m away

20 m away

10 m away

R I V U X G

b

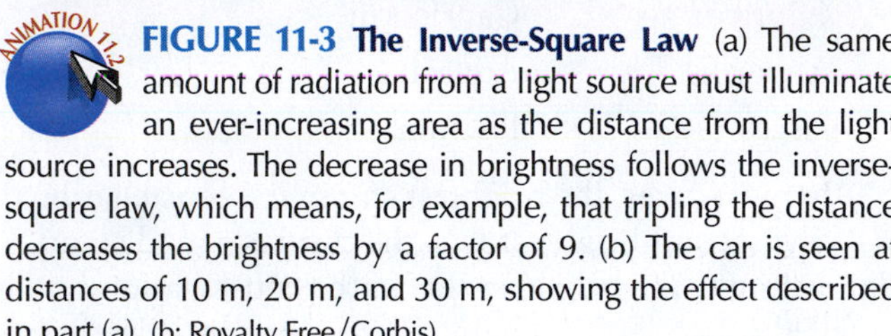

FIGURE 11-3 The Inverse-Square Law (a) The same amount of radiation from a light source must illuminate an ever-increasing area as the distance from the light source increases. The decrease in brightness follows the inverse-square law, which means, for example, that tripling the distance decreases the brightness by a factor of 9. (b) The car is seen at distances of 10 m, 20 m, and 30 m, showing the effect described in part (a). (b: Royalty Free/Corbis)

Returning to our two identical stars, one twice as far away as the other, the farther one's brightness decreases to $(½)^2 = ¼$ the brightness of its closer twin by the time the starlight gets to us.

Absolute magnitudes are usually determined by calculations based on apparent magnitudes. Of course, we cannot just move a star to 10 pc distance and then remeasure its apparent magnitude. However, we can calculate the absolute magnitude of a nearby star: We first measure its apparent magnitude and then we find its distance by measuring its parallax angle. Combining these numbers, as shown in *An Astronomer's Toolbox 11-3,* gives the star's absolute magnitude.

An Astronomer's Toolbox 11-3

The Distance-Magnitude Relationship

The closer a star, the brighter it appears. The inverse-square law leads to a simple equation for absolute magnitude, M. Suppose a star's apparent magnitude is m and its distance from Earth is d (measured in parsecs). Then

$$M = m - 5 \log (d/10)$$

where log stands for the base-10 logarithm. This distance-magnitude relation can be rewritten as

$$m - M = 5 \log d - 5$$

Example: Consider Proxima Centauri, the nearest star to Earth (other than the Sun). By measuring its parallax angle, we know this star is at a distance from Earth of $d = 1.3$ pc. Its apparent magnitude is $m = +11.1$. Therefore, its absolute magnitude is

$$M = 11.1 - 5 \log (1.3/10) = 11.1 - (-4.4) = +15.5$$

Compare! The Sun is an average star with M = +4.8, so Proxima Centauri is an intrinsically dim star. If you know any two of d, m, and M, you can calculate the third variable. For example, if we know a star's absolute and apparent magnitudes, the equation can be used to determine its distance.

Try these questions: A star is observed to have an apparent magnitude $m = +0.268$ and an absolute magnitude M = −0.01. How far from Earth is the star in parsecs and light-years? A star is observed to have an apparent magnitude $m = +1.17$ and is at a distance of 25.1 ly from Earth. What is its absolute magnitude? *(Remember to convert to parsecs first.)* A star is at a distance from Earth of 8.61 ly and has an absolute magnitude of $m = +1.45$. What is its apparent magnitude? You can check your results and identify the stars by referring to Appendix Table E-5 and Table E-6.

(Answers appear at the end of the book.)

Knowing the Sun's true distance and its apparent magnitude, we can use the inverse-square law to determine how bright it would appear at 10 pc. At that distance, it would have an apparent magnitude of $m = +4.8$. Therefore, the absolute magnitude of the Sun is $M = +4.8$.

2 Because absolute magnitudes tell astronomers how bright stars are compared with one another, this information can be used to evaluate models of stellar evolution, which we discuss in the next few chapters. Absolute magnitudes range from roughly $M = -10$ for the brightest stars to $M = +17$ for the dimmest. Although absolute magnitudes give us comparisons between the energy outputs of stars, we also need to know the total energy they release. The total amount of electromagnetic power (energy emitted each second) is called a star's **luminosity.** We saw in Chapter 4 that the greater the luminosity, the brighter the object. Therefore, the smaller or more negative a star's absolute magnitude, the greater its luminosity. For convenience, stellar luminosities are expressed in multiples of the Sun's luminosity, denoted $L_{\odot}$. As we saw earlier, this value is about 3.83×10^{26} W. The intrinsically brightest stars ($M = -10$) have luminosities of $10^6\ L_{\odot}$. In other words, each of these stars has the energy output of a million Suns. The dimmest stars ($M = +17$) have luminosities of $10^{-5}\ L_{\odot}$. We will provide both luminosity and absolute magnitude data about stars in the chapters that follow.

Why can the total energy output of a star be specified equally by its absolute magnitude and its luminosity?

THE TEMPERATURES OF STARS

Armed with knowledge of stellar luminosities, astronomers early in the twentieth century searched for ways to use this information to determine why stars shine. A major step in this process was to find a graph on which physically similar stars are located near each other, because such groupings often provide insight into how objects work. Astronomers found this graph when they plotted either the luminosity or, equivalently, the absolute magnitude versus the stars' surface temperatures. The resulting graph led to our models of both stellar activity and evolution. To find the surface temperatures, we start with a fact that is easily overlooked: Stars are not all the same color.

11-4 A star's color reveals its surface temperature

3 If you look carefully at the stars with your naked eyes, you can see that they have different colors, most commonly red, orange, yellow, white, and blue-white (see the figure that opens this chapter and Figures 11-2 and 11-4a). As we saw for the Sun, stars behave very nearly like blackbodies. Therefore, their colors are determined by their surface temperatures. Recall from Figure 4-2 that different blackbody curves are associated with blackbodies of different temperatures. These curves differ in both their intensities and the locations of their peaks, as described by Wien's law.

Three typical blackbody curves are presented in Figure 11-4. The intensity of light from a relatively cool star peaks at long wavelengths, so the star looks red. The intensity of light from a very hot star peaks at shorter wavelengths, making it look blue. The maximum intensity of a star of intermediate temperature, such as the Sun, is found near the middle of the visible spectrum.

To accurately determine the peaks of stars' blackbody spectra and, hence, their surface temperatures, astronomers need to know which blackbody curve most accurately describes each star. **Photometry** provides this information. In this process, a telescope collects starlight that is then passed through one of a set of colored filters and recorded on a CCD. At least three photometric images are taken of each star through filters that pass different wavelengths of light. The intensity of a star's image is different at different wavelengths. By plotting the three data points and finding which unique blackbody curve they fit on, we can determine that star's surface temperature.

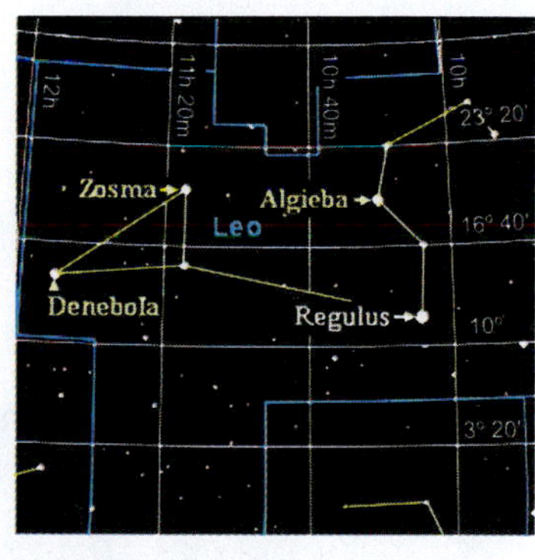

If a star's surface is very hot, perhaps 10,000 K (for example, the hottest star in Figure 11-4b), its radiation is skewed toward the ultraviolet, which makes the star bright as detected through an ultraviolet-passing filter, dimmer through a blue filter, and dimmer still through a yellow filter. Regulus, in Leo, is such a star. If a star is cool, say, 3000 K (for example, the coolest star in Figure 11-4b), its radiation peaks at long visible or even infrared wavelengths, making the star bright through a red filter, dimmer through a yellow filter, and dimmer still through a blue filter. Betelgeuse (Figure 11-2a) and Aldebaran are examples of such cool stars. Humans are blackbody radiators with peak temperatures around 300 K. Our peak is in the infrared.

11-5 A star's spectrum also reveals its surface temperature

Another way to determine a star's surface temperature is from a study of its spectrum, a technique called **stellar spectroscopy** (see Chapter 4). Recall that spectral lines result from the absorption and scattering of starlight by gases in the star's atmosphere, in interstellar space, and in Earth's atmosphere. Astronomers begin by taking the spectrum of a star and then identifying and discarding spectral lines due to interstellar gas and Earth's atmosphere. What

a

FIGURE 11-4 Temperature and Color (a) This beautiful Hubble Space Telescope image shows the variety of colors of stars. (b) These diagrams show the relationship between the color of a star and its surface temperature. The intensity of light emitted by three stars is plotted against wavelength (compare with Figure 4-2). The range of visible wavelengths is indicated. The location of the peak of a each star's intensity curve, relative to the visible-light band, determines the apparent color of its visible light. The insets show stars of about these surface temperatures. Ultraviolet (uv) extends to 10 nm. See Figure 3-6 for more on wavelengths of the spectrum. (a: Hubble Heritage Team/AURA/STScI/NASA; left inset: Andrea Dupree/Harvard-Smithsonian CFA, Ronald Gilliland/STScI, NASA, and ESA; center inset: NSO/AURA/NSF; right inset: Till Credner, Allthesky.com)

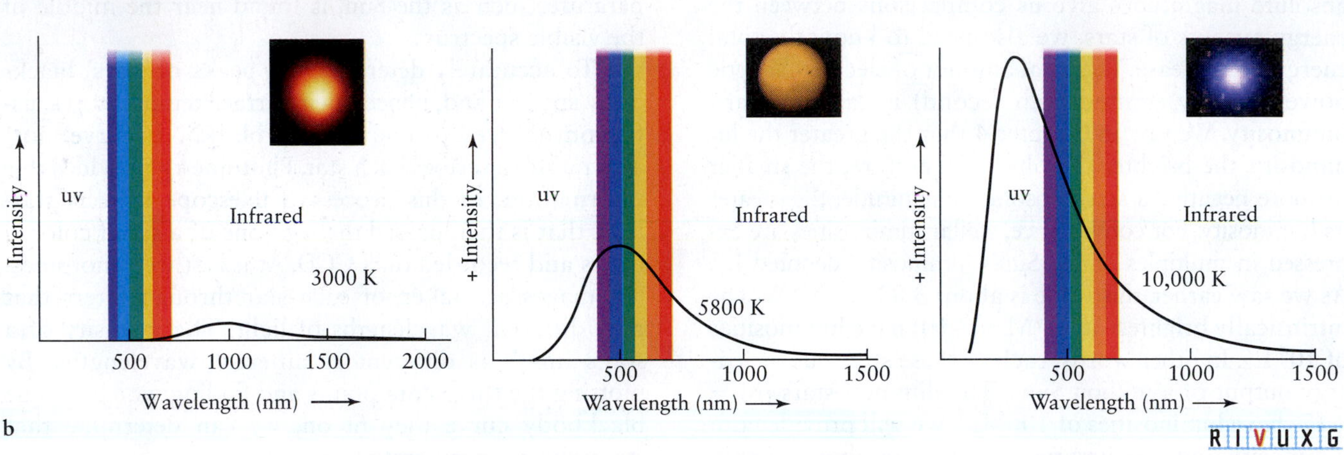

b

A star of which temperature has the most intense blue radiation: 3000 K, 5200 K, or 6800 K?

is left are spectral lines created in the star's atmosphere. These lines provide us with knowledge of the chemical compositions of the outer layers of stars.

At first glance, there seems to be a bewildering variety of stellar spectra, some of which are shown in Figure 11-5. Some stellar spectra show prominent absorption lines of hydrogen. Others exhibit many absorption lines of calcium and iron. Still others are dominated by broad absorption lines created by molecules, such as titanium oxide.

Fortunately, these spectra all hold clues about the nature of stars and, in particular, their surface temperatures. Consider the abundant hydrogen gas found in nearly every star's atmosphere. Although hydrogen accounts for about three-quarters of the mass of a typical star, strong (meaning dark) hydrogen absorption lines do not show up in every star's spectrum. The strength of the absorption lines depends on the star's temperature. In the late 1920s, astronomer Cecilia Payne and physicist Meghnad Saha succeeded in explaining why a star's visible hydrogen spectrum is affected by its surface temperature.

Niels Bohr's model of the hydrogen atom (recall Figure 4-12) shows why. As Bohr proposed in the early 1900s, the visible hydrogen lines, called *Balmer lines,* are produced when photons excite electrons initially in the second, $n = 2$, energy level of hydrogen to a higher energy level. Recall from Section 4-6 that these are labeled H_α, H_β, H_γ, and so on. At 10,000 K, most hydrogen electrons absorb enough energy from photons to normally reside in the $n = 2$ level. When these electrons gain further energy, they move to higher energy levels. Therefore, a star with that surface temperature produces the strongest Balmer lines.

Why do hotter or cooler stars produce weaker (less dark) Balmer absorption lines? Suppose first that the star is much hotter than 10,000 K. High-energy photons that stream through the photosphere completely strip away (ionize) electrons from most of the hydrogen atoms there. Because an ionized hydrogen atom cannot produce spectral lines, a very hot star has very dim hydrogen Balmer lines, even though it contains great quantities of hydrogen. Conversely, if a star is much cooler than 10,000 K, most of the photons that pass through its photosphere possess too little energy to keep many electrons there in the $n = 2$ (first excited state) of the hydrogen atoms, or to boost electrons in the $n = 2$ state to higher energy states. Therefore, these stars also produce dim Balmer lines. To produce strong Balmer lines, a star must be hot enough to excite electrons out of the $n = 2$ state but not hot enough to ionize a significant fraction of the atoms.

As you can see in Figure 11-5, at temperatures much cooler than 10,000 K, the spectral lines of other

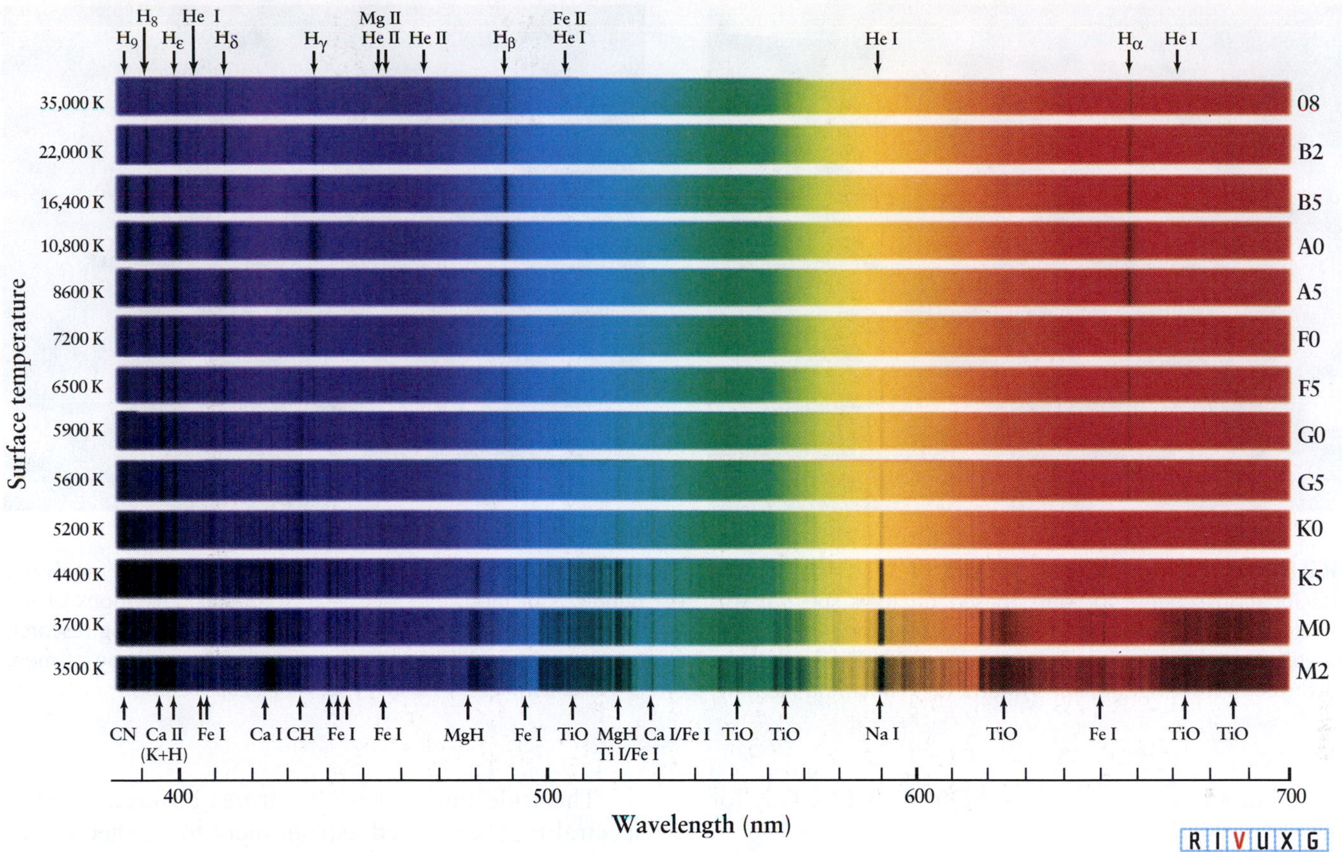

FIGURE 11-5 The Spectra of Stars with Different Surface Temperatures The corresponding spectral types are indicated on the right side of each spectrum. (Note that stars of each spectral type have a range of temperature.) The hydrogen Balmer lines are strongest in stars with surface temperatures of about 10,000 K (called A-type stars). Cooler stars (G- and K-type stars) exhibit numerous atomic lines caused by various elements, indicating temperatures from 4000 to 6000 K. Several of the broad, dark bands in the spectrum of the coolest stars (M-type stars) are caused by titanium oxide (TiO) molecules, which can exist only if the temperature is below about 3700 K. Recall from Section 4-5 that the Roman numeral I after a chemical symbol means that the absorption line is caused by a neutral atom; a numeral II means that the absorption is caused by atoms that have each lost one electron. (R. Bell, University of Maryland, and M. Briley, University of Wisconsin at Oshkosh)

elements dominate a star's spectrum. For example, the spectral line of neutral sodium, Na I, while dim at around 10,000 K, is quite strong at around 3700 K because photons have enough energy to excite sodium atoms without tearing away its electrons. By surveying the relative strengths of a variety of absorption lines, astronomers can determine a star's surface temperature to high precision.

11-6 Stars are classified by their spectra

We have seen that astronomers can determine a star's surface temperature from either the peak of its blackbody or the strength of its various spectral lines. To make use of the diverse stellar spectra, astronomers since the middle of the nineteenth century have grouped similar spectra into classes, or **spectral types.** According to one classification scheme popular in the late 1800s, a star was assigned a letter between A and P, depending on the strength of the Balmer hydrogen lines in the star's spectrum. It was assumed that the strength of these lines was uniquely related to the star's surface temperature, but as we have seen in the preceding section, this belief is unjustified.

In the early 1900s, William Pickering and Williamina Fleming, followed by Annie Jump Cannon, and their colleagues at Harvard Observatory (Figure 11-6) set up the spectral classification scheme we use today. Many of the early A through P categories were dropped because the Balmer lines in a star's spectrum can be weak whether the star is very cool or very hot. The remaining Balmer-based spectral types were thus reordered by stellar surface temperature into the **OBAFGKM sequence.** This sequence is most easily learned with

> Why do very hot stars have weak hydrogen lines, even though they are composed primarily of hydrogen?

a

b

FIGURE 11-6 Classifying the Spectra of Stars The modern classification scheme for stars, based on their spectra, was developed at the Harvard College Observatory in the late nineteenth century. Female astronomers, initially led by Edward C. Pickering (not shown) and (a) Williamina Fleming, standing, and then by (b) Annie Jump Cannon, analyzed hundreds of thousands of spectra. Social conventions of the time prevented most female astronomers from using research telescopes or receiving salaries comparable to those of men. (a: Harvard College Observatory; b: © Bettmann/Corbis)

the aid of a mnemonic, such as "Oh, Be A Fine Guy (or Girl), Kiss Me!"

The hottest stars have surface temperatures of 30,000 to 50,000 K and are classified as O type; their spectra are dominated by He II (singly ionized helium) and Si IV (triply ionized silicon). The coolest stars have surface temperatures of 2500 to 3000 K and are classified as M type. Table 11-1 includes representative examples of each spectral type.

The wide range of temperatures covered by each spectral type prompted astronomers to further subdivide the OBAFGKM temperature sequence. Each spectral type is now broken up into 10 temperature subranges. These 10 finer steps are indicated by adding an integer from 0 (hottest) through 9 (coolest). Thus, an A8 star is hotter than an A9 star, which is hotter than an F0 star, which is hotter than an F1 star, and so on. Test yourself on this: What class of star is just slightly

TABLE 11-1 The Spectral Sequence

Spectral class	Color	Temperature (K)	Spectral lines	Examples
O	Blue-violet	30,000–50,000	Ionized atoms, especially helium	Naos (ζ Puppis), Mintaka (δ Orionis)
B	Blue-white	11,000–30,000	Neutral helium, some hydrogen	Spica (α Virginis), Rigel (β Orionis)
A	White	7500–11,000	Strong hydrogen, some ionized metals	Sirius (α Canis Majoris), Vega (α Lyrae)
F	Yellow-white	5900–7500	Hydrogen and ionized metals, such as calcium and iron	Canopus (α Carinae), Procyon (α Canis Minoris)
G	Yellow	5200–5900	Both neutral and ionized metals, especially ionized calcium	Sun, Capella (α Aurigae)
K	Orange	3900–5200	Neutral metals	Arcturus (α Boötis), Aldebaran (α Tauri)
M	Red-orange	2500–3900	Strong titanium oxide and some neutral calcium	Antares (α Scorpii), Betelgeuse (α Orionis)

cooler than a K9? (The caption for Figure 11-7 has the answer.) The Sun, whose spectrum is dominated by singly ionized metals (especially Fe II and Ca II), is a G2 star.

Insight Into Science

Tolerate Idiosyncrasies In astronomy, the word *metal* applies to all elements other than hydrogen or helium. In chemistry, elements are classified as metals or nonmetals according to their physical and chemical properties. To a chemist, sodium and iron are metals, whereas carbon and oxygen are not.

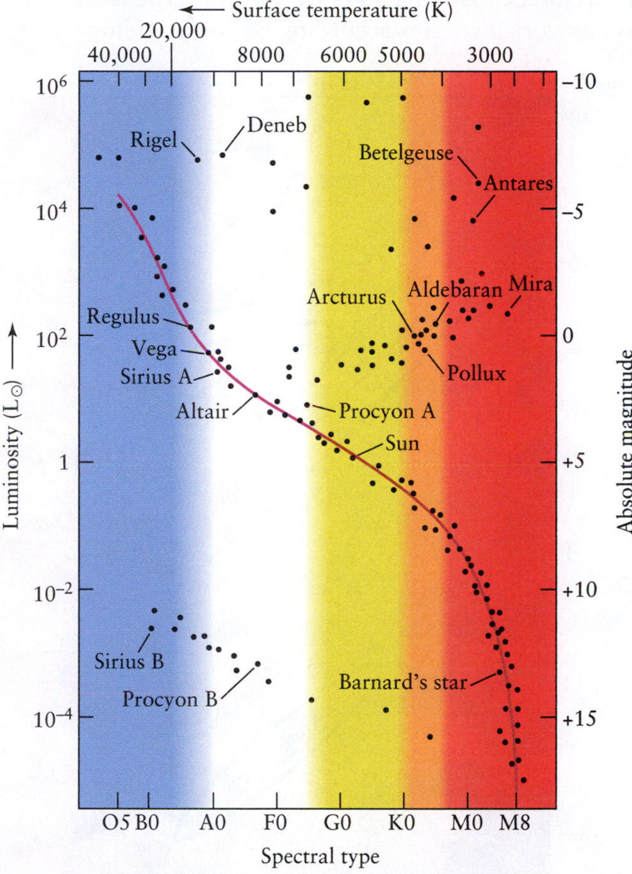

FIGURE 11-7 A Hertzsprung-Russell Diagram On an H-R diagram, the luminosities of stars are plotted against their spectral types. Each dot on this graph represents a star whose luminosity and spectral type have been determined. Some well-known stars are identified. The data points are grouped in just a few regions of the diagram, revealing that luminosity and spectral type are correlated: Main-sequence stars fall along the red curve, giants are to the right, supergiants are on the top, and white dwarfs are below the main sequence. The absolute magnitudes and surface temperatures are listed at the right and top of the graph, respectively. These are sometimes used on H-R diagrams instead of luminosities and spectral types. (*Answer to text question:* An M0 star is the next coolest after a K9.)

TYPES OF STARS

Which star is hottest: F5, B6, or M3?

Keeping in mind that patterns of objects often reveal information about them, early twentieth-century astronomers began searching for relationships between different properties of stars. The relationship between luminosities and surface temperatures turned out to be a key to understanding types of stars. As often happens in such scientific quests, the significance of this relationship was discovered nearly simultaneously by two independent researchers.

11-7 The Hertzsprung-Russell diagram identifies distinct groups of stars

Around 1911, the Danish astronomer Ejnar Hertzsprung noticed that patterns emerge when the luminosities of stars (or their equivalent absolute magnitudes) are plotted against their surface temperatures (or their equivalent spectral types). Luminosity and absolute magnitude indicate the total energy emitted by a star or other body. These two terms each provide quantitative measurements in answer to the question: "How bright is that star?" Within 2 years, the American astronomer Henry Norris Russell independently discovered the same result. Graphs of stellar luminosity or absolute magnitude against surface temperature or spectral type are now known as **Hertzsprung-Russell diagrams, or H-R diagrams.**

The H-R diagram is valuable because it shows that stars do not have random surface temperatures and luminosities; the two factors are correlated. Figure 11-7 is a typical H-R diagram. Each dot represents a star whose luminosity and spectral type have been determined. The surface temperatures are plotted along the top of this figure and the absolute magnitudes along the right side. You can therefore see the equivalence of the temperature and spectral type and the equivalence of the luminosity and absolute magnitude.

Bright stars are near the top of the diagram; dim stars are near the bottom. Contrary to intuition, hot (O and B) stars are toward the left side and cool (M) stars are toward the right. Hertzsprung and Russell made this choice because of the standard sequence OBAFGKM.

Insight Into Science

From Patterns to Models Scientists look at patterns of behavior in related objects as valuable clues to underlying properties and their causes. Guided by this data, they then create mathematical models, which are used to make fresh predictions. For example, astronomers analyze the relationship between the luminosities and surface temperatures of stars as displayed on an H-R diagram to gain insight into the internal activities of stars.

The band of stars in Figure 11-7 stretching diagonally across the H-R diagram and on which a red curve is superimposed represents most of the stars we see in the nighttime sky. Called the **main sequence,** it extends from the hot, bright, bluish stars in the upper left corner of the diagram down to the cool dim stars (called *red dwarfs*) in the lower right corner. Each star on this band is a **main-sequence star.** Just over 91% of the stars surrounding the solar system fall into this category.

Observations reveal that the number of main-sequence stars decreases with increasing surface temperature. Therefore, along the main sequence, the cooler M, K, and G stars are the most common ones and the hot O stars are the rarest. The Sun (spectral type G2, absolute magnitude +4.8) is a main-sequence star.

4 To the right of the main sequence on the H-R diagram is a second major grouping of stars. These stars are bright but cool. From the Stefan-Boltzmann law (see *An Astronomer's Toolbox 4-1*), we know that a cool object radiates much less light from each unit of surface area than does a hot object. To be so luminous, these cool stars must therefore be huge compared to main-sequence stars of the same temperature, so they are called **giant stars.**

Giants are typically 10 to 100 times the radius of the Sun and have surface temperatures between 2000 and 20,000 K. The cooler members of this class of stars (those with surface temperatures between 2000 and 4500 K) are often called **red giants** because they appear reddish in the nighttime sky. Aldebaran in the constellation Taurus and Arcturus in Boötes are examples of red giants that you can easily see with the naked eye.

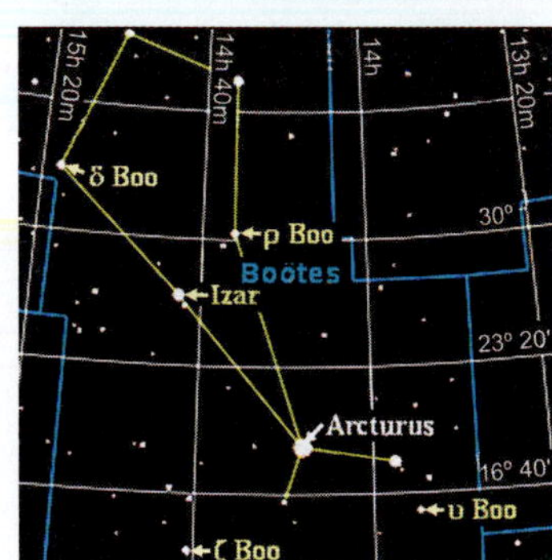

A few rare stars are considerably bigger and more luminous than typical giants. Located along the top of the H-R diagram, these superluminous stars are appropriately called **supergiants.** They can extend in radius up to about 1000 $R_\odot$. Betelgeuse in Orion and Antares in Scorpius are two examples that are visible in the nighttime sky. Together, giants and supergiants comprise less than 1% of the stars in our vicinity.

The remaining 8% of stars in our neighborhood of space fall in a final grouping toward the lower left and bottom of the H-R diagram. As their placement on the diagram shows, these stars are hot, dim, and tiny compared to the Sun. Called **white dwarfs,** we will see that they are actually remnants of stars. White dwarfs are roughly the same size as Earth, and because of their low luminosities, they can be seen only with the aid of a telescope.

For a given spectral class (that is, temperature), a star of which luminosity class—main sequence, giant, or supergiant—is brightest?

5 These results are summarized in Figure 11-8, where the dashed lines indicate the radii of stars. Notice that most main-sequence stars are roughly the same size as the Sun. It is important to keep in mind that, although the H-R diagram is invaluable in helping astronomers *organize* stars by their physical properties (temperature and absolute magnitude), the diagrams do not *explain* the physical mechanisms that produce these characteristics. As we will see shortly, the locations of stars on the H-R diagram provided crucial clues that astrophysicists used in developing and testing their theories of how stars shine.

11-8 Luminosity classes set the stage for understanding stellar evolution

We have seen that a star's surface temperature largely determines which lines are prominent in its spectrum. Therefore, classifying stars by spectral type is the same as categorizing them according to surface temperature.

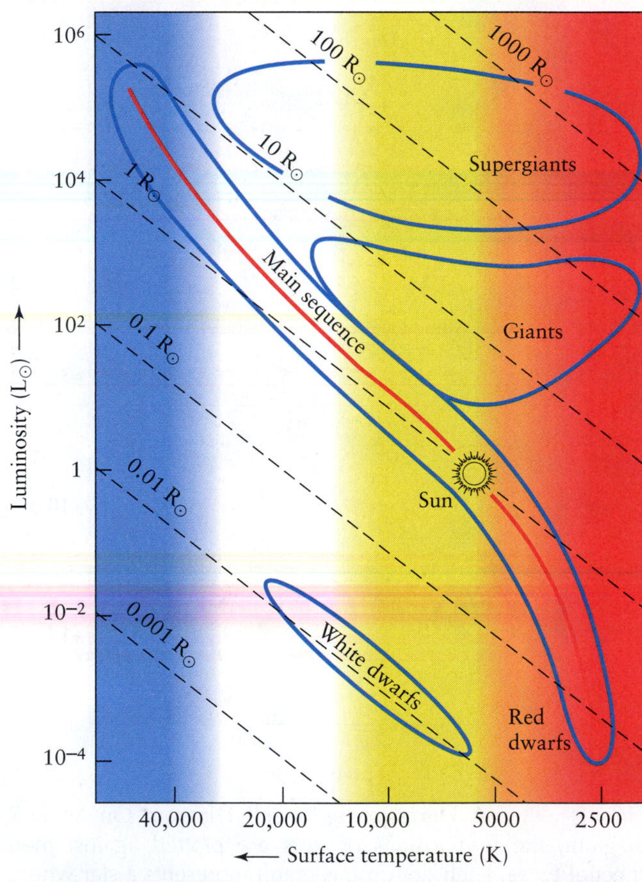

FIGURE 11-8 The Types of Stars and Their Sizes On this H-R diagram, stellar luminosities are plotted against the surface temperatures of stars. The *dashed diagonal lines* indicate stellar radii. For stars of the same radius, hotter stars (corresponding to moving from right to left on the H-R diagram) glow more intensely and are more luminous (corresponding to moving upward on the diagram) than cooler stars. While individual stars are not plotted, we show the regions of the diagram in which main-sequence, giant, supergiant, and white dwarf stars are found. Note that the Sun is intermediate in luminosity, surface temperature, and radius; it is very much a middle-of-the-road star.

However, as you can see in Figure 11-8, stars of the same surface temperature can have different luminosities. For example, a star with a surface temperature of 5800 K could be a white dwarf, a main-sequence star, a giant, or a supergiant.

Astronomers have found subcategories in which to place stars by studying their spectral absorption lines in detail. This refined classification is possible primarily because absorption lines are affected by the density and pressure of the gas in a star's atmosphere, both of which depend on whether the star is a white dwarf, main-sequence star, giant, or supergiant (Figure 11-9).

Based upon the differences in stellar spectra, W. W. Morgan and P. C. Keenan of the Yerkes Observatory developed a system of **luminosity classes** in the 1930s. Luminosity classes Ia and Ib encompass all of the supergiants; giants of various luminosity are assigned classes II, III, and IV; and main-sequence stars are luminosity class V. Figure 11-10 summarizes these results. We will see in Chapter 12 that the different luminosity classes correspond to different stages of stellar evolution. Note on Figure 11-10 that white dwarfs are not given a luminosity class because they are stellar remnants that do not create energy by fusion like the Sun and other stars in the five luminosity classes. We will study white dwarfs further in Chapter 13.

Plotting luminosity classes on the H-R diagram (see Figure 11-10) provides a useful subdivision of star types. In fact, astronomers commonly describe a star by both its spectral type and its luminosity class. The Sun, for example, is a G2 V star. This notation supplies a great deal of information about the star, because its spectral type is correlated with its surface temperature. Thus, an astronomer knows immediately that a G2 V star is a main-sequence star with a luminosity of 1 $L_\odot$ and a surface temperature of around 5800 K. Similarly, knowing that Aldebaran is a K5 III star tells an astronomer that it is a red giant with a luminosity of around 370 $L_\odot$ and a surface temperature of about 4400 K (see Figure 11-7).

The H-R diagram also provides us with enough data to determine a star's radius. Recall from *An Astronomer's Toolbox 4-1* that there is a relationship between a star's luminosity, L, temperature, T, and radius, r: $L = \sigma T^4 4\pi r^2$, where σ (sigma) is a constant. Knowing two properties, the luminosity and temperature, we can solve this equation for the third, namely the radius. Lines of constant stellar radii are drawn on Figure 11-8.

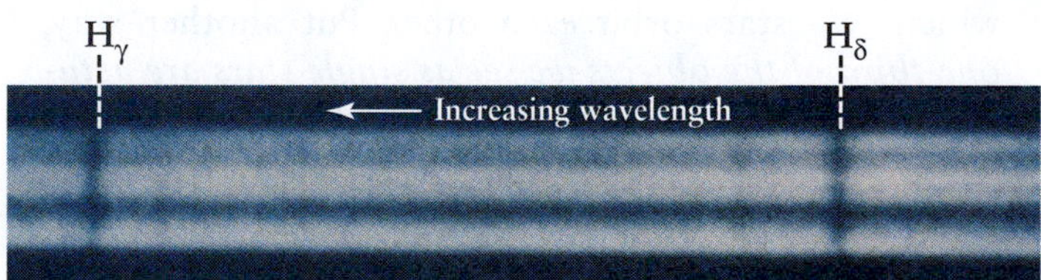

a A supergiant star has a low-density, low-pressure atmosphere: its spectrum has narrow absorption lines

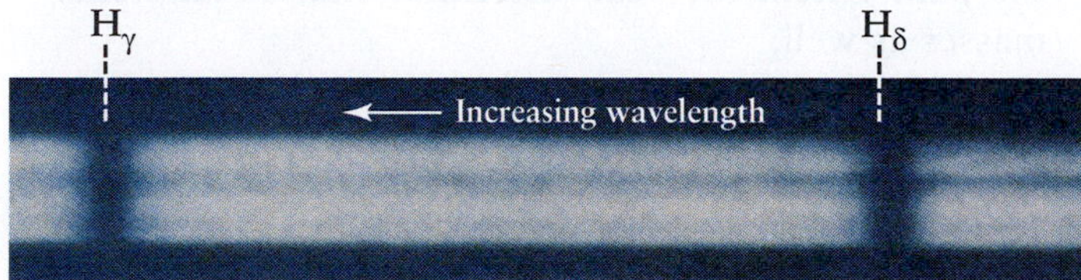

b A main-sequence star has a denser, higher-pressure atmosphere: its spectrum has broad absorption lines

R I V U X G

FIGURE 11-9 Stellar Size and Spectra These spectra are from two stars of the same spectral type (B8) and, hence, the same surface temperature (13,400 K) but different radii and luminosities: (a) the B8 supergiant Rigel (luminosity 58,000 $L_\odot$) in Orion and (b) the B8 main-sequence star Algol (luminosity 100 $L_\odot$) in Perseus. (From W. W. Morgan, P. C. Keenan, and E. Kellman, *An Atlas of Stellar Spectra*)

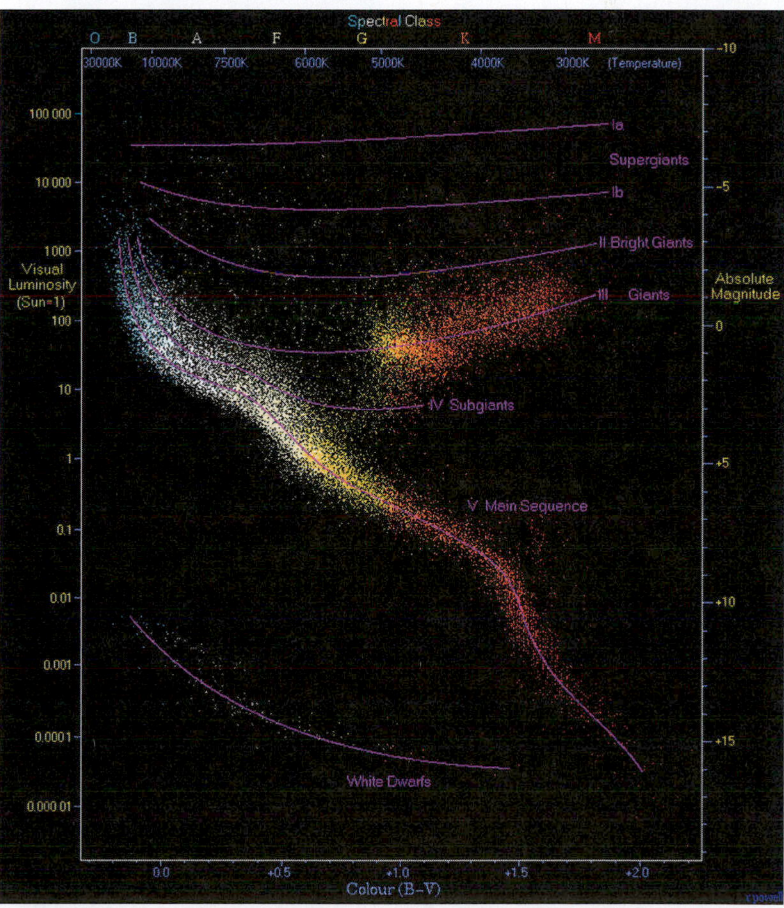

FIGURE 11-10 Luminosity Classes Dividing the H-R diagram into regions, called *luminosity classes,* permits finer distinctions between giants and supergiants. Luminosity classes Ia and Ib encompass the supergiants. Luminosity classes II, III, and IV indicate giants of different brightness. Luminosity class V indicates main-sequence stars. White dwarfs do not have their own luminosity class. (Richard Powell/atlasoftheuniverse.com)

11-9 A star's spectral type and luminosity class provide a second distance-measuring technique

A star's spectral type and luminosity class, combined with information from the H-R diagram, give astronomers information necessary for determining the distances to stars millions of light-years away, far beyond the maximum distance that can be measured using stellar parallax. The process works like this:

1. Astronomers observe a distant star's apparent magnitude and spectrum.

2. From its spectrum, they determine what spectral class the star belongs to (or, equivalently, its surface temperature).

3. The spectrum also reveals the star's luminosity class, that is, whether it is a supergiant, giant, or main-sequence star. Combining the temperature and the luminosity class determines the star's location on the H-R diagram (see Figure 11-10). From this position, the star's approximate absolute magnitude can be read off the diagram.

4. Using the apparent and absolute magnitudes in the distance-magnitude relationship (see *An Astronomer's Toolbox 11-3*), the star's distance can then be calculated.

A main-sequence star of which spectral type—F5, A8, or K0—is largest?

Consider, for example, the star Regulus in the constellation Leo (see the margin chart for Leo on page 357). Regulus's spectrum reveals it to be a B7 V star (a hot, blue, main-sequence star). Placing it on the H-R diagram (see Figure 11-7), we see that its luminosity is $140\ L_{\odot}$ and its absolute magnitude is −0.52. Given the star's apparent magnitude, we can use the distance-magnitude relationship to determine its distance from Earth. Because both the spectral type and luminosity class are obtained spectroscopically, this method of determining distances is called **spectroscopic parallax.** (This name is misleading, however, because no parallax angle is involved.)

Spectroscopic parallax is limited in accuracy because of the spread of stars in each luminosity class—the stars in each class do not fall on a single, narrow line (hence the use of the word *approximate* in the previous numbered process, above). It is also limited in that spectra of distant stars become increasingly hard to determine. As a result, errors of 10% in distance are common using spectroscopic parallax. Accepting such errors, the power of this method is that it can be used for stars at much greater distances than those determined by stellar parallax. Indeed, this method even provides distances to stars in other galaxies tens of millions of light-years away.

What observations must be made to determine a star's distance using spectroscopic parallax?

STELLAR MASSES

The final clue that astronomers needed to understand how stars work is their mass—how much matter they have. If the masses were related to the stellar luminosities and locations on the H-R diagram (and they are), then astronomers would have enough information to start making scientific models of why stars shine. The mass of each star, it turns out, determines the amount of energy it has available to manufacture light and other electromagnetic radiation. Models of how stars generate that radiation, developed in the twentieth century and still being refined today, provide us with further insights into how nature works. With the knowledge gained about thermonuclear fusion in stars, physicists are researching how to control this fusion to generate electricity.

The problem with finding stellar masses is that it cannot be done directly by examining isolated stars. The mass of a star can be determined only by its gravitational effects on other bodies, using Newton's law of gravity. When Newton derived the formula for Kepler's third law, he discovered that for any two objects in orbit, such as a planet around the Sun, a moon around a planet, or two stars around each other, the period of their orbits is related to the sum of their masses (see *An Astronomer's Toolbox 11-4*). You have seen in Section 2-5 that for objects orbiting the Sun, we can ignore the mass of the smaller body, thereby getting Kepler's original third law, as stated in Chapter 2. For stars orbiting each other, we use Newton's full equation, which includes the masses of both bodies.

Fortunately for astronomers, one-third of the stars 6
near our solar system are members of star systems in which two stars orbit each other. Put another way, *one-third of the objects we see as single stars are actually pairs of stars in orbit around each other.* They are so distant that we cannot resolve the individual stars in the pairs and so they appear to us as single objects. However, using telescopes to observe the periods of the orbits and the distances between stars, astronomers can determine the sum of the stellar masses of the pair. Often, they can determine individual stellar masses as well.

11-10 Binary stars provide information about stellar masses

A pair of stars located at nearly the same position in the night sky is called a *double star.* Between 1782 and 1838, William Herschel, his sister Caroline, and his son John catalogued thousands of them. Some double stars are not physically close together and do not orbit each other. These **optical doubles** (sometimes called *apparent binaries*) just happen to lie in the same direction, as seen from Earth. For example, δ Herculis, visible with the

An Astronomer's Toolbox 11-4

Kepler's Third Law and Stellar Masses

The same gravitational force that holds Earth in orbit around the Sun also keeps pairs of stars in orbit around each other. For any pair of orbiting bodies, the orbits are ellipses, and Kepler's third law can be written as

$$M_1 + M_2 = a^3/P^2$$

Here, M_1 and M_2 are the two masses (expressed in solar masses), a is the length of the semimajor axis of the ellipse (in astronomical units), and P is the orbital period (in years). Note that a is also the average separation between the two bodies. Thus, the sum of the masses can be found once we know the orbital period and average separation.

Example: Suppose two stars make up a binary system, with one star's elliptical orbit having a semimajor axis, a, of 4 AU. We find that this star takes 2.5 years to complete one orbit. Then the sum of the stars' masses is

$$M_1 + M_2 = (4)^3/(2.5)^2 = 10.2\ M_\odot$$

The total mass of the system is 10.2 $M_\odot$.

Try these questions: Why can Kepler's third law for a planet orbiting the Sun be written as $P^2 = a^3$? (*Hint:* For the Sun and a planet, the combined mass is extremely close to the Sun's mass alone.) Two identical stars are observed to orbit each other with a period of 2 years at a separation of 2.32 AU. What are their combined masses? How long would it take a 1-$M_\odot$ and a ½-$M_\odot$ star to orbit each other if their separation were 20 AU?

(Answers appear at the end of the book.)

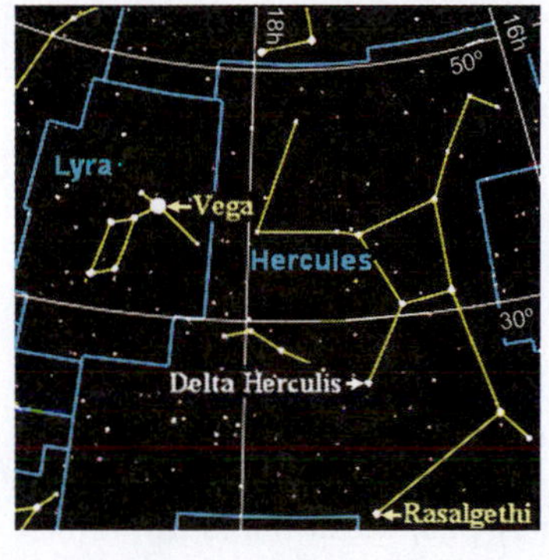

naked eye, is an optical double with a dim star. Through a telescope these stars appear close together, but that is an optical illusion.

Other double stars are true **binary stars**—pairs in which two stars orbit each other. In the case of **visual binaries**, both stars can be seen, using a telescope if necessary (Figure 11-11). Astronomers can plot the orbit of one star around the other in a visual binary.

To see how we determine the masses of stars, consider a visual binary. Newton rewrote Kepler's third law as a relation between the masses of the stars (in solar masses), their orbital period around each other in Earth years (this period is the same for both stars, so it does not matter which star we assume orbits the other), and the average separation between the stars in AU (which equals the semimajor axis of the elliptical orbit):

$$\text{The sum of the masses} = \frac{\text{The cube of the semimajor axis}}{\text{The square of the period}}$$

Thus, by observing the separation between a pair of stars and how long one of them takes to complete its orbit, we can calculate the sum of the masses of the two stars, presented mathematically in *An Astronomer's Toolbox 11-4*.

In some cases, we can also determine the individual masses of the stars. To do this, we need to know the distances of the stars from the center of mass of the pair. As with the Earth-Moon system discussed in Section 6-8, both stars in a binary system move in elliptical orbits around a common point between them. This point, called the **center of mass**, is determined by the masses of the stars and can be understood by analogy to a pair of connected masses whirling on a frictionless surface (see Figure 6-29). One point somewhere between their centers moves in a straight line. This point is the center of mass of the system of masses. Just as the two masses orbit around their center of mass, the two stars in a binary system orbit around their center of mass under the influence of their mutual gravitational attraction (Figure 11-12). The more massive star is closer to the center of mass, just as a heavier child must be closer to the pivot point or fulcrum of a seesaw for it to balance (Figure 11-12b).

We can locate the center of mass of a visual binary by plotting the separate orbits of the two stars, as in Figure 11-11, using the background stars as reference points. The center of mass lies at the common focus of the two elliptical orbits. To get the relative sizes of the orbits, however, we must be able to determine the plane of orbit of the two stars. This calculation is not always possible, which is why we cannot always determine the individual masses of the stars. However, when the stars in a binary system periodically eclipse each other, as seen from Earth, we know that the plane of their orbit must be nearly parallel to our line of sight to them (Figure 11-13a, b). Comparing the relative sizes of the two orbits around the center of mass yields the ratio of the two stars' masses, M_1/M_2.

Eclipsing binaries can be detected even when the stars cannot be resolved as two distinct images in a telescope. The apparent magnitude of the image of an eclipsing binary dims each time one star blocks out part of the other. An astronomer can measure light intensity from

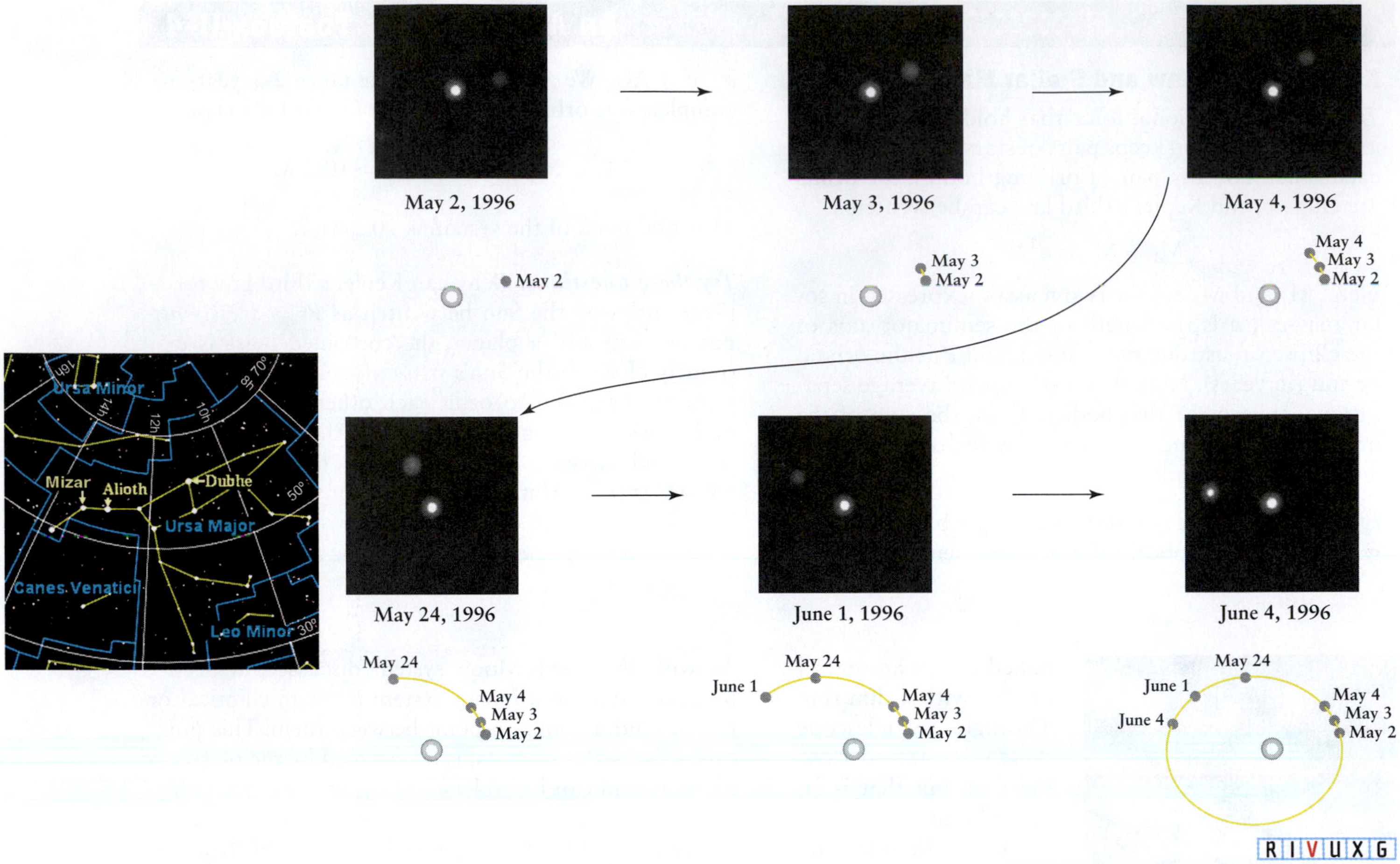

May 2, 1996

May 3, 1996

May 4, 1996

May 2

May 3
May 2

May 4
May 3
May 2

Ursa Minor
Mizar
Alioth
Dubhe
Ursa Major
Canes Venatici
Leo Minor

May 24, 1996

June 1, 1996

June 4, 1996

May 24
May 4
May 3
May 2

June 1
May 24
May 4
May 3
May 2

June 1
May 24
June 4
May 4
May 3
May 2

R I V U X G

FIGURE 11-11 A Binary Star System About one-third of the objects we see as "stars" in our region of the Milky Way Galaxy are actually double stars. Mizar in Ursa Major is a binary system with stars separated by only about 0.01 arcsec. The images and plots show the relative positions of the two stars over nearly half of their orbital period. The orbital motion of the two binary stars around each other is evident. Either star can be considered fixed in making such plots. (Technically, this pair of stars is Mizar A and its dimmer companion. These two are bound to another binary pair, Mizar B and its dimmer companion.) (Navy Prototype Optical Interferometer, Flagstaff, AZ. Courtesy of Dr. Christian A. Hummel)

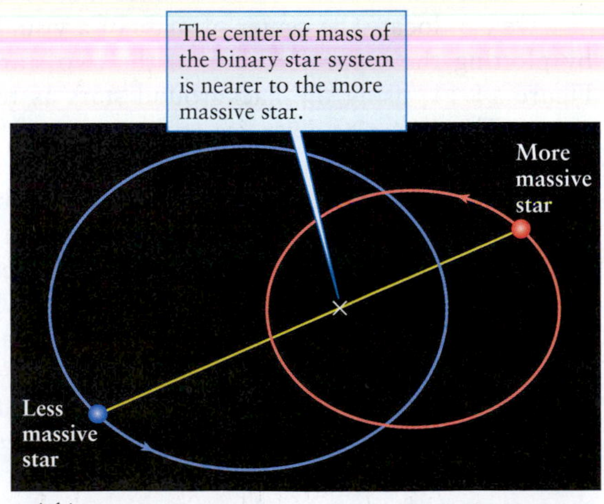

a A binary star system

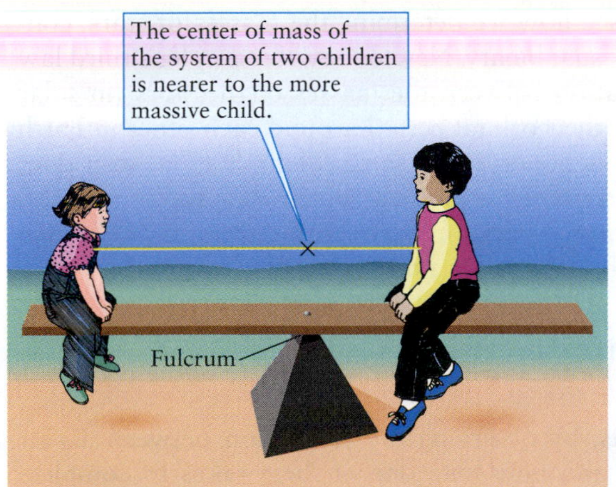

b A "binary system" of two children

ANIMATION 11.3

FIGURE 11-12 Center of Mass of a Binary Star System (a) Two stars move in elliptical orbits around a common center of mass. Although the orbits cross each other, the two stars are always on opposite sides of the center of mass and thus never collide. (b) A seesaw balances if the center of mass of the two children is at the fulcrum. When balanced, the heavier child is always closer to the fulcrum, just as the more massive star is closer to the center of mass of a binary star system.

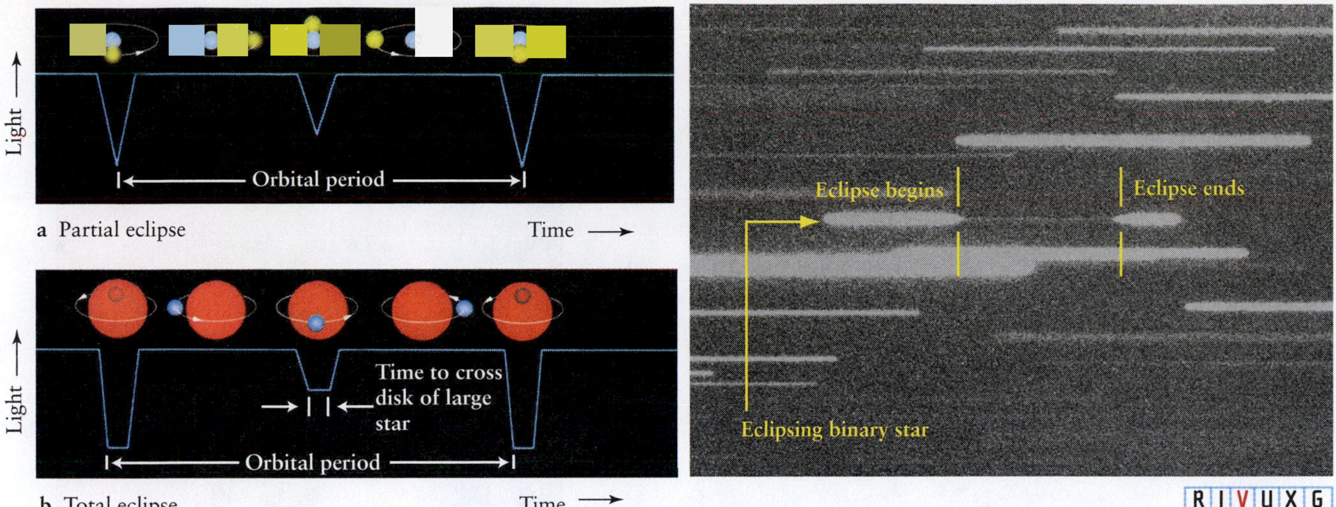

FIGURE 11-13 Representative Light Curves of Eclipsing Binaries The shape of the light curve (*blue*) reveals that the pairs of stars have orbits in planes nearly edge on to our line of sight. It also provides details about the two stars that make up an eclipsing binary. Illustrated here are (a) a partial eclipse and (b) a total eclipse. (c) The binary star NN Serpens, indicated by the arrow, undergoes a total eclipse. The telescope was moved during the exposure so that the sky drifted slowly from left to right. During the 10.5-min eclipse, the dimmer but larger star in the binary system (an M6 V star) passed in front of the more luminous but smaller star (a white dwarf). The binary became so dim that it almost disappeared. (European Southern Observatory)

binaries very accurately as a function of time. From these **light curves,** such as those shown in Figure 11-13, we can see at a glance whether the eclipse is partial, creating a V-shaped trough in the light curve (Figure 11-13a), or total, creating a flat-bottomed trough (Figure 11-13b). These data provide details of how close the plane of their orbit is to being perpendicular to our line of sight.

When we can determine M_1/M_2, we can find the individual masses, because we already know the sum $M_1 + M_2$ from Kepler's third law (that is, two equations and two unknowns yield the values for both unknowns). The individual masses are determined by combining the ratio of the masses and the sum of the masses. The range of stellar masses thus determined extends from 0.08 $M_\odot$ to about 150 $M_\odot$. (Stars with masses up to 265 $M_\odot$ have been observed by other means.)

Light curves can also reveal information about stellar atmospheres. Suppose that one star of a binary is a tiny white dwarf (the size of Earth) and the other is a giant (much larger than the Sun). By observing exactly how the light from the white dwarf is gradually cut off as it begins to move behind the edge of the giant during an eclipse, astronomers can infer the pressure and density in the upper atmosphere of the giant. Such information is invaluable in testing models of stellar structure.

Many binary stars are separated by several astronomical units or more. Other than orbiting each other, these stars behave as though they are isolated; that is, the models we develop in the following chapters for the evolution of isolated stars apply to these binary stars as well. However, there are also **close binary** systems, with only a few stellar diameters separating the stars. Such stars are so close together that the gravity of each one dramatically affects the appearance and evolution of the other. If one member of a close binary is a giant, some of the gas of its outer layers is pulled onto its more compact companion—mass is transferred from one star to the other. We will explore more about such systems in Chapter 12. In addition to binary systems of stars, a few systems composed of three stars bound to each other exist in our neighborhood of space. Polaris, the North Star, for example, is part of a triple system.

To determine the sum of the masses of two stars in a binary star system, what observations of them must be made?

11-11 Main-sequence stars have a relationship between mass and luminosity

As binary star data accumulated, an important trend began to emerge. On the main sequence, the more luminous the star, the more massive it is. This mass-luminosity relation can be conveniently displayed graphically (Figure 11-14a); note that the axes of this figure are logarithmic. The Sun's mass lies in the middle of both logarithmic ranges. The **mass-luminosity relation** demonstrates that *the main sequence on the H-R diagram is a progression in mass as well as in luminosity and surface temperature*. The hot, bright, bluish stars in the upper left corner of the H-R diagram (see Figure 11-14b) are the most massive main-sequence stars. As we move down the sequence, stellar masses decrease until we reach the dim, cool, reddish stars in the lower right corner of the H-R diagram. The correlation of stellar luminosities with mass

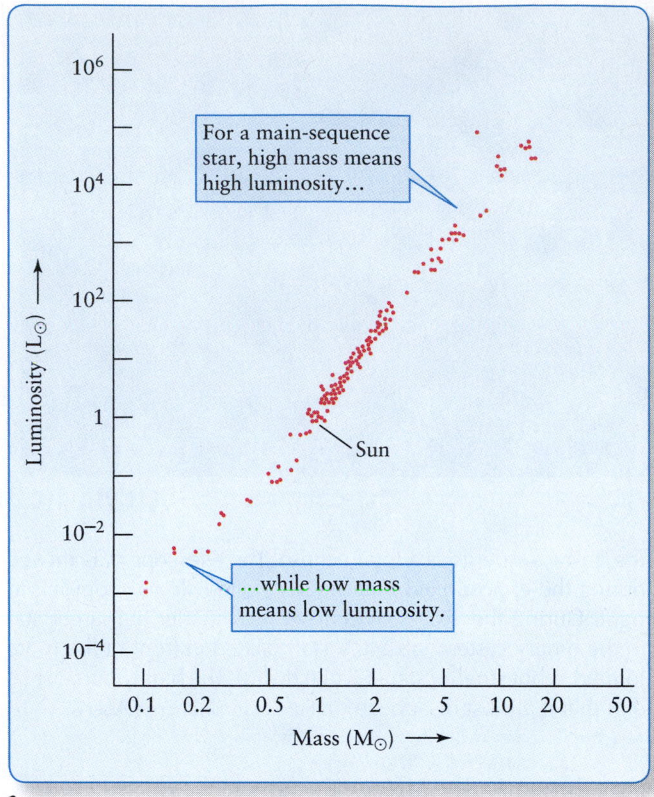

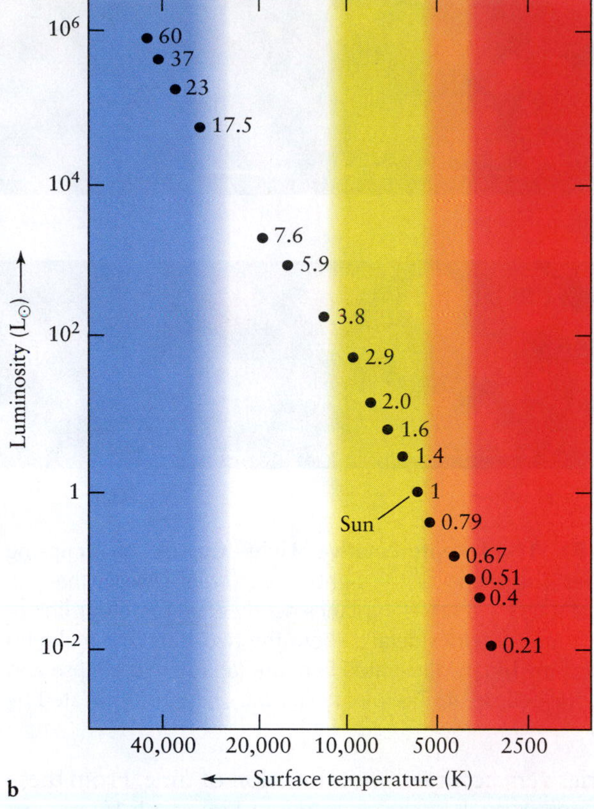

FIGURE 11-14 The Mass-Luminosity Relation (a) For main-sequence stars, mass and luminosity are directly correlated—the more massive a star, the more luminous it is. A main-sequence star of mass 10 $M_\odot$ has roughly 3000 times the Sun's luminosity (3000 $L_\odot$); one with 0.1 $M_\odot$ has a luminosity of only about 0.001 $L_\odot$. To fit the whole sequence on one page, the luminosities and masses are plotted using logarithmic scales. (b) On this H-R diagram, each dot represents a main-sequence star. The number next to each dot is the mass of that star in solar masses ($M_\odot$). As you move up the main sequence from the lower right to the upper left, the mass, luminosity, and surface temperature of main-sequence stars all increase.

For what types of stars does the mass-luminosity relationship not apply?

suggests (correctly, as it turns out) that a star's energy production is closely linked to its mass. Indeed, mass is *the* crucial factor in explaining why stars are as bright as they are and how they evolve.

11-12 The orbital motion of binary stars affects the wavelengths of their spectral lines

Many binary stars are scattered throughout our Milky Way Galaxy, but only those that are nearby or are widely separated can be distinguished as visual binaries. A remote binary often presents the appearance of a single star because even our best telescopes cannot resolve the images of its individual stars. These binary systems can still be detected, however, through spectroscopy.

Spectral analysis yields incongruous spectral lines for some stars. For example, the spectrum of what appears at first to be a single star may include strong absorption lines for both hydrogen (indicating a hot, type A star) and titanium oxide (indicating a cool, type M star). Because a single star cannot display both types of absorption lines prominently, such an observation must be revealing a binary system.

Spectroscopy can also detect the movements of stars orbiting each other because of the Doppler shift in spectral lines. As we saw in Section 4-7, an approaching source of light has shorter wavelengths than if the same source were stationary, and a receding source has longer wavelengths. The amount of the shift is proportional to the speed of the light source: The greater the speed, the greater the shift.

If the two stars in a binary are orbiting at more than a few kilometers per second, they produce spectral lines that shift back and forth regularly. Such stars are called **spectroscopic binaries.** The motions of the stars revolving around their center of mass produce the periodic shifting of the spectral lines. In many spectroscopic binaries, one of the stars is so dim that its spectral lines cannot be detected. Instead, a single set of spectral lines from the star that we can see shifts regularly back and forth. Such a *single-line spectroscopic* *binary* yields less information about its two stars than does a *double-line spectroscopic binary,* in which the lines from both stars are visible. In particular, a double-line binary, in which one star eclipses the other, yields the individual masses of its two stars, whereas a single-line binary cannot provide that information.

Figure 11-15 shows four spectra of a spectroscopic binary system. At Stage 1, star A is moving toward Earth; hence, its spectrum is blueshifted. At the same time, star B is moving away, so its spectrum is redshifted. In the spectrum below the drawing, you can see two sets of spectral lines, each slightly offset from each other. A few days later, at Stage 2, the stars have progressed along their orbits so that star A is moving toward the right and star B toward the left, as seen from Earth. Because neither star is moving toward or away from us, their spectral lines return to their normal positions. Stage 3 shows the opposite spectra from Stage 1 (star A is now redshifted and star B is blueshifted), while Stage 4 shows the same spectrum as Stage 2. Figure 11-16 shows spectra taken of the binary star system κ (kappa) Arietis, demonstrating these shifts.

The shifts in spectral lines can yield significant information about the orbital velocities of the stars in a spectroscopic binary. This information is best displayed as a **radial-velocity curve,** in which radial velocity is graphed over time. In Figure 11-15, the wavy pattern repeats with a period of about 15 days, which is the orbital period of the binary. This pattern is displaced upward from the zero-velocity line by about 12 km/s, which is the overall motion of the binary system away from Earth. Superimposed on this overall recessional motion are the periodic approaches and recessions of the two stars as they orbit around each other.

Combining radial-velocity curves and light curves, we can obtain other useful information from eclipsing binaries. If an eclipsing binary is also a double-line spectroscopic binary, astronomers can calculate the masses, diameters, varied brightness, speeds, and stellar separation of each star. These stars are rare, however, because the orbital planes of most spectroscopic binaries are tilted so that eclipses do not occur, as seen from Earth.

While most of the stars we see in the night sky are single objects or in binary systems, some are in groups

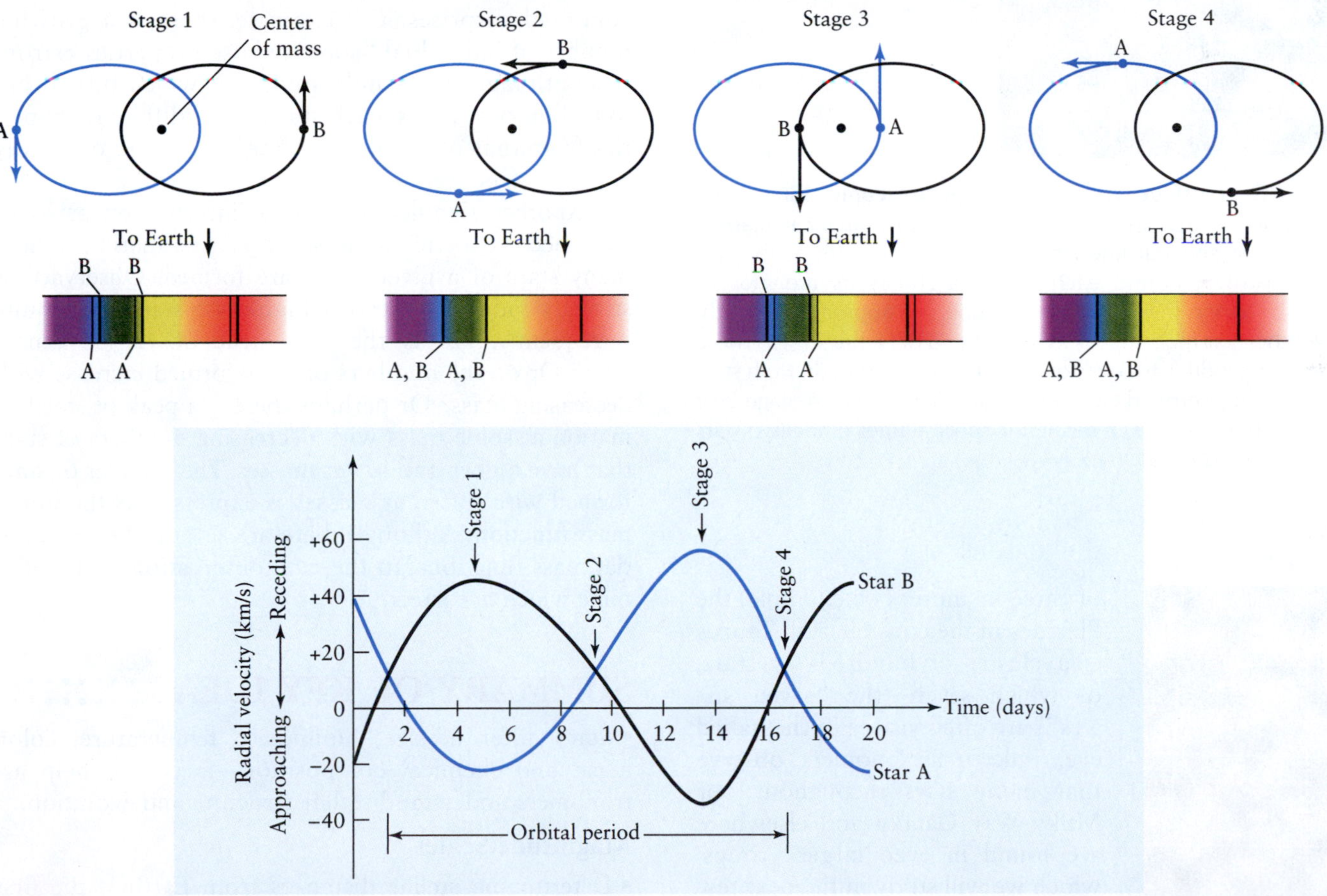

ANIMATION 11.5 **FIGURE 11-15 Spectral-Line Motion in Binary Star Systems** The diagrams at the top indicate the positions and motions of the stars, labeled A and B, relative to Earth. Below each diagram is the spectrum we would observe for these two stars at each stage. The changes in colors (wavelengths) of the spectral lines are due to changes in the stars' Doppler shifts, as seen from Earth. The graph displays the radial-velocity curves of the binary HD 171978. (The HD means that this is a star from the *Henry Draper Catalogue* of stars.) The entire binary is moving away from us at 12 km/s, which is why the pattern of radial velocity curves is displaced upward from the zero-velocity line.

When one of the stars in a spectroscopic binary is moving toward us and the other is receding from us, we see *two* sets of spectral lines due to the Doppler shift.

B A

To Earth

a

b

When both stars are moving perpendicular to our line of sight, there is no Doppler splitting and we see a *single* set of spectral lines.

A

B

To Earth

R I V U X G

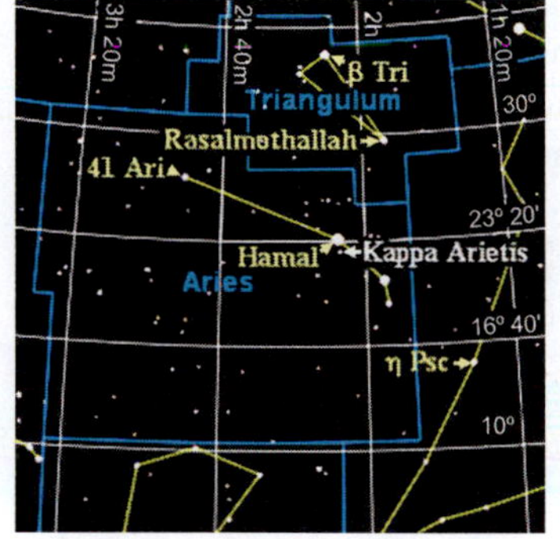

FIGURE 11-16 A Double-Line Spectroscopic Binary The spectrum of the double-line spectroscopic binary κ (kappa) Arietis has spectral lines that shift back and forth as the two stars revolve around each other. (a) The stars are moving parallel to the line of sight, with one star approaching Earth, the other star receding, as in Stage 1 or 3 of Figure 11-15. These motions produce two sets of shifted spectral lines. (b) Both stars are moving perpendicular to our line of sight, as in Stage 2 or 4 of Figure 11-15. As a result, the spectral lines of the two stars have merged. (Lick Observatory)

How do the spectral lines of a spectroscopic binary change as observed from Earth?

of three or more. For example, the Pleiades in the constellation Taurus is a cluster of hundreds of stars, of which seven (the "seven sisters") are often visible to the naked eye. Indeed, astronomers observe that many stars throughout our Milky Way Galaxy and elsewhere are found in even larger groups, which we will study in the next few chapters.

This chapter has provided us with observational data that serve, in large measure, as the basis for the models of stellar activity and evolution presented in the next three chapters. The models make predictions that lead to new observations and refined models. As we proceed through these models, we will take up some further observations driven by these refinements.

11-13 Frontiers yet to be discovered

The legacy of *Hipparcos* is a better understanding of the size of our region of the cosmos. As noted earlier, stellar parallax is the most accurate means of measuring distances to the stars, requiring the fewest assumptions about how stars work and the nature of the material between us and the stars. Therefore, a major goal of space-based observational astronomy is to get more and better parallax measurements for stars at greater distances. To this end, satellites will be sent far from Earth to increase the baseline used to make these measurements (see Figure 11-1) and hence the angular accuracy that is possible.

Classifying stars by luminosity and spectral type is an ongoing process providing an ever-growing body of data about stellar properties. This information yields continual surprises. For example, there is a growing number of individual "stars" whose properties astronomers thought they understood but which turned out to be binary systems, each star having different properties from that of the single "star" they were originally believed to be.

Another significant piece of information astronomers need to better understand star formation is how many stars of a given mass are formed. Observations strongly indicate that very-high-mass stars form quite infrequently, but is the same true of very-low-mass stars? Or do the numbers of stars formed increase with decreasing mass? Or perhaps there is a peak of star formation at some mass with decreasing numbers of stars that have higher and lower masses. The number of stars formed with different masses is expressed as the **initial mass function.** Although there are several theoretic initial mass functions, in the end, observations will determine which is correct.

SUMMARY OF KEY IDEAS

• Stars differ in size, luminosity, temperature, color, mass, and chemical composition—facts that help astronomers understand stellar structure and evolution.

Magnitude Scales

• Determining stellar distances from Earth is the first step to understanding the nature of the stars. Distances to the nearer stars can be determined by stellar parallax, which is the apparent shift of a star's location against the background stars while Earth moves along its orbit around the Sun. The distances to more remote stars are determined using spectroscopic parallax.

• The apparent magnitude of a star, denoted *m*, is a measure of how bright the star appears to Earth-based observers. The absolute magnitude of a star, denoted M, is a measure of the star's true brightness and is directly related to the star's energy output, or luminosity.

• The luminosity of a star is the amount of energy emitted by it each second.

• The absolute magnitude of a star is the apparent magnitude it would have if viewed from a distance of 10 pc. Absolute magnitudes can be calculated from the star's apparent magnitude and distance from Earth.

The Temperatures of Stars

• Stellar temperatures can be determined from stars' colors or stellar spectra.

• Stars are classified into spectral types (O, B, A, F, G, K, and M) based on their spectra or, equivalently, their surface temperatures.

Types of Stars

• The Hertzsprung-Russell (H-R) diagram is a graph on which luminosities of stars are plotted against their spectral types (or, equivalently, their absolute magnitudes are plotted against surface temperatures).

• The H-R diagram reveals the existence of four major groupings of stars: main-sequence stars, giants, supergiants, and white dwarfs.

• The mass-luminosity relation expresses a direct correlation between a main-sequence star's mass and the total energy it emits.

• Distances to stars can be determined using their spectral types and luminosity classes.

Stellar Masses

• Binary stars are fairly common. Those that can be resolved into two distinct star images (even if it takes a telescope to do this) are called visual binaries.

• The masses of the two stars in a binary system can be computed from measurements of the orbital period and orbital dimensions of the system.

• Some binaries can be detected and analyzed, even though the system may be so distant (or the two stars so close together) that the two star images cannot be resolved with a telescope.

• A spectroscopic binary is a system detected from the periodic shift of its spectral lines. This shift is caused by the Doppler effect as the orbits of the stars carry them alternately toward and away from Earth.

• An eclipsing binary is a system whose orbits are viewed nearly edge on from Earth, so that one star periodically eclipses the other. Detailed information about the stars in an eclipsing binary can be obtained by studying the binary's light curve.

A WHAT DID YOU THINK?

1 *How near to us is the closest star other than the Sun?* The closest star, Proxima Centauri, is about 40 trillion km (25 trillion mi) away. Light from there takes about 4 years to reach Earth.

2 *How luminous is the Sun compared with other stars?* The most luminous stars are about a million times brighter, and the least luminous stars are about a hundred thousand times dimmer than the Sun.

3 *What colors are stars, and why do they have these colors?* Stars are found in a wide range of colors, from red through violet as well as white. They have these colors because they have different surface temperatures.

4 *Are brighter stars hotter than dimmer stars?* Not necessarily. Many brighter stars (such as red giants) are cooler but larger than hotter, dimmer stars (such as white dwarfs).

5 *Compared to the Sun, what sizes are other stars?* Stars range from more than 1000 times the Sun's diameter to less than 1/100 the Sun's diameter.

6 *Are most stars isolated from other stars, as the Sun is?* In the vicinity of the Sun, one-third of stars are found in gravitationally bound pairs or larger groups.

Key Terms for Review

Review Questions

The answers to all computational problems, which are preceded by an asterisk (*), appear at the end of the book.

1. Stellar parallax measurements are used in astronomy to determine which of the following properties of stars? **a.** speeds, **b.** rotation rates, **c.** distances, **d.** colors, **e.** temperatures

2. The brightness a star would have if it were at 10 pc from Earth is called its **a.** absolute magnitude, **b.** apparent magnitude, **c.** luminosity, **d.** spectral type, **e.** center of mass

3. Measurements of a binary star system are required to determine what property of stars? **a.** luminosity, **b.** apparent magnitude, **c.** distance, **d.** mass, **e.** temperature

4. A star with which of the following apparent magnitudes appears brightest from Earth? **a.** 6.8, **b.** 3.2, **c.** 0.41, **d.** −0.44, **e.** −1.5

5. A star of what spectral class has the strongest (darkest) H_α line? (*Hint:* See Figure 11-5): **a.** B2, **b.** A0, **c.** A5, **d.** G5, **e.** M0

6. Describe how the parallax method of finding a star's distance is similar to the binocular (two-eye) vision of animals.

7. What is stellar parallax?

8. How do astronomers use stellar parallax to measure the distances to stars?

9. Why do stellar parallax measurements work only with relatively nearby stars?

10. What is the difference between apparent magnitude and absolute magnitude?

11. Briefly describe how you would determine the absolute magnitude of a nearby star.

12. What does a star's luminosity measure?

13. Why is the magnitude scale "backward" from what common sense dictates?

14. Does the star Betelgeuse, whose apparent magnitude is $m = +0.5$, look brighter or dimmer than the star Pollux, whose apparent magnitude is $m = +1.1$?

***15.** Consider two identical stars, with one star 5 times farther away than the other. How much brighter will the closer star appear than the more distant one?

16. How and why is the spectrum of a star related to its surface temperature?

17. What is the primary chemical component of most stars?

18. A star of which spectral type has the strongest Na I absorption lines? At approximately what wavelength is this line normally found? *Hint:* See Figure 11-5.

19. Why does a G2 star have many more absorption lines than a B0 star?

20. Draw an H-R diagram and sketch the regions occupied by main-sequence stars, giants, supergiants, and white dwarfs. Briefly discuss the different ways you could have labeled the axes of your graph.

21. To test your understanding of the H-R diagram, do Interactive Exercise 11.1. You can print out your answers, if required.

22. How can observations of a visual binary lead to information about the masses of its stars?

23. What is a radial-velocity curve? What kinds of stellar systems exhibit such curves?

24. What is the difference between a single-line and a double-line spectroscopic binary?

25. What is meant by the light curve of an eclipsing binary? What sorts of information can be determined from such a light curve?

26. What is the mass-luminosity relation? To what kind of stars does it apply?

27. Refer to Figure 11-7: **a.** Which are the hottest and coolest named stars on the diagram? **b.** Which are the brightest and dimmest named stars on the diagram? **c.** Which are the hottest and coolest named main-sequence stars on the diagram? **d.** Which named stars are white dwarfs? giants? supergiants?

Advanced Questions

28. What is the inverse-square law? Use it to explain why a headlight on a car can appear brighter than a star, even though the headlight emits far less light energy per second.

***29.** Van Maanen's star, named after the Dutch astronomer who discovered it, has a parallax angle of 0.232 arcsec. How far away is the star?

30. Explain how sailors on a ship traveling parallel to a coastline at a known speed can use parallax angle measurements to determine the distance to the shore.

***31.** Suppose that a dim star were located 2 million AU from the Sun. Find **a.** the distance to the star in parsecs and **b.** the parallax angle of the star.

***32.** How many times brighter is a star of apparent magnitude $m = -1$ than a star of apparent magnitude $m = +7$?

33. Sketch the radial-velocity curve of a binary whose stars are moving in nearly circular orbits that are **a.** perpendicular and **b.** parallel to our line of sight.

34. Sketch the light curve of an eclipsing binary whose stars are moving along highly elongated orbits with the major axes of the orbits **a.** pointed toward Earth and **b.** perpendicular to our line of sight.

***35.** **a.** What is the approximate mass of a main-sequence star that is 10,000 times as luminous as the Sun? **b.** What is the approximate luminosity of a main-sequence star whose mass is one-tenth that of the Sun?

***36.** What is the approximate surface temperature of a main-sequence star luminosity 100 times as bright as the Sun?

Discussion Questions

37. Discuss the advantages and disadvantages of measuring stellar parallax from a space telescope at the distance of Jupiter from the Sun compared to the same measurements made from Earth's surface.

38. How does a star's rotation affect the appearance of its spectral lines? *Hint:* Assume we are looking at the star from above its equator. Then, at every instant, half of the spinning star is approaching Earth, while the other half is receding. Consider the resulting Doppler shift in the spectral lines.

What If...

39. All of the stars in our Milky Way Galaxy actually were the same distance from Earth, say, 10 pc? Describe what the night sky might look like. How about the daytime sky? *Hint:* If you are doing this quantitatively, assume for simplicity that all of the stars have the same absolute magnitudes as the Sun and that there are roughly 200 billion stars in the Galaxy.

40. All stellar parallax angles, p, were observed to be increasing? What would that imply about the motions of stars? Is there another observational technique that could be used to confirm this motion?

41. A star's parallax were observed to oscillate—regularly increase in angle and then decrease in angle—over a period of 100 years? What would that imply about the star?

42. The Sun were part of a binary star system of two main-sequence stars and Earth orbited around one of the stars, while the other star just slightly changed Earth's distance of orbit, compared to its orbit today? You might want to discuss such things as climate, tides, impacts from meteoroids and asteroids, and habitability.

43. The Sun were an M-type star, rather than a G-type star? Assuming that Earth orbiting the M-type star had the same composition and orbit distance as it does today, what would be different here?

44. The Sun were a B-type star, rather than a G-type star? Assuming that Earth orbiting the B-type star had the same composition and orbit distance as it does today, what would be different here? We will pick up this question again in Chapter 13 for further insights.

Web Questions

45. Search the Web for the periods of 10 binary star systems. Plot these on a graph of time (on the horizontal axis) versus number of systems. If possible, combine this information with similar data from your classmates. Do you see any patterns in the periods of these star systems?

46. To explore the range of periods of binary star systems, locate on the Web binary star systems with periods of **a.** less than 1 week, **b.** between 1 day and 1 week, **c.** between 1 and 2 years, **d.** between 40 and 50 years, and **e.** more than 400 years.

Observing Projects

47. You can use *Starry Night*™ to examine the 10 brightest stars in Earth's night sky. Select **Favourites > Discovering the Universe > Atlas.** Click on the **Options** tab, expand the **Constellations** list and click on **Boundaries, Labels,** and **Stick Figures (Astronomical)** to display these features. Use the **File** (Windows) or **Starry Night** (Macintosh) menu command to open the **Preferences** dialog window. Ensure that the **Cursor Tracking (HUD)** preferences include **Apparent Magnitude, Distance from Observer,** and **Temperature** in the **Show** list. Before closing the **Preferences** dialog window, it might be helpful to increase the saturation for **Star colour** under the **Brightness/Contrast** preferences. Click on the **Lists** side pane tab, expand the **Observing Lists** layer and click the **10 Brightest Stars** option. Then expand the **List Viewer** layer and select **All Targets** from the **Show** drop-down menu to see a list of the 10 brightest stars in Earth's night sky. Double-click on each of the stars in this list in turn to center the star in the view. For each star in the list, use the HUD to compile a table of these stars that includes each star's apparent magnitude, distance, and temperature. You may also wish to sketch the star's position within its constellation. Alternatively, you may find it helpful to print out relevant star charts around these stars, using *Starry Night*™. **a.** Which is the brightest star in Earth's night sky? What features of this star make it so bright in our sky? **b.** Which of these brightest stars has the highest temperature? What would you expect to be the color of this star compared to others in the list? **c.** Use *Starry Night*™ to determine which of these stars is visible from your location. Click the **Home** button, then the **Stop** button, and finally the **Sunset** button to show the view from your home location today at sunset. Again, it may be helpful to open the **Options** pane and display the constellation **Boundaries, Labels,** and **Stick Figures (Astronomical)** in the **Constellations** list. Open the **Lists** side pane and double-click each entry in the list of the **10 Brightest Stars.** If the star is visible in your sky, the program will center it in the view or alternately suggest a **Best Time** for observing this star. For those stars in the list that are visible from your home location, go outside and observe them in the real sky. See if you can tell which of these stars has the highest temperature on the basis of your conclusion regarding the star's color; check your estimate against the table you compiled in part a. (*Hint:* The colors of stars are not very distinct and a dark sky background is needed in order to distinguish differences in stellar colors.)

48. Locate the stars Betelgeuse and Rigel in Orion. Observe them both with the unaided eye and through a small telescope. Are they the same color? To determine their color(s), it helps to compare them to their neighbors.

49. In this exercise, you can use *Starry Night*™ to look at one example of how stars in our neighborhood are often members of a small group by examining the stars in the vicinity of **Mizar** in the Big Dipper. Select **Favourites > Discovering the**

Universe > Mizar. To help you to find these stars in the real sky, the view indicates the location of Mizar in the Big Dipper asterism. Ensure that the HUD feature is configured to include **Name, Object type,** and **Distance from Sun.** From a dark site, Mizar can be seen in the real sky to have a companion star. Zoom in to a field of view of about 2°, when a second star will move away from Mizar. **a.** What is the name of the fainter star and where is it located with respect to Mizar? **b.** From the HUD, what are the distances of Mizar and its companion from the Sun? **c.** Use the angular separation tool to measure the angular separation and actual physical distance between Mizar and its companion. **d.** Considering this information about Mizar and its companion, do you think that they constitute: i) a visual binary—that is, two stars that can be resolved from Earth and are together in space; ii) a spectroscopic binary—that is, two stars that cannot be resolved from Earth but whose spectra show them to be separate entities; or iii) an optical double, an accidental alignment of two stars at different distances from the Sun? Explain your reasoning in support of this conclusion. (Note: As a comparison, the closest star to the Sun in space is Proxima Centauri, about 4.3 ly away.) **e.** If you zoom in on Mizar, you will note that it consists of two closely spaced stars, Mizar itself and Zeta Ursae Majoris to the southeast of it. Zoom in to a field of view of about 2′. What is the angular separation of these two stars when viewed from Earth? Spectroscopy of these stars reveals that each of them is in fact a close binary pair of stars. Thus, there are five stars in this Mizar group. Most stars in our sky are members of a binary pair or a larger group of stars.

STARRY NIGHT

50. Stars exhibit a wide variety of properties. Astronomers have developed a convenient method for organizing stars into groups by plotting two basic stellar properties, intrinsic brightness and surface temperature, on a graph in which each point represents a single star. This graph is known as a Hertzsprung-Russell diagram (HRD), after its originators. Intrinsic brightness is represented as either luminosity or absolute magnitude. Stars within a particular population are not distributed randomly on this brightness-temperature graph but form into distinct groups that astronomers have related to the evolution of stars with time, the groupings representing particular stages in a star's lifetime. For example, a band of stars across the H-R diagram from high-temperature, high-brightness to low-temperature, low-brightness regions is known as the main sequence and is occupied by stars that are "burning" hydrogen in the nuclear furnaces within their interiors. Another group of stars above the main sequence, the red giants, have evolved beyond this stage and have consumed all of their hydrogen in nuclear fusion reactions to make heavier elements. They are now generating energy by "burning" these heavier elements in different nuclear reactions in their hot, dense cores.

Starry Night™ can display the HRD of all stars in the main view. You can use this facility to investigate the distribution on the HRD of stars in our near space. Select **Favourites > Discovering the Universe > Denver** to display the sky above Colorado. Open the **Status** pane and expand the **H-R Options** layer. Select **Use absolute magnitudes** to represent the star's intrinsic brightness on the y-axis and select **Labels > Kelvin** × **1000** to represent the star's surface temperature on the x-axis. Check the **Labels** option and select **Gridlines, Regions,** and **Main Sequence.** Now, expand the **Hertzsprung-Russell** layer to show the HRD of all of the stars that are currently in the main view. This graphical representation shows the absolute magnitudes of stars increasing upward, as a function of their surface temperatures plotted in an inverse direction, the hottest stars appearing to the left of the diagram. Absolute magnitude is related to the star's luminosity. **a.** Use the hand tool to scroll around the sky. Watch the HRD change as different stars enter and leave the main window. Does the distribution of stars in the HRD change drastically from one part of the sky to another, or are all types of stars approximately equally represented in all directions from Earth? **b.** If you place the cursor over a star, a red dot appears in the HRD at the position for this star. Use this facility to estimate and make a note of the absolute magnitude, M_V, and the surface temperature, T, of the following stars that are labeled in the main window: Altair, Deneb, Enif, 74 Ophiuchi, and 51 Pegasi. If the Sun has an absolute magnitude of about +5 and a temperature of about 6000K, which of these stars most resembles the Sun? **c.** In which region of the HRD do we find this star? **d.** If are similar for the Sun and this star, which other two parameters will necessarily be similar? Explain your reasoning. **e.** Identify the red giant region of the HRD, containing a significant fraction of all the stars in the Sun's near neighborhood. What is the average absolute magnitude of these stars? Compare this value to that of the Sun. How much greater is the average luminosity of these stars than that of the Sun, that is, how much brighter are these stars than the Sun, intrinsically? **f.** Note that most of these red giant stars have significantly lower temperatures than that of the Sun. How then can these cooler stars have significantly higher energy outputs if they are so much cooler than the Sun?

Chapter 12

The Lives of Stars from Birth Through Middle Age

RIVUXG The young super star cluster R136 is located in the Tarantula Nebula in the nearby Large Magellanic Cloud galaxy. These stars, surrounded by glowing interstellar clouds, are only a few million years old. (NASA, ESA, and F. Paresce [INAF-IASF, Bologna, Italy], R. O'Connell [UVA, Charlottesville], and Wide Field Camera 3 Science Oversight Committee)

WHAT DO YOU THINK?

1. **How do stars form?**
2. **Are stars still forming today? If so, where?**
3. **Do more massive stars shine longer than less massive ones? What is your reasoning?**
4. **When stars like the Sun stop fusing hydrogen into helium in their cores, do the stars get smaller or larger?**

Answers to these questions appear in the text beside the corresponding numbers in the margins and at the end of the chapter.

Time changes everything. We grow up, we grow older, and we die. We are most familiar with changes brought about by aging in our own lives and the lives of other people, but everything else in the universe alters with time, too. These changes are due to the forces in nature that cause interactions between matter and energy. For example, over time, exposure to the effects of weather and pollution causes even solid stone to crumble (Figure 12-1).

Gravity provides the energy that enables stars to shine. As we saw in studying the Sun (Chapter 10), gravity compresses stars, thereby heating them and causing fusion to occur. As a result, stars emit huge amounts of radiation. Over time, their chemical compositions, masses, and brightnesses all vary—stars age. Major stages in the life of each star can last from days to hundreds of billions of years. Cumulative lifetimes of stars range from millions to hundreds of billions of years. We will explore their evolution in this and the next chapter.

In this chapter you will discover

- how stars form
- what a stellar "nursery" looks like
- how astronomers use the physical properties of stars to learn about stellar evolution
- the remarkable transformations of older stars into giants
- how the Hertzsprung-Russell (H-R) diagram is your guide to the stellar life cycle
- how pairs of orbiting stars can change each other

Insight Into Science

Beware Poetic License All too often, the names and descriptions we use for nonliving objects seem to give them human qualities. Anthropomorphism—assigning human attributes to nonhuman creatures or even to nonliving objects, like stars—is descriptive but not scientific. In fact, scientists avoid this practice where possible, because it creates unjustified expectations. Nevertheless, anthropomorphism does creep into astronomy. For lack of better terms, astronomers describe stellar change with words related to living things, such as *evolution, birth, aging, maturing, growing old,* and *dying*. This is poetic license of a sort: It piques our interest but muddles our science. As you study stellar evolution, always be mindful of that.

PROTOSTARS AND PRE-MAIN-SEQUENCE STARS

We saw in Chapter 5 that the solar system is believed to have formed from a collapsing, rotating fragment of a cloud of gas and dust some 4.6 billion years ago. This scientific theory requires that interstellar matter existed before then so that the solar system had something from which to form. Taking this reasoning one step further, if such interstellar matter exists today, perhaps star formation continues.

12-1 Gas and dust exist between the stars

Matter does indeed exist between the stars. We can see a relatively small amount of it through visible light telescopes (Figure 12-2a); however, the bulk of it is too cold to be seen optically and requires the use of infrared (Figure 12-2b) and radio telescopes. We call all the matter between stars the **interstellar medium,** and it contains at least 10% of the observed mass in our Galaxy. Observations of the spectra of the interstellar medium reveal that it is composed of gas containing isolated atoms and molecules and tiny pieces of dust. Table 12-1 summarizes the composition of the interstellar medium.

FIGURE 12-1 Everything Ages In less than 60 years (1935 to 1994), corrosive gases in the air caused this statue of George Washington in New York City to erode more rapidly than normal. Even without the significant pollution that humans add to the air, everything in contact with it decays. (a: NYC Parks Photo Archive/Fundamental Photographs; b: Kristen Brochmann/Fundamental Photographs)

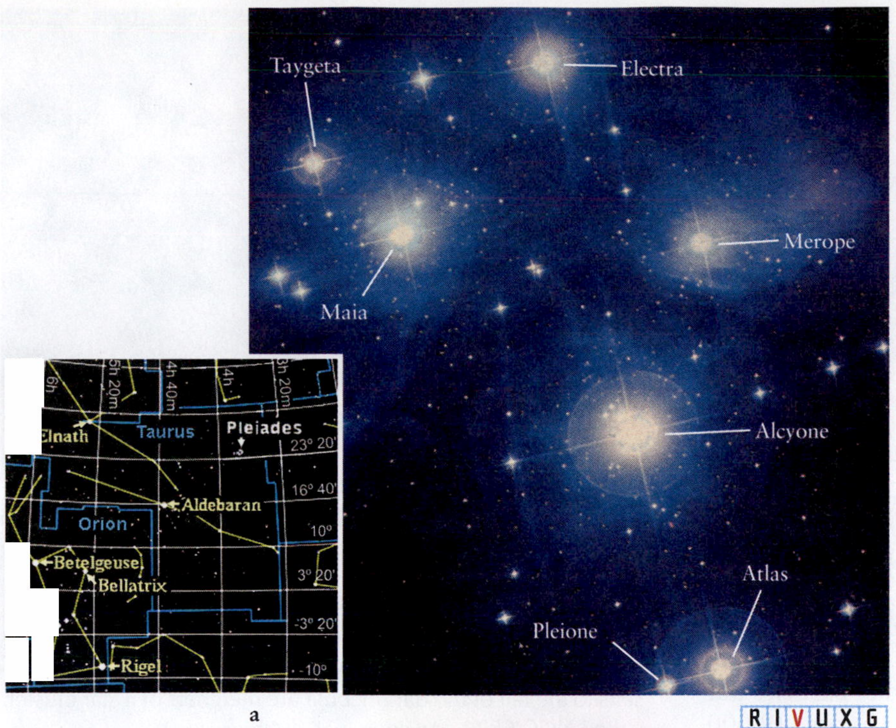

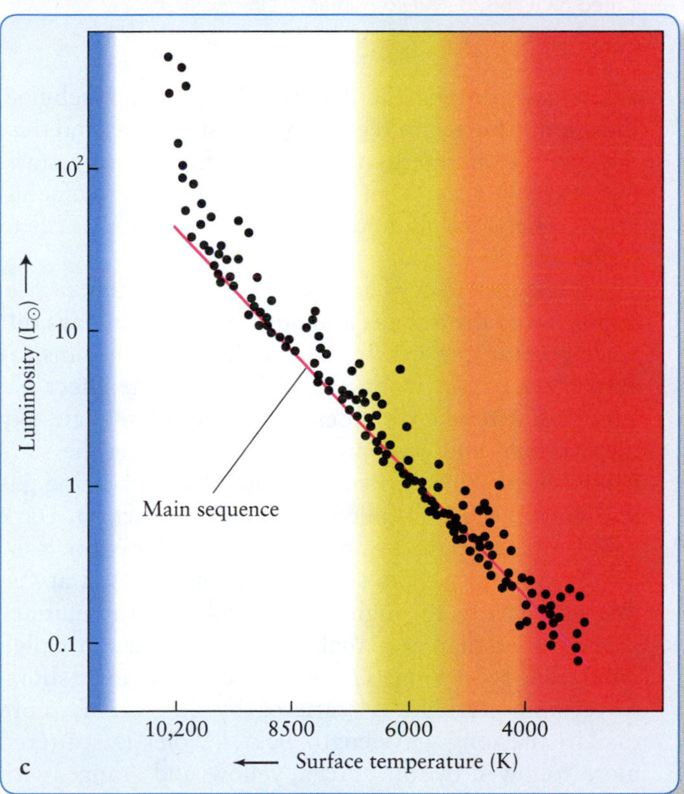

FIGURE 12-2 Stars and the Interstellar Medium (a) This open cluster, called the Pleiades, can easily be seen with the naked eye in the constellation Taurus (the Bull). Pleiades lies about 440 light-years (134 pc) from Earth. The stars are not shedding mass, unlike the stars in Figures 12-15a and 12-23. The blue glow surrounding the stars of the Pleiades is a reflection nebula created as some of the stars' radiation scatters off preexisting dust grains in their vicinity. (b) The same region of the sky in a false-color infrared image taken by the Spitzer Space Telescope. Gases are seen here to exist in more areas than can be detected in visible light. (c) Each dot plotted on this H-R diagram represents a star in the Pleiades whose luminosity and surface temperature have been determined. Note that most of the cool, low-mass stars have arrived at the main sequence, indicating that hydrogen fusion has begun in their cores. The cluster has a diameter of about 5 light-years, is about 100 million years old, and contains about 500 stars. (a: Anglo-Australian Observatory; b: NASA/JPL-Caltech/J. Stauffer [SSC/Caltech])

A variety of molecules is found in interstellar space, including molecular hydrogen (H_2), carbon monoxide (CO), carbon dioxide (CO_2), water (H_2O), ammonia (NH_3), formaldehyde (H_2CO), and the sugar glycoaldehyde ($C_2H_4O_2$), among many others. We can identify these molecules by their unique spectral emissions, just as we can identify elements from atomic spectra.

Table 12-1 Composition of the Interstellar Medium

	Particle number (%)	Mass (%)
Hydrogen (atoms and molecules)	90	74
Helium	9	25
Metals*	1	1

*Metals are all elements except hydrogen and helium.

FIGURE 12-3 A Connection to Interstellar Space The charred layer created by overcooking this beef contains compounds of carbon and hydrogen, called polycyclic aromatic hydrocarbons. These molecules are also found in interstellar clouds. (Dena Digilio Betz)

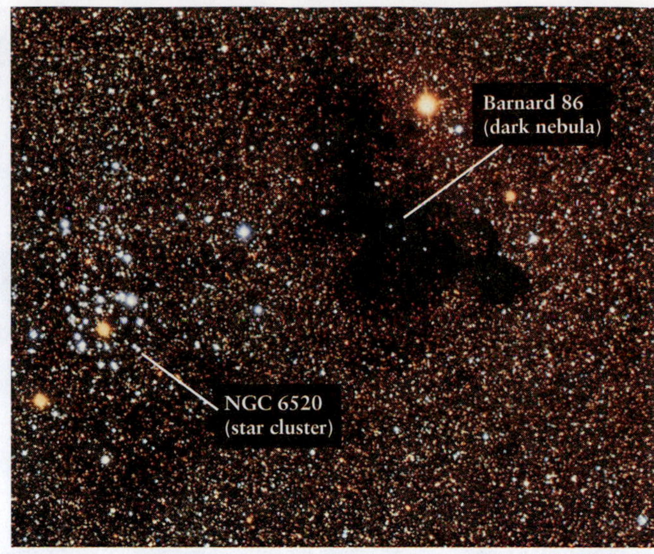

FIGURE 12-4 A Dark Nebula The dark nebula Barnard 86 is located in Sagittarius. It is visible in this photograph simply because it blocks out light from the stars beyond it. The bluish stars to the left of the dark nebula are members of a star cluster, called NGC 6520. (Anglo-Australian Observatory)

The dust found in space also has a variety of structures. Some carbon-based (that is, organic) particles out there are typically 0.005 microns (μm) across (10,000 times smaller than the diameter of a typical human hair). Among these are collections of molecules similar to those found in engine exhaust and burnt meat (Figure 12-3), called *polycyclic aromatic hydrocarbons* (PAHs), which are composed of only carbon and hydrogen atoms. Much larger pieces of space dust have cores of carbon or silicon compounds surrounded with mantles of ice and other materials. These grow to more than 0.30 μm in diameter (nearly 200 times smaller than the diameter of a human hair).

Working with the knowledge that new stars form from the gas and dust of the interstellar medium, astronomers map this matter to identify places to look for young stars. When it can be seen in the visible part of the spectrum, the gas and dust in the interstellar medium glow as a result of scattered light from stars in its vicinity. For example, the interstellar medium is dramatically highlighted by stars in the Pleiades star cluster (Figure 12-2a) located in the constellation Taurus. The bluish haze, starlight scattered by the interstellar gas and dust, is called a **reflection nebula.** A **nebula** (plural, *nebulae*) is a dense region of interstellar gas and dust. Nebulae are often embedded in much larger bodies of gas and dust, called **molecular clouds.** The largest molecular clouds, called **giant molecular clouds,** contain upward of a million solar masses of matter and extend up to about 300 light-years across.

Astronomers identify several other types of nebulae. Most important among these are dark and emission nebulae. **Dark nebulae** (Figure 12-4) are regions of interstellar gas and dust that are sufficiently dense to prevent most of the visible light from behind them getting to us. They look like regions of empty space, whereas they are actually among the densest of interstellar nebulae. **Emission nebulae** are regions of interstellar gas and dust that glow from energy they receive from nearby stars, from exploding stars, and from collisions between nebulae. A variety of dark and emission nebulae are visible in Orion (Figure 12-5).

While the effect of a dark nebula is an extreme case, finding more distant gas and dust in space is always difficult because the nearby interstellar medium dims, or even blocks, light from behind it. The same effect occurs on Earth where thin cloud layers in our night sky dim starlight and even prevent us from seeing the stars behind them. This darkening of light by intervening gas and dust in space is called **interstellar extinction.** Dark nebulae are extreme examples of interstellar extinction.

Even when we can see a star or other object at visible wavelengths through the interstellar medium, it appears redder than it actually is, a phenomenon called **interstellar reddening.** This effect occurs because short-wavelength starlight is scattered by dust grains more than is the long-wavelength light. (Violet is scattered most, followed by blue, green, yellow, and orange, with red scattered least.) Therefore, when we observe a star through gas and dust, we are seeing less short-wavelength light from it than we would if the interstellar medium were not there (Figure 12-6a). The more gas and dust between the objects and us, the redder the objects appear (Figure 12-6b). Interstellar reddening is different from reddening due to the Doppler shift (see Section 4-7). The Doppler shift causes all wavelengths of electromagnetic radiation to lengthen equally, whereas

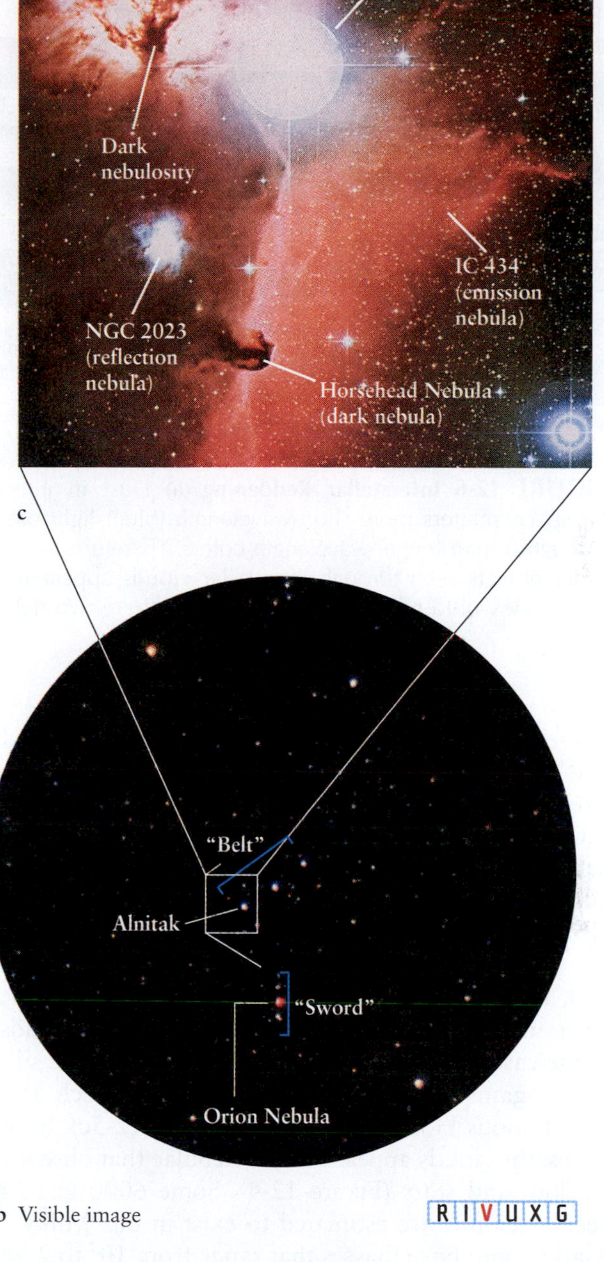

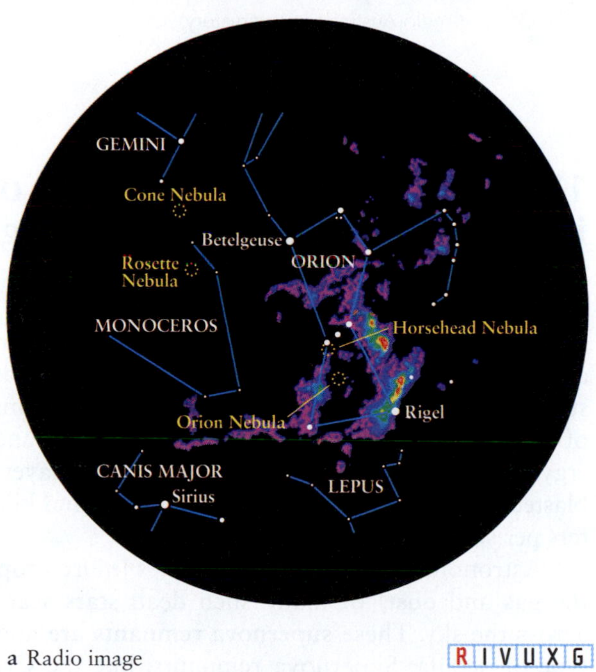

FIGURE 12-5 A Gas- and Dust-Rich Region of Orion (a) This color-coded radio map of a large section of the sky shows the extent of giant molecular clouds in Orion and Monoceros as seen in the radio part of the spectrum. The intensity of carbon monoxide (CO) emission is displayed by colors in the order of the rainbow, from violet for the weakest to red for the strongest. Black indicates no detectable emission. The locations of four prominent star-forming nebulae are indicated on the star chart overlay. Note that the Orion and Horsehead nebulae are sites of intense CO emission, indicating that stars are forming in these regions. (b, c) A variety of nebulae appear in the sky around Alnitak, also called ζ (zeta) Orionis, the easternmost star in the belt of Orion. To the left of Alnitak is a bright, red emission nebula, called NGC 2024. The glowing gases in emission nebulae are excited by ultraviolet radiation from young, massive stars. Dust grains obscure part of NGC 2024, giving the appearance of black streaks, while the distinctively shaped dust cloud, called the Horsehead Nebula, blocks the light from the background nebula IC 434. The Horsehead Nebula is part of a larger complex of dark interstellar matter, seen in the lower left of this image. Above and to the left of the Horsehead Nebula is the reflection nebula NGC 2023, whose dust grains scatter blue light from stars between us and it more effectively than any other color. All of this nebulosity lies about 1600 light-years from Earth, while the star Alnitak is only 815 light-years away from us. NGC refers to the New General Catalog of stars and IC stands for Index Catalogs, two supplements to the NGC. (a: R. Maddalena, M. Morris, J. Moscowitz, and P. Thaddeus; b: Royal Observatory, Edinburgh; c: R. C. Mitchell, Central Washington University)

interstellar reddening, due to the stronger scattering of shorter wavelengths, does not change the wavelengths of the starlight we receive—only their intensities.

We have been able to discover distant stars and clouds of interstellar gas and dust whose visible light is obscured by the nearby interstellar medium because the distant objects emit radio or infrared photons that are scattered relatively little on their way to us. Likewise, we use radio and infrared telescopes to find nearby interstellar gas and dust that, as blackbodies, are too cold to emit much visible light.

Stars form from gas and dust that become **Jeans unstable** (see Section 5-2). Most of this matter is hydrogen in the form of molecules (rather than atoms). However, molecular hydrogen is relatively hard to detect in space. Therefore, radio astronomers often search for cold interstellar gas in the form of carbon monoxide, which emits lots of photons at short radio (microwave)

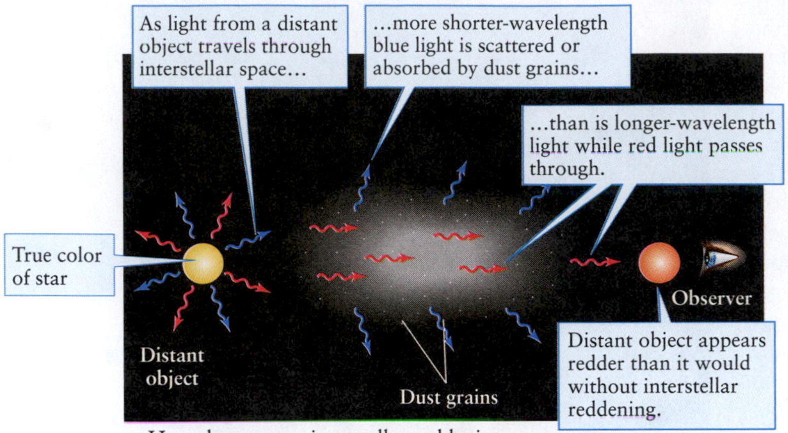

As light from a distant object travels through interstellar space...

...more shorter-wavelength blue light is scattered or absorbed by dust grains...

...than is longer-wavelength light while red light passes through.

True color of star

Distant object

Dust grains

Observer

Distant object appears redder than it would without interstellar reddening.

a How dust causes interstellar reddening

NGC 3576: A closer nebula (9,000 ly from Earth)

NGC 3603: A more distant nebula (20,000 ly from Earth)

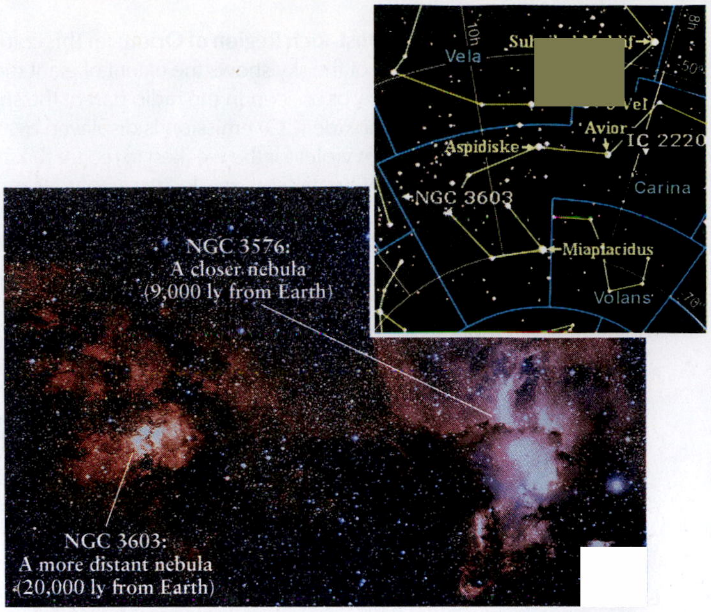

b Reddening depends on distance

R I V U X G

FIGURE 12-6 Interstellar Reddening (a) Dust in interstellar space scatters more short-wavelength (blue) light passing through it than longer-wavelength colors. Therefore, stars and other objects seen through interstellar clouds appear redder than they would otherwise. (b) Light from these two nebulae pass through different amounts of interstellar dust and therefore they appear to have different colors. Because NGC 3603 is farther away, it appears a ruddier shade of red than does NGC 3576. (Anglo-Australian Observatory)

wavelengths of 2.6 and 1.3 mm. Calculations based on the known abundances of elements reveal that there are about 10,000 hydrogen molecules (H_2) for every CO molecule in the interstellar medium. Consequently, wherever astronomers detect strong emission of CO, they deduce that an enormous amount of molecular hydrogen gas must also be present.

In mapping the locations of CO emission (Figure 12-5a), astronomers came to realize that interstellar gas and dust are often concentrated in giant molecular clouds. In some cases, these clouds appear as dark nebulae silhouetted against a glowing background light, such as Orion's famous Horsehead Nebula (Figure 12-5c). In other cases, the clouds appear as dark nebulae that obscure the background stars (Figure 12-4). Some 6000 giant molecular clouds are estimated to exist in our Milky Way Galaxy, and have masses that range from 10^5 to 2×10^6 $M_\odot$ and diameters that range from 50 to 300 light-years. The density inside each of these clouds ranges from 10^2 to 10^5 hydrogen molecules per cubic centimeter—thousands of times greater than the average density of the gas and dust dispersed throughout interstellar space, but some 10^{15} times less dense than the air we breathe. Having located interstellar matter, we will turn to why some of this gas and dust becomes Jeans unstable, and therefore collapses to form new stars (and planets). We explore this activity by expanding on the material in Section 5-2.

The interstellar medium is composed primarily of what kinds of things?

12-2 Supernovae, collisions of interstellar clouds, and starlight trigger new star formation

We will see in detail in Chapter 13 that a supernova is a violent detonation that ends the life cycle of a massive star. The core of the doomed star collapses in a matter of seconds, releasing vast quantities of particles and energy that blow the star apart. The star's outer layers are blasted into space at speeds of several thousand kilometers per second.

Astronomers have found the ashes (more properly, the gas and dust) of many such dead stars scattered across the sky. These **supernova remnants** are another type of nebula. Supernova remnants, like the Cygnus Loop shown in Figure 12-7, have a distinctly arched appearance, as would be expected for an expanding shell of gas. As it passes through the surrounding interstellar medium, the supernova remnant slams into preexisting matter, exciting the electrons in the atoms and molecules there, causing the gases to glow. If the expanding shell of a supernova remnant rams into a giant molecular cloud, it can compress the cloud, thus stimulating star birth in it. As we learned in Chapter 5, there is evidence that such an event happened around the time the solar system formed. For more about seeing nebulae on your own, see *Guided Discovery: Observing the Nebulae.* 1

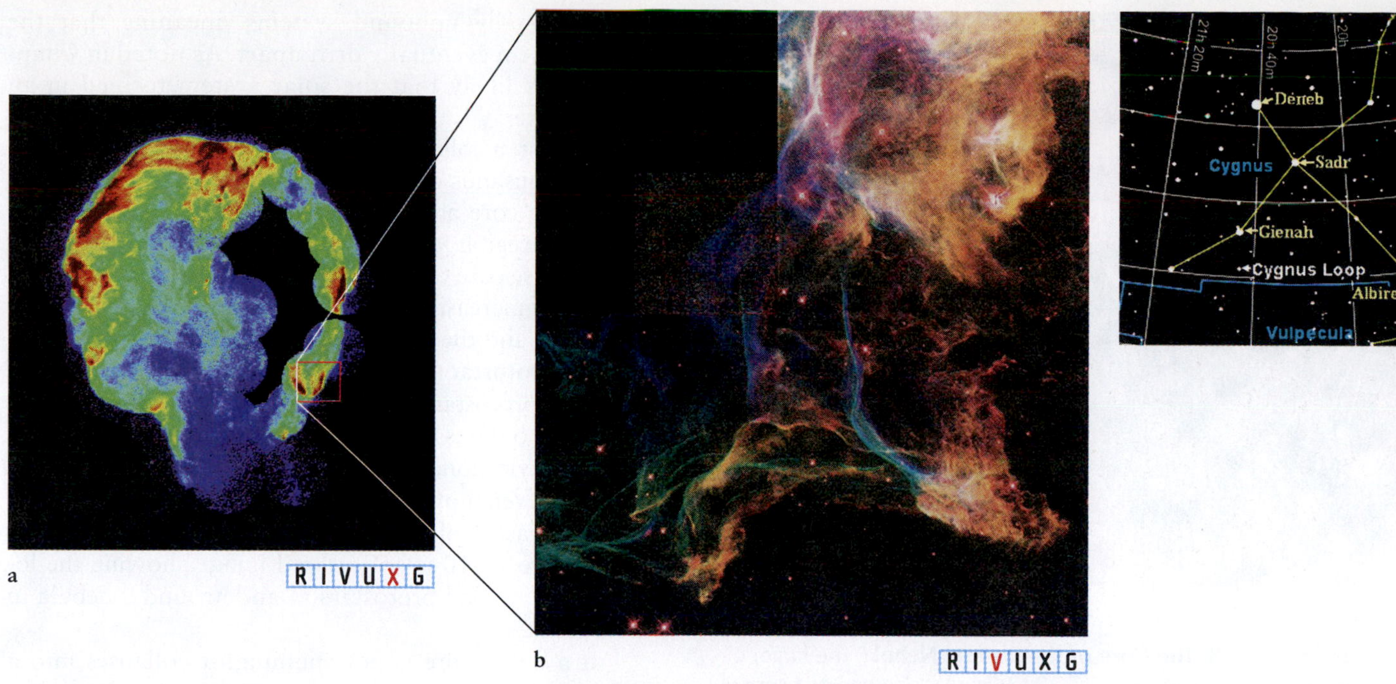

FIGURE 12-7 A Supernova Remnant (a) X-ray image of the Cygnus Loop, the remnant of a supernova that occurred nearly 20,000 years ago. The expanding spherical shell of gas now has a diameter of about 120 light-years. The entire Cygnus Loop has an angular diameter in our sky 6 times wider than the Moon. (b) This visible-light Hubble Space Telescope image of part of the Cygnus Loop shows emission from different atoms false color-coded with blue from oxygen, red from sulfur, and green from hydrogen. (a: Nancy Levenson/NASA; b: Jeff Hester, Arizona State University and NASA)

Guided Discovery

Observing the Nebulae

Distant nebulae—clusters of stars and glowing gases—are among the most impressive objects in the night sky. Binoculars and the naked eye are enough to let you "get your hands dirty" exploring them.

You can observe the Great Nebula of Orion (M42) during winter in the northern hemisphere with the naked eye. In all likelihood, you have seen it dozens of times without knowing it. To locate the Great Nebula, find the constellation Orion (using, for example, the star charts at the end of the book). Locate Orion's belt. Due south of the belt are three stars in a row that make up Orion's sword. Examine the sword very carefully with your naked eye. Do any of the stars in it look at all odd? Now look at them through a pair of binoculars. Which one is different from the others? That one is the Great Nebula in Orion—not a star at all! How does what you see compare with Figure 12-17?

The North America Nebula and the Pelican Nebula in Cygnus are best spotted in autumn. Pick a dark, moonless night and use binoculars rather than a telescope. Higher magnification reveals too small a region of the sky for you to see the entirety of these vast, dim nebulae. To find them, first locate the bright star Deneb on the tail of Cygnus (using, for example, the star chart for Cygnus in this chapter, *Starry Night*™, or the star charts at the end of the book). The North America Nebula is located 3° east of Deneb, whereas the Pelican Nebula is located 2° southeast of it. These angles are both very small, so sweep around the sky east of Deneb. If your binoculars are powerful enough, you should be able to see the outlines that give these nebulae their names.

The constellations of Orion and Monoceros encompass one of the most accessible regions of the sky for studying star formation and the interaction of young stars with the interstellar medium. Figure 12-5 shows a map of this region made with a radio telescope tuned to a wavelength of 2.6 mm, which is emitted by CO. Note the extensive areas covered by giant molecular clouds. Such comprehensive maps of CO emission help astronomers understand how the large-scale structure of the interstellar medium is related to the formation of stars.

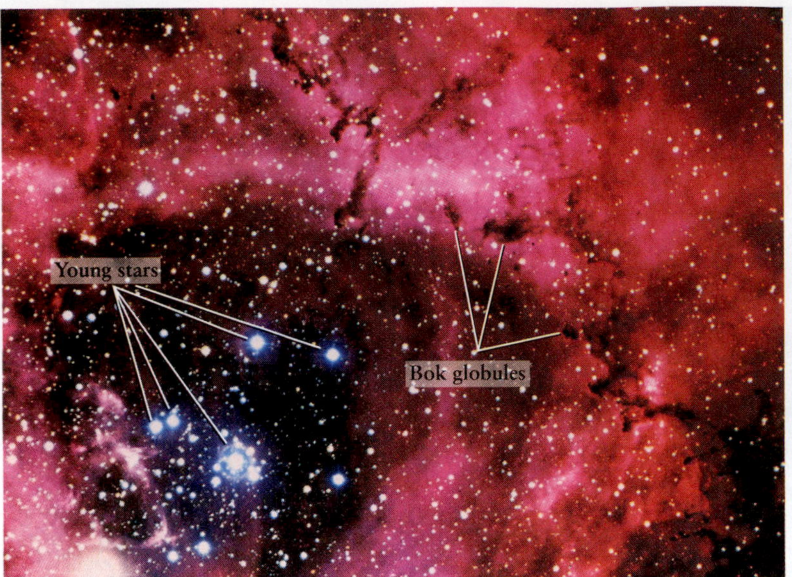

FIGURE 12-8 The Core of the Rosette Nebula The large, circular Rosette Nebula (NGC 2237) is near one end of a sprawling giant molecular cloud in the constellation Monoceros (the Unicorn). Radiation from young, hot stars has blown gas away from the center of this nebula. Some of this gas has become clumped in Bok globules that appear silhouetted against the glowing background gases. New star formation is taking place within these globules. The entire Rosette Nebula has an angular diameter on the sky nearly 3 times that of the Moon, and it lies some 3000 light-years from Earth. (Anglo-Australian Observatory)

A simple collision between two interstellar clouds can also create regions that are Jeans unstable so that they collapse and form new stars. Likewise, radiation from O and B stars, which are especially bright and hot, will ionize (remove electrons from their atomic orbits) the gas that surrounds them, which then moves away and compresses the nearby interstellar medium (Figure 12-8).

2 As small regions of a giant molecular cloud become Jeans unstable and collapse, they become denser, preventing light from behind and inside them from escaping. These darker regions are called **Bok globules** (see the very small dark nebulae in Figure 12-8), named after astronomer Bart Bok, who first studied them in the 1940s. Infrared observations show compact regions of gas and dust, called dense cores, inside Bok globules. These cores are destined to become stars, as their gravity pulls their matter inward.

Often a giant molecular cloud has several hundred or even thousands of dense cores. In that case, hundreds or thousands of stars form together. Such stellar nurseries will become **open clusters** of stars like the Pleiades (see Figure 12-2). Open clusters are gravitationally unbound systems, meaning that the stars in them eventually drift apart. As noted in Chapter 5, it is likely that the solar system formed in an open cluster.

What types of events can initiate the process of star formation?

At first, a collapsing dense core is just a cool, dusty region thousands of times larger than our solar system. The dense core actually collapses from the inside out. The inner region falls in rapidly, leaving the outer layers of the dense core to drift in at a more leisurely rate. This process of increasing mass in the central region is called *accretion,* and the newly forming object at the center is called a **protostar** (Figure 12-9). Although fusion has not begun, a protostar emits energy, some of which comes from the compression and heating of its interior caused by the gravitational force from its growing mass of hot gas. However, most of the energy it releases comes from infalling gases colliding with the surface of the protostar. Figure 12-10 is an infrared image showing the locations of myriad protostars in and around a nebula in Centaurus.

If a dense core is not spinning, it collapses into a sphere, which, ultimately, becomes an isolated star. If a dense core is spinning, it collapses into a disk, which may then condense into two or three stars. Or, if the disk has a low enough mass, it may become a single star with orbiting protoplanets, as we discussed in Chapter 5.

12-3 When a protostar ceases to accumulate mass, it becomes a pre–main-sequence star

Protostars are physically larger than the main-sequence stars into which they are evolving. A protostar of 1 $M_\odot$, for example, has about 5 times the diameter of the Sun. Because of their large sizes, protostars emit great quantities of radiation and gas (analogous to solar wind), and they can be observed as sources of infrared radiation. At this point, they cannot be seen in visible light because they are enshrouded by their outer layer of gas and dust (see Figure 12-9). Much matter is still slowly falling inward from the dense core's outer shell. Eventually, the radiation and particles that flow off the protostar exert enough outward force to halt the infall of this gas and dust. As a result, mass accretion stops, and the protostar becomes a **pre–main-sequence star.**

A pre–main-sequence star contracts slowly, unlike the rapid collapse of a protostar. When the temperature at its core reaches 10^7 K, hydrogen fusion begins. As we saw in Chapter 10, this thermonuclear process releases enormous amounts of energy. The outpouring of energy from hydrogen fusion creates enough pressure inside the pre–main-sequence star to stop its contraction. In the final stages of pre–main-sequence evolution, the outer shell of gas and dust finally dissipates (Figure 12-11; see also Figure 12-8). For the first time, the star is revealed via visible light to the outside universe.

FIGURE 12-9 Protostar in a Bok Globule (a) This visible-light image shows a small dark nebula (equivalently, a Bok globule) called L1014 located in the constellation Cygnus. (b) When viewed in the infrared, a protostar is visible within the nebula. (a: Deep Sky Survey; b: NASA/JPL-Caltech/N. Evans, University of Texas at Austin)

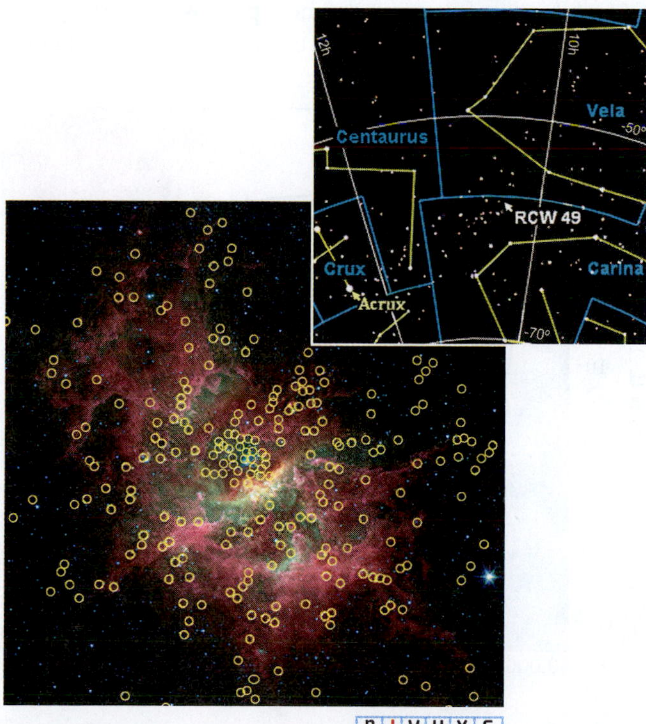

FIGURE 12-10 A Cluster of Protostars Over 300 protostars (yellow circles) were observed in the infrared by the Spitzer Space Telescope. This cluster of newly forming stars is 13,700 light-years away in the constellation Centaurus. The nebula, some of whose gas is being converted into stars, is called RCW 49 and contains more than 2200 stars and protostars. Most of the interior of this nebula is hidden from our eyes by the dust it contains. (NASA/JPL-Caltech/E. Churchwell, University of Wisconsin)

FIGURE 12-11 Pre–Main-Sequence Stars Seen in infrared, the two large bright objects in the center of this image are pre–main-sequence stars. They have recently shed their cocoons of gas and dust, but still have strong stellar winds that create their irregular shapes. The two stars are an optical double, that is, they are not orbiting each other. (Atlas Image courtesy of 2MASS/UMASS/IPAC-Caltech/NASA/NSF)

12-4 The evolutionary track of a pre-main-sequence star depends on its mass

The more massive a pre–main-sequence star is, the more rapidly it begins hydrogen fusion in its core. For example, calculations indicate that a 5 $M_\odot$ pre–main-sequence star starts fusing less than a million years after it first forms from a protostar, whereas a 1 $M_\odot$ pre–main-sequence star takes a few tens of millions of years to begin fusion.

Why can't protostars be observed with visible light telescopes?

Calculations reveal that the minimum temperature required to start normal hydrogen fusion (see *An Astronomer's Toolbox 10-1*) in the core of a star is 10 million K. However, pre–main-sequence stars less massive than 0.08 $M_\odot$ do not have enough gravitational force compressing and heating their cores to ever get this hot. As a result, these small bodies contract to become planetlike orbs of hydrogen and helium, called **brown dwarfs** (Figure 12-12). See *Guided Discovery: Extrasolar Planets and Brown Dwarfs* for further discussion of these important, albeit nonstellar, bodies. To date the lowest mass star that has been observed has 0.091 $M_\odot$. This star, part of a binary system, is only 16% larger than Jupiter.

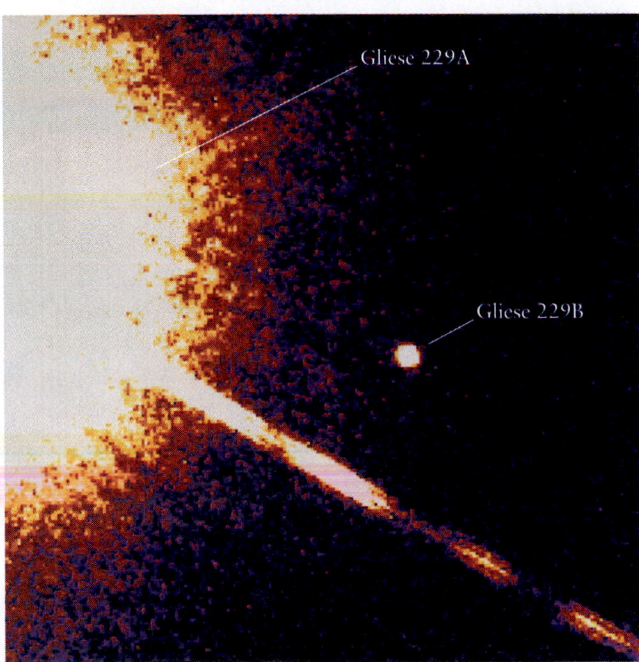

FIGURE 12-12 A Brown Dwarf Located 18 light-years (6 pc) from Earth in the constellation Lepus (the Hare), Gliese 229B was the first confirmed brown dwarf ever observed. With a surface temperature of about 1000 K, its spectrum is similar to that of Jupiter. Gliese 229B is in orbit around a star. The overexposed image of part of its companion, Gliese 229A, appears on the left. The two bodies are separated by about 43 AU. Gliese 229B has from 20 to 50 times the mass of Jupiter, but the brown dwarf is compressed to the same size as Jupiter. The spike of light was produced when Gliese 229A overloaded part of the Hubble Space Telescope's electronics. (S. Kularni, California Institute of Technology; D. Golimowski, Johns Hopkins University; NASA)

Astrophysicists use computers and the equations of stellar structure (described in Section 10-8) to model the evolution of a pre–main-sequence star. By calculating changes in the energy that the contracting star emits, computer simulations can follow its changing position on a Hertzsprung-Russell (H-R) diagram (Figure 12-13). Keep in mind that such an **evolutionary track** represents changes in a star's temperature and luminosity, not its motion in space.

Protostars transform into pre–main-sequence stars as they cross a curve called the **birth line** (see the *blue line* on Figure 12-13). A star's exact location on this curve depends primarily on its mass and, to a much smaller extent, on the amount of metal it contains. (Recall from Section 11-6 that all elements other than hydrogen and helium are considered metals by astronomers.)

Spectroscopic observations of pre–main-sequence stars show that many are vigorously ejecting gas, often in oppositely directed jets, just before they reach the main sequence. Gas-ejecting stars in spectral classes G and cooler (that is, G, K, and M) are called **T Tauri stars** (Figure 12-14), after the first example discovered in the constellation of Taurus. Some astronomers propose that the onset of hydrogen fusion is preceded by vigorous chromospheric

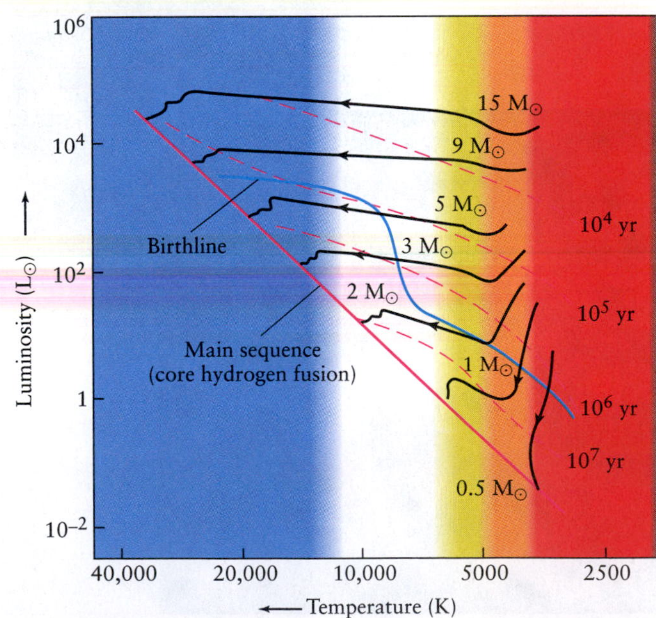

FIGURE 12-13 Pre–Main-Sequence Evolutionary Tracks This H-R diagram shows evolutionary tracks based on models of seven stars having different masses. The dashed lines indicate the stage reached after the indicated number of years of evolution. The birth line, shown in blue, is the location where each protostar stops accreting matter and becomes a pre–main-sequence star. Note that all tracks terminate on the main sequence at points that agree with the mass-luminosity relation (see Figure 11-14a).

Extrasolar Planets and Brown Dwarfs

The lowest mass that an object can have and still maintain the fusion of normal hydrogen into helium, as occurs in the Sun (see *An Astronomer's Toolbox 10-1*), is 0.08 $M_{\odot}$, or about 75 times the mass of Jupiter. Astronomers have discovered hundreds of objects in our Galaxy with less than this mass. Like Jupiter, they are primarily composed of hydrogen and helium, with traces of other elements. Many of them are found in orbit around stars, while some are found as free-floating masses that apparently formed without ever orbiting a star. An intriguing question has arisen: *What should they be called?*

Although normal hydrogen fusion does not occur in them, bodies with between 13 and 75 times Jupiter's mass do fuse deuterium (a rare form of hydrogen) into helium and those with between 60 and 75 times Jupiter's mass also fuse lithium (the element with three protons) into helium. Both of these types of fusion occur very briefly (in cosmic terms) because of the limited supplies of deuterium and lithium in any known object in space. All objects between 13 and 75 times Jupiter's mass are called *brown dwarfs*. Objects with less than 13 times Jupiter's mass that are orbiting stars are *extrasolar planets* or *exoplanets* (see Chapter 5), while free-floating bodies with less than 13 times Jupiter's mass are often called *sub-brown dwarfs*. Bear in mind that these definitions are still undergoing discussion and revision in the astronomy community.

Because they emit relatively little energy compared to stars, extrasolar planets and brown dwarfs are dim and therefore very challenging to observe. Those in orbit are detected by their gravitational or eclipsing effects on the stars they orbit. The first brown dwarf was discovered only in 1994. Named Gliese 229B, it is located in orbit around a star, Gliese 229A (see Figure 12-12), in the constellation Lepus, about 18 ly (6 pc) from Earth. A decade later, an extrasolar planet, 2M1207b, was observed orbiting brown dwarf 2M1207 (Figure 5-15). Hundreds more brown dwarfs have been found, along with more than a dozen sub-brown dwarfs. Many of these are found in active star-forming regions, such as the Orion Nebula (see Figure 12-17) and the rho Ophiuchi cloud (see accompanying figure). Astronomers have also found more than 520 extrasolar planets. In 2002, astronomers observed clouds and storms on a brown dwarf similar to, but probably much larger than, the storms observed on the giant planets in our solar system.

Brown dwarfs of larger mass have the interesting feature that when they fuse deuterium or lithium, the helium they create moves upward, out of the core where it is formed. This helium is replaced with fresh deuterium or lithium fuel to fuse. The upward motion of the helium and downward motion of deuterium and lithium-rich hydrogen are due to convection, and, as a result of this motion, eventually all the deuterium and lithium are consumed. We say that these brown dwarfs are fully convective (for further discussion of fully convective stars, see Section 12-8). This convective behavior is different than we find in the Sun (see Chapter 10), which has a separate core, convective zone, and radiative zone that do not share atoms. Flares have been observed from brown dwarfs. By analogy to the Sun's flares caused by magnetic fields emerging from its surface, astronomers believe that some brown dwarfs rotate and have magnetic fields.

Based on the numbers and locations of the known brown dwarfs and sub-brown dwarfs, astronomers estimate that there may be as many of these bodies in our Milky Way Galaxy as there are stars. Even in these numbers, brown dwarfs do not contribute a substantial amount of mass or gravitational force in the Galaxy because they have such small individual masses.

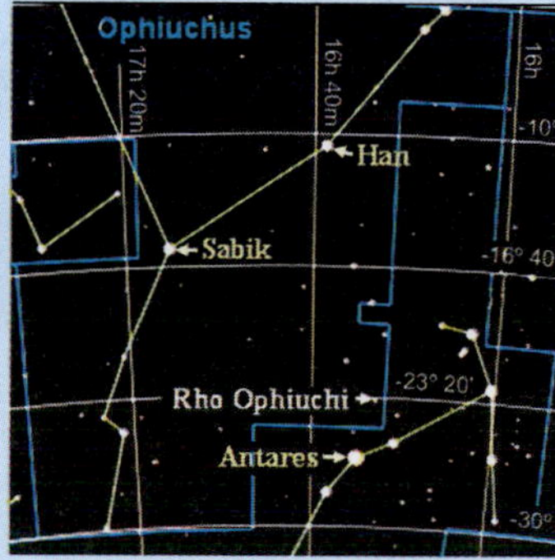

A Stellar Nursery Full of Brown Dwarfs Besides containing more than 100 young stars, the rho Ophiuchi cloud, located 540 ly away in the constellation Ophiuchus, contains at least 30 brown dwarfs. By studying these objects, astronomers expect to learn more about early stellar evolution. This infrared image is color coded, with *red* indicating 7.7-μm radiation and *blue* indicating 14.5-μm radiation. (Infrared Space Observatory, NASA)

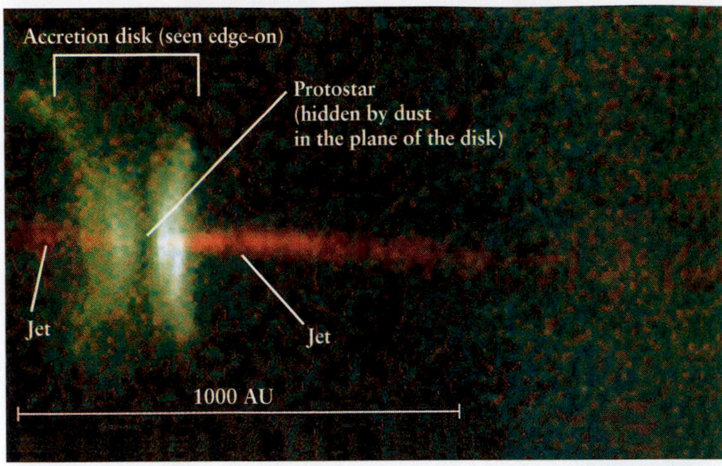

FIGURE 12-14 T Tauri star Protostar HH 30 is emitting gas primarily as two opposed jets. Called a T Tauri star, it is not yet on the main sequence. The star is hidden inside a disk of gas and dust. (C. Burrows, the WFPC-2 Investigation Definition Team, and NASA)

How are T Tauri stars different from main-sequence stars, like the Sun?

activity marked by enormous spicules (see Section 10-2) and flares (see Section 10-6) that propel the star's outermost layers back into space. In fact, an infant star going through its T Tauri stage can lose as much as 0.4 $M_{\odot}$ of matter and also shed its cocoon while still a pre–main-sequence star. In light of all this activity, it is not surprising that observations reveal T Tauri stars to be variable, meaning that they change brightness much more than, say, the Sun does today. We will discuss variable stars further in Sections 12-12 and 12-13.

Pre–main-sequence stars more massive than 2 $M_{\odot}$ become hotter without much change in overall luminosity. The evolutionary tracks of these pre–main-sequence stars thus traverse the H-R diagram nearly horizontally, from right to left (see Figure 12-13). A star more massive than about 7 $M_{\odot}$ has no pre–main-sequence phase at all. Its gravitational compression is so great that it begins to fuse hydrogen in its protostellar phase.

Until 2004, the upper limit to main-sequence stellar mass was thought to be around 120 $M_{\odot}$. This number was based on calculations that show that above this mass, protostars rapidly develop extremely high fusion rates in their cores, which lead to extremely high surface temperatures—temperatures so great that their outer layers are superheated and thereby expelled into interstellar space. This expulsion, in turn, decreases their masses and their temperatures. An example of a very massive star in the process of shedding mass as it settles down onto the main sequence is the Pistol Star (Figure 12-15a). One of the most luminous stars in our Galaxy, the Pistol Star may have started as a protostar with 200 $M_{\odot}$. Observations reveal that every few thousand years it expels shells of gas, and it may have less than 10 $M_{\odot}$ left when the expulsion of matter stops. The entire process of mass loss by very massive stars takes only a few million years.

In 2004, however, astronomers observed a main-sequence star that apparently has between 130 and 150 $M_{\odot}$ (Figure 12-15b). In 2006 they saw the remnants of a star that must have been at least 150 $M_{\odot}$. Astronomers are trying to reconcile the theory of massive star formation with these observations.

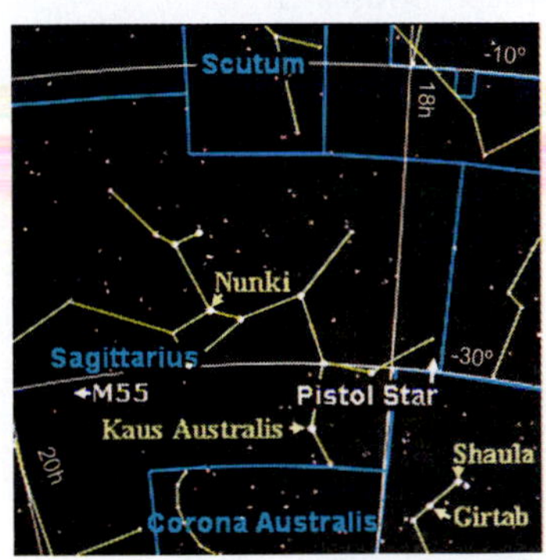

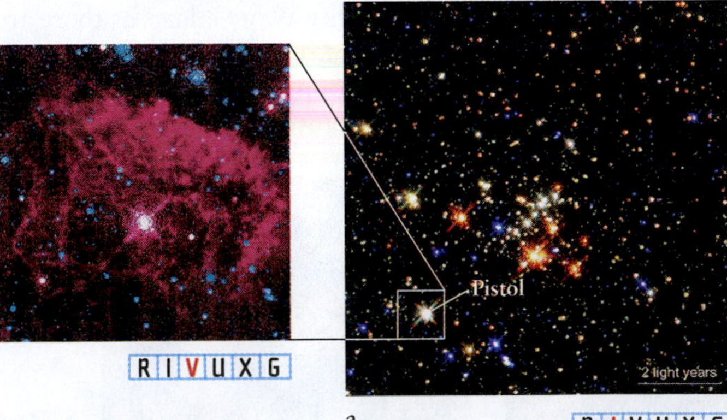

FIGURE 12-15 Mass Loss from a Supermassive Star (a) The Quintuplet Cluster is 25,000 light-years from Earth. (Inset) Within the cluster is one of the brightest known stars, called the Pistol. Astronomers calculate that the Pistol formed nearly 3 million years ago and originally had 100–200 $M_{\odot}$. The structure of the gas cloud suggests the star ejected the gas we see in two episodes, 6000 and 4000 years ago. The gas from any previous ejections is so thinly spread now that we cannot see it. The nebula shown in the inset is more than 4 ly (1.25 pc) across—it would stretch from the Sun nearly to the closest star, Proxima Centauri. The name Pistol was given to the star based on early, low-resolution radio images of its gas, which initially looked like an old-fashioned pistol aimed to the left near the top of the inset. (b) The largest, most massive known star, LBV 1806-20, is 5 million times brighter and apparently some 150 times more massive than the Sun. This drawing shows the star's color and its size compared to the Sun. (a: D. Filger, NASA; b: Image by Dr. Stephen Eikenberry, Meghan Kennedy/University of Florida)

12-5 H II regions harbor young star clusters

We can detect young open clusters of stars from the magnificent glows they create in the nebulae in which they form. Figures 12-5 and 12-16 show examples of these emission nebulae. Because these nebulae are predominantly ionized hydrogen, they are also called **H II regions.** To see why H II regions occur, remember that the most massive pre–main-sequence stars, those of spectral types O and B, are exceptionally hot. Their surface temperatures are typically 15,000 to 35,000 K, causing them to emit vast quantities of ultraviolet radiation. This energetic radiation easily ionizes any surrounding hydrogen gas, thereby creating an H II region. Photons from an O5 star can ionize hydrogen atoms up to 30 light-years away.

H II denotes ionized hydrogen, which has no electrons in orbit that can make transitions and give off or absorb photons; therefore, how can we observe it? While some hydrogen atoms in the H II regions are being knocked apart by ultraviolet photons, some of the free protons and electrons approach each other so closely that they combine to become neutral hydrogen, H I. As these new hydrogen atoms assemble, their electrons return to their ground states ($n = 1$). This downward cascade through each atom's energy levels, releasing photons with each jump between levels, is what makes the nebula glow. Particularly prominent is the transition from $n = 3$ to $n = 2$, which produces H_α photons at 656 nm in the red portion of the visible spectrum (review the emission line spectrum in Figure 4-8c, d). Thus, the nebula around a newborn star cluster often shines with a distinctive reddish hue (see Figure 12-16). Nebulae that have oxygen often look green when seen with the eye through a telescope because this gas has a green emission line at 501 nm. Because the eye is more sensitive to green light than red, the dimmer oxygen emission line appears brighter in our brains than does the H_α line, which shows up well on CCD images, which usually appear red.

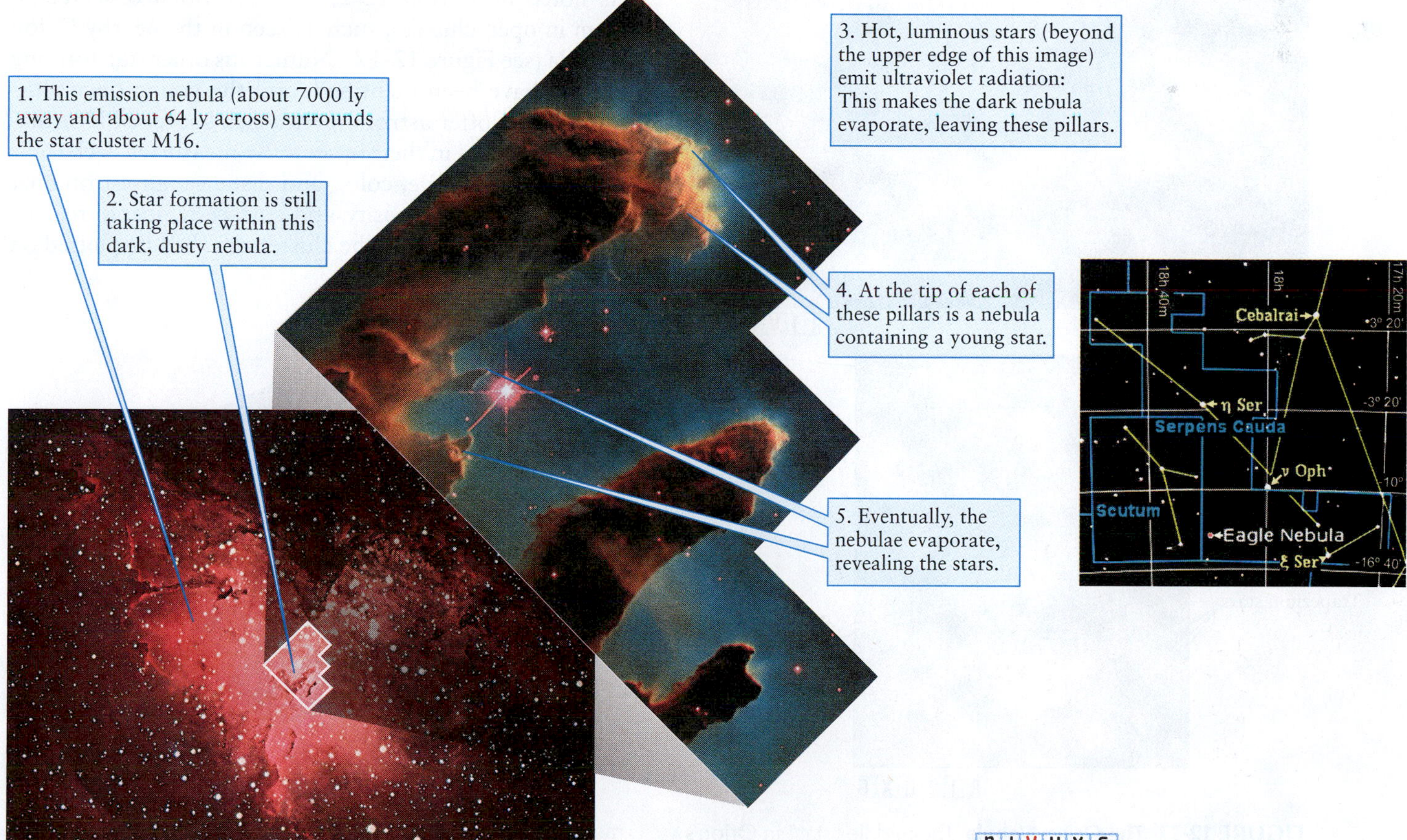

ANIMATION 12.1

FIGURE 12-16 An H II Region This emission nebula, M16, called the Eagle Nebula because of its shape, surrounds a star cluster. It is so numbered because it was the sixteenth object in the Messier Catalogue of astronomical objects. Star formation is presently occurring in M16, which is located 7000 light-years from Earth in the constellation of Serpens Cauda (the Serpent's Tail). Several bright, hot O and B stars are responsible for the ionizing radiation that causes the gases to glow. (Inset) Star formation is occurring inside these dark pillars of gas and dust. Intense ultraviolet radiation from existing massive stars off to the right of this image is evaporating the dense cores in the pillars, thereby prematurely terminating star formation there. Newly revealed stars are visible at the tips of the columns. (Anglo-Australian Observatory; J. Hester and P. Scowen, Arizona State University; NASA)

An H II region is a small, bright "hot spot" in a giant molecular cloud. The collection of hot, bright O and B stars that produces the ionizing ultraviolet radiation is called an **OB association.** The famous Orion Nebula (Figure 12-17) is an example. Four O and B stars in an open cluster called the Trapezium, at the heart of the Orion Nebula, are the primary sources of the ionizing radiation that causes the surrounding gases to glow. The Orion Nebula is embedded in a giant molecular cloud whose mass is estimated at 500,000 $M_{\odot}$.

The OB association that creates an H II region also affects the rest of the giant molecular cloud in which it is imbedded (see Figure 12-17 insets). Detailed models indicate that vigorous stellar winds, along with ionizing ultraviolet radiation from these stars, carve out a cavity in the cloud. Where this outflow is supersonic, it creates a shock wave, like the sonic boom created by fast-flying aircraft (for example, see the shock waves shown in the right inset of Figure 12-17). The shock wave forms along the outer edge of the expanding H II region, compressing hydrogen gas as it passes and thereby stimulating a new round of star birth. As more O and B stars form, they power the expansion of the H II region still farther into the giant molecular cloud. Meanwhile, the older O and B stars left behind begin to disperse (Figure 12-18). In this way, an OB association "eats into" a giant molecular cloud, creating stars in its wake. The inset in Figure 12-18 shows star formation around a single O star.

12-6 Plotting a star cluster on an H-R diagram reveals its age

As noted in Section 12-2, stars are often observed to form in open clusters, such as seen in the nearby Orion Nebula (see Figure 12-17). Numerous other star-forming regions have been identified, and the young open clusters in them offer astronomers a rich source of information about stars in their infancy. By measuring each star's apparent magnitude, color, and distance, an astronomer can deduce its luminosity and surface temperature. The data for all the stars in the cluster can then be plotted on

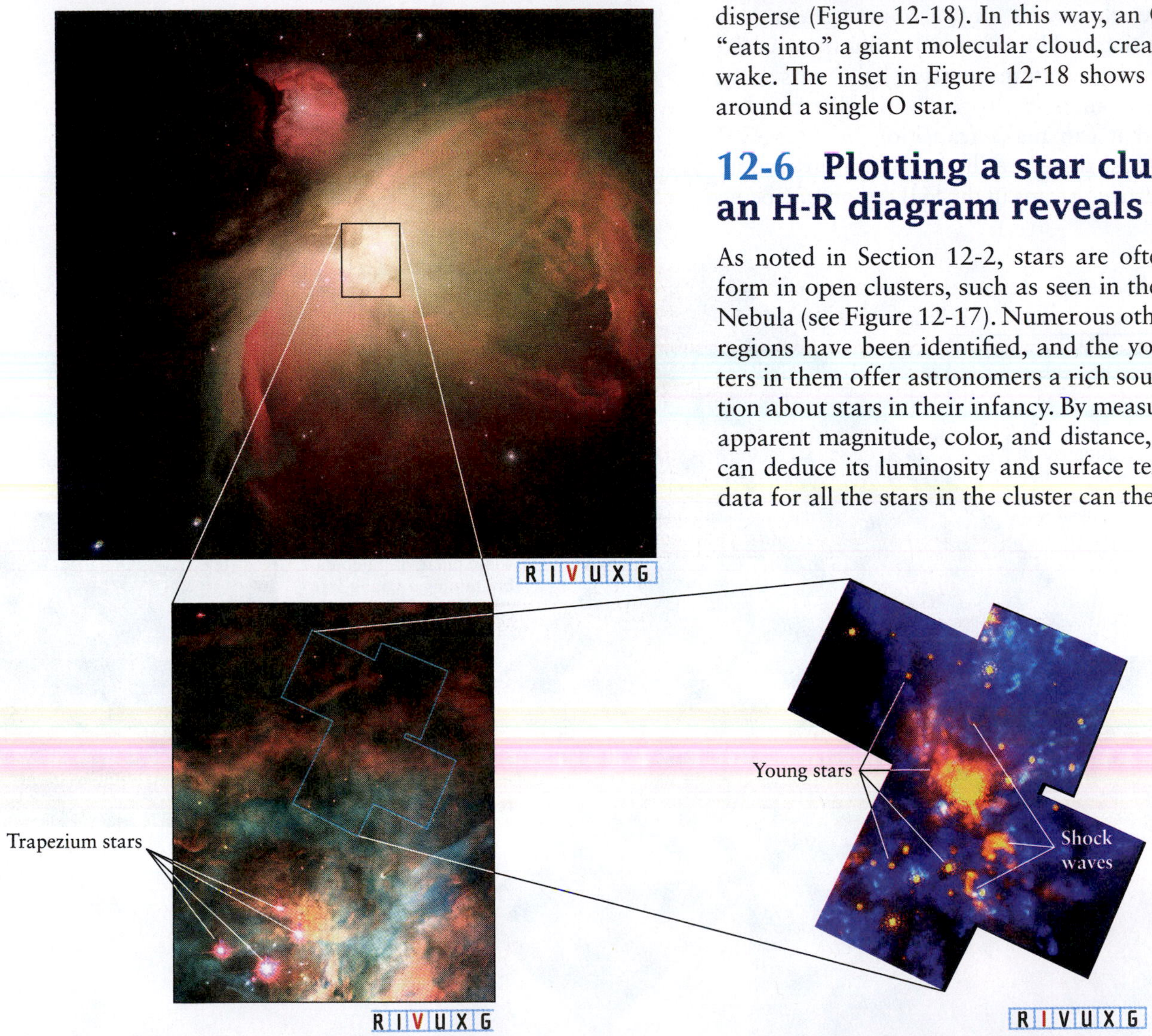

FIGURE 12-17 The Orion Nebula The middle "star" in Orion's sword is actually the Orion Nebula, part of a huge system of interstellar gas and dust in which new stars are now forming. The Orion Nebula is a region visible to the naked eye. It is 1600 ly (490 pc) from Earth and has a diameter of roughly 16 ly (5 pc). This nebula's mass is about 300 $M_{\odot}$. (Left inset) This view at visible wavelengths shows the inner regions of the Orion Nebula. At the lower left are four massive stars, the brightest members of the Trapezium star cluster, which cause the nebula to glow. (Right inset) This view shows that infrared radiation penetrates interstellar dust that absorbs visible photons. Numerous infrared objects, many of which are stars in the early stages of formation, can be seen, along with shock waves caused by matter flowing out of protostars faster than the speed of sound waves in the nebula. Shock waves from the Trapezium stars may have helped trigger the formation of the protostars in this view. (ESO—European Southern Observatory; left inset: C. R. O'Dell, S. K. Wong, and NASA; right inset: R. Thompson, M. Rieke, G. Schneider, S. Stolovy, E. Erickson, D. Axon, and NASA)

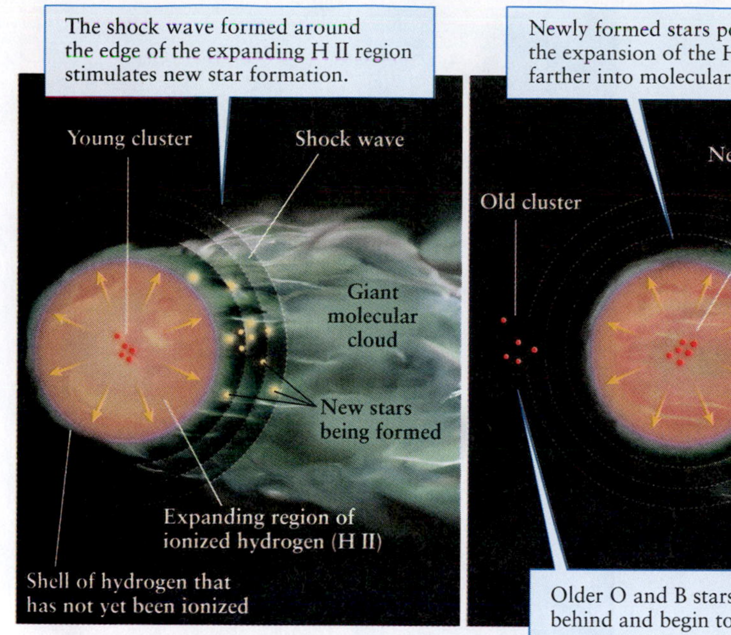

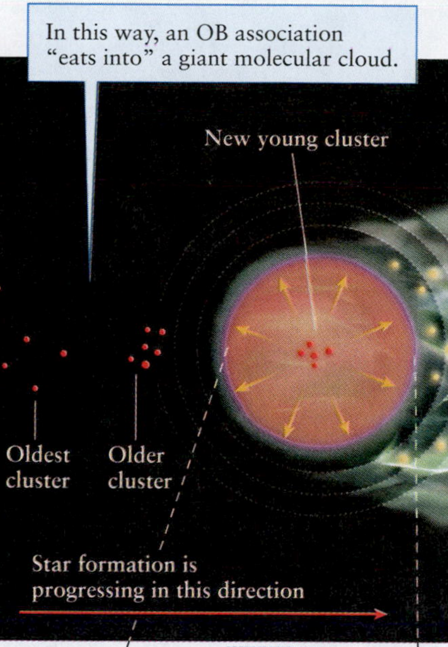

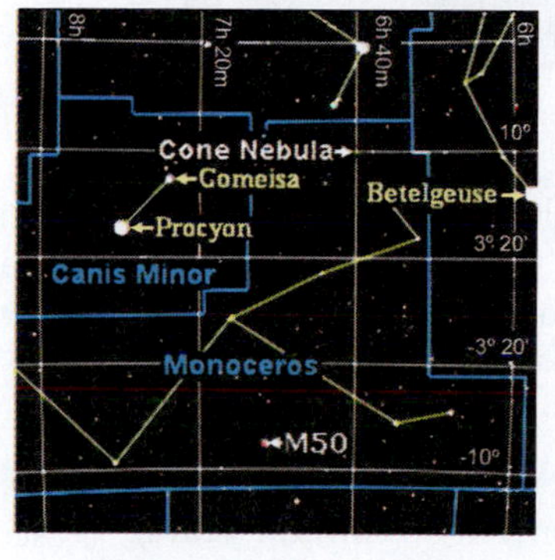

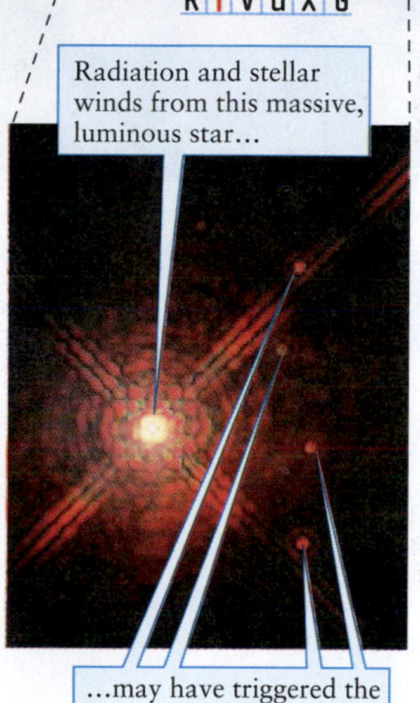

FIGURE 12-18 The Evolution of an OB Association High-speed particles and ultraviolet radiation from young O and B stars produce a shock wave that compresses gas farther into the molecular cloud, stimulating new star formation deeper in the cloud. Meanwhile, older stars are left behind. (Inset) Stars forming around a massive star 2500 ly (770 pc) away in the constellation Monoceros's Cone Nebula. The stars (small dots on the right side of the inset) arrayed around the bright, massive central star are believed to have formed as a result of the central star compressing surrounding gas with high-speed particles and radiation. The younger stars are just 0.04–0.08 light-years from the central star. (Adapted from C. Lada, L. Blitz, and B. Elmegreen; inset: R. Thompson, M. Rieke, G. Schneider, and NASA)

an H-R diagram, as shown for the Pleiades in Figure 12-2 or for the cluster NGC 2264 in Figure 12-19.

Because all of the stars in a cluster begin forming at the same time, and stars with different masses arrive on the main sequence at different times, astronomers can use the H-R diagram to determine the age of a cluster. For example, note that the hottest stars in NGC 2264 lie on the main sequence. These hot stars have surface temperatures around 20,000 K and are extremely bright and massive. Their radiation also causes the surrounding gases to glow. Most of the stars cooler than about 10,000 K have not yet arrived at the main sequence. These less massive stars, which are in the final stages of pre–main-sequence contraction, are just now beginning to ignite thermonuclear reactions at their centers.

From the H-R diagram of a young cluster we can see which are the lowest-mass stars that have already entered the main sequence. Using this information, along with theories of stellar evolution, we can determine the cluster's age. The cluster NGC 2264, for example, is roughly 2 million years old. In contrast, nearly all of the stars in the Pleiades (see Figure 12-2c) have completed their pre–main-sequence stage. That cluster's age is calculated to be about 100 million years, which is how long it takes for the least massive stars to finally begin hydrogen fusion in their cores.

> What are groups of high-mass stars called?

Open clusters dissipate. As noted earlier, open clusters, such as the Pleiades and NGC 2264, possess barely enough mass to hold themselves together. A star moving faster than the average speed for the cluster occasionally escapes. This lowers the total gravitational force of the cluster, making it easier for other stars to leave. Observations indicate that the stars in most open clusters separate from each other and eventually mix with the rest

> Which star arrives on the main sequence first, one that is 0.5 $M_\odot$ or one that is 2 $M_\odot$?

a NGC 2264 R I V U X G

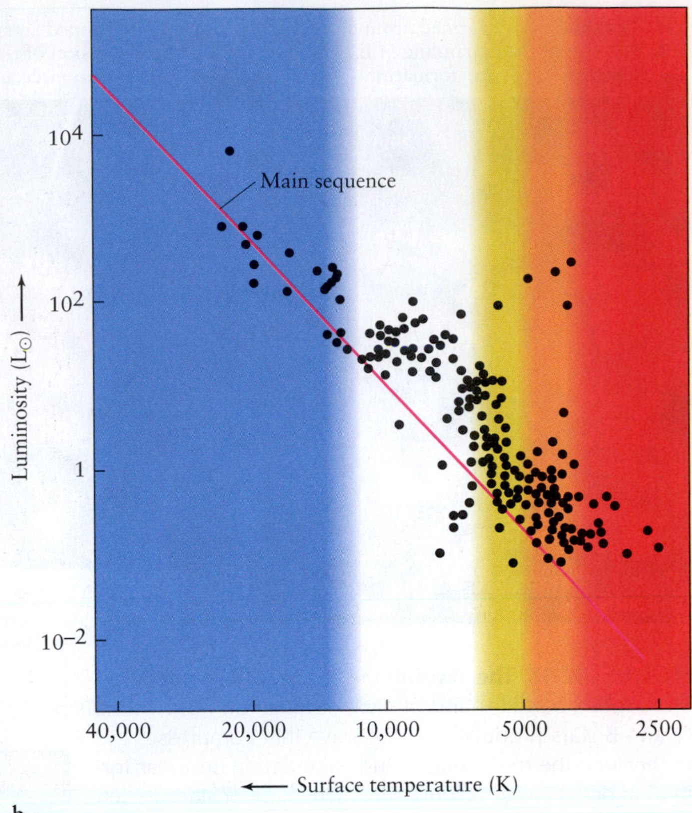

b

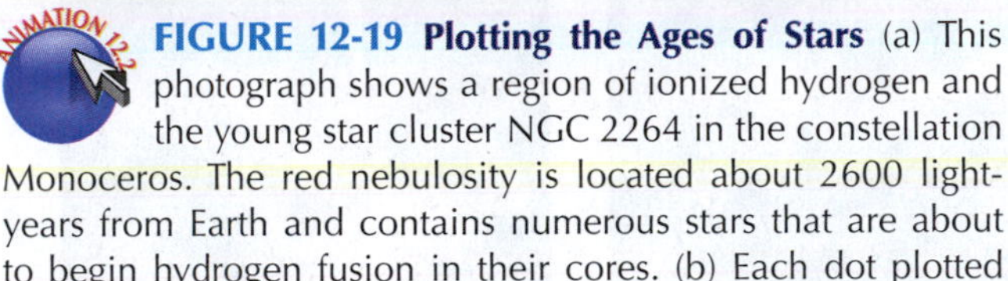

FIGURE 12-19 Plotting the Ages of Stars (a) This photograph shows a region of ionized hydrogen and the young star cluster NGC 2264 in the constellation Monoceros. The red nebulosity is located about 2600 light-years from Earth and contains numerous stars that are about to begin hydrogen fusion in their cores. (b) Each dot plotted on this H-R diagram represents a star in this cluster whose luminosity and surface temperature have been measured. Note that most of the cool, low-mass stars have not yet arrived at the main sequence. Calculations of stellar evolution indicate that this star cluster started forming about 2 million years ago. (a: David Malin/Anglo-Australian Observatory)

of the stars in the Galaxy within 10–50 million years after the cluster began forming, although some open clusters last for a few hundred million years.

MAIN-SEQUENCE AND GIANT STARS

In the first part of this chapter, we saw how stars form from dense cores of gas and dust inside giant molecular clouds. As these cores collapse, the gas and dust in their centers quickly form protostars. Eventually, protostars have pulled in (accreted) most of the mass available to them and they become pre–main-sequence stars that slowly contract. As a pre–main-sequence star evolves, fusion begins in its core.

Recall from Chapter 10 that in our Sun, thermal pressure created by fusion in the core pushes outward, everywhere balancing the inward force of gravity so that the Sun neither expands nor collapses. The Sun is said to be in hydrostatic equilibrium (see Figure 10-21). When each layer of a pre–main-sequence star can finally support all of the layers above it (meaning that collapse ceases), that star also comes into hydrostatic equilibrium, and a main-sequence star is born (see Figure 12-13). *Main-sequence stars are those stars in hydrostatic equilibrium, with nuclear reactions fusing hydrogen into helium in their cores at nearly constant rates.* Figure 12-20 summarizes the star formation process.

12-7 Stars spend most of their lives on the main sequence

The **zero-age main sequence** (ZAMS) is the set of locations on the H-R diagram where pre–main-sequence stars of different masses first become stable objects, neither shrinking nor expanding. This sequence is denoted by the solid red line that appears on several of the H-R diagrams in this chapter. Note that the theoretical evolutionary tracks in Figure 12-13 end at locations along the main sequence that agree with the observed mass-luminosity relation (recall Figure 11-14a): The most massive main-sequence stars are the most luminous, whereas the least massive stars are the least luminous.

The energy emitted by stars is primarily generated 
by thermonuclear fusion (see *An Astronomer's Toolbox 10-1*). More luminous (brighter) stars generate and radiate more energy each second than do dimmer

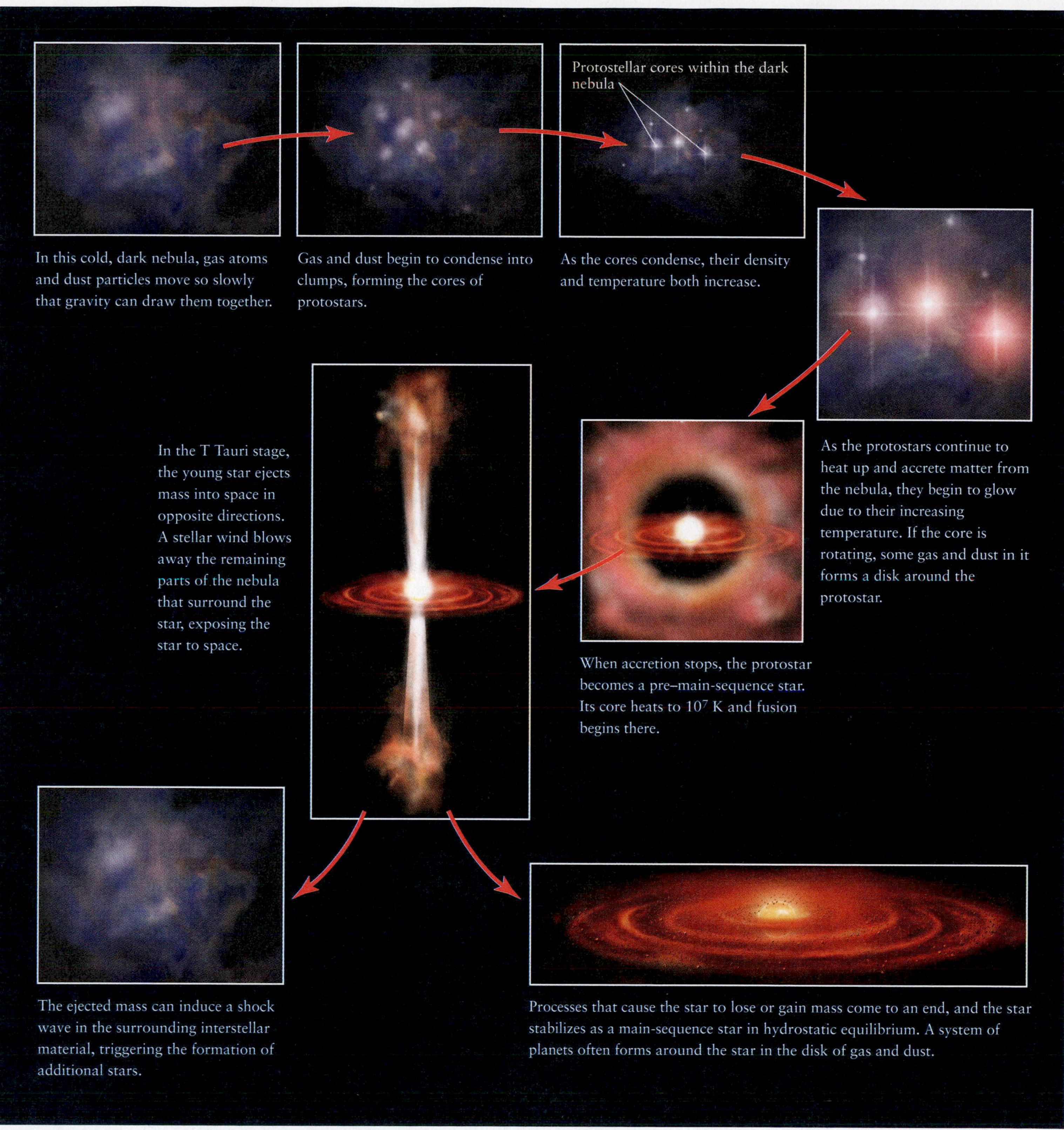

FIGURE 12-20 A Summary of the Star Formation Process This set of drawings takes you through the star formation process for the Sun and other stars with less than about 1.5 solar masses.

stars. Put another way, higher-mass stars are consuming more fuel than are lower-mass stars. This raises the interesting question of whether higher-mass stars have enough extra nuclear fuel to shine longer than lower-mass stars. The answer is *no*. The equations of nuclear fusion reveal that higher-mass stars consume fuel so quickly that these stars actually spend *less* time at each stage of stellar evolution than do lower-mass stars. For example, the tremendous internal pressure created by gravity inside O stars forces them to consume all of their core hydrogen in only a few million years, as shown in Table 12-2. Conversely, stars of very low mass take hundreds of billions of years to convert their cores from hydrogen into helium. Because, as we will see, the universe is roughly 14 billion years old, no main-sequence star with mass less than about 0.75 $M_\odot$ has yet moved into the next stage of stellar evolution.

About how long ago was the Sun a ZAMS star?

The equations further reveal that the conversion of hydrogen into helium in every star's core takes a long time compared to any other stage of its stellar evolution. (These stages are presented in the following sections and the next two chapters.) That is why the vast majority of stars represented on an H-R diagram are located on the main sequence. This raises another intriguing question: Because most stars are on the main sequence, why are most of the stars visible to the naked eye other kinds of stars? The answer will become apparent shortly.

EVOLUTION OF STARS WITH MASSES BETWEEN 0.08 AND 0.4 $M_\odot$

Because stars with mass less than 0.4 $M_\odot$ and stars with mass more than 0.4 $M_\odot$ evolve beyond the ZAMS so differently, we explore their evolutionary stages separately.

12-8 Red dwarfs convert essentially their entire mass into helium

The lowest-mass main-sequence stars, called **red dwarfs,** have masses between 0.08 $M_\odot$ and 0.4$M_\odot$. They have the lowest core temperatures of all stars. Because fusion rates depend critically on temperature, these stars produce the least amount of energy and therefore emit the least energy; they are the dimmest of all main-sequence stars.

Red dwarfs are also different from higher-mass main-sequence stars in that the lower-mass stars transport the helium that they create out of their cores. This hot helium moves upward by convection, while cooler, denser gases from the hydrogen-rich outer layers move downward and into the core (Figure 12-21). This hydrogen is fused into helium, too. As a result, these stars eventually convert their entire mass into helium.

Fusion in the cores of red dwarfs is so slow that they remain on the main sequence for trillions of years. (In comparison, the theory of solar evolution predicts that the Sun's total lifetime on the main sequence will be about 10 billion years.) When a red dwarf has eventually converted its entire mass into helium, fusion will cease inside it because the star does not have enough pressure in its core to heat and fuse that helium into anything else. These helium bodies will then radiate the heat they generated while on the main sequence, becoming cooler and dimmer. They will move down and to the right on the H-R diagram.

We saw in Section 11-11 how there is a relationship between the masses and luminosities of main-sequence stars (for example, see Figure 11-14). Besides being the lowest mass, dimmest main-sequence stars, red dwarfs are also the most common stars in our Galaxy, representing some 85% of all its stars. However, red dwarfs are so dim that astronomers continue to discover more of them even in our astronomical neighborhood. Red dwarfs typically form as isolated bodies surrounded with planets, unlike more massive stars, which are more apt to form

Table 12-2 Main-Sequence Lifetimes

Mass ($M_\odot$)	Surface temperature (K)	Luminosity ($L_\odot$)	Time on main sequence (10^6 years)	Spectral class
25	35,000	80,000	3	O
15	30,000	10,000	15	B
3	11,000	60	500	A
1.5	7000	5	3000	F
1.0 (Sun)	6000	1	10,000	G
0.75	5000	0.5	15,000	K
0.50	4000	0.03	200,000	M

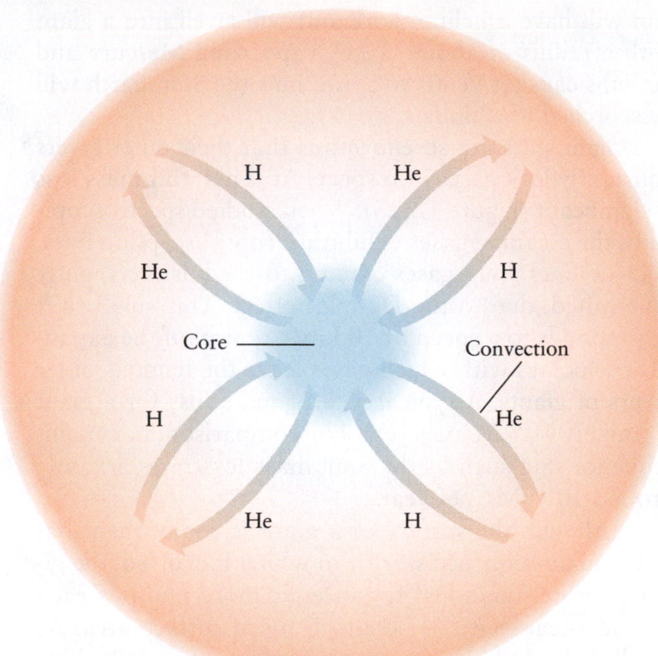

FIGURE 12-21 Fully Convective Star This drawing shows how the helium created in the cores of red dwarfs rises into the outer layers of the star by convection, while the hydrogen from the outer layers descends into the core. This process continues until the entire star is composed of helium.

in binary star systems. Because red dwarfs are the most common stars and are usually single, astronomers have recently concluded that most stars in our Galaxy are isolated, rather than binary star systems, as they thought until just a few years ago. Current calculations indicate that two-thirds of the stars in our Galaxy are isolated.

EARLY AND MIDDLE EVOLUTION OF STARS WITH MORE THAN 0.4 $M_\odot$

The changes in stars with more than $0.4M_\odot$ are much more complicated. In this chapter, we consider what happens to these stars on the main sequence and in the evolutionary stage after it. In the following two chapters, we explore the ends of their lives.

12-9 When core hydrogen fusion slows down, a main-sequence star with mass greater than 0.4 $M_\odot$ becomes a giant

Recall that the main sequence is defined by the fusion of hydrogen into helium in a star's core. Unlike lower-mass stars, stars with more than $0.4M_\odot$ do not transport helium out of their cores by convection. Therefore, the chemical composition of their cores differs from the composition of their outer layers throughout their lives.

Is the name red dwarf appropriate for the stars so named? Why or why not?

When the hydrogen in the core of a main-sequence star with more than 0.4 $M_\odot$ is mostly converted into helium, fusion in the core drastically slows down. The temperature of the core at that time is not high enough to enable its helium to fuse into other elements. However, the star continues to evolve and, contrary to intuition, the end of core hydrogen fusion actually causes these stars to expand in size.

Recall from our discussion of the Sun in Chapter 10 that, while on the main sequence, a star's outer layers are supported by thermal pressure, the energy for which is supplied by fusion in the core. As the rate of this fusion decreases, a star can no longer support the crushing weight of its outer layers, and so the hydrogen-rich gas just outside the core is compressed inward and thereby heated enough to begin fusing into helium (Figure 12-22). This process is called **hydrogen shell fusion** because it occurs in a shell a few thousand kilometers thick surrounding the core. Recall from *An Astronomer's Toolbox 10-1* that the proton-proton chain is the major source of energy inside the Sun. This fusion process also predominates in hydrogen shell fusion.

Why, then, does the star expand to become a giant? 4
The photons created in shell fusion are generated closer to the star's outer layers than are the photons generated in the core. As noted in Section 10-8, photons provide the energy and hence the outward pressure that keeps stars in hydrostatic equilibrium. Photons first created in the

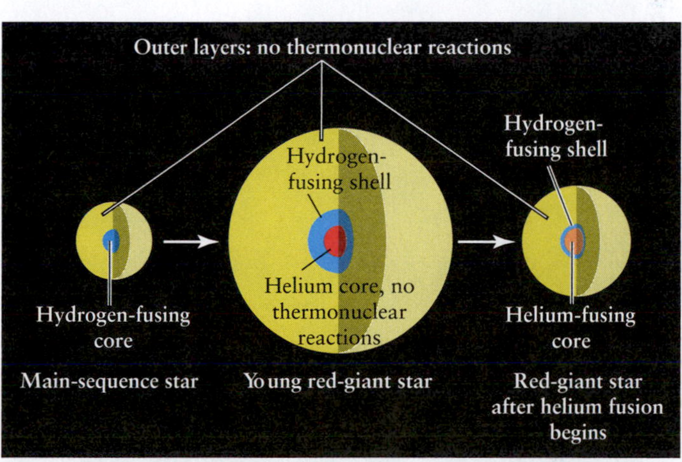

FIGURE 12-22 Evolution of Stars Off the Main Sequence (a) Hydrogen fusion occurs in the cores of main-sequence stars. (b) When the core is converted into helium, fusion there ceases and then begins in a shell that surrounds the core. The star expands into the giant phase. This newly formed helium sinks into the core, which heats up. (c) Eventually, the core reaches 10^8 K, and core helium fusion begins. This activity causes the core to expand, slowing the hydrogen shell fusion and thereby forcing the outer layers of the star to contract.

shell have more energy with which to push the outer layers outward than do photons from the core, which have to hold out more of the star's mass. In other words, the core photons have lost energy by the time they emerge from the core, so they can't push the outer layers out as much as can shell photons. The outer layers of stars with shell fusion therefore expand farther out (the stars become larger) than they were on the main sequence.

Insight Into Science

More Than the Sum of Its Parts We have gotten to the point in science where some computer models that numerically solve complex sets of equations can accurately reproduce phenomena without fully explaining them. For example, computer models of the equations that describe how stars evolve show them expanding into the giant phase. This gives us some confidence that the equations are meaningful representations of reality. However, the models are so complex that astrophysicists are unable to identify the specific physical effects that take place that cause the stars to expand.

Where in the Sun will the energy be generated to expand it into the giant phase?

Far from the fusing shell, the surface gases cool and the temperature of the star's bloated surface decreases between 3000 and 6000 K, depending on the star's total mass. In about 5 billion years, our Sun will have a helium core and will swell into a giant with a radius of about ½ AU, vaporizing Mercury and perhaps causing Venus to spiral into the Sun. Earth will be scorched to a cinder.

Giant stars are so enormous that their outer layers constantly leak gases into space. At times, this mass loss is significant (Figure 12-23). When studied spectroscopically, the escaping gases exhibit narrow absorption lines, and the lines from gases coming toward us are slightly blueshifted, due to the Doppler effect. This small shift corresponds to a speed of 10 km/s, typical of the expansion velocities with which gases leave the tenuous outer layers of giants. A typical rate of mass loss for a giant is roughly 10^{-7} $M_\odot$ per year. For comparison, in a main-sequence star, such as the Sun, mass loss rates are only around 10^{-14} $M_\odot$ per year.

Although the surface of a giant is cooler than that of the main-sequence star from which it evolved, the giant is more luminous: It can emit more photons each second because it has so much more surface area. As a full-fledged giant (Figure 12-24a), our Sun will shine 2000 times more brightly than it does today.

ANIMATION 12.3 **FIGURE 12-23 A Mass-Loss Star** A red giant star is shedding its outer layers, thereby creating this reflection nebula, labeled IC 2220 and called Toby Jug, located in the constellation Carina. The star is embedded inside the nebula and is not visible in this image. (Anglo-Australian Observatory)

12-10 Helium fusion begins at the center of a giant

When a star first becomes a giant, its hydrogen-fusing shell surrounds a small, compact core a few times the diameter of Earth, composed of almost pure helium. At first, no thermonuclear reactions occur in the helium-rich core of a giant because the temperature there is too low to fuse helium nuclei. The hydrogen-fusing shell creates more helium, which settles into the core. The eventual transformation of the core follows two dramatically different routes, depending on the star's total mass. The lower-mass giants, from 0.4 to about 2 solar masses, go down an exceptionally bizarre path.

0.4 to 2 $M_\odot$—Helium Flash The helium atoms in cores of young giant stars with masses between 0.4 and about 2 $M_\odot$ are squeezed by the gravitational force into a crystal-like solid. At the pressures they are under, these atoms are completely ionized, separating into nuclei and electrons. The electrons, distributed between the nuclei, are so closely crowded together that they come under the influence of a phenomenon called the **Pauli exclusion principle,** first formulated in 1925 by the Austrian physicist Wolfgang Pauli. According to this principle, nature does not allow two identical particles to exist in the same place at the same time. As the electrons are pressed closer and closer together, the exclusion principle prevents them from freezing in place between the nuclei. Instead, many of them must vibrate faster and faster so that they do not become "identical," here meaning being in the same place and moving with the same speeds as adjacent electrons.

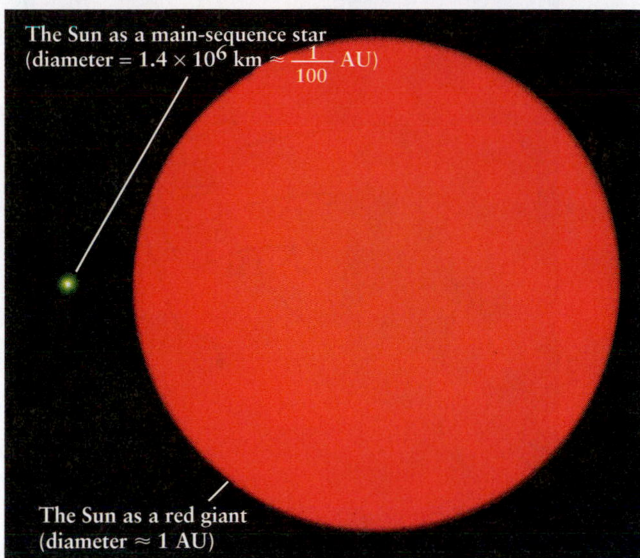

a The Sun today and as a red giant

b Red giant stars in the star cluster M50 RIVUXG

FIGURE 12-24 The Sun Today and as a Giant (a) In about 5 billion years, when the Sun expands to become a giant, its diameter will increase a hundredfold from what it is now, while its core becomes more compact. Today, the Sun's energy is produced in a hydrogen-fusing core whose diameter is about 200,000 km. When the Sun becomes a giant, it will draw its energy from a hydrogen-fusing shell that surrounds a compact helium-rich core. The helium core will have a diameter of only 30,000 km. The Sun's diameter will be about 100 times larger, and it will be about 2000 times more luminous as a giant than it is today. (b) This composite of visible and infrared images shows red giant stars in the open cluster M50 in the constellation of Monoceros (the Unicorn). (T. Credner and S. Kohle, Astronomical Institutes of the University of Bonn)

Astronomers call the electrons in this situation *degenerate*. Degeneracy keeps the electrons apart, provides a pressure in the core, and prevents the core from collapsing. Thus, the helium-rich core of a low-mass giant is supported by **electron degeneracy pressure.** This state provides a greater outward pressure than did the normal pressure in the star before degeneracy developed in its core. The solid core grows from infalling helium and heats up under the gravitational influence of the star's mass.

Here is the weirdest part. The equations predict that, unlike the pressure of an ordinary gas, the pressure inside a degenerate core does not change with temperature. In a normal gas, like the air we breathe, when the temperature increases, the pressure also increases, causing the gas to expand and cool. When a degenerate core's temperature increases, the pressure there does not increase. Without the "safety valve" of increasing pressure, the star's core cannot expand and cool when overheated.

Eventually, the core temperature reaches about 100 million K, at which time **core helium fusion** begins (see Figure 12-22c). This fusion creates energy, so the core's temperature increases. But a degenerate core's pressure is unable to increase. If the core pressure increased, the core would expand and cool. Because the core does not immediately expand, during the first few hours of core helium fusion, both the core temperature and the fusion rate increase dramatically. This situation is called the **helium flash.**

When the core temperature reaches around 3.5×10^8 K, the degenerate helium is forced to transform back into an ordinary gas. Suddenly, the usual safety valve operates, namely that the high temperature increases the core pressure, causing the core to expand and cool, moderating the rate of fusion occurring there within a few hours. The expanding core, in turn, enlarges and cools the hydrogen-fusing shell around it. Fusion in the shell therefore decreases, so the flow of photons from the shell directly to the outer layers of the star drops, causing the outer regions to cool. As they cool, they contract—the star shrinks until it reaches a new hydrostatic equilibrium.

Keep in mind that the helium flash occurs in and acts upon the star's core. It does not cause a sudden flash in brightness. A 0.4-to-2 $M_\odot$ helium-fusing giant is left smaller, dimmer, and hotter than it was before it began fusing helium. Figure 12-25a shows the expected changing brightness of the Sun when going through this process. Figure 12-25b includes evolutionary tracks of two stars in this mass range as they leave the main sequence, move into

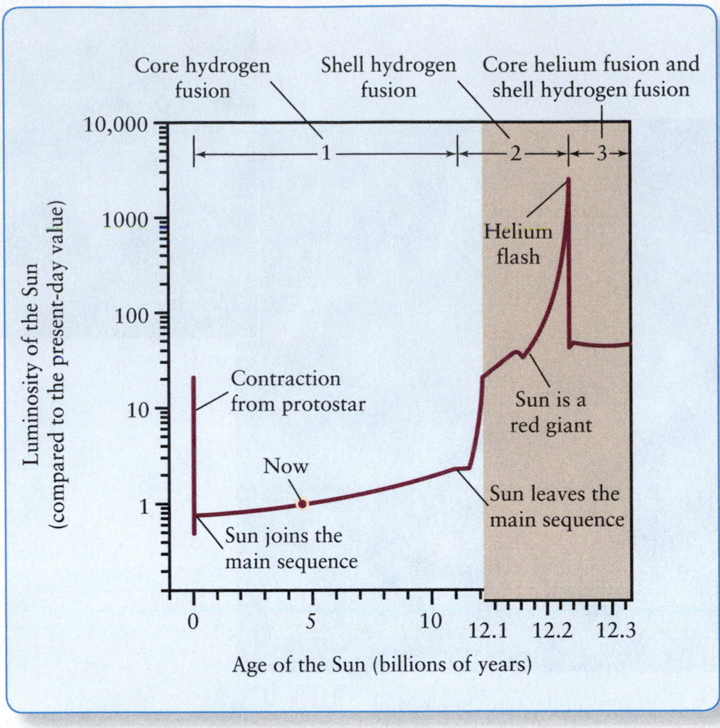

a

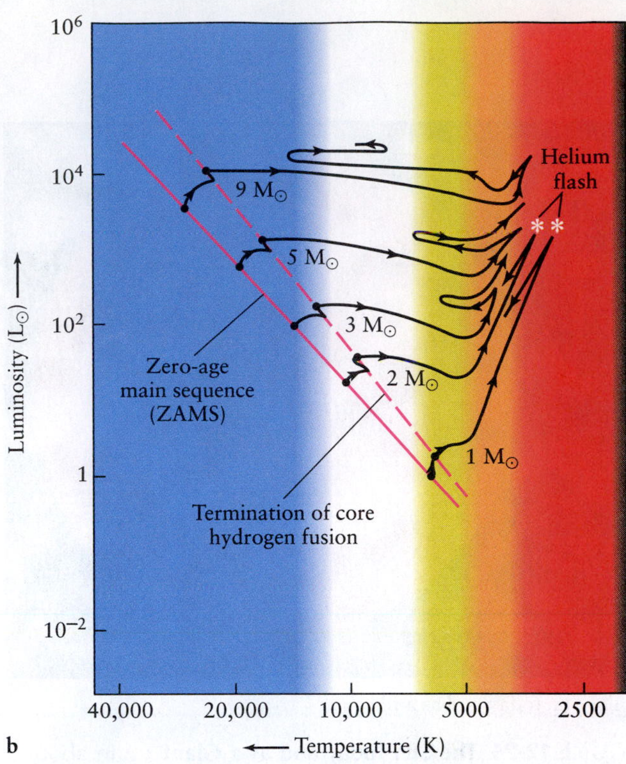

b

FIGURE 12-25 Post–Main-Sequence Evolution (a) The luminosity of the Sun changes as our star evolves. It began as a protostar with decreasing luminosity. On the main sequence today, it gradually brightens. Giant phase evolution occurs more rapidly, with faster and larger changes of luminosity. Note the change in scale of the horizontal axis at 12 billion years. (b) Model-based evolutionary tracks of five stars are shown on this H-R diagram. In the high-mass stars, core helium fusion ignites smoothly where the evolutionary tracks make a sharp turn upward into the giant region of the diagram.

the giant phase, go through the helium flash, and begin to settle back down. We will follow their evolution through the rest of the giant phase after briefly exploring how stars with more than 2 $M_\odot$ proceed off the main sequence.

More than 2 $M_\odot$—No Helium Flash The pressure created by the gravitational force of a star with more than 2 solar masses is enough to smoothly compress and heat its core until it reaches the fusion temperature of helium. In other words, the cores of giants in this mass range never become degenerate. When an ordinary gas, such as the helium that exists in the cores of these stars, is compressed, it heats up; when it expands, it cools down. If the energy production created by helium fusion overheats its core, the core expands, cooling its gases and slowing the rate of thermonuclear reactions. Conversely, if too little energy is being created to support the star's overlying layers, they move inward, compressing the core. The resulting increase in temperature speeds up the thermonuclear reactions and thus increases the energy output, which stops the contraction. Either way, the "safety valve" keeps it from collapsing or exploding.

Will the Sun undergo a helium flash? Why or why not?

Figure 12-25b shows the evolution off the main sequence of three stars in this mass range, which do not undergo helium flash. These stars move up and to the right across the giant region of the H-R diagram, getting larger until core fusion begins, whereupon they contract slightly—less than stars that undergo the helium flash.

12-11 Life in the giant phase has its ups and downs

The dominant process for fusing helium in the cores of giants is called the *triple alpha process*, in which three helium atoms fuse to become a carbon atom. (Early in the twentieth century, particles emitted by radioactive nuclei were called *alpha particles*. They were later identified as helium nuclei.) The triple alpha process works in two steps. First, two helium nuclei fuse to create beryllium. Within 10^{-8} s, a third helium nucleus collides with the short-lived beryllium to create carbon. This process also releases energy and can be summarized as follows:

$$^{4}He + {}^{4}He + {}^{4}He \rightarrow {}^{12}C + \gamma$$

where γ denotes energy emitted as photons. The fusion in giants does not stop there. Some of the carbon thus created can fuse with another helium nucleus to produce oxygen:

$$^{12}C + {}^{4}He \rightarrow {}^{16}O + \gamma$$

As in main-sequence stars, energy is released in the form of gamma rays.

Giant stars fuse helium in their cores for about 10% as long as they spend fusing hydrogen while on the main sequence. For example, the Sun will spend a total of about 10 billion years on the main sequence, after which it will be a giant for about 1 billion years. While helium fusion is occurring in a giant's core, further gravitational contraction of the star's core ceases.

Despite the fact that all stars spend most of their lives on the main sequence, most of the stars visible to the naked eye are giants! This occurs because the relatively few giants in our neighborhood are so luminous that they outshine their neighbors in the night sky.

VARIABLE STARS

After core helium fusion begins, the evolutionary tracks of mature stars move partway back toward the main sequence (right to left on an H-R diagram; Figure 12-25). Their pressure-temperature safety valves are not perfect, so as they shrink, they overheat and core fusion causes them to expand to larger sizes than they would be at equilibrium. Becoming too large, their pressure and temperature decrease and they collapse, overcompressing and thereby overheating themselves again. This process is analogous to dropping a tennis ball, which bounces repeatedly after striking the ground. In other words, a star under these conditions becomes unstable and pulsates. The region on the H-R diagram between the main sequence and the right side of the giant branch, where this occurs, is called the **instability strip** (Figure 12-26). These so-called **variable stars** can be easily identified by their changes in brightness amid a field of stars of constant luminosity.

How many helium atoms does it take to make one oxygen atom?

Lower-mass, post–helium-flash stars pass through the lower end of the instability strip as they move horizontally on the H-R diagram along their evolutionary tracks. These stars become **RR Lyrae variables,** named after the prototype in the constellation of Lyra (the Lyre). RR Lyrae variables all have periods shorter than 1 day. Higher-mass stars that evolve from the same interstellar medium as RR Lyrae pass back and forth through the upper end of the instability strip on the H-R diagram. These stars become Type I Cepheid variables, often simply called *Cepheids*. (There is another whole group of variable stars called Type II Cepheids that formed earlier in the history of the universe from gases with virtually no metals. We discuss them in more detail shortly.)

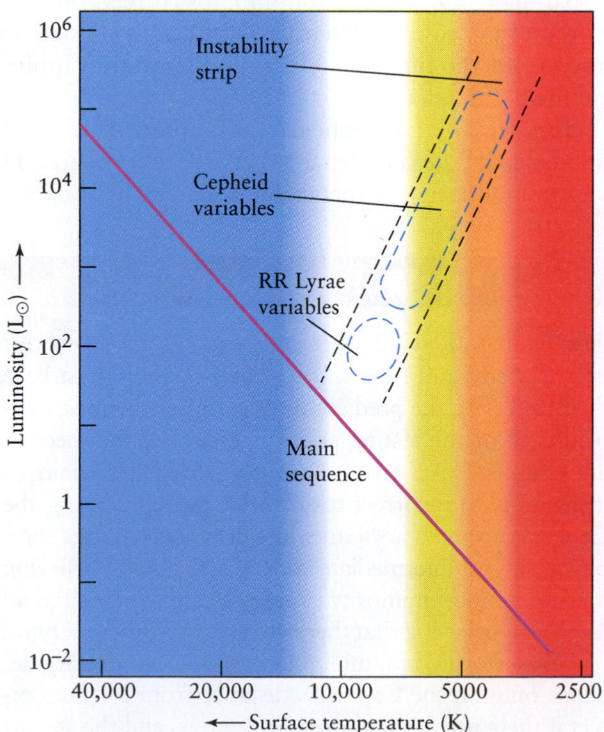

FIGURE 12-26 The Instability Strip The instability strip occupies a region between the main sequence and the giant branch on the H-R diagram. A star passing through this region along its evolutionary track becomes unstable and pulsates.

12-12 A Cepheid pulsates because it is alternately expanding and contracting

A **Cepheid variable** is characterized by the way in which its light output varies—most of these stars follow a pattern of rapid brightening followed by gradual dimming. A Cepheid variable brightens and fades because the star's outer layers cyclically expand and contract. Lines in the spectrum of δ (delta) Cephei shift back and forth with the same 5.4-day period that characterizes its variations in magnitude. According to the Doppler effect measured from this star, these shifts mean that the star's surface is alternately approaching and receding from us.

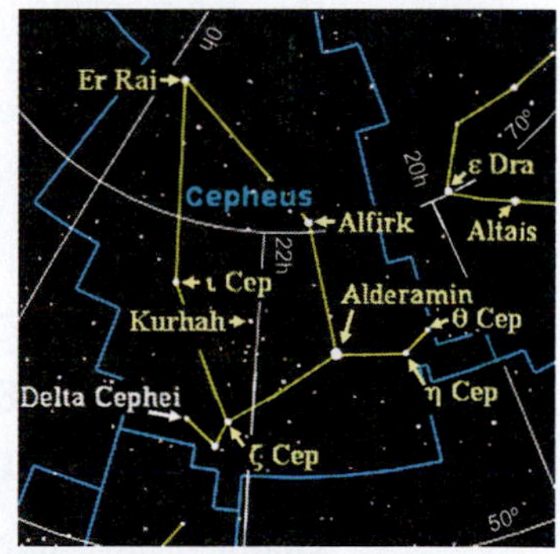

When a Cepheid variable pulsates, its gases alternately heat up and cool down; the surface temperature cycles between about 6300 K and 5000 K. Thus, the characteristic light curve (luminosity versus time curve, as in Figure 11-13) of a Cepheid variable results from changes in both size and surface temperature.

Just as a bouncing ball eventually comes to rest, a pulsating star would soon stop pulsating without some mechanism to maintain its oscillations. Variable stars have a layer of gas rich in partially ionized helium (helium with one electron stripped off). This gas, absorbing a lot of

a

b

c

R I V U X G

d

FIGURE 12-27 Analogy for Cepheid Variability (a) As pressure builds up in this pot, the force on the lid (analogous to a Cepheid's outer layers) increases. (b) When the pressure inside the pot is sufficient, it lifts the lid off (expands the star's outer layers) and thereby allows some of the energy inside to escape. This process cycles (two are shown), as do the luminosity and temperature of Cepheid stars. (a–d: Janet Horton)

How do astronomers observe that Cepheids are changing size?

photons leaving the core, heats, expands, and pushes the outer layers outward, just as steam raises the lid of a pot (Figure 12-27). Eventually, the ionized helium layer is spread so thin that photons pass through it. The ionized helium cools and contracts, and the star's outer layers collapse, compressing the interior gases until the process repeats itself. In our analogy with the pot of boiling water, when enough of the steam inside escapes, the lid falls back and the steam inside builds up again. Variability only occurs when the conditions of temperature and pressure inside the star are appropriate. As a result, giants are variable only when the stars are in the instability strip of an H-R diagram.

12-13 Cepheids enable astronomers to estimate vast distances

Cepheids are important to astronomers because there is a direct relationship between a Cepheid's period of pulsation and its average luminosity. This correlation is called, appropriately enough, the **period-luminosity relation.** We can determine the distance to a Cepheid in four steps. First, we observe its period. Second, we use the period-luminosity relation to determine its luminosity and, hence, its absolute magnitude (recall that the luminosity and absolute magnitudes are directly related to each other). Third, we observe its apparent magnitude. Finally, we use the distance-magnitude relationship in *An Astronomer's Toolbox 11-3* to calculate its distance.

Dim Cepheid variables pulsate rapidly, with periods of 1 to 2 days, and have average brightness variations of a few hundred times our Sun's luminosity. The most luminous Cepheids have the longest periods of all Cepheids, with variations occurring over 100 days and average brightness variations equaling 10,000 $L_{\odot}$. As they are so bright, we can see Cepheids far beyond the boundaries of our Milky Way Galaxy. Because the changes in brightness of Cepheids can be seen at distances where other techniques for measuring distance, such as stellar parallax, fail, the period-luminosity relation plays an important role in determining the overall size and structure of the universe. We will explore this application further in Chapter 16.

The details of a Cepheid's pulsation depend on the abundance of heavy elements in its atmosphere. The average luminosity of metal-rich Cepheids is roughly 4

Insight Into Science

Firm Foundations? To determine things far from our everyday size and time scales, scientists often must base their results on the predictions of complex theories, even before all of the testing and refinement of the theories are complete. An error anywhere along the chain of ideas leads to incorrect results. For example, using the Cepheid variable stars to measure distances requires combining several intermediate concepts, including believing that the period-luminosity relationship applies well to all Cepheids, believing that there are just two different types of Cepheids, and assuming that the peak luminosity depends only on the Cepheid's distance from us (not correct if there are gas and dust between us and the star, as there often are). If any of these or myriad other assumptions is wrong, the calculated distances will be incorrect.

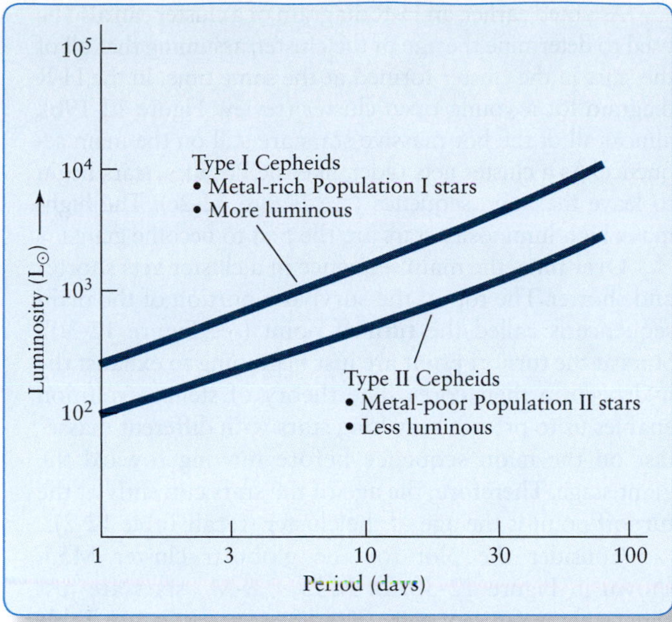

FIGURE 12-28 The Period-Luminosity Relation The period of a Cepheid variable is directly related to its average luminosity: The more luminous the Cepheid, the longer its period and the slower its pulsations. Type I Cepheids (δ Cephei stars) are brighter, more massive, and more metal-rich stars than Type II Cepheids. The greater brightness of the Type I Cepheids is a result of their higher mass. (Adapted from H. C. Arp)

times greater than the average luminosity of metal-poor Cepheids that have the same period. The difference in luminosity is a result of the higher mass of the metal-rich Cepheids compared to the metal-poor ones. Thus, there are two classes: **Type I Cepheids** (also called δ *Cephei stars*), which are the brighter, more massive, metal-rich stars, and **Type II Cepheids,** which are the dimmer, lower mass, metal-poor stars. The period-luminosity relations for both types of variables are shown in Figure 12-28.

12-14 Globular clusters are bound groups of old stars

Many giant stars are found in groups, called **globular clusters,** so named because of their spherical shapes. A typical globular cluster, like the one shown in Figure 12-29, may contain up to a million stars in a volume 200 light-years across. Like open clusters, the stars in globular clusters all formed at about the same time. Unlike open clusters of young stars, however, globular clusters are gravitationally bound groups of old stars from which few stars escape. At least 157 globular clusters in our Galaxy are distributed in a sphere around the center of the Milky Way. Other galaxies also have them spherically distributed around their centers.

Astronomers know that most globular clusters are old because they contain no high-mass main-sequence stars. (The young star cluster Westerlund 1 in our Galaxy and similar clusters elsewhere are apparently exceptions. These are discussed later in this section.) If you measure the luminosity and surface temperature of many stars in an older globular cluster and plot the data on a color-magnitude diagram (Figure 12-30), you will find that the upper half of the main sequence is missing. All of the high-mass main-sequence stars evolved long ago into giants, leaving behind only lower-mass, slowly evolving main-sequence stars still undergoing core hydrogen fusion.

The H-R diagram of a globular cluster typically shows a roughly horizontal grouping of stars to the left of the center portion of the diagram (see Figure 12-30). These hot stars, called **horizontal branch stars,** undergo helium fusion in their cores, and they have luminosities of about 50 $L_\odot$ (see Figure 12-29). Eventually, these stars will move back toward the giant region as core helium fusion and hydrogen shell fusion devour their fuel.

What are typical periods of Cepheid variable stars?

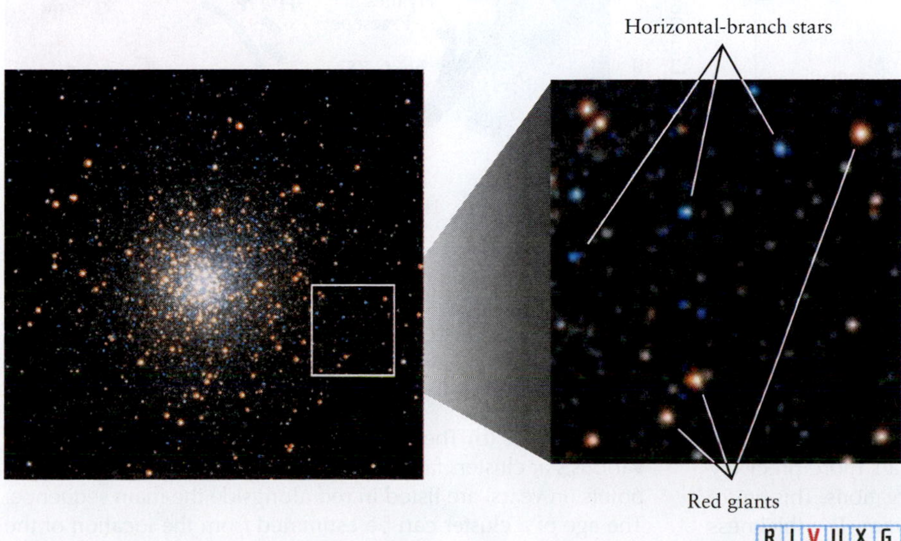

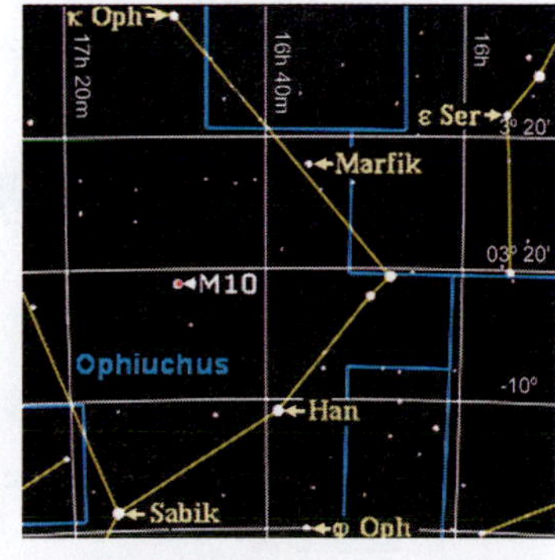

FIGURE 12-29 A Globular Cluster This cluster, M10, is about 85 light-years across and is located in the constellation Ophiuchus (the Serpent Holder), roughly 16,000 light-years from Earth. Most of the stars here are either red giants or blue horizontal-branch stars with both core helium fusion and hydrogen shell fusion. (T. Credner and S. Kohle, Astronomical Institutes of the University of Bonn)

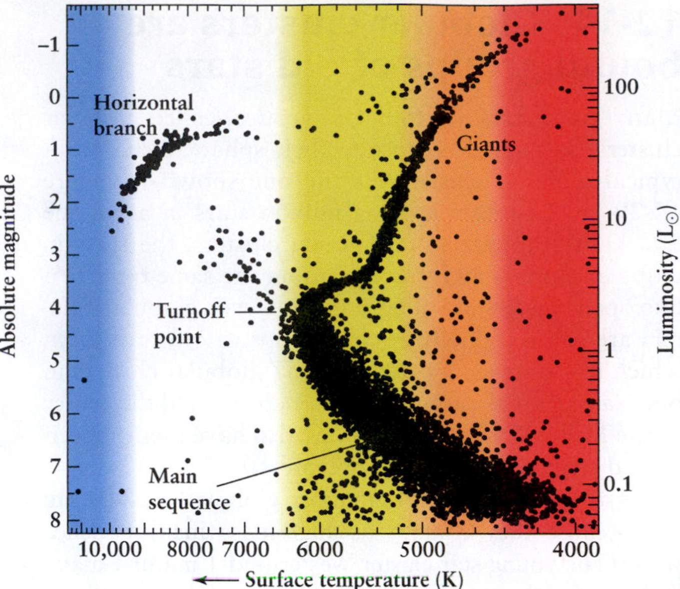

FIGURE 12-30 An H-R Diagram of a Globular Cluster Each dot on this graph represents the absolute magnitude and surface temperature of a star in the globular cluster M55. Note that the upper half of the main sequence is missing. The horizontal branch stars are stars that recently experienced the helium flash in their cores and now exhibit core helium fusion and hydrogen shell fusion.

As noted earlier, an H-R diagram of a cluster can also be used to determine the age of the cluster, assuming that all of the stars in the cluster formed at the same time. In the H-R diagram for a young open cluster (review Figure 12-19b), almost all of the hot massive stars are still on the main sequence. As a cluster gets older, like the Pleiades, stars begin to leave the main sequence (see Figure 12-2c). The high-mass, high-luminosity stars are the first to become giants.

Over time, the main sequence in a cluster gets shorter and shorter. The top of the surviving portion of the main sequence is called the **turnoff point** (see Figure 12-30). Stars at the turnoff point are just beginning to exhaust the hydrogen in their cores. The theory of stellar evolution enables us to predict how long stars with different masses last on the main sequence before moving toward the giant stage. Therefore, the age of the stars currently at the turnoff point is the age of the cluster (recall Table 12-2).

Consider the plot for the globular cluster M55, shown in Figure 12-30. In M55, 0.8-$M_\odot$ stars are just leaving the main sequence. Therefore, according to Table 12-2, the cluster's age is approximately 13.5 billion years.

Using data collected from the *Hipparcos* satellite, Figure 12-31a is a composite diagram of isolated and cluster stars, reflecting the overall structure of the H-R diagram. Using data from stellar evolution theory as presented in Table 12-2, the ages of star clusters can be estimated from the turnoff points, such as those depicted in

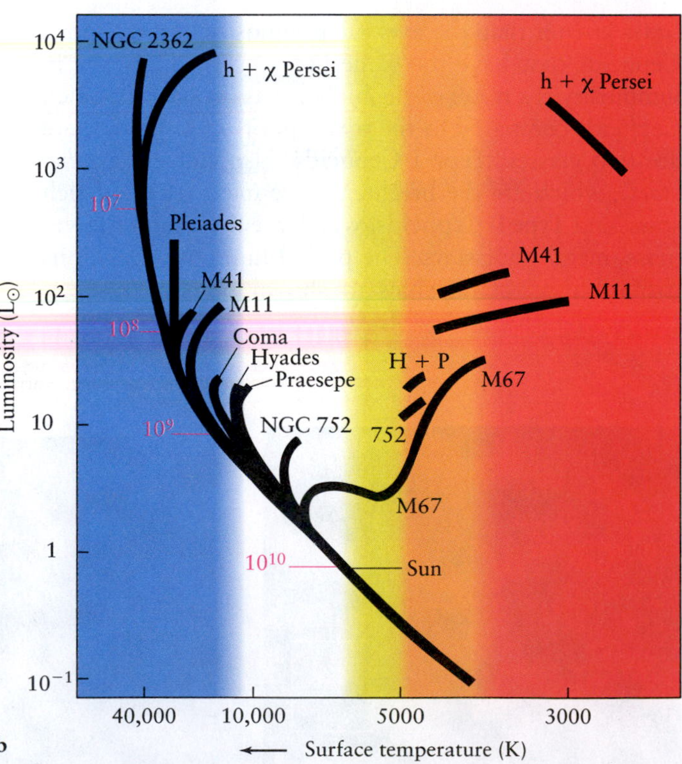

FIGURE 12-31 Structure of the H-R Diagram (a) Data taken by the *Hipparcos* satellite placed 41,453 stars more precisely on the H-R diagram than any previous observations. This figure shows the overall structure of the H-R diagram. The thickness of the main sequence is due in large part to stars of different ages turning off the main sequence at different places, as shown in (b). (b) The black bands indicate where data from various star clusters fall on the H-R diagram. The ages of turnoff points (in years) are listed in red alongside the main sequence. The age of a cluster can be estimated from the location of the turnoff point, where the cluster's most massive stars are just now leaving the main sequence.

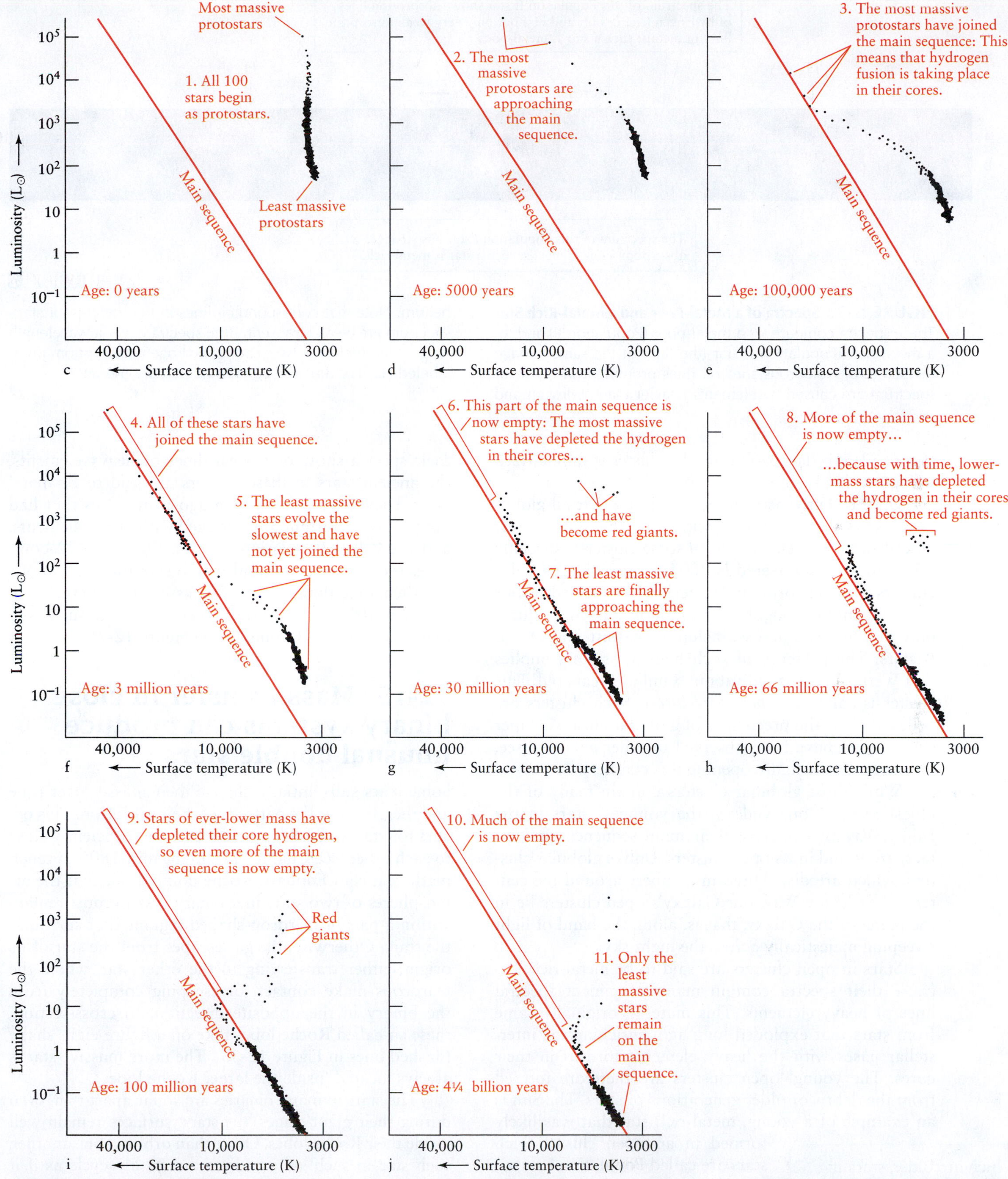

FIGURE 12-31 ***(continued)*** (c–j) Summarizing the evolution of a theoretical cluster of 100 stars, as shown by their locations on the H-R diagram. (In principle, each star's evolution could be followed separately.) After a star passes through the red giant phase, it is deleted from the diagram.

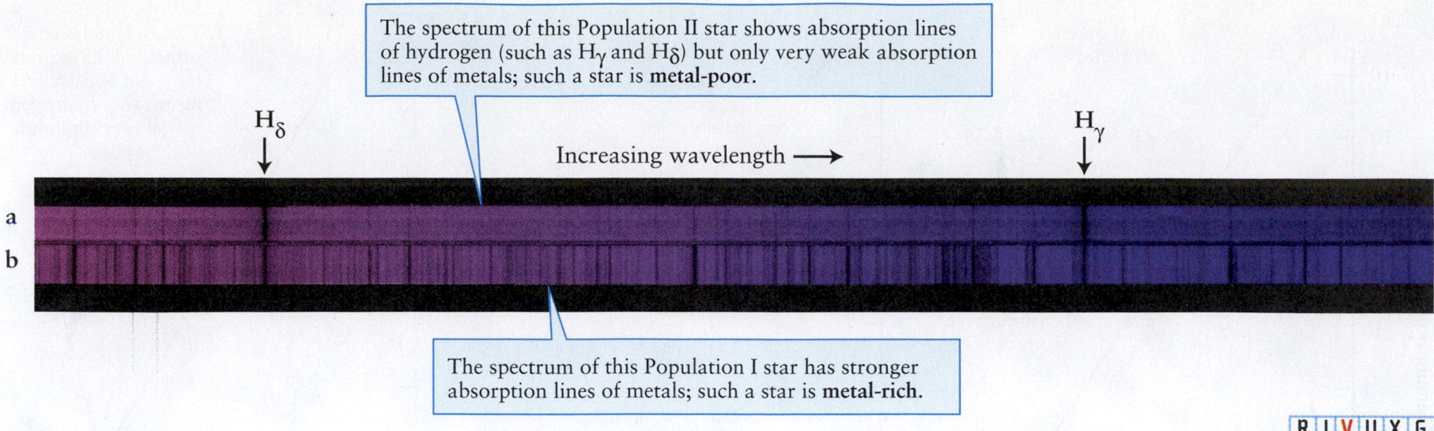

FIGURE 12-32 Spectra of a Metal-Poor and a Metal-Rich Star These spectra compare (a) a metal-poor (Population II) and (b) a metal-rich (Population I) star (the Sun) of the same surface temperature. Numerous spectral lines prominent in the solar spectrum are caused by elements heavier than hydrogen and helium. Note that corresponding lines in the metal-poor star's spectrum are weak or absent. Both spectra cover a wavelength range that includes two strong hydrogen absorption lines, labeled H_γ (410 nm) and H_δ (434 nm). (Lick Observatory)

Figure 12-31b. The evolution of a cluster is summarized in Figure 12-31c–j.

Until recently, astronomers believed that all globular clusters were very old, with turnoff points as just described. However, analysis of some clusters, especially Westerlund 1, discovered in 2004, suggest that globular clusters are still forming. Westerlund 1, located in our Galaxy, contains roughly half a million stars in a volume only 6 light years across, including at least 200 O and B stars. The presence of such high mass stars implies that Westerlund 1 is only about 5 million years old. This cluster is called a *super star cluster*. Such clusters are believed to be the precursors of globular clusters. Super star clusters have been observed in other galaxies (see, for example, the figure opening this chapter).

While most globular clusters contain many of the oldest stars in our Galaxy, the youngest stars in the Milky Way (which have their main sequences still intact) are found in its open clusters. Unlike globular clusters, which are distributed in a sphere around the center of the Milky Way, our Galaxy's open clusters lie in the plane of the Galaxy, that is, along the band of light sweeping majestically across the night sky.

Stars in open clusters are said to be metal-rich, because their spectra contain many prominent spectral lines of heavy elements. This material originally came from stars that exploded long ago, enriching the interstellar gases with the heavy elements formed in their cores. The young, open clusters are therefore formed from the debris of older generations of stars. The Sun is an example of a young, metal-rich star that was likely formed in an open cluster. Such stars are called **Population I stars.**

Which are older, stars in open clusters or stars in globular clusters?

Older globular clusters are generally located above or below the plane of our Galaxy. Because their spectra show only weak lines of heavy elements, the ancient stars in these clusters are said to be metal poor. They were created long ago from gases that had not yet been substantially enriched with heavy elements, and are called **Population II stars.** Figure 12-32 compares the spectra of a Population II star and the Sun. We see, then, that along with their masses, the differences in the chemistries of stars affects their evolution, such as in their variability of Cepheids (see Figure 12-28).

12-15 Mass transfer in close binary systems can produce unusual double stars

Some stars substantially change their masses over time and thereby alter the patterns of their evolution. This occurs for stars in binary systems that are sufficiently close to each other (see Chapter 11). In the mid-1800s, French mathematician Édouard Roche pointed out that the atmospheres of two stars in a binary system must remain within a pair of teardrop-shaped regions that surround the stars. Otherwise, the gas escapes from the star of its origin, either transferring to the other star, where the teardrops make contact, or escaping completely from the binary in the opposite direction. In cross-section, these so-called **Roche lobes** take on a figure-eight shape (dashed lines in Figure 12-33). The more massive star is always located inside the larger Roche lobe.

The stars in many binaries are so far apart that, even during their giant stages, the stars' surfaces remain well inside their Roche lobes. Other than orbiting one another, each star in such systems lives out its life cycle as if it were single and isolated, and the system is referred to as a **detached binary** (see Figure 12-33a). If two stars are relatively close together, however, one star may fill or over-

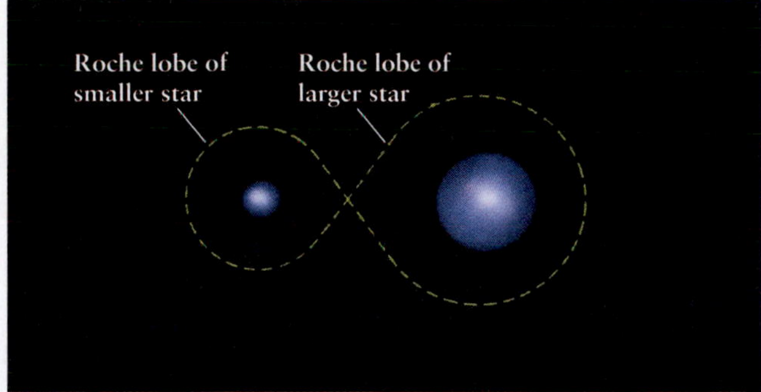

a Detached binary: Neither star fills its Roche lobe.

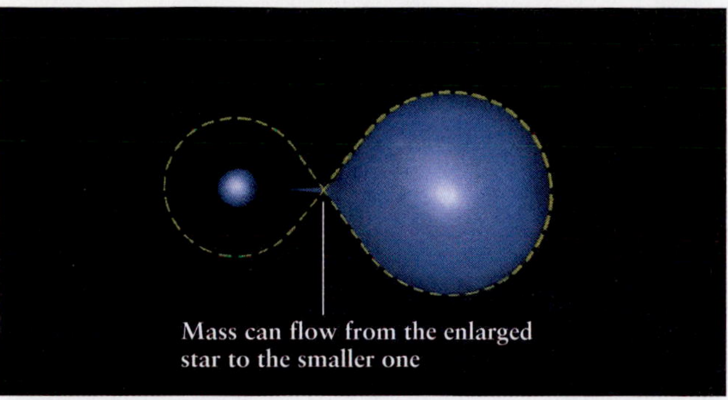

b Semidetached binary: One star fills its Roche lobe.

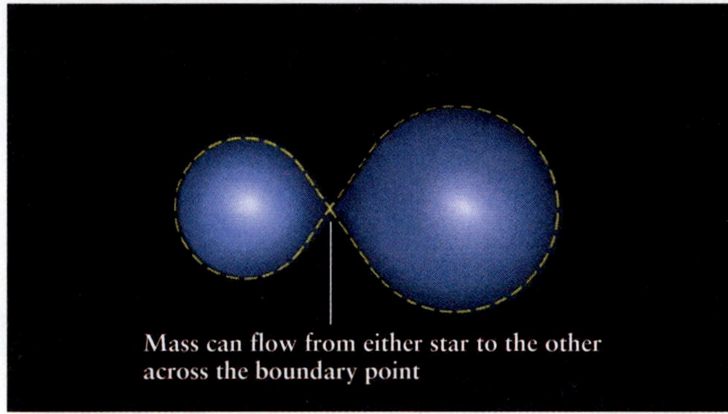

c Contact binary: Both stars fill their Roche lobes.

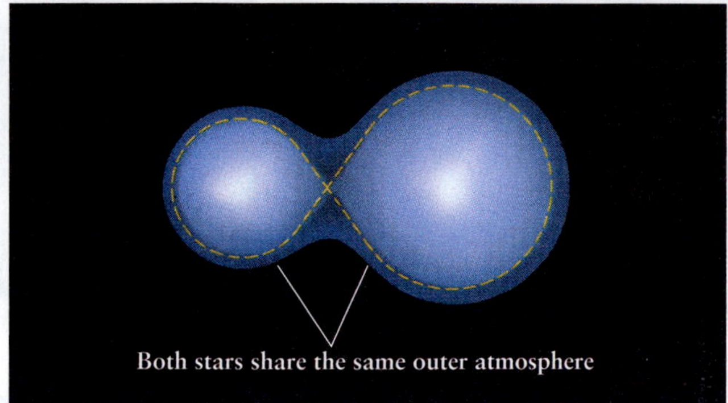

d Overcontact binary: Both stars overflow their Roche lobes.

FIGURE 12-33 Detached, Semidetached, Contact, and Overcontact Binaries (a) In a detached binary, neither star fills its Roche lobe. (b) If one star fills its Roche lobe, the binary is semidetached. Mass transfer is often observed in semidetached binaries. (c) In a contact binary, both stars fill their Roche lobes. (d) The two stars in an overcontact binary both overflow their Roche lobes. The two stars actually share the same outer atmosphere.

flow its Roche lobe as it expands into the giant phase. This system is a **semidetached binary,** and gases then flow across the point where the two Roche lobes touch and fall onto the companion star (Figure 12-33b). When both stars completely fill their Roche lobes, the system is a **contact binary,** because the two stars actually touch and exchange gas (Figure 12-33c). It is quite unlikely, however, that both stars will exactly fill their Roche lobes at the same time. It is more likely that they overflow their lobes, giving rise to a common atmospheric envelope. Such a system is an **overcontact binary** (Figure 12-33d).

Semidetached and contact binaries are easiest to detect if they are also eclipsing binaries (recall Figure 11-13). Their light curves have a distinctly rounded appearance, caused by these tidally distorted egg-shaped stars. The eclipsing binary β (beta) Persei, commonly called Algol (from the Arabic term for "demon"), is a semidetached binary that can easily be seen with the naked eye in the constellation Perseus. From Algol's light curve (Figure 12-34a), astronomers have determined that the binary contains a star that fills its Roche lobe. Sometime in the past, as it expanded and became a giant, this star dumped a significant amount of gas onto its companion.

Mass transfer is still occurring in a semidetached eclipsing binary, called β Lyrae, in the constellation Lyra. Like β Persei, β Lyrae contains a giant that fills its Roche lobe (Figure 12-34b). For many years, astronomers were puzzled that the detached companion star in β Lyrae is severely underluminous, contributing virtually no light at all to the visible radiation coming from the system. Furthermore, the spectrum of β Lyrae contains unusual features, some of which are consistent with gas flowing between the stars and around the system as a whole.

The β Lyrae system was explained in 1963, when the Chinese-American astronomer Su-Shu Huang proposed that the underluminous star in β Lyrae is enveloped in a huge **accretion disk** of gas captured from its bloated companion. The disk is so large and thick that it completely shrouds the secondary star, making it impossible to observe at visible wavelengths. The primary star is overflowing its Roche lobe, with gases streaming onto the disk at the rate of 1 $M_\odot$ per hundred thousand years.

The fate of a semidetached system like β Persei or β Lyrae depends primarily on how fast its stars evolve. If the detached star expands to fill its Roche lobe while the companion star fills its own Roche lobe, then the result is an overcontact binary. An example is W Ursae Majoris, in which two stars share the same photosphere (Figure 12-34c).

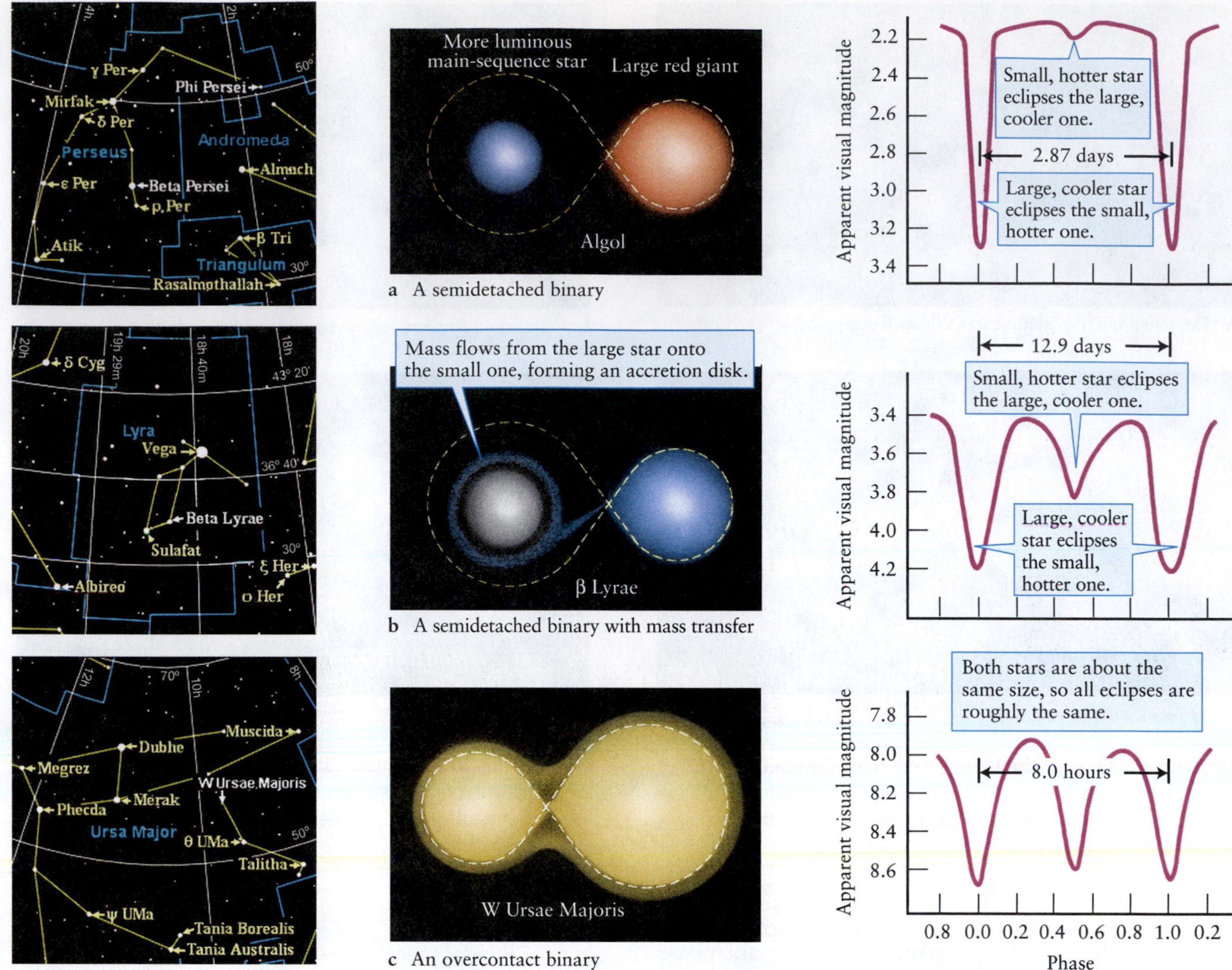

FIGURE 12-34 Three Close Binaries Sketches of and light curves for three eclipsing binaries are shown. The phase denotes the fraction of the orbital period from one primary minimum to the next. (a) Algol, also known as β Persei, is a semidetached binary. The deep eclipse occurs when the giant star (right) blocks the light from the smaller, but more luminous, main-sequence star. (b) β Lyrae is a semidetached binary in which mass transfer has produced an accretion disk that surrounds the detached star. This disk is so thick and opaque that it renders the secondary star almost invisible. (c) W Ursae Majoris is an overcontact binary. Both stars therefore share their outer atmospheres. The short, 8-h period of this binary indicates that the stars are very close to each other.

Theories of the evolution of binary systems that transfer mass reveal that some remarkable transformations are possible, such as the case of φ (phi) Persei. As shown in Figure 12-35, the more massive star in a binary can actually be transformed into the lower-mass star, and vice versa! When stars in such systems gain mass, their increased gravitational force increases their rate of fusion, which increases the rate at which they evolve. Conversely, the stars that lose mass begin evolving more slowly than before their mass loss. In Chapters 13 and 14 we will see that mass transfer produces some of the most extraordinary objects in the sky.

In which type of binary system are the two stars most actively exchanging mass?

12-16 Frontiers yet to be discovered

Stars, whether isolated, in pairs, or in much larger groups, are our primary source of information about the universe. Details of each step of stellar evolution remain to be explored; the pre–main-sequence evolution of stars is being revealed in sources like the Eagle Nebula (see Figure 12-16). These observations provide data with which to compare the theories of early star formation. Over the past two decades, observations have revealed that more stars lose mass in the form of stellar winds while on the main sequence than previously expected. Consequently, we have a lot to learn about why such stars shed mass.

FIGURE 12-35 Mass Exchange Between Close Binary Stars This sequence of drawings shows how close binary stars can initially be isolated but, as they age, grow and exchange mass. Such mass exchange leads to different fates than if the same stars had evolved in isolation.

As noted, we still need to understand the details of why stars swell into the red giant phase. Also, the mechanisms that cause stars to be variable are still under investigation. This latter work has been stimulated recently by the as-yet-unexplained observation that Betelgeuse has shrunk 15% in diameter since 1993. This is a huge change in size, equal to the distance between the Sun and Venus.

SUMMARY OF KEY IDEAS

Protostars and Pre–Main-Sequence Stars

- Enormous, cold clouds of gas and dust, called giant molecular clouds, are scattered about the disk of the Galaxy.
- Star formation begins when gravitational attraction causes clumps of gas and dust, called protostars, to coalesce in Bok globules within a giant molecular cloud. As a protostar contracts, its matter begins to heat and glow. When the contraction slows down, the protostar becomes a pre–main-sequence star. When the pre–main-sequence star's core temperature becomes high enough to begin hydrogen fusion and stop contracting, it becomes a main-sequence star.
- The most massive pre–main-sequence stars take the shortest time to become main-sequence stars (O and B stars).
- In the final stages of pre–main-sequence contraction, when hydrogen fusion is about to begin in the core, the pre–main-sequence star may undergo vigorous chromospheric activity that ejects large amounts of matter into space. G, K, and M stars at this stage are called T Tauri stars.
- A collection of a few hundred or a few thousand newborn stars formed in the plane of the Galaxy is called an open cluster. Stars escape from open clusters, most of which eventually dissipate.

Main-Sequence and Giant Stars

- The Sun has been a main-sequence star for 4.6 billion years and should remain so for about another 5 billion years. Less massive stars than the Sun evolve more slowly and have longer main-sequence lifetimes. More massive stars than the Sun evolve more rapidly and have shorter main-sequence lifetimes.
- Main-sequence stars with mass between 0.08 and 0.4 $M_{\odot}$ convert all of their mass into helium and then stop fusing. Their lifetimes last hundreds of billions of years, so none of these stars has yet left the main sequence.
- Core hydrogen fusion ceases when hydrogen in the core of a main-sequence star with $M > 0.4\ M_{\odot}$ is gone, leaving a core of nearly pure helium surrounded by a

shell where hydrogen fusion continues. Hydrogen shell fusion adds more helium to the star's core, which contracts and becomes hotter. The outer atmosphere expands considerably, and the star becomes a giant.

• When the central temperature of a giant reaches about 100 million K, the thermonuclear process of helium fusion begins. This process converts helium to carbon, then to oxygen. In a massive giant, helium fusion begins gradually. In a less massive giant, it begins suddenly in a process called the helium flash.

• The age of a stellar cluster can be estimated by plotting its stars on an H-R diagram. The upper portion of the main sequence disappears first, because more massive main-sequence stars become giants before low-mass stars do.

• Giants undergo extensive mass loss, sometimes producing shells of ejected material that surround the entire star.

• Relatively young stars are metal-rich (Population I); ancient stars are metal-poor (Population II).

Clusters of stars

• Groups of between a few hundred and a few thousand stars, formed together from a single interstellar cloud in the disk of our Galaxy, are called open clusters.

• Stars in open clusters go their separate ways.

• Groups of hundreds of thousands to millions of stars formed together from a common interstellar cloud are called globular clusters.

• Stars in globular clusters remain bound together.

Variable Stars

• When a star's evolutionary track carries it through a region called the instability strip in the H-R diagram, the star becomes unstable and begins to pulsate.

• RR Lyrae variables are low-mass, pulsating variables with short periods. Cepheid variables are high-mass, pulsating variables exhibiting a regular relationship between the period of pulsation and luminosity.

• Mass can be transferred from one star to another in close binary systems. When this occurs, the evolution of the two stars changes.

A WHAT DID YOU THINK?

1 *How do stars form?* Each star forms from the collective gravitational attraction of a clump of gas and dust inside a giant molecular cloud.

2 *Are stars still forming today? If so, where?* Yes. Astronomers have seen stars that have just arrived on the main sequence, as well as infrared images of gas and dust clouds in the process of forming stars. Most stars in the Milky Way form in giant molecular clouds in the disk of the Galaxy.

3 *Do more massive stars shine longer than less massive ones? What is your reasoning?* No. Lower-mass stars last longer because the lower gravitational force inside them causes fusion to take place at much slower rates compared to the fusion inside higher-mass stars. These latter stars therefore use up their fuel more rapidly than do lower mass stars.

4 *When stars like the Sun stop fusing hydrogen and helium in their cores, do the stars get smaller or larger?* They get larger. Such stars start fusing hydrogen and helium outside their cores. This new fusion, closer to the star's surface, is able to push the star's outer layers out farther than they had been before.

Key Terms for Review

accretion disk, 403
birth line, 384
Bok globule, 382
brown dwarf, 384
Cepheid variable, 397
contact binary, 403
core helium fusion, 395
dark nebulae, 378
detached binary, 402
electron degeneracy pressure, 395
emission nebula, 378
evolutionary track, 384
giant molecular cloud, 378
globular cluster, 399
H II regions, 387
helium flash, 395
horizontal branch star, 399
hydrogen shell fusion, 393
instability strip, 397
interstellar extinction, 378
interstellar medium, 376
interstellar reddening, 378
Jeans unstable, 379
molecular cloud, 378
nebula (*plural* nebulae), 378
OB association, 388
open cluster, 382
overcontact binary, 403
Pauli exclusion principle, 394
period-luminosity relation, 398
Population I star, 402
Population II star, 402
pre–main-sequence star, 382
protostar, 382
red dwarf, 392
reflection nebula, 378
Roche lobe, 402
RR Lyrae variable, 397
semidetached binary, 403
supernova remnant, 380
T Tauri stars, 384
turnoff point, 400
Type I Cepheid, 399
Type II Cepheid, 399
variable stars, 397
zero-age main sequence (ZAMS), 390

Review Questions

1. Consider a star behind a cloud of interstellar gas and dust as seen from our perspective. Which of the following would you see? **a.** The star appears brighter than it would if the cloud were not present, **b.** The star appears to be moving toward us, **c.** The star appears redder than it would if the cloud were not present, **d.** The star would be invisible at all wavelengths, **e.** The star would always appear green.

2. What is the lowest mass that a star can have on the main sequence? **a.** There is no lower limit, **b.** 0.003 $M_\odot$, **c.** 0.08 $M_\odot$, **d.** 0.4 $M_\odot$, **e.** 2.0 $M_\odot$

3. What is the source of energy that enables a main-sequence star to shine? **a.** Friction between its atoms, **b.** Fusion of hydrogen in a shell that surrounds the core, **c.** Fusion of helium in its core, **d.** Fusion of hydrogen in its core, **e.** Burning of gases on its surface

4. What are giant molecular clouds, and what role do they play in star formation?

5. Why are low temperatures necessary for dense cores to form and contract into protostars?

6. Why do thermonuclear reactions not occur on the surface of a main-sequence star?

7. What is an evolutionary track, and how do such tracks help us interpret the H-R diagram?

8. Draw the pre–main-sequence evolutionary track of the Sun on an H-R diagram. Briefly describe what was occurring throughout the solar system at various stages along this track. (You may find it useful to review Chapter 11.)

9. On what grounds are astronomers able to say that the Sun has about 5 billion years remaining in its main-sequence stage?

10. What will happen inside the Sun 5 billion years from now when it begins to evolve into a giant?

11. How is the evolution of a main-sequence star with less than 0.4 $M_\odot$ fundamentally different from that of a main-sequence star with more than 0.4 $M_\odot$?

12. Draw the post–main-sequence evolutionary track of the Sun on an H-R diagram up to the point when the Sun becomes a helium-fusing giant. Briefly describe what might occur throughout the solar system as the Sun undergoes this transition.

13. What does it mean when an astronomer says that a star "moves" from one place to another on an H-R diagram?

14. What is the helium flash and what causes it?

15. Explain how and why the turnoff point on the H-R diagram of a cluster is related to the cluster's age.

16. Why do astronomers believe that most globular clusters are made of old stars?

17. What are Cepheid variables, and how are they related to the instability strip?

18. What occurs in Cepheid stars that is analogous to the vapor raising the lid on a pot of boiling water?

19. What are RR Lyrae variables, and how are they related to the instability strip?

20. What is a Roche lobe, and what is its significance in close binary systems?

21. What are the differences between detached, semi-detached, contact, and overcontact binaries?

Advanced Questions

The answers to all computational problems, which are preceded by an asterisk (*), appear at the end of the book.

22. How is a degenerate gas different from an ordinary gas?

23. If you took a spectrum of a reflection nebula, would you see absorption lines, emission lines, or no lines? Explain your answer.

24. Why is it useful to plot the *apparent* magnitudes of stars in a single cluster on an H-R diagram?

25. What might happen to the massive outer planets when the Sun becomes a giant?

*26. How many 1.5-$M_\odot$ main-sequence stars would it take to equal the luminosity of one 15-$M_\odot$ star?

*27. How many times longer does a 1.5-$M_\odot$ star fuse hydrogen in its core than does a 15-$M_\odot$ star?

28. Why does a shock wave from a supernova produce relatively few high-mass O and B stars compared to the number of low-mass A, F, G, K, and M stars produced?

29. How would you distinguish a newly formed protostar from a giant, given that they occupy the same location on the H-R diagram?

30. What observational consequences would we find in H-R diagrams for star clusters as a result of the universe having a finite age? Could we use these consequences to establish constraints on the possible age of the universe? Explain your answers.

Discussion Questions

31. Discuss the possibility of life-forms and biological processes occurring in giant molecular clouds. In what ways might conditions favor or hinder biological evolution?

32. Is there any evidence that Earth has ever passed through a star-forming region in space?

What If...

33. The solar system passed through a giant molecular cloud? How would this encounter affect Earth and life on it?

34. The Sun began entering its giant phase today? What would happen to Earth and to life on it?

35. Earth were orbiting a 0.5-$M_\odot$ star at a distance of 1 AU? What would be different for Earth and life on it? What effects would moving Earth closer to the lower-mass Sun have?

36. The Sun were a variable star? How would this change life on Earth, and, assuming we could live in

orbit around such a star, how might it change our perspective of the cosmos?

Web Questions

37. To test your understanding of where stars are formed, do Interactive Exercise 12.1 on the Web. You can print out your results, if required.

38. To test your understanding of variable stars, do Interactive Exercise 12.2 on the Web. You can print out your results, if required.

39. To test your understanding of close binary star systems, do Interactive Exercise 12.3 on the Web. You can print out your results, if required.

Observing Projects

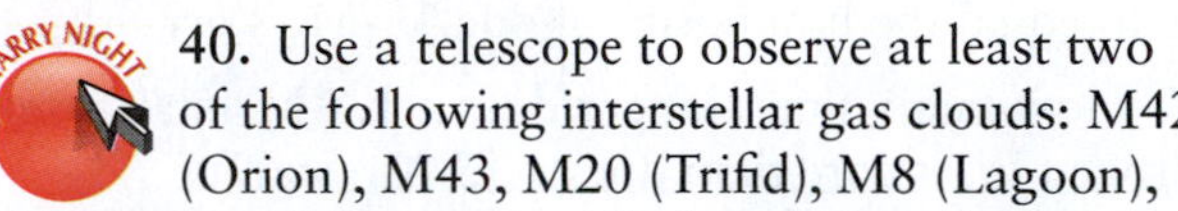

40. Use a telescope to observe at least two of the following interstellar gas clouds: M42 (Orion), M43, M20 (Trifid), M8 (Lagoon), M17 (Omega). You can easily locate them by using the coordinates in the list below, by using the *Starry Night*™ program, or by looking at star charts published in such magazines as *Astronomy* and *Sky & Telescope*. For each nebula, can you identify the stars responsible for the ionizing radiation that causes the nebula to glow? Draw a picture of what you see through the telescope and compare it with a photograph of the object. Which components of the nebula are not visible through your telescope, when compared to photographs in textbooks or the images displayed in *Starry Night*™? Why do you think that these components are not visible through your telescope?

Nebula	Right ascension	Declination
M42 (Orion)	5h 35.4m	−5° 27′
M43	5h 35.6m	−5° 16′
M20 (Trifid)	18h 02.6m	−23°02′
M8 (Lagoon)	18h 03.8m	−24° 23′
M17 (Omega)	18h 20.8m	−16° 11′

41. Several star clusters can be seen relatively easily through a good pair of binoculars. You can locate them with the aid of your *Starry Night*™ program, by using the star charts published in such magazines as *Astronomy* and *Sky & Telescope*, or by using the list of coordinates below. Observe as many of these clusters as you can. Use a telescope, if available, to look at them. Note the overall distribution of stars in each cluster. Can you see any of these clusters with the naked eye? What difference do you note between binocular and telescopic images of individual clusters?

Star cluster	Right ascension (2000)	Declination (2000)
M45 (Pleiades)	3h 46.0m	+24° 22′
Hyades	4h 27.0m	+16° 00′
M44 (Beehive)	8h 40.1m	+19° 45′
Coma	12h 25.0m	+26° 00′
M11	18h 51.1m	−06° 16′

42. Use the *Starry Night*™ program to examine a star-forming region. Select **Favourites > Discovering the Universe > Nebulas in Sagittarius.** The view is centered on the constellation Sagittarius, with several prominent clusters and nebulae labeled, including M20, the Trifid Nebula.
a. In which month is M20 highest in the sky at midnight, where it is best placed for observing with a telescope? Explain how you determined this. **b.** Click the right mouse button (Ctrl-Click on a Macintosh) with the cursor over M20 and select **Magnify** from the contextual menu. Explain why different areas of the nebula have the colors that they have.

43. You can use *Starry Night*™ to examine the Hertsprung-Russell Diagram (HRD) of the stars in a localized cluster of stars that was formed relatively recently in astronomical time, the Pleiades in the constellation of Taurus. Select **Favourites > Discovering the Universe > Pleiades** to display this cluster of young stars in the view. You can display an image of the dust and gas surrounding these stars briefly by zooming in to a field of view of about 4° and toggling on and off **View > Deep Space > Messier Objects.** Click **File > Revert** to return to the wider-field view. Click on the **Status** tab to display the HRD of all stars in this field of view around the Pleiades. Note that this HRD appears to contain stars from all groups and is similar to that of stars in our local neighborhood. However, if you restrict the distance to display only the stars within this localized cluster, a different pattern emerges. Click on the **Distance cut-off** checkbox in the **H-R Options** layer of the **Status** pane to restrict the distance to a range between 300 and 400 ly. **a.** Where on the HRD do you now find the majority of the stars of this cluster? **b.** In view of the existence within this cluster of very hot stars with high output of energy, what does this tell you about the age of this cluster compared to the general population of stars?

WHAT IF... Earth Orbited a 1.5-$M_\odot$ Sun?

Earth is at a perfect distance from a wonderful star. The Sun provides just enough heat so that liquid water, necessary to sustain life, can exist on Earth. Would our planet still be suitable for the evolution of life if the Sun were 1.5 $M_\odot$ rather than 1.0 $M_\odot$? To begin with, we need to know how far from the new Sun, let's call it Sol II, to place Earth.

A Very Sunny Day Sol II's surface temperature would be 8400 K and it would appear blue-white in our sky. Sol II would give off 7 times as much energy per second as our present Sun, due to the combination of a higher temperature and a 20% larger radius than the Sun. The effect of Sol II's increased energy emission would require that Earth be located much farther away from Sol II than our present distance from the Sun. To understand why, consider the increase in infrared (heat) output from Sol II. That extra heat would have the initial effect of increasing the average global temperature on Earth at our present distance by about 10 K (about 20°F). This does not seem like a lot, but that slight increase would quickly boost the atmospheric temperature much higher.

The extra heat from Sol II would also cause more ocean water to evaporate into the atmosphere. Because water is a greenhouse gas (that is, it traps infrared radiation), the air temperature would increase, causing even more water to evaporate from the oceans, which, in turn, would cause the air to heat even more. This vicious cycle, called the *runaway greenhouse effect,* would make Earth's surface so hot and dry that it would be uninhabitable.

By moving Earth about 2.6 times farther away from Sol II, the temperature would become suitable for life. At that distance, the year would be 1249 days long. While moving away from the heat is one thing, moving away from the ultraviolet radiation emitted by Sol II is something else altogether.

Ultraviolet Excess Just by increasing the Sun's mass by 50%, the ultraviolet radiation emitted would be several thousand times stronger. This would occur because the energy output of stars with different surface temperatures varies with wavelength. While the output of visible light would change only slightly, the output of ultraviolet radiation would be vastly greater. Therefore, even though Earth's surface temperature would be suitable farther from Sol II, the flood of ultraviolet radiation would be so strong that the ozone layer would be overwhelmed, and the level of ultraviolet radiation at Earth's surface would be much higher than it is today. This would be so even with the greater concentration of ozone created by the increased ultraviolet from Sol II. (Note that stars both create and destroy ozone in their planets' atmospheres.)

Life would have to evolve greater protection from ultraviolet radiation than it has today. If this were not a great enough challenge, suppose intelligent life-forms were evolving on Earth and orbiting Sol II 4.6 billion years after the solar system formed. They would discover that their star had evolved so rapidly that it was just about to expand into the giant phase!

(John Elk III/Bruce Coleman) R I V U X G

CLOUDY WITH A CHANCE OF STARS

Making a star is no easy thing

Adapted from an article by Erick T. Young

Though the basic idea for how stars form dates back to the eighteenth-century work of Immanuel Kant and Pierre-Simon Laplace, and physicists worked out how stars shine and evolve in the first half of the twentieth century, the birth of stars remains one of the most vibrant topics in astrophysics today.

Four questions, in particular, trouble astronomers. First, if dense, floating clouds of gas and dust (called cores) are the eggs of stars, where are the cosmic chickens? The clouds must come from somewhere, and their formation is not well understood. Second, what causes the core to begin collapsing, making it denser, hotter, and subject to nuclear fusion? Whatever the initiation mechanism is, it determines the rate of star formation and the final masses of stars.

Third, how do embryonic stars affect one another? The standard theory does not describe what happens when stars form in close proximity, as most do—even quite possibly our own Sun. How does growing up in a crowded nursery differ from being an only child?

Fourth, how do massive stars manage to form at all? The standard theory works well for building stars of up to 20 times the Sun's mass, but breaks down for bigger ones. Their tremendous brilliance—combined with blasts of ultraviolet radiation, high-velocity outflows, and supersonic shock waves—should disrupt or blow away clouds before nascent stars could completely form.

[STANDARD THEORY]

A Star Is Born—With Difficulty

The standard theory of star formation neatly explains isolated low- to medium-mass stars but leaves many conceptual gaps.

Star formation begins with a giant molecular cloud, a cold, nebulous mass of gas and dust.

Within the cloud, an especially dense subcloud of gas and dust—known as a core—collapses under its own weight.

The core fragments into multiple stellar embryos. In each, a protostar nucleates and pulls in gas and dust.

PROBLEM #1: Where does the cloud come from? A mixture of material produced in the big bang or ejected from stars must somehow coagulate.

PROBLEM #2: Why does the core collapse? The model does not specify how the balance of forces that stabilizes the cloud is disrupted.

PROBLEM #3: How do the embryos affect one another? The standard theory of star formation treats stars in isolation.

This knowledge is needed to dissect distant galaxies and make sense of planets being discovered beyond our solar system. Star formation underlies almost everything in astronomy, from the rise of galaxies to the genesis of planets. A more sophisticated theory must consider both the initial conditions in the core and the influences of a fledgling star's surroundings and its stellar neighbors.

Swaddled in Dust

The fundamental obstacle to observing star formation is that stars cloak their own birth. Though astronomers can see how the process begins and ends, what happens in the middle is hard to observe. Infrared observations have revealed nascent stars deeply embedded in dust, but have trouble seeing the earliest steps leading from molecular clouds to these protostars. Much of the radiation from those middle steps is far-infrared and submillimeter wavelengths where the astronomer's toolbox is relatively primitive.

In the mid-1990s, astronomers discovered clouds so dense (more than 10,000 atoms per cubic centimeter) that they are opaque even to the thermal infrared wavelengths that usually penetrate dusty regions. These "infrared dark clouds" are much larger (100 to 100,000 times the mass of the Sun) than clouds that had been previously discovered at optical wavelengths. They appear to be the missing link between molecular clouds and protostars. Dark clouds and dense cores could represent the crucial formative stage of stars, determining whether a nascent star becomes massive and dies young, exploding catastrophically, or develops into a smaller star that lives longer.

What Pulled the Trigger?

Astronomers are also making some progress on the second major unresolved problem: What causes a cloud or core to collapse? Something triggers an imbalance between gravity, external pressure, and internal pressure. Collapse begins when gravity begins to dominate.

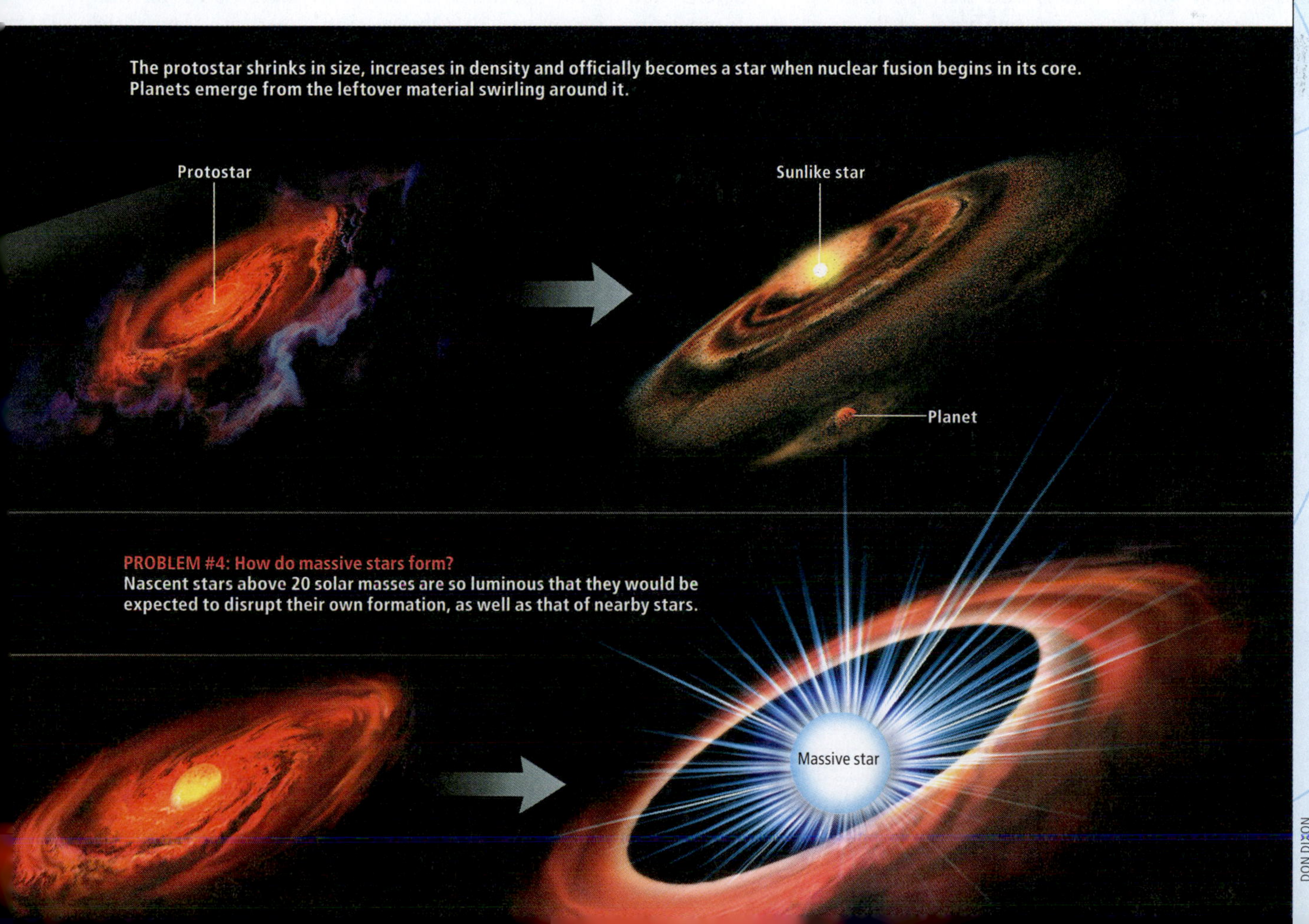

[PROBLEM #1]

The Dark Origins of Interstellar Clouds

Astronomers have gradually identified the stages by which clouds coalesce from diffuse interstellar gas and become progressively denser. The stage immediately prior to protostar formation is represented by so-called infrared dark clouds. Opaque even to infrared light, they show up as black streaks in this image from the Galactic Legacy Infrared Midplane Survey Extraordinaire (GLIMPSE), captured by the Spitzer Space Telescope. Their size and mass are just right for forming stars.

COURTESY OF THE GLIMPSE TEAM/UNIVERSITY OF WISCONSIN–MADISON

Astronomers have mapped molecular clouds, identifying a large number of quiescent, isolated cores in nearby clouds. Some show evidence of slow inward motions and may be making stars. For example, an infrared source in the center of Barnard 335 (in the constellation Aquila) may be an early-stage protostar, suggesting that the balance recently tilted towards collapse.

Other studies suggest that the clouds might be compressed by an external force. Astronomers discovered a striking example of this compression in the W5 region of Cassiopeia. Young protostars are embedded there in dense gas pockets compressed by radiation from an earlier generation of stars. Because compression is a rapid process, these widely-scattered objects must have formed almost simultaneously. But massive stars disrupt their birthplaces, making it difficult to reconstruct the conditions under which they formed. It is also difficult to observe dimmer, lower-mass stars, confounding confirmation that they, too, were formed in synchrony.

Life in a Stellar Nursery

Although questions remain, the standard model explains observations of isolated star-forming cores fairly well. But many, perhaps most, stars form in clusters. The model does not account for how this congested environment affects their birth. Two competing theories have been developed through computer modeling.

The "competitive accretion" theory surmises that many small protostars form, move rapidly through the cloud and compete to collect remaining gas. Some grow large; the smaller ones may be ejected from the cluster, creating stellar runts that roam the galaxy.

In the alternative "turbulent-core model," turbulence within the gas helps to trigger collapse. The relative size of stars is determined by the spectrum of turbulent motions rather than competition for material. Observations seem to favor this model, but in some situations, both turbulence and competitive accretion might operate: the competitive-accretion model may be important in regions of dense star clusters.

The famous Christmas Tree Cluster (NGC 2264) in the constellation Monoceros provides a snapshot of the stages when turbulence or competitive accretion would leave their mark. The youngest stars are clumped tightly and spaced evenly, possibly illustrating dense cores gravitationally collapsing out of a molecular cloud. Though observations support the turbulent model, compact groupings suggest that protostars must be competing, at least on a small scale.

Supersize This Star

Though massive stars are short-lived and rare, they help galaxies evolve. Massive stars inject energy into space—both radiation and mass outflows. When they die, they can explode as supernovae, expelling matter rich in heavy elements. The Milky Way is riddled with bubbles and supernova remnants created by such stars. But the standard theory fails to explain why, once a protostar reaches a certain threshold, pressure exerted by its radiation does not overpower gravity and prevent further growth. Additionally, the winds that such a star generates disperse its natal cloud, further limiting growth—and interfering with formation of nearby stars.

New three-dimensional simulations show that during stellar growth, dense gas regions can alternate with bubbles where

[PROBLEM #2]

The Onset of Collapse

Current models are vague as to how clouds become destabilized and collapse. New Spitzer infrared images reveal that nearby massive stars are often responsible.

In the W5 region of the galaxy, massive stars (which look bluish) have cleared out a cavity in a molecular cloud. On the rim of the cavity are protostars (embedded in whitish and pinkish gas) that are all roughly the same age, indicating that their formation was triggered by the massive stars; other processes would not have been so synchronized.

In the cluster NGC 2068, protostars are lined up like pearls on a string. Though widely scattered, they have formed almost simultaneously, and again the most likely culprit is a group of nearby massive stars.

COURTESY OF NASA, JPL/CALTECH AND HARVARD-SMITHSONIAN CENTER FOR ASTROPHYSICS (*W5*); COURTESY OF ERICK T. YOUNG AND NASA (*NGC 2068*)

[PROBLEM #3]

Life in a Crowded Nursery

Contradicting the assumptions made in the standard model of star formation, newborn stars can interfere with one another's formation. Spitzer has found an example in the Christmas Tree Cluster (NGC 2264), which contains a dense cluster of stars of varying ages. At high resolution, some of the youngest "stars" turn out to be tight groupings of protostars—as many as 10 of them within a radius of 0.1 light-year, close enough to affect one another.

COURTESY OF NASA, JPL/CALTECH AND PAULA S. TEIXEIRA *Harvard-Smithsonian Center for Astrophysics*

starlight streams out. This alternation may counter radiation pressure. Verifying this model will be tricky because it is hard to catch rare massive stars in the act of forming.

There are many possible reasons. Internal pressure might ebb as heat or magnetic fields dissipateNew technology and improved methods will soon help answer questions about star formation. A Boeing 747 that flies above the obscuring water vapor in Earth's atmosphere will observe the far-infrared and submillimeter wavelengths where star formation is easiest to see. At longer wavelengths, the Atacama Large Millimeter Array, 66 high-precision antennae now under construction in the Chilean Andes, will allow mapping of individual protostars in exquisite detail.

With new observations, astronomers hope to trace the complete life cycle of the interstellar medium from atomic clouds to molecular clouds to prestellar cores to stars—and back into diffuse gas. They also hope to observe star-forming disks, trace the infall of material from clouds, and compare the effects of different environments on stellar birth.

These answers will ripple across astrophysics, helping to explain many of the most important aspects of today's universe.

[PROBLEM #4]

Breaking through the Mass Ceiling

Recent computer simulations of star formation show that a massive star is able to reach a seemingly impossible size because it does not grow uniformly. Radiation emitted by the protostar pushes gas away, creating giant voids (bubbles) within the gas, but does not completely choke off the inward flow of gas, because material collects into filaments in the interstices of these voids.

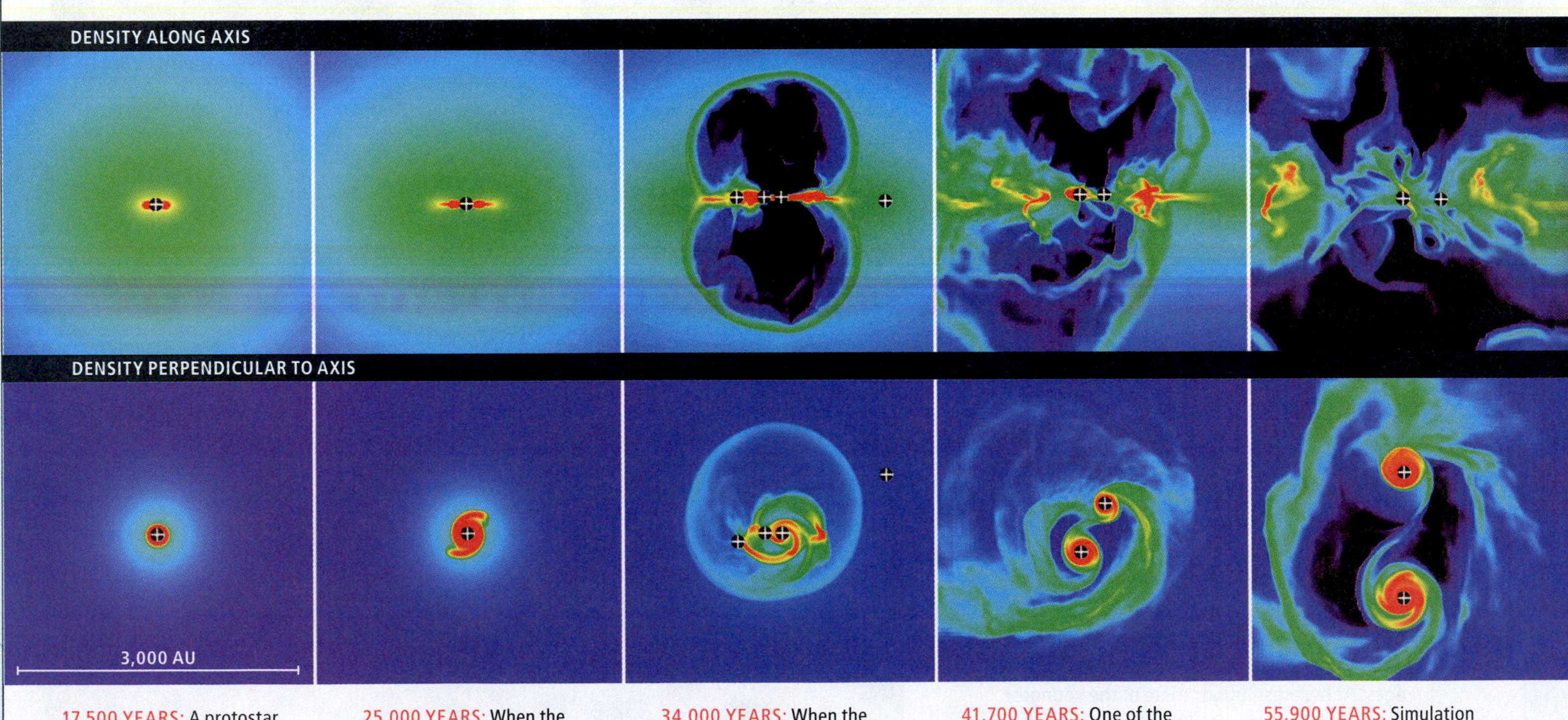

17,500 YEARS: A protostar has formed, and gas falls in nearly uniformly. Gravitational potential energy released by the descent of the gas causes it to glow.

25,000 YEARS: When the protostar has grown to about 11 solar masses, the disk around it becomes gravitationally unstable and develops a spiral shape.

34,000 YEARS: When the protostar exceeds 17 solar masses, radiation pushes gas out, creating bubbles. But gas still flows in around them. Smaller protostars form.

41,700 YEARS: One of the small protostars grows faster than the central one and soon rivals it in size. Accretion is not only uneven in space but also unsteady in time.

55,900 YEARS: Simulation ends as the central star reaches 42 solar masses and its companion 29. Some 28 solar masses of gas remain and will probably fall in eventually.

FROM "THE FORMATION OF MASSIVE STAR SYSTEMS BY ACCRETION," BY MARK R. KRUMHOLZ ET AL., IN *SCIENCE*, VOL 323; JANUARY 15, 2009

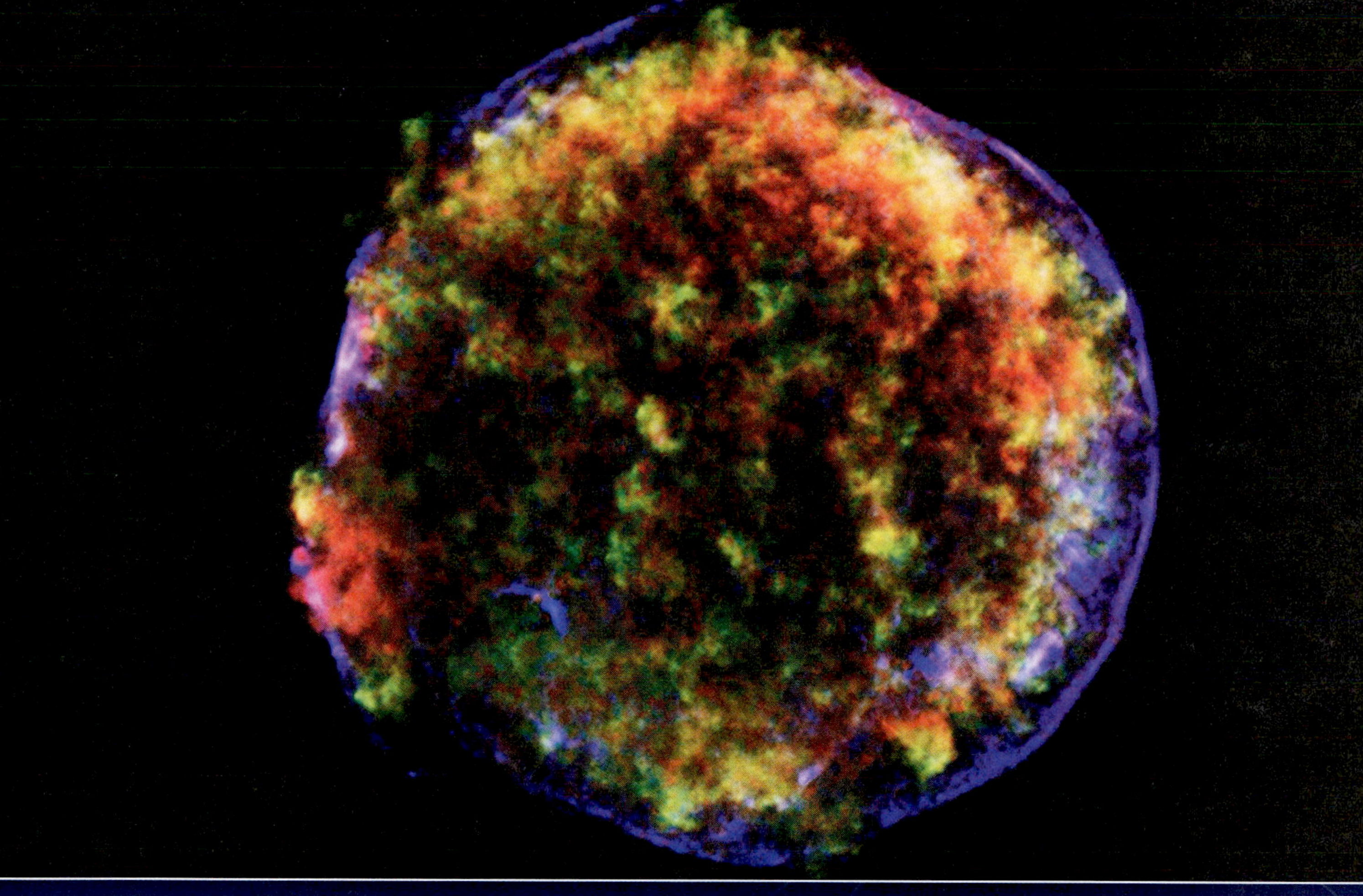

RIVUXG This X-ray image of Tycho's supernova, first seen as a visible light object by Tycho Brahe in 1572, was taken in 2003 by the Chandra X-ray Observatory. Gas and dust with temperatures in the millions of kelvins (shown in red and green) are expanding outward at about 10 million km/h, following a shell of high-energy electrons (blue). (NASA/CXC/Rutgers/J. Warren & J. Hughes et al.)

Chapter 13

The Deaths of Stars

Q WHAT DO YOU THINK?

1. **Will the Sun someday cease to shine brightly? If so, how will this occur?**
2. **What is a nova? How does it differ from a supernova?**
3. **What are the origins of the carbon, silicon, oxygen, iron, uranium, and other heavy elements on Earth?**
4. **What are cosmic rays? Where do they come from?**
5. **What is a pulsar?**

Answers to these questions appear in the text beside the corresponding numbers in the margins and at the end of the chapter.

Most video and computer games have one thing in common: Players move through different levels, each progressively more difficult than the last. The biggest challenge comes at the final level, when you have to face the most powerful or evil entity. Stellar evolution is similar, with stars passing through different stages of stellar activity before they can move on to the next stage. There is one big difference—a game player can win the final level, but a star at the final stage of evolution always ceases to shine with the vigor that it had previously. In the end, stars eject vast quantities of gas and dust into interstellar space. In human terms, they die. In this chapter and in Chapter 14, we learn how the later stages of stellar evolution are significantly different for stars with different masses. Some of them stop evolving by relatively mild emissions of their outer layers, while others have spectacular finales.

In this chapter you will discover

- what happens to stars when core helium fusion ceases
- how heavy elements are created
- the characteristics of the end of stellar evolution
- why some stars go out relatively gently, and others go out with a bang
- the incredible density of the matter in neutron stars and how these objects are observed

LOW-MASS STARS AND PLANETARY NEBULAE

We have seen that red dwarf stars, with masses less than $0.4M_\odot$, never get to the giant phase of stellar evolution. Rather, they convert all their mass to helium, stop fusing, and then just cool off. We will now explore the fates of stars greater than $0.4M_\odot$ as they proceed along through the giant phase and beyond.

In Chapter 12, we discovered that when hydrogen fusion in the shell surrounding the core first begins, the energy from it causes a star to expand and become a giant. During this process, low-mass stars move over to, and ascend, the giant branch on the H-R diagram for the first time (Figure 13-1a; see also Figure 12-25b). As this happens, mass is expelled into space in the form of stellar winds, which reduce the masses of these stars. Then core helium fusion begins, with stars of less than about 2 $M_\odot$ undergoing a core helium flash in the giant stage (Figure 13-1a). Stars with more than 2 $M_\odot$ begin helium fusion more gradually. In either case, after core fusion begins again, stars shrink and move onto the *horizontal branch* (Figure 13-1b). Their cores are eventually converted into carbon and oxygen, helium fusion ceases, and these stars undergo another stage that closely parallels the end of core hydrogen fusion (Figure 13-1c). The destinies of stars depend on their masses, and we have two mass ranges to consider: 0.4 to 8 $M_\odot$ (hereafter, *low-mass* stars), and more than 8 $M_\odot$ (hereafter, *high-mass* stars).

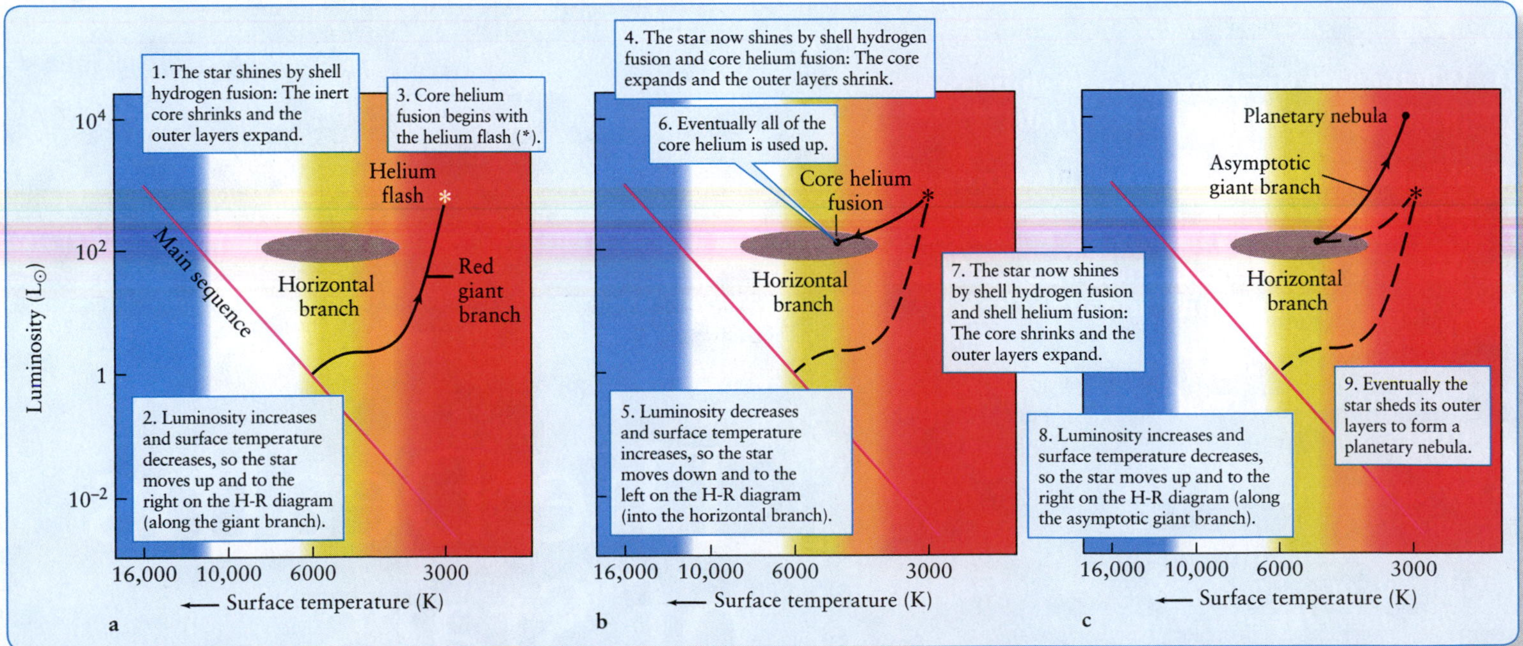

ANIMATION 13.1 **FIGURE 13-1 Post–Main-Sequence Evolution of Low-Mass Stars** (a) A typical evolutionary track on the H-R diagram as a star makes the transition from the main sequence to the giant phase. The asterisk (*) shows the helium flash occurring in a low-mass star. (b) After the helium flash, the star converts its helium core into carbon and oxygen. While doing so, its core reexpands, decreasing shell fusion. As a result, the star's outer layers recontract. (c) After the helium core is completely transformed into carbon and oxygen, the core recollapses, and the outer layers reexpand, powered up the asymptotic giant branch by hydrogen shell fusion and helium shell fusion.

13-1 Low-mass stars become supergiants before expanding into planetary nebulae

Calculations reveal that carbon and oxygen atoms require a temperature of at least 600 million K to fuse. Because the core of a low-mass giant on the horizontal branch only reaches about 200 million K, fusion of these elements in the core does not occur. Photon production in the core therefore decreases when its helium fuel is nearly used up, and the inner regions of the star again contract, compressing and heating the shell of helium-rich gas just outside the core. As a result, **helium shell fusion** begins outside of the core; this shell is itself surrounded by a hydrogen-fusing shell (Figure 13-2). All of this takes place within a volume roughly the size of Earth.

Once helium shell fusion commences, the new outpouring of energy again pushes the outer envelope of the star out. A low-mass star thus leaves the horizontal branch and ascends the H-R diagram for a second, and final, time. Powered by the fusion from two shells, such an **asymptotic giant branch star,** or **AGB star,** becomes brighter than ever before. When the Sun moves along the asymptotic giant branch, it will expand to a radius of about 1 AU, enveloping Earth in its tenuous outer layers (Figure 13-2). An 8-$M_\odot$ star will have a diameter as big as the orbit of Mars and shine with the luminosity of 10^4 $L_\odot$. At the top of the asymptotic giant branch, these stars are now so bright that they are classified as low-temperature, red supergiants (see Figure 13-1c).

AGB stars are destined to self-destruct. Giant stars of all spectral types emit strong stellar winds. Low-mass stars expel their outer layers more slowly than higher-mass stars, but both types of stars reduce their masses significantly from what they were on the main sequence. During their initial ascent up the giant branch, stars can lose as much as 30% of their masses. On the asymptotic giant branch, stars lose even more, often surrounding themselves with thickening cocoons of gas and dust. A star of the Sun's mass loses about 10^{-5} $M_\odot$ per year at this stage, eventually dumping half of its mass back into space.

This mass loss limits the force of gravity available to compress the star's core and the regions of shell fusion. A fine balance is struck: The low-mass AGB star is compressed until its core of carbon and oxygen becomes degenerate, meaning that the electrons provide a growing repulsive force that stops its contraction (as discussed in Section 12-10). However, even in the most massive of these low-mass stars (now much less than their original 8 $M_\odot$ due to their stellar winds), the core temperature cannot reach the 600 million K necessary to fuse carbon or oxygen. Therefore, no further core fusion occurs.

The final stage through which a low-mass star passes begins with a "thermal runaway" (meaning a rapid rise in temperature) in the helium shell, just like the helium flash in its core earlier in its life (described in Section 12-10). This process occurs because the triple alpha process (see Section 12-11) is extremely sensitive to temperature, and, as the temperature goes up slightly, the fusion rate skyrockets. The increase in energy output from the thin helium-fusing shell, called a **helium shell flash,** expands the star, thereby briefly decreasing the temperature in the shell and slowing the rate of fusion.

A star in this mass range undergoes several helium shell flashes as its helium shell thickens. During each flash, the helium shell's energy output jumps a thousandfold. These brief outbursts are separated by relatively quiet intervals lasting about 100,000 years, during which the helium shell gradually becomes thicker. The energy from the increased fusion during the helium shell flashes causes the star's outer layers to expand and, therefore, cool further. Eventually, the outer gases are sufficiently cool so that electrons and ions in them can recombine. This process is exactly the opposite of ionization (discussed in Section 4-5). Whereas ionization requires an electron to absorb a photon, recombination forces an electron to emit a photon.

Just as the collisions of particles create pressure (force acting over an area), the collisions of photons with matter also create pressure. The impacts of photons emitted during recombination, along with the photons from the helium flashes and the photons normally flowing outward from fusion, generate enough pressure to eject more and more of the star's outer layers into space. Eventually, the mass of the star decreases so much that there is not

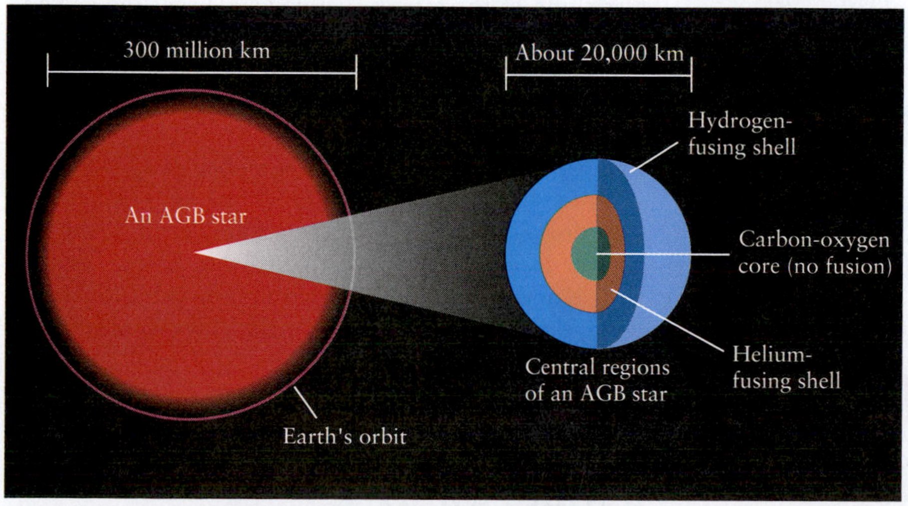

FIGURE 13-2 The Structure of an Old Low-Mass Star Near the end of its life, a low-mass star, like the Sun, travels up the AGB and becomes a red supergiant. (The Sun will eventually be about as large as the diameter of Earth's orbit.) The star's inert core, its hydrogen-fusing shell, and its helium-fusing shell are contained within a volume roughly the size of Earth. The inner layers are not shown to scale here.

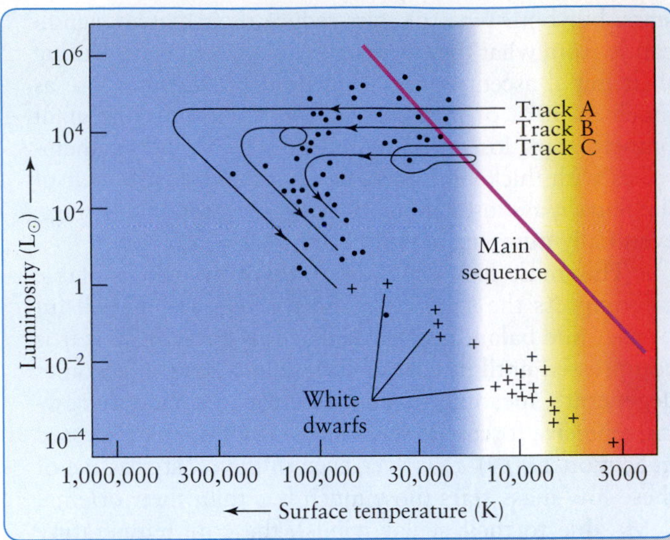

	Mass ($M_\odot$)		
Evolutionary track	Red supergiant	Ejected nebula	White dwarf
A	3.0	1.8	1.2
B	1.5	0.7	0.8
C	0.8	0.2	0.6

FIGURE 13-3 Evolution from Red Supergiants to White Dwarfs The evolutionary tracks of three low-mass supergiants are shown as they eject planetary nebulae. The table gives their masses as supergiants, the amount of mass they lose as planetary nebulae, and their remaining (white dwarf) masses. The loops indicate periods of instability and adjustment for the white dwarfs. The dots on this graph represent the central stars of planetary nebulae whose surface temperatures and luminosities have been determined. The crosses are white dwarfs for which similar data exist.

enough pressure to sustain any shell fusion. As the ejected material expands and cools, some of it condenses to form dust grains. When enough of the gas has left the star for the core to be visible, the expanding dust and gases are considered to be a **planetary nebula** (see Figure 13-1c). As the outer layers of the star are shed, an increasingly hot interior is revealed, and the star moves to the left across the H-R diagram (Figure 13-3). Including the mass lost prior to the planetary nebula phase, low-mass stars lose as much as 80% of their mass by shedding their outer layers.

Insight Into Science

Names Often Don't Change as Understanding Improves Planetary nebulae have *nothing* to do with planets. This term was coined by the astronomer William Herschel in the eighteenth century. When viewed through the telescopes of the day, these glowing objects, often green in appearance, looked like planets similar to Uranus. Astronomers have continued to use this term, but do not be fooled—we now know that planetary nebulae indicate the presence of dying stars, not planets.

Planetary nebulae are quite common in our Galaxy. 1
More than 1800 have been identified, and astronomers estimate that 20,000 to 100,000 planetary nebulae exist in the Milky Way. Indeed, the fate of the Sun is to shed its outer layers as a planetary nebula. The rate at which the gases are removed in a planetary nebula is so slow that they really are not explosive, compared either to our common conception of an explosion on Earth or to the supernova explosions that we will examine shortly.

Because stellar winds from red giants and subsequent planetary nebulae are so plentiful, astronomers estimate that they return a total of about 5 $M_\odot$ per year of gas and dust to the disk of our Galaxy. This mass amounts to about 85% of all matter expelled by all types of stars. This material goes into the formation of new, metal-rich Population I stars, along with their associated planets and debris. Thus, planetary nebulae play an important role in the chemical and physical evolution of the Galaxy.

The outflowing gases ejected in a planetary nebula take a breathtaking variety of shapes when they interact with gases surrounding their stars, with companion stars, and with the stars' magnetic fields, which are often 10 to 100 times stronger than the Sun's field (Figures 13-4 and 13-5). The Hourglass Nebula (Figure 13-5c) appears initially to have shed mass in a doughnut shape around itself. In the star's final death throes, this gas and dust forced the outflow to go in two directions perpendicular to the plane of the doughnut, creating what is called a *bipolar planetary nebula*.

Spectroscopic observations of planetary nebulae show bright emission lines of hydrogen, carbon, neon, magnesium, oxygen, and nitrogen. From the Doppler shifts of these lines, we conclude that the gas and dust are moving outward from dying low-mass stars at speeds of 10 to 30 km/s. A typical planetary nebula moving outward for 10,000 years has a diameter of a few light-years.

By astronomical standards, a planetary nebula is short-lived. After about 50,000 years, its gases spread over distances so far from the cooling central star that its nebulosity simply fades from view. The gases then mingle and mix with the surrounding interstellar medium, enriching it with moderately heavy elements.

13-2 The burned-out core of a low-mass star becomes a white dwarf

The exposed carbon-oxygen cores of low-mass stars are stable objects supported by electron degeneracy pressure. Such burned-out hulks are called white dwarfs, and this is the fate of the Sun's core. Typical white dwarfs are roughly the size of Earth and are covered with a thin coating of hydrogen and helium. The densities of these stellar "corpses" are typically

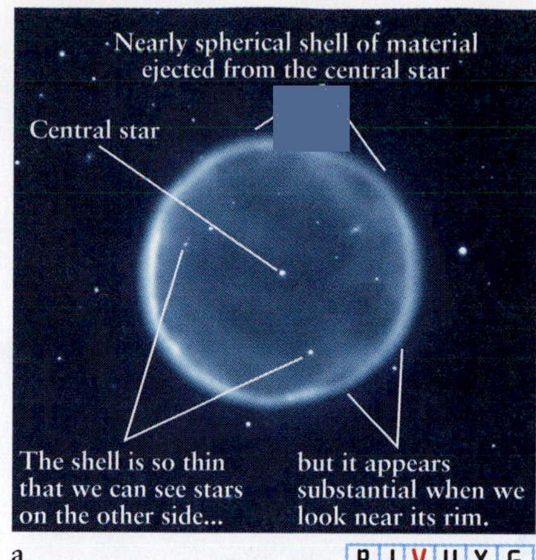

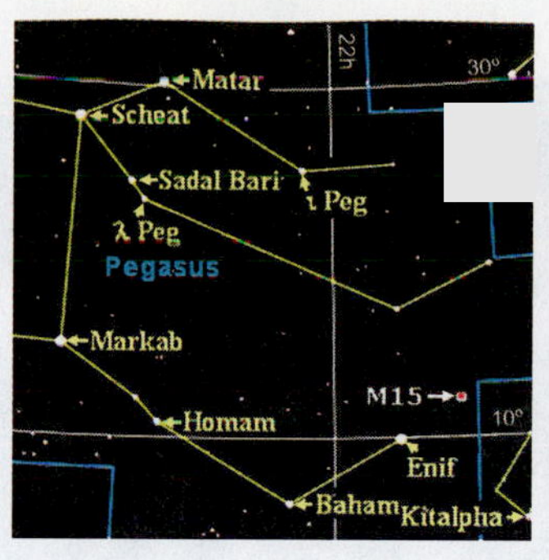

a

R I V U X G

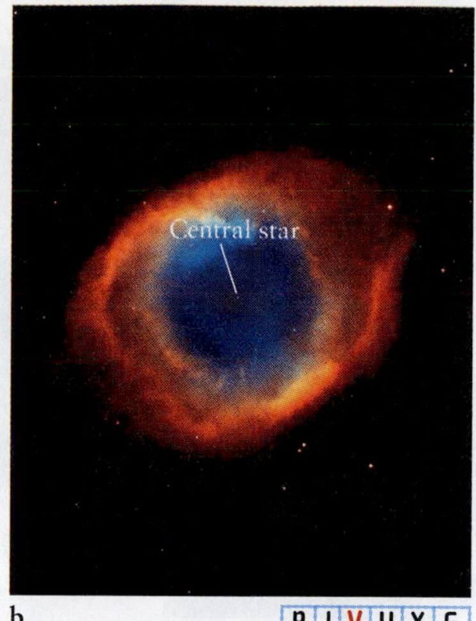

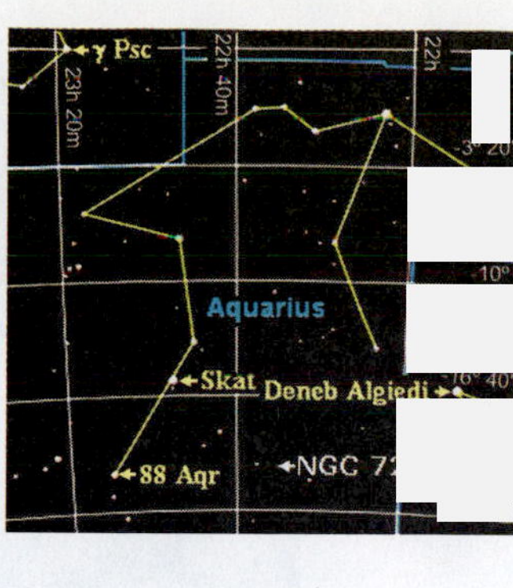

b

R I V U X G

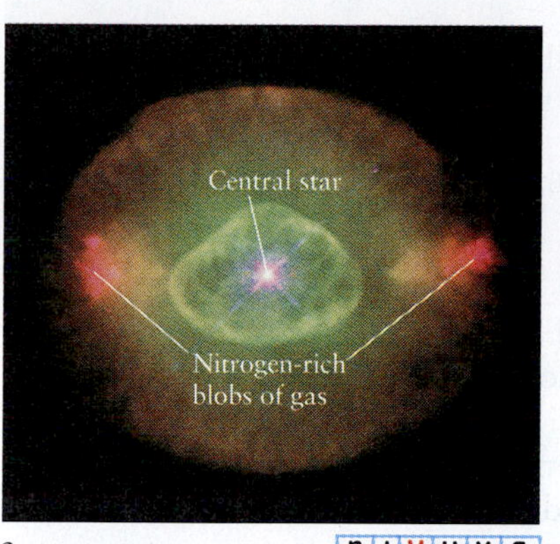

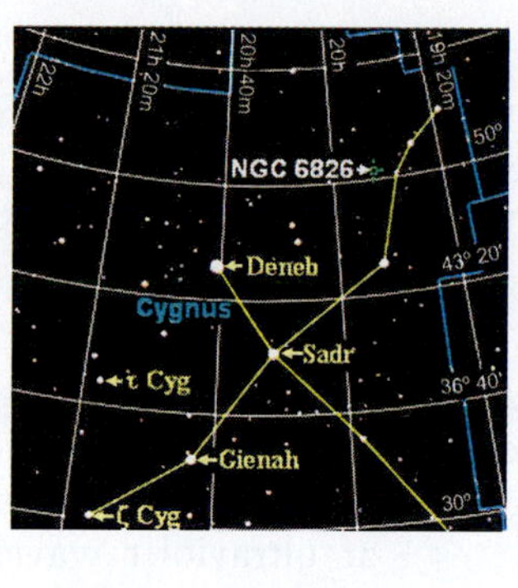

c

R I V U X G

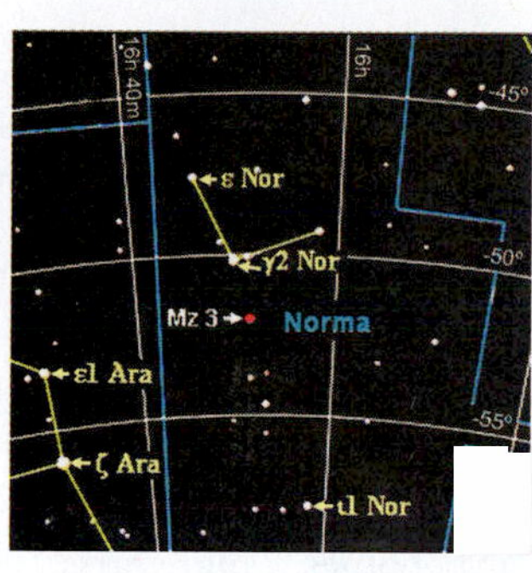

d

R I V U X G

FIGURE 13-4 Some Shapes of Planetary Nebulae The outer shells of dying low-mass stars are ejected in a wonderful variety of patterns. (a) An exceptionally spherical remnant, this shell of expanding gas, in the globular cluster M15 in the constellation Pegasus, is about 7000 ly (2150 pc) away from Earth. (b) The Helix Nebula, NGC 7293, located in the constellation Aquarius, about 700 ly (215 pc) from Earth, has an angular diameter equal to about half that of the full Moon. Red gas is mostly hydrogen and nitrogen, whereas the blue gas is rich in oxygen. (c) NGC 6826 shows, among other features, lobes of nitrogen-rich gas (red). The process by which they were ejected is as yet unknown. This planetary nebula is located in Cygnus. (d) Mz 3 (Menzel 3), in the constellation Norma (the Carpenter's Square), is 3000 ly (900 pc) from Earth. The dying star, creating these bubbles of gas, may be part of a binary system. (a: WIYN/NOAL/NSF; b–d: NASA, NOAO, ESA, The Hubble Helix Nebula Team, M. Meixner/STScI, and T. A. Rector/ NRAO; b: Howard Bons, STScI/Robin Ciardullo, Pennsylvania State University/ NASA; c: AURA/STScI/NASA)

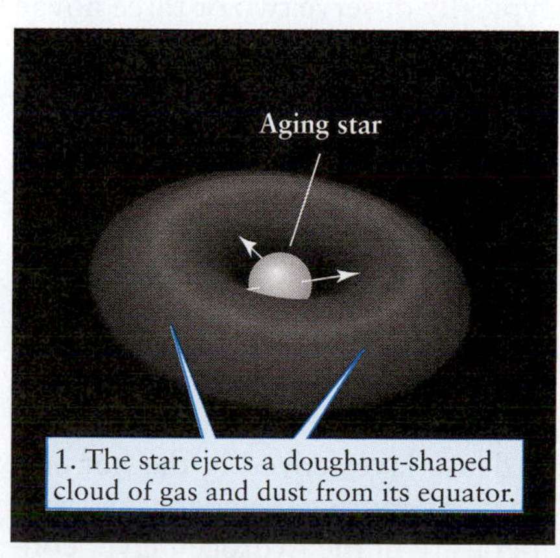

a

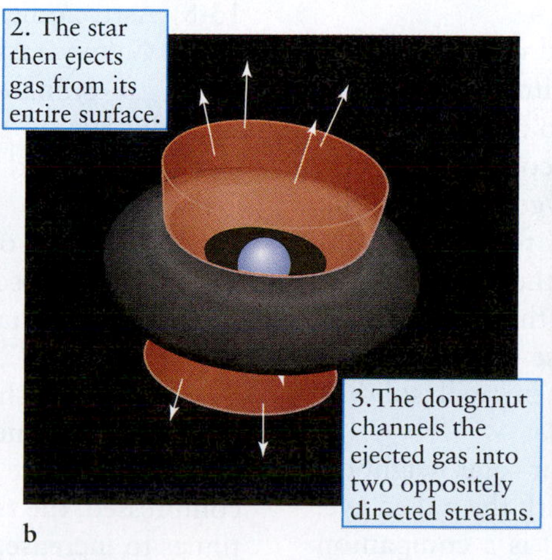

b

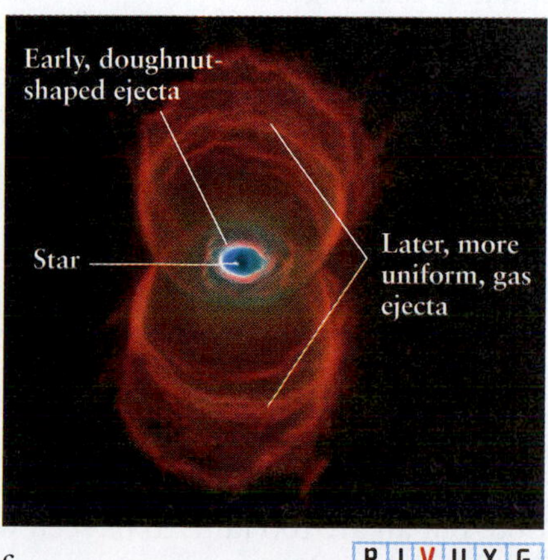

c

R I V U X G

FIGURE 13-5 Formation of a Bipolar Planetary Nebula Bipolar planetary nebulae may form in two steps. Astronomers hypothesize that (a) first, a doughnut-shaped cloud of gas and dust is emitted from the star's equator, (b) followed by outflow that is channeled by the original gas to squirt out perpendicularly to the plane of the doughnut. (c) The Hourglass Nebula appears to be a "textbook" example of such a system. The bright ring is believed to be the doughnut-shaped region of gas lit by energy from the planetary nebula. The Hourglass is located about 8000 ly (2500 pc) from Earth. (c: R. Sahai and J. Trauer, JPL; WFPC-2 Science Team; and NASA)

a

R I V U X G

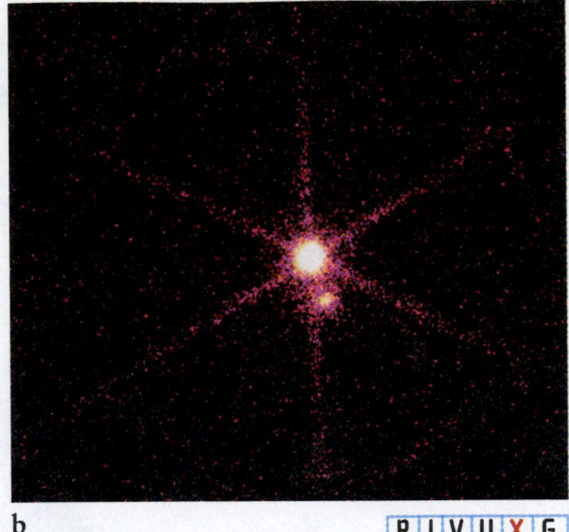

b

R I V U X G

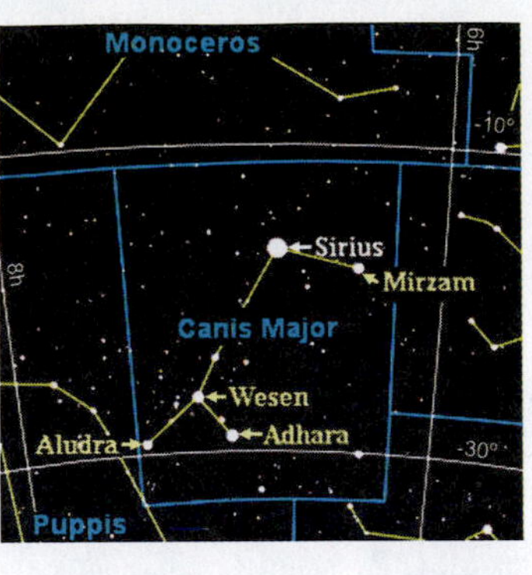

FIGURE 13-6 Sirius and Its White Dwarf Companion (a) Sirius, the brightest-appearing star in the night sky, is actually a double star. The smaller star, Sirius B, is a white dwarf, seen here at the five o'clock position in the glare of Sirius. The spikes and rays around the bright star, Sirius A, are created by optical effects within the telescope. (b) Since Sirius A (11,000 K) and Sirius B (30,000 K) are hot blackbodies, they are strong emitters of X rays. (a: R. B. Minton; b: NASA/SAO/CXC)

What luminosity class will the Sun be in just before it becomes a planetary nebula?

10^9 kg/m^3. In other words, a teaspoonful of white dwarf matter brought to Earth would weigh 5 tons.

Theoretical evolutionary tracks of three burned-out stellar cores are shown in Figure 13-3. These particular white dwarfs evolved from main-sequence stars of between 0.8 and 3.0 $M_\odot$. During the ejection phase, the appearance of these remnants changes rapidly. They appear to race along their evolutionary tracks on the H-R diagram, sometimes executing loops called *thermal pulses* as they temporarily reheat (tracks B and C in Figure 13-3). Finally, as the ejected planetary nebulae fade and the stellar cores cool, the evolutionary tracks of the dying stars take a sharp turn downward toward the white dwarf region of the diagram.

As billions of years pass, an isolated white dwarf radiates its stored energy into space, cooling and decreasing in luminosity. It moves down and to the right on the H-R diagram. The Sun will become a cold, dark, dense sphere of degenerate gases rich in oxygen and carbon, about the size of Earth. We can actually follow the theory a little further. Astronomers predict that after billions of years of radiating away their heat, the interior temperatures of white dwarfs will decrease to about 4000 K, and the carbon and oxygen in them will solidify, transforming them into giant crystals.

Many white dwarfs are found in our solar neighborhood, but all are too faint to be seen with the naked eye. The first white dwarf to be discovered is a companion to the bright star Sirius. The binary nature of Sirius was first deduced in 1844 by the German astronomer Friedrich Bessel, who noticed that the star was moving back and forth, as if orbited by an unseen object. This companion, Sirius B (Figure 13-6), was first glimpsed in 1862. Recent satellite observations at ultraviolet wavelengths—where white dwarfs emit most of their light—demonstrate that the surface temperature of Sirius B is about 30,000 K.

Of what two elements is a white dwarf composed?

13-3 White dwarfs in close binary systems can create powerful explosions

Occasionally, a star in the sky suddenly becomes between 10^4 and 10^6 times brighter than it is normally. This phenomenon is called a **nova.** Its abrupt rise in brightness is followed by a gradual decline that may stretch over several months or more (Figures 13-7 and 13-8). Astronomers typically observe two or three novae in our Galaxy each year. Novae are fairly common, with about 20 estimated to occur in the Milky Way Galaxy annually. We do not see all of them because of the interstellar gas and dust that obscures many parts of the Galaxy.

Painstaking observations strongly suggest that novae occur in close binary systems that contain a white dwarf. The ordinary companion star fills its Roche lobe (recall Figures 12-33, 12-34, and 12-35), so it gradually deposits fresh hydrogen onto the white dwarf. This new mass becomes a dense layer covering the hot surface of the white dwarf. As more gas is deposited and compressed, the temperature in the hydrogen layer continues to increase. Finally, at about 10^7 K, hydrogen fusion ignites throughout the layer, blowing it into interstellar space. This explosion is the nova. After a nova, fusion ceases on the white dwarf. The companion star, however, may retain enough mass to supply a new layer 

a

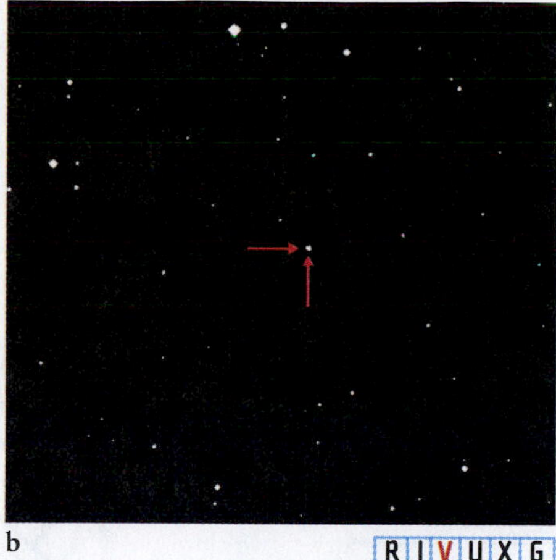

b

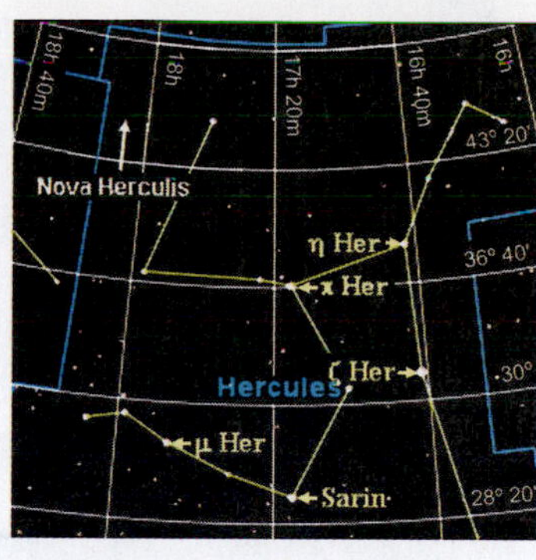

R I V U X G

FIGURE 13-7 Nova Herculis 1934 These two pictures show a nova (a) shortly after peak brightness as a magnitude –3 star and (b) 2 months later, when it had faded to magnitude +12. Novae are named after the constellation and year in which they appear. (UCO/Lick Observatory)

of surface hydrogen, enabling some novae to reoccur. White dwarfs are not damaged by novae initiated on their surfaces.

Recent visible observations by the Hubble Space Telescope and X-ray observations by the Chandra X-ray Observatory reveal that novae are much more complex than astronomers had thought. Rather than smooth shells of expanding gas, they are composed of thousands of clumps of gas. These observations are stimulating astrophysicists to reexamine the details of the nova mechanism described above to try to explain these newly observed effects.

Theoretical models predict that degenerate matter, such as the electrons in a white dwarf, can withstand only a finite amount of pressure, above which the degenerate matter cannot exert enough outward force to prevent the core from collapsing further. This limits the mass of a white dwarf to less than 1.4 $M_{\odot}$. This mass is called the **Chandrasekhar limit,** after the astrophysicist Subrahmanyan Chandrasekhar, who received a Nobel Prize in 1990 for his pioneering theoretical studies of white dwarfs a half-century earlier.

Why are novae not associated with isolated white dwarfs?

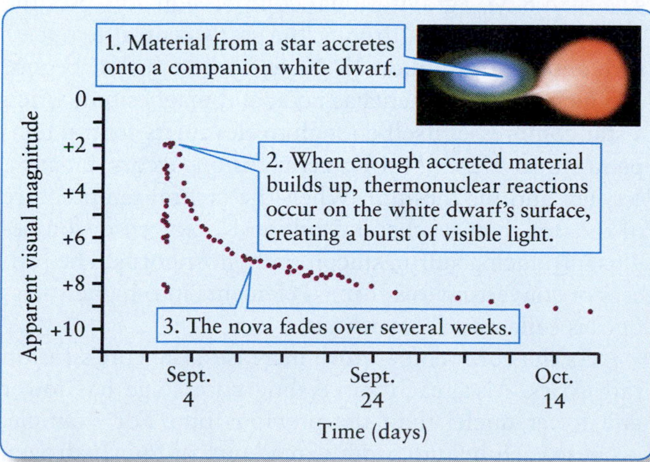

FIGURE 13-8 The Light Curve of a Nova This graph shows the history of Nova Cygni 1975, a nova that was observed to blaze forth in the constellation of Cygnus in September 1975. The rapid rise in magnitude followed by a gradual decline is characteristic of many novae, although some oscillate in intensity as they become dimmer.

13-4 Accreting white dwarfs in close binary systems can also explode as Type Ia supernovae

ANIMATION 13.2

Whereas a nova only removes hydrogen and helium that build up on the surface of a white dwarf, some of these stellar remnants actually blow apart completely. This event is called a **Type Ia supernova,** and it occurs in white dwarfs that are part of semidetached binary systems. To trigger a Type Ia supernova, a swollen giant companion star dumps gas onto a white dwarf that is close to the Chandrasekhar limit. When the added gas causes the white dwarf's mass to cross that limit, the increased pressure deep inside the dwarf enables carbon fusion to begin in the core. In a catastrophic runaway process reminiscent of the helium flash, the rate of carbon fusion skyrockets, and with no outer layers to absorb the energy, the star blows up.

A Type Ia supernova is powered by nuclear energy. What we see is the fallout from a gigantic thermonuclear explosion, which produces a wide array of radioactive isotopes. Especially abundant is an unstable isotope of nickel that decays into a radioactive isotope of cobalt. Most of the electromagnetic display of a Type Ia supernova results directly from the radioactive decay of this nickel into cobalt.

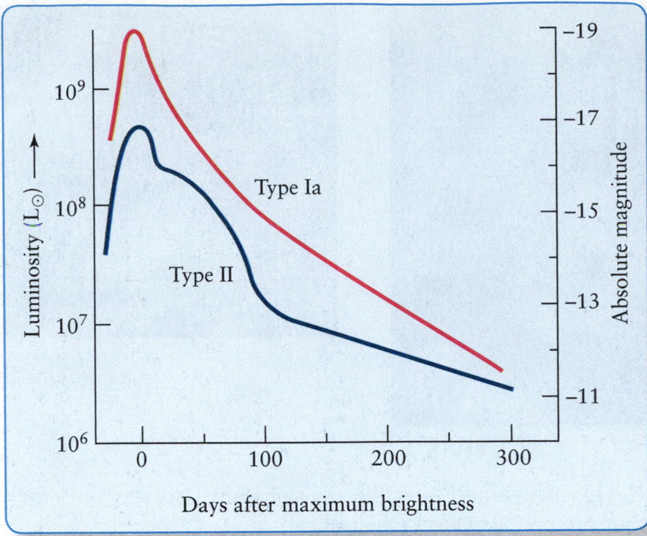

FIGURE 13-9 Supernova Light Curves A Type Ia supernova, which gradually declines in brightness, is caused by an exploding white dwarf in a close binary system. A Type II supernova is caused by the explosive death of a massive star and usually has alternating intervals of steep and gradual declines in brightness.

A Type Ia supernova's spectrum lacks hydrogen lines because the star from which it came was virtually free of this element. The light curve of the supernova, which shows its change of brightness over time, begins with a sudden rise (Figure 13-9). The supernova typically reaches an absolute magnitude of –19 at peak brightness and then declines gradually for more than a year.

Because these supernovae are so bright, they can be seen far beyond the bounds of the Milky Way. Whereas our Galaxy is about 10^5 ly in diameter, Type Ia supernovae can be seen in other galaxies at distances of over 10^9 ly away. Because they all peak at the same absolute magnitude, these supernovae can be used to determine distances to myriad galaxies throughout the universe. This calculation is done in two steps. First, the apparent magnitude of a Type Ia supernova at its peak brightness is observed. Second, because we know its absolute magnitude, we use the distance-magnitude relationship (see *An Astronomer's Toolbox 11-3*) to calculate its distance.

HIGH-MASS STARS AND TYPE II SUPERNOVAE

In main-sequence stars of at least 8 $M_\odot$, sufficient gravitational force compresses and heats the core enough to enable its carbon and oxygen to fuse. This fusion leads to the creation of a host of other elements that are eventually ejected into interstellar space.

13-5 A series of fusion reactions in high-mass stars leads to luminous supergiants

When helium fusion ends in the core of a star more massive than 8 $M_\odot$, gravitational compression collapses the carbon-oxygen core, driving the star's central temperature above 600 million K. Now, carbon fusion begins, producing such elements as neon and magnesium. When a star compresses itself enough to elevate its central temperature to 1.2 billion K, neon fusion occurs, creating oxygen and magnesium. When the central temperature of the star then reaches 1.5 billion K, oxygen fusion begins, producing sulfur, silicon and phosphorus. The process of converting lower-mass elements into higher-mass ones is called **nucleosynthesis.**

As the core temperature increases, the core's fusion rate soars. Also, each succeeding core stage has fewer and fewer nuclei than the previous one. For example, because each helium atom is made up of four hydrogen atoms, only one-quarter as many helium atoms exist in a star's core as there were hydrogen atoms. Combining the higher fusion rate and fewer fusible atoms causes each cycle of core thermonuclear fusion to occur more rapidly than the previous stage. For example, detailed

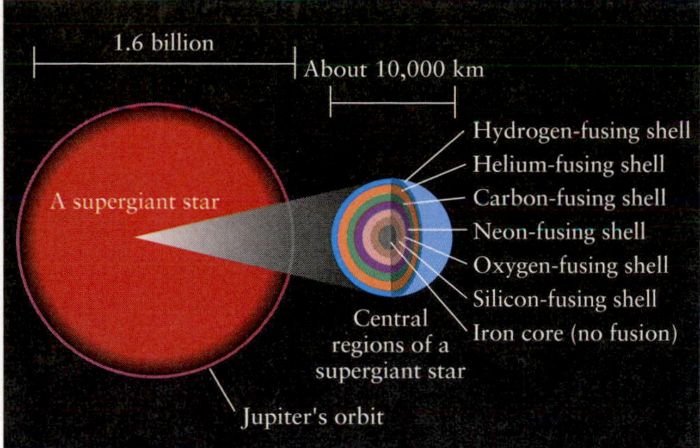

FIGURE 13-10 The Structure of an Old High-Mass Star Near the end of its life, a high-mass star becomes a supergiant with a diameter almost as wide as the orbit of Jupiter. The star's energy comes from six concentric fusing shells, all contained within a volume roughly the same size as Earth.

FIGURE 13-11 Mass Loss by a Supermassive Star Gas and dust ejected by the massive star HD 148937 in the constellation Norma. Orange is H_α, blue is oxygen, and red is sulfur. Located about 4200 ly (1300 pc) away, this star has 40 $M_\odot$ and is more than halfway through its 6-million-year lifespan. (Gemini Observatory/AURA)

calculations for a star that was initially 25 $M_\odot$ on the main sequence reveal that carbon fusion occurs for 600 years, neon fusion for 1 year, and oxygen fusion for only 6 months. After half a year of core oxygen fusion, gravitational compression forces the central temperature up to 2.7 billion K, and silicon fusion begins. This thermonuclear process proceeds so furiously that the entire core supply of silicon in a 25-$M_\odot$ star is converted into iron in 1 day.

Each stage of fusion adds a new shell of matter outside of the core (Figure 13-10), creating a structure that resembles the layers of an enormous onion. Together, the tremendous numbers of fusion-generated photons created in all of these shells, as well as in the core, push the outer layers of the star farther and farther outward. The star expands to become a luminous supergiant, almost as wide across as Jupiter's orbit around the Sun.

High-mass main-sequence stars evolve into luminous supergiant stars that are brighter than 10^5 $L_\odot$, emit winds throughout most of their existence, and have mass-loss rates that exceed those of giants (see Section 12-9). Figure 13-11 shows a supergiant star losing mass. Betelgeuse (see Figure 11-2a), 470 ly away in the constellation of Orion, is another good example of a supergiant experiencing mass loss. Spectroscopic observations show that Betelgeuse is losing mass at the rate of 1.7×10^{-7} $M_\odot$ per year and is surrounded by ejected gas, in a *circumstellar shell*. This huge shell is expanding at 10 km/s, and escaping gases have been detected at distances of 10,000 AU from the star. The expanding circumstellar shell has an overall diameter of one-third of a light-year.

Silicon fusion in high-mass stars involves many types of nuclear reactions, but its major final product is iron. Eventually, the core is converted entirely into iron, which is surrounded by layers of shell fusion that consume the star's remaining reserves of fuel (see Figure 13-10). While the star's enormously bloated atmosphere is nearly as big as the orbit of Jupiter, its entire energy-producing region is again contained in a volume the size of Earth. The buildup of an inert core of iron nuclei and electrons signals the impending violent death of a massive star.

Why do type Ia supernovae not have any hydrogen lines in their spectra?

Where in massive stars are the higher-mass elements formed?

13-6 High-mass stars blow apart in violent supernova explosions

Unlike lighter elements, iron cannot fuel further thermonuclear reactions. The protons and neutrons inside iron nuclei are already so tightly bound together that no further energy can be extracted by fusing still more particles with them. The sequence of fusion stages in the cores of high-mass stars therefore ends.

Because iron atoms do not fuse and emit energy, the electrons in the core must now support the star's outer layers by the strength of electron degeneracy pressure alone. Soon, however, the continued deposition of fresh iron from the silicon-fusing shell causes the core's mass

Table 13-1 Evolutionary Stages of a 25-$M_\odot$ Star

Stage	Central temperature (K)	Central density (kg/m^3)	Duration of stage
Hydrogen fusion	4×10^7	5×10^3	7×10^6 years
Helium fusion	2×10^8	7×10^5	5×10^5 years
Carbon fusion	6×10^8	2×10^8	600 years
Neon fusion	1.2×10^9	4×10^9	1 year
Oxygen fusion	1.5×10^9	1×10^{10}	6 months
Silicon fusion	2.7×10^9	3×10^{10}	1 day
Core collapse	5.4×10^9	3×10^{12}	0.2 second
Core bounce	2.3×10^{10}	4×10^{17}	milliseconds
Supernova explosion	about 10^9	varies	hours

to exceed the Chandrasekhar limit. Electron degeneracy suddenly fails to support the star's enormous weight, and the core collapses.

Any isolated main-sequence star of more than 8 $M_\odot$ develops an iron core as a luminous supergiant. When the core collapses, a rapid series of cataclysms is triggered that tears the star apart in a few seconds. Let us see how this happens in the death of a 25-$M_\odot$ star, according to computer simulations (Table 13-1).

In a 25-$M_\odot$ star, electron degeneracy pressure fails when the density is sufficiently high inside the iron core. The core, only some 3000 km in diameter, then collapses immediately. In roughly $\frac{1}{10}$ s, the central temperature exceeds 5 billion K. Gamma-ray photons associated with this intense heat have so much energy that they begin to break apart the iron nuclei, a process called **photodisintegration.** Although millions of years passed from the time the star arrived on the main sequence with a hydrogen- and helium-filled core until its core became iron, it takes less than a second to convert the core back into elemental protons, neutrons, and electrons!

Within another $\frac{1}{10}$ s, as the density continues to climb, the electrons in the core are forced to combine with the core's protons to produce neutrons, and the process releases a flood of neutrinos. As we saw in Section 10-9, neutrinos have no electric charge and have very little mass, and most pass through Earth or the Sun without interaction. Nevertheless, the matter deep inside a collapsing high-mass star is so fantastically dense that its newly created neutrinos cannot immediately escape from the star's core. Slamming into nearby particles, the neutrinos provide a pressure pushing outward from the core.

According to the calculations, about $\frac{1}{4}$ s after the collapse begins, the density of the entire core reaches 4×10^{17} kg/m^3, which is *nuclear density,* the density at which neutrons and protons are normally packed together inside atomic nuclei. (Compare this with the density of water, 10^3 kg/m^3.) When the neutron-rich material of the core reaches nuclear density, it suddenly stiffens. Thus, the core resists the collapse much more than when it was less dense. The collapse of the core halts so abruptly that it rebounds and begins rushing back out in a process called *core bounce.*

The model predicts that, during this critical stage, the star's unsupported layers of shell-fusing matter are plunging inward at up to 15% of the speed of light. The outward-flowing neutrinos and rebounding core slam into this matter (the red regions in Figure 13-12a). The impact stops the core's rebound while causing the infalling matter to reverse course. In just a fraction of a second, a tremendous volume of matter begins to move back up toward the star's surface. This matter accelerates rapidly as it encounters less and less resistance, and soon it forms an outgoing shock wave. After a few hours, this shock wave reaches the star's surface, lifting the star's outer layers away from the core in a mighty blast. The star becomes a **Type II supernova.**

As the supernova expands, the star's luminosity, which was already some 10^4 to 10^5 $L_\odot$, suddenly increases by another factor of between 10^3 and 10^6. A supernova is often 10% as bright as all of the other stars in its galaxy combined. The brightest supernova that has been observed emitted as much energy as a hundred billion stars as luminous as our Sun!

Many of the elements with more protons than iron 3
are created during supernovas. Recall from Chapter 4 that neutrons can transform into protons by emitting an electron. As the layers of a massive dying star are blasted into space, they are compressed so much by the neutrinos and the shock wave that neutrons are forced into the elements that have been created as the star evolved. Packed with extra neutrons, many of these nuclei are unstable. As a result, neutrons begin emitting electrons, a process that converts these neutrons to protons. Once this occurs,

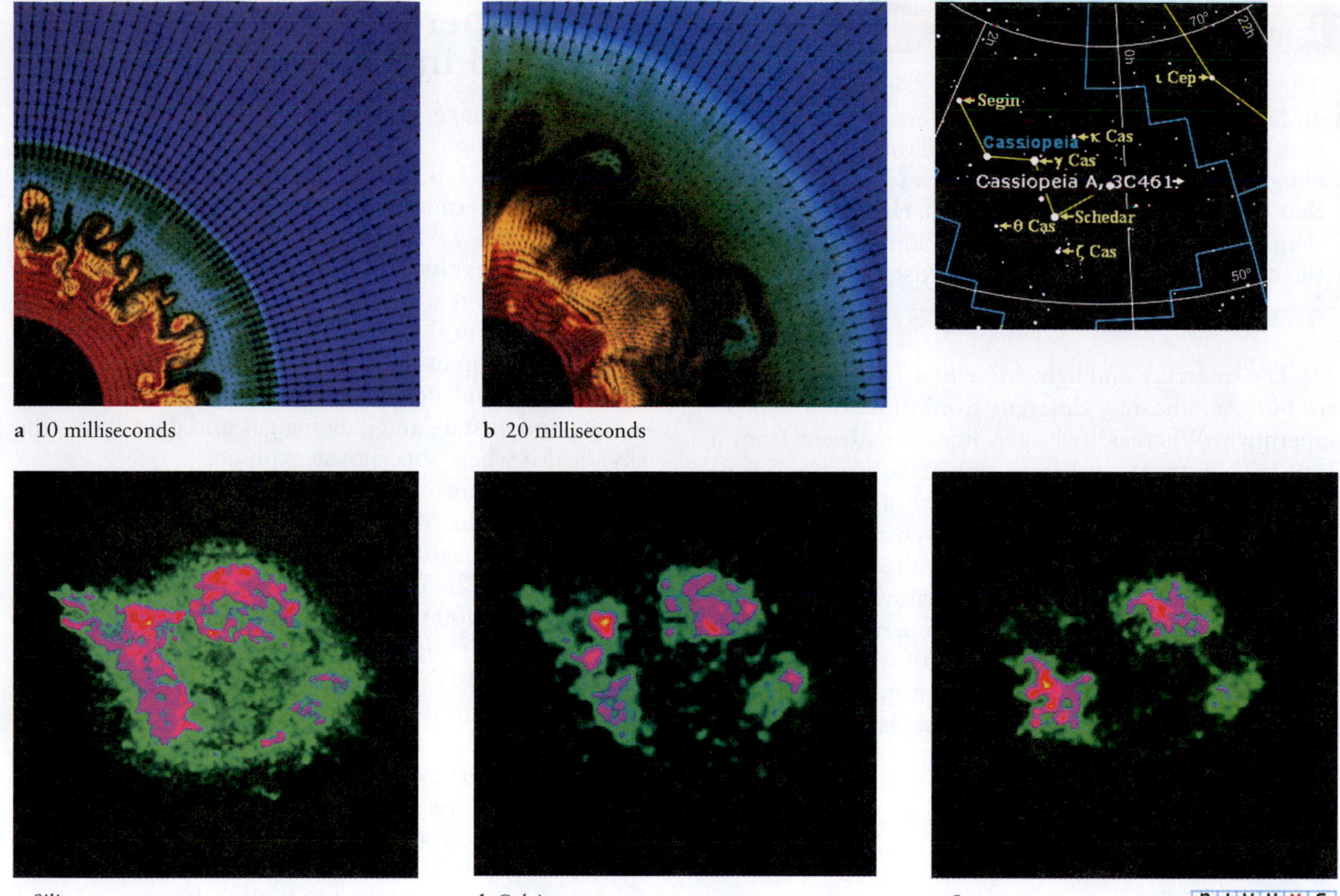

FIGURE 13-12 Supernovae Proceed Irregularly Images (a) and (b) are computer simulations showing the chaotic flow of gas deep inside the star as it begins to explode as a supernova. This uneven flow helps account for the globs of iron and other heavy elements emitted from deep inside, as well as the lopsided distribution of all elements in the supernova remnant, as shown in (c), (d), and (e). These three pictures are X-ray images of supernova remnant Cassiopeia A taken by Chandra at different wavelengths. (a and b: Courtesy of Adam Burrows, University of Arizona, and Bruce Fryxell, NASA/GSFC; c, d, and e: U. Hwang et al., NASA/GSFC)

that nucleus is no longer the element it was a moment before. For example, a neutron-enriched iron atom transforms into cobalt when a neutron transforms to a proton in it. Whereas the neutron-enriched iron was not stable, the cobalt it becomes can be—an atom of cobalt has been created. This process creates many of the heavy elements that exist today. Likewise, the supernova can cause atoms to bond together and create interstellar dust particles (see Sections 5-1 and 5-2). In summary, the products of shell fusion and this final burst of fusion during supernovae are where many of the heavy elements on Earth and other terrestrial planets come from, including some of the atoms in our bodies. We really are made of stardust.

Computer simulations indicate that the onion skins of different elements do not just move out of a star going supernova in lockstep (Figures 13-12a, b). Therefore, the ejected gases are not expected to be uniform shells of matter. Indeed, observations reveal that the material emerging in a supernova comes out quite irregularly. Figure 13c–e show the distributions of some of the elements in the supernova Cassiopeia A (Cas A). Other observations show that some of the dense elements, like iron, ejected in supernovae are actually created deep inside the innermost silicon and oxygen layers of the supernovae and then shot out like bullets or geysers. Other elements are created throughout the expanding stars. However, calculations and observations reveal that supernovae do not create enough of the heaviest elements, like gold, silver, and platinum, to account for the amounts of these elements found on Earth. We will explore their likely origin in Section 13-14.

What role do neutrinos play in Type II supernovae?

Over its lifetime, our model 25-$M_{\odot}$ star ejects more than 20 $M_{\odot}$ of its mass back into space—by stellar winds, by gas ejected during unstable periods, and during its supernova phase. Less massive stars return proportionately less mass to the interstellar medium.

Insight Into Science

The Process of Science The best scientific theories and models also provide explanations about matters that they were not initially designed to study. In modeling stellar evolution and supernovae, the processes that create and eject metals (that is, elements heavier than hydrogen and helium) from stars also provide the elements necessary for life to exist.

The spectrum and light curve of a Type II supernova are both significantly different from those of a Type Ia supernova. Whereas hydrogen lines are absent from a Type Ia's spectrum, they are strong in the spectra of Type II supernovae. Both types begin with a sudden rise in brightness (see Figure 13-9). However, a Type II supernova usually peaks at $M = -17$, a hundred times dimmer than a Type Ia. Whereas Type Ia supernovae decline gradually for more than a year, Type II supernovae alternate between periods of steep and gradual declines in brightness. Type II light curves, therefore, have a steplike appearance.

FIGURE 13-13 The Gum Nebula The Gum Nebula is the largest known supernova remnant. It spans 60° across the sky and is centered roughly on the southern constellation of Vela. The nearest portions of this expanding nebula are only 300 ly from Earth. The supernova explosion occurred about 11,000 years ago, and the remnant now has a diameter of about 2300 ly. Only the central regions of the nebula are shown here. (Royal Observatory, Edinburgh)

13-7 Supernova remnants are observed in many places

Astronomers have seen more than 1000 supernovae. These observations suggest that in a typical galaxy like the Milky Way, Type Ia supernovae occur approximately once every 36 years, whereas Type II supernovae occur about once every 44 years. Thus, there should be about 5 supernovae exploding in our Milky Way Galaxy each century. Vigorous stellar evolution in our Milky Way occurs primarily in the Galaxy's disk, because stars there continue to form in the giant molecular clouds. The disk (which we are in and which we see edge on as the milky-white band of stars and glowing gas and dust across the sky) is also where supernovae explode.

Astronomers find the debris of local supernova explosions in the Milky Way scattered across the Galaxy's disk. A beautiful example is the Cygnus Loop (see Figure 12-7). The star's outer layers were blasted into space 20,000 years ago so violently that they are still traveling at supersonic speeds. As this expanding shell of gas plows through the interstellar medium, it collides with atoms and molecules, making the gases glow.

Many supernova remnants cover sizable fractions of the sky. The largest is the Gum Nebula, with an angular diameter of 60° (its central region is shown in Figure 13-13). This nebula looks so big because it is so close: Its near side is only about 300 ly from Earth. Studies of the Gum Nebula's expansion rate suggest that the supernova exploded about 11,000 years ago. At maximum brilliance, the exploding star was probably as bright as the Moon at first quarter.

The Milky Way's disk is so filled with interstellar gas and dust that we cannot see very far into it. Supernovae are thought to erupt every few decades in remote parts of our Galaxy, but their detonations are hidden from optical telescopes by the interstellar medium. Many supernova remnants in our Galaxy can be detected only at nonvisible wavelengths, ranging from X rays through radio waves. For example, Figure 13-14 shows images of the supernova remnant Cassiopeia A as seen in radio (Figure 13-14a) and X ray (Figure 13-14b). Visible-light photographs of this part of the sky reveal only a few small, faint wisps. Thus, radio searches for supernova remnants are more fruitful than visual searches. Only 24 supernova remnants have been found in visible-light photographs, but more than 3100 remnants in our Galaxy and in others have been discovered by radio astronomers.

From the expansion rate of the nebulosity in Cassiopeia A, astronomers calculate that the light from that supernova explosion first reached Earth about 300 years ago. Although telescopes were in wide use by the late 1600s, no record of the event is known. In fact, the

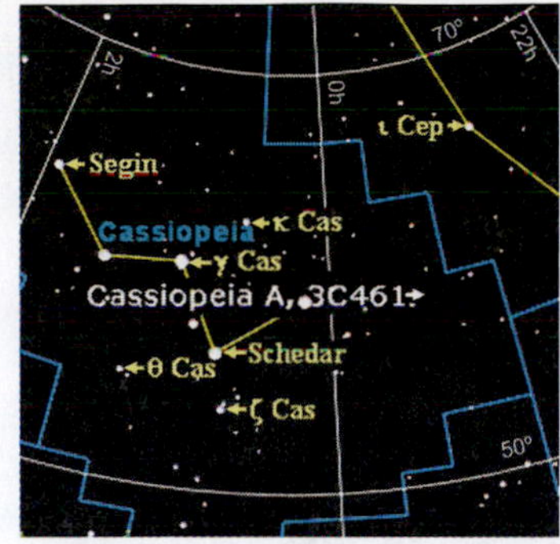

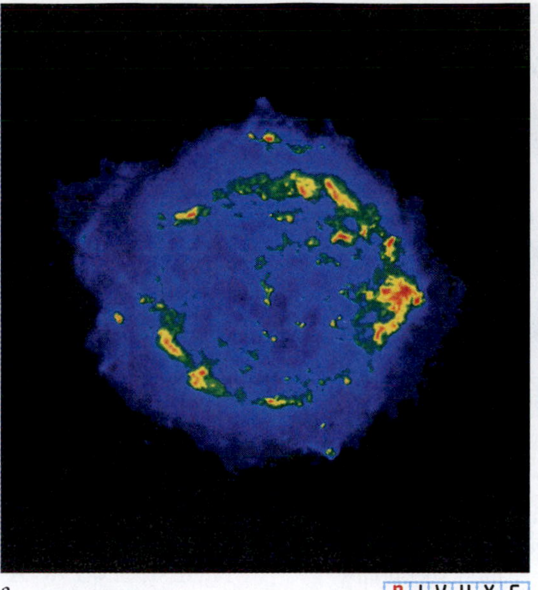
a R I V U X G

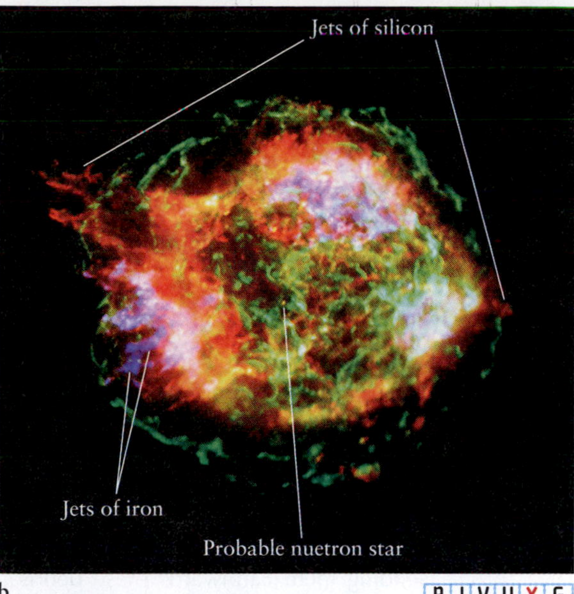

b R I V U X G

ANIMATION 13.4

FIGURE 13-14 Cassiopeia A Supernova remnants, such as Cassiopeia A, are typically strong sources of X rays and radio waves. (a) A radio image produced by the Very Large Array (VLA). (b) A corresponding X-ray picture of Cassiopeia A taken by Chandra. The opposing jets of silicon, probably guided by powerful magnetic fields, were ejected early in the supernova, before the iron-rich jets were released. Radiation from the supernova that produced this nebula first reached Earth 300 years ago. The explosion occurred about 10,000 ly from here. (a: NASA/ CXC/GSFC/U. Hwang et al.; b: Very Large Array)

last supernova of a star near its maximum brightness in our Galaxy, seen by the naked eye, was observed by Johannes Kepler in 1604. In 1572, Tycho Brahe also recorded the sudden appearance of an exceptionally bright star in the sky.

In 2002, astronomers discovered that 2 million years ago a group of bright O and B stars, called the Scorpius-Centaurus OB association, passed within 150 ly of Earth, and that one or more supernovae in this group may have occurred at that time. Such a close explosion could have damaged Earth's ozone layer and caused the extinction of some ocean life at the end of the Pliocene epoch.

Besides the gas and dust enriched with heavy elements, supernovae leave a heritage long after their remnants have dissipated. Much of this gas and dust is recycled into new stars and planets, such as our solar system.

13-8 Cosmic rays are not rays at all

4 In observations connected with supernovae, astronomers have detected particles flying through space at speeds that exceed 90% the speed of light. These high-speed particles are called **cosmic rays** or **primary cosmic rays.** They were named *rays* before their true identity as particles was known, and, as often happens, this misleading name has stuck. Cosmic rays are primarily hydrogen nuclei. A few percent of cosmic rays are electrons and positrons (positively charged electrons). Less than 1% of cosmic rays are nuclei of more-massive, non-hydrogen elements.

> Why might we not be able to see all the supernovae in our Galaxy?

Most primary cosmic rays coming in our direction collide with gas in the atmosphere some 15 km above Earth's surface. This is fortunate, because each cosmic ray packs an enormous amount of energy that potentially could harm living tissue. A collision in the air divides a cosmic ray's energy among several gas particles, which are shoved Earthward. These particles, in turn, often hit other gas atoms, creating a cascade of lower-energy particles that eventually reach Earth's surface as a **cosmic ray shower** (Figure 13-15). These cosmic rays created from atoms in our atmosphere are called **secondary cosmic rays.**

The origins of cosmic rays are just beginning to become clear. At first, many astronomers thought that moderate-energy cosmic rays were emitted directly by supernovae. This theory made sense because these explosions are the best-known source of the energy necessary to give such cosmic rays their tremendous speeds. However, observations by the *Advanced Composition Explorer* satellite reveal that the isotopes of

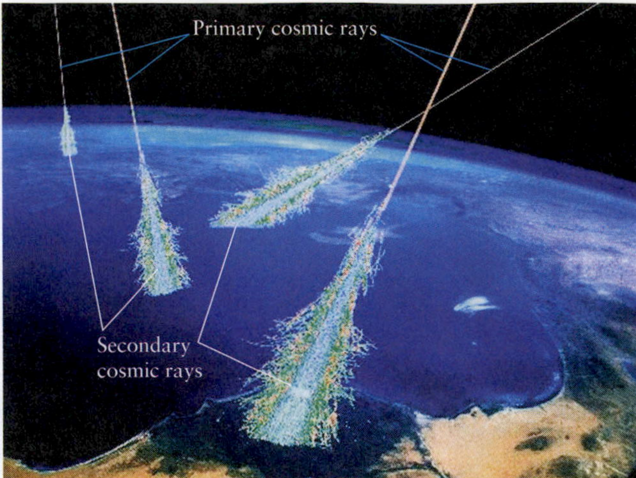

FIGURE 13-15 Cosmic Ray Shower Cosmic rays from space slam into particles in the atmosphere, breaking them up and sending them Earthward. These debris are called secondary cosmic rays. This process of impact and breaking up continues as secondary cosmic rays travel downward, creating a cosmic ray shower, as depicted in this artist's conception of four such events. (Simon Swordy [U. Chicago], NASA)

nickel coming to Earth in the form of cosmic rays are not the isotopes emitted by supernovae. Instead, evidence is mounting that many cosmic rays are created when supernova debris slams into gas already in space, causing some of this preexisting matter to accelerate and become cosmic rays.

Much less common than cosmic rays from supernova remnants are ultrahigh-energy cosmic rays, some of which come from in our Galaxy, while many come from outside of it. Each of these particles has as much kinetic energy as a baseball thrown by a major league pitcher. In 2004, the Pierre Auger Observatory, consisting of 1600 cosmic ray sensors designed to detect the highest energy cosmic rays, was commissioned. Located on the Pampa Amarilla ("Yellow Prairie") in Argentina, it is designed to give astronomers information about the directions from which these cosmic rays come. Observations of their paths suggest that many of them come from galaxies containing supermassive black holes, which we will discuss in Section 14-8. Whether these black holes themselves generate the ultrahigh-energy cosmic rays is still under investigation.

13-9 Supernova 1987A offered a detailed look at a massive star's death

Some supernovae in other galaxies can be seen without a telescope. In 1885, a supernova in the Andromeda Galaxy was just barely visible to the unaided eye. On February 23, 1987, a supernova was observed in the Large Magellanic Cloud, a galaxy near our own Milky Way. The supernova, designated SN 1987A, was the first to be observed that year (Figure 13-16), and it was so bright that it could easily be seen with the naked eye. What made SN 1987A such an exciting discovery was that it gave astronomers a rare opportunity to study the death of a nearby massive star using modern equipment that provided data with which to verify the theory of supernovae described earlier.

At first, SN 1987A reached only a tenth of the luminosity predicted for an exploding massive star. For the next 85 days, it gradually brightened. Then it began dimming, as is characteristic of an ordinary supernova.

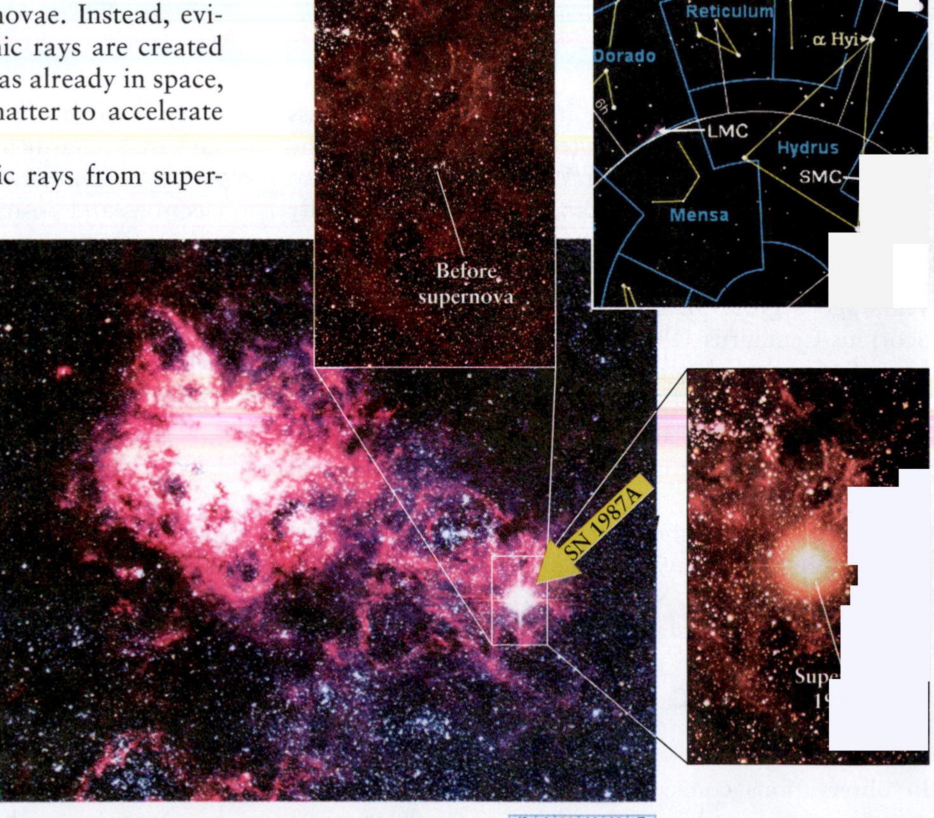

FIGURE 13-16 Supernova 1987A A supernova was discovered in a nearby galaxy called the Large Magellanic Cloud (LMC) in 1987. This photograph shows a portion of the LMC that includes the supernova and a huge H II region, called the Tarantula Nebula. At its maximum brightness, observers at southern latitudes saw the supernova without a telescope. (Insets) The star before and after it exploded. (European Southern Observatory; insets: Anglo-Australian Observatory/ David Malin Images)

Fortunately, the star had been observed before it became a supernova (Figure 13-16 insets). The Large Magellanic Cloud is about 160,000 ly from Earth—near enough to us that many of its stars have been individually observed and catalogued. The star had been identified as a B3 I supergiant. The theory of the evolution of such stars provided an explanation for SN 1987A's unusually slow brightening.

When this star was on the main sequence, its mass was about 20 $M_\odot$, although by the time it exploded, it had shed many solar masses. The evolutionary track for an aging 20-$M_\odot$ star wanders back and forth across the top of the H-R diagram, so the star alternates between being a hot (blue) supergiant and a cool (red) supergiant. The star's size changes significantly as its surface temperature changes. A blue supergiant is only 10 times larger in diameter than the Sun, but a red supergiant of the same luminosity is 1000 times larger. Because the doomed star was a relatively small blue supergiant when it exploded, it reached only a tenth of the brightness that it would have attained had it been a red supergiant at that time.

Ordinary telescopic observations of a supernova explosion can only show us the expanding outer layers of the dying star. No photons created in the core can penetrate through the supernova gases directly to our telescopes. As the outer layers of the supernova thin, however, neutrinos escape directly from the core into space. By detecting these neutrinos and measuring their properties, astronomers can learn many details about the star's collapsing core, especially about core bounce. Several neutrino detectors were operating when the neutrinos from SN 1987A reached Earth.

What are cosmic rays?

Nearly a day before SN 1987A was first observed in the sky, teams of scientists at neutrino detectors in Japan and the United States reported observing Cerenkov flashes (see Section 10-9) from a burst of neutrinos. The Kamiokande II detector in Japan detected 12 neutrinos at about the same time that 8 were found by the IMB (Irvine-Michigan-Brookhaven) detector in a salt mine under Lake Erie. Neutrinos preceded the visible supernova outburst because they escaped from the dying star before the shock wave from the collapsing core reached the star's surface. The particles were detected in Earth's northern hemisphere, where the supernova is always below the horizon, after having passed through Earth. The discovery of these neutrinos provided strong support for the theory that supernovae are caused by the collapse of a star's core, its subsequent bounce, and the flow of neutrinos that help cause the supernova, as described in Section 13-6.

About 3½ years after SN 1987A's detonation was first seen from Earth, astronomers using the Hubble Space Telescope obtained a picture showing several rings of glowing gas around the exploded star (Figure 13-17).

a Supernova 1987A seen in 1996 R I V U X G

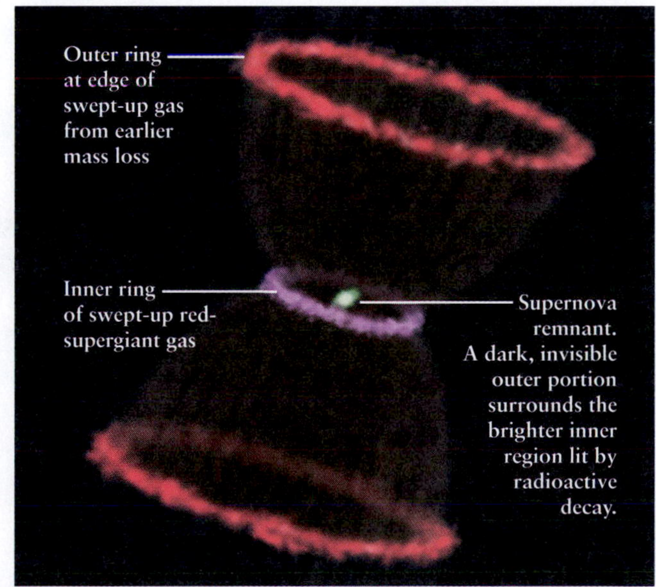

b An explanation of the rings

FIGURE 13-17 Shells of Gas Around SN 1987A (a) Intense radiation from the supernova explosion caused three rings of gas surrounding SN 1987A to glow in this Hubble Space Telescope image. This gas was ejected from the star 20,000 years before the star detonated. All three rings lie in parallel planes. The inner ring is about 1.3 ly across. The white and colored spots are unrelated stars. (b) When the progenitor star of SN 1987A was still a red supergiant, a slowly moving wind from the star filled the surrounding space with a thin gas. When the star contracted into a blue supergiant, it produced a faster-moving stellar wind. The interaction between the fast and slow winds somehow caused gases to pile up along an hourglass-shaped shell surrounding the star. The burst of ultraviolet radiation from the supernova ionized the gas in the rings, causing the rings to glow. The supernova itself, at the center of the hourglass, glows because of energy released from radioactive decay. (Robert P. Kirshner and Peter Challis, Harvard-Smithsonian Center for Astrophysics; STScI)

This gas had been emitted by the star 20,000 years before the supernova, when a hydrogen-rich stellar envelope was ejected by stellar winds from the doomed star, as discussed earlier in this chapter. This gas expanded in an hourglass shape, because it was blocked from expanding around the star's equator, either by a preexisting ring of gas or by the orbit of an as-yet-unseen companion star.

While emitting these gases, the doomed star was a red supergiant. It then shrank and in its final blue giant phase, the same star emitted higher-speed gas that compressed the hourglass gas into three rings—a narrow one around the equator and wider ones at the top and bottom of the hourglass. This relic gas is now being illuminated by photons from the supernova. In 1998, astronomers began to observe gas ejected from the supernova striking the preexisting circumstellar gas. This collision has caused the ring of gas to brighten, and it should eventually illuminate more of the earlier gas that had been ejected.

Why were neutrinos from SN 1987A observed before the light from this event?

SN 1987A provided astronomers with invaluable information, in part because the doomed star had been studied before it exploded and its distance from Earth was known. The supernova was also located in an unobscured part of the sky, and neutrino detectors happened to be operating at the time of the outburst. Astronomers will be monitoring the progress of this supernova for years to come.

NEUTRON STARS AND PULSARS

When stars that initially had between 8 and 25 $M_{\odot}$ explode as supernovae, their expanding cores are stopped by the outer layers into which they slam. The cores recontract and remain intact as highly compressed clumps of neutrons, called **neutron stars.** Isolated neutron stars, resulting from these explosions, are stable stellar remnants with masses between 1.4 $M_{\odot}$ and about 3 $M_{\odot}$.

13-10 The cores of many Type II supernovae become neutron stars

The neutron was discovered during laboratory experiments in 1932. Inspired by the realization that white dwarfs are supported by electron degeneracy pressure, Fritz Zwicky and Walter Baade proposed, within a year, that a similar repulsion between neutrons could support remnant neutron cores. They used the fact that the Pauli exclusion principle also prevents neutrons with identical properties from packing too closely together. When neutrons are pressed together, many of them have to move rapidly (so as not to be identical to their neighbors). This motion provides a pressure, called **neutron degeneracy pressure,** that equations predict is even greater than electron degeneracy pressure (see Section 12-10) and thus allows stellar remnants with masses beyond the Chandrasekhar limit to be stable.

Most scientists ignored Zwicky and Baade's theory for years. After all, a neutron star seemed to be a rather unlikely object. To transform protons and electrons into neutrons, the density in the star would have to be equal to nuclear density, about 10^{17} kg/m^3. Thus, a teaspoon full of neutron star matter brought back to Earth would weigh about 1 billion tons. Put another way, neutron star matter is 200 million times denser than matter found in a white dwarf. Furthermore, an object compacted to nuclear density would be very small. A 2-$M_{\odot}$ neutron star would have a diameter of only 20 km (12 mi) and would fit inside any major city on Earth. The surface gravity on one of these neutron stars would be so strong that the escape velocity would equal one-half the speed of light. All of these conditions appeared to be so outrageous that few astronomers paid any serious attention to the subject of neutron stars—until 1968.

In that year, Jocelyn Bell, then a graduate student in radio astronomy at Cambridge University, noticed that the radio telescope she was using was detecting regular pulses from one particular location in the sky. Careful repetition of the observations demonstrated that the radio pulses were arriving with a regular period of 1.3373011 s. The regularity of this pulsating radio source was so striking that the Cambridge team suspected that they might be detecting signals from an advanced alien civilization, which accounts for the name first assigned to it: LGM1 (short for Little Green Men 1). This possibility was soon discarded as several more of these pulsating radio sources, which soon came to be known as **pulsars,** were discovered across the sky. The broad distribution of pulsars would require incredibly widespread civilizations, inconsistent with what we have found in our searches for alien intelligence (see Chapter 19). In all cases, the periods were extremely regular, ranging between 0.2 and 1.5 s (Figure 13-18).

When the discovery of pulsars was officially announced in early 1968, astronomers around the world

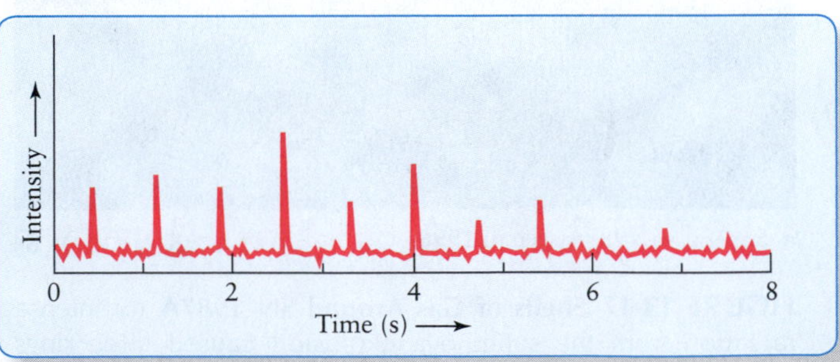

FIGURE 13-18 A Recording of a Pulsar This chart recording shows the intensity of radio emissions from one of the first pulsars to be discovered, PSR 0329+54. Note that some of the pulses are weak and others are strong. Nevertheless, the spacing between pulses is so regular (0.714 s) that it is more precise than most clocks on Earth.

began proposing all sorts of explanations. Many of these theories were bizarre, and arguments raged for months. We have already ruled out alien civilizations. Astronomers therefore looked for some recurring event in a star's life. We know that a star sheds matter as it swells to a giant. Perhaps some stars alternately expand and contract, emitting energy as they pulsate. However, this explanation also fails because any star expanding and contracting as fast as a pulsar emits pulses would explode.

How big is a neutron star?

Late in 1968, all controversy was laid to rest with the discovery of a pulsar in the middle of the Crab Nebula. In A.D. 1054, Native American, Chinese, and possibly other peoples saw and recorded the appearance of a "guest star" in the constellation of Taurus. When we turn a telescope toward this location, we find the Crab Nebula, shown in Figure 13-19. It looks like the residue of an explosion and is, in fact, a supernova remnant. At the

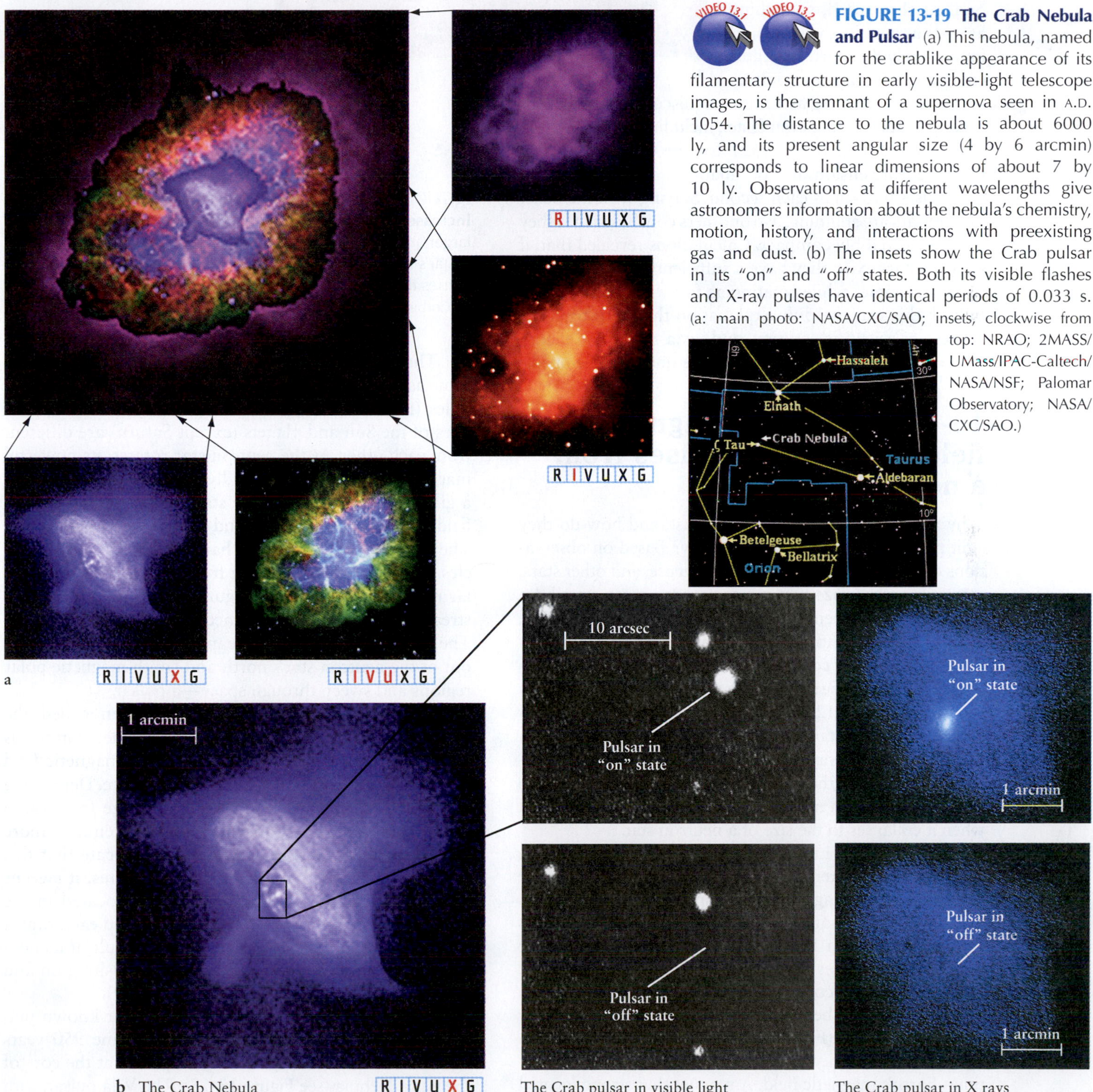

VIDEO 13.1 VIDEO 13.2

FIGURE 13-19 The Crab Nebula and Pulsar (a) This nebula, named for the crablike appearance of its filamentary structure in early visible-light telescope images, is the remnant of a supernova seen in A.D. 1054. The distance to the nebula is about 6000 ly, and its present angular size (4 by 6 arcmin) corresponds to linear dimensions of about 7 by 10 ly. Observations at different wavelengths give astronomers information about the nebula's chemistry, motion, history, and interactions with preexisting gas and dust. (b) The insets show the Crab pulsar in its "on" and "off" states. Both its visible flashes and X-ray pulses have identical periods of 0.033 s. (a: main photo: NASA/CXC/SAO; insets, clockwise from top: NRAO; 2MASS/UMass/IPAC-Caltech/NASA/NSF; Palomar Observatory; NASA/CXC/SAO.)

center of the Crab Nebula is the Crab pulsar (Figure 13-19b).

With expansion and contraction ruled out, the likely scenario is that pulsars are spinning objects. If the Crab pulsar is spinning, it is spinning too fast to be a white dwarf. Its period is 0.033 s, which means that it rotates 30 times each second. At that speed, something as wide as a white dwarf would immediately fly apart. Strengthening the case against pulsars being white dwarfs, astronomers discovered numerous other pulsars rotating even faster, including PSR 1937+21 (PSR for pulsar and 1937+21 for the object's approximate right ascension and declination), a stellar remnant that pulses 640 times each second. Staying with the belief that pulsars are spinning, astronomers concluded that they must be incredibly compact. Calculations revealed that, if neutron stars exist, they have a sufficiently small diameter to remain intact while rotating as fast as pulsars pulsate. Indeed, this was the *only* explanation that withstood the process of scientific scrutiny, and, as a result, the theory of neutron stars described earlier was quickly developed.

FIGURE 13-20 Analogy for How Magnetic Field Strengths Increase When growing, these wheat stalks cover a much larger area than when they are harvested and bound together. A star's magnetic field behaves similarly. The collapsing star carries the field inward, thereby increasing its strength. (a: Corbis; b: Oscar Burriel/Photo Researchers, Inc.)

13-11 A rotating magnetic field explains the pulses from a neutron star

Why are neutron stars rotating so fast, and how do they emit pulses of electromagnetic energy? Based on observations of the rotation of the Sun, Betelgeuse, and other stars, astronomers theorize that many, perhaps most, stars have some angular momentum (rotation). The Sun, for example, takes nearly a full month to rotate once around its axis. But the rotation rate of collapsing stars increases, just as pirouetting ice skaters speed up when they pull in their arms, as shown in Figure 2-12. Recall from Section 2-7 that this occurs because the total amount of angular momentum in an isolated system always remains constant. To conserve angular momentum, the core of a high-mass main-sequence star rotating once a month spins faster than once a second when it collapses to the size of a neutron star.

In addition to rapid rotation, most neutron stars are believed to have intense magnetic fields. In an average star like our Sun, the magnetic field is spread out over millions upon millions of square kilometers just under the star's surface (see Section 10-5). However, when a star of solar dimensions collapses down to a neutron star, its magnetic field becomes very concentrated, in the same way as stalks of wheat in a field become concentrated when you gather them in a bundle (Figure 13-20). The strength of a collapsing star's magnetic field increases to as much as 10^{15} times Earth's magnetic field. A magnet that strong located halfway to the Moon would lift iron rods off Earth.

The axis of rotation of a typical neutron star is not the 5
same as the axis connecting its north and south magnetic poles (Figure 13-21), just as the magnetic and rotation axes of the Sun and planets (except Saturn) are different from each other. As the neutron star rotates, its powerful magnetic field therefore rapidly changes direction. Like a giant electric generator, the star creates intense electric fields, which act on protons and electrons near its surface. The powerful electric fields channel these charged particles, causing them to flow out from the neutron star's polar regions, as sketched in Figure 13-21. As the particles stream along the field, they accelerate and emit energy. The result is two very thin beams of radiation that pour out of the neutron star's north and south magnetic polar regions and sweep through space—a pulsar.

This explanation for pulsars is often called the **lighthouse model.** A rotating, magnetized neutron star is somewhat like a lighthouse beacon. The magnetic field causes nearby gases to accelerate and radiate. Depending on the conditions, this radiation can range from radio waves to gamma rays, and many pulsars emit in more than one spectral range. Calculations indicate that this radiation is beamed, and, as the star rotates, it sweeps around the sky. If Earth happens to be located in the right direction, a brief flash can be observed each time a beam whips past our line of sight. As a result of its neutron star's rotation, the Crab Nebula is flashing on and off 30 times each second (see Figure 13-19).

The Crab pulsar is one of the youngest known pulsars, its creation having been observed some 950 years ago. Also visibly flashing is the Vela pulsar at the core of the Gum Nebula (see Figure 13-13). The Vela pulsar, with a period of 0.089 s, is the slowest pulsar ever detected at

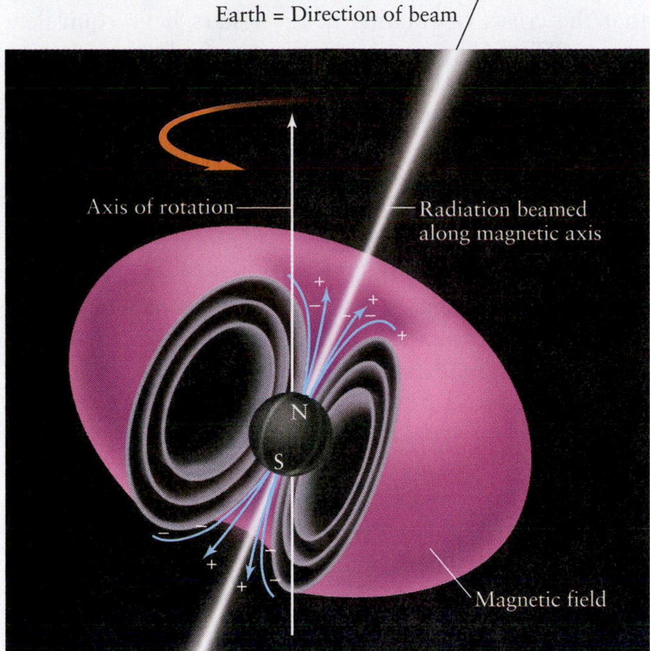

a One of the beams from the rotating neutron star is aimed toward Earth: We detect a pulse of radiation.

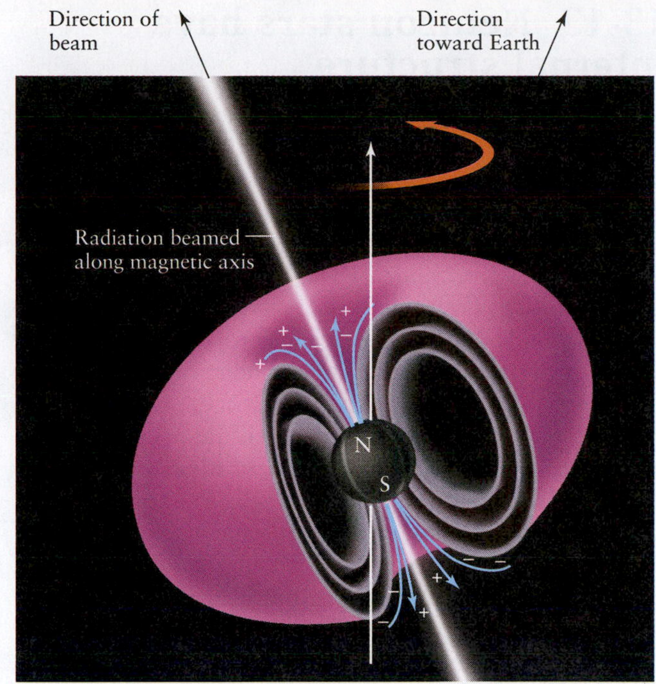

b Half a rotation later, neither beam is aimed toward Earth: We detect that the radiation is "off."

FIGURE 13-21 A Rotating, Magnetized Neutron Star Calculations reveal that many neutron stars rotate rapidly and possess powerful magnetic fields. Charged particles are accelerated near a neutron star's magnetic poles and produce two oppositely directed beams of radiation. As the star rotates (going from *a* to *b* is half a rotation period), the beams sweep around the sky. If Earth happens to lie in the path of a beam *(a,* but not *b),* we see the neutron star as a pulsar.

visual wavelengths. Because they use some of their energy of rotation to create the pulses they emit, *isolated pulsars slow down as they get older*. Only the very youngest pulsars are energetic enough to emit visible flashes along with their radio pulses, and the Vela pulsar was created just 11,000 years ago, which is recent in astronomical terms.

13-12 Rotating neutron stars create other phenomena besides normal pulsars

Pulsars typically have magnetic fields about 10^{12} times stronger than the magnetic field at the surface of Earth. Exceptional circumstances can conspire to raise the field strength a thousand times higher in some pulsars. If the neutron star is very hot (around 10^{11} K) and spinning very fast (100 times a second) when it forms, convection occurs throughout the neutron star, just as it does for red dwarfs (see Section 12-8). This inward and outward motion, combined with the spin, generates extra magnetic fields via the dynamo effect (see Section 6-4). Combining the neutron star's two magnetic fields yields a field 10^{15} times greater than the field at Earth's surface. The resulting object is called a **magnetar,** and its effects are dramatic.

The magnetic fields of a magnetar are so strong that they cause the neutron star's surface to buckle and emit stupendous bursts of X rays and gamma rays, along with photons at other wavelengths, that last for fractions of a second. Such magnetars are called *soft gamma ray repeaters*, or SGRs. ("Soft" means "low energy." Even so, soft gamma rays are very powerful photons.) Interacting with surrounding gases, the magnetic fields of a magnetar rapidly slow the neutron star's rotation rate. After 10^4 years, the body spins only once every 10 s, and the SGR activity subsides. Twenty-one magnetars have been observed since 1998, including one in 2004 that released the brightest flash ever seen from outside our solar system. Its photons scattered off the Moon and lit up Earth's atmosphere.

> If the lighthouse model of pulsars is correct, do we see all nearby pulsars? Why or why not?

In 2006, radio astronomers discovered yet another type of emission from rotating neutron stars. Called *rotating radio transients*, or RRATs, these objects give off bursts of radio and other emissions that last a few thousandths of a second, turn off for minutes or hours, and then spring briefly back to life. The neutron stars involved rotate slowly compared to most pulsars, spinning once every 0.4 to about 10 s. The most promising explanation, at present, is that these are older pulsars or magnetars that have slowed down.

> How do the rotation rates of magnetars change with time?

13-13 Neutron stars have internal structure

Neutron stars are not composed solely of neutrons. Astronomers have developed a theoretical model for the surface and internal structure of neutron stars (Figure 13-22). A neutron star's interior has a radius of about 10 km, with a core of superconducting protons and superfluid neutrons. A *superconductor* is a material in which electricity and heat flow without the system losing energy, whereas a *superfluid* has the strange property that it flows without any friction. Both superconductors and superfluids have been created in the laboratory. Surrounding a neutron star's core is a layer of superfluid neutrons. The surface of the neutron star is a solid, brittle crust of dense nuclei and electrons about ⅓-km thick. The gravitational force of the neutron star is so great at its surface that climbing a bump there just 1 mm high would take more energy than it takes to climb Mount Everest.

Neutron stars may also have atmospheres, as indicated by absorption lines in the spectrum of at least one of them. The absorption lines are believed to be due to scattering of radiation from the neutron star by gases in its atmosphere, just as scattering in the Sun's atmosphere causes absorption lines in its spectrum. In 2005, astronomers found indications that some isolated neutron stars expel energy long after they form. In particular, the Cassiopeia A remnant has regions glowing from radiation deposited on them as recently as 50 years ago (some 275 years after the supernova itself). The cause of this activity is still being determined.

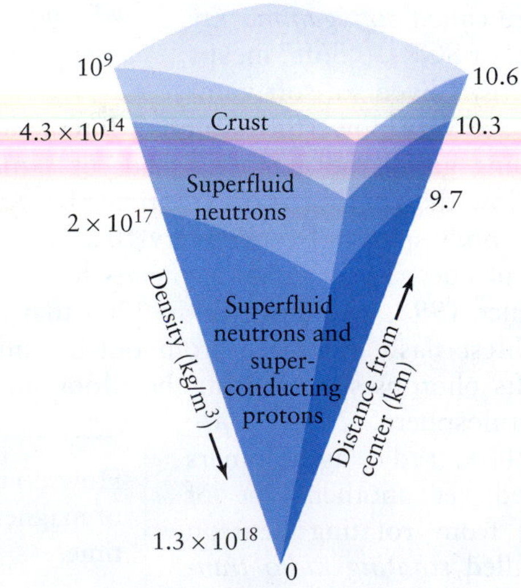

FIGURE 13-22 Model of a Neutron Star's Interior This drawing shows the theoretical model of a 1.4-$M_\odot$ neutron star. The neutron star has a superconducting, superfluid core 9.7 km in radius, surrounded by a 0.6-km-thick mantle of superfluid neutrons. The neutron star's crust is only 0.3 km thick (the length of four football fields) and is composed of heavy nuclei and free electrons. The thicknesses of the layers are not shown to scale.

Just as Earth's crust has earthquakes, calculations show that the crusts of rotating neutron stars have equivalent events. When a rotating neutron star (pulsar) undergoes such a quake, its rotation rate suddenly changes to conserve its angular momentum. This change creates a **glitch** in the rate at which the pulsar is slowing down. Such events have been observed in a number of pulsars (Figure 13-23).

While most pulsars appear to be isolated bodies, more than 90 have been discovered with companions (not necessarily other neutron stars). These pulsars are called *binary pulsars* and are of interest for several reasons. First, many binary pulsars pull mass onto themselves from their companions. This infall causes the pulsar to speed up (like the skater in Figure 2-12 pulling in her arms). As a result of this effect, some pulsars spin nearly a thousand times per second and are called *millisecond pulsars*. Second, some pulsars, like the one labeled PSR 1916+13, have compact massive companions, which are often also neutron stars. Observations of binary pulsars with neutron star companions reveal that the two bodies are spiraling toward each other, as predicted by Einstein's gravitational theory, called *general relativity* (see Sections 2-1 and 14-4 for details).

In 2004, astronomers observed, for the first time, a pair of pulsars in orbit around each other (Figure 13-24). This discovery provides an exciting opportunity to further test the predictions of general relativity. For example, this theory predicts that these two pulsars should be precessing (the directions that their rotation axes point should be changing; see Section 1-8) and that, like the binary pulsars just discussed, they should be spiraling in toward each other. Observations of this system in the next few years will test these predictions.

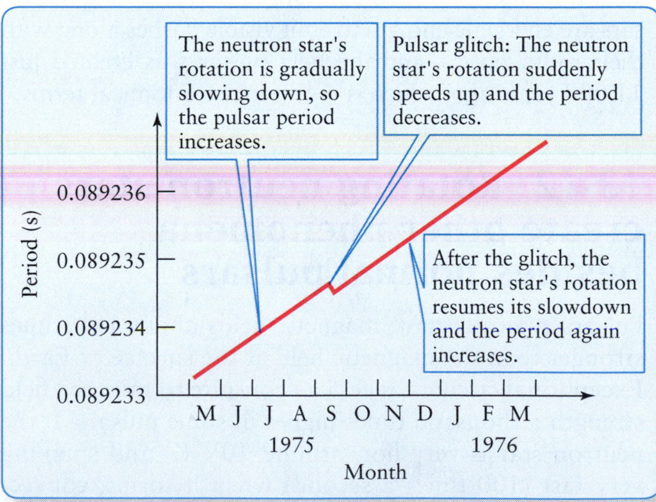

FIGURE 13-23 A Glitch Interrupts the Vela Pulsar's Spindown Rate An isolated pulsar radiates energy, which causes it to slow down. This "spin down" is not always smooth. As it slows down, it becomes more spherical, and so its spinning, solid surface must readjust its shape. Because the surface is brittle, this readjustment is often sudden, like the cracking of glass, which causes the angular momentum of the pulsar to suddenly jump. Such an event, called a glitch and shown here for the Vela pulsar in 1975, changes the pulsar's rotation period.

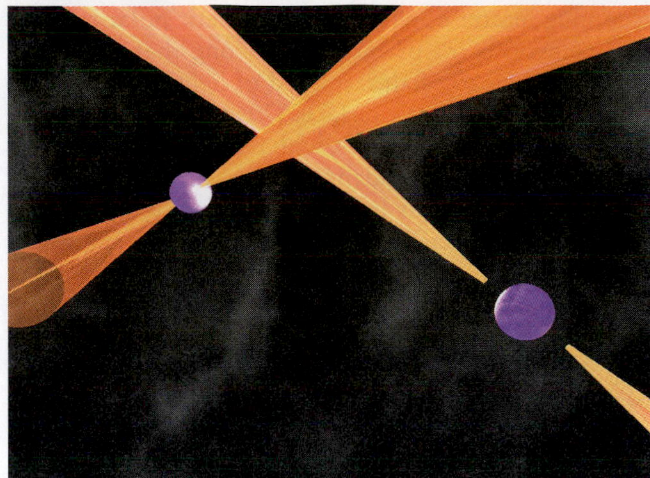

FIGURE 13-24 Double Pulsar This artist's conception shows two pulsars that orbit their center of mass. The double pulsar they represent is called PSR J0737-3039, which is about 1500 ly from Earth in the constellation Puppis. One pulsar has a 23-ms period and the other has a 2.8-s period. The two orbit once every 2.4 h. (Michael Kramer/Jodrell Bank Observatory, University of Manchester)

13-14 Colliding neutron stars may provide some of the heavy elements in the universe

In Sections 13-5 and 13-6 we discussed nucleosynthesis, the process of creating the heavier elements. We saw how a variety of elements up to iron are created inside massive stars before the supernova occurs and are then ejected into space during the explosion. We also saw that even heavier elements are created and ejected during the supernovae themselves. However, models indicate that supernovae do not create enough of the elements heavier than iron to account for the amounts of these elements found in the universe.

Computer simulations of what happens during collisions of two neutron stars predict that, in these impacts, some of the neutrons are splashed into space, where many of them decay back into protons, fuse together, and help form various elements. The interactions that follow create a variety of heavy elements in amounts consistent with observations. For example, formation of an isotope of the element thulium requires neutron densities only available in neutron stars.

13-15 Binary neutron stars create pulsating X-ray sources

Neutron stars in binary systems may also hold the key to another regular pulse from the sky—pulses from X-ray sources. During the 1960s, astronomers obtained tantalizing X-ray views of the sky during short rocket and balloon flights that briefly lifted X-ray detectors above Earth's atmosphere. Several strong X-ray sources were discovered, and each was named after the constellation in which it was located.

Astronomers were so intrigued by these preliminary discoveries that NASA built and launched *Explorer 42*, an X-ray-detecting satellite that could make observations 24 hours a day. The satellite was launched from Kenya (to place it in an orbit above Earth's equator) in 1970, on the seventh anniversary of Kenyan independence. To commemorate the occasion, *Explorer 42* was renamed *Uhuru*, which means "freedom" in Swahili.

Why do glitches change the rotation rates of pulsars?

Uhuru gave us our first comprehensive look at the X-ray sky. As the satellite slowly rotated, its X-ray detectors swept across the heavens. Each time an X-ray source came into view, signals were transmitted to receiving stations on the ground. Before its battery and transmitter failed in early 1973, *Uhuru* had succeeded in locating 339 X-ray sources.

The discovery of pulsars was still fresh in everyone's mind when the *Uhuru* team discovered X-ray pulses coming from Centaurus X-3 in early 1971. Figure 13-25 shows data from one sweep of *Uhuru*'s detectors across Centaurus X-3. The pulses have a regular period of 4.84 s. A few months later, similar pulses were discovered coming from a source called Hercules X-1, which has a period of 1.24 s. Because the periods of these two X-ray sources are so short, astronomers suspected that they had found rapidly rotating neutron stars.

Why does most of the iron on Earth not come from the iron cores of high-mass stars that explode as Type II supernovae?

It soon became clear, however, that systems such as Centaurus X-3 and Hercules X-1 are not ordinary

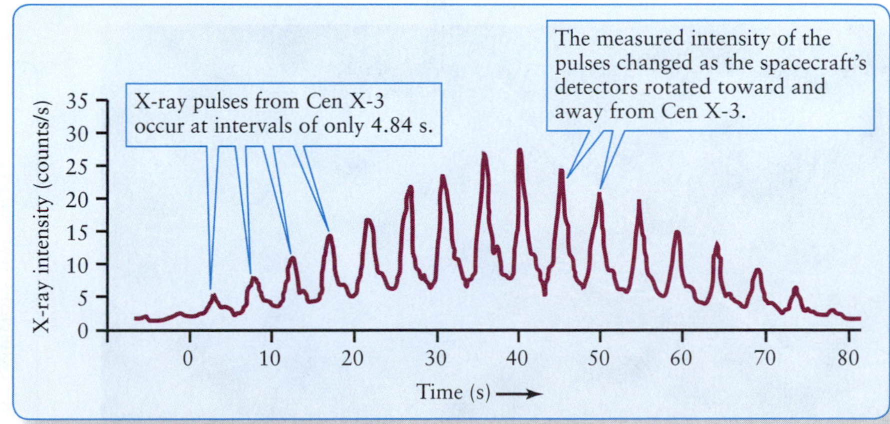

FIGURE 13-25 X-Ray Pulses from Centaurus X-3 This graph shows the intensity of X rays detected by *Uhuru* as Centaurus X-3 moved across the satellite's field of view. The variation in the height of the pulses was a result of the changing orientation of *Uhuru*'s X-ray detectors toward the source as the satellite rotated. The short pulse period suggests that the source is a rotating neutron star.

pulsars like the Crab or Vela pulsars. Every 2087 days, Centaurus X-3 completely turns off for almost 12 h. This fact suggests that Centaurus X-3 is an eclipsing binary (see Figure 11-13), and that the X-ray source takes nearly 12 h to pass behind its companion star.

The case for the binary nature of Hercules X-1 is even more compelling. It has an off state corresponding to a 6-h eclipse every 1.7 days, and careful timing of the X-ray pulses shows a periodic Doppler shift every 1.7 days. This information provides direct evidence of orbital motion around a companion star. When the X-ray source is approaching us, its pulses are separated by slightly less than 1.24 s. When the source is receding from us, slightly more than 1.24 s elapse between the pulses.

Astronomers now realize that systems such as Centaurus X-3 and Hercules X-1 are examples of binary systems in which one object is a neutron star. Because all of these binaries have very short orbital periods, the distance between the ordinary star and the neutron star must be small. This proximity enables the neutron star to capture gas escaping from the companion star.

To explain the pulsations of X-ray sources such as Centaurus X-3 or Hercules X-1, astronomers propose that the ordinary star either fills or nearly fills its Roche lobe. Either way, matter escapes from the star. If the star fills its lobe, as in the case of Hercules X-1, mass loss results from direct overflow through the Roche lobe. If the star's surface lies just inside its lobe, as with Centaurus X-3 (Figure 13-26), a stellar wind carries off the mass. A typical rate of mass loss for the ordinary star in such a pair is approximately 10^{-9} $M_\odot$ per year.

The neutron star in a pulsating X-ray source, like an ordinary pulsar, rotates rapidly and has a powerful magnetic field inclined to the axis of rotation (recall Figure 13-21). As the gas from the companion star falls toward the neutron star, the magnetic field funnels the incoming matter down onto its magnetic polar regions. The neutron star's gravity is so strong that the gas is traveling at nearly half the speed of light by the time it crashes onto the star's surface. This violent impact creates hot spots at both poles, with temperatures of about 10^8 K, that emit abundant X rays with a luminosity nearly 100,000 times brighter than the Sun. Observations of three such X-ray sources in 2005 revealed the impact sites on the neutron stars to range in size from 100 m to 4 km across. As the neutron star rotates, two beams of X rays from its polar caps sweep around the sky. If Earth happens to be in the path of one of the beams, we can observe this pulsating X-ray source. The pulse period is thus equal to the neutron star's rotation period. For example, the neutron star in Hercules X-1 completes one rotation every 1.24 s.

13-16 Neutron stars in binary systems can also emit powerful isolated bursts of X rays

Neutron stars can acquire additional mass from a companion star, which leads to the flaring up of the neutron stars. Beginning in late 1975, astronomers analyzing data from X-ray satellites detected sudden, powerful bursts of radiation. The record of a typical burst is shown in Figure 13-27. The source, called an **X-ray burster,** emits X rays at a constant low level; then, suddenly and without warning, an abrupt increase occurs, followed by a gradual decline. A typical burst lasts for 20 s. Several dozen X-ray bursters have been located, most lying in the plane of the Milky Way Galaxy.

X-ray bursters, like novae, are believed to arise from mass transfer in binary star systems. With a burster, however, the stellar remnant is a neutron star rather than a white dwarf. Gases escaping from the ordinary

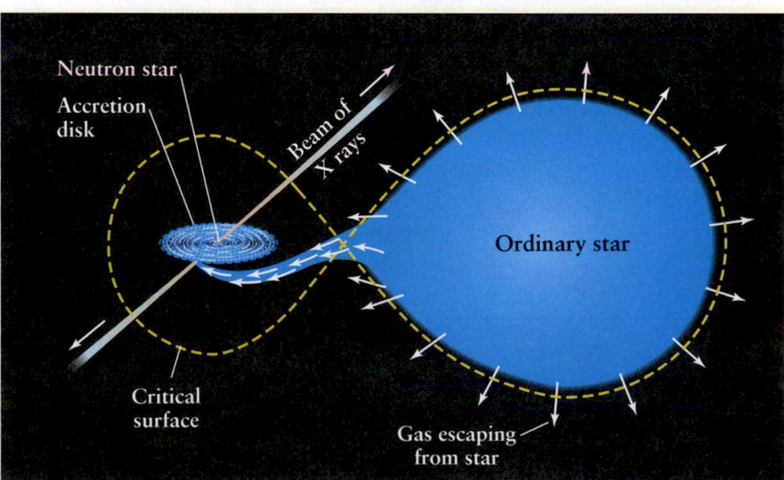

FIGURE 13-26 A Model of a Pulsating X-Ray Source Gas transfers from an ordinary star to the neutron star. The infalling gas is funneled down onto the neutron star's magnetic poles, where it strikes the star with enough energy to create two X-ray-emitting hot spots. As the neutron star spins, beams of X rays from the hot spots sweep around the sky.

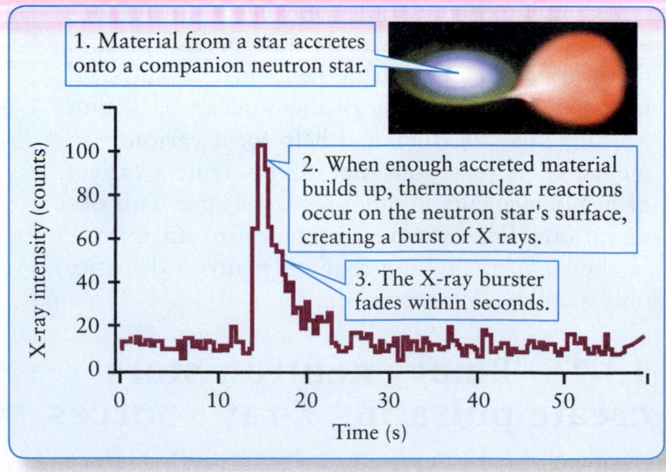

FIGURE 13-27 X Rays from an X-Ray Burster A burster emits X rays with a constant low intensity interspersed with occasional powerful bursts. This burst was recorded in September 1975 by an X-ray telescope that was pointed toward the globular cluster NGC 6624.

companion star fall onto the neutron star. The energy released as this gas crashes down onto the neutron star's surface produces the low-level X rays that are continuously emitted by the burster.

Most of the gas falling onto the neutron star is hydrogen, which becomes compressed against the hot surface of the star by the star's powerful surface gravity. In fact, temperatures and pressures in this accreting layer are so high that the arriving hydrogen is promptly converted by fusion into helium. Constant hydrogen fusion soon produces a layer of helium that covers the entire neutron star. When the helium layer is about 1-m thick, helium fusion ignites explosively, and we observe a sudden burst of X rays. In other words, whereas explosive hydrogen fusion on a white dwarf produces a nova, explosive helium fusion on a neutron star produces an X-ray burster. •

13-17 Smaller, more exotic stellar remnants composed of quarks may exist

In death as well as in life, the mass of a star determines its fate. Just as there is an upper limit to the mass of a white dwarf (the Chandrasekhar limit), there is also an upper limit to the mass of a neutron star, called the *Oppenheimer-Volkov limit.* Above this limit, neutron degeneracy pressure cannot support the overbearing weight of the star's matter pressing in from all sides. The Chandrasekhar limit for a white dwarf is 1.4 $M_{\odot}$, and the Oppenheimer-Volkov limit for a neutron star is about 3 $M_{\odot}$.

According to calculations first done in 1964 and experiments verifying the theory done in 1968, many elementary particles, such as neutrons and protons (but not electrons), are actually composed of smaller particles, called **quarks.** Normally, quarks cannot exist as separate objects—in the cases of neutrons and protons, they clump together in groups of three. However, when the pressure on them is great enough, neutrons can become so compressed that they dissolve into their constituent quarks. As free particles, quarks obey the Pauli exclusion principle, meaning that identical quarks cannot occupy the same place at the same time. So as with electrons and neutrons, quarks create a pressure that resists collapse, thereby creating quark stars. Neutron stars just above the Oppenheimer-Volkov limit may become quark stars. The upper limit for the mass of quark stars is unknown, but it is probably not much greater than the Oppenheimer-Volkov limit.

If the remnant core from a supernova is above the mass limit of neutron stars (and quark stars, if their existence is verified), then there is another fate in store for it. As noted earlier, the gravitational attraction near a neutron star is so strong that the particles escaping from its surface must be moving at roughly half the speed of light. If the stellar corpse has slightly greater than 3 $M_{\odot}$, so much matter becomes crushed into such a small volume that the repulsion created by the Pauli exclusion principle on its neutrons is overcome. The remnant therefore collapses further and the escape velocity exceeds the speed of light. Because nothing can travel faster than light, nothing—not even light—can leave such a collapsed object. The discovery of neutron stars and the calculations regarding the limit to the Pauli exclusion principle inspired astrophysicists to consider seriously one of the most bizarre and fantastic objects ever predicted by modern science—the black hole, which we examine in Chapter 14.

Is an X-ray burster similar to a supernova in the evolution of its source?

Figure 13-28 summarizes the evolutionary history of stars, as discussed in Chapters 12 and 13.

13-18 Frontiers yet to be discovered

The incredible telescopes now in existence or scheduled to go soon are revolutionizing our understanding of the final stages of star formation. In coming years, astronomers hope to pin down the details of how planetary nebulae, novae, supernovae, and stellar remnants behave. Among specific questions to be answered—What are the physical processes that cause planetary nebulae? Why do planetary nebulae have such a variety of shapes? What causes the hourglass shape of so many stellar remnants? Why are novae composed of many blobs of gas? Do quark stars really exist, and, if so, what are their properties?

What is the Oppenheimer-Volkov limit?

SUMMARY OF KEY IDEAS

- Stars with higher masses fuse more elements than do stars with lower masses.
- Stars lose mass via stellar winds throughout their lives.

Low-Mass Stars and Planetary Nebulae

- A low-mass (below 8 $M_{\odot}$) main-sequence star becomes a giant when hydrogen shell fusion begins. It becomes a horizontal-branch star when core helium fusion begins. It enters the asymptotic giant branch and becomes a supergiant when helium shell fusion starts.
- Stellar winds during the thermal pulse phase eject mass from the star's outer layers.
- The burned-out core of a low-mass star becomes a dense carbon-oxygen body, called a white dwarf, with about the same diameter as that of Earth. The maximum mass of a white dwarf (the Chandrasekhar limit) is 1.4 $M_{\odot}$.
- Explosive hydrogen fusion may occur in the surface layer of a white dwarf in some close binary systems, producing sudden increases in luminosity that we call novae.
- An accreting white dwarf in a close binary system can also become a Type Ia supernova when carbon fusion ignites explosively throughout such a degenerate star.

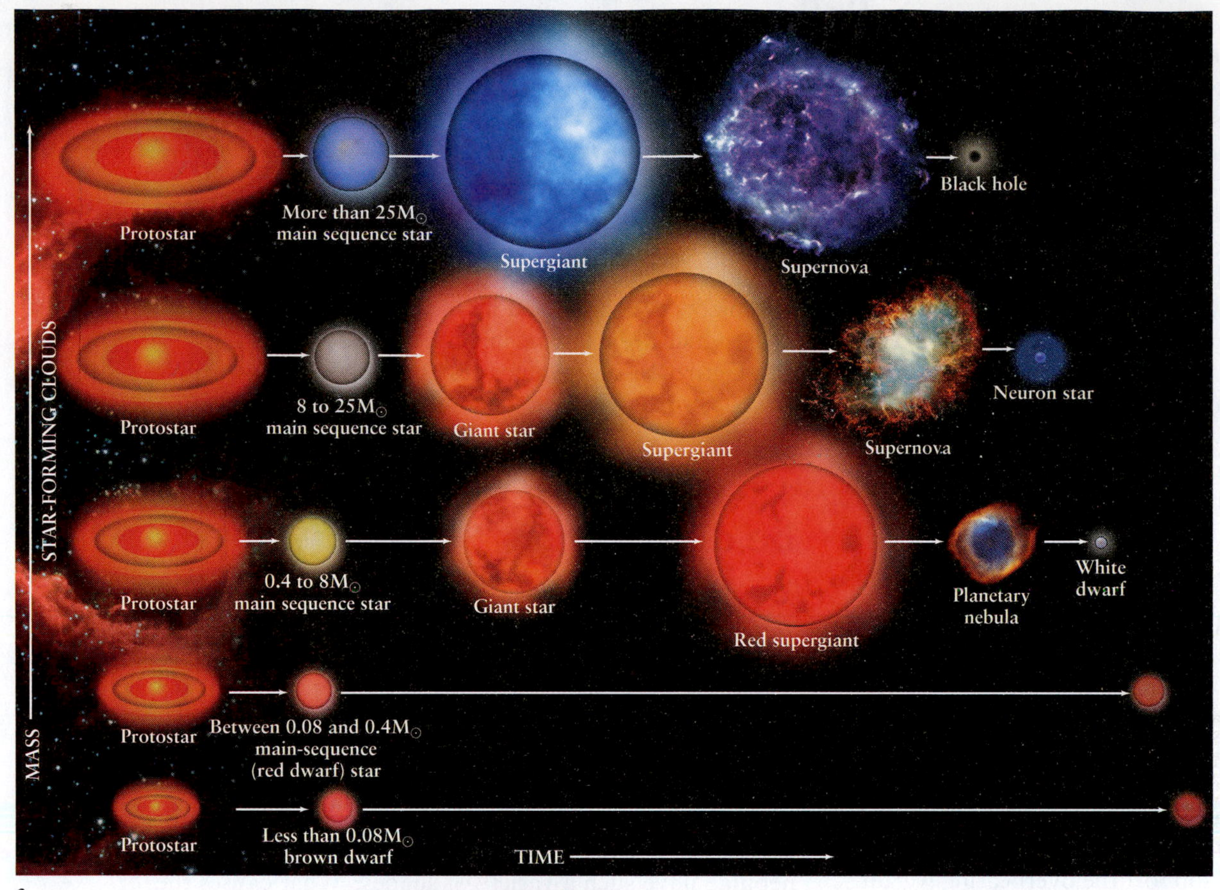

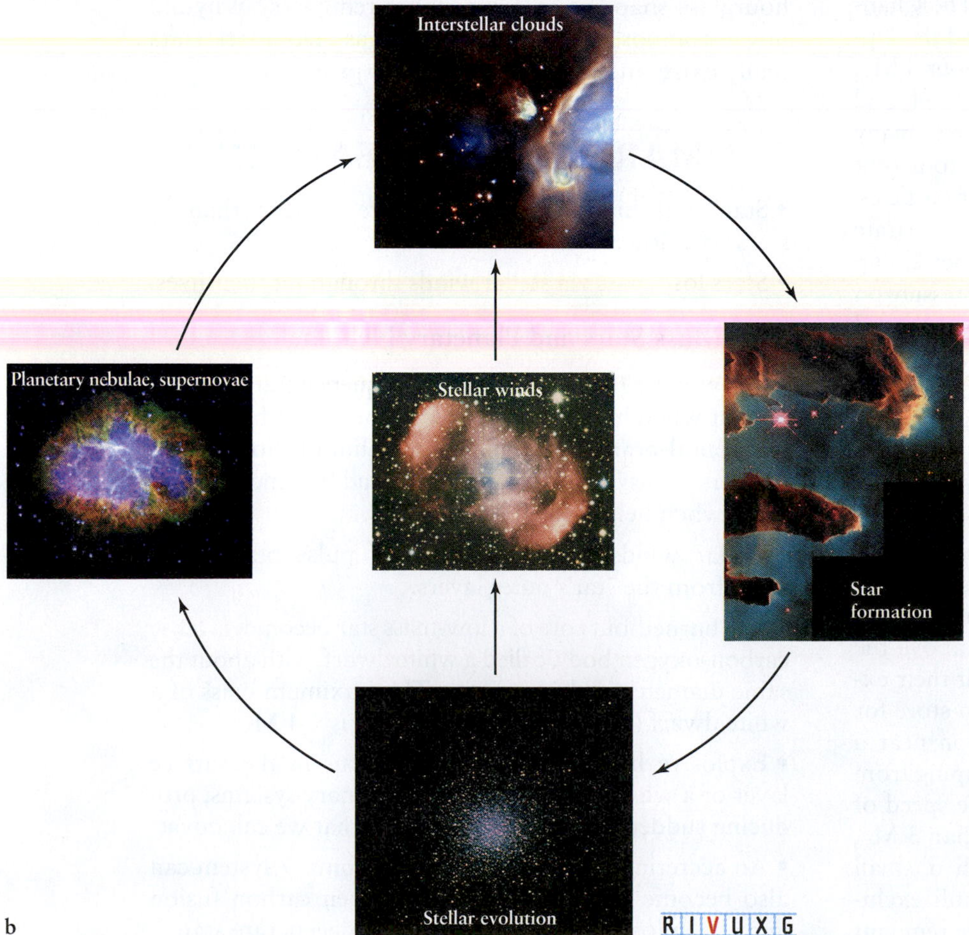

FIGURE 13-28 A Summary of Stellar Evolution (a) The evolution of isolated stars depends primarily on their masses. The higher the mass, the shorter the lifetime. Stars less massive than about 8 $M_\odot$ can eject enough mass to become white dwarfs. High-mass stars can produce Type II supernovae and become neutron stars or black holes. The horizontal (time) axis is not to scale, but the relative lifetimes are accurate. (b) The cycle of stellar evolution is summarized in this figure. (Top inset: Infrared Space Observatory, NASA; right inset: Anglo-Australian Observatory/J. Hester and P. Scowen, Arizona State University/NASA; bottom inset: NASA; left inset: NASA; middle inset: Anglo-Australian Observatory)

High-Mass Stars and Supernovae

- After exhausting its central supply of hydrogen and helium, the core of a high-mass (above 8 $M_\odot$) star undergoes a sequence of other thermonuclear reactions. These reactions include carbon fusion, neon fusion, oxygen fusion, and silicon fusion. This last fusion eventually produces an iron core.
- A high-mass star dies in a supernova explosion that ejects most of the star's matter into space at very high speeds. This Type II supernova is triggered by the gravitational collapse and subsequent bounce of the doomed star's core.
- Neutrinos were detected from Supernova 1987A, which was visible to the naked eye. Its development supported theories of Type II supernovae.

Neutron Stars, Pulsars, and (perhaps) Quark Stars

- The core of a high-mass main-sequence star containing between 8 and 25 $M_\odot$ becomes a neutron star. The cores of slightly more massive stars may become quark stars. A neutron star is a very dense stellar corpse consisting of closely packed neutrons in a sphere roughly 20 km in diameter. The maximum mass of a neutron star, called the Oppenheimer-Volkov limit, is about 3 $M_\odot$.
- A pulsar is a rapidly rotating neutron star with a powerful magnetic field that makes it a source of periodic radio and other electromagnetic pulses. Energy pours out of the polar regions of the neutron star in intense beams that sweep across the sky.
- Some X-ray sources exhibit regular pulses. These objects are believed to be neutron stars in close binary systems with ordinary stars.
- Explosive helium fusion may occur in the surface layer of a companion neutron star, producing a sudden increase in X-ray radiation, called an X-ray burster.

WHAT DID YOU THINK?

1 *Will the Sun someday cease to shine brightly? If so, how will this occur?* Yes. The Sun will shed matter as a planetary nebula in about 6 billion years and then cease nuclear fusion. Its remnant white dwarf will dim over the succeeding billions of years.

2 *What is a nova? How does it differ from a supernova?* A nova is a relatively gentle explosion of hydrogen gas on the surface of a white dwarf in a binary star system. Supernovae, on the other hand, are explosions that cause the nearly complete destruction of massive stars.

3 *What are the origins of the carbon, silicon, oxygen, iron, uranium, and other heavy elements on Earth?* These elements are created during stellar evolution, by supernovae, and by colliding neutron stars.

4 *What are cosmic rays? Where do they come from?* Cosmic rays are high-speed particles (mostly hydrogen and other atomic nuclei) in space. Many of them are thought to have been created as a result of supernovae.

5 *What is a pulsar?* A pulsar is a rotating neutron star in which the magnetic field's axis does not coincide with the rotation axis. The beam of radiation it emits periodically sweeps across our region of space.

Key Terms for Review

asymptotic giant branch (AGB) star, 417
Chandrasekhar limit, 421
cosmic ray, 427
cosmic ray shower, 427
glitch, 434
helium shell flash, 417
helium shell fusion, 417
lighthouse model, 432
magnetar, 433
neutron degeneracy pressure, 430
neutron star, 430
nova (*plural* novae), 420
nucleosynthesis, 422
photodisintegration, 424
planetary nebula, 418
primary cosmic ray, 427
pulsar, 430
quark, 437
secondary cosmic ray, 427
Type Ia supernova, 421
Type II supernova, 424
X-ray burster, 437

Review Questions

1. Will the Sun shed most of its mass, and, if so, what is that event called? **a.** yes, as a planetary nebula, **b.** yes, as a supernova, **c.** yes, as a white dwarf, **d.** yes, as a neutron blast, **e.** no

2. A white dwarf is composed of primarily **a.** neutrons, **b.** hydrogen and helium, **c.** iron, **d.** cosmic rays, **e.** carbon and oxygen.

3. What prevents a neutron star from collapsing? **a.** hydrogen fusion, **b.** friction, **c.** electron degeneracy pressure, **d.** neutron degeneracy pressure, **e.** helium fusion

4. What is the difference between a giant star and a supergiant star?

5. Why is knowing the temperature in a star's core so important in determining which nuclear reactions can occur there?

6. What determines the temperature in the core of a star?

7. What is a planetary nebula, and how does it form?

8. What is the Chandrasekhar limit?

9. What is a neutron star?

10. Compare a white dwarf and a neutron star. Which of these stellar corpses is more common? Why?

11. To test your understanding of the stages of stellar evolution, do Interactive Exercise 13.1. You can print out your results, if required.

12. What is the Oppenheimer-Volkov limit?

13. On an H-R diagram, sketch the evolutionary track that the Sun will follow between the time it leaves the main sequence and when it becomes a white dwarf.

Approximately how much mass will the Sun have when it becomes a white dwarf? Where will the rest of its mass have gone?

14. Why have searches for supernova remnants at visible wavelengths been less fruitful than searches at other wavelengths?

15. To test your understanding of how stars "die," do Interactive Exercise 13.2. You can print out your results, if required.

16. To test your understanding of neutron stars, do Interactive Exercise 13.3. You can print out your results, if required.

17. Why do astronomers believe that pulsars are rapidly rotating neutron stars?

18. To test your understanding of rotating neutron stars, pulsars, and novae, do Interactive Exercise 13.4. You can print out your results, if required.

19. What is the difference between Type Ia and Type II supernovae?

20. Compare a nova with a Type Ia supernova. What do they have in common? How are they different?

21. Compare a nova and an X-ray burster. What do they have in common? How are they different?

22. Describe what X-ray pulsars, pulsating X-ray sources, and X-ray bursters have in common. How are they different manifestations of the same type of astronomical object?

Advanced Questions

The answers to all computational problems, which are preceded by an asterisk (*), appear at the end of the book.

23. What prevents thermonuclear reactions from occurring at the center of a white dwarf? If no thermonuclear reactions take place in its core, why doesn't that body collapse?

24. Suppose you wanted to determine the age of a planetary nebula. What observations would you make, and how would you use the resulting data?

25. Why is the rate of expansion of the gas shell in a planetary nebula often not uniform in all directions?

26. What kinds of stars would you monitor if you wished to observe a supernova explosion from its very beginning? Look up the tabulated lists of the nearest and brightest stars in Appendix Tables E-5 and E-6. Which, if any, of these stars are possible supernova candidates? Explain.

27. To determine the period of a pulsar accurately, astronomers must take Earth's orbital motion around the Sun into account. Explain why.

*28. The distance to the Crab Nebula is about 2000 pc. When did the nebula actually explode?

29. To test your overall understanding of stellar evolution, do Interactive Exercise 13.5. You can print out your results, if required.

Discussion Questions

30. Suppose that you discover a small glowing disk of light while searching the sky with a telescope. How would you determine whether this object was a planetary nebula? What else could this object be?

31. Immediately after the first pulsar was discovered, one explanation offered was that the pulses were signals from an extraterrestrial civilization. Why did astronomers discard this idea?

32. Describe how astronomers can determine whether a supernova at a known distance is Type Ia or Type II, assuming that they can see the supernova from the time it begins to brighten. There are at least two valid answers to this question.

33. Describe how astronomers can determine whether a supernova at an unknown distance is Type Ia or Type II.

What If ...

34. Earth passed through an old supernova remnant? What would happen to Earth and life on it?

35. The Sun were a B-type star, rather than a G-type star? Assuming that Earth orbiting the B-type star had the same composition and orbital distance that it has today, what would be different on Earth? If you answered this question in Chapter 11, you might want to see how the material in this chapter enhanced your previous answer.

36. The Sun were an M-type star, rather than a G-type star? Assuming that Earth orbiting the M-type star had the same composition and orbital distance that it has today, what would be different on Earth?

37. The Sun were expanding into a giant star today? How might we cope with this change?

Web Questions

38. It has been claimed that the Dogon tribe in western Africa has known for thousands of years that Sirius is a binary star. Search the Web for information about these claims. What is their basis? Why are scientists skeptical, and how do they refute these claims?

39. Search the Web for recent information about SN 1987A. Sketch the shape of the supernova remnant. Has a pulsar been detected yet in the center of this supernova remnant? If so, how fast is it spinning? Has the supernova debris thinned out enough to give a clear view of the neutron star?

40. Search the Web for information about SN 1994I, a supernova that occurred in the galaxy M51 (NGC 5194). Why was this supernova unusual? Was it bright enough to have been seen by amateur astronomers?

Observing Projects

41. Planetary nebulae represent the late stages of the evolution of stars whose masses are similar to that of the Sun. They have reached the point in their lives when hydrogen burning within their cores has ceased and a period of activity and reorganization within this core has led to the ejection of their outer layers. They are distributed throughout our galaxy. You can use *Starry Night*™ to explore the distribution of these spectacular objects in our sky and to view several of them individually. Set the view for your home location at some time in the evening with a field of view of about 100°. Open the **Options** pane and expand the **Deep Space** layer. Expand the **NGC-IC Database** list, click in its box to activate the display of the objects in this list and click **Off** all entries except **Planetary Nebula.** Use the hand tool to move around the sky. Note that these nebulae are mostly concentrated around the Milky Way in our sky. If you have access to a telescope, try to locate and observe several of these planetary nebulae, if possible on a clear, moonless night. Some of the more notable planetary nebulae include: Little Dumbbell (M76), NGC 1535, Eskimo, Ghost of Jupiter, Owl (M97), Ring (M57), Blinking Planetary, Dumbbell (M27), Saturn Nebula, and NGC 7662.

If you do not have access to a telescope, use *Starry Night*™ to examine in detail two of these planetary nebulae, M57 (the Ring Nebula) and M27 (the Dumbbell Nebula), and compare their shapes and sizes. Select **Favourites > Discovering the Universe > Atlas** and use the **Find** pane to center upon and magnify these two nebulae in turn. You can compare a ground-based image of M57 with a high-resolution image taken by the Hubble Space Telescope by opening the **Options** pane, expanding the **Deep Space** list and clicking on the **Messier Objects** alone. You can replace this image by a space image by clicking in the **Hubble Images** box and clicking off the **Messier Objects.** Note that the Hubble image is displayed in a different alignment from that of the ground-based image.

For each of these objects, note their **Distance from observer** in the HUD, and then use the angular separation tool to measure the approximate angular radius of each of these nebulae. **a.** How do you account for the difference in the shape of these two planetary nebulae? **b.** What is the nature of the central star in each of these nebulae? **c.** Calculate the physical size of each of these nebulae. (*Hint:* Translate angular size in arcseconds to radians and use the small-angle relationship. 1 radian $= 206{,}265''$; 1 ly $= 9.46 \times 10^{12}$ km.) **d.** Assuming that both of these nebulae have been expanding at the same rate (measured in km/s), which of the stars at the cores of these nebulae reached the end of the active hydrogen-burning phase of its life first (that is, which of these nebulae is the oldest)? **e.** Using an average rate of expansion of 20 km/s for the shell of gas that forms the Ring Nebula (M57), approximately how long ago, in years, did the star that formed this nebula initiate the expansion of its outer atmosphere?

42. Two supernova remnants can be seen through modest telescopes, one in the winter sky and the other in the summer sky. Since both of these are quite faint, you should schedule your observations for a moonless night. The winter sky contains the Crab Nebula, a supernova remnant from a star that exploded and became bright enough to be seen in the daytime for a few days in A.D. 1054. It is located near the star marking the eastern horn of Taurus (the Bull) and can be found using *Starry Night*™ software. Its coordinates are R.A. 5 h 34.5 m and Dec. = +22° 00′. The Veil (or Cirrus) Nebula in the summer sky is so vast that it extends across a significant region of the sky, in contrast to the Crab Nebula, which fits easily into the field of view of the eyepiece of a telescope. The easiest way to find the Veil Nebula is to aim the telescope at the star 52 Cygni (R.A. = 20 h 45.7 m and Dec. = +30° 43′), which is coincident with one of the brightest portions of the nebula in our sky. If you then move the telescope slightly north or south until 52 Cygni is just out of the field of view, you should see giant wisps of glowing gas, the remnants of the explosion of a massive star.

If you do not have access to a telescope, you can examine the images of both of these supernova remnants in *Starry Night*™. Use the **Find** tool to locate the Crab Nebula and adjust **Date** and **Time** to move it high in your sky. Click on the drop-down list to the left of the listing of Crab Nebula and click on **Magnify** to see a composite of images from several sources, including the Hubble Space Telescope and the Chandra spacecraft. Open the **Options** pane, expand the **Deep Space** list and click **Off** all sources except **Messier Objects** to see a visible light image of the nebulous wisps of gas that make up the supernova remnant. Replace this image with the Hubble image to see in more detail the active regions near the center of the nebula. It is here at the center of this nebula that a spinning neutron star, the result of incredible implosion during the supernova explosion, is producing a beam of radiation that passes over Earth many times per second to appear as a rapidly flashing source known as a *pulsar.* This pulsar is instrumental in powering the expansion of the nebula.

Use the **Find** tool to locate, center, and magnify the image of the Veil Nebula. You can magnify images of the East and West portions of this widely spaced nebula separately by using the drop-down lists next to the Veil Nebula (East) and (West) listings to magnify and examine the wisps of gas that make up the remains of this star explosion. You may have to open the **Options** pane and expand the **Deep Space** list to ensure that the **Bright NGC Objects** are displayed.

WHAT IF ... A Supernova Exploded Near Earth?

The Flash What would happen on Earth if a supernova occurred only 50 ly away? (Considering the titanic forces that supernovae release, it should come as no surprise that the high-energy electromagnetic radiation from such an explosion detonating much closer than this distance would immediately kill virtually all life on Earth.) Neutrinos would foreshadow by a few hours the visible flash and pending flood of X rays and gamma rays from the supernova. The doomed star would then grow tremendously luminous, 50 times brighter than the Moon and only 8000 times less bright than the Sun. It would be brighter than the light from all other stars in the night sky combined.

The lethal X rays and gamma rays would be the first causes of death on Earth. Within days, organisms from virtually all species of plants and animals would begin dying of radiation poisoning. Entire radiation-sensitive species might then be annihilated. The radiation would also cause many cancers and other internal diseases over succeeding years, leading to many more plant and animal deaths. At the same time, ultraviolet radiation reaching Earth's surface would cause an astronomical jump in the rates of skin cancer and cataracts.

The first blast of ultraviolet radiation from the supernova would destroy Earth's ozone layer within a matter of days, transforming the ozone primarily into atomic oxygen. With the removal of this protective barrier, ultraviolet radiation from both the supernova and the Sun would saturate Earth's surface. After the intensity of ultraviolet radiation from the supernova diminished, sunlight would begin repairing the ozone layer, eventually returning it to normal levels.

The brightness and emission of all the electromagnetic radiation from the supernova would peak after a month. It would then fade, but the supernova remnant would be visible for millennia as an expanding cloud of gas and dust in the night sky. Fortunately for the life that survived the onslaught of high-energy radiation from the supernova, the remnant cloud would primarily emit harmless visible light.

The Aftermath The most energetic particles from the supernova and its environs, the so-called *cosmic rays,* would also affect Earth's surface. Moving at 90% of the speed of light, cosmic rays would arrive here only 5 to 10 years after the photons. Cosmic ray energies are much higher than those of any particles normally existing on Earth. For example, a typical cosmic ray has enough energy to light a small lightbulb for 1 s; it takes over 600,000 trillion electrons flowing through wiring in your house to do the same thing.

Some cosmic rays would slam into nitrogen or oxygen molecules in the air, shattering the air molecules and creating cosmic ray showers. The highest-energy cosmic rays from the supernova would reach Earth intact, as do the highest-energy cosmic rays from other sources today. These would break up atoms of the objects they strike on Earth's surface, enhancing the earlier biological damage caused by the supernova's electromagnetic radiation.

The bulk of the matter ejected into space by the supernova would travel at speeds of 16,000 km/s (60 million km/h), nearly 20 times slower than the emitted photons. Thus, the bulk of the supernova remnant would take at least 1000 years longer to reach us than did its initial radiation. By the time the bulk of the blast wave reaches the solar system, it would have become so thin and diffuse that it would probably do little damage to Earth's atmosphere. However, we could expect this material to deposit a thin but exotic mix of elements into the upper atmosphere. All of this material would eventually fall to Earth and alter the chemistry of both the ocean and the soil.

The Outcome The damage done to life by both the electromagnetic radiation and cosmic ray particles would disrupt the global food chain. In the oceans, large quantities of plankton and other microscopic organisms at the foundation of the chain would die off. As a result, many of the larger aquatic species that feed on these smaller organisms would starve. Similarly, on the surface, most plant life would wither, and many herbivorous animals would starve as a result. This would, of course, lead to the death and dislocation of animals throughout the food chain. Surviving plants and animals would undergo genetic mutations due to the supernova's radiation. Most of these genetic changes would lead to the immediate death of the altered plant or animal, or their offspring, but a few changes would be beneficial and enable the mutated organisms to thrive.

How would the human race fare after the supernova? Our long-term survival would depend, in part, on whether surviving humans could coexist with surviving plants and animals, and with each other.

WHAT DO YOU THINK?

1. Are black holes empty holes in space? If not, what are they?
2. Does a black hole have a solid surface? If not, what is at its surface?
3. What power or force enables black holes to draw things into them?
4. How close to a black hole do you have to be for its special effects to be apparent?
5. Can you use black holes to travel to different places in the universe?
6. Do black holes last forever? If not, what happens to them?

Answers to these questions appear in the text beside the corresponding numbers in the margins and at the end of the chapter.

This artist's conception shows what we would see looking toward a black hole from just above the hot, gaseous accretion disk spiraling around it. Some of this gas is so hot that it expands and jets outward perpendicular to the accretion disk, guided by magnetic fields. The rest of the gas in the disk spirals into the black hole. (A. Hobart, CXC)

Chapter 14

Black Holes: Matters of Gravity

Stars are formed and held together by gravity. At each stage of stellar development that we have considered, stars are prevented from collapsing further by the outward forces generated by fusion, electromagnetism, and degeneracy pressure. In this chapter we explore the fate of stellar remnants with more than 3 $M_\odot$, in which gravitation "wins" the battle, but eventually loses the war between the inward and outward forces.

When the gravitational force of an object is so great that it overcomes all opposing repulsive forces or pressures (like neutron degeneracy pressure), the object collapses in on itself. Its gravitational attraction then becomes so strong that nothing—not even light—can escape from it. When this happens, the matter and the space around it become a *black hole.*

Black holes inspire awe, fear, and uncertainty. Many people harbor the mistaken belief that black holes are giant vacuum cleaners destined to "suck up" all matter in the universe. Happily, the equations that describe black holes reveal that they are more benign than that, but they are still truly strange.

The idea that gravity could prevent matter from escaping was first put forward in the eighteenth century, but it was not until the twentieth century, when Albert Einstein presented his ideas of *relativity* (how matter and motion affect mass, distance, and time), that the modern concept of a black hole developed. We must first explore his relativity theories in order to understand these exotic objects.

In this chapter you will discover

- that Einstein's theory of general relativity predicts the existence of regions of space and time that are severely distorted by the extremely dense matter they contain
- that space and time are not separate entities
- how black holes arise
- the surprisingly simple theoretical properties of black holes
- that X rays and jets of gas are created near many black holes
- the fate of black holes
- the unsurpassed energy emitted by gamma-ray bursts

THE RELATIVITY THEORIES

It seems reasonable that if I watch you traveling at, say, 100 km/h (about 60 mi/h), my measurements of your mass, the rate at which your watch runs, and the length of your car would be the same as the measurements you make of these things. In fact, the measurements will not be the same. At the beginning of the twentieth century, Einstein began a revolution in science by disregarding his common sense, making a few physical assumptions about nature, and using mathematics to explore their consequences. The resulting equations revealed profound, albeit highly counterintuitive, insights into how nature works. Armed with the equations that Einstein derived, scientists have been exploring and using the consequences of his theories to develop new technology and a deeper understanding of how matter, energy, space, and time interact.

14-1 Special relativity changes our conception of space and time

In 1905, Albert Einstein derived his **theory of special relativity,** a description of how motion affects our measurements of distance, time, and mass. He was guided by two innovative ideas. Although the implications of the theory proved revolutionary, the first notion seems simple:

Your description of physical reality is the same regardless of the (constant) velocity at which you move.

In other words, as long as you are moving in a straight line at a constant speed, you experience the same laws of physics as anyone else moving at any other constant speed and in any other direction. To illustrate this first idea, suppose you were inside a closed boxcar moving smoothly in a straight line at 100 km/h and you dropped a pen from a height of 2 m. You time how long it takes the pen to reach the floor (about 0.64 s). Then the train is stopped and you repeat the same experiment in the station. The time it would take the pen to fall would be identical.

The second idea seems more bizarre:

Regardless of your speed or direction, you always measure the speed of light to be the same.

Suppose that you are in a (very powerful) car moving toward a distant street lamp at 150,000 km/s (93,000 mi/s). In simpler terms, this is 50% the speed of light, denoted $0.5c$ (Figure 14-1a). A friend of yours leaning on the lamppost sees you coming and also sees light from the lamp heading toward you at c. How fast do *you* see the photons of light coming at you? It would seem reasonable that you clock them at $c + 0.5c$, which is the speed of the photons toward you plus your speed toward them, but this is incorrect. You will, in fact, see the photons coming toward you with the same speed, c, that your friend sees them leaving the lamp (Figure 14-1b). You would, however, see that they have a different color than your friend sees. That difference is due to the Doppler shift, as explained in *An Astronomer's Toolbox 4-3.*

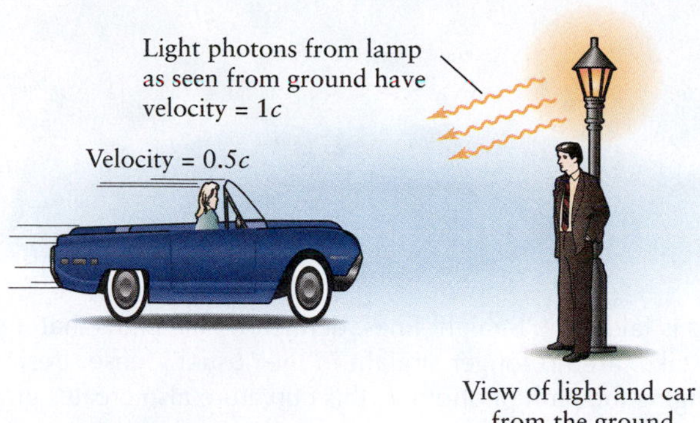

FIGURE 14-1 The Speed of Light Is Constant (a) As seen from the ground, photons of light from the lamp are traveling toward the car with a speed $v = c$ (ignoring the slight decrease in speed due to the presence of the air). (b) As seen from the car, moving toward the lamp at $v = 0.5c$, the photons are also traveling at the speed $v = c$. The difference in color between the light in the two figures is due to the Doppler shift.

Einstein's theory of special relativity expresses these two assumptions mathematically, and the results of these assumptions have been confirmed in innumerable experiments. Perhaps the most well-known result of special relativity is the relationship between matter and energy, $E = mc^2$. The fact that mass is multiplied by the square of a very large number reveals that a little mass creates a lot of energy. This equation implicitly states that, contrary to most daily experience (and what most people are taught as children), matter *can* be converted into energy and energy *can* be converted into matter. This conversion takes place when stars fuse elements together, like the hydrogen fusion that occurs in the Sun today.

Three other fundamental results of special relativity also defy our everyday experience:

1. **The length of an object decreases as its speed increases.** In other words, if a train moves toward or away from you, you would measure its length to be shorter than you would measure its length to be when it is stopped at the station (Figure 14-2). (This result also explains why the car in Figure 14-1 appears too short.) This result is called *length contraction*. However, if you measure the length of the moving train while you are inside it and moving with it, your measurement of its length will be the same as you measured on the ground when it was at rest. The word *relativity* emphasizes the importance of the relative speed between the observer and the measured object.

2. **Clocks that you see as moving run more slowly than do clocks you see at rest.** This result is called *time dilation*. Indeed, the faster a clock moves relative to you, the slower it appears to tick from your perspective. For example, air travelers actually age more slowly than they would if they did not fly (although because of the relatively low speeds of aircraft travel compared to the speed of light, this difference is imperceptibly small). People flying in aircraft do not feel time passing more slowly, however, because their biological activities slow down at the same rate as the clocks around them. Only an observer moving relative to an airplane sees that the clock and the activities of the high-speed travelers aboard the plane have slowed. These connections between motion and clocks mean that space and time cannot be considered as two separate concepts. Relativity requires us to combine them in a single entity, thus creating the concept of **spacetime.**

3. **The mass of an object increases as it moves faster.** The concept of mass is discussed in Sections 2-5 and 2-7. The equations of special relativity reveal that an object moving at the speed of light would have an infinite amount of mass. It is impossible for any object with mass to reach this speed, much less go faster than the speed of light, because an infinite amount of mass is

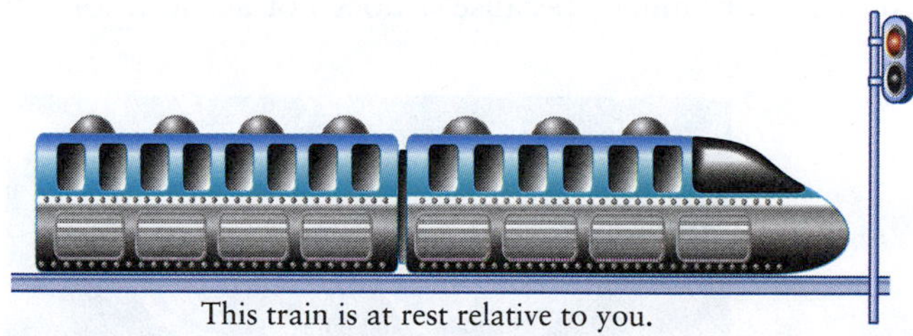

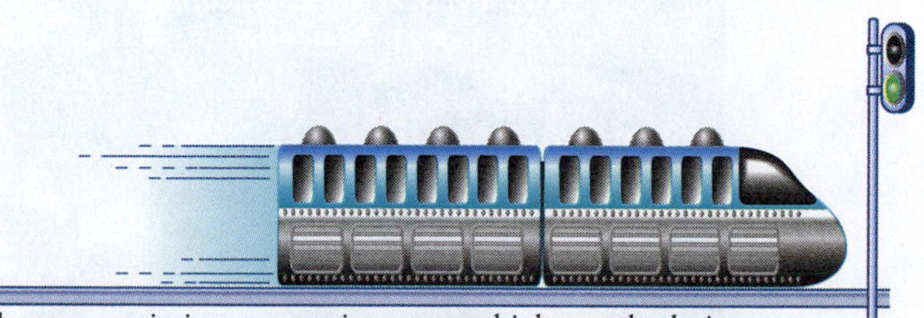

FIGURE 14-2 Movement and Space According to the theory of special relativity, the faster an object moves, the shorter it becomes in its direction of motion as observed by someone not moving with the object. It becomes infinitesimally short as its speed approaches the speed of light. The dimensions perpendicular to the object's motion are unchanged.

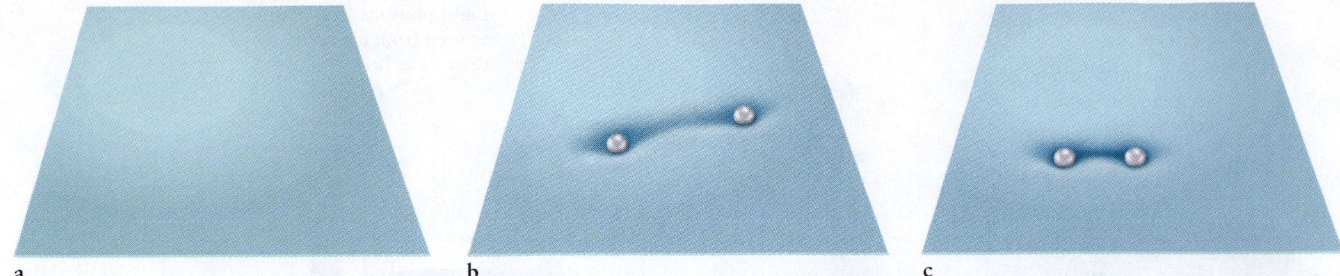

FIGURE 14-3 Curved Spacetime (a) This flat surface represents two dimensions in spacetime. In the absence of any matter in the spacetime, straight lines are straight in our intuitive sense (the sheet is flat). (b) In the presence of matter, spacetime curves, as shown by the curvature of the sheet when mass is laid on it. Straight lines, defined by the paths that light rays take, are no longer straight in the "usual" sense. Besides changing the path of photons, this curvature also creates gravity, which (c) pulls the two masses toward each other.

How much mass, m, would have to be destroyed to create an amount of energy, E?

more than the total mass that exists in the universe. Furthermore, the equation $F = ma$ (F is the force exerted on an object, m is its mass, and a is the acceleration it undergoes; see Section 2-7) reveals that if an object's mass is infinite, the force necessary to accelerate this object must also be infinite. Because the total force available to move matter in the universe is finite, no massive object can be sped up to the speed of light: The speed of light is the universal speed limit.

14-2 General relativity explains how matter warps spacetime, creating gravitational attraction

While special relativity tells us how the mass of an object is related to its (constant) speed, this theory is special (in the sense of limited) because it does not account for the effects of acceleration and gravitation. These effects were incorporated by Einstein in 1915 into a more general theory, called the **theory of general relativity** (or, more simply, general relativity). It describes how spacetime changes shape in the presence of matter (Figure 14-3). The greater the mass, the more the distortion or *curvature*. Furthermore, the curvature of spacetime creates attraction between all pieces of matter in the universe (Figure 14-3c)—this attraction is what we call the *gravitational force*.

Another result of general relativity is that time slows down in the presence of matter. The greater the concentration of matter, the slower clocks tick (Figure 14-4). For example, time passes more slowly for us here on Earth's surface than it would for astronauts on the Moon, which has less mass than Earth.

Newton's laws of motion and his universal law of gravitation are accurate only for objects with relatively small masses, slow velocities compared to the speed of light, and low densities (such as those for objects found on Earth). Newton's laws are also limited to motion suf-

FIGURE 14-4 Time Slows Down near Matter (a) Two clocks in space set at exactly the same time are (b) brought to Earth and the Moon. From a vantage point far from Earth and the Moon, the clock on Earth is ticking more slowly than the clock on the Moon. This occurs because mass slows down the flow of time, and Earth has more mass (and a higher density, which adds to the effect) than the Moon.

ficiently far from large masses (such as the Sun) or high-density objects (such as neutron stars). We can use Newton's laws to accurately predict the paths of projectiles on Earth and the motion of our planet around the Sun, but they do not describe the orbit of Mercury around the Sun very well (because that planet is so close to our star) or two neutron stars that orbit each other. In such cases, general relativity correctly predicts how objects move and how time passes. Closer to home, the accurate determination of location using GPS satellites orbiting Earth requires the use of the equations of general relativity.

14-3 Spacetime affects the behavior of light

Besides affecting the behavior of objects, the curvature of spacetime changes the path and wavelength of light that passes near any matter. (These behaviors are not predicted by Newton's laws.) Light travels along trajectories in space called *geodesics*. You can get an idea of how geodesics work by imagining that you are flying on the shortest possible route from one city to another. Your path would be analogous to a geodesic for a photon, and like that particle in the presence of matter in space, you would actually be following the curve of Earth's surface, rather than going in a "straight line." The first experimental verification of general relativity was its prediction that geodesics are sufficiently curved to cause light from stars behind the Sun to arc around it (Figure 14-5).

Photons that leave the vicinity of a star also lose energy in climbing out of the star's gravitational field. They do not slow down like a bullet fired upward. Rather, they shift to longer wavelengths (Figure 14-6), an effect we see in the spectra of some white dwarfs, whose light appears redder than it would if this effect did not occur. This shift of wavelengths leaving the vicinity of a massive object is called **gravitational redshift.**

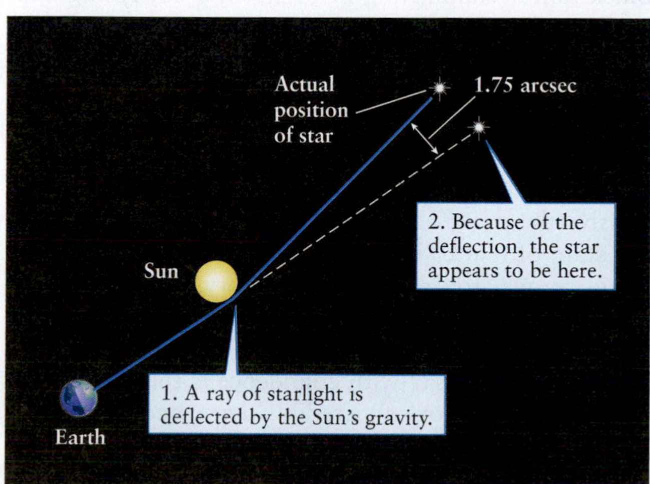

FIGURE 14-5 Curved Spacetime and the Path of Light The warping of space by matter causes light to be deflected. This was the first prediction of general relativity to be confirmed, in 1919. This confirmation came when stars behind the Sun were observed during an eclipse. The star in this drawing was not observed where it was supposed to be, as a result of the Sun's gravity changing the path of its light.

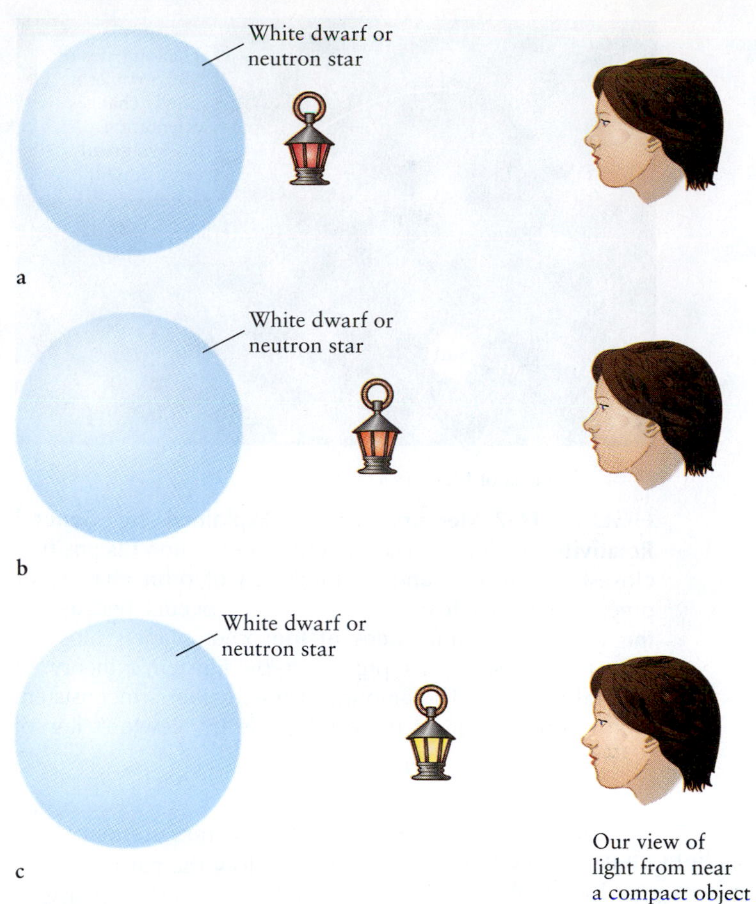

FIGURE 14-6 Gravitational Redshift The color of light from the same object located at different distances from a mass appears different as seen from far away. The photons that leave the vicinity of the massive object lose energy and are therefore redshifted. The closer the light source is to the mass, the redder the light appears, and hence the name gravitational redshift. The same argument applies to light leaving the surfaces of different stars.

> What property of matter does general relativity address that is not included in special relativity?

General relativity has been confirmed again and again, as seen in these observations:

- Light is measurably deflected by the gravitational curving of space due to the presence of matter like stars or entire galaxies containing billions of stars (see Figure 14-5).
- The perihelion position of Mercury as seen from the Sun shifts, or precesses, by 43 arcsec per year more than is predicted by Newtonian gravitational theory (Figure 14-7). Its actual precession is exactly the amount predicted by general relativity.

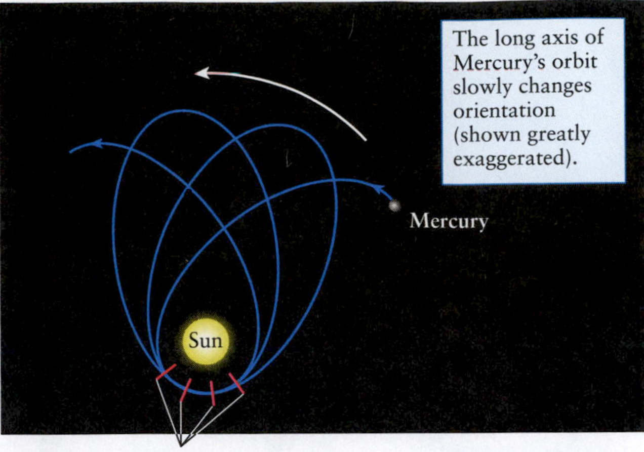

FIGURE 14-7 Mercury's Orbit Explained by General Relativity The location of Mercury's perihelion (its position closest to the Sun) and its long axis of orbit change, or precess, with each orbit. This change occurs because of the gravitational influences of the other planets plus the curvature of space as predicted by Einstein's theory of general relativity. The amount of precession is inconsistent with the prediction of the orbit made by Newton's law of gravity alone.

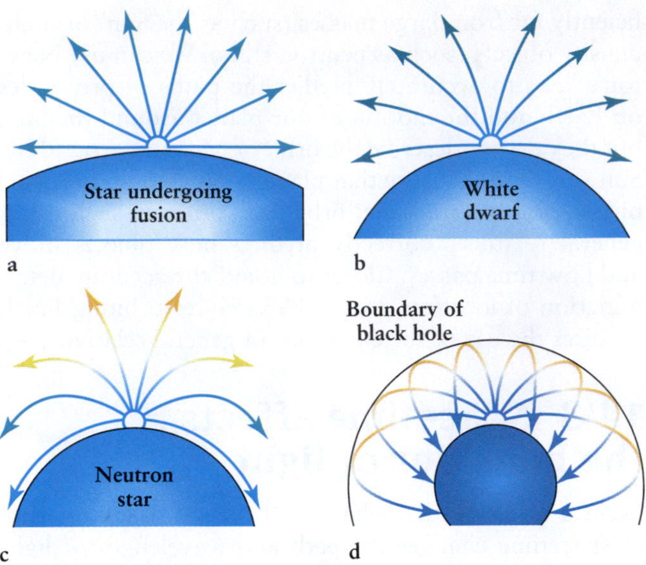

FIGURE 14-8 Trapping of Light by a Black Hole (a) The paths and colors of light rays departing from a main-sequence, giant, or supergiant star are affected very little by the star's gravitational force. (b) Light leaving the vicinity of a white dwarf curves and redshifts more, whereas (c) near a neutron star, some of the photons actually return to the star's surface. (d) Inside a black hole, all light remains trapped. Most photons curve back in, except those that fly straight upward, which become infinitely redshifted, thereby disappearing.

Does light change direction as it goes past Earth? Why or why not?

- The orbits of stars in binary systems follow the paths predicted by general relativity, rather than those predicted by Newtonian gravitation.
- The spectra of stars are observed to have the gravitational redshifts predicted by general relativity (see Figure 14-6).

14-4 General relativity predicts the fate of massive star cores—black holes

Let us now return to stellar evolution and consider what happens when high-mass main-sequence stars (with more than 25 $M_\odot$) explode as supernovae. As with the stars that contain between 8 $M_\odot$ and 25 $M_\odot$, high-mass stars become neutron stars. The neutron-star remnants of these high-mass stars, however, have more than 3 $M_\odot$. This is problematic because above this mass, the gravitational force of the neutron star overcomes the neutron degeneracy pressure created by the interactions among its particles. As a result, such remnant stars collapse further, and no force in nature can stop their infall.

We can use the equations of general relativity to understand the fate of collapsing neutron stars. We just saw that all matter warps the space around itself (see Figures 14-3 and 14-5). When matter gets sufficiently dense, it actually causes space near it to curve so much that it closes in on itself (Figure 14-8). Photons flying outward at an angle from such a collapsing star arc back inward. Photons flying straight upward redshift so much that they lose all their energy (see Section 3-3) and hence cease to exist.

1 If light, the fastest moving of all known things, cannot escape from the vicinity of such dense matter, then nothing can! Such regions out of which no matter or any form of electromagnetic radiation can leave are called **black holes.** Plummeting in on itself, a collapsing neutron star becomes so dense that it ceases to consist of neutrons. General relativity predicts that in creating a black hole, matter compresses to infinite density (and zero volume), a state called a **singularity.** However, we know that general relativity and our other theories of physics are invalid in such a situation. A more comprehensive theory of nature must be developed to explain the state of matter in a black hole's singularity. Efforts are under way to develop such a theory. The best-known nascent explanation is called *superstring theory*, but its mathematics is daunting and much work remains to be done. All we can say with confidence is that the matter in black holes becomes incredibly dense and compact.

INSIDE A BLACK HOLE

The formation of a black hole is complicated, but its nature is surprisingly simple: It contains matter at its center or in a ring, it has a boundary shaped like a sphere, it either has a net electric charge or not, and it either rotates or does not rotate.

14-5 Matter in a black hole becomes much simpler than elsewhere in the universe

2 A black hole is separated from the rest of the universe by a boundary, called its **event horizon.** The event horizon is not like the surface of a solid or liquid body. No matter exists at this location except for the instant it takes infalling mass to cross the event horizon and enter the black hole.

We cannot look inside a black hole because no electromagnetic radiation escapes from it. Our understanding of its structure comes from the equations of general relativity. According to Einstein's theory, the event horizon is a sphere. The distance from the center of the black hole to its event horizon is called the **Schwarzschild radius** (abbreviated R_{Sch}), after the German physicist Karl Schwarzschild, who first determined its properties. *An Astronomer's Toolbox 14-1* shows how to calculate this distance, which depends only on the black hole's mass. The more massive the black hole, the larger its event horizon.

According to the equations of general relativity, when a stellar remnant collapses to a black hole, it loses its internal magnetic field. The field's energy radiates away in the form of electromagnetic radiation and **gravitational radiation.** Emitted as **gravitational waves,** gravitational radiation travels as ripples in the very fabric of spacetime. Gravitational waves are also created when neutron stars or black holes collide or when stars and stellar remnants are in close orbits around each other. (Actually, gravitational radiation is emitted whenever any two things move around each other, such as a pair of dancers. When the moving bodies are less massive than stars or stellar remnants, however, we have no hope of detecting their very weak gravitational radiation with either present or projected technology.)

> Why do neutron stars of more than about 3 $M_\odot$ collapse to form black holes?

Gravitational radiation has not yet been directly detected, unlike, say, visible light that we "see" by its effects on our eyes or on a CCD. However, gravitational radiation has been observed indirectly by its effects on the orbits of some binary star systems. In particular, a pair of neutron stars orbiting each other emit so much energy as gravitational waves that the two bodies spiral toward each other. The change in their orbit is correctly predicted by general relativity. This agreement earned a Nobel Prize in Physics in 1993 for Joseph Taylor and Russell Hulse, who discovered the first binary pulsar in 1974 and measured the changes in its neutron stars' orbits.

The ripples in spacetime created by gravitational waves from stars or stellar remnants are incredibly tiny. On Earth, each meter-wide volume of space changes by less than 10^{-25} m as the waves pass by. Astronomers around the world are building *gravitational wave detectors* to measure these small changes (Figure 14-9). They have been in operation since 2002, and astronomers expect that these detectors or the next generation of them will provide a direct means to observe high-energy gravitational activity in the cosmos.

Characteristics of Black Holes Besides losing its internal magnetic field, matter within a black hole loses almost all traces of its composition and origin. Indeed, it retains only three properties that it had before entering the black hole: its *mass,* its *angular momentum,* and its *electric charge.* Familiar concepts, such as proton, neutron, electron, atom, and molecule, no longer

An Astronomer's Toolbox 14-1

The Sizes of Black Holes

In 1918, Karl Schwarzschild, a German physicist, discovered the first solution of Einstein's equations. His solution describes the nature of nonrotating black holes. According to Einstein's general theory of relativity, the Schwarzschild radius R_{Sch} of any black hole can be found from its mass, *M:*

$$R_{sch} = \frac{2GM}{c^2}$$

where R_{Sch} is measured in meters; M is the black hole's mass in kilograms; c is the speed of light, 3×10^8 m/s^2; and G is the gravitational constant, 6.67×10^{-11} m^3/kg·s^2. Expressing M in terms of solar masses, this equation can be conveniently approximated by

$$R_{Sch} \approx 3\ M_{BH}$$

where M_{BH} is the black hole's mass in solar masses and R_{Sch} is in kilometers.

Example: A 5-$M_\odot$ main-sequence star has a radius of 3×10^6 km, whereas the previous equation reveals that a black hole with the same mass has a 15-km Schwarzschild radius.

Try these questions: A primordial black hole with the mass of Mount Everest would have a Schwarzschild radius of just 1.5×10^{-15} m! What is Mount Everest's mass? What is the Schwarzschild radius of a galactic black hole that contains 3 billion solar masses? A black hole with a Schwarzschild radius equal to the radius of Earth has how much mass?

(Answers appear at the end of the book.)

FIGURE 14-9 Laser Interferometer Gravitational Wave Observatory (LIGO) Located in Hanford, Washington, this is one of several gravitational wave (colloquially, gravity wave) detectors around the world. It has two perpendicular arms, each 4 km long. Gravity waves that pass the detector cause unequal changes in the lengths of the arms. These changes are detected by lasers inside each arm. (LIGO Laboratory)

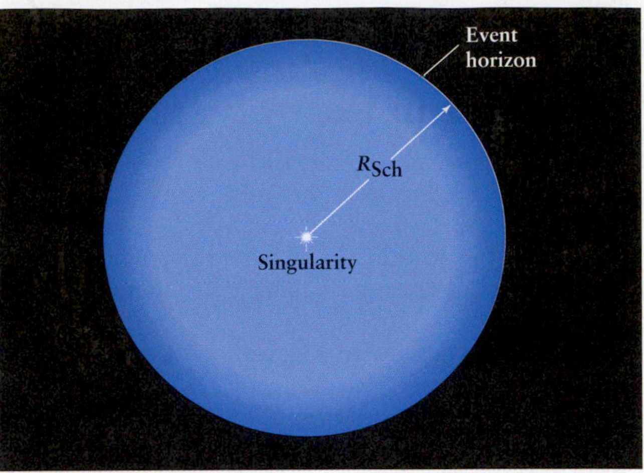

FIGURE 14-10 Structure of a Schwarzschild (Nonrotating) Black Hole A nonrotating black hole has only two notable features: its singularity and its boundary. Its mass, called a singularity because it is so dense, collects at its center. The spherical boundary between the black hole and the outside universe is called the event horizon. The distance from the center to the event horizon is the Schwarzschild radius, R_{Sch}. There is no solid, liquid, or gas surface at the event horizon. In fact, except for its location at the boundary of the black hole, an event horizon lacks any features at all.

apply. In addition, because few large bodies appear to have a net charge, it is also doubtful that black holes do. We therefore predict that there are only two different types of black holes: those that rotate and those that do not.

Types of Black Holes If the mass creating a black hole is not rotating, the black hole that is formed does not rotate either. We call nonrotating black holes **Schwarzschild black holes** (Figure 14-10). General relativity predicts that all the mass in such a black hole collapses to a point of infinite density at its center, the singularity mentioned earlier. The rest of the volume from the event horizon to the singularity of a Schwarzschild black hole is empty space.

When the matter that creates a black hole possesses angular momentum, that matter collapses to a ring-shaped singularity located inside the black hole between its center and the event horizon (Figure 14-11). Such rotating black holes are called **Kerr black holes,** in honor of the New Zealand mathematician Roy Kerr, who first calculated their structure in 1963. Once again, the black hole is empty except for the singularity. Most Kerr black holes should be spinning thousands of times every second, even faster than the pulsars that we studied in Chapter 13. The black hole with the highest known rotation rate, discovered in 2006, rotates a thousand times per second, consistent with this theory.

Unlike Schwarzschild black holes, equations indicate that Kerr black holes possess a doughnut-shaped region directly *outside* of their event horizons in which

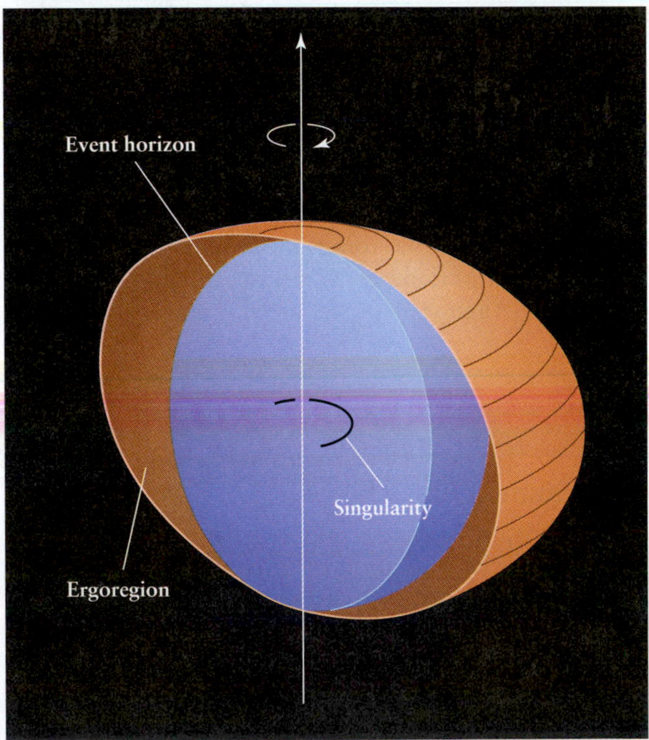

FIGURE 14-11 Structure of a Kerr (Rotating) Black Hole Rotating black holes are only slightly more complex than nonrotating ones. The singularity of a Kerr black hole is located in an infinitely thin ring around the center of the hole. The ring appears as an arc in this cutaway drawing. The event horizon is again a spherical surface. There is also a doughnut-shaped region, called the ergoregion, just outside the event horizon, in which nothing can remain at rest. Space in the ergoregion is being curved or pulled around by the rotating black hole.

FIGURE 14-12 Swirling Space in an Ergoregion Just as the chocolate in this blender is being dragged by the spinning blade, so, too, is space dragged by a spinning (Kerr) black hole. (Jack Andersen/Food Pix)

objects cannot remain at rest without falling into the black hole. Called **ergoregions,** they are volumes of spacetime that the rotating black hole drags around, like so much batter in a blender (Figure 14-12). In 1997, matter orbiting a black hole was observed to behave in a fashion consistent with the existence of an ergoregion surrounding the hole. If it is moving fast enough, an object that enters the ergoregion can again fly out of it; however, if an object stops in the ergoregion, it must fall into the black hole. In the first decade of the twenty-first century, astronomers began observing spectra of atoms orbiting in the ergoregions of particularly massive black holes. The spectra showed the effects of gravitational redshift and Doppler shift predicted by Einstein's general relativity for matter orbiting in these regions.

Where is the singularity of a rotating black hole?

14-6 Falling into a black hole is an infinite voyage

Imagine being in a spacecraft orbiting only 1000 3
Schwarzschild radii (15,000 km) from an isolated 5-$M_\odot$ black hole (see *An Astronomer's Toolbox 14-1* for the equation of the Schwarzschild radius). You are held in orbit by the black hole's gravitational force. Even at that short distance, the only effect the black hole has on you is its gravitational attraction. Only when you get very, very close to the event horizon do bizarre things begin to happen. To investigate these changes, you send a cube-shaped probe with a clock toward the black hole, with the same side of the cube always facing "downward" toward the black hole. The probe emits a blue glow so that you can follow its progress. What happens to the cube as it approaches the black hole?

From the time you launch the probe until it reaches 4
about 100 Schwarzschild radii (1500 km), you see the probe descend as if it were falling toward a planet or moon (Figure 14-13a). Time measured by the probe's clock slows down only slightly. At 100 Schwarzschild radii, however, the probe begins to respond significantly to the powerful tidal effect from the black hole. The face of the probe closest to the event horizon receives

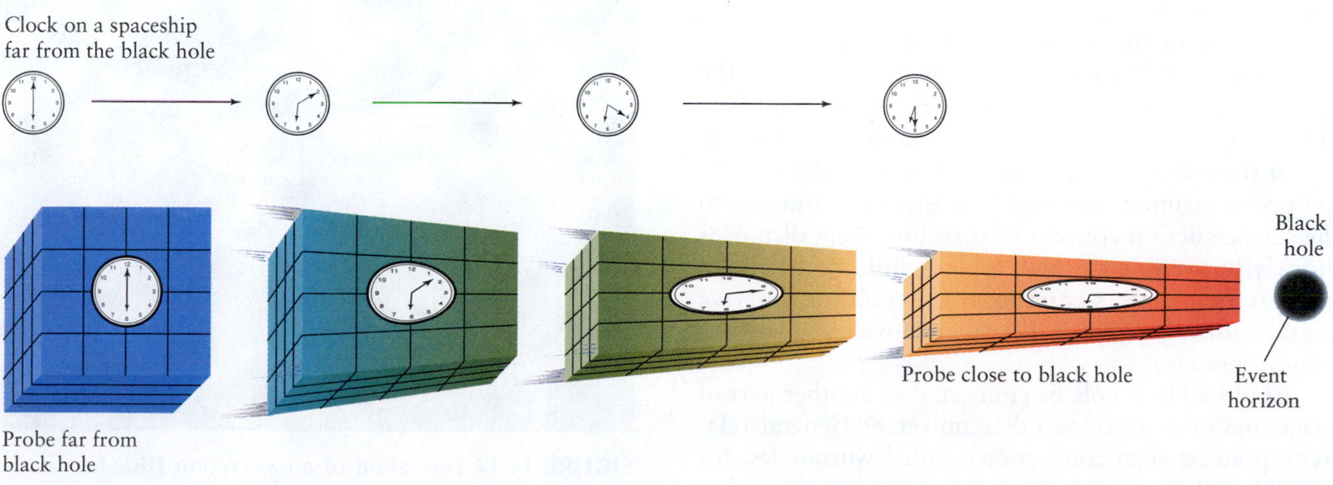

FIGURE 14-13 Effect of a Black Hole's Tidal Force on Infalling Matter (a) A cube-shaped probe 1500 km from a 5-$M_\odot$ black hole. (b, c, d) Near the Schwarzschild radius, the probe is pulled long and thin by the difference in the gravitational forces felt by its different sides. This tidal effect is a greatly magnified version of the Moon's gravitational force on Earth. The probe changes color as its photons undergo extreme gravitational redshift and time slows down on the probe, as seen from far away.

perceptibly more gravitational pull from the hole than its farther-away parts, and it begins to stretch apart.

By the time the probe comes within a few Schwarzschild radii of the event horizon, the tidal forces on it are so great that it violently elongates. The part of the probe closest to the black hole accelerates downward and away from the rest of the probe. Furthermore, the sides of the probe are drawn together: They are falling in straight lines toward a common center. The net gravitational effect of moving close to the event horizon is for the probe to be pulled long and thin. From a practical perspective, this means that the probe would be violently torn apart, because it is not composed of perfectly elastic material.

As the probe nears the black hole, the blue photons leaving it must give up more and more energy to escape the increasing gravitational force. However, unlike a projectile fired upward, photons cannot slow down. Rather, they lose energy by increasing their wavelengths (see *An Astronomer's Toolbox 3-1*). This is another example of the gravitational redshift predicted by general relativity. The closer the probe gets to the event horizon, the more its light is redshifted—first to green, then yellow, then orange, then red, then infrared, and, finally, radio waves (see Figure 14-13b, c, d).

How would a 1-$M_{\odot}$ black hole 1 AU from Earth affect our planet?

Stranger still is the black hole's effect on time. General relativity predicts that when the probe approaches within a few Schwarzschild radii of the black hole, its infall rate will slow down as seen from far away. Also, signals from the probe show you that its clocks are running much more slowly than they did when it left your spacecraft. Time dilation becomes so great near the event horizon that the probe will appear to hover above it and its clocks will stop.

Anyone in the probe would observe something else altogether: They see the probe actually cross the event horizon and continue falling toward the black hole's singularity in a normal period of time, according to their own watch. Pulled apart by tidal effects, the probe disintegrates as it falls inward. Contrary to the science fiction concept of traveling great distances quickly by passing through a black hole, calculations indicate that objects entering them could not survive passage through, even if there were a way to come out somewhere else.

5 Could a black hole be connected to another part of spacetime or even some other universe? General relativity predicts such connections, called **wormholes,** for Kerr black holes, but astrophysicists are skeptical that the equations are correct in this regard. Their conviction is called **cosmic censorship:** Nothing can leave a local region of space that contains a singularity (that is, a black hole).

EVIDENCE FOR BLACK HOLES

14-7 Several binary star systems contain black holes

Black holes are more than fine points of relativity theory: They are real. Their presence has been observed by their effects on the orbits of other stars and on gas and dust near them, and more of them are being located all of the time. To find evidence for black holes, we look first to binary star systems.

The technique for detecting black holes formed from collapsing stars is based on the interaction between the black hole and its binary companion. When one star in a close binary becomes a black hole, its gravitational attraction pulls off some of its companion's atmosphere. However, such black holes have diameters of only a few kilometers, so there is not enough room for all that gas to fall straight in. Rather, the infalling gas swirls into the black hole like water going down a bathtub drain (Figure 14-14). The gas waiting

FIGURE 14-14 Formation of an Accretion Disk Just as the water in this photograph swirls around waiting to get down the drain, the matter pulled toward a black hole spirals inward. Angular momentum of the infalling gas and dust causes them to form an accretion disk around the hole. (Chris Collins/Corbis)

FIGURE 14-15 X Rays Generated by Accretion of Matter Near a Black Hole Stellar-remnant black holes, such as Cygnus X-1, LMC X-3, V404 Cygni, and probably A0620-00, are detected in close binary star systems. This drawing (of the Cygnus X-1 system) shows how gas from the 30-$M_{\odot}$ companion star, HDE 226868, transfers to the black hole, which has at least 11 $M_{\odot}$. This process creates an accretion disk. As the gas spirals inward, friction and compression heat it so much that the gas emits X rays, which astronomers can detect. (Courtesy of D. Norton, Science Graphics)

Table 14-1 Stellar Black Holes in the Milky Way

X-ray source name	Mass* of companion	Mass* of black hole
Cygnus X-1	24–42	11–21
V404 Cygni	~0.6	10–15
V461 Sgr	~10	5.5–8
GS 2000+25	~0.7	6–14
H 1705-250	0.3–0.6	6.4–6.9
GRO J1655-40	2.34	7.02
A 0620-00	0.2–0.7	5–10
GS 1124-T68	0.5–0.8	4.2–6.5
GRO J042+32	~0.3	6–14
4U 1543-47	~2.5	2.7–7.5

*Solar masses

R. Blandford and N. Gehrels, "Revisiting the Black Hole," *Physics Today*, June 1999.

to fall in forms an **accretion disk,** a disk of gas and dust spiraling in toward the black hole. Magnetic fields in the disk help pull the debris inward. Calculations reveal that this gas is compressed and thereby heated so much from the collisions of its particles that it gives off X rays (Figure 14-15). Thus, if a visible star has a sufficiently tiny, sufficiently massive X-ray–emitting companion, we have located a black hole. See *Guided Discovery: Identifying Stellar-Remnant Black Holes* for details on how this is done.

To date, in the Milky Way Galaxy alone, at least 10 stellar-remnant black holes in binary star systems have been identified (Table 14-1). There is also growing evidence that black holes can collide with each other or with different types of objects, such as neutron stars.

14-8 Other black holes range in mass up to billions of solar masses

Based on the equations of general relativity, the fate of massive neutron stars led as early as 1939 to the idea of black holes. Since then, calculations have supported at least three other types of black holes:

Why is the gas in an accretion disk heated?

1. Early in the life of the universe, black holes could have formed from the condensation of vast amounts of gas and also from the collisions of stars during the process of galaxy formation, thereby creating **supermassive black holes,** each with millions or billions of solar masses.

2. We saw in Section 12-2 that stars form in clusters. If a cluster forms with enough stars concentrated in a sufficiently small volume of space, collisions between stars in the center of the cluster may lead to the formation of a black hole with between a few hundred and a few thousand solar masses. Such objects are called **intermediate-mass black holes.**

3. Black holes could have formed during the explosive beginning of the universe as tiny amounts of matter were compressed sufficiently to form **primordial black holes.** These bodies would have masses ranging from a few grams to the mass of a planet.

Supermassive Black Holes In May 1994, the Hubble Space Telescope obtained compelling evidence for a black hole at the center of the galaxy M87. In the nucleus of M87 is a tiny, bright source of light. Spectra showed that nearby gas and stars are orbiting it extremely rapidly. They can be held in place only if the bright object contains some 3 *billion* solar masses (Figure 14-16). Given that the source's size is only slightly larger than the solar system, it can only be a black hole.

Identifying Stellar-Remnant Black Holes

Shortly after the *Uhuru* X-ray satellite was launched in the early 1970s, astronomers found a promising black-hole candidate—an X-ray source called Cygnus X-1. This source is highly variable and irregular. Its strong X-ray emission flickers on timescales as short as a hundredth of a second. If different parts of an X-ray source grew bright at different times, its emission would be a continuous stream. For Cygnus X-1 to flicker, the entire star must brighten and dim as a unit. Therefore, light must have time to travel across Cygnus X-1 between pulses. Because light travels 3×10^8 km/s, it travels 3000 km in the "flicker time" of a hundredth of a second. This value means that for all of Cygnus X-1 to brighten and darken simultaneously, it must be about 3000 km across, smaller in diameter than Earth.

Cygnus X-1 occasionally emits radio radiation, and, in 1971, radio astronomers succeeded in associating Cygnus X-1 with the visible star HDE 226868 (Figure GD 14-1), a B0 supergiant with a surface temperature of about 31,000 K. Because such stars do not emit significant X rays, HDE 226868 alone cannot be the Cygnus X-1 X-ray source. Spectroscopic observations soon showed that the lines in the spectrum of HDE 226868 shift back and forth within a period of 5.6 days. This behavior is characteristic of a single-line spectroscopic binary (see Section 11-12), and HDE 226868's companion is too dim to produce its own set of spectral lines. The clear implication is that HDE 226868 and Cygnus X-1 are the two components of a binary star system.

The B0 supergiant HDE 226868's mass is estimated at about 30 $M_\odot$, like other B0 supergiants. As a result, Cygnus X-1 must have at least 11 $M_\odot$; otherwise, it would not exert enough gravitational pull to make the B0 star wobble by the amount deduced from the periodic Doppler shift of its spectral lines. Cygnus X-1 cannot be a white dwarf or a neutron star, because its mass is too large for either of these objects. The only remaining possibility is that it must be a fully collapsed star—a black hole.

In the early 1980s, a similar binary system was identified in the nearby galaxy called the *Large Magellanic Cloud*. The X-ray source, called LMC X-3, exhibits rapid fluctuations, just like those of Cygnus X-1. LMC X-3 orbits a B3 main-sequence star every 1.7 days. From its orbital data, astronomers conclude that the mass of LMC X-3 is about 10 $M_\odot$, which would make it a black hole.

Another black-hole candidate is a spectroscopic binary in the constellation Monoceros that contains the flickering X-ray source A0620-00. The visible companion of A0620-00 is an orange-colored dwarf star. The low-mass, main-sequence star of spectral type K orbits the X-ray source every 7.75 h. From orbital data, astronomers conclude that the mass of A0620-00 must be greater than 5 $M_\odot$, probably about 9 $M_\odot$.

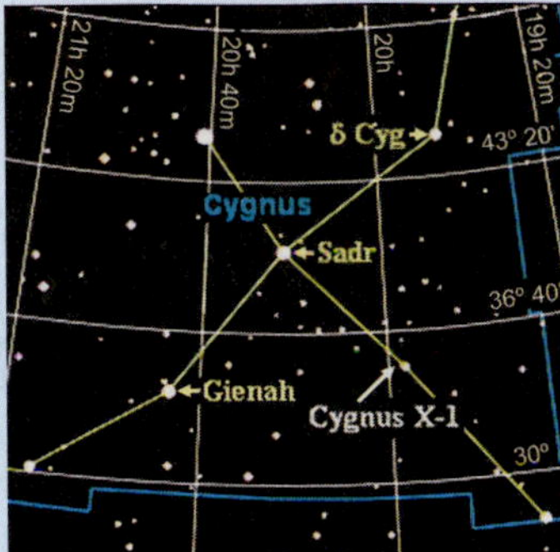

FIGURE GD 14-1 HDE 226868 This star is the visual companion of the X-ray source Cygnus X-1. This binary system is located about 8000 light-years from Earth and contains a black hole of at least 11 $M_\odot$ in orbit with HDE 226868, a B0 blue supergiant star. The photograph was taken with the 200-in. telescope at Palomar Observatory on Palomar Mountain, north of San Diego. The slightly dimmer star above is an optical double that is not part of the binary system. (J. Kristian, Carnegie Observatories)

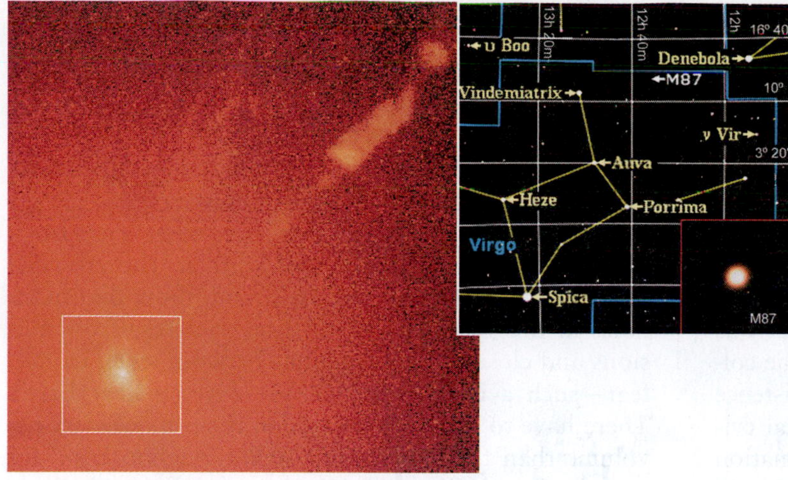

FIGURE 14-16 Supermassive Black Hole The bright region in the center of galaxy M87 has stars and gas held in tight orbits by a black hole. M87's bright nucleus (center of the region in the white box) is only about the size of the solar system but it pulls on the nearby stars with so much force that astronomers calculate that it is a 3×10^9-$M_\odot$ black hole. One of the bright jets of gas shooting out perpendicular to the black hole's accretion disk is visible at the upper right on this image. (Holland Ford, STScI/Johns Hopkins University; Richard Harms, Applied Research Corp.; Zlatan Tsvetanov, Arthur Davidsen, and Gerard Kriss at Johns Hopkins University; Ralph Bohlin and George Hartig at STScI; Linda Dressel and Ajay K. Kochhar, Applied Research Corp., Landover, MD; and Bruce Margon, University of Washington, Seattle)

Since 1994, black holes in the centers of many galaxies have been identified by their X-ray emissions and gravitational effects on surrounding gas and stars. For example, a distinct, frisbee-shaped accretion disk around the supermassive black hole in NGC 7052 was seen by the Hubble Space Telescope in 1998 (Figure 14-17). A black hole containing several million solar masses has even been found at the center of our Milky Way, only 26,000 light-years from Earth. Indeed, evidence for central, supermassive black holes has been found in most of the several dozen nearest galaxies.

Earlier in this book we studied how tidal effects of one body on another occur in several situations, including the Earth-Moon system and the heating of Io and Europa by Jupiter. In 2004, astronomers observed several supermassive black holes in other galaxies absorbing parts of passing stars. These black holes created such high tides on those stars that the stars were pulled apart. Some of each star's mass was then drawn into the black hole. These discoveries were made because infalling gas from the stars was rapidly heated to millions of kelvins, causing the gas to emit a burst of X rays that was observed by

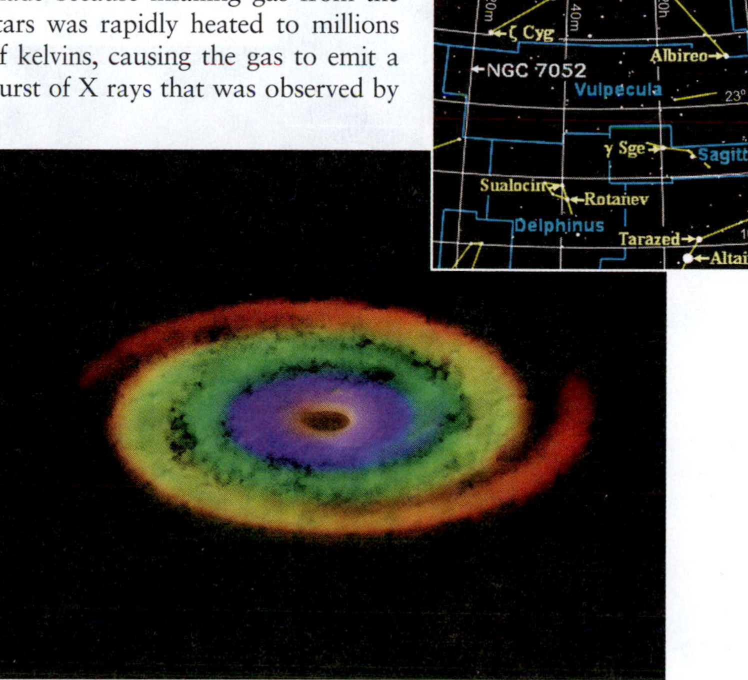

FIGURE 14-17 Accretion Disk Around a Supermassive Black Hole (a) Swirling around a 3×10^8-$M_\odot$ black hole in the center of the galaxy NGC 7052, this disk of gas and dust is 3700 ly across. The gas is cascading into the black hole, which will consume it all over the next few billion years. The black hole appears bright because of light emitted by the hot, accreting gas outside its event horizon. NGC 7052 is 191 million ly from Earth in the constellation Vulpecula. (b) This drawing shows how the gases spiraling inward in an accretion disk heat up as they approach the black hole. Color coding follows Wien's law: red (coolest), followed by orange, yellow, green, blue, and violet (hottest). (R. P. van der Marel, STScI/F. C. van den Bosch, University of Washington/NASA; b: NASA/CXC/SAO)

orbiting telescopes. Subsequent observations also reveal that supermassive black holes can consume vast quantities of interstellar gas, sometimes stripping enough gas from a galaxy to prevent it from forming new generations of stars.

As we will explore in more detail in Chapter 18, galaxies were created from condensing gas and stars in the early universe. Our technology is now providing us with observations of that epoch of the universe, and the formation process of supermassive black holes in the centers of galaxies is now being studied. These black holes apparently formed, in part, from the gas that was condensing to create galaxies and, in part, from the collisions of stars and black holes that came into existence shortly after the beginning of time. Observational evidence indicates that the formation of supermassive black holes is still ongoing. We will explore this further in Chapter 16.

Where are supermassive black holes located?

Intermediate-Mass Black Holes At the end of the twentieth century, astronomers began discovering objects that appear to be black holes with between 10^2 and 10^5 $M_\odot$. Again, these objects were identified as potential black holes by the intensity and spectra of the X rays that they emit and by the orbits of stars around them. For example, near the center of the galaxy M82, astronomers found what appears to be a 500–1000 $M_\odot$ black hole (Figure 14-18), while in the galaxy NGC 1313, astronomers have observed what appears to be two intermediate-mass black holes, each with 200–500 $M_\odot$. Most intermediate-mass black holes are observed in globular clusters.

In support of the belief that these objects are black holes, computer simulations of stars in the crowded regions, where such objects are found, show that black holes in this mass range can form as a result of collisions and close tidal interactions between stars. Do not fear—such a black hole will not develop near Earth. There have to be at least a million times more stars per volume than there are in our stellar neighborhood for such frequent stellar collisions or near misses to occur.

Primordial Black Holes Even more exotic black holes may have formed along with the universe itself. The British astrophysicist Stephen Hawking has proposed that the Big Bang explosion from which most astronomers believe the universe emerged may have been chaotic and powerful enough to have compressed tiny knots of matter into primordial black holes. Their masses may have ranged from a few grams to greater than the mass of Earth. Astronomers have not yet observed evidence of primordial black holes, although that does not mean they do not necessarily exist—just that we cannot yet detect them.

FIGURE 14-18 An Intermediate-Mass Black Hole M82 is an unusual galaxy in the constellation Ursa Major. The inset shows an image of the central region of M82 from the Chandra X-ray Observatory. The bright, compact X-ray source shown varies in its light output over a period of months. The properties of this source suggest that it is a black hole of roughly 500 $M_\odot$. (Subaru Telescope, National Astronomical Observatory of Japan; inset: NASA/SAO/CXC)

14-9 Black holes and neutron stars in binary systems often create jets of gas

We saw in Section 13-11 how beams of radiation are emitted by neutron stars. Similarly, many neutron stars and black holes emit pairs of gas jets shooting out in opposite directions.

Consider a black hole (neutron stars behave analogously) in a binary system. If the companion star is still fusing material, then its outer layers can be pulled off, as discussed in Section 12-15. The infall of this matter toward the compact companion is too rapid for all of the mass to enter the event horizon immediately. The resulting accretion disk around the black hole is the key to explaining the presence of the jets.

As the disk mass spirals down toward the event horizon, this gas is compressed into a smaller volume. Such compression heats the gas, which, in turn, causes its pressure to increase. As a result of the increased pressure, the gas expands, forming a doughnut-shaped region around the black hole (Figure 14-19). As the gas starts the final plunge toward the black hole, its temperature skyrockets to tens or even hundreds of millions of kelvins. At such temperatures, the pressure is so great that much of the infalling gas expands and, finding little resistance perpendicular to the plane of the accretion disk, it squirts out as two jets (Figure 14-19). These jets are prevented from spreading out by a magnetic field created as the hot gases orbit in the accretion disk and by pressure from surrounding gas.

Whereas neutron stars and stellar-mass black holes create jets as a result of being in binary star systems, supermassive black holes have so much gravitational attraction that they pull huge quantities of nearby interstellar gas and dust into orbit around themselves without needing a binary companion. Figure 14-17a shows such a disk, and Figure 14-16 shows one of the jets emitted by the supermassive black hole in M87. We will explore more about jets created by black holes in Chapter 17.

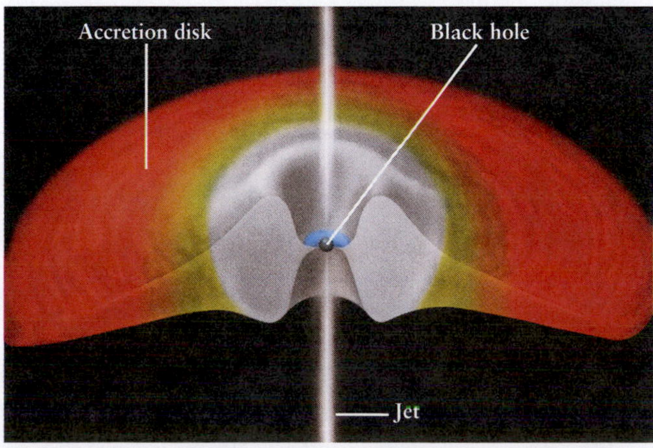

FIGURE 14-19 Jets Created by a Black Hole in a Binary System Some of the matter spiraling inward in the accretion disk around a black hole is superheated and redirected outward to produce two powerful jets of particles that travel at close to the speed of light. The companion star is off to one side of this drawing.

GAMMA-RAY BURSTS

Some black holes, neutron stars, and the supernovae that accompany their formation are apparently the sources of the most energetic and enigmatic energy emissions known in the universe. Called **gamma-ray bursts,** these pulses of gamma rays (the shortest-wavelength electromagnetic radiation) emit more energy than supernovae, from even smaller volumes of space.

Why doesn't all the gas in a black hole's accretion disk enter the black hole?

14-10 Gamma-ray bursts are the most powerful explosions in the known universe

Because atomic bomb blasts emit pulses of gamma rays, the U.S. military put the gamma-ray-detecting *Vela* satellites in orbit around Earth to monitor illegal nuclear explosions in the 1960s. In 1973, astronomers were told that since 1967, these satellites had been detecting bursts of gamma rays from objects in space. This news intrigued the astronomical community because, at that time, there was no known mechanism to emit such radiation.

Gamma-ray bursts each last between a few milliseconds and about 1000 s (just over 16 min). There appear to be at least three types of such bursts. One group typically lasts for a few tenths of a second. The second group typically lasts for about 40 s. The third group appears to have properties of both of the first two groups, hence it is called a *hybrid gamma-ray burst.* Unlike X-ray bursts (see Section 13-16), each gamma-ray burst occurs only once.

In 2004, astronomers detected the first remnant of a gamma-ray burst that occurred in the Milky Way. In 2008, they observed a gamma-ray burst 7.5 billion light-years away that also emitted enough visible light to be (just barely) visible to the naked eye here on Earth. For comparison, it was as bright as the Triangulum galaxy, which contains some 30 billion stars and is only 3 million light-years away. More than 5000 gamma-ray bursts have been observed, and more are being discovered at a rate of about one per day. Plotting their locations on the celestial sphere (Figure 14-20), astronomers have discovered that the shorter-lived bursts occur slightly more commonly in the plane of the Milky Way Galaxy than elsewhere, while the longer-lived ones occur randomly over the sky.

Since 1997, optical and infrared "afterglows" of many gamma-ray bursts have been detected (Figure 14-21). These often persist for days, which allow astronomers time to study their spectra in detail. Such observations reveal that the light has passed through

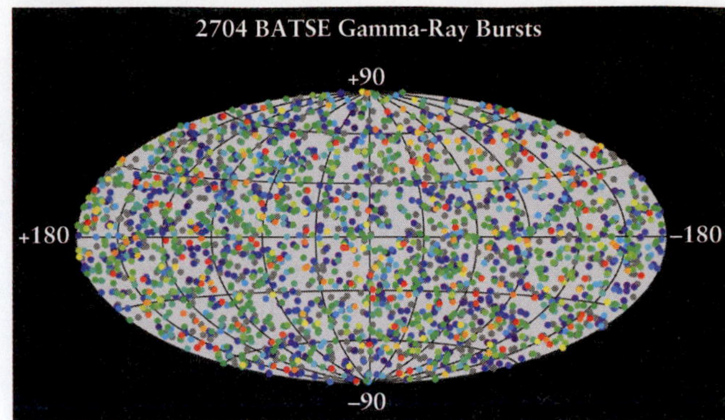

FIGURE 14-20 The Most Powerful Known Bursts Gamma-ray bursts have been observed everywhere in the sky, indicating that, unlike X-ray bursters, most do not originate in the disk of the Milky Way Galaxy. This map of the entire sky "unfolded" onto the page shows 2704 bursts detected by the Burst and Transient Source Experiment (BATSE) aboard the Compton Gamma-Ray Observatory. (The colors indicate the brightness of the bursts; they are coded brightest in red, dimmest in violet. Gray dots indicate incomplete information about the burst strength.) (NASA)

intergalactic gas clouds. Combining this discovery with observations of the bursts' spectra and redshifts reveals that the longer-lasting bursts typically originate more than 6 billion light-years away, far outside our Galaxy. The most distant gamma-ray burst observed so far occurred 13.1 billion light-years away. Many of the shorter bursts occur in more nearby galaxies. Taking their distance into account, a typical long-lived gamma-ray burst emits as much energy in 100 s as the Sun will emit over the 10 billion years that it will be on the main sequence. Gamma-ray burst afterglows emit as much as 2.5 million times as much visible light as does the brightest supernova ever observed.

Deep sky observations of some gamma-ray bursts have begun to show galaxies in the same locations that the bursts were discovered. As Figure 14-22 reveals, the bursts do not appear in the centers of their host galaxies. Likewise, the bursts in the plane of the Milky Way are not located near the galactic nucleus. These observations indicate that bursts are not associated with the supermassive black holes in the galactic cores. Research suggests that gamma-ray bursts are instead associated with stellar cataclysms.

In 2003, observations by NASA's *High-Energy Transient Explorer* observed a 30-s gamma-ray burst that was seen to occur at the same place as a particularly powerful supernova, called a *hypernova,* that occurs in **Wolf-Rayet stars.** These are rotating stars of at least 20 $M_\odot$ with strong magnetic fields and stellar winds. The bursts associated with these stars occur when gamma rays and a lot of gas are funneled out along the rotation axis of the star. If we are looking down the axis from Earth, the gamma rays create the burst, and the gas, traveling more slowly, creates the subsequent afterglow. Observations reveal that hypernovae and their associated gamma-ray bursts are extremely messy affairs, with powerful aftershocks, rather than a single, smooth outburst. Because gamma-ray bursts associated with supernovae occur near the beginning of these events, the bursts enable astronomers to watch the supernovae unfold, providing further insights into these explosions and the mechanism by which they generate gamma-ray bursts.

Other gamma-ray bursts occur when black holes collide, when a black hole swallows a neutron star, or when a pair of neutron stars collide and form a black hole. Observations made in 2005 confirm calculations predicting that the collision of such pairs of objects each lead to a rapid, intense burst of gamma rays.

To better understand gamma-ray bursts, astronomers have constructed telescopes for use both on Earth

22 seconds

48 seconds

73 seconds

R I V U X G

FIGURE 14-21 Gamma-Ray Bursts In 1999, astronomers photographed the visible-light counterpart of a 100-s gamma-ray burst some 9 billion light-years away in the constellation Boötes. The times indicated are after the burst began. (Carl Akerlof, University of Michigan, Los Alamos National Laboratory, Lawrence Livermore National Laboratory)

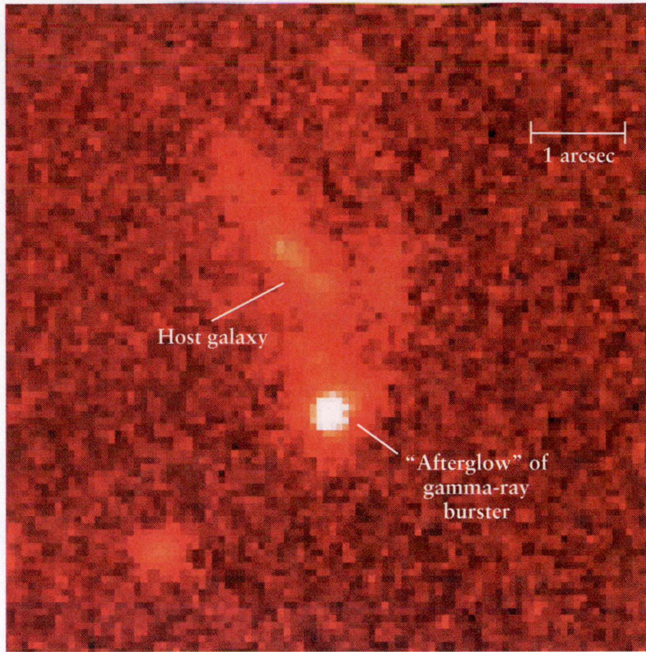

FIGURE 14-22 The Host Galaxy of a Gamma-Ray Burst The Hubble Space Telescope recorded this visible-light image in 1999, 16 days after a gamma-ray burst was observed at this location. The image (color-coded to show contrast, rather than true color) reveals a faint galaxy that is presumed to be the home of the gamma-ray burst. The galaxy actually has a bright blue color, indicating the presence of many young stars. (Andrew Fruchter, STScI; and NASA)

and in orbit that are able to rapidly slew (change direction) to look in the direction in which these events occur in space. These telescopes, including the *Swift Gamma-Ray Burst Mission* and the *High Energy Transient Explorer 2*, see these bursts and provide new insights into their origins and dynamics. *Swift* carries not only a gamma-ray telescope to locate the bursts, but also separate X-ray and optical telescopes to study their afterglows.

14-11 Black holes evaporate

In exploring the fate of black holes, astrophysicists find that, once again, these objects confound common sense—they evaporate! With a black hole cloaked behind the event horizon and its mass collapsed into a singularity, it must seem that there is no way of getting mass from the black hole back out into the universe again. No way, that is, until you recall that mass and energy are two sides of the same coin. What if there was a way of converting the black hole's mass into a form of energy that *could* get out, such as gravitational energy? The transformation does occur, according to an idea first proposed by Stephen Hawking.

Black holes convert their mass into energy by a process called **virtual particle** production. If things in this chapter aren't weird enough, consider this: The laws of nature allow pairs of particles, called virtual particles, to spontaneously appear. Each pair consists of a particle and its antiparticle—such as a proton and an antiproton, an electron and an antielectron (positron), or a pair of photons. The particles in each pair are identical, except for having opposite electric charges, and they annihilate each other and normally disappear without a trace within a very short period of time, typically 10^{-21} s.

We know that virtual particles actually flash into existence spontaneously throughout the universe because some have been made into real particles in the laboratory. This creation is done using high-speed particles in particle accelerators that slam into the pair of virtual particles and push them apart before they annihilate each other. The mass or energy given to the virtual particles in making them real comes from the laboratory particles, which lose an equal amount of energy. Calculations reveal that billions and billions of pairs of virtual particles are seething in and out of existence in your body (and every other equivalent volume of space) every second.

How long do gamma-ray bursts last?

Now, suppose two virtual particles form *just outside* the event horizon of a black hole. They would experience an exceptionally powerful gravitational field. If one particle is created slightly farther from the hole than its companion, the two virtual particles feel a tidal force from the black hole's tremendous gravitational pull. If the gravitational tidal force is strong enough, it can pull the two virtual particles apart before they can annihilate each other. This separation makes them real (Figure 14-23).

To conserve momentum, at least one of the newly 6
formed real particles always falls into the black hole.

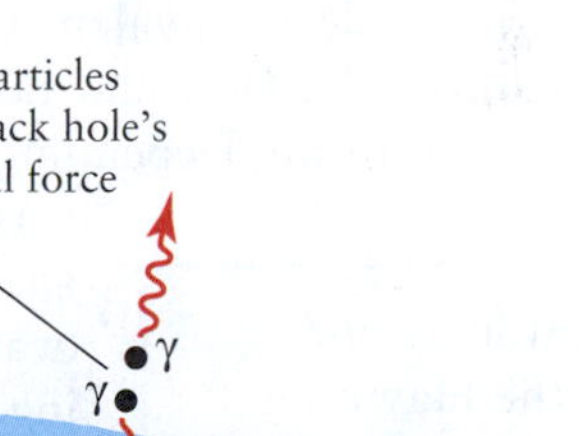

FIGURE 14-23 Evaporation of a Black Hole Throughout the universe, pairs of virtual particles spontaneously appear and disappear so quickly that they do not violate any laws of nature. The tidal force just outside of the event horizon of a black hole is strong enough to tear apart two virtual particles that appear there before they destroy each other. The gravitational energy that goes into separating them makes them real and, therefore, permanent. At least one of each pair of newly created particles falls into the black hole. Sometimes the other particle escapes into the universe. Because the gravitational energy used to create the particles came from the black hole, the hole loses mass and shrinks, eventually evaporating completely. Here we see just a few particles in the making: an electron (e^-) and a positron (e^+) and a pair of photons (γ).

But the other particle will sometimes have enough kinetic energy to escape from the vicinity of the black hole. This latter particle flies free into space, and the black hole has effectively emitted energy equal to $E = mc^2$, where m is the mass of the freed particle and c is the speed of light. Alternatively, the black hole may lose the energy due to one of two photons created, $E = hc/\lambda$, where λ is the wavelength of the photon and h is Planck's constant.

Where the virtual particles become real is a temporary void in space outside the event horizon. The black hole fills the void with energy that comes from its mass. This energy is transmitted outside the event horizon as gravitational radiation to replace the energy taken away by the newly formed particles. This is called the **Hawking process,** and the particles leaving the vicinity of the black hole are called *Hawking radiation* (even though many of the particles are not electromagnetic radiation).

The time it takes a black hole to completely evaporate by the Hawking process increases with the mass of the hole. More massive black holes have lower tidal forces at their event horizons than do less massive and, therefore, smaller-diameter, black holes. Therefore, the *rate* at which virtual particles are converted into real particles around bigger black holes is actually slower than it is around smaller ones. The faster the real particles are produced, the faster they wear down the mass of the black hole; thus, lower-mass black holes evaporate more rapidly. Indeed, calculations indicate that in its final moments of evaporation, a black hole should create particles so quickly that it should appear to explode.

A stellar black hole of 5 $M_\odot$ would take more than 10^{62} years to evaporate (much longer than the present age of the universe), whereas a 10^{10}-kg primordial black hole (equivalent to the mass of Mount Everest; Figure 14-24) would take only about 15 billion years to evaporate. Depending on the precise age of the universe, a black hole of this size and age should be in the final throes of evaporating. Astronomers are trying to identify events in space that correspond to the violent, particle-generating deaths of primordial black holes.

Why does at least one particle in the Hawking process always fall into the black hole?

14-12 Frontiers yet to be discovered

Black holes take us to the limits of science. They represent a state of matter that we are not yet fully capable of explaining. This state of affairs makes black holes an especially intriguing field of study. Understanding the nature of black hole singularities is a major research goal. Numerous other frontiers exist in this realm. For example, by directly observing gravitational radiation, we will have a new tool for exploring violent activity in the universe. We still need to understand the details of black hole formation, both from stars and by other means. We have yet to understand in detail why mid-mass and high-mass black holes have the masses they have.

FIGURE 14-24 Mount Everest Primordial black holes, formed at the beginning of time, may have had masses similar to that of Mount Everest, as shown here. Calculations indicate that the universe is old enough for such black holes to have evaporated. Their final particle production rate is so high that they should look as though they are exploding. (Galen Rowell/Corbis)

If black holes merge, as is now expected, the details of these events and their effects on the rest of the universe have yet to be understood. It also remains to be shown whether or not the cosmic censorship theorem is correct. If not, the implications of the existence of wormholes need to be examined in more depth. Among the more intriguing issues surrounding black holes is whether primordial black holes exist. Although the evaporation process has not yet been seen, we should soon have the technology to discover it, if it is occurring.

Gamma-ray bursts still require much explanation. How does so much energy get transformed so quickly into electromagnetic radiation in these events? Why do some supernovae emit most of their energy as gamma rays, while others emit most as less energetic electromagnetic photons?

SUMMARY OF KEY IDEAS

- If a stellar corpse is more massive than about 3 $M_\odot$, gravitational compression overcomes neutron degeneracy and forces it to collapse further and become a black hole.
- A black hole is an object so dense that the escape velocity from it exceeds the speed of light.

The Relativity Theories

- Special relativity reveals that space and time are intimately connected and change with an observer's relative motion.

• As seen by observers moving more slowly, the faster an object moves, the slower time passes for it (time dilation) and the shorter it becomes (length contraction).

• According to general relativity, mass causes space to curve and time to slow down. These effects are significant only near large masses or compact objects.

Inside a Black Hole

• The event horizon of a black hole is a spherical boundary where the escape velocity equals the speed of light. No matter or electromagnetic radiation can escape from inside the event horizon. The distance from the center of the black hole to the event horizon is called the Schwarzschild radius.

• The matter inside a black hole collapses to a singularity. The singularity for nonrotating matter is a point at the center of the black hole. For rotating matter, the singularity is a ring inside the event horizon.

• Matter inside a black hole has only three physical properties: mass, angular momentum, and electric charge.

• Nonrotating black holes are called Schwarzschild black holes. Rotating black holes are called Kerr black holes. The event horizon of a Kerr black hole is surrounded by an ergoregion in which all matter must constantly move to avoid being pulled into the black hole.

• Matter that approaches a black hole's event horizon is stretched and torn by the extreme tidal forces generated by the black hole, light from the matter is redshifted, and time slows down.

• Black holes can evaporate by the Hawking process, in which virtual particles near the black hole become real. These transformations of virtual particles into real ones decrease the mass of a black hole until, eventually, it disappears.

Evidence of Black Holes

• Observations indicate that some binary star systems harbor black holes. In such systems, gases captured by the black hole from the companion star heat up and emit detectable X rays and jets of gas.

• Supermassive black holes exist in the centers of many galaxies. Intermediate-mass black holes appear to exist in globular clusters of stars. Very low mass (primordial) black holes may have formed at the beginning of the universe.

Gamma-Ray Bursts

• Gamma-ray bursts are events believed to be caused by some supernovae and by the collisions of dense objects, such as neutron stars or black holes. Some occur in the Milky Way and nearby galaxies, whereas many others occur billions of light-years away from Earth.

• Typical gamma-ray bursts occur for a few tens of seconds and emit more energy than the Sun will radiate over its entire 10-billion-year lifetime.

A WHAT DID YOU THINK?

1 *Are black holes empty holes in space? If not, what are they?* No. Black holes contain highly compressed matter—they are not empty.

2 *Does a black hole have a solid surface? If not, what is at its surface?* No. The surface of a black hole, called the *event horizon,* is empty space. No stationary matter exists there.

3 *What power or force enables black holes to draw things into them?* The only force that pulls things in is the gravitational attraction of the matter and energy in the black hole.

4 *How close to a black hole do you have to be for its special effects to be apparent?* About 100 times the Schwarzschild radius.

5 *Can you use black holes to travel to different places in the universe?* No. Most astronomers believe that the wormholes predicted by general relativity do not exist.

6 *Do black holes last forever? If not, what happens to them?* No. Black holes evaporate.

Key Terms for Review

accretion disk, 453
black hole, 448
cosmic censorship, 452
ergoregion, 451
event horizon, 449
gamma-ray burst, 457
gravitational radiation, 449
gravitational redshift, 447
gravitational wave, 449
Hawking process, 460
intermediate-mass black hole, 453
Kerr black hole, 450
primordial black hole, 453
Schwarzschild black hole, 450
Schwarzschild radius, 449
singularity, 448
spacetime, 445
supermassive black hole, 453
theory of general relativity, 446
theory of special relativity, 444
virtual particle, 459
wormhole, 452
Wolf-Rayet stars, 458

Review Questions

1. Which property, if any, of normal matter ceases to exist in a black hole? **a.** mass, **b.** chemical composition, **c.** angular momentum, **d.** electric charge, **e.** all of these properties exist in a black hole

2. Supermassive black holes are found in which of the following locations? **a.** in the centers of galaxies, **b.** in globular clusters, **c.** in open (or galactic) clusters, **d.** between galaxies, **e.** in orbit with a single star

3. Which feature is found with Kerr black holes but not Schwarzschild black holes? **a.** a singularity, **b.** an event horizon, **c.** gravitational redshift of photons outside of the black hole, **d.** an ergoregion, **e.** warping of nearby spacetime

4. Under what conditions do all outward pressures on a collapsing star fail to stop its inward motion?

5. In what way is a black hole blacker than black ink or a black piece of paper?

6. If the Sun suddenly became a black hole, how would Earth's orbit be affected?

7. What is cosmic censorship?

8. What are the differences between rotating and non-rotating black holes?

9. Why are all of the observed stellar-remnant black-hole candidates members of close binary systems?

10. If light cannot escape from a black hole, how can we detect X rays from such an object?

Advanced Questions

The answers to all computational problems, which are preceded by an asterisk (*), appear at the end of the book.

***11.** What is the Schwarzschild radius of a black hole, measured in kilometers, containing 3 $M_{\odot}$? 30 $M_{\odot}$?

12. If more massive stars evolve and die before less massive ones, why do some black-hole candidates have lower masses than their stellar companions?

13. Under what circumstances might a neutron star in a binary star system become a black hole?

14. Which type of black hole, nonrotating or rotating, do science fiction writers use (implicitly) in sending spaceships from one place to another through the hole? Why would the other type not be suitable?

***15.** You are standing in a train car at rest in the station, as in Figure 14-2. You measure the car's length to be 10 m long. The train speeds up to the considerable speed of 0.95c with you still inside the car. How long will you measure the car to be now? Explain how you got your answer.

What If ...

16. A black hole of 5 $M_{\odot}$ passed by Earth at, say, Neptune's distance from the Sun? What would happen to our planet's orbit and to life here? *Hint:* The Sun's gravitational force is about 180 times greater than the maximum effect that black hole would have on us. The Sun's gravitational force on Earth is about 17,000 times stronger than Jupiter's gravitational force on us, for comparison.

17. A primordial black hole with the mass of our Moon approached, passed through, and exited Earth? What might happen to our planet and to life here? *Hints:* You may want to calculate the black hole's Schwarzschild radius. Also, specify whether this is a "high speed" or "low speed" event.

18. A primordial black hole exploded nearby? What would astronomers observe?

Web Questions

19. To test your understanding of black hole structure, do Interactive Exercise 14.1 on the Web. You can print out your answers, if required.

20. Search the Web for information about the stellar-mass black-hole candidate named V4641 Sgr. In what ways does it resemble other black-hole candidates, such as Cygnus X-1 and V404 Cygni? In what ways is it different and more dramatic? How do astronomers currently explain why V4641 Sgr is different?

Observing Projects

21. Use the *Starry Night*™ program to investigate the X-ray source Cygnus X-1, a suspected black hole. This region of space is one of the brightest in our sky at X-ray wavelengths. Click the **Home** button in the toolbar and then use the **Find** pane to center the field of view on Cygnus X-1. If Cygnus X-1 is below the horizon, allow the program to reset the time to when it can best be seen. Click the checkbox to the left of the listing for Cygnus X-1 to apply a label to this object. Use the button bar or **Zoom** controls to set the field of view to 100 degrees. **a.** Use the time controls in the toolbar to determine when Cygnus X-1 rises and sets on today's date from your location. **b. Zoom in** until you can see an object at the location indicated by the label. What apparent magnitude and radius does *Starry Night*™ give for this object? (*Hint:* You can obtain this information from the HUD or by using the **Show Info** command from the contextual menu for this object.) Keeping in mind that the object that gives rise to this X-ray source is a black hole, to what must this apparent magnitude and radius refer? Explain.

PART IV

Understanding the Universe

R I V U X G

The Sombrero Galaxy (also designated M104, the 104th galaxy listed in the Messier catalog) is 29 million ly from us. By combining infrared, optical, and X-ray observations, we can gain insights into its disk of stars, gas, and dust, along with its central region, and the hot gas surrounding it. (X-ray: NASA/UMass/Q. D. Wang et al.; optical: NASA/STScI/AURA/Hubble Heritage; infrared: NASA/JPL-Caltech/Univ. AZ/R. Kennicutt/SINGS Team)

AN ASTRONOMER'S ALMANAC

964 Andromeda Galaxy discovered by Abd al-Rahman al-Sufi.

1517–1519 Ferdinand Magellan is the first European to record existence of Large and Small Magellanic Cloud galaxies.

1755 Milky Way proposed by Immanuel Kant to be only one among many galaxies.

1785 William Herschel calculates, incorrectly, the location of the solar system within the Milky Way.

Photography, Radio

1845 William Parsons (Lord Rosse) observes and sketches spiral-shaped nebulae later shown to be external galaxies.

1912–1915 Henrietta Leavitt discovers the period–luminosity relationship for Cepheid Type I stars. Einstein's general theory of relativity predicts a dynamic universe.

1914–1922 Redshift observed in spectra of galaxies taken by Vesto Slipher.

1917–1918 Harlow Shapley observes that the Milky Way galaxy is surrounded by a spherical distribution of globular clusters not centered on Earth; determines, essentially correctly, the location of the solar system in the Milky Way and the location of Galactic center.

1922 Alexander Friedmann discovers expanding universe solution to Einstein's equations of general relativity.

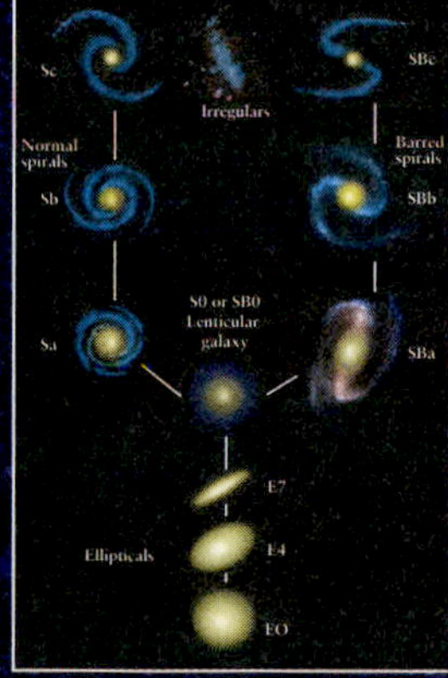

1924–1929 Edwin Hubble uses Cepheid variable stars to prove that nebulae are galaxies external to the Milky Way; proposes "tuning fork" classification scheme for galaxies based on their shapes; determines the relationship between the distance to superclusters of galaxies and their speed away from the Milky Way.

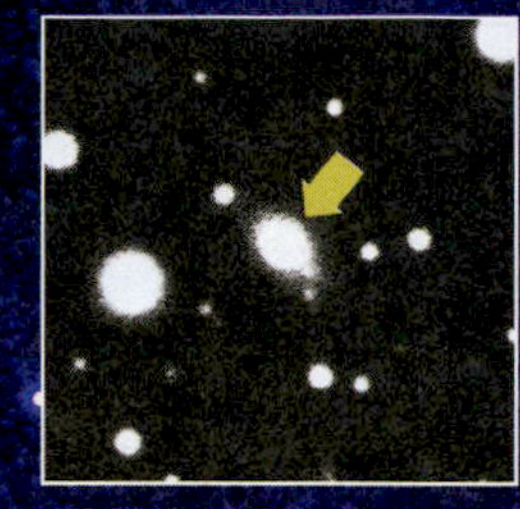

1927–1929 Bertil Lindblad calculates the solar system's motion through the Milky Way. BL Lac discovered.

Television, Radar

1930–1936 Robert Trumpler establishes presence of interstellar dust. Karl Jansky detects intense radio waves from direction of Galactic center. Grote Reber builds first radio telescope. Georges Lemaître proposes that Big Bang occurred.

1943 First Seyfert galaxy detected by Carl Seyfert.

1944 Hendrik van de Hulst discovers that neutral hydrogen emits 21-cm radiation.

Transistors

1948 Ralph Alpher, Hans Bethe, and George Gamow predict the cosmic microwave background radiation.

THE UNIVERSE

1951 Walter Baade and Rudolph Minkowski discover optical counterpart to Cygnus A. Orion and Perseus arms of Milky Way discovered by William Morgan.

1952 Simple experiment by Stanley Miller and Harold Urey yields first organic molecules created in a laboratory.

1959 Third Cambridge radio catalog published.

Lasers, Sputnik

1960–1961 Frank Drake conducts an early search for extraterrestrial intelligence using a radio telescope; derives an equation to calculate the likely number of extraterrestrial civilizations.

1965–1968 Arno Penzias and Robert Wilson discover the cosmic microwave background radiation. Donald Lynden-Bell proposes black holes as engines for quasars and other active galaxies.

1973–1978 Jerry Ostriker and James Peebles discover that the visible mass of a galaxy is not sufficient to keep it together. Brent Tully and Richard Fisher discover a relationship between galactic luminosity and stellar velocities. Vera Rubin and Kent Ford show that rotation curves of spiral galaxies are flatter than predicted by Newtonian gravitation.

Microprocessor, CCD

1980–1981 Alan Guth proposes inflationary phase of the evolution of the universe. Robert Kirshner, August Oemler, Paul Schecter, and Stephen Shectman detect void in the distribution of galaxies in the constellation Boötes.

1989 Black holes discovered at centers of Andromeda and its companion galaxy, M32.

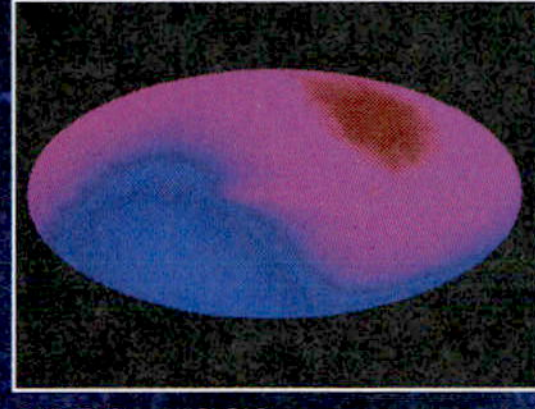

1990–1992 COBE spacecraft confirms blackbody nature of the cosmic microwave background and discovers anisotropy in the cosmic background radiation.

Internet

1998 Type Ia supernova observations reveal that the universe is accelerating outward.

2003 Hubble constant, H_0, determined to be 71 ±0.4 km/s/Mpc.

2005 Spitzer Space Telescope confirms Milky Way is a barred spiral galaxy.

WHAT DO YOU THINK?

1. What is the shape of the Milky Way Galaxy?
2. Where is our solar system located in the Milky Way Galaxy?
3. Is the Sun moving through the Milky Way Galaxy and, if so, about how fast?

Answers to these questions appear in the text beside the corresponding numbers in the margins and at the end of the chapter.

R I V U X G The center of the Milky Way Galaxy, as seen by the Chandra X-ray Observatory. The false colors reveal extremely hot gases that surround black holes, neutron stars, and white dwarfs. The image covers an area of space approximately 300 by 800 ly across. The supermassive black hole at the center of our Galaxy is enshrouded by the hot, white gas cloud in the center of the image. (NASA/UMass/D. Wang et al.).

Chapter 15

The Milky Way Galaxy

By the beginning of the twentieth century, astronomers knew that we live in a forest of stars and interstellar matter. Just as nearby trees and leaves hide what lies behind them, the stars and interstellar clouds we can see block light from more distant objects. A discussion ensued back then as to whether there is only one forest, culminating in the inconclusive Shapley–Curtis debate on whether or not the Milky Way is the entire universe. Observations presented in 1924 finally enabled astronomers to step outside our forest and see clearly that there are others, separated from ours by vast regions of relatively empty space. Our forest is called the **Milky Way Galaxy.** Observations have since revealed that the other forests, also called *galaxies,* number in the billions. By the end of the twentieth century, astronomers knew that the galaxies that fill the universe are clustered together, rather than randomly dotting the cosmic landscape. They had also discovered that these clusters of galaxies are themselves gathered together in "superclusters."

The more stars, nebulae, galaxies, and other features that astronomers observed in the cosmos, the more evident it became that something else dwelled in each galaxy that they could not see. Wraithlike, this "dark matter" interacted with everything else without showing itself. As the twenty-first century dawned, astronomers started getting glimpses of these wraiths, revealing that there are several kinds of them. Nevertheless, other than that it interacts gravitationally with matter that we can observe, we still do not know the nature of most of this dark matter.

In this part of the book, we explore the galaxies, clusters of galaxies, superclusters of galaxies, and the nature of the space that connects them, along with the evolution of the universe. We begin by studying our Milky Way. For those fortunate enough to live away from bright outdoor lights and air pollution, the Milky Way appears as a veiled band overlaid with the glow of individual stars. Centuries of observations have firmly established that our solar system is part of an enormous assemblage of hundreds of billions of stars, along with gas, dust, and other matter, all held together by their mutual gravitational attraction. Most of the stars in our Galaxy are located in a disk that looks from its edge like a flying saucer in an old science fiction movie (Figure 15-1a). The inner part of the disk is filled with stars, some of which form a bar (Figure 15-1b). Spiral arms swirl out from the ends of the bar, and, within these arms, new stars form from the debris of earlier generations of stars. The Galaxy's remaining stars are located in a two-shell, spherical halo that surrounds the disk. (Note: The word "Galaxy" when used alone, as here, is capitalized only when it refers to our Milky Way.)

In the following two chapters, we will examine the rest of the galaxies and the supermassive black holes that lie at the hearts of most of them. Then, like biologists who study the origins and evolution of life, we consider what astronomers know about the formation, evolution, and fate of the universe. The book ends with a discussion of astrobiology—the interrelationship between life and astronomy.

a

b

FIGURE 15-1 Schematic Diagrams of the Milky Way (a) This edge-on view shows the Milky Way's disk, containing most of its stars, gas, and dust, and its halo, containing many old stars. Individual stars in the halo are too dim to be visible on this scale, so the bright regions in the halo represent clusters of stars. (b) Our Galaxy has two major arms and several shorter arm segments, all spiraling out from the ends of a bar of stars and gas that passes through the Galaxy's center. The bar's existence and the presence of two major arms were confirmed by the Spitzer Space Telescope. (b: NASA/JPL-Caltech/R. Hurt [SSC])

In this chapter you will discover

- the Milky Way Galaxy—billions of stars along with gas and dust bound together by mutual gravitational attraction
- the structure of our Milky Way Galaxy
- Earth's location in the Milky Way
- how interstellar gas and dust enable star formation to continue
- that observations reveal the presence of significant mass in the Milky Way that astronomers have yet to identify
- that there is a massive black hole at the center of our Galaxy

DEFINING THE MILKY WAY

Prior to the twentieth century, astronomers did not know the large-scale distribution of stars and other matter in the universe. Throughout history, most people, including many astronomers, believed that the Milky Way contains all of the stars in the cosmos. In other words, they thought that the "Galaxy" and the "universe" were the same thing. We begin the study of galaxies by learning how that belief changed.

15-1 Studies of Cepheid variable stars revealed that the Milky Way is only one of many galaxies

The concept that our Galaxy is but one of many was put forth in 1755, when the German philosopher Immanuel Kant suggested that vast collections of stars lie far beyond the confines of the Milky Way. Less than a century later, the Irish astronomer William Parsons observed the structure of some of those "island universes" proposed by Kant. Parsons was the third earl of Rosse in Ireland. He was rich, he liked machines, and he was fascinated by astronomy. Accordingly, he set about building gigantic telescopes. In February 1845, his pièce de résistance was finished. This telescope's massive mirror measured 1.8 m (6 ft) in diameter and was mounted at one end of an 18-m (60-ft) tube controlled by cables, straps, pulleys, and cranes (Figure 15-2a). For many years, this triumph of nineteenth-century engineering enjoyed the distinction of being the largest telescope in the world.

a

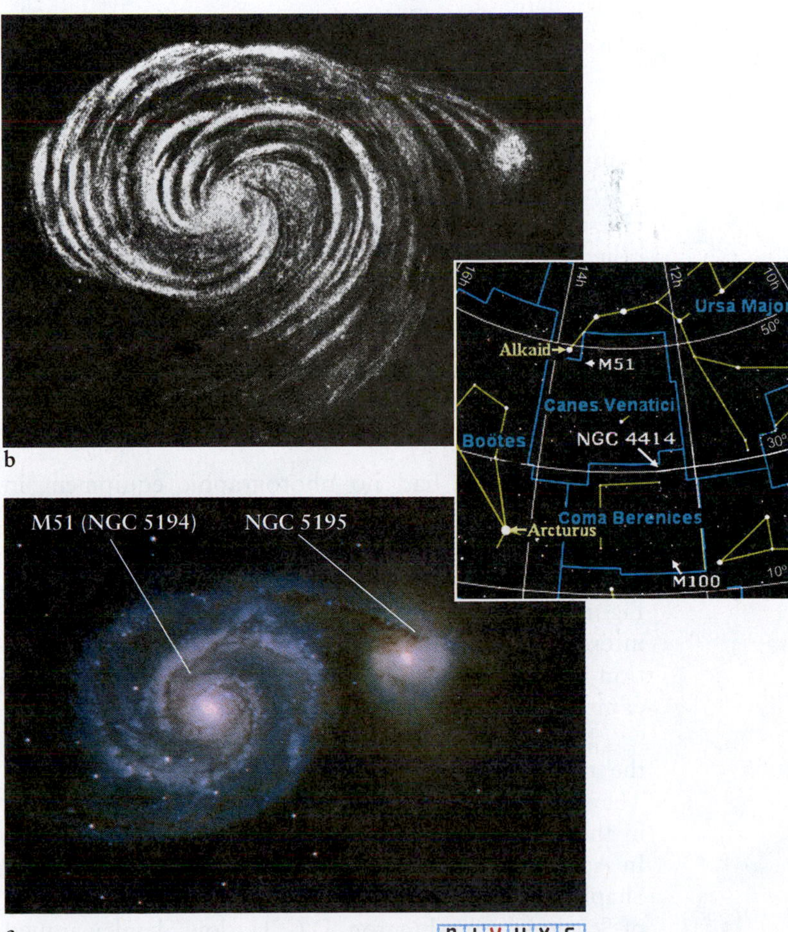

b

c

FIGURE 15-2 High-Tech Telescope of the Mid-Nineteenth Century (a) Built in 1845, this structure housed a 1.8-m-diameter telescope, the largest of its day. The improved resolution it provided over other telescopes was similar to the improvement that the Hubble Space Telescope provided over Earthbound optical instruments when it was launched. The telescope, as shown here, was restored to its original state in 1996–1998. (b) Using his telescope, Lord Rosse made this sketch of the spiral structure of the galaxy M51 and its companion galaxy NGC 5195. (c) A modern photograph of M51 (also called NGC 5194) and NGC 5195. The spiral galaxy M51 in the constellation of Canes Venatici is known as the Whirlpool Galaxy because of its distinctive appearance. The two galaxies are about 20 million ly from Earth. (a: Birr Castle Demesne; b: Lund Humphries; c: NOAO)

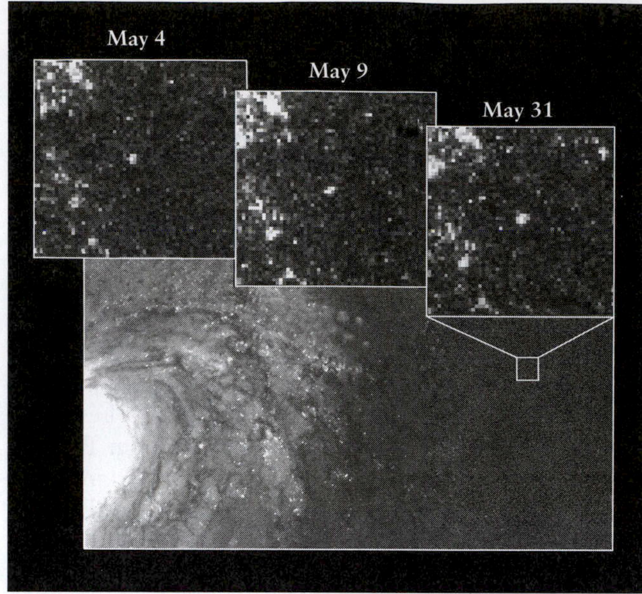

FIGURE 15-3 A Cepheid Variable Star in Galaxy M100 One of the most reliable ways to determine the distance to moderately remote galaxies is to locate Cepheid variable stars in them, as discussed in the text. The distance of 50 million ly (15.2 Mpc) from Earth to the galaxy M100 in the constellation Coma Berenices was determined using Cepheids. Insets: The Cepheid in this view, one of 20 located to date in M100, is shown at different stages in its brightness cycle, which recurs over several weeks. (Dr. Wendy L. Freedman, Observatories of the Carnegie Institution of Washington; NASA)

With this new telescope, Lord Rosse examined many of the glowing interstellar clouds previously discovered and catalogued by the Herschels. William Herschel, his sister Caroline, and his son John, among others, discovered and recorded details of many fuzzy-looking astronomical objects, called **nebulae** (singular, **nebula**). With the high resolution provided by his telescope, Lord Rosse observed that some of these nebulae have a distinct spiral structure. A particularly good example is M51, also called the *Whirlpool Galaxy* or NGC 5194.

Lord Rosse had no photographic equipment in 1845, so he made drawings of what he observed. Figure 15-2b is his drawing of M51. Views like this inspired him to echo Kant's proposal of island universes. Figure 15-2c shows a modern photograph of M51. It is interesting to note the differences between the perception and interpretation of astronomical objects and a camera's recording of them.

Most astronomers of Rosse's day did not agree with the notion of island universes outside of our Galaxy. They thought that the Milky Way contained all stars in the universe—that the Milky Way *was* the universe. In April 1920, a formal discussion, now known as the **Shapley–Curtis debate,** was held at the National Academy of Sciences in Washington, D.C. Harlow Shapley argued that the spiral nebulae are relatively small, nearby objects scattered around our Galaxy. Heber D. Curtis championed the island universe theory, arguing that each of these spiral nebulae is a separate rotating system of stars, much like our own Galaxy. While the Shapley–Curtis debate focused scientific attention on the size of the universe, nothing was decided, because no one had any firm evidence to demonstrate exactly how far away the spiral nebulae were. Astronomers desperately needed to devise a way to measure the distances to the nebulae. A young teacher and former basketball player originally from Missouri, who moved to Chicago to study astronomy, finally achieved this goal. His name was Edwin Hubble.

In 1923, Hubble took an historic photograph of M31, then called the Andromeda Nebula. It was one of the spiral nebulae around which controversy raged. On the photographic plate he discovered what first appeared to be a nova. Referring to previous plates of that region, he soon realized that the object was actually a Cepheid variable star. As we saw in Section 12-12, these pulsating stars vary in brightness periodically. Further scrutiny over the next several months revealed many other Cepheids. Figure 15-3 shows a Cepheid in the galaxy M100, at different stages of brightness.

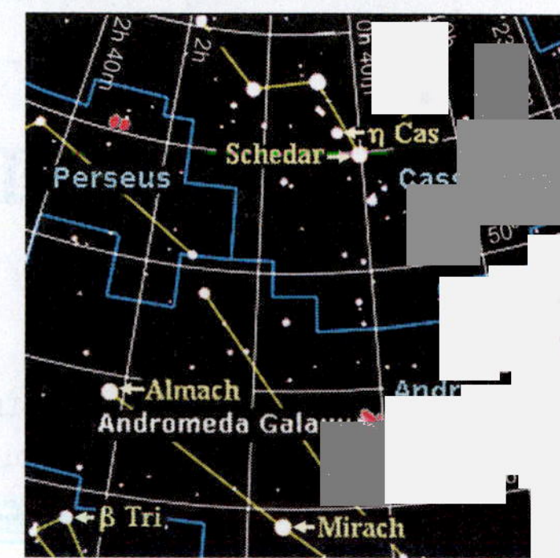

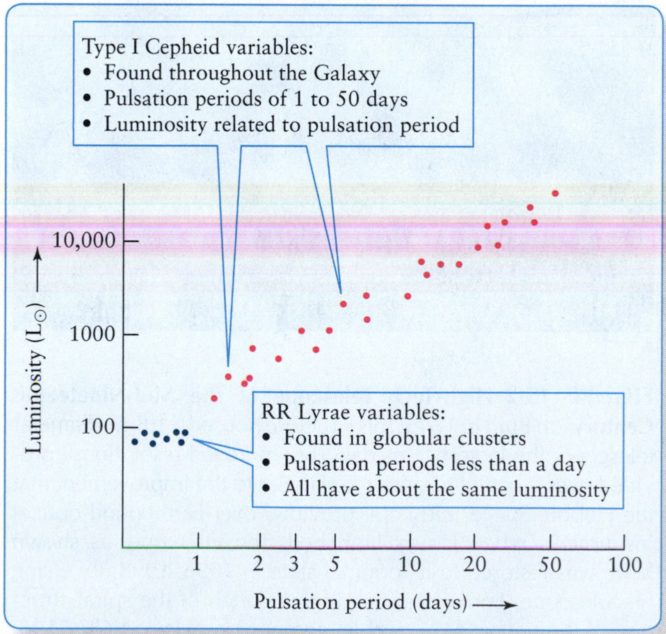

FIGURE 15-4 The Period-Luminosity Relation This graph shows the relationship between the periods and average luminosities of classical (Type I) Cepheid variables and the closely related RR Lyrae stars (discussed in Chapter 12). Each dot represents a Cepheid or RR Lyrae whose luminosity and period have been measured.

Only a decade before, in 1912, the American astronomer Henrietta Leavitt had published an important study of Cepheid variables. Leavitt studied many Cepheids in the Small Magellanic Cloud, then also believed to be a nebula, but now known to be a galaxy passing very close to the Milky Way. Leavitt's study led her to the period-luminosity relation for Type I Cepheids (see Section 12-13). Leavitt established that a direct relationship exists between a Cepheid's luminosity (or absolute magnitude) and its period of oscillation, which we saw in Figure 12-28 and which is presented in more detail in Figure 15-4. By observing the star's period and apparent magnitude, its distance can be calculated, as described in *An Astronomer's Toolbox 15-1*.

An Astronomer's Toolbox 15-1

Cepheids and Supernovae as Indicators of Distance

Because their periods are directly linked to their luminosities, Cepheid variables are among the most reliable tools that astronomers have for determining the distances to relatively nearby galaxies. To this day, astronomers use this link—much as Hubble did back in the 1920s—to measure intergalactic distances. More recently, they have begun to use Type Ia supernovae, which are far more luminous and thus can be seen much farther away, to determine the distances to more remote galaxies.

Example: In 1992, a team of astronomers used Cepheid variables in a galaxy called IC 4182 to deduce that galaxy's distance from Earth. They used the Hubble Space Telescope on 20 separate occasions to record images of the stars in IC 4182. By comparing these images, the astronomers could pick out which stars vary in brightness. In this way, they discovered 27 Cepheids in IC 4182. Using their observations, they determined the (changing) apparent magnitudes of the Cepheid variables and plotted their light curves. The Hubble Space Telescope is particularly well-suited for studies of this kind, because its extraordinary angular resolution makes it possible to pick out individual stars at great distances. One such Cepheid has a period of 42.0 days and an average apparent magnitude (m) of +22.0. (See Section 11-2 for an explanation of the apparent magnitude scale.) By comparison, the dimmest star you can see with the naked eye has $m = +6$; this Cepheid in IC 4182 appears less than one-millionth as bright. The star's spectrum shows that it is a metal-rich Type I Cepheid variable.

According to the period-luminosity relation shown in Figure 15-4, such a Type I Cepheid with a period of 42.0 days has an average luminosity of 33,000 $L_\odot$. In other words, this Cepheid has an average absolute magnitude (M) of −6.5 (compared to M = +4.8 for the Sun). Hence, the difference between the Cepheid's apparent and absolute magnitudes, called its **distance modulus,** is

$$m - \mathrm{M} = (+22.0) - (-6.5) = 22.0 + 6.5 = 28.5$$

From *An Astronomer's Toolbox 11-3*, we see that the distance modulus of a star is related to its distance in parsecs (d) by

$$m - \mathrm{M} = 5 \log d - 5$$

where "log" is the base 10 logarithm. This equation can be rewritten as

$$d = 10^{(m - \mathrm{M} + 5)/5} \text{ parsecs}$$

Inserting the value for the distance modulus in this equation, we can obtain the distance to the Cepheid variable and, hence, the distance to the galaxy of which that star is part:

$$\begin{aligned} d &= 10^{(28.5 + 5)/5} \text{ parsecs} \\ &= 10^{6.7} \text{ parsecs} = 5.0 \times 10^6 \text{ parsecs} \end{aligned}$$

The galaxy is 5 Mpc (1 Mpc = 1 megaparsec = 10^6 parsecs), or 16 million (1.6×10^7) ly, from Earth.

Astronomers are interested in IC 4182 because a Type Ia supernova was observed there in 1937. Type Ia supernovae are exploding white dwarfs that all reach the same maximum brightness at the peak of their outbursts (see Section 13-4). Once astronomers knew the peak absolute magnitudes of Type Ia supernovae, they could use these supernovae as distance indicators. Because the distance to IC 4182 is known from its Cepheids, the 1937 observations of the supernova in that galaxy allow us to calibrate Type Ia supernovae as distance indicators. At maximum brightness, the 1937 supernova reached an apparent magnitude of $m = +8.6$. Because the distance modulus of the galaxy (m – M) is 28.5, we see that when a Type Ia supernova is at maximum brightness, its absolute magnitude is

$$\mathrm{M} = m - (m - \mathrm{M}) = 8.6 - 28.5 = -19.9$$

Whenever astronomers find a Type Ia supernova in a remote galaxy, they can combine this absolute magnitude with the observed maximum apparent magnitude to get the galaxy's distance modulus, from which the galaxy's distance can be easily calculated (just as we did above for the Cepheids in IC 4182). This technique has been used to determine the distances to galaxies hundreds of millions of parsecs away.

Try these questions: At what distance in parsecs is a star with distance modulus 20? Epsilon (ε) Indi has an apparent magnitude of +4.7 and distance of 3.6 pc. What is its absolute magnitude? What would the Sun's apparent magnitude and distance modulus be as seen from 100 pc away? Its absolute magnitude is +4.83.

(Answers appear at the end of the book.)

Referring to Figure 15-4, how would the brightness of a Cepheid variable with peak luminosity of 1000 $L_\odot$ change if it were observed every 5 days?

Thanks to this work, Hubble knew that he could use the characteristics of the fluctuating light of Cepheid variable stars to help him calculate the distance to M31 and put the Shapley–Curtis debate to rest once and for all. His observations of Type I Cepheids led him to determine that M31 is some 2.2 million ly *beyond* the Milky Way. This distance proves that M31 is not an open or globular cluster in our Galaxy, but rather an enormous separate stellar system—a separate galaxy. M31, now called the *Andromeda Galaxy,* is the most distant object in the universe that can be seen with the naked eye. Similar calculations have been done for all galaxies in which Cepheids can be observed. The distances to even more remote galaxies have been determined from observations of Type Ia supernovae in them (see *An Astronomer's Toolbox 15-1*).

Insight Into Science

Room for Debate Lacking definitive data, competent scientists can develop and believe strikingly different explanations for the same observations. At the time of the Shapley–Curtis debate, the observations allowed both points of view. It was only with the advent of the distance measurement technique used by Hubble that the spiral nebulae were definitely shown to lie outside of the Galaxy.

Hubble's results, which he presented at the end of 1924, did settle the Shapley–Curtis debate. The universe was recognized to be far larger and populated with far bigger objects than most astronomers had imagined. Hubble had discovered the realm of the galaxies. Today, we know that the universe contains myriad galaxies, of which the Milky Way is just one. Like the Milky Way, each **galaxy** is a grouping of millions, billions, or even trillions of stars, along with gas, dust, and matter in other forms, all gravitationally bound together.

We now apply Hubble's method to find Earth's place in the Milky Way Galaxy. Recall that the Sun's proximity to us makes it our best-understood star. It might seem that the nearness of the stars and clouds in the Milky Way would make it the best-understood galaxy. However, the clouds of gas and dust that surround the solar system make it very challenging for astronomers to survey completely the distant parts of the Galaxy, which are only now coming into focus.

THE STRUCTURE OF OUR GALAXY

Because the band of the Milky Way completely encircles us, astronomers long ago suspected that the Sun and all of the stars that we see are part of it. In the 1780s, William Herschel took the first steps toward mapping our Galaxy's structure. He attempted to deduce the Sun's location in the Galaxy by counting the number of stars in 683 regions of the sky. He reasoned that the greatest density of stars should be seen toward the Galaxy's center and a lesser density seen toward the edge. However, Herschel found roughly the same density of stars all along the Milky Way. He therefore concluded that we are at the center of the Galaxy.

Herschel was wrong: Earth has no privileged place in the Milky Way. The Sun is about 26,000 ly (8000 pc) from the Galaxy's center, the **galactic nucleus.** Herschel's physical understanding of the cosmos was incomplete, so he misinterpreted his observations and thus came to an incorrect conclusion.

15-2 Cepheid variables help us locate our Galaxy's center

While studying star clusters in the 1930s, R. J. Trumpler discovered the reason for Herschel's mistake. Herschel did not know about interstellar gas and dust, which affected his counts of the stars. Trumpler noticed that remote star clusters appear dimmer than would be expected just from their distance alone. Something must be blocking starlight on its way toward Earth. He correctly concluded that interstellar space is not a perfect vacuum. Instead, it contains dust that absorbs light from distant stars. Great patches of this dust are clearly visible in wide-angle photographs (Figure 15-5). Like the stars, this dust is concentrated in the plane of the Galaxy.

R I V U X G

FIGURE 15-5 Our Galaxy This wide-angle photograph spans half the Milky Way. The center of the Galaxy is in the constellation Sagittarius, in the middle of this photograph. The dark lines and blotches are caused by hundreds of interstellar clouds of gas and dust that obscure the light from background stars, rather than by a lack of stars. (Dirk Hoppe)

This interstellar dust almost completely obscures from view visible light emanating from the center of our Galaxy. Visual photons from there are mostly absorbed or scattered before they reach us. Therefore, Herschel was seeing only nearby stars, and he measured apparent magnitudes that were dimmer than they would have been had there been no interstellar dust. Without adjusting for its effects, he concluded that the stars were farther away than they really are. He also had no idea of the true size of the Galaxy and could not see the vast number of stars located in the general direction of the galactic center that are hidden by the dust.

Insight Into Science

A Little Knowledge Incomplete information often leads to incorrect interpretation of data and, therefore, to incorrect conclusions. Herschel's lack of knowledge about the matter in interstellar space prevented him from correctly interpreting the distribution of stars that surround Earth and, thus, led to his inaccurate conclusion about the position of the Sun within the Galaxy.

Because interstellar dust is concentrated in the plane of the Galaxy, the absorption of starlight is strongest in those parts of the sky covered by the Milky Way. Above or below the plane of the Galaxy, our view is relatively unobscured. Knowledge of our true position in the Galaxy eventually came from observations of globular clusters (see Section 12-14). Shapley used the period-luminosity relation for variable stars to determine the distances to the then-known 93 globular clusters in the sky. (More than 150 are known today.) From their directions and distances, he mapped out the distribution of these clusters in three-dimensional space. By 1917, Shapley had discovered that the globular clusters are located in a spherical distribution centered not on Earth but on a point in the Milky Way toward the constellation Sagittarius. Figure 15-6 shows two globular clusters in a relatively clear part of the sky in that direction. Shapley then made a bold conjecture: *The globular clusters orbit the center of the Milky Way, which is located in Sagittarius.* His pioneering research has since been observationally verified. Earth is not at the center of the Galaxy.

15-3 Nonvisible observations help map the galactic disk

To see into the dust-filled plane of the Milky Way, astronomers use radio wave, infrared, X-ray, and gamma-ray telescopes (see Figure 3-37). These wavelengths are scattered much less by the interstellar gas and dust located throughout the Galaxy's disk than are visible or ultraviolet wavelengths. Observations of the distant parts of the Galaxy were first made using radio telescopes. Radio waves penetrate Earth's atmosphere, so we can observe them anywhere that we can build a radio telescope. (Recall that infrared observations must be made at high altitudes or in space, and that X-ray and gamma-ray observations are almost always made from space.)

Detecting the radio emission directly from interstellar hydrogen—by far the most abundant element in the universe—is a primary means of mapping the Galaxy. Unfortunately, the major transitions of electrons in the hydrogen atom (see Figure 4-11) produce photons at ultraviolet and visible wavelengths that do not penetrate the interstellar medium. How, then, can radio telescopes directly detect all of this hydrogen? The answer lies in atomic physics.

In addition to mass and charge, particles such as protons and electrons possess a tiny amount of angular momentum, commonly called **spin.** According to the laws of quantum mechanics, the electron and proton in a hydrogen atom can spin only in either parallel or opposite directions (Figure 15-7); they can have no other spin orientations. If the electron in a hydrogen atom flips from one orientation to the other, the atom must gain or lose a tiny amount of energy. In particular,

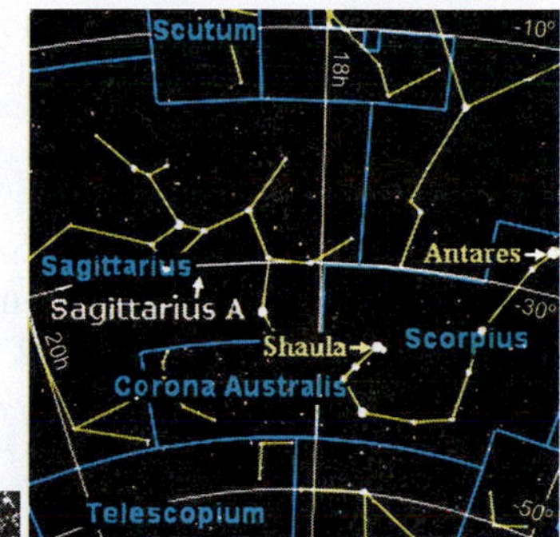

FIGURE 15-6 A View Toward the Galactic Center More than a million stars in the disk of our Galaxy fill this view, which covers a relatively clear window just 4° south of the galactic nucleus in Sagittarius. Beyond the disk stars you can see two prominent globular clusters. Although most regions of the sky toward Sagittarius are thick with dust, very little obscuring matter appears in this tiny section of the sky. (Harvard Observatory)

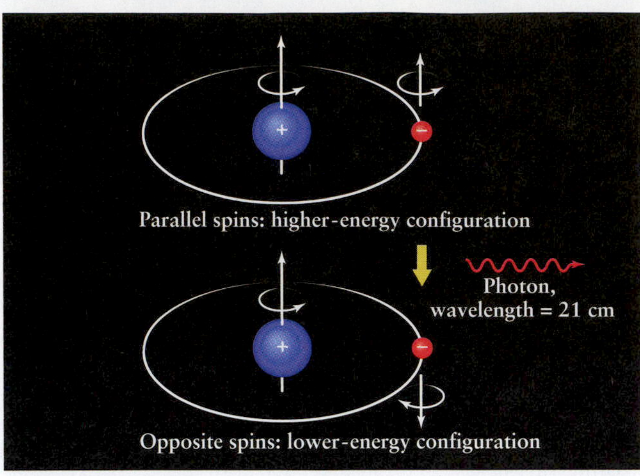

FIGURE 15-7 Electron Spin and the Hydrogen Atom Due to their spins, electrons and protons both act as tiny magnets. When an electron and the proton it orbits are spinning in the same direction, their energy is higher than when they are spinning in opposite directions. When the electron flips from the higher-energy to the lower-energy configuration, the atom loses a tiny amount of energy that is radiated as a radio photon with a wavelength of 21 cm.

when flipping from parallel to opposite spins, the atom simultaneously emits a low-energy radio photon whose wavelength is 21 cm. This flip happens rarely in each atom, so it is only because the Galaxy has vast quantities of interstellar hydrogen gas that it can be detected at all. In 1951, a team of astronomers first succeeded in detecting the faint hiss of 21-cm radio static from spin flips.

The detection of **21-cm radio radiation** was a major breakthrough in mapping the **disk** of the Galaxy. To see why, suppose that you aim your radio telescope across the Galaxy, as sketched in Figure 15-8. Your radio receiver picks up 21-cm emission from hydrogen clouds at points 1, 2, 3, and 4. However, the radio waves from these various clouds are Doppler shifted (see *An Astronomer's Toolbox 4-3*) by slightly different amounts because they have different radial velocities (motion toward or away from Earth). Because these radio waves from gas clouds in different parts of the Galaxy arrive at our radio telescopes with slightly different wavelengths as a result of being Doppler shifted, it is possible to identify which radio signals come from which gas clouds and thus to produce an initial map of the Galaxy, such as that shown in Figure 15-9a.

Our radio map reveals numerous arched lanes of neutral hydrogen gas. If this were the overall structure of the Galaxy, then the Milky Way would appear unlike any other observed galaxy (see Chapter 16 for typical images of other galaxies). Indeed, the other disk-shaped galaxies we observe have spiral arms. Different observations were needed to improve our understanding of the Galaxy's disk features. Note that photographs of a barred spiral galaxy (Figure 15-10) show arms outlined by "spiral tracers"—bright, Population I stars and emission nebulae. As we saw in Chapter 12, these features indicate active star formation. If the Milky Way is spiral, then another useful way to further chart its structure and show that it has **spiral arms** is to map the locations of star-forming complexes filled with H II regions, giant molecular clouds, and massive, hot, young stars in groups, called *OB associations*.

What situation(s) on Earth are analogous to the obscuration caused by interstellar gas and dust?

1 Dust absorption limits the range of visual observations in the plane of the Galaxy to less than 24,000 ly from Earth. However, astronomers can use visible observations of nearby bright OB associations and associated H II regions to plot the spiral arms near the Sun (see Figure 15-9a inset). Radio observations of hydrogen and carbon monoxide molecules (discussed in Section 12-1) have been used to chart more remote star-forming regions of the Galaxy. Taken together, all of these observations indicate that our Galaxy has about 200 billion stars located in and between two major spiral arms and several minor arms (see Figure 15-9b). (Each arm is named after the constellation in which it is centered, as seen from Earth.) We will explore the origin of the spiral arms in Chapter 16, when we present more evidence about the cause of spiral structure from observations of other galaxies. Recent observations by the Spitzer Space

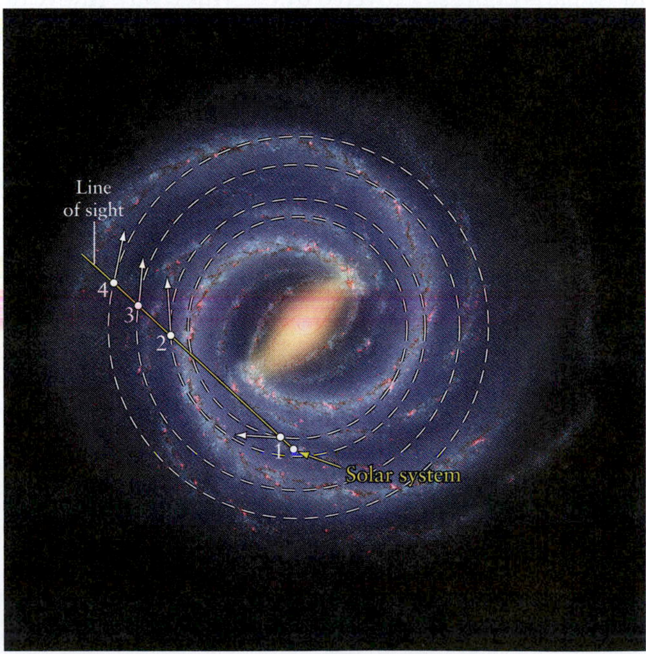

FIGURE 15-8 A Technique for Mapping the Galaxy Hydrogen clouds at different locations along our line of sight are moving around the center of the Galaxy at different speeds. The component of their motion away from us varies with their distance from the solar system. Radio waves from the various gas clouds, therefore, exhibit slightly different Doppler shifts, permitting astronomers to sort out the gas clouds and map the Galaxy. (NASA/JPL-Caltech/R. Hurt [SSC])

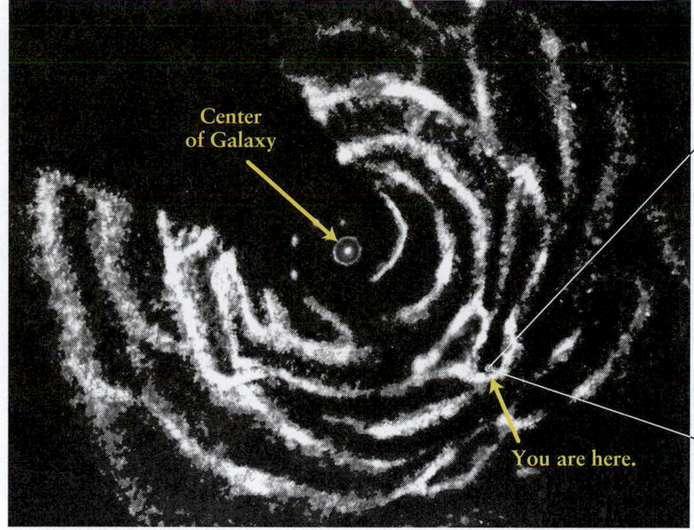

a

R I V U X G

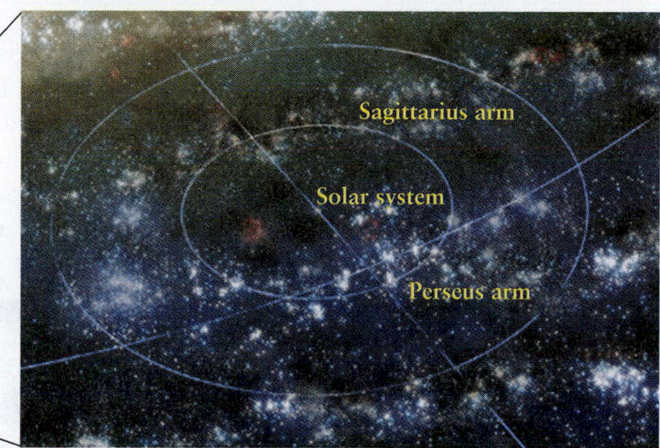

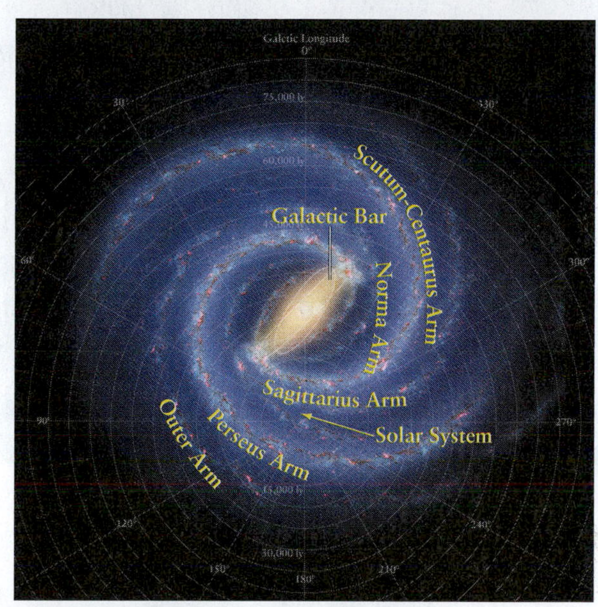

b

FIGURE 15-9 A Map of the Galaxy (a) This map, based on radio telescope surveys of 21-cm radiation, shows the distribution of hydrogen gas in a face-on view of the Galaxy. This view just hints at spiral structure. The galactic nucleus is marked with a dot surrounded by a circle. Details in the large, blank, wedge-shaped region toward the upper left of the map are unknown, because gas in this part of the sky is moving perpendicular to our line of sight and thus does not exhibit a detectable Doppler shift. Inset: This drawing, based on visible-light data, shows that our solar system lies between the two major arms of the Milky Way Galaxy. (b) This drawing labels the spiral arms in the Milky Way. (a: Courtesy of G. Westerhout; inset: National Geographic; b: NASA/JPL-Caltech/R. Hart [SSC])

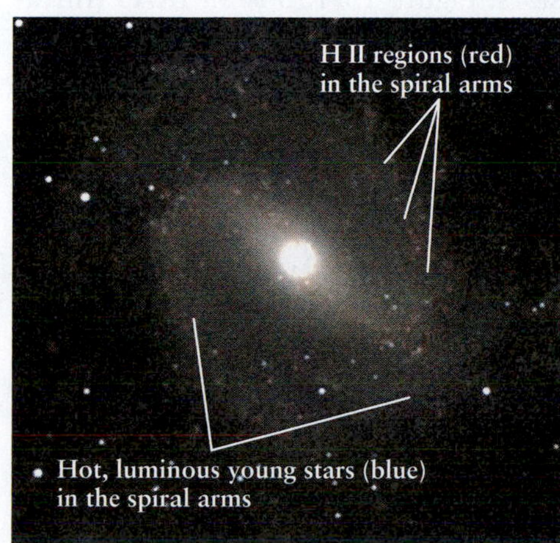

a Visible-light view of M83

R I V U X G

b 21-cm radio view of M83

R I V U X G

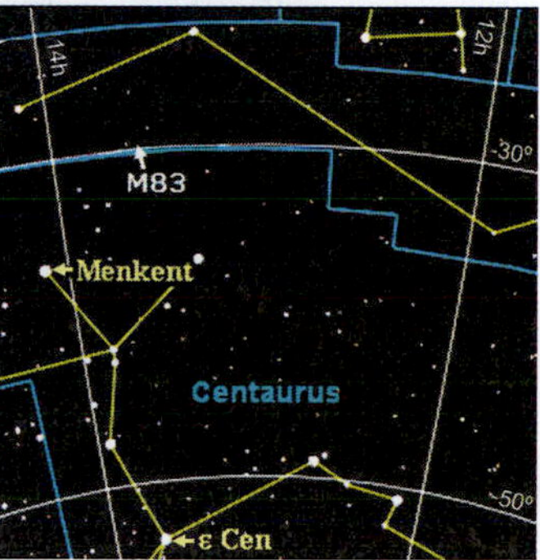

FIGURE 15-10 Two Views of a Barred Spiral Galaxy The galaxy M83 is in the southern constellation of Centaurus, about 12 million ly from Earth. (a) At visible wavelengths, spiral arms are clearly illuminated by young stars and glowing H II regions. (b) A radio view at 21-cm wavelength shows the emission from neutral hydrogen gas. Note that the spiral arms are more clearly demarcated by visible stars and H II regions than by 21-cm radio emission. (a: S. Van Dyk/IPAC; b: VLA, NRAO)

Telescope confirm previous observations that a bar of stars and gas crosses the center of the Galaxy, also shown in Figure 15-9b.

2 The observable disk of our Galaxy is about 100,000 ly in diameter and about 2000 ly thick (Figure 15-11a). Its two major spiral arms are designated the Perseus arm and the Scutum-Centaurus arm (see Figure 15-9b). There are also several less pronounced arms, including the Sagittarius arm, the Normal arm, the Orion spur, and the Outer arm. The solar system is located between the arms, and is closest to the Orion spur. On the side toward the galactic center is the Sagittarius arm, which stargazers in the northern hemisphere see in the summer when they look at the portion of the Milky Way stretching across Scorpius and Sagittarius (see Figure 15-5). Directed away from the galactic center is the Perseus arm, which is visible in the northern hemisphere in the winter.

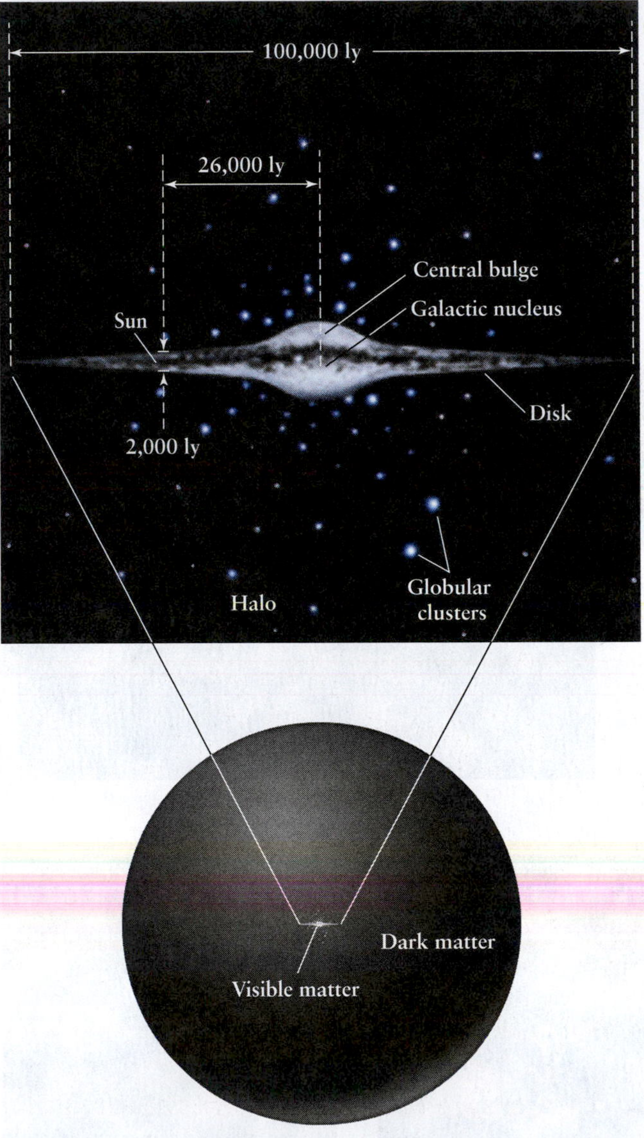

FIGURE 15-11 Our Galaxy As seen from the side, three major visible components of our Galaxy are a thin disk, a central bulge, and a two-part halo system. As noted earlier, there is also a central bar. The visible Galaxy's diameter is about 100,000 ly, and the Sun is about 26,000 ly from the galactic center. The disk contains gas and dust along with Population I (young, metal-rich) stars. The halo is composed almost exclusively of Population II (old, metal-poor) stars. Inset: The visible matter in our Galaxy fills only a small volume compared to the distribution of dark matter, whose composition is presently unknown. Dark matter's presence is felt by its gravitational effect on visible matter.

Observations reveal that the Galaxy's arms spiral out from a bar of stars, gas, and dust (Figure 15-9b) running through a flattened sphere of stars, called the **central bulge** (also called the *nuclear bulge*), that is about 12,000 ly in diameter. The central bulge is also seen in Figure 15-12, a wide-angle infrared image of the Galaxy taken by the COBE satellite. The central bulge is centered on the galactic nucleus 26,000 ly away from us. The stars, gas, and dust in the bar move one way down it and then back the other way.

15-4 The galactic nucleus is an active, crowded place

If you lived on a planet near the center of the Galaxy, which is called the galactic nucleus, you would see a million stars as bright as Betelgeuse appears from Earth today. The total intensity of starlight from all those nearby stars would be equivalent to 200 of our full Moons. Night would never really fall. Stranger still, the Galaxy around you would be filled with intense activity.

Figure 15-13 shows three infrared views that look toward the nucleus of the Galaxy. Figure 15-13a is a wide-angle view covering a 50° segment of the Milky Way through Sagittarius and Scorpius. The prominent band across this image is a thin layer of dust in the plane of the Galaxy. The numerous knots and blobs along the dust layer are interstellar clouds heated by young O and B stars. Figure 15-13b is an IRAS (Infrared Astronomical Satellite) view of the galactic center. Numerous streamers of dust (blue) surround it. The strongest infrared emission (white) comes from **Sagittarius A** (often abbreviated Sgr A), which is also a grouping of several powerful sources of radio waves. One of these sources, called Sagittarius A* (pronounced "A-star"), is believed to be the galactic nucleus. Figure 15-13c shows stars within 1 ly of Sagittarius A*, with resolution of 0.02 ly.

Radio observations add more to the picture of the galactic center. Some of the most detailed radio images of it come from the Very Large Array (VLA). Figure 15-14a is a wide-angle view of Sagittarius A and surrounding features, including arcs of gas, at least three supernovae remnants, and localized radio sources. Huge filaments, such as the one labeled "Arc," lie perpendicular to the plane of the Galaxy and stretch 200 ly northward of the galactic disk, then abruptly arch southward toward Sagittarius A. The orderly arrangement of these filaments suggests that a magnetic

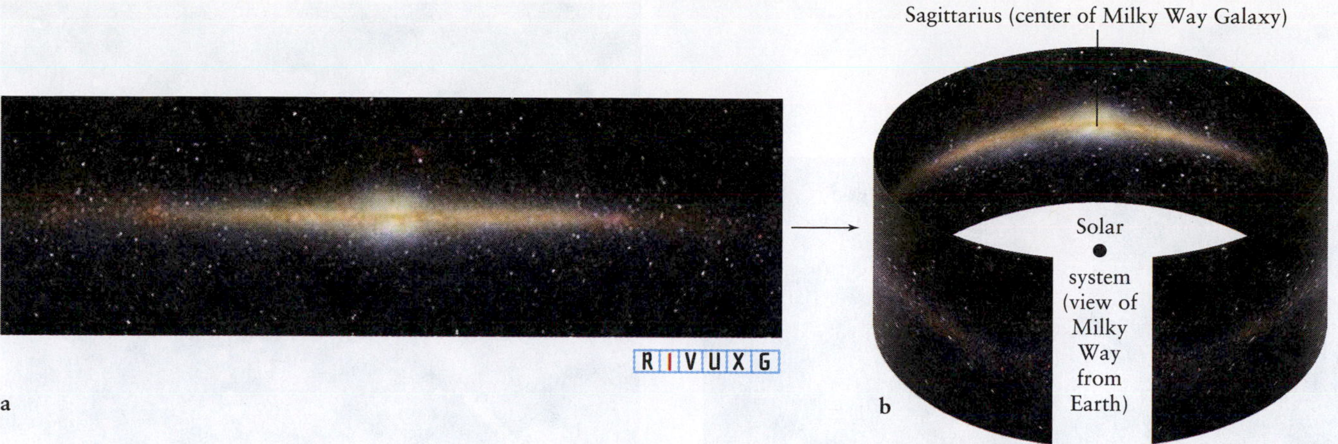

FIGURE 15-12 Infrared View of the Milky Way (a) Taken by the COBE satellite in 1997, this infrared image shows the disk and central bulge of our Galaxy, as they would be seen from outside the Galaxy. Most of the sources scattered above and below the disk are nearby stars. Stars appear white, whereas interstellar dust appears orange. Note that the dust that obscures light from more distant stars in Figure 15-5 is quite bright in this infrared image. (b) Because we are embedded in it, the Galaxy appears wrapped around us. (The COBE Project, DIRBE, NASA)

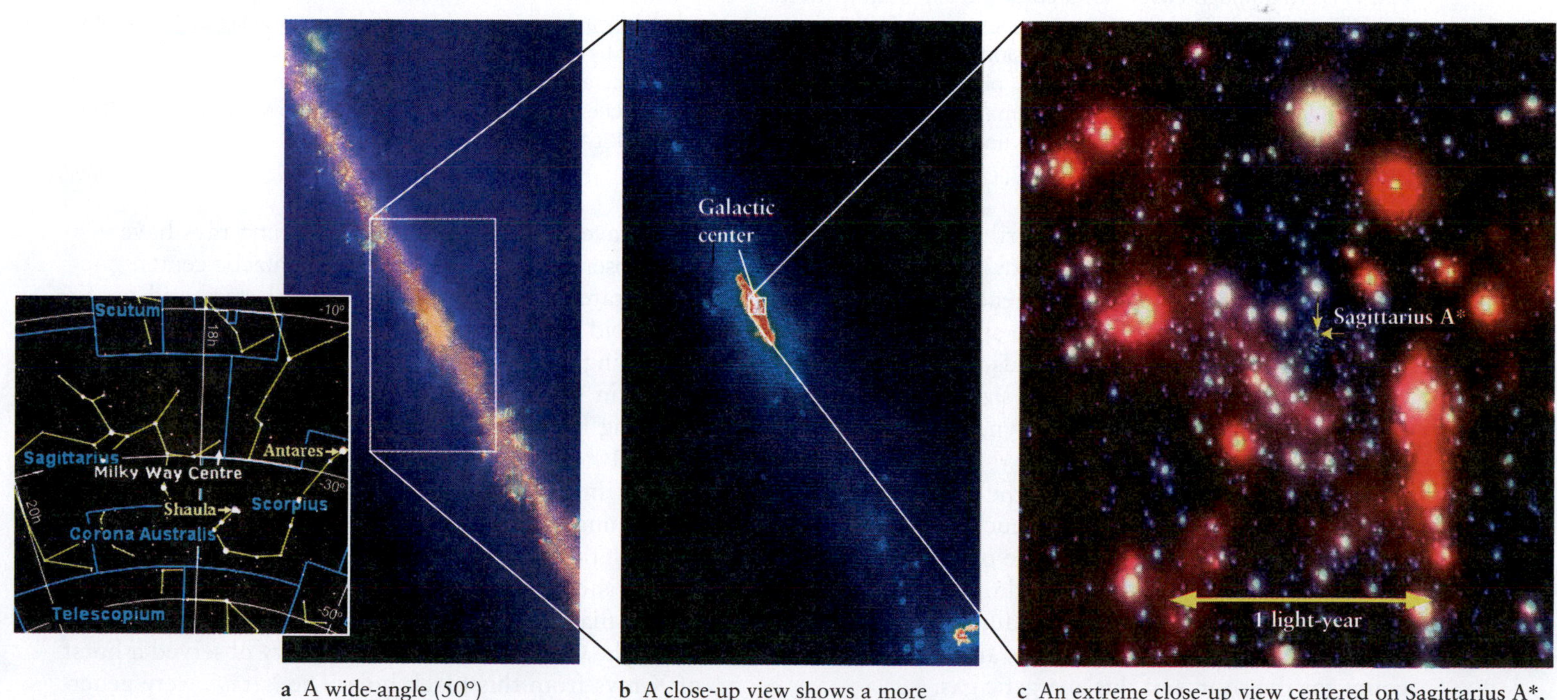

a A wide-angle (50°) infrared view.

b A close-up view shows a more luminous region at the galactic center.

c An extreme close-up view centered on Sagittarius A*, a radio source at the very center of the Milky Way Galaxy, shows hundreds of stars within 1 ly (0.3 pc).

FIGURE 15-13 The Galactic Center (a) This wide-angle view at infrared wavelengths shows a 50° segment of the Milky Way centered on the nucleus of the Galaxy. Black represents the dimmest regions of infrared emission, with blue the next strongest, followed by yellow and red; white represents the strongest emission. The prominent band diagonally across this photograph is a layer of dust in the plane of the Galaxy. Numerous knots and blobs along the plane of the Galaxy are interstellar clouds of gas and dust heated by nearby stars. (b) This close-up infrared view of the galactic center covers the area outlined by the white rectangle in (a). (c) This infrared image shows about 300 of the brightest stars less than 1 ly from Sagittarius A*, which is at the center of the picture. The distribution of stars and their observed motions around the galactic center imply a very high density (about a million solar masses per cubic light-year) of less luminous stars. (a, b: NASA; c: R. Schödel et al., MPE/ESO)

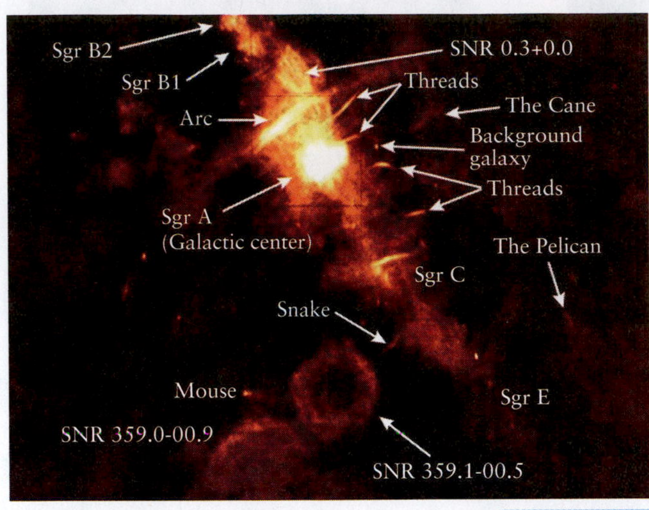

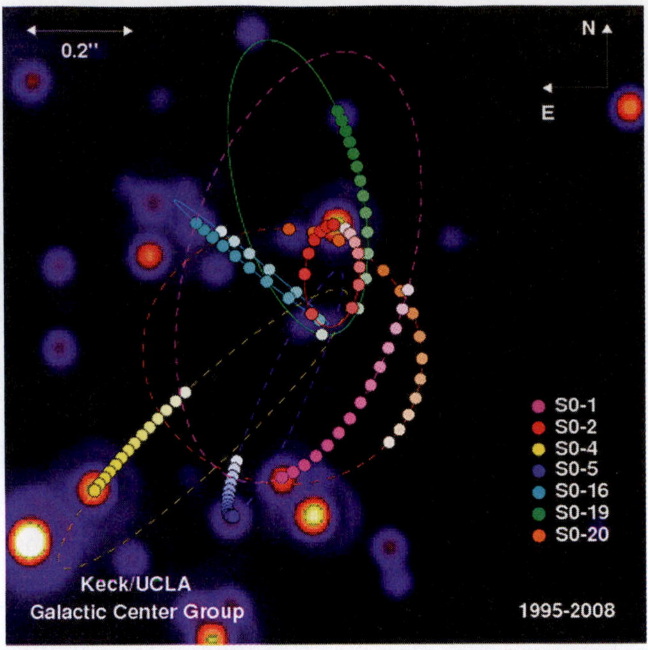

FIGURE 15-14 Two Views of the Galactic Nucleus (a) A radio image taken at the VLA of the galactic nucleus and environs. This image covers an area of the sky 8 times wider than the Moon. SNR means supernova remnant. The numbers following each SNR are its right ascension and declination. The Sgr (Sagittarius) features are radio-bright objects. (b) The colored dots superimposed on this infrared image show the motion of seven stars in the vicinity of the unseen massive object (denoted by the yellow star) at the position of the radio source Sagittarius A*, part of Sgr A in (a). The orbits were measured over a 14-year period. This plot indicates that the stars are held in orbit by a 4×10^6-$M_\odot$ black hole. (a: Naval Research Laboratory produced by N. E. Kassim, D. S. Briggs, T. J. W. Lazio, T. N. LaRosa, J. Imamura, and S. D. Hyman. Originally from the NRAO Very Large Array. Courtesy of A. Pedlar, K. Anantharamiah, M. Gross, and R. Ekers; b: Keck/UCLA Galactic Center Group)

field may be controlling the distribution and flow of ionized gas, just as magnetic fields on the Sun funnel such gas to create solar prominences. This resulting radio emission, produced by high-speed electrons that spiral around these magnetic fields, is called **synchrotron radiation.** Despite its small size, Sagittarius A is one of the brightest sources of synchrotron radiation in the entire sky.

X-ray observations taken by the Chandra telescope in 2004 reveal that the galactic nucleus is also bathed in ultrahot gas with a temperature of 100 million K. If this gas is more than a single outburst from an as-yet-unknown source, it must continually be replenished to remain as hot as it is. Astronomers are still trying to account for the source of this energetic gas.

If the center of our Galaxy is not active and bizarre enough, recent gamma-ray observations reveal positrons being ejected from that region. (Recall that positrons have positive electric charges but are otherwise identical to electrons.) The source of these positrons is still unknown. Positrons can be detected because when a positron and an electron collide, they annihilate each other. Their mass is converted into energy as gamma rays with well-defined wavelengths. These special gamma rays have also been observed emanating from the galactic center.

The presence of supernova remnants at the center of our Galaxy implies what other activity is occurring in that region?

Infrared observations also reveal stars and gas in very rapid orbits around Sagittarius A* (Figure 15-14b). Something massive must be holding this high-speed matter in such tight orbits around the galactic nucleus. Using Kepler's third law, astronomers calculate that 4.3×10^6 $M_\odot$ is needed to prevent the stars and gas from flying off into interstellar space. The observed broadening of spectral lines further suggests that an object with that mass in a volume only the size of the solar system is concentrated at Sagittarius A*. Astronomers believe that the object is a supermassive black hole. In 2001, the Chandra X-ray Observatory observed a burst of X rays from this black hole. The X rays were generated by gas heating up as it fell into it. The bigger a black hole, the longer such a burst of X rays will last. The duration of the observed blast indicated that the black hole is no wider than 1 AU across, consistent with theoretical calculations.

As we saw in Chapter 14, extraordinary activity is also occurring in the nuclei of many other galaxies, implying the presence of supermassive black holes at their centers as well. Astronomers are actively studying these regions in an effort to understand the complex, intriguing events that are happening there.

15-5 Our Galaxy's disk is surrounded by a two-shell spherical halo of stars and other matter

As mentioned at the beginning of this chapter, stars have been observed in a spherical distribution, called the **halo**, which consists of two concentric shells centered on the galactic nucleus and extending far beyond the disk (see Figure 15-11). The inner halo was originally discovered because of the globular clusters that it contains. Looking out of the plane of the Milky Way's disk and between globular clusters, we see apparently unobstructed views of distant galaxies, such as the Whirlpool Galaxy observed by Lord Rosse (see Section 15-1). Most of the globular clusters orbit within about 100,000 ly of the nucleus, about twice the distance to the edge of the visible disk in which we reside.

Appearances can be deceiving. Although the brightness of the globular clusters suggests that they contain most of the stars in the halo, it turns out that about 99% of the halo's stars are isolated *halo field stars* spread all through the halo. The localized concentrations of stars in globular clusters account for only 1% of the halo's stars. It is worth noting, however, that some globular clusters are observed to contain intermediate-mass black holes, which contribute significantly to their total mass.

The globular clusters and field stars in the inner halo orbit in the same general sense that the stars in the disk orbit. The stars in the outer shell of the halo, discovered in 2007, have very little metal, implying that they are among the oldest stars in the universe. Furthermore, these latter stars orbit in the opposite sense to the motion of the disk and inner halo stars. These big differences between the two components of the halo strongly suggest that they formed from different components of the early universe. Understanding this formation process will help astronomers develop a theory of how the Milky Way developed.

The various components of the Milky Way Galaxy, including the two shells of the halo, overlap and intersect each other. For example, the disk slices through the central bulge, and globular clusters and halo field stars periodically pass through the plane of the disk. Figure 15-15 shows the shapes of the orbits of typical central bulge, disk, and halo stars and clusters. Our solar system is presently moving at an angle of 25° from the plane of the Galaxy's disk, as illustrated by the red path in the figure. We will move above the plane about 230 ly and then return, crossing the plane every 33 million years.

Since 1994, astronomers have observed 23 small galaxies that orbit in our Galaxy's halo. The Sagittarius Dwarf and the Canis Major Dwarf (Figure 15-16) are named after the constellations in which their central regions lie. They have both spread widely through the halo and periodically pass through the disk of the Milky Way. These two galaxies are much smaller in size and mass than the Milky Way and are losing stars due to the disrupting influence of our Galaxy's strong gravitational attraction. Within the next 100 million years, they will cease to exist, their stars becoming part of the Milky Way. This process of a bigger galaxy consuming a smaller one is called **galactic cannibalism.** The most recently discovered dwarf, called the Ursa Major Dwarf spheroidal galaxy, was found in 2005. Computer simulations of galactic cannibalism performed in 2010 suggest that most of the halo field stars orbiting in the Milky Way are remnants of smaller galaxies.

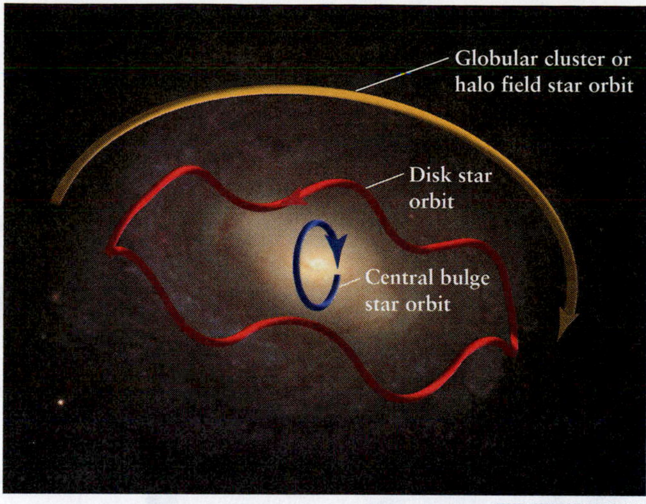

FIGURE 15-15 Orbits of Stars in Our Galaxy This disk galaxy, M58, looks very similar to what the Milky Way Galaxy would look like from far away. The colored arrows show typical orbits of stars in the central bulge (*blue*), disk (*red*), and halo (*yellow*). Interstellar clouds, clusters, and other objects in the various components have similar orbits. (NOAO/AURA/NSF)

15-6 The Galaxy is rotating

Our solar system is moving at 878,000 km/h (500,000 mi/h) 3
around the center of our Galaxy. Just as the orbital motion of the planets keeps them from falling into the Sun, the motion of the stars and interstellar clouds around the galactic center keeps these bodies apart. If the stars and clouds in our Galaxy were not in orbit, their mutual gravitational forces would have caused them to fall together and form one massive black hole billions of years ago. (Indeed, such infall of gas clouds early in the life of the universe is believed to be the origin of the supermassive black holes located at the centers of the Milky Way and other galaxies.) Just as detecting the positions of the stars and clouds has been difficult, so too has been measuring their orbital motions.

Observations show that the Sun's distance from the center of the Galaxy varies little throughout its orbit. About 80% of the stars in our neighborhood have

a

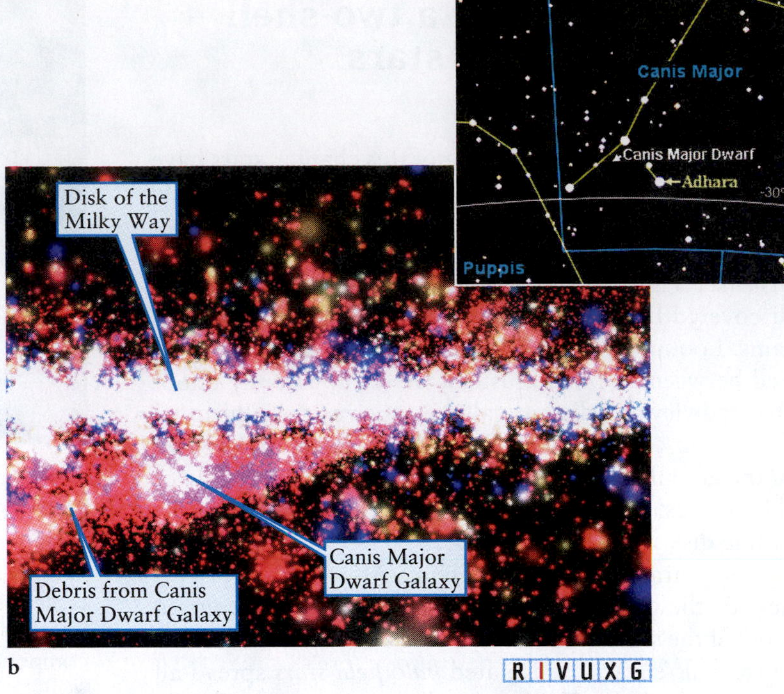

b

FIGURE 15-16 The Nearest Galaxy (a) The Canis Major Dwarf Galaxy is a dwarf elliptical galaxy that lies some 25,000 ly from the Milky Way. This infrared radiation–based image shows the Milky Way's spiral arms, as well as the distribution of stars being stripped from the Canis Major Dwarf Galaxy by our Galaxy's gravitational tidal force. Containing only about 1 billion stars, the Canis Major Dwarf Galaxy will be completely pulled apart within the next 100 million years or so by the Milky Way. (b) View from Earth of the Canis Major Dwarf Galaxy and its path of debris. (a: R. Ibata/Strasbourg Observatory, ULP et al., 2MASS, NASA; b: Nicola Martin & Rodrigo Ibata, Observatoire de Strasbourg, 2003)

similar-shaped orbits. The other 20% have orbits that are either taking them outward from the center or inward toward it, including at least 16 stars moving outward with enough speed to leave the Galaxy forever. The fastest of these exiting, so-called *hypervelocity stars* is moving outward with a speed of at 2.7 million km/h.

Radio observations of 21-cm radiation from hydrogen gas provide important clues about our Galaxy's overall rotation. By measuring Doppler shifts, astronomers can determine the speed of objects toward or away from us across the Galaxy. These observations clearly indicate that our Galaxy does not rotate like a rigid body (Figure 15-17a), but rather exhibits *differential rotation,* meaning that stars at different distances from the galactic center orbit the Galaxy at different rates (Figure 15-17b).

To understand differential rotation, we focus on those stars with nearly circular orbits. Because of the stars' differential rotation in the Galaxy, the Sun is like a car on a circular freeway with the fast lane on one side and the slow lane on the other. As sketched in Figure 15-17b, stars in the fast lane (closer to the center of the Galaxy) are passing the Sun and thus appear from our vantage point to be moving in one direction, while stars in the slow lane (farther from the center of the Galaxy than our solar system) are being overtaken by the Sun and therefore appear to be moving in the opposite direction. This is like the retrograde motion of the planets discussed in Section 2-2.

What effect, which also occurs in the solar system, is responsible for the removal of many stars from smaller galaxies orbiting the Milky Way?

Unfortunately, like the 21-cm observations, studying the motion of nearby stars and gas reveals only how fast they are moving relative to the Sun. To get a complete picture of the Galaxy's rotation, we must find out how fast the Sun itself is orbiting the center of the Galaxy. The Swedish astronomer Bertil Lindblad proposed a method of computing this speed. He noted that globular clusters do not move in the orderly pattern shown in Figure 15-17. Different globular clusters in the halo of our Galaxy orbit in different planes, and so they do not participate in the organized rotation of the objects in the Galaxy's disk. However, the combined velocities of the globular clusters around the center of the Galaxy must average to zero or else they would drift, en masse, relative to the rest of the Galaxy. Using the motions of globular clusters as a reference, astronomers calculated the speed of the Sun's orbit around the galactic center given earlier.

Knowing the Sun's speed and its distance from the galactic center, astronomers can calculate the Sun's orbital period. Traveling at 878,000 km/h, our Sun takes

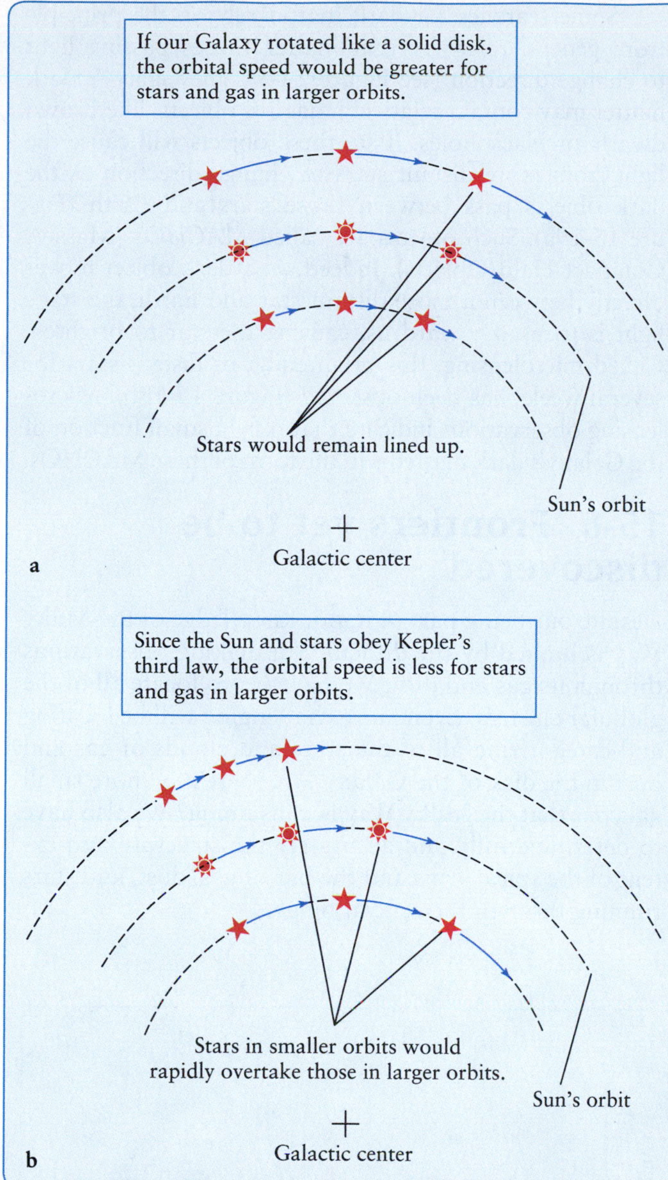

FIGURE 15-17 Differential Rotation of the Galaxy (a) If all stars in the Galaxy had the same angular speed, they would orbit in lockstep. (b) However, stars at different distances from the galactic center have different angular speeds. Stars and clouds farther from the center take longer to go around the Galaxy than do stars closer to the center. As a result, stars closer to the Galaxy's center than the Sun are overtaking the solar system, whereas stars farther from the center are lagging behind us.

about 230 million years to complete one trip around the Galaxy. When the solar system last passed our present location in the Milky Way, early dinosaurs of the Triassic period roamed Earth.

By combining the true speed of the Sun with the relative speed of the stars around us, as measured by radio astronomers, we can also determine the actual orbital speeds of the stars. This computation gives us the **rotation curve** of the Galaxy, a graph that shows the

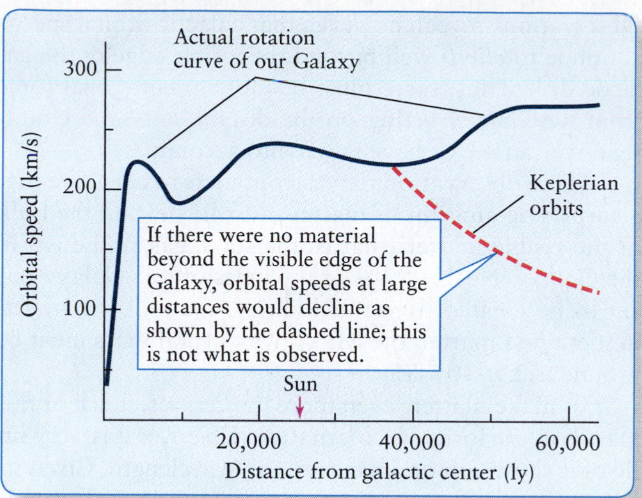

FIGURE 15-18 The Galaxy's Rotation Curve The blue curve shows the orbital speeds of stars and gas in the Galaxy, and the dashed red curve shows Keplerian orbits that would be caused by the gravitational force from all the known objects in the Galaxy. Because the data (blue curve) do not show any such decline, there is, apparently, an abundance of dark matter that extends to great distances from the galactic center. This additional mass gives the outer stars higher speeds than they would have otherwise.

orbital speeds of stars and interstellar clouds at various distances from the center of the Milky Way (Figure 15-18).

Knowing the Sun's velocity around the Galaxy from the rotation curve, we can use Kepler's third law to estimate that the mass of the Galaxy that lies between us and the galactic nucleus is about 110 billion $M_{\odot}$.

The complete trip of the solar system around the Galaxy is analogous to what period in Earth's motion in the solar system?

MYSTERIES AT THE GALACTIC FRINGES

Because the band of the Milky Way extends all around Earth, we know that stars and other matter exist beyond the Sun's orbit. In recent years, astronomers have been astonished to discover just how much matter lies farther from the center of our Galaxy than we are.

15-7 Most of the matter in the Galaxy has not yet been identified

According to Kepler's third law, out beyond the bulk of the Galaxy's visible disk, the orbital speeds of stars should decrease with distance from the Galaxy's center (see the dashed line in Figure 15-18), just as the orbital speeds of the planets decrease with increasing distance from the Sun (which contains the bulk of the solar system's mass).

Observations reveal, however, that galactic orbital speeds continue to *climb* well beyond the visible edge of the galactic disk. Thus, there must be more gravitational force from the Galaxy acting on the distant stars and clouds than we can see or have taken into account.

These observations led astronomers to calculate that a surprising amount of matter must lie beyond the bulk of the visible matter (that is, the stars, gas, and dust) in the Galaxy. Nearly 90% of the mass of our Galaxy has yet to be located. If we include the mass of the as-yet-unidentified matter, the Milky Way's total mass must be around $1.2 \times 10^{12} M_\odot$.

To make matters even more mysterious, much of this mass is in the form of **dark matter:** Whatever it is, very little of it shows up on images at any wavelength. Given its gravitational effect on the stars and gas in the Galaxy that we can detect, astronomers deduce that this dark matter is spherically distributed all around the Galaxy in a massive halo extending at least 12 times farther from the galactic nucleus than we are, far beyond the halo defined by the visible globular clusters and halo field stars (see inset in Figure 15-11). Small fractions of the dark matter are composed of neutrinos and white dwarfs, but the nature of the rest of it—whether black holes, gas, Jupiterlike bodies, brown dwarfs, or something far more exotic—has yet to be determined. The unidentified matter in our Galaxy is sometimes called the **missing mass.** This is a poor name because the matter is not missing; we just do not yet know its location or nature.

Some searches for dark matter rely on the principle from general relativity that matter causes passing light to change direction (see Figure 14-5). The Galaxy's dark matter may contain relatively massive objects, like brown dwarfs or black holes. If so, these objects will cause the light from more distant stars to change direction as the dark objects pass between those stars and Earth (Figure 15-19a). Such objects are called *MACHOs* (Massive Compact Halo Objects). Indeed, as a dark object moves directly between a more distant star and Earth, the star's light is focused toward us, causing the star to brighten. Called **microlensing,** this brightening of distant stars for several weeks has been observed (Figure 15-19b). Microlensing observations indicate that only a small fraction of the Galaxy's dark matter is in the form of these MACHOs.

15-8 Frontiers yet to be discovered

Despite our being part of it, our knowledge of the Milky Way is limited by the difficulty in making observations through its gas and dust. We have yet to locate all of the globular clusters. Even more challenging will be locating and categorizing all of the stars and clouds of gas and dust in the disk of the Galaxy. Are there any more small galaxies that the Milky Way is consuming? We also have to determine fully and accurately the structure and extent of the spiral arms and the bar of gas, dust, and stars running through the central bulge.

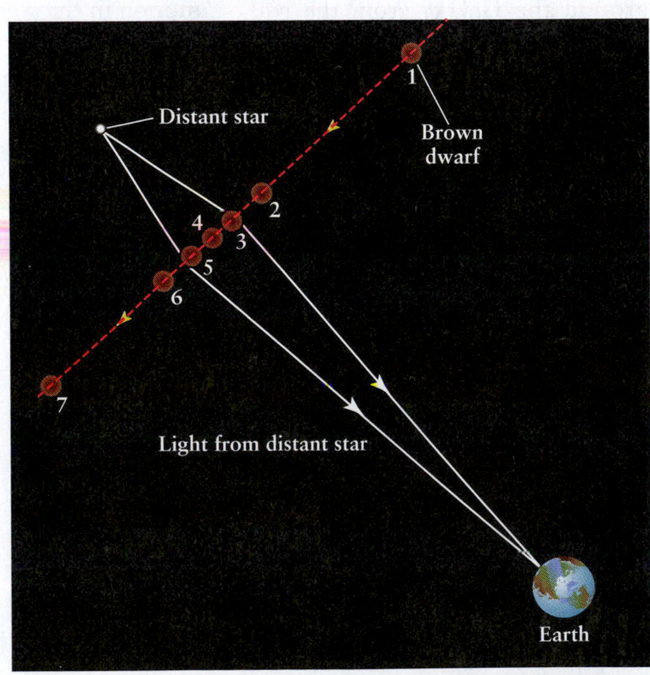

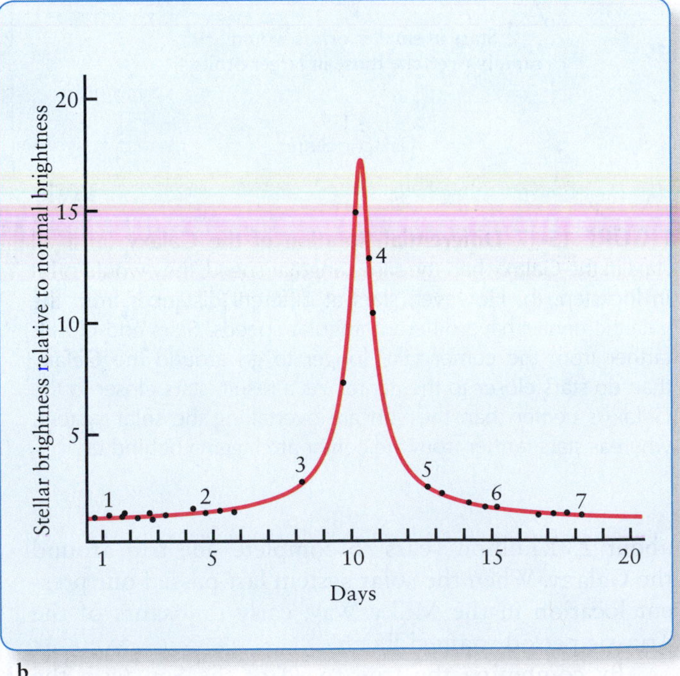

FIGURE 15-19 Microlensing by Dark Matter in the Galactic Halo (a) Gravitational fields cause light to change direction. A white dwarf, brown dwarf, or black hole in the Galaxy's halo passing between Earth and a more distant star will focus the starlight in our direction, making distant objects appear brighter than they are normally. (b) The light curve of the gravitational microlensing of light from a star in the Galaxy's central bulge by an intervening object.

Our observations of the Galaxy's nucleus hint at its remarkable activity that has yet to be explained. What is heating the 100-million-K gas? What is creating the positrons there, whose resulting gamma rays we observe? In its farthest reaches, what else is in the Galaxy's halo that creates all of the gravitational force we detect, but whose source is not yet observed? In other words, what is the dark matter? Another major unsolved mystery is the formation process of our Galaxy. Did stars form first? Did the supermassive black hole form and then pull gas and dust into orbit, which then formed stars? Did smaller galaxies form and then coalesce into the Milky Way as it is today? Astronomers are presently working to answer all of these questions.

SUMMARY OF KEY IDEAS

Discovering the Milky Way

- A century ago, astronomers were divided on whether or not the Milky Way Galaxy and the universe were the same thing.
- The Shapley–Curtis debate was the first major public discussion between astronomers as to whether the Milky Way contains all the stars in the universe.
- Cepheid variable stars are important in determining the distance to other galaxies.
- Edwin Hubble proved that there are other galaxies far outside of the Milky Way.

The Structure of Our Galaxy

- Our Galaxy has a disk about 100,000 ly in diameter and about 2000 ly thick, with a high concentration of interstellar dust and gas. It contains around 200 billion stars.
- Interstellar dust obscures our view into the plane of the galactic disk at visual wavelengths. However, hydrogen clouds can be detected beyond this dust by the 21-cm radio waves emitted by changes in the relative spins of electrons and protons in the clouds, as well as by other nonvisible emissions.
- The center, or galactic nucleus, has been studied at gamma-ray, X-ray, infrared, and radio wavelengths, which pass readily through intervening interstellar dust and H II regions that illuminate the spiral arms. These observations have revealed the dynamic nature of the galactic nucleus, but much about it remains unexplained.
- A supermassive black hole of about 4.3×10^6 $M_\odot$ exists in the galactic nucleus.
- The galactic nucleus of the Milky Way is surrounded by a flattened sphere of stars, called the central bulge, through which a bar of stars and gas extends.
- A disk with two bright arms of stars, gas, and dust spirals out from the ends of the bar in the galactic central bulge.
- Young OB associations, H II regions, and molecular clouds in the galactic disk outline huge spiral arms where stars are forming.
- The Sun is located about 26,000 ly from the galactic nucleus, between the spiral arms. The Sun moves in its orbit at a speed of about 878,000 km/h and takes about 230 million years to complete one orbit around the center of the Galaxy.
- The entire Galaxy is surrounded by two halos of matter. The inner halo includes a spherical distribution of globular clusters and field stars, as well as large amounts of dark matter. It orbits in the same general direction as the disk. The outer halo is composed of dark matter and very old stars, which have retrograde orbits.

Microlensing is similar to the physics of which: reflecting telescopes or refracting telescopes?

WHAT DID YOU THINK?

1 *What is the shape of the Milky Way Galaxy?* The Milky Way is a barred spiral galaxy. A bar of stars, gas, and dust runs through its central region. It has two major spiral arms, several minor arms, and is surrounded by a complex spherical halo system of stars and dark matter.

2 *Where is our solar system located in the Milky Way Galaxy?* The solar system is between the Sagittarius and Perseus spiral arms, about 26,000 ly from the center of the Galaxy (about halfway out to the visible edge of the galactic disk).

3 *Is the Sun moving through the Milky Way Galaxy and, if so, about how fast?* Yes. The Sun orbits the center of the Milky Way Galaxy at a speed of 878,000 km/h.

Key Terms for Review

central bulge, 476
dark matter (missing mass), 482
disk (of a galaxy), 474
distance modulus, 471
galactic cannibalism, 479
galactic nucleus, 472
galaxy, 472
halo (of a galaxy), 479
microlensing, 482
Milky Way Galaxy, 468
nebula (*plural*, nebulae), 470
rotation curve (of a galaxy), 481
Sagittarius A, 476
Shapley–Curtis debate, 470
spin (of an electron or proton), 473
spiral arm, 474
synchrotron radiation, 478
21-cm radio radiation, 474

Review Questions

The answers to computational problems, which are preceded by an asterisk (*), appear at the back of the book.

1. Where in the Galaxy is the solar system located? **a.** in the nucleus, **b.** in the halo, **c.** in a spiral arm, **d.** between two spiral arms, **e.** in the central bulge

2. What is located in the nucleus of the Galaxy? **a.** a globular cluster, **b.** a spiral arm, **c.** a black hole, **d.** the solar system, **e.** a MACHO

3. Which statement about the Milky Way Galaxy is correct? **a.** Our Galaxy is but one of many galaxies. **b.** Our Galaxy contains all stars in the universe. **c.** All stars in our Galaxy take the same time to complete one orbit. **d.** Most stars in our Galaxy are in the central bulge. **e.** None of the stars in our Galaxy move.

4. What was the Shapley–Curtis debate all about? Was a winner declared at the end of the debate? Whose ideas turned out to be correct?

5. How did Edwin Hubble prove that M31 is not a nebula in our Milky Way Galaxy?

6. As seen in the northern hemisphere, why is the Milky Way far more prominent in July than in December? Planetarium software, such as *Starry Night*™, may be useful in answering this question.

7. What observations led Harlow Shapley to conclude that we are not at the center of the Galaxy?

8. Explain why globular clusters spend more time in the galactic halo than in the plane of the disk, even though their eccentric orbits take them across the disk of the Galaxy.

9. How do hydrogen atoms generate 21-cm radiation? What do astronomers learn about our Galaxy from observations of that radiation?

10. Why do astronomers believe that vast quantities of dark matter surround our Galaxy?

11. What is synchrotron radiation?

12. Why are there no massive O and B stars in globular clusters?

13. What evidence indicates that a supermassive black hole is located at the center of our Galaxy?

***14.** Approximately how many times has the solar system orbited the center of the Galaxy since the Sun and planets were formed?

Advanced Questions

15. Why do astronomers not detect 21-cm radiation from the hydrogen in giant molecular clouds?

16. Determine the order of redshifts in Figure 15-8 from highest blueshift to lowest.

17. Describe the rotation curve you would get if the Galaxy rotated like a rigid body.

18. Compare the apparent distribution of open clusters, which contain young stars, with the distribution of globular clusters in the Milky Way. Why are open clusters also referred to as *galactic clusters?*

***19.** The visible disk of the Galaxy is about 100,000 ly in diameter and 2000 ly thick. If about five supernovae explode in the Galaxy each century, how often on average would you expect to see a supernova within 1000 ly of the Sun?

20. Give reasons for the rapid rise in the Galaxy's rotation curve (see Figure 15-18) at distances close to the galactic center.

What If ...

21. The solar system were located in a globular cluster 26,000 ly above the plane of the Galaxy's disk? Make a drawing of what the Milky Way might look like from that location.

22. The solar system were located at the edge of the visible galactic disk? How would the Milky Way appear to us?

23. The solar system were located at the center of the Milky Way Galaxy? What would we see, and how would things be different on Earth?

24. A globular cluster were crossing the disk of the Milky Way in our vicinity right now? What would we see, and how might things be different on Earth?

Web Questions

25. To test your understanding of our Galaxy's rotation, do Interactive Exercise 15.1 on the Web. You can print out your results, if required.

26. Search the Web for information on the Shapley–Curtis debate. Who proposed the debate in the first place? Briefly outline the relevant scientific beliefs held by the two men. What points did they agree upon and what did they disagree on? Why was the debate considered inconclusive? Support your argument with examples, if possible.

27. To test your understanding of the Milky Way's structure, do Interactive Exercise 15.2 on the Web. You can print out your results, if required.

28. Clouds of positrons (see Section 15-4 and *An Astronomer's Toolbox 10-1*) have been discovered near the center of the Galaxy. Search the Web for information about this discovery. When was it made? How were the clouds discovered? What is thought to be the origin of these particles? What happens when ordinary electrons and positrons collide?

Observing Projects

STARRY NIGHT

29. Use the *Starry Night*™ program to observe the Milky Way. **a.** Select **Favourites > Discovering the Universe > Milky Way.** This view shows the entire celestial sphere from the center of a transparent Earth. The brightness of the Milky Way has been enhanced. Use the hand tool to look at different parts of the Milky Way. Can you identify the direction toward the galactic nucleus? This is the direction in which the Milky Way appears broadest. To check your identification, use the **Find** pane to center the field of view on the constellation Sagittarius. **b.** Click the **Home** button in the toolbar to reset the program to display the view from your home location at the current time and date. Click the **Stop** button to stop time flow and then set the local time to midnight (12:00:00 A.M.) and the date to January 1, then February 1, and so on. In which month is the galactic nucleus highest in the sky at midnight, so that it is most easily seen?

STARRY NIGHT

30. Use *Starry Night*™ to determine the angle between the equatorial coordinate system and the plane of the Milky Way. You will need a protractor for this project. Select **Favourites > Discovering the Universe > Milky Way.** Select **View > Celestial Guides > Equator** from the menu to display a red line in the view that represents the celestial equator, the projection of Earth's equator on the sky. Next, select **View > Galactic Guides > Equator** to draw a blue line on the screen that represents the plane of the Milky Way Galaxy. If necessary, use the hand tool to drag the screen around until you see the Milky Way crossing the celestial equator with the celestial equator horizontal. **a.** Use a protractor to directly measure the angle between the celestial equator and the central plane of the Milky Way represented by the galactic equator, being very careful not to damage your computer screen. Alternatively, select **File > Export As Image...** from the menu and print a copy of this saved image on which to measure the angle between the celestial equator and the plane of the Milky Way. Right-click (Ctrl-click on a Macintosh) on the point at which the galactic equator intersects the celestial equator. The contextual menu indicates the celestial coordinates of this point in the sky. **b.** What is the right ascension of the point in the sky at which the Milky Way intersects the celestial equator? Now rotate the celestial sphere by 12 h around the celestial equator. Is the Milky Way again crossing the celestial equator? **c.** Measure the angle between the two planes again at this intersection. How does this compare with your earlier measurement? Explain.

STARRY NIGHT

31. Use the *Starry Night*™ program to measure the dimensions of the Milky Way Galaxy. Select **Favourites > Discovering the Universe > Milky Way Galaxy** to display a face-on view of the Milky Way Galaxy from a distance of 0.128 Mly from Earth with a field of view 130° wide. The Sun is labelled, and is seen directly below the center of the Galaxy in this view. **a.** Use the angular separation tool to measure the angular separation between the Sun and the center of the Galaxy as seen from this vantage point, and find the corresponding distance in light years (ly). (Note: Both values are shown in the display beside the line drawn by the angular separation tool.) Convert the angular separation to a decimal number in degrees ($1° = 60' = 3600''$) and then calculate the scale of the image of the Galaxy in ly/degree. Be careful not to change the zoom or elevation in the view after calculating this scale. **b.** Use the location scroller to view the Galaxy edge on and oriented vertically on the screen, with the Sun still below the center. To do this, place the location scroller tool at the center of the right-hand edge of the screen, hold down the mouse button (on a two-button mouse, hold down the left mouse button), and move the location scroller directly to the left, toward the center of the Galaxy. Then use the angular separation tool to measure the angular separation of the Sun from the center of the Galaxy. This value should be approximately the same as you found in part a. **c.** Use the angular separation tool to find the total angular diameter of the Milky Way Galaxy as seen from this viewpoint, measured from one end to the other of this edge-on view. Convert this value to a decimal number in degrees, and then use the scale that you calculated in part a to find the diameter of the Galaxy in light years. **d.** In a similar fashion, use this viewpoint to measure the distance in degrees and light years for the following quantities: the diameter (in the plane of the Galaxy) of the central bulge, the thickness (perpendicular to the plane of the Galaxy) of the central bulge, and the thickness (perpendicular to the plane of the Galaxy) of the disk of the Galaxy at the location of the Sun.

WHAT DO YOU THINK?

1 **Are most of the stars in spiral galaxies located in their spiral arms?**

2 **Do all galaxies have spiral arms?**

3 **Are galaxies isolated objects?**

4 **Is the universe contracting, unchanging in size, or expanding?**

Answers to these questions appear in the text beside the corresponding numbers in the margins and at the end of the chapter.

RIVUXG The Starburst Galaxy M82 shows a frenzy of star formation in its central regions and jets of gas perpendicular to the galaxy. Both activities are believed to have been initiated by passage of the galaxy near its neighbor, M81. (X-ray: NASA/CXC/JHU/D.Strickland; optical: NASA/ESA/STScI/AURA/The Hubble Heritage Team; infrared: NASA/JPL-Caltech/U. of Arizona/C. Engelbracht)

Chapter 16

Galaxies

We live in a galactic suburb, far from the center of the action. The Sun is just one of about 200 billion stars in the Milky Way, and our Galaxy is but one of between 50 billion and one trillion galaxies in the visible universe. In studying galaxies, we have entered the realm of truly cosmic systems of stars, gas, and dust. Although nearby stars are separated from us by light-years (trillions of miles), neighboring galaxies are hundreds of thousands of times farther away, and the most distant galaxies are at least 13.2 billion light-years from Earth.

Finding patterns is a powerful tool built into the human psyche. Early in the twentieth century, astronomers observed that galaxies appear in surprisingly few shapes. The classification scheme Edwin Hubble developed is where we begin our study of the large-scale structure of the universe—but that is only the beginning. By the mid-twentieth century, clusters of galaxies were discovered, followed by clusters of clusters (now called *superclusters*) of galaxies. Between these ensembles of stars, gas, and dust are vast regions of space relatively devoid of matter. Through establishing these patterns, astronomers now know the large-scale structure of visible matter in the universe, which we also present in this chapter.

In this chapter you will discover

- how galaxies are categorized by their shapes
- the processes that produce galaxies of different shapes
- that galaxies are found in clusters that contain huge amounts of dark matter
- why clusters of galaxies form in superclusters
- how some galaxies merge and others devour their neighbors
- that the universe is changing size

TYPES OF GALAXIES

Despite the incredible number of galaxies, there is surprising consistency in their overall shapes. Edwin Hubble began cataloging their appearance in the 1920s. The **Hubble classification** of galaxies—spirals, barred spirals, ellipticals, irregulars, and their subclasses—is still used today.

16-1 The winding of a spiral galaxy's arms is correlated to the size of its central bulge

Spiral galaxies without bars (often called *normal spiral galaxies*) are characterized by a central bulge and arched lanes of stars and glowing interstellar clouds, which appear as arms that spiral out from the bulge. Observations of many spiral galaxies reveal that the more tightly wound the spiral, the larger the central bulge.

As shown in Figure 16-1, spirals with tightly wound spiral arms (and fat central bulges) are called *Sa* (for spiral type a) *galaxies*. Those with moderately wound spiral arms (and a moderate central bulge) are *Sb galaxies*. Loosely wound spirals (with small central bulges) are *Sc galaxies*. A typical spiral galaxy contains about 100 billion stars and measures nearly 10^5 light-years in diameter.

The closest spiral galaxy to the Milky Way is the Andromeda Galaxy, M31 (Figure 16-2). M31 is visible to the naked eye as a fuzzy blob in the constellation of Andromeda. The disk of this galaxy is tilted as seen from Earth, typical of many spiral galaxies. Some are tilted so much that their spiral arms are not evident. We can still classify these galaxies as specific spiral types by observing the sizes of their central bulges. For example, M104 (Figure 16-3a) has a huge central bulge. It must therefore be an Sa galaxy with tightly wound arms. An Sb

a Sa (NGC 1357) R I V U X G

b Sb (M81) R I V U X G

c Sc (NGC 4321) R I V U X G

FIGURE 16-1 Spiral Galaxies (Nearly Face-on Views) Edwin Hubble classified spiral galaxies according to the tightness of their spiral arms and the sizes of their central bulges. Sa galaxies have the largest central bulges and the most tightly wound spiral arms, whereas Sc galaxies have the smallest central bulges and the least tightly wound arms. (a: NASA/Hubble Space Institute; b: Giovanni Benintende; c: Anglo-Australian Observatory)

FIGURE 16-2 Andromeda (M31) Andromeda is a beautiful spiral galaxy and the only galaxy easily visible to the naked eye from Earth's northern hemisphere. It has dim, red giant stars (not visible here) extending half a million light-years from its nucleus. Without a telescope, Andromeda appears to be a fuzzy blob in the constellation of the same name. Located only 2.5 Mly (0.77 Mpc) from us, Andromeda is gravitationally bound to the Milky Way, and it covers an area in the sky roughly 5 times as large as the full Moon. Two other galaxies, M32 and M110, are also labeled on this photograph. The points of light that pepper the image are stars in our Galaxy. (Bill and Sally Fletcher/Tom Stack and Associates)

galaxy (Figure 16-3b) has a smaller central bulge. The tiny central bulge of an Sc galaxy (Figure 16-3c) is hardly noticeable at all in an edge-on view.

Besides the degree of winding, the overall appearance of individual spiral arms typically has one of two structures. In some galaxies, called *flocculent spirals* (from the word meaning "fleecy"), the spiral arms are broad, fuzzy, chaotic, and poorly defined (Figure 16-4a). Other galaxies, called *grand-design spirals,* exhibit beautiful arching arms outlined by brilliant H II regions and

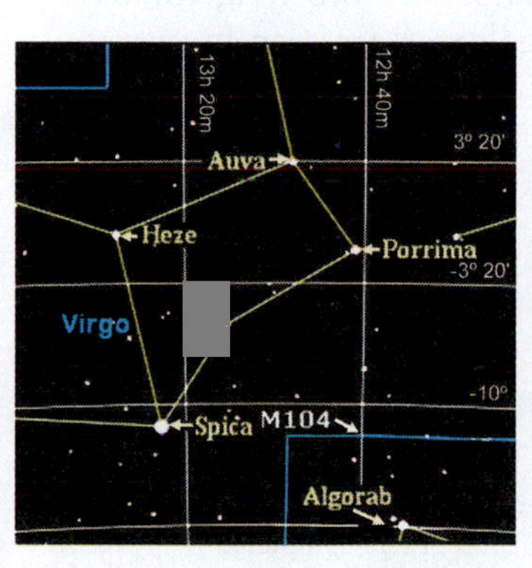

a M104; an Sa galaxy

b NGC 891: an Sb galaxy

c NGC 5907: an Sc galaxy

FIGURE 16-3 Spiral Galaxies Seen Nearly Edge-on from the Milky Way (a) Because of its large central bulge, this galaxy (called the Sombrero Galaxy) is classified as an Sa. If we could see it face-on, the spiral arms would be tightly wound around a voluminous bulge. (b) Note the smaller central bulge in this Sb galaxy. (c) At visible wavelengths, interstellar dust obscures the relatively insignificant central bulge of this Sc galaxy. (a: European Southern Observatory; b: © Malin/IAC/RGO; c: Brand Ehrhorn/Adam Block/NOAO/AURA/NSF)

a M33: a spiral galaxy with flocculent spiral arms R I V U X G

Grand design spiral arms

b M74: a grand-design spiral galaxy R I V U X G

FIGURE 16-4 Variety in Spiral Arms The differences in spiral galaxies suggest that at least two mechanisms create spiral arms. (a) This flocculent spiral galaxy has fuzzy, poorly defined spiral arms. (b) This grand-design spiral galaxy has well-defined spiral arms. (a: NASA; b: Gemini Observatory/AURA)

OB associations. In these galaxies, the spiral arms are thin, delicate, graceful, and well defined (Figure 16-4b).

At first glance, it may seem that the shapes of spiral galaxies are created by the orbital motions of strings of stars, with the stars closer to the center of the galaxy orbiting more rapidly and thereby leading the outer edges of the spirals. But, in this case, common sense leads us astray. Although the stars closer to the center of the galaxy do orbit faster than those farther away, spiral arms cannot be orbiting strings of stars. If they were, the spiral arms should "wind up" from the inside out (Figure 16-5), wrapping themselves tightly around the nucleus. While this action would take too long for us to observe in a single galaxy, galaxies are seen in just a few states of "winding" (see Figure 16-1), which is not consistent with this model.

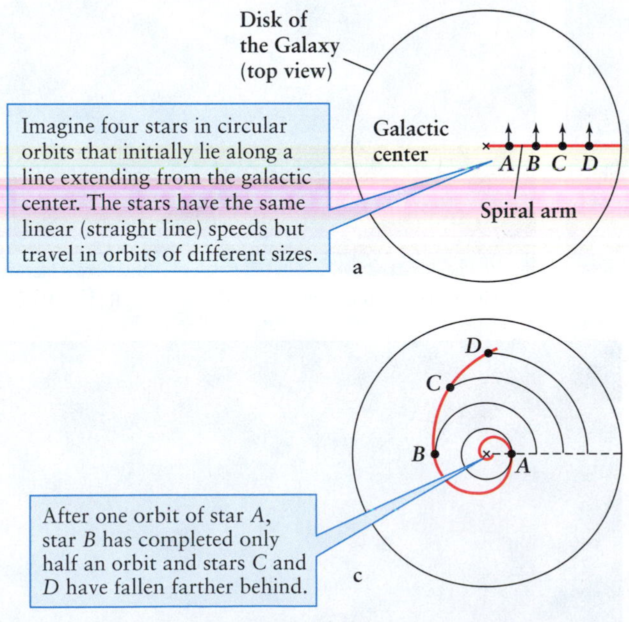

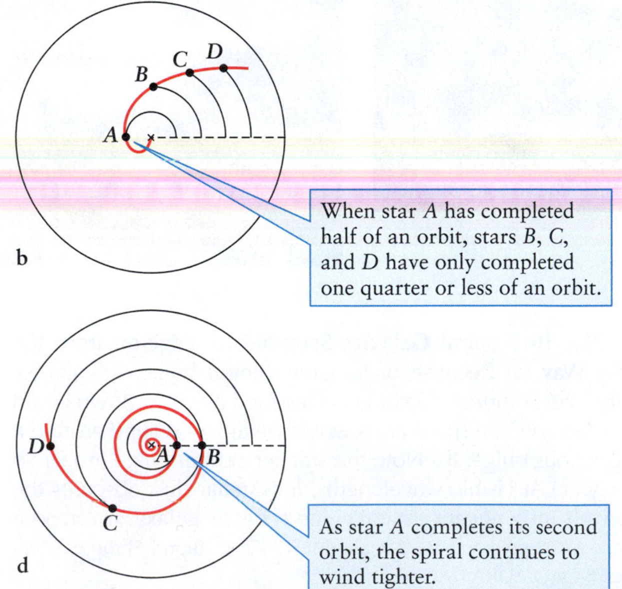

FIGURE 16-5 The Winding Dilemma The rotation curve of the disk stars in our Galaxy indicates that most of them have the same linear (straight line) speed. Those stars farther from the center take longer to go around because they have a greater distance to travel at the same speed than stars closer to the center of the Galaxy, which orbit in smaller circles. All four dots in these drawings have circular orbits at the same linear speed. Think of these dots as (a) a few stars initially in a straight line. As time goes on, the outer stars are left behind, creating (b–d) a spiral shape that becomes more and more tightly wound. Such tightening is not observed in our Galaxy or in other galaxies.

If star motions caused spirals, then after a few galactic rotations, the spiral structure should disappear altogether. Why, then, do we observe spiral arms in our Galaxy and many others? This *winding dilemma* led astronomers to two explanations, one for flocculent spirals and the other for grand-design spirals.

16-2 Explosions create flocculent spirals, and waves create grand-design spirals

Flocculent Spiral Galaxies Consider a disk galaxy so young that it has not yet formed spiral arms. Stars begin forming in one of its giant molecular clouds. The radiation and winds from these stars compress the nearby interstellar gas, triggering the formation of additional stars. Moreover, massive stars quickly explode as supernovae and produce shock waves, which further compress surrounding gases. Shock waves are very powerful compressions of the gas, like the sonic booms created by lightning, the snap of a whip, or a supersonic aircraft. Thus, as new stars trigger the birth of still other stars, the star-forming region grows, a process called *self-propagating star formation.*

The continuing birth of stars accounts for spiral arms in flocculent spiral galaxies. The galaxy's differential rotation (inner regions orbit faster than outer regions) drags the inner edge of each young region ahead of its outer edge, spreading the star-forming region into a spiral arm. This region of the galaxy is bright because it is highlighted by its brilliant O and B stars, and by the nebulae that these stars cause to glow. However, high-mass O and B stars explode relatively quickly, and each spiral arm dims before it winds up (that is, there is no winding dilemma here). As old arms fade, new star-forming regions appear and create new spiral arms.

Comparing it to the galaxies in Figure 16-1 or 16-3, what is the spiral classification of M74, shown in Figure 16-4b?

Where self-propagating star formation is the only process creating spiral arms in a galaxy, bits and pieces of these arms appear constantly, only to disappear as the bright, massive stars in them die. This process creates chaotic spiral arms, such as those observed in flocculent galaxies (see Figure 16-4a) but not the smooth spirals of grand-design galaxies. Thus, self-propagating star formation cannot be the whole story.

Grand-Design Spiral Galaxies Our search for the mechanism that generates grand-design spirals begins in a pond. If you throw a rock into the water, you create ripples. As you know from experience, the ripples move outward in concentric rings from where the rock struck (Figure 16-6a). But suppose the pond water is rotating when you throw in a rock. Now what shape do the ripples have? According to calculations done by astronomer Bertil Lindblad in the 1920s, ripples created in a rotating disk-shaped system of liquid (the pond) or gas and dust (a disk galaxy) will be spiral. These ripples are called **spiral density waves.**

Russian scientists in the 1970s tested Lindblad's model. Lacking the computer power to create simulations of galaxies, they threw pebbles into pie pans of water rotating on phonograph turntables. The resulting water wave patterns were indeed spirals, as drawn in Figure 16-6b.

a

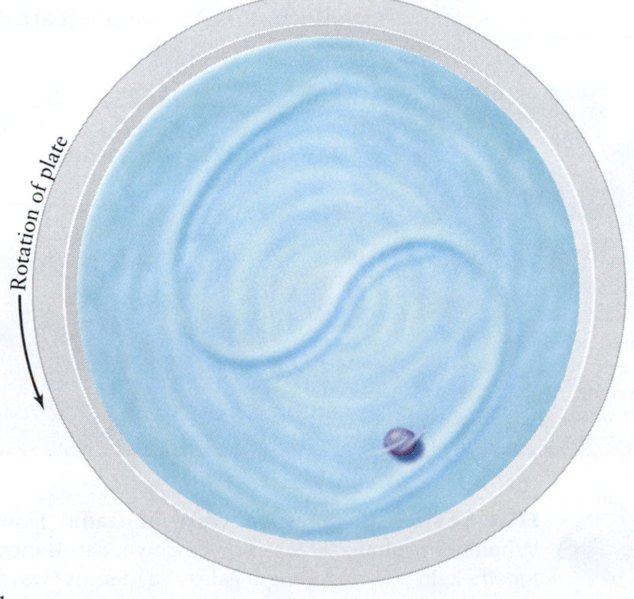

b

FIGURE 16-6 Ripples in Water (a) The usual circular ripples expanding from the place where a rock was thrown into the water. (b) Ripples in rotating water creates spiral arms, as do ripples in the gas and dust of a disk galaxy. (a: © Royalty Free/Corbis)

Furthermore, spiral density waves never wind up. In the case of the gas and dust in a galaxy, these waves orbit in a rigid spiral pattern at about half the average speed of the interstellar medium through which they move.

In the mid-1960s, two American astrophysicists looked carefully at how galactic spiral density waves traveling through the disk of a galaxy lead to spiral arms. C. C. Lin and Frank Shu argued that density waves cause interstellar gas and dust to pile up temporarily. This piling-up happens because, unlike water waves, which move up and down as they travel along, spiral density waves in the gas and dust of a galaxy are compressional waves, just like sound. As a result, they cause interstellar gas and dust to become denser as this material passes through the waves.

Lin and Shu demonstrated that a spiral arm is a galactic traffic jam. Imagine a slow moving truck on a busy freeway. The cars normally cruise at 100 km/h (about 65 mi/h), but the truck causes a bottleneck. The cars slow down temporarily to avoid hitting other cars and the slowly moving truck. As seen from the air, there is a noticeable congestion of cars around the truck (Figure 16-7). An individual car spends only a few moments in the moving traffic jam before resuming its usual speed, but the traffic jam itself lasts all day long.

The vehicles in the traffic jam represent interstellar gas and dust entering a spiral arm. The truck represents a small part of a spiral density wave. Like the cars catching up to and passing the truck, the interstellar medium sweeps through the more slowly moving spiral density waves. The waves compress some of this interstellar gas and dust, which contract to form new, metal-rich stars. We see spiral arms because they contain bright O and B stars and copious quantities of dust and gas that these stars illuminate. The sprawling dust lanes in Figure 16-4b attest to the passage of that material through a spiral density wave. While all of this occurs, the density wave maintains its shape, and, overall, the spiral arms in these galaxies are better defined than in the flocculent galaxies discussed earlier.

> Fill in the blanks: If analogous events occurred on Earth, we would call self-propagating star formation a ________ reaction or a ________ effect.

A few high-mass stars form in most open clusters, 1
such as those created in spiral density waves. So why are spiral galaxies not eventually filled with enough of these bright stars to wash out the spiral arms and to make the disk uniformly bright? The answer is that the massive stars highlighting each density wave are short lived and they explode before they finish passing through it. The remaining, longer-lived, lower-mass stars, like our Sun, fill the space between the spiral arms without emitting as much light (see Figure 16-4b). As pronounced as spiral arms may appear, they contain only 5% more stars than are found between them.

It takes an enormous amount of energy from the spiral density waves to compress interstellar gas and dust. After a billion years or so, even spiral density waves would begin to fade away. Some driving mechanisms must keep them going, akin to throwing more rocks in the pond to maintain its ripples. We are not yet completely certain what those mechanisms are, but one likely explanation for reinvigorating the galactic spiral structure is the passage of a nearby companion galaxy. As this nearby galaxy periodically passes close by, its gravitational attraction pulls on the gas, stars, and dust of the spiral galaxy to generate new density waves. Indeed, grand-design galaxies are usually found in the presence of a companion galaxy.

Although most spiral galaxies have two arms, a sizable minority has more. Astronomers have not yet established why the numbers of arms vary.

Because new stars are formed from interstellar gas and dust, it is plausible that galaxies with different amounts of this material will have different star formation rates and therefore different overall structures. Infrared and radio observations reveal that Sa galaxies typically contain about 4% gas and dust, whereas Sb galaxies contain 8%, and Sc galaxies contain 25%. Continually improving computer simulations of galaxies reveals more and more of the details of how various concentrations of gas and dust affect galactic structure. Figure 16-8 summarizes the activity in a spiral galaxy.

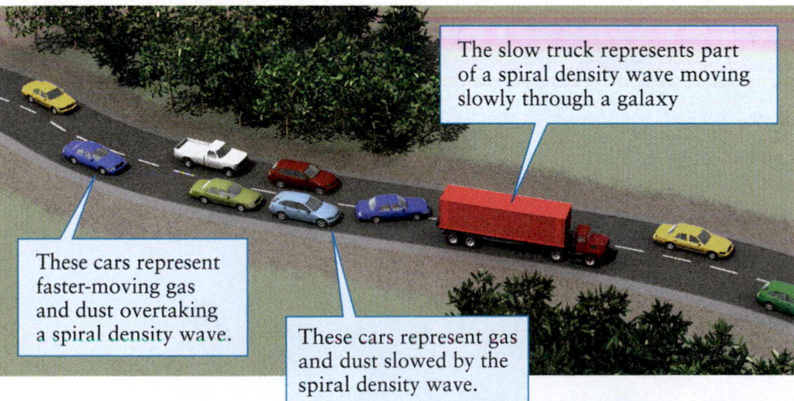

ANIMATION 16.1

FIGURE 16-7 Compression Wave in Traffic Flow When normal traffic flow is slowed down, cars bunch together. In a grand-design galaxy, a density wave moves through the stars and gas. The wave is merely a region of slightly denser matter, which, in turn, creates more gravitational force. This force compresses the gas and enhances star formation, which highlights the spiral density wave.

16-3 Bars of stars run through the central bulges of barred spiral galaxies

The Milky Way is a **barred spiral,** a spiral galaxy with a bar of stars crossing through the central bulge. Such bars are composed of stars and gas that basically flow first one way down the bar and then turn around and

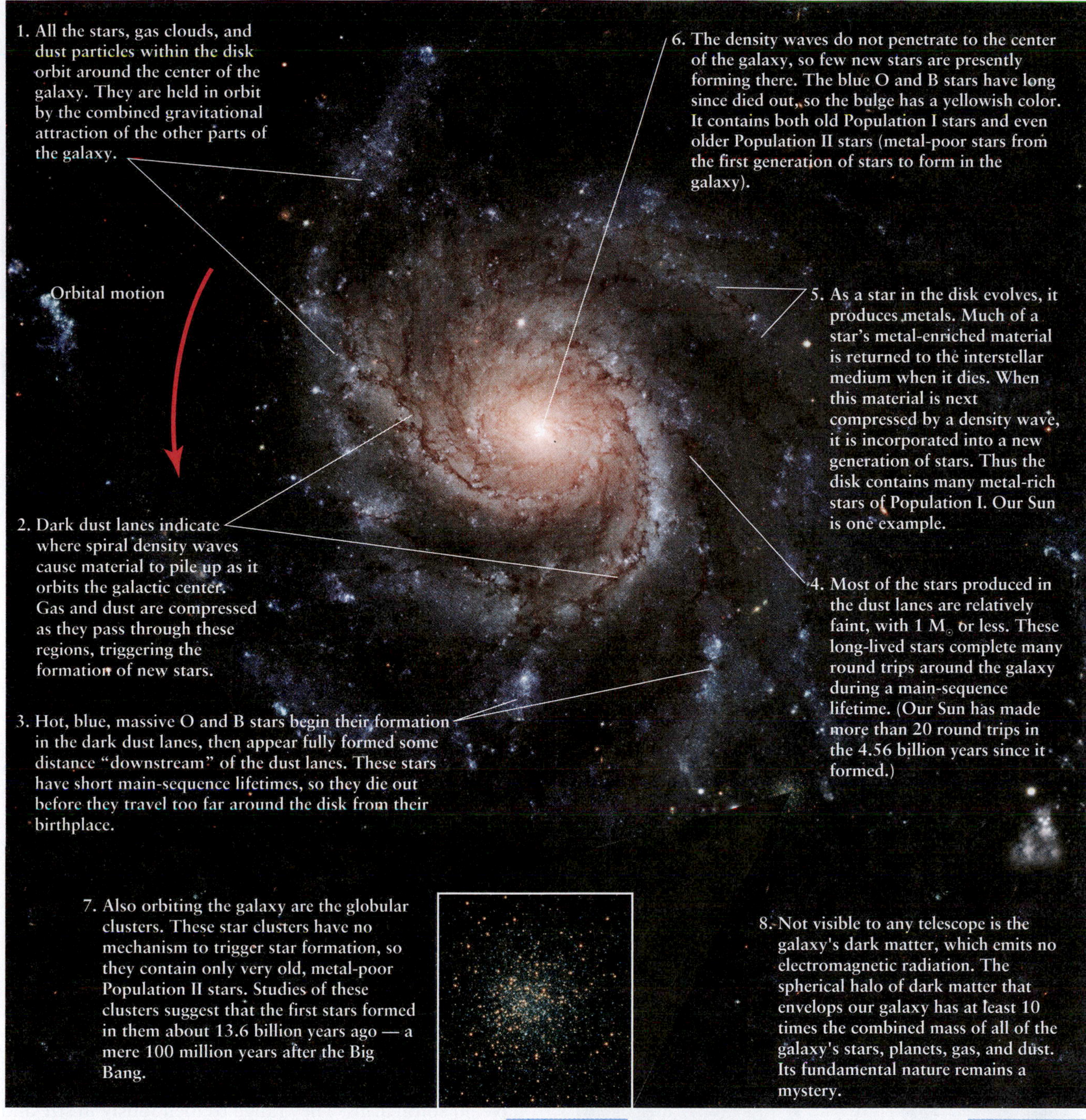

FIGURE 16-8 Dynamics of a Grand-Design Spiral Galaxy This figure summarizes the activities taking place in a grand-design spiral galaxy. (Image of spiral galaxy M101: NASA and ESA; image of globular cluster M3: S. Kafka and K. Honeycutt, Indiana University/WIYN/NOAO/NSF)

flow the other way. Bars form in spiral galaxies that have less total mass than the normal or "unbarred" spirals described above. Indeed, the higher total mass of normal spirals prevents bars from forming. About one-third of all spiral galaxies are barred spirals, the remainder being the normal spirals. The two main arms in barred spirals typically extend from the ends of the bar rather than from the central bulge itself.

Edwin Hubble found that the winding of the spiral arms in barred spirals also correlates with the size of the central bulge (Figure 16-9), just as for normal spirals. An *SBa* (for spiral, barred, type *a*) *galaxy* has a large central bulge (and tightly wound spiral arms). Likewise, a barred

Why do you think that spiral arms do not extend farther inward than the bar in barred spiral galaxies?

a M58: an SBa galaxy

b M83: an SBb galaxy

c NGC 1365: an SBc galaxy

FIGURE 16-9 Barred Spiral Galaxies As with spiral galaxies, Edwin Hubble classified barred spirals according to the tightness of their spiral arms (which correlates with the sizes of their central bulges). (a) SBa galaxies have the most tightly wound spirals and largest central bulges, (b) SBb galaxies have moderately tight spirals and medium-sized central bulges, and (c) SBc galaxies have the least tightly wound spirals and the smallest central bulges. (a: Johan H. Knapen and Nik Szymanek, University of Hertfordshire; b: European Southern Observatory; c: Jean-Charles Cuillandre/CFHT/Photo Researchers, Inc.)

spiral with moderately wound spiral arms (and a moderate central bulge) is an *SBb galaxy,* and an *SBc galaxy* has loosely wound spiral arms (and a tiny central bulge). The Milky Way is classified as an SBab, which has properties between those of an SBa and an SBb. Sa and SBa are sometimes called *early type galaxies,* while Sc and SBc are called *late type galaxies.*

The motions of the stars in many nearby galaxies have been measured from Doppler shifts of their spectra. In all spiral and barred spiral galaxies discovered to date, except one, the arms trail around behind as the galaxies rotate. These galaxies are thus called **trailing-arm spirals.** The exception is the galaxy NGC 4622. This galaxy has the points of its spirals leading the motion of the arms. It is a *leading-arm spiral.* Calculations predict that the leading arms are unstable and that this galaxy will eventually switch and become a trailing-arm spiral.

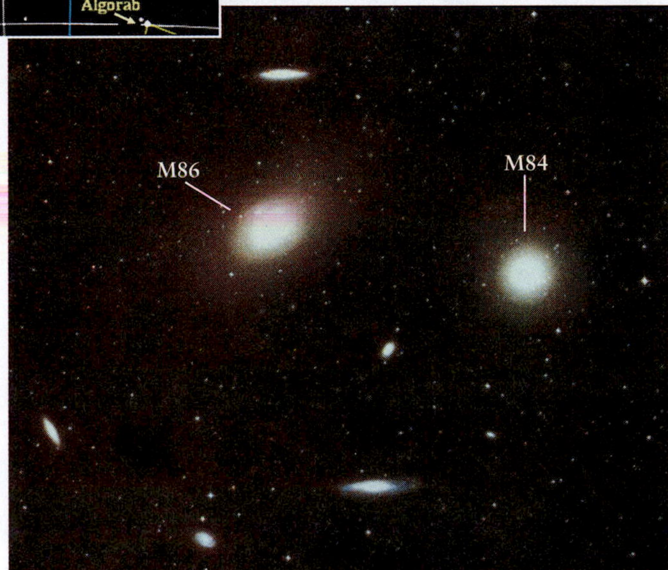

FIGURE 16-10 Giant Elliptical Galaxies The Virgo cluster is a rich, sprawling collection of more than 2000 galaxies about 50 million light-years from Earth. Only the center of this huge cluster appears in this photograph. The two largest galaxies in the cluster are the giant elliptical galaxies M84 and M86. (Royal Observatory, Edinburgh)

16-4 Elliptical galaxies display a wide variety of sizes and masses

Elliptical galaxies, named for their distinctive shapes, 2
have no spiral arms. They range tremendously in size and mass—from the biggest to the smallest galaxies in the universe. Figure 16-10 includes two *giant elliptical galaxies,* M84 and M86, that form part of a cluster of galaxies in the constellation Virgo. These enormous giant elliptical galaxies are each about 2 million light-years in diameter, 20 times the diameter of the Milky Way Galaxy.

Giant ellipticals, containing some 10 trillion solar masses, are rare compared to other types of galaxies, whereas *dwarf elliptical galaxies* are extremely common. Dwarf ellipticals are only a fraction of the size of an average elliptical galaxy and contain so few stars—only a few million—that we see these galaxies as nearly transparent. Because you can actually see straight through the center of a dwarf elliptical galaxy and out

Leo 1: an E4 galaxy R I V U X G

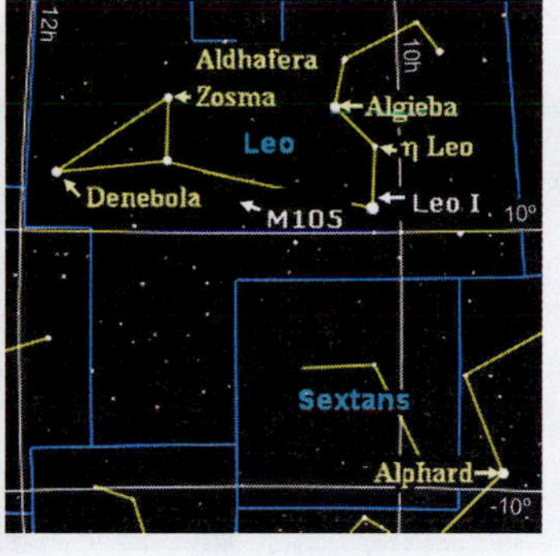

FIGURE 16-11 A Dwarf Elliptical Galaxy This nearby E4 galaxy, called Leo I, is about 600,000 light-years from Earth. It is only 3000 light-years in diameter and is so sparsely populated with stars that you can see right through its center. Leo I is a satellite galaxy of the Milky Way. (NASA)

the other side (Figure 16-11), distant ones are hard to detect. In 1999 astronomers discovered the existence of dwarf galaxies that are much more compact and have stars orbiting much closer together than traditional dwarf elliptical galaxies. Whereas traditional dwarf elliptical galaxies are a few thousand light-years in diameter, *ultra compact dwarfs* are a few hundred light-years across. Ultra compact dwarfs represent groups of stars intermediate in size and number between globular clusters (see Section 12-14) and traditional dwarf galaxies. Many more dwarf and ultra dwarf ellipticals undoubtedly exist than have been identified, and the uncertainty in their numbers is a major reason that we do not know the total number of galaxies in the visible universe.

Hubble subdivided elliptical galaxies according to how round or oval they look. The roundest elliptical galaxies are called *E0 galaxies,* whereas the most elongated are *E7 galaxies.* These latter are about 3 times longer than they are wide. Elliptical galaxies of intermediate elongation are numbered E1 to E6 (Figure 16-12).

The Hubble scheme classifies galaxies solely by their appearance from Earth. But whenever we observe anything in the sky, we are seeing a two-dimensional view of a three-dimensional object. In the case of elliptical galaxies, we observe length and width, but we have no way of knowing anything about the third dimension of depth. What looks like an E0 galaxy (basically circular) might actually be egg-shaped when viewed from another angle. Conversely, an elongated E7 galaxy might look circular when viewed face-on.

Elliptical galaxies look far less dramatic than their spiral and barred spiral cousins because they contain relatively little interstellar gas and dust. For example, the only gas visible in M105 (Figure 16-12a), is the dark line running from upper left to lower right. All the orange images are stars. Because stars form in interstellar clouds, relatively few stars should be forming in ellipticals compared to the numbers forming in spirals. Observations of spectra confirm the hypothesis that elliptical galaxies contain primarily Population II, low-mass,

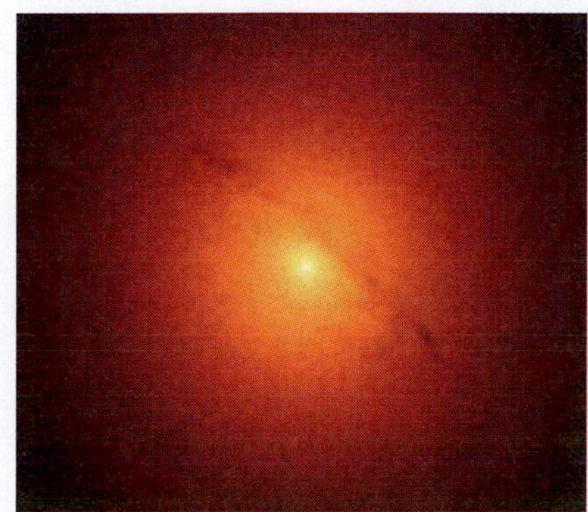

a E0 (M105)

b E3 (NGC 4406)

c E6 (NGC 3377) R I V U X G

FIGURE 16-12 Elliptical Galaxies Hubble classified elliptical galaxies according to how round or elongated they appear. An E0 galaxy is round; a very elongated elliptical galaxy is an E7. Three examples are shown. (a and c: Karl Gebhardt [University of Michigan], Tod Lauer [NOAO], NASA; b: Jean-Charles Cuillandre, Hawaiian Starlight, CFHT)

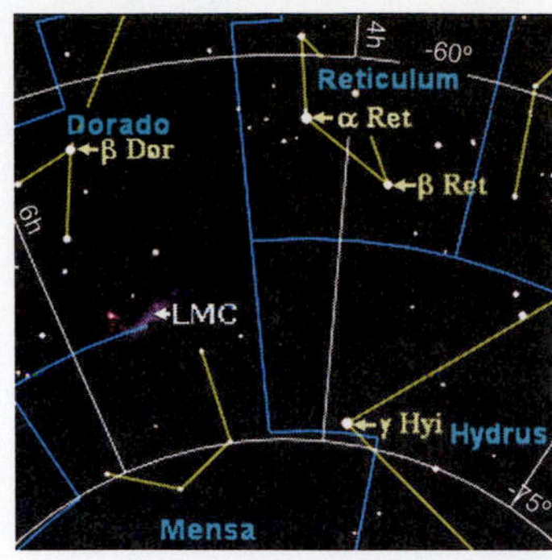

a Large Magellanic Cloud, an Irr I galaxy

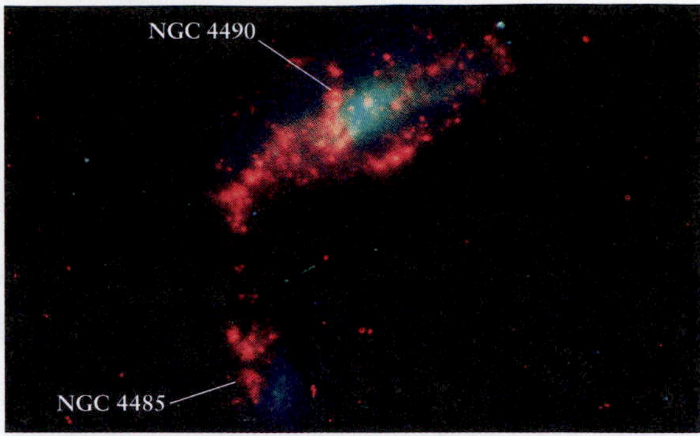

b NGC 4485 (Irr II) and NGC 4490 (Sc) galaxies

R I V U X G

FIGURE 16-13 Irregular Galaxies (a) At a distance of only 179,000 light-years, the Large Magellanic Cloud (LMC), an Irr I irregular galaxy, is passing close to our Milky Way Galaxy. About 62,000 light-years across, the LMC spans 22° across the sky, about 44 times the angular size of the full Moon. Note the huge H II region (called the Tarantula Nebula or 30 Doradus). Its diameter of 800 light-years and mass of 5 million solar masses make it the largest known H II region. (b) The small irregular (Irr II) galaxy NGC 4485 interacts with the highly distorted Sc galaxy NGC 4490, also called the Cocoon Galaxy. This pair is located in the constellation Canes Venatici. (a: Anglo-Australian Observatory; b: M. Altmann/Hoher Observatory)

> Instead of being classified an E0, what other classification could M105 in Figure 16-12 possibly have?

long-lived stars. However, billions of years ago, many ellipticals contained bright, massive stars that have since exploded as supernovae. We know this because the Chandra X-ray Observatory has discovered large numbers of remnant black holes and neutron stars in several of these galaxies.

FIGURE 16-14 Hubble's Tuning Fork Diagram Hubble summarized his classification scheme for galaxies with this tuning fork diagram. Elliptical galaxies are classified by how oval they appear, whereas spirals and barred spirals are classified by the sizes of their central bulges and the correlated winding of their spiral arms. An S0 or SB0 galaxy, also called a lenticular galaxy, is an intermediate type between ellipticals and spirals. It has a disk but no spiral arms.

16-5 Galaxies without global structure are called irregular

Edwin Hubble found some galaxies that cannot be classified as spirals, barred spirals, or ellipticals. He called these **irregular galaxies.** They are generally rich in interstellar gas, dust, and both young and old stars. Irregular galaxies with numerous OB associations and only hints of organized structure are denoted *Irr I.* The nearby Large Magellanic Cloud (LMC) (Figure 16-13a) and the Small Magellanic Cloud (SMC) are examples of Irr I galaxies. (Although they are among the closest galaxies to us, neither of them orbits the Milky Way.) Both can be seen with the naked eye from southern latitudes.

Occasionally, irregulars are observed that appear highly distorted and completely asymmetrical, as though created by collisions between galaxies or by violent activity in their nuclei. These galaxies are denoted *Irr II,* as shown by NGC 4485 (Figure 16-13b). Irregular galaxies are typically smaller and less massive than spirals, containing between about 10^8 and 3×10^{10} $M_\odot$.

16-6 Hubble presented spiral and elliptical galaxies in a tuning fork-shaped diagram

Edwin Hubble connected the three regularly shaped types of galaxies—spirals, barred spirals, and ellipticals—in a diagram shaped like a tuning fork (Figure 16-14). According to his scheme, S0 or SB0 galaxies, called **lenticulars** (lens-shaped), are an intermediate type between ellipticals and the two kinds of spirals.

Table 16-1 Some Properties of Galaxies

	Spiral (S) and barred spiral (SB) galaxies	Elliptical galaxies (E)	Irregular galaxies (Irr)
Mass ($M_{\odot}$)	10^9 to 4×10^{11}	10^7 to 10^{13}	10^8 to 3×10^{10}
Luminosity ($L_{\odot}$)	10^8 to 2×10^{10}	3×10^5 to 10^{11}	10^7 to 10^9
Diameter (ly)	1.6×10^4 to 8×10^5	3×10^3 to 6.5×10^5	3×10^3 to 3×10^4
Stellar populations	Disk: young Population I central bulge; halo: Population II and old Population I	Population II and old Population I	Mostly Population I
Percentage of observed galaxies	77%	*20%	3%

*This percentage does not include dwarf elliptical galaxies that are as yet too dim and distant to detect. Hence, the actual percentage of galaxies that are ellipticals is likely to be higher than shown here.

Although they often look somewhat like ellipticals, lenticular galaxies have both a central bulge and a disk, like spiral galaxies, but they lack spiral arms. For want of any better scheme, the irregular galaxies are sometimes placed between the ends of the tuning fork tines of the Hubble diagram.

The idea that one type of galaxy may change into another type has been in and out of favor with astronomers for decades. Indeed, such transformations were part of Hubble's motivation in relating the different types of galaxies as he did. As we will see later in this chapter and in Chapter 18, extensive observations by the Hubble Space Telescope and by ground-based telescopes reveal that interactions between galaxies sometimes do lead to changes in their structures. For example, when two disk galaxies merge they often morph into a giant elliptical galaxy. However, it appears that most galaxies remain unchanged in overall shape once they are fully formed. The properties of the different types of galaxies are summarized in Table 16-1.

16-7 Galaxies built up in size over time

In Chapter 18, we will explore the observational evidence for the formation and early evolution of galaxies. Nevertheless, it is worth summarizing here what astronomers know about why the galaxies have the structures we observe. With the exception of dwarf ellipticals, galaxies formed from smaller ensembles of stars, gas, and dust. The underlying dark matter that clumped together in the young universe drew this matter together, often assisted by supermassive black holes formed early in the life of the universe. These black holes became the nuclei of the galaxies. The more massive the black hole, the larger a disk galaxy's central bulge. Once formed, about three-quarters of all galaxies have maintained their structures, whereas the remaining quarter have changed (for example, from spiral to elliptical).

The Large Magellanic Cloud is visible to the naked eye in the southern hemisphere. From Figure 16-13, what terrestrial things do you think it could be mistaken for?

Normal (unbarred) spiral galaxies formed when swirling eddies of gas-rich matter collided and formed a disk centered on an especially massive black hole and lots of dark matter. When the amount of dark matter and the mass of the central black hole were too low, the matter in the central regions of the disk had bar-shaped orbits, creating barred spiral galaxies.

When the collision of galaxy-forming gases led to especially rapid star formation, the resulting galaxies were elliptical or irregular. Giant elliptical galaxies were (and still are) formed from the collisions of two spiral galaxies (of any type).

Galaxy collisions do not just occur in pairs. Stephan's quintet, discovered in 1877 by the French astronomer Edouard Stephan, is a group of five galaxies bound together. Four of its members are undergoing multiple collisions with each other. They will eventually merge to form a single giant elliptical galaxy. This is not a unique interaction: Collisions of other groups of three or four galaxies have been observed since then.

What physical property of matter discussed in Chapter 2, besides gravitation, is essential in forming all spiral galaxies?

CLUSTERS AND SUPERCLUSTERS

As we consider the vastness of galaxies, it is hard to imagine that structures this huge orbit each other in groups. But they do.

16-8 Galaxies occur in clusters, which occur in larger clumps called superclusters

3 Galaxies are not scattered randomly throughout the universe, but rather they are grouped together in **clusters.** The Hercules cluster (Figure 16-15), so called because it is located in the constellation of Hercules, is typical. Observations of galactic motion indicate that members of a cluster of galaxies are gravitationally bound together: The galaxies in a cluster orbit each other and, occasionally, even collide. Calculations based on observations of the Andromeda Galaxy's velocity strongly suggest that it and our Milky Way will collide in a few billion years.

FIGURE 16-15 A Cluster of Galaxies This group of galaxies, called the Hercules cluster, is about 650 million light-years from Earth. Both elliptical and spiral galaxies within this cluster can be easily identified. (NOAO)

Clusters of galaxies are themselves grouped together in huge associations, called **superclusters.** A typical supercluster contains dozens of individual clusters spread over a volume typically 150 million light-years across. Figure 16-16 shows the distribution of superclusters in our vicinity of the universe. The nearer ones, out to the Virgo cluster, are members of our *Local Supercluster.* Observations indicate that most superclusters are not gravitationally bound units—most clusters in each supercluster are drifting away from most of the other clusters in that same supercluster. Furthermore, the superclusters are all moving away from one another.

How superclusters are distributed in space started to become clearer in the early 1980s, when astronomers began systematic surveys of the positions of observable objects in the nearby universe. Completed in 2005, the first

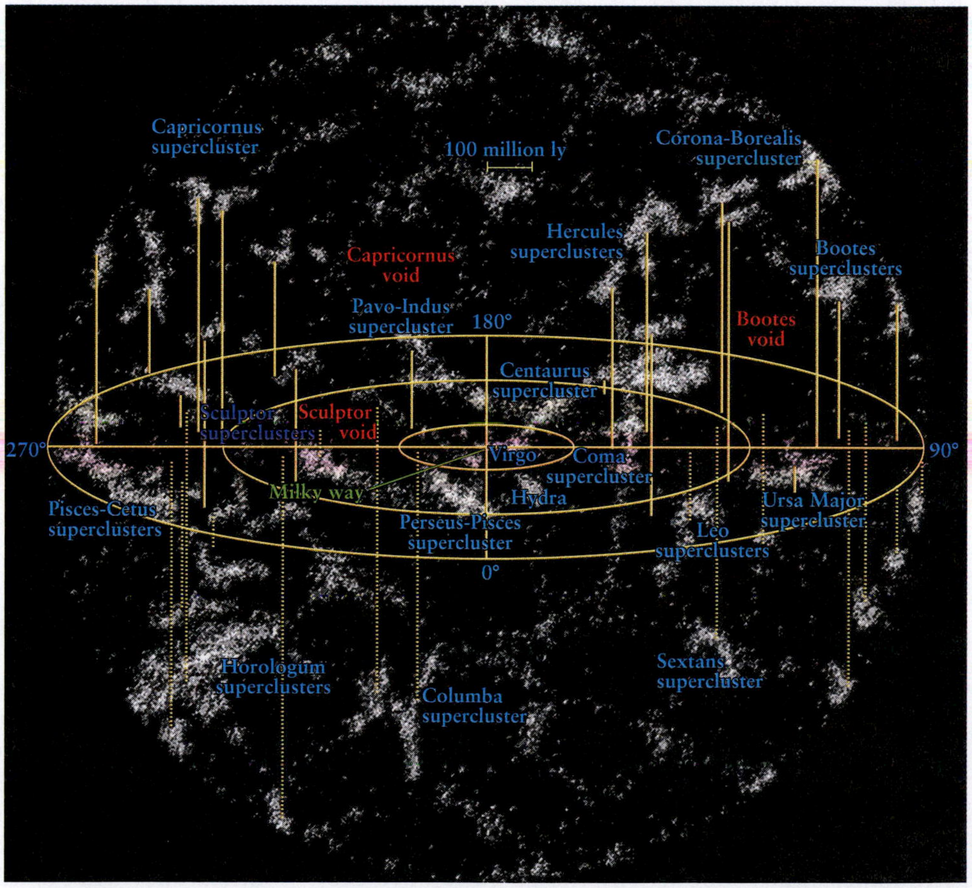

FIGURE 16-16 Superclusters in Our Neighborhood This diagram shows the distances and relative positions of superclusters within 950 million light-years of Earth. Note also the labeling of some of the voids, which are large, relatively empty regions between superclusters. (Kirk Korista)

Astronomers observe very hot (10^7-K) gas between clusters of galaxies. Why does this gas not appear in Figure 16-15?

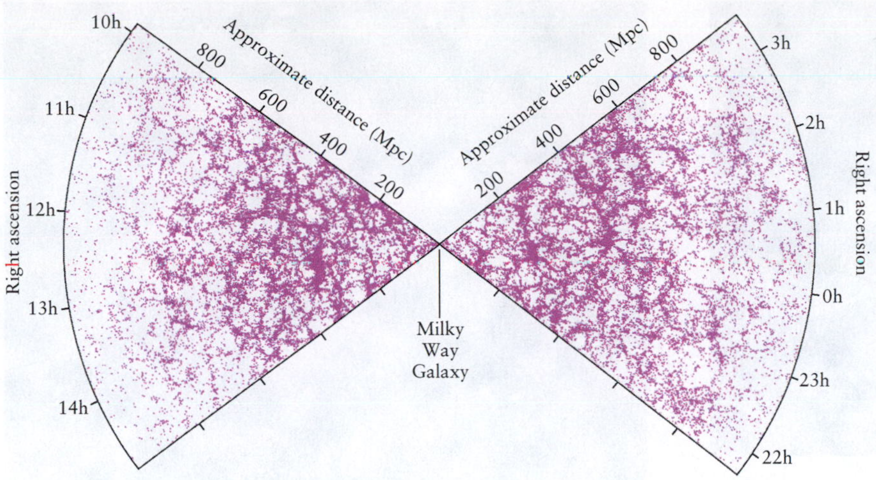

FIGURE 16-17 Structure in the Universe This map shows the distribution of 62,559 galaxies in two wedges extending in opposite directions from Earth out to distances of 3.25 billion light-years. For an explanation of right ascension (r.a.), see Section 1-3. Note the prominent voids surrounded by thin areas full of galaxies. (Courtesy of the 2dF Galaxy Redshift Survey Team)

Sloan Digital Sky Survey observed nearly 200 million galaxies, stars, and other celestial objects, and measured the redshifts of more than a million of them. This survey, made in the northern hemisphere, covered about one-quarter of the sky and gives us a three-dimensional map of 100 times the volume of space as we had mapped before. Another set of observations, called the Two Degree Field (2dF) Survey, studied regions of both the northern and southern hemispheres.

These data reveal that most superclusters are located on the boundaries between enormous bubble-like *voids* in which few galaxies are observed (Figures 16-16 and 16-17). These voids are roughly spherical and typically measure between 100 million and 400 million light-years in diameter. They are not completely empty, however. Observations reveal hydrogen clouds in some of them, while others may be subdivided by strings of dim galaxies.

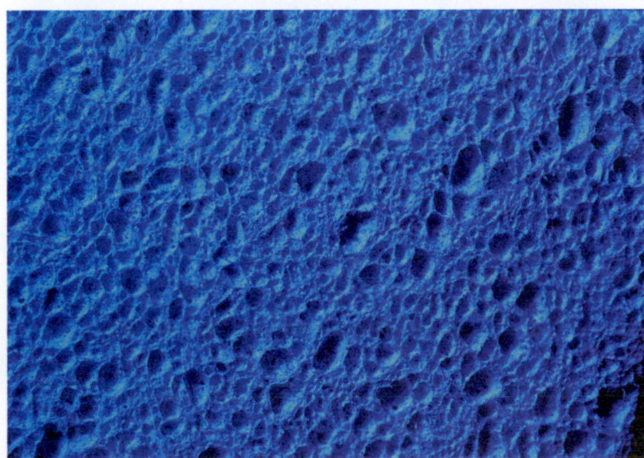

FIGURE 16-18 Foamy Structure of the Universe A sponge that recreates the distribution of bright clusters of galaxies throughout the universe. The empty spaces in the foam are analogous to the voids found throughout the universe. The spongy regions are analogous to the locations of most of the superclusters of galaxies. (Image Source/Super Stock)

The distribution of clusters of galaxies throughout the universe appears to have the structure of a sponge (Figure 16-18) or soap bubbles, with the galaxies located where the sponge or soapy material is, surrounded by voids. Astronomers believe that this spongy pattern contains important clues about conditions in the early universe that have not yet been fully fathomed.

16-9 Clusters of galaxies may appear densely or sparsely populated and regular or irregular in shape

A cluster of galaxies is said to be either a **poor cluster** or a **rich cluster**, depending on whether it contains less than or more than a thousand galaxies. For example, the Milky Way Galaxy, the Andromeda Galaxy, and the Large and Small Magellanic Clouds belong to a poor cluster called the **Local Group.** The Local Group contains roughly 40 galaxies, over a third of which are dwarf ellipticals. Figure 16-19 shows a map covering most of the Local Group. The Virgo cluster is the nearest rich cluster, with more than 2000 galaxies.

New galaxies in the Local Group continue to be found. These discoveries happen primarily because astronomers are developing techniques to observe objects that lie in the same plane as, but beyond, the Milky Way. In addition to the 1994 discovery of the Sagittarius Dwarf Galaxy, the dwarf galaxy Antlia (Figure 16-20) was first observed in 1997 and the Canis Major Dwarf (see Figure 15-16) was found in 2003. The Canis Major Dwarf is only about 25,000 light-years from the Milky Way.

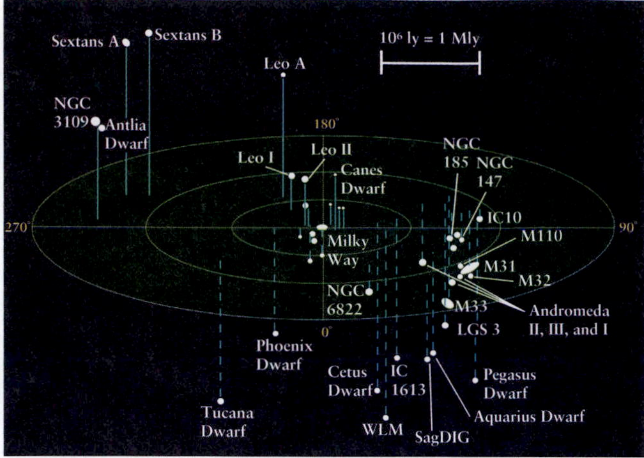

FIGURE 16-19 The Local Group Our Galaxy belongs to a poor, irregular cluster that consists of about 40 galaxies, called the Local Group. This map shows the distribution of about three-quarters of the galaxies. The Milky Way and Andromeda galaxies are the largest and most massive galaxies in the Local Group. Andromeda (M31) and the Milky Way are each surrounded by a dozen satellite galaxies. The recently discovered Canis Major Dwarf Galaxy is the Milky Way's nearest known neighbor.

R I V U X G

FIGURE 16-20 A Recently Discovered Member of the Local Group The galaxy Antlia was first detected in 1997. It lies about 3 million light-years away, outside the region depicted in Figure 16-19. This galaxy contains only about a million stars. (M. J. Irwin, Royal Greenwich Observatory, and A. B. Whiting and G. K. T. Hau, Institute of Astronomy, Cambridge University)

Astronomers also categorize clusters of galaxies according to their shapes. A **regular cluster** is distinctly spherical, with a marked concentration of galaxies at its center. Numerous gravitational interactions over billions of years have spread the galaxies into this distribution. In contrast, galaxies in an **irregular cluster** are more randomly scattered about a sprawling region of the sky. The Virgo cluster is irregular.

The nearest example of a rich, regular cluster is the Coma cluster, located about 300 million light-years from us in the constellation of Coma Berenices (Figure 16-21). Despite its great distance, more than 1000 bright galaxies within it are easily visible from Earth.

What are two reasons that new galaxies are still being discovered in the Local Group?

Rich, regular clusters, like the Coma cluster, contain mostly elliptical and lenticular galaxies. Only 15% of the Coma cluster's presently known galaxies are spirals and irregulars. It is likely that the Coma cluster contains many thousands of as-yet-undetected dwarf ellipticals—maybe as many as 10,000 galaxies overall. Irregular clusters, such as the Virgo cluster and the Hercules cluster (see Figure 16-15), display a more even mixture of galaxy types. Two-thirds of the 200 brightest galaxies in the Hercules cluster are spirals, nearly a fifth are ellipticals, and the rest are irregular.

16-10 Galaxies in a cluster can collide and combine

Occasionally, two or more galaxies in a cluster or two or more galaxies from neighboring clusters collide. Sometimes these galaxies pass through one another (Figure 16-22) and sometimes they merge. In 2002, the Hubble Space Telescope observed four galaxies colliding simultaneously. There is so much space between stars that the probability of two stars crashing into each other during galactic collisions is extremely small. On the other hand, the galaxies' huge clouds of interstellar gas and dust are so large that they do collide, slamming into each other and producing strong shock waves. The colliding interstellar clouds are stopped in their tracks.

A violent collision can strip both galaxies of their interstellar gas and dust. The energy associated with the collision heats this gas and dust to extremely high temperatures. This process is a major source of the hot **intergalactic gas** often observed in rich, regular clusters. The Chandra X-ray Observatory has observed intergalactic gas with temperatures between 300,000 and 100 million K.

Another source of intergalactic gas is gas escaping from dwarf galaxies. This gas is ejected by supernovae of stars created in a burst of star formation when these galaxies formed. Whereas our Galaxy's gravitational force traps supernova remnants in the disk, this gas escapes into intergalactic space from the relatively weak gravitational attraction of dwarf galaxies.

Less violent collisions and near misses between galaxies allow the compressed interstellar gas and dust to cool sufficiently to have new stars form from it. These collisions can also stimulate galaxywide star formation, which may account for the **starburst galaxies** that blaze with the light of numerous newborn stars. These galaxies

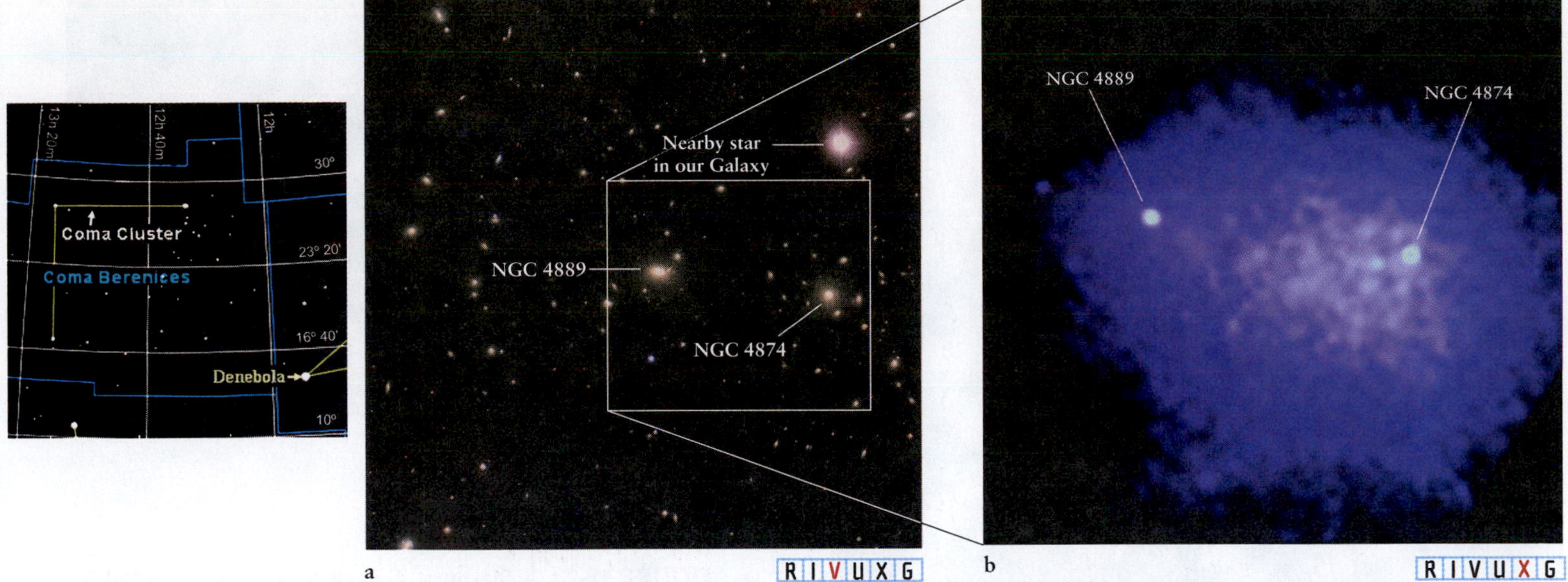

FIGURE 16-21 The Coma Cluster (a) This rich, regular cluster, containing thousands of galaxies, is about 300 million light-years from Earth. (b) Regular clusters are composed mostly of elliptical and lenticular galaxies and are sources of X rays. This Chandra image shows Coma's central region, which is 1.5 million light-years across. The gas cloud emitting most of these X rays has a temperature of 100 million K. (a: O. Lopez-Cruz and I. Shelton, NOAO/AURA/NSF; b: NASA/CXC/SAO/A. Vikhlinin et al.)

are characterized by bright centers surrounded by clouds of warm interstellar dust, indicating a recent, vigorous episode of star birth (Figure 16-23). Their warm dust is so abundant that starburst galaxies are among the most luminous objects in the universe at infrared wavelengths.

Observational evidence indicates that globular clusters are formed during collisions of interstellar gas and dust. While most globular clusters are over 10 billion years old, some galaxies also contain very young clusters and, like the Milky Way, have some very dense star-forming regions that are believed to be precursors

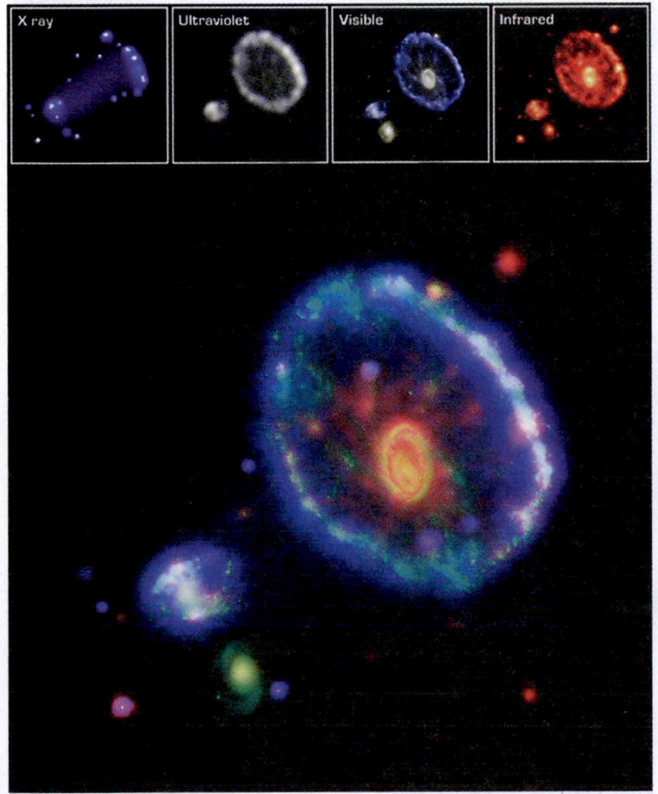

b

R I V U X G

FIGURE 16-22 Galaxies with Rings (a) A composite image of the Cartwheel Galaxy. This ring-shaped assemblage 500 million light-years from Earth is likely the result of one galaxy, probably the blue-white one below it at the eight o'clock position, having passed through the middle of the larger one. Astronomers suspect that the passage created a circular density wave in the Cartwheel that stimulated a burst of star formation, creating many bright blue and white stars. Ultraviolet is in blue, visible light in green, infrared in red, and X-ray in violet. (b) Infrared image of the Andromeda Galaxy. The ring of hot dust indicates star formation, probably caused by the passage of another galaxy through Andromeda. The fact that the ring is disturbed suggests that yet another galaxy had a close interaction with Andromeda. (a: NASA/JPL-Caltech/P. Appleton [SSC/Caltech]; b (infrared): NASA/JPL-Caltech/K. Gordon [Univ. of Arizona]; b (visible): NOAO/AURA/NSF)

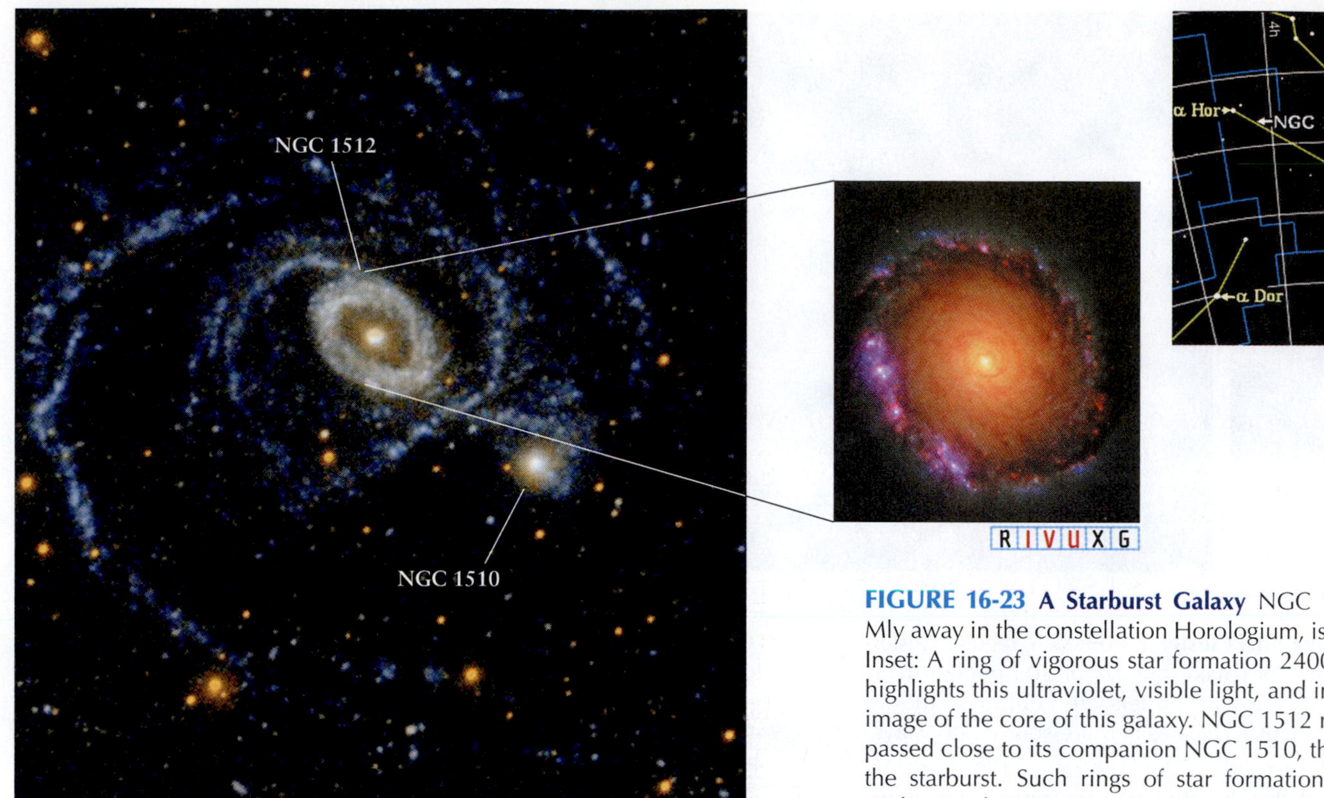

FIGURE 16-23 A Starburst Galaxy NGC 1512, located 30 Mly away in the constellation Horologium, is 70,000 ly across. Inset: A ring of vigorous star formation 2400 light-years wide highlights this ultraviolet, visible light, and infrared composite image of the core of this galaxy. NGC 1512 may have recently passed close to its companion NGC 1510, thereby stimulating the starburst. Such rings of star formation are common in starburst galaxies. (NASA/JPL/Caltech)

of new globular clusters. Whether those in our Galaxy formed as a result of an early collision of the Milky Way with another galaxy has yet to be determined.

The Irr II starburst galaxy M82 is one member of a nearby cluster of about a dozen galaxies that include the beautiful spiral galaxy M81 and a fainter compan-

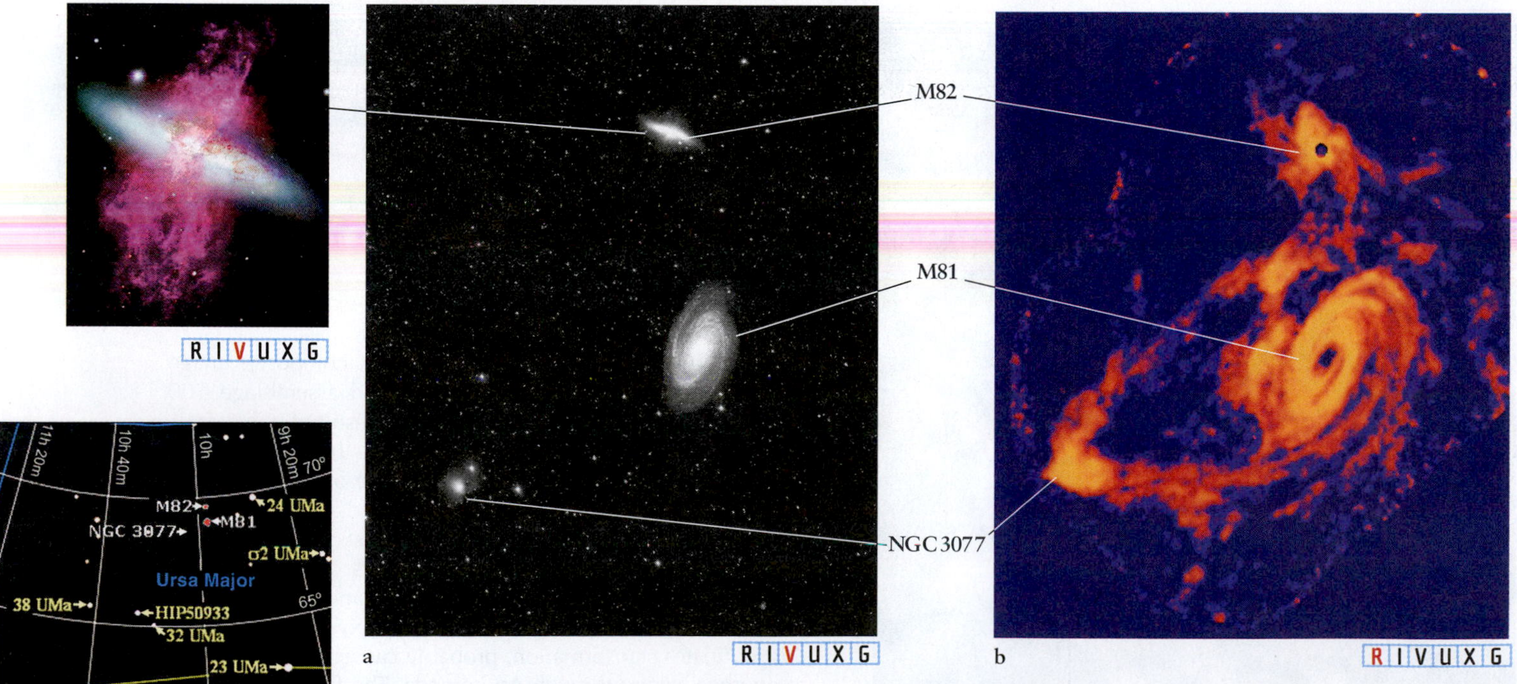

FIGURE 16-24 The M81 Group The Irr II starburst galaxy M82 (see also the chapter-opening image) is in a nearby cluster of about a dozen galaxies, including the spectacular spiral M81. Several of the galaxies in this cluster are connected by streamers of hydrogen gas. (a) The three brightest galaxies in the cloud at visual wavelengths. The inset on the left shows large volumes of hydrogen gas, in red, being ejected from M82. (b) This radio image, created from data taken by the Very Large Array, shows the streamers of hydrogen gas that connect the bright galaxies and also several dim ones, seen as regions of bright orange here. (a: Palomar Sky Survey; b: M. S. Yun, VLA, and Harvard)

ion, called NGC 3077 (Figure 16-24a). A radio survey of that region of the sky revealed enormous streams of hydrogen gas connecting these and other, smaller galaxies in that cluster (Figure 16-24b). The loops and twists in these streamers suggest that the three galaxies have had several close encounters over the ages, most recently about 600 million years ago. Gravitational interactions between colliding galaxies can also result in thousands of stars being hurled out into intergalactic space along huge, arching streams, as seen in Figure 16-25a.

In rich clusters, astronomers observe near misses between galaxies (Figure 16-25b). If galaxies are surrounded by extended halos of dim stars, these near misses strip the galaxies of these outlying stars. In this way, a loosely dispersed sea of dim stars populates the space between galaxies in a cluster.

Similarly, after galaxies collide, some interior stars are flung far and wide, scattering material into intergalactic space. Such isolated stars have been observed by the Hubble Space Telescope. However, other stars slow down, often causing the remnants of the galaxies to

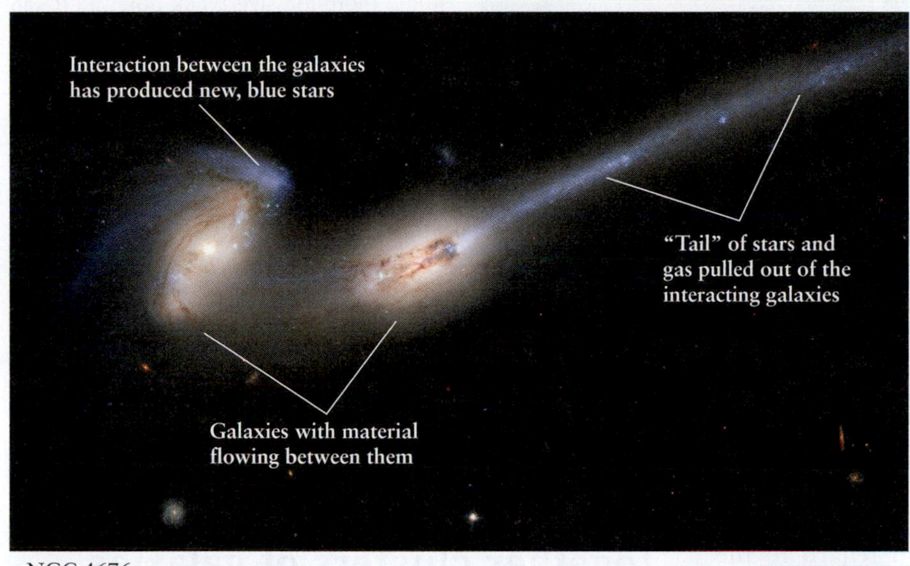

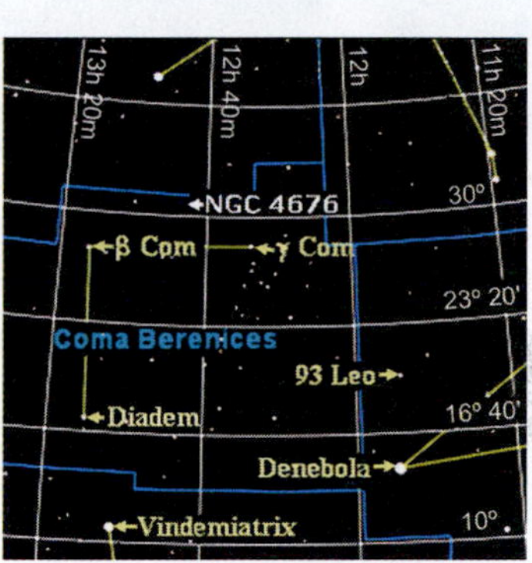

a NGC 4676 R I V U X G

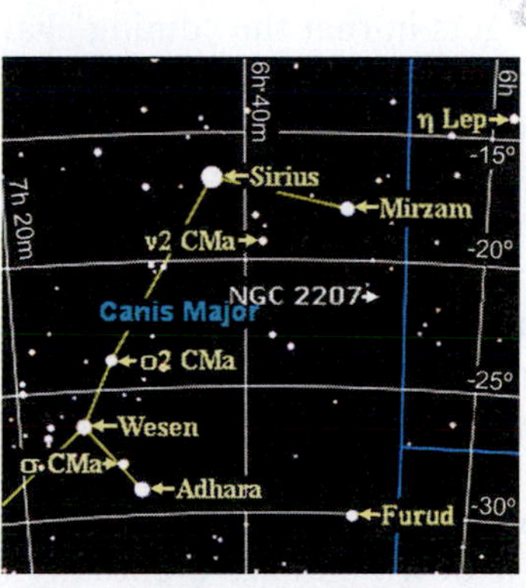

b IC 2163 (*left*) and NGC2207 R I V U X G

FIGURE 16-25 Interacting and Colliding Galaxies (a) Pairs of colliding galaxies often exhibit long "antennae" of stars ejected by the collision. This particular system is known as NGC 4676 or "the Mice" (because of its tails of stars and gas). It is 300 million light-years from Earth in the constellation Coma Berenices. The collision has stimulated a firestorm of new star formation, as can be seen in the bright blue regions. Mass can also be seen flowing between the two galaxies, which will eventually merge. (b) These two galaxies, NGC 2207 (right) and IC 2163, are orbiting and tidally distorting each other. Their most recent close encounter occurred 40 million years ago when the two were perpendicular to each other and about 1 galactic diameter apart. Computer simulations indicate that they should eventually coalesce. (a: NASA, H. Ford/JHU, G. Illingworth/UCSC/Lick, M. Clampin/STScI, G. Hartig/STScI, The ACS Science Team, and ESA; b: NASA)

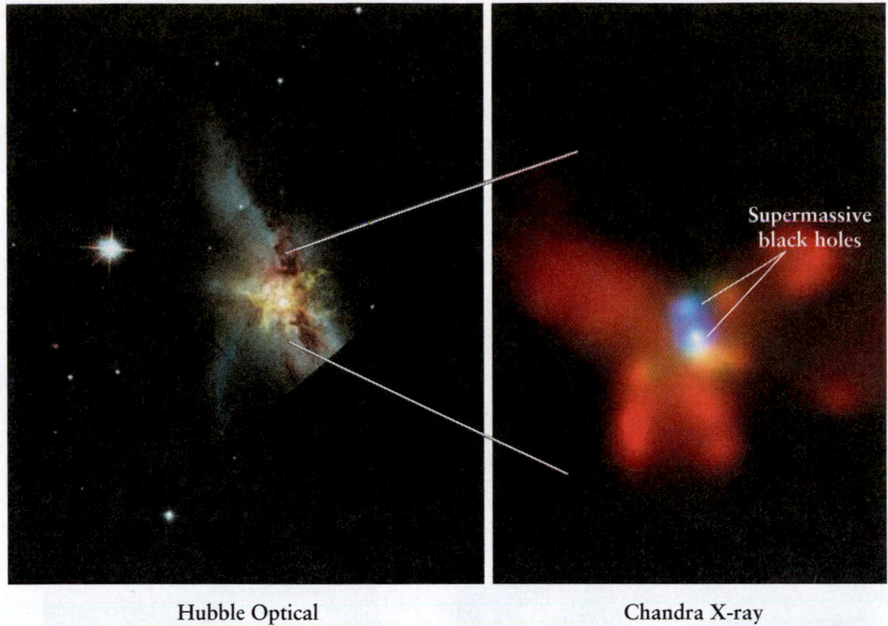

FIGURE 16-26 Merging Galaxies This contorted object, NGC 6240, in the constellation Ophiuchus is the result of two spiral galaxies in the process of merging. The widespread blue-green area reveals that the collision between the two galaxies has triggered an immense burst of star formation. Inset: The Chandra X-ray Observatory shows that at the heart of this system are two supermassive black holes, one from each of the original galaxies. Within a few hundred million years, these black holes are expected to merge into a single more-massive black hole. (R. P. van der Marel/J. Gersgen/STScI/NASA; inset: S. Komossa/G. Hasinger/MPE et al., CXC/NASA)

merge. Several dramatic examples of **galactic mergers** have been discovered, such as the galaxy known as NGC 6240 (Figure 16-26). The supermassive black holes in the nuclei of the original two galaxies (see Section 14-8) have not yet merged (see Figure 16-26 inset), but are expected to do so within a few hundred million years.

As mentioned in Chapter 15 with regard to the Milky Way tearing apart the Canis Major Dwarf Galaxy, astronomers also speak of galactic cannibalism, which occurs when a large galaxy captures and "devours" a smaller one. Cannibalism differs from mergers in that the "dining" galaxy is significantly bigger than its "dinner," whereas merging galaxies are about the same size.

Many astronomers suspect that giant ellipticals are the product of galactic mergers and cannibalism, including spiral-spiral collisions. Observations reveal that giant galaxies typically occupy the centers of rich clusters. Once formed, giant ellipticals can continue to grow because smaller galaxies are often located around them (see Figure 16-10). As they pass through the extended halo of a giant elliptical, these smaller galaxies slow down and are eventually consumed by the larger galaxy.

Figure 16-27, a computer simulation, shows a large, disk-shaped galaxy devouring a small satellite galaxy. The large galaxy consists of 90% stars (blue) and 10% gas (white) by mass. It is surrounded by a halo of dark matter whose mass is about 3.3 times that of the disk. The satellite galaxy, which has a tenth of the mass of the large galaxy, contains only stars (yellow). Initially, the satellite is in circular orbit about the large galaxy. Note that spiral arms appear in the large galaxy as the collision proceeds. Two billion years elapse as the satellite spirals in toward the core of the large galaxy. Although much material is stripped from the satellite, most of its stars plunge into the central region of the large galaxy.

16-11 Dark matter helps hold together clusters of galaxies

What keeps the rapidly moving galaxies together in clusters? There must be sufficient matter to provide the necessary gravitational force to bind galaxies together in clusters and, in some cases, clusters into gravitationally bound superclusters. However, *no cluster of galaxies contains enough visible matter to stay bound together.* Most of the matter binding clusters of galaxies is dark matter (matter that we cannot yet directly detect). This must be true because, otherwise, the galaxies would have long ago wandered apart in random directions, and the clusters would no longer exist. Recall that a similar problem arose in Chapter 15 regarding dark matter in individual galaxies. Analyses of the total visible mass demonstrate that the total mass needed to bind together a typical rich cluster is 10 times greater than the mass of material that shows up on visible-light images. Furthermore, observations of the distribution of galaxies in clusters have determined that the dark matter usually has the same distribution as the galaxies in the cluster, and that each galaxy corresponds to a concentration of dark matter.

Quite a bit of evidence has accumulated to support the idea of extended dark matter halos surrounding galaxies. Many galaxies have rotation curves similar to

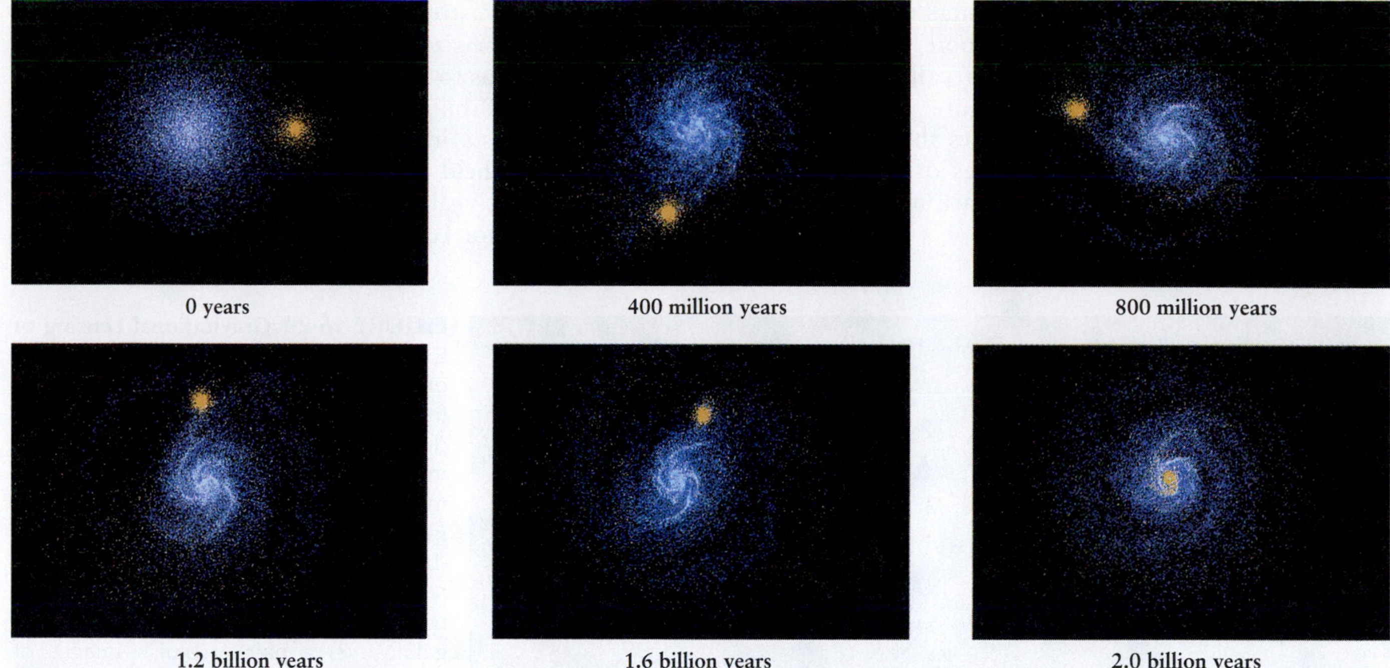

FIGURE 16-27 Simulated Galactic Cannibalism This computer simulation shows a small galaxy (yellow stars) being devoured by a larger, disk-shaped one (blue stars, white gas). Note how spiral arms are generated in the disk galaxy by its interaction with the satellite galaxy. (Lars Hernquist, Institute for Advanced Study with simulations performed at the Pittsburgh Supercomputing Center)

those of our Milky Way (recall Figure 15-18). These rotation curves remain remarkably flat out to surprisingly great distances from the galaxies' centers. For example, Figure 16-28 shows the rotation curves of four spiral galaxies. In all cases, the orbital speed is fairly constant. According to Kepler's third law, we should see a decline in orbital speed toward the outer portions of a galaxy. In most cases, we do not, which implies that we still have not detected the true edge of these and many similar galaxies. As with the Milky Way Galaxy, astronomers conclude that a considerable amount of dark matter must extend well beyond the visible portions of most galaxies.

Astronomers using X-ray telescopes have discovered that some of the dark matter is actually very hot intergalactic gas. Satellite observations have revealed X rays pouring from the space between galaxies in rich clusters that does not emit discernable visible light. This X-ray radiation is emitted by substantial amounts of hot intergalactic gas at temperatures between 10 million and 100 million K. Figure 16-21b shows this hot gas in the center of the Coma cluster. Observations by the Chandra X-ray Observatory indicate that this intergalactic gas contains twice as much mass as exists in all of that cluster's galaxies. Because its physical nature has now been identified, this intergalactic matter is no longer considered to be "dark."

Nature assists astronomers with visualizing dark matter by the way its mass gravitationally warps space, as described by Einstein's general theory of relativity (see Section 14-2). We saw in Chapter 15 how the same effect occurs on small scales here in the Milky Way. The local distortion is called *microlensing,* and when dark matter and the visible matter in entire gal-

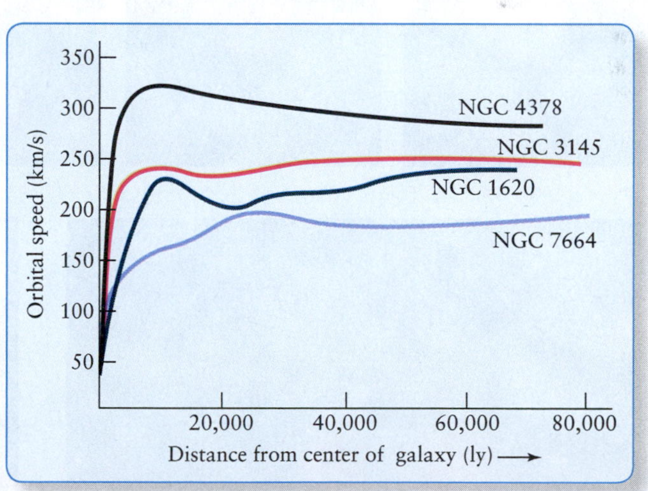

FIGURE 16-28 The Rotation Curves of Four Spiral Galaxies This graph shows how the orbital speed of material in the disks of four spiral galaxies varies with the distance from the center of each galaxy. If most of each galaxy's mass were concentrated near its center these curves would fall off at large distances. But these and many other galaxies have flat rotation curves that do not fall off. This indicates the presence of extended halos of dark matter. See Figure 15-18 to compare these to the Milky Way's rotation curve.

After the Milky Way collides with the Andromeda Galaxy, in what type of galaxy will our descendants likely live?

axies or clusters of galaxies focus light from more distant galaxies, the effect is called **gravitational lensing** (Figure 16-29a). Figure 16-29b shows the gravitationally lensed images of distant galaxies caused by dark matter and closer galaxies.

In 2000, astronomers observed an intergalactic cloud of gas the size of a galaxy. Observations revealed that for that gas to remain bound together, it must contain about 100 billion solar masses of matter. Because no galaxy exists in that space, astronomers conclude that the gas is held together by that much dark matter.

Sometimes visible matter and dark matter separate, shown in Figure 16-29c. Presumably the result of two

Apparent location

Actual location

Apparent location

Massive galaxy and/or dark matter

a

b

R I V U X G

FIGURE 16-29 Gravitational Lensing of Extremely Distant Galaxies (a) Schematic of how a gravitational lens works. Light from the distant object changes direction due to the gravitational attraction of the intervening galaxy and underlying dark matter. The more distant galaxy appears in different places than it actually is. (b) Three examples of gravitational lensing: (1) The blue ring is a galaxy that has been lensed by the redder elliptical galaxy; (2) a pair of bluish images of the same object lensed symmetrically by the brighter, redder galaxy between them; and (3) the lensed object appears as a blue arc under the gravitational influence of the group of four galaxies. (c) This cluster contains over a thousand galaxies. The red shows the cluster's hot gas, while the blue is its dark matter, determined by gravitational lensing. (d) A mechanism proposed to explain the separation of gas and dark matter in clusters like (c). (b: NASA, ESA, A. Bloton [Harvard-Smithsonian CfA] and the SLACS Team; c: X-ray: NASA/CXC/CfA/M. Markevitch et al.; visible: NASA/STScI; Magellan/U. Arizona/D. Clew et al.; lensing map: NASA/STScI; SO FW; Magellan/U. Arizona/D. Clowe et al.; d: NASA/CXC/M. Weiss)

c Composite image of galaxy cluster 1E0657-56 showing visible galaxies, X-ray emitting gas (red) and dark matter (blue). R I V U X G

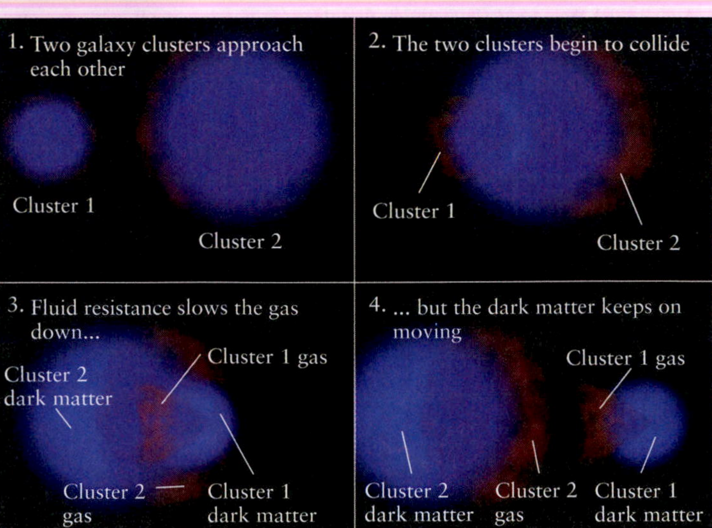

d A model of how the gas and dark matter in 1E0657-56 could have become separated clusters of galaxies colliding (Figure 16-29d). The details of the mechanism by which this separation occurs is still under investigation. In 2007, astronomers observed distant galaxies undergoing gravitational lensing without the presence of an intervening visible galaxy. They deduced that the effect was caused by just the presence of dark matter.

The amount of gravitational lensing by dark matter alone or by dark matter and a galaxy or by a cluster of galaxies provides astronomers with more than just enlarged, distorted images of distant objects. It tells us how much mass is contained in the dark matter and galaxy or cluster that does the lensing. The greater the mass doing the lensing, the greater the effect on the distant object. The visible matter in the galaxies accounts for only about 10% to 20% of the gravitational lensing that occurs. Thus, dark matter accounts for 80% to 90% of the mass in the universe.

SUPERCLUSTERS IN MOTION

Whenever an astronomer finds an object in the sky, one of the first tasks is to determine its composition. As we saw in Chapters 3 and 4, that means attaching a spectrograph to a telescope and recording the object's spectrum. As long ago as 1914, the American astronomer Vesto Slipher, working at the Lowell Observatory in Arizona, took spectra of "spiral nebulae," now known to be spiral galaxies. He was surprised to discover that the spectral lines of 11 of the 15 spiral nebulae that he studied were substantially redshifted. These redshifts indicate that they are moving away from the Milky Way at significant speeds. This result left scientists with a major puzzle: Is the universe expanding?

16-12 The redshifts of superclusters indicate that the universe is indeed expanding

During the 1920s, Edwin Hubble and Milton Humason recorded the spectra of many galaxies with the 100-in. telescope on Mount Wilson, confirming that most galaxies are rapidly receding from the Milky Way. This recession implies that the universe *is* expanding, although, as we will see in a moment, it does not imply that the Milky Way is at the center of it!

4 Using the Doppler effect (recall *An Astronomer's Toolbox 4-3*), Hubble calculated the speed at which the galaxies are moving away from us. Using techniques such as determining the brightness of Cepheid variables (see Section 15-1), he also estimated the distances to several of these galaxies. He found a direct correlation between the distance to a galaxy and the size of its redshift: *Galaxies in distant clusters and superclusters are moving away from us more rapidly than galaxies in nearby clusters and superclusters.* Figure 16-30 shows this correlation for five elliptical galaxies. This recessional motion pervades the universe and is now called the **Hubble flow.**

If the dark matter in Figure 16-29c suddenly vanished, what would we see of the clusters of galaxies behind it?

The separation of clusters in the same supercluster is slower than the general Hubble flow of the universe because these clusters are relatively close to one another, so their gravitational attractions slow each other down.

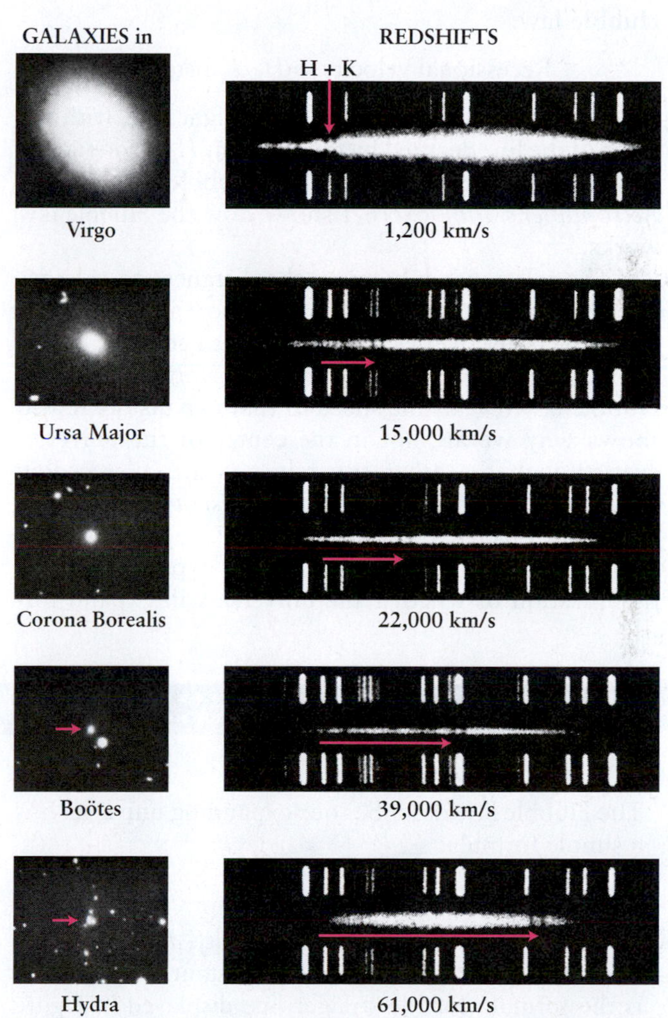

FIGURE 16-30 Five Galaxies and Their Spectra The photographs of these five elliptical galaxies were all taken at the same magnification. They are labeled according to the constellation in which each galaxy is located. The spectrum of each galaxy is the hazy band between the comparison spectra at the top and bottom of each plate. In all five cases, the so-called H and K lines of calcium are seen. The recessional velocity (calculated from the Doppler shifts of the H and K lines) appears below each spectrum. Note that the fainter—and thus more distant—a galaxy is, the greater is its redshift. (Carnegie Observatories)

Keep in mind that the Hubble flow does not occur for galaxies in any given cluster, because all of those galaxies are gravitationally bound to each another. To recap: The Hubble flow applies to the separation of superclusters from one another and many clusters of galaxies from one another.

When Hubble plotted the redshift data on a graph of distance versus velocity, he found that the points lie nearly along a straight line. Figure 16-31 shows the relationship between the distances to galaxies and their recessional motion, including many of whose distances were measured using the apparent magnitude of Type Ia supernovae. The information in this figure can be stated as a formula, called the **Hubble law:**

$$\text{Recessional velocity} = H_0 \times \text{distance}$$

This equation simply describes a straight line, with the slope of the line denoted by the constant H_0 (pronounced "H naught"), commonly called the **Hubble constant.** *An Astronomer's Toolbox 16-1* shows how the Hubble law works.

The relationship between the distances to galaxies and their redshifts was one of the most important astronomical discoveries of the twentieth century. It tells us that we are living in an expanding universe, and the Hubble law reveals the speed of that expansion. It also shows why we are not in the center of the universe, even though the other superclusters are all receding from us, as discussed in *Guided Discovery: The Expanding Universe.*

The discovery that the universe is expanding raised the question of whether the universe will expand forever or whether the gravitational attraction between all of its parts is enough to cause it some day to stop and recollapse. Imagine trying to escape forever from Earth. A normal jump might raise you up ⅓ m or so, but if you had a sufficiently powerful cannon to shoot you upward fast enough, you could escape Earth's gravitational attraction completely and enter interplanetary space. To determine whether the superclusters are ex-

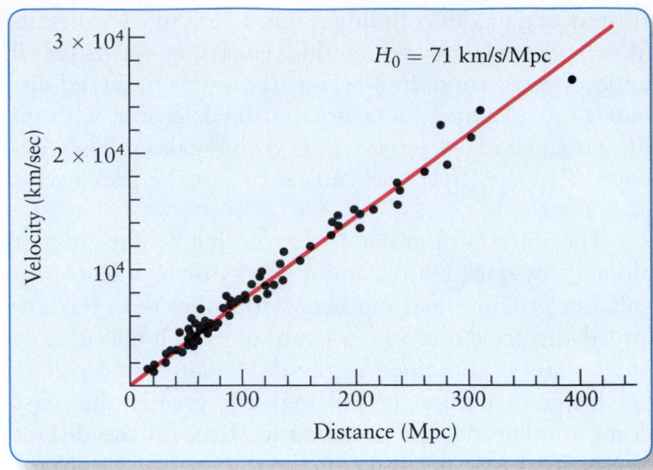

ANIMATION 16.2

FIGURE 16-31 The Hubble Law The distances and recessional velocities of distant galaxies are plotted on this graph. The straight line is the "best fit" for the data. This linear relationship between distance and speed is called the Hubble law. For historical reasons, distances between galaxies, clusters of galaxies, and superclusters of galaxies are usually given in megaparsecs (Mpc) rather than millions of light-years.

An Astronomer's Toolbox 16-1

The Hubble Law

The Hubble law describes our expanding universe. It is a simple formula:

$$v = H_0 \times d,$$

where v is a galaxy's recessional velocity, d is its distance from Earth, and H_0 is the Hubble constant. The equation is the formula for the straight line displayed in Figure 16-31, and the Hubble constant is the slope of this line.

As Figure 16-31 shows, we usually measure a galaxy's velocity in kilometers per second and its distance in megaparsecs. You will see in Section 16-13 that astronomers have determined the Hubble constant to be $H_0 \approx 71$ km/s/Mpc. If you have distances in millions of light-years (Mly), you can convert H_0 using 1.00 Mpc = 3.26 Mly.

Example: A galaxy 1 Mpc from us is moving away at a rate of

$$v = (71 \text{ km/s/Mpc}) \times (1 \text{ Mpc}) = 71 \text{ km/s}$$

A galaxy 2 Mpc away is therefore receding at 142 km/s, and so on. A galaxy located 100 million parsecs from Earth rushes away from us with a speed of 7100 km/s or 16 million mi/h!

Compare! While a supercluster of galaxies 10 Mpc away recedes at 710 km/s, the solar system orbits the center of our Galaxy at a comparable 230 km/s.

Try these questions: At what speed is a galaxy observed to be 8 billion light-years away moving from us? How far away from us is a galaxy observed to be moving away at 216 km/s? If H_0 were 50 km/s/Mpc, would a galaxy 4 billion light-years away be moving away from us faster or slower than the same galaxy with $H_0 = 71$ km/s/Mpc?

(Answers appear at the end of the book.)

Guided Discovery

The Expanding Universe

To understand the universe, scientists build models—mathematical representations of the world around us. In developing these pictures, it often helps to begin with a simple analogy. Because the expansion of the universe is hard to visualize, imagine for a moment that batter for a chocolate chip cake is floating in an oven in the International Space Station. As the cake bakes, it expands and the chocolate chips move farther apart. The chips do not get larger and do not move through the batter. Each chocolate chip remains at rest in its own little bit of cake and is carried along as the cake spreads out.

Now, think of each chocolate chip as a supercluster of galaxies and the batter between the chocolate chips as the rest of spacetime. As the universe expands, the distance between widely separated superclusters of galaxies grows larger and larger. The expansion of the universe is the expansion of spacetime.

Suppose you were on one particular chocolate chip, say the chocolate chip labeled 3 in the oven at the left in the figure. You would see chocolate chips 2 and 4 moving away from you with equal velocities. Every 10 min, chocolate chips 2 and 4 get 1 cm farther away as the cake expands, as shown in the oven on the right. In that same time interval, chocolate chips 1 and 5 move twice as far away from you as chocolate chips 2 and 4. To go twice as far away in the same time, chocolate chips 1 and 5 must be moving twice as fast as the closer chocolate chips 2 and 4. This is exactly what Hubble's law of cosmological expansion (recessional velocity $= H_0 \times$ distance) says: Double the distance and you double the velocity.

To see why Earth need not occupy the center of the universe, put yourself on chocolate chip 4. You see chocolate chips 3 and 5 moving away at the same rate—in fact, they recede at the same rate as you saw chocolate chips 2 and 4 move away when you were on chocolate chip 3. From chocolate chip 4, chocolate chips 2 and 6 are moving away twice as fast as chocolate chips 3 and 5. From any chocolate chip, you will see all of the other chocolate chips moving away. Similarly, no matter where we are in the universe, we see all of the other superclusters receding from us in the same manner.

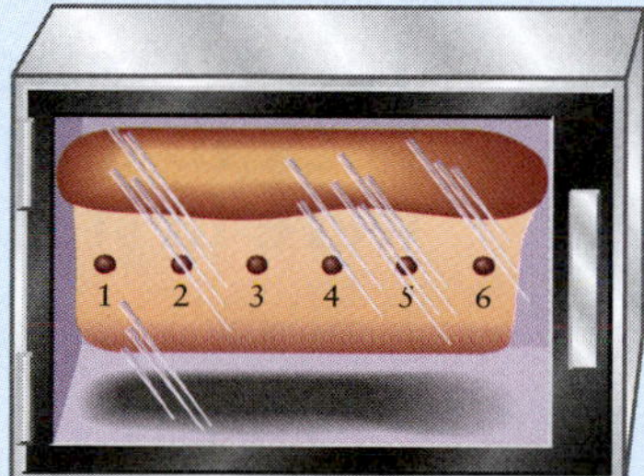

FIGURE GD16-1 The Expanding Chocolate Chip Cake Analogy The expanding universe can be compared to a chocolate chip cake baking and expanding in the International Space Station. Just as all of the chocolate chips move apart as the cake rises, all of the superclusters of galaxies recede from each other as the universe expands.

panding away from one another fast enough to escape their mutual gravitation forever, astronomers have worked hard to determine the exact value of H_0. We explore that here and pick up the question of the fate of the universe in Chapter 18.

Insight Into Science

Power in Numbers Much of the power of science comes from expressing systematic data in mathematical form. By themselves, the spectra of galaxies give a qualitative picture of galactic motion, and the pattern of their recessional velocities suggests that a large-scale property of the universe is in effect here. When the data are converted into the equation, called the *Hubble law,* it becomes clear that the universe is expanding, and we can begin to look for the reason behind this expansion.

16-13 Different techniques determine the expansion of the universe at different distances from Earth

To determine the Hubble constant, astronomers must measure the redshifts and distances to many galaxies. Although redshift measurements from spectra can be quite precise, it is very difficult to accurately measure the distances to remote galaxies. We cannot use the method of stellar parallax at the distances of galaxies. Recall from Section 11-1 that only the distances to the stars in our Galaxy within about 150 pc (500 ly) can be determined very precisely this way.

If the Hubble constant were 3 times its present value, how much slower or faster would superclusters be moving apart?

Distances to other galaxies are determined from observations of the apparent magnitudes of objects in them or apparent

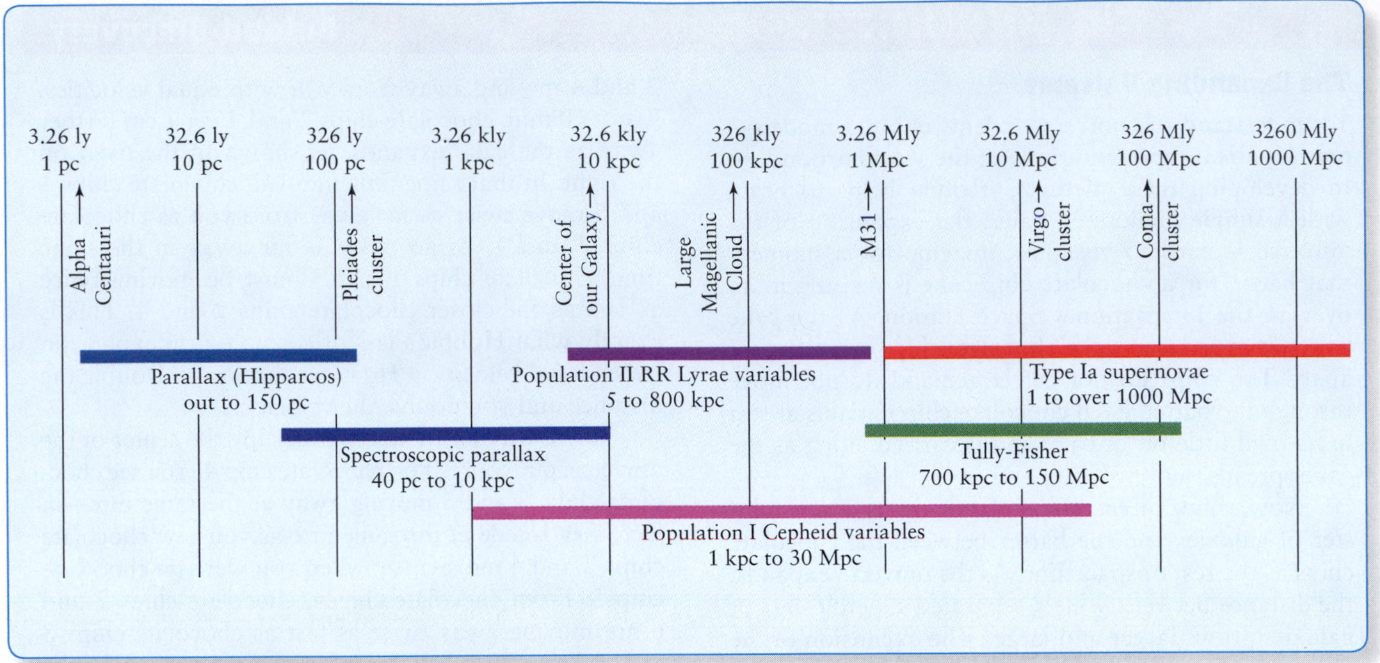

FIGURE 16-32 Techniques for Measuring Cosmological Distances Astronomers use different methods to determine different distances in the universe. All of the methods shown here are discussed in the text.

magnitudes of entire galaxies. These observations are compared to known absolute magnitudes for similar objects. Having the apparent magnitude and the absolute magnitude enables astronomers to calculate distances, as discussed in *An Astronomer's Toolbox 11-3*. Astronomers use the term **standard candle** to denote any object whose absolute magnitude is known. The various techniques used to determine large-scale distances are presented in *Guided Discovery: The Tully-Fisher Relation and Other Distance-Measuring Techniques*. Figure 16-32 summarizes the distances for which all of the measuring techniques described here are valid. Including data from all of the methods yields a distance-velocity graph, as shown in Figure 16-31. Using this diagram, astronomers have calculated that the Hubble constant is

$$H_0 = 71 \pm 4 \text{ km/s/Mpc}$$

The plus or minus 4 km/s/Mpc indicates the possible range for H_0, when taking into account all of the errors that may still remain in the observations and calculations. A significant source of error is the effect of intergalactic gas that causes the standard candles to appear dimmer than they would if the gas were not present. Astronomers must determine the amounts of both intergalactic and interstellar gas and take their effects into account, something that is often hard to do.

What are a few of the techniques that we use to measure distances on Earth?

16-14 Astronomers are looking back to a time when galaxies were first forming

The Hubble Space Telescope has been looking deeper and deeper into the universe. In 1998, working in concert with the Keck I telescope, Hubble observed galaxies 8 Bly (where Bly denotes a billion light-years) away, many of them in the process of merging (Figure 16-33). The same year it saw galaxies nearly 12 Bly away (Figure 16-33b). In 2004, galaxies up to 13.2 Bly away were observed. We are seeing these latter objects as they were less than a billion years after the universe came into existence. Although many of the early galaxies are not as well formed as nearby galaxies, the existence of galaxy-sized collections of stars so far back in time is an important clue to the processes of star and galaxy formation that occurred in the early universe.

The observational data now being accumulated from the period of time when the universe was less than a few billion years old are helping astronomers develop and test theories of the evolution of the early universe. We will explore this point further in Chapter 18.

16-15 Frontiers yet to be discovered

The farther out in the universe that we explore, the more there is to discover. In the realm of the galaxies, we have yet to understand fully how the different types

The Tully-Fisher Relation and Other Distance-Measuring Techniques

Spectroscopic parallax (see Section 11-9) provides distances for stars up to 10 kpc (33 kly, where kly denotes 1000 light-years). This distance is still within the Milky Way. To determine distances to other galaxies, we need sources in them that become very bright compared to the luminosity of normal stars, and whose absolute magnitudes are well known. Using the brightnesses of RR Lyrae variable stars (see Section 12-11), astronomers can measure distances to the Large and Small Magellanic Clouds, about 100 kpc (330 kly). To measure the Hubble flow, we must determine distances to much more distant galaxies.

Cepheid variable stars (see Section 15-2 and Figure 15-4) can now be observed out to 30 Mpc (100 Mly, where Mly denotes a million light-years) from Earth. The distances to galaxies in this nearby volume of space can thus be determined from the Cepheid period-luminosity law. Plotting the distances to galaxies versus their recessional velocity for the small volume of space extending out 30 Mpc from Earth (shown on the left side of the graph in Figure 16-31) enabled astronomers to get a first estimate for H_0 from the slope of the line through these data. This result gives $H_0 \approx 75$ km/s/Mpc. Errors in both distance and recessional velocities led to inaccuracies in the slope of this curve.

Beyond 30 Mpc, even the brightest Cepheid variables, which have absolute magnitudes of about -6, are not visible with current technology. In the 1970s, the astronomers Brent Tully and Richard Fisher developed another method for determining distances. They discovered that the width of the hydrogen 21-cm emission line of a spiral galaxy (discussed in Chapter 15) is related to the galaxy's absolute magnitude: *The broader the line, the brighter the galaxy.* The connection goes like this: Interstellar gas orbits in galaxies, as we have seen. Different hydrogen gas clouds in different places in the same galaxy have different speeds toward or away from us, and so we observe a variety of slightly different wavelengths from this gas, centered around 21 cm. This combination of emissions makes the 21-cm emission line appear "broad."

The **Tully-Fisher relation** says that the greater the galaxy's mass, the greater its luminosity and the more rapidly the stars and gas orbit in it. The faster it rotates, the greater the range of speeds and hence the greater the Doppler shifts we see from it. Measuring the width of the 21-cm line, therefore, tells us the galaxy's mass. Assuming that stars with different masses, and hence luminosities, occur with the same frequencies in each galaxy, the galaxy's mass tells us its *absolute* magnitude. Combining this with observations of the galaxy's *apparent* magnitude allows us to calculate its distance from us using the distance-magnitude relationship (see *An Astronomer's Toolbox 11-3*).

Because line widths can be measured quite accurately, astronomers can use the Tully-Fisher relation to determine the luminosities (absolute magnitudes) of many spiral galaxies and thus their distances from Earth.

The key method for determining the distances to the most remote galaxies is to measure the brightness of Type Ia supernova explosions within them. These supernovae all reach an absolute magnitude of -19 at the peak of their outbursts (see figure below). Using their observed apparent magnitudes and the distance modulus equation (see *An Astronomer's Toolbox 15-1*), their distances can be calculated. With this method, astronomers in 2001 measured a supernova nearly 3 billion parsecs (10 billion light-years) away, and more distant Type Ia supernovae have been measured since then. Measurements of distances to galaxies using the Tully-Fisher relation and Type Ia supernovae added data to Figure 16-31 and yield a more accurate Hubble constant of 71 km/s/Mpc.

FIGURE GD16-2 Two Supernovae in NGC 664 In 1997 the rare occurrence of two supernovae in the same galaxy at the same time was observed in the spiral galaxy NGC 664, located about 300 Mly (90 Mpc) from Earth. Supernovae observed in remote galaxies are important standard candles used by astronomers to determine the distances to these faraway objects. The two supernovae overlap each other, as shown. The upper, yellow-orange supernova was observed to occur 2 months before the hotter, blue one, which was observed to occur less than 2 weeks before this image was made and had not yet achieved maximum brightness. (Perry Berlind and Peter Garnavich, Harvard Smithsonian Center for Astrophysics)

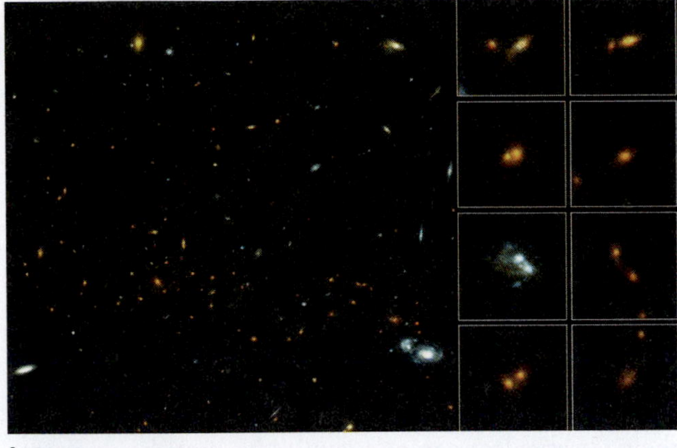

a

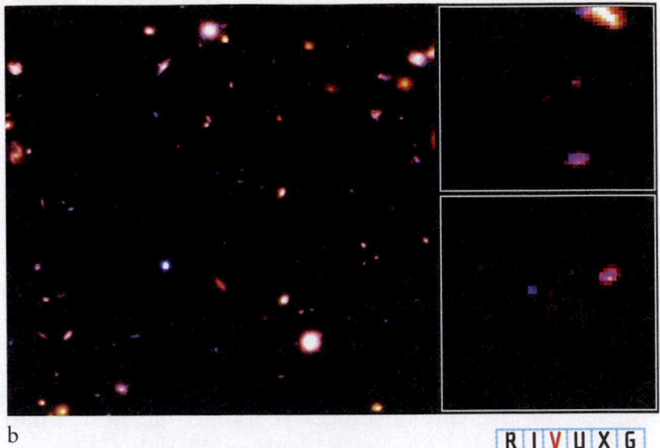

b

FIGURE 16-33 Distant Galaxies (a) The young cluster of galaxies MS1054-03, shown on the left, contains many orbiting pairs of galaxies, as well as remnants of recent galaxy collisions. Several of these systems are shown at the right. This cluster is located 8 billion light-years away from Earth. (b) This image of more than 300 spiral, elliptical, and irregular galaxies contains several galaxies that are an estimated 12 billion light-years from Earth. Two of the most distant galaxies are shown in the images on the right, in red, at the centers of the pictures. (a, b: P. Van Dokkum, Uner of Granengen, ESA, and NASA)

of galaxies form. Under what conditions did the older globular clusters form? How long does it take interacting galaxies to combine? Do supermassive black holes form before, simultaneously with, or after galaxies come into existence? Moving into the realm of the global structure of the universe, what is the overall distribution of clusters and superclusters of galaxies in the universe? Are most of them really found on spongelike or soap-bubblelike surfaces and, if so, what does that structure tell us about the underlying distribution of dark matter and other properties of the cosmos? Perhaps most intriguing, what is dark matter?

In what way are telescopes time machines?

SUMMARY OF KEY IDEAS

Types of Galaxies

• The Hubble classification system groups galaxies by their shapes into four major types: spiral, barred spiral, elliptical, and irregular.

• The arms of spiral and barred spiral galaxies are sites of active star formation.

• According to the theory of self-propagating star formation, spiral arms of flocculent galaxies are caused by the births and deaths of stars over extended regions of a galaxy. Differential rotation of a galaxy stretches the star-forming regions into elongated arches of stars and nebulae that we see as spiral arms.

• According to the spiral density wave theory, spiral arms of grand-design galaxies are caused by density waves. The gravitational field of a spiral density wave compresses the interstellar clouds that pass through it, thereby triggering the formation of stars, including OB associations, which highlight the arms.

• Elliptical galaxies contain much less interstellar gas and dust than do spiral galaxies; little star formation occurs in elliptical galaxies.

• Irregular galaxies are rich in gas and dust, and star formation occurs in them.

• Lenticular galaxies are disk galaxies without spiral arms.

Clusters and Superclusters

• Galaxies group into clusters rather than being randomly scattered through the universe.

• A rich cluster contains at least a thousand galaxies; a poor cluster may contain only a few dozen up to a thousand galaxies. A regular cluster has a nearly spherical shape with a central concentration of galaxies; in an irregular cluster, the distribution of galaxies is asymmetrical.

• Our Galaxy is a member of a poor, irregular cluster, called the Local Group.

• Rich, regular clusters contain mostly elliptical and lenticular galaxies; irregular clusters contain more spiral and irregular galaxies. Giant elliptical galaxies are often found near the centers of rich clusters.

• Each galaxy is held together with the aid of dark matter.

• No cluster of galaxies has an observable mass large enough to account for the observed motions of its galaxies; a large amount of unobserved mass must be present between the galaxies.

• Hot intergalactic gases emit X rays in rich clusters.

• When two galaxies collide, their stars initially pass each other, but their interstellar gas and dust collide violently, either causing gas and dust to be stripped from the galaxies or triggering prolific star formation. The gravitational effects of a galactic collision can cast stars out of their galaxies into intergalactic space.

• Galactic mergers occur. A large galaxy in a rich cluster may also grow steadily through galactic cannibalism.

Superclusters in Motion

• A simple linear relationship exists between the distance from Earth to galaxies in other superclusters and the redshifts of those galaxies (a measure of the speed at which they are receding from us). This relationship is the Hubble law: Recessional velocity $= H_0 \times$ distance, where H_0 is the Hubble constant.

• Astronomers use standard candles—Cepheid variables, the brightest supergiants, globular clusters, H II regions, supernovae in a galaxy, and the Tully-Fisher relation—to calculate intergalactic distances. Because of difficulties in measuring the distances to remote galaxies, the value of the Hubble constant, H_0, is not known with complete certainty.

WHAT DID YOU THINK?

1 *Are most of the stars in spiral galaxies located in their spiral arms?* No. The spiral arms contain only 5% more stars than the regions between the arms.

2 *Do all galaxies have spiral arms?* No. Galaxies may be either spiral, barred spiral, elliptical, or irregular. Only spirals and barred spirals have arms.

3 *Are galaxies isolated objects?* No. Galaxies are grouped in clusters, and clusters are grouped in superclusters.

4 *Is the universe contracting, unchanging in size, or expanding?* The universe is expanding.

Key Terms for Review

barred spiral galaxy, 492
cluster (of galaxies), 498
elliptical galaxy, 494
galactic merger, 504
gravitational lensing, 506
Hubble classification, 488
Hubble constant, 508
Hubble flow, 507
Hubble law, 508
intergalactic gas, 500
irregular cluster (of galaxies), 500
irregular galaxy, 496
lenticular galaxy, 496
Local Group, 499
poor cluster (of galaxies), 499
regular cluster (of galaxies), 500
rich cluster (of galaxies), 499
spiral density wave, 491
spiral galaxy, 488
standard candle, 510
starburst galaxy, 500
supercluster (of galaxies), 498
trailing-arm spiral galaxy, 494
Tully-Fisher relation, 511

Review Questions

1. What are the most massive galaxies in the universe? **a.** normal spirals, **b.** barred spirals, **c.** giant ellipticals **d.** dwarf ellipticals, **e.** irregulars

2. Spiral density waves are directly responsible for which of the following? **a.** flocculent spirals, **b.** grand-design spirals, **c.** supernovae, **d.** the collisions between galaxies, **e.** galactic cannibalism

3. Which of the following statements about the motion of galaxies is correct? **a.** All galaxies are moving apart. **b.** Superclusters of galaxies are all moving apart. **c.** Superclusters of galaxies are all moving toward each other. **d.** The Milky Way Galaxy is at the center of the universe. **e.** All clusters of galaxies in each supercluster are moving toward each other.

4. What is the common name for the sonic boom created by lightning?

5. In spiral galaxies, what spectral classes of stars are only found in spiral arms?

6. What is the Hubble classification scheme? Which category includes the biggest galaxies? Into which category do the smallest galaxies fall? Which type of galaxy is the most common?

7. In which Hubble types of galaxies are new stars most commonly forming? Describe the observational evidence that supports your answer.

8. What is the difference between a flocculent spiral galaxy and a grand-design spiral galaxy?

9. Briefly describe how the theory of self-propagating star formation accounts for the existence of spiral arms in some spiral galaxies.

10. Briefly describe how the spiral density wave theory accounts for the existence of spiral arms in some spiral galaxies.

11. How is it possible that galaxies in our Local Group still remain to be discovered? In what part of the sky are these galaxies located? What sorts of observations might reveal these galaxies?

12. Can any galaxies besides our own be seen with the naked eye? If so, which ones?

13. What is the difference between a rich cluster of galaxies and a poor one? What is the difference between a regular cluster of galaxies and an irregular one?

14. How can a collision between galaxies produce a starburst galaxy?

15. Why do astronomers believe that considerable quantities of dark matter must exist in clusters of galaxies?

16. Explain why the dark matter in galaxy clusters cannot be neutral hydrogen.

17. What is the Hubble law?

18. Some galaxies in the Local Group exhibit blueshifted spectral lines. Why are these blueshifts not violations of the Hubble law?

19. What is a standard candle? Why are standard candles important to astronomers trying to measure the Hubble constant?

20. What kinds of stars would you expect to find populating the space between galaxies in a cluster?

Advanced Questions

The answers to all computational problems, which are preceded by an asterisk (*), appear at the end of the book.

*21. Suppose a spectrum of a distant galaxy showed that its redshift corresponds to a speed of 22,000 km/s. How far away is the galaxy in Mpc?

*22. A cluster of galaxies in the southern constellation of Pavo (the Peacock) is located 100 Mpc from Earth. How fast, on average, are galaxies in this cluster receding from us? Why do different galaxies in the cluster show different velocities as measured from Earth?

23. How would you determine what fraction of a distant galaxy's redshift is caused by the galaxy's orbital motion around the center of mass of its cluster?

Discussion Questions

24. Discuss the advantages and disadvantages (for example, reliability, repeatability, etc.) of using the various standard candles to determine extragalactic distances.

25. Discuss whether the various Hubble types of galaxies actually represent some sort of evolutionary sequence.

26. Discuss the sorts of phenomena that can occur when galaxies collide. Do you think that such collisions can change the Hubble type of a galaxy? Explain your answers.

27. From what you know about stellar evolution, the interstellar medium, and spiral density wave theory, explain the appearance and structure of the spiral arms of grand-design spiral galaxies.

What If...

28. The solar system was located in an active star-forming region in a spiral arm, rather than on the edge of the Orion arm? How would the solar system be different?

29. The Milky Way collided with the Andromeda Galaxy? What might we experience on Earth? (This merger will occur in the distant future.)

30. No distant galaxies showed any redshift or blueshift? Discuss what that would imply about the universe.

31. All distant galaxies showed a blueshift, rather than a redshift? Discuss what that would imply about the universe.

Web Questions

32. To test your understanding of the global properties of galaxies, do Interactive Exercise 16.1 on the Web. You can print out your answers, if required.

33. To test your understanding of the local group of galaxies, do Interactive Exercise 16.2 on the Web. You can print out your answers, if required.

34. To test your understanding of the various types of galaxies, do Interactive Exercise 16.3 on the Web. You can print out your answers, if required.

35. To test your understanding of star formation in the spiral density wave model, do Interactive Exercise 16.4 on the Web. You can print out your results, if required.

Observing Projects

36. If you have access to a telescope with an aperture of at least 30 cm (12 in.) in a dark location, observe as many as possible of the spiral galaxies shown in the list of selected "Spiral Galaxies and Interacting Galaxies" in Appendix E. Many of these galaxies are members of the Virgo cluster, which can be seen from the northern hemisphere most conveniently during the spring. Because all galaxies are quite faint, be sure to schedule your observations for a Moonless night. The best view is obtained when a galaxy is high in the sky. While at the eyepiece, make a sketch of what you see. Can you distinguish any spiral structure in these galaxies?

Use *Starry Night*™ to examine each of the listed spiral galaxies in turn. If your telescope observations were successful, you can compare your sketches with the appearance of each of these spiral galaxies. If no telescope was available, you can still use *Starry Night*™ to examine these galaxies in detail. Click on **Home** and use the **Find** pane to locate and magnify each galaxy in turn. In each case, note whether the spiral is edge-on or face-on to observers upon Earth, and consider particularly the source of the light in the spiral arms to determine whether stars or diffuse nebulae are the major contributors to this light.

37. Use the same session at the telescope and the same procedure as in the previous exercise to observe as many as possible of the interacting galaxies listed under "Spiral Galaxies and Interacting Galaxies" in Appendix E. In particular, attempt to make sketches of what you see while at the eyepiece. Can you distinguish hints of interplay between the galaxies?

Use *Starry Night*™ to examine each of the listed galaxies in turn. If your telescope observations were successful, you can compare your sketches with the appearance of each of these interacting galaxies. If no telescope was available, you can still use *Starry Night*™ to examine these intriguing galaxy collisions. Click on **Home** and use the **Find** pane to locate each galaxy in turn. In each case, look for evidence of collision and write a comment on how this collision manifests itself in the resulting shapes of the colliding galaxies. Can you see the difference between head-on and more distant collisions?

38. The galaxies M81 and M82 (Figure 16-24) appear to be near to each other and interacting, as seen on photographs and radio images taken from the Earth. In this exercise, you can use *Starry Night*™ to determine just how close they are in space. Open the **Favourites** pane and select **Discovering the Universe > M81 and 82** from this pane. The view, covering a field only 1° wide, shows both M81, also called Bode's Galaxy and M82, also called the Cigar Galaxy. Note the difference in appearance between these two galaxies. Use the angular measurement tool to determine the angular distance separating these galaxies, as seen from Earth and note the distance to M81, as shown by the **HUD** when the cursor is placed over this galaxy. Compare this separation to the apparent size of the Full Moon (about 30′ in diameter). To find the actual physical distance separating these two galaxies, use the small-angles equation. (Note: In this equation, if a distance d at a distance D away from observer is seen to subtend a small angle θ, with both d and D measured in the same units, then θ in units of radians is simply $\theta = d/D$.) Translate the measured angle into units of radians (1 radian = 57.3°, 1° = 60′.) and multiply this resulting angle by the distance to M81. Are these two galaxies as close to each other as the Milky Way and the Large Magellanic Cloud, which are about 0.195 Mly apart?

39. To give you experience in interpreting and classifying the images of galaxies, you can use *Starry Night*™ to visit a variety of galaxies and determine whether they are spiral, barred spiral, or irregular. Click on **Home** to see the sky from your home location and then use the **Find** pane to visit each of the galaxies listed below in turn. For each object, type its name in the search box of the **Find** pane and press the **Enter** key. Be careful—include spaces, where indicated, when typing in object names. To go to the galaxy without slewing, click the spacebar. Use the **Zoom** buttons to examine each galaxy in detail and then classify it as a spiral (S), barred spiral (SB), or irregular (Irr). You do not need to worry about the subclasses in this question.

IC 4182	IC 5152	M33	M58	M74
M81	M82	M83	M94	M109
NGC 1232	NGC 1313	NGC 1365	LMC	SMC

40. Clusters of galaxies contain different numbers and distributions of galaxies. In this exercise you can use *Starry Night*™ to compare a few of these groupings. You can start by looking at one of the largest galaxy clusters, the Virgo cluster. Open the **Favourites** pane and select **Discovering the Universe > Virgo Cluster-Milky Way** from this pane. You are looking at this group of galaxies from a distance of about 66 Mly from the Sun. In this view, the Milky Way Galaxy and the Virgo cluster are separated by a huge void in space. Use the location scroller to move around the Virgo cluster and consider its overall shape and its relationship to neighboring galaxies. What is the general shape of the Virgo cluster? Zoom in towards this cluster until individual galaxies are shown and again use the location scroller to help you to identify each classification of galaxy—elliptical, spiral, and irregular—within the group. Click on **File > Revert** to return to the original view and identify several other groups of galaxies. Select one or two clusters in turn, move the cursor over a galaxy within the selected group to open the object contextual menu and select the **Highlight** option to identify this group. You can select the **Centre** option to move the selected cluster to the center of the view and examine the cluster's extent across space. Again, use the location scroller and zoom in to examine this cluster from various viewpoints. Describe the distributions of their galaxies within the cluster, compared to that in the Virgo Cluster. For example, what are their shapes and relative sizes compared to Virgo and to each other? See if you can recognize the walls of galaxies surrounding large voids in space that link these concentrated regions of galaxies.

41. In this exercise, you can use *Starry Night*™ to view and examine the Great Northern Wall of Galaxies. Select **Favourites >**

Discovering the Universe > Great Northern Wall to view a wide region of space from about 500 Mly away from the Milky Way. The galaxies in this Great Wall are highlighted in yellow and the image is centered on the Milky Way. Select the location scroller from the cursor selection list and use it to move around the Milky Way and study the Great Wall from different viewpoints. You will see the Great Wall of Galaxies move around the Milky Way. Study the distribution of galaxies in the Great Wall as you move around in space and write down an explanation for this name in terms of its shape and structure.

WHAT IF... The Solar System Were Located Closer to the Center of the Galaxy?

It is a clear, Moonless night. Stars by the thousands twinkle serenely against the ebony darkness of space. Overhead the soft white span of the Milky Way catches your attention, and you try to see individual stars in its glowing haze.

Most of the 6000 stars visible to the naked eye throughout the year are within 300 light-years of Earth. The rest of the Galaxy's 200 billion or so stars are too dim or too obscured by interstellar gas and dust to be easily observed. What if the solar system were one-third of its present distance from the center of our Galaxy? At that distance, we would still be in the realm of the spiral arms, extremely close to the central bulge, and none of the stars we see now would be visible.

Population Explosion Out in the galactic suburbs, where we live today, the population density is about 1 star per 300 cubic light-years. The displaced Earth would be surrounded by nearly 5 times as many stars, or 1 per 60 cubic light-years. Also, the solar system would pass much more frequently through the dust-rich spiral arms of the Galaxy. Scattering starlight, this interstellar matter would glow as diaphanous wisps throughout our nighttime sky. Furthermore, whenever the solar system was actually in a spiral arm, several nearby stars would be of the high-mass, high-luminosity variety. The combined light from these stars and the shimmering clouds would be so great that for millions of years at a time night would never fall.

Evolution Interruption Recall from Chapter 13 that high-mass stars evolve much more rapidly than average-mass stars like the Sun. If we were closer to the center of the Milky Way and frequently passing through the spiral arms of the Galaxy, massive stars would explode near us much more frequently than they do today. These nearby supernovae, occurring within 50 light-years of Earth, would deposit lethal radiation, damage life, and cause mass extinctions. As a result, the direction of biological evolution would change more frequently.

At Earth's current location, the closest star, Proxima Centauri, is more than 4 light-years away. Would Earth be in danger of colliding with a star if the solar system were closer to the center of the Galaxy? Several stars would certainly be much closer than Proxima Centauri is now. However, stars are so small compared to the vastness of a galaxy that the likelihood of a collision would still approach nil.

Much more likely would be the passage of a star so close to the Sun that Earth's orbit would be disturbed. If Earth's orbit became, say, more elliptical, the change of seasons would be noticeably affected. With the seasonal effect of Earth's tilt compounded by greater changes in distance between Earth and the Sun, one hemisphere of Earth would suffer much more extreme temperatures than it does today, while the other would see less variation. This would affect the evolution of life, of course, and the distribution of life-forms on the planet.

Close Encounters Earth-evolved life at our new location in the Galaxy would be more likely to encounter sentient beings on planets that orbit nearby stars. Today, after a century of broadcasting radio and television signals, a 200-light-year-diameter sphere of space centered on Earth is filled with such signals. There are about 6300 stars in that sphere. At our new location there would be 31,500 stars in the same volume and the probability of many more stars with life-supporting planets that orbit them. Perhaps one of the reasons that life on Earth has been able to evolve so long unaffected by other intelligent life is that the solar system lies near the fringe of the Galaxy.

THE GHOSTS OF GALAXIES PAST

Strangely moving stars may be the remnants of past galaxies devoured by our Milky Way

Adapted from an article by Rodrigo Ibata and Brad Gibson

The stars in the night sky reside in our own galaxy, the Milky Way. It is natural to think of them as native suns, born in the Milky Way. But some, like Arcturus, the second brightest northern star, move differently and have a different chemical composition from most stars in our Galaxy. The origin of these atypical stars has been debated since the 1960s. Did the gravity of our galaxy's spiral arms force these stars into oddball orbits or are they immigrants, formed in regions beyond the Milky Way from material that was never part of it?

Astronomers have found that Arcturus and many other stars are immigrants or "kidnap victims" that were born into smaller galaxies and then captured, plundered and assimilated by the Milky Way. Over time, our galaxy may have vanquished hundreds of neighbors, with their former inhabitants now intermingling with Milky Way "natives." By observing these stars, astronomers can reconstruct the violent history of our galaxy and probe the nature of the unseen dark matter that governs its existence.

Space Invaders

Stellar immigrants line up in long streams that often lead back to a star cluster or one of the Milky Way's satellite galaxies—presumably their home or what remains of it. The Milky Way's gravitational pull deforms a satellite galaxy or star cluster, literally lifting off stars and creating a stream. Eventually, this stream-like arrangement is lost as stars disperse. In some cases, what is streaming away is not stars, but gas. This process builds up our own galaxy.

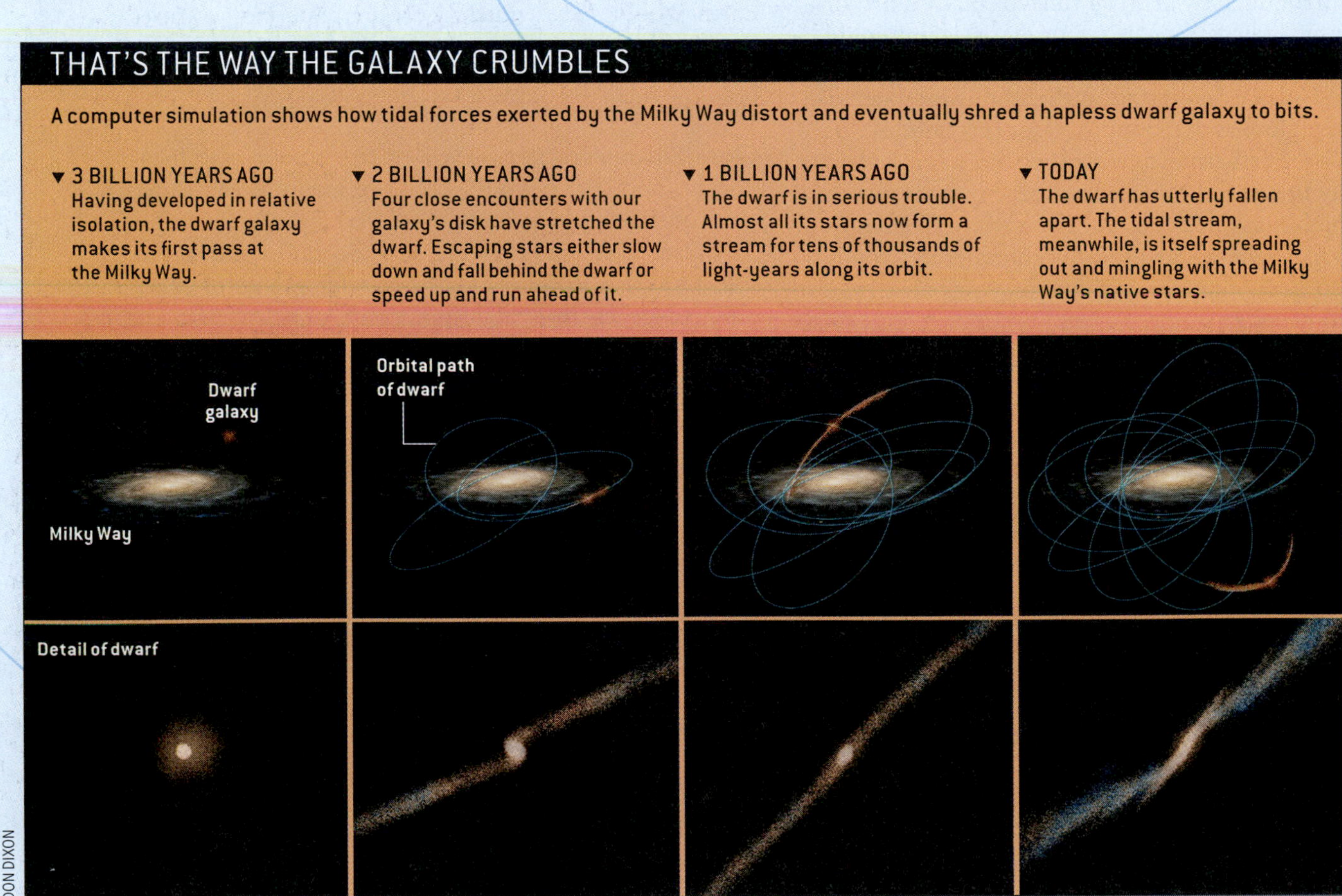

A Work in Progress

These observations have revolutionized theoretical understanding of galaxy formation. Astronomers once thought that all galaxies experienced early, runaway growth, settling quickly into their present form. Today, researchers think that only dwarf galaxies (up to a billion stars in mass) went through an abrupt formation period. Large galaxies, like the Milky Way (about one trillion stars) formed later through the accretion or merging of dwarfs—a process that continues to this day.

Astronomers are now probing the next questions regarding the chemical makeup of these ancient galactic building blocks—and the distribution of dark matter

Forensic astronomers are trying to identify stars that are currently migrating to the Galaxy, which ones migrated in the past—and where they came from. Most stars are born in groups of several thousand to tens of thousands that emerge from the same chemically unique gas cloud. Researchers hope that "chemical fingerprinting" will identify stars of common origin. They may one day be able to identify stars formed from the same gas cloud as the Sun but now spread across the Galaxy.

From 2011 through 2020, Gaia, the world's most technologically ambitious space telescope, will aim to map the three-dimensional structure of our Galaxy, obtaining precise positional and velocity measurements for one billion stars, nearly one percent of the galaxy. The satellite is also intended to measure the chemical compositions of several million stars.

A decade ago, when we discovered the stellar stream of the Sagittarius dwarf galaxy, many of our colleagues considered it a mere curiosity with no broader significance. But it soon became the poster child for the Milky Way's tangled history of merging and accreting—processes now thought to be the main drivers of galaxy formation and evolution. Assimilated galaxies have brought new stars, gas and dark matter and have triggered waves of star formation. These immigrants have kept our galaxy a vibrant place.

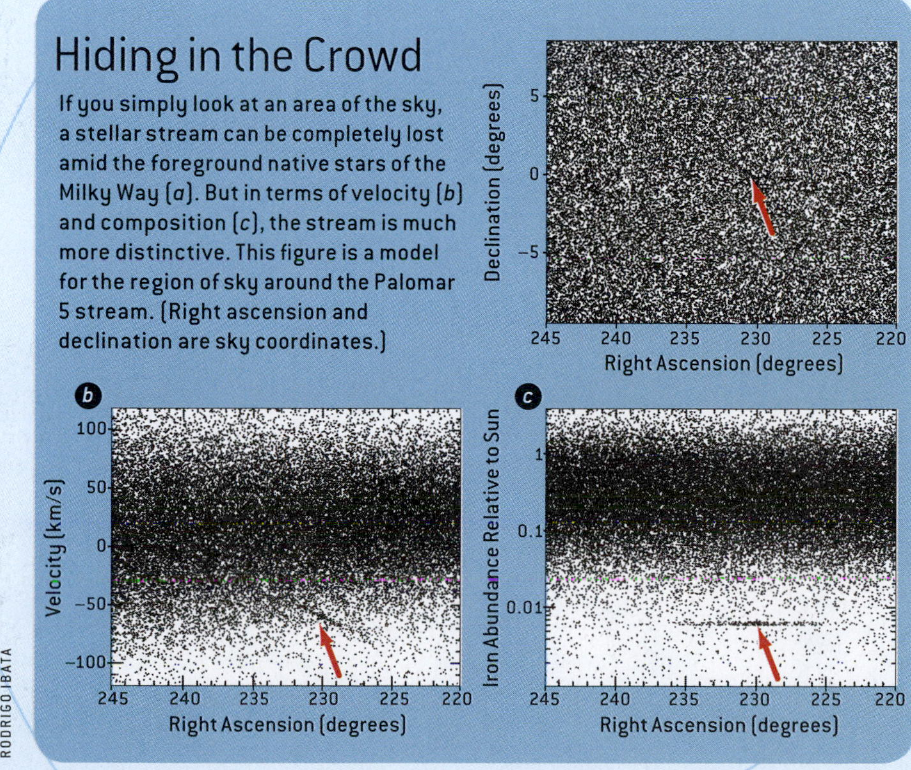

Hiding in the Crowd

If you simply look at an area of the sky, a stellar stream can be completely lost amid the foreground native stars of the Milky Way (*a*). But in terms of velocity (*b*) and composition (*c*), the stream is much more distinctive. This figure is a model for the region of sky around the Palomar 5 stream. (Right ascension and declination are sky coordinates.)

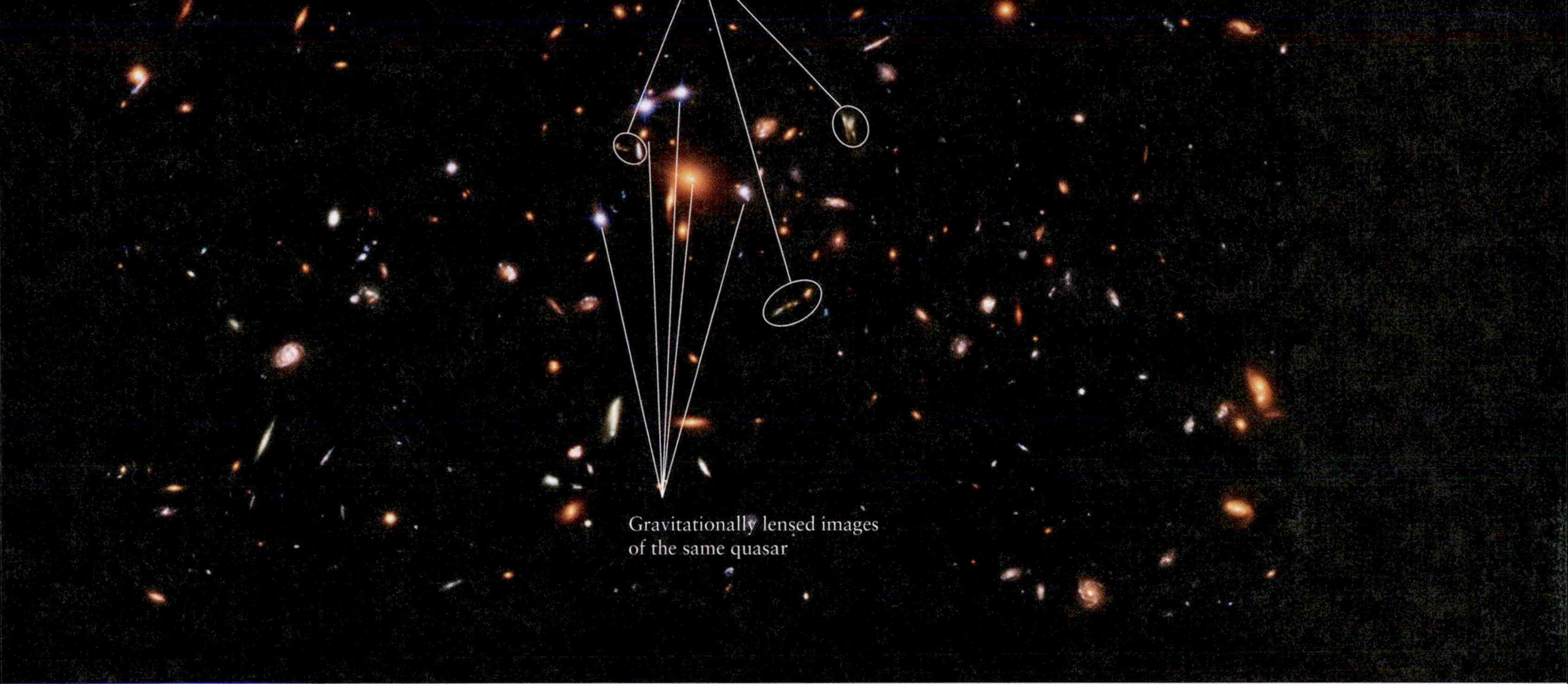

Q WHAT DO YOU THINK?

1. What does "quasar" stand for?
2. What do quasars look like?
3. Where do quasars get their energy?

Answers to these questions appear in the text beside the corresponding numbers in the margins and at the end of the chapter.

RIVUXG Gravitational lensing by galaxies in the intervening cluster brings five images of a distant quasar in our direction. (The fifth image is behind the large, orange galaxy.) The cluster of galaxies is 7 billion ly away, whereas the quasar is 10 billion ly distant. Three images of the same galaxy are also visible. (ESA, NASA, K. Sharon [Tel Aviv University], and E. Ofek [Caltech])

Chapter 17

Quasars and Other Active Galaxies

In this chapter we look at active galaxies—energy sources of almost unimaginable power. While the short-lived outputs of supernovae boggle the mind, they represent miniscule amounts of energy compared to quasars, Seyfert galaxies, radio galaxies, and BL Lac objects. Quasars, for example, emit more energy each second than the Sun does in 200 years, and they continue to do so for millions of years. Evidence shows that most galaxies undergo periods of similar activity. Some truly remarkable activity must occur deep within quasars and other ultra-high energy sources to make them so luminous.

In this chapter you will discover

- bright and unusual objects called active galaxies
- distant, luminous quasars
- the unusual spectra and small volumes of quasars
- the extremely powerful BL Lac objects
- supermassive black holes that serve as central engines for radio galaxies, quasars, Seyfert galaxies, and BL Lac objects

QUASARS

Our knowledge of quasars began in an amateur astronomer's backyard. Grote Reber built the first radio telescope in 1936 behind his home in Illinois, opening the realm of nonvisual astronomy. By 1944, Reber had detected strong radio emissions from sources in the constellations Sagittarius, Cassiopeia, and Cygnus. Two of these sources, Sagittarius A (Sgr A) and Cassiopeia A (Cas A), are in our Galaxy. The first is the galactic nucleus (see Section 15-4), and the second is a supernova remnant (see Chapter 13). However, Reber's third source, called Cygnus A (Cyg A), proved hard to categorize (Figure 17-1). Others quickly refined Reber's observations, but the mystery only deepened in 1954, when Walter Baade and Rudolph Minkowski, using the 200-in. optical telescope on Mount Palomar, discovered a strange-looking galaxy at the position of Cygnus A (Figure 17-1 inset).

The galaxy associated with Cyg A is very dim. Nevertheless, Baade and Minkowski managed to photograph its spectrum. They detected a redshift that corresponds to a speed of 14,000 km/s. According to the Hubble law, this speed indicates that Cyg A lies 635 million ly (194 Mpc) from Earth.

Because Cyg A is one of the brightest radio sources in the sky, its enormous distance intrigued astronomers. Although barely visible through the giant optical telescope at Palomar, Cyg A's radio waves can be picked up by amateur astronomers with backyard equipment. Its radio energy output must therefore be colossal. In fact, Cyg A shines with a radio luminosity 10^7 times as bright as the radio emission from an entire ordinary galaxy, such as Andromeda. The object that creates the Cyg A radio emissions has to be something extraordinary.

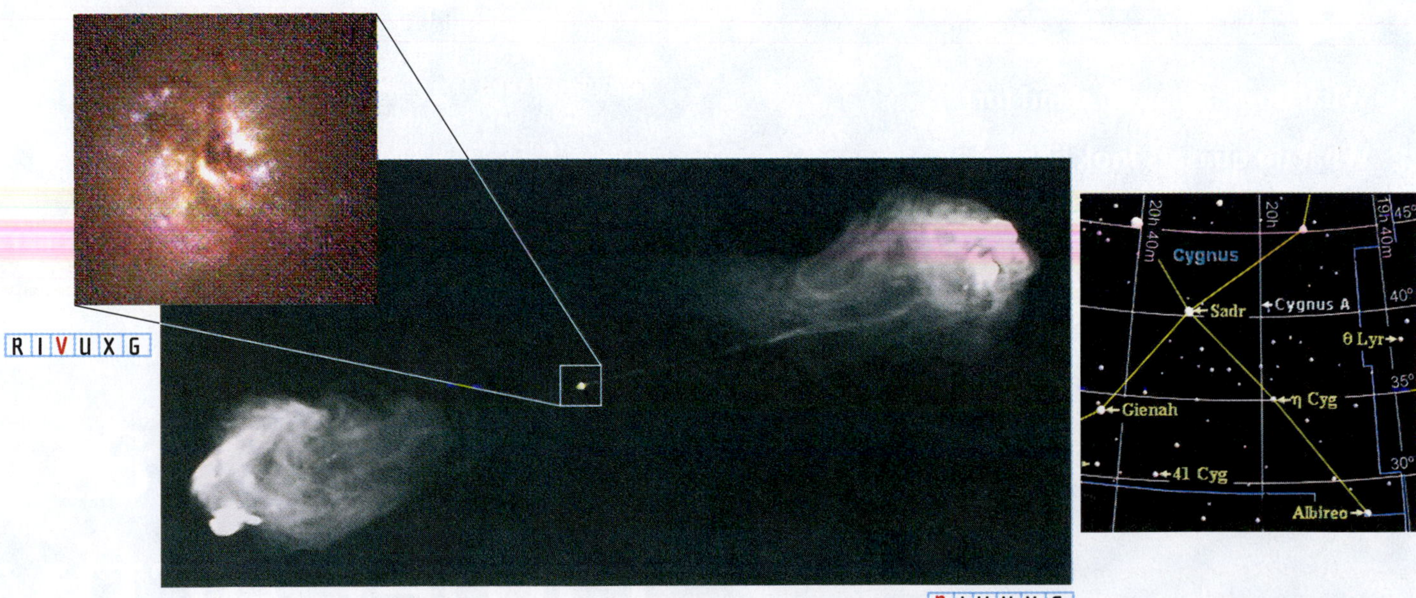

FIGURE 17-1 Cygnus A (3C 405) Radio image produced from observations made at the Very Large Array. Most of the radio emissions from Cygnus A come from the radio lobes located on either side of the peculiar galaxy seen in the inset, a Hubble Space Telescope image. Each of the two radio lobes extend about 160,000 ly from the optical galaxy and contain a brilliant, condensed region of radio emission. Inset: At the heart of this system of gas lies a strange-looking galaxy that has a redshift that corresponds to a recessional speed of 5% of the speed of light. According to the Hubble law, Cygnus A is therefore 635 million ly from Earth. Because Cygnus A is one of the brightest radio sources in the sky, this remote galaxy's energy output must be enormous. (R. A. Perley, J. W. Dreher, J. J. Cowan, NRAO; inset: William C. Keel, Robert Fosbury)

17-1 Quasars look like stars but have huge redshifts

Cygnus A is not the only powerful radio source in the far-distant sky. Starting in the 1950s, radio astronomers created long lists of radio sources. One of the most famous lists, the *Third Cambridge Catalogue,* was published in 1959. (The first two catalogues were filled with inaccuracies.) Even today, astronomers often refer to its 471 radio sources by their "3C numbers." Cyg A, for example, is designated 3C 405, because it is the 405th source in the Cambridge list. Because of the extraordinary luminosity of Cyg A, astronomers were eager to learn whether any other sources in the 3C catalog had similar properties.

One interesting case was 3C 48. In 1960, Allan Sandage used the Palomar telescope to discover a "star" at the location of this radio source (Figure 17-2). Recall that stars are blackbodies whose peak intensity is typically visible light and whose radio emission is much less intense. Because ordinary stars are not strong sources of radio emission, 3C 48 had to be something unusual. Indeed, its spectrum showed a series of emission lines that, initially, no one could identify. Although 3C 48 was clearly an oddball, many astronomers thought it was just another strange star in our Galaxy.

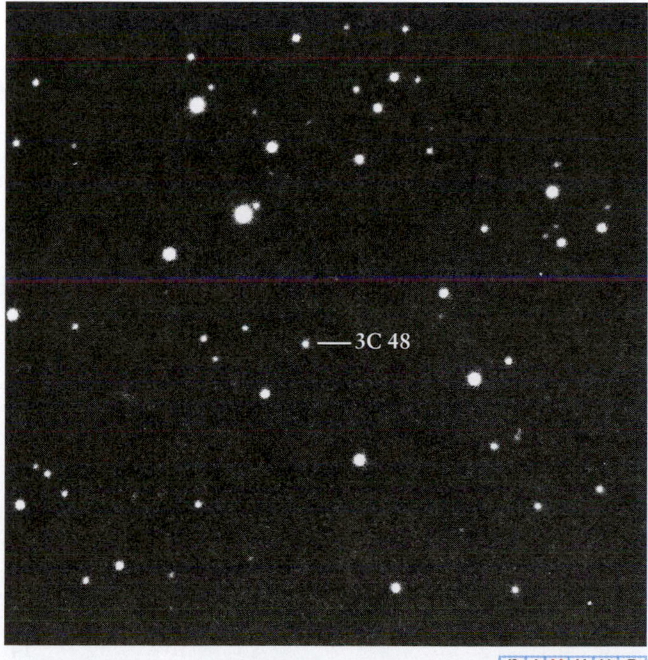

FIGURE 17-2 Quasar 3C 48 For several years, astronomers erroneously believed that this object was simply a peculiar, nearby star that happened to emit radio waves. Actually, the redshift of this starlike object is so great that, according to the Hubble law, it must be roughly 4 billion ly away. (Alex G. Smith, Rosemary Hill Observatory, University of Florida)

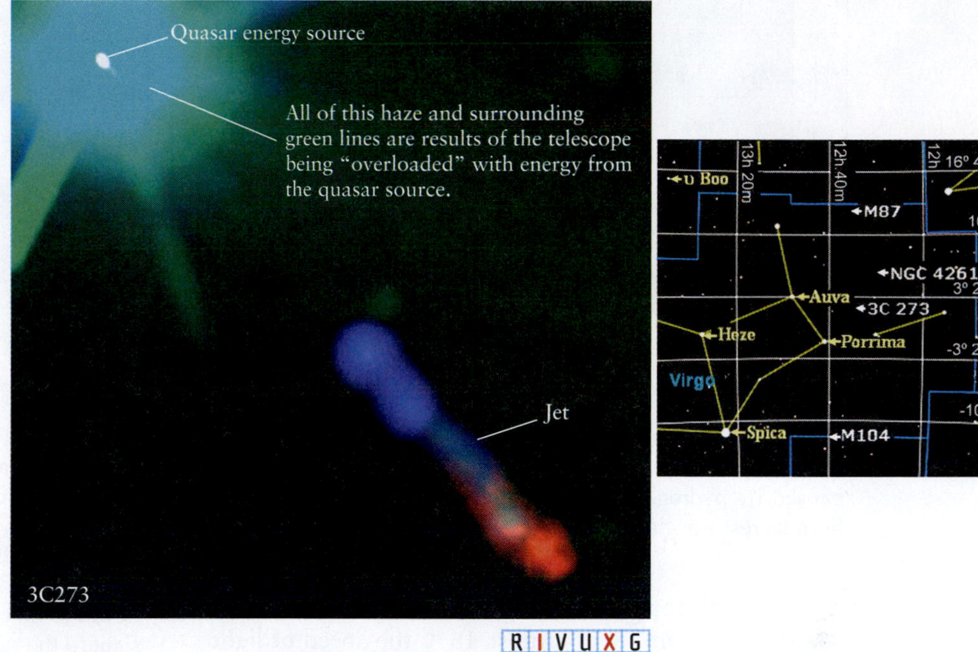

FIGURE 17-3 Quasar 3C 273 This combined X-ray and infrared view shows the starlike object associated with the radio source 3C 273 and the luminous jet it has created. The jet is also visible in the radio and visible parts of the spectrum. By 1963, astronomers determined that the redshift of this quasar is so great that, according to the Hubble law, it is nearly 2 billion ly from Earth. (S. Jester, D. E. Harris, H. L. Marshall, K. Meisenheimer, H. J. Roser, and R. Perley)

Another such "star," called 3C 273, was discovered in 1962. Like 3C 48, this object (Figure 17-3) emits a series of bright spectral emission lines that no one could then identify. These spectral lines are brighter than the background radiation at other wavelengths, called the *continuum* (recall Kirchhoff's laws in Section 4-4).

A breakthrough finally came in 1963, when Maarten Schmidt at the California Institute of Technology identified four of the brightest spectral lines of 3C 273 as four spectral lines of hydrogen. However, these emission lines from 3C 273 are found at much longer wavelengths than the usual wavelengths of these lines. Schmidt concluded that the hydrogen lines are subjected to a substantial redshift. Furthermore, the intense emission lines mean that something unusual is heating the gas.

Spectra for stars in our Galaxy exhibit comparatively small Doppler shifts, because these stars cannot move extremely fast relative to the Sun without soon escaping from the Galaxy. Schmidt thus concluded that 3C 273 is not a nearby star after all. Pursuing this conclusion,

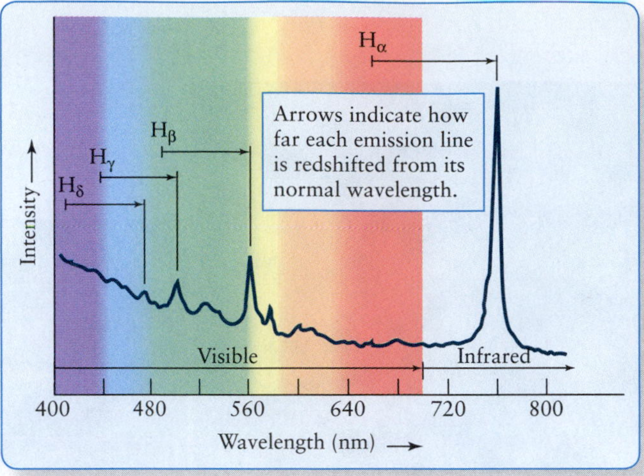

FIGURE 17-4 Spectra of 3C 273 The visible and infrared spectra of 3C 273 are dominated by four bright emission lines caused by hydrogen. This radiation is redshifted nearly 16% from its rest wavelengths.

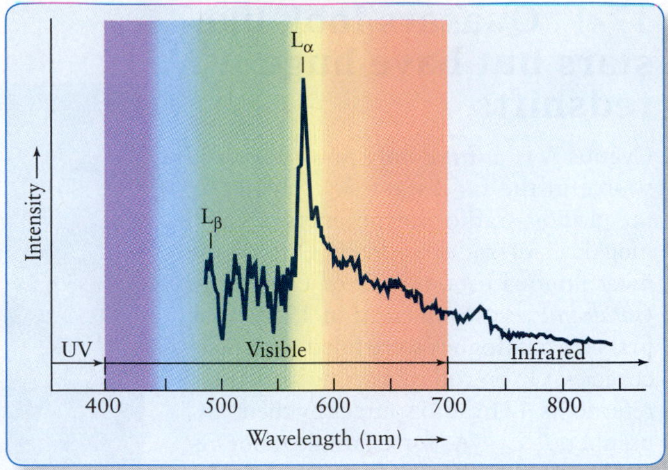

FIGURE 17-5 Spectrum of a High-Redshift Quasar The light from this quasar, known as PKS 2000-330, is so highly redshifted that spectral emission lines normally found in the far-ultraviolet (L_α and L_β) are seen at visible wavelengths. Note the many deep absorption lines on the short-wavelength side of L_α. These lines, collectively called the Lyman-alpha forest, are believed to be created by remote clouds of gas along our line of sight to the quasar. Hydrogen in these clouds absorbs photons from the quasar at wavelengths less redshifted than the quasar's L_α emission line.

he promptly found that its redshift corresponds to a speed away from us of almost 16% the speed of light. According to the Hubble law, this huge redshift implies an impressive distance to 3C 273 of roughly 2 billion light-years (Bly).

Figure 17-4 shows the spectrum of 3C 273, which looks nothing like a star's blackbody spectrum with absorption lines. Remember from Chapter 4 that a spectrograph records energy intensity at different wavelengths. The emission lines appear as peaks, and absorption lines appear as valleys. The emission lines are caused by excited gas atoms that emit radiation at specific wavelengths. Figure 17-5 shows the spectrum of an object so far away that the expansion of the universe (see Section 16-12) gives it a redshift of 92% of the speed of light.

Inspired by Schmidt's success, astronomers looked again at the spectral lines of 3C 48. Sure enough, the emission lines corresponded to hydrogen but with a redshift that corresponds to a velocity of nearly one-third the speed of light. Therefore, 3C 48 must be nearly twice as far away as 3C 273, or about 4 billion ly from Earth, assuming that the Hubble constant, H_0, is 71 km/s/Mpc.

1 Because of their starlike appearances and strong radio emissions, 3C 48 and 3C 273 were dubbed **quasi-**
2 **stellar radio sources,** a term soon shortened to **quasars.** Related objects, called **quasi-stellar objects (QSOs),** also look like stars but have strong energy outputs in all different parts of the electromagnetic spectrum, with most being strongest in the infrared. More than 200,000 quasars and QSOs have been discovered. In this book, we will use the word *quasars* to cover QSOs, too.

How are the spectra of quasars different from the spectra of stars?

All quasars have redshifts of at least 0.06, or 6%, of the speed of light. The upper limit corresponds to speeds away from Earth of greater than 90% of the speed of light. From the Hubble law, it follows that the distances to these high-redshift quasars are typically in the range of 10 billion to more than 13 billion ly. Plotting their distances, astronomers found that quasars began forming early in the life of the universe (Figure 17-6), their number peaking about 2 billion years after it came

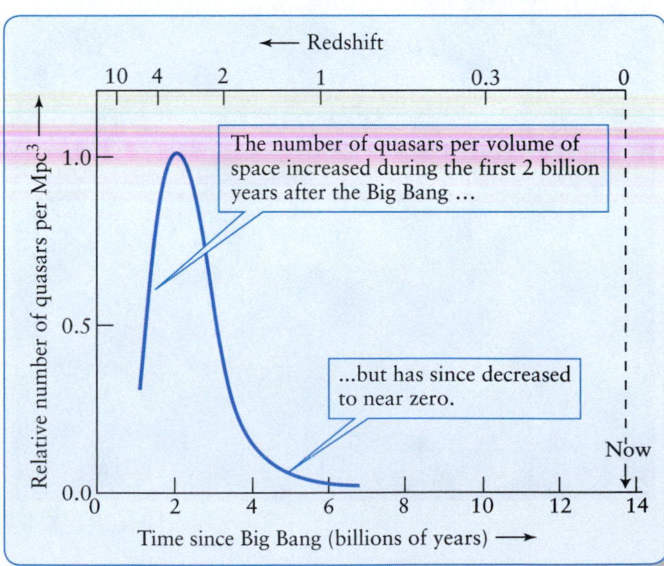

FIGURE 17-6 History of Quasar Formation The greater the redshift of a quasar, the farther it is from Earth and the farther back in time we are seeing it. By observing the number of quasars found at different redshifts, astronomers can calculate how the density of quasars has changed over time.

into existence. Their number dropped to near zero about 7 billion years ago. The reason for this evolution will become clear shortly.

17-2 A quasar emits a huge amount of energy from a small volume

Galaxies are big and bright. A typical large galaxy, like our own Milky Way, contains hundreds of billions of stars and shines with the luminosity of 10 billion Suns. The largest and most luminous galaxies, the giant ellipticals, are only 10 times brighter. But beyond 8 billion ly from Earth, even the brightest galaxies are too faint to be easily detected. Most ordinary galaxies are too dim to be detected at half that distance. If quasars can easily be seen more than 13 billion ly away, they must be far more luminous than galaxies. Indeed, a typical quasar is 100 times brighter than our Milky Way.

In the mid-1960s, several astronomers discovered that quasars fluctuate in brightness. Going back over old images, they found that some newly identified quasars had actually been photographed in the past but had not been considered as anything special because they looked like stars. One photograph of 3C 273, for example, dates back to 1887. By carefully examining these old images, as well as recent ones, astronomers could see that *quasars occasionally flare up.* Prominent outbursts from quasar 3C 279 reached Earth around 1937 and 1943 (Figure 17-7). During these events, the luminosity of 3C 279 increased by a factor of at least 25. At peak periods it was emitting at least 10,000 times as much energy as the entire Milky Way.

Fluctuations and Their Timescales Suppose that somewhere on a star or other spherical object an event occurred that began making the object more luminous and then returned it to normal brightness with a total variation or cycle time of, for example, 1 min. If that fluctuation in luminosity were traveling at the speed of light, then it would take half a minute to cross the object and complete the process of brightening it, and another half a minute to reverse, so that the object would return to its original brightness 1 min after the event began. If the same fluctuation traveled more slowly, then the object would have to be smaller than half a light-minute across for it to return to its original state after 1 min. Therefore, the object could not be more than half a light-minute across (ignoring the curve of a sphere). In this way, astronomers can calculate the maximum size an object can have from the timescale of its fluctuation.

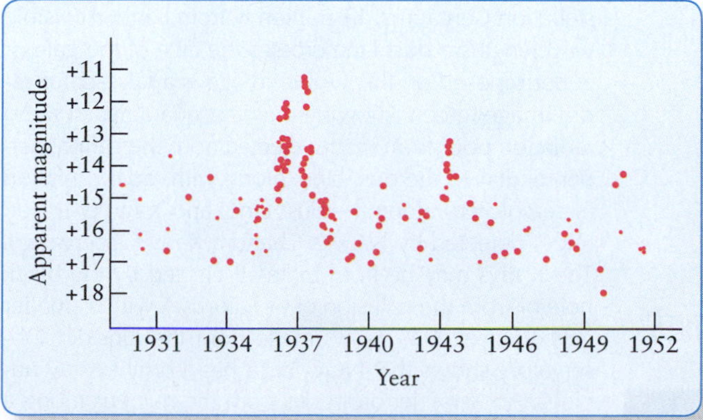

FIGURE 17-7 Brightness of 3C 279 This graph shows variations in the brightness of the quasar 3C 279. Note the especially large outburst observed in 1937. These data were obtained by carefully examining old photographic plates in the files of the Harvard College Observatory.

Many quasars vary in brightness over only a few months, weeks, or days. In fact, X-ray observations reveal large variations in as little as 3 h. This rapid flickering means that the source of the quasar's energy must be quite small by galactic standards. The energy-emitting region of a typical quasar—the "powerhouse" that blazes with the luminosity of 100 galaxies—is less than 1 light-day in diameter. Something must be producing the luminosity of 100 galaxies from a volume with approximately the same diameter as our solar system. Before considering their energy source, let us examine other objects that will turn out to be powered by the same "engine," as it is called by astronomers.

Which can fluctuate faster—an object 1 million km in diameter or an object 1 billion km in diameter?

OTHER ACTIVE GALAXIES

Quasars are but one kind of very powerful energy source in space. The others include Seyfert galaxies, radio galaxies, double-radio sources, and BL Lacertae objects. Located at the centers of galaxies, these powerhouses are collectively called **active galactic nuclei,** or **AGN.** The galaxies that contain AGN are called **active galaxies.** Active galaxies have unusually bright, starlike (that is, a pinpoint as seen from Earth) nuclei; their spectra do not look anything like a star's blackbody spectrum; they have strong emission lines; many are highly variable, meaning that their brightness levels change in different parts of the electromagnetic spectrum; they have jets of electrons and protons and beams of radiation emanating from their cores like the emissions observed for 3C 273; and most are more luminous than ordinary galaxies.

17-3 Active galaxies can be either spiral or elliptical

Seyfert Galaxies Carl Seyfert at the Mount Wilson Observatory in California discovered the first active galaxies in 1943 while surveying spiral galaxies. Now called **Seyfert galaxies,** these spirals reveal exceptionally bright, starlike nuclei and strong emission lines in their

FIGURE 17-8 Seyfert Galaxy NGC 1566 This Sc galaxy is a Seyfert galaxy some 50 Mly (16 Mpc) from Earth in the southern constellation Dorado (the Goldfish). The nucleus of this galaxy is a strong source of radiation whose spectrum shows emission lines of highly ionized atoms. (NASA/JPL-Caltech/R. Kennicutt [University of Arizona] and the SINGS Team)

spectra. For example, the rich spectrum of NGC 4151 has many prominent emission lines. Some of these lines are produced by iron atoms with a dozen or more electrons stripped away, indicating that NGC 4151 contains some extremely hot gas. Individual Seyfert galaxies also vary in brightness. For example, at times the magnitude of NGC 4151 changes over the course of a few days.

Another example of a Seyfert galaxy is NGC 1566 (Figure 17-8). At infrared wavelengths, this galaxy shines with the brilliance of 10^{11} Suns. This extraordinary luminosity has been observed to vary by as much as 7×10^9 $L_\odot$ over only a few weeks. In other words, the infrared power output of the nucleus of NGC 1566 increases and decreases by an amount nearly equal to the total luminosity of our entire Galaxy.

Many more Seyfert galaxies have been discovered in recent years. Approximately 10% of the most luminous galaxies in the sky are Seyferts. Some of the brightest Seyfert galaxies shine as brightly as faint quasars, with point sources in their centers. These are labeled Type 1 Seyferts. Seyfert galaxies whose centers are obscured by large amounts of gas and dust are Type 2 Seyferts.

Radio Galaxies Some elliptical galaxies, called **radio galaxies** because of their strong radio emission, are like lower-energy quasars. Although their energy source

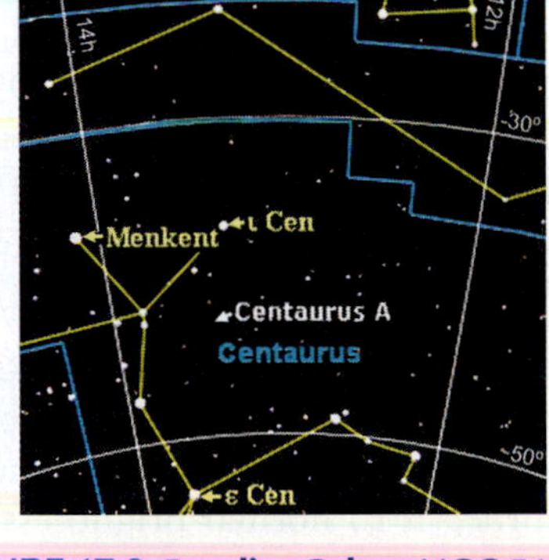

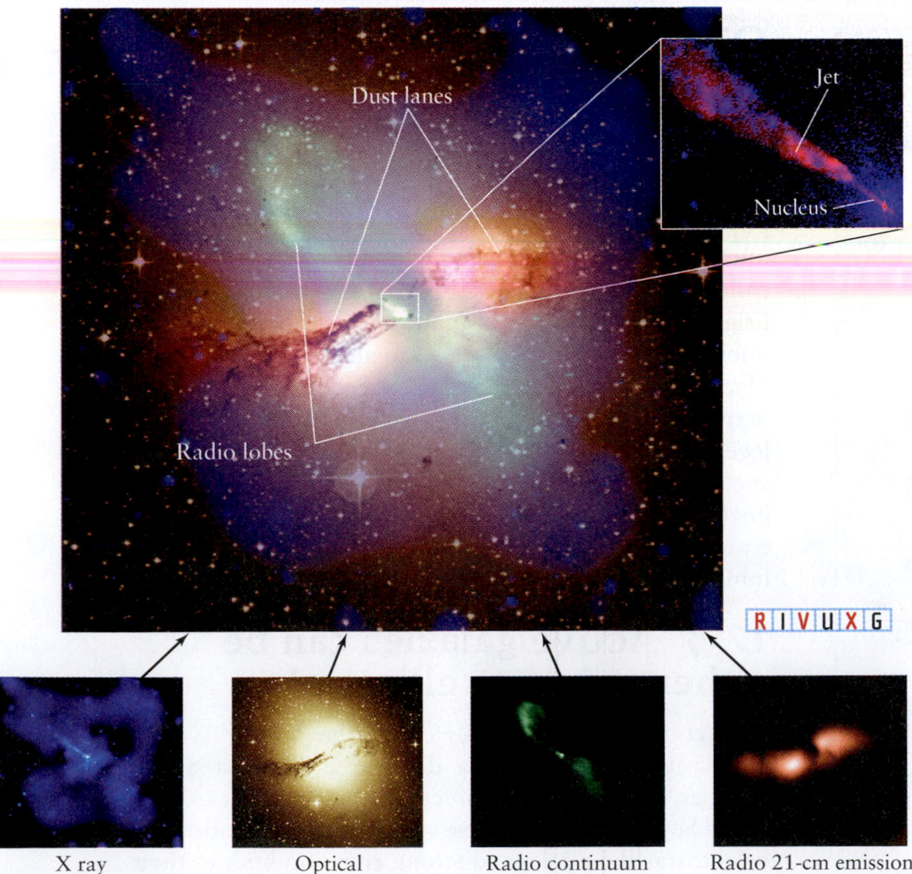

FIGURE 17-9 Peculiar Galaxy NGC 5128 (Centaurus A) This extraordinary radio galaxy is located in the constellation Centaurus, 11 million ly from Earth. At visible wavelengths a dust lane crosses the face of the galaxy. Superimposed on this visible image is a false-color radio image (green) showing that vast quantities of radio radiation pour from matter ejected from the galaxy perpendicular to the dust lane, along with radio emission (rose-colored) along the dust lane, and X-ray emission (blue) detected by NASA's Chandra X-ray Observatory. The X rays may be from material ejected by the black hole or from the collision of Centaurus A with a smaller galaxy. Inset: This X-ray image from the Einstein Observatory shows that NGC 5128 has a bright X-ray nucleus. An X-ray jet protrudes from the nucleus along a direction perpendicular to the galaxy's dust lane. (X ray: NASA/CXC/M. Karovska et al.; radio 21-cm: NRAO/VLA/J. Van Gorkom/Schminovich et al.; radio continuum: NRAO/VLA/J. Conden et al.; optical: Digitized Sky Survey U.K. Schmidt Image/STScI; inset: X ray: NASA/CXC/Bristol U./M. Hardcastle; radio: RAO/VLA/Bristol U./M. Hardcastle)

is very small, like that of a quasar, the resulting radio emissions of these active galaxies are spread over very large areas, often hundreds of thousands of light-years across.

Some radio galaxies are accompanied by unusual optical features; namely, the galaxies look as though they are exploding, which they are not. Such galaxies are therefore called **peculiar galaxies** (denoted *pec*). *Peculiar* is not the classification of a galaxy shape, like *spiral, barred spiral,* or *elliptical.* Any of the Hubble classes of galaxies (see Chapter 16) can be peculiar.

Figure 17-9 is a combined X-ray, optical, and radio image of the peculiar elliptical galaxy NGC 5128 (Centaurus A) in the southern constellation of Centaurus. The dark region that cuts the elliptical galaxy in two is actually a region of very thick dust through which little light from the galaxy behind it can penetrate. The galaxy is ejecting large volumes of hot X ray–emitting gas, as well as jets of gas in two **radio lobes** shown in this figure. NGC 5128 is classified as an E0 (pec). The E0 denotes a circular elliptical galaxy, while (pec) means that this galaxy has the features of a peculiar galaxy.

Detailed observations of Centaurus A reveal that both its radio and X-ray jets (see Figure 17-9 inset) originate in the galaxy's nucleus. Particles and energy stream out of the galaxy's nucleus toward the radio lobes at nearly the speed of light. Such galaxies, with two radio lobes, are now called **double-radio sources.** By 1970, radio astronomers had discovered dozens of other double radio sources, usually in giant elliptical galaxies. Double-radio sources are among the brightest radio objects in the universe.

All double-radio sources appear to have a central engine that ejects electrons, protons, and magnetic fields outward along two oppositely directed jets. After traveling many thousands or even millions of light-years, this material slows down, allowing charged particles and magnetic fields to produce the radio radiation that we detect. This type of radio emission, called *synchrotron radiation* (see Section 15-4), occurs whenever energetic electrons, protons, or ions move in a spiral path within a magnetic field.

The idea that a double-radio source emits jets of particles is supported by the existence of **head-tail sources,** so named because each such source appears to have a region of concentrated radio emission (the head) with two tails of gas that stream from it in opposite directions and then sweep back. NGC 1265, an active elliptical galaxy in the Perseus cluster of galaxies, shows these properties. NGC 1265 is known to be moving at a high speed (2500 km/s) relative to the Perseus cluster as a whole. In a radio map (Figure 17-10), its emission has a distinctly windswept appearance.

Protons and electrons ejected in the two jets from NGC 1265 are deflected by the galaxy's passage through the sparse intergalactic medium. The head-tail source leaves behind a trail of particles, like the trail of smoke that pours from a rapidly moving steam train.

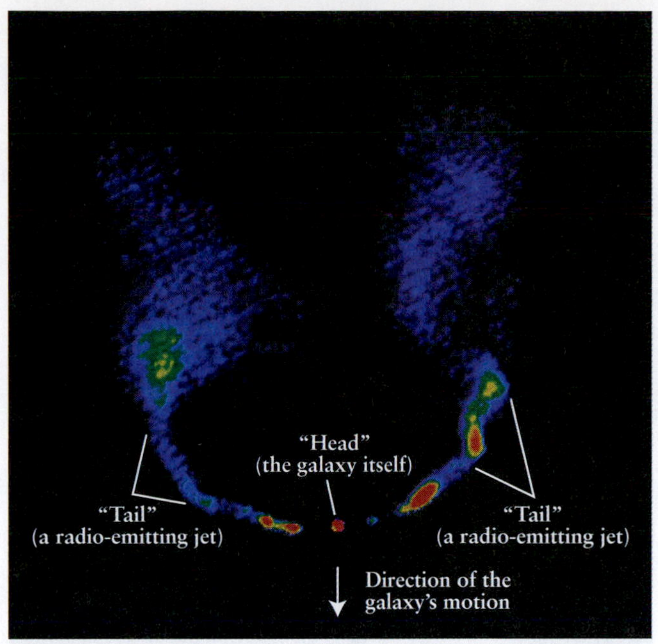

FIGURE 17-10 Head-Tail Source NGC 1265 This active elliptical galaxy is moving at a high speed through the intergalactic medium. Because of this motion, the two tail jets trail the galaxy at its head, giving this radio source a distinctly windswept appearance. (NRAO)

What Hubble types of galaxies can be peculiar?

The giant elliptical galaxy M87 is a relatively weak radio galaxy, emitting only 1% as much energy as the Milky Way. Conversely, the distant, peculiar, elliptical galaxy Cygnus A (Figure 17-1) is a powerful radio galaxy that emits 40 times as much energy as does our Galaxy. Pairs of orbiting black holes and accompanying double-radio sources have also been observed, such as 3C 75 (Figure 17-11). Table 17-1 summarizes the relative power emitted by the Sun, the Milky Way Galaxy, quasars, Seyfert galaxies, and radio galaxies.

BL Lacertae Objects and Blazars Although radio galaxies emit less energy from their nuclei than do quasars, there are also objects of the same size that from our vantage point emit even more energy than quasars. These

Table 17-1 Galaxy and Quasar Luminosities

Object	Luminosity (watts)
Sun	4×10^{26}
Milky Way Galaxy	10^{37}
Seyfert galaxies	$10^{36}–10^{38}$
Radio galaxies	$10^{36}–10^{38}$
Quasars	$10^{38}–10^{42}$

FIGURE 17-11 Binary Head-Tail Source This combined radio and X-ray image of 3C 75 shows the head-tail sources emanating from supermassive black holes in a pair of galaxies that are in the process of merging. The black holes are separated by 25,000 ly and are 300 million ly away from Earth. (X ray: NASA/CXC/D. Hudson, T. Reiprich et al. [AIFA])

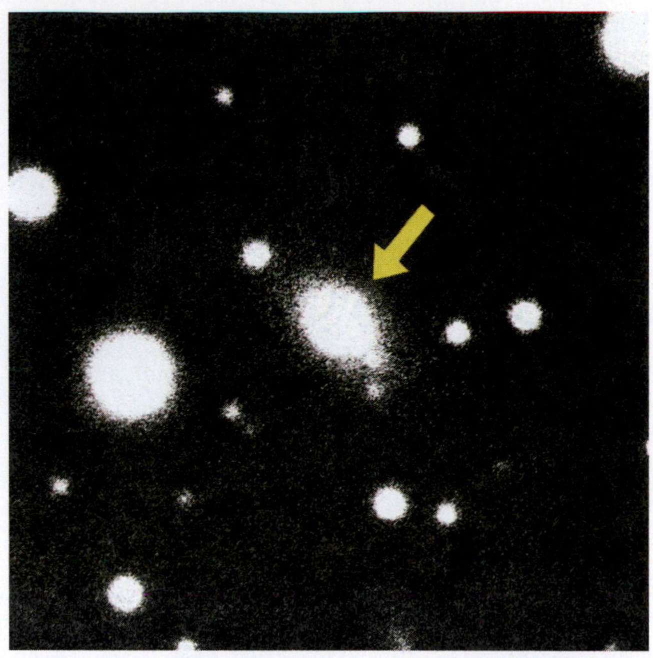

FIGURE 17-12 BL Lacertae This photograph shows fuzz around BL Lacertae (arrow). The redshift of this fuzz indicates that BL Lacertae is about 900 Mly (280 Mpc) from Earth. BL Lac objects appear to be giant elliptical galaxies with bright quasarlike nuclei, much as Seyfert galaxies are spiral galaxies with quasarlike nuclei. BL Lac objects contain much less gas and dust than do Seyfert galaxies. (T. D. Kinman, NOAO/AURA)

are the **BL Lacertae objects.** The name comes from their prototype, BL Lacertae (also called **BL Lac**) in the constellation Lacerta (the Lizard). BL Lac (Figure 17-12) was discovered in 1929, when it was mistaken for a variable star, largely because its brightness varies by a factor of 15 in only a few months. A BL Lac's most intriguing characteristic is an unusually weak set of spectral lines. In other words, its spectrum is not as easily identified as those of stars or interstellar clouds. Nevertheless, the spectra enable astronomers to determine the redshifts of these objects.

BL Lac objects are believed to be elliptical galaxies with bright quasars at their centers, much as Seyfert galaxies are spiral galaxies with quasarlike centers. BL Lac objects change intensities—they are variable. Those that have periods of a day or less are also called **blazars.** In 2004, a blazar with 10 billion solar masses located 12.5 billion ly away was discovered. It is intriguing that this incredibly massive object formed only a billion years after the universe formed.

SUPERMASSIVE ENGINES

Quasars, BL Lac objects, Seyfert galaxies, and radio galaxies are excellent examples of how astronomers can use one mechanism to explain an apparently wide variety of objects in space. In 1968, the British astronomer Donald Lynden-Bell suggested that the gravitational field of a supermassive black hole (see Chapter 14) could cause enormous releases of energy for quasars and other active galaxies.

17-4 Supermassive black holes exist at the centers of most galaxies

As we saw in Chapter 14, black holes with a wide range of masses are being discovered throughout the universe. Supermassive black holes (those exceeding a million solar masses or so) have been detected in a rapidly growing number of galaxies, including our Milky Way (see Chapter 16). To emphasize the similarities in active galaxies, we consider four further examples: M31, M32, M87, and M104.

The Andromeda Galaxy (M31) is the second largest, second most massive galaxy in the Local Group (after the Milky Way). In the mid-1980s, several astronomers made careful spectroscopic observations of the core of M31. Using measurements of Doppler shifts, they determined that stars within 50 ly of the galactic core are orbiting the nucleus at exceptionally high speeds, which suggests that a massive object is located at the galaxy's center. Without the gravity of such an object to keep the

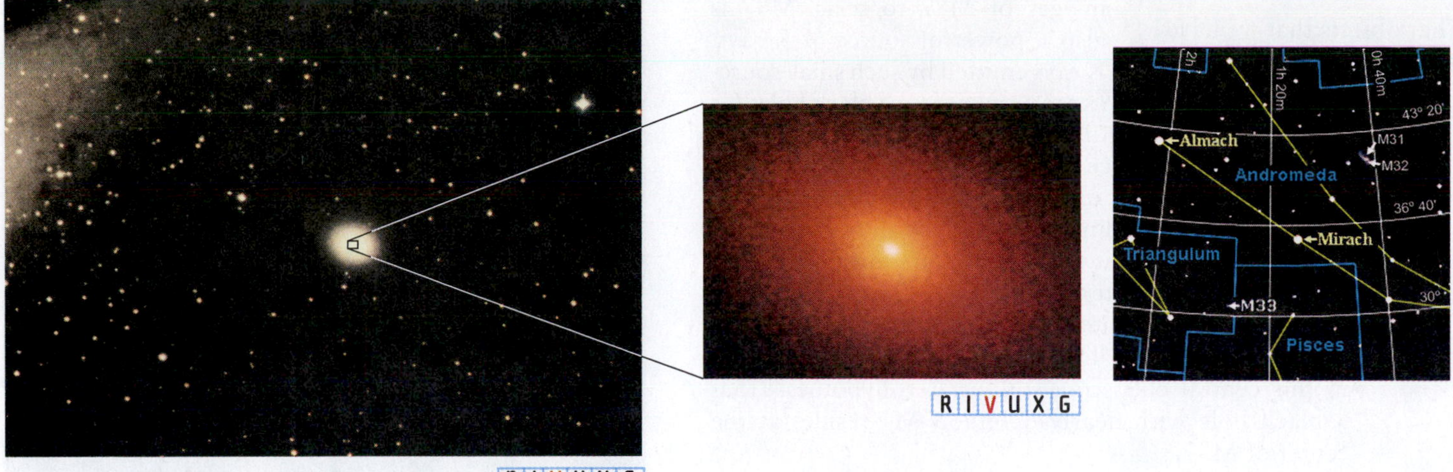

FIGURE 17-13 Elliptical Galaxy M32 This small galaxy is a satellite of M31, a portion of which is seen at the left of this wide-angle photograph. Both galaxies are roughly 2.5 million ly from Earth. Inset: High-resolution image from the Hubble Space Telescope that shows the center of M32. Note the concentration of stars at the nucleus of the galaxy. The nucleus is only 175 ly across. (Palomar Observatory; inset: NASA, ESA)

stars in their high-speed orbits, they would have escaped from the core region long ago. From such observations, astronomers estimate the mass of the central object in M31 to be about 50 million solar masses. That much matter confined to such a small volume strongly suggests that a supermassive black hole is located there.

Near M31 is a small elliptical galaxy, called M32 (Figure 17-13). High-resolution spectroscopy indicates that stars close to the center of M32 are also orbiting this galaxy's nucleus at unusually high speeds, which is explained by the presence of a supermassive black hole there, too. A picture taken by the Hubble Space Telescope (Figure 17-13 inset) shows that the concentration of stars at the core of M32 is truly remarkable. Its number density of stars is more than 100 million times greater than the density of stars in the Sun's neighborhood. The concentration of stars and their high speeds strongly support the belief that a supermassive black hole also exists at the center of M32.

One early candidate galaxy for study by the Hubble Space Telescope was the giant elliptical galaxy M87 (Figure 17-14). Located some 50 million ly from Earth, M87 is an active galaxy that has long been recognized as unusual. In 1918, Heber Curtis at the Lick Observatory reported that the center of M87 has "a curious straight ray… apparently connected with the nucleus by a thin line of matter." Figure 17-14 shows radio and visible

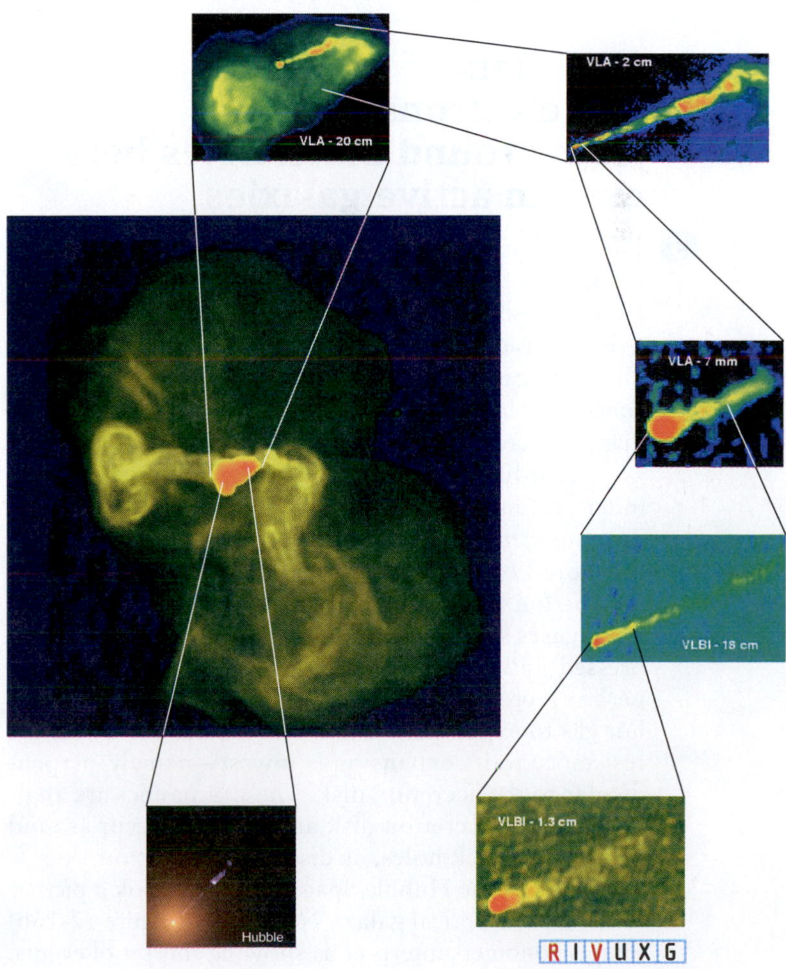

FIGURE 17-14 Giant Elliptical Galaxy M87 M87 is located near the center of the sprawling, rich Virgo cluster, which is about 50 million ly from Earth. Embedded in this radio image of gas is the galaxy M87 from which the gas has been ejected (bottom inset). Images at different radio and visible wavelengths reveal a variety of details about the structure of the jets of gas. M87's extraordinarily bright nucleus and the gas jets result from a 3-billion-$M_\odot$ black hole, whose gravity causes huge amounts of gas and an enormous number of stars to crowd around it. (NASA and the Hubble Heritage Team [STScI/AURA]; Frazer Owen [NRAO], John Biretta [STScI] and colleagues)

What objects that we have studied are millions or billions of solar masses?

images of M87 to scale. M87 is also a powerful source of X rays. X rays emitted by such small sources are a signature of black holes (as discussed in Section 14-7).

The Hubble Space Telescope image of M87 in Figure 17-14 shows an exceptionally bright, starlike nucleus and a surrounding disk of gas with trailing spiral arms. To produce this fiery glow, stars must be packed so tightly at the center of M87 that their density is at least 300 times greater than that normally found at the centers of giant ellipticals. The motions of the stars in this central clustering support the hypothesis that a black hole with nearly 3 billion $M_\odot$ resides at the center of M87.

Even edge-on spiral galaxies are revealing their innermost secrets. John Kormendy used a 3.6-m telescope on Mauna Kea to examine the core of M104 (Figure 17-15) spectroscopically. Once again, high-speed gas orbiting the galaxy's nucleus was found. These observations suggest that the center of this galaxy is dominated by a supermassive black hole that contains a billion solar masses. Similar spectroscopic observations of the edge-on disk galaxy NGC 3115 also reveal evidence for a billion-solar-mass black hole at its core.

What makes it so hard for us to see deep into elliptical galaxies?

FIGURE 17-15 Sombrero Galaxy (M104) This spiral galaxy in Virgo is nearly edge on to our Earth-based view. Spectroscopic observations indicate that a billion-solar-mass black hole is located at the galaxy's center. You can see the bright region in the galaxy's center created by stars and gas that orbit the black hole. (X ray: NASA/UMass/Q.D.Wang et al.; optical: NASA/STScI/AURA/Hubble Heritage; infrared: NASA/JPL-Caltech/U. of Arizona/R. Kennicutt/SINGS Team)

17-5 Jets of protons and electrons ejected from around black holes help explain active galaxies

3 The gravitational energy associated with supermassive black holes at the centers of galaxies creates huge jets of gas comprised of protons and electrons. To see how this works, consider a black hole at the center of a young galaxy filled with gas and dust. Because the centers of galaxies are congested places, a black hole located there will capture a massive accretion disk of this debris (Figure 17-16a).

According to Kepler's third law, the material in the inner regions of such a disk orbits the most rapidly. The inner matter, therefore, constantly interacts with the more slowly moving gases in the outer regions, and the friction between them heats all of the gas. As energized gases spiral toward the black hole, they are compressed and heated to millions of degrees. This temperature creates great pressure, causing some of the hot gas to expand and eventually squirt out where the resistance to its expansion is lowest—namely, perpendicular to the accretion disk. These dynamics are analogous to the accretion disk and jets that occur around solar-mass black holes, as discussed in Section 14-9.

In 1995, the Hubble Space Telescope took a picture of the giant elliptical galaxy NGC 4261 (Figure 17-16b) that astronomers interpret as showing this set of events. A disk of gas and dust about 800 ly in diameter is seen orbiting a supermassive black hole. The speed of the material in this disk indicates that it is held in orbit by a 1.2-billion-$M_\odot$ object. The Figure 17-16b inset is a Hubble image of the galaxy's nucleus. The dark, horizontal oval in the picture is our oblique view of the accretion disk, and ground-based radio and optical observations show a double-lobed structure (Figure 17-16b) that is bisected by the oval.

What confines the ejected gas to narrow jets? At first, the gas still falling toward the black hole prevents the jets from spreading. To see why, consider water spraying out of the nozzle of a hose. The stream of water broadens and fans out through a wide angle in the air. A similar process would occur around these black holes if the infalling gas were not there. If the nozzle is placed in a swimming pool, however, the stream of water will not fan out as much at first (Figure 17-17). Similarly, as the two jets of hot gas leave the vicinity of the black hole, they must blast their way through the gas that is still crowding inward. Passage through this material causes the jets to become narrow, concentrated beams. Then, magnetic fields created in the hot, swirling accretion disk spiral around the jets of gas and help keep the ejecta in columns (Figure 17-18).

After traveling for hundreds of thousands of years or more, the jets of gas interact with enough preexisting

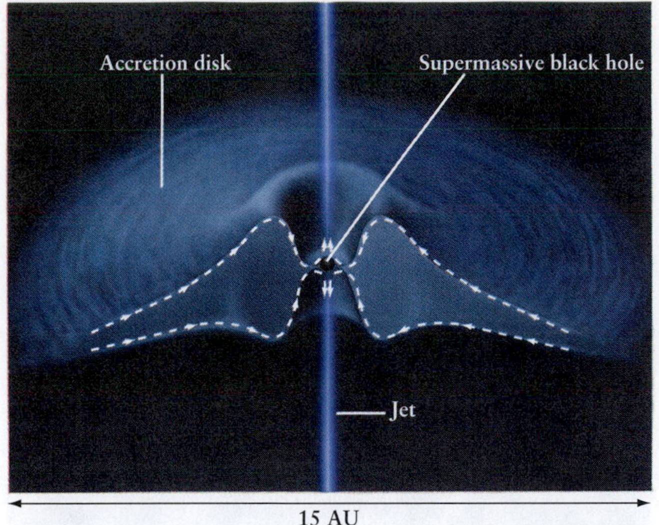

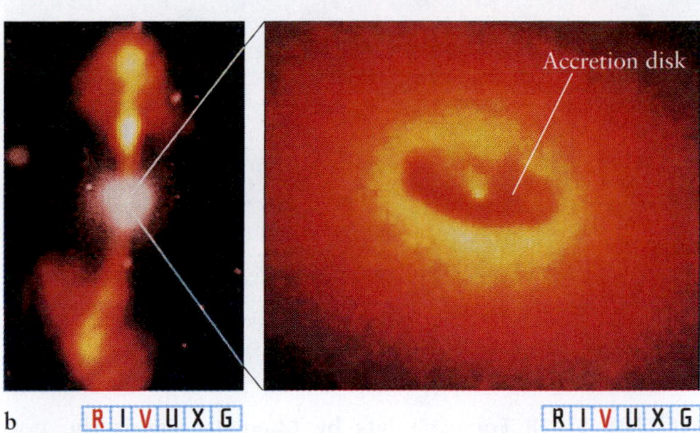

FIGURE 17-16 Supermassive Black Holes as Engines for Galactic Activity (a) In the accretion disk around a supermassive black hole, inswirling gas heats and expands. Pulled inward, compressed, and heated further, some of it is eventually expelled perpendicular to the disk in two jets. (b) The giant elliptical galaxy NGC 4261 is a double-radio source located in the Virgo cluster, about 100 million ly from Earth. An optical photograph of the galaxy (white) is combined with a radio image (orange and yellow) to show both the visible galaxy, which does not emit much radio energy, and its jets, which do. Inset: This Hubble Space Telescope image of the nucleus of NGC 4261 shows a disk of gas and dust about 800 ly (250 pc) in diameter, orbiting a supermassive black hole. (b: NASA; inset: ESA)

interstellar and intergalactic gas to be stopped, thereby forming the lobes and other features. In 2003, astronomers using the Chandra X-ray Observatory observed gas jetting out from a supermassive black hole in a galaxy at the center of the Perseus cluster of galaxies and striking gas that resides between galaxies. This *intracluster* gas became compressed by the jet and generated concentric rings or shells of sound waves 35,000 ly in wavelength. Put another way, the sound created by the jet is a B-flat that is 57 octaves below middle C.

This "rocket nozzle" model explains not only double-radio sources but also the jets and beams we see protruding from other active galaxies. We can now tie together all of the various objects discussed in this chapter. *The major difference between the different types of active galaxies is the angle at which we view the central engine.* As Figure 17-19 shows, an observer sees a radio galaxy or a double-radio source when the accretion disk is viewed nearly edge on, because the jets are nearly in the plane of the sky. At a steeper angle, a quasar is seen. If one of the jets is aimed almost directly at Earth, the galaxy is a BL Lac object. The Seyferts, active spiral galaxies, are distinguished by whether we can see the bright core (Type 1) or not (Type 2). Their placements in Figure 17-19 are still under investigation.

FIGURE 17-17 Focusing Jets by Pressure (a) If a high-speed jet of gas or liquid encounters little pressure (from the surrounding air, in this image), then it will spread out. (b) If the jet encounters high pressure, such as occurs when it enters water, then it will maintain longer its shape as a column. (a: Comstock Images/Getty Images; b: Fundamental Photographs, New York)

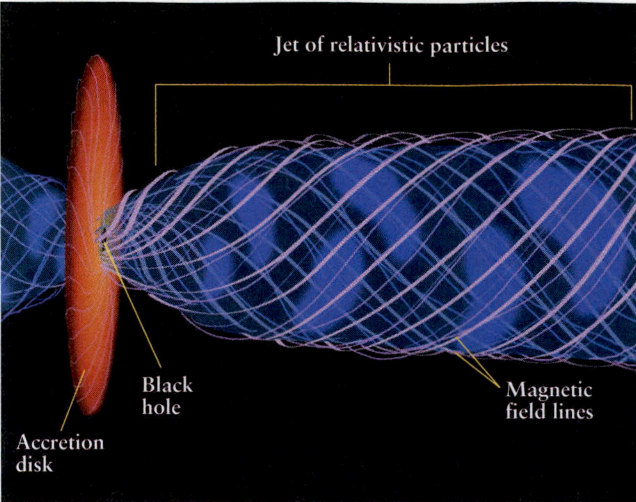

FIGURE 17-18 Focusing Jets by Magnetic Fields The hot, ionized accretion disk around the black hole rotates and creates a magnetic field that is twisted into spring-shaped spirals above and below the disk. Some of the accretion disk's gas falling toward the black hole is overheated and squirted at high speeds into the two tubes created by the magnetic fields. The fields keep the gas traveling directly outward from above and below the disk, thus creating the two jets.

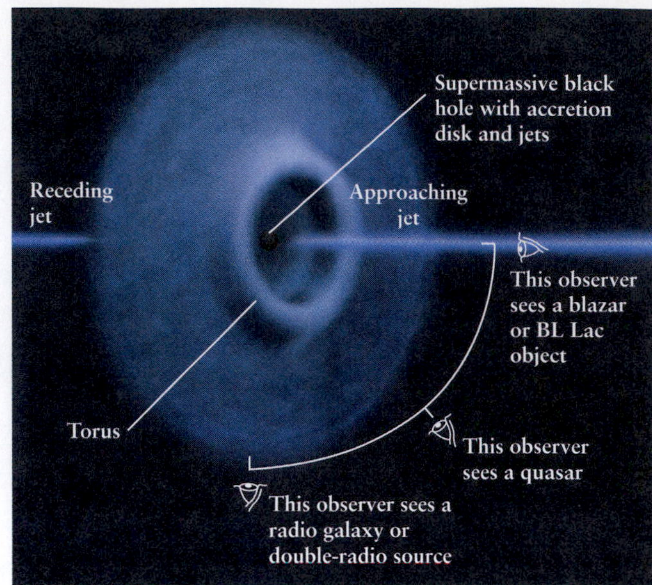

ANIMATION 17.3

FIGURE 17-19 Orientation of the Central Engine and Its Jets BL Lacertae objects, quasars, and double-radio sources appear to be the same type of object viewed from different directions. If one of the jets is aimed almost directly at Earth, we see a BL Lac object. If the jet is somewhat tilted to our line of sight, we see a quasar. If it is tilted farther, we see an active galaxy. If the jets are nearly perpendicular to our line of sight, we see a double-radio source. The central region of the system is shown in Figures 17-16a and 17-18.

There are many active galaxies that we *should* see but do not. These objects have such thick and dusty accretion disks that little visible light or X-rays get through them. We cannot see the energy sources, for example, of quasars using visible light or X-ray telescopes in such situations. However, infrared radiation should pass through the gas and dust. Thus, in 2005, astronomers used the Spitzer Space Telescope to study regions of galaxies in which quasars should exist. In just one tiny patch of sky they found 21 quasars. Extrapolating to the total volume of space that contains quasars, they expect "stealth" active galaxies to exist by the millions or more.

The other major factor in the observed properties of active galaxies and their evolutionary history is the amount of material available to make the accretion disk. Overlap in the brightness levels of different types of active galaxies occurs because central black holes and their accretion disks have different masses. The intensities of active galaxy emissions change as the amount of gas and dust in their accretion disks varies.

When supermassive black holes are young, they are surrounded by vast amounts of gas, which create massive accretion disks. Over time, much of this gas either enters the black hole or jets away from the disk, as just discussed. When an accretion disk is depleted, that active galaxy becomes more and more quiet. Because the galaxies we observe close to us are about the same age as the Milky Way, over 13 billion years old, most have small or no accretion disks around the massive black holes in their nuclei. Thus, we do not see many active galaxies close to us. However, when galaxies collide, gas can be available to enable quiet galaxies to once again become active.

Insight Into Science

Occam's Razor Revisited It is possible to create *separate* theories for quasars, BL Lac objects, and other active galaxies. However, scientists prefer a comprehensive single theory that explains as many things as possible, because it incorporates common properties of the various phenomena. This search for simple, more powerful explanations is the central engine of progress in science.

17-6 Gravity focuses light from quasars

The radiation emitted by all quasars is subject to the gravitational lensing described by Einstein's theory of general relativity (see Sections 14-3 and 16-11). Like the light from stars behind the Sun, the light from quasars is deflected as it travels past other galaxies or intergalactic gas and dark matter toward us. As we saw in Section 16-11, this gravitational lensing can lead to our receiving several images of objects. Figure 17-20 shows four images of a quasar in a configuration called an **Einstein cross.** If the background object is exactly behind an intervening quasar, the light is actually focused as a ring, called an **Einstein ring** (see Figure 16-29b), rather than as several separate images.

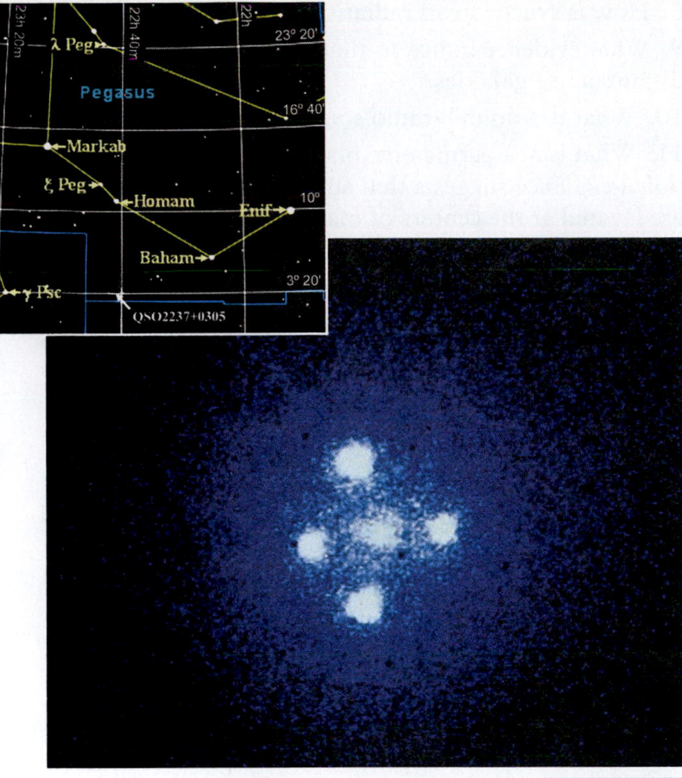

FIGURE 17-20 Gravitational Lensing of Quasars Image from the Hubble Space Telescope that shows the gravitational lensing of a quasar in the constellation of Pegasus. The quasar, about 8 billion ly from Earth, is seen as four separate images that surround a galaxy that is only 400 million ly away. This pattern is called an Einstein cross. The diffuse image at the center of the Einstein cross is the core of the intervening galaxy. The physical effect that creates these multiple images is the same as that seen for galaxies, as depicted in Figure 16-29. (NASA/ESO)

As you know, light travels at a finite speed. For example, light from the Sun takes about 8 min to reach Earth. We see the Sun as it was 8 min ago. The farther an object is from us, the longer it takes light and other radiation from it to reach Earth. We see a galaxy 1 billion ly away as it was a billion years ago. When we look at quasars, we are seeing objects as they appeared when the universe was younger still. In the last few years, astronomers using the Sloan Digital Sky Survey have discovered quasars at a distance corresponding to when the universe was only about 800 million years old.

17-7 Frontiers yet to be discovered

Quasars, active galaxies, BL Lac objects, and double-radio sources are rich areas for observational and theoretical research. Why do different quasars emit most strongly in different parts of the spectrum? Are quasars an early phase of most galaxies, including the Milky Way, as many astronomers are coming to believe? What causes quasar variability? Do we have the proper explanations that blazars, quasars, and radio galaxies are caused by supermassive black holes seen from different orientations? How did supermassive black holes form so quickly in the life of the universe? New discoveries in this realm are being made nearly every week.

> What other astronomical objects have been observed after their light passed through a gravitational lens?

SUMMARY OF KEY IDEAS

- The development of radio astronomy in the late 1940s led to the discovery of very powerful and extremely distant energy sources.

Quasars and Other Active Galaxies

- An active galaxy is an extremely luminous galaxy that has one or more unusual features: an unusually bright, starlike nucleus; strong emission lines in its spectrum; rapid variations in luminosity; and jets or beams of radiation that emanate from its core. Active galaxies include quasars, Seyfert galaxies, radio galaxies, double-radio sources, and BL Lacertae objects.
- A quasar, or quasi-stellar radio source, is an object that looks like a star but has a huge redshift. This redshift corresponds to a distance of billions of light-years from Earth, according to the Hubble law.
- To be seen from Earth, a quasar must be very luminous, typically about 100 times brighter than an ordinary galaxy. Relatively rapid fluctuations in the brightness levels of some quasars indicate that they cannot be much larger than the diameter of our solar system.
- An active spiral galaxy with a bright, starlike nucleus and strong emission lines in its spectrum is categorized as a Seyfert galaxy.
- An active elliptical galaxy is called a radio galaxy. It has a bright nucleus and a pair of radio-bright jets that stream out in opposite directions.
- BL Lacertae (BL Lac) objects (some of which are called blazars) have bright nuclei whose cores show relatively rapid variations in luminosity.
- Double-radio sources contain active galactic nuclei located between two characteristic radio lobes. A head-tail radio source shows evidence of jets of high-speed particles that emerge from an active galaxy.

Supermassive Central Engines

- Many galaxies contain huge concentrations of matter at their centers.
- Some matter that spirals in toward a supermassive black hole is squeezed into two oppositely directed beams that carry particles and energy into intergalactic space.

• The energy sources from quasars, Seyfert galaxies, BL Lac objects, radio galaxies, and double-radio sources are probably matter ejected from the accretion disks that surround supermassive black holes at the centers of galaxies.

WHAT DID YOU THINK?

1 *What does "quasar" stand for?* Quasi-stellar radio source.

2 *What do quasars look like?* They look like stars, but they emit much more energy than any star.

3 *Where do quasars get their energy?* A quasar is thought to be powered by a supermassive black hole with millions or billions of solar masses at the center of a galaxy.

Key Terms for Review

active galactic nuclei (AGN), 525
active galaxy, 525
blazar, 528
BL Lacertae (BL Lac) object, 528
double-radio source, 527
Einstein cross, 532
Einstein ring, 532
head-tail source, 527
peculiar galaxy (pec), 527
quasar (quasi-stellar radio source), 524
quasi-stellar object (QSO), 524
radio galaxy, 526
radio lobe, 527
Seyfert galaxy, 525

Review Questions

1. What is a double-radio source seen along one axis of a jet called? **a.** a BL Lac object, **b.** a quasar, **c.** a binary star system, **d.** a double-radio source, **e.** a pulsar

2. What two things does the engine of a quasar contain? **a.** a supermassive black hole and a binary companion, **b.** a stellar-mass black hole and a binary companion, **c.** a supermassive black hole and an accretion disk, **d.** a stellar-mass black hole and an accretion disk, **e.** a supergiant star and an accretion disk

3. Suppose you suspected a certain object in the sky to be a quasar. What sort of observations would you perform to confirm your hypothesis?

4. Explain why astronomers do not use any of the standard candles described in Chapter 16 to determine the distances to quasars.

5. Explain how the rate of variability of a source of light can be used to place an upper limit on the size of the source.

6. What is an active galaxy? List the different kinds of active galaxies. How do they differ from one another?

7. Why do astronomers believe that the energy-producing region of a quasar is very small?

8. How is synchrotron radiation produced?

9. What evidence indicates that quasars are extremely distant active galaxies?

10. What is a double-radio source?

11. What is a supermassive black hole? What observational evidence suggests that supermassive black holes are located at the centers of many galaxies?

12. Why do many astronomers believe that the engine at the center of a quasar is a supermassive black hole surrounded by an accretion disk?

13. How does the orientation of the jets that emanate from the center of a galaxy (relative to our line of sight) relate to the type of active galaxy that we observe?

Advanced Questions

14. In the 1960s, some astronomers suggested that quasars might be compact objects ejected at high speeds from the centers of nearby ordinary galaxies. Why does the absence of blueshifted quasars disprove this hypothesis?

15. When quasars were first discovered, many astronomers were optimistic that these extremely luminous objects could be used to probe distant regions of the universe. For example, it was hoped that quasars would provide high-redshift data from which the Hubble constant could be more accurately determined. Why have these hopes not been realized?

Discussion Questions

16. Explore the belief that quasars, double-radio sources, and giant elliptical galaxies represent an evolutionary sequence.

17. Some quasars show several sets of absorption lines whose redshifts are less than the redshift of the quasars' emission lines. For example, the quasar PKS 0237-23 has five sets of absorption lines, all with redshifts somewhat less than the redshift of the quasar's emission lines. Propose an explanation for these sets of absorption lines.

What If . . .

18. Earth passed through a jet emitted by a radio galaxy? What would we see and what might happen to Earth?

19. A jet of gas suddenly appeared across the night sky? What might it indicate has recently happened?

20. Sirius suddenly became ring-shaped for a few hours and then returned to normal? What might cause such an event?

Web Question

21. To test your understanding of active galaxies, do Interactive Exercise 17.1 on the Web. You can print out your answers, if required.

WHAT DO YOU THINK?

1. What is the universe?
2. Did the universe have a beginning?
3. Into what is the universe expanding?
4. How strong is gravity compared to the other forces in nature?
5. Will the universe last forever?

Answers to these questions appear in the text beside the corresponding numbers in the margins and at the end of the chapter.

Bursts of star formation in the early universe, depicted by science artist Adolf Schaller. (Adolf Schaller, STScI/NASA/K. Lanzetta, SUNY)

Chapter 18
Cosmology

For millennia, our ancestors speculated about whether the universe has existed forever or whether it began some finite time in the past. Did God create the universe, or was its creation an event that required no divine intervention? Science has yet to provide an answer to this question. But astronomers and other scientists *have* accumulated a great deal of knowledge about what has happened to the universe beginning a tiny fraction of a second after it began, and what its fate will be. Although the overall theory of cosmic evolution is not complete, its major elements are consistent with observations.

In this chapter we will examine the distribution of matter on the largest scales in the universe and how this matter is moving and changing. Using this information, we will explore the Big Bang model of how the universe has evolved from its earliest moments, when it was smaller than an atom. We will see how it grew and created matter as we know it, forming and evolving large-scale systems, like galaxies and groupings of galaxies. Finally, we will consider the fate of the universe.

In this chapter you will discover

- cosmology, which seeks to explain scientifically how the universe began, how it evolves, and its fate
- the best cosmological theory that we have for the evolution of the universe—the Big Bang
- how astronomers trace the emergence of matter and the formation of galaxies
- how astronomers explain the overall structure of the universe
- our understanding of the fate of the universe

THE BIG BANG

1 The **universe** consists of all matter, energy, and spacetime that we can ever detect or that will ever be able to affect us. (We use this definition because there may be matter and energy in other dimensions or matter and energy that are moving away from us so quickly that their influence will never reach us.) So far in this text, we have explored matter on size scales from atoms to superclusters of galaxies. We also learned in Section 16-12 that the superclusters are all moving away from one another, implying that the universe is expanding. In this chapter we take the observational evidence from the rest of the book and use it to explore the Big Bang theory of cosmology. **Cosmology** is the study of the large-scale structure and evolution of the universe.

18-1 General relativity predicts an expanding (or contracting) universe

Modern cosmology almost began in 1915, when Einstein published his theory of general relativity. To his surprise and dismay, the relativity equations predicted that the universe is not static: They indicated that it should be either expanding (which it is) or contracting. But Einstein was not ready for what the equations were telling him.

At the time that Einstein published the theory of general relativity, the existence of galaxies and clusters and superclusters of galaxies had not yet been established and the 1925 discovery of the Hubble flow (Section 16-12) was a decade away. The prediction of a changing universe flew in the face of the then-widely accepted belief in an infinite, static universe, a concept promoted by Isaac Newton more than two centuries earlier. Newton believed that each star is fixed in place and held under the influence of a uniform gravitational pull from every part of the cosmos. If the stars were not uniformly distributed, he argued, one region would have more mass than another. The denser region's gravity would then attract other stars, causing them to further clump together. Because he did not observe this clumping, Newton concluded that the stars in the universe must be distributed uniformly over an infinite space.

The apparently static universe and the prevailing philosophy that the universe had existed forever made Einstein doubt the implications of his own theory, so he missed the opportunity to propose that we live in a changing universe. Instead, he adjusted his elegant equations to yield a static, finite cosmos. He did this by adding a repulsive (outward-pushing) term, called the **cosmological constant,** to his equations so that gravity's normal attractive force would be counterbalanced and the universe would be static. After observations revealed that the universe is expanding, Einstein said that adding the cosmological constant was the biggest blunder of his career.

Although the value of the cosmological constant that Einstein inserted was wrong, the concept of such a constant may be correct. Observations since 1997 indicate that the universe is not just expanding but actually *accelerating* outward. This acceleration means that there must be an outward pressure that more than counteracts the effects of normal gravitation, which is trying to slow the universe's expansion. One of the two current theories that can explain this acceleration is the presence of a cosmological constant that creates the outward pressure. We discuss these two theories further in Section 18-15.

18-2 The expansion of the universe creates a Dopplerlike redshift

Edwin Hubble is credited with discovering that we live in an **expanding universe** (see Section 16-12). The redshifts of clusters and superclusters of galaxies that

Hubble found moving away from us appear to be produced by the Doppler effect, but they actually are not. Recall that the normal Doppler shift is caused by an object moving toward or away from us through *fixed* spacetime (see Section 4-7). However, using Einstein's theory of general relativity, we find that spacetime, the fabric of the universe, is *not fixed* but is actually expanding. This expansion is what is carrying the superclusters away from each other, and, in many cases, carrying clusters in a given supercluster away from one another. (The gravitational force that holds objects like planets and stars and entire galaxies together is so strong that these bodies and systems are not being pulled apart. The expansion of space just acts to increase the separation between large groups of galaxies.)

The redshift that Hubble observed, caused by the expansion of the universe, is properly called the **cosmological redshift.** In other words, the photons that we observe from galaxies in other superclusters are all redshifted because space is expanding. To understand why, consider a wave drawn on a rubber band (Figure 18-1). The wave has an unstretched wavelength (see Section 3-1) of λ_0. As the rubber band stretches, the wavelength increases ($\lambda > \lambda_0$). Now imagine a photon coming toward us from a distant galaxy. As the photon travels through space, space is expanding, and, like stretching the rubber band, this expansion stretches the photon's wavelength. When the photon reaches our eyes, we see a drawn-out wavelength—the photon is redshifted. The longer the photon's journey, the more its wavelength is stretched by the expansion of the universe. Therefore, astronomers observe larger redshifts in photons from relatively distant galaxies than in photons from relatively nearby galaxies.

What are the three kinds of redshifts presented in this book?

As we saw in Section 16-12, Hubble discovered the linear relationship between the distances to galaxies in other superclusters and the redshifts of those galaxies' spectral lines. For example, a galaxy twice as far from Earth as another galaxy has twice the cosmological redshift of the closer one. The normal Doppler shift has the same relationship. As a result of the two effects being described by the same equations, the normal Doppler shift and the cosmological redshift predict the same relationship between redshift and motion—except for the most distant galaxies and quasars, where effects of relativity must be taken into account. Working with relatively nearby galaxies, Hubble was fully justified in using the Doppler equation to calculate the recession of galaxies.

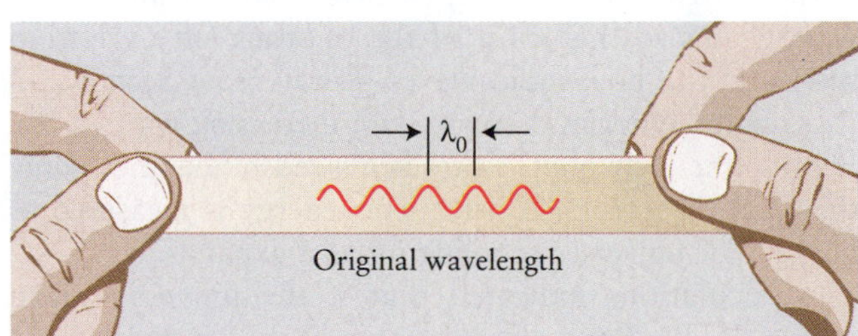

a A wave drawn on a rubber band ...

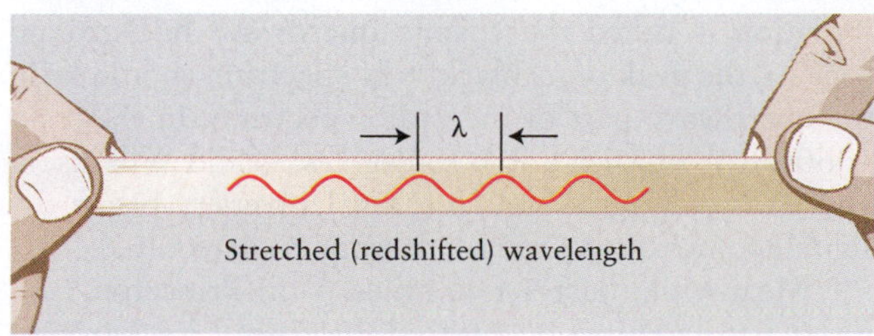

b ... increases in wavelength as the rubber band is stretched.

FIGURE 18-1 Cosmological Redshift Just as the waves drawn on this rubber band are stretched along with the rubber band, so too are the wavelengths of photons stretched as the universe expands.

18-3 The Hubble constant is related to the age of the universe

Hubble's law gives us a way to estimate the age of the universe. Imagine watching a movie of two superclusters receding from each other. If we then run the film backward, time runs in reverse, and we observe the superclusters coming together. The time it would take for them to collide is the time since they were last combined as parts of the same matter. *Assuming that the universe began expanding when it came into existence and that it has always expanded at a constant rate,* we can use a simple equation to estimate the age of the universe:

$$\text{Time since the Big Bang} = \frac{\text{separation distance}}{\text{recessional velocity}}$$

Recall that Hubble found the relationship

$$\text{Recessional velocity} = H_0 \times \text{separation distance}$$

which we can rewrite as

$$H_0 = \frac{\text{recessional velocity}}{\text{separation distance}}$$

Comparing the first and last equations here (see *An Astronomer's Toolbox 18-1*), we see that Hubble's constant is the reciprocal of the time since the universe began. Using a Hubble constant of 71 km/s/Mpc, we find:

$$1/H_0 \approx 1/71 \text{ km/s/Mpc} \approx 13.7 \text{ billion years}$$

As of 2010, the most accurate measurement is 13.75 ± 0.12 billion years. In what follows, we will simply use an age of 13.7 billion years.

An Astronomer's Toolbox 18-1

H_0 and the Age of the Universe

The Hubble constant may not look like a measure of time, but it is. All we need to do to see this is a little unit conversion. Our procedure will be to remove all distance units from H_0 and then invert it. Recall that we are using $H_0 = 71$ km/s/Mpc, which has different units of distance in the numerator and denominator. As we saw in Chapter 16, this mix of units makes it easy to determine how fast a galaxy is receding in kilometers per second once its distance in megaparsecs is known.

Example: A galaxy 10 Mpc away is receding at 71 km/s/Mpc $\times$ 10 Mpc $=$ 710 km/s.

We know that 1 pc $= 3.09 \times 10^{13}$ km, and so 1 Mpc $= 3.09 \times 10^{19}$ km. Because there are 3.156×10^7 s in a year, we get the conversion factors we need:

$$\frac{1}{H_0} = \frac{1}{71 \text{ km/s/Mpc}} \times (3.09 \times 10^{19} \text{ km/Mpc})$$

$$\times \frac{1}{3.156 \times 10^7 \text{ s/years}} = 13.7 \times 10^9 \text{ years}$$

If the universe had always been expanding at the same speed, then $1/H_0$ would be the age of the universe. However, for the first several billion years of its existence, the expansion of the universe was slowing down, and it has been speeding up since.

Decades of observations by many astronomers have confirmed that all distant galaxies are moving away from Earth. Not only that, *the Hubble law relating velocities and distances is the same in all directions*. The Hubble constant, H_0, is the same no matter where you aim your telescope. The fact that the recession rates are the same in all directions is the condition called *isotropy*. The universe is isotropic.

The general expansion away from us seems to suggest that Earth is at the center of the universe. After all, where else could we be if all of the distant galaxies are moving away from us? In fact, the answer is that we could be practically anywhere in the universe! To understand why, see *Guided Discovery: The Expanding Universe* (Chapter 16).

Try these questions: How old would the universe be if Hubble's constant were half its actual value? Twice its actual value?

(Answers appear at the end of the book.)

18-4 Remnants of the Big Bang have been detected

2 3 In 1927, the Belgian astrophysicist and Catholic priest Georges Lemaître proposed that, as a consequence of the equations of Einstein's theory of general relativity, the universe is expanding. Using Hubble's subsequent observations of an expanding universe, Lemaître logically concluded that the superclusters must have expanded from a much smaller volume. He proposed that the universe began as an extraordinarily hot and dense *primordial atom* of energy. Just as energy created in an explosion causes debris to expand outward, the primordial energy created at the beginning of time caused the universe to expand. This initial event is now called the **Big Bang.** Space and time (that is, spacetime; see Chapter 14) did not exist before that moment. Rather than expanding into preexisting spacetime, the Big Bang explosion created spacetime, and that spacetime (in which we live) has been expanding ever since. The Big Bang led to the formation of all spacetime, matter, and energy. As noted earlier in this chapter, science does not yet have an explanation for why the Big Bang occurred.

If the universe were twice as old as it is now, how would the Hubble constant compare to the value it has today?

Shortly after World War II, the astrophysicists Ralph Alpher and George Gamow proposed that a fraction of a second after the Big Bang, the universe was an incredibly hot blackbody—in excess of 10^{32} K. As heat is a measure of how many photons there are and how energetic they are, the early universe must therefore have been filled with intensely high-energy electromagnetic radiation. As the universe expanded, the cosmological redshift stretched these gamma rays to longer and longer wavelengths so that most of the photons left over from this event are now radio waves. Recall from Section 3-3 that photon energies decrease with increasing wavelength. Because the early photons are being redshifted, the energy they each have is decreasing. This energy is measured as heat, so the universe is cooling off as it expands.

Calculations indicated that if the universe began with a hot Big Bang, the remnants of that energy should still fill all space today, giving the entire universe a temperature only a few kelvins above absolute zero. This radiation is called the **cosmic microwave background** because the peak of its blackbody spectrum should lie in the microwave part of the radio spectrum. In the early 1960s, Robert Dicke, P. J. E. Peebles, David Wilkinson, and their colleagues at Princeton University began designing a microwave radio telescope to detect it.

Meanwhile, just a few miles from Princeton, two physicists had already detected the cosmic background radiation. Arno Penzias and Robert Wilson of Bell Laboratories were working on a new horn antenna designed to relay telephone calls to Earth-orbiting communications satellites (Figure 18-2). However, Penzias and Wilson were mystified. No matter where in the sky

FIGURE 18-2 Bell Labs Horn Antenna This Bell Laboratories horn antenna at Holmdel, New Jersey, was used by Arno Penzias (right) and Robert Wilson in 1965 to detect the cosmic microwave background. (Lucent Technologies, Bell Laboratories)

they pointed their antenna, they detected a faint background noise. All efforts to eliminate this background noise failed, even the careful removal of static noise-generating pigeon droppings from the antenna.

Thanks to a colleague, they learned of the then-theoretical cosmic microwave background and the work of Dicke, Peebles, and Wilkinson in trying to locate it. Communicating with the Princeton astronomers, Penzias and Wilson presented their finding and thus were able to claim the first detection—their annoying noise was, in fact, the remnant energy of the Big Bang. The cosmic microwave background is required by the Big Bang theory but is neither required nor predicted by any other competing cosmological theory. Detection of the cosmic microwave background is the principal reason why the Big Bang is accepted by astronomers as the correct cosmological theory.

What other objects that we have studied in this book have (nearly) blackbody spectra?

Precise measurements of the cosmic microwave background were first made by the *Cosmic Background Explorer* (COBE) satellite, which operated between 1989 and 1994. The radiation has since been measured more precisely by other equipment, such as the *Wilkinson Microwave Anisotropy Probe* (WMAP) and the balloon-carried *BOOMERANG* (Figure 18-3). The data

a

b

FIGURE 18-3 In Search of Primordial Photons (a) The *Wilkinson Microwave Anisotropy Probe* (WMAP) satellite, launched in 2001, improved upon the measurements of the spectrum and angular distribution of the cosmic microwave background taken by the COBE satellite. (b) The balloon-carried telescope BOOMERANG orbited above Antarctica for 10 days, collecting data used to resolve the cosmic microwave background, with 10 times higher resolution than that of COBE. All of these experiments found local temperature variations across the sky but no overall deviation from a blackbody spectrum. (a: NASA/WMAP Science Team; b: The BOOMERANG Group, University of California, Santa Barbara)

Insight Into Science

Consider Plausible Alternatives Until They Are Shown to Be Inconsistent with Observations Sir Fred Hoyle, who aptly named the Big Bang in 1949, was one of the very few astrophysicists who never believed that it occurred. Hoyle, Herman Bondi, and Thomas Gold proposed the *steady-state theory* in which the universe has existed forever, is continuously expanding, and, as it expands, new matter is created that takes the place of the receding matter. Where this new matter might come from is unknown, but the laws of physics do not necessarily exclude its creation. Today, astronomers accept the Big Bang theory rather than the steady-state theory because cosmic microwave background observations are consistent with the Big Bang theory, but not with the steady-state theory.

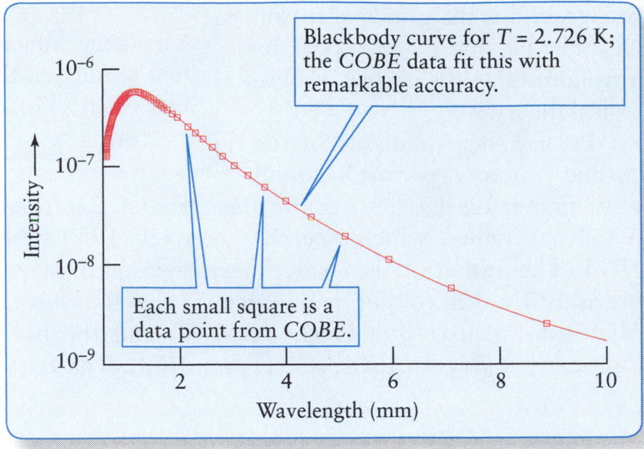

FIGURE 18-4 Spectrum of the Cosmic Microwave Background The little squares on this graph are COBE's measurements of the brightness of the cosmic microwave background plotted against wavelength. To a remarkably high degree of accuracy, the data fall along a blackbody curve for 2.73 K. The peak of the curve is at a wavelength of 1.1 mm, in accordance with Wien's law. (Courtesy of E. Cheng; NASA COBE Science Team)

taken by COBE and shown in Figure 18-4 reveal that, as predicted, this ancient radiation has the spectrum of a blackbody with a temperature of approximately 2.73 K (referred to as the *3-degree background radiation*). The observations by COBE earned the 2006 Nobel Prize in Physics for astrophysicists John C. Mather and George F. Smoot.

The photons of the 3-degree background radiation are not the only remnant of the Big Bang that astronomers have found. In 2008, the neutrinos created along with the photons in the Big Bang were indirectly observed by WMAP. (Recall from Section 10-9 that neutrinos are also created today in the nuclear fusion occurring in the Sun and other stars.) The discovery of these *cosmic neutrinos* was made by comparing the predicted effects of those neutrinos on the young universe with observations of the earliest observable remnants of the Big Bang, about which we will have more to say shortly. The theory of these neutrinos and the observations of their predicted effects agree to high precision. The numbers of neutrinos in the *cosmic neutrino background* is impressive: Several million of them pass through your body every second. There are similar numbers of neutrinos in similar volumes of space everywhere in the universe.

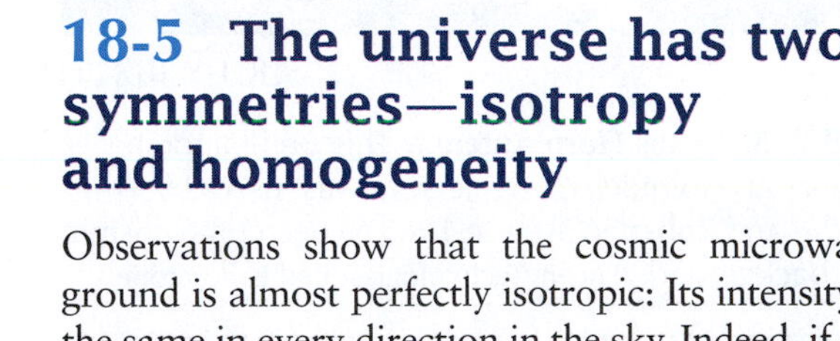

18-5 The universe has two symmetries—isotropy and homogeneity

Observations show that the cosmic microwave background is almost perfectly isotropic: Its intensity is nearly the same in every direction in the sky. Indeed, if you could look at the cosmic microwave background from all directions in the entire universe with the naked eye, it would appear as uniform as a well-painted wall. However, when extremely high resolution measurements are made, tiny differences in the temperature of the universe are observed in different directions. One difference is caused by our motion through the cosmic microwave background. Figure 18-5 shows a map of the microwave sky. It is very, very slightly warmer than average in the direction of the constellation Leo and slightly cooler in the opposite direction, toward Aquarius. The observed temperature differences shown in Figure 18-5 mean that the solar system is moving toward Leo at a speed of 390 km/s. These observations can be understood as follows: As we move toward Leo, our motion through the cosmic microwave background radiation creates a normal Doppler shift that causes photons from that direction to appear to have shorter wavelengths—radiation from the direction of Leo is blueshifted, whereas radiation from the opposite direction, Aquarius, is redshifted because we are moving away from that region (Figure 18-6). Furthermore, a photon's wavelength determines its energy (see Section 3-3). The energy of the photons then

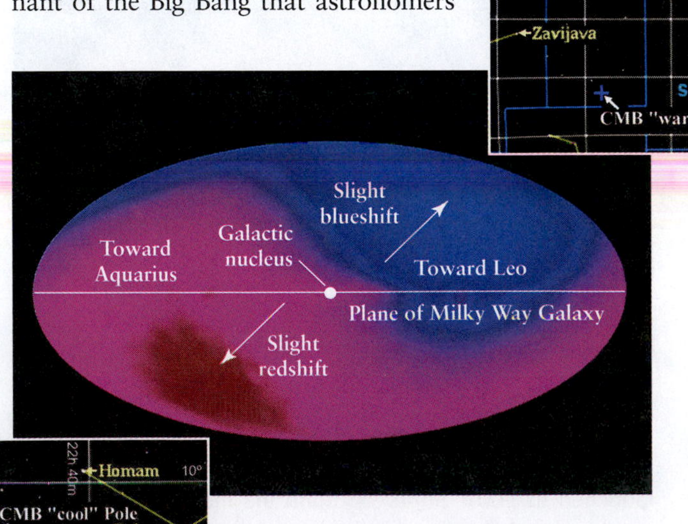

FIGURE 18-5 Microwave Sky This map of the microwave sky was produced from data taken by instruments on board COBE. The galactic center is in the middle of the map, and the plane of the Milky Way runs horizontally across the map. Color indicates Doppler shift and temperature: Blue is where the microwave background is blueshifted and appears warmer, while red is where it is redshifted and appears cooler. This Doppler shift across the sky is caused by Earth's motion through the microwave background. The resulting variation in background temperature is quite small, only 0.0033 K above the average radiation temperature of 2.726 K (which we have rounded to 2.73 K in the text). (NASA)

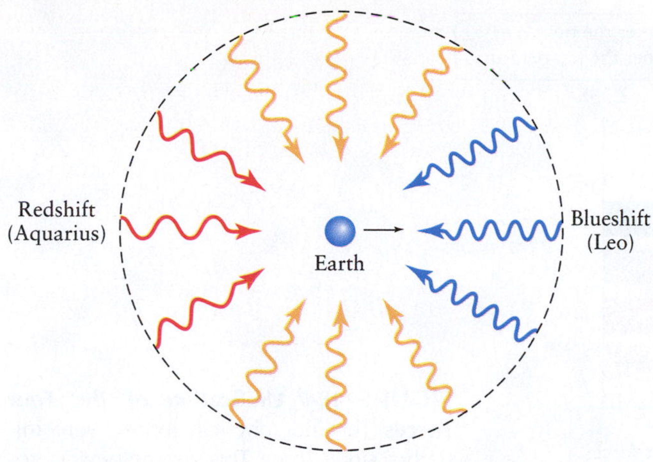

FIGURE 18-6 Our Motion Through the Microwave Background Because of the Doppler effect, we detect shorter wavelengths in the microwave background and a higher temperature of radiation in that part of the sky toward which we are moving. This part of the sky is the area shown in blue in Figure 18-5. In the opposite part of the sky, shown in red in Figure 18-5, the microwave radiation has longer wavelengths and a cooler temperature.

determines the temperature that we measure for that part of the sky. Observing slightly shorter-wavelength photons in the direction of Leo means that space in that direction appears slightly warmer, because of our motion, than does space in other directions.

Taking into account the known velocity of the Sun around the center of our Galaxy, astronomers calculate that the entire Milky Way Galaxy is moving relative to the cosmic microwave background at 600 km/s—some 1.3 million mi/h—in the general direction of the Centaurus cluster. The gravitational attraction of four nearby clusters of galaxies, most significantly the *Great Attractor,* and an enormous supercluster called the *Shapley concentration,* is believed to be pulling us in that direction. Subtracting the motion of Earth, the Sun, the Milky Way, and also the Local Cluster with respect to the microwave background, the average intensity of radiation is found to be the same in all directions (that is, isotropic) to about 1 part in 100,000.

Although the existence of the cosmic background radiation provides compelling support for the Big Bang theory, its near-isotropy was, until recently, a major problem. The issue was that the original Big Bang theory did not require space to be isotropic—after accounting for Earth's motion through the cosmos, the temperature in one direction *could* have been many degrees different than the temperature in any other direction, but it is not.

Isotropy is not limited to the blackbody radiation observed throughout the universe. It is also found on sufficiently large scales when exploring the numbers of galaxies in different directions. Astronomers have counted the numbers of galaxies at different distances from us and in a variety of directions. The number of galaxies stays roughly constant, allowing, as we saw in Section 16-8, for the distribution of most galaxies on the boundaries of sponge-shaped regions or bubbles throughout the universe. Because of this *uniformity with distance,* we say that the universe is also homogeneous. Isotropy and **homogeneity** must be explained by any viable theory of cosmology, such as the Big Bang.

The need to explain isotropy and homogeneity, among other things, has led to numerous refinements in the Big Bang theory. As a result, this theory now provides an accurate scenario for the evolution of the universe from a tiny fraction of a second after it formed and onward.

A BRIEF HISTORY OF SPACETIME, MATTER, ENERGY, AND EVERYTHING

The composition of the universe during its first few minutes of existence was profoundly different from what it is today. To understand why, we need to expand briefly on the nature of the four known physical forces: gravity, electromagnetism, and the strong and weak nuclear forces.

18-6 All physical forces in nature were initially unified

Only the effects of electromagnetism and gravity can extend over infinite distances. The electromagnetic force holds electrons in orbit around the nuclei in atoms, which allows normal matter to exist. However, over large volumes of space, the net effects of electromagnetism cancel out because a negative electric charge exists for every positive charge and a south magnetic pole exists for every north magnetic pole.

What object other than Earth could meaningfully have been used at the center of Figure 18-6?

Although gravity is the weakest of all four forces, it is the only force whose effect is both infinite in extent and attractive for all matter and energy. Those characteristics are why gravity determines the evolutionary behavior of stars, the orbits of planets, the existence of galaxies, and many other large-scale phenomena. Indeed, the attractive force of gravity dominates the universe at astronomical distances. (Despite all of this, we will learn shortly that, under certain conditions, gravity may have a repulsive component.)

In contrast to gravity and electromagnetism, both the strong and the weak nuclear forces act over extremely short ranges. Their influences extend only over atomic nuclei, distances less than about 10^{-15} m. The **strong nuclear force** holds protons and neutrons together. Without this force, nuclei would disintegrate because of the electromagnetic repulsion of their positively charged

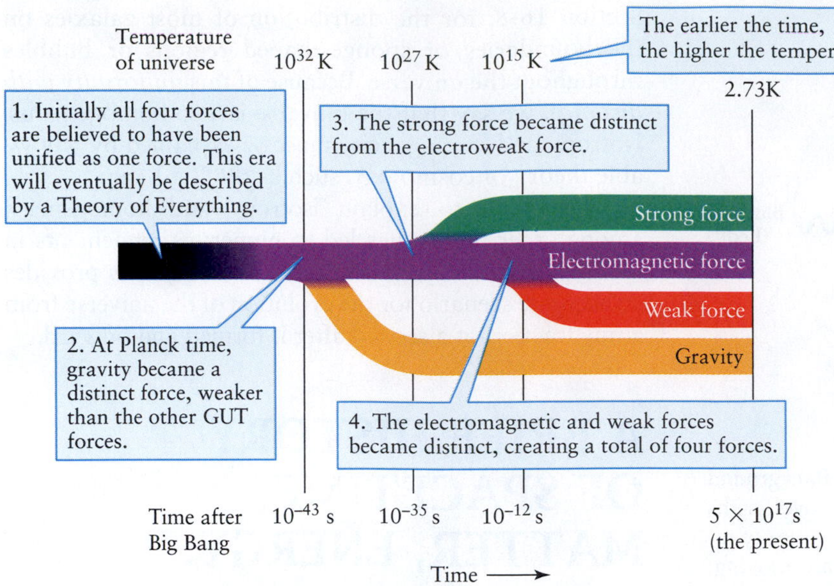

FIGURE 18-7 Unification of the Four Forces The four physical forces were initially a single force. This symmetry was broken as the universe expanded and cooled. This figure shows the time and temperature of the universe when the forces separated from each other.

protons. Thus, the strong nuclear force overpowers the electromagnetic force inside nuclei.

The **weak nuclear force** is at work in certain kinds of radioactive decay, such as the transformation of a neutron into a proton. Protons and neutrons are composed of more basic particles, called **quarks.** A proton is composed of two "up" quarks and one "down" quark, whereas a neutron is made up of two "down" quarks and one "up" quark. The weak nuclear force is at play whenever a quark changes from one variety to another. For example, in one kind of radioactive decay, a neutron transforms into a proton as a result of one of the neutron's "down" quarks changing into an "up" quark.

To examine details of the physical forces, scientists use particle accelerators that hurl high-speed particles at targets. High speed is equivalent to high energy and therefore high temperature. In such experiments, physicists find that *at sufficiently high temperatures, the different forces begin to behave the same way*. For example, at extremely high temperatures, the electromagnetic force, which works over all distances under "normal" circumstances, and the weak force, which only works over very short distances under the same "normal" circumstances, become identical. They are no longer separate forces, but become a single force called the *electroweak force*. The experiments verifying this were done at the CERN particle accelerator in Europe in the 1980s.

Theories have also been developed that describe conditions under which the electromagnetic, weak, and strong forces all have the same strength (that is, when they act as one). These **Grand Unified Theories** (or **GUTs**) describe how the energies or, equivalently, temperatures at which this unification of forces occurs are greater than can ever be achieved in any laboratory on Earth. But recall from the earlier discussion that when the universe was young, it was very hot—hot enough for this unifying of forces to occur. Detailed calculations reveal that in the first 10^{-35} s of existence, the temperature of the universe was more than high enough for the electromagnetic, weak, and strong forces to have the same strength.

The relationship between gravitation and the other three forces is still uncertain. Scientists have not yet been able to include gravity's effects in a single theory with the weak, strong, and electromagnetic forces. Efforts to create such a theory are under way, under the general name **Theories of Everything.** Mathematical formulations of such theories, including loop quantum gravity and **superstring theories,** are being developed, although these theories do not yet make predictions that scientists have been able to test (putting them on the outskirts of science). Assuming gravity will be joined to the other forces, we include its expected effects here. Figure 18-7 shows that, during the beginning moments of the universe (from 0 to 10^{-43} s, called the *Planck era*), the four forces are believed to have all been unified—only one physical force existed in nature. However, as it assumes a Theory of Everything force exists, this assumption remains to be proven.

Insight Into Science

Different Realms of Science Are Working Together Because physicists who study elementary particles have no hope of building accelerators powerful enough to test the unified theories of fundamental forces described here, they have joined with astrophysicists and astronomers to use the entire universe as a large laboratory. If there is a way to see, even indirectly through remnants of events, what happened during the earliest moments of the universe, they could use those observations to test the theory. Amazingly, such observations can be, and are being, made.

18-7 Equations explain the evolution of the universe, even before matter and energy, as we know them, existed

The earliest time at which our current equations can explain the behavior of the universe is about 10^{-43} s after the Big Bang. This is called the **Planck time,** in honor of Max Planck, who helped derive some of the fundamental quantum physics that was used to calculate this time. (The time before this moment, but still after the Big Bang, is sometimes referred to as the **Planck era.**) At the Planck time, gravity is thought to have become a separate force, leaving electromagnetism and the weak and strong nuclear forces still united as the GUT force. Which of the present GUT models, if any, is correct, remains to be seen.

The Planck time corresponded to when the currently observable universe was much smaller than the size of an atom today. Keep in mind that the universe was not (and is not) expanding into preexisting spacetime. Rather, the expansion was (and is still) creating spacetime. Therefore, you could not stick a thermometer into the early universe from "outside" to take its temperature. However, we can calculate the total amount of energy in the universe at that early time, and it corresponds to a temperature then of about 10^{32} K. There were no separate particles and electromagnetic radiation then, just gravitation and the GUT force, so matter as we now know it, did not exist.

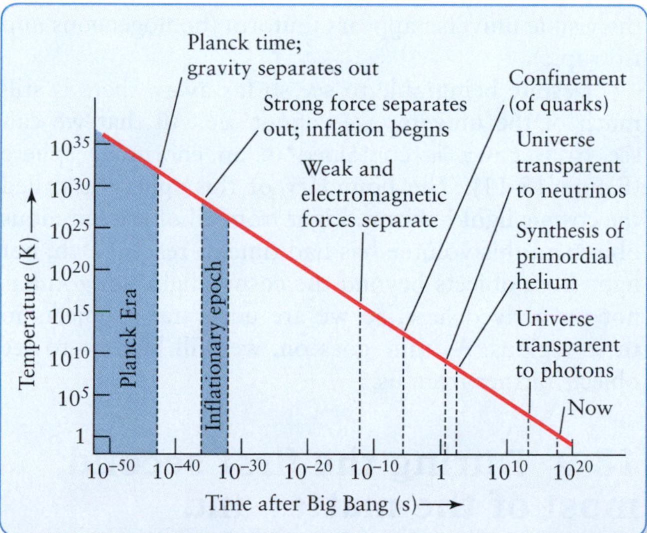

FIGURE 18-8 Early History of the Universe Current theory holds that, as the universe cooled, the four forces separated from their initially unified state. The inflationary epoch lasted from 10^{-36} s to 10^{-32} s after the Big Bang. Quarks became confined together, thereby creating neutrons and protons 10^{-6} s after the Big Bang. The universe became transparent to light (that is, photons decoupled from matter) when the universe was about 1.6×10^{13} s (380,000 years) old. The physics of the Planck era is presently unknown.

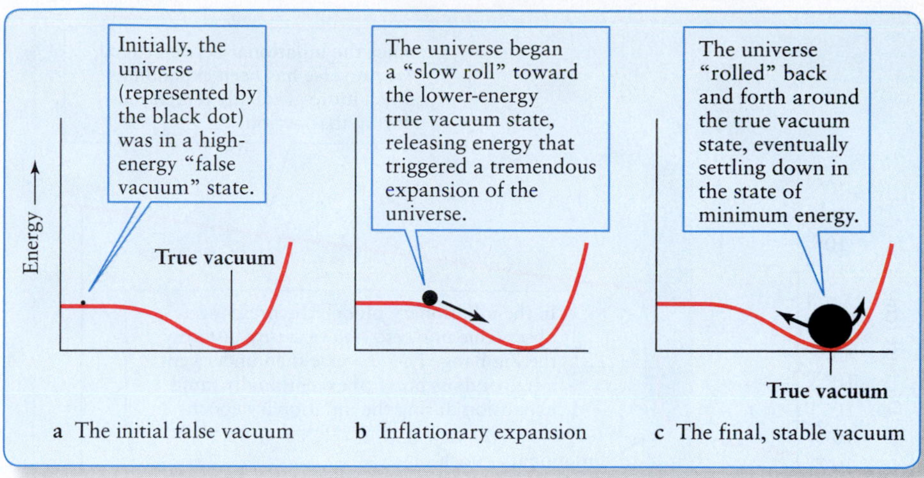

FIGURE 18-9 Cause of Inflation (a) The universe formed in an unstable energy state that (b) began to transition to a stable configuration (c). This transition provided the energy that caused inflation.

For the first 10^{-35} s of the universe's existence (the blink of an eye takes about one-tenth of a second), calculations indicate that the universe expanded at the initial rate and that its temperature dropped to 10^{27} K (Figure 18-8). (For comparison, room temperature is about 300 K.) By then, the energy in each volume of the universe had decreased to the point at which the strong nuclear force could no longer remain unified with the electromagnetic and weak nuclear forces. At that instant, the strong nuclear force emerged (see Figure 18-7), distinct from the other two forces, which remained together as the electroweak force. For a short period, then, there were three forces in nature: gravitation, the strong nuclear force, and the electroweak force.

> Which fundamental force is essential in the process of nuclear fusion?

Many astronomers hypothesize that before the strong force decoupled from the electroweak force, the universe was in an unstable state, called a *false vacuum.* It was unstable in the same sense as you would be if you woke up one day to find yourself on a greased, half-meter wide shelf on the face of a cliff thousands of meters above the cliff bottom. The slightest move and you slide off and fall. Analogous to the transition you make by falling off the shelf, the young universe is believed to have suddenly made a transition (Figure 18-9) into a stable *true vacuum* state. The universe's transition to the true vacuum state caused its expansion to momentarily accelerate.

As the universe began moving toward the true vacuum state, it was already expanding. The transition caused

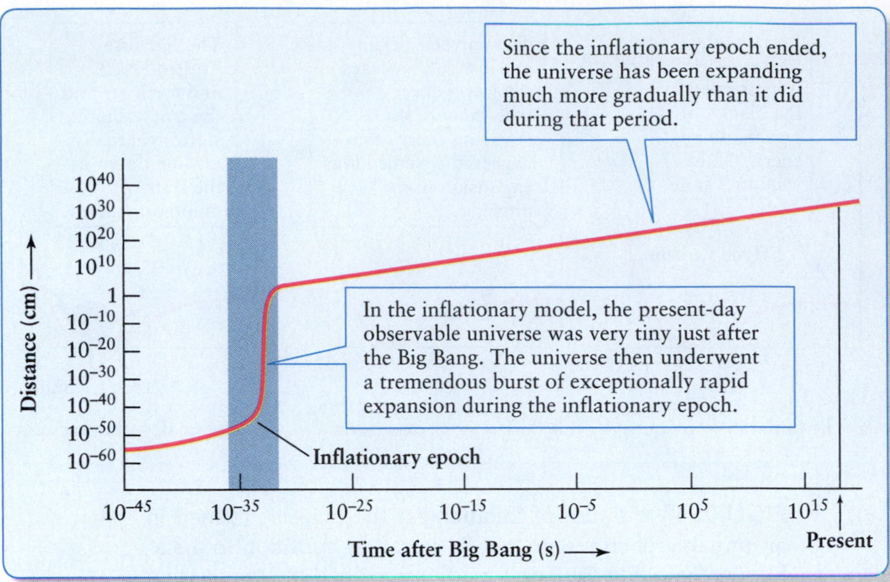

FIGURE 18-10 Observable Universe Before and After Inflation Shortly after the Big Bang, the universe expanded by a factor of about 10^{50} due to inflation. This growth in the size of the presently observable universe occurred in a very brief time (shaded interval).

Is the space in a sphere 3 m around you isotropic?

the expansion rate to increase dramatically. From about 10^{-36} s until about 10^{-32} s after the Big Bang, the universe ballooned outward about 10^{50} times larger (Figure 18-10). The currently observable universe expanded to about the size of a soccer ball. This huge rate of expansion of the universe is termed **inflation.** Shortly after inflation ended, at about 10^{-12} s after the Big Bang, the weak and electromagnetic forces became separate, heralding the first formation of elementary particles: quarks, electrons, photons, neutrinos, and their antiparticles.

18-8 Homogeneity and isotropy are results of inflation

Because disorder is more likely to occur in nature than order, astronomers theorize that the matter and energy in the early universe were distributed chaotically and expanded turbulently. That is also why an explosion on Earth expands turbulently, with lots of swirling gases, rather than expanding smoothly, like the surface of a balloon as it is being inflated. As a result, some places in the early universe had higher temperatures than others. If the universe had never undergone inflation, but had continued expanding only at the rate calculated for its motion at the Planck time, today we should be able to see different parts of the universe with significantly different temperatures. However, as we saw in Figure 18-5, the temperatures of every region of the universe are the same to within a very tiny fraction of 1 K. This situation is called the **isotropy problem** or **horizon problem.** Realizing how uniform the universe's temperature is, astronomers have hypothesized that inflation very early in its existence smoothed out the universe that we can see.

During the so-called **inflationary epoch,** the universe became so large that today we can see only a tiny portion of it. We are limited in what we can see by the speed of light. Recall that light takes time to travel from astronomical objects to Earth. We see the Sun, for example, as it was about 8 minutes ago. The farther we look into space, the further back in time we are seeing. Because the universe came into existence about 13.7 billion years ago, we cannot see objects that are farther away than more than 13.7 billion light-years away from Earth (see the caveat about this number in the caption to Figure 18-11).

At a distance of 13.7 billion light-years, we see objects like galaxies and clusters of galaxies as they were just being formed. But because of inflation, the matter comprising these apparently distant objects was once extremely close to us. At that early time, before inflation, these objects had the same temperature, pressure, and density as the matter and energy in our nearest neighborhood of the cosmos. As the universe inflated, that nearby matter and energy became much more distant, but maintained those properties. Therefore, the most distant reaches of space we can see are composed of matter and energy with the same characteristics as in our present neighborhood of the universe. That is why the visible universe appears uniform (homogeneous and isotropic).

Despite being able to see so far away, there is still much of the universe we cannot see. All that we can see from Earth is contained in an enormous sphere (Figure 18-11). The boundary of this sphere is called the **cosmic light horizon.** Light from all of the luminous objects in this volume has had time to reach Earth, but light from objects beyond the cosmic light horizon has not yet arrived here, so we are unaware of objects in these regions. As time goes on, we will be able to see objects farther from us.

18-9 During the first second, most of the matter and antimatter in the universe annihilated each other

When inflation ended, the universe resumed the relatively leisurely expansion initiated by the Big Bang (see Figure 18-10) that continues today as the Hubble flow (see Section 16-12). Arriving at the time 10^{-12} s, when the temperature of the universe had dropped to 10^{15} K, the electromagnetic force separated from the weak

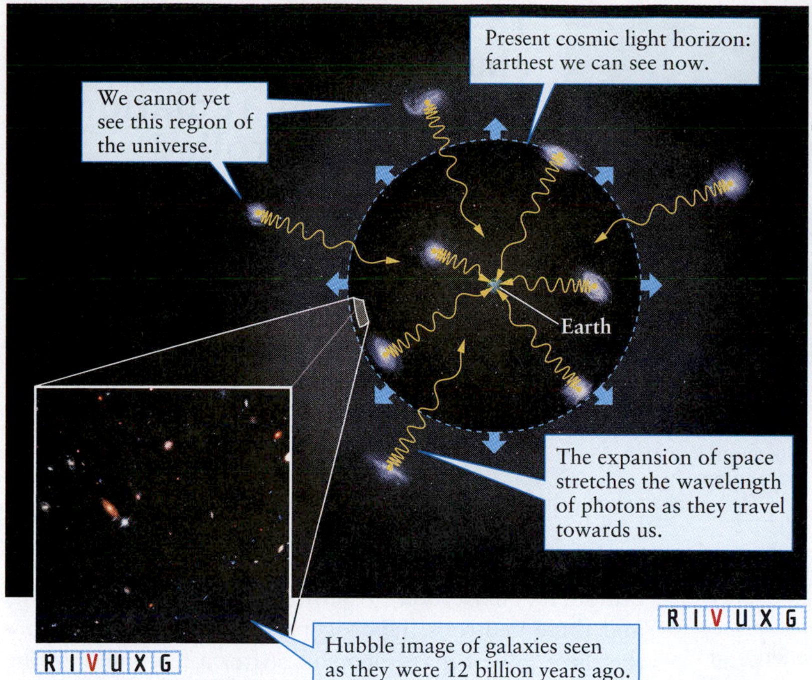

FIGURE 18-11 Observable Universe This diagram shows why we only see part of the entire universe. As time passes, this volume grows, meaning that light from more distant galaxies reaches us. The farthest galaxies we see (inset) are as they were within a few hundred million years after the Big Bang. These galaxies, formed at the same time as the Milky Way, appear young because the light from their beginnings is just now reaching us. While the light from the most distant galaxies we see was traveling toward us, the universe has been growing. Therefore, objects that appear 13 billion light-years away from us are actually about 3 times farther away today. Put another way, if we could see it all as it is today, rather than as it was when photons from distant objects started their journeys, the visible universe would be nearly 46 billion light-years in radius. Inset: This image shows some of the most distant galaxies we have seen. (Robert Williams and the Hubble Deep Field Team, STScI, and NASA)

nuclear force (as noted in Section 18-7; see also Figures 18-7 and 18-8). It is also likely that dark matter formed by this time, but because we know virtually nothing of its nature, we can only acknowledge its existence. From that moment on, all four forces interacted with particles essentially as they do today. At 10^{-12} s, the universe's matter was primarily the individual building blocks of protons and neutrons—namely quarks, antiquarks, and the other elementary particles: neutrinos, antineutrinos, electrons, and positrons.

The next significant event is calculated to have occurred at 10^{-6} s, when the universe's temperature dropped to 10^{13} K. Just as it is energetically favorable for boiling water to turn into gas, the temperature of the universe was so high before this time that it was energetically favorable for quarks to exist as isolated particles, rather than to combine in twos and threes to create the particles that exist in the universe today. We know they were initially isolated because similar seas of quarks were recreated under equivalent conditions in the laboratory in 2001 by smashing dense atoms together. After 10^{-6} s, a period appropriately called **confinement**, the universe was sufficiently cool so that quarks could finally stick together to form individual protons, neutrons, and their antiparticles. Figure 18-8 summarizes the connections between particle physics and cosmology.

Equations show that early in the first second, the universe was so hot that virtually all photons were gamma rays possessing incredibly high energies. These energies were high enough so that photons could be transformed into matter and antimatter, according to Einstein's equation $E = mc^2$. This process is the reverse of what we saw in Section 10-7, where we discussed how matter is converted into energy in nuclear fusion. To make a particle of mass m, you need a photon with energy E at least as great as mc^2, where c is the speed of light.

Why is inflation a necessary part of the Big Bang theory of cosmology?

The creation of matter from energy is routinely observed in laboratory experiments that involve high-energy gamma rays. A highly energetic photon colliding with an atomic nucleus can create a pair of particles (Figure 18-12a). This process, called **pair production,** always creates one ordinary particle and its antiparticle. For example, if one of the particles is an electron, the other is an antielectron, also called a positron. (Recall from Section 14-11 that similar pairs of particles are also believed to be produced near a black hole.)

A positron has the same mass as an electron but the opposite charge. When an electron and a positron meet, they annihilate each other and their energy is converted into photons (Figure 18-12b). Theory predicts that at times earlier than 1 second in the evolution of the universe, photon collisions created vast numbers of pairs of particles. Consequently, shortly after the Big Bang, all space was chock full of protons, neutrons, electrons, neutrinos, their antiparticles, and dark matter, all immersed in a phenomenally hot bath of high-energy photons.

As the universe expanded, the temperature declined until after the first second, photons could no longer create pairs of particles and antiparticles.

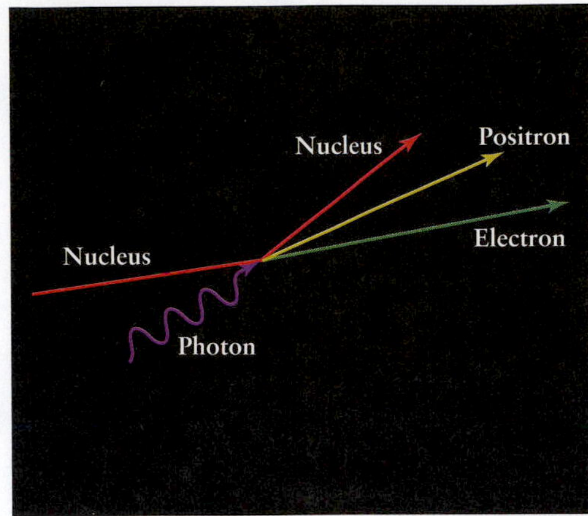

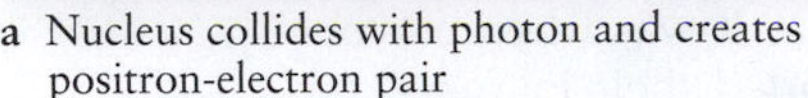

a Nucleus collides with photon and creates positron-electron pair

Time ⟶

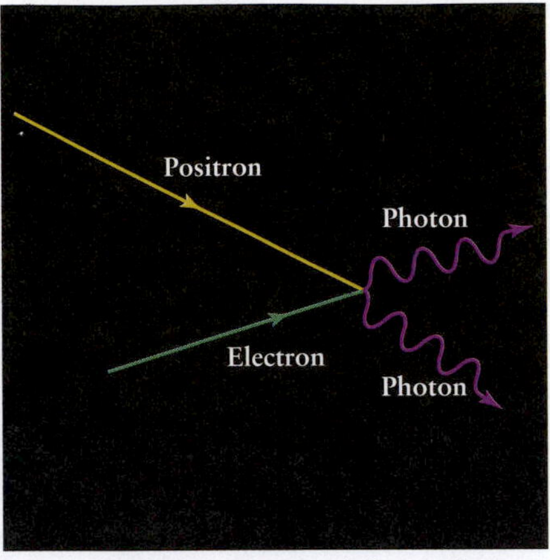

b Positron and electron annihilate each other and create two photons

Time ⟶

FIGURE 18-12 Pair Production and Annihilation (a) A particle and an antiparticle can be created when a high-energy photon collides with a nucleus. (b) Conversely, a particle and an antiparticle can annihilate each other and emit energy in the form of gamma rays. (The processes are more complex and are only summarized in these drawings.)

Particles and antiparticles continued to collide, however, annihilating each other and converting their masses back into high-energy photons, but nothing could replenish the dwindling supply of particles and antiparticles.

If all particles had annihilated their antiparticles in the early universe, no matter would be left at all and we would not be here. And yet, here we are and astronomers observe very few antiparticles in the universe. Because particles and antiparticles are always created or destroyed in pairs, why aren't they found in equal numbers in the universe today, or rather, why didn't they all destroy each other shortly after they were created?

Physicists theorize that the symmetry or equality between the number of particles and the number of antiparticles in the early universe was broken by the weak force—a *symmetry breaking*. This effect has been observed on Earth in high-energy particle accelerators over the past half century. Due to symmetry breaking, the number of particles was actually slightly greater than the number of antiparticles in the early universe. For every billion antiprotons, perhaps a billion plus one protons formed; for every billion antielectrons, a billion plus one electrons formed. Virtually all of the antiparticles created in the first second of time annihilated normal particles shortly thereafter. The remaining particles in the universe today are the slight excess of normal matter created as a result of this symmetry breaking.

18-10 The universe changed from being controlled by radiation to being controlled by matter

As the universe expanded and cooled further, the remaining protons and neutrons began colliding and fusing together, a process called **primordial nucleosynthesis.** For the most part, the nuclei with two or more particles were quickly separated again by the gamma rays in which they were bathed, a process called nuclear fission. However, during the first 3 min, enough helium and lithium atoms were created (and not subsequently split apart) to form most of the free helium and at least 10% of the trace element lithium that exist today. A lithium atom has three protons and three neutrons. Hydrogen, helium, and lithium are the three lowest-mass elements. After a few minutes, the universe was too cool to allow fusion to create more massive elements, such as carbon and oxygen. All elements other than the three lowest-mass ones formed later, due to stellar evolution, as discussed in Chapters 10, 12, and 13.

Insight Into Science

Share Information Although the evolution of the universe is a truly large-scale event, the activity in the first 3 min was nevertheless intimately connected to the creation and properties of elementary particles, the smallest objects in the universe. Physicists who study elementary particles and astrophysicists who study cosmology now regularly hold meetings to compare notes and connect their fields.

Calculations indicate that for tens of thousands of years, photons dominated the behavior of the universe in two important ways. One effect of the photons was to contribute to the gravitational attraction in the universe. This contradicts intuition, because photons are massless and mass is what we normally associate with causing gravitational attraction. Nevertheless, the gravitational contribution of photons to the universe is a tested prediction of general relativity. For about 30,000 years after the Big Bang, the energy density, and hence the gravitational effect, of photons was greater than the gravitational effect of matter. This situation was called the **radiation-dominated universe.**

The major effect of the radiation-dominated universe was that photons prevented matter from collapsing together and forming stars and galaxies. As the

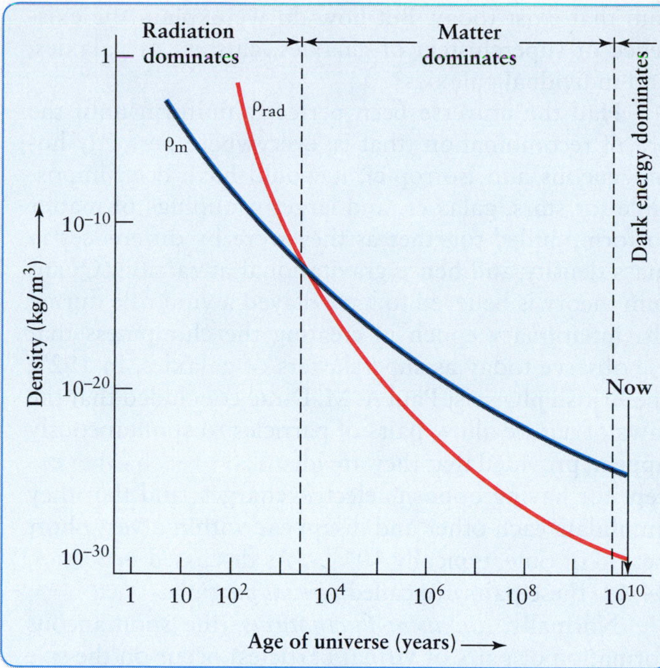

FIGURE 18-13 Evolution of Density For approximately 30,000 years after the Big Bang, the gravitational effects from photons (ρ_{rad}, shown in red) exceeded the attraction of all the matter in the universe (ρ_m, shown in blue). This early period is said to have been radiation-dominated. Later, however, continued expansion of the universe caused ρ_{rad} to become less than ρ_m, at which time the universe became matter-dominated.

universe expanded, the gravitational effects of both photons and matter decreased as they became more spread out. The effect of photons decreased faster than that of matter (Figure 18-13) because the photons lost energy, and hence gravitational influence, due to the cosmological redshift, but particles with mass did not similarly lose mass or energy. After about 30,000 years, matter came to dominate the gravitational behavior of the universe, a situation that has persisted ever since. As a result, we live in a **matter-dominated universe** today. It was only after the universe became matter-dominated that gas could respond to gravitational interactions and begin clumping together where the force of gravity was sufficiently strong (an issue we explore further in a moment).

Photons dominated the early universe in another way for about the first 380,000 years. The photons back then could only travel short distances before they collided with particles. These interactions prevented electrons from settling into orbit around nuclei—as soon as electrons did so, photons struck them and kicked them out of orbit, a process called *ionization* (see Section 4-5). During that epoch, the universe was completely filled with a shimmering expanse of high-energy photons colliding vigorously with nuclei and electrons. This state of matter, in which nuclei and electrons are separated from each other, is called a *plasma*. In such an environment you would see a glow from local photons, but because of their vigorous interactions with matter you would not be able to see photons from even a few meters or yards away (Figure 18-14a). The term **primordial fireball** describes the universe during this time.

By around 380,000 years, the universe was so large that the cosmological redshift had stretched photon wavelengths (see Figure 18-1) and therefore decreased photons' energies to the point where they could no longer ionize hydrogen. This stage is called **decoupling,** meaning that the massive particles and the photons were effectively disconnected from each other. Electrons began orbiting nuclei, thereby creating neutral atoms (Figure 18-14b). The space between nuclei became so

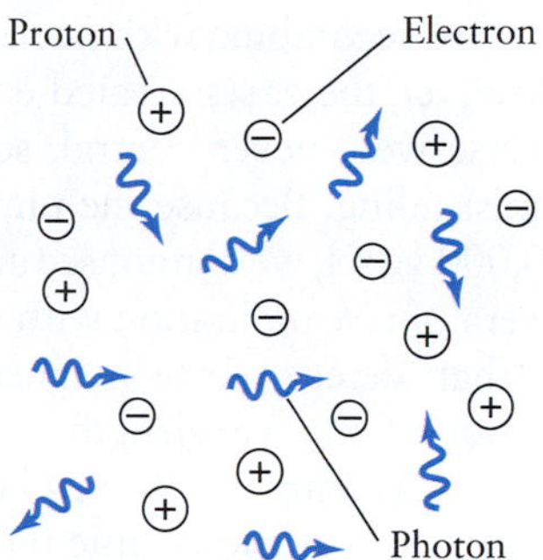

a Before recombination:
- Temperatures were so high that electrons and protons could not combine to form hydrogen atoms.
- The universe was opaque: Photons underwent frequent collisions with electrons.
- Matter and radiation were at the same temperature.

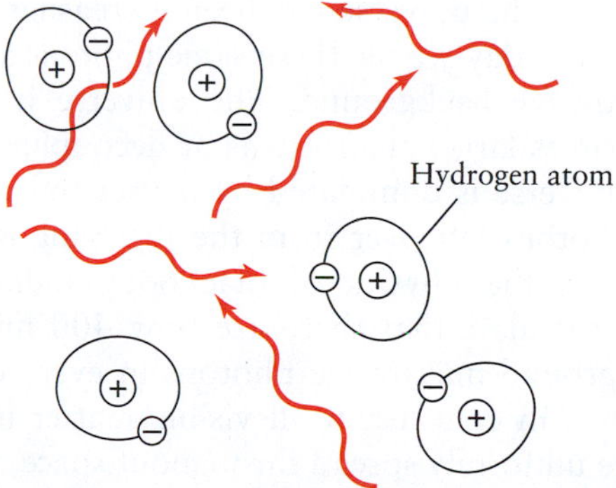

b After recombination:
- Temperatures became low enough for hydrogen atoms to form.
- The universe became transparent: Collisions between photons and atoms became infrequent.
- Matter and radiation were no longer at the same temperature.

FIGURE 18-14 Era of Recombination (a) Before recombination, the energies of photons in the cosmic background were high enough to prevent protons and electrons from forming hydrogen atoms. (b) As soon as the energy of the background radiation became too low to ionize hydrogen, neutral atoms came into existence.

How, if at all, would the universe be different, if antiparticles had exceeded particles in the symmetry breaking described earlier?

large that most primordial photons now passed by these particles, rather than colliding with them. Without these collisions to scatter photons, the universe ceased being hazy. It became clear, as it is today, enabling electromagnetic radiation to travel long distances unimpeded and, important from an astronomer's point of view, enabling us to see far into the universe today.

This dramatic period, when the universe transformed from being opaque to transparent, took about 50,000 years to complete and is referred to as the **era of recombination,** referring to nuclei and electrons combining. There are many instances, both in space and in the laboratory, where electrons are stripped from atoms. After such ionization occurs, electrons often recombine with nuclei to recreate neutral atoms. However, the gases created early in the evolution of the universe were never neutral, so the "re" in recombination is misleading. Because the universe was opaque in its first 380,000 years, we cannot see any further into the past than the era of recombination with telescopes that detect electromagnetic radiation of any wavelength.

We say that "most" of the helium in the universe was formed in primordial nucleosynthesis. Where was/is the rest of it formed?

We can use Wien's law to calculate that the cosmic background radiation had a temperature of 3000 K at the era of recombination. Such a blackbody has a peak microwave wavelength of 0.001 mm. The temperature history of the universe is graphed in Figure 18-8.

Since the primordial gamma-ray photons formed, the expansion of the universe has been increasing their wavelengths, so today we see these same photons as the cosmic microwave background. The universe is now about 1000 times larger than it was at decoupling. Although the universe is dominated by matter today, the number of photons left over from the Big Bang is still immense. From the physics of blackbody radiation, astronomers calculate that there are now 400 million cosmic background microwave photons in every cubic meter of space. In contrast, if all visible matter in the universe were uniformly spread throughout space, there would be roughly one hydrogen atom in every 3 cubic meters of space. In other words, photons continue to outnumber atoms by more than a billion to one. The microwave background contains the most ancient photons we expect ever to be able to observe.

18-11 Galaxies formed from huge clouds of primordial gas

The Big Bang theory explains the evolution of spacetime, matter, and energy since the Planck era. It accounts for the hydrogen, most of the helium, and some of the lithium that exist today. But how do we explain the existence of superclusters of galaxies, clusters of galaxies, and individual galaxies?

Had the universe been perfectly uniform until the era of recombination (that is, everywhere perfectly homogeneous and isotropic), it would have been impossible for stars, galaxies, and larger groupings of matter to form, pulled together as they were by differences in mass density and hence gravitational attraction. Quantum theory is believed to have played a vital role during the inflationary epoch in creating the clumpiness that we observe today as superclusters of galaxies. In 1927, the British physicist Paul A. M. Dirac concluded that the laws of nature allow pairs of particles to spontaneously appear, provided that they are identical to each other except for having opposite electric charges, and that they annihilate each other and disappear within a very short period of time, typically 10^{-21} s. As discussed in Section 14-11, these pairs are called *virtual particles*.

Normally, *quantum fluctuations* (the spontaneous formation of pairs of virtual particles) occur on the size scales of pairs of particles. However, the quantum fluctuations on microscopic scales that were occurring during the inflationary period were stretched by the inflation until they were large enough to affect the spacetime around them. Those particles created tiny regions of slightly higher temperature, density, and pressure in the inflating universe, like little firecrackers going off all over the place. Also like firecrackers, those events sent out sound waves that traveled through the rapidly expanding universe. These sound waves are believed to have led to the formation of superclusters of galaxies. Here is how that may have taken place.

Sound waves are waves that compress and rarefy the medium through which they travel. The rate at which these compressions and rarefactions in the air enter your ears each second, for example, determines the pitches of the sound you hear. As quantum fluctuations stretched out in the early universe, they compressed the surrounding medium, sending out spherical sound waves, which traveled as long as the gas in the young universe could support them. The speed of these sound waves was impressive by everyday standards. Sound in our atmosphere travels at about 350 m/s (1100 ft/s). The sound waves generated in the inflating universe traveled at 1.75×10^5 m/s.

The sound waves created at the end of inflation in the early universe traveled out through the plasma of the universe until decoupling, 380,000 years later. At that time, when photons no longer scattered quickly off the gas in the universe, the waves stopped moving.

Because those sound waves were compression and rarefactions of the gas in the young universe, when they stopped traveling they established shells of higher density gas (compressions) and interior regions of lower density (rarefactions). The slightly denser and slightly more rarefied regions of the universe that existed at decoupling were also slightly warmer and slightly cooler, respectively, than the average gas in the cosmos. Although those differences in temperature were incredibly small, they have been detected by careful analysis of the data from the COBE and WMAP satellites, which observed the cosmic microwave background temperature at the time of decoupling. Their data (Figure 18-15a) show that these variations in temperature were only about 1 part in 100,000, implying that the average density of matter in the early universe varied by a similar amount. Such tiny differences in density were nevertheless sufficient to start the process of supercluster, cluster, and galaxy formation.

Astrophysicists have plotted the angular sizes of the peaks (warm regions) created by the acoustic (sound) waves, as shown in Figure 18-15b. Their sizes are clustered around 1.2° in angle (Figure 18-15b). In other words, the bubbles of sound generated by the quantum fluctuations spread out to that angular size, as seen from Earth some 13.7 billion light-years away. Allowing for the expansion of the universe since decoupling, those bubbles are about 960 million light-years in diameter today. This size is consistent with predictions of their properties made by the Russian physicist Andrei Sakharov in 1965. Like the sound generated by a violin, there are overtones (higher pitches) in the peaks of the cosmic microwave background. These overtones have different amplitudes. Their relative amplitudes tell us the relative amounts of visible matter and dark matter in the universe, from which we conclude that there is about 5 times as much dark matter as there is visible matter.

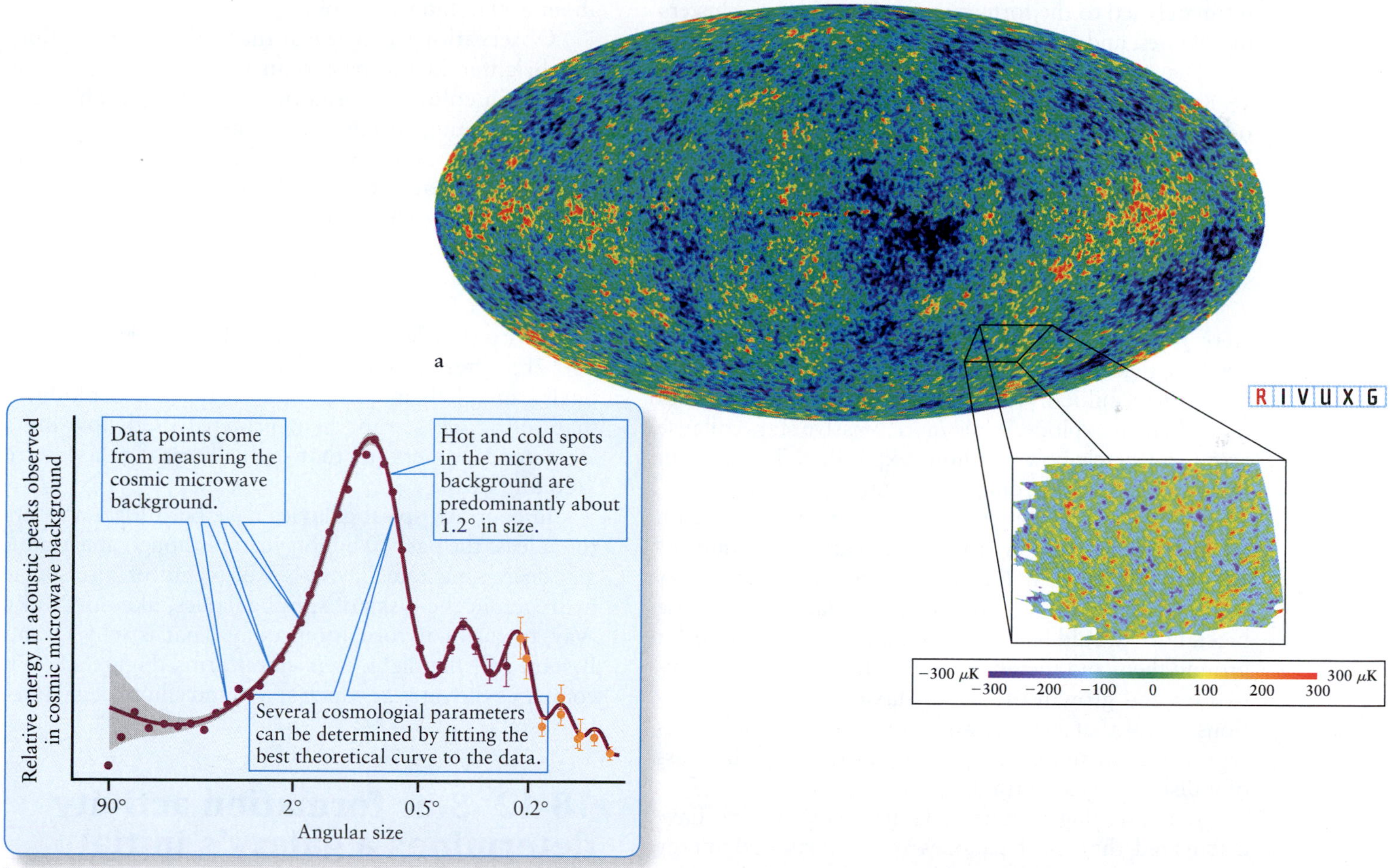

FIGURE 18-15 Structure of the Early Universe (a) This microwave map of the entire sky, produced from data taken by the *Wilkinson Microwave Anisotropy Probe* (WMAP), shows temperature variations in the cosmic microwave background. Red regions are about 0.00003 K warmer than the average temperature of 2.73 K; blue regions are about 0.00003 K cooler than the average. Inset: These tiny temperature fluctuations, observed by BOOMERANG, are related to the large-scale structure of the universe today, indicating where superclusters and voids grew. The radiation detected to make this map is from a time 380,000 years after the Big Bang. (b) Acoustic peaks show the sizes of the hot spots on the inset map in (a), along with overtones that provide information about the kinds of matter in the universe. (a: NASA/WMAP Science Team; inset: NSF/NASA; b: NASA/WMAP Science Team)

We can now outline the scenario for how matter clumped in the early universe. When the universe was radiation-dominated, the particles in it were moving too fast to be pulled together and form structures, like stars, galaxies, clusters, or superclusters. During that time, sound waves from quantum fluctuations created growing shells of higher density gases, surrounding regions of lower density gases. Following decoupling, those shells stopped growing. Virtually all matter in the early universe consisted of hydrogen, helium, and trace amounts of lithium gas. It took time for the gas to cool sufficiently to start forming stars, a period called the **dark ages.** (Despite the name of that era, the hydrogen in the dark ages was emitting 21-cm photons, as discussed in Section 15-7. These photons have now been redshifted by the Hubble flow to tens of meters in length, which is where astronomers are looking for them.) Gravitational attraction in the gases on the shells ultimately led to the formation of superclusters, clusters of galaxies, and galaxies.

The voids seen throughout the cosmos (see Section 16-8) are the relatively empty spaces surrounded by shells of superclusters. The diameters of some of the voids have been measured. They are all 1 billion light-years or smaller, consistent with the 960 million light-year diameters of the shells on which superclusters formed. That not all voids are the same size comes from the fact that many shells of sound crossed each other before decoupling.

Computer simulations indicate that the first stars were typically 100–500 $M_\odot$, a million times or more brighter than the Sun, and that they lived for no more than a few million years. Stars with between 100 and 250 $M_\odot$ then explode, while more massive stars collapse and form black holes without exploding. The stars in this early cohort that did explode provided the first metals in the universe (other than the primordial lithium). During the first billion years, it appears that clumps of gas that contained millions or billions of solar masses also collapsed to create supermassive black holes. These black holes would have attracted other matter into orbit around them and thereby served as the seeds for the formation and growth of some galaxies. Recent observations reveal that massive and supermassive black holes typically grow until they have about 0.2% of the mass of a disk galaxy's central bulge.

By observing remote galaxies, astronomers have discovered that most galaxies initially emitted energy we associate with quasars and other active galaxies. This implies that most galaxies have supermassive black holes at their centers. The earliest quasars dating back to when the universe was only 900 million years old provide information about when the first stars formed. Those quasars have spectral lines created by iron and magnesium. In order for those elements to be in those quasars, the iron and magnesium had to be created in stars that formed 700 million years earlier (then exploded and put the metals into the interstellar medium). These metals were then attracted by the black hole. This scenario means that the first stars existed within 200 million years of the Big Bang (Figure 18-16a), thus defining the end of the dark ages.

The earliest generation of stars are called *Population III* stars. By 600 million years after the Big Bang, galaxies were beginning to coalesce (Figure 18-16b). We also have observational evidence that galaxies were in clusters within 2 billion years of the Big Bang (Figure 18-16c).

From about 600 million years to about 6 billion years after it formed, the universe underwent heavy star formation. Galaxies grew in size during this period, some continuously, others suddenly and quickly. Star formation decreased as the amount of interstellar hydrogen diminished. One observed galaxy, with about 8 times the mass in stars as the Milky Way has, reached its full star-forming potential when the universe was only about 800 million years old.

Observations also reveal that galaxies were bluer and brighter in the past than they are today. These changes in color and brightness suggest a high abundance of young, bright, hot, massive stars in newly formed galaxies (Figure 18-17a). As galaxies age, these blue O and B stars become supergiants and eventually die. Therefore, galaxies grow somewhat redder and dimmer. This is especially true of those elliptical galaxies that formed early on. They appear to form nearly all of their stars in one vigorous burst of activity that lasts for about a billion years (Figure 18-17b), after which star formation diminishes drastically and their massive stars evolve and explode so that today we see in them primarily red, low-mass stars. Astronomers say that such elliptical galaxies are "red and dead."

In contrast, spiral galaxies have been forming stars for at least the past 10 billion years, although at a gradually decreasing rate. There is still plenty of interstellar hydrogen in the disks of spiral galaxies, like our Milky Way, to fuel star formation today. That is why O and B stars still highlight their spiral arms. Figure 18-17b compares the rates at which spiral and elliptical galaxies form stars.

18-12 Star formation activity determines a galaxy's initial structure

Imagine a developing galaxy, called a *protogalaxy,* forming from a cloud of gas. Theory proposes that the rate of star formation determines whether this protogalaxy becomes a spiral or an elliptical galaxy. If stars form slowly enough, then the gas surrounding them has plenty of time to settle by collision with other infalling gas into a flattened disk, just like the early

a Early bursts of star formation

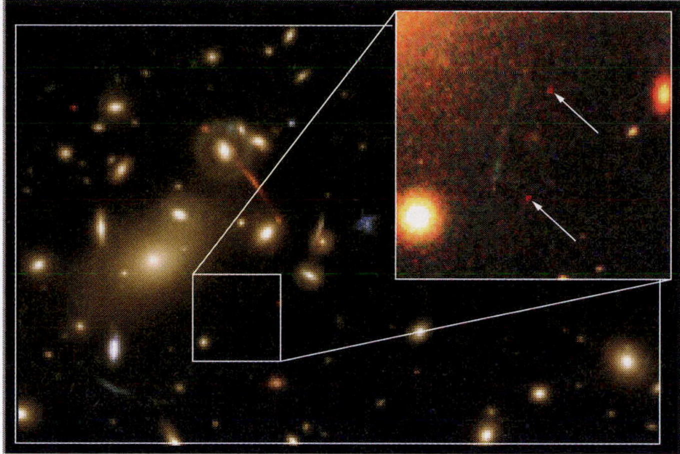

b Arrows indicate galaxies beginning to form 13.4 billion years ago

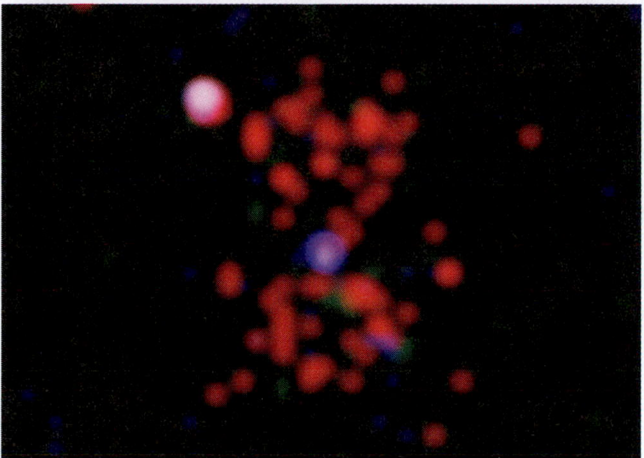

c Massive cluster of galaxies formed 11.2 billion years ago

FIGURE 18-16 Galaxies Forming by Combining Smaller Units (a) This painting indicates how astronomers visualize the burst of star formation that occurred within a few hundred million years after the Big Bang. The arcs and irregular circles represent interstellar gas illuminated by supernovae. (b) Using the Hubble and Keck telescopes, astronomers discovered two groups of stars (arrows) 13.4 billion light-years away that are believed to be protogalaxies, from which bigger galaxies grew. These protogalaxies were discovered because they were enlarged by the gravitational lensing of an intervening cluster of galaxies. (c) The Chandra X-ray Observatory imaged gravitationally bound gas around the distant galaxy 3C 294. The X-ray emission from this gas is the signature of an extremely massive cluster of galaxies, in this case, at a distance of about 11.2 billion light-years from us. (a: Adolf Schaller, STScI/NASA/K. Lanzetta, SUNY; b: Richard Ellis (Caltech) and Jean-Paul Kneib (Observatorie Midi-Pyrenees, France), NASA, ESA; c: NASA)

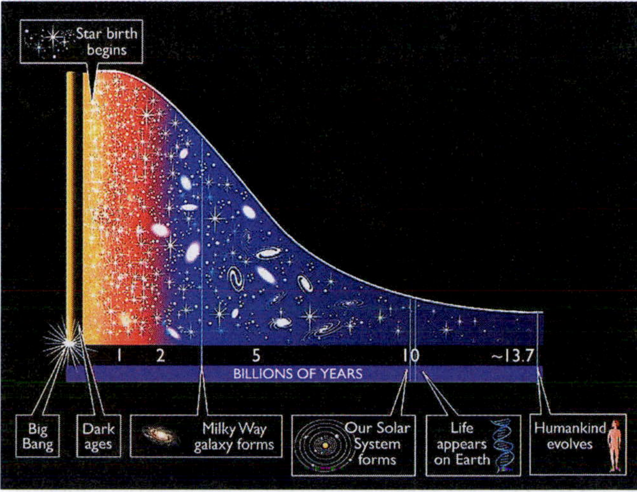

a

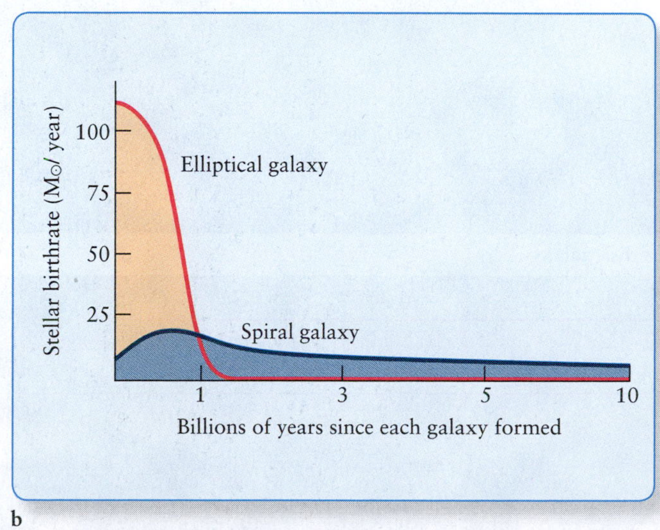

b

FIGURE 18-17 Stellar Birth Rates (a) This figure shows that star formation started quickly in the life of the universe and has been tapering off ever since. (b) Most of the stars in an elliptical galaxy are created in a brief burst of star formation when the galaxy is very young. In spiral galaxies, stars form at a more leisurely pace that extends over billions of years.

If the first elements were hydrogen, helium, and lithium, where did Population II stars in globular clusters and elsewhere get their other metals?

solar system. Star formation continues because the protogalactic disk contains an ample supply of hydrogen, and a spiral or lenticular (disk-shaped but without spiral arms) galaxy is created. If, however, the initial stellar birthrate is high in the protogalaxy, the theory predicts that virtually all pregalactic gas is used up in the creation of stars before a disk can form. In this case, an elliptical galaxy is created. Figure 18-18 depicts these contrasting sequences of events.

Not all galaxies maintain their initial structure over the evolution of the universe. As we discussed in Section 16-10, some galaxies collide, and these collisions can change a galaxy's structure from spiral, for example, to elliptical. Indeed, astronomers who use extremely long exposures, called Hubble Deep Field images (for example, see the inset on Figure 18-11), have observed elliptical galaxies during the first few billion years of the universe's existence that are far less uniform in color than are closer ellipticals. They observed blue stars in the young ellipticals consistent with the merger of spirals and a resulting burst of star formation, as well as lots of lower-mass (yellow and red) stars. Our understanding of galactic formation and evolution is far from complete.

Determining the location and nature of the dark matter in the universe (see Sections 15-7 and 16-11) is still of paramount importance in understanding the cosmos. The observable stars, gas, and dust in a galaxy or cluster of galaxies account for only about 10% to 20% of each object's mass. (Recall that this observable matter does not have enough mass to hold galaxies or clusters of galaxies together.) We have very little idea of what the remaining 80% to 90% of each galaxy or cluster of galaxies is composed of. However, some progress in understanding the distribution of this dark matter is being made. In 2002, astronomers who map large numbers of galaxies and use computer simulations of the effects of dark matter determined that on the scales of clusters of galaxies, the locations of the galaxies often coincide with the concentrations of dark matter, while voids between clusters coincide with voids of dark matter. It is likely that the gravitational force of the dark matter caused gas to concentrate in the same regions, thereby stimulating formation of supermassive black holes and galaxies. In 2007, using images of galaxies that were focused by a gravitational lens of dark matter, astronomers were able to plot the locations of this dark matter (Figure 18-19).

Sequence of events →

a Formation of a disk galaxy

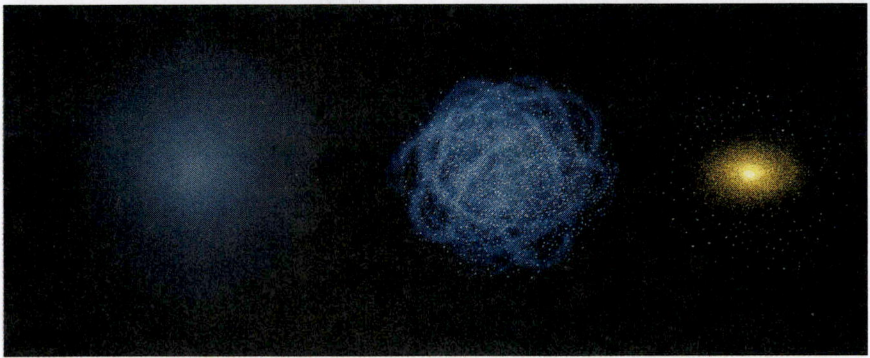

b Formation of an elliptical galaxy

FIGURE 18-18 Creation of Spiral and Elliptical Galaxies A galaxy begins as a huge cloud of primordial gas that collapses gravitationally. (a) If the rate of star birth is low, then much of the gas collapses to form a disk, and a spiral galaxy is created. (b) If the rate of star birth is high, then the gas is converted into stars before a disk can form, resulting in an elliptical galaxy.

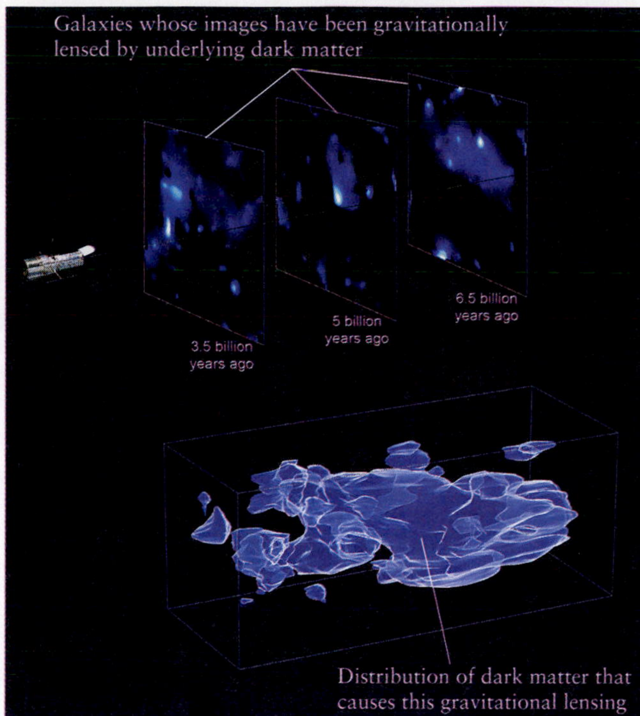

FIGURE 18-19 Mapping Dark Matter (Top) The Hubble Space Telescope observed that galaxies in the same direction, but at different distances from Earth, undergo different amounts of gravitational lensing. (Bottom) Much of this effect is due to dark matter. By subtracting out the lensing effects of intervening galaxies, the distorted shapes of the galaxies at various distances enable astronomers to determine the distribution of dark matter. (NASA, ESA, and R. Massey [CIT])

THE FATE OF THE UNIVERSE

We now turn to the future. Will the universe last forever? Or will it someday stop expanding and collapse?

18-13 The average density of matter is one factor that determines the future of the universe

As superclusters move apart in the expanding universe, their mutual gravitational attractions act on one another and thereby slow the rate at which they separate. If that were all that mattered in determining the fate of the universe, you could simply locate and add up all of the visible and (presently) dark matter in existence and see if its gravitational force is enough to eventually stop the expansion.

This process is analogous to what engineers do in calculating, for example, how high a cannonball shot upward from the surface of Earth will travel. Earth's gravitational force slows the ball's ascent. If the cannonball's speed upward is less than the escape velocity from Earth's gravity (about 11 km/s straight up), it will fall back to Earth. If the ball's speed exceeds the escape velocity, it will leave Earth and continue outward forever, despite the relentless pull of Earth's gravity. On the boundary between these two scenarios is the situation when the cannonball's speed equals the escape velocity. In that case, the cannonball will just barely escape falling back to Earth, slowing forever, and come to rest an infinite distance away.

If dark matter did not exist, would any gravitational lensing occur in the universe?

By analogy with the cannonball fired from Earth, it would seem that if the universe is expanding too slowly to overcome the mutual gravitational attraction of its parts, it should stop expanding and someday collapse. If it is expanding at exactly its escape velocity, it should expand until coming to a stop an infinite time in the future. Or, if it is expanding fast enough, it should slow down, but continue to expand forever. However, none of these options is correct!

The laws of physics pertaining to the evolution of the universe, as spelled out in the theory of general relativity and in recent observations of distant objects, reveal that reality is more complex than this simple analogy. Just when astronomers were getting comfortable with the idea that the gravitational force is always attractive, general relativity once again showed that reality ignores our common sense by demonstrating that gravity can be repulsive.

We now know two additional effects that must be considered: the effect of matter on the shape of the universe and the presence of a repulsive gravitational force, named **dark energy** *(not to be confused with dark matter)* by cosmologist Michael Turner. We begin by considering the effect that matter and energy have on the shape of the universe.

18-14 The overall shape of spacetime affects the future of the universe

During the 1920s, Alexandre Friedmann in Russia, Georges Lemaître in Belgium, Willem de Sitter in the Netherlands, and Einstein himself applied the theory of general relativity to the expanding universe. General relativity predicts that the presence of matter curves the fabric of spacetime, as we saw in Chapters 14, 15, and 17, in the form of gravitational lensing and the distortion of spacetime around black holes. Similarly, the presence of energy also curves spacetime—recall that matter and energy are related by $E = mc^2$.

Only three possibilities exist for the overall shape of the universe. These possibilities are determined by the amount of mass and energy the universe contains and how fast it is expanding. For example, imagine shining two powerful laser beams out into space. Suppose that we can align these two beams so that they are perfectly parallel as they leave Earth. Suppose, further, that nothing gets in the way of these two beams. We follow them across the spacetime whose shape we wish to determine. The light beams will begin to diverge due to the expanding universe carrying them apart. This effect happens regardless of any properties of the spacetime, and, because we are not interested in the effect of expansion right now, we compensate for it (that is, we ignore it) in what follows. The three possibilities for the paths of the laser beams, due to the actual curvature of the universe, are

What, if any, objects have we given enough velocity to escape forever from Earth?

1. Two beams of light, starting out parallel, gradually get closer and closer together as they move across the universe, eventually intersecting at some enormous distance from Earth (Figure 18-20a). In this case, we say that space is *positively curved*. Analogously, the lines of constant longitude on Earth's surface are parallel at the equator but intersect at the poles. Thus, if the universe has this effect on light, the shape of the universe is like the surface of a sphere: Space is spherical and the universe has positive curvature. This shape would happen if the universe had such high mass and energy densities that the mass and energy literally curved space back in on itself. In the absence of any outward-pushing force, such as would be supplied by a sufficiently large cosmological constant (see Section 18-1), such a universe does not have enough energy to keep expanding forever. It would someday stop expanding and thereafter collapse.

It is easiest to understand such a universe by visualizing a two-dimensional version of it as being the surface of a balloon on which you must remain. In such a universe, you could, in principle, keep going in a straight line (which is a curve on the balloon's surface) and eventually end up back where you started. Like the surface of a balloon, a three-dimensional universe with positive curvature has no outside or center and is said to be a **closed universe.**

2. Two beams of light remain parallel regardless of how far they travel (Figure 18-20b). In this case, space is not curved. That is, space is flat and it can extend without limit. This is the structure that meets our commonsense belief.

3. Two initially parallel beams of light gradually diverge farther and farther apart as they move across the universe (Figure 18-20c). This is what would happen if the energy of the Big Bang was sufficiently great to assure that the universe is going to expand forever. In this case, the universe has negative curvature. A horse's saddle is a good example of a negatively curved or *hyperbolic* surface. Initially parallel lines drawn on a saddle always diverge. Thus, in a negatively curved universe, we would describe space as hyperbolic. A hyperbolic universe extends without limit and is called an **open universe.** Both the flat and the hyperbolic universes are open and will grow larger forever.

Most astronomers have come to believe that the universe is flat. The reason for this stems from the observations that the universe is homogeneous and isotropic. The only mechanism we have at present to explain these properties of the matter and energy distribution in space is inflation, as discussed earlier. However, the equations predict that if inflation occurred it stretched the volume of the universe and the matter and energy in it so much the that universe must be very nearly flat.

Telescopic observations strongly support the belief that the universe is flat. This conclusion comes from examining the sizes of the regions of slightly higher and lower temperature in the cosmic microwave background (see Figure 18-15) and comparing them to the sizes predicted for a flat universe. The observations are consistent with the theoretical variations. Furthermore, if the universe had positive curvature, light from the hot regions of the early universe would be curved and thereby focused, creating bigger, brighter images than are observed. Figure 18-21 summarizes the effects of space curvature on observations of the cosmic microwave background.

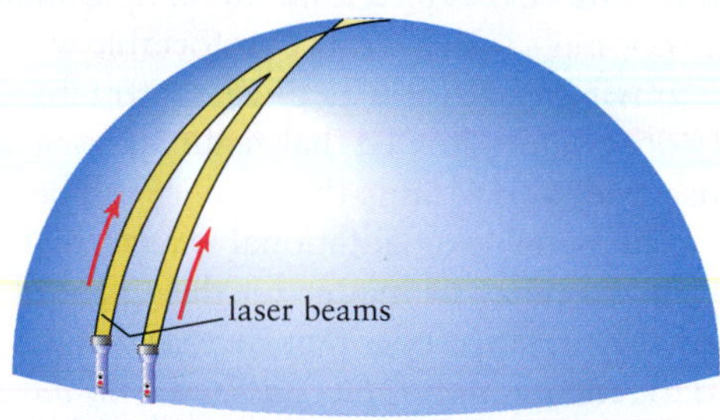

a Parallel light beams converge

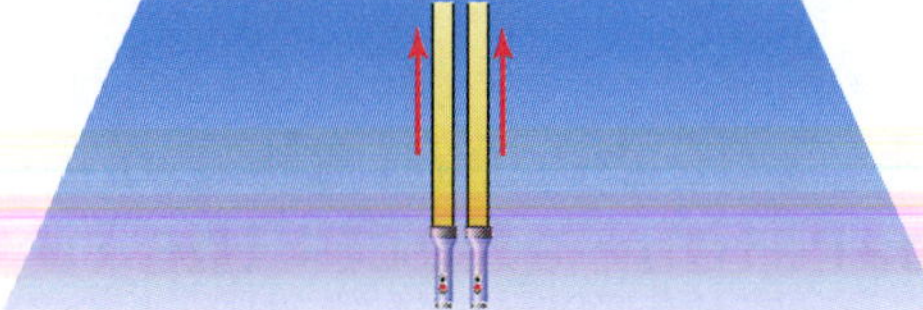

b Parallel light beams remain parallel

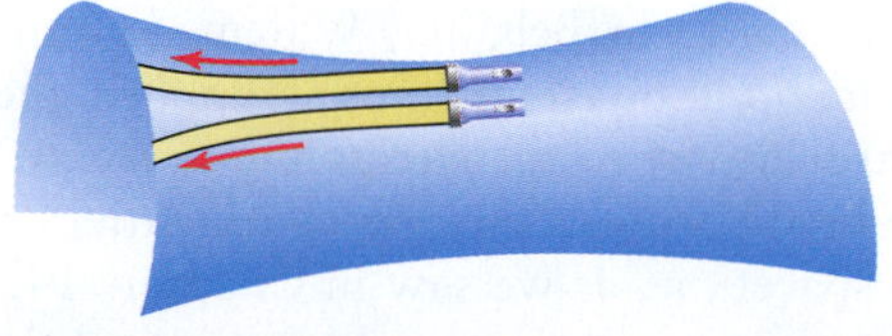

c Parallel light beams diverge

FIGURE 18-20 Possible Geometries of the Universe The shape of space (represented here as two-dimensional for ease of visualization) is determined by the matter and energy contained in the universe. The curvature is either (a) positive, (b) zero, or (c) negative, depending on whether the average matter and energy density throughout space is greater than, equal to, or less than a critical value. The lines on each curve are initially parallel. They converge, remain parallel, or diverge, depending on the curvature of space.

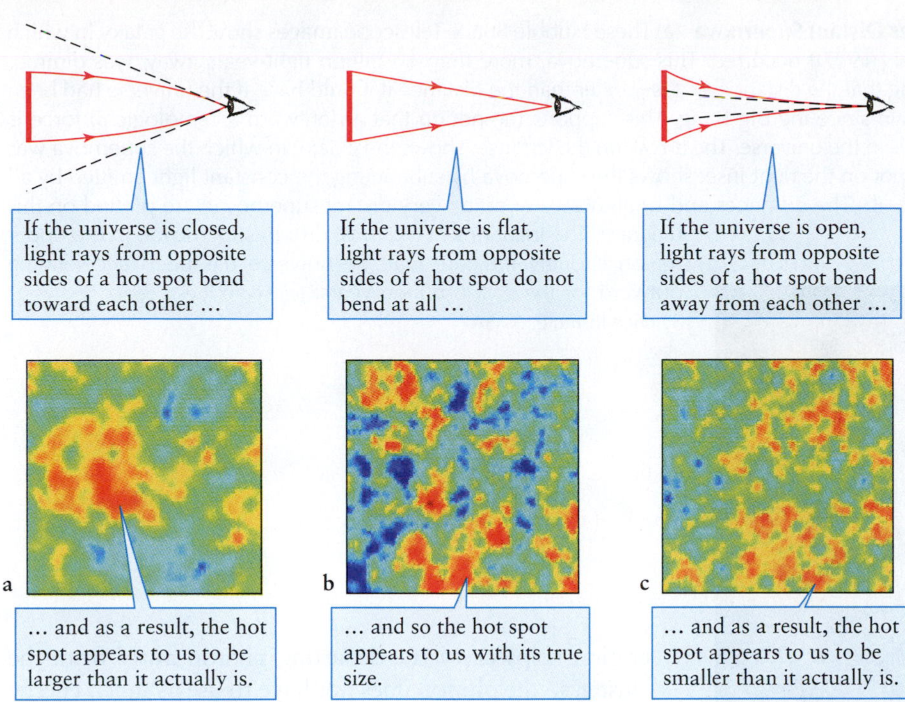

FIGURE 18-21 Cosmic Microwave Background and the Curvature of Space Temperature variations in the early universe appear as "hot spots" in the cosmic microwave background. The apparent sizes of these spots depend on the curvature of space. (a) In a closed universe with positive curvature, light rays from opposite sides of a hot spot bend toward each other. Hence, the hot spot appears larger than it actually is (dashed lines). (b) The light rays do not bend in a flat universe. (c) In an open universe, light rays bend apart. The dashed lines show that a hot spot would appear smaller than its actual size. (The BOOMERANG Group, University of California, Santa Barbara)

18-15 Dark energy is causing the universe to accelerate outward

Until the past few years, there was a major inconsistency between the distribution of observed matter and energy in the universe and the flatness of space. Just as there is not enough visible matter to account for galaxies and clusters of galaxies that remain as bound systems, there is not enough observed matter and energy to account for a flat universe. Visible matter accounts for only 4% of the required mass. The cosmic microwave background photons add only 0.005% of the required gravitational effects needed for flatness, and calculations of the mass necessary to keep galaxies and clusters bound, combined with gravitational lensing by dark matter of distant galaxies and quasars, reveal that the dark matter known to exist accounts for only about 23% of the required mass. Therefore, the universe has only about 27% of the required mass and energy necessary to make it flat. Allowing for the errors that still exist in all of these observations, the possible range of mass and energy from all matter and photons in the universe is still between only 20% and 40% of that required for flatness. Yet flat it certainly appears to be.

By the mid-1990s, combined evidence of the microwave background and large-scale structure had forced astronomers to the conclusion that in order for the universe to be flat, the remaining energy must be a form of a dark energy with the remarkable feature that it causes the universe to accelerate outward. This energy is considered "dark" because we don't yet know what it is or where it originates. Corroborating evidence for the existence of such dark energy was found in the recent observations of Type Ia supernovae in extremely distant galaxies (Figure 18-22).

The light curves (see Section 13-4) for Type Ia supernovae (explosions of white dwarfs in binary star systems) are very well known. Studies show that these supernovae in nearby galaxies behave similarly to those observed in the Milky Way, regardless of their environment. Because they are so bright, such supernovae are excellent standard candles for determining distances to objects billions of light-years away. Assuming that supernovae in distant galaxies also behave similarly to those in our Galaxy (an assumption that is now undergoing careful scrutiny), observations first made in 1998 revealed that the supernovae in distant galaxies appear dimmer than they would if the universe had been continually decelerating or even expanding at a constant speed. In other words, the universe is now expanding faster than it was, so the distant galaxies are farther away than they otherwise would have been if the expansion were slowing or even continuing at a constant rate. The universe is *accelerating* outward. Figure 18-23 summarizes the major stages of cosmic evolution discussed in this chapter.

There must be some kind of repulsive force acting to increase the rate at which superclusters separate today. Astronomers hypothesize that this is due to some kind of dark energy that has a repulsive gravitational effect, as mentioned earlier. Confirmation that dark energy exists came in 2003, when astronomers observed light from very distant galaxies passing through more nearby clusters of galaxies on its way toward Earth. As the distant light moved toward a cluster on its way to us, that light was gravitationally blueshifted, meaning the light gained energy as it was pulled toward the cluster. If there were no dark energy in a flat universe, then, as that light left the vicinity of the cluster, it would have been redshifted (lost energy) by exactly the same amount and come to us with the same wavelengths it would have had if it had never passed through the cluster.

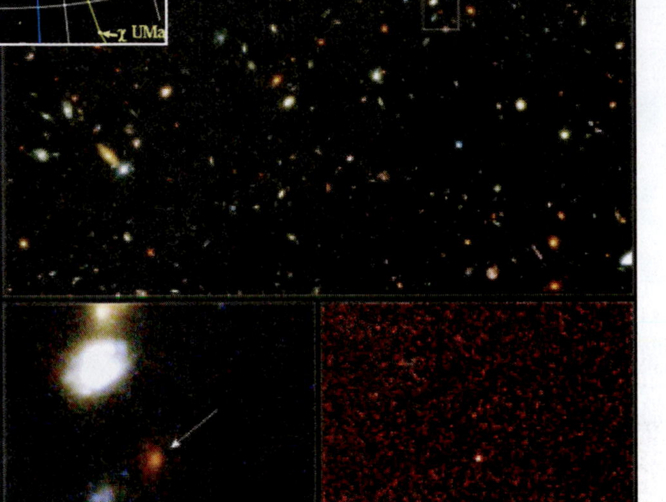

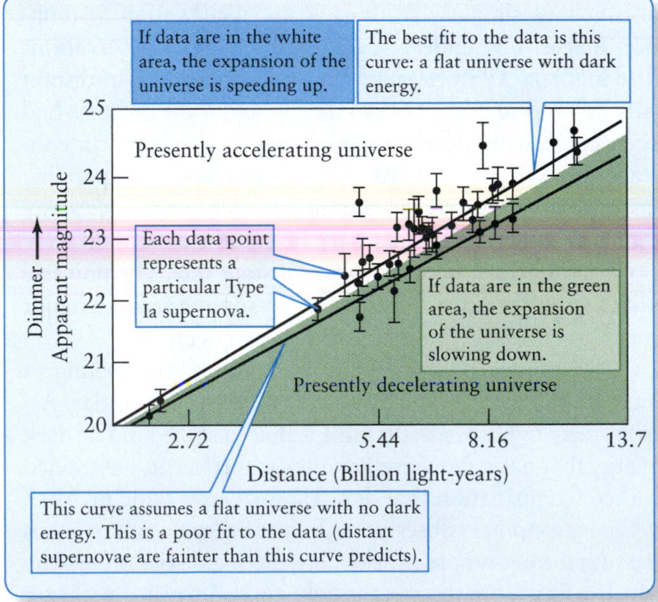

FIGURE 18-22 Dimmer Distant Supernova (a) These Hubble Space Telescope images show the galaxy in which the Type Ia supernova SN 1997ff occurred. This supernova, more than 10 billion light-years away, was dimmer than expected, indicating that the distance to it is greater than the distance it would have if the universe had been continually slowing down since the Big Bang. This supports the notion that an outward (cosmological) force is acting over vast distances in the universe. The arrow on the left inset shows the galaxy in which the supernova was discovered. The bright spot on the right inset shows the supernova by subtracting the constant light emitted by all the other nearby objects. (b) The distances and brightnesses of many very distant supernovae are plotted on this diagram. The locations of the most distant supernovae in the upper region strongly indicate that the universe has been accelerating outward for the past 6 billion years. (a: Adam Riess, Space Telescope Science Institute, NASA)

However, if dark energy exists, then the universe is accelerating outward (moving outward faster and faster). So the universe would have expanded more during the time that the light was leaving the cluster than when that light was first moving toward the cluster. The universe's acceleration helps carry the departing photon away from the cluster, so the photon does not have to use as much energy leaving the cluster as it gained traveling toward it. Under those conditions, the light would have gained more energy from the cluster's gravitational attraction while traveling toward the cluster than the light would have lost to that gravitational attraction as it was moving away from the cluster. In other words, the amount of blueshift that the light underwent falling into the cluster would be greater than the amount of redshift that it underwent leaving the vicinity of the cluster. This was precisely what astronomers discovered: The light from distant galaxies undergoes a net blueshift when it passes through clusters of galaxies on its way to us—the universe is accelerating outward.

Cosmological Constant There are presently two viable theoretical explanations for dark energy. Let us first consider what would happen if the cosmological constant that Einstein introduced and then rejected actually did exist. As Einstein had intended, it would provide a repulsive force to nature. The energy associated with that force would appear in the vacuum of space. The cosmological constant form of vacuum energy has the property of contributing a repulsive gravitational force that competes with the normal attractive gravitational force from matter and radiation that slows down the expansion. Which gravitational force wins depends on whether there is more vacuum energy or more matter and radiation.

As the universe expands, the average density of matter and energy decreases. In other words, on average there is less matter and energy in each cubic meter of space every second than there was the second before. However, the density of energy created by the cosmological constant is unchanged—as the universe expands the amount of vacuum energy in each cubic meter of space remains constant. If the universe is still expanding when the vacuum energy exceeds the matter and energy per unit volume, then the repulsive gravitational force

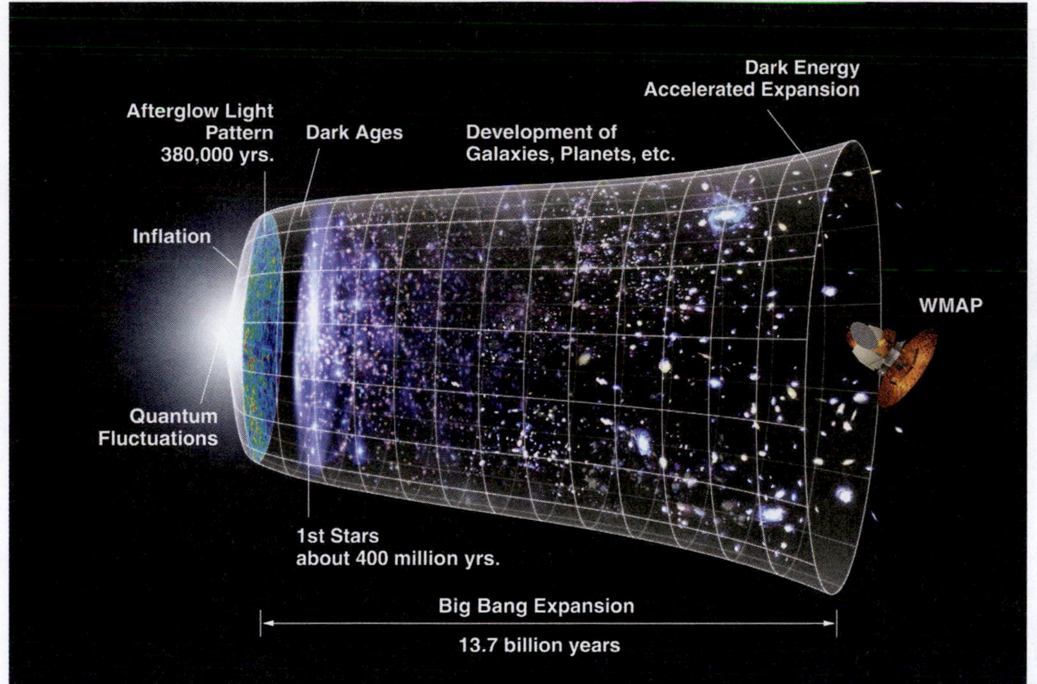

FIGURE 18-23 Big Picture of the Evolution of the Universe This figure shows our current thinking about the evolution of star and galaxy formation in the early universe, as well as the present day acceleration of the universe's expansion. (NASA/WMAP Science Team)

would dominate and force the universe to start accelerating outward, as is observed.

The concern that some astronomers express about the cosmological constant is that the repulsive force it creates has just the right strength to have allowed the universe to slow down for several billion years and then to slowly cause it to accelerate outward, as seen today. To explain this apparent coincidence would require fine-tuning the ratio of the vacuum energy to the matter and radiation to a remarkably high degree of precision. Scientists do not yet have a good reason why the repulsion created by a cosmological constant is not, say, a hundred times greater, in which case structure throughout the universe, such as stars and galaxies, would not have formed yet, or a billion times less, in which case the repulsive effects of the cosmological constant would be negligible today.

Quintessence Despite these concerns, observations of distant Type Ia supernovae suggest that the outward force acting on the universe is nearly constant, consistent with the dark energy being created by the physics we characterized with the cosmological constant. However, these observations do not yet rule out the major competing theory of dark energy, called **quintessence.** Scientists are exploring a variety of mathematical descriptions of quintessence. Quintessence differs from the vacuum energy associated with the cosmological constant in that the energy of quintessence is not constant, but changes slowly with the expansion of the universe and can also be changed by the flow of energy and matter through the universe. Because it can change, quintessence need not have been fine-tuned to a strength that makes the universe flat and creates just the slight acceleration we are seeing. Instead, it could have started out with some arbitrary strength and then been adjusted by physical properties of the universe to match the growth of the universe.

As the universe grew, the energy in quintessence would eventually come to dominate the gravitational energy from matter. If this took place a few billion years ago, quintessence would have begun exerting a negative gravitational force and so the universe would have begun accelerating.

The unexpected discovery that the universe is accel- 5
erating outward completely alters astronomers' perspectives about the fate of the cosmos. This discovery tells us that *the universe will expand forever*. At the same time, it neatly solves the mystery of why the universe is flat, as required by the occurrence of inflation in the early universe. Dark energy provides the remaining 72% of the energy needed to account for the universe being flat. The contributions of all of its major components to the mass and energy in the universe are shown in Figure 18-24, with details listed in Appendix E-12.

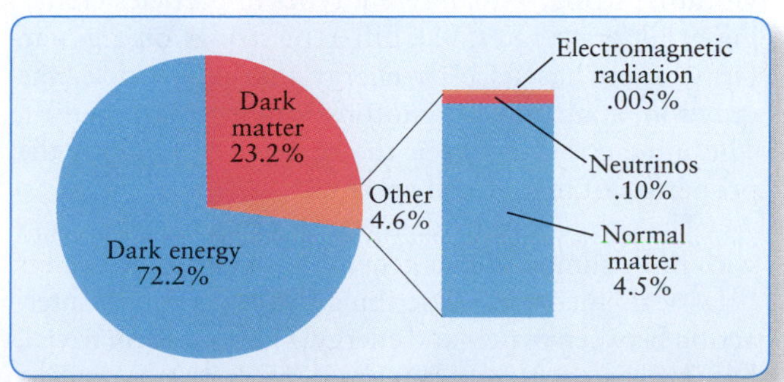

FIGURE 18-24 Percentages of the Major Components of the Universe.

We have seen that our present understanding of the laws of nature is incomplete, especially in that our theories do not explain the nature of the universe during the Planck era (or why the universe began in the first place). Driven by these challenges, scientists have begun to develop superstring theories that may resolve these issues. The *Guided Discovery: Superstring Theory and M-Theory* is intended to give you a brief introduction to them and what they now predict.

Guided Discovery

Superstring Theory and M-Theory

To combine gravitation and the other three fundamental forces in nature into one comprehensive *Theory of Everything,* scientists have had to consider a universe that contains more than the four dimensions we know about today (three of space and one of time). The theories that mathematically describe this new formulation of the universe are called *superstring theories*. There are basically five such theories, which allow all of the particles we have been studying—such as quarks, protons, neutrons, electrons, and photons—to exist.

Spacetime in superstring theories has 10 dimensions, of which 6 are everywhere rolled up into such tiny volumes that we cannot detect them directly. The other 4 dimensions are our normal spacetime. Superstring's more general spacetime carries with it properties that allow scientists to combine all four forces in nature into one set of equations.

The difficulty in reconciling quantum mechanics (describing the weak, strong, and electromagnetic forces) and general relativity (describing gravitation) is that the three forces in quantum mechanics are quantized, whereas general relativity is not. In other words, the weak, strong, and electromagnetic forces are transmitted by particles. For example, the quanta of electromagnetism are photons. Gravity, as described by general relativity, is based on a smooth and continuous, rather than quantized, force. Specifically, the distortion of spacetime by matter and energy creates the gravitational force.

Superstring theories begin with a different assumption about all particles and their interactions than do either quantum mechanics or general relativity. The new theories assert that each particle is actually a tiny vibrating string, with different types of particles vibrating at different rates, like different strings on a guitar. Gravitation has its own energy-sharing particle, the graviton, analogous to photons for electromagnetism. The interactions between the strings create all of the properties of matter and energy.

The predictions made by superstring theories begin with the assumption that general relativity is the correct "classical" theory for describing the gravitational interaction between matter and energy. This may seem trivial, but, because general relativity today correctly predicts everything in its realm of validity, a more comprehensive (superstring) theory needs to keep that accuracy, or the larger theory is wrong. Some additional predictions of superstring theories include the following:

- The universe cannot have positive curvature (which it does not, as we have seen).
- Some of the clumpy structure that we see as superclusters of galaxies could have been caused by the effects of superstring activity during inflation (with the rest due to the expansion of quantum fluctuations).
- Spacetime may not be entirely smooth. It may have structural defects, like a flawed diamond or ice that has broken into abutting chunks. These defects in spacetime would appear as one-dimensional "cosmic strings" with great density. Their attraction would pull normal matter around them, creating strings of galaxies. These cosmic strings have not yet been found.
- Some of the dark matter may be particles predicted by string theory.
- Most versions of superstring theory include a cosmological constant, but there is no underlying reason yet known for the value of the cosmological constant that may exist today.
- The speed of light is the same for all photons. If different wavelengths of light from the same event arrived at different times, then superstring (and relativity) theories would be wrong.

You may have found the idea of five superstring theories of the universe to be four too many. So do scientists who study string theory. American physicist Ed Witten has shown that in 11 dimensions, all five string theories are equivalent. The one 11-dimensional theory of strings is called M-theory.

Superstring and M-theories are, so far, consistent with observations, but it remains to be seen they will continue to maintain consistency with future observations and, very importantly, if they will make predictions that can be tested. Without being able to do that, they will remain truly elegant mathematical formalisms, but not science.

18-16 Frontiers yet to be discovered

Some of the biggest unanswered questions about the cosmos follow from the discoveries presented in this chapter. What caused the Big Bang? What happened during the Planck time before 10^{-43} s? Is our universe the only one, or are there others, perhaps with profoundly different physical properties than ours, that were created at the same time but that are inaccessible to us? What caused the universe to come into existence in a false vacuum? What are the details of the formation processes of galaxies? Why was gas in the early universe not eventually transformed into systems of stars a million times bigger or smaller than galaxies? Did black holes grow with the evolution of the galaxies, as is currently believed? What are the origins of the dark energy? Which, if either, of the current candidates for dark energy—the cosmological constant or quintessence—is correct? One of the things that makes this time in human existence so fascinating is that it is likely we will have answers to most of these questions within your lifetime.

SUMMARY OF KEY IDEAS

The Big Bang

• Astronomers believe that the universe began as an exceedingly dense cosmic singularity that expanded explosively in an event called the Big Bang. The Hubble law describes the ongoing expansion of the universe and the rate at which superclusters of galaxies move apart.

• The observable universe extends about 13.7 billion light-years in every direction from Earth to what is called the cosmic light horizon. We cannot see any objects that may exist beyond the cosmic light horizon because light from these objects has not had enough time to reach us.

• According to the theory of inflation, early in its existence, the universe expanded very rapidly for a short period, spreading matter that was originally far from our location (and hence at different temperatures and densities) throughout a volume of the universe so large that we cannot yet observe it. The observable universe today is thus a growing volume of space containing matter and radiation that was in close contact with our matter and radiation during the first instant after the Big Bang (and hence at the same temperature, pressure, and density). Inflation explains the isotropic and homogeneous appearance of the universe.

A Brief History of Spacetime, Matter, Energy, and Everything

• Four basic forces—gravity, electromagnetism, the strong nuclear force, and the weak nuclear force—explain the interactions observed in the universe.

• According to current theory, all four forces were identical just after the Big Bang. At the end of the Planck time (about 10^{-43} s after the Big Bang), gravity became a separate force. A short time later, the strong nuclear force became a distinct force. A final separation created the electromagnetic force and the weak nuclear force.

• Before the Planck time, the universe was so dense that known laws of physics did not describe the behavior of spacetime, matter, and energy back then.

• In its first 30,000 years, the universe was radiation-dominated, during which time photons prevented matter from forming clumps. Then it was matter-dominated, during which time superclusters and smaller clumps of matter formed. Today it is dark-energy–dominated. Dark energy of some sort supplies a repulsive gravitational force that causes superclusters to accelerate away from each other.

• During the first 380,000 years of the universe, matter and energy formed an opaque plasma, called the primordial fireball. Cosmic microwave background radiation is the greatly redshifted remnant of the universe as it existed about 380,000 years after the Big Bang.

• About 380,000 years after the Big Bang, spacetime expansion caused the temperature of the universe to fall below 3000 K, allowing protons and electrons to combine and thereby form neutral hydrogen atoms. This period is called the era of recombination. The universe became transparent during the era of recombination, with the photons that existed back then still traveling through space today. In other words, the microwave background radiation is composed of the oldest photons in the universe.

• Clusters of galaxies and individual galaxies formed from pieces of enormous hydrogen and helium clouds, each of which became a separate supercluster of galaxies.

• All of the superclusters and some of the clusters of galaxies within each supercluster are moving away from one another.

• Supermassive black holes appear to have "seeded" the formation of most galaxies.

• During the matter-dominated era, structure formed in the universe. As the universe goes farther into the dark-energy–dominated era, the large-scale structure of superclusters of galaxies will fade away.

The Fate of the Universe

• The average density of matter and dark energy in the universe determines the curvature of space and the ultimate fate of the universe.

• Observations show that the universe is flat and that the cosmic microwave background is almost perfectly isotropic, resulting from a brief period of very rapid expansion (the inflationary epoch) in the very early universe.

• The universe is accelerating outward and it will expand forever.

A WHAT DID YOU THINK?

1 *What is the universe?* It is all of the matter, energy, and spacetime that will ever be detectable from Earth or that will ever affect us.

2 *Did the universe have a beginning?* Yes. It occurred about 13.7 billion years ago, in an event called the Big Bang.

3 *Into what is the universe expanding?* Nothing. The Big Bang created space and time (spacetime), as well as all matter and energy in the universe. Spacetime is expanding to accommodate the expansion of the universe.

4 *How strong is gravity compared to the other forces in nature?* Gravity is by far the weakest force.

5 *Will the universe last forever?* Current observations support the belief that the universe will last forever.

Key Terms for Review

Big Bang, 538
closed universe, 554
confinement, 545
cosmic light horizon, 544
cosmic microwave background, 538
cosmological constant, 536
cosmological redshift, 537
cosmology, 536
dark ages, 550
dark energy, 553
decoupling, 547
era of recombination, 548
expanding universe, 536
Grand Unified Theory (GUT), 542
homogeneity, 541
inflation, 544
inflationary epoch, 544
isotropy, 541
isotropy problem (horizon problem), 544
matter-dominated universe, 547
open universe, 554
pair production, 545
Planck era, 543
Planck time, 543
primordial fireball, 547
primordial nucleosynthesis, 546
quark, 542
quintessence, 557
radiation-dominated universe, 546
strong nuclear force, 541
superstring theories, 542
Theories of Everything, 542
universe, 536
weak nuclear force, 542

Review Questions

1. Which force in nature is believed to have formed second? **a.** gravity, **b.** electromagnetic force, **c.** weak force, **d.** strong force, **e.** all formed at the same time

2. Inflation most directly explains which of the following? **a.** Why the universe is homogeneous **b.** Why the universe is expanding **c.** Why the universe is going to last forever **d.** Why the universe has a background temperature **e.** Why the universe has particles

3. What does it mean when astronomers say that we live in an expanding universe?

4. Explain the difference between a Doppler redshift and a cosmological redshift.

5. In what ways are the fate of the universe, the shape of the universe, and the average density of the universe related?

6. Assuming that the universe will expand forever, what will eventually become of the microwave background radiation?

7. What does it mean to say that the universe is dark energy–dominated? When was the universe radiation-dominated? When was it matter-dominated? How did radiation domination show itself?

8. Explain the difference between an electron and a positron.

9. Where do astronomers believe most of the photons in the cosmic microwave background originated?

10. Give examples of the actions or roles of each of the four basic physical forces in the universe.

11. What is the observational evidence for the **a.** Big Bang, **b.** inflationary epoch, and **c.** confinement of quarks?

Advanced Questions

12. Explain why the detection of cosmic microwave background radiation was a major blow to the steady-state theory.

13. Some so-called "creation scientists" claim that the universe came into existence 6000 years ago. What is Hubble's constant for such a cosmos? Is this a reasonable number? Explain your answers.

Discussion Questions

14. Discuss the implications that science cannot yet tell us what caused the Big Bang or what, if anything, existed before the Big Bang occurred.

15. Explain why gravitational attraction has dominated the behavior of the universe until recently and why the dark energy determines the fate of the universe.

What If...

16. The universe were destined to collapse and the collapse was under way? What would be different in space and on Earth under those conditions?

17. Our solar system formed very early in the evolution of our Galaxy, when the universe was just 2 billion years old? What would be different in space and on Earth?

18. Our solar system formed much later in the evolution of the universe than it actually did? How would observations of stars and galaxies be different than they are now?

19. Our solar system formed with the first generation of stars? What would be different about the solar system? Would Earth exist as an inhabitable world? Why or why not?

Web Question

20. Structure of the Early Universe. Search the Web for information about the *Planck* spacecraft, which is mapping the universe. In what ways is *Planck* an improvement over the WMAP mission? What new insights do scientists hope to gain about the cosmos from *Planck*? What have they already learned from it?

Observing Project

STARRY NIGHT **21.** In an attempt to explore the far reaches of the Universe, the Hubble Space Telescope (HST) took long-exposure images of very dark regions of space that appear to contain no bright stars or galaxies. These images, known as the Hubble Deep Field and Hubble Ultra Deep Field images, reveal very rich fields of faint and very distant galaxies. The light now arriving at Earth from some of these galaxies has traveled for over 13 billion years and was collected by the HST at a rate of photons per minute! Thus, the light was emitted very early in the life of the universe, only a few hundred million years after the Big Bang. You can examine and measure these two images.

a. In *Starry Night*™, open the **Favourites** pane and select **Discovering the Universe > Atlas.** Open the **Find** pane, ensure that the Query (Q) box is empty and click on the **Down** (▼) arrow in this box to display a list of image sources. Click on **Hubble Images** and double-click on **Hubble Deep Field** to center the view on this dark region of space. Note its position with respect to the Big Dipper. (Note: If you cannot identify this region of the northern sky, click on **View > Constellations > Asterisms** and **View > Constellations > Labels.** Remove these indicators after you have identified the region.) Zoom in to a field of view about 3° wide and note that the region still appears to be devoid of objects. Select **View > Deep Space > Hubble Images** from the menu to turn on the display of Hubble images in the **View** and reveal what this remarkable telescope discovered in this seemingly empty patch of sky. Zoom in further until the **Hubble Deep Field** (HDF) fills the view. One-quarter of the full HDF, with dimensions of $1.15' \times 1.15'$, is displayed in *Starry Night*™. All of the objects that appear on this long-exposure image are galaxies containing millions of stars. Examine this image carefully and attempt to identify each kind of galaxy—spiral, barred spiral, elliptical, and irregular—in this field. Choose five or six of the largest galaxies in this field, record their shapes and galaxy types, and use the angular separation tool to measure carefully and record their angular dimensions.

b. Select **File > Revert** from the menu. Return to the **Find** pane and the list of **Hubble Images** and click on **Hubble Ultra Deep Field** (HUDF) to center the view of this "dark" region of the sky. Zoom in on this region and note that, even at a field of view as small as 2°, no objects can be seen in the position of this long-exposure image. Now, select **View > Deep Space > Hubble Images** from the menu to reveal the image that the Hubble Space Telescope obtained of this region of the sky. Zoom in further until the full HUDF, with dimensions of $3.3' \times 3.3'$, fills the field of view to see this rich field of faint and very distant galaxies. Examine this image carefully and attempt to identify each kind of galaxy—spiral, barred spiral, elliptical, and irregular—in this field. Again, select five or six of the largest galaxies in this field, record their shapes and galaxy types, and use the angular separation tool to measure their dimensions.

c. Consider the mix of different kinds of galaxies and assess whether the proportions of different kinds are the same in these two images. Compare the angular sizes of the largest galaxies in these two images.

THE FIRST FEW MICROSECONDS

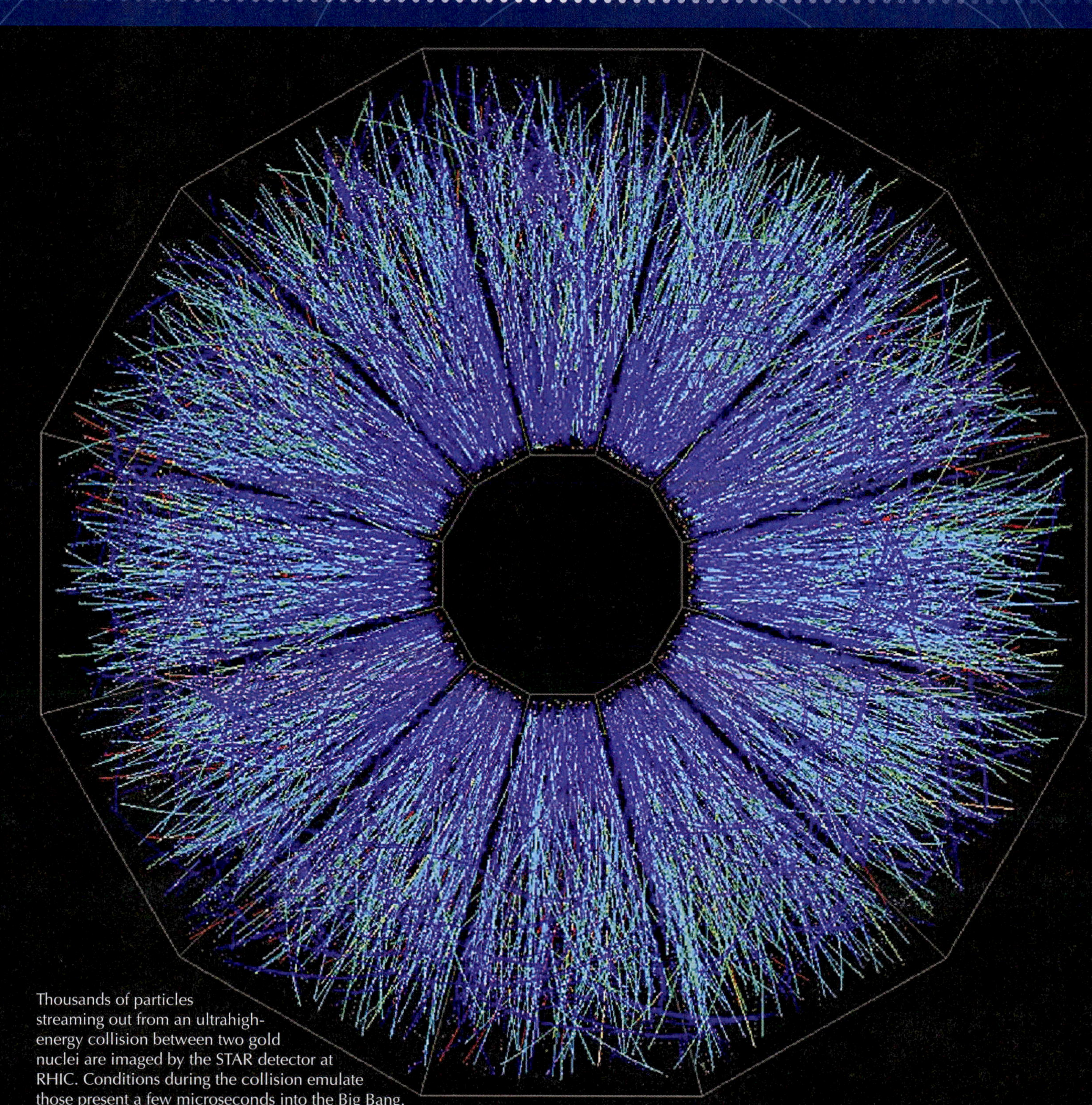

Thousands of particles streaming out from an ultrahigh-energy collision between two gold nuclei are imaged by the STAR detector at RHIC. Conditions during the collision emulate those present a few microseconds into the Big Bang.

In recent experiments, physicists have replicated conditions of the infant universe—with startling results

Adapted from an article by Michael Riordan and William A. Zajc

Since 2001, scientists have been using a powerful atom smasher to mimic conditions that existed at the birth of the universe. Inside Brookhaven National Laboratory's Relativistic Heavy Ion Collider (RHIC), two beams made of gold nuclei travel in opposite directions at nearly the speed of light, generating bursts of matter and energy when they collide head-on.

During the first few microseconds of the Big Bang, matter was an ultrahot, superdense brew of subatomic particles called quarks and gluons flying around and crashing into one another. A sprinkling of electrons, photons, and other elementary particles seasoned the soup. This mixture measured trillions of degrees, more than 100,000 times hotter than the Sun's core.

But the temperature plummeted as the cosmos expanded, just like gas cools today when it expands rapidly. As the quarks and gluons slowed down, some began to briefly stick together. After nearly 10 microseconds had elapsed, strong forces shackled the quarks and gluons together, locking them permanently within heavy nuclear particles like protons and neutrons that are collectively called "hadrons." Studying this phase transition into hadrons is key for scientists trying to understand how the universe evolved—and the fundamental forces involved.

The RHIC Experiments

The protons and neutrons that form the nuclei of every atom today are like subatomic prison cells where quarks are chained forever. Even in violent collisions, when the quarks seem on the verge of breaking out, new "walls" form to keep them confined. No one has ever witnessed a solitary quark drifting alone through a particle detector. But researchers can liberate quarks and gluons by smashing heavy nuclei together in "mini bangs." Inside RHIC, two strings of 870 superconducting magnets cooled by tons of liquid helium steer particle beams around two interlaced 3.8-km rings. The beams clash at four points where these rings cross.

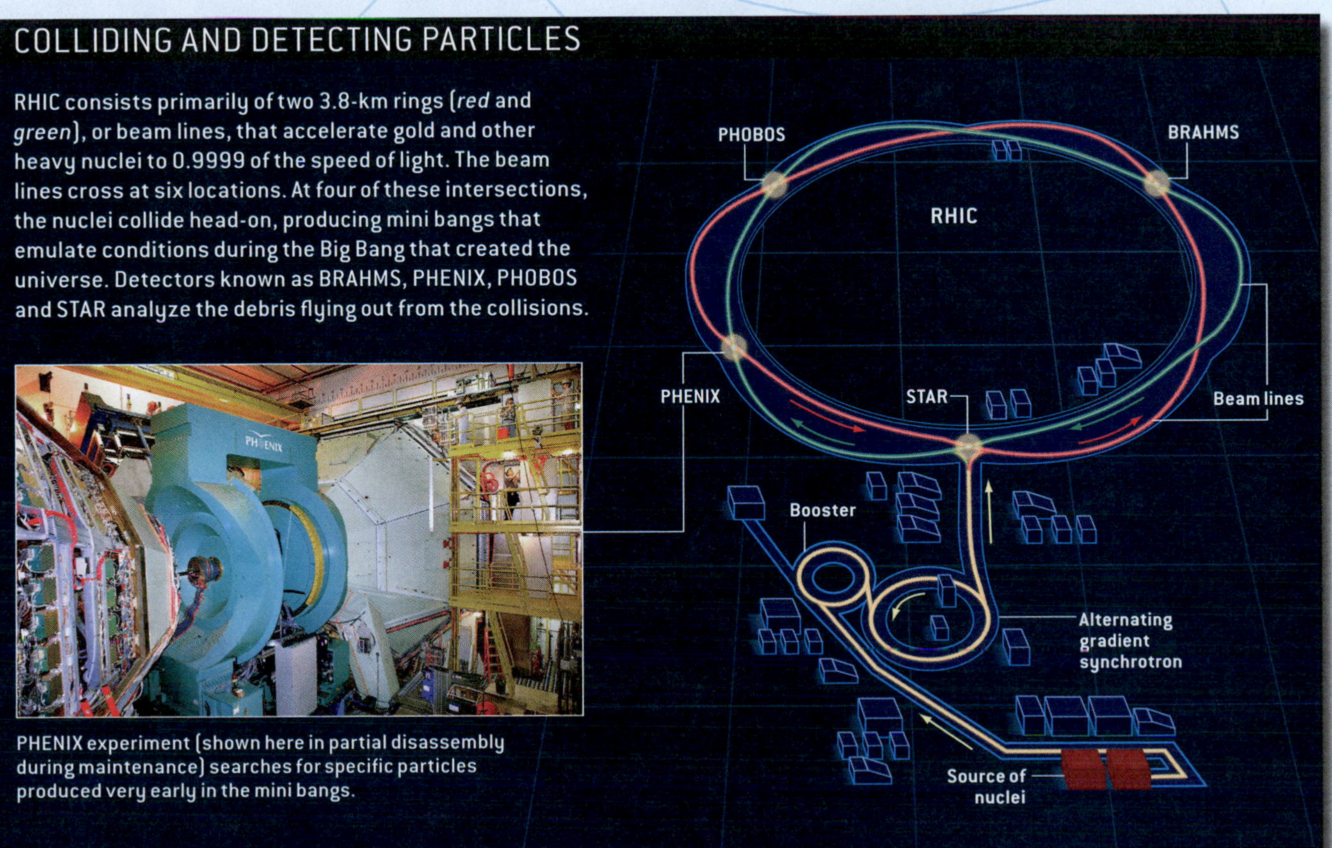

COLLIDING AND DETECTING PARTICLES

RHIC consists primarily of two 3.8-km rings (*red* and *green*), or beam lines, that accelerate gold and other heavy nuclei to 0.9999 of the speed of light. The beam lines cross at six locations. At four of these intersections, the nuclei collide head-on, producing mini bangs that emulate conditions during the Big Bang that created the universe. Detectors known as BRAHMS, PHENIX, PHOBOS and STAR analyze the debris flying out from the collisions.

PHENIX experiment (shown here in partial disassembly during maintenance) searches for specific particles produced very early in the mini bangs.

BROOKHAVEN NATIONAL LABORATORY/RHIC COLLABORATION (*photograph*); LUCY READING-IKKANDA (*illustration*)

Physicists squeeze the highest possible energies into the smallest possible volume to recreate the enormous temperatures, pressures, and densities of those first few microseconds. They use extremely dense nuggets of matter, atomic nuclei from metals like gold and lead: A thimbleful of these nuclei would weigh 300 million tons.

When two gold nuclei smash head-on inside RHIC, they produce a microscopic fireball consisting of thousands of particles, allowing researchers to study their collective properties. Both the nuclei and their constituent protons and neutrons melt, and many quarks, antiquarks (antimatter opposites of the quarks), and gluons are created from the available energy. Typically, more than 5000 elementary particles are briefly liberated. The pressure generated at the moment of collision is a whopping 1030 times atmospheric pressure.

But about 50 trillionths of a trillionth of a second later, all the quarks, antiquarks, and gluons recombine into hadrons that then explode outward. Sophisticated particle detectors record as much information as possible about the subatomic debris.

A Perfect Surprise

The picture emerging from these experiments is both consistent and surprising. The quarks and gluons break out of confinement and behave collectively, if fleetingly. Though scientists expected this exotic substance to be similar to the plasma inside a lightning bolt, it acts more like a liquid than a gas.

The energy densities created in head-on collisions between gold nuclei are about 100 times those of the nuclei themselves—because of relativity. Both nuclei flatten into ultrathin disks of protons and neutrons just before they meet, so all their energy

A MINI BANG FROM START TO FINISH

RHIC generates conditions similar to the first few microseconds of the big bang by slamming together gold nuclei at nearly the speed of light. Each collision, or mini bang, goes through a series of stages, briefly producing an expanding fireball of gluons (*green*), quarks and antiquarks. The quarks and antiquarks are mostly of the up, down and strange species (*blue*), with only a few of the heavier charm and bottom species (*red*). The fireball ultimately blows apart in the form of hadrons (*silver*), which are detected along with photons and other decay products. Scientists deduce the physical properties of the quark-gluon medium from the properties of these detected particles.

Gold nuclei traveling at 0.9999 of the speed of light are flattened by relativistic effects.

The particles of the nuclei collide and pass one another, leaving a highly excited region of quarks and gluons in their wake.

The quark-gluon plasma is fully formed and at maximum temperature after 0.7×10^{-23} second.

Quarks and gluons are locked inside protons and neutrons

Quarks and gluons are freed from protons and neutrons but interact strongly with their neighbors

Heavier charm and bottom quarks are formed in quark-antiquark pairs early in the fireball

Photons are emitted throughout the collision aftermath but most copiously early on

Photon

LUCY READING-IKKANDA

is crammed into a very tiny volume at the moment of impact. Then, particles dart in every direction, bashing into one another. Physicists estimate that the resulting energy density is at least 15 times what is needed to free the quarks and gluons.

Evidence for the rapid formation of such a hot, dense liquid comes from a phenomenon called jet quenching. When two protons collide at high energy, some of their quarks and gluons meet nearly head-on, blasting narrow, back-to-back sprays of hadrons (called jets) in opposite directions. Their behavior mimics what would occur within a dense liquid state of matter.

Indications of liquidlike behavior of the quark-gluon medium is also seen in the phenomenon of elliptic flow. In collisions that occur slightly off-center—which is often the case—the hadrons that emerge reach particle detectors arranged in an elliptical pattern. This behavior indicates substantial pressure within the medium and shows that the quarks and gluons behaved collectively before reverting back into hadrons. They were acting like a liquid—that is, not a gas. From a gas, the hadrons would have emerged uniformly in all directions.

This liquid behavior must mean these particles interact strongly during their moments of liberation. The decrease in the strength of their interactions is overwhelmed by a dramatic increase in the number of newly liberated particles. It is as though the prisoners escaped from their cells only to find themselves in a jail-yard crush, jostling with all the other escapees. This conflicts with the theoretical picture of this medium as a weakly interacting gas. And the elliptical asymmetry suggests this surprising liquid flows with almost no viscosity and therefore little friction. The hottest, densest matter ever encountered is probably the most perfect liquid ever observed.

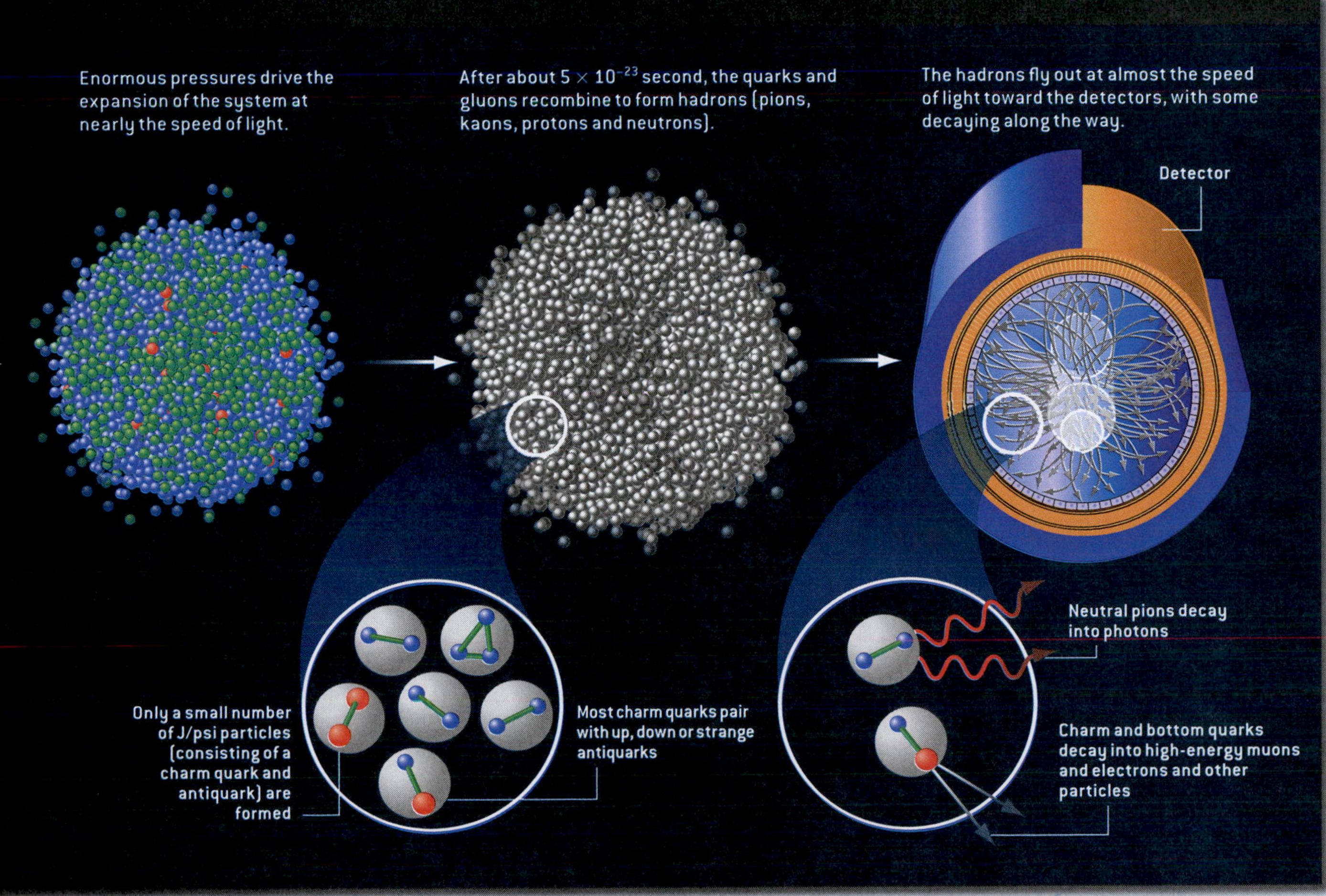

Future Challenges

Data from the RHIC experiments is forcing theorists to reconsider some cherished ideas about matter in the early universe. It is also raising new questions. Experimenters are now studying the different kinds of quarks emerging from the mini bangs. When quarks were originally predicted in 1964, they were thought to occur in three versions: up, down, and strange. In RHIC collisions, these three quark species and their antiquarks are created abundantly and in roughly equal numbers. Two heavier quarks, dubbed charm and bottom, turned up in the 1970s. Because much more energy is required to create heavy quarks, they appear earlier in the mini bangs (at the point where when energy densities are higher) and less frequently. Therefore, they are valuable tracers of flow patterns and other properties that develop early in the evolution of a mini bang.

Researchers are also trying to determine the temperature of the quark-gluon fluid by using its own light, just as astronomers measure a distant star's temperature from its light spectrum emission. A hot broth of these particles should shine briefly, like a flash of lightning, because it emits high-energy photons that escape the medium unscathed. But measuring this spectrum has proved challenging because many other photons are generated by the decay of hadrons.

Experiments at the Large Hadron Collider (LHC) near Geneva are now observing collisions of lead nuclei. The mini bangs produced by the LHC will reach several times the energy density of RHIC collisions, and the temperatures should surpass 10 trillion degrees. This work will allow physicists to simulate and study conditions during the very first microsecond of the Big Bang.

The overriding question is whether the liquidlike behavior witnessed at RHIC will persist at the higher temperatures and densities at the LHC. Some theorists project that the force between quarks will become weak and that the quark-gluon plasma will finally start behaving like a gas, as originally expected. Others maintain that the quarks and gluons should remain tightly coupled in their liquid embrace. We await the verdict of experiment, which may bring other surprises

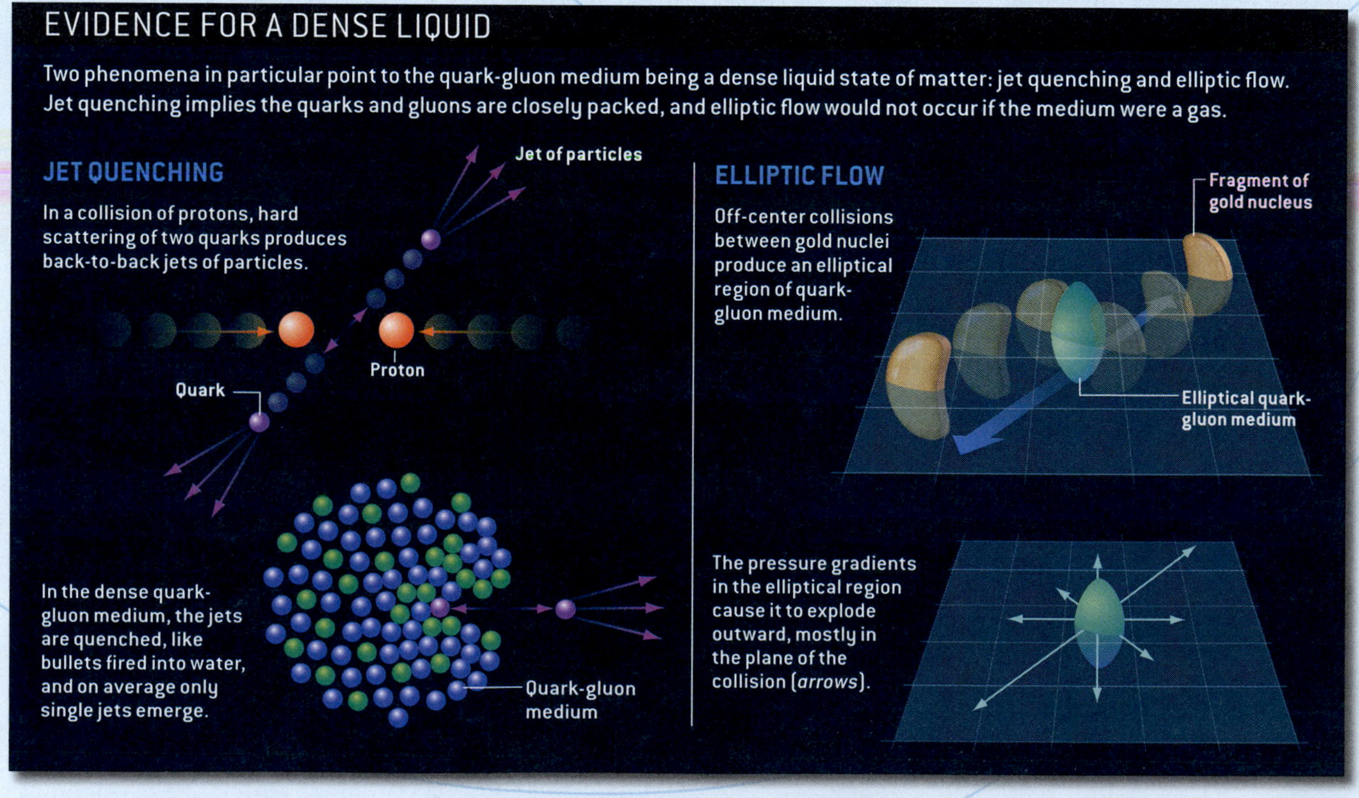

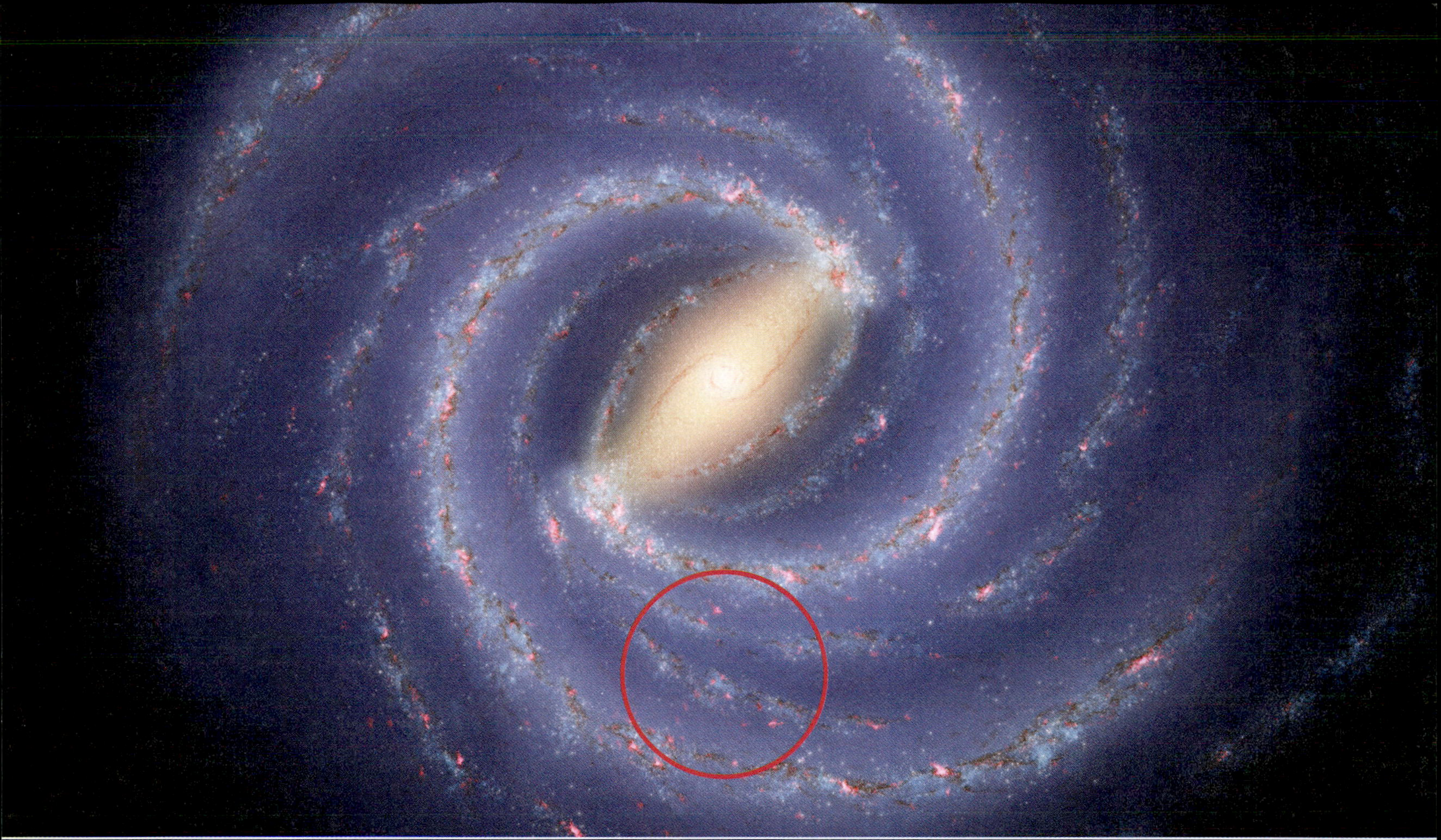

Centered on Earth, the red circle shows the region in which astronomers are searching for extraterrestrial intelligence. (NASA/Space Telescope Science Institute)

Chapter 19
Astrobiology

WHAT DO YOU THINK?

1. Why is water so important to the formation of life?
2. What element is uniquely suited to be the foundation of life as we know it, and why?
3. How do astronomers search for extraterrestrial intelligence?
4. Have astronomers located any extraterrestrial civilizations?
5. If advanced alien civilizations exist, is there any way they might know of our existence?

Answers to these questions appear in the text beside the corresponding numbers in the margins and at the end of the chapter.

If, as we said in the first chapter of this book, the starry night sky inspires us to look beyond ourselves, the study of astronomy often motivates us to contemplate the formation of Earth, the nature of the stars, and even the creation of the universe. Perhaps two of the most compelling questions in this realm of nature are how life began here on Earth and whether there is life elsewhere in the universe.

Scientists are searching for the origins of life on Earth by doing experiments in hopes of recreating steps in that process. In 2010, for example, a simple artificial cell was fabricated. A key part of these experiments involves trying to determine the physical environment of Earth and the materials available at the time when life began.

Understanding the origins of life may be helped forward by discovering primitive life elsewhere in our solar system, such as inside Mars, Europa, Ganymede, and, possibly, other moons of the solar system. Astronomers have begun looking for it, developing life-seeking space missions to other worlds, including two that went to Mars in the 1970s (Figure 19-1; see also Section 7-13).

The search for advanced extraterrestrial life has gone in several directions, starting right here on Earth. Conceivably, advanced civilizations from elsewhere have sent spacecraft to visit. In fact, some people believe that they have seen extraterrestrials or, at least, their spacecraft, commonly called *unidentified flying objects* (UFOs). This discussion leads to the question of whether life on Earth was "seeded" intentionally or accidentally by alien visitors. If Earth was visited, that event is possible, but Occam's razor requires us to explore simpler avenues for the beginning of life here.

Are there other advanced civilizations elsewhere in the cosmos? Some people find comfort in the belief that Earth is the only home for life. Others look at the vastness of the universe and the size and diversity of the objects in the cosmos and thereby find reason to expect there may be other worlds suitable for sustaining life. Some scientists are now undertaking high-tech searches for distant extraterrestrial life, trying to answer this question of whether or not we are alone in the universe or whether others are out there, perhaps searching just as we are, looking for company in the cosmos.

FIGURE 19-1 ***Viking* Mars Lander** Astronomer and renowned science popularizer Carl Sagan poses by a model of the *Viking* lander. This image was taken in Death Valley, California, where the background creates the feel of a Martian landscape. Sagan was instrumental in choosing some of the experiments flown on the *Viking* spacecraft. (NASA Jet Propulsion Laboratory)

In this chapter you will discover

- what qualities scientists believe a world must have in order to support life
- why many scientists are open to the possibility that primitive life exists elsewhere in the solar system
- how scientists estimate the number of planets orbiting other stars that could support complex life
- how scientists search for life beyond our own solar system—and the results of those searches
- how we are trying to communicate with extraterrestrial life

19-1 Astrobiology connects the cosmos and the origins of life

The amazing advances in all realms of science that began in the seventeenth century and accelerated to tremendous heights in the last few years have finally enabled scientists to meaningfully explore issues surrounding the beginnings of life on Earth. **Astrobiology** combines all of the natural sciences in an effort to understand the formation, evolution, and future of life here and throughout space. As we will explore shortly, efforts to replicate the conditions under which life formed on Earth have been carried out since the 1950s. The discovery of strong evidence that liquid water, essential for life as we know it, exists in several worlds in the solar system has added fuel to the astrobiology effort.

Astrobiology as a unified discipline received a big boost in 1996, when a meteorite from Mars, labeled ALH84001, was opened to reveal organic-looking structures that *could* have been formed by life on Mars. This possibility sparked the interest of many people, including those who fund NASA and other organizations worldwide. Thousands of scientists now work in this field.

Finding other potential homes for life in the solar system is one facet of astrobiology. Finding homes beyond the solar system is another. The number of known extrasolar planets has passed 520. Although most are Jupiterlike (that is, hydrogen-rich bodies), Earthlike planets such as Gliese 581 c and the yet-to-be confirmed Gliese 581 g are now being discovered. Some of these planets may have liquid water on their surfaces.

The search for advanced civilizations is also under way. It began on Earth, in the form of people who believe they have seen aliens or, more commonly, their UFO spacecraft. Beliefs are one thing, scientific evidence is something else. Despite all of the alleged UFO sightings and personal encounters that have been reported, no one has produced a single piece of physical evidence to convince scientists that an intelligent, extraterrestrial life-form has *ever* visited Earth. Furthermore, despite all of the wonderful science-fiction books, movies, and TV shows, there is no known way for spacecraft to travel faster than the speed of light. Therefore, unless they come from planets around the few stars within a few light-years of Earth, aliens would have to travel for centuries, or even millenia, to span the void of space between their worlds and ours. After all that travel time, it would seem illogical for space visitors not to land here.

*S*earches for *e*xtra*t*errestrial *i*ntelligence (**SETI**) in space began in 1960, when the American astronomer Frank Drake carried out Project Ozma (named after the land in L. Frank Baum's Oz series of books). Drake used a radio telescope at the National Radio Astronomy Observatory in West Virginia to "listen" for radio signals from civilizations that might have existed around two Sunlike stars: τ (tau) Ceti and ε (epsilon) Eridani. He found nothing.

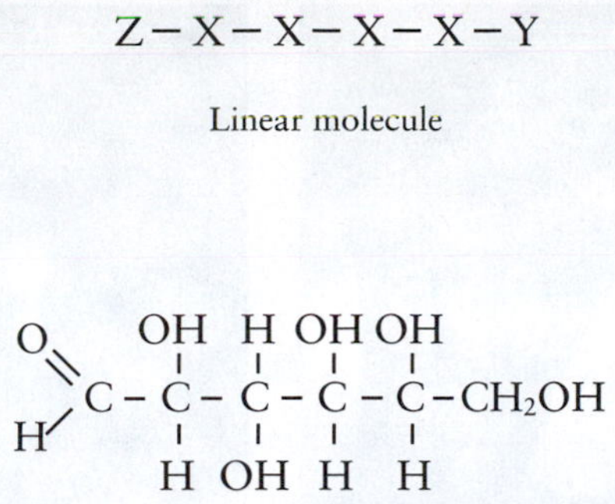

Z–X–X–X–X–Y

a Linear molecule

```
O     OH  H  OH OH
 \\    |   |   |   |
  C – C – C – C – C – CH2OH
 /     |   |   |   |
H      H  OH   H   H
```

b Nonlinear molecule: Glucose

FIGURE 19-2 Creating Complex Molecules (a) Atoms (denoted by capital letters) that can bond strongly to only two other atoms can make linear chains, as depicted here. However, when such atoms bond to atoms that can only make one bond, denoted here as Y and Z, the chain stops. In no case can atoms X, Y, and Z combine to create nonlinear chains other than loops. (b) When an atom, like carbon (C), can share electrons with more than two other atoms (in carbon's case, with four atoms), then the complex, nonlinear chains essential for life can form. Chains of carbon atoms form the backbone of organic molecules. For example, glucose, with carbon, oxygen (O), and hydrogen (H), is a nonlinear molecule that serves as a nutrient for many life-forms; it is a sugar. The lines indicate bonds between atoms.

19-2 The existence of life depends on chemical and physical properties of matter

1 To build a physical foundation for astrobiology, let us begin by considering some of the fundamental requirements for the existence of life as we know it. First is the presence of liquid water. Water, neither acidic nor basic, enables many kinds of atoms and molecules dissolved in it to bond to one another or separate from one another in large numbers. While water molecules combine with some atoms and molecules, water also allows many interactions to occur in it without chemical interference. Such diverse activity is essential for the creation and evolution of the complex molecules and systems of molecules that function in living organisms. Given all of these factors, biologists believe that life on Earth began to develop in primordial bodies of water.

The second requirement is an element that can bond strongly, but not too strongly, with at least three other atoms to allow for complex molecules necessary for life. Atoms able to make only two bonds can form chains (Figure 19-2a) or loops of atoms, but cannot create complex structures (Figure 19-2b) that contain atoms of other elements, as is necessary for making the variety of molecules essential for life.

Five elements can bond covalently (that is, by sharing electrons) with three or more elements: boron (B), carbon (C), nitrogen (N), silicon (Si), and phosphorus (P). Bonding by donating or receiving electrons (called *ionic bonding*) creates bonds that are unsuitable for the formation of the complex, flexible, rapidly-changing molecules that life requires. Carbon is unique among the five elements that can make at least three covalent bonds in that its bonds are flexible yet strong. As counterexamples, silicon-silicon bonds come apart under the slightest disturbance; silicon-oxygen bonds create gels and liquids that are very hard to alter (Figure 19-3a); and silicon-oxygen-oxygen bonds are so strong that they create rocks (Figure 19-3b). The other three elements that might serve as the backbone of life—boron, nitrogen, and phosphorus—have similar bonding problems. Despite the enormous range of conditions under which terrestrial life exists, all of it is based on the unique properties of carbon.

How would most scientists respond to the statement: "Aliens might be here, but we just haven't seen them yet?"

Carbon atoms form chemical bonds that can combine to create especially long, complex molecules. These molecules can be further linked in elaborate chains, lattices, and fibers. It is for these reasons that carbon-based compounds, called **organic molecules**, are the stuff of life. Indeed, biologists have already identified more than 6 million carbon compounds. 2

The variety and stability of living organisms depend on the complex, self-regulating, chemical reactions of organic molecules. Happily, carbon is also among the most

a

b

FIGURE 19-3 Non-Carbon Organic Molecules? When any element other than carbon that can make three or more covalent bonds combines, it makes compounds that are too soft, too hard, too reactive, or too inert to be useful in supporting life. Consider silicon. (a) The silicon-oxygen pair that creates the backbone of silicone is too inert to allow such molecules to react rapidly and thereby serve as organic molecules. Furthermore, these bonds produce gel or liquid compounds, as shown. (b) When the backbone is silicon-oxygen-oxygen, the bonds are rigid, as in this quartz rock. (a: Richard Megna/Fundamental Photographs, New York; b: Mark A. Schneider/Visuals Unlimited)

abundant elements in the universe. The unique versatility and abundance of carbon imply that extraterrestrial biology would also be based on organic chemistry.

Third, a life-supporting environment must also have a variety of other elements essential for developing life. Other constituents of organic molecules, such as hydrogen, oxygen, nitrogen, and sulfur, are plentiful in our Galaxy. All of the organic molecules necessary to form life may develop naturally in the oceans of Earth and similar planets. Nevertheless, some astronomers have proposed that after planets came into existence, organic material from space rained down on these worlds, thereby expediting the evolution of life. This belief is supported by the fact that spectra taken of comets in our solar system reveal that they contain an assortment of organic compounds.

Further evidence for the existence of organic molecules in interplanetary space was collected by, among others, the spacecraft *Giotto* as it passed close to the nucleus of Halley's comet. In 1997, NASA launched the *Stardust* mission, which rendezvoused with Comet Wild 2 in January 2004, collecting dust and other debris from the comet. The spacecraft returned safely to Earth in January 2006. It revealed the presence of organic molecules that had been released from the comet, supporting the belief that comets may have helped seed compounds useful in the formation of life on Earth.

Evidence of extraterrestrial organic molecules also comes from certain meteorites called *carbonaceous chondrites* (Figure 19-4). Many of the carbonaceous chondrites that have fallen to Earth contain a variety of organic substances. As noted in Section 9-11, carbonaceous chondrites are ancient meteorites dating from the formation of the solar system. If they contained organic molecules while in space, it seems reasonable to conclude that, even from their earliest days, the planets have been continually bombarded with organic compounds.

In interstellar clouds, carbon atoms have combined with other elements to produce an impressive variety of organic compounds. Since the 1960s, radio astronomers have detected telltale microwave emission lines from interstellar clouds that help identify more than 80 of these carbon-based compounds. Examples include polycyclic aromatic hydrocarbons (PAHs), a type of molecule composed of just carbon and hydrogen atoms, as well as ethyl alcohol (CH_3CH_2OH), formaldehyde (H_2CO), acetic

a

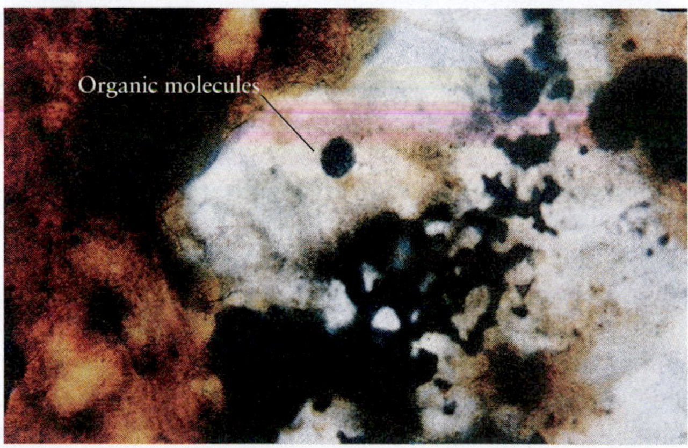

b

FIGURE 19-4 Carbonaceous Chondrite (a) Carbonaceous chondrites are meteorites that date back to the formation of the solar system. This sample is a piece of the Allende meteorite, a large carbonaceous chondrite that fell in Mexico in 1969. (b) Chemical analysis of newly fallen carbonaceous chondrites discloses that they are rich in organic compounds. (a: From the collection of Ronald A. Oriti; b: Harvard Smithsonian Center for Astrophysics) acid ($HCOOCH_3$), and acetaldehyde (CH_3CHO). In 2003, scientists detected the amino acid glycine in space. In 2005, PAHs were observed more than 9 billion ly away. These were the first observations of complex organic molecules from that far in the past. Because stars and planets condense from interstellar gas and dust enriched in these elements, these compounds can help provide the building blocks of life elsewhere in the Galaxy and, indeed, in many other galaxies throughout the universe. Indeed, in 2004, astronomers using the Spitzer Space Telescope detected some of the ingredients for life, including methanol, water, and carbon dioxide, in orbit around newly forming stars.

Interstellar space is not the only source of organic material. In a classic experiment performed in 1952, the American chemists Stanley Miller and Harold Urey demonstrated that simple chemicals can combine on liquid-water–bearing planets to form organic compounds. Hoping to recreate conditions on primitive Earth and in its atmosphere, as then understood, Miller and Urey mixed hydrogen, ammonia, methane, and water vapor in a closed container. They then subjected their mixture to an electric arc to simulate lightning bolts. At the end of a week, the inside of the container had become coated with a reddish-brown substance rich in compounds essential to life (Figure 19-5).

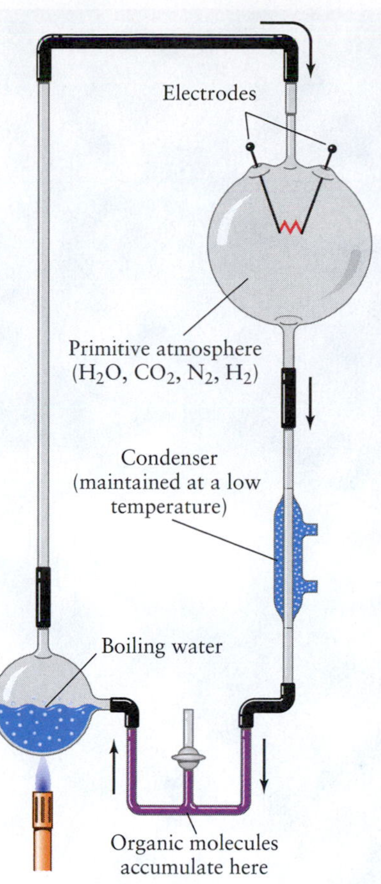

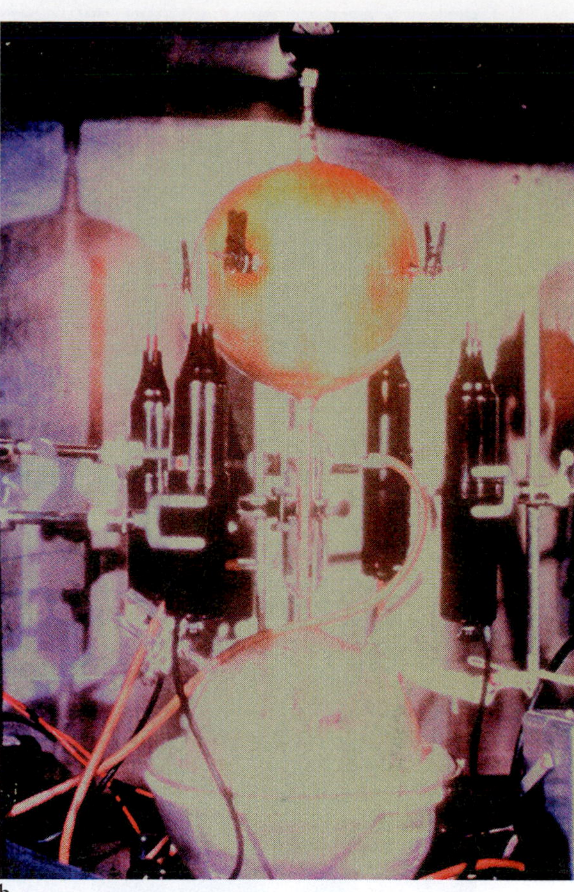

FIGURE 19-5 Miller-Urey Experiment Updated (a) Modern versions of this classic experiment prove that numerous organic compounds important to life can be synthesized from gases that were present in Earth's primordial atmosphere. This experiment supports the hypothesis that life on Earth arose as a result of ordinary chemical reactions. (b) This photograph shows a slime. (b: NASA)

Since that time, geochemists have come to realize that Earth's early atmosphere was more likely to have been a mixture of carbon dioxide, nitrogen, and hydrogen, along with water vapor "outgassed" from volcanoes and deposited by comet impacts. Modern versions of the Miller-Urey experiment have therefore used these common gases, and, once again, organic compounds were produced.

Keep in mind that scientists have not yet created life in a test tube. Biologists have yet to figure out, among other things, how simple organic molecules gathered into cells and developed systems for self-replication.

The fourth issue surrounding the existence of life elsewhere relates to physical properties of planets and life—the environments supporting life-forms must have suitable temperature, radiation, and other factors. The fundamental factor here is that water must be liquid on (or possibly in) a world for life to evolve there. The region around a star or planet where this happens is called the **habitable zone**. Earth is in the Sun's habitable zone, while Europa is in Jupiter's.

If a planet is too close to a star, the temperature and deadly ultraviolet radiation levels will be too high for life to evolve very far. If the planet is too far away, it will be too cold to have liquid water. If a planet's environment is otherwise hostile—such as having too much radiation, being too windy, or being too seismically active—life may either not be able to spread or may quickly become extinct. If the star is too massive, it will explode before advanced life on any of its planets has a chance to evolve very far. If the star is too small, the planet would have to be very close to it to be warm enough for life, but then tidal forces from the star would lock the planet in synchronous rotation, making most of its surface either too hot (daytime side) or too cold (nighttime side) for life to flourish. Even a world in synchronous rotation with a suitable temperature on the star-facing side is very unlikely to support life, because water in its atmosphere will drift to the night side, permanently freeze, and the star-lit side will eventually become arid. These issues are summarized in Figure 19-6.

Synchronous rotation occurred in what other situation that we explored earlier in this book?

Habitable Zones for Life

Intelligent civilizations in our Milky Way Galaxy can evolve only in a certain region called the galactic habitable zone. In that zone, a suitable planet must lie within the planetary habitable zone of its parent star. (From C. H. Lineweaver, Y. Fenner, and B. K. Gibson)

Galactic habitable zone

Too close to the center of the Milky Way Galaxy:
- The distances between stars are small, so there can be close encounters between stars that would disrupt a planetary system.
- There are also frequent outbursts of potentially lethal radiation from supernovae and from the supermassive black hole at the very center of the Galaxy.

Too far from the center of the Milky Way Galaxy:
- Stars are deficient in elements heavier than hydrogen and helium, so they lack both the materials needed to form Earthlike planets and the chemical substances required for life as we know it.

Planetary habitable zone

The Star:
- Must have a mass that is neither too large or too small.
- If the star's mass is too large, it will use up its hydrogen fuel so rapidly that it will move off the main sequence before life can evolve on any of its planets.
- If the star's mass is too small, the habitable planet would be so close that it would be in synchronous rotation. Water on the star-lit side would vaporize and become locked up as ice on the permanently dark side.

The Neighborhood:
- There needs to be one or more large Jovian planets whose gravitational forces will clear away comets and meteors.

The Planet:
- Must be a terrestrial planet with a solid surface.
- Must have enough mass to provide the gravity needed to retain an atmosphere and oceans.
- Must be at a comfortable distance from the star so that water can be liquid on its surface.
- Must be in a stable, nearly circular orbit. (A highly elliptical orbit would cause excessively large temperature swings as the planet moved toward and away from the star.)

FIGURE 19-6 Zone For Habitable Planets This figure summarizes the locations in the Galaxy and in orbit around stars where habitable planets might be found. Earth, of course, is in such a location.

19-3 Evidence is mounting that life might exist elsewhere in our solar system

Despite all of these potential limitations, scientists have strong evidence that everything does not have to be optimal for life to form. Geologists and biologists have discovered life on Earth in some incredibly challenging environments, such as on the ocean floor, in a lake far under the Antarctic ice pack, deep inside our planet's crust, and even in hot geothermal vents (Figure 19-7). It therefore seems reasonable to believe that life could have originated off Earth under similarly challenging conditions. Scientists consider at least four places in our solar system as possible habitats for life past or present. They are Jupiter's moons Europa, Ganymede, and Callisto, and the planet Mars.

Astronomers have discovered (see Chapter 7) that Mars once had surface liquid water and that it may still have underground bodies of liquid water. We saw in Sections 8-5 through 8-7 that Europa, Ganymede, and Callisto are also likely to have liquid water under their frozen surfaces. These four bodies, then, are candidates for having evolved simple life-forms. Indeed, we saw in Chapter 7 what may be fossilized bacterial life from Mars (see Figure 7-39).

Confirmation that life evolved on any of these worlds would demonstrate that the formation of complex molecules and structures is possible off Earth. However, given the limited chemical resources available on other worlds in our solar system, these locations are probably unsuitable for the evolution of complex life, like ourselves, that is self-aware and able to create advanced civilizations.

19-4 Searches for advanced civilizations try to detect their radio signals

We have not yet discovered Earthlike planets orbiting Sunlike stars. The closest confirmed case was the discovery in 2007 of a 5-Earth-mass planet orbiting the M3 star Gliese 581, which has about a third the Sun's mass. Furthermore, we saw in Section 5-11 that many stars have Jupiter-mass planets that orbit very close to them. The presence of massive planets near stars would make it very hard for those stars to support habitable planets because the gravitational tugs from the massive planets would pull Earthlike planets into highly elliptical orbits. During parts of such orbits, terrestrial worlds would be overheated and overradiated by their stars, whereas during other parts of their orbits, the planets would freeze. However, not all Sunlike stars have close-orbiting massive planets.

Billions of Sunlike stars in our Galaxy remain to be studied, and most astronomers and many other people expect that we will eventually discover habitable planets around some of them. This belief helps motivate SETI.

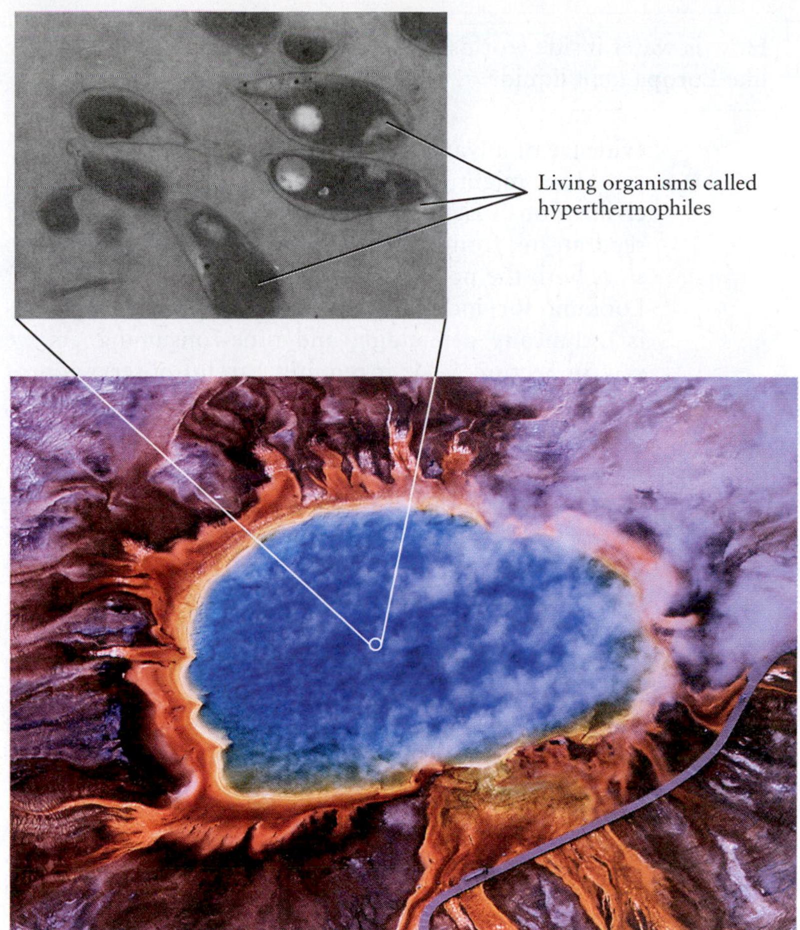

a Grand prismatic spring, Yellowstone National Park

b

R I V U X G

FIGURE 19-7 Hyperthermophiles (a) These microscopic thermophiles (heat-loving organisms) live in water that is between 80°C and 100°C (85°F–140°F). (b) Tube worms (light-green tubes) with hemoglobin-rich red plumes. They reside around black smokers—vents in the ocean bottom that are in the same temperature range as the hot springs shown in (a). These vents are over 3 km (2 mi) under water. (a: Jim Peaco; July 2001 Yellowstone National Park Image by NPS Photo; inset: NASA; b: Fisheries and Oceans of Canada)

How is water inside worlds like Europa kept liquid?

Unlike the fossil evidence for life on Mars, the effort to locate life outside of the solar system is based on searching for high-tech evidence of advanced civilizations.

3 How might we ascertain whether extraterrestrial civilizations exist, given the tremendous distances that separate us from other stars and the huge number of stars with the potential to support life-bearing planets? Looking for individual, habitable, extrasolar planets is technically demanding and time-consuming. As we saw in Section 5-11, it requires careful observation of data available only in the past 25 years. Furthermore, the discovery of a planet does not imply that it harbors life. SETI astronomers therefore need a way to bypass the daunting search for habitable planets. Starting in the 1950s and 1960s, they identified a promising approach to searching directly for extraterrestrial intelligence—they listen for radio transmissions from distant civilizations.

As we have seen, radio waves can travel immense distances without being significantly altered by the interstellar medium. Because they penetrate gas and dust, radio waves are a logical choice for interstellar communication. Even if alien civilizations are not trying to communicate with us, they probably use radio waves in their own technology. But at what frequencies should we look to maximize the odds of detecting alien signals?

We could try random frequencies in the hopes of hearing something—anything. But the number of directions in which to look and frequencies to check are both daunting. We need to find some way to improve the odds. It turns out that some radio wavelengths travel farther without being absorbed by interstellar gas and dust, or without competing with noise from other sources, than do other radio wavelengths. The SETI pioneer Bernard Oliver pointed out that a range of relatively clear frequencies exists in the neighborhood of the microwave emission lines of hydrogen and the hydroxyl radical (OH) (Figure 19-8). This region of the microwave part of the radio spectrum is called the **water hole**, a humorous reference to the H and OH lines being so close together. If extraterrestrial beings are purposefully sending messages into space (to communicate with spacecraft or other worlds), it seems reasonable that they would choose to transmit on these frequencies.

Even if only a few alien civilizations are scattered across the Galaxy, we have the technology to detect radio transmissions from them. Since Project Ozma was carried out, the number of organized searches for signals from extraterrestrials approaches 100. SETI searches use radio telescopes (Figure 19-9) along with sophisticated electronic equipment and powerful computers to conduct both a targeted search and an all-sky survey. Although NASA has created astrobiology centers to study issues regarding life in the solar system and Galaxy, most of the SETI programs in operation today are privately funded. Several programs are actively listening and analyzing enormous volumes of data.

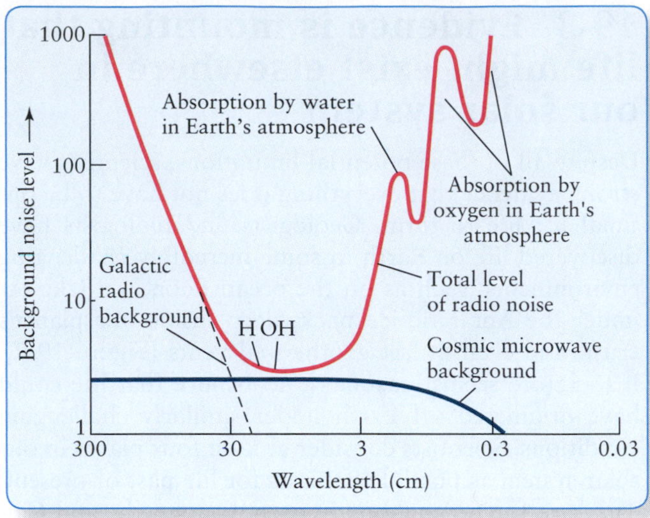

FIGURE 19-8 Water Hole The so-called water hole is a range of radio wavelengths from about 3 to 30 cm that happens to have relatively little cosmic noise. Some scientists suggest that this noise-free region would be well-suited for interstellar communication.

Indeed, so much data are available to analyze that one SETI project, called SETI@home (http://setiathome.berkeley.edu), has enlisted the help of home-computer

FIGURE 19-9 Radio Telescope Used for SETI The Arecibo Observatory's radio telescope, with a diameter of 305 m (1000 ft) is the largest single-aperture telescope in the world. It is located in Arecibo, Puerto Rico. In the past decade, it has been used in an all-sky survey, along with an antenna located in the Mojave Desert in California, to search for extraterrestrial intelligence. In 1996, an antenna in Canberra, Australia, joined the network. (Courtesy of the NAIC-Arecibo Observatory, a facility of the NSF)

users. SETI researchers provide actual data from radio telescopes to be analyzed, along with a data-analysis program. While the home computer's screensaver is on, the program runs, the data are analyzed, and the results are sent back to the researchers to be combined with data from other home-computer users. The processed data are replaced with fresh data to be analyzed. Another SETI program uses an array of small, inexpensive TV satellite antennas to look for signals from space.

4 Figure 19-10 shows the region of the Galaxy from which radio signals similar to the kinds of transmissions we humans make could be detected. Occasionally, a SETI project does detect powerful or unusual signals from space. However, because none of these signals are ever repeated, SETI researchers do not yet believe they have discovered any extraterrestrial civilization.

The detection of a message from an alien civilization would be one of the greatest events in human history. Communicating with aliens could dramatically change the course of civilization on Earth through the sharing of scientific information or an awakening of social or humanistic enlightenment. In only a few years, our industry and social structure could advance centuries into the future. The effects of such change would touch every person on Earth.

FIGURE 19-10 Region of the Search for Extraterrestrial Intelligence The red circle, centered on the solar system, shows the region of the Milky Way in which SETI searches can reasonably expect to detect radio emissions from alien civilizations. This reasoning assumes that the signals are similar in nature to the kinds of radio emissions we generate on Earth. (NASA/Space Telescope Science Institute)

19-5 The Drake equation: How many civilizations are likely to exist in the Milky Way?

Just how many planets throughout our Galaxy are likely to harbor complex life? The first person to tackle this question quantitatively was Frank Drake. He proposed that the number of technologically advanced civilizations in the Galaxy (designated by the letter N) could be estimated by what is now called the **Drake equation:**

$$N = R^* f_p n_e f_l f_i f_c L$$

in which

R^* = rate at which Sunlike stars form in the Galaxy

f_p = fraction of Sunlike stars that have planets

n_e = number of planets per solar-type star system suitable for life

f_l = fraction of those habitable planets on which life actually arises

f_i = fraction of those life-forms that evolve into intelligent species

f_c = fraction of those species that develop adequate technology and then choose to send messages out into space

L = lifetime of that technologically advanced civilization

> If we detect incidental, rather than directed, signals from alien civilizations, what information might they contain?

The Drake equation expresses quantitatively the number of extraterrestrial civilizations as a product of terms, some of which can be estimated from what we know about stars and stellar evolution. For instance, thanks to new evidence for extrasolar planets, astronomers can now hope to determine the first two terms, R^* and f_p, by observation. We should probably exclude stars larger than about 1.5 $M_\odot$, because they have main-sequence lifetimes shorter than the time it took for intelligent life to develop on Earth—some 3.8 to 4.0 billion years. If that period is typical of the time needed to evolve higher life-forms, then a massive star would become a giant or even explode as a supernova before self-aware creatures could evolve on any of its planets.

Although low-mass stars have much longer lifetimes, they are less well-suited than stars in the range of 1 $M_\odot$ for supporting life on their planets because they are so cool. As noted earlier, only planets very near a low-mass star would be sufficiently warm for water to be a liquid. But, as also mentioned earlier, a planet that close would become tidally coupled to its star, developing synchronous rotation. One side would have continuous daylight, leading to the evaporation of oceans, while the other side would be in perpetual, frigid darkness. The only place that life could survive on such planets is in the narrow ring at the boundary between day and night. Such a small fraction of the land available greatly reduces the likelihood that life would be able to evolve to the complexity necessary for technological civilizations to develop.

As "ideal" life-supporting stars, this leaves main-sequence stars with masses near those of the Sun. These stars are of spectral types between F5 and M0. Based on statistical studies of star formation in the Milky Way, astronomers calculate that roughly one of these Sunlike stars forms in the Galaxy each year, yielding a value of $R^* = 1$ per year.

We learned in Sections 5-2 through 5-7 that the planets in our solar system formed in conjunction with the birth of the Sun. We have also seen evidence that similar processes of planetary formation may be commonplace around isolated stars. Many astronomers, therefore, assign f_p a value of 1, meaning they believe it is likely that most Sunlike stars have planets.

Unfortunately, the rest of the factors in the Drake equation are very hypothetical. The chances that a planetary system has an Earthlike world are not known. Were we to consider our own solar system as representative, we could put n_e at 1. Let us be more conservative, however, and suppose that 1 in 10 solar-type stars is orbited by a habitable planet, making $n_e = 0.1$.

From what we know about the evolution of life on Earth, we might assume that, given appropriate conditions, the development of life is a certainty, which would make $f_l = 1$. This value is, of course, an area of intense interest to biologists. For the sake of argument, we might also assume that evolution naturally leads to the development of intelligence (a conjecture that is hotly debated) and also make $f_i = 1$. It is anyone's guess as to whether these intelligent extraterrestrial beings would attempt communication with other civilizations in the Galaxy, but, if we assume they all would, f_c would also be put at 1.

The last variable, L, the longevity of technological civilization, is the most uncertain of all—it cannot be tested! Looking at our own example, we see a planet whose atmosphere and oceans are increasingly polluted, potentially destroying the food chain. When we add in how close we have come to destroying ourselves with weapons of mass destruction, it may be that we humans are among the lucky few technological civilizations to squeak through its first years. In other words, L may be as short as 100 years. Putting all of these numbers together, we arrive at

$$N = 1 \times 1 \times 0.1 \times 1 \times 1 \times 1 \times 100 = 10$$

With what two numbers used in the Drake equation calculation presented previously do you disagree most? Using your numbers, how many civilizations do you estimate exist?

Therefore, out of the hundreds of billions of stars in the Galaxy, there may be only 10 civilizations technologically advanced enough to communicate with us. Of course, the numbers used here are just estimates. A wide range of values has been proposed for the terms in the Drake equation, and these various numbers produce vastly different estimates of N, the total number. Increasing the average lifetime of advanced civilizations significantly increases N. Some scientists argue that there is exactly one advanced civilization in the Galaxy and that we are it. Others speculate that there may be tens of millions of planets inhabited by intelligent creatures. Although we do not know yet, science enables us to home in on the number.

19-6 Humans have been sending signals into space for more than a century

We have been doing more than just listening passively for 5
voices from the cosmos. Astronomers have intentionally broadcast well-focused signals through radio telescopes toward star systems likely to harbor advanced life, including a 1974 "hello" message aimed at the globular cluster M13 (Figure 19-11a) and a 2008 broadcast of the Beatles' song "Across the Universe." Unintentional broadcasts into space have occurred for longer than that. Since 1895, when the first radio transmission was made by the Italian engineer Guglielmo Marconi, we have been sending messages about ourselves into space.

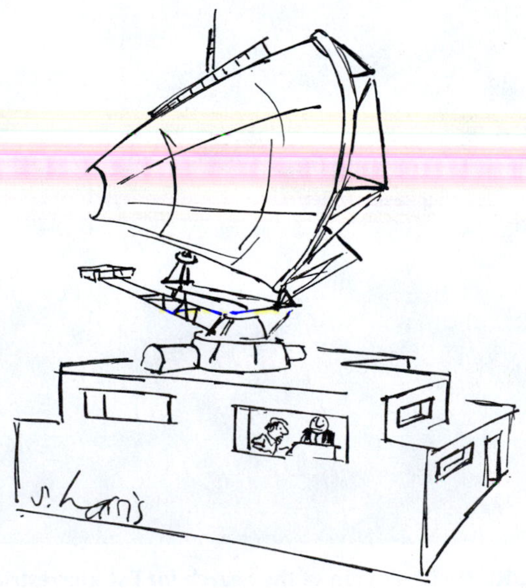

"AS I READ IT, WE'RE RECEIVING A MESSAGE FROM OUTER SPACE TELLING US TO STOP BOMBARDING THEM WITH UNINTELLIGIBLE MESSAGES."

© Sidney Harris

a

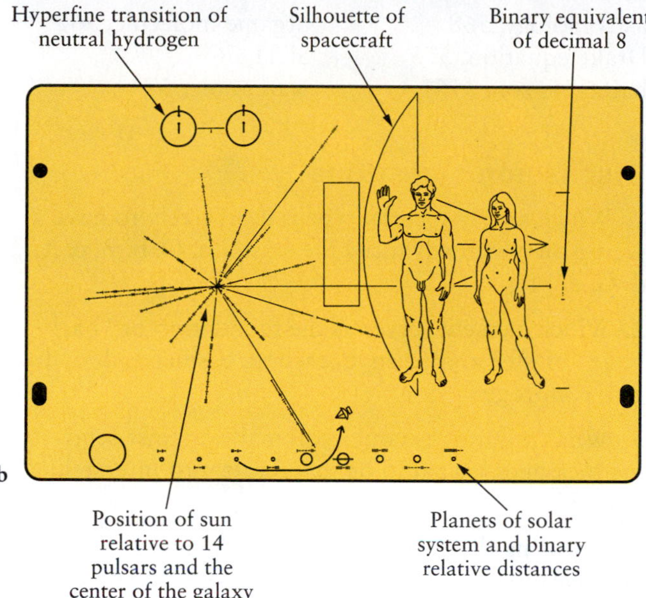

b

c

R I V U X G

FIGURE 19-11 Memorabilia in Space (a) Humans have beamed radio signals into space, hoping that the message will someday be intercepted by an alien civilization. This is a visual version of the signal sent in 1974 from the Arecibo radio telescope toward the globular cluster M13. The *Pioneer* and *Voyager* spacecraft, now in interstellar space, also carry messages from Earth. (b) The plaques on *Pioneer 10* and *Pioneer 11* provide information about where humans are, what we look like, and some of the science we know. (c) Images and sounds sent on *Voyager 1* and *Voyager 2* were stored on phonographic records, long before DVDs were even a twinkle in an engineer's eye. There are also instructions for playing the record, which contains information about our biology, our technology, and our knowledge base. Each record also contains the sounds of children's voices. It is remotely possible that another race might someday discover the spacecraft. (a: Frank Drake/UCSC, et al., Arecibo Observatory/Cornell/NAIC: b and c: NASA)

The spherical region within some 115 ly of the solar system is now filled with signals from radio and television shows. If other advanced civilizations fall within that sphere, they could very well be listening to radio programs or watching television shows from decades past, trying to make sense of our species.

Four spacecraft also carry information about the human race. Launched in 1972 and 1973, respectively, the *Pioneer 10* and *Pioneer 11* spacecraft carry plaques about us (Figure 19-11b). They are now making their way into interstellar space. In 1977, the two *Voyager* spacecraft were sent toward the outer solar system. They carry old-fashioned phonograph records encoded with human images and voices. Because space is so vast and the spacecraft are so small, the likelihood of their being discovered is truly remote. And yet millions, perhaps billions, of years from now, one of the spacecraft just might reach another race of intelligent creatures. From its path and the information on board, those creatures might be able to determine where the *Pioneer* and *Voyager* spacecraft came from. If their travel budgets are sufficiently large, they might even decide to come and visit us. At the least, they could send us radio messages. Will our descendants be here to receive them?

What negative consequence could arise from an alien civilization discovering a *Pioneer* spacecraft?

19-7 Frontiers yet to be discovered

We still have much to learn about astrobiology. Myriad steps in the formation of life on Earth remain to be understood. Although the evidence for liquid water once having existed on Mars is overwhelming, we still need to confirm whether life ever existed there. Does life exist on Mars, Europa, Ganymede, or Callisto today? Astronomers are still working to refine their ideas of how organic compounds formed in space. A paramount frontier is detecting life elsewhere in the universe.

SUMMARY OF KEY IDEAS

- The chemical building blocks of life exist throughout the Milky Way Galaxy.
- Organic molecules and water have been discovered in interstellar clouds, in some meteorites, in comets, and in newly forming star and planet systems.
- Astronomers are using radio telescopes to search for signals from other self-aware life in the Galaxy. This effort is called the search for extraterrestrial intelligence, or SETI. SETI is primarily done at frequencies where radio waves pass most easily through the interstellar medium. So far, these searches have not detected any life outside Earth.
- The Drake equation is used to estimate the number of technologically advanced civilizations in the Galaxy whose radio transmissions we might discover. Estimates of this number vary from one to millions.
- Everyday radio and television transmissions from Earth, along with intentional broadcasts into space, may be detected by other life-forms.

A WHAT DID YOU THINK?

1. *Why is water so important to the formation of life?* Water allows many interactions between atoms and molecules dissolved in it.

2. *What element is uniquely suited to be the foundation of life as we know it, and why?* Carbon is the only atom that can bond flexibly (but not too flexibly) with three or more other atoms.

3. *How do astronomers search for extraterrestrial intelligence?* They search for radio signals from other advanced civilizations.

4. *Have astronomers located any extraterrestrial civilizations?* No. Extraterrestrial civilizations have not yet been discovered.

5. *If advanced alien civilizations exist, is there any way they might know of our existence?* Yes. They might detect our intentional radio broadcasts into space, our everyday radio and television broadcasts, or, possibly, one of our spacecraft traveling in interstellar space.

Key Terms for Review

astrobiology, 568
Drake equation, 575
habitable zone, 571
organic molecule, 569
SETI, 569
water hole, 574

Discussion Questions

1. Which stellar spectral type is most likely to have a planet on which advanced life exists? **a.** O, **b.** B, **c.** A, **d.** G, **e.** M

2. Which element serves as the foundation or "backbone" of life? **a.** oxygen, **b.** carbon, **c.** silicon, **d.** hydrogen, **e.** nitrogen

3. What arguments could you make against sending messages via radio or spacecraft into interstellar space?

4. Try using the Drake equation with values that you find reasonable. How many civilizations do you estimate there are in our Galaxy?

5. Discuss the possible biological effects of our probes that visit other life-sustaining worlds.

6. What social effects might probes from other worlds have on us?

7. List some of the pros and cons in the argument that alien spacecraft have visited Earth.

What If ...

8. Our planet were the moon of a much larger, Jupiter-like planet that orbits the Sun at 1 AU? What might Earth and life inhabiting it be like?

9. Our Moon were Earth-sized and also had life on it?

10. We discover living organisms in Europa's subsurface oceans? What precautions should we take in the search for such life, and why?

11. We develop life in a Miller-Urey–like experiment? Explore some of the implications that this event would have.

Web Questions

12. What information would you have put on the record sent out on the *Voyager* spacecraft? Compare your answer to the results listed in the Web link associated with that record.

13. To test your understanding of the Miller-Urey experiment, do Interactive Exercise 19.1 on the Web. You can print out your answers, if requested.

14. To test your understanding of the Drake equation, do Interactive Exercise 19.2 on the Web. You can print out your answers, if requested.

On other worlds, plants could be red, blue, even black

Adapted from an article by Nancy Y. Kiang

Beyond the solar system, astronomers have discovered more than 520 extrasolar planets orbiting other stars. The world's space agencies are developing telescopes that will search for signs of life on Earth-sized planets by observing their light spectra.

Scientists are looking for physical and chemical signs of fundamental life processes: "biosignatures." Photosynthesis could produce very conspicuous biosignatures, as it does on Earth. Our first firm fossil evidence for photosynthesis dates to about 3.4 billion years ago.

Photosynthesis is adapted to the spectrum of light that reaches organisms. This spectrum is determined by the parent star's radiation spectrum. Light of any color, from violet through near-infra-

Filtering Starlight

The color of plants depends on the spectrum of the star's light, which astronomers can easily observe, and filtering of light by air and water, which the author and her colleagues have simulated based on the likely atmospheric composition and life's own effects.

STAR TYPE: M (mature)
MASS*: 0.2
LUMINOSITY*: 0.0044
LIFETIME: 500 billion years
ORBIT OF MODELED PLANET: 0.07 AU
*Relative to Sun

STAR TYPE: M (young)
MASS*: 0.5
LUMINOSITY*: 0.023
LIFETIME: Flaring: 1 billion years
Total: 200 billion years
ORBIT OF MODELED PLANET: 0.16 AU

Photosynthetic pigments absorb different ranges of wavelengths. All land plants on earth rely on chlorophyll a and b and a mixture of carotenoid pigments. Algae and cyanobacteria use different pigments called phycoerythrin and phycocyanin.

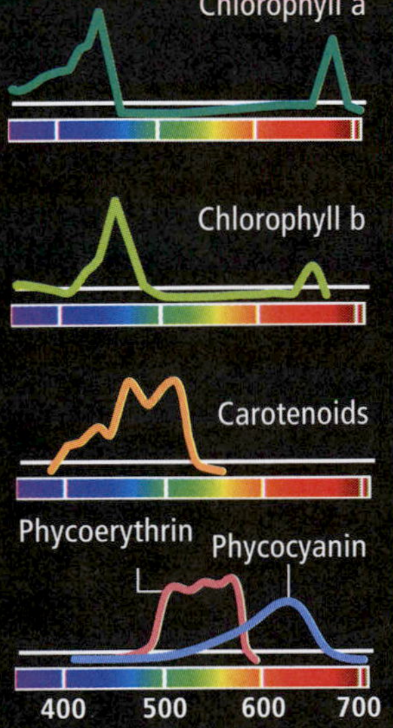

STARLIGHT
Before entering the atmosphere, starlight has a distinctive spectrum. The overall shape is determined by the surface temperature of the star, with a few dips produced by absorption in the star's own atmosphere.

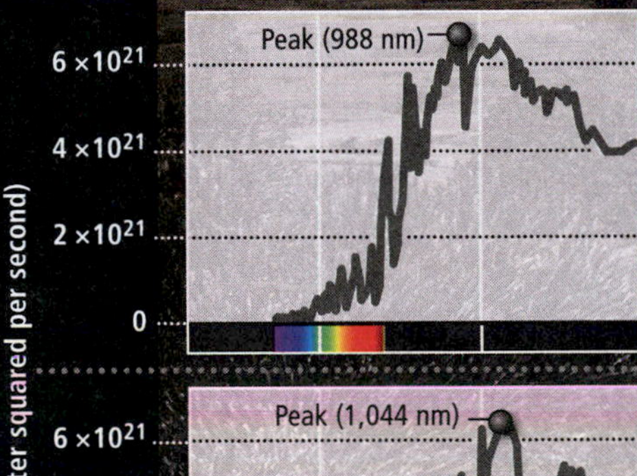

SURFACE
Atmospheric gases absorb the starlight unevenly, shifting its peak color and introducing absorption bands—wavelengths that are screened out. These bands are best known for Earth (the G-star case).

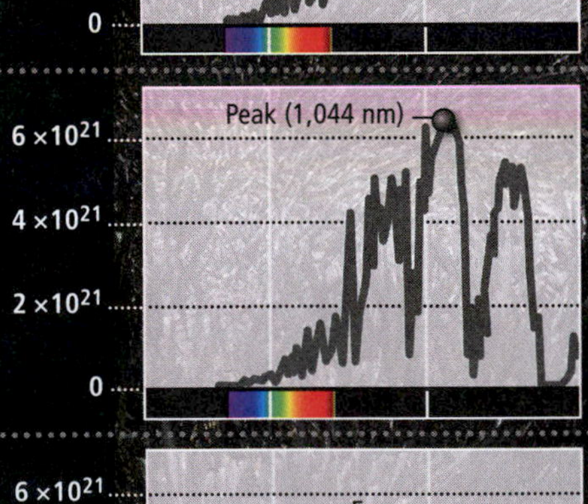

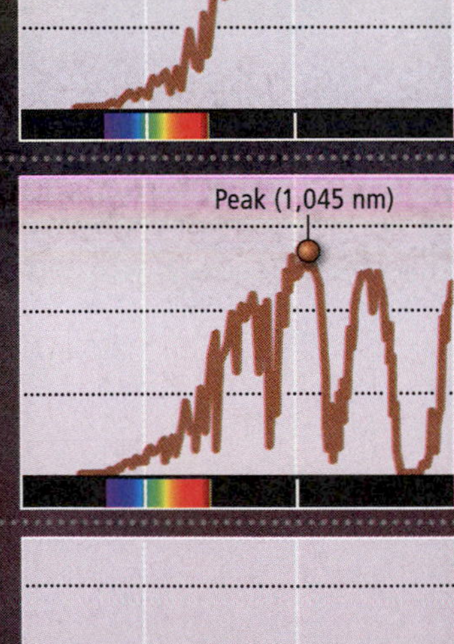

UNDERWATER
Water tends to transmit blue light and absorb red and infrared light. The graphs shown here are for water depths of 5 and 60 cm. (The mature M-star case is for a low-oxygen atmosphere.)

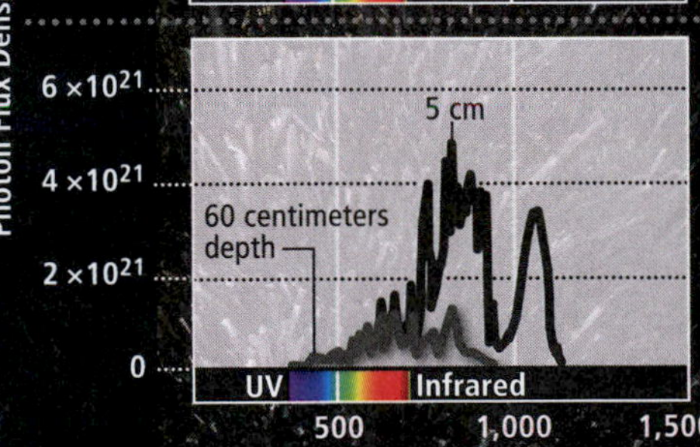

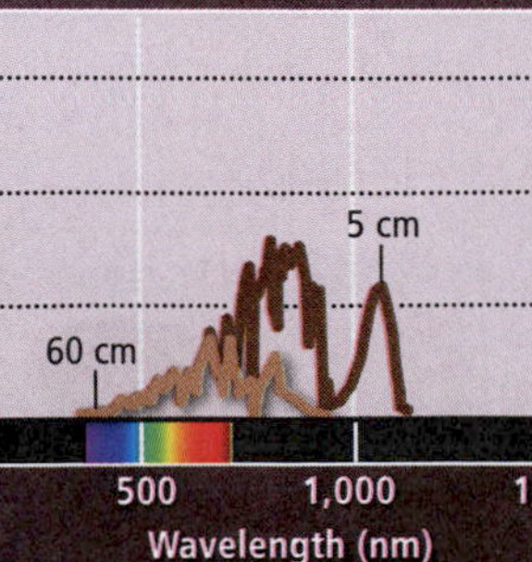

red, could power photosynthesis. That spectrum would also determine plant color—yellow, green, red, even black. Around stars hotter and bluer than our Sun, plants would tend to absorb blue light and could look green to yellow to red. Around cooler stars such as red dwarfs, planets receive less visible light, so plants might try to absorb as much of it as possible, making them look black.

Photosynthetic pigments in other solar systems must satisfy the same rules as on Earth: Pigments tend to absorb photons that are either the most abundant, the shortest available wavelength (most energetic), or the longest available wavelength (where the reaction center absorbs).

Finding life on other planets—abundant life, not just fossils or microbes eking out a meager living under extreme conditions—is a fast-approaching reality. Our understanding of photosynthesis will be key to designing research missions and interpreting their data.

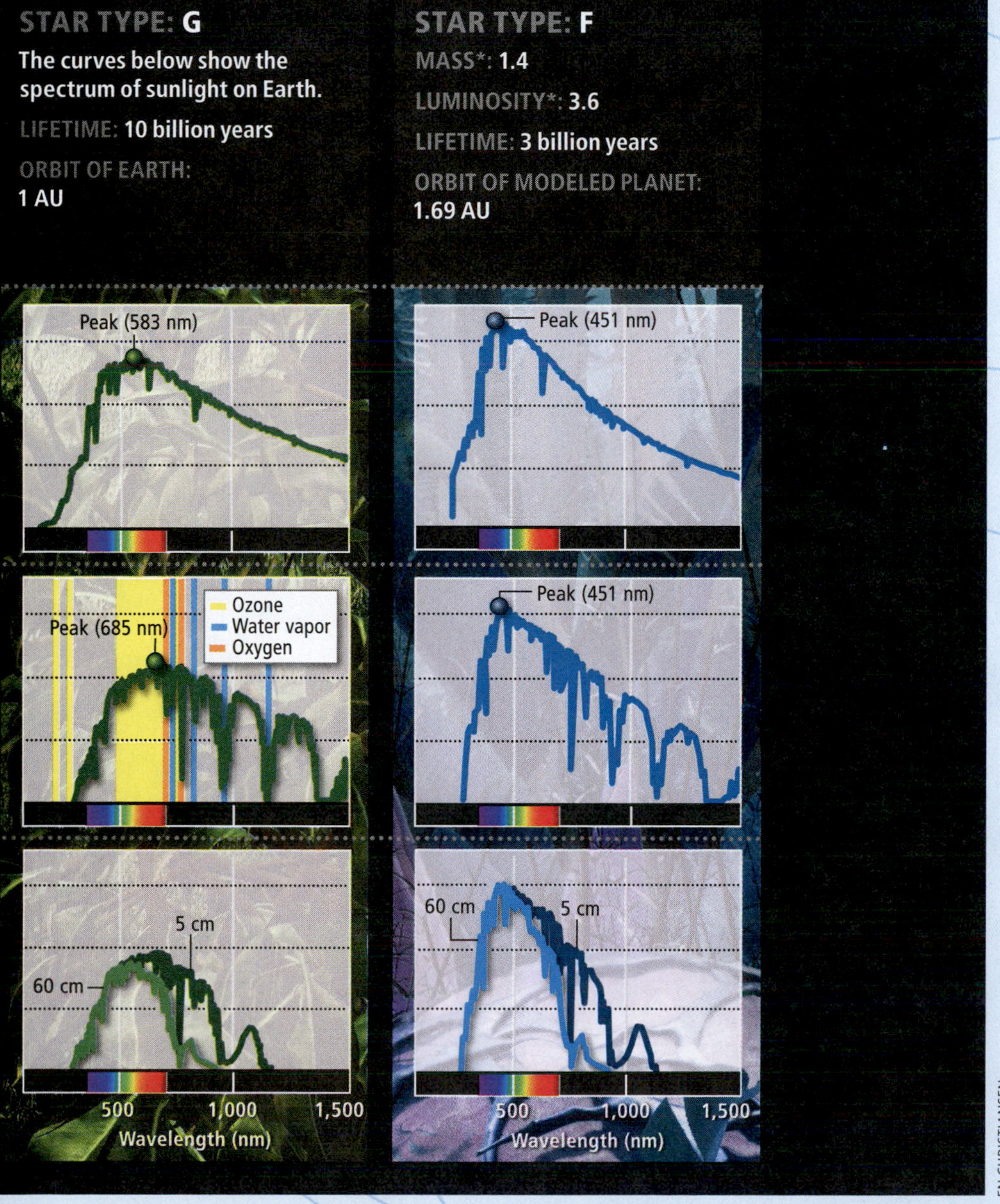

THE COLOR OF PLANTS ON OTHER WORLDS (continued)

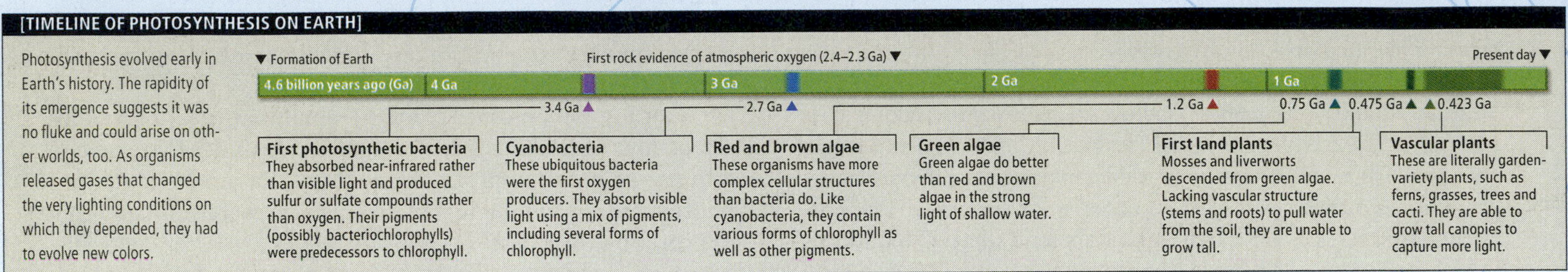

DETAIL OF HYPOTHETYCAL F-STAR FOLIAGE

KENN BROWN AND CHRIS WREN *Mondolithic Studios*

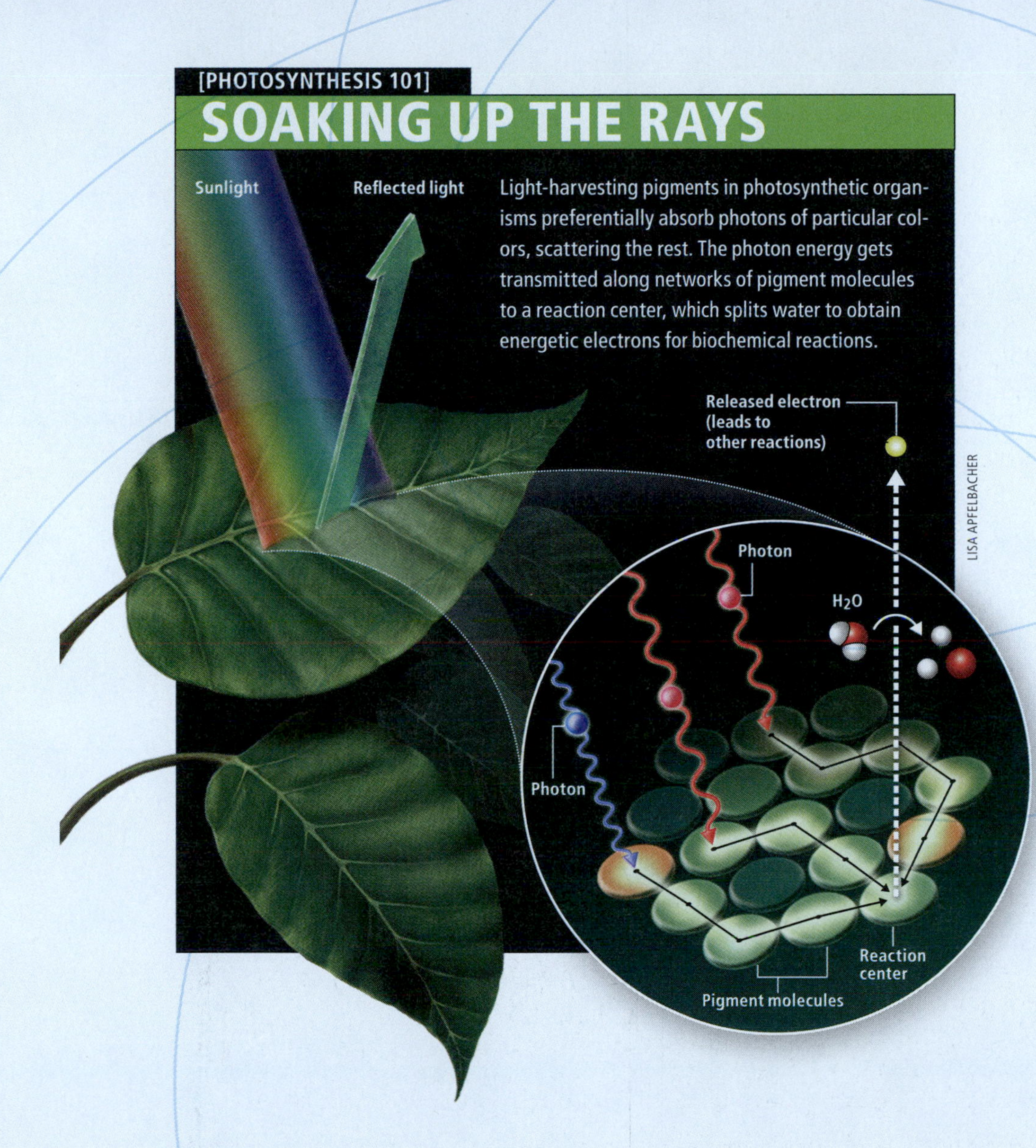
[PHOTOSYNTHESIS 101]
SOAKING UP THE RAYS
Sunlight
Reflected light
Light-harvesting pigments in photosynthetic organisms preferentially absorb photons of particular colors, scattering the rest. The photon energy gets transmitted along networks of pigment molecules to a reaction center, which splits water to obtain energetic electrons for biochemical reactions.
Released electron (leads to other reactions)
Photon
H_2O
Photon
Reaction center
Pigment molecules
LISA APFELBACHER

Appendix A: Powers-of-Ten Notation

Astronomy is a science of extremes. As we examine various cosmic environments, we find an astonishing range of conditions—from the incredibly hot, dense centers of stars to the frigid, near-perfect vacuum of interstellar space. To describe such divergent conditions accurately, we need a wide range of both large and small numbers. Astronomers avoid such confusing terms as "a million billion billion" (1,000,000,000,000,000,000,000,000) by using a standard shorthand system. All the cumbersome zeros that accompany such a large number are consolidated into one term consisting of 10 followed by an *exponent*, which is written as a superscript and called the **power of ten**. The exponent merely indicates how many zeros you would need to write out the long form of the number. Thus,

$$10^0 = 1$$

$$10^1 = 10$$

$$10^2 = 100$$

$$10^3 = 1000$$

$$10^4 = 10{,}000$$

and so forth. Equivalently, the exponent tells you how many tens must be multiplied together to yield the desired number. For example, ten thousand can be written as 10^4 ("ten to the fourth") because $10^4 = 10 \times 10 \times 10 \times 10 = 10{,}000$. Similarly, 273,000,000 can be written as 2.73×10^8.

In scientific notation, numbers are written as a figure between 1 and 10 multiplied by the appropriate power of 10. The distance between Earth and the Sun, for example, can be written as 1.5×10^8 km. Once you get used to it, you will find this notation more convenient than writing "150,000,000 kilometers" or "one hundred and fifty million kilometers."

This powers-of-ten system can also be applied to numbers that are less than 1 by using a minus sign in front of the exponent. A negative exponent tells you that the location of the decimal point is as follows:

$$10^0 = 1.0$$

$$10^{-1} = 0.1$$

$$10^{-2} = 0.01$$

$$10^{-3} = 0.001$$

$$10^{-4} = 0.0001$$

and so forth. For example, the diameter of a hydrogen atom approximately is 1.1×10^{-8} cm. That is more convenient than saying "0.000000011 centimeter" or "11 billionths of a centimeter." Similarly, 0.000728 equals 7.28×10^{-4}.

Using the powers-of-ten shorthand, one can write large or small numbers like these compactly:

$$3{,}416{,}000 = 3.416 \times 10^6$$

$$0.000000807 = 8.07 \times 10^{-7}$$

Because powers-of-ten notation bypasses all the cumbersome zeros, a wide range of circumstances can be numerically described conveniently:

$$\text{one thousand} = 10^3$$

$$\text{one million} = 10^6$$

$$\text{one billion} = 10^9$$

$$\text{one trillion} = 10^{12}$$

and also

$$\text{one thousandth} = 10^{-3} = 0.001$$

$$\text{one millionth} = 10^{-6} = 0.000001$$

$$\text{one billionth} = 10^{-9} = 0.000000001$$

$$\text{one trillionth} = 10^{-12} = 0.000000000001$$

Try these questions: Write 3,141,000,000 and 0.0000000031831 in scientific notation. Write 2.718282×10^{10} and 3.67879×10^{-11} in standard notation.
(Answers appear at the end of the book.)

Appendix B: Guidelines for Solving Math Problems and Reading Graphs

Astronomy relies on mathematics. Although we have distilled the concepts into words, you can gain further insights into the distances, intensities, and other astronomical quantities from the equations presented in this book. In addition, much scientific information is presented in the form of graphs, a technique that presents data very compactly and often provides a good way to see the trends that the data reveal. This appendix thus provides guidance on three things: setting up problems to be solved analytically (using equations), solving algebra equations, and reading graphs.

Setting Up and Solving Analytical Problems

At the end of most chapters, there are questions, indicated by an asterisk (*), that require mathematical solutions. You will need to translate the question into an equation or two in order to solve for some value. This is best done systematically. In what follows, we will work a question, appropriate to Chapter 2: How fast does a meteoroid (small piece of rocky space debris) of mass 5 kg (kilograms) move if it has a kinetic energy of 10 J ($1\ \text{J} = 1\ \text{kg} \times \text{m}^2/\text{s}^2$)? This approach should work for most of the questions in the book.

Step 1. *Write down all the information you are given in terms of the variable names used in the book.* For example, denoting mass by m and kinetic energy by KE, write: mass $m = 5$ kg, kinetic energy $KE = 10$ J.

Step 2. *Identify and write down the thing you are trying to find in terms of its variable name.* In this case: speed, $v = ?$.

Step 3. *Find (usually) one or (sometimes) two equations that are needed to solve for the unknown.* All the equations presented in the book are listed in Appendix C. For our problem, use the kinetic energy equation $KE = \frac{1}{2}mv^2$, because this equation has the unknown, v, along with only variables and constants that are given, namely KE and m. (As in this case, you will often need only one equation, but sometimes you will have to solve first one equation and then another to get the answer. As an example, you might be given the information above but asked for the value of a momentum, p, of the meteoroid. The equation for momentum, also in Chapter 2, is $p = mv$. First you would solve for the speed, as we are now doing, and then use that v in this last equation.)

Step 4. *Manipulate the equation(s) until you have the unknown variable on one side and all the other variables and constants on the other side.* Variables can be added, subtracted, multiplied, and divided in the same way that numbers are combined. For our example, we want to find v, so we start with $KE = \frac{1}{2}mv^2$, multiply by 2, and divide by m. This gives $v^2 = 2\ KE/m$. Because we want v, we now take the square root of both sides or $v = \sqrt{2KE/m}$.

Step 5. *Make sure that all the units match up.* Units on both sides of the equation must be identical. Period. If you have one value in kilometers (km) and another value in meters (m), then the result you will get by combining numbers is not meaningful. If you end up with inconsistent units on opposite sides of an equation, convert one unit to the other. For example, change kilometers into meters or vice versa. In this case, use the fact that $1\ \text{km} = 10^3\ \text{m}$.

Step 6: *Plug in all the numbers and solve for the unknown.* In our case, $v = \sqrt{2(10\ \text{J}/5\ \text{kg})}$ or $v = 2$ m/s.

Reading Graphs

The graphs you will encounter in this book are compact ways of displaying patterns of information relating two variables, like the temperature and luminosity of stars or the temperature of an atmosphere at different altitudes. The relevant values of one of the variables are presented along the horizontal or x axis and the relevant values of the other variable are presented along the vertical or y axis. The word "relevant" here indicates that often graphs do not start at 0. Consider three examples. First is data presented in Chapter 7 concerning the temperature of Venus's atmosphere at different altitudes (**Figure 1**).

The horizontal axis of the graph indicates the temperature in kelvins (K). (Unlike degrees Celsius or degrees Fahrenheit, temperature using the Kelvin scale is simply noted as kelvins.)

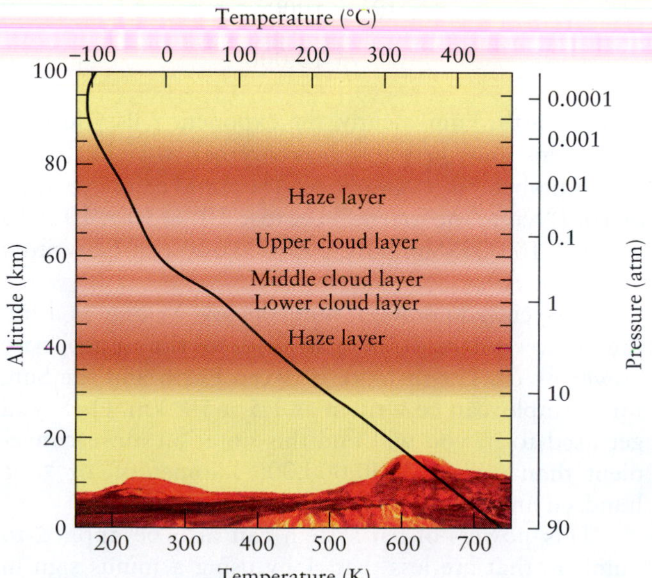

Figure 1 Linear Graph

The vertical axis denotes the altitude above Venus's surface in kilometers (km). Note that both axes are always labeled with a name (temperature or altitude, here) and units (K or km, respectively). Bear in mind that some variables in graphs you will see in this book increase to the right or upward (these are more common), but some variables will increase in the opposite directions.

To read a graph, note that a value on the horizontal axis is transferred directly upward through the graph. For example, all points on the blue line in Figure 1 (which extends upward from 400 K) have a temperature of 400 K. Equivalently, the value given on the vertical axis is transferred to all points horizontally across from this value. All the points on the black line in Figure 1 are at an altitude of about 43 km above Venus's surface. We have interpolated between 40 and 45 to get this value (**Figure 2**).

A curve or a set of points on the graph presents the *relationship* between the variable represented on the horizontal axis and the variable on the vertical axis. Each point on a curve or each separate point relates the two variables. Choose a point, say the dot on Figure 1. It represents the temperature at a certain altitude above Venus's surface. To find the temperature at that point, you slide directly down (along the blue line in this example) from the point and read the value of the horizontal variable under it. To find the altitude for that point, you slide directly over to the side (along the black line) and read the value of the vertical variable there. In our example, sliding along the blue line leads to 400 K on the temperature line. Therefore, this point corresponds to a temperature of 400 K. Moving horizontally from the point, you encounter, by interpolation, the 43 km indicator. Combining the data, you conclude that *the temperature of Venus's atmosphere 43 km above its surface is 400 K.*

The red curve in Figure 1 provides the relationship between altitude and temperature for a wide range of locations above Venus in a representation that is much more informative than a table of heights and temperatures. Specifically, this curve shows you the trend of temperature with altitude.

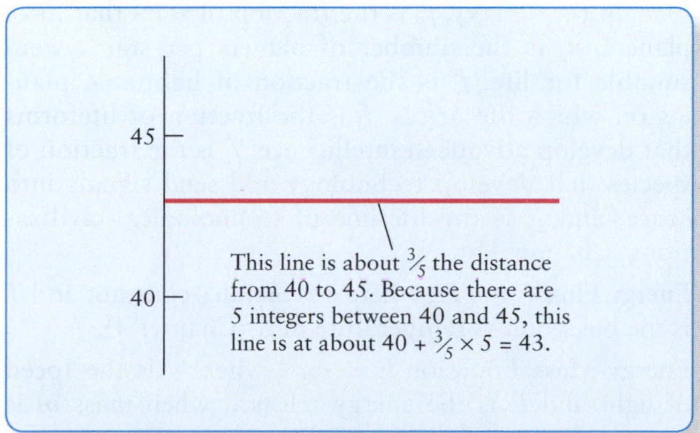

Figure 2 Interpolation

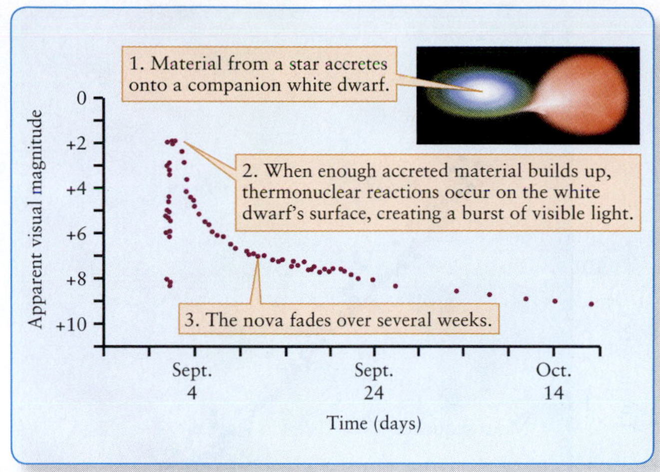

Figure 3 The Brightness of a Nova

Try these questions: What is the temperature at 20 km? What is the altitude at which the temperature is 300 K? For most of this graph, what is the general *trend* of the temperature with height? (Answers appear at the end of the book.)

Sometimes, the known information is not a curve, but rather a set of points, as in our second example (**Figure 3**), from Chapter 13. In this case, the graph connects the apparent magnitude of a nova (how bright it appears to be as seen from Earth regardless of its distance or other factors) and time. Each dot indicates how bright the nova (an explosion on the surface of certain stars) was at different times. For example, the peak brightness of the nova was an apparent magnitude of about +2 and it occurred on September 2. Noting that time passes to the right, you can immediately see that the trend of the nova's brightness is to increase rapidly and decrease more slowly.

Try these questions: What is the apparent magnitude on September 24? October 9? On what two days was the apparent magnitude +6? (Answers appear at the end of the book.)

The graphs so far have shown variables that change uniformly along the axes. For example, the distance on Figure 1 from 300 K to 400 K is the same as the distance from 400 K to 500 K, and so on. Many graphs you will encounter have variables that do not change uniformly (that is, linearly) along the axes. That means that the change in value going along each axis varies—there is not the same amount of change per centimeter along the axis. In **Figure 4**, from Chapter 12, for example , the luminosity (total energy emitted per second) increases upward logarithmically, while the temperature decreases going from left to right in a more complex, nonuniform way.

The purpose of logarithmic and other nonlinear axes is to present in compact form data that varies very widely in values. For example, the dimmest star represented in Figure 4 is just less than 0.1 times as luminous

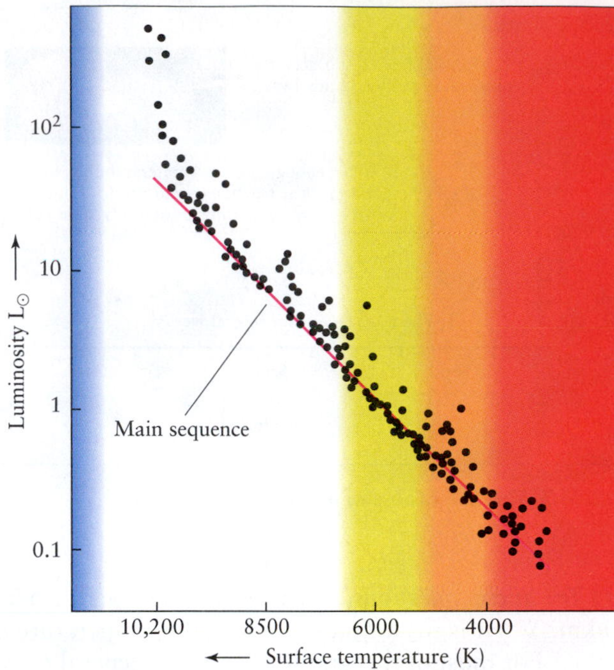

Figure 4 Logarithmic Graph

as the Sun, while the brightest star is nearly 1000 times as luminous. The process of getting information from logarithmic graphs is the same as linear graphs. You must just be careful not to think of values as doubling or tripling when you go over or up two or three intervals. Figure 5 shows how a logarithmic scale varies over one decade of values. The same numbering intervals apply for any decade of values, for example, 1 to 10 or 10^5 to 10^6. As you can see, the numbers bunch up near the highest value, so you need to interpolate these graphs more carefully than linear graphs.

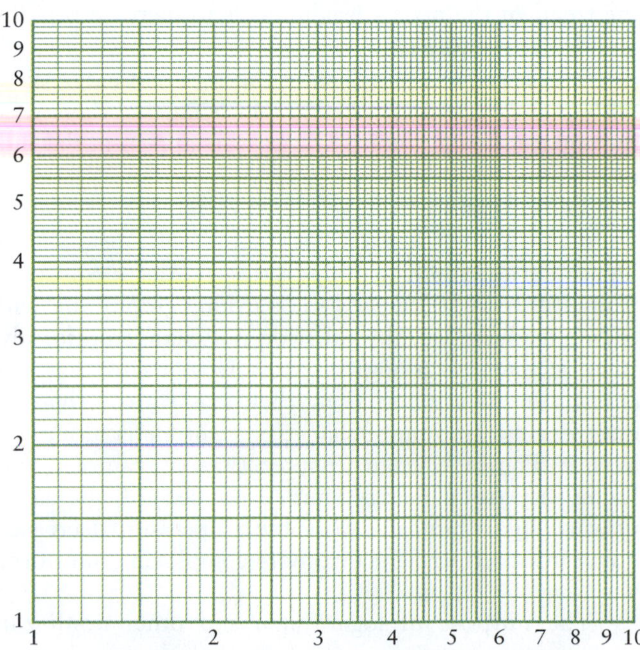

Figure 5 Logarithmic Scale

Referring to Figure 4, the luminosity of a star with surface temperature of 4000 K is about 0.1 $L_\odot$. Note, also, that some graphs are linear on one axis and logarithmic on the other axis.

Try these questions: Approximately what is the luminosity of a star with surface temperature 8500 K? What is the surface temperature of a star with the same luminosity as the Sun (1 $L_\odot$)?
(Answers appear at the end of the book.)

Appendix C: Key Formulas

Angular Momentum $L = I\omega$, where L is the angular momentum, I is the object's moment of inertia, and ω is the object's angular velocity (Chapter 2).

Area of a Circle $A = \pi d^2/4$, where π is approximately 3.14, and d is the diameter of the circle (Chapter 3).

Average Density $\rho = m/V$, where ρ is the average density, m is the total mass, and V is the volume of the object (Chapter 5).

Distance from Parallax $d = 1/p$ where d is the distance in parsecs and p is the parallax angle in arcseconds, or $d_{ly} = 3.26/p$, where d_{ly} is the distance in light-years (Chapter 11).

Distance-Magnitude Relationship $M = m - 5 \log(d/10)$, where M is the star's absolute magnitude, m is its apparent magnitude, and d is its distance in parsecs (Chapter 11).

Doppler Shift $\Delta\lambda/\lambda_o = v/c$, where $\Delta\lambda$ is the change in wavelength $(\lambda - \lambda_0)$, λ is the observed wavelength, λ_0 is the wavelength the object emits as seen by someone not moving toward or away from it, v is the speed of the object toward or away from the observer, and c is the speed of light (Chapter 4).

Drake Equation $N = R^* f_p n_e f_l f_i f_c L$, where N is the number of advanced civilizations in our Galaxy estimated by the equation, R^* is the rate at which solar-type stars form in the Galaxy, f_p is the fraction of stars that have planets, n_e is the number of planets per star system suitable for life, f_l is the fraction of habitable planets on which life arises, f_i is the fraction of lifeforms that develop advanced intelligence, f_c is the fraction of species that develop technology and send signals into space, and L is the lifetime of technological civilizations (Chapter 19).

Energy Flux $F = \sigma T^4$, where σ is Stefan's constant and T is the blackbody's temperature in K (Chapter 4).

Energy-Mass Equation $E = mc^2$, where c is the speed of light and E is the energy released when mass m is converted to energy (Chapter 10).

Gravitational Force $F = G(m_1m_2/r^2)$, where G is the gravitational constant, m_1 and m_2 are the masses of the two interacting objects, and r is the distance between them (Chapter 2).

Gravitational Potential Energy (always true) $PE = GmM/r$, where PE is the potential energy, G is the gravitational constant, m is the mass of the object whose gravitational potential energy you are measuring, M is the mass of the object creating the gravitational potential, and r is the distance between the centers of the two (Chapter 2).

Gravitational Potential Energy (near Earth) $PE = mgh$, where m is the mass of the object whose gravitational potential energy you are measuring, $g = 9.8$ m/s^2, and h is the height of the object above Earth's surface (Chapter 2).

Kepler's Third Law $P^2 = a^3$, where P is the sidereal period in Earth years and a is its semimajor axis in astronomical units (Chapter 2).

Kinetic Energy $KE = \frac{1}{2}\,mv^2 = p^2/2m = L^2/2I$, where KE is its kinetic energy, m is its mass, v is its speed, p is the momentum, L is the angular momentum, and I is the object's moment of inertia. The first two equalities apply if the object is moving in a straight line. The last equality applies if the object is revolving or rotating (Chapter 2).

Luminosity $L = F \cdot 4\,\pi r^2$, where L is the luminosity (total energy per second) emitted, F is the energy flux, and r is the radius of the object (Chapter 4).

Magnification $M = f_o/f_e$, where f_o is the focal length of the primary mirror or objective lens and f_e is the focal length of the eyepiece (Chapter 3).

Momentum $\mathbf{p} = m\mathbf{v}$, where $\mathbf{p}$ is the momentum, m is the mass, and $\mathbf{v}$ is the velocity (Chapter 2).

Newton's Force Law $\mathbf{F} = m\mathbf{a}$, where F is the force acting on an object, m is its mass, and a is its acceleration ($\mathbf{F}$ and $\mathbf{a}$ are in boldface to denote that they act in some direction or other) (Chapter 2).

Newton's Version of Kepler's Third Law for Binary Star Systems $M_1 + M_2 = a^3/P^2$, where M_1 and M_2 are the masses of the stars, a is the average distance between them, and P is the period of their orbit in years (Chapter 11).

Photon Energy $E = hc/\lambda$, where E is the photon's energy, h is Planck's constant, c is the speed of light, and λ is the wavelength of the light (Chapter 3).

Pressure $P = F/A$, where P is the pressure, F is the force, and A is the area over which the force acts (Chapter 6).

Recessional Velocity of Galaxies $v = H_0 \times d$, where v is the recessional velocity in kilometers per second, H_0 is the Hubble constant, and d is the distance to the galaxy in megaparsecs (Chapter 16).

Schwarzschild Radius of a Black Hole $R_{Sch} = 2\ GM/c^2$, where R_{Sch} is the Schwarzschild radius in meters, G is the gravitational constant, M is the mass of the black hole in kilograms, and c is the speed of light (Chapter 14).

Size of a Distant Object $D = R \times \tan\theta$, where D is the physical diameter of an object, R is the distance to it, and $\tan\theta$ is the tangent of the angle it makes in the sky (Chapter 1).

Wien's law $\lambda_{max} = 2.9 \times 10^3/T$, where λ_{max} is the peak wavelength of the blackbody and T is the blackbody's temperature in kelvins (Chapter 4).

Work $W = Fd$, where W is the work, F is the force acting, and d is the distance moved in the direction that the force acts (Chapter 2).

Appendix D: Temperature Scales

Three temperature scales are in common use. Throughout most of the world, temperatures are expressed in degrees Celsius (°C), named in honor of the Swedish astronomer Anders Celsius, who proposed it in 1742. The **Celsius temperature scale** (also known as the "centigrade scale") is based on the behavior of water, which freezes at 0°C and boils at 100°C at sea level on Earth.

Scientists usually prefer the **Kelvin scale,** named after the British physicist Lord Kelvin (William Thomson), who made many important contributions to our knowledge about heat and temperature. On the Kelvin temperature scale, water freezes at 273 K and boils at 373 K. Note that we do not use the degree symbol with the Kelvin temperature scale.

Because water must be heated by 100 K or 100°C to go from its freezing point to its boiling point, you can see that the size of a kelvin is the same as the size of a degree Celsius. When considering temperature *changes*, measurements in kelvins and in degrees Celsius lead to the same number.

A temperature expressed in kelvins is always equal to the temperature in degrees Celsius plus 273. Scientists prefer the Kelvin scale because it is closely related to the physical meaning of temperature. All substances are made of atoms, which are very tiny (a typical atom has a diameter of about 10^{-10} m) and constantly in motion. The temperature of a substance is directly related to the average speed of its atoms. If something is hot, its atoms are moving at high speeds. If a substance is cold, its atoms are moving much more slowly.

The coldest possible temperature is the temperature at which atoms move as slowly as possible (they can never quite stop completely). This minimum possible temperature, called *absolute zero*, is the starting point for the Kelvin scale. Absolute zero is 0 K, or −273°C. Because it is impossible for anything to be colder than 0 K, there are no negative temperatures on the Kelvin scale.

In the United States, many people still use the now outdated Fahrenheit scale, which expresses temperature in degrees Fahrenheit (°F). When the German physicist Gabriel Fahrenheit introduced this scale in the early 1700s, he intended 0°F to represent the coldest temperature then achievable (with a mixture of ice and saltwater) and 100°F to represent the temperature of a healthy human body. On the Fahrenheit scale, water freezes at 32°F and boils at 212°F. Because there are 180 degrees Fahrenheit between the freezing and boiling points of water, a degree Fahrenheit is only 100/180 (= 5/9) the size of the other scales.

The following equation converts from degrees Fahrenheit to degrees Celsius:

$$T_C = 5/9\ (T_F - 32)$$

To convert from Celsius to Fahrenheit, a simple rearrangement of terms gives the relationship

$$T_F = 9/5\ T_C + 32$$

where T_F is the temperature in degrees Fahrenheit and T_C is the temperature in degrees Celsius.

For example, consider a typical room temperature of 68°F. Using the first equation, we can convert this measurement to the Celsius scale as follows:

$$T_C = 5/9\ (68 - 32) = 20°C$$

To arrive at the Kelvin scale, we simply add 273 degrees to the value in degrees Celsius. Thus, 68°F = 20°C = 293 K. The accompanying figure displays the relationships among these three temperature scales.

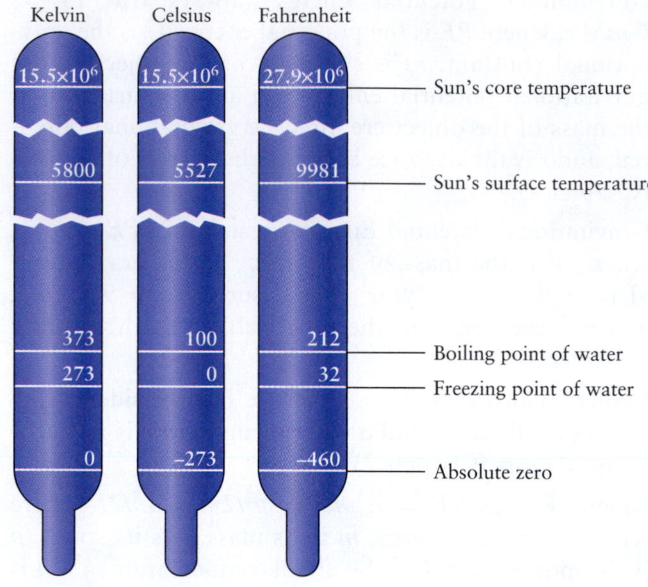

Try these questions: The Sun's surface temperature is about 5800 K. What is its temperature in Celsius and Fahrenheit? The temperature of empty space is about 3 K. What is its temperature in Celsius and Fahrenheit? (Answers appear at the end of the book.)

Appendix E: Data Tables

Table E-1 The Planets: Orbital Data

Planet	Semimajor axis (AU)	Semimajor axis (10^6 km)	Sidereal period (year)	Sidereal period (day)	Synodic period (day)	Mean orbital speed (km/s)	Orbital eccentricity	Inclination of orbit to ecliptic (°)
Mercury	0.3871	57.9	0.2408	87.97	115.88	47.9	0.206	7.00
Venus	0.7233	108.2	0.6152	224.70	583.92	35.0	0.007	3.39
Earth	1.0000	149.6	1.0000	365.26	—	29.8	0.017	0.00
Mars	1.5237	227.9	1.8808	686.98	779.94	24.1	0.093	1.85
Jupiter	5.2034	778.6	11.862	4,332.6	398.9	13.1	0.048	1.31
Saturn	9.5371	1433.5	29.457	10,759	378.1	9.7	0.054	2.48
Uranus	19.1913	2872.5	84.01	30,685	369.7	6.8	0.047	0.77
Neptune	30.0690	4495.1	164.79	60,189	367.5	5.4	0.009	1.77

TABLE E-2 The Planets: Physical Data

Planet	Equatorial diameter (km)	Equatorial diameter (Earth = 1)	Mass (kg)	Mass (Earth = 1)	Mean density (kg/m^3)	Rotation period* (days)	Inclination of equator to orbit (°)	Surface gravity (Earth = 1)	Albedo	Escape speed (km/s)
Mercury	4,879	0.383	3.302×10^{23}	0.055	5427	58.646	173.4	0.38	0.106	4.3
Venus	12,104	0.949	4.869×10^{24}	0.815	5243	243.02^R	177.4	0.91	0.65	10.4
Earth	12,756	1.000	5.974×10^{24}	1.000	5515	0.996	23.45	1.000	0.37	11.2
Mars	6,794	0.533	6.419×10^{23}	0.107	3933	1.025	25.19	0.38	0.15	5.0
Jupiter	142,984	11.209	1.899×10^{27}	317.83	1326	0.413	3.13	2.5	0.52	59.5
Saturn	120,536	9.449	5.685×10^{26}	95.16	687	0.445	26.73	1.1	0.47	35.5
Uranus	51,118	4.007	8.683×10^{25}	14.54	1270	0.717^R	97.77	0.91	0.51	21.3
Neptune	49,528	3.883	1.024×10^{26}	17.147	1638	0.671	28.32	1.1	0.4	23.5

* For Jupiter, Saturn, Uranus, and Neptune, the internal rotation period is given. A superscript *R* means that the rotation is retrograde (opposite the planet's orbital motion).

TABLE E-3 Satellites of the Planets

Planet	Satellite	Discoverers of planet	Average distance from center (km)	Orbital (sidereal) period* (days)	Orbital eccentricity	Diameter of satellite (km)	Mass (kg)
EARTH	Moon	—	384,400	27.322	0.0549	3476	7.349×10^{22}
MARS	Phobos	Hall (1877)	9,378	0.319	0.02	28 × 23 × 2	1.1×10^{16}
	Deimos	Hall (1877)	23,459	1.262	0.00	16 × 12 × 10	2.4×10^{15}
JUPITER	Metis	Synott (1980)	127,960	0.2948	0.00	44	1×10^{17}
	Adrastea	Jewitt et al. (1979)	128,980	0.2983	0.00	24 × 16 × 20	1.9×10^{16}
	Amalthea	Barnard (1892)	181,300	0.4982	0.00	270 × 170 × 150	7.5×10^{18}
	Thebe	Synott (1979)	221,900	0.6745	0.02	98	8×10^{17}
	Io	Galileo (1610)	421,600	1.769	0.00	3643	8.93×10^{22}
	Europa	Galileo (1610)	670,900	3.551	0.01	3138	4.80×10^{22}
	Ganymede	Galileo (1610)	1,070,000	7.155	0.00	5268	1.48×10^{23}
	Callisto	Galileo (1610)	1,883,000	16.689	0.01	4821	1.08×10^{23}
	Themisto	Kowal (1975)	7,435,000	130.02	0.24	8	(?)
	Leda	Kowal (1974)	11,094,000	238.72	0.15	18	6×10^{15}
	Himalia	Perrine (1904)	11,480,000	250.57	0.16	170	9.5×10^{18}
	Lysithea	Nicholson (1938)	11,720,000	259.22	0.11	38	8×10^{16}
	Elara	Perrine (1905)	11,737,000	259.65	0.21	80	8×10^{17}
	S/2000/J11	Sheppard et al. (2000)	12,654,000	287	0.25	4	(?)
	Euporie	Sheppard et al. (2001)	19,017,000	553.1^R	0.16	2	(?)
	Chaldene	Sheppard et al. (2000)	20,375,000	723.8^R	0.24	4	(?)
	Iocaste	Sheppard et al. (2000)	20,733,000	632^R	0.22	5	(?)
	Kale	Sheppard et al. (2001)	20,804,000	721^R	0.48	2	(?)
	Orthosie	Sheppard et al. (2001)	20,876,000	623^R	0.27	2	(?)
	Thyone	Sheppard et al. (2001)	20,876,000	632^R	0.30	4	(?)
	Euanthe	Sheppard et al. (2001)	20,947,000	620^R	0.18	3	(?)
	Harpalyke	Sheppard et al. (2000)	21,019,000	623^R	0.23	4	(?)
	Praxidike	Sheppard et al. (2000)	21,162,000	625^R	0.22	7	(?)
	Ananke	Nicholson (1951)	21,200,000	631^R	0.17	28	4×10^{16}
	Hermippe	Sheppard et al. (2001)	21,376,000	632^R	0.25	4	(?)
	Taygete	Sheppard et al. (2000)	21,734,000	732^R	0.25	5	(?)
	Erinome	Sheppard et al. (2000)	21,948,000	728^R	0.27	3	(?)
	Carme	Nicholson (1938)	22,600,000	692^R	0.21	46	9×10^{16}
	Isonoe	Sheppard et al. (2000)	22,806,000	726	0.26	4	(?)
	Pasithee	Sheppard et al. (2001)	22,949,000	715^R	0.29	2	(?)
	Eurydome	Sheppard et al. (2001)	23,378,000	721^R	0.35	3	(?)
	Aitne	Sheppard et al. (2001)	23,449,000	741^R	0.29	3	(?)
	Pasiphae	Melotte (1908)	23,500,000	735^R	0.38	36	2×10^{17}
	Megaclite	Sheppard et al. (2000)	23,521,000	753^R	0.43	5	(?)
	Sponde	Sheppard et al. (2001)	23,592,000	749^R	0.45	2	(?)

TABLE E-3 Satellites of the Planets (continued)

Planet	Satellite	Discoverers of planet	Average distance from center (km)	Orbital (sidereal) period* (days)	Orbital eccentricity	Diameter of satellite (km)	Mass (kg)
	Sinope	Nicholson (1914)	23,700,000	758^R	0.28	28	8×10^{16}
	Autonoe	Sheppard et al. (2001)	23,979,000	765^R	0.42	4	(?)
	Kalyke	Sheppard et al. (2000)	24,164,000	743^R	0.24	5	(?)
	Callirrhoe	Scotti et al. (1999)	24,200,000	759^R	0.28	8	(?)
SATURN	Pan	Showalter (1990)	133,570	0.573	0.00	20	(?)
	Atlas	Terrile (1980)	137,670	0.602	0	40 × 30 × 30	8×10^{17}
	Prometheus	Collins et al. (1980)	139,353	0.613	0.00	140 × 80 × 100	8×10^{17}
	Pandora	Collins et al. (1980)	141,700	0.629	0.00	110 × 70 × 100	2×10^{17}
	Epimetheus	Walker (1966)	151,422	0.694	0.01	140 × 100 × 100	5×10^{17}
	Janus	Dolfuss (1966)	151,472	0.695	0.01	220 × 160 × 200	2×10^{18}
	Mimas	Herschel (1789)	185,520	0.942	0.02	392	3.8×10^{19}
	Enceladus	Herschel (1789)	238,020	1.370	0.00	444	7.3×10^{19}
	Tethys	Cassini (1684)	294,660	1.888	0.00	1050	6.3×10^{20}
	Calypso	Pascu et al. (1980)	294,660	1.888	0 (?)	30 × 20 × 25	8×10^{17}
	Telesto	Smith et al. (1980)	294,660	1.888	0 (?)	24	8×10^{17}
	Dione	Cassini (1684)	377,400	2.737	0.00	1120	1.1×10^{21}
	Helene	Laques et al. (1980)	377,400	2.737	0.01	40 × 30 × 30	8×10^{17}
	Rhea	Cassini (1672)	527,040	4.518	0.00	1528	2.3×10^{21}
	Titan	Huygens (1655)	1,221,830	15.945	0.03	5150	1.3×10^{23}
	Hyperion	Bond et al. (1848)	1,481,100	21.277	0.10	410 × 260 × 220	8×10^{17}
	Iapetus	Cassini (1671)	3,561,300	79.330	0.03	1436	1.6×10^{21}
	Phoebe	Pickering (1898)	12,952,000	550.56^R	0.16	220	4×10^{17}
URANUS	Cordelia	*Voyager 2* (1986)	49,752	0.335	0.00	26	8×10^{17}
	Ophelia	*Voyager 2* (1986)	53,763	0.376	0.01	30	8×10^{17}
	Bianca	*Voyager 2* (1986)	59,166	0.435	0.00	42	8×10^{17}
	Cressida	*Voyager 2* (1986)	61,767	0.464	0.00	66	8×10^{17}
	Desdemona	*Voyager 2* (1986)	62,658	0.474	0.00	54	8×10^{17}
	Juliet	*Voyager 2* (1986)	64,358	0.493	0.00	84	8×10^{17}
	Portia	*Voyager 2* (1986)	66,097	0.513	0.00	108	8×10^{17}
	Rosalind	*Voyager 2* (1986)	69,927	0.558	0.00	54	8×10^{17}
	Mab	Showalter et al. (2003)	97,734	.(?)	0 (?)	26	(?)
	Belinda	*Voyager 2* (1986)	75,236	0.624	0.00	66	8×10^{17}
	Perdita	*Voyager 2* (1986)	76,420	0.638	0 (?)	20	(?)
	Puck	*Voyager 2* (1986)	86,004	0.762	0.00	144	8×10^{17}
	Cupid	Showalter et al. (2003)	74,800	.(?)	0 (?)	10	(?)
	Miranda	Kuiper (1948)	129,872	1.413	0.00	–470	6.6×10^{19}
	Ariel	Lassell (1851)	190,945	2.520	0.00	–1160	1.4×10^{21}
	Umbriel	Lassell (1851)	265,998	4.144	0.01	1169	1.2×10^{21}
	Titania	Herschel (1787)	436,298	8.706	0.00	1578	3.5×10^{21}

TABLE E-3 Satellites of the Planets (continued)

Planet	Satellite	Discoverers of planet	Average distance from center (km)	Orbital (sidereal) period* (days)	Orbital eccentricity	Diameter of satellite (km)	Mass (kg)
	Oberon	Herschel (1787)	583,519	13.463	0.00	1523	3.0×10^{21}
	Caliban	Gladman et al. (1997)	7,169,000	579^R	0.16	80	8×10^{17}
	Sycorax	Gladman et al. (1997)	12,213,000	1284^R	0.52	160	8×10^{17}
	Stephano	Kavelaars et al. (1999)	>7,979,000	677^R	0.23	32	8×10^{17}
	Prospero	Kavelaars et al. (1999)	>16,665,000	1993^R	0.43	50	8×10^{17}
	Setebos	Kavelaars et al. (1999)	>17,879,000	2194^R	0.59	47	8×10^{17}
NEPTUNE	Naiad	*Voyager 2* (1989)	48,230	0.294	0.00	58	8×10^{17}
	Thalassa	*Voyager 2* (1989)	50,070	0.311	0.00	80	8×10^{17}
	Despina	*Voyager 2* (1989)	52,530	0.335	0.00	148	8×10^{17}
	Galatea	*Voyager 2* (1989)	61,950	0.429	0.05	178	8×10^{17}
	Larissa	*Voyager 2* (1989)	73,550	0.553	0.00	−190	8×10^{17}
	Proteus	*Voyager 2* (1989)	117,650	1.122	0.00	−415	8×10^{17}
	Triton	Lassell (1846)	354,760	5.877^R	0.00	2704	2.14×10^{22}
	Nereid	Kuiper (1949)	5,513,400	360.1	0.75	340	2×10^{19}

* A superscript R means that the satellite orbits in a retrograde direction (opposite to the planet's rotation).

TABLE E-4 Dwarf Planets

	Pluto	Ceres	Eris
Year of Discovery	1930	1801	2003
Semimajor axis (AU)	39.4817	2.766	67.668
Semimajor axis (10^6 km)	5906.4	413.7	10210
Sidereal period (year)	247.7	4.599	557
Sidereal period (day)	90470	1680	2.03×10^5
Mean orbital speed (km/s)	4.67	17.88	3.44
Orbital eccentricity	0.249	0.080	0.442
Inclination of orbit to ecliptic (°)	17.14	10.59	44.19
Equatorial diameter (km)	2390	941	2400
Equatorial diameter (Earth = 1)	0.19	0.074	0.19
Mass (kg)	1.3×10^{22}	9.5×10^{20}	1.7×10^{22}
Mass (Earth = 1)	2.2×10^{-3}	1.6×10^{-4}	2.8×10^{-3}
Mean density (kg/m^3)	2030	2080	2100
Rotation period (days)	6.388^R	0.3781	?
Inclination of equator to orbit (°)	122.5	4	?
Surface gravity (Earth =1)	0.06	0.028	0.07
Escape velocity	1.2	0.51	1.3
Location	Kuiper Belt (TNO)	Asteroid Belt	TNO
Number of satellites	3	0	1

TABLE E-5 The Nearest Stars

Name*	Parallax (arcsec)	Distance (ly)	Spectral type	Radial velocity** (km/s)	Proper motion (arcsec/year)	Apparent visual magnitude	Absolute visual magnitude	Luminosity (Sun = 1)
Sun			G2 V			−26.7	+4.85	1.00
Proxima Centauri	0.769	4.22	M5.5 V	−22	3.853	+11.09	+15.53	8.2×10^{-4}
Alpha Centauri A	0.747	4.40	G2 V	−25	3.710	−0.01	+4.38	1.77
Alpha Centauri B	0.747	4.40	K0 V	−21	3.724	+1.34	+5.71	0.55
Barnard's Star	0.547	5.94	M4 V	−111	10.358	+9.53	+13.22	3.6×10^{-3}
Wolf 359	0.419	7.80	M6 V	+13	4.696	+13.44	+16.6	3.5×10^{-4}
Lalande 21185	0.393	8.32	M2 V	−84	4.802	+7.47	+10.44	0.023
L 726-8 A	0.374	8.56	M5.5 V	+29	3.368	+12.54	+15.4	9.4×10^{-4}
L 726-8 B	0.374	8.56	M6 V	+32	3.368	+12.99	+15.9	5.6×10^{-4}
Sirius A	0.380	8.61	A1 V	−9	1.339	−1.43	+1.47	26.1
Sirius B	0.380	8.61	white dwarf	−9	1.339	+8.44	+11.34	2.4×10^{-3}
Ross 154	0.337	9.71	M3.5 V	−12	0.666	+10.43	+13.07	4.1×10^{-3}
Ross 248	0.316	10.32	M5.5 V	−78	1.617	+12.29	+14.8	1.5×10^{-3}
Epsilon Eridani	0.310	10.49	K2 V	+17	0.977	+3.73	+6.19	0.40
Lacaille 9352	0.304	10.73	M1.5 V	+10	6.896	+7.34	+9.75	0.051
Ross 128	0.299	10.87	M4 V	−31	1.361	+11.13	+13.51	2.9×10^{-3}
L 789-6	0.294	11.09	M5 V	−60	3.259	+12.33	+14.7	1.3×10^{-3}
61 Cygni A	0.286	11.36	K5 V	−65	5.281	+5.21	+7.49	0.16
61 Cygni B	0.286	11.44	K7 V	−64	5.172	+6.03	+8.31	0.095
Procyon A	0.286	11.40	F5 IV–V	−4	1.259	+0.38	+2.66	7.73
Procyon B	0.286	11.40	white dwarf	−4	1.259	+10.7	+12.98	5.5×10^{-4}
BD +59° 1915 A	0.281	11.61	M3 V	−1	2.238	+8.94	+11.18	0.020
BD +59° 1915 B	0.281	11.61	M3.5 V	+1	2.313	+9.70	+11.97	0.010
Groombridge 34 A	0.281	11.65	M1.5 V	+12	2.918	+8.08	+10.32	0.030
Groombridge 34 B	0.281	11.65	M3.5 V	+11	2.918	+11.06	+13.3	3.1×10^{-3}
Epsilon Indi	0.276	11.82	K5 V	−40	4.704	+4.69	+6.89	0.27
GJ 1111	0.276	11.82	M6.5 V	−5	1.290	+14.78	+16.98	2.7×10^{-4}
Tau Ceti	0.274	11.90	G8 V	−17	1.922	+3.49	+5.68	0.62
GJ 1061	0.272	12.08	M5.5 V	−20	0.826	+13.09	+15.26	1.0×10^{-3}
L 725-32	0.269	12.12	M4.5 V	+28	1.372	+12.10	+14.25	1.7×10^{-3}
BD +05° 1668	0.263	12.40	M3.5 V	+18	3.738	+9.84	+11.94	0.011
Kapteyn's star	0.255	12.79	M1.5 V	+246	8.670	+8.84	+10.87	0.013
Lacaille 8760	0.253	12.89	M0 V	+28	3.455	+6.67	+8.69	0.094
Krüger 60 A	0.248	13.05	M3 V	−33	0.990	+9.79	+11.76	0.010
Krüger 60 B	0.248	13.05	M4 V	−32	0.990	+11.41	+13.38	3.4×10^{-3}

* Stars that are components of binary systems are labeled A and B.

** A positive radial velocity means the star is receding; a negative radial velocity means the star is approaching.

Compiled from the *Hipparcos General Catalogue* and from data reported by the Research Consortium on Nearby Stars. The table lists all known stars within 4.00 parsecs (13.05 light-years).

TABLE E-6 The Visually Brightest Stars

Name	Designation	Distance (ly)	Spectral type	Radial velocity* (km/s)	Proper motion (arcsec/year)	Apparent visual magnitude	Apparent visual brightness** (Sirius = 1)	Absolute visual magnitude	Luminosity (Sun = 1)
Sirius A	α CMa A	8.61	A1 V	−9	1.339	−1.43	1.000	+1.47	26.1
Canopus	α Car	313	F0 I	+21	0.031	−0.62	0.470	−5.53	1.4×10^4
Arcturus	α Boo	36.7	K2 III	−5	2.279	−0.05	0.278	−0.31	190
Rigil Kentaurus	α Cen A	4.4	G2 V	−25	3.71	−0.01	0.268	+4.38	1.77
Vega	α Lyr	25.3	A0 V	−14	0.035	+0.03	0.258	+0.58	61.9
Capella	α Aur	42.2	G8 III	+30	0.434	+0.08	0.247	−0.48	180
Rigel	β Ori A	773	B8 Ia	+21	0.002	+0.18	0.225	−6.69	7.0×10^5
Procyon	α CMi A	11.4	F5 IV–V	−4	1.259	+0.38	0.184	+2.66	7.73
Achernar	α Eri	144	B3 IV	+19	0.097	+0.45	0.175	−2.77	5250
Betelgeuse	α Ori	427	M2 Iab	+21	0.029	+0.45	0.175	−5.14	4.1×10^4
Hadar	β Cen	525	B1 II	−12	0.042	+0.61	0.151	−5.42	8.6×10^4
Altair	α Aql	16.8	A7 IV–V	−26	0.661	+0.77	0.132	+2.2	11.8
Aldebaran	α Tau A	65.1	K5 III	+54	0.199	+0.87	0.119	−0.63	370
Spica	α Vir	262	B1 V	+1	0.053	+0.98	0.108	−3.55	2.5×10^4
Antares	α Sco A	604	M1 Ib	−3	0.025	+1.06	0.100	−5.28	3.7×10^4
Pollux	β Gem	33.7	K0 III	+3	0.627	+1.16	0.091	+1.09	46.6
Fomalhaut	α PsA	25.1	A3 V	+7	0.368	+1.17	0.090	+1.74	18.9
Deneb	α Cyg	3230	A2 Ia	−5	0.002	+1.25	0.084	−8.73	3.2×10^5
Mimosa	β Cru	353	B0.5 III	+20	0.05	+1.25	0.084	−3.92	3.4×10^4
Regulus	α Leo A	77.5	B7 V	+4	0.249	+1.36	0.076	−0.52	331

Data in this table was compiled from the Hipparcos General Catalogue.

* A positive radial velocity means the star is receding; a negative radial velocity means the star is approaching.

** This is the ratio of the star's apparent brightness to that of Sirius, the brightest star in the night sky.

Note: Acrux, or α Cru (the brightest star in Crux, the Southern Cross), appears to the naked eye as a star of apparent magnitude +0.87, the same as Aldebaran, but it does not appear in this table because Acrux is actually a binary star system. The blue-white component stars of this binary system have apparent magnitudes of +1.4 and +1.9, and thus they are dimmer than any of the stars listed here.

TABLE E-7 The Constellations

Name	Meaning	R.A.	Dec.	Genitive*	Abbreviation
Andromeda	proper name; princess	1	+40	Andromedae	And
Antlia	air pump	10	−35	Antliae	Ant
Apus	bee	16	−75	Apodis	Aps
Aquarius[1,2]	waterman	22	−10	Aquarii	Aqr
Aquila	eagle	20	+15	Aquilae	Aql
Ara	altar	17	−55	Arae	Ara
Aries[2]	ram	3	+20	Arietis	Ari
Auriga	charioteer	6	+40	Aurigae	Aur
Boötes	proper name; herdsman, wagoner	15	+30	Boötis	Book
Caelum	engraving tool	5	−40	Caeli	Cae
Camelopardalis	giraffe	6	+70	Camelopardalis	Cam
Cancer[2]	crab	8.5	+15	Cancri	Cnc
Canes Venatici	hunting dogs	13	+40	Canum Venaticorum	CVn
Canis Major	larger dog	7	−20	Canis Majoris	CMa
Canis Minor	smaller dog	8	+5	Canis Minoris	CMi
Capricornus[1,2]	water-goat	21	−20	Capricornii	Cap
Carina	keel	9	−60	Carinae	Car
Cassiopeia	proper name; queen	1	+60	Cassiopeiae	Cas
Centaurus	centaur	13	−45	Centauri	Cen
Cepheus	proper name; king	22	+65	Cephei	Cep
Cetus	whale	2	−10	Ceti	Cet
Chamaeleon	chameleon	10	−80	Chamaeleontis	Cha
Circinus	compasses	15	−65	Circini	Cir
Columba	dove	6	−35	Columbae	Col
Coma Berenices	Berenice's hair	13	+20	Comae Berenices	Com
Corona Australis[3]	southern crown	19	+40	Coronae Australis	CrA
Corona Borealis[4]	northern crown	16	+30	Coronae Borealis	CrB
Corvus[5]	crow, raven	12	−20	Corvi	Crv
Crater	cup	11	−15	Crateris	Crt
Crux[6]	southern cross	12	−60	Crucis	Cru
Cygnus	swan	21	+40	Cygni	Cyg
Delphinus[1]	Dolphin	21	+10	Delphini	Del
Dorado[7]	swordfish	6	−55	Doradus	Dor
Draco[8]	dragon	15	+60	Draconis	Dra
Equuleus	little horse	21	+10	Equulei	Equ
Eridanus	proper name; river	4	−30	Eridani	Eri
Fornax	furnace	3	−30	Fornacis	For
Gemini[2]	twins	7	+20	Geminorum	Gem
Grus	crane	22	−45	Gruis	Gru
Hercules[9]	proper name; hero	17	+30	Herculis	Her

TABLE E-7 The Constellations (continued)

Name	Meaning	R.A.	Dec.	Genitive*	Abbreviation
Horologium	clock	3	−55	Horologii	Hor
Hydra	water serpant	12	−25	Hydrae	Hya
Hydrus	water snake	2	−70	Hydri	Hyi
Indus	Indian	22	−70	Indi	Ind
Lacerta	lizard	22	+45	Lacertae	Lac
Leo[2]	lion	11	+15	Leonis	Leo
Leo Minor	smaller lion	10	+35	Leonis Minoris	LMi
Lepus	hare	6	−20	Leporis	Lep
Libra[2,10]	scales	15	−15	Librae	Lib
Lupus	wolf	15	−45	Lupi	Lup
Lynx	lynx	8	+45	Lyncis	Lyn
Lyra[5]	lyre	19	+35	Lyrae	Lyr
Microscopium	microscope	21	−40	Microscopii	Mic
Monoceros	unicorn	7	0	Monocerotis	Mon
Mensa	table	6	−75	Mensae	Men
Musca (Australis)[11]	(southern) fly	12	−70	Muscae	Mus
Norma	square	16	−50	Normae	Nor
Octans[12]	octant	—	−90	Octantis	Oct
Ophiuchus[13]	serpent-bearer	17	0	Ophiuchi	Oph
Orion	proper name; hunter, giant	6	0	Orionis	Ori
Pavo	peacock	20	−70	Pavonis	Pav
Pegasus	proper name; winged horse	23	+20	Pegasi	Peg
Perseus	proper name; hero	3	+45	Persei	Per
Phoenix	phoenix	1	−50	Phoenicis	Phe
Pictor	easel	6	−55	Pictoris	Pic
Pices[1,2]	fishes	1	+10	Piscium	Psc
Piscis Austrinus[1]	southern fish	22	−30	Piscis Austrini	PsA
Puppis	stern	8	−30	Puppis	Pup
Pyxis	compass	9	−30	Pyxidis	Pyx
Reticulum	net	4	−60	Reticuli	Ret
Sagitta	arrow	20	+20	Sagittae	Sge
Sagittarius[2,14]	archer	19	−25	Sagittarii	Sgr
Scorpius[2]	scorpion	17	−30	Scorpii	Sco
Sculptor[15]	sculptor's workshop	1	−30	Sculptoris	Scl
Scutum[16]	shield	19	−10	Scuti	Sct
Serpens[13]	serpent	17	0	Serpentis	Ser
Sextans	sextant	10	0	Sextantis	Sex
Taurus[2]	bull	5	+20	Tauri	Tau
Telescopium	telescope	19	−50	Telescopii	Tel
Triangulum	triangle	2	+30	Trianguli	Tri
Triangulum Australe	southern triangle	16	−65	Trianguli Australis	TrA

TABLE E-7 The Constellations (continued)

Name	Meaning	R.A.	Dec.	Genitive*	Abbreviation
Tucana[17]	toucan	0	−65	Tucanae	Tuc
Ursa Major	larger bear	11	+60	Ursae Majoris	UMa
Ursa Minor[18]	smaller bear	16	+80	Ursae Minoris	UMi
Vela	sails	10	−45	Velorum	Vel
Virgo[2]	Virgin	13	0	Virginis	Vir
Volans	flying fish	8	−70	Volantis	Vol
Vulpecula	fox	20	+25	Vulpeculae	Vul

*Genitive is the grammatical case denoting possession. For example, astronomers denote the brightest or α (alpha) star in Orion (Betelgeuse) as α Orionis.

[1]Constellations of the area of the sky known as the wet quarter for its many watery images.

[2]A zodiac constellation.

[3]Sometimes considered as Sagittarius's crown.

[4]Ariadne's crown.

[5]Corvus was Orpheus's companion, Lyra his harp.

[6]Originally a part of Centaurus.

[7]Contains the Large Magellanic Cloud and the south ecliptic pole.

[8]Contains the north ecliptic pole.

[9]One of the oldest constellations known.

[10]Originally the claws of Scorpius.

[11]Originally named Musca Australis to distinguish it from Musca Borealis, the northern fly, which is now defunct; "Australis" is now dropped.

[12]Contains the south celestial pole.

[13]Ophiucus is identified with the physician Aesculapius, and Serpens with the caduceus.

[14]Contains the galactic center.

[15]Originally named by Lacaille l'Atelier du Sculpteur (in Latin, Apparatus Sculptoris); now known simply as Sculptor. Contains the south galactic pole.

[16]Shield of the Polish hero John Sobieski.

[17]Contains the Small Magellanic Cloud.

[18]Contains the north celestial pole.

TABLE E-8 Some Useful Astronomical Quantities

Astronomical unit:	$1\ \text{AU} = 1.496 \times 10^{11}$ m
Light-year:	$1\ \text{ly} = 9.461 \times 10^{15}$ m $= 63{,}240$ AU
Parsec:	$1\ \text{pc} = 3.086 \times 10^{16}$ m $= 3.262$ ly
Solar mass:	$1\ M_{\odot} = 1.989 \times 10^{30}$ kg
Solar radius:	$1\ R_{\odot} = 6.960 \times 10^{8}$ m
Solar luminosity:	$1\ L_{\odot} = 3.827 \times 10^{26}$ W
Earth's mass:	$1\ M_{\oplus} = 5.974 \times 10^{24}$ kg
Earth's equatorial radius:	$1\ R_{\oplus} = 6.378 \times 10^{6}$ m
Moon's mass:	$1\ M_{\text{Moon}} = 7.349 \times 10^{22}$ kg
Moon's equatorial radius:	$1\ R_{\text{Moon}} = 1.738 \times 10^{6}$ m

TABLE E-9 Some Useful Physical Constants

Speed of light:	$c = 2.998 \times 10^{8}$ m/s
Gravitational constant:	$G = 6.668 \times 10^{-11}$ N m^2 kg^{-2}
Planck constant:	$h = 6.626 \times 10^{-34}$ J s $= 4.136 \times 10^{-15}$ eV s
Boltzmann constant:	$k = 1.380 \times 10^{-23}$ J K^{-1} $= 8.617 \times 10^{-5}$ eV K^{-1}
Stefan–Boltzmann constant:	$\sigma = 5.670 \times 10^{-8}$ W m^{-2} K^{-4}
Mass of electron:	$m_e = 9.109 \times 10^{-31}$ kg
Mass of 1 H atom:	$m_H = 1.673 \times 10^{-27}$ kg

TABLE E-10 Common Conversions between U.S. Customary and Metric Units

1 inch = 2.54 centimeters (cm)

1 cm = 0.394 inch (in)

1 yard = 0.914 meter (m)

1 meter = 1.09 yards = 39.37 inches

1 mile = 1.61 kilometers (km)

1 km = 0.621 mile (mi)

TABLE E-11 Spiral Galaxies and Interacting Galaxies

Spiral galaxy	R.A.	Decl.	Hubble type	Interacting galaxies	R.A.	Decl.
M31 (NGC 224)	0^h 42.7^m	+41° 16′	Sb	M51 (NGC 5194)	13^h 29.9^m	+47° 12′
M58 (NGC 4579)	12 37.7	+11 49	Sb	NGC 5195	13 30.0	+47 16
M61 (NGC 4303)	12 21.9	+ 4 28	Sc	M65 (NGC 3623)	11 18.9	+13 05
M63 (NGC 5055)	13 15.8	+42 02	Sb	M66 (NGC 3627)	11 20.2	+12 59
M64 (NGC 4826)	12 56.7	+21 41	Sb	M81 (NGC 3031)	9 55.6	+69 04
M74 (NGC 628)	1 36.7	+15 47	Sc	M82 (NGC 3034)	9 55.8	+69 41
M83 (NGC 5236)	13 37.0	−29 52	Sc	M95 (NGC 3351)	10 44.0	+11 42
M88 (NGC 4501)	12 32.0	+14 25	Sb	M96 (NGC 3368)	10 46.8	+11 49
M90 (NGC 4569)	12 36.8	+13 10	Sb	M105 (NGC 3379)	10 47.8	+12 35
M94 (NGC 4736)	12 50.9	+41 07	Sb			
M98 (NGC 4192)	12 13.8	+14 54	Sb			
M99 (NGC 4254)	12 18.8	+14 25	Sc			
M100 (NGC 4321)	12 22.9	+15 49	Sc			
M101 (NGC 5457)	14 03.2	+54 21	Sc			
M104 (NGC 4594)	12 40.0	−11 37	Sa			
M108 (NGC 3556)	11 11.5	+55 40	Sc			

TABLE E-12 Mass and Energy Inventory for the Universe

	Individual contribution	Section total
Dark matter and dark energy contributions		0.954 ± 0.003*
Dark energy	0.72 ± 0.03	
Dark matter	0.23 ± 0.03	
Primeval graviational radiation	$\leq 10^{-10}$	
Contributions from Big Bang era		0.0010 ± 0.0005
Electromagnetic radiation	$10^{-4.3 \pm 0.000001}$	
Neutrinos	$10^{-2.9 \pm 0.1}$	
Normal particle (baryon) rest mass		0.045 ± 0.003
Charged particles (plasma) between stars and galaxies	0.0418 ± 0.003	
Main sequence stars in elliptical galaxies and nuclear bulges	0.0015 ± 0.0004	
Neutral hydrogen & helium	0.00062 ± 0.00010	
Main sequence stars in galactic disks and in irregular galaxies	0.00055 ± 0.00014	
White dwarfs	0.00036 ± 0.00008	
Molecular gas	0.00016 ± 0.00006	
Substellar objects	0.00014 ± 0.00007	
Black holes	0.00007 ± 0.00002	
Neutron stars	0.00005 ± 0.00002	
Planets	$10^{-6 \pm 0.1}$	

*The numbers after the plus or minus (±) symbol indicate the possible errors in the given numbers.
This table lists the major contributions to the mass and energy of the universe.

Appendix F: Periodic Table of the Elements

1	2	3	4	5	6	7	8	9	10	11	12	13	14	15	16	17	18
1 H Hydrogen																	2 He Helium
3 Li Lithium	4 Be Beryllium											5 B Boron	6 C Carbon	7 N Nitrogen	8 O Oxygen	9 F Fluorine	10 Ne Neon
11 Na Sodium	12 Mg Magnesium											13 Al Aluminum	14 Si Silicon	15 P Phosphorus	16 S Sulfur	17 Cl Chlorine	18 Ar Argon
19 K Potassium	20 Ca Calcium	21 Sc Scandium	22 Ti Titanium	23 V Vanadium	24 Cr Chromium	25 Mn Manganese	26 Fe Iron	27 Co Cobalt	28 Ni Nickel	29 Cu Copper	30 Zn Zinc	31 Ga Gallium	32 Ge Germanium	33 As Arsenic	34 Se Selenium	35 Br Bromine	36 Kr Kryton
37 Rb Rubidium	38 Sr Strontium	39 Y Yttrium	40 Zr Zirconium	41 Nb Niobium	42 Mo Molybdenum	43 Tc Technetium	44 Ru Ruthenium	45 Rh Rhodium	46 Pd Palladium	47 Ag Silver	48 Cd Cadmium	49 In Indium	50 Sn Tin	51 Sb Antimony	52 Te Tellurium	53 I Iodine	54 Xe Xenon
55 Cs Cesium	56 Ba Barium	57 La Lanthanum	72 Hf Hafnium	73 Ta Tantaium	74 W Tungsten	75 Re Rhenium	76 Os Osmium	77 Ir Iridium	78 Pt Platinum	79 Au Gold	80 Hg Mercury	81 Tl Thallium	82 Pb Lead	83 Bi Bismuth	84 Po Polonium	85 At Astatine	86 Rn Radon
87 Fr Francium	88 Ra Radium	89 Ac Actinium	104 Rf Rutherfordium	105 Db Dubnium	106 Sg Seaborgium	107 Bh Bohrium	108 Hs Hassium	109 Mt Meitnerium	110 Ds Darmstadium	111 Rg Roentgenium	112 Cn Copernicium	113 Uut Ununtrium	114 Uug Ununguadium	115 Uup Ununpentium	116 Uuh Ununhexium	117 Uus Ununseptium	118 Uuo Ununoctium

58 Ce Cerium	59 Pr Praseodymium	60 Nd Neodymium	61 Pm Promethium	62 Sm Samarium	63 Eu Europium	64 Gd Gadolinium	65 Tb Terbium	66 Dy Dysprosium	67 Ho Holmium	68 Er Erbium	69 Tm Thulium	70 Yb Ytterbium	71 Lu Lutetium
90 Th Thorium	91 Pa Protactinium	92 U Uranium	93 Np Neptunium	94 Pu Plutonium	95 Am Americium	96 Cm Curium	97 Bk Berkelium	98 Cf Californium	99 Es Einsteinium	100 Fm Fermium	101 Md Mendelevium	102 No Nobelium	103 Lr Lawrencium

Appendix G: Largest Optical Telescopes in the World

Aperture (m)	Name	Location	Altitude (m)
10.4	Gran Telescopio Cararias	La Palma, Canary Islands, Spain	2400
10	Keck I	Mauna Kea, Hawaii	4123
10	Keck II	Mauna Kea, Hawaii	4123
10	SALT	Sutherland, South Africa	1759
9.2	Hobby-Eberly	Mt. Fowlkes, Texas	2072
8.4 × 2	Large Binocular Telescope	Mt. Graham, Arizona	3170
8.3	Subaru	Mauna Kea, Hawaii	4100
8.2	Antu	Cerro Paranal, Chile	2635
8.2	Kueyen	Cerro Paranal, Chile	2635
8.2	Melipal	Cerro Paranal, Chile	2635
8.2	Tepun	Cerro Paranal, Chile	2635
8.1	Gemini North (Gillett)	Mauna Kea, Hawaii	4100
8.1	Gemini South	Cerro Pachon, Chile	2737
6.5	MMT	Mt. Hopkins, Arizona	2600
6.5	Walter Baade	La Serena, Chile	2282
6.5	Landon Clay	La Serena, Chile	2282
6	Bolshoi Teleskop Azimutalnyi	Nizhny Arkhyz, Russia	2070
6	LZT	British Columbia, Canada	395
5	Hale	Palomar Mtn., California	1900
4.2	William Herschel	La Palma, Canary Islands, Spain	2400
4.2	SOAR	Cerro Pachon, Chile	2738
4.2	LAMOST	Xinglong Station, China	950
4	Victor Blanco	Cerro Tololo, Chile	2200
3.9	Anglo-Australian	Coonabarabran, Australia	1100
3.8	Mayall	Kitt Peak, Arizona	2100
3.8	UKIRT	Mauna Kea, Hawaii	4200
3.7	AEOS	Maui, Hawaii	3058
3.6	"360"	Cerro La Silla, Chile	2400
3.6	Canada-France-Hawaii	Mauna Kea, Hawaii	4200
3.6	Telescopio Nazionale Galileo	La Palma, Canary Islands, Spain	2387
3.5	MPI-CAHA	Calar Alto, Spain	2200
3.5	New Techonology	Cerro La Silla, Chile	2400
3.5	ARC	Apache Point, New Mexico	2788
3.5	WIYN	Kitt Peak, Arizona	2100
3.5	Starfire	Kirtland Air Force Base, New Mexico	1900
3	Shane	Mt. Hamilton, California	1300
3	NASA IRTF	Mauna Kea, Hawaii	4160
2.7	Harlan Smith	Mt. Locke, Texas	2100
2.6	BAO	Byurakan, Armenia	1405
2.6	Shajn	Crimea, Ukraine	600

(continued)

Aperture (m)	Name	Location	Altitude (m)
2.5	Hooker	Mt. Wilson, California	1700
2.5	Isaac Newton	La Palma, Canary Islands, Spain	2382
2.5	Nordic Optical	La Palma, Canary Islands, Spain	2382
2.5	duPont	La Serena, Chile	2282
2.5	Sloan Digital Sky Survey	Apache Point, New Mexico	2788
2.4	Hiltner	Kitt Peak, Arizona	2100
2.4	Lijiang	Lijiang City, China	3250
2.4	Hubble Space Telescope	Low Earth orbit	6×10^5
2.3	WIRO	Jelm Mtn., Wyoming	2900
2.3	ANU	Coonabarabran, Australia	1100
2.3	Bok	Kitt Peak, Arizona	2100
2.3	Vainu Bappu	Kavalur, India	700
2.3	Aristarchos	Mt. Helmos, Greece	2340
2.2	ESO-MPI	Cerro La Silla, Chile	2335
2.2	MPI-CAHA	Calar Alto, Spain	2200
2.2	UH	Mauna Kea, Hawaii	4200
2.1	Kitt Peak 2.1 meter	Kitt Peak, Arizona	2100
2.1	Otto Struve	Davis Mountains, Texas	2070
2.1	UNAM	San Pedro, Mexico	2800
2.1	Jorge Sahade	El Leoncito, Argentina	2552
2	Himalayan Chandra	Hanle, India	4517
2	Alfred Jensch Teleskop	Tautenburg, Germany	
2	Carl Zeiss Jena	Azerbaijan	
2	Ondrejov	Ondrejov, Czech Republic	
2	RCC	Chepelare, Bulgaria	
2	Bernard Lyot	Pic du Midi, France	2877
2	Faulkes Telescope North	Haleakala, Maui, Hawaii	3050
2	Faulkes Telescope South	Siding Springs, Australia	3822
2	MAGNUM	Haleakala, Maui, Hawaii	3058
1.0 (× 6)	CHARA Array	Mt. Wilson, California	1700

GLOSSARY

A ring The outermost ring of Saturn visible from Earth; it is located just beyond Cassini's division.

absolute magnitude The apparent magnitude that a star would have if it were 10 parsecs from Earth.

absorption line A dark line in a continuous spectrum created when photons of a certain energy are absorbed by atoms or molecules.

absorption line spectrum Dark lines superimposed on a continuous spectrum.

acceleration A change in the direction or magnitude of a velocity.

accretion The gradual accumulation of matter by an astronomical body, usually caused by gravity.

accretion disk An orbiting disk of matter spiraling in toward a star or black hole.

active galactic nuclei (AGN) Supermassive black holes in the cores of some galaxies that emit particles and radiation which, when viewed from different angles, create Seyfert galaxies, radio galaxies, double radio sources, BL Lacertae objects, and quasars.

active galaxy A very luminous galaxy, often containing an active galactic nucleus.

active optics A system that adjusts a reflecting telescope in response to changes in temperature and shape of the mount; it helps optimize an image.

adaptive optics Primary telescope mirrors that are continuously and automatically adjusted to compensate for the distortion of starlight due to the motion of Earth's atmosphere.

AGB star *See* **asymptotic giant branch (AGB) star.**

albedo The fraction of sunlight that a planet, asteroid, or satellite scatters directly back into space.

amino acids A class of chemical compounds that are the building blocks of proteins.

angle The opening between two straight lines that meet at a point.

angular diameter (angular size) The arc angle across an object.

angular momentum A measure of how much energy an object has stored in its rotation and/or revolution.

angular resolution (resolution) The angular size of the smallest detail of an astronomical object that can be distinguished with a telescope.

annular eclipse An eclipse of the Sun in which the Moon is too distant to cover the Sun completely so that a ring of sunlight is seen around the Moon at mid-eclipse.

anorthosite A light-colored rock found throughout the lunar highlands and in some very old mountains on Earth.

aphelion The point in its orbit where a planet or other solar system body is farthest from the Sun.

Apollo asteroid An asteroid that is sometimes closer to the Sun than Earth is.

apparent magnitude A measure of the brightness of light from a star or other object as seen from Earth.

arc angle The measurement of the angle between two objects or two parts of the same object.

asteroid Any of the rocky objects larger than a few hundred meters in diameter (and not classified as a planet or moon) that orbits the Sun.

asteroid belt A 1½-astronomical-unit-wide region between the orbits of Mars and Jupiter in which most of the asteroids are found.

astrobiology The study of life in the universe.

astronomical unit (AU) The average distance between the Earth and the Sun: 1.5×10^8 km = 93 million mi.

asymptotic giant branch (AGB) star A red giant star that has completed core helium fusion and has re-expanded for a second time.

atom The smallest particle of an element that has the properties characterizing that element.

atomic number The number of protons in the nucleus of an atom.

autumnal equinox The intersection of the ecliptic and the celestial equator where the Sun crosses the equator moving from north to south. The beginning of autumn (around September 23).

average density The mass of an object divided by its volume.

B ring The brightest of the three rings of Saturn visible from Earth; it lies just inside the Cassini division.

barred spiral galaxy A spiral galaxy in which the spiral arms begin from the ends of a bar running through the nuclear bulge.

belt asteroid An asteroid whose orbit lies in the asteroid belt.

belts (of Jupiter) Dark, reddish bands in Jupiter's cloud cover.

Big Bang An explosion that took place roughly 15 billion years ago, creating all space, time, matter, and energy in which the universe emerged.

binary star Two stars revolving about each other; a double star.

birth line A line on the Hertzsprung-Russell diagram corresponding to where stars with different masses transform from protostars to pre–main-sequence stars.

BL Lacertae (BL Lac) object A type of active galaxy; a blazar.

black hole An object whose gravity is so strong that the escape velocity from it exceeds the speed of light.

blackbody A hypothetical perfect radiator that absorbs and reemits all radiation falling upon it.

blackbody curve The curve obtained when the intensity of radiation from a blackbody at a particular temperature is plotted against wavelength.

blazar A BL Lacertae object.

blueshift A shift of all spectral features toward shorter wavelengths; the Doppler shift of light from an approaching source.

Bok globule A small, roundish dark nebula in which stars are forming.

brown dwarf Any of the planetlike bodies with less than 0.08 $M_{\odot}$ and more than about 13 $M_{Jupiter}$; such bodies do not have enough mass to sustain fusion in their cores.

C ring The faint, inner portion of Saturn's main ring system.

caldera The crater at the summit of a volcano.

capture theory The idea that the Moon was created at a different location in the solar system and subsequently captured by Earth's gravity.

carbonaceous chondrites A class of extremely ancient, carbon-rich meteorites.

Cassegrain focus An optical arrangement in a reflecting telescope in which light rays are reflected by a secondary mirror through a hole in the primary mirror.

Cassini division A prominent gap between Saturn's A and B rings discovered in 1675 by J. D. Cassini.

celestial equator A great circle on the celestial sphere 90° from the celestial poles.

celestial sphere A hypothetical sphere of very large radius centered on the observer; the apparent sphere of the night sky.

center of mass The point around which a rigid system is perfectly balanced in a gravitational field; also, the point in space around which mutually orbiting bodies have elliptical orbits.

central bulge A flattened sphere of stars centered on a spiral galaxy's nucleus, extending out to the vicinity of the spiral arms.

Cepheid variable One of two types of yellow, supergiant, pulsating stars.

Cerenkov radiation Radiation produced by particles traveling through a substance faster than light can.

Chandrasekhar limit The maximum mass of a white dwarf, about 1.4 $M_{\odot}$.

charge-coupled device (CCD) A type of solid-state silicon wafer designed to detect photons.

chondrites Stony meteorites that have never been melted or otherwise altered since they formed. They contain a variety of debris, including spherical droplets called chondrules.

chromatic aberration An optical property whereby different colors of light passing through a lens are focused at different distances from it.

chromosphere The layer in the solar atmosphere between the photosphere and the corona.

circumpolar stars All the stars that never set at a given latitude; all the stars between Polaris and the northern horizon.

close binary A binary star whose members are separated by a few stellar diameters.

closed universe A universe that contains enough matter to cause it to recollapse. It is finite in extent and has no "outside."

cluster (of galaxies) A collection of a few hundred to a few thousand galaxies bound by gravity.

cocreation theory The theory that the Moon formed simultaneously with Earth and in orbit around it.

collision-ejection theory The theory that the Moon was created by the impact of a planet-sized object with Earth; presently considered the most plausible theory of the Moon's formation.

coma (of a comet) The nearly spherical, diffuse gas surrounding the nucleus of a comet near the Sun.

comet A small body of ice and dust in orbit about the Sun. While passing near the Sun, a comet's vaporized ices give rise to a coma, tails, and a hydrogen envelope.

configuration (of a planet) A particular geometric arrangement of Earth, a planet, and the Sun.

confinement The moment shortly after the Big Bang when quarks bound together to form particles like protons and neutrons.

conjunction The alignment of two bodies in the solar system so that they appear in the same part of the sky as seen from Earth.

conservation of angular momentum The law of physics stating that the total amount of angular momentum in an isolated system remains constant.

conservation of linear momentum If the sum of the external forces on a system remains zero, the total linear momentum of the system remains constant.

constellation Any of the 88 contiguous regions that cover the entire celestial sphere, including all the objects in each region; also, a configuration of stars often named after an object, a person, or an animal.

contact binary A close binary system in which both stars fill or overflow their Roche lobes.

continental drift The gradual movement of the continents over the surface of Earth due to plate tectonics.

continuous spectrum (continuum) A spectrum of light over a range of wavelengths without any spectral lines.

convection The transfer of energy by moving currents of fluid or gas containing that energy.

convective zone A layer in a star where energy is transported outward by means of convection; also known as the *convective envelope* or *convection zone*.

core The central portion of any astronomical object.

core (of the Sun) The central quarter of the Sun's radius, in which hydrogen is fused into helium, releasing the energy that enables the Sun to shine.

core helium fusion The fusion of helium to form carbon and oxygen at the center of a star.

corona The Sun's outer atmosphere.

coronal hole A dark region of the Sun's inner corona as seen at X-ray wavelengths.

coronal mass ejection Large volumes of high-energy gas released from the Sun's corona.

cosmic censorship The belief that the only connection between a black hole and the universe is the black hole's event horizon.

cosmic light horizon A sphere, centered on Earth, whose radius equals the distance traveled by light since the Big Bang.

cosmic microwave background Photons from every part of the sky with a blackbody spectrum at 2.73 K; the cooled-off radiation from the primordial fireball that originally filled all space.

cosmic rays (primary cosmic rays) High-speed particles traveling through space.

cosmic ray shower (see also the related **secondary cosmic rays**) Groups of particles from Earth's atmosphere propelled Earthward by the impact of a cosmic ray.

cosmological constant A number sometimes inserted in the equations of general relativity that represents a pressure that opposes gravity throughout the universe.

cosmological redshift An increase in wavelength from distant galaxies and quasars caused by the expansion of the universe.

cosmology The study of the formation, organization, and evolution of the universe.

coudé focus A reflecting telescope in which a series of mirrors direct light to a remote focus away from the moving parts of the telescope.

crater A circular depression on a celestial body caused by the impact of a meteoroid, asteroid, or comet or by a volcano.

crust The solid surface layer of some astronomical bodies, including the terrestrial planets, the moons, the asteroids, and some stellar remnants.

dark ages The age of the universe between the time of decoupling and the first burst of star formation.

dark energy A repulsive gravitational effect that is causing the universe to accelerate outward.

dark matter The as-yet-undetected matter in the universe that is underluminous and probably quite different from ordinary matter.

dark nebula A cloud of interstellar gas and dust that obscures the light of more distant stars.

declination (dec) The coordinate on the celestial sphere exactly analogous to latitude on Earth; measured north and south of the celestial equator.

decoupling The epoch in the early universe when electrons and ions first combined to create stable atoms; the time when electromagnetic radiation ceased to dominate over matter.

degree (°) A unit of angular measure or a temperature measure.

dense core Any of the regions of interstellar gas clouds that are slightly denser than normal and destined to collapse to form one or a few stars.

detached binary A binary system in which the surfaces of both stars are inside their Roche lobes.

differential rotation The rotation of a nonrigid object in which parts at different latitudes or different radial distances move at different speeds.

diffraction grating An optical device consisting of closely spaced lines ruled on a piece of glass that is used like a prism to disperse light into a spectrum.

direct motion The gradual, eastward apparent motion of a planet against the background stars as seen from Earth.

disk (of a galaxy) A flattened assemblage of stars, gas, and dust in a spiral galaxy like the Milky Way.

distance modulus The difference between the apparent and absolute magnitudes of an object.

diurnal motion Cyclic motion with a 1-day period.

Doppler shift The change in wavelength of radiation due to relative motion between the source and the observer along the line of sight.

double radio source An extragalactic radio source characterized by two large lobes of radio emission, often located on either side of an active galaxy.

Drake equation A mathematical equation used to estimate the number of extraterrestrial civilizations that may exist in our Galaxy.

dust devil Whirlwind found in dry or desert areas on both Earth and Mars.

dust tail A comet tail caused by dust particles escaping from the comet's nucleus.

dwarf planet A celestial body that is in orbit around the Sun and has sufficient mass for its self-gravity to pull the body into a nearly spherical shape, but does not have enough gravity to clear its orbital neighborhood of all the small debris orbiting there.

dynamo theory The generation of a magnetic field by circulating electric charges.

eclipse path The track of the tip of the Moon's shadow along Earth's surface during a total or annular solar eclipse.

eclipsing binary A double star system in which stars periodically pass in front of each other as seen from Earth.

ecliptic The annual path of the Sun on the celestial sphere; the plane of Earth's orbit around the Sun.

Einstein cross The appearance of four images of the same galaxy or quasar due to gravitational lensing by an intervening galaxy.

Einstein ring The circular or arc-shaped image of a distant galaxy or quasar created by gravitational lensing by an intervening galaxy.

ejecta blanket The ring of material surrounding a crater that was ejected during the crater-forming impact.

electromagnetic force The interaction between charged particles, the second of four fundamental forces in nature.

electromagnetic radiation Radiation consisting of oscillating electric and magnetic fields, namely gamma rays, X rays, visible light, ultraviolet and infrared radiation, and radio waves.

electromagnetic spectrum The entire array of electromagnetic radiation.

electron A negatively charged subatomic particle usually found in orbit about the nucleus of an atom.

electron degeneracy pressure A powerful pressure produced by repulsion of closely packed (degenerate) electrons.

element A substance that cannot be decomposed by chemical means into simpler substances. Every atom of the same element contains the same number of protons.

ellipse A closed curve obtained by cutting completely through a circular cone with a plane; the shape of planetary orbits.

elliptical galaxy A galaxy with an elliptical shape, little interstellar matter, and no spiral arms.

elongation The angle between a planet and the Sun as seen from Earth.

emission line A bright line of electromagnetic radiation.

emission line spectrum A spectrum that contains only bright emission lines.

emission nebula A glowing gaseous nebula whose light comes from fluorescence caused by a nearby star.

Encke division A thin gap in Saturn's A ring, possibly first seen by J. F. Encke in 1838.

energy flux The amount of energy emitted from each square meter of an object's surface per second.

equinox Either of the two days of the year when the Sun crosses the celestial equator and is therefore directly over Earth's equator; *see also* **autumnal equinox** *and* **vernal equinox.**

era of recombination The time, roughly 500,000 years after the Big Bang, when the universe became transparent.

ergoregion The region of space immediately outside the event horizon of a rotating black hole where it is impossible to remain at rest.

event horizon The location around a black hole where the escape velocity equals the speed of light; the boundary of a black hole.

evolutionary track On the Hertzsprung-Russell diagram, the path followed by a point representing an evolving star.

excited state The orbit of an electron with energy greater than the lowest energy orbit (or state) available to that election.

expanding universe The motion of the superclusters of galaxies away from each other.

eyepiece lens A magnifying lens used to view the image produced at the focus of a telescope.

F ring A thin ring just beyond the outer edge of Saturn's main ring system.

filament A dark curve seen above the Sun's photosphere that is the top view of a solar prominence.

fission theory The theory that the Moon formed from matter flung from Earth because the planet was rotating extremely fast.

flare *See* **solar flare.**

focal length The distance from a lens or concave mirror to where converging light rays meet.

focal plane The plane at the focal length of a lens or concave mirror on which an extended object is focused.

focal point The place at the focal length where light rays from a point object (that is, one that is too distant or tiny to resolve) are converged by a lens or concave mirror.

focus (*plural* **foci**) **(of an ellipse)** The two points inside an ellipse the sum of whose distances from any point on the ellipse is constant.

force That which can change the momentum of an object.

frequency The number of peaks or troughs of a wave that pass a fixed point each second. Also, the number of complete vibrations or oscillations per second.

galactic cannibalism A collision between two galaxies of unequal mass and size in which the smaller galaxy is absorbed by the larger galaxy.

galactic merger A collision and subsequent merger of two roughly equal-sized galaxies.

galactic nucleus The center of a galaxy; the center of the Milky Way Galaxy.

galaxy A large assemblage of stars, gas, and dust bound together by their mutual gravitational attraction.

Galilean moon (Galilean satellite) Any one of the four large moons of Jupiter (Callisto, Europa, Ganymede, Io) that is visible from Earth through a small telescope.

gamma ray The most energetic form of electromagnetic radiation.

gamma-ray burst A short burst of gamma rays; the sources of the bursts are outside our Galaxy.

gas (ion) tail The relatively straight tail of a comet produced by the solar wind acting on ions in a comet's coma.

giant molecular cloud A large interstellar cloud of cool gas and dust in a galaxy.

giant star A star whose diameter is roughly 10 to 100 times that of the Sun.

glitch A sudden speedup in the period of a pulsar.

globular cluster A large spherical cluster of gravitationally bound stars usually found in the outlying regions of a galaxy.

Grand Unified Theory (GUT) A theory that describes and explains the four physical forces.

granules Lightly colored convection features about 1000 km in diameter seen constantly in the solar photosphere.

gravitation The tendency of all matter to attract all other matter.

gravitational lensing The distortion of the appearance of an object by a source of gravity between it and the observer.

gravitational radiation *See* **gravitational waves.**

gravitational redshift The redshift of photons leaving the gravitational field of any massive object, such as a star or black hole.

gravitational waves Ripples in the overall geometry of space produced by nonspherical moving objects.

gravity *See* **gravitation.**

Great Dark Spot A large, dark, oval-shaped storm that used to be in Neptune's southern hemisphere.

Great Red Spot A large, red-orange, oval-shaped storm in Jupiter's southern hemisphere.

greenhouse effect The trapping of infrared radiation near a planet's surface by the planet's atmosphere.

ground state The lowest energy level of an atom.

H II region A region of ionized hydrogen in interstellar space.

habitable zone The region around any star wherein water can exist in liquid form and, hence, life as we know it can conceivably exist.

halo (of a galaxy) A spherical distribution of globular clusters, isolated stars, and possibly dark matter that surrounds a galaxy.

Hawking process The formation of real particles from virtual ones just outside a black hole's event horizon; the means by which black holes evaporate.

head-tail source A radio galaxy whose radio emission is deflected from the galaxy.

heliocentric cosmology A theory of the formation and evolution of the solar system with the Sun at the center.

helioseismology The study of vibrations of the solar surface.

helium flash The explosive ignition of helium fusion in the core of a low-mass, giant star.

helium shell flash The explosive ignition of helium fusion in a thin shell surrounding the core of a low-mass star.

helium shell fusion Helium fusion that occurs in a thin shell surrounding the core of a star.

Hertzsprung-Russell (H-R) diagram A plot of the absolute magnitude or luminosity of stars versus their surface temperatures or spectral classes.

highlands Heavily cratered, mountainous regions of the lunar surface.

homogeneity The property of the universe being smooth or uniform as measured over suitably large distance intervals.

horizon problem The difficulty in explaining why seemingly disconnected regions of the universe have the same temperature.

horizontal-branch stars A group of post–helium-flash stars near the main sequence on the Hertzsprung-Russell diagram of a typical globular cluster.

hot-spot volcanism The creation of volcanoes on a planet's surface caused by a reservoir of hot magma in the planet's mantle under a thin part of the crust.

Hubble classification A system of classifying galaxies according to their appearance into one of four broad categories: spirals, barred spirals, ellipticals, and irregulars.

Hubble constant (H_0) The constant of proportionality in the relation between the recessional velocities of remote galaxies and their distances; the correct value will determine the age of the universe.

Hubble flow The recession of the galaxies caused by the expansion of the universe.

Hubble law The relationship that states that the redshifts of remote galaxies are directly proportional to their distances from Earth.

hydrocarbon A molecule based on hydrogen and carbon.

hydrogen envelope An extremely large, tenuous sphere of hydrogen gas surrounding the head of a comet.

hydrogen fusion The thermonuclear fusion of hydrogen to produce helium.

hydrogen shell fusion Hydrogen fusion that occurs in a thin shell surrounding the core of a star.

hydrostatic equilibrium A balance between the weight of a layer in a star and the pressure that supports it.

hyperbola An open curve obtained by cutting a cone with a plane.

impact breccia A rock consisting of various fragments cemented together by the impact of a meteoroid.

impact crater A crater on the surface of a planet or moon produced by the impact of an asteroid, meteoroid, or comet.

inferior conjunction The configuration when Mercury or Venus is directly between the Sun and Earth.

inflation A sudden expansion of space.

inflationary epoch A brief period shortly after the Big Bang during which the scale of the universe increased very rapidly.

infrared radiation Electromagnetic radiation of a wavelength longer than visible light but shorter than radio waves.

initial mass function The numbers of stars on the main sequence at all different masses.

instability strip A region on the Hertzsprung-Russell diagram occupied by pulsating stars.

interferometry A method of increasing resolving power by combining electromagnetic radiation obtained by two or more telescopes.

intergalactic gas Gas located between the galaxies within a cluster of galaxies.

intermediate-mass black hole A black hole with a mass between a few hundred and few thousand solar masses.

interstellar extinction The dimming of starlight as it passes through the interstellar medium.

interstellar medium Interstellar gas and dust.

interstellar reddening The reddening of starlight passing through the interstellar medium resulting from the scattering of short wavelength light more than long wavelength light.

inverse-square law The gravitational attraction between two objects and the apparent brightness of a light source are both inversely proportional to the square of its distance.

ion An atom that has become electrically charged due to the loss or addition of one or more electrons.

ionization The process by which an atom loses or gains electrons.

ionosphere (thermosphere) Region of Earth's atmosphere, above the mesosphere, in which sunlight ionizes many atoms.

iron meteorite A meteorite composed primarily of iron with an admixture of nickel; also called an *iron*.

irregular cluster (of galaxies) An unevenly distributed group of galaxies bound together by their mutual gravitational attraction.

irregular galaxy An asymmetrical galaxy having neither spiral arms nor an elliptical shape.

isotopes Atoms that all have the same number of protons (atomic number) but different numbers of neutrons. Their nuclear properties often differ greatly.

isotropy The fact that the average number of galaxies at different distances from Earth is the same in all directions; also, the fact that the temperature of the cosmic microwave background is essentially the same in all directions.

isotropy problem *See* **horizon problem.**

Jeans instability The condition under which gravitational forces overcome thermal forces to cause part of an interstellar cloud to collapse and form stars and planets.

Kepler's laws Three statements, formulated by Johannes Kepler, that describe the motions of the planets.

Kerr black hole Any rotating, uncharged black hole.

kinetic energy The energy an object has as a result of its motion.

Kirchhoff's laws Three statements formulated by Gustav Kirchhoff describing what physical conditions produce each type of spectra.

Kirkwood gaps Gaps in the spacing of asteroid orbits discovered by Daniel Kirkwood caused by gravitational attractions of planets.

Kuiper belt A doughnut-shaped ring of space around the Sun beyond Neptune that contains many frozen comet bodies, some of which are occasionally deflected toward the inner solar system.

Kuiper belt object (KBO) Space debris orbiting the Sun beyond Neptune in the Kuiper belt. The debris includes small solar system bodies like Pluto, as well as myriad small rocky and icy objects, including many comet nuclei.

law of equal areas (Kepler's second law) The physical law that a line joining a planet and the Sun sweeps out equal areas in equal intervals of time.

law of inertia (Newton's first law of motion) The physical law that an object will stay at rest or move at a constant speed in a fixed direction unless acted upon by an outside force.

law of universal gravitation Newton's law of gravitation, which describes how the gravitational force between two bodies depends on their masses and separation.

lenticular galaxy A disk-shaped galaxy without spiral arms.

light curve A graph that displays variations in the brightness of a star or other astronomical object over time.

light-gathering power A measure of how much light a telescope intercepts and brings to a focus.

light-year (ly) The distance that light travels in a vacuum in 1 year.

lighthouse model The explanation that a pulsar pulses by rotating and funneling energy outward via magnetic fields that are not aligned with the rotation axis.

limb (of the Sun) The apparent edge of the Sun as seen in the sky.

limb darkening The phenomenon whereby the Sun is darker near its limb than near the center of its disk.

line of nodes The line along which the plane of the Moon's orbit intersects the plane of the ecliptic.

liquid metallic hydrogen A metallike form of hydrogen that is produced under extreme pressure.

Local Group The cluster of about 40 galaxies of which our own Galaxy is a member.

long-period comet A comet that takes tens of thousands of years or more to orbit the Sun once.

luminosity The rate at which electromagnetic radiation is emitted from a star or other object.

luminosity class The classification of a star of a given spectral type according to its luminosity and density; the classes are supergiant, bright giant, giant, subgiant, and main sequence.

lunar eclipse An eclipse during which Earth blocks light that would have struck the Moon.

lunar phases The names given to the apparent shapes of the Moon as seen from Earth.

magnetar Especially hot, rapidly rotating neutron stars whose motion helps generate extra-strong magnetic fields.

magnetic dynamo A theory that explains phenomena of the solar cycle as a result of periodic winding and unwinding of the Sun's magnetic field in the solar atmosphere.

magnification The number of times larger in angular diameter an object appears through a telescope than when it is seen by the naked eye.

main sequence A grouping of stars on the Hertzsprung-Russell diagram extending diagonally across the graph from the hottest, brightest stars to the dimmest, coolest stars.

main-sequence star A star, fusing hydrogen to helium in its core, whose surface temperature and luminosity place it on the main sequence on the Hertzsprung-Russell diagram.

mantle That portion of a terrestrial planet located between its crust and core.

mare (*plural* **maria**) Latin for "sea"; a large, relatively crater-free plain on the Moon.

mare basalt Dark, solidified lava that covers the lunar maria.

mascons Regions of high-density matter near the surface of the Moon.

mass A measure of the total amount of material in an object.

mass-luminosity relation The direct relationship between the masses and luminosities of main-sequence stars.

matter-dominated universe A universe in which the radiation field that fills all space is unable to prevent the existence of neutral atoms.

mesosphere The layer in Earth's atmosphere above the stratosphere.

metals All elements except hydrogen and helium.

meteor The streak of light seen when any space debris vaporizes in Earth's atmosphere; a "shooting star."

meteor shower Frequent meteors that seem to originate from a common point in the sky.

meteorite A fragment of space debris that has survived passage through Earth's atmosphere.

meteoroid A small rock in interplanetary space.

microlensing The gravitational focusing of light from a distant star by a closer object to give a brighter image of the star.

Milky Way Galaxy The galaxy in which our solar system resides.

missing mass *See* **dark matter.**

model A hypothesis that has withstood observational or experimental tests.

molecular clouds Nebulae that are often embedded in much larger bodies of gas and dust.

molecule Two or more atoms bonded together.

moment of inertia A measure of the inertial resistance of an object to changes in the object's rotational motion about the axis.

momentum A measure of the inertia of an object; an object's mass multiplied by its velocity.

moons (natural satellites) Bodies that orbit larger objects, which in turn orbit stars.

neap tide The least change from high to low tide during a day; it occurs during the first and third quarter phases of the Moon.

nebula (*plural* **nebulae**) A cloud of interstellar gas and dust.

neutrino A subatomic particle, with no electric charge and little mass, that is important in many nuclear reactions and in supernovae.

neutron A nuclear particle with no electric charge and with a mass nearly equal to that of the proton.

neutron degeneracy pressure A powerful pressure produced by degenerate neutrons.

neutron star A very compact, dense stellar remnant composed almost entirely of neutrons.

Newton's laws of motion Newton's equations that describe the motion of matter as a result of forces acting on it.

Newtonian reflector An optical arrangement in a reflecting telescope in which a small, flat mirror reflects converging light rays to a focus on one side of the telescope tube.

Nice model A theory that describes the formation of the planets and other objects orbiting the Sun, with Jupiter and the other giant planets forming first, followed by the inner planets, asteroid belt, Kuiper belt, and Oort comet cloud.

north celestial pole The location on the celestial sphere directly above Earth's northern rotation pole.

northern lights (aurora borealis) Light radiated by atoms and ions in Earth's upper atmosphere due to high-energy particles from the Sun and seen mostly in the northern polar regions.

northern vastness (northern lowlands) Relatively young, crater-free terrain in the northern hemisphere of Mars.

nova (*plural* **novae**) A star in a binary system that experiences a sudden outburst of radiant energy, temporarily increasing its luminosity by a factor of between 104 and 106.

nucleosynthesis The formation, by fusion, of higher mass elements from lower mass ones.

nucleus (of a comet) A collection of ices and dust that constitute the solid part of a comet.

nucleus (of an atom) The massive part of an atom, composed of protons and neutrons; electrons surround a nucleus.

OB association An unbound group of very young, massive stars predominantly of spectral types O and B.

OBAFGKM sequence The sequence of stellar spectral classifications from hottest to coolest stars.

objective lens The principal lens of a refracting telescope.

Occam's razor The principle of choosing the simplest scientific theory that correctly explains any phenomenon.

occultation The eclipsing of an astronomical object other than the Moon or Sun by another astronomical body.

Oort cloud A hollow spherical region of the solar system beyond the Kuiper belt where most comets are believed to spend most of their time.

open cluster A loosely bound group of young stars in the disk of the galaxy; a galactic cluster.

open universe A universe with a hyperbolic shape; lacks the mass necessary to someday stop expanding and recollapse. It will expand forever.

opposition The configuration of a planet when it is at an elongation of 180° and thus appears opposite the Sun in the sky.

optical double A pair of stars that appear to be near each other but are unbound and at very different distances from Earth.

orbital inclination The tilt or angle of an object's orbital plane around the Sun compared to the ecliptic.

organic molecule A carbon-based compound.

overcontact binary A close binary system in which the two stars share a common atmosphere.

ozone layer The lower stratosphere, where most of the ozone in the air exists.

pair production The creation of a particle and an antiparticle from energetic photons.

parabola An open curve formed by cutting a circular cone at an angle parallel to the sides of the cone.

parallax The apparent displacement of an object relative to more distant objects caused by viewing it from different locations.

parsec (pc) A unit of distance equal to 3.26 lightyears.

partial eclipse A lunar or solar eclipse in which the eclipsed object does not appear completely covered.

Pauli exclusion principle A principle of quantum mechanics that states that two identical particles cannot simultaneously have the same position and momentum.

peculiar galaxy (pec) Any Hubble class of galaxy that appears to be blowing apart.

penumbra The portion of a shadow in which only part of the light source is covered by the shadowmaking body.

penumbral eclipse A lunar eclipse in which the Moon passes only through Earth's penumbra.

perihelion The point in its orbit where a planet is nearest the Sun.

period-luminosity relation A relationship between the period and average luminosity of a pulsating star.

periodic table A listing of the chemical elements according to their properties; created by D. Mendeleev.

photodisintegration The breakup of nuclei in the core of a massive star due to the effects of energetic gamma rays.

photometry The measurement of light intensities.

photon A discrete unit of electromagnetic energy.

photosphere The region in the solar atmosphere from which most of the visible light escapes into space.

pixel A contraction of the term "picture element"; usually refers to one square of a grid into which the light-sensitive component of a charge-coupled device is divided.

plage A bright spot on the Sun believed to be associated with an emerging magnetic field.

Planck era Time from the Big Bang until the Planck time (10^{-43} s).

Planck time The earliest time, about 10^{-43} s after the Big Bang, for which science has equations describing the universe. At that time all four forces in nature today (gravity, electromagnetism, weak, and strong) behaved as one force.

Planck's law A relationship between the energy carried by a photon and its wavelength.

planet An object orbiting a star that is held together by its own gravitational force in a nearly spherical shape, that is able to clear its neighborhood of debris, and is not the moon (or satellite) of a larger orbiting body.

planetary differentiation The process early in the life of each planet whereby denser elements sank inward and lighter ones rose.

planetary nebula A luminous shell of gas ejected from an old, low-mass star.

planetesimal Primordial asteroidlike object from which the planets accreted.

plasma A hot, ionized gas.

plate tectonics The motions of large segments (plates) of Earth's surface caused by convective motions in the underlying mantle.

polymer A long molecule composed of many smaller molecules.

poor cluster (of galaxies) A cluster of galaxies with only a few members.

Population I star A star, such as the Sun, whose spectrum exhibits spectral lines of many elements heavier than helium; a metal-rich star.

Population II star A star whose spectrum exhibits comparatively few spectral lines of elements heavier than helium; a metal-poor star.

positron An electron with a positive rather than negative electric charge; an antielectron.

potential energy The energy stored in an object as a result of its location in space.

pre–main-sequence star The stage of star formation just before the main sequence; it involves slow contraction of the young star.

precession (of Earth) A slow, conical motion of Earth's axis of rotation caused by the gravitational pull of the Moon and Sun on Earth's equatorial bulge.

precession of the equinoxes The slow westward motion of the equinoxes along the ecliptic because of Earth's precession.

primary cosmic rays (cosmic rays) High-speed particles traveling through space.

primary mirror The large, concave, light-gathering mirror in a reflecting telescope, analogous to the objective lens on a refracting telescope.

prime focus The point in a reflecting telescope where the primary mirror focuses light.

primordial black hole A relatively low-mass black hole hypothetically formed at the beginning of the universe.

primordial fireball The extremely hot gas that filled the universe immediately following the Big Bang.

primordial nucleosynthesis The transformation by fusion of protons and electrons into hydrogen isotopes, helium, and some lithium in the first few minutes of the existence of the universe.

prograde orbit An orbit of a moon or satellite around a planet that is in the same direction as the planet's rotation.

prominence Flamelike protrusion seen near the limb of the Sun and extending into the solar corona. The side view of a filament.

proper motion The change in the location of a star on the celestial sphere.

proton A heavy, positively charged nuclear particle.

protoplanetary disk (proplyd) A disk of material encircling a protostar or a newborn star.

protostar The earliest stage of a star's life before fusion commences and when gas is rapidly falling onto it.

protosun The Sun prior to the time when hydrogen fusion began in its core.

pulsar A pulsating source associated with a rapidly rotating neutron star with an off-axis magnetic field.

quantum mechanics The branch of physics dealing with the structure and behavior of atoms and their interactions with each other and with light.

quark A particle that is a building block of the heavy nuclear particles such as protons and neutrons.

quasar (quasi-stellar radio source) A starlike object with a very large redshift.

quasi-stellar object (QSO) A quasar.

quintessence One of the explanations of the dark energy causing the universe to accelerate outward.

radial velocity That portion of an object's velocity parallel to the line of sight.

radial-velocity curve A plot showing the variation of radial velocity with time for a binary star or variable star.

radiation (photon) pressure The transfer of momentum carried by radiation to an object on which the radiation falls.

radiation-dominated universe The time at the beginning of the universe when the electromagnetic radiation prevented ions and electrons from combining to make neutral atoms.

radiative zone A region inside a star where energy is transported outward by the movement of photons through a gas from a hot location to a cooler one.

radio galaxy A galaxy that emits an unusually large amount of radio waves.

radio lobes Vast regions of radio emission on opposite sides of a radio galaxy.

radio telescope A telescope designed to detect radio waves.

radio wave Long-wavelength electromagnetic radiation.

radioactive Unstable atomic nuclei that naturally decompose by spontaneously emitting particles.

red dwarf A low-mass main-sequence star.

red giant A large, cool star of high luminosity.

redshift The shifting to longer wavelengths of the light from remote galaxies and quasars; the Doppler shift of light from any receding source.

reflecting telescope (reflector) A telescope in which the principal light-gathering component is a concave mirror.

reflection The rebounding of light rays off a smooth surface.

reflection nebula A comparatively dense cloud of gas and dust in interstellar space that is illuminated by a star between it and Earth.

refracting telescope (refractor) A telescope in which the principal light-gathering component is a lens.

refraction The bending of light rays when they pass from one transparent medium to another.

refractor *See* **refracting telescope.**

regolith The powdery, lifeless material on the surface of a moon or planet.

regular cluster (of galaxies) An evenly distributed group of galaxies bound together by mutual gravitational attraction.

resonance The large response of an object to a small periodic gravitational tug from another object.

retrograde motion The occasional backward (that is, westward) apparent motion of a planet against the background stars as seen from Earth. Retrograde motion is an optical illusion.

retrograde orbit The orbit of a moon or satellite around a planet that is in the direction opposite to the planet's rotation.

retrograde rotation The rotation of a planet opposite to its direction of revolution around the Sun. Only Pluto, Uranus, and Venus have retrograde rotation.

revolution The orbit of one body about another.

rich cluster (of galaxies) A cluster of galaxies with many members.

right ascension (r.a.) The celestial coordinate analogous to longitude on Earth and measured around the celestial equator from the vernal equinox.

rille A winding crack or depression in the lunar surface caused by the collapse of a solidified lava tube.

ringlet Any one of numerous, closely spaced, thin bands of particles in planetary ring systems.

Roche limit The shortest distance from a planet or other object at which a second object can be held together by its own gravitational forces.

Roche lobe The teardrop-shaped regions around each star in a binary star system inside of which gas is gravitationally bound to that star.

rotation The spinning of a body about an axis passing through it.

rotation curve (of a galaxy) A graph showing how the orbital speed of material in a galaxy depends on the distance from the galaxy's center.

RR Lyrae variable A type of pulsating star with a period less than 1 day.

Sagittarius A The strong radio source associated with the nucleus of the Milky Way Galaxy.

scarp A cliff on Mercury believed to have formed when the planet cooled and shrank.

Schmidt corrector plate A specially shaped lens used with spherical mirrors that corrects for spherical aberration and provides an especially wide field of view.

Schwarzschild black hole Any nonrotating, uncharged black hole.

Schwarzschild radius The distance from the center to the event horizon in any black hole.

scientific method The method of doing science based on observation, experimentation, and the formulation of hypotheses (theories) that can be tested.

scientific notation The style of writing large and small numbers using powers of ten.

scientific theory An idea about the natural world that is subject to verification and refinement.

seafloor spreading The process whereby magma upwelling along rifts in the ocean floor causes adjacent segments of Earth's crust to separate.

secondary cosmic rays (cosmic ray shower) Particles from Earth's atmosphere given high speeds Earthward by cosmic rays from space.

secondary mirror A relatively small mirror used in reflecting telescopes to guide the light out the side or bottom of the telescope.

seeing disk The size that a star appears to have on a photographic or charge-coupled-device image as a result of the changing refraction of the starlight passing through Earth's atmosphere.

seismic waves Vibrations traveling through or around an astronomical body usually associated with earthquakelike phenomena.

seismograph A device used to record and measure seismic waves, such as those produced by earthquakes.

semidetached binary A close binary system in which one star fills or is overflowing its Roche lobe.

semimajor axis (of an ellipse) Half of the longest dimension of an ellipse.

SETI The search for extraterrestrial intelligence.

Seyfert galaxy A spiral galaxy with a bright nucleus whose spectrum exhibits emission lines.

Shapley–Curtis debate An inconclusive debate between Harlow Shapley and Heber Curtis in 1920 about whether certain nebulae were beyond the Milky Way.

shepherd satellite (moon) A small satellite whose gravitational tug is responsible for maintaining a sharply defined ring of matter around a planet such as Saturn or Uranus.

short-period comet A comet that orbits the Sun in the vicinity of the planets, thereby reappearing with tails every 200 years or less.

sidereal month The period of the Moon's revolution about Earth measured with respect to the Moon's location among the stars; 27⅓ Earth days.

sidereal period The orbital period of one object about another measured with respect to the stars.

singularity A place of infinite curvature of spacetime in a black hole.

small solar system bodies (SSSBs) All objects in the solar system that are not planets, dwarf planets, or moons.

snow line The distance from the Sun beyond which ices stayed frozen in the early solar system.

solar corona The Sun's outer atmosphere.

solar cycle A 22-year cycle during which the Sun's magnetic field reverses its polarity twice.

solar day From noontime to the next noontime; for Earth it is 24 hours.

solar eclipse An eclipse during which the Moon blocks the Sun.

solar flare A violent eruption on the Sun's surface.

solar luminosity ($L_\odot$) The total energy emitted by the Sun each second.

solar model A set of equations that describe the internal structure and energy generation of the Sun.

solar nebula The cloud of gas and dust from which the Sun and the rest of the solar system formed.

solar system The Sun, planets, their satellites, asteroids, comets, and related objects that orbit the Sun.

solar wind An outward flow of particles (mostly electrons and protons) from the Sun.

south celestial pole The location on the celestial sphere directly above Earth's south rotation pole.

southern highlands Older, cratered terrain in the Martian southern hemisphere.

southern lights (aurora australis) Light radiated by atoms and ions in Earth's upper atmosphere due to high-energy particles from the Sun; seen mostly in the southern polar regions.

spacetime The concept from special relativity that space and time are both essential in describing the position, motion, and action of any object or event.

spectral analysis The identification of chemicals by the appearance of their spectra.

spectral type A classification of stars according to the appearance of their spectra.

spectrograph A device for photographing a spectrum.

spectroscope A device for directly viewing a spectrum.

spectroscopic binary A double star whose binary nature can be deduced from the periodic Doppler shifting of lines in its spectrum.

spectroscopic parallax A method of determining a star's distance from Earth by measuring its surface temperature, luminosity, and apparent magnitude.

spectrum (*plural* **spectra**) The result of electromagnetic radiation passing through a prism or grating so that different wavelengths are separated.

spherical aberration An optical property whereby different portions of a spherical lens or spherical, concave mirror have slightly different focal lengths, thereby producing a fuzzy image.

spicule A narrow jet of rising gas in the solar chromosphere.

spin (of an electron or proton) A small, well-defined amount of angular momentum possessed by electrons, protons, and other particles.

spiral arms Lanes of interstellar gas, dust, and young stars that wind outward in a plane from the central regions of some galaxies.

spiral density wave A spiral-shaped pressure wave that orbits the disk of a spiral galaxy and induces new star formation.

spiral galaxy A flattened, rotating galaxy with pinwheel-like spiral arms winding outward from the galaxy's nuclear bulge.

spoke A moving dark region of Saturn's rings.

spring tide The greatest daily difference between high tide and low tide, occurring when the Moon is new or full.

stable Lagrange points Locations throughout the solar system where the gravitational forces from the Sun and a planet keep space debris trapped.

standard candle An object whose known luminosity can be used to deduce the distance to a galaxy.

starburst galaxy A galaxy where there is an exceptionally high rate of star formation.

Stefan-Boltzmann law The relationship stating that an object emits energy at a rate proportional to the fourth power of its temperature, in Kelvins.

stellar evolution The changes in size, luminosity, temperature, and chemical composition that occur as a star ages.

stellar parallax The apparent shift in a nearby star's position on the celestial sphere resulting from Earth's orbit around the Sun.

stellar spectroscopy The study of the properties of stars encoded in their spectra.

stony meteorite A meteorite composed of rock with very little iron; also called a *stone*.

stony-iron meteorite A meteorite composed of roughly equal amounts of rock and iron.

stratosphere The second layer in Earth's atmosphere, directly above the troposphere.

strong nuclear force The force that binds protons and neutrons together in nuclei.

summer solstice The point on the ecliptic where the Sun is farthest north of the celestial equator; the day with the largest number of daylight hours in the northern hemisphere, around June 21.

sunspot A temporary cool region in the solar photosphere created by protruding magnetic fields.

sunspot maximum The time during the solar cycle when the number of sunspots is greatest.

sunspot minimum The time during the solar cycle when the number of sunspots is minimum.

supercluster (of galaxies) A gravitationally bound collection of many clusters of galaxies.

supergiant A star of very high luminosity.

supergranule A large convective cell in the Sun's chromosphere containing many granules.

superior conjunction The configuration when Mercury or Venus is on the opposite side of the Sun from Earth.

supermassive black hole A black hole whose mass exceeds 1000 solar masses.

supernova remnant A nebula left over after a supernova detonates.

superstring theories A set of theories that hope to describe the nature of spacetime and matter at a more fundamental level than is presently possible.

synchronous rotation The condition when a moon's rotation rate and revolution rate are equal or when a planet's rotation rate equals its moon's revolution rate.

synchrotron radiation The radiation emitted by charged particles moving through a magnetic field; nonthermal radiation.

synodic month (lunar month) The period of revolution of the Moon with respect to the Sun; the length of one cycle of lunar phases; 29½ Earth days.

synodic period The interval between successive occurrences of the same configuration of a planet as seen from Earth.

T Tauri stars Young, variable pre–main-sequence stars associated with interstellar matter that show erratic changes in luminosity.

terminator The line dividing day and night on the surface of any body orbiting the Sun; the line of sunset or sunrise.

terrestrial planet Any of the planets Mercury, Venus, Earth, or Mars; a planet with a composition and density similar to that of Earth.

Theories of Everything Theories under development that comprehensively explain all four fundamental forces in nature.

theory *See* **scientific theory.**

theory of general relativity A description of spacetime formulated by Einstein explaining how gravity affects the geometry of space and the flow of time.

theory of special relativity A description of mechanics and electromagnetic theory formulated by Einstein according to which measurements of distance, time, and mass are affected by the observer's motion.

thermonuclear fusion A reaction in which the nuclei of atoms are fused together at a high temperature.

3-to-2 spin-orbit coupling The rotation of Mercury, which makes three complete rotations on its axis for every two complete orbits around the Sun.

time zone One of 24 divisions of Earth's surface separated by 15° along lines of constant longitude (with allowances for some political boundaries).

total eclipse A solar eclipse during which the Sun is completely hidden by the Moon, or a lunar eclipse during which the Moon is completely immersed in Earth's umbra.

trailing-arm spiral galaxy A galaxy with its spiral arms pointing away from the direction of rotation, characteristic of all but one known spiral galaxies.

trans-Neptunian objects (TNOs) Objects orbiting the Sun beyond the orbit of Neptune.

transition (of an electron) The change in energy and orbit of an electron around an atom or molecule.

transition zone Region between the Sun's chromosphere and corona where the temperature skyrockets to about 1 million K.

transverse velocity The portion of an object's velocity perpendicular to our line of sight to it.

Trojan asteroid One of several asteroids at stable Lagrange points that share Jupiter's orbit about the Sun.

troposphere The lowest level of Earth's atmosphere.

Tully-Fisher relation A correlation between the width of the 21-centimeter line of a spiral galaxy and its absolute magnitude.

turnoff point The location of the brightest main-sequence stars on the Hertzsprung-Russell diagram of a globular cluster.

21-cm radiation Radio emission from a hydrogen atom caused by the flip of the electron's spin orientation.

twinkling The apparent change in a star's brightness, position, or color due to the motion of gases in the Earth's atmosphere.

Type I Cepheid A Population I Cepheid variable star found in the disks of spiral galaxies.

Type Ia supernova A supernova occurring after a white dwarf accretes enough mass from a companion star to exceed the Chandrasekhar limit.

Type II Cepheid A Population II Cepheid variable star found in elliptical galaxies and in the halos of disk galaxies that is 1.5 magnitudes dimmer than a Type I Cepheid.

Type II supernova A supernova occurring after a massive star's core is converted to iron.

ultraviolet (UV) radiation Electromagnetic radiation of wavelengths shorter than those of visible light but longer than those of X rays.

umbra The central, completely dark portion of a shadow.

universal constant of gravitation The constant of proportionality in Newton's law of gravitation, usually denoted G.

universe All space along with all the matter and radiation in space.

Van Allen radiation belts Two flattened, doughnut-shaped regions around Earth where many charged particles (mostly protons and electrons) are trapped by Earth's magnetic field.

variable star A star whose luminosity varies.

velocity A quantity that specifies both direction and speed of an object.

vernal equinox The point on the ecliptic where the Sun crosses the celestial equator from south to north; the beginning of spring, around March 21.

very-long-baseline interferometry (VLBI) A method of connecting widely separated radio telescopes to make observations of very high resolution.

virtual particle A particle and its antiparticle, created simultaneously in pairs and which quickly disappear without a trace.

visual binary A double star in which the two components can be resolved through a telescope.

water hole The part of the electromagnetic spectrum at a few thousand megahertz where there is very little background noise from space.

wavelength (λ) The distance between two successive peaks in a wave.

weak nuclear force A nuclear interaction involved in certain kinds of radioactive decay.

weight The force with which a body presses down on the surface of a world such as Earth.

white dwarf A low-mass stellar remnant that has exhausted all its thermonuclear fuel and contracted to a size roughly equal to the size of Earth.

Widmanstätten patterns Crystalline structure seen inside iron meteorites.

Wien's law The relationship that the dominant wavelength of radiation emitted by a blackbody varies inversely with its temperature.

winter solstice The point on the ecliptic where the Sun is farthest south of the celestial equator; fewest hours of daylight in the northern hemisphere, around December 22.

Wolf-Rayet stars Rotating stars of at least 20 $M_\odot$ with strong magnetic fields and stellar winds.

work Change in an object's energy as a result of a force being applied to it.

wormhole A hypothetical connecting passage between black holes and other places in the universe.

X rays Electromagnetic radiation whose wavelength is between that of ultraviolet light and gamma rays.

X-ray burster A neutron star in a binary star system that accretes mass, undergoes thermonuclear fusion on its surface, and therefore emits short bursts of X rays.

Zeeman effect A splitting of spectral lines in the presence of a magnetic field.

zenith The point on the celestial sphere directly overhead.

zero-age main sequence (ZAMS) The positions of stars on the Hertzsprung-Russell diagram that have just begun to fuse hydrogen in their cores.

zodiac A band of 13 constellations around the sky through which the Sun appears to move throughout the year.

zones (on Jupiter) Light-colored bands in Jupiter's cloud cover.

ANSWERS TO MARGIN QUESTIONS AND COMPUTATIONAL QUESTIONS

Chapter 1

MARGIN QUESTION: The planets move through constellations, while the stars do not. What two possible properties of the planets could allow them to do that? ANSWER: Two explanations for why the planets move through the constellations, but the stars do not are: 1. The planets are closer than the stars and the difference in distances allows us to see the planets move. (correct explanation) 2. The planets are about the same distances as the stars, but the planets are moving faster across the sky. (incorrect explanation)

MARGIN QUESTION: Explain why Figure 1-13 must have been taken facing west. *Hint*: Examine Figure 1-17. ANSWER: In the northern hemisphere, stars move southward (to the right) as they rise and northward (to the right) as they set. If this figure were taken in the morning, the stars would be moving in the wrong direction.

MARGIN QUESTION: Why are some of the paths in Figure 1-28 wider than others? ANSWER: The angle that the shadow strikes Earth determines how wide the eclipse path is. This angle varies with latitude, time from noon, and distance of the Moon from Earth. For example, near the equator, near noon, and with the Moon relatively far from Earth, the shadow is thinnest, while near the poles, especially far from noon, and when the Moon is closest to Earth, the shadow is particularly wide.

43. 1 more sidereal month than synodic month

53. $R_\odot \approx 6.53 \times 10^8$ m

Toolbox 1-1: ~0.53°. This is the same angle as the Moon makes in the sky; ~0.27°; ~1.1°

Chapter 2

MARGIN QUESTION: We saw in Chapter 1 that the Moon's orbit around Earth is not circular. Where in its orbit is the Moon moving fastest and where is it moving slowest? ANSWER: By Kepler's second law, the Moon is moving fastest where it is closest to Earth (perigee) and slowest where it is farthest away (apogee).

MARGIN QUESTION: Sitting in a moving car, how can you experimentally verify that your body has inertia? ANSWER: To show that your body has inertia, drive a car and then rapidly put on the brakes. You will feel yourself forced forward against the seatbelt. Without inertia, you would slow down at the same rate as the car does without having to be restrained by the seatbelt.

17. (a) 8.3 pc = 27.1 ly;
(b) 6.52 ly = 2.0 pc;
(c) 8459 AU = 1.26×10^{12} km;
(d) 2.7×10^3 Mpc = 2.7×10^6 kiloparsec

20. 5.2 square AU in 1995; 26 square AU in 5 years

21. a = 100 AU; maximum distance is almost 200 AU

22. 2.8 years

29. At 2 AU, 1 years = 2.8 present years

30. At ½ AU, 1 year = 0.35 present years

31. $0.01 \times$ present gravitational attraction

32. same acceleration; same length of the year

Toolbox 2-1: ~2.48×10^{13} mi, ~3.99×10^{13} km

Toolbox 2-2: Energy increases 9-fold; energy decreases to ¼; 2000 J; 0 J; change its moment of inertia, I, and its angular velocity ω.

Toolbox 2-3: Your weight in pounds; same; 9.8 m/s² *or* 32 ft/s²; 9×10^{21} N; It is one-quarter the present force.

Chapter 3

MARGIN QUESTION: The speed of sound is about 0.34 km/s (0.21 mi/s). How can you use this and the information in this section to determine the distance you are from a lightning strike? ANSWER: To determine the distance to a lightning strike, measure the time interval between the lightning and subsequent thunder. Since you see the lightning virtually instantaneously after it strikes, while the thunder travels at the much slower "speed of sound," the distance to the strike is just the speed of sound times the time between the lightning and thunder clap.

MARGIN QUESTION: Where in Figure 3-9c does the light from the primary mirror actually come to a focal point? ANSWER: The focal point of the primary mirror on Figure 3-9c is where the two light rays reflecting off the primary mirror meet to the left of the secondary mirror.

MARGIN QUESTION: Why do the human eye and brain clear the images that they receive many times per second? ANSWER: The brain clears images it receives in order to detect motion. Without clearing what we see very often, the brain would get saturated with light and would be unable to detect changes, i.e., motion.
MARGIN QUESTION: Where in a typical house would infrared detectors indicate most activity? ANSWER: At home, infrared detectors are useful for detecting things in the kitchen, in the furnace room, in fireplaces, and wherever else there are heat sources.

4. 9 times more

17. 1/25

18. Palomar gathers 10^6 times more light than the human eye.

20. (a) 222×; (b) 100×; (c) 36×

Toolbox 3-1: 406 nm; violet; 203 nm; assuming green at 500 nm, $E = 4.0 \times 10^{-19}$ J

Chapter 4

MARGIN QUESTION: From what we have just discussed, what do you think causes fireworks to have their distinctive colors? ANSWER: Transitions of electrons in the elements and molecules used in the fireworks emit specific wavelengths, giving fireworks their distinctive colors.
MARGIN QUESTION: Which Balmer line in Figure 4-11 is H_α? ANSWER: H_α on Figure 4-11 is the transition from $n = 3$ to $n = 2$. This transition's energy is least (the resulting photon have the longest wavelength) of all Balmer transitions.
MARGIN QUESTION: Does a police siren approaching you sound higher or lower in pitch than the siren at rest relative to you? ANSWER: An approaching siren has a higher pitch than does the same siren at rest relative to you.

15. 7½ times

16. 238 nm

17. 6520 K

18. approaching; 13 km/s

19. receding; 21 km/s

20. about 90,000 km/s

24. $\lambda = 500$ nm (same as now), 4× brighter than now

28. Rigel, no; Deneb, no; Arcturus, yes; Vega, no; Betelgeuse, no

Toolbox 4-1: ~5270 K; ~9700 nm; infrared
Toolbox 4-2: 1/8 kg; ~0.35 kg; ~17½
Toolbox 4-3: 3×10^4 km/s *or* 0.1c; $\sim 3.3 \times 10^{-4}$; −0.1

Chapter 5

MARGIN QUESTION: Which elements on Earth may have been unchanged since the universe began? ANSWER: Much of the hydrogen and helium, and some of the lithium on Earth formed shortly after the universe began. All other elements here were definitely formed from these building blocks since then.
MARGIN QUESTION: Why is Earth's albedo continually changing? ANSWER: The surface and atmospheric features of Earth, including cloud cover, ocean surface activity, snow, ice, and vegetation cover change, thereby changing the amount of light scattered back into space (i.e., the albedo).

Chapter 6

MARGIN QUESTION: An increase in the vegetation on Earth will have what effect on the carbon dioxide level in the atmosphere? ANSWER: Increased vegetation decreases the carbon dioxide level in the atmosphere by converting some of the CO_2 (along with water) into oxygen and glucose.
MARGIN QUESTION: What happens to the Van Allen belts when Earth's magnetic field is flipping? ANSWER: When the Van Allen belts are flipping, Earth's magnetic field and the resulting Van Allen belts vanish, allowing more high energy particles to enter Earth's atmosphere than occurs today.
MARGIN QUESTION: Why haven't the ocean tides on Earth put our planet into synchronous rotation with respect to the Moon? ANSWER: Earth's angular momentum is too great for it to have been slowed down enough to be in synchronous rotation with the Moon today.

Chapter 7

MARGIN QUESTION: If Mercury were struck by a large planetesimal, why would this collision not produce a moon, as happened when Earth was struck early in its history? ANSWER: Two possible reasons the large impact on Mercury didn't create a moon there are that the impacting body hit head on, putting debris in all directions, but not in orbit, and that the impacting body was composed of low-density material that didn't have enough mass to cause an ejection.
MARGIN QUESTION: Why isn't Mercury in synchronous rotation with respect to the Sun? ANSWER: Mercury is not in synchronous orbit around the Sun because the planet's orbit is too elliptical to keep the same face toward the Sun.
MARGIN QUESTION: Why does little cratering occur on Venus today, even compared to the present low rate of cratering on our Moon? ANSWER: Venus's thick atmosphere vaporizes virtually every object heading toward the planet's surface, preventing cratering from occurring there today.
MARGIN QUESTION: Why do you think it is easier to stand up in a dust devil on Mars than in the equivalent-speed event here on Earth? ANSWER: The air on Mars is less dense than the air on Earth, so the air pressure on Mars is lower and winds of equivalent speed have less impact there than on Earth.
MARGIN QUESTION: Would our Moon have to be closer or farther away to orbit in the same direction, but rise in the west? ANSWER: By Kepler's third law, our Moon would have to be closer to Earth for it to rise in the west.

Chapter 8

MARGIN QUESTION: What creates most of the heat inside Io that causes it to have volcanoes and geysers? ANSWER: Friction created by tidal distortion (rubbing of rock on rock) in Io creates the heat necessary to form volcanoes and geysers there.
MARGIN QUESTION: What two effects cause Saturn's belt and zone system? ANSWER: Convection of Saturn's outer layers and the planet's rotation cause that planet to have a belt and zone system.
MARGIN QUESTION: What evidence do astronomers have that Enceladus has liquid water in its interior? ANSWER: Ice floes on the surface and water vapor in the atmosphere strongly indicate that there is liquid water inside Enceladus.
MARGIN QUESTION: Why do Uranus's rings remain in orbit? ANSWER: Uranus's rings are held in orbit by two shepherd moons, Cordelia and Ophelia.
MARGIN QUESTION: Is our Moon inside or outside Earth's Roche limit? ANSWER: Our Moon is outside Earth's Roche limit. Otherwise, the Moon would have been pulled apart and made into a ring system around Earth.

Chapter 9

MARGIN QUESTION: What objects are classified today as planets? What objects are classified as dwarf planets? ANSWERS: Planets are Mercury, Venus, Earth, Mars, Jupiter, Saturn, Uranus, and Neptune. Dwarf planets are Pluto, Ceres, Eris, Haumea, and Makemake.
MARGIN QUESTION: Why do astronomers doubt that the asteroid belt was once made up of a single planet? ANSWER: The total mass of all the objects in the asteroid belt combined is much less than that of any planet.
MARGIN QUESTION: What evidence do we have that Comet Shoemaker-Levy 9 was not one solid chunk of rock surrounded by ice? ANSWER: Shoemaker-Levy 9 broke into numerous pieces under the relatively weak tidal influence of Jupiter. A solid comet would not have been so easily pulled apart.
MARGIN QUESTION: Why are stony-iron meteorites so rare compared to stony or iron meteorites? ANSWER: Stony-iron meteorites come from the boundaries between the rocky and metal parts of asteroids. These are very thin layers; hence, the meteorites from them are rare.

29. 5.0×10^{14} tons

Chapter 10

MARGIN QUESTION: Which of the Sun's three atmospheric layers is coolest? Which is densest? ANSWER: A portion of the Sun's chromosphere is the coolest atmospheric layer, while its photosphere is the densest part.
MARGIN QUESTION: Why doesn't the Sun collapse under the influence of its own enormous gravitational attraction? ANSWER: The Sun doesn't collapse because some of the energy it generates in its core pushes outward, counteracting the inward force of gravity.
MARGIN QUESTION: Why did the earlier neutrino detectors not detect the predicted number of neutrinos from the Sun? ANSWER: The earlier neutrino detectors were only sensitive to one type of neutrino. Since some of the Sun's neutrinos transformed from one type to another before reaching Earth, the early detectors did now detect all of them.

5. next maximum in 2005, next minimum in 2012
16. 1400 kg/m^3
17. 4.8%
18. 500 nm = visible light; 58 nm = ultraviolet; 1.9 nm = X ray
32. $\approx 25d$, equator
Toolbox 10-1: 6×10^{17} helium atoms; 2.25×10^{14} J

Chapter 11

MARGIN QUESTION: For which object—the Moon, Mars, or the star Sirius—is the parallax angle smallest from Earth?
ANSWER: The parallax angle for Sirius is smaller than that of Mars and our Moon because Sirius is farthest away of the three.
MARGIN QUESTION: A main-sequence star of which spectral type—F5, A8, or K0—is largest? ANSWER: Hotter main sequence stars are larger than cooler main sequence stars. Therefore, the A8 main sequence stars are larger than F5 or K0 main sequence stars.
MARGIN QUESTION: For what types of stars does the mass-luminosity relationship not apply? ANSWER: The mass-luminosity relationship does not apply for any type of star other than those on the main sequence.
MARGIN QUESTION: How do the spectral lines of a spectroscopic binary change as observed from Earth? ANSWER: The spectral lines of a spectroscopic binary Doppler shift in opposite directions, as seen from Earth unless the two stars orbit perpendicular to our line of sight. In the latter case, the spectral lines do not change at all.

15. 25 times brighter
29. 4.3 pc
31. (a) 9.7 pc;
(b) 0.10 arcsec
32. 1585 times brighter
35. (a) $\approx 10\ M_\odot$;
(b) $\approx 10^{-3}\ L_\odot$
36. $\approx$ 10,000–15,000 K
Toolbox 11-1: $\sim 4.85 \times 10^{-6}$ pc; $3.26 \times 10^{-2\prime\prime}$; $\sim$2.6 pc *or* $\sim$8.47 ly; $\sim$237 pc; $\sim 4.22 \times 10^{-3\prime\prime}$
Toolbox 11-2: 40 times; 631 times; 5.5
Toolbox 11-3: $\sim$11.4 pc, $\sim$37.1 ly; + 1.74 (Fomalhaut); −1.44 (Sirius A)
Toolbox 11-4: We can ignore M_2 (planet's mass) as tiny compared to the Sun's mass, M_1. Since the Sun's mass in these units is 1, the equation becomes $1 = a^3/P^2$, or $P^2 = a^3$; $\sim 1.56\ M_\odot$; 73 years

Chapter 12

MARGIN QUESTION: How are T Tauri stars different from main-sequence stars, like the Sun? ANSWER: Unlike solar-mass main-sequence stars, T Tauri stars are emitting large amounts of gas and dust, and they are changing brightness significantly.
MARGIN QUESTION: Which star arrives on the main sequence first, one that is 0.5 $M_{\odot}$ or one that is 2 $M_{\odot}$? ANSWER: Higher-mass stars arrive on the main sequence more quickly than lower-mass stars, so a 2-$M_{\odot}$ star would arrive there more rapidly than a 0.5-$M_{\odot}$ star.
MARGIN QUESTION: Will the Sun undergo a helium flash? Why or why not? ANSWER: The Sun will undergo a helium flash because its core will remain degenerate as it expands into the giant phase.
MARGIN QUESTION: How many helium atoms does it take to make one oxygen atom? ANSWER: It takes four helium atoms to make one oxygen atom.
MARGIN QUESTION: How do astronomers observe that Cepheids are changing size? ANSWER: Their cyclic changes in brightness and Doppler shift indicate that Cepheids are changing size.
26. 2000
27. 200 times longer

Chapter 13

MARGIN QUESTION: Why do type Ia supernovae not have any hydrogen lines in their spectra? ANSWER: Type Ia supernovae lack hydrogen lines because the star that is exploding was composed primarily of carbon and oxygen.
MARGIN QUESTION: Why were neutrinos from SN 1987A observed before the light from this event? ANSWER: Neutrinos were observed from SN 1987A before light from this event because the neutrinos were released by the dying star before the visible light was generated by the blast.
MARGIN QUESTION: If the lighthouse model of pulsars is correct, do we see all the nearby pulsars? Why or why not? ANSWER: We do not see the nearby pulsars whose beams do not sweep in our direction.
MARGIN QUESTION: Why do glitches change the rotation rates of pulsars? ANSWER: Glitches change the rotation rates of a pulsar by redistributing its mass. To conserve its angular momentum, the pulsar's rotation rate must also change.
28. about 7470 years ago

Chapter 14

MARGIN QUESTION: How much mass, m, would have to be destroyed in order to create an amount of energy, E? ANSWER: By Einstein's equation $E = mc^2$, to create energy, E, requires mass E/c^2.
MARGIN QUESTION: How would a 1-$M_{\odot}$ black hole 1 AU from Earth affect our planet? ANSWER: A black hole with the same mass as the Sun at the same distance from Earth as the Sun would have the same gravitational attraction on Earth as the Sun. The black hole would have no other effects on Earth.
MARGIN QUESTION: Why does at least one particle in the Hawking process always fall into the black hole? ANSWER: At least one particle must enter the black hole in the Hawking process to conserve the linear momentum of the pair of particles that the process creates.
11. 8.9 km, 89 km
15. 10 m
Toolbox 14-1: $\sim 1.0 \times 10^{12}$ kg; 9×10^{9} km; $\sim 2130\ M_{\odot}$

Chapter 15

MARGIN QUESTION: Referring to Figure 15-4, how would the brightness of a Cepheid variable with peak luminosity of 1000 $L_{\odot}$ change if it were observed every 5 days? ANSWER: A Cepheid variable with peak luminosity of 1000 $L_{\odot}$ has a period of 5 days, so if it were observed every 5 days, its luminosity would appear unchanged from one observation to the next.
MARGIN QUESTION: The presence of supernova remnants at the center of our Galaxy implies what other activity is occurring in that region? ANSWER: Since stars that become supernovae, and the remnants they leave, have short lives in astronomical terms, supernova remnants at the center of our Galaxy imply that new star formation must be occurring there.
14. 20 times
19. about once every 3750 years
Toolbox 15-1: 1×10^{5} pc; ~ 6.92; $M = 9.83$; $m - M = 5$

Chapter 16

MARGIN QUESTION: Comparing it to the galaxies in Figure 16-1, what is the spiral classification of M74, shown in Figure 16-4b? ANSWER: M74 is an Sc galaxy.
MARGIN QUESTION: The Large Magellanic Cloud is visible to the naked eye in the southern hemisphere. From Figure 16-13, what terrestrial things do you think it could be mistaken for? ANSWER: The Large Magellanic Cloud is sometimes mistaken for part of the Milky Way or for a cloud in the night sky.
MARGIN QUESTION: If the dark matter in Figure 16-29c suddenly vanished, what would we see of the clusters of galaxies behind it? ANSWER: If the dark matter between us and a cluster of galaxies suddenly vanished, the shapes we would see of many of those galaxies would change (since the dark matter lenses the light from them).
MARGIN QUESTION: If the Hubble constant were 3 times its present value, how much slower or faster would superclusters be moving apart? ANSWER: If the Hubble constant were 3 times its present value, superclusters would be moving apart 3 times faster.

MARGIN QUESTION: In what way are telescopes time machines? ANSWER: The light entering telescopes travels at a finite speed, so what we see through them are events and objects as they were some time in the past.

21. 306 Mpc

22. 7200 km/s

Toolbox 16-1: $v = 1.74 \times 105$ km/s; $d = 3.0$ Mpc; slower

Chapter 17

MARGIN QUESTION: How are the spectra of quasars different from the spectra of stars? ANSWER: Quasar spectra have emission lines, whereas stellar spectra have absorption lines.

MARGIN QUESTION: What other astronomical objects have been observed after their light passed through a gravitational lens? ANSWER: Besides quasars, astronomers have observed galaxies that have been gravitationally lensed.

Chapter 18

MARGIN QUESTION: If the universe were twice as old as it is now, how would the Hubble constant compare to the value it has today? ANSWER: Since the inverse of the Hubble constant is its age, if the universe were twice as old, the Hubble constant would be half as large as it is today.

MARGIN QUESTION: If dark matter did not exist, would any gravitational lensing occur in the universe? ANSWER: Even without dark matter, gravitational lensing would occur due to the gravitational effects of superclusters of galaxies, clusters of galaxies, individual galaxies, and other mass distributions.

Toolbox 18-1: 27.6×10^9 years; 6.9×10^9 years

Chapter 19

MARGIN QUESTION: Synchronous rotation occurred in what other situation that we explored earlier in this book? ANSWER: Examples of synchronous rotation that we have considered are the orbits of most of the moons in the solar system around their respective planets, including our Moon orbiting Earth, and the synchronous rotation of Pluto with respect to its moon Charon.

MARGIN QUESTION: What negative consequence could arise from an alien civilization discovering a Pioneer spacecraft? ANSWER: If an alien civilization discovered a Pioneer spacecraft, it could trace the spacecraft back to Earth. Visits from this alien civilization might introduce diseases for which we have no defense, or they might lead to interplanetary war if we couldn't get along with our new neighbors.

Appendix A

3.141×10^9, 3.1831×10^{-9}; 27182820000, 0.0000000000367879

Appendix B

580 K, 55 km, temperature decreases with height; +8, +10, September 1, September 6; $8L_\odot$, 6000 K

Appendix D

5527°C, 9980°F; –270°C, –454°F

INDEX

Page numbers in *italics* indicate figures.

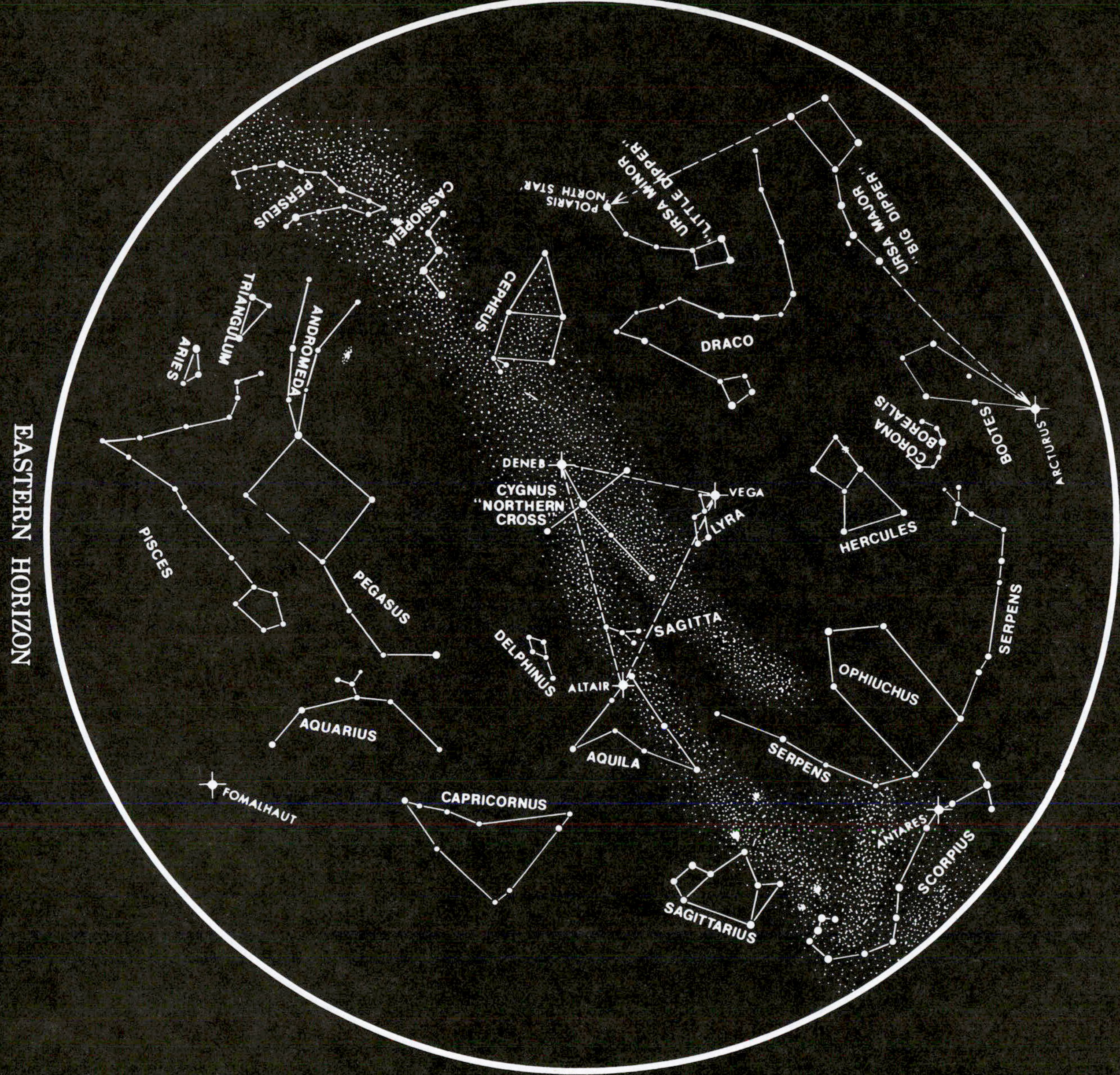

SOUTHERN HORIZON

THE NIGHT SKY IN SEPTEMBER

Chart time (Daylight Savings Time):

11 pm . . . First of September
10 pm . . . Middle of September
9 pm . . . Last of September

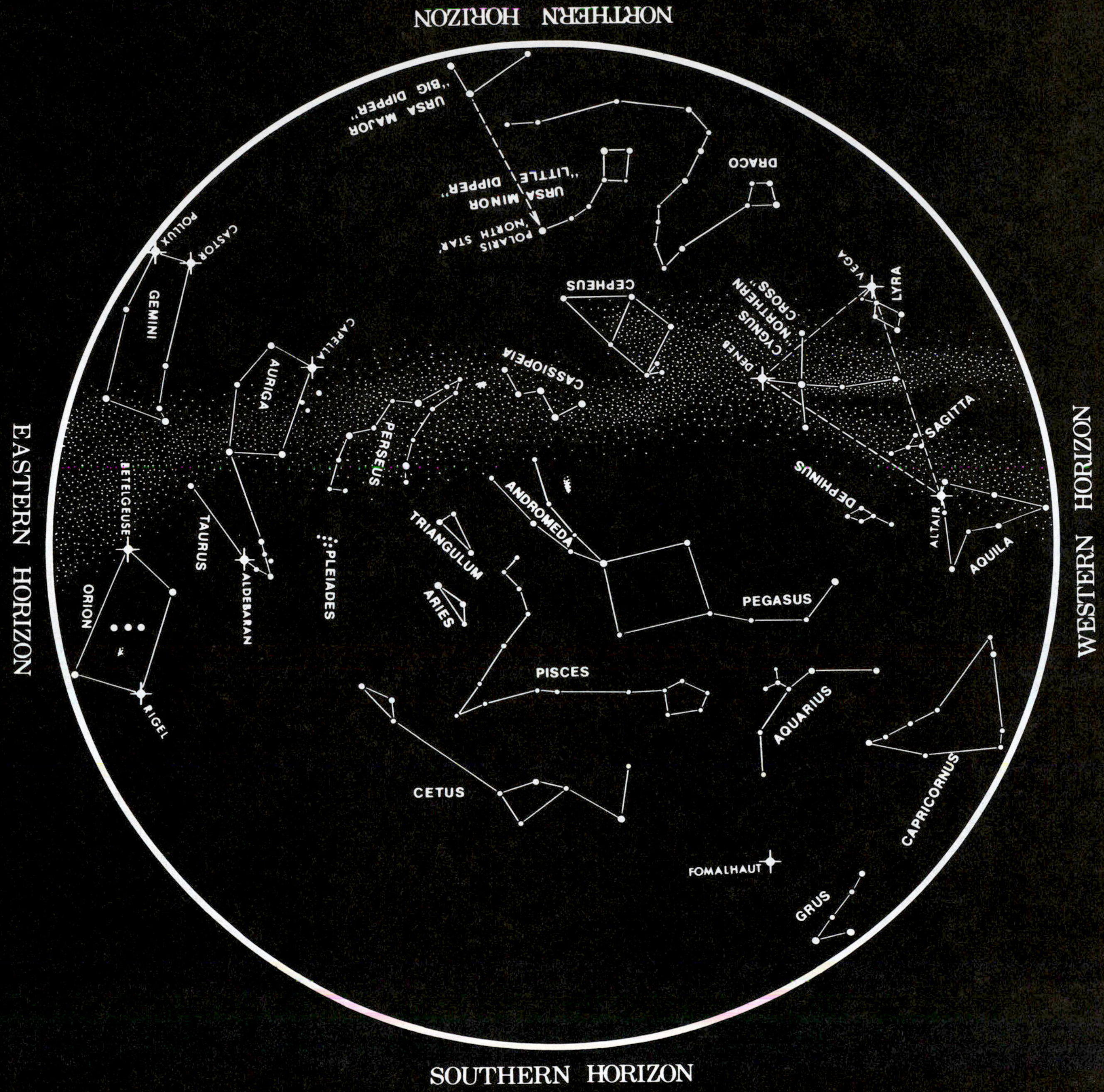

NORTHERN HORIZON

EASTERN HORIZON

WESTERN HORIZON

SOUTHERN HORIZON

URSA MAJOR "BIG DIPPER"
URSA MINOR "LITTLE DIPPER"
POLARIS "NORTH STAR"
DRACO
CEPHEUS
CYGNUS "NORTHERN CROSS"
DENEB
VEGA
LYRA
CASSIOPEIA
CAPELLA
AURIGA
POLLUX
CASTOR
GEMINI
PERSEUS
SAGITTA
DELPHINUS
ALTAIR
AQUILA
ANDROMEDA
TRIANGULUM
PLEIADES
TAURUS
ALDEBARAN
BETELGEUSE
ORION
RIGEL
ARIES
PEGASUS
PISCES
AQUARIUS
CAPRICORNUS
CETUS
FOMALHAUT
GRUS

THE NIGHT SKY IN NOVEMBER

Chart time (Local Standard Time):

10 pm . . . First of November
9 pm . . . Middle of November
8 pm . . . Last of November

NORTHERN HORIZON

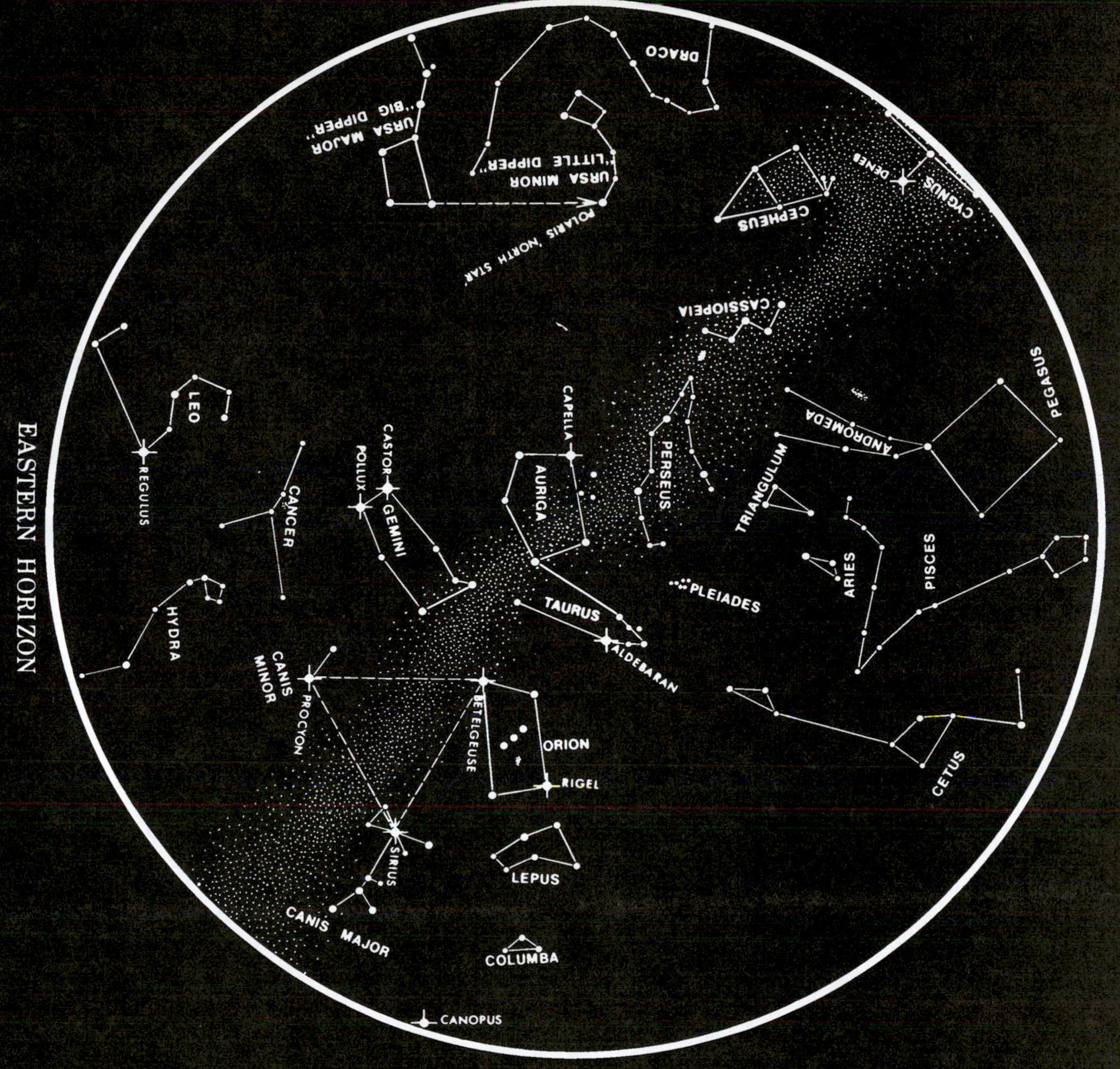

EASTERN HORIZON

WESTERN HORIZON

SOUTHERN HORIZON

THE NIGHT SKY IN JANUARY

Chart time (Local Standard Time):

10 pm . . . First of January
9 pm . . . Middle of January
8 pm . . . Last of January

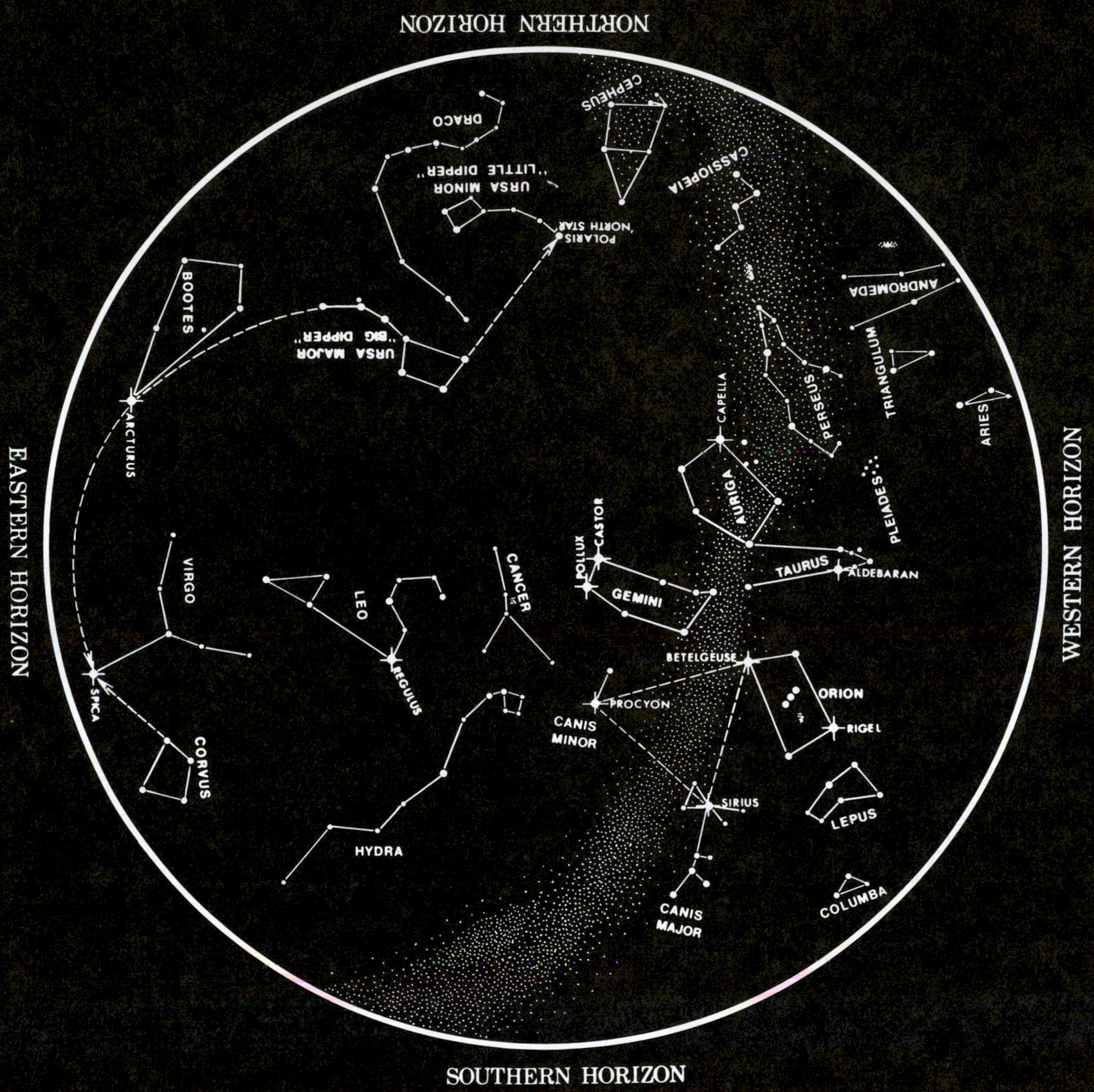

NORTHERN HORIZON

EASTERN HORIZON

WESTERN HORIZON

SOUTHERN HORIZON

THE NIGHT SKY IN MARCH

Chart time (Local Standard Time):

10 pm... First of March
9 pm... Middle of March
8 pm... Last of March

NORTHERN HORIZON

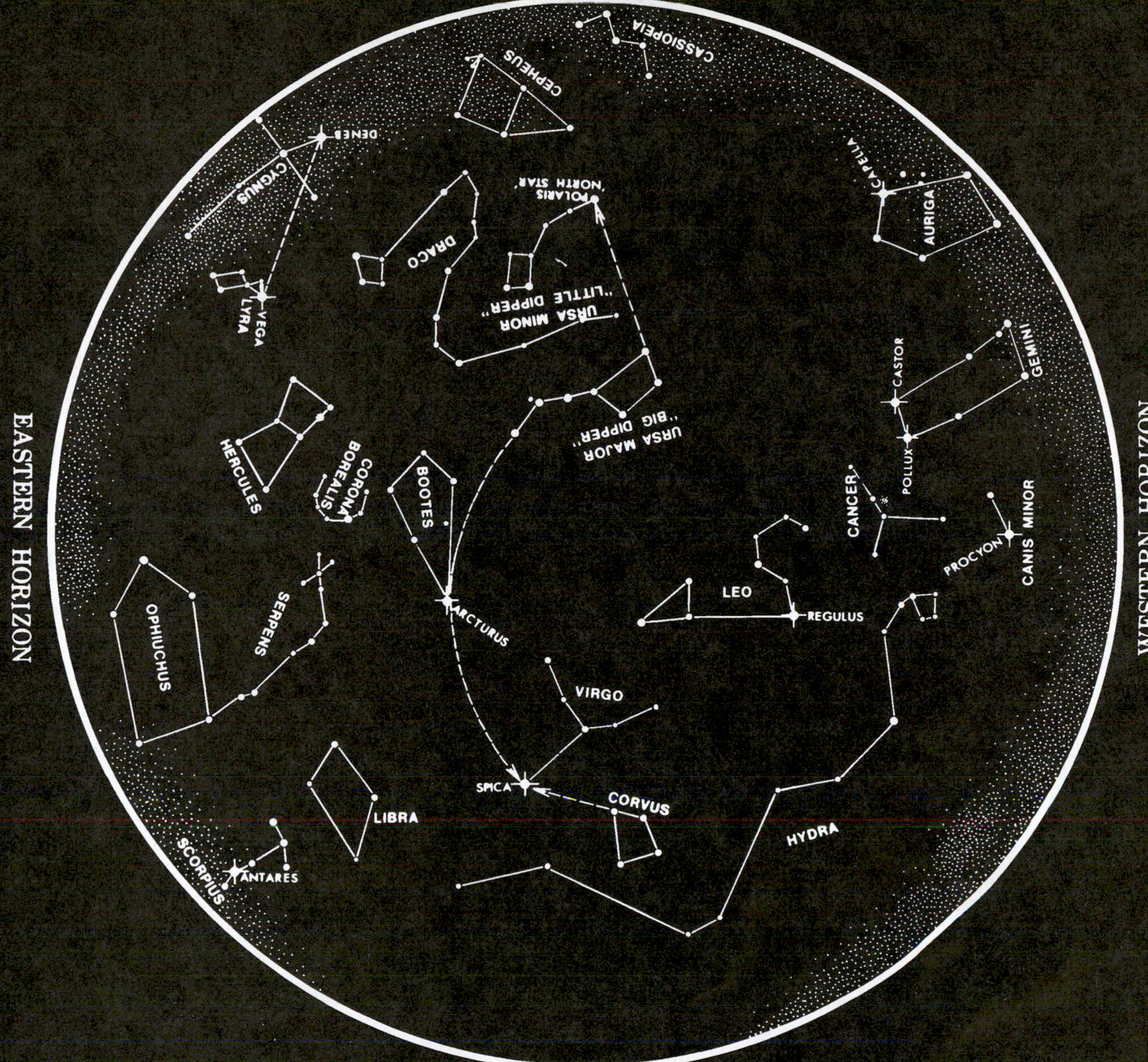

SOUTHERN HORIZON

THE NIGHT SKY IN MAY

Chart time (Daylight Savings Time):

11 pm . . . First of May
10 pm . . . Middle of May
9 pm . . . Last of May

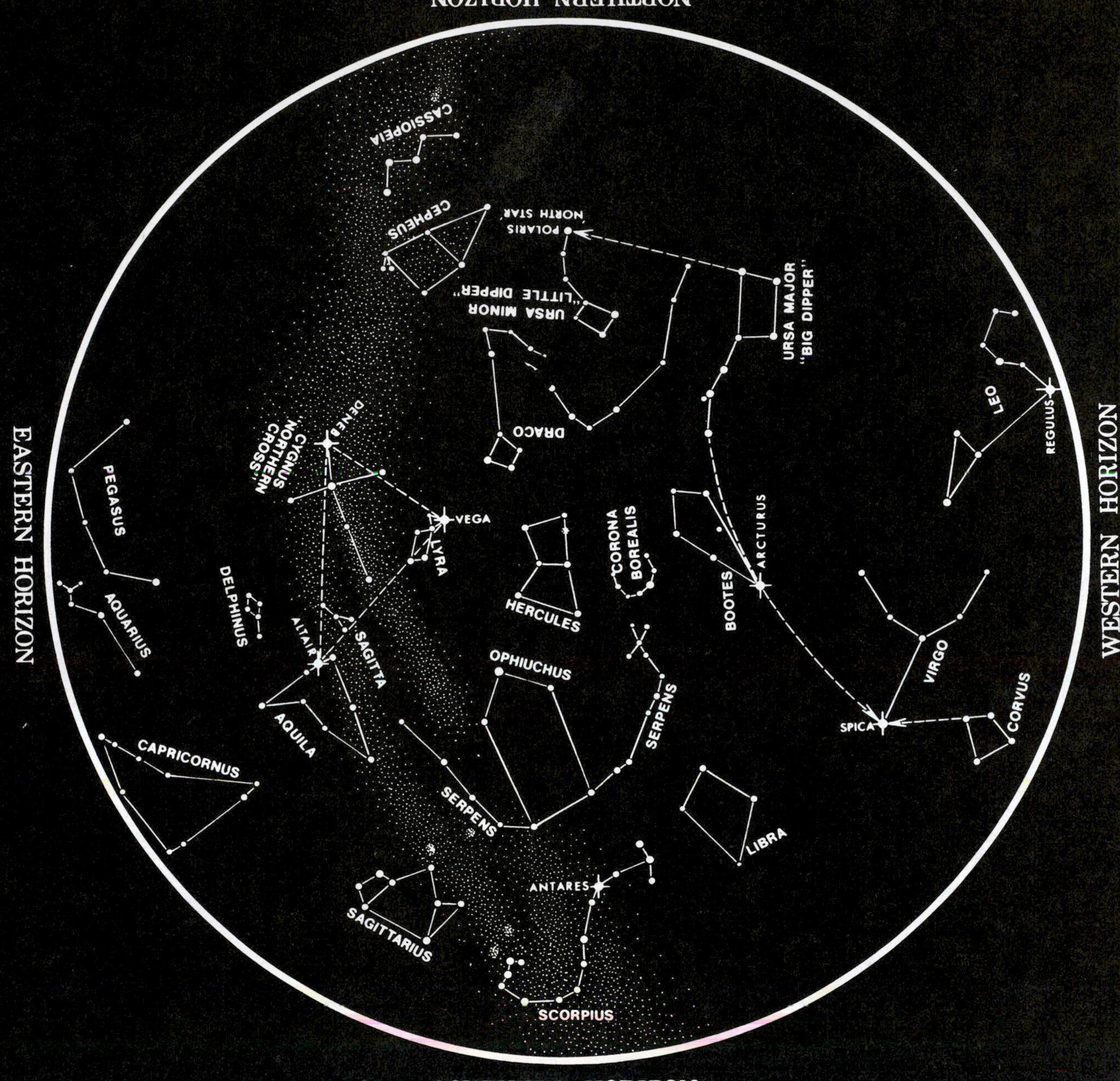

THE NIGHT SKY IN JULY

Chart time (Daylight Savings Time):

11 pm . . . First of July
10 pm . . . Middle of July
9 pm . . . Last of July